Communications in Computer and Information Science 723

More information about this series at http://www.springer.com/series/7899

María Valdés Hernández
Víctor González-Castro (Eds.)

Medical Image Understanding and Analysis

21st Annual Conference, MIUA 2017
Edinburgh, UK, July 11–13, 2017
Proceedings

Editors
María Valdés Hernández
Department of Neuroimaging Sciences
University of Edinburgh
Edinburgh
UK

Víctor González-Castro
Universidad de León
León
Spain

ISSN 1865-0929 ISSN 1865-0937 (electronic)
Communications in Computer and Information Science
ISBN 978-3-319-60963-8 ISBN 978-3-319-60964-5 (eBook)
DOI 10.1007/978-3-319-60964-5

Library of Congress Control Number: 2017944214

Printed on acid-free paper

This Springer imprint is published by Springer Nature
The registered company is Springer International Publishing AG
The registered company address is: Gewerbestrasse 11, 6330 Cham, Switzerland

Preface

This volume comprises the proceedings of the 21st edition of the Medical Image Understanding and Analysis (MIUA) Conference, an annual forum organized in the United Kingdom for communicating research progress within the community interested in biomedical image analysis. Its goals are the dissemination and discussion of research in medical image analysis to encourage the growth and to raise the profile of this multi-disciplinary field that has an ever-increasing real-world applicability. The conference constitutes an excellent opportunity to network, generate new ideas, establish new collaborations, learn about and discuss different topics, listen to speakers of international reputation, and present and show medical image analysis tools.

This year's edition was organized by The Row Fogo Centre for Research in Ageing and Dementia in conjunction with Edinburgh Imaging (http://www.ed.ac.uk/clinical-sciences/edinburgh-imaging) at The University of Edinburgh, in partnership with the Scottish Imaging Network A Platform for Scientific Excellence (SINAPSE) (http://www.sinapse.ac.uk/), the *Journal of Imaging* (http://www.mdpi.com/journal/jimaging), and the EPSRC Medical Image Analysis Network (MedIAN) (http://www.ibme.ox.ac.uk/MedIAN) and MathWorks (https://uk.mathworks.com/); it was supported by the British Machine Vision Association (BMVA), Toshiba Medical Visualisation Systems Europe, General Electric, Bayer, Siemens, Edinburgh Imaging, Optos, Holoxica Ltd., AnalyzeDirect and NVIDIA.

The number and level of submissions of this year's edition were unprecedented. In all, 105 technical papers and 22 abstracts showing clinical applications of image-processing techniques (the latter not considered for inclusion in this volume) were revised by an expert team of 93 reviewers from British(38), Spanish(25), French(6), Italian(11), Swedish(5), Greek(2), New Zealand(1), American(3), German(1), Dutch(1), Bangladeshi(1) and Swiss(1) institutions. Each of the 127 submissions were reviewed by two to four members of the Program Committee. Based on their ranking and recommendations, 18 of 105 papers were accepted, 65 of 105 papers were provisionally accepted pending minor revisions and suggestions, and 22 of 105 papers were rejected for publication in this volume. From the papers rejected, 15 were considered to need a major revision, and their authors were invited to address the reviewers' comments and submit their revised work to the *Journal of Imaging* (conference partner), to which the reviews will be provided for facilitating the new review process. After a second round of revision, we are including in this volume 82 full papers. We hope you agree with us that they all show high quality and represent a step forward in the medical image analysis field.

We thank all members of the MIUA 2017 Organizing, Program, and Steering Committees and, particularly, all those who supported MIUA 2017 by submitting papers and attending the meeting. We thank Professor Joanna Wardlaw, Head of the Academic Hub and Centre that organized the conference, for her welcome words.

We also thank our speakers Professors Sir Michael Brady, Daniel Rueckert, Ingela Nyström, and Jinah Park, and Dr. Constantino Carlos Reyes Aldasoro and Konstantinos Kamnitsas for sharing their success, knowledge, and experiences. We hope you enjoy the proceedings of MIUA 2017.

May 2017

Maria Valdes Hernandez
Víctor González-Castro

Organization

Program Chairs

Maria Valdes Hernandez	University of Edinburgh, UK
Víctor González-Castro	Universidad de León, Spain

MIUA Steering Committee

Yan Chen	Loughborough University, UK
Bill Crum	King's College London, UK
Alastair Gale	Loughborough University, UK
Víctor González-Castro	Universidad de León, Spain
Tryphon Lambrou	University of Lincoln, UK
Stephen McKenna	University of Dundee, UK
Nasir Rajpoot	University of Warwick, UK
Constantino Carlos Reyes-Aldasoro	City, University of London, UK
Greg Slabaugh	City, University of London, UK
Maria Valdes Hernandez	University of Edinburgh, UK
Xianghua Xie	Swansea University, UK
Xujiong Ye	University of Lincoln, UK
Reyer Zwiggelaar	Aberystwyth University, UK

Local Organizing Committee

Devasuda Anblagan	University of Edinburgh, UK
Lucia Ballerini	University of Edinburgh, UK
Kristin Flegal	University of Glasgow/SINAPSE, UK
Anne Grant	University of Edinburgh, UK
Taku Komura	University of Edinburgh, UK
Tom MacGillivray	University of Edinburgh, UK
Ian Marshall	University of Edinburgh, UK
Cyril Pernet	University of Edinburgh, UK
Amos Storkey	University of Edinburgh, UK
Emanuele Trucco	University of Dundee, UK
Edwin van Beek	Clinical Research Imaging Centre, University of Edinburgh, UK

Technical Program Committee

Jose Luis Alba-Castro	University of Vigo, Spain
Enrique Alegre	Universidad de León, Spain

Luciano Alparone	University of Florence, Italy
Devasuda Anblagan	University of Edinburgh, UK
Paul Armitage	University of Sheffield, UK
Juan Arribas	University of Valladolid, Spain
Lucia Ballerini	University of Edinburgh, UK
Isabelle Bloch	Telecom ParisTech, France
Leonardo Bocchi	University of Florence, Italy
Gunilla Borgefors	Centre for Image Analysis, Uppsala University, Sweden
Larbi Boubchir	University of Paris 8, France
Gloria Bueno	Universidad de Castilla-La Mancha, Spain
Maria J. Carreira	Universidade de Santiago de Compostela, Spain
Alessandro Cavinato	École Polytechnique Fédérale de Lausanne, Switzerland
Yan Chen	Loughborough University, UK
Samuel Danso	University of Edinburgh, UK
Johan Debayle	École Nationale Supérieure des Mines de Saint-Étienne, France
David Alexander Dickie	University of Edinburgh, UK
Javier Escudero	University of Edinburgh, UK
María del Milagro Fernández Carrobles	Universidad de Castilla-La Mancha, Spain
Laura Fernández-Robles	Universidad de León, Spain
Samuele Fiorini	Università degli studi di Genova, Italy
Francesco Fontanella	Università di Cassino e del Lazio meridionale, Italy
Alejandro Frangi	University of Sheffield, UK
Alastair Gale	Loughborough University, UK
María García	University of Valladolid, Spain
Andrea Garzelli	University of Siena, Italy
Yann Gavet	École Nationale Supérieure des Mines de Saint-Etienne, France
Andrea Giachetti	University of Verona, Italy
Calum Gray	University of Edinburgh, UK
Manuel Graña	Basque Country University (UPV/EHU), Spain
Enrico Grisan	University of Padova, Italy
Juan M. Górriz	University of Granada, Spain
Taku Komura	University of Edinburgh, UK
Violet Kovacheva	Institute of Cancer Research, UK
Tryphon Lambrou	University of Lincoln, UK
Joakim Lindblad	Mathematical Institute of the Serbian Academy of Sciences and Arts, Serbia
Ruggiero Lovreglio	University of Auckland, New Zealand
Evelyne Lutton	INRA, France
Tom MacGillivray	University of Edinburgh, UK
Sasan Mahmoodi	University of Southampton, UK
Filip Malmberg	Uppsala University, Sweden
Ian Marshall	University of Edinburgh, UK
Stephen McKenna	University of Dundee, UK

Bjoern Menze	Technical University of Munich, Germany
Mariofanna Milanova	University of Arkansas at Little Rock, USA
Carmel Moran	University of Edinburgh, UK
Philip Morrow	Ulster University, UK
Susana Munoz Maniega	University of Edinburgh, UK
Henning Müller	HES-SO, Switzerland
Bill Nailon	University of Edinburgh/NHS Lothian, UK
Mark Nixon	University of Southampton, UK
Jorge Novo Buján	University of A Coruña, Spain
Marcos Ortega	University of A Coruña, Spain
Gonzalo Pajares	University Complutense of Madrid, Spain
Pietro Pala	University of Florence, Italy
Georgios Papanastasiou	University of Edinburgh, UK
Roberto Paredes	Universidad Politécnica de Valencia, Spain
Alejandro Pazos Sierra	Universidade da Coruña, Spain
Enrico Pellegrini	University of Edinburgh, UK
Cyril Pernet	University of Edinburgh, UK
Jean-Charles Pinoli	École Nationale Supérieure des Mines, France
Ian Poole	TMVS, UK
Kashif Rajpoot	University of Birmingham, UK
Nasir Rajpoot	University of Warwick, UK
Javier Ramírez	University of Granada, Spain
Constantino Carlos Reyes-Aldasoro	City, University of London, UK
Scott Semple	University of Edinburgh, UK
Korsuk Sirinukunwattana	Harvard Medical School, USA
Greg Slabaugh	City, University London, UK
Natasa Sladoje	Centre for Image Analysis, Uppsala University, Sweden
Robin Strand	Centre for Image Analysis, Uppsala University, Sweden
Pablo G. Tahoces	Universidade de Santiago de Compostela, Spain
Emanuele Trucco	University of Dundee, UK
Maria Valdes Hernandez	University of Edinburgh, UK
Cesar Veiga	Instituto de Investigación Sanitaria Galicia Sur - IISGS, Spain
Xianghua Xie	Swansea University, UK
Xujiong Ye	University of Lincoln, UK
Xenophon Zabulis	Foundation for Research and Technology, Greece
Matteo Zanotto	Istituto Italiano di Tecnologia, Italy
Reyer Zwiggelaar	Aberystwyth University, UK

Contents

Image Enhancement and Alignment

Modeling and Ssegmentation of Preclinical, Body and Histological Imaging

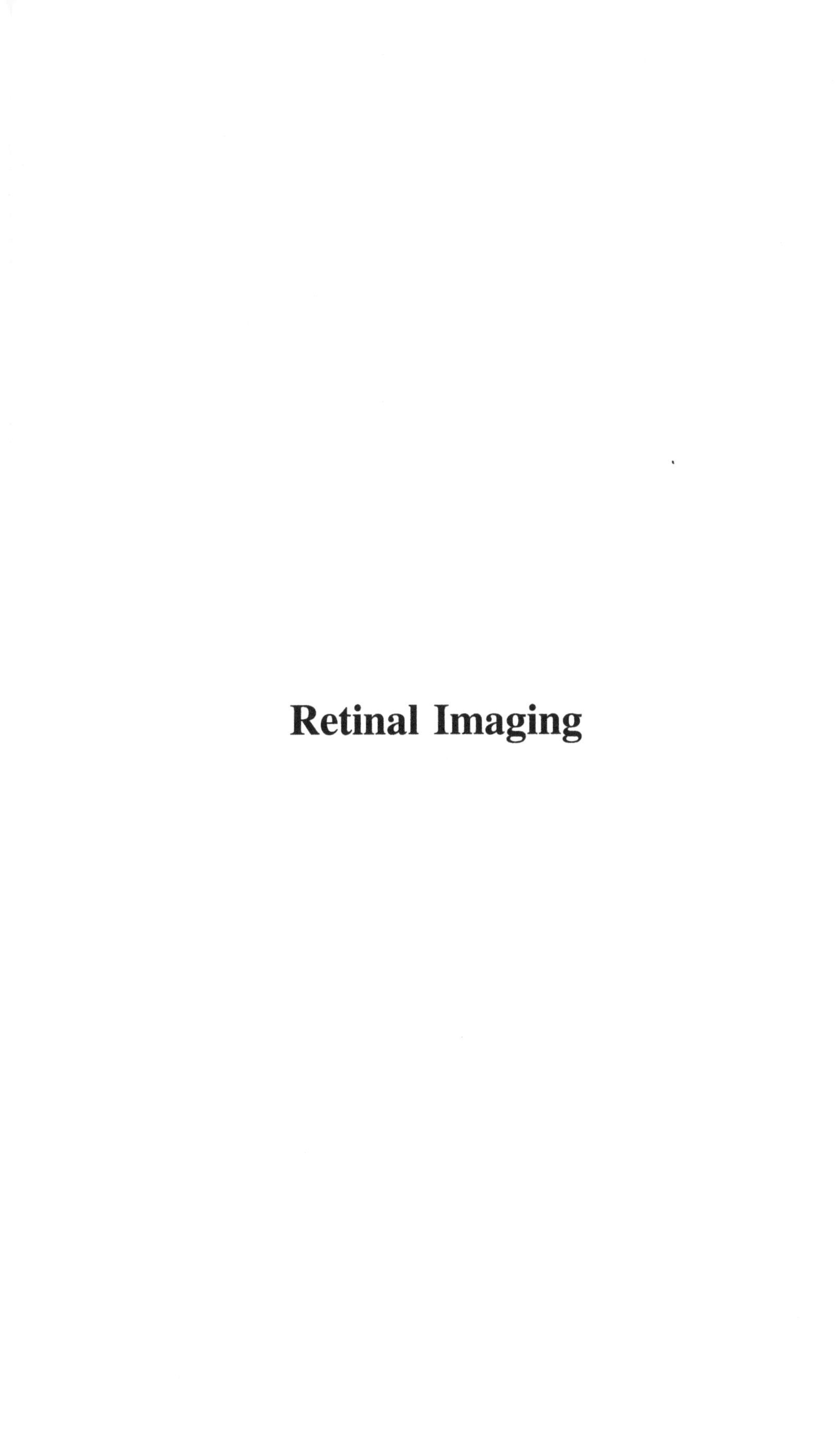

Retinal Imaging

End-to-End Learning of a Conditional Random Field for Intra-retinal Layer Segmentation in Optical Coherence Tomography

Arunava Chakravarty(✉) and Jayanthi Sivaswamy

Center for Visual Information Technology, IIIT Hyderabad, Hyderabad, India
arunava.chakravarty@research.iiit.ac.in, jsivaswamy@iiit.ac.in

Abstract. Intra-retinal layer segmentation of Optical Coherence Tomography images is critical in the assessment of ocular diseases. Existing Energy minimization based methods employ handcrafted cost terms to define their energy and are not robust to the presence of abnormalities. We propose a novel, Linearly Parameterized, Conditional Random Field (LP-CRF) model whose energy is learnt from a set of training images in an end-to-end manner. The proposed LP-CRF comprises two convolution filter banks to capture the appearance of each tissue region and boundary, the relative weights of the shape priors and an additional term based on the appearance similarity of the adjacent boundary points. All the energy terms are jointly learnt using the Structured Support Vector Machine. The proposed method segments all retinal boundaries in a single step. Our method was evaluated on 107 Normal and 220 AMD B-scan images and found to outperform three publicly available OCT segmentation software. The average unsigned boundary localization error is 1.52 ± 0.29 pixels for segmentation of 8 boundaries on the Normal dataset and 1.9 ± 0.65 pixels for 3 boundaries on the combined AMD and Normal dataset establishing the robustness of the proposed method.

Keywords: CRF · SSVM · OCT · Segmentation

1 Introduction

Optical Coherence Tomography (OCT) images of retina provide a cross-sectional view of the intra-retinal tissue which is composed of 7 adjacent layers [2], separated by 8 boundaries as depicted in Fig. 1a. The boundaries ordered from the top to bottom are the: (i) Inner Limiting Membrane (ILM) separating the Nerve Fiber Layer (NFL) from vitreous, (ii) *NFL/GCL* boundary separating NFL from the Ganglion Cell and Inner Plexicon layer (GCL-IPL), (iii) *IPL/INL* separating GCL-IPL from the Inner Nuclear Layer (INL), (iv) *INL/OPL* separating INL from the Outer Plexiform Layer (OPL), (v) *OPL/ONL* separating OPL from the Outer Nuclear and Inner Segment (ONL-IS) layer, (vi) *IS/OS* separating ONL-IS from the Outer Segment (OS) layer (vii) RPE_{in} separating OS from the Retinal Pigment Epithelium (RPE) layer and finally the (viii) RPE_{out} boundary separating RPE from the choroid.

M. Valdés Hernández and V. González-Castro (Eds.): MIUA 2017, CCIS 723, pp. 3–14, 2017.
DOI: 10.1007/978-3-319-60964-5_1

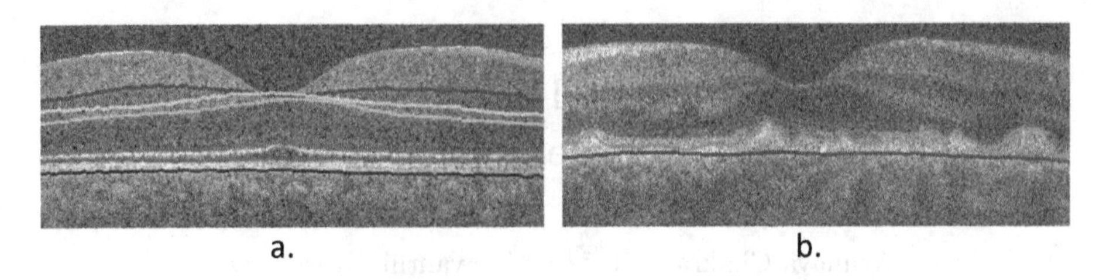

Fig. 1. Retinal layer boundaries in OCT B-scans of (a) Healthy retina, listed from top to bottom: ILM(Red), NFL/GCL(Green), IPL/INL(blue), INL/OPL(yellow), OPL/ONL(cyan), IS/OS(magenta), RPE_{in}(pink) and RPE_{out}(purple); (b) Retina with AMD: ILM(Red), RPE_{in}(Green) and RPE_{out}(Blue). (Color figure online)

Accurate segmentation of these layers is necessary to quantify the morphological changes in the retinal tissue and detect ocular diseases such as Age Related Macular Degeneration (AMD) [1]. In AMD, the drusen deposits in the RPE layer lead to irregularities and undulations in the RPE_{in} boundary as depicted in Fig. 1b. Currently, commercial OCT systems are equipped to segment only two or three layers and often fail for images of poor quality or/and with lesions.

Early work on OCT layer segmentation focussed on the segmentation of few (2–4) prominent layers. These methods used peak, valley and signed gradient analysis on the intensity profiles for each A-scan (column) in the OCT slice followed by regularization across adjacent A-scans [5]. Their performance suffered due to the lack of strong boundaries and overlapping intensities between the adjacent layers, presence of speckle noise and vessel shadows.

To overcome these challenges, deformable model based methods were proposed that incorporated shape priors in addition to the boundary and regional appearance of the layers. Constraints on the thickness and smoothness of layers were imposed in [9] while in [11] the parallelism between the adjacent layer boundaries was enforced. A circular arc based shape prior was employed in [14]. However, these methods are sensitive to the initial contour initialization leading to high convergence time or entrapment of the curve in the local minima.

Graph based methods have been explored in [1,2] where the layers in each OCT slice is segmented sequentially using the Dijkstra's shortest path algorithm. In [4], an energy composed of multiple cost terms was defined to capture the appearance and shape priors for each layer. The energy minimization was formulated as a Minimum Cost Closed Set (MCCS) problem on a geometric graph. Each cost term in the energy had to be *handcrafted* manually and then combined using empirically determined relative weights. These methods were initially designed for the segmentation of healthy images and require complex disease-specific modifications in [1,12] to handle abnormalities.

Thus, the goal of obtaining an optimal energy function for the joint multi-layer segmentation of OCT images remains an open problem. The minima of such an energy must correspond to the desired boundaries across a large set of images and should be applicable for both healthy and abnormal cases without the need for disease specific modifications. We address this problem with a novel *linearly parameterized* Conditional Random Field (LP-CRF) formulation whose parameters are learnt in a data-driven, end-to-end manner with a Structured Support Vector Machine (SSVM) [3]. The convolution filters used to capture the appearance of each layer region and boundary as well as the relative weights of the appearance and shape prior based cost terms are implicitly modelled within the LP-CRF model. As a result our method doesnot require any explicit feature extraction or tunable weights. The contributions of this paper are:

- We eliminate the need to handcraft the energy by learning it from a set of training images in an end-to-end manner. The proposed method outperforms the existing methods [4] with similar but handcrafted energy cost terms.
- Our CRF formulation seamlessly incorporates both hard and soft constraints on shape priors in a single pairwise edge between each neighbor which is difficult to implement in the MCCS approach [4,13], requiring additional directed edges in the graph construction.
- We jointly segment all the layers in a single optimization step in contrast to the existing methods that need to handle each layer differently.
- Our method is able to learn a single energy function for both Normal as well as AMD cases for a three layer boundary segmentation problem.
- The proposed method has been shown to be robust to data acquired from multiple centres with different scanners and at different resolutions.

2 Method

An overview of the proposed method is presented in Fig. 2a. The input image is standardized by extracting the region of interest (ROI), flattening the retinal tissue and removing speckle noise (see Sect. 2.1). The task of joint extraction of the multiple layer boundaries is posed as a CRF based Energy Minimization (see Sect. 2.2). The total CRF energy is learnt in an end-to-end manner by employing SSVM (see Sect. 2.3). This involves parameterization of the energy as a linear function LP-CRF (see Sect. 2.4). During testing, a CRF is constructed from the standardized image using the learnt parameters. Thereafter, the optimal labelling of the CRF is evaluated and brought back to the original image space by reversing the image flattening and ROI extraction operations.

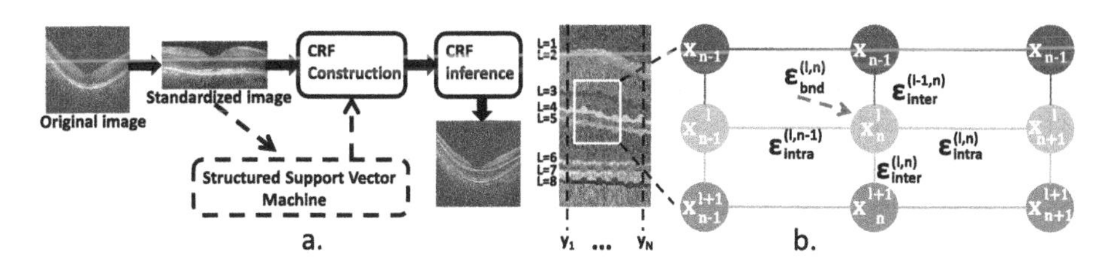

Fig. 2. a. Overview of the proposed method, training is represented with dotted lines. b. Graphical model of the proposed CRF formulation.

2.1 Image Preprocessing

The curvature of the retinal surface is flattened by using the method reported in [2]. Each image column is shifted by an offset obtained by fitting a quadratic polynomial on a rough estimate of the RPE_{out} boundary. Next, the ROI is extracted by cropping out the dark (intensity values close to 0) background regions at the top and bottom of the image. The ROI is estimated as the rectangular region encompassing the largest connected component obtained by thresholding the input image at 0.3 after scaling the intensity to [0,1]. To reduce holes due to the dark layers within the ROI, a large Gaussian filter with σ empirically set to 9 was employed prior to thresholding. Thereafter, the speckle reducing anisotropic diffusion [15] is applied. The ROI is resized to 190×600 to handle the variations in image resolution and an intensity standardization scheme based on [10] is applied to handle the inter and intra-scanner intensity variations.

2.2 Problem Formulation as a CRF

The joint multi-layer segmentation problem seeks to extract L layer boundaries in an OCT image I of size $X \times Y$. Each boundary is labeled as $1 \leq l \leq L$ ordered from the top to bottom and represented by N points with coordinates $(x_{l,n}, y_n)$, obtained by uniformly sampling the image columns at y_n positions. The height along the X-axis where the lth boundary passes through the column y_n is denoted by $x_{l,n}$. Thus, each $x_{l,n}$ is a discrete random variable that can take a value from the label set $\Omega = \{1 \leq i \leq X, i \in Z^+\}$. The set of random variables $X = \{x_n^l | 1 \leq l \leq L, 1 \leq n \leq N\}$ defines a random field and $\mathbf{x} \in \Omega^{L \times N}$ denotes a feasible *labelling* of X obtained by assigning a label from Ω to each $x_{l,n}$.

A unary boundary cost $\varepsilon_{bnd}^l(x_{l,n})$ is defined for each $x_{l,n}$ to capture the likelihood of the lth boundary to pass through $(x_{l,n}, y_n)$ given the appearance of the OCT image around that location. Moreover, based on the Markovian assumption, the label of each $x_{l,n}$ is also considered to be dependent on its immediate neighbors $x_{l,n+1}$ on the same boundary and $x_{l+1,n}$ on the adjacent $(l+1)^{th}$ boundary which are captured by the pairwise Intra-layer cost $\varepsilon_{intra}^{l,n}(x_{l,n}, x_{l,n+1})$ and the pairwise Inter-layer cost $\varepsilon_{inter}^{l,n}(x_{l,n}, x_{l+1,n})$ terms respectively. An undirected graphical representation of the CRF for a *4-neighborhood* is depicted in Fig. 2b. The smoothness and similarity in the appearance of the adjacent points on a boundary is captured by $\varepsilon_{intra}^{l,n}(x_{l,n}, x_{l,n+1})$ while $\varepsilon_{inter}^{l,n}(x_{l,n}, x_{l+1,n})$ captures the tissue appearance and layer thickness priors between the l and $(l+1)^{th}$ layer. Thus, all the three cost terms are dependent on the observed image I. To simplify the notation, the input arguments for the cost terms have been omitted in the rest of the paper and simply represented by ε_{bnd}^l, $\varepsilon_{intra}^{l,n}$ and $\varepsilon_{inter}^{l,n}$ respectively. Thus, the CRF energy for I is defined as

$$E(\mathbf{x}, I) = \sum_{l=1}^{L}\sum_{n=1}^{N} \varepsilon_{bnd}^{l} + \sum_{l=1}^{L}\sum_{n=1}^{N-1} \varepsilon_{intra}^{l,n} + \sum_{l=1}^{L-1}\sum_{n=1}^{N} \varepsilon_{inter}^{l,n}$$
$$= E_{bnd}(\mathbf{x}, I) + E_{intra}(\mathbf{x}, I) + E_{inter}(\mathbf{x}, I), \quad (1)$$

where $E_{bnd}(\mathbf{x}, I)$, $E_{intra}(\mathbf{x}, I)$ and $E_{inter}(\mathbf{x}, I)$ are the sum of all the unary, intra-layer and the inter-layer cost terms in the entire CRF respectively. The labelling $\mathbf{x}$ that *maximizes* $E(\mathbf{x}, I)$ for an image I corresponds to the desired segmentation. During implementation, the max CRF inference in Eq. 1 is converted into a minimization problem by taking the negative of all the unary and pairwise cost terms, and solved using the Sequential-Tree Reweighted Message Passing algorithm [6].

Our objective is to parameterize $E(\mathbf{x}, I)$ by a set of parameters $\boldsymbol{\theta}$ which can be learnt from a set of training images. Next, we look at the problem of learning $\boldsymbol{\theta}$ in Sect. 2.3 followed by an appropriate definition of $E_\theta(\mathbf{x}, I)$ in Sect. 2.4.

2.3 The Structured Support Vector Machine Formulation

Let $\{I^{(k)}, \mathbf{x}^{(k)}\}_{k=1}^{K}$ denote a set of K training OCT image slices $I^{(k)}$ with corresponding ground truth (GT) labelling $\mathbf{x}^{(k)}$. Given a feasible labelling $\mathbf{x}$ for $I^{(k)}$, we define a loss function $\Delta(\mathbf{x}^{(k)}, \mathbf{x}) = \sum_{l=1}^{L}\sum_{n=1}^{N} \mid x_{l,n}^{(k)} - x_{l,n} \mid$ as sum of the unsigned distances between the corresponding boundary points across all layers.

The energy function $E_\theta(\mathbf{x}, I)$ must map all possible labellings $\mathbf{x}$ for each image I to energy values such that the correct labelling corresponds to the highest energy, i.e., $\mathbf{x}^{(k)} = argmax_{\mathbf{x}} E_\theta(\mathbf{x}, I^{(k)}), \forall k$. Moreover, E_θ must tend to assign higher energy scores to labellings with a lower loss Δ.

SSVM [3] is an extension of the Support Vector Machines that can be applied to solve the above problem under the assumption that E_θ is a linear function i.e., $E_\theta(\mathbf{x}, I) = \boldsymbol{\theta}^\top.F(\mathbf{x}, I)$ where $F(\mathbf{x}, I)$ is known as the joint feature function. It imposes $L2$ regularization on $\boldsymbol{\theta}$ to improve the generalization on unseen images by solving the following optimization,

$$\min_{\theta,\xi\geq 0} \quad \frac{\lambda}{2} \mid\mid \boldsymbol{\theta} \mid\mid^2 + \frac{1}{M}\sum_{k=1}^{M} \xi_k$$
$$\text{s.t.} \quad \boldsymbol{\theta}^\top.(F^{(k)}(\mathbf{x}^{(k)}) - (F^{(k)}(\bar{\mathbf{x}}^{(k)})) \geq \Delta(\mathbf{x}^{(k)}, \bar{\mathbf{x}}^{(k)}) - \xi_k \quad \forall k, \forall \bar{\mathbf{x}}^{(k)} \in \mathcal{Y}^{(k)}, \quad (2)$$

where ξ_k are the slack variables and λ is the regularization parameter which was fixed to 10^{-4} in our experiments. The constraints ensure that for each image, the GT labelling $\mathbf{x}^{(k)}$ has a higher score than all incorrect labellings $\bar{\mathbf{x}}^{(k)}$ by a margin scaled by the loss $\Delta(\mathbf{x}^{(k)}, \bar{\mathbf{x}}^{(k)})$. Here $\mathcal{Y}^{(k)}$ denotes the combinatorially large set of all incorrect labellings for I_k resulting in an extremely large $(\sum_k |\mathcal{Y}^{(k)}|)$ number of constraints making it difficult to be solved directly. In this work, we employed the Block Co-ordinate Frank Wolfe Algorithm [7] to solve Eq. 2 which replaces the $|\mathcal{Y}^{(k)}|$ constraints for each image I_k by the *most violating constraint*. We refer the readers to [7] for more details. The method used to find the most violating constraint (known as the max oracle) is problem specific and involves solving the following optimization problem: $y_k^* = argmax_x(\Delta(\mathbf{x}^{(k)}, \bar{\mathbf{x}}^{(k)}) + E_\theta(\mathbf{x}, I))$. Since in our case, $\Delta(\mathbf{x}^{(k)}, \bar{\mathbf{x}}^{(k)})$ is separable at each x_n^l, the max oracle is defined similar to the CRF inference problem in Eq. 1. with an additional term $\mid x_n^l - \bar{x}_n^l \mid$ added to the unary terms for each x_n^l.

2.4 Linear Parameterization of CRF Energy

We define the individual cost terms in Eq. 1 as the linear functions denoted by, $E_{bnd}(\mathbf{x}, I) = \mathbf{w}_{bnd}^\top . F_{bnd}(\mathbf{x}, I)$, $E_{intra}(\mathbf{x}, I) = \mathbf{w}_{intra}^\top . F_{intra}(\mathbf{x}, I)$ and $E_{inter}(\mathbf{x}, I) = \mathbf{w}_{inter}^\top . F_{inter}(\mathbf{x}, I)$ respectively. Then, the net energy is given by $E_\theta(\mathbf{x}, I) = \boldsymbol{\theta}^\top . F(\mathbf{x}, I)$ where $F(\mathbf{x}, I) = [F_{bnd}^\top \, F_{intra}^\top \, F_{inter}^\top]^\top$ and $\boldsymbol{\theta}^\top = [\mathbf{w}_{bnd}^\top \, \mathbf{w}_{intra}^\top \, \mathbf{w}_{inter}^\top]$. The details of the individual cost terms are discussed below.

Boundary Cost: For each layer boundary l, we aim to learn a $p \times p$ convolution filter $\mathbf{u}_l$ which would have a high response only at the pixels lying on the lth boundary. Let $\mathbf{I_{l,n}}$ represent a $p \times p$ image patch centered at $(x_{l,n}, y_n)$. Both $\mathbf{u}_l$ and $\mathbf{I_{l,n}}$ are linearly indexed to $p^2 \times 1$ vectors so that their dot product gives the filter response at $x_{l,n}$. Thus, the boundary cost for each $x_{l,n}$ is defined as $\varepsilon_{bnd}^l(x_n^l) = \mathbf{u}_l^\top . \mathbf{I_{l,n}}$ and the total Boundary cost over the entire CRF is

$$E_{bnd}(\mathbf{x}, I) = \sum_{l=1}^{L} \sum_{n=1}^{N} \mathbf{u}_l^\top . \mathbf{I_{l,n}} = \sum_{l=1}^{L} \mathbf{u}_l^\top \sum_{n=1}^{N} \mathbf{I_{l,n}} = \mathbf{w}_{bnd}^\top . F_{bnd}(\mathbf{x}, I), \quad (3)$$

where $\mathbf{w}_{bnd}^\top = [\mathbf{u}_1^\top ... \mathbf{u}_L^\top]$ and $F_{bnd}(\mathbf{x}, I) = [\sum_{n=1}^{N} \mathbf{I_{l,n}} ... \sum_{n=1}^{N} \mathbf{I_{l,n}}]^\top$.

Pairwise Intra-layer Energy: The interaction between the adjacent points on the lth boundary consists of a shape and an appearance term. The shape prior $d_{intra}^{l,n}(x_{l,n}, x_{l,n+1}) = \exp\{\frac{-1}{2}.(\frac{(x_{l,n+1} - x_{l,n}) - \mu_{intra}^{l,n}}{\sigma_{intra}^{l,n}})^2\}$ preserves the boundary smoothness by penalizing large deviations of the signed height gradient $(x_{l,n+1} - x_{l,n})$ from the Gaussian distributions whose mean $\mu_{intra}^{l,n}$ and standard deviation $\sigma_{intra}^{l,n}$ are separately learnt from the GT of the training images for each layer l, and column y_n.

Emphasising the shape prior can lead to gross segmentation errors in the presence of abnormalities such as AMD which affect the boundary smoothness. Hence, an additional appearance-based term is introduced to favour the similarity in the appearance of the adjacent boundary points. $S(x_{l,n}, x_{l,n+1}) = 1 - min \sum_{k=1}^{255} min\{h_k(x_{l,n}, y_n), h_k(x_{l,n+1}, y_{n+1})\}$ is a histogram intersection based similarity measure between the normalized histograms h_k of $p \times p$ image patches centered at $(x_{l,n}, y_n)$ and $(x_{l,n+1}, y_{n+1})$ respectively.

The pairwise intra-layer energy is modelled as a linear combination of the shape and appearance terms, $\varepsilon_{intra}^{l,n} = \alpha_l . d_{intra}^{l,n}(x_{l,n}, x_{n+1}^l) + \beta_l . S(x_{l,n}, x_{l,n+1})$, where α_l and β_l are relative weight coefficients learnt automatically in an end-to end manner during training. Therefore, the total Intra-layer pairwise energy is

$$\begin{aligned} E_{intra}(\mathbf{x}, I) &= \sum_{l=1}^{L} \sum_{n=1}^{N-1} (\alpha_l . d_{intra}^{l,n}(x_{l,n}, x_{l,n+1}) + \beta_l . S(x_{l,n}, x_{l,n+1})) \\ &= \sum_{l=1}^{L} \alpha_l . (\sum_{n=1}^{N-1} d_{intra}^{l,n}(x_{l,n}, x_{l,n+1})) + \sum_{l=1}^{L} \beta_l . (\sum_{n=1}^{N-1} S(x_{l,n}, x_{l,n+1})) \\ &= \mathbf{w}_{intra}^{\top} . F_{intra}(\mathbf{x}, I). \end{aligned} \tag{4}$$

Here, $E_{intra}(\mathbf{x}, I)$ is linearized by taking $\mathbf{w}_{intra}^{\top} = [\alpha_1 \alpha_2 ... \alpha_L \beta_1 \beta_2 ... \beta_L]$ and $F_{intra}(\mathbf{x}) = [d^1 \, d^2 \dots d^L \, S^1 \, S^2 \dots S^L]^{\top}$, where $d^i = \sum_{n=1}^{N-1} d_{intra}^{i,n}(x_{i,n}, x_{i,n+1})$ and $S^i = \sum_{n=1}^{N-1} S(x_{i,n}, x_{i,n+1})$ respectively.

Pairwise Inter-layer Energy: It captures the interaction between the adjacent layer boundaries by employing a shape and an appearance based cost. The shape prior $d_{inter}^{l,n}$ penalizes the deviation of the layer thickness $(x_{l+1,n} - x_{l,n})$ from apriori learnt Gaussian distributions with mean $\mu_{inter}^{l,n}$ and standard deviation $\sigma_{inter}^{l,n}$ for each column y_n. Moreover, hard constraints are also imposed on the minimum T_{mn}^{l} and maximum T_{mx}^{l} layer thickness of each layer l by assigning ∞ to the infeasible labellings. Therefore,

$$d_{inter}^{l,n} = \begin{cases} \exp\{\frac{-1}{2} . (\frac{(x_n^{l+1} - x_n^{l}) - \mu_{inter}^{l,n}}{\sigma_{inter}^{l,n}})^2\}, & if \; T_{mn}^{l} \le (x_{l+1,n} - x_{l,n}) \le T_{mx}^{l} \\ \infty, \; otherwise. \end{cases} \tag{5}$$

The parameters $\mu_{inter}^{l,n}$, $\sigma_{inter}^{l,n}$, T_{mn}^{l} and T_{mx}^{l} are learnt for each layer from the GT of training images. $T_{mn}^{l} > 0$ ensures that the layer boundaries donot intersect.

Let R_l denote the tissue region between the l and $(l+1)^{th}$ boundary. To capture the appearance of each R_l, a $q \times q$ convolution filter $\mathbf{v}_l$ is defined for each of the $L-1$ regions such that $\mathbf{v}_l$ has a high *average response* given by $\frac{1}{|x_{l+1,n} - x_{l,n}|}(\sum_{j=x_{l,n}}^{x_{l+1,n}} \mathbf{v}_l^{\top} . \mathbf{I_{j,n}})$ in each column y_n within the layer R_l.

$\varepsilon_{inter}^{l,n}$ is defined as a linear combination of the appearance and shape cost terms, $\varepsilon_{inter}^{l,n} = \frac{1}{|x_{l+1,n} - x_{l,n}|}(\sum_{j=x_{l,n}}^{x_{l+1,n}} \mathbf{v}_l^{\top} . \mathbf{I_{j,n}}) + \gamma_l . d_{inter}^{l,n}(x_{l,n}, x_{l+1,n})$, where γ_l provides a relative weight between the two terms. Both $\mathbf{v}_l$ and γ_l are learnt in an end-to-end manner. Therefore, the total Inter-layer pairwise Energy is

$$\begin{aligned} E_{inter}(\mathbf{x}, I) &= \sum_{l=1}^{L-1} \sum_{n=1}^{N} \left\{ \frac{1}{| \, x_{l+1,n} - x_{l,n} \, |} \left(\sum_{j=x_{l,n}}^{x_{l+1,n}} \mathbf{v}_l^{\top} . \mathbf{I_{j,n}} \right) + \gamma_l . d_{inter}^{l,n} \right\} \\ &= \sum_{l=1}^{L-1} \mathbf{v}_l^{\top} \left\{ \sum_{n=1}^{N} \frac{1}{| \, x_{l+1,n} - x_{l,n} \, |} \sum_{j=x_{l,n}}^{x_{l+1,n}} \mathbf{I_{j,n}} \right\} + \sum_{l=1}^{L-1} \gamma_l \left\{ \sum_{n=1}^{N} d_{inter}^{l,n} \right\} \\ &= \mathbf{w}_{inter}^{\top} . F_{inter}(\mathbf{x}, I) \end{aligned} \tag{6}$$

Here, $E_{inter}(\mathbf{x}, I)$ is linearized by taking $\mathbf{w}_{inter}^{\top} = [\mathbf{v}_1 \mathbf{v}_2 ... \mathbf{v}_{L-1} \gamma_1 \gamma_2 ... \gamma_{L-1}]$ and $F_{inter}(\mathbf{x}) = [\mathbf{r^1} \, \mathbf{r^2} \dots \mathbf{r^{L-1}} \, t^1 \, t^2 \dots t^{L-1}]^{\top}$, where $t^i = \sum_{n=1}^{N} d_{inter}^{i,n}(x_{i,n}, x_{i+1,n})$ and $\mathbf{r^i} = \sum_{n=1}^{N} \frac{1}{|x_{i+1,n} - x_{i,n}|} \sum_{j=x_{i,n}}^{x_{i+1,n}} \mathbf{I_{j,n}}$ respectively.

3 Results

Dataset: The proposed method was evaluated on the macular OCT B-scans of both Normal and AMD subjects using two datasets, *Normal* dataset [2] and *AMD* dataset [1] kindly provided by Chiu et al.

The *Normal* dataset consists of 107 B-scans obtained from 10 OCT volumes. The B-scans in 5 volumes are of size 300×800 pixels with a pixel resolution of $(3.23, 6.7, 67)$ µm/pixel along the axial, lateral and azimuthal directions. The *AMD* dataset is characterized by the presence of drusen and geographic atrophy. It consists of 220 B-scans sampled from 20 OCT volumes collected from 4 different clinics. All B-scans were of size 512×1000 pixels with pixel resolution varying in the range of (3.06–3.24, 6.50–6.60, 65–69.8) $\mu m/pixel$.

In both datasets, eleven B-scans are linearly sampled from the OCT volumes such that the sixth B-scan is centered at the fovea. The GT comprises of manual markings by a senior grader. While GT for all 8 layer boundaries (see Fig. 1a.) are available for the *Normal* dataset, GT of only 3 boundaries (see Fig. 1b.) are available for the *AMD* dataset.

Evaluation Metrics and Benchmarking: The accuracy of the proposed method was analyzed in terms of the boundary localization error (BLE) for each layer boundary, the Dice coefficient (DC) and the retinal Layer Thickness Error (LTE) for the segmented tissue regions. BLE is defined as the average (signed or unsigned) distance in pixels between the extracted boundary and the GT along each column (A-scan) in the image. DC measures the extent of overlap between the extracted and GT tissue regions. LTE is defined as the average absolute difference in the thickness (in pixels) between the extracted and GT tissue region across each column in the image. While DC provides a global measure of the accuracy, LTE is more sensitive to localized inaccuracies at each column. Ideally, BLE, LTE should be 0 and DC should be 1.

The proposed method was benchmarked against the results obtained from three publicly available OCT segmentation software: (A) CASEREL[1] based on [2], (B) the Iowa Reference Algorithm (IRA) based on [4] and (C) OCTSEG[2] based on [8]. Each software extracts different number of layers and the RPE_{in} boundary which is critical for AMD detection is only available in IRA. Additionally, the INL/OPL boundary is also unavailable in OCTSEG.

Experimental Setup: The experiments were carried out in two settings. In *Experiment-1*, the proposed method was evaluated on the task of the joint segmentation of all the 8 layer boundaries on the *Normal* dataset. In *Experiment-2*, a single CRF model was learnt on the combined *Normal* and *AMD* dataset for extraction of the three layers, ILM, RPE_{in} and RPE_{out} as the GT of only these three boundaries were available for the *AMD* dataset. In both experiments, y_n was sampled 4 pixels apart for computational tractability and the intermediate boundary points were obtained using b-spline interpolation. The size of both $\mathbf{u}_l$

[1] http://pangyuteng.github.io/caserel/.

[2] https://www5.cs.fau.de/research/software/octseg/.

and $\mathbf{v}_l$ filters were fixed to 11 × 11. A Matlab implementation of our method took around 9 seconds to process each B-scan on a i7 processor with 8 GB RAM.

Experiment-1: The proposed method was evaluated using a five-fold cross-validation on the *Normal* dataset. In each fold, the B-scans from two OCT volumes were used for testing while the B-scans from the remaining 8 volumes were used for training. The boundary and region based metrics for each layer are provided in Tables 1 and 2, respectively. The proposed method consistently outperforms other methods for all the 8 boundaries showing a 45% improvement in the average unsigned BLE (1.52/0.29) across all layers compared to IRA (2.80/0.91). This indicates the advantage of learning the energy in contrast to the handcrafted cost terms used in IRA based on [4]. Moreover, while IRA performs the segmentation in 2 steps, namely, the outer (1, 7, 8) layer boundaries followed by the inner layers (2–6), the proposed method extracts all the

Table 1. (mean/std.) of BLE (in pixels) for 8 layer boundaries on the Normal dataset

	ILM	NFL/GCL	IPL/INL	INL/OPL	OPL/ONL	IS/OS	RPE_{in}	RPE_{out}
UnsignedError								
CASEREL	1.31/0.40	2.83/2.59	5.02/2.67	5.42/2.59	5.37/4.71	1.72/0.81	—	1.70/0.60
OCTSEG	1.87/3.19	6.56/2.92	3.89/3.92	—	3.48/2.99	1.11/1.49	—	1.11/1.69
IRA	3.10/0.99	3.39/1.89	2.40/1.15	4.06/1.30	3.33/1.21	1.08/0.48	2.19/0.84	2.84/1.06
Proposed	1.09/0.28	1.66/0.64	1.51/0.47	1.68/0.55	1.95/0.81	1.15/0.85	1.47/0.75	1.67/0.76
SignedError								
CASEREL	−1.02/0.59	1.36/3.14	−3.24/3.96	−3.00/4.26	−3.71/5.68	−0.98/1.04	—	0.13/1.03
OCTSEG	−0.36/2.97	0.42/5.32	2.73/1.54	—	1.54/3.35	−0.22/1.42	—	0.01/1.78
IRA	−2.99/1.04	−2.88/2.29	−2.10/1.36	−3.98/1.39	−2.54/1.98	0.72/0.70	0.90/2.02	−2.77/1.10
Proposed	−0.00/0.73	0.17/1.23	0.00/0.97	0.10/1.30	−0.06/1.63	0.28/1.03	0.01/1.37	−0.21/1.52

Table 2. (mean/std.) of LTE and Dice for 7 tissue regions on the Normal dataset. Ground Truth mean layer thickness is reported to provide a context for the LTE.

	NFL	GCL-IPL	INL	OPL	ONL-IS	OS	RPE
GT layer thickness							
	8.23	22.63	11.29	8.21	25.45	8.13	9.33
Avg. layer thickness err.							
CASEREL	3.31/2.84	5.52/2.51	2.33/0.62	5.03/1.52	5.23/4.30	—	—
OCTSEG	6.84/2.74	5.71/1.83	—	—	3.36/2.61	—	—
IRA	2.39/1.00	2.44/0.94	2.54/0.91	2.67/1.29	3.85/1.25	2.12/0.88	3.87/1.44
Proposed	1.97/0.67	2.06/0.59	1.85/0.53	2.32/0.96	2.26/1.08	1.73/0.77	1.83/0.85
Dice coeff.							
CASEREL	0.79/0.13	0.82/0.09	0.60/0.19	0.60/0.16	0.88/0.08	—	—
OCTSEG	0.61/0.14	0.79/0.13	—	—	0.91/0.06	—	—
IRA	0.68/0.11	0.88/0.07	0.71/0.11	0.63/0.11	0.92/0.03	0.82/0.06	0.70/0.07
Proposed	0.84/0.06	0.93/0.02	0.87/0.03	0.80./0.07	0.94/0.02	0.86/0.07	0.85/0.06

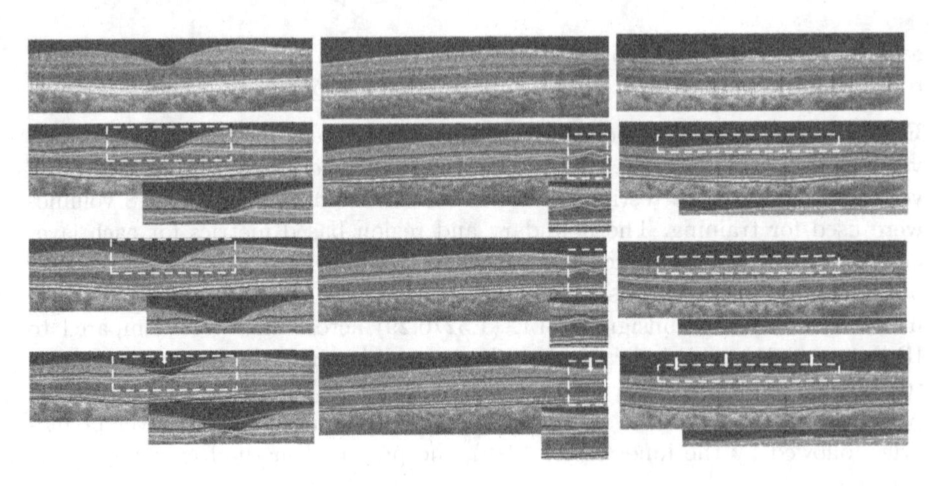

Fig. 3. Qualitative results for 7 layer segmentation on 3 normal images (each column). 1st row: input image, 2nd row: GT, 3rd row: proposed method, 4th row: IRA software.

8 boundaries in a single step, yet significantly outperforming IRA on the inner (*INL/OPL*, *OPL/ONL*) layers. Sample qualitative results are provided in Fig. 3. The proposed method was also compared against the manual markings provided by a second human expert on a subset of 28 images. The average unsigned BLE (1.60/0.71) across all 8 layers for this subset was found to be within the inter-observer variance of the 2nd expert (1.91/0.88).

Experiment-2: The results of a five-fold cross-validation of our method on the combined *Normal* and *AMD* dataset is reported in Table 3. In each fold, B-scans from 2 Normal and 4 AMD OCT volumes were used for testing while the B-scans

Table 3. Performance (mean/std.) for 3 layer boundaries on the combined dataset.

	Unsigned BLE (pixels)			LTE (pixels)	
	ILM	RPE_{in}	RPE_{out}	ILM-RPE_{in}	RPE_{in}-RPE_{out}
AMD dataset alone					
CASEREL	1.31/1.25	—	2.93/3.55	—	—
OCTSEG	5.20/8.09	—	3.35/3.13	—	—
IRA	3.75/5.22	4.45/5.45	3.51/5.44	5.48/2.64	4.47/2.34
Proposed	1.25/0.35	2.19/1.13	2.85/2.28	2.56/1.11	3.45/1.58
AMD + Normal dataset					
CASEREL	1.31/1.06	—	2.54/3.01	—	—
OCTSEG	4.07/7.01	—	2.75/2.85	—	—
IRA	3.53/4.30	3.68/4.59	3.28/4.48	5.07/2.43	4.27/2.10
Proposed	1.19/0.34	2.04/1.07	2.47/2.00	2.33/1.05	3.07/1.51

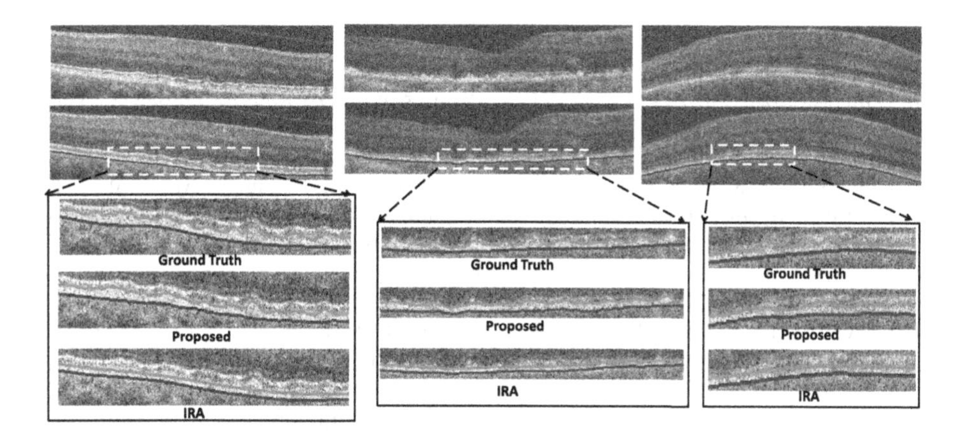

Fig. 4. Qualitative results on 3 AMD images (each column). 1st row: input image, 2nd row: segmentation using the proposed method. The cropped region is magnified for comparison with the GT and the segmentation of the IRA software.

of the remaining OCT volumes were used to learn the CRF energy. Among the available software, CASEREL is fine-tuned for optimal peformance on the AMD images. OCT-SEG and IRA cannot be trained on the combined dataset as they are unsupervised and employ handcrafted cost terms. In contrast, our method is trainable in an end-to-end manner and can be easily adapted to different abnormalities. A *single* model was learnt on the combined dataset.

CASEREL performs best for the ILM and RPE_{out} boundaries while the proposed method outperforms it marginally. Though, RPE_{in} boundary is most affected by AMD, currently, only IRA provides its segmentation and the proposed method outperforms it by $\frac{3.68-2.04}{3.68} \approx 44\%$ on the unsigned BLE. Qualitative results depicted in Fig. 4 indicate that the proposed method is sensitive to the irregularities in the RPE_{in} boundary in the presence of drusen.

4 Conclusion

Motivated by the need to learn an optimal energy function for intra-retinal layer segmentation in OCT images, we presented a LP-CRF model whose parameters are learnt using SSVM in an *end-to-end* manner. The proposed method outperforms the existing methods on the tasks of 8 layer segmentation in Normal images and 3 layer segmentation on a combined dataset of Normal and AMD images. The max-margin property of SSVM ensures good generalization of the learnt energy on unseen test set and the proposed method was found to be robust to variations in image scanner and resolution. The improvement in performance over the IRA software (that employs a similar energy function) demonstrates the effectiveness of learning the energy over handcrafting. While currently, the proposed method has been evaluated on macular OCT scans, future work will include an extension of the method to handle peripapillary OCT.

References

1. Chiu, S.J., Izatt, J.A., O'Connell, R.V., Winter, K.P., Toth, C.A., Farsiu, S.: Validated automatic segmentation of AMD pathology including drusen and geographic atrophy in SD-OCT images. IOVS **53**(1), 53–61 (2012)
2. Chiu, S.J., Li, X.T., Nicholas, P., Toth, C.A., Izatt, J.A., Farsiu, S.: Automatic segmentation of seven retinal layers in SDOCT images congruent with expert manual segmentation. Opt. Exp. **18**(18), 19413–19428 (2010)
3. Finley, T., Joachims, T.: Training structural SVMs when exact inference is intractable. In: ICML, pp. 304–311 (2008)
4. Garvin, M.K., Abramoff, M.D., Wu, X., Russell, S.R., Burns, T.L., Sonka, M.: Automated 3-D intraretinal layer segmentation of macular spectral-domain optical coherence tomography images. IEEE TMI **28**(9), 1436–1447 (2009)
5. Ishikawa, H., Stein, D.M., Wollstein, G., Beaton, S., Fujimoto, J.G., Schuman, J.S.: Macular segmentation with optical coherence tomography. IOVS **46**(6), 2012–2017 (2005)
6. Kolmogorov, V.: Convergent tree-reweighted message passing for energy minimization. IEEE PAMI **28**(10), 1568–1583 (2006)
7. Lacoste-Julien, S., Jaggi, M., Schmidt, M., Pletscher, P.: Block-coordinate frank-wolfe optimization for structural SVMs. In: ICML, pp. 53–61 (2013)
8. Mayer, M., Hornegger, J., Mardin, C.Y., Tornow, R.P.: Retinal nerve fiber layer segmentation on FD-OCT scans of normal subjects and glaucoma patients. Opt. Exp. **1**(5), 1358–1383 (2010)
9. Novosel, J., Thepass, G., Lemij, H.G., de Boer, J.F., Vermeer, K.A., van Vliet, L.J.: Loosely coupled level sets for simultaneous 3D retinal layer segmentation in optical coherence tomography. Med. Image Anal. **26**(1), 146–158 (2015)
10. Nyul, L.G., Udupa, J.K., Zhang, X.: New variants of a method of MRI scale standardization. IEEE TMI **19**(2), 143–150 (2000)
11. Rossant, F., Bloch, I., Ghorbel, I., Paques, M.: Parallel double snakes. Application to the segmentation of retinal layers in 2D-OCT for pathological subjects. Pattern Recogn. **48**(12), 3857–3870 (2015)
12. Shi, F., Chen, X., Zhao, H., Zhu, W., Xiang, D., Gao, E., Sonka, M., Chen, H.: Automated 3-D retinal layer segmentation of macular optical coherence tomography images with serous pigment epithelial detachments. IEEE TMI **34**(2), 441–452 (2015)
13. Song, Q., Bai, J., Garvin, M.K., Sonka, M., Buatti, J.M., Wu, X.: Optimal multiple surface segmentation with shape and context priors. IEEE TMI **32**(2), 376–386 (2013)
14. Yazdanpanah, A., Hamarneh, G., Smith, B.R., Sarunic, M.V.: Segmentation of intra-retinal layers from optical coherence tomography images using an active contour approach. IEEE TMI **30**(2), 484–496 (2011)
15. Yu, Y., Acton, S.T.: Speckle reducing anisotropic diffusion. IEEE TIP **11**(11), 1260–1270 (2002)

Superpixel-Based Line Operator for Retinal Blood Vessel Segmentation

Tong Na[1,2], Yitian Zhao[2(✉)], Yifan Zhao[3], and Yue Liu[2]

[1] Georgetown Preparatory School, North Bethesda 20852, USA
[2] Beijing Engineering Research Center of Mixed Reality and Advanced Display, School of Optoelectronics, Beijing Institute of Technology, Beijing, China
yitian.zhao@bit.edu.cn
[3] EPSRC Centre for Innovative Manufacturing in Through-life Engineering Services, Cranfield University, Cranfield, UK

Abstract. Automated detection of retinal blood vessels plays an important role in advancing the understanding of the mechanism, diagnosis and treatment of cardiovascular disease and many systemic diseases. Here, we propose a new framework for precisely segmenting vasculatures. The proposed framework consists of two steps. Inspired by the Retinex theory, a non-local total variation model is introduced to address the challenges posed by intensity inhomogeneities and relatively poor contrast. For better generalizability and segmentation performance, a superpixel based line operator is proposed as to distinguish between lines and the edges, and thus allows more tolerance in the position of the respective contours. The results on three public datasets show superior performance to its competitors, implying its potential for wider applications.

Keywords: Vessel · Segmentation · Total variation · Retinex · Superpixel · Line operator

1 Introduction

The accurate detection of retinal vessels is essential for many clinical applications to support early detection, diagnosis and optimal treatment. Manual annotation of vascular structure is an exhausting task for graders, and computer-aided automatic/semi automatic vascular detection methods can significantly reduce the amount of time. However, many factors cause inaccuracy in vessel segmentation, including poor contrast, noise and pathologies such as micro-aneurysms, hemorrhages, and exudate.

Over the past two decades, a tremendous amount of vessel segmentation methods have been developed for different types of medical images. Numerous fully automated, semi-automated methods have been proposed, as evidenced by extensive reviews [1–3]. In general, all established automated segmentation methods may be categorized as either supervised segmentation [4–9] or unsupervised segmentation [3,10–15] regarding the overall system design and architecture.

M. Valdés Hernández and V. González-Castro (Eds.): MIUA 2017, CCIS 723, pp. 15–26, 2017.
DOI: 10.1007/978-3-319-60964-5_2

Unsupervised segmentation refers to methods that achieve the segmentation of blood vessels without using training data or explicitly using any supervised classification techniques [16]. This category includes most segmentation techniques in the literature, such as active contour models [13,22], wavelets [14], line operator [10] and our new framework, as described in this paper. In contrast, supervised methods [4–8,17,28] require a manually annotated set of training images for classifying a pixel either as vessel or non-vessel. Most of these methods in supervised catergory use Support Vector Machine, AdaBoost, Neural Networks, Conditional Random Field, etc.

However, the computer-aided vessel segmentation have yet to completely solve the challenging problems, such as posed by the high degree of anatomical variation across the population, and to the increasing complexity of the surrounding tissue and varying scales of vessels within an image. Moreover, artifacts during image acquisition, such as noise, poor contrast and low resolution, exacerbate this problem.

In this paper, we proposed a novel vessel segmentation framework. It comprises two main phases: a non-local total variation regularized intensity inhomogeneity correction, and superpixel based line operator segmentation model. The contributions of this work may be summarized as three folds: **(1)** A Retinex-based inhomogeneity correction method is introduced to normalize the imbalance illumination. When it is extended to vessel image intensity inhomogeneity correction, it has showed good performance and facilitates the subsequent processes. **(2)** The sensitivity for the detection of vessels is significantly improved after the superpixel adapted to the line operator. **(3)** The proposed segmentation framework achieves the best performance in the comparison studies on three publically available datasets.

2 Method

In this section, we describe the proposed method for the extraction of vessels by using Retinex-based inhomogeneity correction and superpixel enabled line operator segmentation. The main steps of our approach are illustrated in Fig. 1.

2.1 Retinex-Based Inhomogeneity Correction

The retinal images acquired with a fundus camera sometimes have poor contrast due to too strong or too low illumination conditions, it usually inherited from image acquisition. To this end, an inhomogeneity correction method is proposed to handle these problems in this paper.

The Retinex theory has been successfully adopted to computer vision field [18], in order to remove unfavorable illumination effects from images to improve their quality and contrast. The Retinex theory shows that any given image $\mathbf{I}$ can be modeled as a component-wise multiplication of two components, the reflectance $\mathbf{R}$ and the illumination $\mathbf{L}$: $\mathbf{I} = \mathbf{L} * \mathbf{R}$. Typically, $\mathbf{R}$ reveals the reflectance of the object of interest more objectively, and can thus be regarded

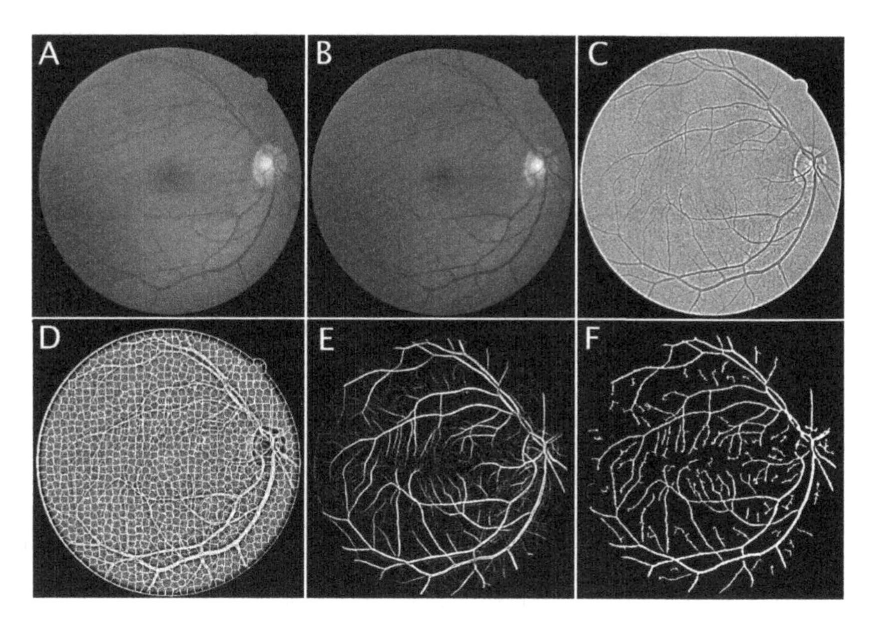

Fig. 1. Overview of the main steps of our method: (A) A random selected color fundus image; (B) The green channel of (A); (C) Results after applying Retinex on (B); (D) Superpixelized results of (C); (E) Vessel response of the proposed method; (F) Segmentation result by the proposed method.

as the enhanced image **I**. A look-up-table log operation can transfer this multiplication into an addition, resulting in $i = \log(\mathbf{I}) = \log(\mathbf{L}) + \log(\mathbf{R}) = l + r$ [18]. Clearly, the recovery of l or r is an ill-posed inverse image decomposition problem.

In this paper, a non-local total variation (TV) regularized model supporting the Retinex theory is adopted. It is very effective that the TV regularizer in recovering edges of images [19]. Such phenomenon coincides with the partial differential equation based Retinex method: the reflectance corresponds to the sharp details in the image and the illumination is spatially smooth. The non-local TV regularized model can be formulated as an energy minimization problem as:

$$\mathbf{R} = \arg\min_{l}\{t \int_{\Omega} |\nabla_w l| + \frac{1}{2}|\nabla(l - i)|_2^2\}, \tag{1}$$

where $l \leq i$. Here, $\int_{\Omega} |\nabla_w l|$ indicates the regularization term, and it is able to find the sharp details. $|\nabla(l - i)|_2^2$ is L_2 term of the gradient of the illumination, it ensures to smooth the illumination. t is the parameter to balance two terms. Ω is the support of the image. For a given image, the non-local weight between pixel $\mathbf{x}$ and $\mathbf{y}$ can be defined as

$$w(\mathbf{x}, \mathbf{y}) = \exp\left\{\frac{-K * (l(\mathbf{x}) - l(\mathbf{y}))^2}{2h^2}\right\}, \tag{2}$$

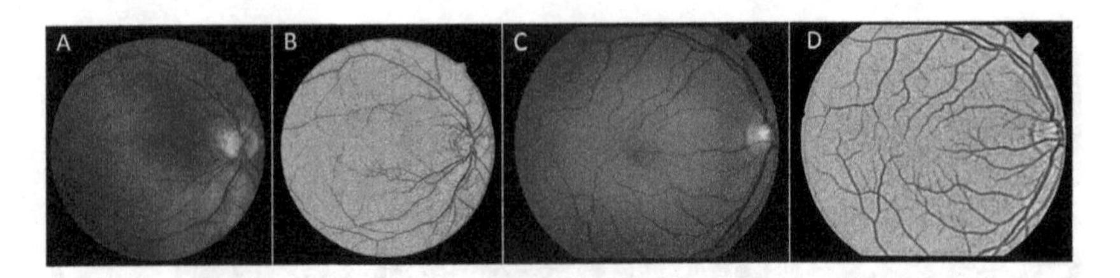

Fig. 2. Illustrative results of image enhancement by using non-local total variation based Retinex approach. (A) and (C): The green channel of two random selected color fundus image. (B) and (D): Results after applying Retinex on (A) and (C).

where K is the Gaussian kernel, and h is the control parameter. The non-local gradient operator at pixel $\mathbf{x}$ can be defined by the yielded non-local weights, as the vector of all partial difference $\nabla_w l(\mathbf{x}, \cdot)$:

$$\nabla_w l(\mathbf{x}, \mathbf{y}) = (l(\mathbf{y}) - l(\mathbf{x}))\sqrt{w(\mathbf{x}, \mathbf{y})}, \forall \mathbf{y} \in \Omega. \tag{3}$$

Hence, the non-local TV regularizer can be defined as

$$\int_{\Omega} |\nabla_w l| = \int_{\Omega} (\int_{\Omega} (l(\mathbf{y}) - l(\mathbf{x}))^2 w(\mathbf{x}, \mathbf{y})\, \mathrm{d}\mathbf{y})^{\frac{1}{2}}\, \mathrm{d}\mathbf{x}. \tag{4}$$

Figure 2 shows two enhanced results produced by applying the non-local TV based Retinex model. It has successfully corrected the contrast between vessels and background well, as well as the region of optic disc. In consequence, the vessels are more easily identifiable.

2.2 Superpixel-Based Line Operator

The basic line operator considers 12 angles, and the angular resolution is 15 degree. The largest average grey level $\mathbf{L}$ is found, which the pixel lies on a line passing through the target pixel. Then the line strength of the pixel is defined as

$$\mathbf{S}(i) = \mathbf{L}(i) - \mathbf{N}(i), \tag{5}$$

where $\mathbf{N}(i)$, is the average grey-level of a square window, centered on the target pixel i, with edge length equal to μ. The winning line is aligned within a vessel if the line strength is large, while the line strength is lower if the line is partially overlapped. In general, the length μ is empirically chosen, such as 15 in [10], and 5 in [20].

However, it usually has varying scales of vessels within an image, and a single value of μ always yield imbalance responses on the vessels. Therefore, in order to achieve better segmentation performance, in this work we applied a modified line operator on the superpixel generated patches rather than on entire image, in particular in regions with low signal noise ratio. The length μ was set to be half of the minimum object length of corresponding superpixel.

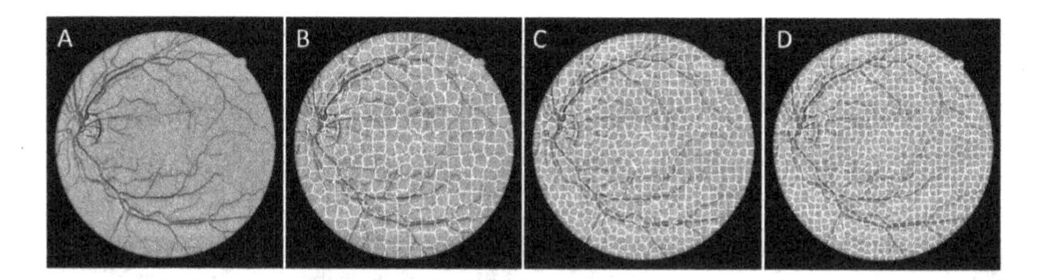

Fig. 3. Illustration of different superpixel numbers generated on an example image: (A) The green channel of a random selected color fundus image; (B) 400 superpixels; (C) 800 superpixels; (D) 1200 superpixels.

To achieve this, we first generate the superpixel upon the vesselness map. The SLIC superpixel algorithm [21] is adapted to replace the rigid structure of the pixel grid. The SLIC is a k-means clustering based method, and is able to assign each pixel to a superpixel according to their intensities and spatial locations. The superpixel clustering procedure starts with the generation of initial cluster centers. Then a distance measure D to cluster centers for all pixels is defined, aims to associate to their nearest cluster centers. The Euclidean distance (d_c) and spatial distance (d_s) are used to define this measure:

$$D = \sqrt{d_c^2 + (\frac{d_s}{S})^2 m^2}, \tag{6}$$

where $S = \sqrt{N/k}$ is the grid interval. k is the desired superpixel number and N is the total number of pixels. m indicates a parameter to balance the weighting of intensity and coordinates. Figure 3 shows an example of superpixel representation, with 400, 800, and 1200 superpixels, respectively.

Let $\mathcal{P}_t \in T$ be a viable local representation as a superpixel t $(t = 1, 2, \cdots, T)$, and let I indicate the input image. The line strength of the pixel in superixel $\mathcal{P}$ is defined as $\mathbf{S}_{\mathcal{P}_t}(i) = \mathbf{L}_{\mathcal{P}_t}(i) - \mathbf{N}_{\mathcal{P}_t}(i)$. In practice, the line path is hard to be exactly matched the pixel grid, hereby, the line and region averages at arbitrary orientations are obtained by using nearest neighbour interpolation instead of bi-linear interpolation.

Multiscale analysis is also performed in this framework. The line strength of the pixel under multi-level superpixel is defined as

$$\mathbf{S}(i) = \frac{1}{P} \sum_{p=1}^{P} \mathbf{S}(i)(\mathcal{P}_t^p | i \in \mathcal{P}_t^p). \tag{7}$$

where P indicates the levels of superpixels that the input image is segmented to. Parameter tuning for optimal numbers of superpixels and levels (P and M) will be discussed in Sect. 4.2. The second column of Fig. 4 demonstrate the final vessel responses of the proposed method. In order to extract the vessel from the response map, our previous proposed infinite perimeter active contour with hybrid region (IPACHR) method [23] is employed for its good performance. The

IPACHR uses an infinite perimeter active contour model for its effectiveness in detecting vessels with irregular and oscillatory boundaries. For more details, we refer readers to the original paper [23]. The third column of Fig. 4 depict the segmentation results.

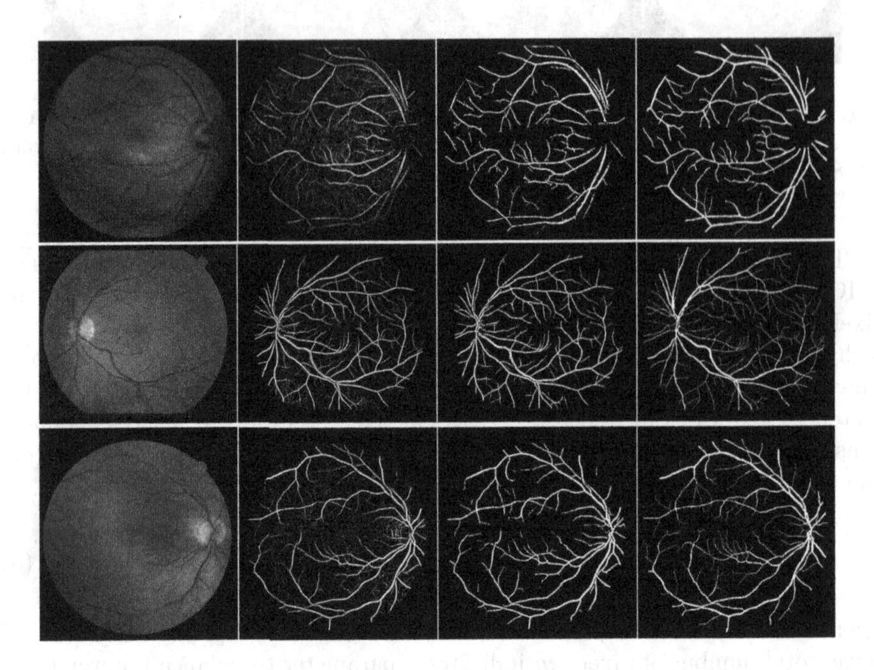

Fig. 4. Examples of vessel segmentation by the proposed method on 3 datasets. From left to right: green channel of random selected color fundus images, results after super-pixel enabled line operator, automated segmentation results, and manual annotations.

3 Datasets and Evaluation Metrics

Three publically available retinal datasets are used in this work to evaluate the proposed segmentation framework: STARE[1], DRIVE[2], and a newly released dataset IOSTAR[3]. The image resolutions of these datasets are 565×584, 700×605, and 1024×1024, respectively.

The segmentation performance is measured by sensitivity se, specificity sp, and accuracy acc. They are defined as $se = \frac{tp}{tp+fn}, sp = \frac{tn}{fp+tn}, acc = \frac{tp+tn}{tp+fp+tn+fn}$, respectively. Here, true positive tp is the count of pixels marked as vessel pixels in both the segmented image and its ground truth. Similarly,

[1] http://www.ces.clemson.edu/~ahoover/stare/.
[2] http://www.isi.uu.nl/Research/Databases/DRIVE/.
[3] http://www.retinacheck.org.

false positive fp identifies the number of incorrectly identified vessel pixels; true negative tn is the number of correctly identified non-vessel pixels; false negative fn indicates the number of incorrectly identified non-vessel pixels. In general, reporting the se and sp obtained at highest acc is a common way in the retinal image segmentation. However, it is possible to produces imbalanced results where a higher sp is favored since vessel has relatively lower amount than background. In such a case, acc will be skewed by the dominant classes. Consequently, in order to evaluate the performance of the proposed vessel segmentation method, the receiving operator characteristics (ROC) curve is computed with true positive ratio versus the false positive ratio. The area under the ROC curve (AUC) is calculated to quantify the performance of the segmentation, since it has the ability to reflect the trade-offs between the sensitivity and specificity.

4 Experimental Results

In this experiment, the green channel of the color fundus images were used for vessel segmentation. Figure 4 illustrates examples of vessel detection performance on three datasets, and manual annotation from observer 2 of the DRIVE and STARE dataset were used as groundtruth.

To reveal the relative performance of our proposed method, we compared it with several existing state-of-the-art vessel detection methods on the most popular datasets: DRIVE and STARE. The results are shown in Table 1, and the chosen methods have been ordered by the category the methods belonging to: the most recent seven supervised methods [4–8,17,28], and nine unsupervised segmentation methods [3,10,14,15,23–27]. Overall, our framework yields state-of-the-art performance and outperforms most methods reported in most of the quality metrics used, as it took into account the global features through the Retinex analysis and the local features through the superpixel-based line operator, therefore, more fine vessels may be detected. For accurate analysis of the DRIVE dataset, the proposed method yields the highest sensitivity, accuracy, and AUC among unsupervised method, and only the sensitivity is lower than the supervised method proposed by Orlando et al. [17]. Note, to the best knowledge of the authors, only Zhang et al. [27] has tested their segmentation method on IOSTAR dataset. In consequence, we only compared with the performance obtained by [27] in the bottom of Table 1, and is by no means exhaustive. In contrast, our method has better performance in terms of all metrics.

Furthermore, three state-of-the-art vessel enhancement methods were employed for comparison purposes. These methods were: isotropic undecimated wavelet filter [14], local phase filter [29] and Combination Of Shifted Filter Responses (BCOSFIRE) [15]. In the interests of reproducibility, the recommended parameters in the literature were used in the experiments. In Fig. 5, we show examples of applying different enhancement methods on a representative patch with multiple vascular bifurcations, curvature changes, intensity inhomogeneity on large vessel and low intensities on tiny vessels. Overall, the proposed method is not only able to detect the vessel regions, but also has the

Table 1. Performance of different segmentation methods, in terms of sensitivity (*se*), specificity (*sp*), accuracy (*acc*) area under the curve (*AUC*), on the **DRIVE**, **STARE** datasets, and **IOSTAR**.

Method	DRIVE				STARE			
	se	sp	acc	AUC	se	sp	acc	AUC
Second observer	0.776	0.972	0.947	0.874	0.895	0.938	0.934	0.917
Supervised methods								
Staal [4]	-	-	0.946	0.952	-	-	0.951	0.961
Soares [5]	0.733	0.782	0.946	0.961	0.721	0.975	0.948	0.967
Lupascu [6]	0.720	-	0.959	0.956	-	-	-	-
You [7]	0.741	0.975	0.943	-	0.726	0.975	0.949	-
Marin [8]	0.706	0.980	0.945	0.959	0.694	0.981	0.952	0.977
Li [28]	0.757	0.982	0.953	0.974	0.773	0.984	0.963	0.988
Orlando [17]	0.789	0.968	-	-	0.768	0.974	-	-
Unsupervised methods								
Ricci [10]	-	-	0.963	0.960	-	-	0.968	0.965
Palomera-Perez [24]	0.660	0.961	0.922	-	0.779	0.940	0.924	-
Fraz [3]	0.715	0.976	0.943	-	0.731	0.968	0.944	-
Bankhead [14]	0.703	0.971	0.937	-	0.758	0.950	0.932	-
Zhao [23]	0.742	0.982	0.954	0.862	0.780	0.978	0.956	0.874
Yin [25]	0.725	0.979	0.940	-	0.854	0.942	0.933	-
Roychowdhury [26]	0.740	0.978	0.949	0.967	0.732	0.984	0.956	0.967
Azzopardi [15]	0.766	0.970	0.944	0.961	0.772	0.970	0.950	0.956
Zhang [27]	0.747	0.976	0.947	0.952	0.768	0.976	0.955	0.961
Proposed method	**0.768***	**0.970**	**0.954***	**0.970***	**0.781**	**0.977**	**0.957***	**0.968***
	IOSTAR							
Zhang [27]	0.755	0.974	0.951	0.962				
Proposed method	0.761	0.975	0.955	0.964				

ability to suppress noise and artifacts. In other words, the results obtained by the proposed method seem more pleasing: stronger enhancement results on tiny vessels, better responses on bifurcations/crossovers, and higher uniformity on intensity inhomogeneity.

4.1 The Effectiveness of Superpixel and Retinex

In this section, the effectiveness of line operator enabled with superpixel and Retinex based image enhancement are validated individually.

Figure 6 demonstrates the segmentation results obtained by the proposed models with and without superpixel enabled. It can be observed from Fig. 6(C) that superpixel contributes significantly to the final performance - more tiny vessels have been detected, and helps to improve the sensitivity of the vessel segmentation. This observation is also confirmed by the ROC curves over three

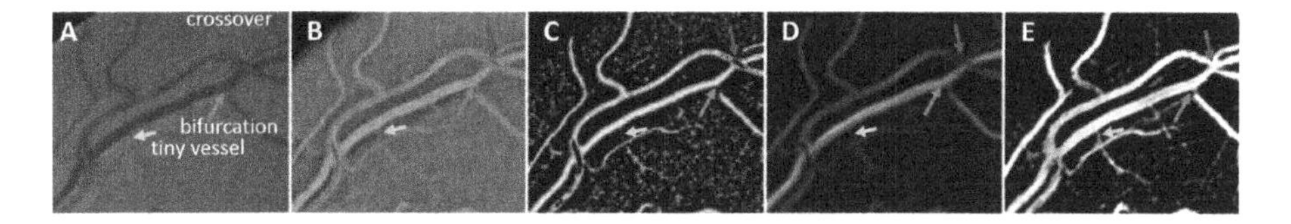

Fig. 5. A comparative study with other enhancement techniques on a selected region with tiny vessel (yellow arrow), bifurcation (green arrow), and crossover (red arrow). (A) The green channel of a selected region of a color fundus image. (B) isotropic undecimated wavelet filter. (C) Local phase. (D) BCOSFIRE. (E) Proposed method. (Color figure online)

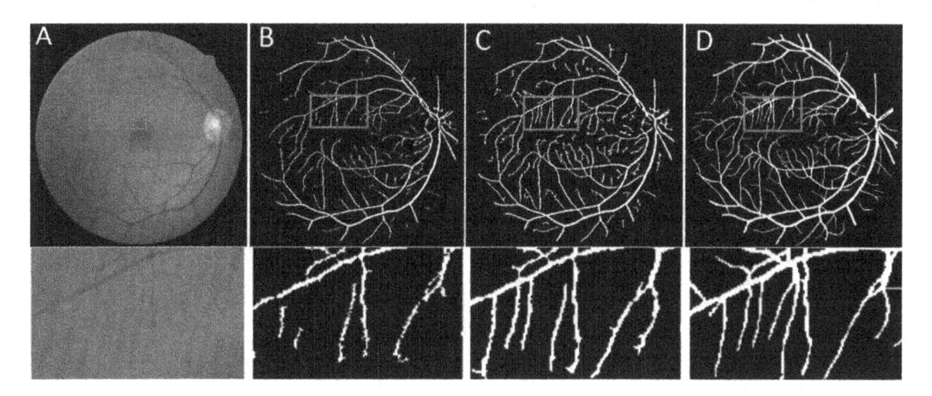

Fig. 6. Segmentation results of the proposed method, and the snapshot of selected region with small vessels. (A) Original image. (B) Segmentation result without superpixel applied. (C) Segmentation result with superpixel applied. (D) Groundtruth. (Color figure online)

different datasets, as illustrated at Fig. 7 (red line). Most existing line operator based segmentation approaches have a certain edge length μ, such as 15 pixels in [10], and 5 pixels in [20]. In this work, the edge length is self-adapting, and it is more sensitive to capture the varying scales of vessels within an image, and this leads to higher se, acc, and AUC.

In addition, the ROC curves of the proposed method with or without Retinex enhancement applied are illustrated at Fig. 7 (green line). Overall, Retinex process affects the final performance significantly, since the optic disk and foveal area always have inhomogeneous intensities, and these inhomogeneities were corrected after Retinex applied. In contrast, the segmentation performances were relative poorer in dataset STARE and IOSTAR than DRIVE when without Retinex applied. That is because STARE and IOSTAR dataset contain some images with pathologies, e.g. presents bright lesions or exudates, blurring vessel, and these features cause more false detections (lower sp). While the proposed Retinex method is capable to normalize these regions to a similar level with the background, and increase the contast between the vessels and background, as thus to avoid the false detection (higher sp), and raise the sensitivity score.

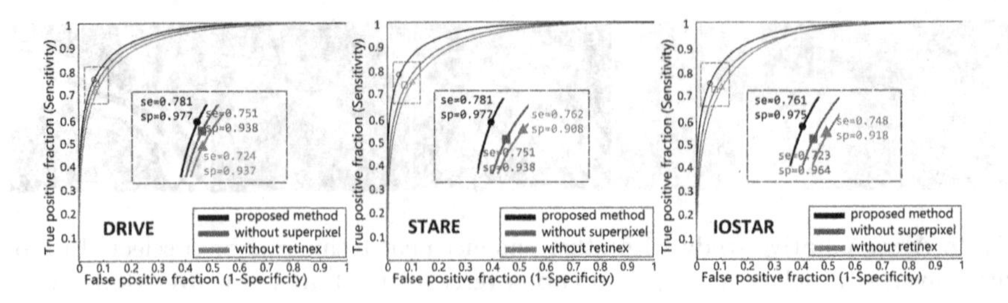

Fig. 7. The ROC curves of the proposed framework with and without the Retinex enhancement applied and superpixel-based the line operator applied over three different datasets respectively. (The reader is referred to the color version of this figure.)

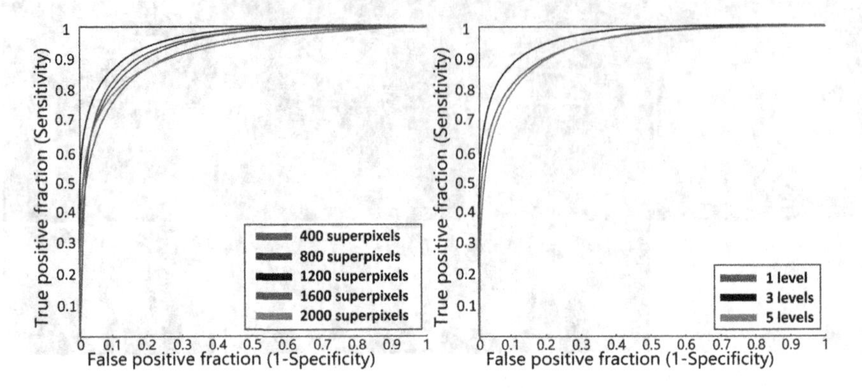

Fig. 8. The ROC curves of the proposed method with (left) different numbers of superpixels: 400,800, 1200, 1600, and 2000; (right) different numbers of levels, after setting the optimal number of the superpixels to 1200. (The reader is referred to the color version of this figure.)

4.2 Parameters Tuning

In this section, we experimentally investigate the suitable numbers of superpixels M and levels of superpixel partition P. It is known that too large number of superpixel leads to false detection, and on the contrary, too few superpixels result in a loss of the edge information of the vessel [30]. To this end, in this experiment, the numbers of superpixels were set to be successively 400, 800, 1200, 1600, and 2000. The left column of Fig. 8 shows the ROC curve of the proposed method under these numbers, and the proposed method achieves the best result when the superpixel number is 1200. As aforementioned at Sect. 2.2, multiscale analysis was used to detect vessel more precisely when an image contains varying scales of vessels. The right column of Fig. 8 shows the segmentation performance under different superpixel levels when the number of superpixels was set to 1200, and it can clearly be seen that the proposed method yields the best performance when the number of levels is 3. In consequence, the number of superpixels at the other two levels are $\frac{1}{3} \times 1200 = 400$, and $\frac{1}{3} \times 1200 = 800$.

5 Conclusions

In this paper, we have presented a new framework for vessel segmentation, which exploits the advantages of non-local total variation based Retinex model for intensity inhomogeneity correction, and superpixel-based line operator for vessel segmentation. Quantitative evaluations on publically-available datasets showed that, compared to established methods, the proposed method achieves competitive vessel segmentation performance. In particular, it shows better performance in handling small, bifurcation, and crossover vessels, even in the case of poor contrast. It has the potential to become a powerful tool for quantitative analysis of vasculature for the management of a wide range of diseases.

Acknowledgments. This work was supported by National Science Foundation Program of China (61601029, 61602322), China Association for Science and Technology (2016QNRC001), and National Key Research and Development Program of China (2016YFB0401202).

References

1. Kirbas, C., Quek, F.: A review of vessel extraction techniques and algorithms. ACM Comput. Surv. **36**, 81–121 (2004)
2. Lesagea, D., Funka-Leaa, G.: A review of 3D vessel lumen segmentation techniques: models, features and extraction schemes. Med. Image Anal. **13**, 819–845 (2009)
3. Fraz, M.M., Remagnino, P., Hoppe, A., Uyyanonvara, B., Rudnicka, A.R., Owen, C.G., Barman, S.A.: Blood vessel segmentation methodologies in retinal images - a survey. Comput. Meth. Prog. Bio. **108**, 407–433 (2012)
4. Staal, J., Abramoff, M.D., Niemeijer, M., Viergever, M.A., van Ginneken, B.: Ridge-based vessel segmentation in color images of the retina. IEEE Trans. Med. Imaging **23**, 501–509 (2004)
5. Soares, J., Cree, M.: Retinal vessel segmentation using the 2D Gabor wavelet and supervised classification. IEEE Trans. Med. Imaging **25**, 1214–1222 (2006)
6. Lupascu, C.A., Tegolo, D., Trucco, E.: FABC: retinal vessel segmentation using AdaBoost. IEEE Trans. Inf. Technol. Biomed. **14**, 1267–1274 (2010)
7. You, X., Peng, Q., Yuan, Y., Cheung, Y., Lei, J.: Segmentation of retinal blood vessels using the radial projection and semi-supervised approach. Pattern Recogn. **44**, 2314–2324 (2011)
8. Marin, D., Aquino, A., Gegundez-Arias, M.E., Bravo, J.M.: A new supervised method for blood vessel segmentation in retinal images by using gray-level and moment invariants-based features. IEEE Trans. Med. Imaging **30**, 146–158 (2011)
9. Wang, Y., Ji, G., Lin, P., Trucco, E.: Retinal vessel segmentation using multi-wavelet kernels and multiscale hierarchical decomposition. Pattern Recogn. **46**, 2117–2133 (2013)
10. Ricci, E., Perfetti, R.: Retinal blood vessel segmentation using line operators and support vector classification. IEEE Trans. Med. Imaging **26**, 1357–1365 (2007)
11. Mendonça, A., Campilho, A.C.: Segmentation of retinal blood vessels by combining the detection of centerlines and morphological reconstruction. IEEE Trans. Med. Imaging **25**, 1200–1213 (2007)

12. Martinez-Perez, M., Hughes, A., Thom, S.A., Bharath, A.A., Parker, K.H.: Segmentation of blood vessels from red-free and fluorescein retinal images. Med. Image Anal. **11**, 47–61 (2007)
13. Al-Diri, B., Hunter, A., Steel, D.: An active contour model for segmenting and measuring retinal vessels. IEEE Trans. Med. Imaging **28**, 1488–1497 (2009)
14. Bankhead, P., McGeown, J., Curtis, T.: Fast retinal vessel detection and measurement using wavelets and edge location refinement. PLoS ONE **7**, e32435 (2009)
15. Azzopardi, G., Strisciuglio, N., Vento, M., Petkov, N.: Trainable COSFIRE filters for vessel delineation with application to retinal images. Med. Image Anal. **19**, 46–57 (2015)
16. Lathen, G., Jonasson, J., Borga, M.: Blood vessel segmentation using multi-scale quadrature filtering. Pattern Recogn. Lett. **31**, 762–767 (2010)
17. Orlandp, J., Prokofyeva, E., Blaschko, M.: A discriminatively trained fully connected conditional random field model for blood vessel segmentation in fundus images. IEEE Trans. Biomed. Eng. **64**, 16–27 (2017)
18. Elad, M.: Retinex by two bilateral filters. Scale Space PDE Methods Comput. Vis. **3459**, 217–229 (2005)
19. Ng, M.K., Wang, W.: A total variation model for retinex. SIAM J. Imaging Sci. **4**, 345–365 (2011)
20. Zwiggelaar, R., Astley, S., Boggis, C., Taylor, C.: Linear structures in mammographic images: detection and classification. IEEE Trans. Med. Imaging **23**, 1077–1086 (2004)
21. Achanta, R., Shaji, A., Smith, K., Lucchi, A., Fua, P.: Slic superpixels compared to state-of-the-art superpixel methods. IEEE Trans. Pattern Anal. Mach. Intell. **34**, 2274–2282 (2012)
22. Zhao, Y., Zhao, J., Yang, J., Liu, Y., Zhao, Y., Zheng, Y., Xia, L, Wang, Y.: Saliency driven vasculature segmentation with infinite perimeter active contour model. Neurocomputing (2017). http://dx.doi.org/10.1016/j.neucom.2016.07.077
23. Zhao, Y., Rada, L., Chen, K., Zheng, Y.: Automated vessel segmentation using infinite perimeter active contour model with hybrid region information with application to retinal images. IEEE Trans. Med. Imaging **34**, 1797–1807 (2015)
24. Palomera-Prez, M., Martinez-Perez, M., Bentez-Prez, H., Ortega-Arjona, J.L.: Parallel multiscale feature extraction and region growing: application in retinal blood vessel detection. IEEE Trans. Inf. Technol. Biomed. **14**, 500–506 (2010)
25. Yin, Y., Adel, M., Bourennane, S.: Retinal vessel segmentation using a probabilistic tracking method. Pattern Recogn. **45**, 1235–1244 (2012)
26. Roychowdhury, S., Koozekanani, D., Parhi, K.: Iterative vessel segmentation of fundus images. IEEE Trans. Biomed. Eng. **62**(7), 1738–1749 (2015)
27. Zhang, J., Dashtbozorg, B., Bekkers, E., Pluim, P., Duits, B., Romeny, R.: Robust retinal vessel segmentation via locally adaptive derivative frames in orientation scores. IEEE Trans. Med. Imaging **35**(12), 2631–2644 (2016)
28. Li, Q., Feng, B., Xie, L., Liang, P., Zhang, H., Wang, T.: A crossmodality learning approach for vessel segmentation in retinal images. IEEE Trans. Med. Imaging **35**(1), 109–118 (2016)
29. Zhao, Y., Liu, Y., Zheng, Y.: Retinal vessel segmentation: an efficient graph cut approach with retinex and local phase. PLoS ONE **10**, e0122332 (2015)
30. Zhao, Y., Zheng, Y., Liu, Y., Yang, J., Zhao, Y., Chen, D., Wang, Y.: Intensity and compactness enabled saliency estimation for leakage detection in diabetic and malarial retinopathy. IEEE Trans. Med. Imaging **36**, 51–63 (2017)

Automatic Detection and Identification of Retinal Vessel Junctions in Colour Fundus Photography

Harry Pratt[1(✉)], Bryan M. Williams[1], Jae Ku[1,3], Frans Coenen[2], and Yalin Zheng[1,3]

[1] Department of Eye and Vision Science, Institute of Ageing and Chronic Disease, University of Liverpool, Liverpool L7 8TX, England
{h.pratt,yzheng}@liverpool.ac.uk
[2] Department of Computer Science, University of Liverpool, Liverpool L69 3BX, England
[3] St Paul's Eye Unit, Liverpool Royal University Hospital, Liverpool L7 8XP, England
http://pcwww.liverpool.ac.uk/~yzheng

Abstract. The quantitative analysis of retinal blood vessels is important for the management of vascular disease and tackling problems such as locating blood clots. Such tasks are hampered by the inability to accurately trace back problems along vessels to the source. This is due to the unresolved challenge of distinguishing automatically between vessel branchings and vessel crossings. In this paper, we present a new technique for tackling this challenging problem by developing a convolutional neural network approach for first locating vessel junctions and then classifying them as either branchings or crossings. We achieve a high accuracy of 94% for junction detection and 88% for classification. Combined with work in segmentation, this method has the potential to facilitate automated localisation of blood clots and other disease symptoms leading to improved management of eye disease through aiding or replacing a clinicians diagnosis.

Keywords: Convolutional neural networks · Retinal imaging · Retinal vessels fundus photography · Vessel classification

1 Introduction

Vascular conditions present a challenging public health problem. They are often life-threatening and damage to blood vessels can lead to significant complications such diabetes, hypertension and stroke. The retina is the only inner organ which can be directly imaged and also serve as a window for the diagnosis of systematic diseases such as cerebral malaria, stroke, dementia and cardiovascular diseases [10]. It is therefore of great importance to better understand and be able to manage such conditions. It is also significant that pathologies can affect veins and

M. Valdés Hernández and V. González-Castro (Eds.): MIUA 2017, CCIS 723, pp. 27–37, 2017.
DOI: 10.1007/978-3-319-60964-5_3

arteries differently. For example, in diabetic retinopathy, there is veinus beading. With the availability of imaging techniques such as colour fundus photography, fundus angiography and optical coherence tomography angiography, there has been a significant need for automated vessel analysis techniques [14,15].

There has been a considerable amount of work in recent years aimed at the effective segmentation of retinal blood vessels in fundus photography, which is a fundamental step for a vessel analysis system. Work such as [2,14,15] has been able to achieve increasingly improved segmentation of retinal vessels. However, a significant remaining challenge is to distinguish between vessel branchings and vessel crossings, where one blood vessel passes over another but does not connect to it. This is important for tracking vessels, separating veins from arteries and with occlusions.

When a blood clot needs to be located, we must be able to trace back along the vessel. The current inability to accurately identify vessel crossings in vessel segmentations hinders this. It is also important to monitor progress after vein and artery occlusions; being able to identify and distinguish vessel crossings and branchings facilitates this. Automating the detection and classification of vessel junctions also allows us to aid clinicians in detecting vascular abnormalities.

In this paper, we present a new hierarchical approach to first automatically determine the locations of blood vessel junctions in colour fundus images and then distinguish between vessel branchings and crossings. We employ an available segmentation of the vessel structure, although an automatic segmentation procedure could be incorporated, to identify points along blood vessels. We then develop a convolutional neural network which is trained on expert annotated data to identify vessel junctions. The same network architecture is then used and trained to learn new convolution filters to distinguish between vessel branchings and crossings. This results in a method which is capable of identifying and classifying vessel junctions without user intervention.

For applications in image analysis and classification, Convolutional Neural Networks (CNNs), a branch of deep learning, has achieved state of the art results for many problems. The 1970's saw the introduction of network architectures being used to analyse image data [4]. These had useful applications and allowed challenging tasks such as handwritten character recognition [3] to be achieved. Decades later, there were several breakthroughs in neural networks that lead to vast improvements in their implementation such as the introduction of dropout [13] and rectified linear units [11]. These theoretical enhancements and the accompanying increase in computing power through graphical processor units (GPUs) meant that CNNs became viable for more complex image recognition problems. Presently, large CNNs are used to successfully tackle highly complex image recognition tasks with many object classes to an impressive standard. CNNs are used in many of the current state-of-the-art image classification tasks including medical imaging. Hence, we use this method combined with expert segmented fundus images and skeletonisation [5,12] to detect and classify vessel junctions within fundus images.

There are many different architectures for neural networks. Recently residual networks have achieved impressive results on the highly competitive competition of ImageNet detection, ImageNet localisation, COCO detection, and COCO segmentation [6]. They were then widely used in the following 2016 ImageNet competition due to impressive performance on general large data sets of small images such as the MNIST [9] dataset for handwritten digits 0–9 and CIFAR-10 [8] a dataset of 10 classes of color images. This makes the network ideal for our patch based method. Hence, the Res18 network structure containing 18 residual layers is used in the CNNs throughout this paper.

The rest of this paper is organised as follows. In Sect. 2, we present our new method for locating and identifying crossings and branchings of retinal vessels, in Sect. 3 we present our experimental results and in Sects. 4 and 5 we discuss this work and present our conclusions.

2 Methods

The images used to implement out method are from the Digital Retinal Images for Vessel Extraction (DRIVE) database with manual segmentations [14]. The images in the DRIVE dataset were obtained from a diabetic retinopathy screening program in The Netherlands. The images were acquired using a Canon CR5 non-mydriatic 3CCD camera with a 45° field of view (FOV) using 8 bits per color plane at 768 by 584 pixels.

Our framework consists of identifying patches of fundus images $z(\mathbf{x})$ and identifying those patches which include junctions, and then distinguishing the type of junction located. We make use of available vessel segmentations given as binary functions $\phi(\mathbf{x})$ defined on the domain $\mathbb{R} \times \mathbb{R}$. In practice, we are dealing with the discrete counterparts $\mathbf{z}$ and ϕ of the image and segmentation function respectively defined over the discrete domain $\Omega \subset \mathbb{Z} \times \mathbb{Z}$.

2.1 Skeletonisation and Patch Extraction

We consider patches of the fundus images centred along the segmented vessels. In order to restrict the number of patches for training to a manageable amount and reduce bias, we aim to reduce to segmentation of the vessels to a skeleton and consider regions centred only on these points. We achieve this by performing a skeletonisation of the level set function $\phi(\mathbf{x})$ for each image. We convolve the level set function with the kernels

$$\kappa_1^j(\alpha^1) = r_j \begin{pmatrix} 0 & 0 & 0 \\ \alpha_1^1 & 1 & \alpha_2^1 \\ 1 & 1 & 1 \end{pmatrix}, \quad \kappa_2^j(\alpha^2) = r_j \begin{pmatrix} \alpha_1^2 & 0 & 0 \\ 1 & 1 & 0 \\ \alpha_2^2 & 1 & \alpha_3^2 \end{pmatrix}, \tag{1}$$

where r_j denotes rotation of the matrix by a multiple j of $\pi/2$ radians and $\alpha^1 = (\alpha_1^1, \alpha_2^1)^\top \in \Psi^2$, $\alpha^2 = (\alpha_1^2, \alpha_2^2, \alpha_3^2)^\top \in \Psi^3$ where $\Psi = \mathbb{Z} \cap [0, 1]$. We thin the segmentation of the vessels by removing the points which are centred on regions matching the above filters. That is, we set such points as background points. We achieve this by iterating

$$\varphi^{\ell+1} = \mathcal{F}_{i,j}\left(\varphi^{\ell}\right), \qquad \ell = 0, 1, \ldots, \qquad \varphi_0(\mathbf{x}) = \phi(\mathbf{x})$$

$$\mathcal{F}_{i,j}(\varphi) = \varphi - 1 + H\left(\left(\kappa_i^j(\alpha^i) * \varphi - \sum \kappa_i^j(\alpha^i)\right)^2\right), \tag{2}$$

beginning with $l_0^1 = 0$ and cycling through $i \in \{1,2\}$, $j \in \{0,1,2,3\}$ (Fig. 1).

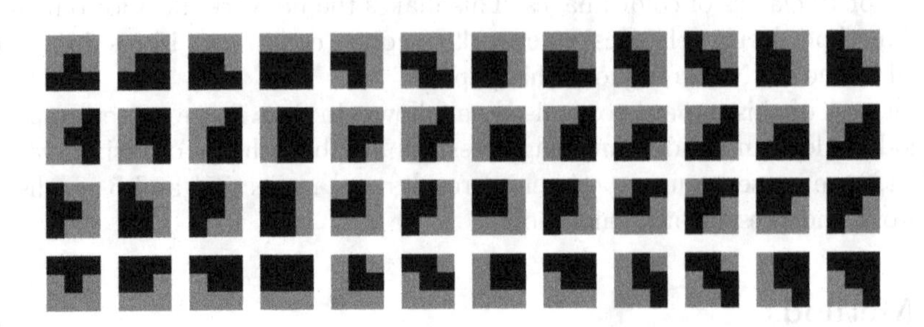

Fig. 1. Kernel functions for skeletonisation

Following this, we extract the patches by cropping the image $z(\mathbf{x})$ to 21×21 pixel windows $\Theta_{\mathbf{p}}$ centred on points $\mathbf{p}$ in the set Υ of points considered the foreground of the skeletonised vessel map. The patch size was selected so that junctions and branches in the vessel would fit within one patch. The patches are given by

$$\Theta_{\mathbf{p}} = \{\mathbf{q} \in \Omega \mid |\mathbf{p} - \mathbf{q}| \leq 10\}, \qquad \mathbf{p} \in \Upsilon = \{\mathbf{p} \in \Omega \mid \varphi(\mathbf{p}) = 1\}.$$

In the training stage, these patches Θ of the images in the training set are used to train the neural network to identify whether a branching or crossing is contained in the image patch. In the test stage, the trained CNN classifies the patches accordingly. This step is described below.

2.2 Junction Identification - CNN $\mathbb{C}_1$

To identify the vessel junctions within the patches created we train our CNN on a high-end graphics processer unit (GPU). The large random access memory of the Nvidia K40c means that we were able to train on the whole dataset of patches at once. The Nvidia K40c contains 2880 CUDA cores and comes with the Nvidia CUDA Deep Neural Network library (cuDNN) for GPU learning. The deep learning package Keras [1] was used alongside the Theano machine learning back end to implement the network. After training, the feed forward process of the CNN can classify the patches produced from a single image in under a second.

We use the Res18 network architecture [6] as deep levels of convolution were required to distinguish the vessel junction type in our small patches. The residual layers incorporate activation, batch normalisation, convolutional, dense and maxpooling layers. We also use L^2 regularisation to improve weight training. There were approximately 100,000 patches for training and 30,000 for testing in the junction identification problem. The classes were weighted as a ratio of junction to background due to the fact that junctions in the training and testing patches were sparse at a ratio of 1:39. The network was trained using Adam stochastic optimisation for backpropegation [7]. The network was trained to classify the patches in to a binary classification of either vessel junction or background. Gaussian initialisation was used within the network to reduce initial training time. The loss function used to optimise was the widely used categorical cross-entropy function. Training was undertaken until reduction of the loss plateaued to obtain optimal results (Fig. 2).

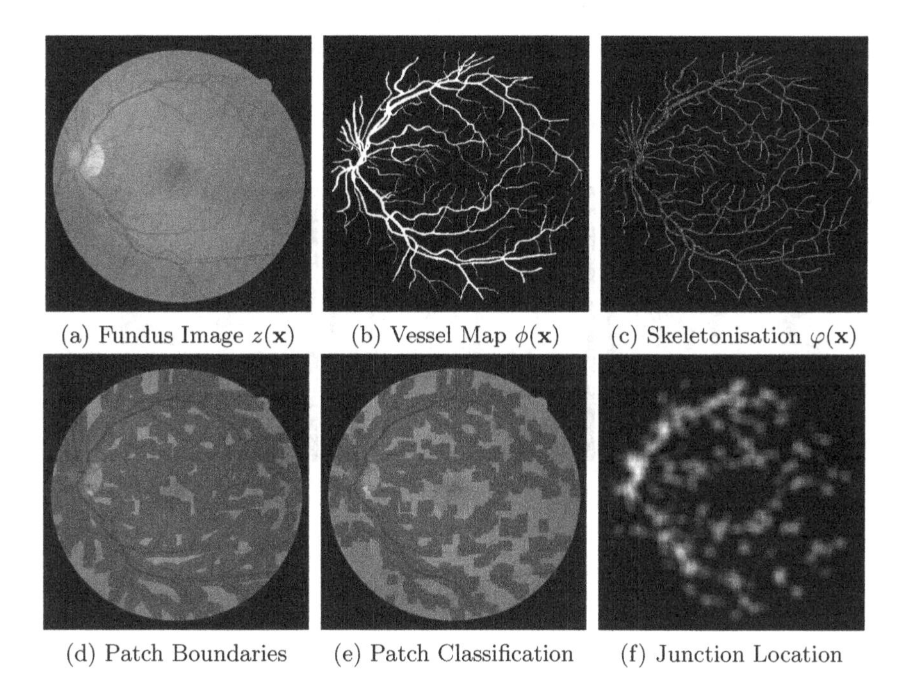
(a) Fundus Image $z(\mathbf{x})$ (b) Vessel Map $\phi(\mathbf{x})$ (c) Skeletonisation $\varphi(\mathbf{x})$

(d) Patch Boundaries (e) Patch Classification (f) Junction Location

Fig. 2. Example of first part of algorithm: locating junctions.

2.3 Locate the Centres

Following the neural network classification, which tell us if a branching or crossing is contained within a patch, we aim to find the locations of the points. We achieve this by forming the cumulative sum image

$$t(\mathbf{q}) = \sum_{\mathbf{p} \in \Upsilon} s^{\mathbf{p}}(\mathbf{q}, l), \qquad s^{\mathbf{p}}(\mathbf{q}, l) = \begin{cases} l^1_{\mathbf{p}} \text{ if } q \in \Theta(\mathbf{p}) \\ 0 \text{ otherwise} \end{cases} \tag{3}$$

and taking the local maxima $\mathbf{r} \in \Upsilon$ as points of interest. We then aim to determine whether points are at crossings or branchings.

2.4 Junction Classification $\mathbb{C}_2$

We extract the patches $\Theta(\mathbf{r})$ and use these to train a neural network to distinguish between crossing and branchings. The second neural network was trained with the Res18 architecture, like the first. Using a relatively small training set of patches, as from our images the majority of patches did not contain junctions, we trained our network in similar fashion to that used in the previous step. Weighted classes were introduced again to cater for the imbalance, in that images from the branching class were substantially more prominent than that of the cross class.

Depending on the patch method there were around 800–2500 patches containing a junction that were used for training. In all methods there were approximately twice as many junction patches containing branching vessels compared to patches containing vessels crossing. Training was performed until a plateau in the reduction of the loss function was reached indicating no further improvement (Fig. 3).

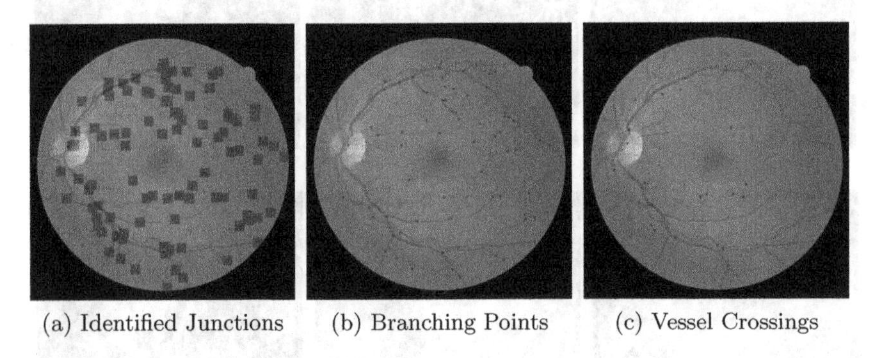

(a) Identified Junctions (b) Branching Points (c) Vessel Crossings

Fig. 3. Example of second part of algorithm: classifying junctions as branchings and crossings.

3 Results

3.1 Overall

We test the ability of our algorithm using 40 images from the DRIVE database with manual segmentations [14]. This data was split to provide 30 images for training the neural networks, leaving 10 for testing. While this may seem a small number for a machine learning approach, is should be noted that the number of patches generated numbered more than 100,000 providing sufficient training data (Fig. 4).

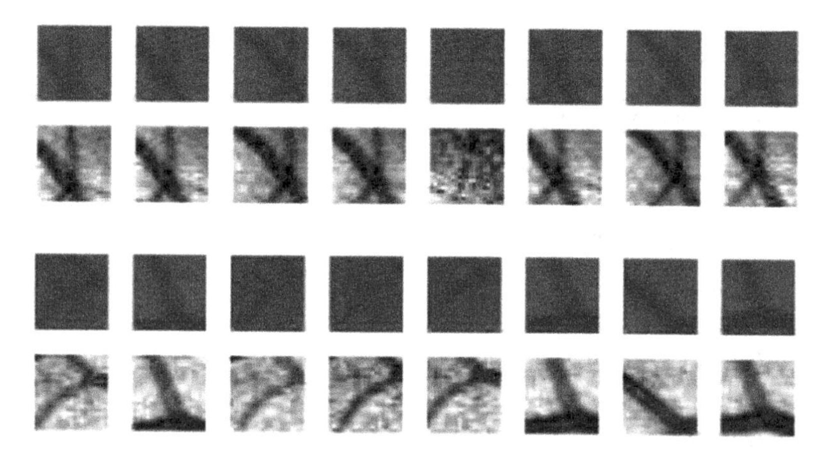

Fig. 4. Example of $\mathbb{C}_2$ input. Rows 1 and 2 (resp. 3 and 4): training patches with crossings (resp. branchings) and their enhanced counterparts for presentation. The neural networks were able to achieve good results using the patches without enhancement.

4 Discussion

We have produced a method that can learn to detect and classify vessel junctions using a very small dataset of 40 fundus images that had been manually classified for junctions and their type. Using the CNN $\mathbb{C}_1$, we managed to detect the junctions to an impressive detection accuracy of over 94% due in part to the relatively large amount of patches containing junctions. Along with the skeletonisation, our deep learning classification $\mathbb{C}_2$ for vessel type gave us an accuracy of 88%. Increasing the size of our dataset would allow better distinction in the classification of the vessel junctions. It is worth noting that junction type training was undertaken on a couple of thousand patches and tested on around 800. Through training on more images the model could be fine tuned to refine the filters and increase identification accuracy (Fig. 5).

The current algorithm works well for images which have been manually segmented but this time-consuming task could be further extended to incorporate automatic segmentation techniques [15]. A further very useful extension would be to automatically determine whether the artery or vein is in front with arteriovenous crossings along with consideration of intra and inter-observer variability. In order to better identify and classify junctions with other nearby junctions, it would be useful to consider extending our method to a multi-scale approach (Fig. 6).

Algorithm 1. $l_p \leftarrow \mathbb{A}\left(z(\mathbf{x}), \phi(\mathbf{x}),\right)$

1: Skeletonisation of vessel map
2: Set $\varphi^0(\mathbf{x}) = \phi(\mathbf{x})$
3: **for** $\ell \leftarrow 0 : maxit$ **do**
4:     **for** $i \leftarrow 1 : 2$ **do**
5:         **for** $j \leftarrow 0 : 3$ **do**
6:             $\varphi^{\ell+1} \leftarrow \mathcal{F}_{i,j}\left(\varphi^{\ell}\right)$ using equation (2)
7:         **end for**
8:     **end for**
9: **end for**
10: Extract and classify the patches
11: Set $\Upsilon = \{\mathbf{p} \in \Omega \mid \varphi(\mathbf{p}) = 1\}$
12: **for** $\mathbf{p} \in \Upsilon$ **do**
13:     $\Theta_{\mathbf{p}} = \{\mathbf{q} \in \Omega \mid |\mathbf{p} - \mathbf{q}| \leq 10\}$
14:     $l^1_{\mathbf{p}} = \mathbb{C}_1\left(\Theta_{\mathbf{p}}\right)$
15: **end for**
16: Calculate the cumulative sum image $t(\mathbf{q})$ using (3) and determine the set of points of interest P.
17: $\forall \mathbf{p} \in P$, extract the patches $\Theta_{\mathbf{p}} = \{\mathbf{q} \in \Omega \mid |\mathbf{p} - \mathbf{q}| \leq 10\}$
18: Classify the extracted patches to obtain $l^2_{\mathbf{p}} = \mathbb{C}_2\left(\Theta_{\mathbf{p}}\right)$

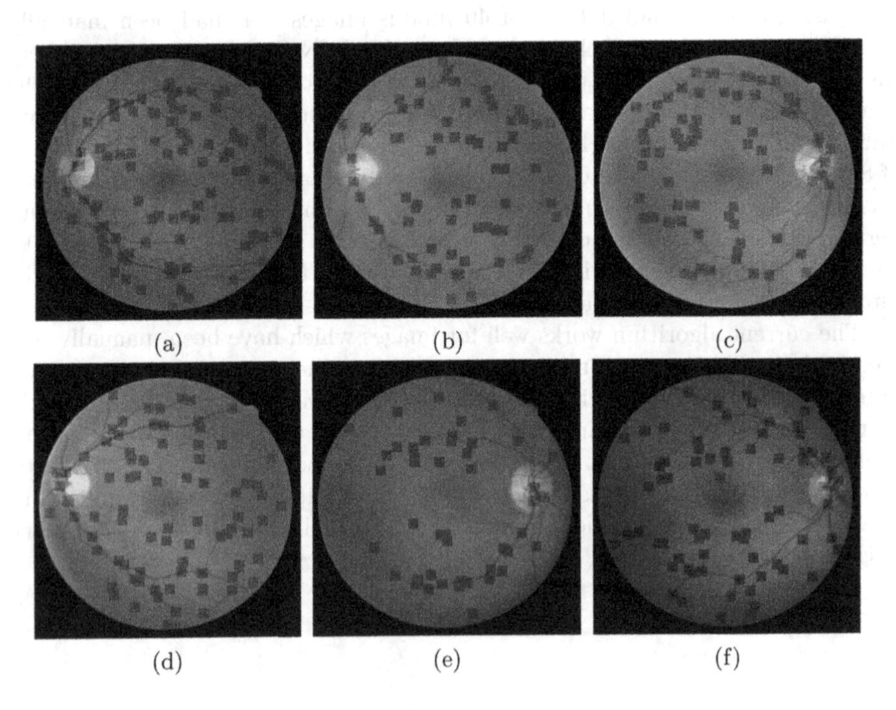

Fig. 5. Example of identifying junctions in fundus images.

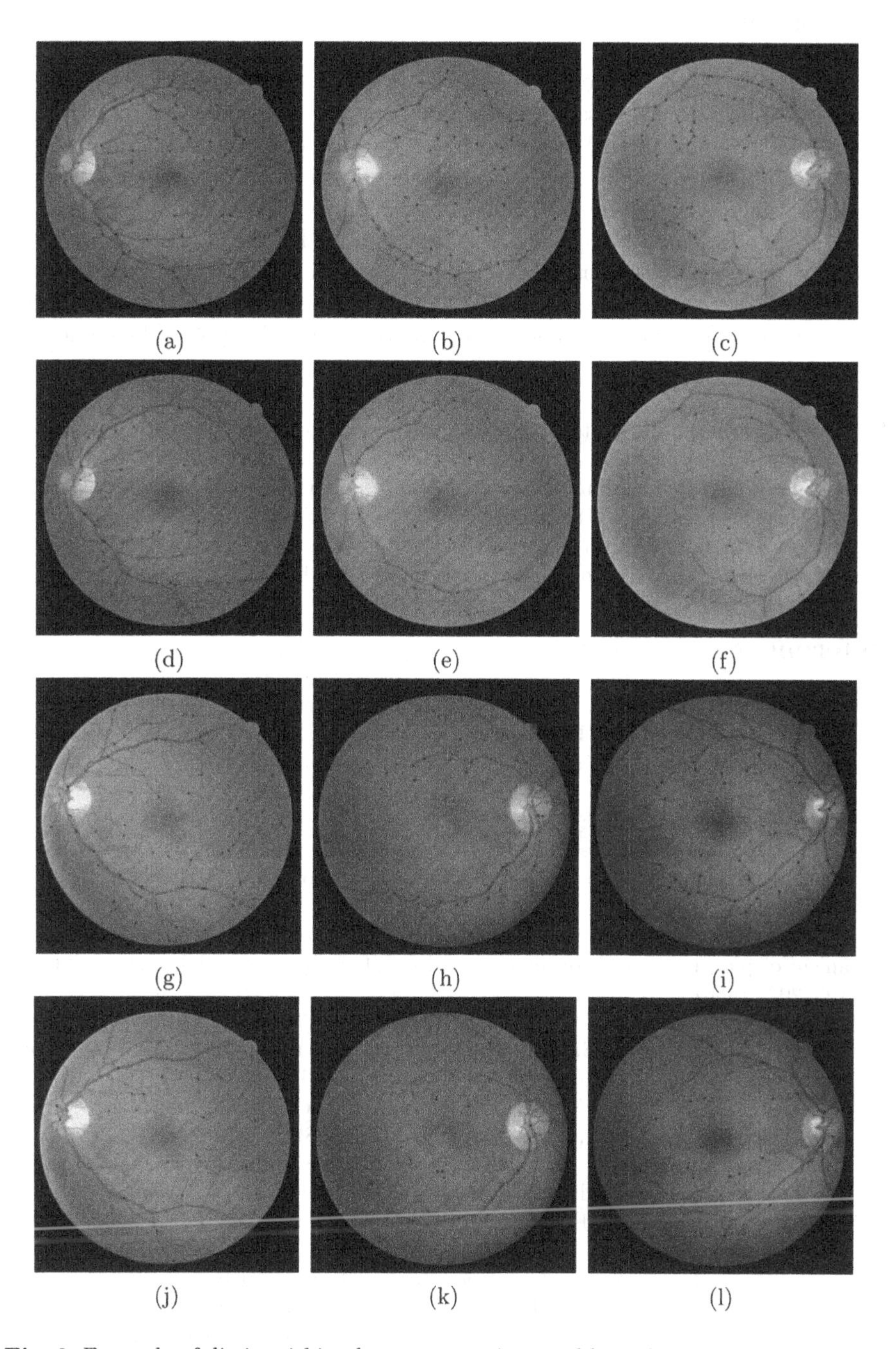

Fig. 6. Example of distinguishing between crossings and branchings in fundus images. In each column, rows one and three show branchings and rows two and four show the crossings for the respective examples.

5 Conclusion

The challenging task of detecting and classifying vessel junctions in fundus images is shown to be possible using our method. The ability to expand on this method to make the detection both quicker and more accurate than manual classification is possible. These preliminary results demonstrate that the overall framework including the deep learning approach proposed is a viable technique to accurately find and identifying vessel junctions with little training data. More extensive testing of this framework could be undertaken to assess the transferability of these results to different size images and different datasets. However, there is no reason why this framework would not be directly applicable to another dataset.

Acknowledgement. H. Pratt acknowledges PhD funding from Fight for Sight charity. This project is funded in part by the National Institute for Health Research's i4i Programme. This paper summarises independent research funded by the National Institute for Health Research (NIHR) under its i4i Programme (Grant Reference Number II-LA-0813-20005). The views expressed are those of the authors and not necessarily those of the NHS, the NIHR or the Department of Health.

References

1. Chollet, F.: Keras (2015). https://github.com/fchollet/keras
2. Chutatape, O., Zheng, L., Krishnan, S.M.: Retinal blood vessel detection and tracking by matched Gaussian and Kalman filters. In: Proceedings of the 20th Annual International Conference of the IEEE Engineering in Medicine and Biology Society, vol. 6, pp. 3144–3149. IEEE (1998)
3. Cun, Y.L., Boser, B., Denker, J.S., Howard, R.E., Habbard, W., Jackel, L.D., Henderson, D.: Advances in Neural Information Processing Systems 2, pp. 396–404. Citeseer (1990)
4. Fukushima, K.: Neocognitron: a self-organizing neural network model for a mechanism of pattern recognition unaffected by shift in position. Biol. Cybern. **36**(4), 193–202 (1980)
5. Gonzalez, R., Wintz, P.: Digital Image Processing (1977)
6. He, K., Zhang, X., Ren, S., Sun, J.: Deep residual learning for image recognition. CoRR abs/1512.03385 (2015). http://arxiv.org/abs/1512.03385
7. Kingma, D.P., Ba, J.: Adam: a method for stochastic optimization. CoRR abs/1412.6980 (2014). http://arxiv.org/abs/1412.6980
8. Krizhevsky, A.: Learning multiple layers of features from tiny images. https://www.cs.toronto.edu/kriz/learning-features-2009-TR.pdf
9. LeCun, Y., Cortes, C.: MNIST handwritten digit database (2010). http://yann.lecun.com/exdb/mnist/
10. MacGillivray, T., Trucco, E., Cameron, J., Dhillon, B., Houston, J., Van Beek, E.: Retinal imaging as a source of biomarkers for diagnosis, characterization and prognosis of chronic illness or long-term conditions. Br. J. Radiol. **87**(1040), 20130832 (2014)
11. Nair, V., Hinton, G.E.: Rectified linear units improve restricted Boltzmann machines. In: Proceedings of the 27th International Conference on Machine Learning (ICML 2010), pp. 807–814 (2010)

12. Sonka, M., Hlavac, V., Boyle, R.: Image Processing, Analysis, and Machine Vision. Cengage Learning (2014)
13. Srivastava, N., Hinton, G., Krizhevsky, A., Sutskever, I., Salakhutdinov, R.: Dropout: a simple way to prevent neural networks from overfitting. J. Mach. Learn. Res. **15**(1), 1929–1958 (2014)
14. Staal, J., Abràmoff, M.D., Niemeijer, M., Viergever, M.A., Van Ginneken, B.: Ridge-based vessel segmentation in color images of the retina. IEEE Trans. Med. Imaging **23**(4), 501–509 (2004)
15. Zhao, Y., Rada, L., Chen, K., Harding, S.P., Zheng, Y.: Automated vessel segmentation using infinite perimeter active contour model with hybrid region information with application to retinal images. IEEE Trans. Med. Imaging **34**(9), 1797–1807 (2015)

Fast Optic Disc Segmentation in Retinal Images Using Polar Transform

Muhammad Nauman Zahoor and Muhammad Moazam Fraz(✉)

School of Electrical Engineering and Computer Science,
National University of Sciences and Technology, Islamabad, Pakistan
{14mscsmzahoor,moazam.fraz}@seecs.edu.pk

Abstract. Glaucoma is one among major causes of blindness. Early detection of glaucoma through automated retinal image analysis helps in preventing vision loss. Optic Disc segmentation from retinal images is considered as the preliminary step in developing the diagnostic tool for early Glaucoma detection. A novel hierarchical technique for optic disc localization and segmentation on retinal fundus images is presented in this paper. Retinal vasculature and pathologies are delineate and removed by using morphological operations as preprocessing steps. Circular Hough transform is used to localize the optic disc. Region of interest is calculated and a novel polar transform based adaptive thresholding is performed to obtain the precise boundary of optic disc. The methodology has shown considerable improvement over existing methods in terms of accuracy and processing time. The algorithm is evaluated on a number of publicly available retinal image sets which includes MESSIDOR, DIA-RETDB1, DRIONS-DB, HRF, DRIVE and RIM-ONE, with average spatial overlap approximately 85%.

Keywords: Optic disc · Retinal image analysis · Polar transform · Segmentation · Glaucoma

1 Introduction

The automated analysis of human retinal images has been widely used for early detection, screening and treatment planning of various retinal, ophthalmic and systemic disease. Retinal image analysis had been increasingly used in large scale screening programs and population based studies for early detection of glaucoma and diabetic retinopathy [1]. The analysis of change in morphological attributes of retinal anatomical structures including Optic Disc (OD), Optic Cup, retinal vasculature and pathologies triggers timely detection and treatment of glaucoma while it is still in early stage [2]. Early detection and hence treatment can help in prognosis of vision loss. The precise segmentation of OD is the preliminary step in development of computer assisted automated system for glaucoma screening. OD is distinguished as a variable sized slightly yellowish elliptical area in the retinal images. The ganglion cell axons leave the eye, forming an optic nerve which transmits the visual information to the brain. The OD has two prominent regions; central brighter region andthe peripheral region containing OD cup which is also termed as the neuro-retinal rim [1]. The Optic Disc and other anatomical features in retinal image is shown in Fig. 1.

M. Valdés Hernández and V. González-Castro (Eds.): MIUA 2017, CCIS 723, pp. 38–49, 2017.
DOI: 10.1007/978-3-319-60964-5_4

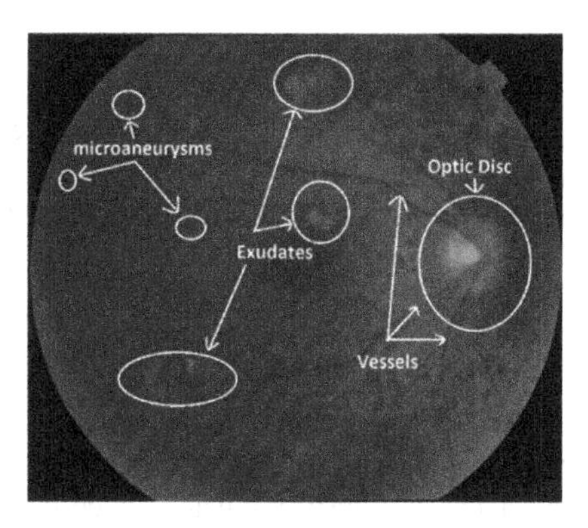

Fig. 1. Retinal features on fundus image

Glaucoma is characterized by the change in color, shape and depth of OD.The presence of parapapillary atrophy produces bright regions around the OD rim distorting the elliptical shape [1]. Moreover, the progression in optic nerve fiber damage causes the structural changes in OD, optic nerve head and nerve fiber layer which results in an increase in optic cup to disc ratio (CDR). The CDR can be accessed by estimating the diameter and the area of OD, the area of rim andoptic cup diameter. The accurate and fast OD segmentation and analysis is the first step towards the development of computer assisted diagnostic system for glaucoma screening in large population based studies.

Retinal pathologies which includes exudates, hemorrhages and other bright lesions, if present, may look similar in appearance to the shape of OD, thus may result in false detection. Hence the robust localization and segmentation of OD is a challenging task because of various factors like presence of pathologies and other anatomical structures, uneven illumination, variability in luminosity and contrast during image acquisition. Moreover, the presence of incoming blood vessels make the OD boundary appeared as non-distinctive fused boundary.

This paper presents a new hybrid approach for fast and accurate localization and segmentation of OD based on Circular Hough Transform (CHT) [3] and Polar Transformation algorithm [4]. The Polar Transformation algorithm has been widely used in many application areas of image segmentation, but has not been applied within the framework of retinal image analysis. To the limit of our knowledge, the Polar Transform (PT) algorithm has been utilized for the first time in localizing and segmenting OD in retinal images. Experimental evaluation shows that this method is computationally fast in processing, robust to the variation in image contrast and illumination, and comparable with the state of the art methodologies in terms of quantitative performance metrics.

It's worth mentioning that the proposed methodology is aimed at contributing to the development of computer assisted system for glaucoma screening. There are other published methods of OD segmentation available in literature but this methodology is computationally fast, produces higher accuracy, robustness and tolerant to vast variety of images as shown by the use of various retinal fundus datasets made up of diseased and healthy images. This all make it suitable for integration with glaucoma detection system.

The organization of the paper is as follows. A comprehensive review OD detection and segmentation methodologies is presented in Sect. 2.The proposed methodology is explained in Sect. 3. The materials and the performance metrics used to evaluate the proposed methodology are illustrated in Sect. 4. The evaluation results and comparative analysis with other methods ispresented in Sect. 5. Section 6 concludes the paper.

2 Related Work

There are a number of methodologies available in literature for Optic Disc Segmentation. Some preliminary work involves the segmentation by modeling the OD as an elliptical or a circular object [5] and performing the shape-based template matching. The irregularity in the shape of OD, the presence of the pathological structures and the difference in the multiple views of OD are the challenges faced by the shape based modeling methodologies. Active contour models had been utilized for OD segmentation. Contours were initialized manually as well as automatically and the energy term; derived from the gradient of image, caused the deformation of the contour. The OD boundary was detected by employing the gradient vector flow based active contour model [6]. The energy minimization was achieved by the using pre-processed images or by applying the constraint on the OD segmentation results to elliptical or circular region [7]. The snake model [8] has also been proposed to improve the OD segmentation in the presence of vessel occlusion. Watershed transformation has also been used to detect the OD contour [9]. Geometric active contour model in combination with maximum local variance has been proposed by Kande et al. [10]. Optic Disc was detected by Staporet et al. [11] using geodesic reconstruction by dilation and extracted Optic Disc using Mathematical Morphology. Lupascu et al. [12] approximated Optic Disc boundary by using texture information and applying a regression based technique to obtain the best circle that defines the OD boundary. Detection and Extraction of Optic Disc was done without assuming any pre-defined shape using mathematical morphology techniques by Welfer et al. [13]. Basit and Fraz [14] used marker controlled watershed transform for Optic Disc segmentation. After application of morphological operations and smoothing filters, gradient magnitude image was modified using Internal and external markers and watershed transformation was applied for OD boundary extraction. A combination of morphological template based technique followed by CHT, was used by Aquino et al. [15]. Principal Component Analysis based mathematical modelling was proposed by Morales et al. [16]. Abdullah et al. [17] had used morphological operation and grow cut algorithm for approximating the OD boundary.

3 Proposed Methodology

This work presents an automated Optic Disc localization and segmentation technique that is able to detect OD without using any initial vascular information and template knowledge. For the segmentation of OD boundary, the first step is to approximate the OD center. Optic Disc appears as the brightest spot in the retinal fundus images but the presence of artifacts can create multiple bright spots. Pathologies in fundus images can take shape of OD while actual OD could lose its brightness. As shown in Fig. 3. The shape of the OD varies from circular to elliptical. This information about the shape of the OD can be used for the detection of OD. The proposed methodology preprocesses the image to remove vessels and enhance the OD boundary using morphological operations. CHT is used for OD localization. Spatial to polar transform is applied to convert circular region of interest into rectangular. Adaptive thresholding is applied to obtain the OD boundary. Flow chart of the proposed methodology is presented in Fig. 2.

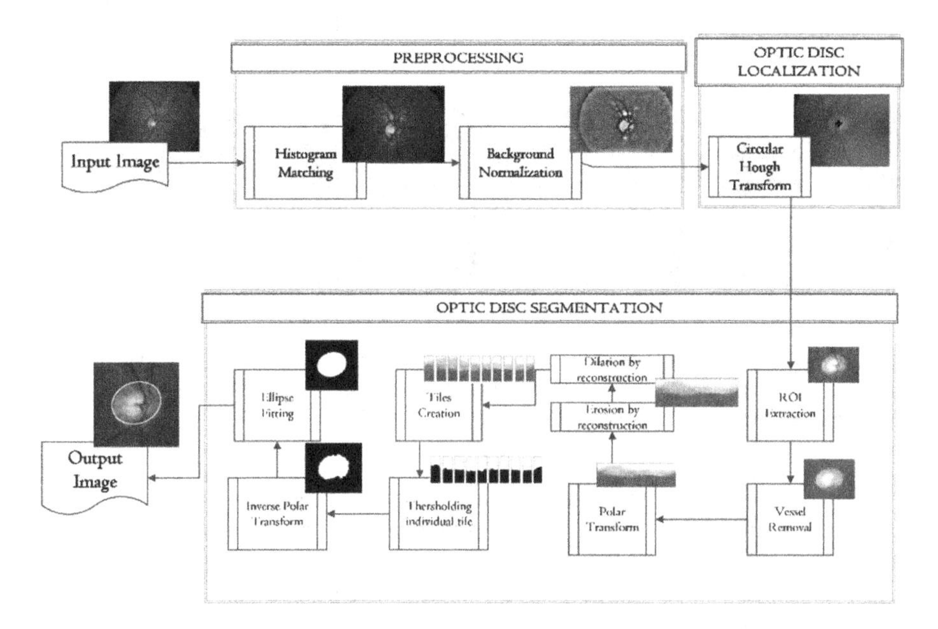

Fig. 2. Proposed methodology workflow

3.1 Preprocessing

Varying conditions during image capture, noise, and uneven illumination and contrast variations are the added challenges of automated optic disc detection and segmentation. Figure 3 shows retinal images under different illumination conditions and affected with pathologies. In order to handle these images autonomously, preprocessing has to be applied.

Histogram Matching [18] has been applied to for normalizing the image variations. The histogram of properly illuminated image (Fig. 4(a)) is taken as reference and the other images' histograms are matched with it which resulted in normalized illumination and color tone. Red channel was chosen as it contains most information of the Optic Disc. Background normalization is then performed by subtracting the image with the estimate of background. The estimate of background I_{bg} is calculated by filtering the image with a large arithmetic mean kernel such that the filtered image doesn't contain

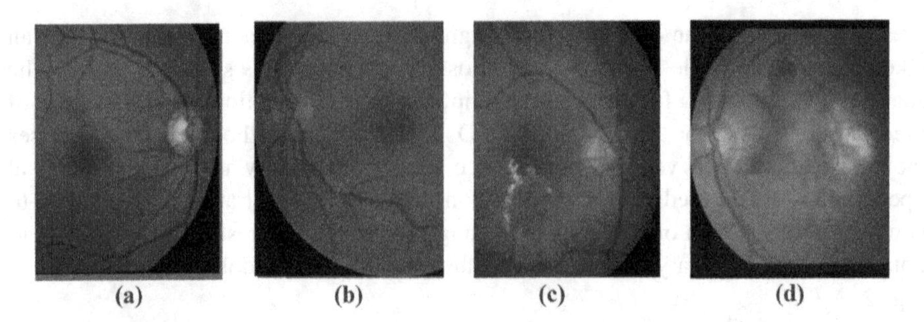

Fig. 3. Different modalities of optic disc (a) OD is bright and visible. (b) OD is not bright. (c) OD is bright together with pathologies. (d) OD with pathologies.

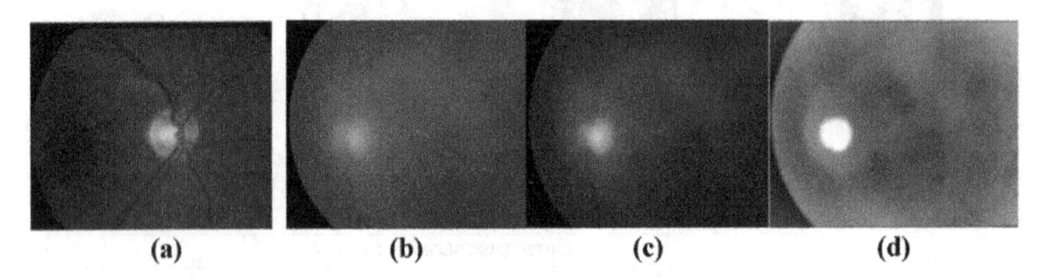

Fig. 4. Preprocessing. (a) Reference image to be matched with. (b) Images under different illumination and reflectance. (c) Image after histogram matching. (d) Red channel background normalized image. (Color figure online)

any visible structures. Original image is also morphologically opened using a 'disk' shaped structuring element with a size *1/100th* the size of the original image to obtain I_{open}. The size *1/100th* of actual image was chosen after manual experimentation on various image databases and keeping the overall best performing size. Background normalized image $I_{normalized}$ is the difference of opened image and the background estimate, as shown in Eq. (1).

$$I_{normalized} = I_{open} - I_{bg} \tag{1}$$

Figure 4 shows the results of histogram matching and background normalization.

3.2 Optic Disc Localization

OD localization from preprocessed images is done by applying CHT which is an extension of Hough transform [19] which detects circular objects in the image. This step requires a radius search range of min radius and a max radius. Max and min radius are approximated to be *1/30th* to *1/10th* of the image width. This radius range covers the Optic Discs in the various image databases and disregards the effects of pathologies in the images. The output of the transform contains all the circles that are present in the given range. The circle with the highest score is kept. Figure 5 shows detected OD.

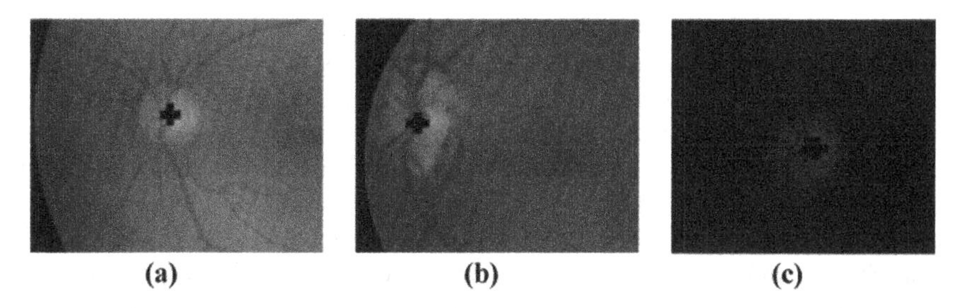

Fig. 5. Optic disc detection results via circular hough transform on normal and pathological images under challenging illumination settings

From the results it is clear that the OD localization technique along with preprocessing methodology is robust and is even able to localize OD under challenging illumination conditions and in the presence of pathologies.

3.3 Optic Disc Segmentation

For the precise segmentation of the optic disc, a region of interest (ROI) is extracted from the original image "*I_orig*". The size of "*I_roi*" from "*I_orig*" is calculated as described by following equation.

$$roi_{size} = r + buff \tag{2}$$

Where, r is the radius of the circle approximated by the CHT and "*buff*" is the number of extra pixels that are not part of the OD but surrounds it. ROI is centered on the circle center approximated by the CHT. ROI contains optic disc pixels surrounded by non-od pixels. The next task is to find precise boundary between OD pixels and non OD pixels. Direct thresholding techniques does not yield good results as the gray level distribution of the OD and non-OD regions is not uniform and applying a global threshold fails. Applying a local threshold on a neighborhood of pixels also does not return good results because of the circular nature of the OD. To overcome these issues, a novel OD segmentation technique is proposed that makes use of Polar Transform (PT) [18].

Polar transform can be defined as a 2D coordinate system where every point is calculated using distance from a reference point and an angle from a reference direction. Polar transform has been used a lot in automated segmentation of iris from image as is done by [20]. To the limit of our knowledge, the PT algorithm has been utilized for the first time in localizing and segmenting OD in retinal images.

For OD segmentation, theROI image is calculated and ROI's pixel coordinates are converted from Cartesian to polar coordinates by applying Polar Transform with the origin point being the center of the ROI image. Due to this transformation, the OD is now straightened. Next this straightened OD is divide in to 10 equal sized sub-tiles. Morphological erosion by reconstruction [18] is applied on each tile followed by morphological dilation by reconstruction. At this step, since precise boundary of the optic disc is needed, morphological opening and closing is avoided as this would also remove the structures in the retina that are to be used as the end boundary points to distinguish between optic disc pixels and the rest of the image. Opening by reconstruction preserves the shape of the components.

After application of opening by reconstruction, each tile is then thresholded using adaptive thresholding. If the output tile is successfully thresholded in to two regions, it is forwarded to the next step as is. If not, then a blank tile (all black) is forwarded. The tiles are then combined and Polar to Cartesian transformation is applied. Ellipse fitting is then performed using ellipse equation in which the boundary obtained via thresholding is used to draw an ellipse over it. This gives the precise OD boundary. All steps of OD segmentation are graphically illustrated in Fig. 6.

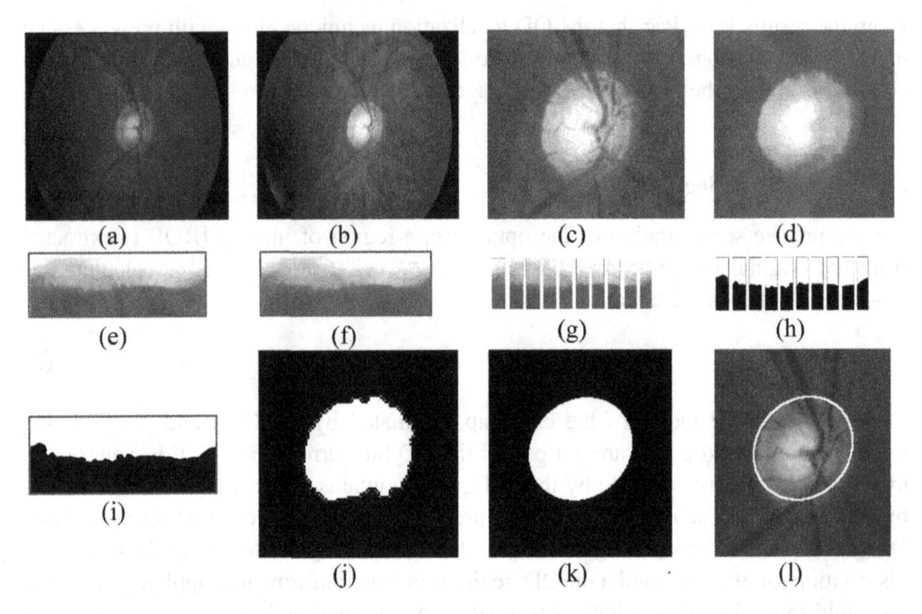

Fig. 6. Optic disc segmentation (a) Original image. (b) Red channel. (c) Region of interest ROI. (d) Vessels removed. (e) Cartesian to polar transform applied on ROI. (f) Morphological operations (erosion by reconstruction followed by dilation by reconstruction). (g) Segregation into tiles (h) Threshold application on individual tiles. (i) Thresholded tiles (j) Conversion from polar to cartesian. (k) Ellipse fitting. (l) Result overlay on the image.

4 Materials

The methodology is evaluated on following publically available datasets shown in Table 1 and a dataset received from a local hospital (Shifa international hospital).

Ground truths of OD boundary are available with RIM-ONE, HRF and Messidor databases whereas the images of DIARETDB1, DRIVE and DRIONS-DB were hand labeled by ophthalmic experts from the Armed forces institute of Ophthalmology, Rawalpindi by [17].

Table 1. Retinal image databases wih count of images as healthy/pathological

	Database	Total images	Healthy	Diabetic retinopathy	Glaucomatous
1	RIM-ONE [21]	118	78	–	40
2	HRF [22]	45	15	14	15
3	MESSIDOR [23]	1200	540	660	–
4	DIARETDB1 [24]	89	05	84	–
5	DRIONS-DB [25]	110	06	83*	21
6	DRIVE [26]	40	33	07	–
7	Shifa database [17]	111	19	92	–

* Images with hypertensive retinopathy

Performance Measures

Pixel based comparison of ground truths and the OD segmentation obtained by the proposed methodology is presented. The four possible outcomes of pixel classification are illustrated in Table 2. Based on these outcomes, the calculated performance metrics are shown in Table 3.

Table 2. Classification of OD pixels

	Algorithm predicted pixel $\in$ OD	Algorithm predicted pixel $\notin$ OD
Actual pixel $\in$ OD	True Positive (TP)	False Positive (FP)
Actual pixel $\notin$ OD	False Negative (FN)	True Negative (TN)

Table 3. OD segmentation performance metrics

Measure	Description
SN	TP/(TP + FN)
SP	TN/(TN + FP)
Accuracy	(TP + TN)/(TP + FP + TN + FN)
Precision	TP/(TP + FP)
F1 Score	2TP/(2TP + FP + FN)
Overlap	(Predicted OD $\cap$ Ground Truth OD)/(Predicted OD $\cup$ Ground Truth OD)

5 Results

The optic disc segmentation results are shown in Fig. 7, where the OD boundary detected by the proposed methodology is shown in white marking. The quantitative performance metrics based on pixel-wise comparison of ground truths and the segmentation results obtained from the algorithm are shown in Table 4. The comparison with other methods are shown in Table 5. It is worth mentioning that the proposed methodology outperforms other techniques and uses less average time per image. The average time per image reported in Table 5 has been taken from the respective

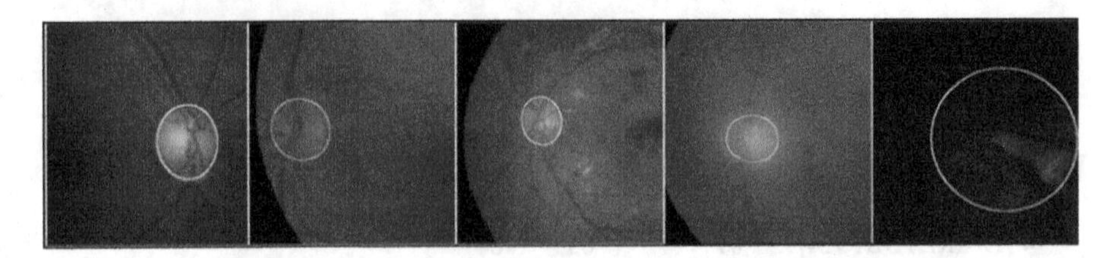

Fig. 7. Results of OD segmentation on various dataset, contain images with uneven illumination conditions and affected with pathologies.

Table 4. OD segmentation quantitative performance measures

Database	Accuracy	Sensitivity	Specificity	Precision	F1 Score	Overlap
DRIONS-DB	0.9986	0.9384	0.9994	0.9463	0.9378	0.8862
DIARETDB1	0.9937	0.9706	0.9949	0.8991	0.9306	0.8734
HRF	0.9774	0.9233	0.9892	0.9448	0.9282	0.8686
SHIFA	0.9963	0.8908	0.9996	0.9873	0.9332	0.8788
MESSIDOR	0.9918	0.8891	0.9973	0.9537	0.9039	0.8441
RIM-ONE	0.9750	0.9112	0.9807	0.8310	0.8491	0.7480
DRIVE	0.9980	0.8309	0.9993	0.9136	0.8500	0.7561

Table 5. OD segmentation comparison with other techniques

DIARETDB1	Overlap	Sensitivity	Specificity	Accuracy	Time (in seconds)
Sopharak et al. [27]	0.294	0.4603	0.9994		
Kande et al. [10]	0.334	0.8808	0.9878		120.5
Seo et al. [28]	0.353	0.6103	0.9987		15.63
Walter et al. [29]	0.369	0.6569	0.9993		308.5
Lupascu et al. [12]	0.309	0.6848	0.9969		
Stapor et al. [11]	0.340	0.8498	0.9964		59.72
Basit and Fraz [14]	0.546	0.7347	0.9944		
Welfer et al. [13]	0.391	0.6341	0.9983		57.16
Abdullah et al. [17]	0.851	0.851	0.9984	0.9772	40
Proposed methodology	**0.874**	**0.9706**	**0.9949**	**0.9937**	**1.3**
DRIONS-DB					
Morales et al. [16]				0.9934	
Abdullah et al. [17]	0.851	0.8508	0.9966	0.9989	43.2
Proposed methodology	**0.886**	**0.9384**	**0.9994**	**0.9986**	**1.6**
MESSIDOR					
Morales et al. [16]				0.9949	
Abdullah et al. [17]	0.879	0.8954	0.9995	0.9989	71.3

(*continued*)

Table 5. (*continued*)

DIARETDB1	Overlap	Sensitivity	Specificity	Accuracy	Time (in seconds)
Proposed methodology	**0.844**	**0.8891**	**0.9973**	**0.9918**	**1.8**
DRIVE					
Sopharak et al. [27]	0.168	0.2104	0.9993		14.92
Kande et al. [10]	0.296	0.6999	0.9888		111.7
Seo et al. [28]	0.310	0.5029	0.9983		7.23
Walter et al. [29]	0.293	0.4988	0.9981		219.6
Salazar-Gonzalez et al. [30]		0.7512	0.9684	0.9412	
Lupascu et al. [12]	0.309	0.7768	0.9968		
Stapor et al. [11]	0.334	0.7368	0.9920		43
Basit and Fraz [14]	0.618	0.8921	0.9921		
Welfer et al. [13]	0.394	0.7357	0.9982		53.65
Abdullah et al. [17]	0.786	0.8188	0.9966	0.9672	59.2
Proposed methodology	**0.756**	**0.8309**	**0.9993**	**0.9980**	**1.6**

publications. We understand that hardware configuration should be taken into account for time comparison benchmarking with other methodologies, which we will do in the future.

6 Discussion and conclusion

Fast and precise segmentation of OD in varying imaging conditions is the preliminary step in development of automated system for glaucoma detection. The correct and precise segmentation of the optic disc will increase the correct diagnosis. A novel optic detection and segmentation methodology has been presented in this paper. The methodology has used hierarchical combination of morphological operations and CHT for the Optic Disc localization. This gives better results as relying only on the "Optic Disc as the brightest spot" technique that is used a lot for the detection purposes can give false results in presence of pathologies and poor illumination setting of optic disc. After detection, a novel polar transform based adaptive thresholding is applied. Based on OD localization, the circular ROI is extracted followed by the application of Polar transform for converting the ROI into distinct rectangular tiles. The tiles are adaptively thresholded to get the precise OD boundary. Inverse polar transform is then applied to get the OD shape back. To the best of our knowledge, the polar transformation has been applied for the first time for the segmentation of Optic Disc.

Evaluation was performed on publically available datasets that are MESSIDOR, DRIVE, HRF, DIARETDB1, RIM-ONE, DRIONS-DB and a dataset received from a local hospital. The outcome shows that the methodology is computationally efficient and performs well even in varying illumination setting and contrast changes and in the presence of pathologies in the image. Optic Disc Segmentation Algorithm was able to achieve an average spatial overlap of 84.4% for MESSIDOR, 87.3% for DIRETDB1, 88.6% for DRIONS-DB, 86.9% for HRF, 75.6% for Drive and 74.8% for RIM-ONE. It has been observed that the proposed methodology segments the OD in images affected with pathologies and under varying illumination in less amount of time without compromising on accuracy.

We have already developed an automated software titled QUARTZ [2], for extraction of quantifiable parameters from retinal vessel morphology. The epidemiologists and other medical/statistical experts can evaluate the association of retinal vessel abnormalities with the biomarkers of systemic diseases. In future, we aim to incorporate the Glaucoma detection module in the aforementioned software system. With the improvement in time efficiency, the proposed methodology can be used as a building block in developing an automated system for early detection of Glaucoma.

References

1. Abramoff, M.D., Garvin, M.K., Sonka, M.: Retinal imaging and image analysis. IEEE Rev. Biomed. Eng. **3**, 169–208 (2010)
2. Fraz, M.M., et al.: QUARTZ: quantitative analysis of retinal vessel topology and size – an automated system for quantification of retinal vessels morphology. Expert Syst. Appl. **42** (20), 7221–7234 (2015)
3. Illingworth, J., Kittler, J.: The adaptive hough transform. IEEE Trans. Pattern Anal. Mach. Intell. **PAMI-9**(5), 690–698 (1987)
4. Luo, X., Liang, T., Wang, W.: Static image segmentation using polar space transformation technique. In: Wong, W.E., Zhu, T. (eds.) Computer Engineering and Networking. LNEE, vol. 277, pp. 533–540. Springer, Cham (2014). doi:10.1007/978-3-319-01766-2_61
5. Osareh, A., et al.: Colour morphology and snakes for optic disc localisation. In: The 6th Medical Image Understanding and Analysis Conference. BMVA Press (2002)
6. Lee, S., Brady, M.: Optic disk boundary detection. In: Mowforth, P. (ed.) BMVC91, pp. 359–362. Springer, London (1991)
7. Lowell, J., et al.: Optic nerve head segmentation. IEEE Trans. Med. Imaging **23**(2), 256–264 (2004)
8. Xu, J., et al.: Optic disk feature extraction via modified deformable model technique for glaucoma analysis. Pattern Recogn. **40**(7), 2063–2076 (2007)
9. Walter, T., Klein, J.-C.: Segmentation of color fundus images of the human retina: detection of the optic disc and the vascular tree using morphological techniques. In: Crespo, J., Maojo, V., Martin, F. (eds.) ISMDA 2001. LNCS, vol. 2199, pp. 282–287. Springer, Heidelberg (2001). doi:10.1007/3-540-45497-7_43
10. Kande, G.B., Subbaiah, P.V., Savithri, T.S.: Segmentation of exudates and optic disk in retinal images. In: Sixth Indian Conference on Computer Vision, Graphics and Image Processing, ICVGIP 2008. IEEE (2008)
11. Stąpor, K., Świtonski, A., Chrastek, R., Michelson, G.: Segmentation of fundus eye images using methods of mathematical morphology for glaucoma diagnosis. In: Bubak, M., Albada, G.D., Sloot, P.M.A., Dongarra, J. (eds.) ICCS 2004. LNCS, vol. 3039, pp. 41–48. Springer, Heidelberg (2004). doi:10.1007/978-3-540-25944-2_6

12. Lupascu, C.A., Tegolo, D., Di Rosa, L.: Automated detection of optic disc location in retinal images. In: 21st IEEE International Symposium on Computer-Based Medical Systems, CBMS 2008. IEEE (2008)
13. Welfer, D., Scharcanski, J., Marinho, D.R.: A morphologic two-stage approach for automated optic disk detection in color eye fundus images. Pattern Recogn. Lett. **34**(5), 476–485 (2013)
14. Basit, A., Fraz, M.M.: Optic disc detection and boundary extraction in retinal images. Appl. Opt. **54**(11), 3440–3447 (2015)
15. Aquino, A., Gegúndez-Arias, M.E., Marín, D.: Detecting the optic disc boundary in digital fundus images using morphological, edge detection, and feature extraction techniques. IEEE Trans. Med. Imaging **29**(11), 1860–1869 (2010)
16. Morales, S., et al.: Automatic detection of optic disc based on PCA and mathematical morphology. IEEE Trans. Med. Imaging **32**(4), 786–796 (2013)
17. Abdullah, M., Fraz, M.M., Barman, S.A.: Localization and segmentation of optic disc in retinal images using circular Hough transform and grow-cut algorithm. PeerJ **4**, e2003 (2016)
18. Gonzalez, R.C., Woods, R.E.: Digital Image Processing, 3rd edn. Prentice-Hall Inc., Upper Saddle (2006)
19. Hough, P.: Method and means for recognizing complex patterns. Google Patents (1962)
20. Luengo-Oroz, M.A., Faure, E., Angulo, J.: Robust iris segmentation on uncalibrated noisy images using mathematical morphology. Image Vis. Comput. **28**(2), 278–284 (2010)
21. Fumero, F., et al.: RIM-ONE: an open retinal image database for optic nerve evaluation. In: 2011 24th International Symposium on Computer-Based Medical Systems (CBMS). IEEE (2011)
22. Odstrcilik, J., et al.: Retinal vessel segmentation by improved matched filtering: evaluation on a new high-resolution fundus image database. IET Image Proc. **7**(4), 373–383 (2013)
23. Decencière, E., et al.: Feedback on a publicly distributed image database: the Messidor database. Image Anal. Stereol. **33**(3), 231–234 (2014)
24. Diaretdb, M.: DiaRetDB1: Diabetic retinopathy database and evaluation protocol (2009)
25. Carmona, E.J., et al.: Identification of the optic nerve head with genetic algorithms. Artif. Intell. Med. **43**(3), 243–259 (2008)
26. Staal, J., et al.: Ridge-based vessel segmentation in color images of the retina. IEEE Trans. Med. Imaging **23**(4), 501–509 (2004)
27. Sopharak, A., et al.: Automatic detection of diabetic retinopathy exudates from non-dilated retinal images using mathematical morphology methods. Comput. Med. Imaging Graph. **32** (8), 720–727 (2008)
28. Seo, J., et al.: Measurement of ocular torsion using digital fundus image. In: 26th Annual International Conference of the IEEE Engineering in Medicine and Biology Society, IEMBS 2004. IEEE (2004)
29. Walter, T., et al.: A contribution of image processing to the diagnosis of diabetic retinopathy-detection of exudates in color fundus images of the human retina. IEEE Trans. Med. Imaging **21**(10), 1236–1243 (2002)
30. Salazar-Gonzalez, A., et al.: Segmentation of the blood vessels and optic disk in retinal images. IEEE J. Biomed. Health Inf. **18**(6), 1874–1886 (2014)

A Novel Technique for Splat Generation and Patch Level Prediction in Diabetic Retinopathy

I. Syed Muhammedh Ajwahir[1(✉)], Kumar Rajamani[2], and S. Ibrahim Sadhar[3]

[1] Indian Institute of Technology, Madras, Chennai, India
ajwahir@gmail.com

[2] Robert Bosch Engineering and Business Solutions Ltd., Bengaluru, India
kumartr@gmail.com

[3] Al-Azhar College of Engineering and Technology, Idukki, India
isadhar@gmail.com

Abstract. Diabetic Retinopathy (DR) is vision threatening and can be prevented with early diagnosis and treatment. This can be achieved with the regular screening of patients known to have Diabetes for 5 years or more. Once detected with DR, it is important for doctors to maintain the progress of the disease down the line. This includes identification and marking of DR features in the fundus images. Manual marking of DR features like exudates and hemorrhages is tedious and error prone job for opthalmologists. Detection of DR is widely done with fundus imaging technique. To help aid ophthalmologists, the DR features in fundus images can be automatically marked using machine learning algorithms [1]. In this paper, a novel and generalized method for segmenting the fundus images is proposed. With our approach, retinal color images are partitioned into non-overlapping segments covering the entire image. Each segment, i.e., splat, contains pixels with similar color and spatial location. A novel method for automated generalised generation of splat is presented in this paper and further marking of the diseased splats is also proposed. The proposed method is tested on DIARETDB1 dataset, achieving accuracies of 87.7% for exudate patches, 84.6% for hemorrhage patches and 80.7% for normal patches.

Keywords: Segmentation · Retinal image · Classification · Diabetic retinopathy · Haemorrhage · Exudate · Vessel segmentation · Support vector machine

1 Introduction

Diabetic retinopathy (DR) is a diabetes mellitus which affects the small blood vessels in the retina. Hence, it is absolutely essential that patient should regularly examine eyes for early diagnosis and treatment to prevent vision loss from DR. One of the easiest ways to diagnose DR is by analyzing the fundus images. For this reason, it is necessary to capture fundus images and identify the DR by

M. Valdés Hernández and V. González-Castro (Eds.): MIUA 2017, CCIS 723, pp. 50–59, 2017.
DOI: 10.1007/978-3-319-60964-5_5

appropriate image processing techniques. As the acquisition of fundus images is inexpensive, non-invasive, and easy to perform, we prefer image based analysis.

1.1 Retinal Image Segmentation

Image segmentation is a common practice in analyzing and identifying retinal images. Plenty of approaches exist in literature to segment the haemorrhage portion of a retinal image. Out of the many available techniques, two successful methods have been utilized in order to achieve the goal of exactly segmenting the required region. They are regular and irregular segmentation. The conventional image over segmentation on a regular grid is a regular image segmentation technique where superpixels are generated [2]. But super pixels are roughly homogeneous in size and shape. Another regular segmentation algorithm called SLIC (Simple Linear Iterative Clustering) that clusters pixels in the combined five-dimensional color and image plane space was proposed by Achanta et al. [3]. The regular technique is not suitable for segmenting retinal images because the segmented regions are regular in size. However, typical retinal image has different size and shape of haemorrhage. Hence, there is a need for irregular segmentation technique for accurately capturing the haemorrhage region of a retinal image.

An irregular splat-based image segmentation method which divides images into irregular grids was proposed by Tang et al. [4]. In this technique splats are created by over-segmenting images using watershed [5] or toboggan algorithms [6]. The challenge involved in segmenting the haemorrhage portion of the retinal image lies on completely segmenting the entire haemorrhage portion. Moreover, detecting all the haemorrhages in an image is a critical requirement for subsequent processing. We propose a new gradient based approach which utilizes watershed algorithm to accurately segment the retinal image. The proposed method takes care of segmenting the haemorrhage portion fully without loosing small or large haemorrhage regions.

1.2 Outline

Details of the proposed framework of image segmentation which utilizes an automated gradient threshold based approach is discussed in Sect. 2. Detailed description and algorithm steps of the proposed technique including the preprocessing stages are discussed in this Section. In Sect. 3, feature extraction for identifying exudate splats, hemorrhage splats and normal ones in DR affected fundus images are presented. Experimental results and quantitative analysis are presented in Sect. 4. Section 5 concludes the paper.

2 Proposed Framework for Segmentation of Splats Based on Automated Gradient Threshold

The proposed algorithm aims at segmenting the retinal image such that every homogeneous part is clustered into a single segment. Disconnected homogeneous regions need to be clustered as independent segments. As the blood vessels and optic disc are having high gradient, they will be captured in any gradient based approach. Hence, before computing the gradient, the high frequency content of the blood vessels and optic disc need to be suppressed. Moreover, the fundus images might be corrupted with noises and non uniform illumination. So, different preprocessing stages need to be introduced in the proposed algorithm before proceeding with the gradient based approach.

2.1 Blood Vessel Detection

The major requirement in our segmentation is that the blood vessels should not be detected as splats. Hence, an accurate extraction and removal of retinal blood vessels is necessary. As the first step, we completely identify the retinal vessels using Matched Filter (MF) utilizing First order Derivative of Gaussian (FDoG) operator [7]. The MF-FDoG is composed of a zero-mean Gaussian function, and the FDoG operator. The MF filter has the Gaussian distribution and is defined as given in Eq. 1

$$f(x,y) = \frac{1}{\sqrt{2\pi}\sigma}\exp\left(-\frac{x^2}{2\sigma^2}\right) - m, \text{for}|x| \leq t.\sigma, |y| \leq L/2 \tag{1}$$

The FDoG filter is defined in Eq. 2

$$f(x,y) = \frac{x}{\sqrt{2\pi}\sigma^3}\exp\left(-\frac{x^2}{2\sigma^2}\right), \text{for}|x| \leq t.\sigma, |y| \leq L/2 \tag{2}$$

where σ is the scale factor, m is the mean and L is the length of the neighborhood along the y-axis. The principle of MF detection is that, the cross-section of a vessel can be modeled as a Gaussian function, a series of Gaussian-shaped filters. However, the MF will have strong responses not only to vessels but also to non-vessel edges, for example, the edges of bright blobs and red lesions in retinal images. Therefore, after thresholding the response image appropriately, many false detections can result. The vessel edges can be better identified in the presence of non vessel region by using the MF-FDoG than by using the MF. Extracted vessels of the original image of Fig. 1(a) is shown in Fig. 1(b).

2.2 Optic Disc Detection

Optic disc is identified using the adaptive mean thresholding technique. The red channel of the image is converted to one dimensional signals in horizontal and vertical projected bins by computing the row-wise and column-wise mean. The location of optic disk has peaks in the projected signals. The threshold is chosen as in Eq. 3

$$T = \min(\max(H), \max(V)) \tag{3}$$

where H and V are horizontal and vertical profiles computed as above. The red channel of the image is then converted to binary image using the threshold above [8]. Using this fact that optic disk has the highest variance, the exudate regions that gets segmented out are rejected and the OD location is determined. Then a uniform high gradient is induced in the green channel of the image so that OD gets segmented out.

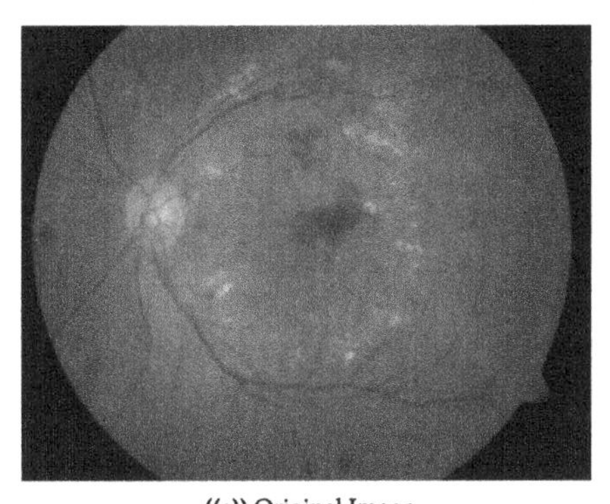

((a)) Original Image

((b)) Vessel segmented out using MF-FDoG

Fig. 1. Vessel extraction

2.3 Contrast Stretching

The next stage of preprocessing is to increase the contrast and enhance the feature as the fundus images might be corrupted with noises and non uniform illumination. We choose the green channel as proposed by [4] since it clearly exhibits the red color features of hemorrhages when compared to other channels. After that, Contrast Limited Adaptive Histogram Equalization (CLAHE) is applied to the image to increase its contrast. CLAHE will divide the image into smaller tiles. On each of these tiles, equalization is applied which will expose the image with more uniformly distributed contrast. Also CLAHE emphasizes some hidden features in the image.

2.4 Automated Gradient Threshold

The proposed algorithm to segment splats of similar color and spatial location is as follows. This works by computing the gradient magnitudes of the contrast enhanced image at a range of scales. The proposed framework also aims at extracting the required background [7]. We define a scale-of-interest (SOI) as $s \in s_1, s_2, s_3, ..., s_n$ and compute the maximum of the gradient magnitudes $|\nabla I(x, y; s)|$ over this SOI as given in Eq. 4.

$$\sqrt{I_x(x,y;s)^2 + I_y(x,y;s)^2} = \sqrt{\left[\frac{\partial}{\partial x}(G_s * I(x,y))\right]^2 + \left[\frac{\partial}{\partial y}(G_s * I(x,y))\right]^2}$$
$$= \sqrt{\left[\frac{\partial G_s}{\partial x} * I(x,y)\right]^2 + \left[\frac{\partial G_s}{\partial y} * I(x,y)\right]^2} \quad (4)$$

where $*$ is the convolution operation. We choose different variances for obtaining the gradient images. Having obtained all the gradient images for every pixel, we choose the maximum gradient value among these gradient images aggregated over SOI as described in Eq. 4. This gradient image is thresholded with a small threshold value. The segmentation is done by watershed algorithm by applying the gradient thresholded image as the input. We fix the number of segments in the output image based on the imaging process and the size of the image by adjusting the gradient threshold iteratively in such a way that the required number of segments in the output of the watershed segmentation algorithm is less than one as shown in Algorithm 1. However, we need to consider the segments of significant sizes only as the small segments are not of interest though they are bounded by homogeneous region. The small segments are likely to appear because of lighting conditions while capturing the retinal image. These very small regions are removed and merged to the nearest splat.

Algorithm 1. Splat generation

```
1: function SPLAT(gradientimage)
2:     segThres ← Number of segments
3:     gradThresh ← small gradient threshold
4:     top:
5:     nseg ← Number of segments after watershed
6:     if segThres > nseg then
7:         gradThresh ← gradThresh + 1
8:         loop:
9:         if gradImage(i)(j) > gradThresh then
10:            newgradImage(i)(j) ← gradImage(i)(j)
11:            t ← i+1
12:            j ← j+1
13:            goto loop.
14:        gradImage = newgradImage
15:        goto top.
16:    return requiredImage ← watershed(gradImage)
```

The image is partitioned into non-overlapping splats of similar intensity resulting in an irregular grid covering the entire image [4]. Splat-based representation is an image re-sampling strategy which maximizes the diversity of training samples by retaining all important samples [4].

3 Classification

The individual splats in an image are labeled as either exudate, hemorrhage or normal. This is done using the ground truth information available for DIARETDB1 database. The images in DIARETDB1 were manually assigned into catogories reperesenting the progressive state of retinopathy: normal, mild, moderate and severe non-proliferative, and profilerative. In this data set a confidence level on ground truth is set as $conf_GT = 0.75$ as discussed in [9]. The ground truth confidence level represent the certainty of the decision that a marked finding is correct.

3.1 Feature Extraction

Features for individual splats are extracted which are helpful for efficient classification of the splat images as detailed below.

1. Minimum, maximum, mean and standard deviation of intensity values in a splat in RGB, HSV, XYZ and LAB color space are computed. To accommodate variations in different color space, we normalize the intensities in each splat.
2. Histogram of 15 bins in RGB, HSV, XYZ and LAB color space is taken.
3. Histogram of Gradient (HoG) for a particular splat extracted, 30 features are considered [10].
4. Scale Invariant Feature Transform (SIFT) features are extracted in Red, Blue, and Green channels splat-wise [11].
5. Local Binary Pattern (LBP) features are extracted for Red, Blue, and Green channels separately [12].

All the extracted features and descriptions are tabulated in Table 1. Hence, a total of 816 features extracted for each of the spalt which are used for classification.

Table 1. Feature details

Feature	Number	Color space
Minimum and maximum	$2 \times 3 \times 4$	RGB, HSV, XYZ and LBP
Mean and standard deviation	$2 \times 3 \times 4$	RGB, HSV, XYZ and LBP
Histogram	15×3	RGB
Histogram of Gradient (HoG)	10×3	RGB
SIFT	128×3	RGB
LBP	58×3	RGB

3.2 Support Vector Model

The features extracted as detailed in Table 1 are used in support vector model to identify the disease affected patches in fundus images. A least squares Support Vector Machine (SVM) with Radial basis function (RBF) kernel 5 is used for the classification [13].

$$K(x, x') = exp(-\gamma * ||x' - x||^2) \tag{5}$$

where x, x′ are two feature vectors and γ is a free parameter.

4 Experimental Results

The proposed algorithm was implemented and the effectiveness of the technique was demonstrated in DiaretDB1 database images [9]. The main reason for choosing the DiaretDB1 database is that it can be used to benchmark diabetic retinopathy detection methods. Moreover, it has a well defined testing protocol and the ground truth images of haemorrhage region of a retinal image. The database is publicly available for benchmarking diagnosis algorithms [9]. The database consists of 89 colour fundus images which were taken in the Kuopio university hospital. The images were selected by the medical experts, and captured with the same 50° field-of-view digital fundus camera with different imaging settings. The data correspond to a practical situation, and can be used to evaluate the general performance of diagnostic methods. The images correspond to commonly used imaging conditions encountered in hospitals. This data set is referred to as "calibration level 1 fundus images" [9].

A typical fundus image of size 1152×1500 from DiaretDB1 [9] database was taken as shown in Fig. 2(a) which consists of haemorrhage portion and blood vessels. The original image was resized to 807×1050 in order to demonstrate the working of the proposed algorithm. It is seen that as discussed in Sect. 2, the edges of the blood vessels were also captured, as expected. Having understood the need for blood vessel suppression, we employ MF-FDoG filter as described by Eqs. 1 and 2 to detect the same. Almost all the blood vessels were detected by MF-FDoG filer. However, it is found that in addition to exact detection of the vessels, some artifacts also appear. Knowing the basic properties of vessels, like size and shape, we remove the artifacts by using the size filter and shape filter aiming at extract recovery of blood vessels. The size filter chooses the objects within the given range of size. The shape filter identifies the objects with minimum shape deviation from the given object.

Finally, we choose the variances $\sigma = 0.5$ to 6.5 in steps of 1, to obtain the gradient images. For every pixel, the maximum gradient value among these gradient images aggregated over SOI was computed. The number of segments in the output image was fixed as 800. The gradient threshold was adjusted to obtain the number of segments in the output of the watershed segmentation algorithm is 9 as generally used. This is obtained after removing the smaller segments. The final output image after segmentation is shown in Figs. 2(b) and (c). The classification results are shown in Figure 2(d) in which Green color markings are exudates and Blue color markings are hemorrhages. It is seen from the output image that all the hemorrhages have been identified fully retaining its size and shape, However, some non hemorrhage have been segmented which can easily be removed based on some specific features of hemorrhage.

The total number of splats for exudates, haemorrhage and normal are counted and the minimum of these counts is taken as reference for train data, test data split. 70% of the data is considered for training and the classifier model is built for each of the classes using Support Vector Machine (SVM) with RBF kernel. Libsvm [14] was used for building the model. 30% of the data is considered for testing against the model built. It is tested on 30 DR images with around 400 normal splats, 70 hemorrhage splat and 120 exudate splats. The 70–30 train test split is done randomly and the trained model is built each time for 10 runs. The 30% data is tested against the model for 10 runs and the performance is

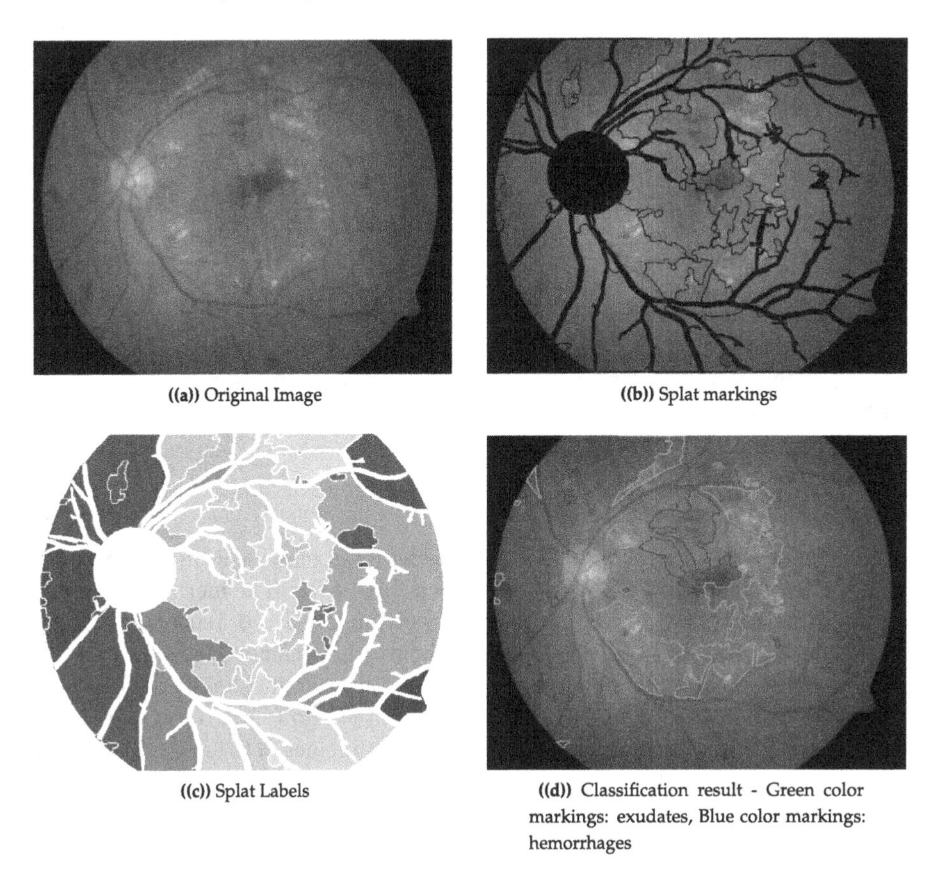

((a)) Original Image

((b)) Splat markings

((c)) Splat Labels

((d)) Classification result - Green color markings: exudates, Blue color markings: hemorrhages

Fig. 2. Multi-class classification (Color figure online)

averaged from these 10 runs. The splat is tested and the performance of the proposed method is given in the Table 2.

Hatanaka et al. [15] proposed a methodology to detect hemorrhage using density analysis, rule-based method and 3 Mahalanobis distance classifiers. They obtained sensitivity and specificity of 80% and 80% respectively.

Acharya et al. [16] used morphological image processing to detect various lesions. First, an image with blood vessels was extracted by using 'ball' shaped structuring elements, in addition to morphological operations. Then, other image with the vessels as well as hemorrhages was extracted using the same technique but slightly increased the ball size. The final detection was obtained by subtracting the image with vessels alone from the image with vessels as well as hemorrhage. The sensitivity and specificity obtained were 82% and 86% respectively.

Sinthanayothin et al. [17] used the concept of region growing in their research but in a very different way. First, the main features of the retinal image such as optic disc, fovea, blood vessels etc. were defined. Then Moat Operator was used

Table 2. Feature details

	Accuracy	Sensitivity	Specificity
Exudate	87.771	87.444	87.822
Hemorrhage	84.649	84.737	83.241
Normal	80.783	81.103	82.536

to define the features of hemorrhage followed by the application of adaptive intensity thresholding. The candidates then undergo the process of recursive region growing segmentation. They obtained sensitivity and specificity of 77.5% and 88.7% respectively.

5 Conclusion

In this paper an effective and efficient method for classification of Diabetic Retinopathy lesions in color fundus images was presented. The method is unique in a way that threshold is chosen by the algorithm itself and hence same algorithms can be used for segmenting hard exudates and hemorrhages as well. In the existing literature, different techniques and different threshold were used for lesion marking. The performance of the algorithm is also found to be very good and can be generalized for segmentation of similar images. The limitation of the proposed technique is that, if the blood vessel detection is not accurate because of high gradients, some portion of it will be captured as splat.

References

1. Walter, T., Klein, J.C., Massin, P., Erginay, A.: A contribution of image processing to the diagnosis of diabetic retinopathy-detection of exudates in color fundus images of the human retina. IEEE Trans. Med. Imaging **21**, 1236–1243 (2002)
2. Ren, X., Malik, J.: Learning a classification model for segmentation. In: Proceedings of 9th International Conference on Computer Vision 2003, vol. 1, pp. 10–17 (2003)
3. Achanta, R., Shaji, A., Smith, K., Lucchi, A., Fua, P., Süsstrunk, S.: SLIC superpixels compared to state-of-the-art superpixel methods. IEEE Trans. Pattern Anal. Mach. Intell. **34**, 2274–2282 (2012). A previous version of this article was published as a EPFL Technical report in 2010: http://infoscience.epfl.ch/record/149300. Supplementary material can be found at: http://ivrg.epfl.ch/research/superpixels
4. Tang, L., Niemeijer, M., Reinhardt, J., Garvin, M., Abramoff, M.: Splat feature classification with application to retinal hemorrhage detection in fundus images. IEEE Trans. Med. Imaging **32**, 364–375 (2013)
5. Roerdink, J.B., Meijster, A.: The watershed transform: definitions, algorithms and parallelization strategies. Fundam. Inf. **41**, 187–228 (2000)
6. Fairfield, J.: Toboggan contrast enhancement for contrast segmentation. In: 1990 Proceedings of 10th International Conference on Pattern Recognition, vol. 1, pp. 712–716 (1990)

7. Zhang, B., Zhang, L., Zhang, L., Karray, F.: Retinal vessel extraction by matched filter with first-order derivative of Gaussian. Comput. Biol. Med. **40**, 438–445 (2010)
8. Oktoeberza, K.Z.W., Nugroho, H.A., Adji, T.B.: Optic disc segmentation based on red channel retinal fundus images. In: Intan, R., Chi, C.H., Palit, H., Santoso, L. (eds.) Intelligence in the Era of Big Data, pp. 348–359. Springer, Heidelberg (2015)
9. Kauppi, T., Kalesnykiene, V., Kamarainen, J.-K., Lensu, L., Sorri, I., Raninen, A., Voutilainen, R., Uusitalo, H., Kalviainen, H., Pietila, J.: The DIARETDB1 diabetic retinopathy database and evaluation protocol. In: Rajpoot, N.M., Bhalerao, A.H. (eds.) Proceedings of the British Machine Conference, pp. 15.1–15.10. BMVA Press, September 2007. doi:10.5244/C.21.15
10. Felzenszwalb, P., McAllester, D., Ramanan, D.: A discriminatively trained, multiscale, deformable part model. In: IEEE Conference on Computer Vision and Pattern Recognition, CVPR 2008, pp. 1–8 (2008)
11. Ke, Y., Sukthankar, R.: PCA-SIFT: a more distinctive representation for local image descriptors. In: Proceedings of the 2004 IEEE Computer Society Conference on Computer Vision and Pattern Recognition, CVPR 2004, vol. 2, pp. II-506–II-513 (2004)
12. Ojala, T., Pietikainen, M., Maenpaa, T.: Multiresolution gray-scale and rotation invariant texture classification with local binary patterns. IEEE Trans. Pattern Anal. Mach. Intell. **24**, 971–987 (2002)
13. Suykens, J., Vandewalle, J.: Least squares support vector machine classifiers. Neural Process. Lett. **9**, 293–300 (1999)
14. Chang, C.C., Lin, C.J.: LIBSVM: a library for support vector machines. ACM Trans. Intell. Syst. Technol. **2**, 27:1–27:27 (2011). http://www.csie.ntu.edu.tw/~cjlin/libsvm
15. Hatanaka, Y., Nakagawa, T., Hayashi, Y., Hara, T., Fujita, H.: Improvement of automated detection method of hemorrhages in fundus images. In: 2008 30th Annual International Conference of the IEEE Engineering in Medicine and Biology Society, pp. 5429–5432 (2008)
16. Acharya, U.R., Lim, C.M., Ng, E.Y.K., Chee, C., Tamura, T.: Computer-based detection of diabetes retinopathy stages using digital fundus images. Proc. Inst. Mech. Eng. Part H: J. Eng. Med. **223**, 545–553 (2009). PMID: 19623908
17. Sinthanayothin, C., Boyce, J.F., Williamson, T.H., Cook, H.L., Mensah, E., Lal, S., Usher, D.: Automated detection of diabetic retinopathy on digital fundus images. Diabet. Med. **19**, 105–112 (2002)

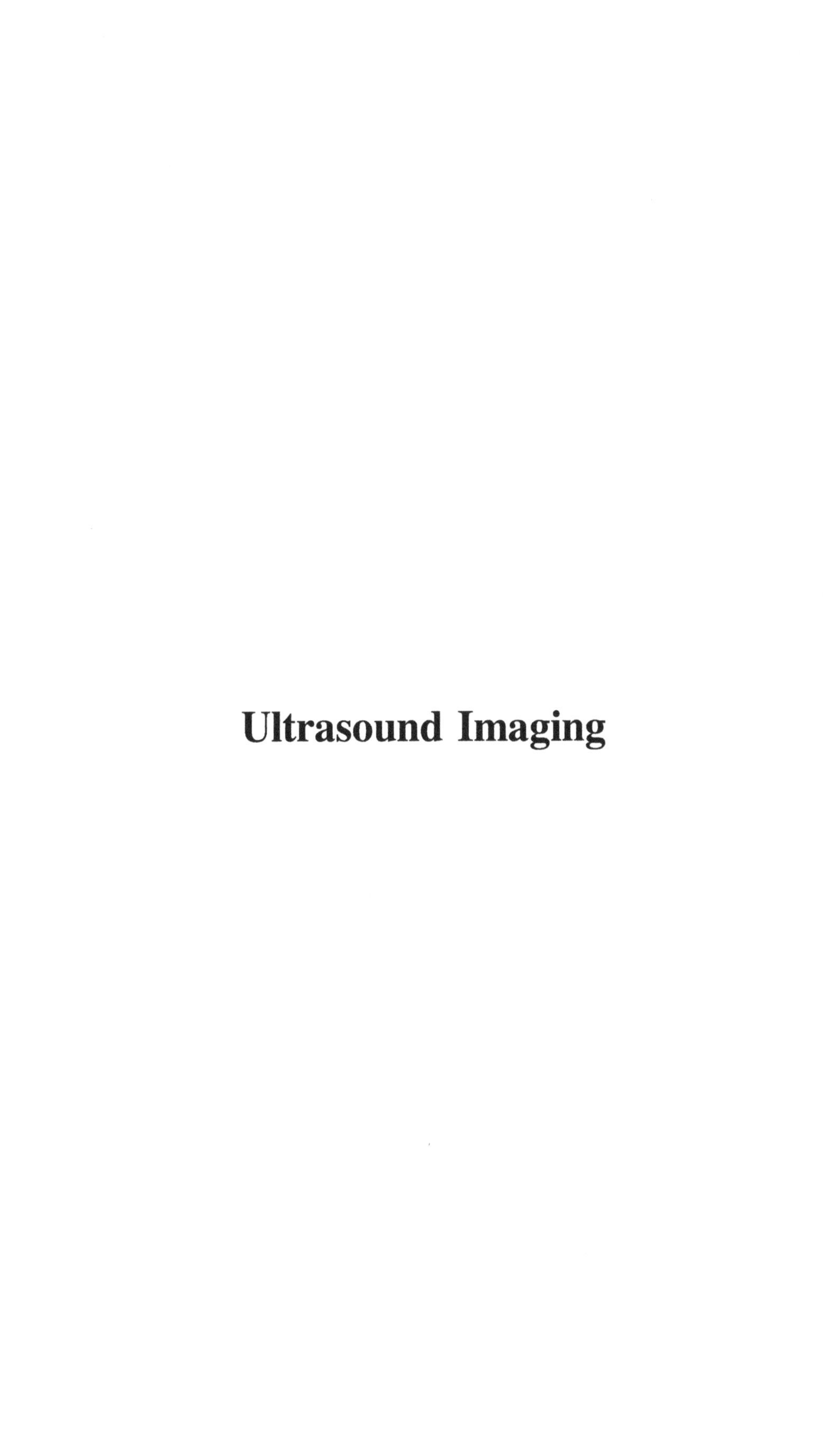

Ultrasound Imaging

Deep Residual Networks for Quantification of Muscle Fiber Orientation and Curvature from Ultrasound Images

Ryan Cunningham(✉), Peter Harding, and Ian Loram

Manchester Metropolitan University, Manchester, England, UK
ryan.cunningham@mmu.ac.uk

Abstract. This paper concerns fully automatic and objective measurement of human skeletal muscle fiber orientation directly from standard b-mode ultrasound images using deep residual (ResNet) and convolutional neural networks (CNN). Fiber orientation and length is related with active and passive states of force production within muscle. There is currently no non-invasive way to measure force directly from muscle. Measurement of forces and other contractile parameters like muscle length change, thickness, and tendon length is not only important for understanding healthy muscle, but such information has contributed to understanding, diagnosis, monitoring, targeting and treatment of diseases ranging from myositis to stroke and motor neurone disease (MND). We applied well established deep learning methods to ultrasound data recorded from 19 healthy participants (5 female, ages: 30 ± 7.7) and achieved state of the art accuracy in predicting fiber orientation directly from ultrasound images of the calf muscles. First we used a previously developed segmentation technique to extract a region of interest within the gastrocnemius muscle. Then we asked an expert to annotate the main line of fiber orientation in 4×4 partitions of 400 normalized images. A linear model was then applied to the annotations to regulate and recover the orientation field for each image. Then we applied a CNN and a ResNet to predict the fiber orientation in each image. With leave one participant out cross-validation and dropout as a regulariser, we were able to demonstrate state of the art performance, recovering the fiber orientation with an average error of just 2°.

Keywords: Ultrasound · Muscle analysis · Residual networks · Convolutional networks · Wavelet · Fiber orientation

1 Introduction

This paper presents a novel fully automatic computational approach to non-invasively measuring human skeletal muscle architectural parameters directly from ultrasound images. In recent years, ultrasound has become a valuable and ubiquitous clinical and research tool for understanding the changes which take place within muscle in ageing, disease, atrophy, and exercise. Ultrasound has been proposed (Harding et al. 2016) as a non-invasive alternative to intramuscular electromyography (iEMG) for measuring twitch frequency, useful for early the diagnosis of motor neurone disease (MND).

M. Valdés Hernnádez and V. González-Castro (Eds.): MIUA 2017, CCIS 723, pp. 63–73, 2017.
DOI: 10.1007/978-3-319-60964-5_6

Ultrasound has also recently demonstrated application to rehabilitative biofeedback (Loram et al. 2017). Other computational techniques have been developed for muscle-ultrasound analysis which would allow estimation of changes in muscle length during contraction (Loram et al. 2006), and changes in fiber orientation and length (Darby et al. 2013; Namburete et al. 2011; Rana et al. 2009). Muscle fiber orientation is one of the main identifying features of muscle state (Lieber and Fridén 2000). Without highly invasive methods, the measurement of force from muscle is currently not possible (Barry and Ahmed 1986; Finni et al. 1998; Finni et al. 2000; Gregor et al. 1987; Holden et al. 1994; Komi 1996; Komi et al. 1987; Lewis et al. 1982). Muscle fiber orientation, curvature and length is known to change with changes in force within the muscle (Herbert et al. 1995; Narici et al. 1996).

Broadly there are two main approaches to extracting fiber orientation/curvature automatically from ultrasound. The first approach is feature tracking, proposed by Darby and others (Darby et al. 2013). Feature tracking can be robust for small movements (i.e. where the difference in texure appearance of features between frames is similar) (Loram et al. 2006), however drift due to noise is inevitable, and further, severe drift can occur where for large movements, which cause disparity between texture patches (Loram et al. 2006; Yeung et al. 1998). Drift can be lessened using methods such as the Kanade Lucas-Tomasi (Shi and Tomasi 1994; Tomasi 1991) pyramidal method of tracking (Darby et al. 2012), but not solved since features can completely disappear from view as they move outside the ultrasound image plane causing feature loss (Darby et al. 2012). Darby et al. (2013) address the drift problem using a Bayesian particle filtering framework which regularizes the tracking based on a Gaussian process (GP) model of fiber shape. Their approach was validated on synthetic data with known curvatures, and evaluation with real data in the form of comparisons to existing techniques. Although this is an important contribution, the authors conclude that their method is outperformed in some scenarios by existing methods on synthetic data, and produces physiologically unrealistic fiber length values on real data (one test person/trial).

The second approach can be described as feature engineering and extraction, as proposed by Rana et al. (2009). Feature engineering can be a powerful approach to image analysis if a good model of the features can be constructed. In the case of muscle fibers within an ultrasound image, they appear as dark parallel lines (with slight curvature) at particular angles (typically 5–30°) within the muscle belly. These dark lines are surrounded by connective tissues which appear bright. The difficulty is that there are other structures which exhibit these features such as blood vessels, image artifacts and muscle boundaries. Rana and others model these high-contrast tube-like/vessel structures using orientated anisotropic wavelets. Due to noise and contrast variations within the image, after manual region of interest (ROI) identification they process the image with a multi-scale vessel enhancement filtering technique (Frangi et al. 1998) which enhances tubular structures in the image. After filtering they convolve orientated wavelets (0–90°) with the image, and for each pixel the maximum convolution gives the orientation of the fiber at that pixel. Their technique was evaluated on synthetic images with known orientations, and real images with operator identification of true fiber orientation. For synthetic images they report accuracy less than 0.02° (mean absolute error) for synthetic images, and they do not quantify the error on real images, instead they report statistically significant differences between the wavelet method and the operator identification of fiber orientation (one test participant/trial during cycling).

A third approach which has thus far not been considered is to learn features which predict fiber orientation directly from data. In recent years, with the advent of graphics processing unit (GPU) computing and advancements in machine learning algorithms, such as the introduction of the rectified linear unit, dropout, and residual networks for very deep learning (Krizhevsky et al. 2012b; He et al. 2015; Hinton, 2014; Krizhevsky et al. 2012b; LeCun et al. 1990; Nair and Hinton 2010), image understanding and analysis is more feasible for challenging images such as medical ultrasound images. As static analysis has a clear and demonstrable advantage over dynamic analysis (tracking), we propose to learn to extract fiber orientation directly from data with manual identifications of fiber orientations. We hypothesize a machine learning approach will provide an improvement over feature engineering (Rana et al. 2009) because a convolutional neural network (CNN) can learn more complex convolution filters, which are more descriptive and discriminative of fibers and other non-fibrous structures in the image, over a filter-bank technique. One problem with a machine learning approach, as highlighted by Rana et al. (2009), is that manual identification of fiber orientation is subjective and difficult, and since supervised machine learning methods typically require large volumes of accurately labeled data, the CNN may not have the required information to create an accurate generalizable solution. We address this problem by regulating the operator identification process using multiple annotations and linear models. We compare a standard 7-layer deep (not counting 4 max-pooling layers) CNN with a 27-layer *very* deep (not counting 4 max-pooling layers) residual neural network (ResNet), both with and without dropout for CNN and ResNet. Results are compared against the wavelet approach on real images undergoing joint rotations and contractions.

2 Methods

2.1 Data Collection

Ultrasound data were recorded from 19 participants (5 female, ages: 30 $\pm$ 7.7) during dynamic standing tasks. Participants stood upright on a programmable/controllable foot pedal system during 3 tasks while strapped to a backboard. During the tasks we recorded calf muscle (*medial gastrocnemius*: MG) activation using electromyography (EMG), ankle joint angle and joint torque, all at 1000 Hz, and ultrasound of the MG at 25 Hz (AlokaSSD-5000 PHD, 7.5 MHz). The three tasks were designed to observe distinct muscle mechanical changes during isolated joint angle changes via sinusoidal modulation of the pedal system (resulting in passive changes in the length of the calf muscles), isolated contraction via pushing down of the toes onto the pedals while the pedals were fixed in a neutral position (resulting in active changes in length of the calf muscles), and combined joint angle changes with contractions (resulting in a mixture of passive and active changes in calf muscle length). Simulink (Matlab, R2013a, The-MathWorks Inc., Natick, MA) was used to interface with the lab equipment (pedal system and EMG), and for video synchronization a hardware trigger was used to initiate recording at the start of each trial.

2.2 Automatic Identification of Region of Interest (Muscle Segmentation)

For segmentation and ROI identification we used an already established technique previously developed by our group (Cunningham et al. 2015) for segmenting skeletal muscle within ultrasound. The technique involved manual annotation of a number of training images (100) and construction of a principal components model of shape. The shape model is then used to define a mean ultrasound image by warping each texture in the space to the mean texture (after shape alignment via Procrustes analysis). Then a dictionary of textures and associated shapes is created from the mean texture and shape using the same method by warping the mean texture to equidistant points in the component model orthogonal to the axes of the n largest components (representing 90% of the shape variance). To segment a new image the dictionary is used to define an initial segmentation, and then a modified Active Shape Model is iteratively used to refine and complete the segmentation. The segmentation of the gastrocnemius muscle allowed extraction of a rectangular ROI (496×128) equidistant from its boundaries and orthogonal to the main axis of the muscle (linear least squares fit over the upper and lower boundary). For the purposes of this paper, generalization performance is not of primary concern since we only need accurate segmentation to evaluate the fiber orientation methods we have developed here. However, accuracy was evaluated on a leave one participant out basis and the results are presented in Sect. 3.

2.3 Regulated Expert Identification of Fiber Orientation

After segmentation and region extraction, an expert was asked to identify regional fiber orientation in 400 normalized (128×128) training images and 50 validation and 50 test images. Since there is regional variation in orientation due to curvature and other factors, each image was divided into a 4×4 grid of 32×32 sub-images, and then the expert identified the main line of fiber orientation in each sub-image. In many cases the expert could not identify fibers. In these cases the expert was allowed to mark the sub-image as undefined. Following manual identification of fiber orientation, linear models were fit to the annotation of each individual image,

$$\mathbf{w} = \left(\mathrm{X}^{\mathrm{T}}\mathrm{X}\right)^{-1}\mathrm{X}^{\mathrm{T}}\mathbf{y},$$

where X is a matrix in which the columns are a constant 1 (bias term) and the horizontal and vertical center coordinates of each sub-image (not including sub-images marked as undefined by the expert) relative to the full image, $\mathbf{y}$ is a row vector containing the corresponding angle of the expert-defined line of fiber orientation for each pair of coordinates, and $\mathbf{w}$ therefore is a vector of coefficients which can multiply an arbitrary pair of coordinates in the space, recovering a linear estimate of local fiber orientation,

$$\theta = \mathrm{X}\mathbf{w}.$$

The linear model is used not only to reconstruct the orientations which the expert could not identify, but also to regulate the expert's inconsistencies and variations in challenging images since physiologically we expect fibers to follow similar trajectories (Fig. 1).

Fig. 1. Regulated expert identification of regional fiber orientation. Left: expert identification of main fiber orientation. Middle: reconstruction after linear model (blue dashed) of original identification (green solid). Right: Line trace of linear fit over entire image, revealing regional curvature over the muscle. (Color figure online)

2.4 Vessel Enhancement and Wavelet Method

After segmentation and ROI identification and before wavelet analysis all images were processed at full resolution (496 × 128) with the vessel enhancement filter in the same way as (Rana et al. 2009) to enhance the appearance of fibers (and other tubular structures) in the image. We used an existing implementation of (Frangi et al. 1998). Parameters were optimized empirically on the training set to obtain processed images similar to those presented by Rana et al. (2009).

Following vessel enhancement all processed images were convolved at full resolution (496 × 128) with a set of orientated (41 × 41) Gabor wavelets,

$$G(x,y) = -exp\left(\frac{(x-k-1)^2(y-k-1)^2}{-dk}\right)\cos\left(\frac{2\pi[(x-k-1)\cos\alpha - (y-k-1)\sin\alpha]}{f}\right),$$

where k is the kernel size, d is a damping term, α is the orientation, and f is the spatial frequency. All parameters were set as originally described in (Rana et al. 2009). Wavelets were constructed in the range of 0°–90° in 1° increments. Following convolution with the filter bank, the maximum wavelet convolution at each pixel was stored. From the list of maximum convolutions, we then removed any convolution values less than one standard deviation from the mean. At that point we applied a linear model to the location of the pixels and wavelet orientations given by the maximum convolutions. This was done as in Sect. 2.3 to resample the orientations at points comparable with the expert identifications and to regulate noise/error (Figs. 2 and 3).

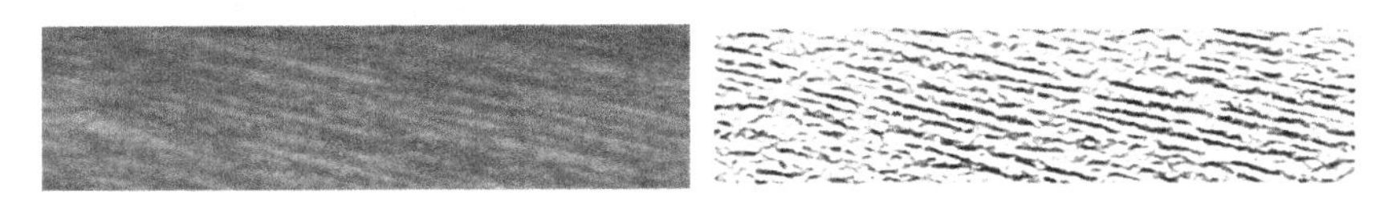

Fig. 2. Vessel filter. This figure shows how a noisy ultrasound ROI image (left) can be enhanced (right) by the vesselness filter prior to being convolved with the wavelet filter bank.

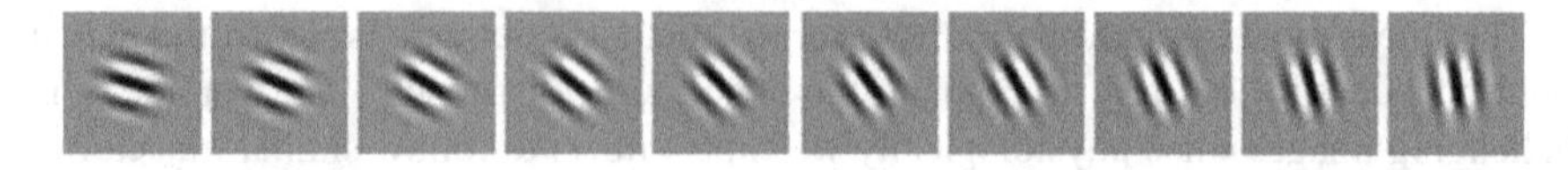

Fig. 3. Wavelet filter bank. The figure shows 10 selected orientated wavelets ranging from 17° to 80° in increments of 7°.

2.5 Deep Learning Method

After segmentation and ROI identification, due to concern over model training time all images were resampled (bilinear interpolation) to a standard size of (128×128). We then trained a CNN and a ResNet, first without dropout, and then with dropout in the fully connected layer. The architecture we chose was a simple one due to concern over training time, yet we empirically confirmed that it was complex enough to learn the training set. The input to each net was the resampled images, and the output was 16 linear regression units, one for each fiber orientation of the subsampled images derived from the linear model of the expert identifications.

Each net had rectified linear units (ReLU) in the hidden layers, and 4 max pooling layers. In between pooling layers the number of convolutional filters in the CNN from input to output was 25-*p*-36-*p*-49-*p*-100-*p*-196 (where *p* denotes max pooling layer) and one fully connected layer of 1024 Leaky ReLUs ($\alpha = 1e - 1$). The ResNet had the same architecture with the main difference being 5 times the number of convolutional layers in between max pooling layers, and there were identity shortcut connections used in alternating consecutive layers (2 shortcut connections per set of 5 convolutional layers). We did not use projection shortcuts since identity shortcuts were adequate for our purposes. Both CNN and ResNet used a convolution stride of 1×1, and the same max pooling dimensions (from input to output: $2 \times 2 - 4 \times 4 - 2 \times 2 - 4 \times 4$). For the CNN the filter sizes were all 3×3 with the exception of the input layer which was 7×7. For the ResNet the filter sizes were 3×3 throughout. Initial weights were drawn from a normal distribution, normalized by the product of the size of fan-in and the dropout coefficient, $w_i = N(0,1)\sqrt{\frac{2}{n\frac{1}{1-p}}}$.

For training, both nets used a learning rate of $5e - 4$ with a momentum of $9e - 1$ and a small amount of L2 weight decay ($1e - 3$), and no learning rate decay or adaptive learning rate was used. Models were trained with ($p = 5e - 1$) and without dropout ($p = 0$) for 250 full passes over the training set with a batch size of 1 (i.e. online stochastic gradient descent). For testing, the mean square error (MSE) was monitored on the validation and testing sets and the model with the lowest validation error was selected as the solution, at which point the testing error was used to estimate generalized performance on unseen data. We also present root MSE for performance evaluation in degree units. Finally, we quantify the linear predictive power (R^2) of each approach for predicting the measured ankle joint torque during all three trials as described in Sect. 2.1.

3 Results

Images are presented along with predictions of the regional fiber orientation in the testing participant for the three approached under investigation (see Fig. 4). All three methods clearly produce realistic estimates of the fiber direction; however the wavelet method demonstrates some fairly substantial deviations from the ground truth (expert annotations). RMSE measures report accuracy in general for all methods comfortable below 2°, with the ResNet approaching errors of 1° in the training and test cases (see Fig. 5). The most important result came from the R^2 linear prediction of measured joint torque during the combined, active, and passive trials in the test participant. Both CNN and ResNet clearly demonstrate greater predictive power over the wavelet method, despite using a lower resolution image (128 × 128 vs 496 × 128). Further, the ResNet method showed much less deviation from ground truth as confirmed by MSE and RMSE standard deviation (see Table 1) and visual assessment (see Fig. 4).

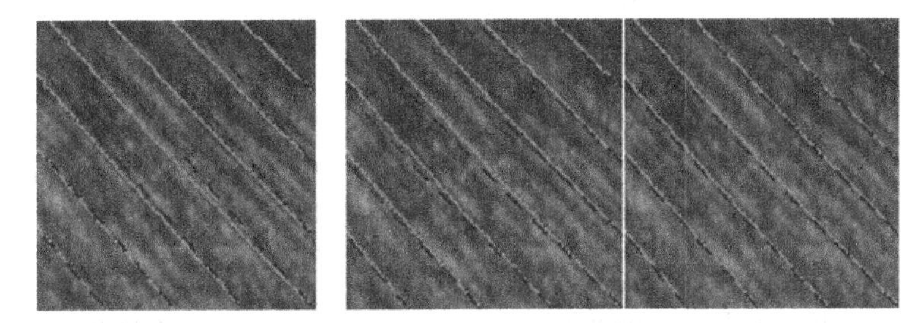

Fig. 4. Method comparison on held out test participant (left: wavelet method, middle: CNN. right: ResNet). The blue dashed line is the prediction of the fiber orientation from each of the methods, whereas the green solid line is the regulated expert label. Notice the clear improvement of the ResNet over the CNN and the wavelet-based method. (Color figure online)

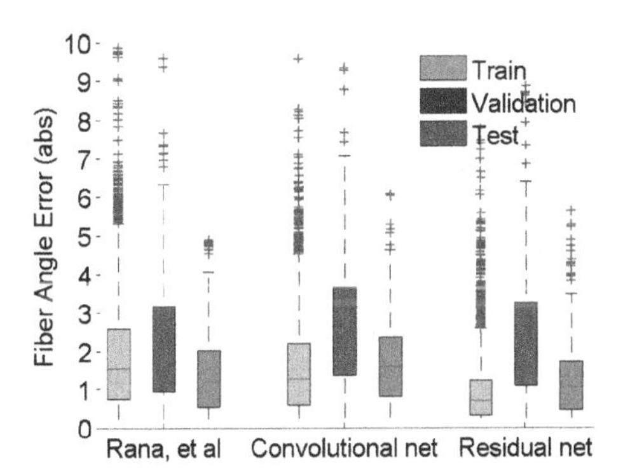

Fig. 5. Absolute fiber angle prediction error. This figure summarizes the errors of the three approaches in real values (angle in degrees).

Table 1. Summary of results. This table sumarises the results of all three methods for each of the labeled data sets. The predictive power is also given (R^2) for the testing participant. In general the convolutional network and wavelet methods perform similarly. Conversely, the residual network reports a clear improvement with generally less deviation and lower error over all cases. The discriminating factor is in the predictive power, where both deep learning methods give a marked improvement over Rana et al.

Method	Data set	RMSE	MSE	R^2
Rana et al. (2009)	Training	$2.30° \pm 3.46°$	0.84 ± 1.22	0.3837
	Validation	$2.76° \pm 3.44°$	0.77 ± 1.07	
	Testing	$1.72° \pm 1.98°$	0.56 ± 0.99	
Convolutional net	Training	$2.00° \pm 3.50°$	0.60 ± 0.95	0.4222
	Validation	$3.08° \pm 3.44°$	0.91 ± 1.08	
	Testing	$1.94° \pm 2.08°$	0.70 ± 1.35	
Residual net*	Training	$1.25° \pm 2.50°$ *	0.24 ± 0.49 *	0.5897*
	Validation	$2.70° \pm 3.03°$ *	0.72 ± 0.93 *	
	Testing	$1.53° \pm 1.88°$ *	0.47 ± 1.08 *	

4 Discussion

The combination of segmentation (Cunningham et al. 2015) with deep learning provides a general framework for extracting physiological information directly from ultrasound images of human skeletal muscle. Comparison with an existing state of the art wavelet convolution method reveals a new benchmark set by deep residual networks (see Table 1). The convolutional network was comparable in accuracy to the wavelet method and in some cases proved more robust, yet it did not demonstrate the gain in performance that the residual network did. The benefits of deep residual networks are clear for classification problems (He et al. 2015), however the work presented here contributes evidence of success in the domain of multiple regression. All of the methods presented here provide the necessary information to quantify fiber orientation which consequently provides the information required to estimate fiber length and pennation angles, which are common features of interest within biomechanics and medical analysis.

Although labels were easy to acquire (c. 8 h of annotation), the sorts of volumes of data typically used to train deep networks was not feasible for this study. Although the results presented here are competitive with the state of the art – *if not a significant improvement* – we would argue strongly that those results could certainly be improved perhaps by an order of magnitude if the size of the labeled data set could be increased by hundreds or thousands. Due to concern over the said small data set, state of the art regularization techniques were used to improve model generalization: dropout, early stopping, weight sharing (convolutions), max-pooling, and identity connections (residual learning). The success of dropout as a regulariser is well established in classification problems (Krizhevsky et al. 2012a), however we our results evidence success in the domain of multiple regression. We show that dropout behaves as expected and continues to minimize error while the nets without dropout diverge before reaching the same level of performance as the nets with dropout.

The previous state of the art used engineered wavelets based on prior knowledge of fiber size and expected orientation, and these wavelets were convolved with the image (after vesselness filtering). This approach (feature engineering) is becoming an out-dated concept where there exists an abundance of data, since deep learning methods can learn exactly what sort of filters would extract only the relevant information and discard all of the irrelevant information by learning directly from data.

This preliminary work did not investigate many of the available free parameters available to deep learning methods. Preventing over-fitting was our primary concern after learning the training data, yet for this preliminary investigation we only used early stopping with and without dropout. Dropout demonstrably works well even without parameterization, although we must highlight that there is potential to use dropout in multiple layers and with different frequencies, all of which would need fully investigating on a validation set. We used some weight decay as is standard practice to help convergence; however we acknowledge that this parameter could also be tuned to improve generalization. We did investigate different architectures in terms of max pooling configuration and number of filters per layer, but we did not attempt to cross-validate all of the different models due to concern over training time – we chose a reasonably sized model with standard pooling configurations and then relied on dropout and early stopping. We note that investigations of different architectures would become more important when the project scales up in terms of increasing the size of the training set.

5 Conclusions

This paper has presented a robust and repeatable method for extracting fiber orientations directly from ultrasound images of human skeletal muscle. A clear improvement has been demonstrated over a widely used and well-established wavelet-based method (Rana et al. 2009). In this paper deep learning principals were applied to automatically learn discriminative features of muscle fibrous tissue from a relatively small data set of 400 labeled images. In order to provide robust and accurate labels a novel annotation method was introduced in which linear models are used to regulate annotations and recover a fiber orientation field. That development led to a successful extension of the wavelet-based method which facilitated comparison with the deep learning methods we have introduced here. The wavelet-based method previously was unable to recover robust estimations of regional fiber orientation, and the application of linear models allowed recovery of that information as evidenced (see Fig. 5 and Table 1). This paper provides further evidence that very deep networks (ResNets) can help performance and generalisation, even when there is not an abundance of labeled data available. With additional data we propose that this project could easily be extended more successfully and this preliminary muscle analysis step could very likely form part of a skeletal muscle analysis system which accurately predicts the passive and active muscle forces non-invasively directly from single ultrasound images and sequences. Such a contribution could enable early diagnoses of diseases such as MND, and would enable personalized musculoskeletal medical diagnosis, monitoring, treatment targeting, and care.

References

Barry, D., Ahmed, A.M.: Design and performance of a modified buckle transducer for the measurement of ligament tension. J. Biomech. Eng. **108**(2), 149–152 (1986). https://doi.org/10.1115/1.3138594

Cunningham, R., Harding, P., Loram, I.: Real-time ultrasound segmentation, analysis and visualization of deep cervical muscle structure. Trans. Med. Imaging **36**(2), 653–665 (2015). https://doi.org/10.1109/TMI.2016.2623819

Darby, J., Hodson-Tole, E.F., Costen, N., Loram, I.D.: Automated regional analysis of B-mode ultrasound images of skeletal muscle movement. J. Appl. Physiol. **112**(2), 313–327 (2012). https://doi.org/10.1152/japplphysiol.00701.2011

Darby, J., Li, B., Costen, N., Loram, I., Hodson-Tole, E.: Estimating skeletal muscle fascicle curvature from B-mode ultrasound image sequences. IEEE Trans. Biomed. Eng. **60**(7), 1935–1945 (2013). https://doi.org/10.1109/TBME.2013.2245328

Finni, T., Komi, P.V., Lukkariniemi, J.: Achilles tendon loading during walking: application of a novel optic fiber technique. Eur. J. Appl. Physiol. Occup. Physiol. **77**(3), 289–291 (1998). https://doi.org/10.1007/s004210050335

Finni, T., Komi, P.V., Lepola, V.: In vivo human triceps surae and quadriceps femoris muscle function in a squat jump and counter. Eur. J. Appl. Physiol. **83**, 416–426 (2000). https://doi.org/10.1007/s004210000289

Frangi, A.F., Niessen, W.J., Vincken, K.L., Viergever, M.A.: Multiscale vessel enhancement filtering. In: Wells, W.M., Colchester, A., Delp, S. (eds.) MICCAI 1998. LNCS, vol. 1496, pp. 130–137. Springer, Heidelberg (1998). doi:10.1007/BFb0056195. https://doi.org/10.1016/j.media.2004.08.001

Gregor, R.J., Komi, P.V., Järvinen, M.: Achilles tendon forces during cycling. Int. J. Sports Med. **8**(Suppl. 1), 9–14 (1987). https://doi.org/10.1055/s-2008-1025698

Harding, P.J., Loram, I.D., Combes, N., Hodson-Tole, E.F.: Ultrasound-based detection of fasciculations in healthy and diseased muscles. IEEE Trans. Biomed. Eng. **63**(3), 512–518 (2016). https://doi.org/10.1109/TBME.2015.2465168

He, K., Zhang, X., Ren, S., Sun, J.: Deep residual learning for image recognition. Arxiv.Org **7**(3), 171–180 (2015). https://doi.org/10.3389/fpsyg.2013.00124

Herbert, R.D., Gandevia, S.C., Herbert, R.D.: Changes in pennation with joint angle and muscle torque: in vivo measurements in human brachialis muscle. J. Physiol. **484**(Pt2), 523–532 (1995). https://doi.org/10.1113/jphysiol.1995.sp020683

Hinton, G.: Dropout: a simple way to prevent neural networks from overfitting. J. Mach. Learn. Res. **15**, 1929–1958 (2014)

Holden, J.P., Grood, E.S., Korvick, D.L., Cummings, J.F., Butler, D.L., Bylski-Austrow, D.I.: In vivo forces in the anterior cruciate ligament: direct measurements during walking and trotting in a quadruped. J. Biomech. **27**(5), 517–526 (1994). https://doi.org/10.1016/0021-9290(94)90063-9

Komi, P.V.: Optic fibre as a transducer of tendomuscular forces. Eur. J. Appl. Physiol. Occup. Physiol. **72**(3), 278–280 (1996). https://doi.org/10.1007/BF00838652

Komi, P.V., Salonen, M., Jarvinen, M., Kokko, O.: In vivo registration of Achilles tendon forces in man. I. Methodological development. Int. J. Sports Med. **8**(Suppl. 1), 3–8 (1987). https://doi.org/10.1055/s-2008-1025697

Krizhevsky, A., Sutskever, I., Hinton, G. E.: ImageNet classification with deep convolutional neural networks. Adv. Neural Inf. Process. Syst. **25**, 1–9 (2012a). https://doi.org/10.1117/12.2176558

Krizhevsky, A., Sutskever, I., Hinton, G.: Imagenet classification with deep convolutional neural networks. In: Advances in Neural Information Processing Systems, pp. 1097–1105 (2012b)

LeCun, Y., Boser, B., Denker, J.S., Henderson, D., Howard, R.E., Hubbard, W., Jackel, L.D.: Handwritten digit recognition with a back-propagation network. In: Advances in Neural Information Processing Systems, pp. 396–404 (1990). https://doi.org/10.1111/dsu.12130

Lewis, J.L., Lew, W.D., Schmidt, J.: A note on the application and evaluation of the buckle transducer for the knee ligament force measurement. J. Biomech. Eng. **104**(2), 125–128 (1982). http://www.ncbi.nlm.nih.gov/pubmed/7078126

Lieber, R.L., Fridén, J.: Functional and clinical significance of skeletal muscle architecture. Muscle Nerve **23**(11), 1647–1666 (2000). https://doi.org/10.1002/1097-4598(200011)23:11<1647:AID-MUS1>3.0.CO;2-M

Loram, I., Bate, B., Harding, P., Cunningham, R., Loram, A.: Proactive selective inhibition targeted at the neck muscles: this proximal constraint facilitates learning and regulates global control. IEEE Trans. Neural Syst. Rehabil. Eng. **25**(4), 357–369 (2017). https://doi.org/10.1109/TNSRE.2016.2641024

Loram, I., Maganaris, C., Lakie, M.: Use of ultrasound to make noninvasive in vivo measurement of continuous changes in human muscle contractile length. J. Appl. Physiol. **100**, 1311–1323 (2006). https://doi.org/10.1152/japplphysiol.01229.2005

Nair, V., Hinton, G.E.: Rectified linear units improve restricted Boltzmann machines. Proc. Int. Conf. Mach. Learn. **3**, 807–814 (2010). https://doi.org/10.1.1.165.6419

Namburete, A.I.L., Rana, M., Wakeling, J.M.: Computational methods for quantifying in vivo muscle fascicle curvature from ultrasound images. J. Biomech. **44**(14), 2538–2543 (2011). https://doi.org/10.1016/j.jbiomech.2011.07.017

Narici, M.V., Binzoni, T., Hiltbrand, E., Fasel, J., Terrier, F., Cerretelli, P.: In vivo human gastrocnemius architecture with changing joint angle at rest and during graded isometric contraction. J. Physiol. **496**(1), 287–297 (1996). https://doi.org/10.1113/jphysiol.1996.sp021685

Rana, M., Hamarneh, G., Wakeling, J.M.: Automated tracking of muscle fascicle orientation in B-mode ultrasound images. J. Biomech. **42**(13), 2068–2073 (2009). https://doi.org/10.1016/j.jbiomech.2009.06.003

Shi, J., Tomasi, C.: Good features to track. In: Proceedings of IEEE Computer Society Conference on Computer Vision and Pattern Recognition, CVPR 1994, pp. 593–600 (1994). https://doi.org/10.1109/cvpr.1994.323794

Tomasi, C.: Detection and tracking of point features. School of Computer Science, Carnegie Mellon University, **91**(April), 1–22 (1991). https://doi.org/10.1016/S0031-3203(03)00234-6

Yeung, F., Levinson, S.F., Parker, K.J.: Multilevel and motion model-based ultrasonic speckle tracking algorithms. Ultrasound Med. Biol. **24**(3), 427–441 (1998). https://doi.org/10.1016/S0301-5629(97)00281-0

Modelling, Speckle Simulation and Quality Evaluation of Synthetic Ultrasound Images

Prerna Singh[1(✉)], Ramakrishnan Mukundan[1], and Rex de Ryke[2]

[1] Department of Computer Science and Software Engineering, University of Canterbury, Christchurch, New Zealand
prerna.singh@pg.canterbury.ac.nz, mukundan@canterbury.ac.nz

[2] Radiology Services, Christchurch District Health Board, Christchurch, New Zealand
Rex.DeRyke@cdhb.health.nz

Abstract. Speckle noise reduction is an important area of research in the field of ultrasound image processing. Several algorithms for speckle noise characterization and analysis have been recently proposed in the area. Synthetic ultrasound images can play a key role in noise evaluation methods as they can be used to generate a variety of speckle noise models under different interpolation and sampling schemes, and can also provide valuable ground truth data for estimating the accuracy of the chosen methods. However, not much work has been done in the area of modelling synthetic ultrasound images, and in simulating speckle noise generation to get images that are as close as possible to real ultrasound images. This paper discusses these aspects, presents novel algorithms for speckle simulation and modelling based on three sampling schemes, and also evaluates the quality of the outputs using image quality metrics. Detailed experimental analysis including both quantitative and subjective assessments are also presented.

Keywords: Ultrasound image analysis · Speckle simulation · Speckle noise reduction · Synthetic ultrasound images · Image quality assessment

1 Introduction

Ultrasound images are known to have poor signal-to-noise ratio, yet they are low cost, non-invasive techniques in diagnostic radiology and hence extensively used in clinical applications. Several new ultrasound image analysis algorithms are currently being researched for noise reduction [1–3], segmentation [4], registration and volume reconstruction [5]. Online ultrasound image databases are now becoming increasingly available and this has greatly benefitted researchers in obtaining reference images for testing and evaluating algorithms [5–7].

The speckle noise in ultrasound images degrades the fine details and edge definitions, and limits the contrast resolution by making it difficult to detect small and low contrast lesions in the body. Therefore, algorithms for ultrasound image filtering and

M. Valdés Hernández and V. González-Castro (Eds.): MIUA 2017, CCIS 723, pp. 74–85, 2017.
DOI: 10.1007/978-3-319-60964-5_7

analysis primarily focus on the characteristics of speckle noise and try to minimize its effects on image interpretation [8]. To analyse the effectiveness or accuracy of speckle reduction techniques, it is necessary to add controlled noise to ideal noiseless images [2]. In the absence of such noiseless ground truth images, researchers commonly use standard non-ultrasound test images (e.g. Lena, Mandrill etc.), model speckle noise on those images and perform algorithm evaluation. This paper addresses the need for generating accurate synthetic models of ultrasound image formation for applications in speckle noise analysis. A synthetic ultrasound image can be sampled using a configuration of points that correspond to either linear or sector scan modes of ultrasound imaging, and interpolated later after generating speckle noise at the sampled points to obtain visually realistic effects. Synthetic images can therefore be used to generate simulated ultrasound images with a wide range of image and noise characteristics useful for filtering methods and noise analysis.

Statistical and empirical methods of generating speckle lack realism owing to the lack of image modeling. There are only very limited algorithms reported in literature for speckle simulation based on image acquisition modeling. Perreault and Auclair-Fortier [9] proposed an efficient simulation model of ultrasound images based on a radial-polar configuration of sampling points and a speckle noise simulation algorithm. We extend their work by considering different types of sampling and interpolation schemes and by performing detailed experimental analysis to compare their effectiveness in producing realistic speckle simulation. Their work used images of Lena and Barbara for generating the simulated images with speckle noise. However, for generating highly realistic synthetic models, we require images that clearly show the anatomical features present in an ultrasound image without the noise content. To achieve this goal, we used an artist to render the features based on three reference ultrasound images, and used this as our base image.

A very important aspect of synthetic image modelling algorithms is quality assessment. In the proposed method, the base synthetic image is modified as outlined above using the acquisition model, speckle noise simulation and interpolation of the sampled points. To the authors knowledge, no prior work has been reported on image quality assessment of ultrasound images using spatial frequency measure (SFM), and spectral activity measure (SAM) metrics. In this paper, the quality of the generated outputs is compared with that of real ultrasound images using these image quality metrics.

The paper presents the complete framework for the development of synthetic ultrasound images including the set of processes in both simulation and evaluation stages. Each stage incorporates a wide range of parametric variations and options, allowing the user to generate synthetic images with varying levels of sampling, interpolation and noise characteristics. This paper is organized as follows: The next section gives a brief outline of the images used and the methods in the processing pipeline. Section 3 describes the simulation model in detail. Section 4 presents the methods used in the evaluation stage. Section 5 presents experimental results and their evaluations, and Sect. 6 gives a summary of the work presented in the paper and outlines future directions.

2 Materials and Methods

All experimental work presented in the paper are based on images derived from three reference ultrasound images sourced from the online ultrasound image gallery [7]. These are ultrasound scans of the liver, and have very similar image features, intensity distribution and noise content. Three reference images were used because ultrasound images could present variations in texture, image quality and speckle noise content as shown by the SFM and SAM values in Table 1 later in Sect. 5. For comparison, the images and their histograms are presented in Fig. 1.

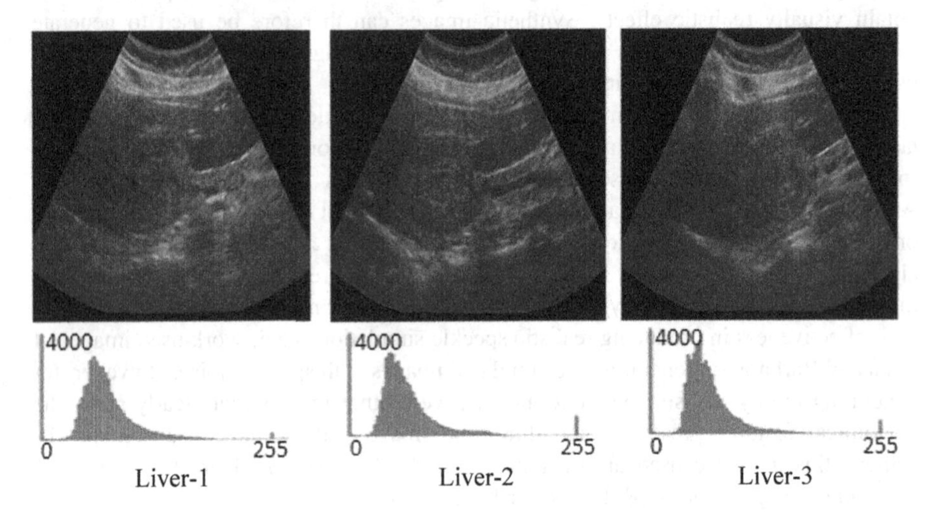

Fig. 1. Reference ultrasound images [7] used in our work, and their histograms.

The three reference images in Fig. 1 were used by an artist to sketch the image features which formed our base synthetic image (Fig. 2). The histogram of the synthetic image bears similarity with those of the reference images.

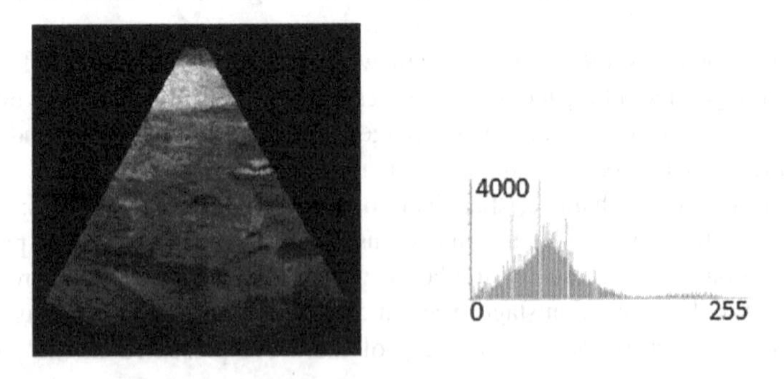

Fig. 2. Artist rendered synthetic image and its histogram.

The main methods used in the speckle simulation modelling and evaluation pipeline are depicted in Fig. 3. Within the simulation model, the synthetic image is first sampled based on an acquisition model, speckle noise is then generated at the sampled points, and an interpolation algorithm used to fill the sector scan region. The evaluation model uses image quality metrics computed for the output are then compared with those of the reference ultrasound images for a quantitative assessment of the quality of the final synthetic images. A subjective evaluation is also performed using expert sonographers.

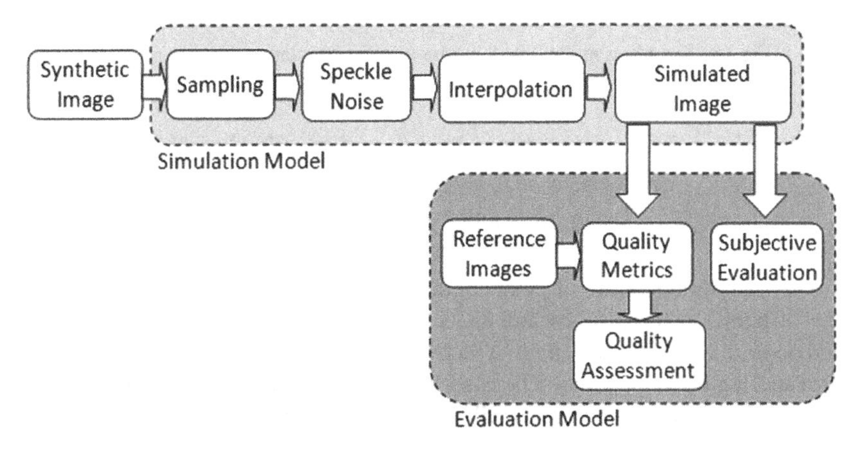

Fig. 3. The simulation and evaluation stages of the processing pipeline.

The processing stages within the simulation and evaluation models are further elaborated in the following sections.

3 The Simulation Model

The first stage of the sampling model is the method that generates a set of points at a coarse spatial resolution. The configuration of points models the loss of resolution of the ultrasound image due to pulse length, and also the scanning mode (sector or linear). One of the original contributions in this field is the paper by Perreault and Auclair-Fortier [9], where a radial-polar sampling model was introduced. We extend their work and propose three types of sampling methods called radial-polar, radial-uniform, and uniform grid. The first two are closely related to sector scan, while the third corresponds to a sampling in linear orthogonal directions (Fig. 4).

In Fig. 4, the sector angle is denoted by Φ, and the extent of the sector is given by radial distances d_{min} and d_{max}. The image width is denoted by w. We also denote the total number of divisions along each radial line (axial resolution) by m, and the number of division of the sector angle (lateral resolution) by n. The Cartesian coordinates of the sampled points for radial-polar sampling are given by

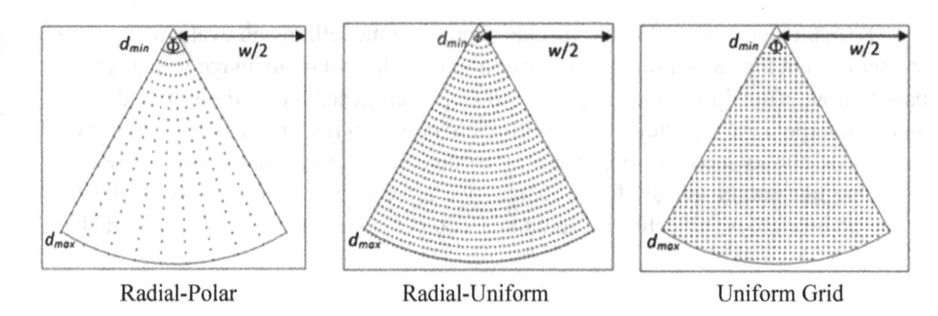

Fig. 4. Sampling models that can be used in simulating speckle noise.

$$\begin{aligned} d_j = d_{\min} + j(d_{\max} - d_{\min})/(m-1); \quad \theta_i &= (3\pi - \Phi)/2 + i\Phi/(n-1) \\ x(i,j) = d_j \cos\theta_i + w/2; \quad y(i,j) &= -d_j \sin\theta_i; \\ i = 0..(n-1); \quad j &= 0..(m-1) \end{aligned} \tag{1}$$

The non-uniform spacing of points in the radial-polar sampling method causes the density of points to increase towards the sector's apex. The radial-uniform sampling method uses a constant arc length Δ between points along each arc to generate a uniform spacing between points. The equations for this sampling model are same as in Eq. (1) except that the polar angle θ will now depend on both i and j as shown below.

$$\theta_{ij} = (3\pi - \Phi)/2 + i\Delta/d_j \tag{2}$$

The uniform grid is the simplest sampling model corresponding to a rectangular arrangement of uniformly spaced points with a constant distance δ between points. If a sector scan region is required, the points outside the region are clipped using the line equations of the two bounding edges. Using Eq. (3), if $f(x, y, \theta_{min}) > 0$ or $f(x, y, \theta_{max}) < 0$, the point (x, y) is outside the sector region.

$$\begin{aligned} \theta_{min} = (3\pi - \Phi)/2; \quad \theta_{max} &= (3\pi - \Phi)/2 \\ f(x, y, \theta) &= (x - w/2)\sin\theta + y\cos\theta \end{aligned} \tag{3}$$

More details and implementation aspects of the above three models are given in [10]. For speckle simulation, we use the method given in [7]. Their model is based on a complex distribution of incoherent phasors (u, v) given by a two-dimensional Gaussian function g_σ. The complex amplitude of each pixel is initialized with the square-root of the sampled intensity value. The number of incoherent phasors $M(x, y)$ at each pixel (x, y) is set as the value of a random number under a uniform distribution within a pre-specified range $[a, b]$. The incoherent phasors are generated and added M times to both the real and imaginary components of the complex value at each pixel. The noisy intensity value is then given by the amplitude of the complex number.

After generating speckle noise at the sampled points, we use an interpolation method to fill the empty space left by the sampling step. In general, the interpolated value at a specified coordinate (x, y) of an image I is computed by grouping the sample values at neighboring pixels (l, m) using the following formula [11]:

$$I(x,y) = \sum_{l,m \in Z} \phi(x-l, y-m)\, I(l,m) \tag{4}$$

where, $\varphi()$ denotes a two-dimensional interpolation/synthesis function that provides the weights of the linear combination of sampled intensity values. Commonly used interpolation methods are B-Spline and cubic Hermite [11, 12]. In [9], the authors used an interpolation scheme using the Lanczos-3 kernel [13, 17].

4 The Evaluation Model

One of the key requirements in the analysis of image modelling and simulation algorithms that use synthetic data is image quality assessment. Image quality metrics are also extensively used in the evaluation of compression and noise filtering algorithms [14]. In this paper, we use the following three quality measures: *E*, *SFM* and *SAM*.

Entropy: It measures the degree of randomness in an image, and is defined as

$$E = -\sum_{j} P_j \log_2 P_j \tag{5}$$

where, P_j is the probability associated with gray level j, and is usually computed as the ratio of the histogram value of the intensity j to the total number of pixels.

Spatial Frequency Measurement (SFM) is a way to measure the overall activity level in an image. SFM is expressed as,

$$SFM = \sqrt{g_R + g_C} \tag{6}$$

where, g_R and g_C denote the mean pixel-level intensity gradients along rows and columns evaluated on an image of size $M \times N$ pixels as given below [16]:

$$\begin{aligned} g_R &= \sqrt{\frac{1}{MN}\sum_{i=0}^{M-1}\sum_{j=1}^{N-1}\left(I(i,j) - I(i,j-1)\right)^2} \\ g_C &= \sqrt{\frac{1}{MN}\sum_{i=1}^{M-1}\sum_{j=0}^{N-1}\left(I(i,j) - I(i-1,j)\right)^2} \end{aligned} \tag{7}$$

The spectral activity measure (SAM) is a measure of image predictability (higher values indicating higher predictability). For an image of size $M \times N$ pixels, it is defined in terms of the Discrete Fourier Transform (DFT) coefficients of the image in the frequency domain as follows:

$$SAM = \frac{\frac{1}{M.N}\sum_{j=0}^{M-1}\sum_{k=0}^{N-1}|F(j,k)|^2}{\left[\sum_{j=0}^{M-1}\sum_{k=0}^{N-1}|F(j,k)|^2\right]^{\frac{1}{M.N}}} \tag{8}$$

where, $F(j, k)$ denotes the DFT coefficient at position (j, k) [16].

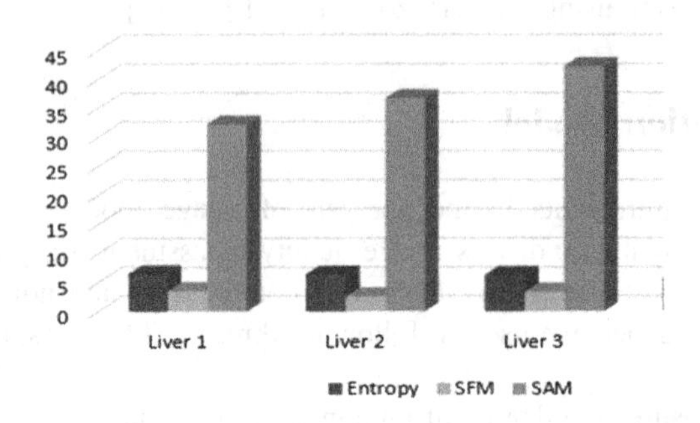

Fig. 5. Entropy, SFM, SAM values of the reference images.

Other traditionally used objective measures such as mean square error (MSE), peak signal-to-noise ratio (PSNR) and speckle index (SI) [15] are useful only in the context of noise reduction and filtering algorithms. For example, the SI value measures the level of residual speckle noise in an image, and is therefore not a useful measure in a synthetic image modelling application.

The values of the entropy, SFM, SAM computed for the three reference images (Liver-1, Liver-2, Liver-3 given in Fig. 1) are shown below in Fig. 5. From the SFM and SAM values, it can be seen that the reference images have higher predictability and less details, as is common in ultrasound images.

In addition to using the above objective measures, we also propose to use subjective assessment of quality by clinical experts in our evaluation model. The importance of subjective evaluation in image quality assessments is emphasized in [16]. The outputs of the speckle simulation stage were assessed by clinical experts, and their subjective evaluations are discussed in the next section.

5 Experimental Results and Analysis

The proposed framework provides allows several options and parametric variations in each stage of the pipeline. As seen in Sect. 3, the three sampling methods and three interpolation schemes themselves give nine possible combinations. Each sampling scheme has its own set of parameters that can be varied over a wide range of values.

The speckle noise generation algorithm also has a set of statistical parameters governing the noise distribution. Due to limitation of space, only a few sample outputs are presented in this section. Further, subtle variations in image or noise characteristics cannot be clearly perceived when images are reduced to fit within a small space.

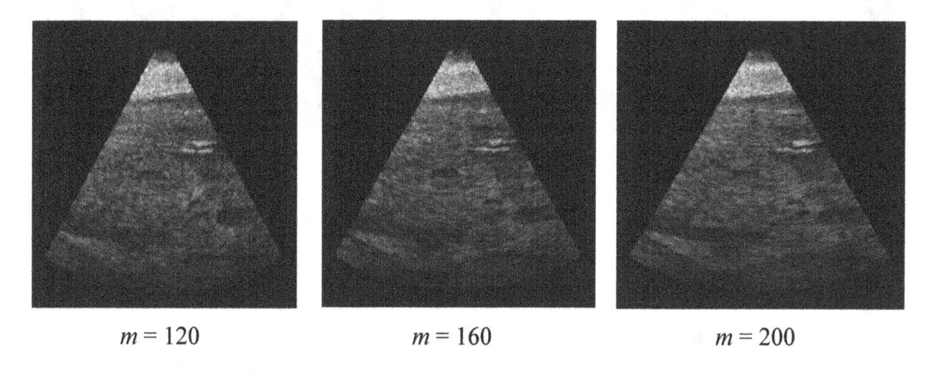

Fig. 6. Effect of changing axial resolution in radial-polar sampling.

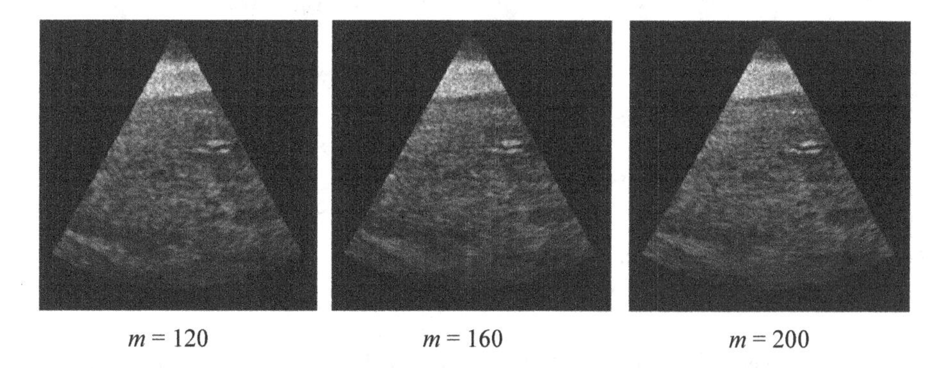

Fig. 7. Effect of changing axial resolution in radial-uniform sampling.

The first row of Fig. 6 shows the variations when the axial resolution m is increased in radial-polar sampling, keeping the lateral resolution fixed at $n = 40$. The interpolation used was Lanczos-3 [17].

Similar results for radial uniform sampling are shown in Fig. 7.

Some of the commonly found artifacts in simulated images when values of certain parameters become large are shown in Fig. 8. In Fig. 8(a), a large value for m results in a dense, overlapping set of points along beam directions resulting in smoothing/merging of pixels. A similar effect is seen when both n and m are large (Fig. 8(b)). When the σ value is large in the speckle generation function, the image becomes too grainy with loss of fine details, as in Fig. 8(c).

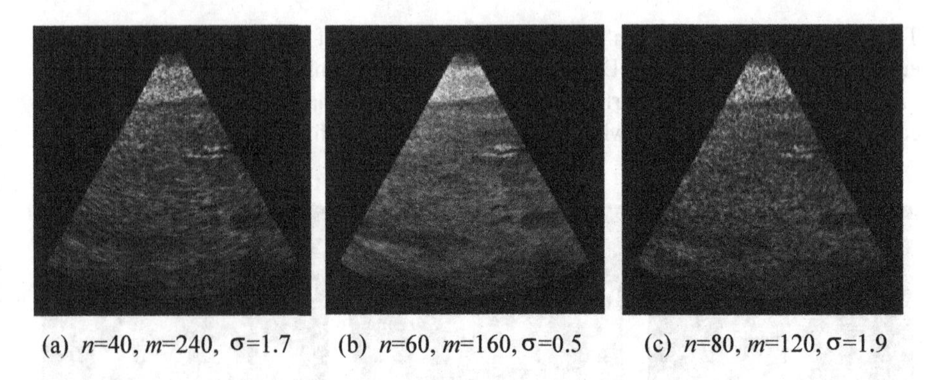

(a) n=40, m=240, σ=1.7 (b) n=60, m=160, σ=0.5 (c) n=80, m=120, σ=1.9

Fig. 8. Image artifacts produced by large values of sampling and noise parameters.

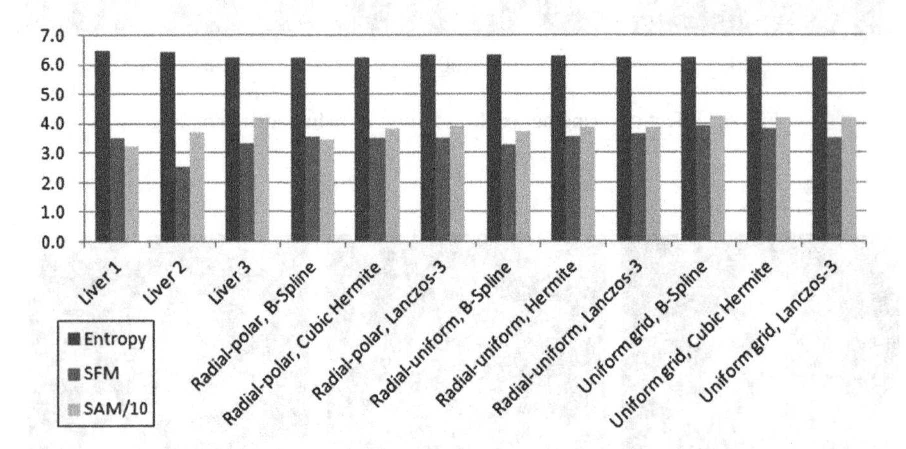

Fig. 9. A comparison of entropy, SFM and SAM values computed for the reference and synthetic images (SAM values scaled by a factor of 10).

Figure 9 gives a comparison of entropy, SFM and SAM values computed for the reference images and also the simulated images generated using various combinations of sampling and interpolation methods. The SAM values have been scaled by a factor of 10 to get a nearly uniform range of values for all three metrics. An important aspect to be considered while computing quality metrics is that the background pixels outside the sector region must be excluded from the computation.

Figure 9 shows that the SFM and SAM values of the generated synthetic images are similar to that of the reference images on an average sense, which points to the fact that the synthetic image is visually similar to a real ultrasound image. The radial-polar sampling scheme with Lanczos-3 interpolation gave SFM values that are closest to the reference value. The SAM values showed larger range of variations. However, the values of all synthetic images were between the minimum and maximum of reference values. The uniform sampling method provided SAM values that are closest to that of the reference image. Four simulated synthetic ultrasound images which gave values

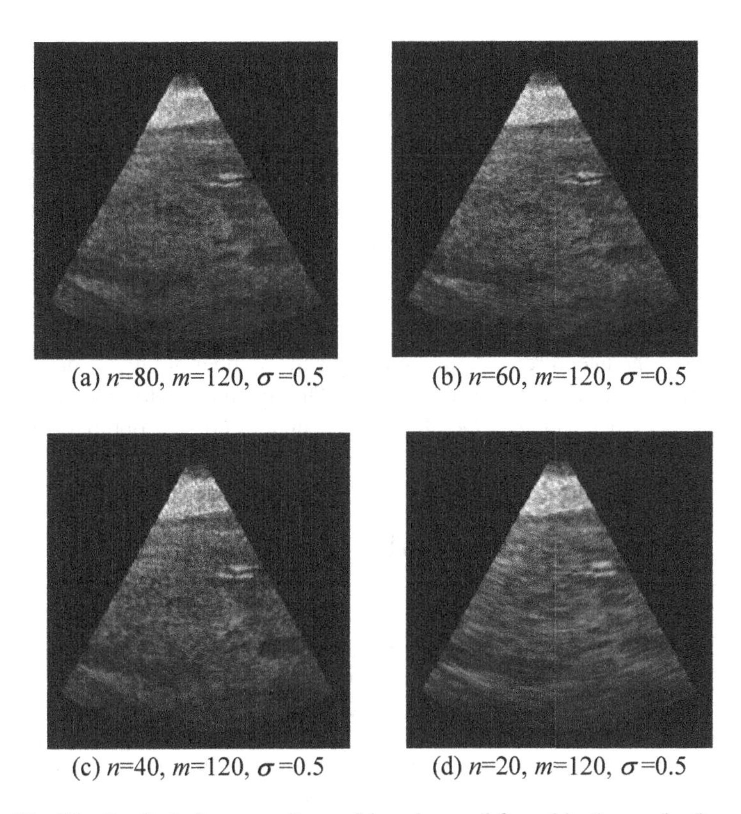

(a) n=80, m=120, σ=0.5 (b) n=60, m=120, σ=0.5

(c) n=40, m=120, σ=0.5 (d) n=20, m=120, σ=0.5

Fig. 10. Synthetic images with speckle noise used for subjective evaluation.

Table 1. Mean subjective evaluation scores assigned by clinical experts.

Image	Figure 10(a)	Figure 10(b)	Figure 10(c)	Figure 10(d)
Mean score	4	3	3	1

closest to the reference values were shown to clinical experts for subjective evaluation. The images and their parameters are shown in Fig. 10.

The subjective evaluation was performed by three experts. They based their evaluation on key visual features such as contrast, grayscale variations, texture and graininess. The images were scored on a scale from 1 to 5. There was a general agreement among the assessors on image quality and how closely the simulated images resembled real ultrasound images. The mean subjective scores are given in Table 1.

As the lateral resolution parameter n is varied from 20 to 80, the smoothing effect due to interpolation is significantly reduced, the graininess improved and image features became more clearly visible, which is important from a clinician's perspective. The assessors also observed that the images have gray-scale variation and graininess closely resembling real ultrasound images.

Overall, based on both quantitative and subjective evaluations, radial-polar sampling method with parameter values $n = 80$, $m = 120$ and speckle simulation with $\sigma = 0.5$ with linear and Lanczos-3 interpolation gave the most accurate simulation of real ultrasound images, where the image size used was 256×256 pixels. In general, for an image of size $N \times N$ pixels, the optimal value m of axial resolution (the number of pixels per beam) depends on the value of N. Our experiments by varying N have shown that the optimal value of m varies proportional to N as follows:

$$m = \text{floor}\ (0.56N - 23), \quad 250 \leq N \leq 500. \tag{9}$$

The value of the lateral resolution n represents the number of ultrasound beams and it specifies the subdivisions of the sector angle. The optimal subdivision was found to be approximately 1 beam/degree, *i.e.*, if the sector angle is 60°, the optimal value of n is also around 60.

6 Conclusions and Future Work

This paper has presented the complete algorithmic framework for generating realistic and simulated ultrasound images incorporating image acquisition models, speckle noise formation processes and image interpolation schemes. These processes within the simulation model allows users to vary a wide range of parameters that control the image and noise formation processes. The simulated images with speckle noise could be used to evaluate noise filtering methods as ground truth data (the corresponding synthetic images without noise) are readily available.

The paper has introduced three sampling schemes, *viz.*, radial-polar, radial-uniform and uniform grid sampling methods. These methods together with the speckle simulation model and the interpolation scheme formed the simulation model of the processing pipeline. In the evaluation model, objective assessment of image quality was performed using entropy, SFM and SAM metrics. A subjective evaluation by clinical experts was also performed.

Experimental analysis shows that the synthetic image with simulated speckle noise has visual characteristics and image features very similar to real ultrasound images. The evaluation study helped to pick the best set of parameters that accurately modelled real ultrasound images, from a very large set of values.

Future work is directed towards more accurate content-specific modelling of speckle noise considering the regions present in the image, which will require additional processing such as region identification or segmentation.

Acknowledgments. We acknowledge the help extended by Dr. Khadijah Hajee Abdoula, Victoria Hospital, Quatre Bornes, Mauritius and Dr. Vivek Aggarwal, Thyroid Clinic, New Delhi, India, in providing their expert advice, valuable inputs and subjective evaluation of synthetic images produced in this research.

References

1. Zhang, J., Cui, W., Wu, L., Lin, G., Cheng, Y.: A novel algorithm based on wavelet-trilateral filter for de-noising medical ultrasound images. In: Control and Decision Conference (2016). doi:10.1109/CCDC.2016.7531648
2. Malutan, R., Terebes, R., Germain, C., Borda, M., Cislariu, M.: Speckle noise removal in ultrasound images using sparse code shrinkage. In: IEEE International Conference on E-Health and Bioengineering (2015). doi:10.1109/EHB.2015.7391394
3. Le, T.: Adaptive noise reduction in ultrasound imaging. In: IEEE Symposium on Signal Processing in Medicine and Biology (2014). doi:10.1109/SPMB.2014.7002961
4. Zang, X., Bascom, R., Gilbert, C., Toth, J., Higgins, W.: Methods for 2D and 3D endobronchial ultrasound image segmentation. IEEE Trans. Biomed. Eng. **63**(7), 1426–1439 (2016). doi:10.1109/TBME.2015.2494838
5. Cortes, C., Kabongo, L., Macia, I., Ruiz, O.E., Florez, J.: Ultrasound image dataset for image analysis algorithms evaluation. In: Chen, Y.-W., Toro, C., Tanaka, S., Howlett, R.J., Jain, L.C. (eds.) Innovation in Medicine and Healthcare 2015. SIST, vol. 45, pp. 447–457. Springer, Cham (2016). doi:10.1007/978-3-319-23024-5_41
6. Telmed Ultrasound Medical Systems. http://www.pcultrasound.com/products/products_usimg/index.html
7. Ultrasound Image Gallery. http://www.ultrasound-images.com/
8. Loizou, C.P., Pattichis, C.S.: Despeckle Filtering for Ultrasound Imaging and Video, Volume 1: Algorithms and Software. Morgan & Claypool, San Rafael (2015)
9. Perreault, C., Auclair-Fortier, M.F.: Speckle simulation based on B-mode echographic image acquisition model. In: 4th Canadian Conference on Computer and Robot Vision, pp, 379–386 (2007). doi:10.1109/CRV.2007.61
10. Singh, P., Mukundan, R., de Ryke, R.: Synthetic models of ultrasound image formation for speckle noise simulation and analysis. In: International Conference on Signals and Systems (ICSigSys-2017) (2017)
11. Kai-yu, L., Wen-dong, W., Kai-wen, Z., Wen-bo, L., Gui-li, X.: The application of B-spline based interpolation in real-time image enlarging processing. In: 2nd International Conference on Systems and Informatics, pp. 823–827 (2014). doi:10.1109/ICSAI.2014.7009398
12. Goceri, E., Lomenie, N.: Interpolation approaches and spline based resampling for MR images. In: 5th International Symposium on Health Informatics and Bioinformatics (2010). doi:10.1109/HIBIT.2010.5478891
13. Somawirata, I.K., Uchimura, K., Koutaki, G.: Image enlargement using adaptive manipulation interpolation kernel based on local image data. In: IEEE International Conference on Signal Processing, Communication and Computing, pp. 474–478 (2012). doi:10.1109/ICSPCC.2012.6335692
14. Xia, Z.W., Li, Q., Wang, Q.: Quality metrics of simulated intensity images of coherent ladar. In: International Conference on Optoelectronics and Microelectronics (2012). doi:10.1109/ICoOM.2012.6316255
15. Mirza, S., Kumar, R., Shakher, C.: Study of various preprocessing schemes and wavelet filters for speckle noise reduction in digital speckle pattern interferometric fringes. Opt. Eng. **44**(4) (2005). doi:10.1117/1.1886749
16. Grgic, S., Grgic, M., Mrak, M.: Reliability of objective picture quality measures. J. Elec. Eng. **55**(1–2), 3–10 (2004). http://citeseerx.ist.psu.edu/viewdoc/versions?doi=10.1.1.138.6936
17. Burger, W., Burge, M.J.: Digital Image Processing: An Algorithmic Introduction Using Java. Springer, Heidelberg (2008). doi:10.1007/978-1-84628-968-2

Multi-level Trainable Segmentation for Measuring Gestational and Yolk Sacs from Ultrasound Images

Dheyaa Ahmed Ibrahim(✉), Hisham Al-Assam, Sabah Jassim, and Hongbo Du

Department of Applied Computing, University of Buckingham, Buckingham, UK
Dheyaa.Ibrahim@buckingham.ac.uk

Abstract. As a non-hazardous and non-invasive approach to medical diagnostic imaging, ultrasound serves as an ideal candidate for tracking and monitoring pregnancy development. One critical assessment during the first trimester of the pregnancy is the size measurements of the Gestation Sac (GS) and the Yolk Sac (YS) from ultrasound images. Such measurements tend to give a strong indication on the viability of the pregnancy. This paper proposes a novel multi-level trainable segmentation method to achieve three objectives in the following order: (1) segmenting and measuring the GS, (2) automatically identifying the stage of pregnancy, and (3) segmenting and measuring the YS. The first level segmentation employs a trainable segmentation technique based on the histogram of oriented gradients to segment the GS and estimate its size. This is then followed by an automatic identification of the pregnancy stage based on histogram analysis of the content of the segmented GS. The second level segmentation is used after that to detect the YS and extract its relevant size measurements. A trained neural network classifier is employed to perform the segmentation at both levels. The effectiveness of the proposed solution has been evaluated by comparing the automatic size measurements of the GS and YS against the ones obtained gynaecologist. Experimental results on 199 ultrasound images demonstrate the effectiveness of the proposal in producing accurate measurements as well as identifying the correct stage of pregnancy.

Keywords: Ultrasound image segmentation · Gestational sac · Yolk sac · Trainable segmentation · Pregnancy stage identification

1 Introduction

Medical imaging has unparalleled value in clinical analysis and plays a critical role in the development of effective medical interventions for the treatment of health conditions [1]. Numerous imaging devices have been constructed for diagnostic purposes, and each is characterized by the way in which it draws on different methods in the composition of images. Diagnostic sonography, as an example, operates by transmitting high-frequency sound waves and registering its reflection through a transducer [2]. Given its high-degree of safety and non-invasive nature, ultrasound scanning is routinely employed during pregnancy to monitor growth and health status of foetus.

M. Valdés Hernández and V. González-Castro (Eds.): MIUA 2017, CCIS 723, pp. 86–97, 2017.
DOI: 10.1007/978-3-319-60964-5_8

Statistical evidence reveals that a range of complications occur during pregnancy, the most prevalent of which is miscarriage. Miscarriages are quite common. In the UK, the figure reaches to nearly a quarter of a million each year [3, 4]. As reported in [3], around 20% of all pregnancies are miscarried prior to 24 weeks, and the majority of these occur in the first trimester (namely, within 12 weeks after conception). Gestational assessments conducted over the course of the first trimester tend to focus on the confirmation of fundamental details regarding the pregnancy – namely, whether it has taken place and additionally how many pregnancies there are – and other details addressed include the location of the gestation sac (GS) and the embryo's health status. An assessment of the dimensions of the GS and the Yolk Sac (YS) provides information regarding the probable gestational age of an early pregnancy, and such assessments are also valuable in diagnosing the extent to which the pregnancy is viable. In cases where the Mean Sac Diameter (MSD) of an empty GS greater than 25 mm is observed,the pregnancy will most likely end up as a miscarriage [5]. The value of MSD is generated from the 3 diameters of the GS in combination with the 3 diameters measured in the sagittal and transverse planes [6]. The same process of scan is repeated at the later stage when the yolk sac is starting to grow to further check the status of the pregnancy. As detailed in [5], another indicator of miscarriage is a smaller GS than anticipated based on the gestational age when compared to the last menstrual period. All the descriptions above indicate the necessity of obtaining accurately measured GS and YS sizes.

It is critical to acknowledge that the quality of diagnostic accuracy based on manual measurements can be affected by inconsistency among the measurements obtained by different gynaecologists and even the same gynaecologist, known as inter- and intra-observer variabilities [7]. Therefore, the potential for positive impacts on the result from the provision of automated tools for the enhancement of diagnostic accuracy is clear. Specifically, a computer-based framework to segment and classify ultrasound images could effectively enhance decision-making in pregnancy diagnostics. Additionally, one of the most beneficial prospects associated with the development of this framework would be saving time and resources. Nevertheless, it must be acknowledged developing an automated diagnostic mechanism based on ultrasound images for miscarriage detection is not a straightforward task.

This paper outlines an automatic system for a trainable multi-level segmentation of the GS and the YS based on neural network classifier followed by the measurement of the MSD. After a first level segmentation of the GS, histogram analysis is used to identify the pregnancy stages by establishing whether the GS is empty or not (i.e. it contains the YS). A second level of segmentation is then applied to detect the YS based on the Hough transform followed by producing the MSD. The experimental results upon a dataset of 199 images confirm that the proposed framework on the one hand yields the automatic measurements for the GS and the YS in terms of MSD very close to the gynaecologist measurements without inter- and intra-observer variabilities, and on the other hand achieves an accuracy of pregnancy stage identification around 97.48%.

The rest of the paper is organized as follows: Sect. 2 sets the scene by briefly describing the medical background and the literature review. In Sect. 3 we introduce our approach for identify the pregnancy stage and extract the GS and YS followed by

MSD measurements. Then, Sect. 4 discusses the experiment result. Finally, Sect. 5 concludes the paper and outlines several possible extensions of our approach.

2 Background and Related Work

A regular pregnancy takes 40 weeks (±2 weeks). The first indicator of pregnancy is the absence of the menstrual period. Pregnancy tests, which are conventionally conducted using urine sample, capitalize on the presence or absence of Human Chorionic Gonadotropin (hCG) [8]. In situations where pregnancy tests of this kind yield positive results, an initial scan is conducted to identify the GS's age and, based on this, to calculate the birth date. As being observable in Fig. 1, the structural anatomical elements of early pregnancy are as follows: **(i)** Amniotic sac is a bag of clear fluid inside the uterus where the foetus starts to develop and grow [9]; **(ii)** GS is a structure that surrounds an embryo, in the very early stages of pregnancy (see Fig. 1(b)); **(iii)** Yolk sac is the ring-shaped structure identified within the gestational sac (see Fig. 1(c)) [9]. Heartbeat inside embryo will be seen along with the yolk sac. Failure to identify fatal heartbeat is a sign of abnormal pregnancy which may lead to miscarriage later [9]. Figure 1(d) shows the embryo attached to the YS inside the GS.

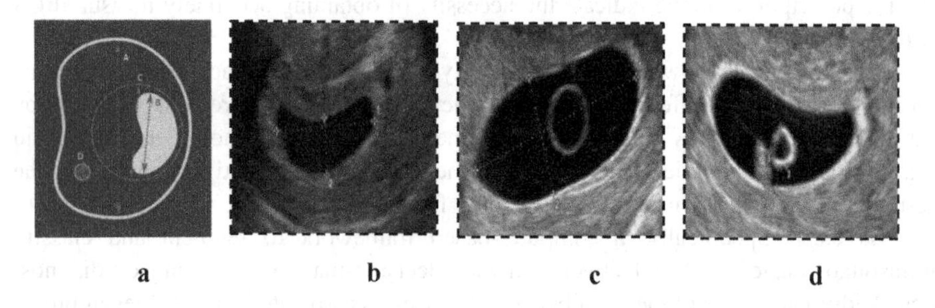

Fig. 1. Examples shows the ultrasound images of a very beginning of pregnancy until developing the embryo (a) The anatomical structures of the early pregnancy A: Gestational sac (GS), B: Crown rump length (CRL) of embryo, C: Amniotic sac and D: Yolk sac (b) Gestational Sac (c) YS within GS (d) embryo attached with YS within GS

The initial structural element that is observable inside the GS is the YS, and an observation of this serves as a confirmation of an intrauterine pregnancy. At the point where the MSD of the GS is 5–6 mm, the YS can be observed by employing transvaginal ultrasound, and it must always be visualized in cases where the MSD of the GS is 8 mm or higher [10].

Early miscarriages are defined as taking place prior to 12 weeks (namely, in the first trimester of pregnancy), while late miscarriages arise in the period from 13–24 weeks. Such phenomena emerge as a consequence of the discontinuation of embryonic development and, in combination with this, the simultaneous discontinuation in the regular development and growth of the GS.

When pregnancy tests yield positive results, it is conventionally the case that the initial scan will be conducted once 12 weeks have elapsed since conception. Scans are provided at an earlier date in certain circumstances: for example, where vaginal bleeding is reported etc. [10]. The assessment of pregnancy within these first 12 weeks is critical for practitioners because it facilitates the evaluation of the foetus (in terms of development, growth, and health status) and the estimation of a birth date [11, 12].

As mentioned previously, the first measurable sign of an early pregnancy are the geometric characteristics of the GS. The first feature calculated is the MSD. Different studies consider that miscarriage should be declared based on different cut-off values for MSD within the range of 13–25 mm [13, 14]. As displayed below, the most up-to-date limits for miscarriage diagnosis are as follows (Miscarriage identification cut-offs according to the NICE guideline):

- Mean gestational sac diameter (MSD) of ≥ 25 mm with no obvious YS
- Mean sac diameter of (GS and YS) of ≥ 25 and no embryo defined in YS

The segmentation of ultrasound images is crucial for the identification of regions of interest (ROI), one of which, for example, is the GS. Nevertheless, the speckle noise found in ultrasound images (speckles being the cause of the noisy and textured backdrop which results from the reflection and scattering of sound waves inside target tissue) is a chief contributor to the decrease in segmentation accuracy.

Several academic initiatives have directed their efforts towards the issue of ultrasound medical image analysis in the recent 10 years. In a particularly notable study, the researchers outlined an automated system for the measurement of GS dimensions [15]. This system relied on a variety of methods to generate accurate figures for GS age and birth date. Other prominent researchers formulated a new way in which to assess the MSD of the GS using 2-dimensional ultrasound video [16]. The novel method relied on the following series of elements: training, detecting, indexing, and measuring. The chief contributions of these studies stem from their provision of novel frameworks by which the detection of miscarriage can be facilitated in an automated manner; moreover, the studies draw on novel indicators to reinforce diagnostic decision-making. Nevertheless, despite the valuable nature of these studies, a fully-automated miscarriage indicator identification framework which employs ultrasound images is still lacking; as previously alluded to, this is a product of the way in which noise complicates the attempt to automatically segment ROIs, thereby meaning it must be conducted manually or semi-automatically [3, 4]. Hence, the present author is alert to the fact that this paper represents the first attempt to outline a fully-automatic system for the identification of pregnancy stage and segmentation of GS with YS.

3 The Proposed Method

Figure 2 provides an overview of the system for the automatic identification of the pregnancy stage as well as detecting and measuring the GS and the YS from a static B-mode image. The framework consists of a sequence of steps starting from segmenting and measuring the GS, followed by automatically identifying the stage of pregnancy, and finally segmenting and measuring the YS. Each step of the framework will be explained in details in the following sub-sections.

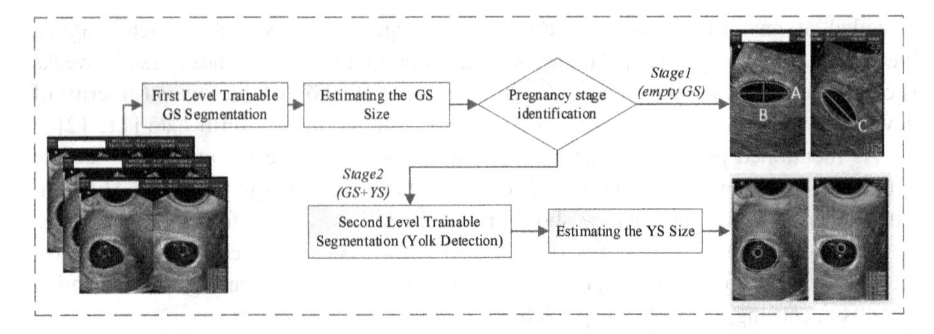

Fig. 2. General framework for the automatic segmentation

3.1 First Level Trainable GS Segmentation

The aim of the first level segmentation is to isolate the GS from the rest of the image. The process starts by first selecting a certain number of images as training images. From each training image, a number of samples, i.e. a small window of square regions of a certain size (e.g. 3 × 3), are taken from the inside of the GS (Class 1) and outside the GS (Class 2). For each window of either class, a set of HOG features (to be described in later) are extracted. The labelled feature vectors collected from all samples are then used to train a neural network classifier. Once training is complete, each image from the testing set is used as an input for the segmentation algorithm. The algorithm scans an image pixel by pixel. For each pixel, a small square region of the same window size with the pixel as the centre is constructed, and the HOG feature of the region is extracted and classified by the trained neural network as being inside the GS or outside GS. If the region is inside a GS, the central pixel is also labelled as inside the GS; otherwise outside the GS. Once all pixels of the image are labelled, the region of the "inside pixels" is taken as the segmented GS.

The central problematic element associated with this stage of the process stems from the fact that upon the application of the trainable segmentation, binary objects remain which resemble the sac. Consequently, it is necessary for a fully automated system to first detect these objects and consequently remove them. This process of filtration takes place in accordance with the non-sac objects' characteristics such as circularity, area, and the greyscale mean [2].

The histogram of oriented gradients (HOG) is a feature descriptor that facilitates the identification of objects in digital images. HOG was formulated for the quantification of gradient orientation occurrences in localised image sections, and have been well documented in [17, 18]. The process of HOG feature extraction involves taking a window around the pixels called cells. The mask [−1,0, 1] is used to computing image gradients. In our adaptation of this extraction method (see Fig. 3), for orientation binning, we directly used the gradient at each image locations for the corresponding orientations. The orientation cells are chosen in the range of 0–180° with 9 bins. For better invariance to illumination, shadowing, etc., contrast-normalization of the local histogram is applied.

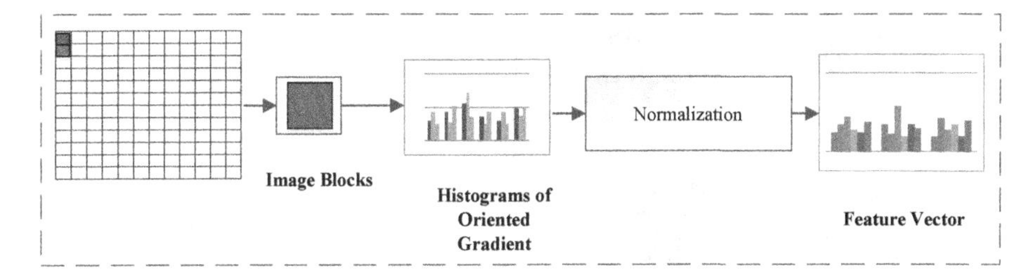

Fig. 3. HOG process

3.2 Estimating the GS Size

As explained earlier, each GS is viewed in two perpendicular planes. The GS usually seems to be more elliptical at the early stages, therefore, our system aims to locate the best-fitted ellipse for the segmented GS for each plane. The region props function in Matlab is utilized to fit an ellipse to the GS. Four parameters can be returned by this process. These are major axes, minor axes, Centroid and Orientation. The GS is presumably has an ellipsoidal shape in 3D, the three main dimensions of the ellipsoid has generally been calculated by the minor and major dimension from the sagittal plane and the major dimension from the transverse plane. The average of the three Dimension is taken as the MSD.

3.3 Histogram Analysis to Identify the Pregnancy Stage

This stage aims to establish if the GS is empty or not i.e. whether it contains a YS inside. To do this, first, the binary image resulted from the GS segmentation step is used as a mask to locate the pixels inside the GS from the original image. Histogram analysis is applied to those pixels to determine whether theYS is present within the GS or not by following the process outlined in Fig. 4.

Consider that the grey-level histogram matches with a GS, f (x, y), constituted of light objects (the YS's border) superimposed on a dark backdrop (the GS). Furthermore, the configuration is organized such that the pixels of the object and backdrop display grayscale levels are categorized into a pair of dominant modes. If this is the case, the immediately discernible method by which the border of the YS can be estimated from the GS involves the selection of a threshold (T) that facilitates the separation of these modes. Consequently, a pixel (x, y) according to which f (x, y) > T (based on bin frequency) can be designated as a YS border.

Under certain circumstances, when the frequency is high within GS, it does not necessarily mean there is a YS: the frequency simply serves to denote small noise objects. Consequently, for each bin with frequency greater than T, pixels of the intensity represented by the bin are located in the GS image (see Fig. 4). A post-processing check is conducted to see if the pixels form an object. If not, the GS is considered as empty (stage1); otherwise, the GS contains YS (stage2).

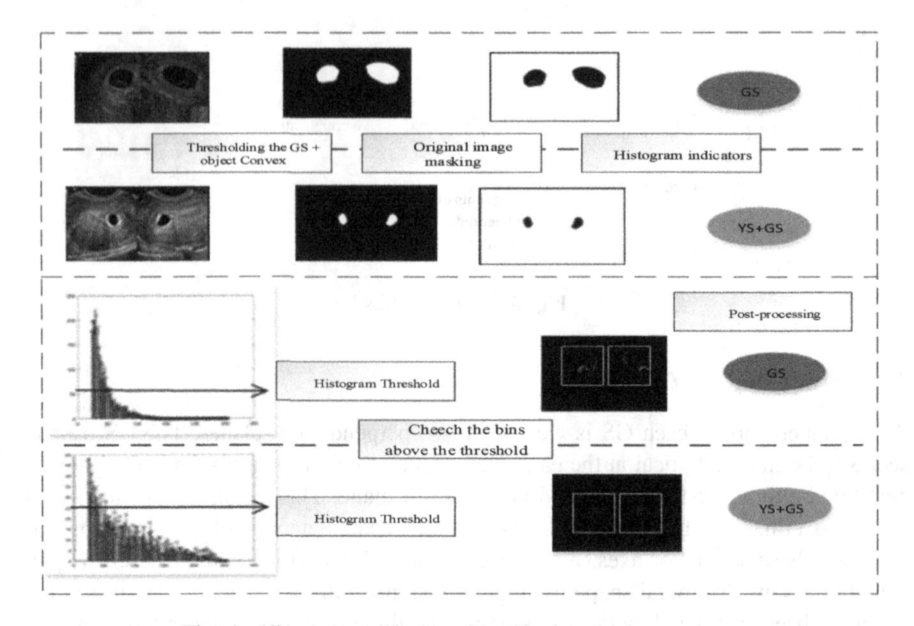

Fig. 4. Histogram analysis to identify the pregnancy stages

3.4 Second Level Trainable Segmentation: Estimating the YS Border

Once the histogram analysis of the area inside the GS establishes the existence of the YS, a second level trainable segmentation is employed at this stage to estimate the YS border. The main challenge here is that in certain scenarios, the YS's border is vague and unclear, which means that techniques based on the pixel intensity thresholding cannot work satisfactorily. The machine learning based approach has more promises in such situations.

A pixel feature (in accordance with the pixel neighbourhood) is used to identify the border pixels. The training phase is implemented by selecting window of size N × N for each yolk boundary and non-boundary in image (see Fig. 5). The testing phase will process only the pixels which are inside the GS sac using the mask from the GS segmentation stage. The output of this stage is a binary GS include the GS with white colour and the YS border with black colour.

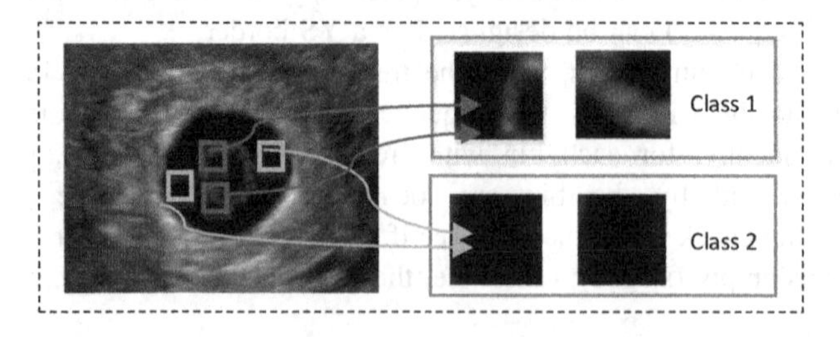

Fig. 5. Select samples for the learning phase

Once the second level trainable segmentation estimates the YS border. A convex hull is employed to facilitate the delineation of the YS from the GS. The rationale for this is of three folds: (i) it facilitates the maintenance of the entire GS and the assessment of the MSD; (ii) it allows just the YS's border to be retained for the Hough transform stage; and (iii) reduces the frequency of objects in conjunction with the likelihood of object circularity, thereby capturing the YS. All the objects inside the convex will be as input to the circle Hough transform to detect the best circle next.

3.5 Locating YS and Measuring its Size

Based on the assumption that a circle exists in (x and y)space, the image coordinates, and that the YS's perimeter points have been acquired, a circle in the (a and b) parameter space corresponds to every point on the circle's perimeter. The (a and b) space constitutes a circle accumulation of the input image for a specified radius (r) (see Fig. 6). The critical piece of information when attempting to identify a circle is the radius, and this is because it determines circle dimensions in the (a and b) space. In cases where a circle constructed in the (a and b) space is not identical in size to the initial circle's radius, the former will not contact at a single location. Hence, a suitable radius means that the constructed circles will contact at a single location, and this constitutes the centre circle identified in the image coordinates. Assuming the YS has acircle shape in 3D, the two principal axes of the circle can be estimated by the radius (R1) of the circle from the sagittal plane and the radius (R2) of the circle from the trans-verse plane. The ((2*R1 + R1)/3) will be taken as the MSD.

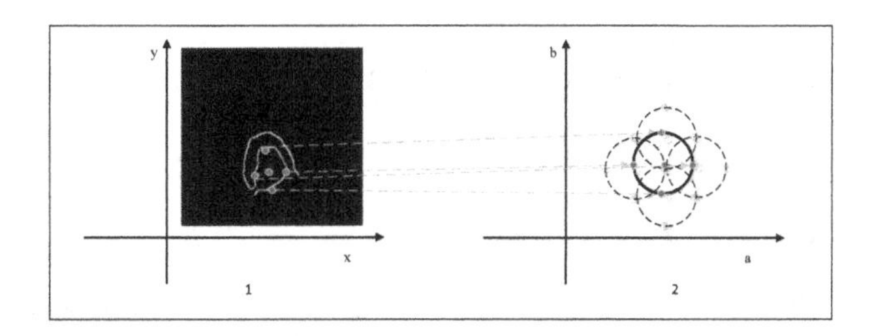

Fig. 6. (1) Circle located on parameter plane *x*,*y*, (2) transform in space *a*,*b*

4 Experiments and Results

4.1 Dataset

The dataset consists of 199 ultrasound images, 184 of which are of an empty GS (157 PUV and 27 miscarriage). The remaining 15 are for GSs with YSs inside. The images were acquired in 3 dataset batches, with the first constituted of 94 images; the second constituted of 90; and the third constituted of 15 with GS and YS. The images are obtained by the Early Pregnancy Unit, Imperial College Healthcare Trust, London. The visualisation of the GS and YS for each image was conducted from a pair of viewpoints:

namely, the sagittal and transverse planes. The images were accompanied with manually-obtained diameters (D_1, D_2, and D_3) together with the resulting MSDs in addition to diagnostic outcomes.

4.2 Experimental Protocol

A 2-layer backpropagation neural network has been used in each of the two segmentation levels. In the first level, the network was trained with 100 sample regions obtained from 8 training images. In the second level of segmentation, the network was trained with 60 sample regions obtained from 3 images due to the limited number of images in batch 3.

It is important to recognize that the features extracted and the quality of segmentation are directly influenced by the block size, i.e. the size of the window around the pixel. As shown in Fig. 7, when the block size is small, more pixels outside the real region of interest are considered as "inside" pixels. This will results in the creation of more irrelevant objects outside the real object of interest, increasing the level of difficulty for removing those objects later (Fig. 7(b)). As the block size increases, fewer pixels outside the object of interest would be confused as "inside" pixels, and hence fewer irrelevant objects are created (Fig. 7(c, d)). However, when the block size gets too large, the possibility that a block contains pixels of both "inside" and "outside" also increases, resulting imprecise classification of the pixel class (Fig. 7(d)). As such, to ensure high-quality segmentation, the block size must be determined carefully. Based on our observation, the block size of 5 × 5 seems optimal.

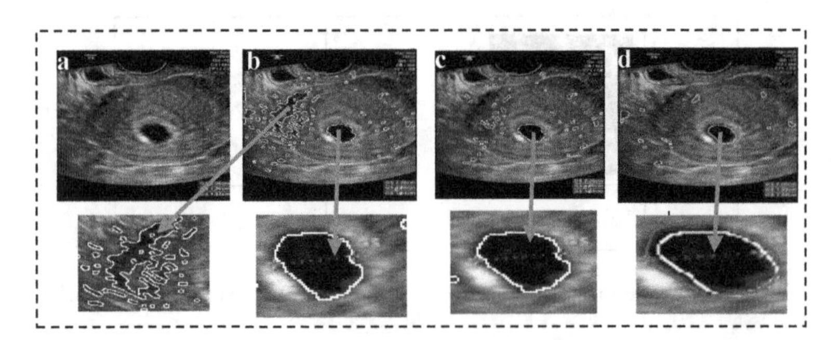

Fig. 7. Trainable segmentation result, (a) original image, (b) segmented image with window size 3*3, (c) segmented image with window size 5*5, (b) segmented image with window size 7*7. The wait line represent the automatic segmentation and the red color represent the manual segmentation. (Color figure online)

The effectiveness of the proposed method was evaluated by conducting examinations at the following phases: (i) evaluate the pregnancy stage identification accuracy based on the segmentation of GS; (ii) assess the segmentation precision by comparing the automatic and manual measurement results of MSD.

4.3 Results

Pregnancy Stages Identification

We are interested in the effectiveness of histogram thresholding and the histogram thresholding followed by post-processing in the proposed method as explained in Sect. 3.3. Two experiments were conducted: one for using the thresholding alone, and the other for using thresholding followed by post-processing. Figure 8 shows that an overall average accuracy of 84.42% with stage1 identification (empty GS) of 83.69% and stage2 identification (GS &YS) of 100% for can be achieved in the first set of the experiment (Histogram threshold). By contrast, an overall accuracy of 97.48% with stage1 identification of 97.28% and with stage2 identification of 100% was achieved in the second experiment as (Histogram threshold with post-processing). The results confirm that the second method approved that the threshold of the histogram is not enough to estimate if there is a YS or not.

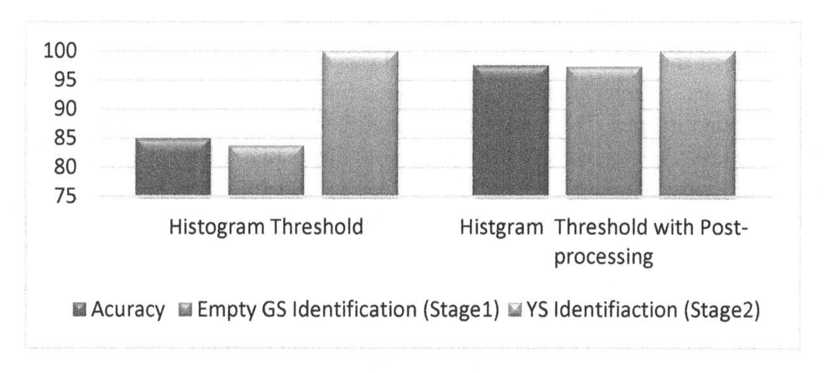

Fig. 8. Pregnancy stage identification

Segmentation Result (Manual vs. Automatic Measurements)

The manual calculation of the proportion of successful GS segmentations was conducted with the human visual system (HVS). Regarding batches 1 and 2, it was noted that 153/184 GS images were segmented which is 83.15%.
Additionally, out of the 15 images in the third batch, GS were segmented precisely, which is 100%. Furthermore, the YS segmentation was successful for 14 out of the 15 images, i.e. 93.3%. To facilitate the provision of a suitable system assessment regarding the subsequent phases, each image associated with unsuccessful segmentation of GS was not included.

Automatic MSD measurements were considered in relation to manual measurements to assess the degree to which the proposed system was effective. Figure 9 provides an indication of the performance. Figure 9a illustrates the result of batch 1 and batch 2 (Empty GS). Figure 9b represents the result for batch three where the points inside the red ellipse represent the measurement for YS and the other points represents the GS measurements. In view of the approximately 45° regression line, it can be concluded that the automatic measurements are very close to the manual measurements. It can also be observed that there is no apparent systematic bias of the automatic measurements.

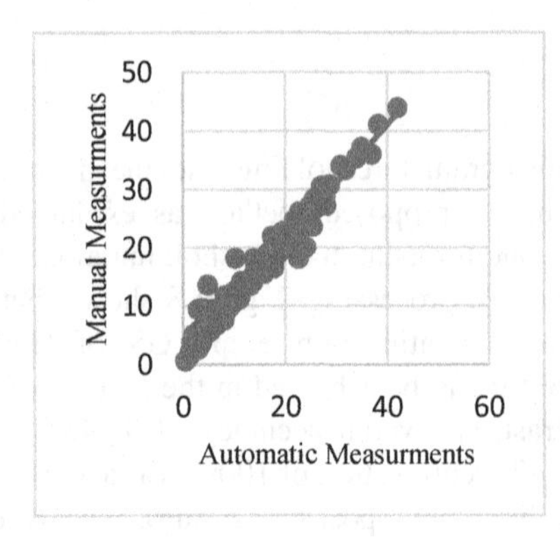

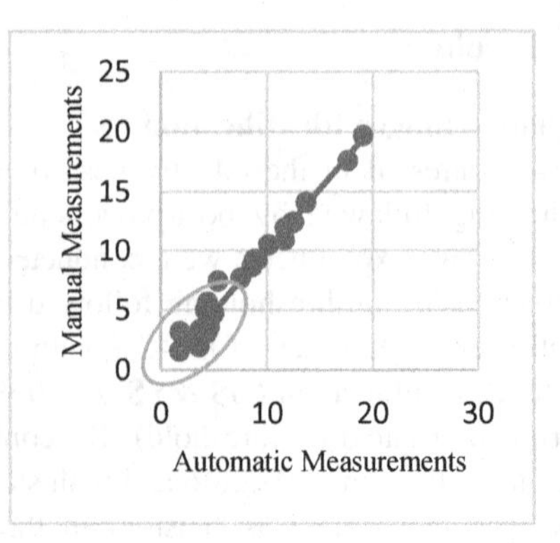

Fig. 9. Comparison manual and automatic MSD measurements. (a) Empty GS (b) GS + YS

5 Conclusion

The purpose of this study was to outline and evaluate a new approach to automatic identification the pregnancy stage and segmentation of the GS and YS from B-mode static ultrasound images. This paper argued that the segmentation of such images is a challenging task due to overlap between colour intensities of the GS and YS with surrounding tissues.

The proposed automatic system overcome these difficulties by formulating a multi-level framework: (1) first-level trainable GS segmentation to locate the GS sac; (2) using histogram properties of the GS to identify the pregnancy stage; (3) second level of trainable segmentation is used to segment and measure the YS. Finally, the MSD for the YS and the GS were extracted. Experiments showed that the automatic measurements were very close to the manuals ones extracted by domain experts which confirms the viability of the proposal.

Our future work includes extending the test of our method on more images, improving the efficiency of the proposed method, and investigate the effectiveness of alternative image-based indicators such as those extracted from the border of the GS.

Acknowledgements. Many thanks to all the collaborators involved in this work. Department of Early Pregnancy, Imperial College, Professor Tom Bourne and Dr. Jessica Farren for their help in preparing the images and all important information related to the datasets used in this study.

References

1. Lee, L.K., Liew, S.C.: A survey of medical image processing tools. In: 2015 4th International Conference on Software Engineering and Computer Systems (ICSECS) (2015)
2. Ibrahim, D.A., Al-Assam, H., Du, H., Farren, J., Al-karawi, D., Bourne, T., Jassim, S.: Automatic segmentation and measurements of gestational sac using static B-mode ultrasound images. In: SPIE Commercial + Scientific Sensing and Imaging (2016)

3. Khazendar, S., Al-Assam, H., Bourne, T., Jassim, S.A.: Automatic identification of early miscarriage based on multiple features extracted from ultrasound images. In: MIUA (2014)
4. Preisler, J., Kopeika, J., Ismail, L., Vathanan, V., Farren, J., Abdallah, Y., Battacharjee, P., Van Holsbeke, C., Bottomley, C., Gould, D., et al.: Defining safe criteria to diagnose miscarriage: prospective observational multicentre study (2015)
5. Ectopic Pregnancy and Miscarriage: Diagnosis and Initial Management. National Institute for Health and Care Excellence (NICE) (2012). Accessed 2015
6. Pexsters, A., Luts, J., Van Schoubroeck, D., Bottomley, C., Van Calster, B., Van Huffel, S., Abdallah, Y., D'Hooghe, T., Lees, C., Timmerman, D., et al.: Clinical implications of intra-and interobserver reproducibility of transvaginal sonographic measurement of gestational sac and crown–rump length at 6–9 weeks' gestation. Ultrasound Obstet. Gynecol. **38**(5), 510–515 (2011)
7. Doubilet, P.M., Benson, C.B., Bourne, T., Blaivas, M.: Diagnostic criteria for nonviable pregnancy early in the first trimester. N. Engl. J. Med. **369**(15), 1443–1451 (2013)
8. Creighton University Medical Center, Omaha Nebraska. Diagnostic and Interventional Radiology Department/Ultrasound of Early Pregnancy. Creighton University Medical Center (2013). http://web.archive.org/web/20070814054851/http:/radiology.creighton.edu/pregnancy.htm#section4
9. Uideline, L.: Ultrasound evaluation of first trimester pregnancy complications. J. Obstet. Gynaecol. Can. **27**(6), 581–585 (2005)
10. Khazendar, S., Farren, J., Al-Assam, H., Du, H., Sayasneh, A., Bourne, T., Jassim, S.: Automatic Identification of Miscarriage Cases Supported by Decision Strength Using Ultrasound Images of the Gestational Sac (2015)
11. Geirsson, R., Busby-Earle, R.: Certain dates may not provide a reliable estimate of gestational age. BJOG: Int. J. Obstet. Gynaecol. **98**(1), 108–109 (1991)
12. Kaur, A., Kaur, A.: Transvaginal ultrasonography in first trimester of pregnancy and its comparison with transabdominal ultrasonography. J. Pharm. Bioallied Sci. **3**(3), 329 (2011)
13. Bourne, T.: Why greater emphasis must be given to getting the diagnosis right: the example of miscarriage. Australas. J. Ultrasound Med. **19**(1), 3–5 (2016)
14. Levi, C., Lyons, E., Lindsay, D.: Ultrasound in the first trimester of pregnancy. Radiol. Clin. North Am. **28**(1), 19–38 (1990)
15. Chakkarwar, V., Joshi, M.S., Revankar, P.S., et al.: Automated analysis of gestational sac in medical image processing, pp. 304–309 (2010)
16. Zhang, L., Chen, S., Li, S., Wang, T.: Automatic measurement of early gestational sac diameters from one scan session. In: SPIE Medical Imaging, p. 796342. International Society for Optics and Photonics (2011)
17. Dalal, N., Triggs, B.: Histograms of oriented gradients for human detection. In: 2005 IEEE Computer Society Conference on Computer Vision and Pattern Recognition (CVPR 2005) (2005)
18. Patil, M.S.S., Junnarkar, M.A., Gore, M.D.: Study of texture representation techniques. image **3**(3), 267–274 (2014)

Weakly Supervised Learning of Placental Ultrasound Images with Residual Networks

Huan Qi[1](✉), Sally Collins[2], and Alison Noble[1]

[1] Institute of Biomedical Engineering (IBME), University of Oxford, Oxford, UK
huan.qi@eng.ox.ac.uk

[2] Nuffield Department of Obstetrics and Gynaecology, University of Oxford, Oxford, UK

Abstract. Accurate classification and localization of anatomical structures in images is a precursor for fully automatic image-based diagnosis of placental abnormalities. For placental ultrasound images, typically acquired in clinical screening and risk assessment clinics, these structures can have quite indistinct boundaries and low contrast, and image-level interpretation is a challenging and time-consuming task even for experienced clinicians. In this paper, we propose an automatic classification model for anatomy recognition in placental ultrasound images. We employ deep residual networks to effectively learn discriminative features in an end-to-end fashion. Experimental results on a large placental ultrasound image database (10,808 distinct 2D image patches from 60 placental ultrasound volumes) demonstrate that the proposed network architecture design achieves a very high recognition accuracy (0.086 top-1 error rate) and provides good localization for complex anatomical structures around the placenta in a weakly supervised fashion. To our knowledge this is the first successful demonstration of multi-structure detection in placental ultrasound images.

1 Introduction

Ultrasonography is a low-cost, non-invasive and non-radiative technique used worldwide for clinical assessment of the human placenta. Expertise is required to both acquire placental ultrasound images and to perform clinical diagnosis from them. These images are particularly challenging for automated biomedical image analysis as the contrast between the textured areas of interest is often low.

Abnormally invasive placentation (AIP) is a general term that covers conditions where the human placenta adheres to the uterus in an invasive fashion. Various diagnostic criteria based on placental ultrasound imaging have been reported or suggested in the literature to characterise this condition [3]. The general approach is to first detect and localise anatomical structures such as the placenta itself, the utero-placental interface and the myometrium within grayscale ultrasound images (B-Mode). Vascular examination using Doppler ultrasound imaging can provide further evidence to support diagnosis (analysis of Doppler

M. Valdés Hernández and V. González-Castro (Eds.): MIUA 2017, CCIS 723, pp. 98–108, 2017.
DOI: 10.1007/978-3-319-60964-5_9

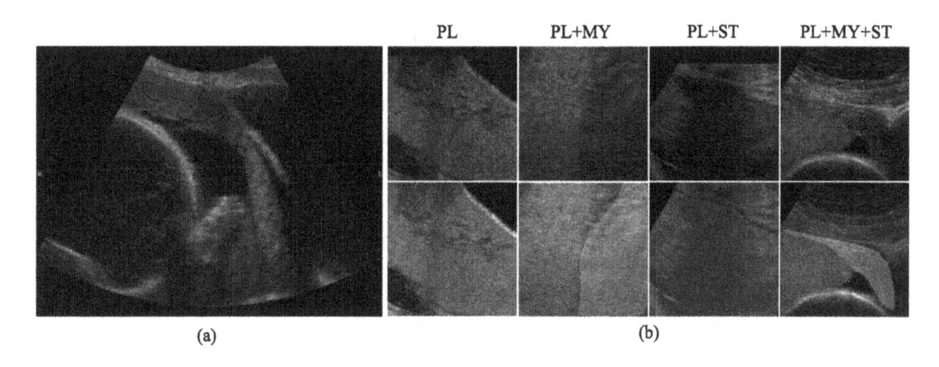

Fig. 1. (a) A placental ultrasound image taken from the *sagittal* plane. (b) Samples from four image categories cropped from sagittal planes, the bottom row shows the reference segmentation mask of the follow anatomical structures: placenta (PL), subcutaneous tissue (ST) and myometrium (MY). Please note that all segmentation masks that appear in this paper are used solely for illustration purpose rather than training models. (Color figure online)

is beyond the scope of the current paper). However, interpretation of the criteria by sonographers is quite inclined to subjectivity [4]. Moreover, manual search for visual evidence among sequences of 2D or 3D placental ultrasound data is sometimes too time-consuming to be considered in clinical workflow.

The contributions in this paper are twofold. First, we propose a deep convolutional neural network (CNN) model for describing anatomical structures present in a 2D placental ultrasound image. This image-level model achieves accurate classification (0.086 top-1 error rate) of the four multi-anatomical structure combinations typically observed in a 2D placental image (as illustrated in Fig. 1), namely (1) placenta only (*PL*); (2) placenta and myometrium (*PL+MY*); (3) placenta and subcutaneous tissue (*PL+ST*); (4) placenta, myometrium and subcutaneous tissue (*PL+MY+ST*). Second, we show that the proposed model achieves good localization of anatomical structures (placenta, myometrium, subcutaneous tissue) based on our multi-structure classification formulation. This is achieved by incorporating a global average pooling (GAP) layer before the fully-connected layer. Thus we demonstrate that image-level classification suffices for localization of anatomical structures in a weakly supervised fashion without any additional training.

2 Related Work

Weakly-Supervised Object Localization: CNN-based weakly-supervised object localization has been a popular research topic in computer vision in recent years and applications are starting to appear in the medical image analysis literature [10], though not to our knowledge for placental ultrasound image analysis. It relies only on image-level labels, rather than annotations in a fully-supervised setting (e.g. manually-annotated bounding box or dense pixel-level annotation), to learn from cluttered scenes with multiple objects. It is of great research interest to develop weakly-supervised localization models that perform comparably to its fully-supervised counterparts due to the fact that the former saves a considerable amount of time in annotation and is less prone to subjectivity. Recent work has further demonstrated that CNNs originally trained for image classification can be used to localize objects via analysis of representative features across layers [1,11,13,16]. For instance, Simonya *et al.* proposed a visualization technique by computing the gradient of the class score with respect to the input image [13]. The resulting saliency map pinpoints the location of objects correlated with the class label. Oquad *et al.* proposed a method to transfer mid-level image representations and explicitly search for high-score regions [11]. Zhou *et al.* recently proposed class activation mapping (CAM) to localize regions with discriminative features in an end-to-end fashion [16]. In general, weakly-supervised localization relies only on image-level classification, which is a desirable property that makes object localization a preferable by-product of classification without additional training.

3 Learning to Classify and Localize with Residual Units

Problem Formulation: Our approach is built on an observation that there are four local anatomical scenarios, which clinicians observe in routine placenta scans, namely *PL, PL+MY, PL+ST, PL+MY+ST*. Thus we have designed a CNN to distinguish between these classes. Further, since the placenta (PL) is shared in all classes it acts as a distractor for localization. To be discriminative, the other three categories are forced to activate their unique regions, which can then be visualized by CAM as is described later. First, however, we describe the general CNN architecture we use.

Deep Residual Learning for Placental Ultrasound Images: Deep residual networks (Res-Net [7,8]) have shown impressive representative ability and good convergence behaviours in recent large-scale natural image classification tasks (e.g. ImageNet ILSVRC 2015 [12]), yielding state-of-the-art performance. In a recent work [8], a simple and effective identity mapping structure was proposed to enable smooth information propagation through the entire network. In addition, large-scale data experiments reveal that the full pre-activation residual unit, as shown in the top-left corner of Fig. 2, consistently outperforms the original design by putting batch normalization [9] and rectified linear unit (ReLU) before convolution. This network design modification has been found to accelerate learning and improves global regularization. In general, a Res-Net typically contains a number of basic residual units. Each unit performs the following computation: $\mathbf{x}_{l+1} = \mathbf{x}_l + \mathcal{F}(\mathbf{x}_l, \mathcal{W}_l)$, where $\mathbf{x}_l$ refers to the input feature to the l-th residual unit and $\mathcal{W}_l$ is a set of weights and biases associated with the l-th residual unit. Here $\mathcal{F}$ denotes the residual function which is learnt with respect to the input feature $\mathbf{x}_l$. Such a design allows a recursive derivation:

$$\mathbf{x}_L = \mathbf{x}_l + \sum_{i=l}^{L-1} \mathcal{F}(\mathbf{x}_i, \mathcal{W}_i)$$

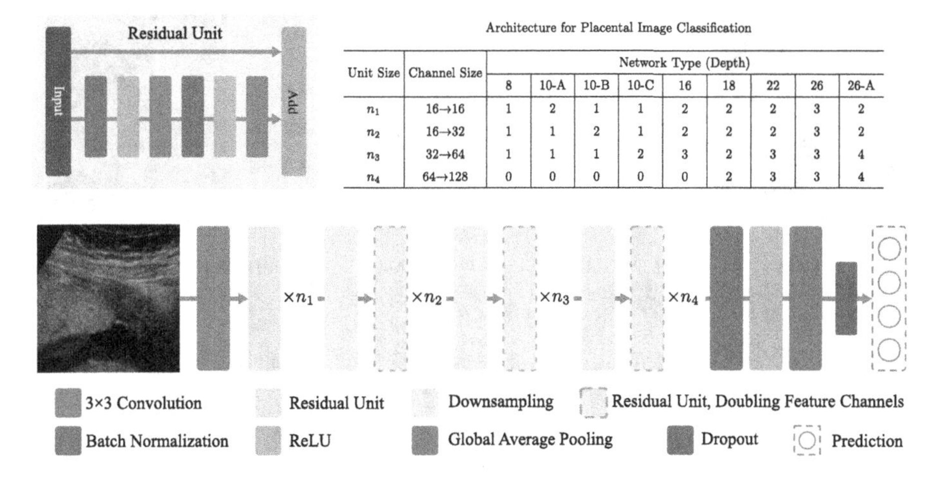

Architecture for Placental Image Classification

Unit Size	Channel Size	Network Type (Depth)								
		8	10-A	10-B	10-C	16	18	22	26	26-A
n_1	16→16	1	2	1	1	2	2	2	3	2
n_2	16→32	1	1	2	1	2	2	2	3	2
n_3	32→64	1	1	1	2	3	2	3	3	4
n_4	64→128	0	0	0	0	0	2	3	3	4

Fig. 2. Proposed deep residual network architecture with different sizes of residual units, as listed in the table.

for any deeper unit L and any shallow unit l, which implies the smooth information propagation through the network.

We adopted the pre-activation residual unit in [8] and designed a series of multi-layer Res-Nets with various representative abilities. The table in Fig. 2 shows different network architectures. In general, the model contains four groups of residual modules, each of which further consists of n_i stacked residual units for $i = 1, 2, 3, 4$. Here n_i is an architecture hyper-parameter that controls the entire depth D of the Res-Net, where $D = 2(n_1 + n_2 + n_3) + 2$ denotes the number of convolutional layer. We follow the principle adopted in recent recognition and segmentation researches [2,7,14] to employ a small convolutional kernel of size 3×3 for all the convolutional layers in Res-Nets. At the *beginning* of the second, third and fourth residual modules, convolutional layers with stride 2 are used to downsample the feature map. Meanwhile, the convolution doubles the feature channel size (also by two), yielding the change of channel sizes: $16 \rightarrow 32 \rightarrow 64 \rightarrow 128$. It is followed by a global average pooling (GAP) layer and a fully-connected layer to generate the final prediction. To increase regularization, we use a dropout layer [15] with dropout probability of $p = 0.3$, which demonstrates a good regularization performance across various architectures according to experiments. The use of GAP is described in the following subsection to boost discriminative localization. The Res-Nets are trained in an end-to-end fashion by stochastic gradient descent with a momentum of 0.9. We tested different combinations of hyper-parameter n_i and report results in the following section.

Global Average Pooling (GAP) for Localization: As shown in [16], the use of GAP encourages the network to identify the *extent* of the object, rather

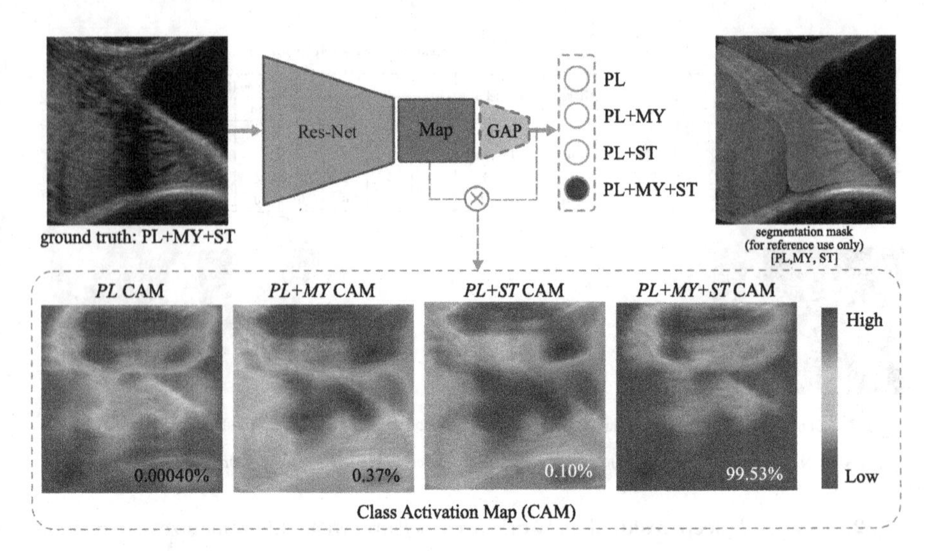

Fig. 3. Proposed pipeline for placental ultrasound image detection and localization. Given an unseen image patch, the network can predict its label, which indicates the multi-anatomical structure combination within. By computing the class activation maps for each category, the network can provide reasonable localization on the detected anatomical structures.

than pinpointing the object on a *specific location* as global max pooling (GMP) does [11]. It is intuitive that this global averaging operation should boost identification of a local discriminative region in order to reach a lower global loss, while a global maximum operation only influences the maximal value of a feature map. We take advantage of GAP to generate CAM [16]. Each class has its corresponding CAM which visualizes discriminative image regions used by the network to identify this specific class. As displayed in the bottom of Fig. 3, the CAM for class i is generated by computing a weighted sum of all feature maps from the rectified activation of the last convolutional layer. Here the weights refer to the corresponding weight vector $\mathcal{W}_i$ learnt in the fully-connected layer. A simple up-sampling would suffice to map the CAM heat-map (28×28) back to the input size (224×224).

For $PL{+}MY$, its CAM is expected to visualize the myometrium region (MY). For the same reason, $PL{+}ST$ CAM would illustrate the region of subcutaneous tissue (ST) and $PL{+}MY{+}ST$ would ideally highlight the joint region of myometrium and subcutaneous tissue. There are potentially many ways to formulate this classification problem. The most intuitive way is probably to build a *multi-label* learning model by training a group of one-verses-all binary classifiers to identify these anatomical structures respectively. However, this type of model suffers from over-fitting and generalization problems in our data experiments. One possible explanation is that there are strong correlations among these anatomical structures (e.g. myometrium is almost always co-localized with placenta), thus it may not be appropriate to model them separately. Moreover,

Table 1. Statistics of placental ultrasound image dataset

	Training	Val	Test	Total
PL	2,764	711	835	4,310
PL+MY	1,064	283	342	1,689
PL+ST	1,834	422	590	2,846
PL+MY+ST	1,256	312	395	1,963
Total	6,918	1,728	2,162	10,808

the multi-label learning does not contribute to the generation of CAM due to the removal of softmax normalization. In this paper we formulate the problem with the intention to both achieve high classification accuracy and boost weakly supervised localization. As shown in Fig. 4, experimental results confirm the validity of our formulation. CAM demonstrates reasonable localization ability for the corresponding anatomical structures. More details will be discussed in the following sections. In this section, we present the two major parts of the proposed model for placental ultrasound image classification and anatomy localization. An overview of our pipeline is given in Fig. 3. We exploit residual units, GAP and CAM to classify ultrasound images and to localize corresponding anatomical structures.

4 Experiments

To evaluate our proposed method, we conducted data experiments on a placental ultrasound image dataset, which was collected as part of a large placenta clinical study. Classification performance of different CNN models are presented. Results of the weakly-supervised localization are also displayed. All the images used in this work are obtained from the *sagittal* plane, annotated by H.Q. under the guidance of S.C., who is a consultant obstetrician and subspecialist in maternal and fetal medicine of John Radcliffe Hospital in Oxford. This method is implemented in Torch 7 [6] on a 64-bit Ubuntu 15.04 machine with a NVIDIA graphics card.

Dataset: All placental ultrasound data ($N = 60$) used in this experiment were obtained as part of a large obstetrics research project [5]. Written informed consent was obtained with local research ethics approval. Static, transabdominal 3D gray-scale ultrasound volumes of the placental bed were obtained according to a predefined protocol with the participant in a semi-recumbent position and a full bladder using a 3D curved array abdominal transducer on a GE Voluson E8 machine. Each 3D volume was then sliced along the sagittal plane into 2D placental ultrasound images, as shown in Fig. 1(a). We then randomly cropped 224×224 image patches from these *sagittal* planes and formed a training and testing dataset by annotating the patches into the four categories described in Sect. 3. In total, the dataset contains 10,808 placental image patches, which is

Table 2. Classification performance of various architectures

Network type	Mean error (val., %)	Mean error (test, %)	Class error (test, %)			
			PL	PL+MY	PL+ST	PL+MY+ST
8	18.23	17.81	13.05	26.61	20.85	15.70
10-A	14.53	14.29	9.82	29.53	12.71	12.91
10-B	12.85	13.55	8.74	30.41	11.36	12.41
10-C	10.82	11.29	7.07	23.98	9.15	12.41
16	11.40	10.78	8.02	22.22	9.15	9.11
18	9.95	9.20	8.62	19.59	**6.10**	**6.08**
22	8.56	8.74	8.26	**14.91**	6.44	7.85
26	9.84	8.88	**6.71**	15.21	8.64	8.35
26-A	**8.08**	**8.60**	7.43	15.21	7.46	7.09

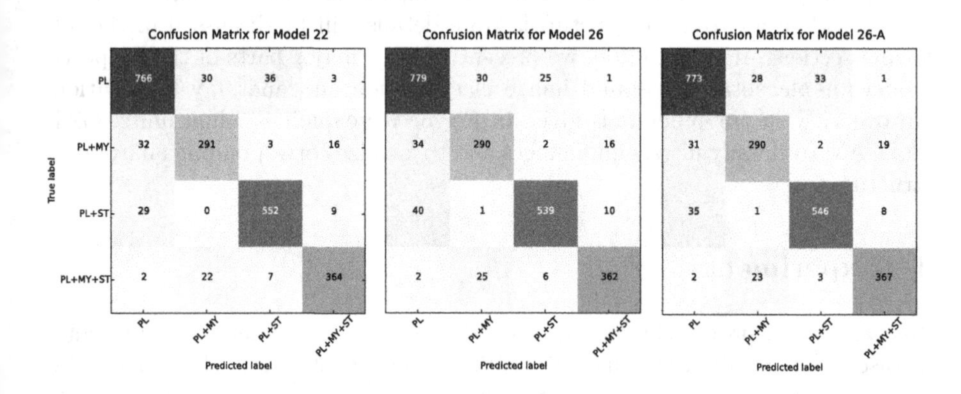

Fig. 4. Confusion matrices in the test stage for best three models.

then randomly divided into a training set (64%), a validation set (16%) and a test set (20%), as shown in Table 1. Here the validation set is used to tune CNN hyper-parameters such as the learning rate, weight decay and architecture parameters $\{n_i\}$.

Evaluation Metrics: We used top-1 error rate to evaluate the classification performance of our proposed method. Experimental results are presented in Table 2. For reference, we also list the top-1 error rate for the validation set and the classification error for individual image categories during the test. It is worth noting that *PL+MY* presents the worst classification performance as expected. In placental ultrasound imaging, it is generally difficult to identify the myometrium for various reasons. First, it may not exist at all due to certain placental abnormality (such as AIP). Second, it appears but the texture or intensity is not sufficiently discriminative to be identified. Third, ultrasound signal dropout may hinder a clear visualization. We also present confusion matrices in the test stage for our best three models in Fig. 4. They all suffer from the same problem of identifying *PL+MY*.

Architecture Hyper-Parameter: In Sect. 3, we introduced four architecture hyper-parameters n_1, n_2, n_3, n_4, corresponding to the number of stacked residual units in each residual module. By altering these hyper-parameters, we can investigate how network depth casts impact on the generalization ability of our classification problem at different abstraction levels. As shown in Fig. 2, nine architectures were evaluated. Here *A,B,C* denotes variants for models of the same depth. Referring to classification performance in Table 2 we see that: (1) more residual units should be put in deeper modules that have larger channel sizes, as demonstrated by the better performance of *10-C* and *26-A* compared to their counterparts of the same depth; (2) the identity mapping structure of Res-Net indeed boosts the propagation of information, yielding better performance for deeper networks without causing *degradation* problems described in [7].

Weakly-Supervised Localization: Here we show results of the weakly supervised localization of placental anatomical structures based on the learnt classification model (model *26-A*). An input image was first classified into one of the four categories. After this, we generated its CAM for the predicted category, which highlights the discriminative regions of this image. Some results are shown in Fig. 5 for each category, where we also provide softmax scores as well as segmentation masks for illustration. For example in the first triple set, the input image is correctly classified as *PL* with a score of 0.8673. CAM heat-map for *PL* highlights the approximate position of the placenta, as verified by the reference segmentation mask. We also present some counterexamples in the bottom of Fig. 5, which are either mis-classified (the correct class is labelled in bold font), mis-localized or both. Good weakly-supervised localization tends to be achieved based on an accurate classification.

Discussions: In the CAM heat-maps of Fig. 5, we observe that the network appears to use texture as well as boundary information as discriminative features. For example, the *PL* CAM *hot zone* typically covers a partition of the placenta as well as the placenta-background boundaries. Similarly, the *PL+MY* CAM *hot zone* covers a part of the myometrium and the myometrium-placenta boundaries. The *PL+ST* CAM *hot zone* contains the subcutaneous tissue as well as the tissue-placenta interface. Finally, the *PL+MY+ST* CAM *hot zone* contains regions of both myometrium and subcutaneous tissue, as well as their interface. This is observed across the test set. Joint analysis of *PL+MY+ST* CAM and *PL+MY* CAM will be carried out in the future, which may provide some insight to help further refine weakly-supervised localization of the myometrium since they both contain a partition of the anatomical structure.

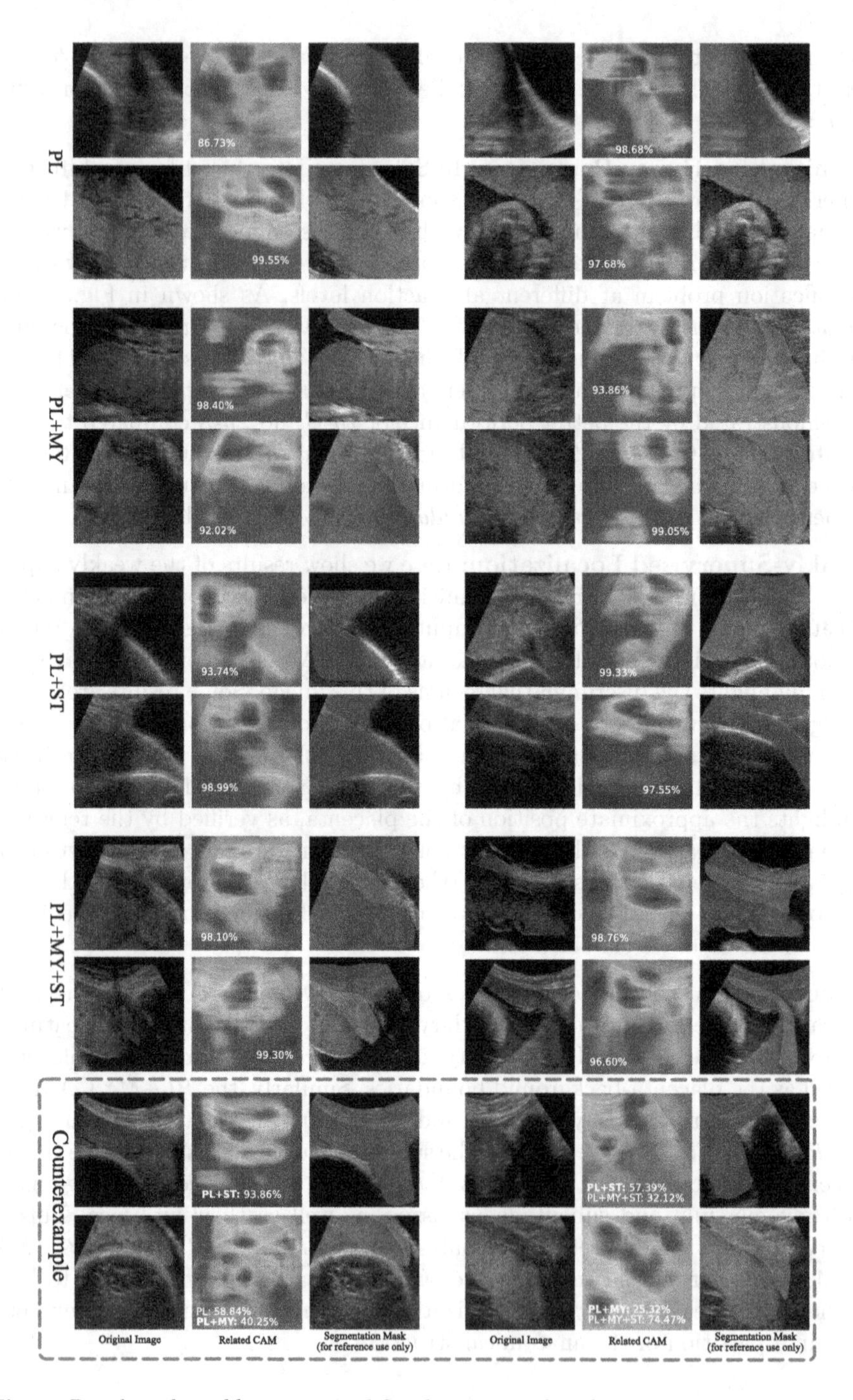

Fig. 5. Results of weakly supervised localization with reference segmentation mask of the follow anatomical structures: placenta (PL), subcutaneous tissue (ST) and myometrium (MY), all images are from the test set. (Color figure online)

5 Conclusion

In this paper, we have formulated automatic placental ultrasound image structure detection and localization as a multi-structure classification problem. The proposed model is based on deep residual networks. Experimental results show good detection accuracy using our approach. Moreover, we demonstrate that reasonable localization of placental anatomical structures can be achieved, without explicit training to perform localization.

References

1. Bazzani, L., Bergamo, A., Anguelov, D., Torresani, L.: Self-taught object localization with deep networks. In: 2016 IEEE Winter Conference on Applications of Computer Vision (WACV), pp. 1–9. IEEE (2016)
2. Chen, H., Dou, Q., Yu, L., Heng, P.A.: Voxresnet: deep voxelwise residual networks for volumetric brain segmentation. arXiv preprint arXiv:1608.05895 (2016)
3. Collins, S.L., Ashcroft, A., Braun, T., Calda, P., Langhoff-Roos, J., Morel, O., Stefanovic, V., Tutschek, B., Chantraine, F.: Proposal for standardized ultrasound descriptors of abnormally invasive placenta (AIP). Ultrasound Obstet. Gynecol. **47**(3), 271–275 (2016)
4. Collins, S.L., Stevenson, G.N., Al-Khan, A., Illsley, N.P., Impey, L., Pappas, L., Zamudio, S.: Three-dimensional power doppler ultrasonography for diagnosing abnormally invasive placenta and quantifying the risk. Obstet. Gynecol. **126**(3), 645–653 (2015)
5. Collins, S., Stevenson, G., Noble, J., Impey, L., Welsh, A.: Influence of power doppler gain setting on virtual organ computer-aided analysis indices in vivo: can use of the individual sub-noise gain level optimize information? Ultrasound Obstet. Gynecol. **40**(1), 75–80 (2012)
6. Collobert, R., Kavukcuoglu, K., Farabet, C.: Torch7: A matlab-like environment for machine learning. In: BigLearn, NIPS Workshop (2011)
7. He, K., Zhang, X., Ren, S., Sun, J.: Deep residual learning for image recognition. In: Proceedings of the IEEE Conference on Computer Vision and Pattern Recognition, pp. 770–778 (2016)
8. He, K., Zhang, X., Ren, S., Sun, J.: Identity mappings in deep residual networks. In: Leibe, B., Matas, J., Sebe, N., Welling, M. (eds.) ECCV 2016. LNCS, vol. 9908, pp. 630–645. Springer, Cham (2016). doi:10.1007/978-3-319-46493-0_38
9. Ioffe, S., Szegedy, C.: Batch normalization: accelerating deep network training by reducing internal covariate shift. arXiv preprint arXiv:1502.03167 (2015)
10. Kamnitsas, K., Ledig, C., Newcombe, V.F., Simpson, J.P., Kane, A.D., Menon, D.K., Rueckert, D., Glocker, B.: Efficient multi-scale 3D CNN with fully connected CRF for accurate brain lesion segmentation. Med. Image Anal. **36**, 61–78 (2017)
11. Oquab, M., Bottou, L., Laptev, I., Sivic, J.: Is object localization for free?-weakly-supervised learning with convolutional neural networks. In: Proceedings of the IEEE Conference on Computer Vision and Pattern Recognition, pp. 685–694 (2015)
12. Russakovsky, O., Deng, J., Su, H., Krause, J., Satheesh, S., Ma, S., Huang, Z., Karpathy, A., Khosla, A., Bernstein, M., Berg, A.C., Fei-Fei, L.: ImageNet large scale visual recognition challenge. Int. J. Comput. Vis. (IJCV) **115**(3), 211–252 (2015)

13. Simonyan, K., Vedaldi, A., Zisserman, A.: Deep inside convolutional networks: Visualising image classification models and saliency maps. arXiv preprint arXiv:1312.6034 (2013)
14. Simonyan, K., Zisserman, A.: Very deep convolutional networks for large-scale image recognition. arXiv preprint arXiv:1409.1556 (2014)
15. Srivastava, N., Hinton, G.E., Krizhevsky, A., Sutskever, I., Salakhutdinov, R.: Dropout: a simple way to prevent neural networks from overfitting. J. Mach. Learn. Res. **15**(1), 1929–1958 (2014)
16. Zhou, B., Khosla, A., Lapedriza, A., Oliva, A., Torralba, A.: Learning deep features for discriminative localization. In: The IEEE Conference on Computer Vision and Pattern Recognition (CVPR), June 2016

Edge Aware Geometric Filter for Ultrasound Image Enhancement

Deepak Mishra[1(✉)], Santanu Chaudhury[1,2], Mukul Sarkar[1], and Arvinder Singh Soin[3]

[1] Indian Institute of Technology Delhi, New Delhi, India
deemishra21@gmail.com
[2] Central Electronics Engineering Research Institute, Pilani, India
[3] Medanta Hospital, Gurgaon, India

Abstract. Despeckling of ultrasound images is essential for subsequent computational analysis. In this paper, an edge aware geometric filter (GF) is proposed for speckle reduction. The behaviour of conventional GF is approximated using commonly used functions like unit step. These approximations help in identifying the natural relationship between GF and other existing spatially adaptive filters. Subsequently, the modifications in GF framework are proposed to take the advantage of edge characteristics. The proposed filter requires almost no parameter tuning and provides good quality outputs for synthetic as well as real ultrasound images. It is compared with the state-of-the-art speckle reducing filters. Improvements of 10.46% and 42% are noticed in mean square error and figure of merit, respectively.

Keywords: Ultrasound image · Speckle · Geometric filter

1 Introduction

Ultrasound (US) images contain granular patterns generated from the constructive and destructive interferences of the backscattered US pulse echoes. These patterns are collectively known as speckle [1,2,6,11,17]. Despeckling of US images depends on subsequent applications. Some applications like tissue characterization, require the speckle to be selectively removed only from the regions of low clinical significance like blood [18]. However, this paper focuses on enhancing the US images to make them suitable for applications involving automatic computational analysis like segmentation. Therefore, the speckle is removed from the entire image, irrespective of the underlying region.

Several speckle reducing filters are reported in literature for US image enhancement [5,12]. Spatially adaptive filters, for example Lee filter [16] and Kuan filter [15], are some of the well known traditional filters. More advanced filters are based on anisotropic diffusion [3,19–21]. These filters prohibit filtering across the edges to preserve object boundaries. Apart form these, wavelet transform based filters [22] and non-local mean (NLM) filters [10] are also popular.

M. Valdés Hernández and V. González-Castro (Eds.): MIUA 2017, CCIS 723, pp. 109–120, 2017.
DOI: 10.1007/978-3-319-60964-5_10

In contrast to the local averaging filters [4], the NLM filters reduce speckle by using a weighted average of non-local image regions. Although, all these filters provide good outputs, their performance is highly sensitive to the tuning of several implementation parameters. It increases the complexity and also leads to catastrophic results in case of inefficient parameter tuning.

In this paper a despeckling filter is proposed which has its roots in the old concepts of geometric filtering [7]. Geometric filter (GF) [7] is a well known speckle reducing filter. Although it aggressively suppresses speckle to quickly reach a stable output, the amount of speckle reduction is not as good as the more advanced diffusion based filters [11]. The key difference is that the GF is unable to use the edge characteristics of the image. In this paper the behaviour of GF is approximated using the function commonly used for signal analysis, the unit step and signum function. This helps us in identification of the natural relationship between GF and spatially adaptive filters like Lee filter. Which in turn enables us to take the advantage of the edge characteristics in GF framework. The proposed edge aware GF requires almost no parameter tuning. It is compared with the existing state-of-the-art filters and a competitive performance is observed.

The paper is organized as follows. Section 2 presents the approximation of the behaviour of GF and its relationship with the spatially adaptive filters. The section also contains details about the proposed filer. Section 3 includes experimental results which are compared with the existing filters. Finally conclusions are presented in Sect. 4.

2 Proposed Filter

This section provides details about the relationship between GF and other existing spatially adaptive filters. Later in the section the edge aware GF is proposed which eliminates the limitations of the conventional GF to provide efficient speckle reduction.

2.1 Geometric Filter (GF)

The GF considers lines of pixels in all four directions, East to West, North to South, North-east to South-west and North-west to South-east [8,9]. These lines which represent gray level profiles, are used to create binary images. The complementary convex hull algorithm is applied on these binary images to make the gray level profile smooth. Accordingly, each pixel in the image is filtered in all four directions to increase the homogeneity among the neighbourhoods. Filtering operation of GF can be easily described using certain rules. Let us assume $x \in \mathbb{R}^2$ represents a pixel and x_1 and x_2 are two neighbours of x lying opposite to each other. Further, f_x is the intensity value of pixel x in the image $f : \Omega \to \mathbb{R}_+, \Omega \subset \mathbb{R}^2$. The intensity value is adjusted based on following rules:

$$\text{if} \quad f_{x_1} \geq f_x + 2 \qquad\qquad \text{then} \quad f_x = f_x + 1 \tag{1a}$$
$$\text{if} \quad f_{x_1} > f_x \quad \text{and} \quad f_x \leq f_{x_2} \qquad \text{then} \quad f_x = f_x + 1 \tag{1b}$$
$$\text{if} \quad f_{x_2} > f_x \quad \text{and} \quad f_x \leq f_{x_1} \qquad \text{then} \quad f_x = f_x + 1 \tag{1c}$$
$$\text{if} \quad f_{x_2} \geq f_x + 2 \qquad\qquad \text{then} \quad f_x = f_x + 1 \tag{1d}$$
$$\text{if} \quad f_{x_1} \leq f_x - 2 \qquad\qquad \text{then} \quad f_x = f_x - 1 \tag{1e}$$
$$\text{if} \quad f_{x_1} < f_x \quad \text{and} \quad f_x \geq f_{x_2} \qquad \text{then} \quad f_x = f_x - 1 \tag{1f}$$
$$\text{if} \quad f_{x_2} < f_x \quad \text{and} \quad f_x \geq f_{x_1} \qquad \text{then} \quad f_x = f_x - 1 \tag{1g}$$
$$\text{if} \quad f_{x_2} \leq f_x - 2 \qquad\qquad \text{then} \quad f_x = f_x - 1 \tag{1h}$$

The filtered image is obtained by iterative application of GF. The time complexity of GF is very low however the filter has some limitations. It works on the assumption that the noise appears as peaks and valleys of the same width in all regions. Considering the spatially correlated nature of speckle, this assumption is a critical assumption which limits the performance of the filter. Further, the GF fails to take the advantage of edge characteristics. Apart from inefficient filtering, it sometimes also results in tempered object boundaries.

2.2 A Different Look to Geometric Filter (GF)

The rules governing the mechanism of GF can be converted into mathematical expression using the standard unit step ($u(.)$) and signum ($sgn(.)$) functions. $u(.)$ can be used to express the rule (1a) as:

$$f_x = f_x + u\left(f_{x_1} - f_x - 2\right) \tag{2}$$

Similar equation can be written for rule (1e) to obtain a combined expression for both the rules, (1a) and (1e), as:

$$f_x = f_x + u\left(f_{x_1} - f_x - 2\right) - u\left(f_x - f_{x_1} - 2\right) \tag{3}$$

As we know, $sgn(k) = u(k) - u(-k)$, (3) can be rewritten as:

$$f_x = f_x + sgn\left(f_{x_1} - f_x\right) \tag{4}$$

where the factor of 2 from the last two terms of (3) is ignored to get a concise expression. Ignoring the factor would require to use stopping criteria like insignificant intensity changes in successive iterations. Similar to (4), an expression for the rules (1d) and (1h) can be written as:

$$f_x = f_x + sgn\left(f_{x_2} - f_x\right) \tag{5}$$

Further the rules (1b) and (1f) can be approximated as:

$$f_x = f_x + \frac{1}{2}\left(sgn\left(f_{x_1} - f_x\right) + sgn\left(f_{x_2} - f_x\right)\right) \tag{6}$$

and similarly the rules (1c) and (1g). Combining these, all the rules can be represented using a single expression as:

$$f_x = f_x + 2 \times sgn\left(f_{x_1} - f_x\right) + 2 \times sgn\left(f_{x_2} - f_x\right) \tag{7}$$

Similar expressions can be obtained for all four directions comprising all the eight neighbours, which would lead us to the following consolidated expression of the filter:

$$f_x^{t+1} = f_x^t + \sum_{x_1 \in \chi} m \times sgn\left(f_{x_1}^t - f_x^t\right) \tag{8}$$

where χ represent the neighbours of x and $m = 2$ is a constant. The superscript t denotes the time or iteration. Equation (8) is an approximated version of GF which is used to establish its relationship with other filters.

2.3 Relationship Between GF and Lee Filter

Mathematical expression of the Lee filter is given as:

$$f_x = \mu_g + k_x \times (g_x - \mu_g) \tag{9}$$

where g and f represent the noisy and filtered images, respectively. μ_g is the local mean and k is a weight factor calculated using local and noise statistics. Equation (9) can be rewritten as:

$$f_x = g_x + (1 - k_x)(\mu_g - g_x) \tag{10}$$

As we know, $\mu_g = \frac{1}{n(\chi)} \sum_{x_1 \in \chi} g_{x_1}$, where $n(\chi)$ is the cardinality of neighbourhood. Thus, (10) can be rewritten as:

$$f_x = g_x + \frac{(1 - k_x)}{n(\chi)} \sum_{x_1 \in \chi} (g_{x_1} - g_x) \tag{11}$$

We can replace g with f^t to reflect iterative filtering and also rewrite (11) as:

$$f_x^{t+1} = f_x^t + \frac{(1 - k_x)}{n(\chi)} \sum_{x_1 \in \chi} |f_{x_1}^t - f_x^t|.sgn\left(f_{x_1}^t - f_x^t\right) \tag{12}$$

where $|.|$ gives the magnitude. If we convert m from (8) into a variable and factorized it into two factors m_1 and m_2 as:

$$m_1 = \frac{(1 - k_x)}{n(\chi)}; \quad m_2 = |f_{x_1}^t - f_x^t| \tag{13}$$

we can see that (12) boils down to (8). This establishes the relationship between GF and Lee filter. Also this shows that the behaviour of GF can be easily adjusted with the careful selection of m_1 and m_2.

2.4 Proposed Filter

Although GF has a notion of direction, it is unable to differentiate between an edge pixel (a pixel lying on an edge) and other pixels. It gives equal weightage to all the neighbouring pixels during filtering. However, for an edge pixel, the neighbours lying in the direction of edge always have a natural dependency. Therefore, these neighbours should have higher weightage during the intensity adjustments. This is the key idea of the proposed filter which helps in increasing the homogeneity along the direction of edges and in turn reducing the speckle. The simplest way to implement the proposed idea is to define m_2 as:

$$m_2 = |\pi_{\overrightarrow{xx_1}}(\overrightarrow{\rho})| \tag{14}$$

where $\overrightarrow{\rho}$ is a unit vector in the direction of the edge passing through the pixel x. The $\pi_{\overrightarrow{xx_1}}(\overrightarrow{\rho})$ gives scalar projection of $\overrightarrow{\rho}$ onto the unit vector $\overrightarrow{xx_1}$. The $m_2 \in [0,1]$ attains the maximum value when the dependency of $\overrightarrow{xx_1}$ on the edge is maximum. $\overrightarrow{\rho}$ can be obtained using any edge indicator. In this work gradients are used. Let us say i and j represent the co-ordinate axis of the image f, then the gradient ∇f is given as:

$$\nabla f = (\partial f_i, \partial f_j) \tag{15}$$

where ∂f_i and ∂f_j are partial derivatives of f in the directions i and j, respectively. The $\overrightarrow{\rho}$ can be defined using ∇f as:

$$\overrightarrow{\rho} = \frac{\overline{\nabla f}}{|\overline{\nabla f}|} \tag{16}$$

where $\overline{\nabla f} = (-\partial f_j, \partial f_i)$ gives a vector parallel to the edge. Now, with the definition of m_2 in (14), we can take $m_1 = 1$ without the loss of generality. Accordingly the expression of the proposed filter is given as:

$$f_x^{t+1} = f_x^t + \sum_{x_1 \in \chi} \left| \pi_{\overrightarrow{xx_1}} \left(\frac{\overline{\nabla f^t}}{|\overline{\nabla f^t}|} \right) \right| \times sgn\left(f_{x_1}^t - f_x^t\right) \tag{17}$$

The proposed definition of m_2 helps in filtering the image according to the edge characteristics. However, our objective is not just to remove speckle but also to enhance the object boundaries. Therefore, a simple modification in the definition of m_2 is done as: $m_2 = |\pi_{\overrightarrow{xx_1}}(\overrightarrow{\rho})| - 0.5$. This negative offset of 0.5 helps in increasing the difference between the pixels belonging to different regions and in turn enhances the object boundaries.

3 Experimental Results and Discussion

The proposed filter is tested on synthetic and real US images. The filtering performance is compared with Lee [16], GF [7], DPAD (detail preserving anisotropic diffusion) [3], OBNLM (optimized Bayesian non-local mean filter) [10] and SBF (squeeze box filter) [23]. The structural similarity measure index (SSIM) [24], figure of merit (FoM) [20] and commonly used mean square error (MSE) are used as evaluation metrics. The mathematical expressions of the evaluation metrics are as follows:

SSIM: it is a measure of structural similarity between the noisy and filtered images:

$$\mathrm{SSIM} = \frac{\left(2\mu_f \mu_{\hat{f}} + c_1\right)\left(2\sigma_f \sigma_{\hat{f}} + c_2\right)}{\left(\mu_f^2 + \mu_{\hat{f}}^2 + c_1\right)\left(\sigma_f^2 + \sigma_{\hat{f}}^2 + c_2\right)} \tag{18}$$

where f is the filtered image and $\hat{f}$ is the noise free reference image. μ_f and σ_f^2 are mean and variances of the image f. c_1 and c_2 are two small constants added to provide stability. The value of SSIM varies between 0 and 1, where 1 represents identical images in terms of structural similarity.

FoM: it is also known as Pratt's figure of merit and calculated as:

$$\mathrm{FoM} = \frac{1}{max\left\{N, N_{ideal}\right\}} \sum_{i=1}^{N} \frac{1}{1 + d_i \alpha} \tag{19}$$

where N and N_{ideal} represent the number of edge pixels obtained from filtered and reference images using the Canny edge detector. The d_i is the Euclidean distance between an edge pixel obtained from filtered image and nearest edge pixel from reference image, and α is a constant set as 1/9. The value of FoM also varies between 0 and 1, where 1 represents the perfect edge preservation.

MSE: it is the measure of intensity differences between the filtered and reference images.

$$\mathrm{MSE} = \frac{1}{N \times N} \sum_{x=1}^{N \times N} \left(\hat{f}_x - f_x\right)^2 \tag{20}$$

where $N \times N$ represent the total number of pixels in the considered image. Low values of MSE are desirable.

The implementation parameters of different filters require tuning to result in the best values of the evaluation metrics. For parameter tuning, initially any one parameter, like the number of iterations (itr), is fixed and rest of the parameters for example step size (dt) for DPAD, smoothing parameter (h) for OBNLM and localization parameter (σ_g) for SBF are varied over a very fine grid. The parameters associated with image regions for example window size (W), patch size (α) and search area (M) are varied over an integer grid. Later, the previously fixed parameter is varied to get the complete set of optimized parameters. In contrast to the existing filters, the proposed filter requires only the tuning of itr, which can also be eliminated by using any stopping criteria.

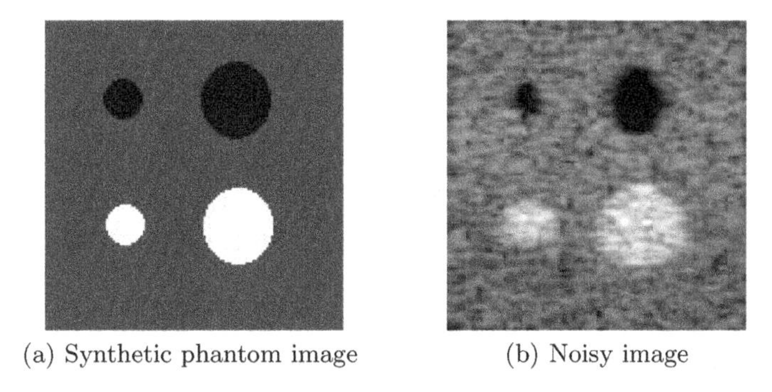

(a) Synthetic phantom image (b) Noisy image

Fig. 1. Synthesized phantom and simulated noisy image.

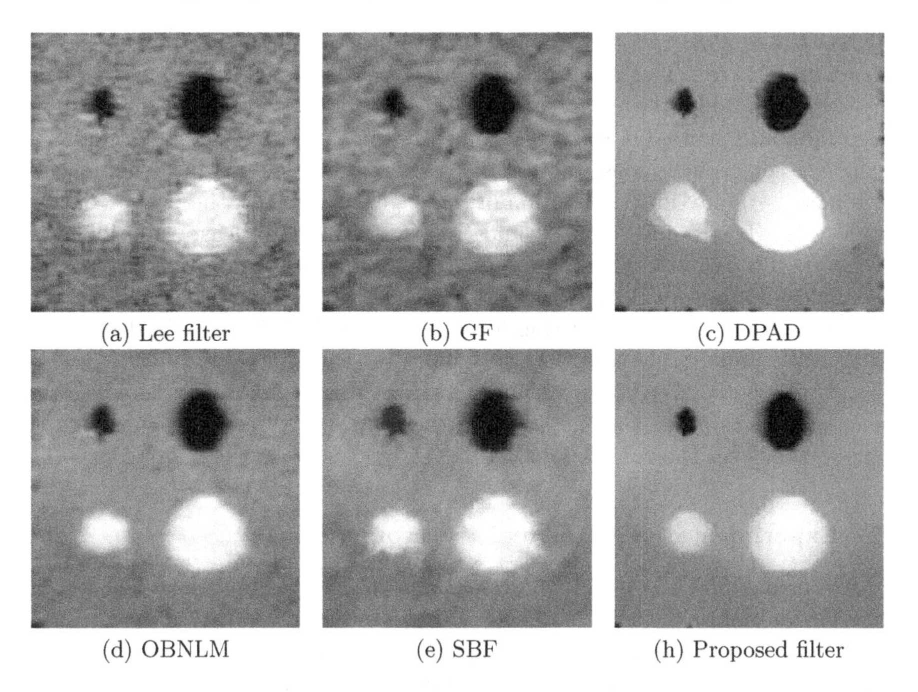

(a) Lee filter (b) GF (c) DPAD

(d) OBNLM (e) SBF (h) Proposed filter

Fig. 2. Filtered outputs of phantom image obtained using (a)–(h) Lee, GF, DPAD, OBNLM, SBF, and the proposed filter.

3.1 Experiment with Synthetic Phantom Image

Figure 1(a) shows the synthetic phantom image containing four cysts. The noisy US image, simulated using Field-II simulator [13,14], is shown in Fig. 1(b). Parameters of the considered filters are optimized to result in the least value of MSE. The optimized parameter values along with the MSE, FoM and SSIM values are listed in Table 1. Since the noisy image is in log compressed domain, the synthetic phantom image is also converted into log compressed domain and used as the reference. The filtered outputs are shown in Fig. 2.

Table 1. Optimized values of the parameters for phantom image and the observed values of evaluation metrics

Filter	Parameters	MSE	FoM	SSIM
Noisy input	-	3000	0.156	0.185
Lee	$W = 13$	1451	0.169	0.344
GF	$itr = 5$	1717	0.161	0.436
DPAD	$dt = 0.1$, $W = 2$, $itr = 706$	956	0.333	0.770
OBNLM	$M = 13$, $\alpha = 3$, $h = 1.5$	1136	0.189	0.689
SBF	$\sigma_g = 0.01$, $W = 3$, $itr = 206$	996	0.276	0.708
Proposed	itr = 157	**856**	**0.473**	**0.800**

Over the uniform intensity regions the gradients are influenced by stochastic behaviour of noise. The inconsistency in the gradients of neighbouring pixels helps the proposed filter to increase homogeneity in every direction, resulting in the reduction of noise. On the other hand, the pixels near the object boundaries or the edge pixels have consistent gradient orientation, therefore, the homogeneity is increased only in the direction of the edge. As a result, the proposed filter provides the best values of FoM and SSIM. Further, the least value of MSE reflects its filtering ability. The proposed filter shows 42% improvement in FoM and 10.46% improvement in MSE, as compared to the best result achieved by existing filters. Among the existing filters, DPAD and SBF result in decent values of MSE and SSIM, however, the quality of their filtered outputs (Fig. 2) is relatively poor. The object boundaries are highly distorted and irregular. An important thing to note here is that the proposed filter shows considerable improvement over the conventional GF.

3.2 Experimentation with Real Ultrasound Image

A real cardiac US image is acquired using GE Healthcare Vivid US system with full consent of the subject. The evaluation metrics used in synthetic image experiment require noiseless reference image. Therefore, a different metric known as contrast to noise ratio (CNR) is used here. It is defined as:

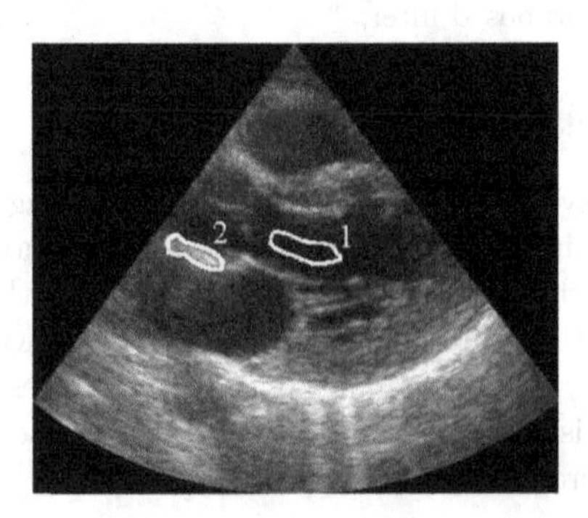

Fig. 3. Cardiac US image with marked regions

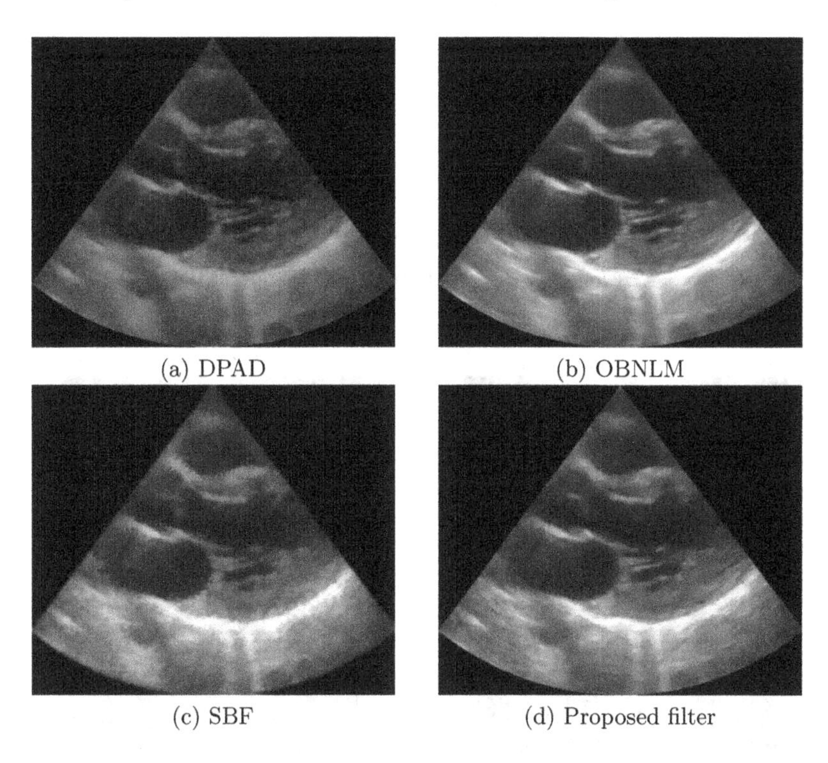

(a) DPAD (b) OBNLM (c) SBF (d) Proposed filter

Fig. 4. Filtered images obtained using (a)–(d) DPAD, OBNLM, SBF, and the proposed filter.

$$\text{CNR} = \frac{|\mu_{C_1} - \mu_{C_2}|}{\sqrt{\sigma^2_{C_1} + \sigma^2_{C_2}}} \tag{21}$$

where μ_{C_r} and $\sigma^2_{C_r}$ are the mean and variance of the class C_r representing object or background. High value of CNR represents the high contrast. CNR measurement requires two regions of similar sizes representing different classes, tissue and background. Two such regions are identified in the image considered for experiment, shown in Fig. 3. Region 1 represents blood chamber and region 2 contains septum wall. The filtered images are shown in Fig. 4. The optimized parameter values are listed in Table 2 along with the measured CNR. Lee filter and GF are not considered due to their poor performances in synthetic phantom image experiment. SBF results in the best value of CNR, however the quality of filtered output is poor. Surprisingly the OBNLM provides better image quality despite having the lowest CNR value. On the other hand, the proposed filter gives balanced output. It results in the second best value of CNR and also provides good quality image.

The proposed filter is further tested on a real US liver images obtained from a publically available database[1]. The liver images and the corresponding filtered outputs obtained using the proposed filter are shown in Fig. 5. It is clear that the noise is efficiently removed while preserving the important structural details.

[1] http://www.ultrasoundcases.info/.

Table 2. Optimized values of the implementation parameters for the real cardiac US image shown in Fig. 3.

Filter	Parameters	CNR
Noisy input	-	2.02
DPAD	$dt = 0.01$, $W = 6$, $It = 65$	2.23
OBNLM	$M = 9$, $\alpha = 7$, $h = 1.2$	2.18
SBF	$\sigma_g = 0.01$, $W = 5$, $It = 59$	**2.33**
Proposed	$itr = 49$	2.27

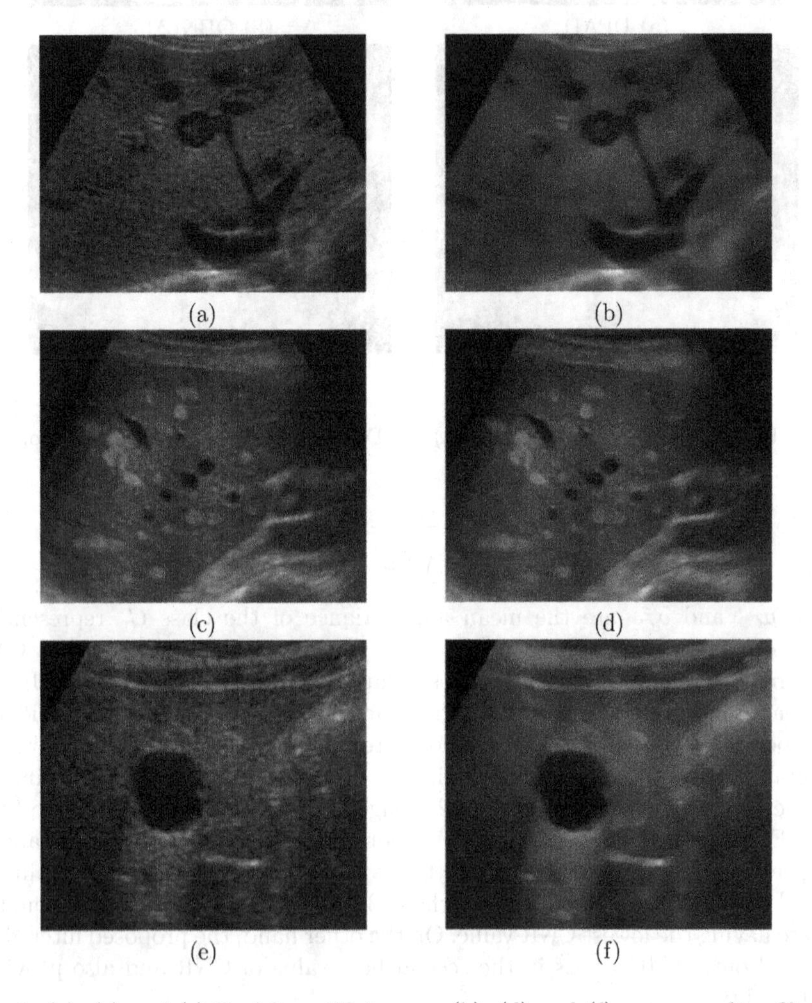

Fig. 5. (a), (c) and (e) Real liver US images, (b), (d) and (f) corresponding filtered outputs obtained using the proposed filter.

4 Conclusions

In this paper an edge aware GF is proposed. The filter uses edge characteristics inside the GF framework. It allows the filter to iteratively remove speckle while preserving the structural details. The filter requires almost no parameter tuning which is a big challenge faced by the existing filters. The filter provides best results for MSE, FoM and SSIM and shows competitive performance in terms of CNR. The obtained filtered images show that the proposed filter is appropriate for US image enhancement.

Acknowledgments. We would like to acknowledge AKAB Healthcare Pvt. Ltd. for funding.

References

1. Kang, J., Lee, J., Yoo, Y.: A new feature-enhanced speckle reduction method based on multiscale analysis for ultrasound B-mode imaging. IEEE Trans. Biomed. Eng. **63**(6), 1178–1191 (2015)
2. Ortiz, S.H.C., Chiu, T., Fox, M.D.: Ultrasound image enhancement: a review. Biomed. Sig. Process. Control **7**(5), 419–428 (2012)
3. Aja-Fernández, S., Alberola-López, C.: On the estimation of the coefficient of variation for anisotropic diffusion speckle filtering. IEEE Trans. Image Process. **15**(9), 2694–2701 (2006)
4. Balocco, S., Gatta, C., Pujol, O., Mauri, J., Radeva, P.: SRBF: speckle reducing bilateral filtering. Ultrasound Med. Biol. **36**(8), 1353–1363 (2010)
5. Zhang, J., Wang, C., Cheng, Y.: Comparison of despeckle filters for breast ultrasound images. Circ. Syst. Sig. Process. **34**(1), 185–208 (2015)
6. Burckhardt, C.B.: Speckle in ultrasound B-mode scans. IEEE Trans. Sonics Ultrason. **25**(1), 1–6 (1978)
7. Crimmins, T.R.: Geometric filter for speckle reduction. Appl. Opt. **24**(10), 1438–1443 (1985)
8. Busse, L.J., Crimmins, T.R., Fienup, J.R.: A model based approach to improve the performance of the geometric filtering speckle reduction algorithm. IEEE Ultrasonics Symp. **1**, 1353–1356 (1995)
9. Alparone, L., Garzelli, A.: Decimated geometric filter for edge-preserving smoothing of non-white image noise. Pattern Recognit. Lett. **19**(1), 89–96 (1998)
10. Coupé, P., Hellier, P., Kervrann, C., Barillot, C.: Nonlocal means-based speckle filtering for ultrasound images. IEEE Trans. Image Process. **18**(10), 2221–2229 (2009)
11. Finn, S., Glavin, M., Jones, E.: Echocardiographic speckle reduction comparison. IEEE Trans. Ultrason. Ferroelect. Freq. Control **58**(1), 82–101 (2011)
12. Gupta, N., Swamy, M.N., Plotkin, E.: Despeckling of medical ultrasound images using data and rate adaptive lossy compression. IEEE Trans. Med. Imaging **24**(6), 743–754 (2005)
13. Jensen, J.A.: FIELD: a program for simulating ultrasound systems. In: 10th Nordic-Baltic Conference on Biomedical Imaging, vol. 4, no. 1, pp. 351–353 (1996)
14. Jensen, J.A., Svendsen, N.B.: Calculation of pressure fields from arbitrarily shaped, apodized, and excited ultrasound transducers. IEEE Trans. Ultrason. Ferroelect. Freq. Control **39**(2), 262–267 (1992)

15. Kuan, D.A., Sawchuk, A.L., Strand, T.I., Chavel, P.: Adaptive restoration of images with speckle. IEEE Trans. Acoust. Speech Sig. Process. **35**(3), 373–383 (1987)
16. Lee, J.S.: Speckle analysis and smoothing of synthetic aperture radar images. Comput. Graph. Image Process. **17**(1), 24–32 (1981)
17. Loizou, C.P., Pattichis, C.S., Christodoulou, C.I., Istepanian, R.S., Pantziaris, M., Nicolaides, A.: Comparative evaluation of despeckle filtering in ultrasound imaging of the carotid artery. IEEE Trans. Ultrason. Ferroelect. Freq. Control **52**(10), 1653–1669 (2005)
18. Ramos-Llordén, G., Vegas-Sánchez-Ferrero, G., Martin-Fernandez, M., Alberola-López, C., Aja-Fernández, S.: Anisotropic diffusion filter with memory based on speckle statistics for ultrasound Images. IEEE Trans. Image Process. **24**(1), 345–358 (2015)
19. Vegas-Sanchez-Ferrero, G., Aja-Fernandez, S., Martin-Fernandez, M., Frangi, A.F., Palencia, C.: Probabilistic-driven oriented speckle reducing anisotropic diffusion with application to cardiac ultrasonic images. In: Jiang, T., Navab, N., Pluim, J.P.W., Viergever, M.A. (eds.) MICCAI 2010. LNCS, vol. 6361, pp. 518–525. Springer, Heidelberg (2010). doi:10.1007/978-3-642-15705-9_63
20. Yu, Y., Acton, S.T.: Speckle reducing anisotropic diffusion. IEEE Trans. Image Process. **11**(11), 1260–1270 (2002)
21. Krissian, K., Westin, C.F., Kikinis, R., Vosburgh, K.G.: Oriented speckle reducing anisotropic diffusion. IEEE Trans. Image Process. **16**(5), 1412–1424 (2007)
22. Yue, Y., Croitoru, M.M., Bidani, A., Zwischenberger, J.B., Clark, J.W.: Nonlinear multiscale wavelet diffusion for speckle suppression and edge enhancement in ultrasound images. IEEE Trans. Med. Imaging **25**(3), 297–311 (2006)
23. Tay, P.C., Garson, C.D., Acton, S.T., Hossack, J.A.: Ultrasound despeckling for contrast enhancement. IEEE Trans. Image Process. **19**(7), 1847–1860 (2010)
24. Wang, Z., Bovik, A.C., Sheikh, H.R., Simoncelli, E.P.: Image quality assessment: from error visibility to structural similarity. IEEE Trans. Image Process. **1**(4), 600–612 (2004)

Cardiovascular Imaging

Tissues Classification of the Cardiovascular System Using Texture Descriptors

Claudia Mazo[1(✉)], Enrique Alegre[2], Maria Trujillo[1], and Víctor González-Castro[2]

[1] Computer and Systems Engineering School, University of Valle, Cali, Colombia
claudia.mazo@correounivalle.edu.co
[2] Industrial and Informatics Engineering School, University of León, León, Spain

Abstract. In this paper, we present an approach to automatically classify tissues of the cardiovascular system using texture information. Additionally, this process makes possible to identify some cardiovascular organs, since some tissues belong to muscles associated to those, i.e. identifying the tissue makes possible to identify the organ. We have assessed rotation invariant Local Binary Patterns (LBPri) and Haralick features to describe the content of histological images. We also assessed Random Forest (RF) and Linear Discriminant Analysis (LDA) for the classification of these descriptors. The tissues were classified into four classes: (i) cardiac muscle of the heart, (ii) smooth muscle of the elastic artery, (iii) loose connective tissue, and (iv) smooth muscle of the large vein and the elastic artery. The experimental validation is conducted with a set of 2400 blocks of 100 × 100 pixels each. The classifier was assessed using a 10-fold cross-validation. The best AUCs (0.9875, 0.9994 and 0.9711 for the cardiac muscle of the heart, the smooth muscle of muscular artery, the smooth muscle of the large vein and the elastic artery classes, respectively) are achieved by LBPri and RF.

Keywords: Fundamental tissues · Histology images · Image processing · Organs of the cardiovascular system · Automatic classification

1 Introduction

The human body is composed of four basic types of tissues: epithelium, connective tissue, muscular tissue and nervous tissue, which vary in their composition and function. Currently, the recognition of healthy tissues and organs is carried out by histology experts and there are no automatic systems able to identify them. An automatic classification platform of histological images of the cardiovascular system would be helpful for educational purposes [1]. It would promote self-learning to on-campus students and also facilitate on-line learning to external or remote students [2] requiring low social and economic investment to train qualified professionals. Furthermore, it could be used for automatic labelling and sorting of large repositories of histology images that are available in hospitals or distributed

M. Valdés Hernández and V. González-Castro (Eds.): MIUA 2017, CCIS 723, pp. 123–132, 2017.
DOI: 10.1007/978-3-319-60964-5_11

through different storage devices of histologists. Thus, subjectivity, time, costs, difficulty and impracticality issues of the manual annotation of histology images may be solved by using automatic classification systems [3]. In this paper, we present an approach to automatically identify tissues and thereafter, indirectly, recognise organs of the cardiovascular system using rotation invariant Local Binary Pattern (LBPri) texture features and the Random Forest (RF) classifier.

Texture descriptors are used to describe the content of medical images [4] and histological images [5,6]. Some histological image studies focused on cell [7,8], tissue and organ recognition [9,10]. For instance, in [11] a statistical model for histological image classification is presented using texture features of Multi-channel Gabor. The probability distribution of texture patterns of each category is approximated by a finite Gaussian mixture model in order to identify ten organs—adrenal gland, the heart, kidney, liver, lung, pancreas, spleen, testis, thyroid and uterus. The validation was performed using 778 histological images. This approach yields an accuracy between 44% and 93% which varies depending on the organ being identified. In particular, the approach yields an accuracy of 80% for the heart, which is the only organ of the cardiovascular system included in the study.

A similar proposal is found in [12], which aims to identify five organs of the gastrointestinal tract—oesophagus, stomach, small intestine, large intestine and anus—by means of a 2D stochastic method for the semantic analysis of the content of histological images. The proposed method is based on the classification of 64×64 pixels blocks. From each block a 25-dimensional feature vector is obtained by concatenating the total Gabor energy and the mean grey value. The model yields an accuracy between 59% and 82% depending on the organ being identified. However, this model has high computational complexity and the evaluation was conducted using a limited dataset of images, i.e. 40 histological images per region.

We have done preliminary advances in the segmentation and the classification of fundamental tissues [13]. In [14] coating epithelial tissue is recognised and classified into flat, cubic and cylindrical in histological images acquired using a 40x lens. We used the K-means algorithm, a sphericity measure and spatial information, obtaining a sensitivity of 0,85. In addition, in [10] loose connective and muscle tissues are recognised by clustering the Structure Tensor, the red and the green colour channels. Expert evaluations scored with an average of 4.85 the identification of loose connective tissue and an average of 4.82 the identification of the muscle tissue, out of 5. However, this approach does not differentiate among different muscle types.

The computer-aided recognition of fundamental tissues and organs presents several problems due to the hard boundary among fundamental tissues, the lack of a panoramic view of the entire sample and the low definition in image areas such as at the edges of the tissues and organs. Moreover, due to the importance and complexity of the appearances of fundamental tissues and organs, their automatic classification remains an open problem.

In this paper, we present an approach to automatically classify the fundamental tissues of the cardiovascular system. We used the LBPri [15] texture descriptor to represent the image content. The RF [16] classifier is used to identify cardiac muscle of the heart, loose connective tissue (i.e., veins, arteries and the heart), smooth muscle of muscular artery, and smooth muscle of the large vein and the elastic artery. Additionally, we outperform our previous results [10,14] using more general and robust computer vision and machine learning methods, which allows us to recognise organs through the same process.

The rest of the paper is structured as follows: the problem statement is presented in Sect. 2. Section 3 explains the proposed approach to automatic classification of fundamental tissues and organs. We present and discuss the dataset, the experiments and the results in Sect. 4. Finally, in Sect. 5 some conclusions are presented.

2 Problem Statement

Different tissues have different roles and functions. For instance, epithelium forms the coverings of surfaces of the body and it has different purposes, e.g. protection, adsorption, excretion, secretion, filtration, and sensory reception. Connective tissue protects, supports, and binds together parts of the body. Muscular tissue produces movement. Nervous tissue receives stimuli and conduct impulses.

Automatic recognition of tissues is based on visual features. Although tissues have unique patterns that make them identifiable from others, their appearances may be different within the same organ, due to changes in the capture zone, cut or sample. In addition, colour is not a reliable feature in histological images since it depends on the applied stain. Nevertheless, the observation of spatial patterns suggests that texture may be a relevant characteristic for tissue recognition. Figure 1 illustrates tissue blocks of 100×100 pixels from different organs. It can

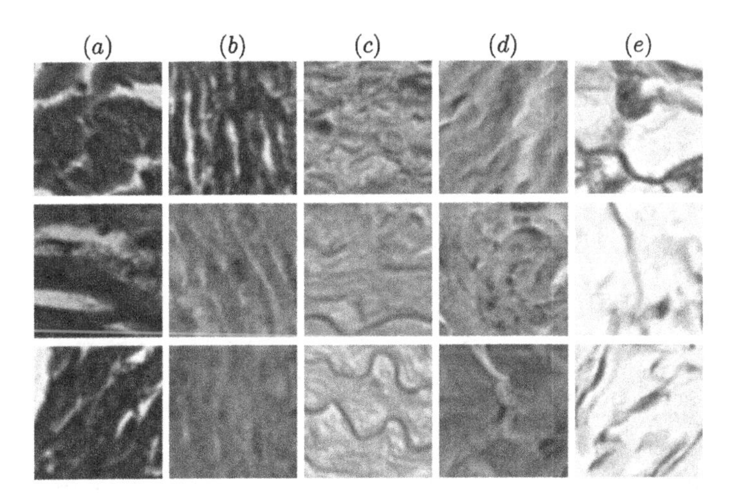

Fig. 1. Examples of blocks of the cardiac muscle tissue of the heart (a), smooth muscle tissue of the muscular artery (b), smooth muscle tissue of the elastic artery (c), smooth muscle tissue of the large vein (d) and loose connective tissue (e)

be observed that there are some intra-class similarity and inter-class difference among the blocks of the same tissue.

The classification of fundamental tissues is affected by the existing variability among blocks of the same tissue, which makes it a challenging and open problem.

3 Classification of Fundamental Tissues

The proposed approach is carried out in three steps: initially, an image is divided into blocks—an image capture protocol was designed to obtain a set of images with the same characteristics and to reduce errors in the automatic classification. Then, relevant texture features are extracted from the blocks and, finally, these descriptors are used to classify them. A general outline of our proposal is shown in Fig. 2. As it is depicted, first of all, blocks of 100×100 pixels are obtained from histological images. In this dataset, a block contains only one type of tissue. Afterwards, textural features are extracted from the blocks to obtain relevant information from them. After evaluating different texture descriptors, we propose to use LBPri to efficiently and robustly represent local micro-patterns. Thus, each block is represented by a vector of 36 features. Finally, Random Forest

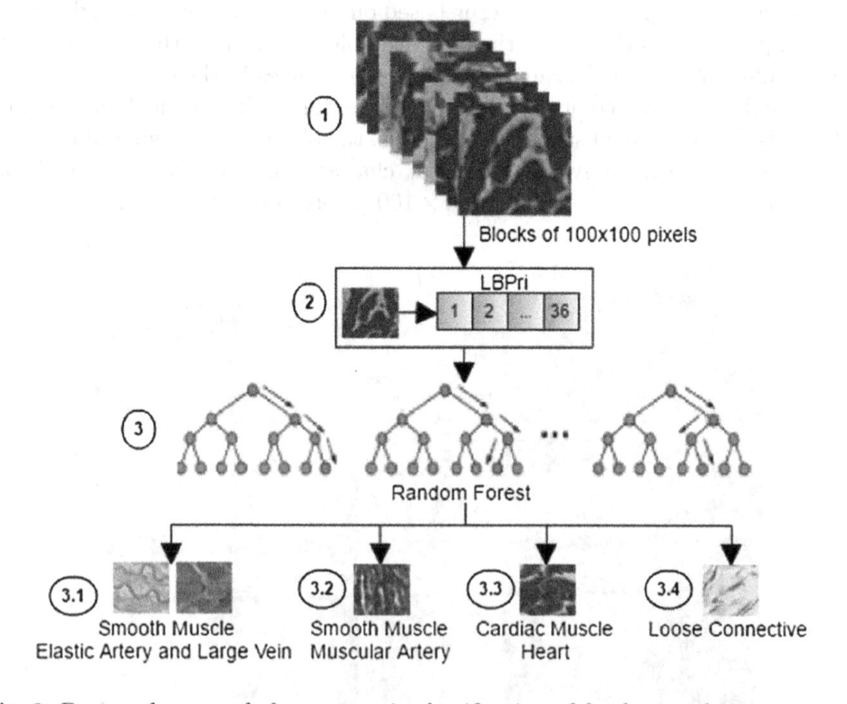

Fig. 2. Proposed approach for automatic classification of fundamental tissues associated with an organ: (1) image block of size 100×100 pixels. (2) LBPri histogram calculated per block. (3) Classification using RF. (3.1), (3.2), (3.3) and (3.4) blocks classified.

(RF) [16] is used as a classifier, after comparing its performance with that of the Linear Discriminant Analysis (LDA) [17]. RF is used to classify blocks as belonging to one of the four considered classes: (i) smooth muscle of the large vein and the elastic artery, (ii) smooth muscle of muscular artery, (iii) cardiac muscle of the heart and (iv) loose connective tissue (3.1, 3.2, 3.3 and 3.4 in Fig. 2, respectively). The smooth muscle of the large vein and the elastic artery were put in the same class taking into account similarities in the composition of the tunica media and in the muscle tissue, which main difference is the thickness of the layer. These similarities are reflected in texture features even when are recognised manually by an expert. The details of the complete process are presented in the rest of this section.

3.1 Tissues Description

The proposed approach uses texture descriptors to extract information about composition and characteristics of fundamental tissue morphology using a block-based recognition algorithm. These features are thereupon used for tissue recognition.

Histological images have particular characteristics depending on each fundamental tissue. Different texture descriptors were assessed for tissue identification and classification. LBP features and its variants have been commonly used as texture descriptors in recent years [18]. Haralick features have been also used in some biomedical applications [19,20]. In this paper, we assessed the rotation invariant Local Binary Patterns (LBPri) and some of the Haralick features (specifically contrast, homogeneity, angular second moment, correlation, entropy, and first and second correlation measures) The parameters *radius* and *number of neighbours* in the LBPri were set to 1 and 8, respectively – i.e. $LBP_{1,8}^{ri}$ (for simplicity, this term will be referred to as LBPri henceforth) – and Haralick features were calculated with distance equal to 3 using 4 directions – i.e. 0°, 45°, 90° and 135°. The feature vectors calculated with LBPri and Haralick have 36 and 48 values, respectively.

3.2 Tissues and Organs Classification

The proposed approach uses machine learning algorithms to recognise fundamental tissues and, in some cases, the organs are also identified. The classification is done for the following four classes: (i) cardiac muscle of the heart, (ii) loose connective tissue (i.e., vein, arteries and the heart), (iii) smooth muscle of the muscular artery, and (iv) smooth muscle of the large vein and the elastic artery. The RF using different parameters and the LDA were compared in order to select the classifier. The RF is an ensemble of decision trees, where the number of trees has a significant effect on the accuracy of the resulting model [16]. The LDA uses a linear combination of features that compute the directions, which represent the axes that maximise the separation of two or more classes [17].

4 Experiments and Results

In this section we analyze the results obtained using the descriptors evaluated with RF. Then, we used the best texture descriptor, according to the previous results, to compare performance with another classifier such as LDA. We use the Receiver Operating Characteristic (ROC) curves for each descriptor we use the area under such curves (AUC) as the quantitative measurement of the classifications. In the rest of the Section, "Muscular" represents the smooth muscle of the muscular artery, "Heart" represents the cardiac muscle of the heart, "Connective" represents the loose connective tissue (i.e. veins, arteries and the heart), and "Elastic-Vein" represents the smooth muscle of the elastic artery and large vein.

4.1 Experimental Setup

Tissue samples were stained with Hematoxylin and Eosin and Masson's trichrome using a laboratory protocol to control the process [21]. The image capture protocol was defined by taking into account some characteristics such as the microscope configuration, software configuration or sample manipulation in order to reduce errors in the automatic classification. We used a dataset of 600 blocks per class (i.e. 2400 blocks), which belong to images obtained from tissue samples of different organs and people and acquired using a $10x$ objective. The histological images were acquired with a Leica *DM750-M* microscope with a resolution of 2048×1536 pixels and stored in *PNG* format. We have made the dataset publicly available at http://biscar.univalle.edu.co/?page_id=1003. Algorithms were implemented in *C++* and *MATLAB*, using the *CImg* library in a *8-cores* CPU and *8*GB *RAM* computer. We used a 10-fold cross-validation in the classification experiments.

4.2 Texture Descriptors Assessment

We have assessed the descriptors LBPri and Haralick features (specifically contrast, angular second moment, homogeneity, correlation, entropy, first and second correlation measures) both on their own. We present the obtained results using RF classifier in Fig. 3, for each set of features per class (one vs all).

LBPri was the best descriptor with the AUCs (0.9986, 0.9952, 0.9453, 0.921) for smooth muscle of the muscular artery, cardiac muscle of the heart, loose connective tissue, and smooth muscle of the elastic artery and large vein, respectively. LBPri presents the highest True Positive Rates, indicating good classification results globally. The worst AUCs are achieved by Haralick features in all cases, which means LBPri is the most effective descriptor for classify blocks of histological images among both local pattern descriptors. This is because Haralick only considers four neighbourhood to distance equal to three while the LBPri descriptor considers eight neighbourhood with radius one, more local information. LBPri is simple, effective, and robust and is proving to be a powerful discriminator in many medical image classification problems.

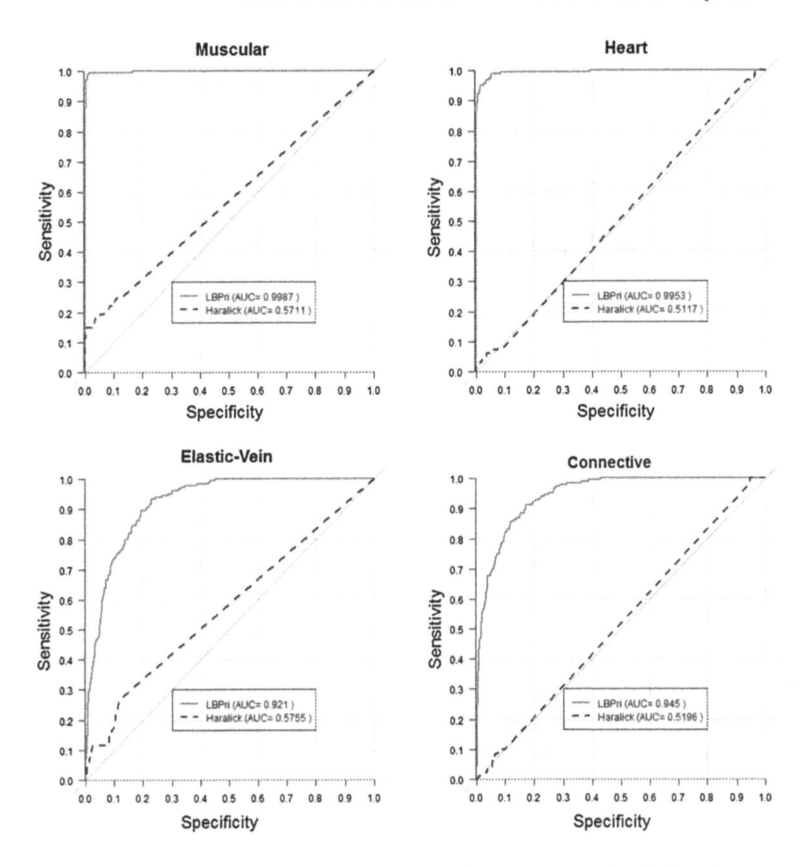

Fig. 3. ROC curves corresponding to the classification of each of the classes vs. the others using RF.

4.3 Evaluation of Different Classification Algorithms

We also compared the Random Forests (RF) and Linear Discriminant Analysis (LDA) using the texture descriptors that performed best in Sect. 4.2, i.e. LBPri. We assessed different parameters in the RF. Figure 4 depicts ROC curves and AUC values of testing performance comparisons in a multi-class classification.

As shown in Fig. 4, a direct comparison of the classification performances shows that the RF has the best classification results globally. The best results are achieved by RF for the cardiac muscle of the heart, the smooth muscle of muscular artery, the smooth muscle of the large vein and the elastic artery classes with AUCs of 0.9875, 0.9994 and 0.9711 respectively. LDA performed better in classifying loose connective tissue, achieving an AUC of 0.9897. Finally, we decided to use the RF with $Deep = 2$ and $MaxTrees = 100$ since they yielded the best performance.

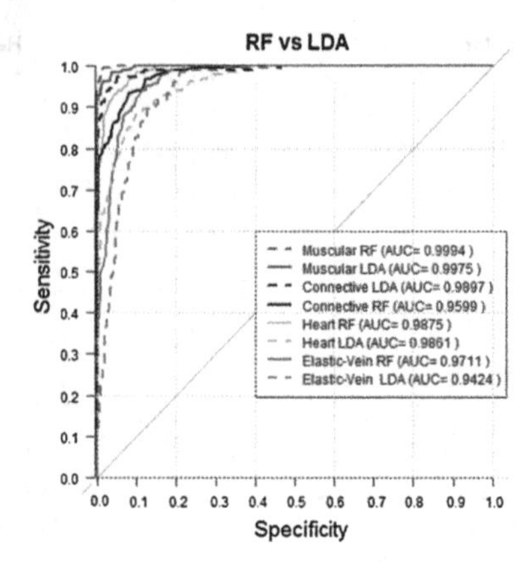

Fig. 4. Comparative evaluation using ROC curves classification of RF and LDA for each class.

5 Conclusions

In this paper, we presented a method that makes possible the recognition of the cardiac muscle of the heart, loose connective tissue (i.e. veins, arteries and the heart) and smooth muscle of the muscular artery, the large vein and the elastic artery with AUC greater than 0.90.

We have determined that, among those assessed, the best texture descriptor is the LBPri. In addition we have evaluated two different classifiers, being the best a RF with $Deep = 2$ and $MaxTrees = 100$ yielded AUC above 0.90.

Using the recognised tissues from a block, we were able to identify some organs. For instance, when cardiac muscle tissue is recognised, we know with no doubt that the specific image contains at least a partial view of tissue form the heart. In a similar way, the correct classification of smooth muscle tissue allows for the recognition of the presence in the image of muscular arteries, veins and large elastic arteries.

Taking into account these results, it is possible to recognise fundamental tissues from the cardiovascular system and also some of the organs associated to these tissues by extracting the appropriate texture feature with the selection of the proper classifier.

This is important for educational purposes: this method may be used in a system to promote self-learning for on-campus students and also facilitate on-line learning to external or remote students. As a result, it would make possible to train qualified professionals low social and economic investments.

In future works, we will extend this proposal by working in the following four lines: (i) to improve the classification method by exploring new classifiers such as Support Vector Machine (SVM), (ii) the classification of small blocks could be used to recognise every part of a whole histological image, (iii) to improve the recognition process by means of a histological ontology which would detect and correct the wrong classifications of the blocks, and (iv) to add new cardiovascular tissue patterns with pathologies, and even from other organs.

Acknowledgements. This work has been supported by COLCIENCIAS and Asociación Universitaria Iberoamericana de Postgrado, AUIP. We thank Liliana Salazar, M.Sc., for providing insight and expertise that greatly assisted the research.

References

1. Izet, M.: E-learning as new method of medical education. Acta Informatica Medica **16**(2), 102–117 (2008). http://dx.doi.org/10.5455/aim.2008.16.102-117
2. Ruiz, J., Mintzer, M., Leipzig, R.: The impact of e-learning in medical education. Acad. Med. **81**(3), 207–212 (2006)
3. Hernadez, A.I., Porta, S.M., Miralles, M., Garca, B.F., Bolmar, F.: La cuanticacin de la variabilidad en las observaciones clnicas. Med. Clin. 424–429 (1990). http://www.ncbi.nlm.nih.gov/pubmed/2082114?dopt=Abstract
4. Nanni, L., Lumini, A., Brahnam, S.: Local binary patterns variants as texture descriptors for medical image analysis. Artif. Intell. Med. **49**(2), 117–125 (2010). doi:10.1016/j.artmed.2010.02.006
5. Herve, N., Servais, A., Thervet, E., Olivo-Marin, J.-C., Meas-Yedid, V.: Statistical color texture descriptors for histological images analysis. In: 2011 IEEE International Symposium on Biomedical Imaging: From Nano to Macro, pp. 724–727 (2011). doi:10.1109/ISBI.2011.5872508
6. Ojansivu, V., Linder, N., Rahtu, E., Pietikinen, M., Lundin, M., Joen-Suu, H., Lundin, J.: Automated classification of breast cancer morphology in histopathological images. Diagn. Pathol. **8**(Suppl. 1), S29 (2013)
7. Mazo, C., Trujillo, M., Salazar, L.: An automatic segmentation approach of epithelial cells nuclei. In: Alvarez, L., Mejail, M., Gomez, L., Jacobo, J. (eds.) CIARP 2012. LNCS, vol. 7441, pp. 567–574. Springer, Heidelberg (2012). doi:10.1007/978-3-642-33275-3_70
8. Nanni, L., Paci, M., dos Santos, F.C., Skottman, H., Juuti-Uusitalo, K., Hyttinen, J.: Texture descriptors ensembles enable image-based classification of maturation of human stem cell-derived retinal pigmented epithelium. PLoS ONE **11**(2), e0149399 (2016). doi:10.1371/journal.pone.0149399
9. Diamond, J., Anderson, N.H., Bartels, P.H., Montironi, R., Hamilton, P.W.: The use of morphological characteristics and texture analysis in the identification of tissue composition in prostatic neoplasia. Hum. Pathol. **35**(9), 1121–1131 (2004)
10. Mazo, C., Trujillo, M., Salazar, L.: Identifying loose connective and muscle tissues on histology images. In: Ruiz-Shulcloper, J., Sanniti di Baja, G. (eds.) CIARP 2013. LNCS, vol. 8259, pp. 174–180. Springer, Heidelberg (2013). doi:10.1007/978-3-642-41827-3_26522
11. Zhao, D., Chen, Y., Correa, N.: Statistical categorization of human histological images. In: IEEE International Conference on Image Processing, ICIP 2005, vol. 3, pp. 628–631 (2005). doi:10.1109/ICIP.2005.1530470
12. Yu, F., Ip, H., Horace, H.S.: Semantic content analysis and annotation of histological images. Comput. Biol. Med. **38**(6), 635–649 (2008). doi:10.1016/j.compbiomed.2008.02.004
13. Boya, J.: Atlas de Histología y Organografía Microscópica. Editorial Medica Panamericana S.A., Madrid (2011)

14. Mazo, C., Trujillo, M., Salazar, L.: Automatic classication of coating epithelial tissue. In: Bayro-Corrochano, E., Hancock, E. (eds.) CIARP 2014. LNCS, vol. 8827, pp. 311–318. Springer, Cham (2014). doi:10.1007/978-3-319-12568-8_38
15. Pietikinen, M., Ojala, T., Xu, Z.: Rotation-invariant texture classication using feature distributions. Pattern Recogn. **33**, 43–52 (2000)
16. Bader-El-Den, M.: Self-adaptive heterogeneous random forest. In: 2014 IEEE/ACS 11th International Conference on Computer Systems and Applications (AICCSA), pp. 640–646 (2014). doi:10.1109/AICCSA.2014.7073259
17. Ghassabeh, Y.A., Rudzicz, F., Moghaddam, H.A.: Fast incremental LDA feature extraction. Pattern Recogn. **48**(6), 1999–2012 (2015). doi:10.1016/j.patcog.2014.12.012
18. Kylberg, G., Sintorn, I.-M.: Evaluation of noise robustness for local binary pattern descriptors in texture classification. EURASIP J. Image Video Process. **2013**, 17 (2013). http://dblp.uni-trier.de/db/journals/ejivp/ejivp2013.html#KylbergS13
19. Canada, B.A., Thomas, G.K., Cheng, K.C., Wang, J.Z., Liu, Y.: Towards efficient automated characterization of irregular histology images via transformation to frieze-like patterns. In: CIVR, pp. 581–590. ACM (2008)
20. Oliveira, D.L., Nascimento, M.Z., Neves, L.A., Batista, V.R., Godoy, M.F., Jacomini, R.S., Duarte, Y.A., Arruda, P.F., Neto, D.S.: Automatic classification of prostate stromal tissue in histological images using Haralick descriptors and local binary patterns. In: Journal of Physics: Conference Series, vol. 490, no. 1 (2013). http://stacks.iop.org/1742-6596/490/i=1/a=012151
21. Alturkistani, H.A., Tashkandi, F.M., Mohammedsaleh, Z.M.: Histological stains: a literature review and case study. Glob. J. Health Sci. **8**(3), 72–79 (2016). http://doi.org/10.5539/gjhs.v8n3p72

Multidimensional Assessments of Abdominal Aortic Aneurysms by Magnetic Resonance Against Ultrasound Diameter Measurements

G. Papanastasiou[1(✉)], V. González-Castro[2], C.D. Gray[1], R.O. Forsythe[1,3], Y. Sourgia-Koutraki[1,3], N. Mitchard[4], D.E. Newby[1,3], and S.I.K. Semple[1,3]

[1] Clinical Research Imaging Centre, University of Edinburgh, Edinburgh, UK
g.papanas@ed.ac.uk
[2] Department of Electric Systems and Automatics Engineering, University of León, León, Spain
[3] Centre for Cardiovascular Science, University of Edinburgh, Edinburgh, UK
[4] Department of Radiology, Royal Infirmary of Edinburgh, Edinburgh, UK

Abstract. Volumetric and 3D tortuosity, curvature and asymmetry measurements were assessed for the first time from magnetic resonance (MR) data against ultrasound (US)-derived diameters, in a pilot cohort of 26 patients with abdominal aortic aneurysm (AAA) disease. We demonstrated that MR multidimensional analysis can better detect morphologic aneurysmal differences over time compared to diameters extracted from US imaging, which is the current clinical standard method for assessing AAA progression. Our 3D descriptors correlated more strongly with MR volumes, compared to US diameters at baseline and over time. Following further assessments in a larger patient cohort, this multidimensional analysis may enhance rupture risk prediction and improve clinical decision making.

Keywords: Abdominal aortic aneurysm · Magnetic resonance imaging · Ultrasound · Multidimensional assessments

1 Introduction

Abdominal aortic aneurysm (AAA) disease is an important cause of morbidity that affects up to 5% of men aged 65–74 years in Western Europe, with a mortality rate that can be greater than 80% when rupture occurs [1]. In the clinical setting, decision making must take into account significant risks associated with elective AAA repair, with a 30-day mortality rate of up to 5% [2]. It is therefore important, that for the clinical management of a patient with AAA, the risks of elective repair must be weighed against the risk of rupture [1, 2].

The current guidelines suggest that the risk of rupture in an AAA increases with its size [3]. Elective repair should be considered when the AAA exceeds 5.5 cm in anteroposterior diameter, or expansion exceeds 0.7 cm per six months, or 1 cm per year [4]. Despite these guidelines, up to 1 in 5 ruptured AAAs occur in small

M. Valdés Hernández and V. González-Castro (Eds.): MIUA 2017, CCIS 723, pp. 133–143, 2017.
DOI: 10.1007/978-3-319-60964-5_12

aneurysms (<5.5 cm), whilst there are many patients under surveillance with aneurysms far exceeding 5.5 cm in diameter that do not rupture [2, 4].

Ultrasound (US) is the reference standard technique for assessing AAA diameter growth rates accurately, with no radiation exposure and at low cost. However, other than visual morphological assessments, it is not possible to provide additional quantitative information about AAA volume and morphology [5]. Computed tomography (CT) data analysis has recently showed that AAA volume may better predict aneurysm growth rate [6], whilst correlated more strongly with increasing estimated biomechanical rupture risk, than maximum diameter [3]. These data suggest that aneurysms exhibit non-linear, discontinuous growth patterns and hence, morphological changes beyond US-derived diameter growth rates must be assessed [3, 5, 6].

Magnetic resonance (MR) imaging is widely accepted as a diagnostic imaging tool for cardiovascular diseases but its utility in monitoring volumetric as well as asymmetry AAA changes has not been rigorously examined [7]. Furthermore, although some AAA diameter and volume [3, 6], 2D tortuosity indices [8, 9] and asymmetry [10] have been assessed (using CT imaging), more advanced multidimensional AAA morphological changes such as 3D tortuosity and curvature have not been extensively investigated.

We assessed growth rates and correlations between AAA volumes derived from MR images versus US-derived diameters in a pilot cohort of patients. We also assessed for the first time multidimensional 3D tortuosity, curvature and asymmetry estimated from MR imaging versus US-derived diameters and MR-derived volumes in this pilot cohort.

2 Methods

2.1 Study Population

26 patients with asymptomatic AAA (diameter 38–56 mm on ultrasound examination) and with aneurysmal expansions <1 cm per year (24 males, 2 females), were retrospectively selected from the MA3RS study database (full cohort N = 350) [11]. After informed consent, MR imaging was acquired in all patients. AAA volumes and multidimensional 3D geometrical descriptors (3D tortuosity, curvature, asymmetry) were extracted from the MR images and comparisons as well as correlations with clinical standard US-derived diameters were assessed.

2.2 Ultrasound and MR Acquisition

Standard US imaging was carried out in an accredited clinical vascular science laboratory. Participants underwent US evaluation of the maximum anteroposterior diameter of the abdominal aorta by an expert ultrasonographer using a 3.5 MHz linear array transducer.

MRI scans were performed on a 3T Magnetom Verio scanner (Siemens, Erlangen, Germany), as previously described [11]. T2-weighted multi-slice Half Fourier Acquisition Single Shot Turbo Spin Echo (HASTE) sequences were used to identify the extent of the aneurysm and anatomical data were obtained using a respiratory-gated T2-weighted turbo spin echo sequence. US and MR protocols were implemented for baseline and 1-year follow up imaging.

2.3 Image and 3D Geometrical Analysis

Maximum anteroposterior diameters (US_Diam) were extracted from the US images by an expert clinician. AAAs from MR images were manually segmented by an experienced user, using dedicated segmentation software (SliceOmatic, TomoVision, Canada). Regions or interest (ROIs) for the lumen, thrombus and aortic wall were manually defined across all slices for which abdominal aortas were MR visible (Fig. 1a, 11–25 slices depending on the length of abdominal aortas). MR data were anatomically landmarked by an expert clinician and matching slices from the same anatomical areas for both baseline and 1-year follow up data were analysed for each subject.

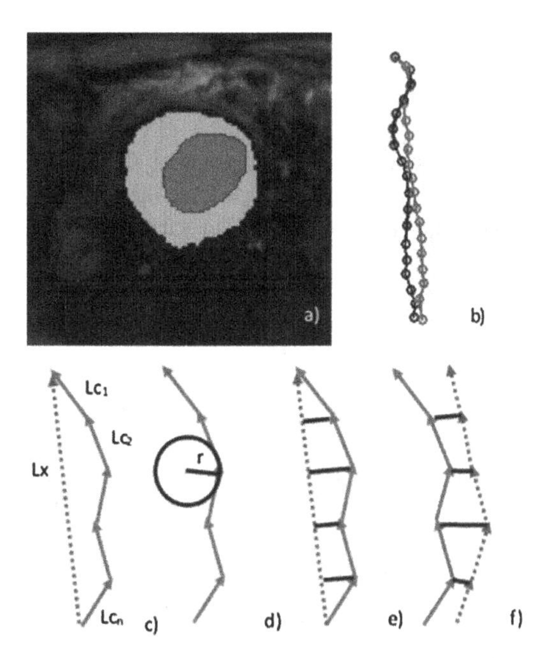

Fig. 1. (a) Lumen (red), thrombus (green) and outer wall (blue) segments shown from a patient with AAA disease at baseline. (b) Centrelines for Lum (lumen, with red) and Com (combined = lumen + thrombus + outer wall, with blue) regions of interest, from the same patient data set. (c) Tortuosity DM algorithm applied across 6 example centroids, where Lc_{1-n} and Lx (dot blue line) are the 3D vectors connecting consecutive centroids and end points in centrelines, respectively. (d) Curvature measured by fitting osculating circles in 3D space. (e) and (f) two different types of vessel asymmetry metrics (in c–f. the solid red centrelines represent the Lum centreline, in f. the dot blue line shows the Com centreline). (Color figure online)

AAA volumes and 3D geometrical descriptors were calculated using in-house software developed in Matlab (Mathworks, USA). From the SliceOmatic data, the centroids for the lumen ROIs (Lum) and the combined ROIs (Com = lumen + thrombus + aortic wall) were derived for all MR slices across all subjects (Fig. 1b) and were then used to derive 3D tortuosity, curvature and asymmetry measurements. AAA volumes and 3D geometrical descriptors were calculated for Lum and Com ROIs and centrelines (Fig. 1b, each centreline connects consecutive centroids) respectively. AAA volumes for Lum and Com ROIs are described as Vol_Lum, Vol_Com respectively.

For 3D tortuosity measurements, the distance measure (DM) algorithm was originally developed for retinal 2D data analysis, as previously described [12]. To adapt the DM algorithm for 3D tortuosity measurements, the lengths of the 3D vectors (3D Euclidean distances) connecting consecutive centroids were calculated and summed across both the Lum and Com centrelines (Fig. 1c, Lc = Lc1 + Lc2 + …Lcn). Furthermore, the length of the 3D vector connecting the two end points of each Lum and Com centreline was measured (Lx). DM tortuosity algorithm was finally calculated in the 3D space for both Lum and Com centrelines (abbreviated as Tor_Lum, Tor_Com respectively) by using:

$$DM = \frac{L_c}{L_x} \tag{1}$$

The curvature was measured for both Lum and Com centrelines (Cur_Lum, Cur_Com respectively). To devise 3D curvature measurements, singular value decomposition was used to fit osculating circles at each centroid in 3D space (see Eq. 2 and Fig. 1d). For the singular value decomposition, the following factorisation was used:

$$A = U\Sigma V^T \tag{2}$$

where A is an 3 × 3 matrix, in which each column represents a 3D centroid point across the Lum and Com centrelines. In matrix A, the second column corresponds to the local 3D centroid point in which the osculating circle was fitted, whilst the first and third columns describe the 3D centroid points before and after the local 3D centroid point, across each centreline. U and V represent orthogonal matrices describing the left and right singular vectors respectively, with V^T denoting the transpose of V. Σ denotes the diagonal matrix in decreasing order. Following singular value decomposition, the first and second unit orthogonal vectors in U were extracted and multiplied by the matrix A. The product of this multiplication defined two 3D unit vectors which were subsequently used to fit an osculating circle for each local 3D centroid point.

The radius of each osculating circle (r) is also the radius of the curvature (κ), whilst curvature equals the reciprocal of radius r (see also Eq. 3) [13]. The smaller the radius r of the circle, the higher the curvature is and vice versa. (note: curvature is 0 for a straight line because the radius tends to infinity).

$$\kappa = \frac{1}{r} \tag{3}$$

Two types of asymmetry measurements were derived. The first type has been previously described by Doyle et al. [10] and is based on calculating the perpendicular distance from each centroid to the straight line connecting the AAA end points (Fig. 1e). This asymmetry type was calculated for both the Lum and Com centrelines (Ass_Lum, Ass_Com respectively). The second type of asymmetry is introduced by the authors, defined as centroid asymmetry (Ass_Cen). The lengths of the 3D vectors connecting paired (corresponding to the same slice) Lum and Com centroids were calculated, summed across all centroids and normalised to the Com centreline length for each AAA (Fig. 1f). In a vessel with no aneurysmal expansion, the lumen and outer wall are approaching concentricity and Ass_Cen tends to 0.

2.4 Statistical Analysis

Statistical analysis was performed using R software (R Foundation for statistical computing, Vienna, Austria). A non-linear growth model (NLGM) was used to assess growth (for AAA expansion measurements: diameter, volumes, asymmetry) and change rates (for 3D tortuosity, curvature), between baseline and 1-year follow-up data [3, 6]:

$$NLGM = (e^{12 \cdot L} - 1) \cdot 100 \frac{\%}{year} \tag{4}$$

with L being a logarithmic growth factor:

$$L = \frac{1}{t} \cdot \ln\left(\frac{M_{follow-up}}{M_{baseline}}\right) \tag{5}$$

in which M is an US or MR estimate and t is the time (in months) between baseline and follow-up measurements. The Shapiro-Wilk test was used to assess normality of the data and comparisons were assessed using paired and unpaired Student's t-tests as appropriate. Box and whisker plots investigated growth/change rates across all US and MR measurements. Linear regression and Pearson's correlation coefficients were used to investigate correlations between MR AAA volumes versus US-derived diameters, and between MR 3D geometrical descriptors versus diameters and volumes (two-sided P value <0.05 were considered significant).

3 Results

26 patients (2 females, 24 males) with AAA were included in the analysis. The median time between baseline and follow up US and MR imaging was 12 months (interquartile range (IQR), 11–13 months). MR imaging was acquired within 1-week from the day of US imaging. The mean (SD) US-derived AAA diameter was 45(4) mm and 47(5) mm, at baseline and follow up respectively.

3.1 Growth and Change Rates

Increased growth rates were observed for all parameters describing AAA expansion (Fig. 2). For US_Diam, median diameter growth rate was 3.32%/year (IQR were 0.00 to 8.56%/year). For Vol_Lum and Vol_Com, median volume growth rates were 7.18 and 8.23%/year (IQRs were 0.38 to 18.54, 4.59 to 13.67) respectively. The growth rates for Vol_Com showed a significant difference versus US_Diam ($P < 0.05$). For 10 patients who showed no diameter increases, Vol_Com showed a significant growth rate increase ($P < 0.001$).

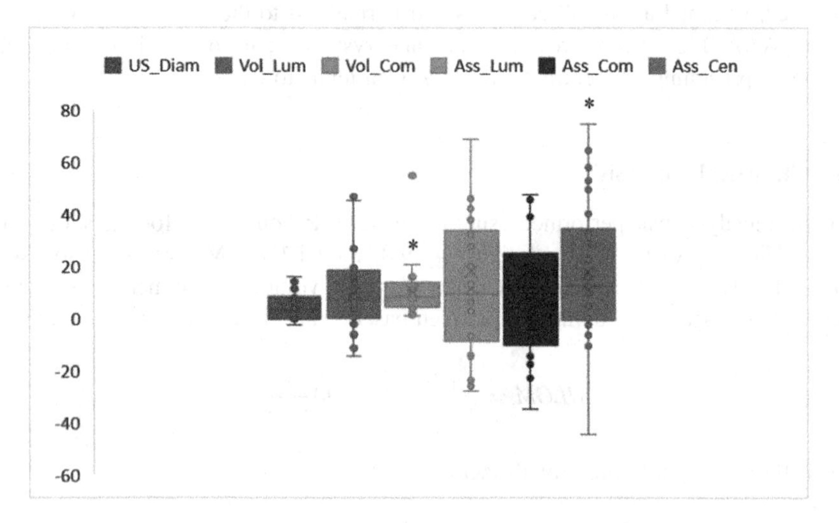

Fig. 2. Growth rates for US and MR measurements demonstrating aneurysmal expansions across all subjects (US_Diam: ultrasound-derived diameters, Vol_Lum: MR-derived volumes of the lumen, Vol_Com: volumes of the combined regions of interest, Ass_Lum: asymmetry for the lumen centreline, Ass_Com: asymmetry for the combined centreline, Ass_Cen: centroid asymmetry). Significant growth rate increases in patients with no diameter changes are indicated with *.

Median growth rates for Ass_Lum and Ass_Com were 9.52 and 5.71%/year (IQRs were −8.60 to 33.90, −9.92 to 24.90%/year) respectively. Ass_Cen demonstrated the largest growth rate (median was 12.74%/year; IQR was −0.40 to 34.52%/year, Fig. 2). The growth rate for Ass_Cen showed a significant difference versus US_Diam ($P < 0.05$). For 10 patients who showed no diameter increases, Ass_Cen showed a significant growth rate increase ($P < 0.01$). No other significant differences were observed between growth rates.

Change rates for 3D tortuosity and curvature were not significantly increased (all median values < US_Diam median).

3.2 Correlations at Baseline

Initially, correlations across all measurements were examined at baseline. Vol_Lum and Vol_Com showed significant, strong correlations with US_Diam (Fig. 3a–b). A significant, strong correlation was also observed between Vol_Com and Ass_Cen (Fig. 3c, all P < 0.001). There was also a moderate but significant correlation between US_Diam and Ass_Cen (r = 0.46, P < 0.05).

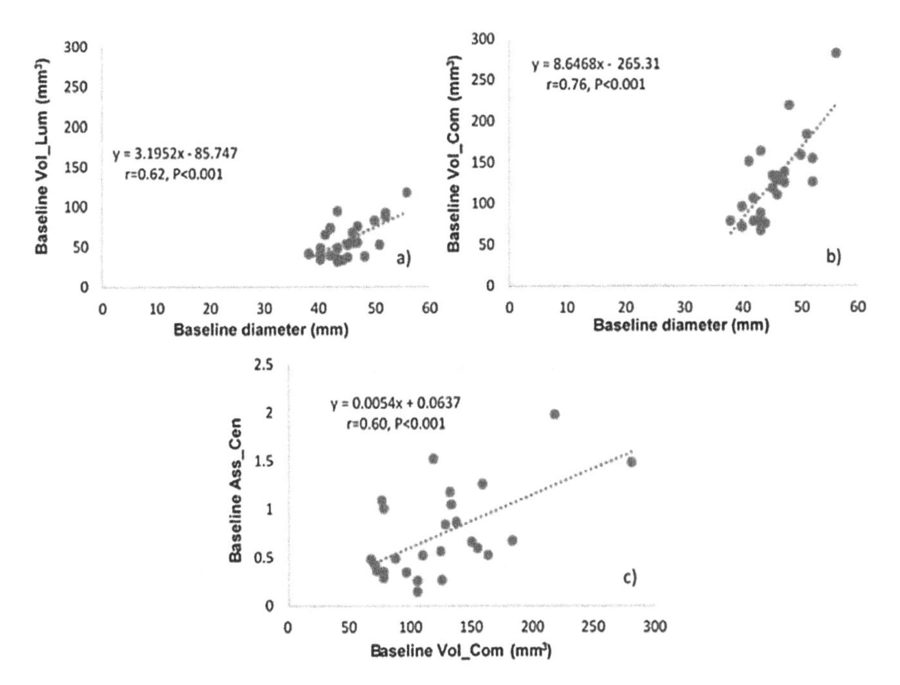

Fig. 3. Significant, strong correlations shown at baseline (Vol_Lum: MR-derived volumes of the lumen, Vol_Com: volumes of the combined regions of interest, Ass_Cen: centroid asymmetry).

All 3D geometrical descriptors (tortuosity, curvature, asymmetry) for both Lum and Com centrelines correlated positively (r = 0.51–0.59, P < 0.001), with Tor_Com versus Ass_Com reaching the highest correlation (r = 0.83, P < 0.001).

3.3 Correlations Between Growth and Change Rates

Subsequently, correlation coefficients were investigated between growth rates for US derived diameters and MR derived volumes versus change rates for all 3D geometrical descriptors. The strongest correlation was observed between Vol_Com and Cur_Com (r = 0.63, P < 0.001, Fig. 3). There were also moderate but significant correlations between US_Diam and Cur_Lum and between Vol_Com and Tor_Com (r = 0.45, 0.50 respectively, P < 0.05).

Significant, strong correlations were finally observed between change rates of 3D geometrical descriptors. Specifically, between Tor_Lum and Cur_Lum (r = 0.60), Tor_Com and Cur_Com (r = 0.60), Tor_Com and Ass_Com (r = 0.71, all P < 0.01). All other correlations at both baseline and 1-year follow-up growth/change rates demonstrated weaker non-significant relationships.

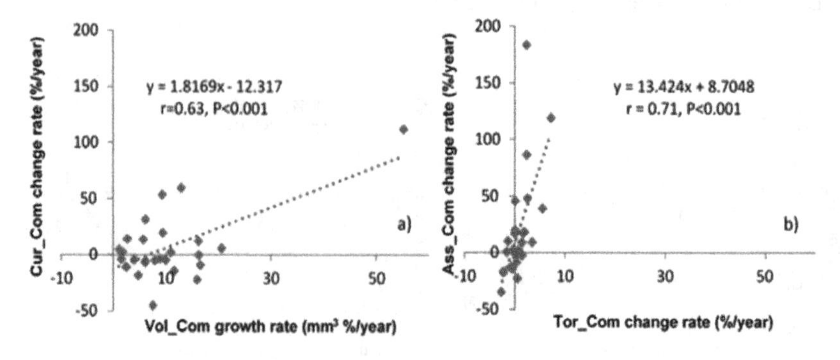

Fig. 4. Significant, strong correlations for growth/change rates. Vol_Com: volume, Cur_Com: curvature, Tor_Com: tortuosity, Ass_Com: asymmetry, of the combined regions of interest (Com).

4 Discussion

We investigated growth rates for US diameters as well as MR volumes and asymmetries derived from a pilot cohort of AAA patients. MR-derived volumes and centroid asymmetries demonstrated significant growth rate increases in patients with small asymptomatic aneurysms and low US diameter increases. We also assessed all 3D geometrical descriptors versus US diameters and MR volumes and demonstrated significant correlations both at baseline and over time (between baseline and 1-year).

4.1 Growth Rates of AAA Expansion Measures

This is the first study assessing MR derived volumes and asymmetries against US reference standard diameter measurements. The AAA structural integrity can be compromised by fast AAA growth and may increase the risk of AAA rupture [6, 14, 15]. Monitoring the growth rate of small AAAs can therefore be crucial for decision making in surveillance programs. In our analysis, we showed that MR-derived volumes may better detect morphologic aneurysmal differences over time compared to US diameters, the current clinical standard method for monitoring AAA changes. This is in agreement with a previous CT study suggesting that volume can better predict AAA growth rate than diameter [3].

Similarly, we demonstrated that centroid asymmetry, our novel 3D asymmetry descriptor, can better represent morphologic aneurysmal changes compared to US diameters. Centroid asymmetry can identify morphological differences beyond diameter and longitudinal growths across the AAA (accounted for by volumetric analysis), describing both areas of thrombus increases and aortic lumen dislocations inside the AAA over time. Moreover, centroid asymmetry demonstrated increased growth rate compared to asymmetry of the combined centreline (Fig. 2), which has previously been suggested as a potentially useful adjunct to diameter using CT data analysis [10].

In our pilot cohort, we also showed for the first time significant, strong correlations between MR-derived volumes and US diameters at baseline (Fig. 3a–b), whilst demonstrated that baseline centroid asymmetry correlated with both US diameters and MR volume. The above correlations at baseline were not observed when the respective growth rates were assessed, which can suggest that volume and centroid asymmetry may follow different growth patterns compared to maximum US diameters (Fig. 2). These results can suggest that MR volume and centroid asymmetry estimates may be able to detect morphological abnormalities at baseline as well as changes over time, and improve decision making in surveillance programs.

4.2 Further Assessment of 3D Geometrical Descriptors

Our analysis demonstrated that changes in the curvature and tortuosity are associated with diameter and volume increases. The strongest correlation was observed between MR volume and curvature of the combined ROIs. Although there was a significant correlation between US diameters and curvature of the lumen, curvature and tortuosity correlated more strongly with MR volume of the combined ROIs. Significant, strong correlations were also shown between change rates of 3D geometrical descriptors (tortuosity and curvature of the lumen, tortuosity and curvature, tortuosity and asymmetry of the combined ROIs). To our knowledge, this is the first study assessing 3D tortuosity and curvature algorithms in AAAs using medical imaging data. Previous studies have mainly assessed 2D tortuosity indices [8, 9], or have focused in biomechanical analysis using peak wall stress [3, 10, 16, 17]. In the context of biomechanical analysis, studies have suggested that AAAs predominantly dilate (thrombotic expansion) in the anterior plane due to spinal cord constraints in the posterior direction and thus, peak wall stress is frequently increased in areas of the posterior wall [3, 16, 17]. Molacek et al. investigated the elastic properties of AAAs by performing a distensibility study using CT data and suggested that thrombotic morphologies may in some cases act protectively against the risk of rupture [18]. Further investigation of our 3D descriptors may provide additional quantitative information to biomechanical analysis, for which cases thrombus morphologies may act as risk [16, 17] or protective factors [18] of rupture.

In conclusion, we showed that MR multidimensional analysis can better detect morphologic aneurysmal differences over time compared to US diameters. In addition, our 3D descriptors correlated more strongly with MR volumes, compared to US diameters at baseline and over time. Further evaluation is underway in our larger patient cohort (N = 350) to further assess our 3D geometrical descriptors and evaluate their ability in predicting rupture risk and disease progression.

References

1. Ashton, H.A., Buxton, M.J., Day, N.E., et al., Multicentre Aneurysm Screening Study Group: The Multicentre Aneurysm Screening Study (MASS) into the effect of abdominal aortic aneurysm screening on mortality in men: a randomised controlled trial. Lancet **360**, 1531–9153 (2002)
2. Forsythe, R.O., Newby, D.E., Robson, J.M.J., et al.: Monitoring the biological activity of abdominal aortic aneurysms beyond ultrasound. Heart **102**(11), 817–824 (2016)
3. Liljeqvist, M.L., Hultgren, R., Gasser, T.C., et al.: Volume growth of abdominal aortic aneurysms correlates with baseline volume and increasing finite element analysis-derived rupture risk. J. Vasc. Surg. **63**, 1434–1442 (2016)
4. Nicholls, S.C., Gardner, J.B., Meissner, M.H., et al.: Rupture in small abdominal aortic aneurysms. J. Vasc. Surg. **28**, 884–888 (1998)
5. Chaikof, E.L., Brewster, D.C., Dalman, R.L., et al.: The care of patients with an abdominal aortic aneurysm: the society for vascular surgery practice guidelines. J. Vasc. Surg. **50**, S2–S49 (2009)
6. Martufi, G., Auer, M., Roy, J., et al.: Multidimensional growth measurements of abdominal aortic aneurysms. J. Vasc. Surg. **58**(3), 748–755 (2013)
7. Buijs, R.V.C., Willems, T.P., Tio, R.A., et al.: Current state of experimental imaging modalities for risk assessment of abdominal aortic aneurysm. J. Vasc. Surg. **57**, 851–859 (2013)
8. Pappu, S., Dardik, A., Tagare, H., et al.: Beyond fusiform and saccular: a novel quantitative tortuosity index may help classify aneurysm shape and predict aneurysm rupture potential. Ann. Vasc. Surg. **22**(1), 88–97 (2008)
9. Fillinger, M.F., Racusin, J., Baker, R.K., et al.: Anatomic characteristics of ruptured abdominal aortic aneurysm on conventional CT scans: implications for rupture risk. J. Vasc. Surg. **39**(6), 1243–1252 (2004)
10. Doyle, B.J., Callanan, A., Burke, P.E., et al.: Vessel asymmetry as an additional diagnostic tool in the assessment of abdominal aortic aneurysms. J. Vasc. Surg. **49**(2), 443–454 (2009)
11. McBride, O.M.B., Berry, C., Burns, P., et al.: MRI using ultrasmall superparamagnetic particles of iron oxide in patients under surveillance for abdominal aortic aneurysms to predict rupture or surgical repair: MRI for abdominal aortic aneurysms to predict rupture or surgery-the MA^{3}RS study. Open Heart **2**, e000190 (2015)
12. Lisowska, A., Annunziata, R., Loh, G.K., et al.: An experimental assessment of five indices of retinal vessel tortuosity with the RET-TORT public dataset. In: Proceedings of 36th Annul International Conference of IEEE Engineering in Medicine and Biology Society, pp. 5414–5417 (2014)
13. Lewiner, T., Gomes Jr., J.D., Lopes, H., et al.: Curvature and torsion estimators based on parametric curve fitting. Comput. Graph. **29**, 641–655 (2005)
14. Freestone, T., Turner, R.J., Coady, A., et al.: Inflammation and matrix metalloproteinases in the enlarging abdominal aortic aneurysm. Arterioscler. Thromb. Vasc. Biol. **15**, 1145–1151 (1995)
15. Anidjar, S., Dobrin, P.B., Eichorst, M., et al.: Correlation of inflammatory infiltrate with the enlargement of experimental aortic aneurysm. J. Vasc. Surg. **16**, 139–147 (1992)
16. Venkatasubramaniam, A.K., Mehta, T., Fagan, M.J., et al.: A comparative study of aortic wall stress using finite element analysis for ruptured and non-ruptured abdominal aortic aneurysm. Eur. J. Vasc. Endovasc. Surg. **28**, 168–176 (2004)

17. Fillinger, M.F., Marra, S.P., Raghavan, M.L., et al.: Prediction of rupture risk in abdominal aortic aneurysm during observation: wall stress versus diameter. J. Vasc. Surg. **37**, 724–732 (2003)
18. Molacek Baxa, J., Houdek, K., Treska, V., et al.: Assessment of abdominal aortic aneurysm wall distensibility with electrocardiography-gated computed tomography. Ann. Vasc. Surg. **25**, 1036–1042 (2011)

Comparison of Automatic Vessel Segmentation Techniques for Whole Body Magnetic Resonance Angiography with Limited Ground Truth Data

Andrew McNeil[1(✉)], Giulio Degano[1], Ian Poole[2], Graeme Houston[3], and Emanuele Trucco[1]

[1] CVIP, Computing, School of Science and Engineering, University of Dundee, Dundee, UK
aymcneil@dundee.ac.uk
[2] Toshiba Medical Visualization Systems Europe, Edinburgh, UK
[3] School of Medicine, Ninewells Hospital and Medical School, Dundee, UK

Abstract. This work is part of a project aimed at automatically detecting vascular disease in whole body magnetic resonance angiograms (WBMRA). Here we present a comparison of four techniques for automatic artery segmentation in WBMRA data volumes; active contours, two "vesselness" filter approaches (the Frangi filter and Optimally Oriented Flux (OOF)) and a convolutional neural network (Convnet) trained for voxel-wise classification. Their performance was assessed on three manually segmented WBMRA datasets, comparing the maximum Dice Similarity Coefficient (DSC) achieved by each method. Our results show that, in the presence of limited training data, OOF performs best for our three patients, achieving a mean DSC of 0.71 across all patients. By comparison, the 3D Convnet achieved a mean DSC of 0.63. We discuss the potential reasons for these differences, and the implications it has for the automated segmentation of arteries in large WBMRA datasets, where ground truth data is often limited and there are currently no pre-trained 3D Convnet models available, requiring models to be trained from scratch. To the best of our knowledge this is the first comparison of these automated vessel segmentation techniques for WBMRA data, and the first quantitative results of applying a Convnet to vessel segmentation in WBMRA, for which no public sets of manually annotated vascular networks currently exist.

1 Introduction

Contrast-enhanced Whole-Body Magnetic Resonance Angiography (WBMRA) is performed by injecting a contrast agent and acquiring images using an MRI scanner as it passes through the arteries of interest [14]. This technique generates high contrast in the lumen (the channel where the blood is flowing), providing a non-invasive, comprehensive imaging method for assessing cardiovascular disease (CVD) throughout the entire body [19]. Analysing these large datasets is

M. Valdés Hernández and V. González-Castro (Eds.): MIUA 2017, CCIS 723, pp. 144–155, 2017.
DOI: 10.1007/978-3-319-60964-5_13

very labour-intensive however, and thus there is a great need for automated, quantitative analysis tools to help stage the disease from these scans.

The first stage of any such system is to locate and segment the arteries of interest. The segmentation of vascular structures is a common task to many medical applications [3, 8, 13], and is a fundamental step in the the quantification of pathologies such as stenoses.

Many vessel segmentation techniques have been proposed in the literature, as explored in [8, 12]. In this work we examine three commonly used techniques—active contours and two "vesselness" filters—comparing their results against a more recent approach using a convolutional neural network (Convnet), structured as a voxel-wise binary classifier following the network structures explored in [21].

2 Materials and Methods

2.1 Patient Data and Ground Truth

The data used in this study consists of three whole-body datasets, each of which are split into four "stations"; station one comprised the head and neck, station 2 the thorax and abdomen, station 3 the pelvis and thighs, and station 4 the feet (see Fig. 1). These were acquired at Ninewells Hospital in Dundee, UK, using a 3.0 Tesla MRI scanner (Siemens Magnetom Trio).

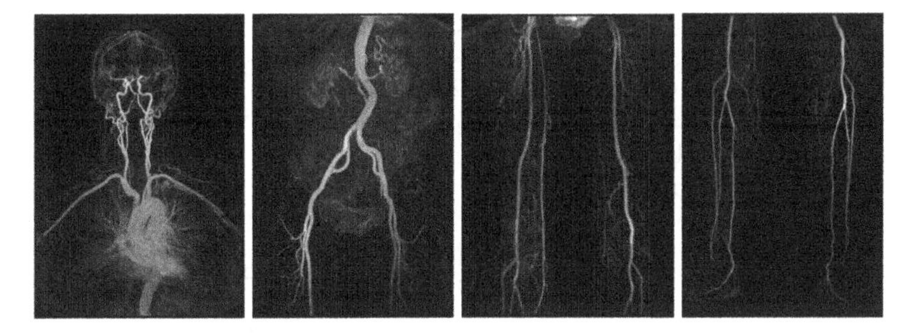

Fig. 1. Maximum intensity projections of the four stations of patient 1, shown after digital subtraction of the pre-contrast from the post-contrast volumes.

The arteries in each individual station were manually segmented by a trained observer using the software package "3DSlicer" [4]. The following pre-processing steps were applied to aid visualisation during the manual segmentation; firstly all volumes were scaled in one direction due to varying slice thicknesses, giving an isotropic voxel size of 0.98 mm^3 (the raw slice thicknesses varied from 0.98 mm–1.3 mm). Next the pre-contrast volume was registered to the post-contrast volume using the mutual information similarity measure. Subtracting the pre-contrast from the post-contrast volume was done to suppress static tissues. To aid with visualisation, an intensity equalisation step was applied in the axial direction, ensuring a consistent vessel intensity across the entirety of each volume. A final artefact correction step was applied, masking border voxels to remove MR artefacts and tissues not covered by the pre-contrast volume.

2.2 Active Contours

Segmentation using active contours, where an initial curve is evolved using a cost function depending on local gradients (external forces) and shape constraints (internal forces), was first proposed in [7] and has been successfully applied to many segmentation problems [1,12]. In general, active contour methods are based on image gradient, detecting edges and generating well-defined boundaries on which to evaluate the internal and external energy.

The level set model tries to solve the optimisation problem by embedding the active contour as a constant set (zero level) in a function ϕ that evolves in time with speed S.

For our comparative study we chose the classic Chan-Vese model [1]. It has been applied to the segmentation of objects whose edges are not well defined by the gradient, and has a well defined implementation for 3D segmentation as described in [25].

The Chan-Vese model is formulated as a "mean-curvature flow"-like evolving active contour, where the stopping term depends not on the gradient of the image, as in classical active contour models, but is instead related to a particular segmentation of the image [1].

For 3D data, we define the bounded domain $\Omega \in \mathbb{R}^n$ (in our case $n = 3$), and the bounded image function $I : \Omega \rightarrow \mathbb{R}$. Ω can be divided into a set of connected domains by a curve C by $\Omega - C = \cup_{i \in I} \Omega_i$. We then define two different regions $R_1 = \cup_{i \in I_1} \Omega_i$ and $R_2 = \cup_{i \in I_2} \Omega_i$ that represent the object support and the background support respectively. The final energy functional $E(\phi, \mu_1, \mu_2)$ is then given by

$$E(\phi, \mu_1, \mu_2) = \lambda_1 \int_{\Omega} (I - \mu_1)^2 H(\phi) d\Omega + \lambda_2 \int_{\Omega} (I - \mu_2)^2 (1 - H(\phi)) d\Omega + \alpha \int_{\Omega} H(\phi) d\Omega + \beta \int_{\Omega} | \bigtriangledown H(\phi)| d\Omega \tag{1}$$

where μ_1 and μ_2 represent the mean value of the object support region and background support region of image I respectively, and $H(\phi)$ is the Heaviside function. Here, the first two terms measure the variations inside and outside the active contour, the third term measures the area inside the contour and the fourth term measures the length of the contour [25].

2.3 Vessel Enhancement Filters

Frangi Filter. This method of enhancing vessel-like structures is based on calculate the local curvature by analysing the Hessian function, so as to extract the main directions in which the local structure can be decomposed [5].

To derive the "vesselness" function we first define λ_k as being the eigenvalue with the k-th smallest magnitude, i.e. $(|\lambda_1| \leq |\lambda_2| \leq |\lambda_3|)$. Therefore, for an ideal tubular structure in a 3D image

$$|\lambda_1| \approx 0, \quad |\lambda_1| \ll |\lambda_2|, \quad |\lambda_2| \approx |\lambda_3| \tag{2}$$

In other words, the curvature should be large in the two directions (around the circumference of the vessel), and very small along the length of the vessel.

The final vesselness function to be evaluated was defined in [5] as being

$$V_0(s) = \begin{cases} 0 & \text{if } \lambda_2 > 0 \text{ or } \lambda_3 > 0 \\ \left(1 - \exp\left(-\frac{R_A^2}{2\alpha^2}\right)\right) \exp\left(-\frac{R_B^2}{2\beta^2}\right) \left(1 - \exp\left(-\frac{S^2}{2c^2}\right)\right) & \text{otherwise} \end{cases} \tag{3}$$

where $R_A = |\lambda_2|/|\lambda_3|$, $R_B = |\lambda_1|/\sqrt{|\lambda_2 \lambda_3|}$, $S = \sqrt{\lambda_1^2 + \lambda_2^2 + \lambda_3^2}$. Here, α, β and c are thresholds which control the sensitivity of the filter to the measures R_A R_B and S.

For the application of enhancing vessels in a 3D angiographic dataset, the vesselness measure in Eq. (3) is analysed at different scales, s, corresponding to the sigma of the Gaussian kernel used in the construction of the Hessian matrix. It logically follows that the response of the filter will be maximised at the scale which approximately matches the size of the vessel in that region. Therefore, the final estimate of vesselness is obtained by integrating the vesselness measure provided by the filter response at different scales,

$$V_0(\gamma) = \max_{s_{min} \leq s \leq s_{max}} V_0(s, \gamma) \tag{4}$$

where s_{min} and s_{max} are the minimum and maximum scales at which relevant structures are expected to be found, chosen so that they cover the range of relevant vessel widths. The vesselness map can then be thresholded to provide a binary vessel tree.

Optimally Oriented Flux. The "optimally oriented flux" filter, first published in [10], evaluates a scalar measure of the flux flowing through a spherical surface. Before computing this value, directional information is extracted by projecting the gradient along "optimal" axes, and the flux measure then evaluated. For each voxel a sphere with variable radius is built, centred on the voxel, which produces an "OOF response" when touching an object edge. If the voxel is inside the curvilinear structure the response will be positive, otherwise it will be negative.

The outwardly oriented flux along the direction $\hat{\rho}$ is firstly computed by projecting the gradient of the image $\mathbf{v}$ along $\hat{\rho}$, with the flux then evaluated through the spherical region S_r with radius r using the definition

$$f(\mathbf{x}; \mathbf{r}, \hat{\rho}) = \int_{\delta Sr} ((\mathbf{v}(\mathbf{x} + \mathbf{h}) \cdot \hat{\rho})\hat{\rho}) \cdot \hat{n} dA \tag{5}$$

where dA is the infinitesimal area of S_r, $\hat{n}$ is the unit normal to the surface at position $h = r\hat{n}$.

As before, the goal is to obtain the principal eigenvalues for each voxel. Inside the vessel, when the local spherical region with surface S_r touches the boundaries of the object $\mathbf{v}$ is aligned opposite to the direction of $\hat{n}$, therefore the eigenvalues $\lambda_1 \leq \lambda_2 \ll 0$. The gradient of the image will be perpendicular to the direction of the curvilinear structure, with a value of $\lambda_3 \approx 0$. In the case where the voxel is in the background, $\mathbf{v}$ will have the same direction as $\hat{n}$, and therefore $\lambda_3 \gg 0$.

To obtain the maximum response to the OOF while changing the radius r, we evaluate of the geometric mean of the eigenvalues, as

$$M(\mathbf{x}; s) = \begin{cases} \sqrt{|\lambda_1(\mathbf{x}, s)\lambda_2(\mathbf{x}, s)|} & \lambda_1(\mathbf{x}, s) \leq \lambda_2(\mathbf{x}, s) < 0 \\ 0 & \text{otherwise} \end{cases} \tag{6}$$

where s represents the scale factor. Similar to the Frangi approach, evaluating the maximum response over an appropriate range of scales generates the final map, which can be thresholded to produce a final segmentation.

2.4 Convolutional Neural Network

In recent years, deep Convnet approaches have been driving advances in many computer vision tasks, such as image classification [9,21] and image segmentation [18,20]. Many network models have been developed for these tasks, and it is a very active area of research [11]. The network structure we chose was inspired by those explored in [21], and recently applied to segmentation tasks in MRI [2,15,17]. To the best of our knowledge, this is the reported results of applying a Convnet to vessel segmentation in WBMRA, for which no public sets of manually annotated vascular networks currently exist.

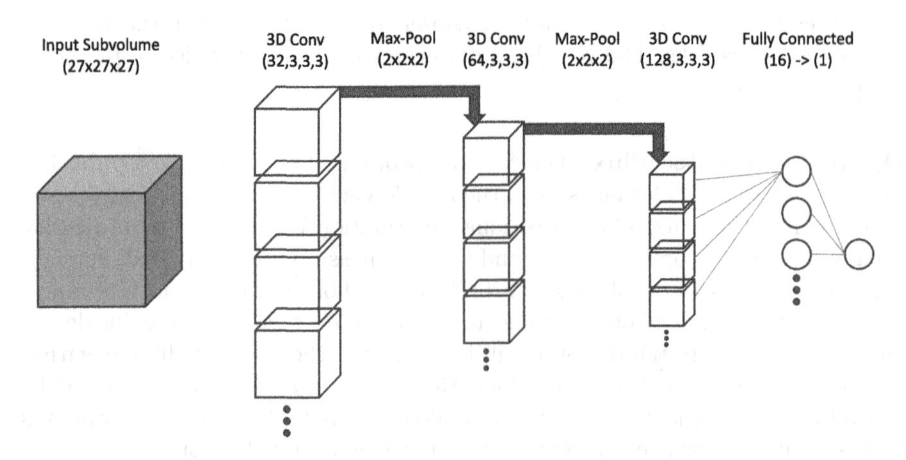

Fig. 2. Structure of the 3D Convnet segmentation network. All layer activation functions were "ReLU" except the final output node, which was "sigmoid". The "Adam" optimiser was used during training, with "binary cross-entropy" selected as the loss function

The final network structure is shown in Fig. 2, consisting of five layers; 3 sets of convolutional and max-pooling layers, followed by two fully connected layers. The output node of the final layer gives a single binary output of vessel/non-vessel for the central voxel of the input patch.

Our network was implemented using Keras v1.1.0 and Theano v0.8.2. All layer activation functions were "ReLU" except the final output node, which was "sigmoid". The "Adam" optimiser was used during training, with "binary cross-entropy" selected as the loss function [6].

While this approach of a voxel-wise classifier has been shown to be less computationally efficient than a fully convolutional network [20], it allowed fine control over dataset balancing for our limited amount of ground truth data, and the fully connected layer gave additional flexibility to the network without increasing the required input volume size, which is inherent to the operation of convolutional layers.

2.5 Comparison Metric

There are many metrics used for evaluating the quality of segmentation in medical images [24]. For our data, we have selected the Dice Similarity Coefficeint (DSC—also referred to as the F1-Measure). This is given by

$$DSC = \frac{2|X \cap Y|}{|X| + |Y|} \tag{7}$$

where $|X|$ is the number of all the vessel voxels in the segmentation obtained by the tested method and $|Y|$ is the number of all the vessel voxels in the ground truth.

2.6 Pre-processing and Parameter Optimisation

Pre-processing. As can be seen in Fig. 1, the subtraction of the pre-contrast from the post-contrast volume still leaves some tissues and non-arterial structures behind, particularly in stations 1 and 2. The most problematic of these are the lungs in station 1, which contain vessels which were not included in the manual ground truth. For this reason, the small region around the lungs and heart were masked out in both the original volumes and the ground truth data, excluding this area from our analyses.

Another artefact which remained after subtraction was the variation of lumen intensity along the length of vessels. These may arise due to poor timing of the contrast agent during acquisition, or by inhomogeneities in the magnetic field (such as surface coil artefacts). A simple procedure was followed to correct these variations in each station, in the form of a local intensity normalisation.

First we make the assumption that each axial slice contains an artery, and they are the brightest objects present (which holds true for all regions except for slices above the head and below the feet—these slices were simple masked to zero after the procedure was applied). We then applied a 7-slice sliding window

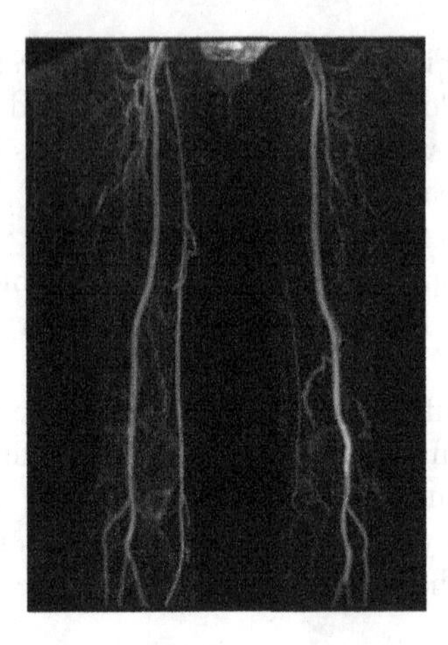
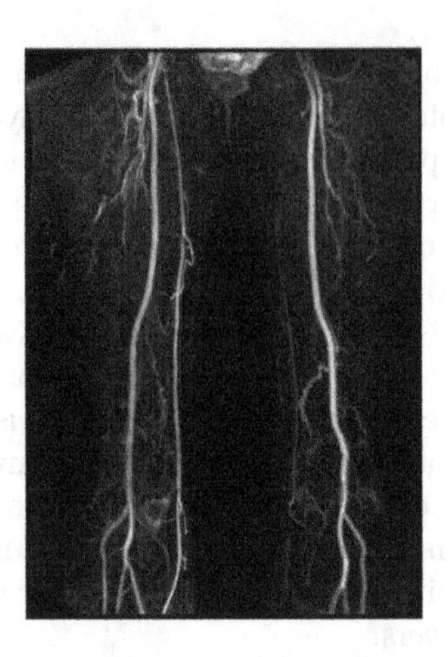

Fig. 3. Results of intensity equalisation on station 3 of patient 1. The MIP of the raw volume is show on the left, and the equalised volume on the right.

axially, in which the local vessel intensity was estimated from its histogram by choosing the highest frequency bin above 70% of the maximum intensity, with this value corresponding to the vessel intensity estimate for the central slice. Once calculated for the entire volume, Gaussian smoothing of the values was applied and then each slice divided by it's corresponding estimate. An example of the results of this processing is show in Fig. 3.

Active Contours. For the active contour method, the Toolbox implementation provided by [25] was used. Values for β, Δt, λ were fixed using a grid search optimisation procedure across all patients. In our case, the optimal values were found to be 0.08 for the smoothing weight term, 0.0002 for the image weight term, and 2.72 for the time step.

The final step was the initialisation of ϕ_0. This choice was critical as it affects the time and the speed of the evolution of the curve. So again under the hypothesis that the highest intensity voxels belong to the vessels, we took a set of seed points with high grey levels as ϕ_0. To keep the process completely automatic we used Otsu's method for generating thresholds from grey-level histograms [16].

For each station we generated 10 thresholds, which served as 10 different sets of seed points. The active contour method was then applied using the above parameters, and the highest Dice score recorded.

Enhancement Filters. The optimal parameters for the enhancement filters are shown in Table 1. These were optimised for each station across all patients using a grid search, with a fixed segmentation threshold.

The final segmentations acquired by calculating the vesselness map using the parameters in Table 1, then 20 thresholds automatically calculated using Otsu's

Table 1. Enhancement filter parameters. The scale factor and radii values are written in the form *minimum:step:maxiumum*.

Station number	SF/Radii	α (Frangi)	β (Frangi)	σ (OOF)
1	1:1:15	0.5	0.5	0.4
2	1:1:10	0.5	0.5	0.5
3	3:0.25:5	0.5	0.5	0.5
4	1:0.5:4	0.5	0.5	0.4

method [16]. The highest Dice score achieved from all 20 segmentation maps was then recorded.

3D Convolutional Neural Network. A number of network structures were explored during optimisation of the network structure. Inspired by models discussed in [20,21], we trained models consisting of 2–6 convolutional layers with 16–128 $3 \times 3 \times 3$ kernels, 1–3 max-pooling layers, and 1–2 fully-connected layers. To help combat overfitting, l_2 weight regularisation was used for each convolutional layer [6], and 20% dropout used on the fully connected layers [22]. All layer weights were initialised from a scaled Gaussian distribution.

A single network was trained for each station, with training patches extracted from two patients and the trained model applied to the held-out third patient in a 3-fold cross-validation setup.

The cubic patches were varied in size according to the network structure used, based on ensuring that the deepest layer still received a patch large enough to perform meaningful calculations on. A minimum side length of 15 voxels was needed to capture the thickest vessels, leading to a side length range of 15–50 voxels for the network structures we explored. For our final network, a patch size of $27 \times 27 \times 27$ was found to be optimal.

Finally, the number of training patches was chosen to maximise the available data. For each station, the minimum number of ground truth vessel voxels across all 3 patients was calculated, and this used as the number of positive samples to be extracted from each patient.

The data was balanced by extracting an equal amount of background samples. The position of the background samples were weighted to have two-thirds from regions within 5 voxels of a vessel and one third sampled randomly from the rest of the volume. This was found to improve the networks tendency to over estimate the diameter of the vessels when the background patches were sampled completely at random.

The total number of training samples used for each station were 136000 for station 1, 160000 for station 2, 28000 for station 3, and finally 24000 for station 4. During training, 5% of the training was data held out for validation, and the best network weights saved as those giving the highest validation accuracy score after 20 epochs (the network performance was found to typically converge after 8–12 epochs).

The networks were trained using an Nvidia Titan X Pascal GPU, with training times of between 3–4 h for each model (depending on the station and number of training samples used).

3 Results

The results of applying the four automated strategies described in Sect. 2 to all stations and patients are shown in Table 2, with the corresponding segmentation results of patient 1 shown in Fig. 4.

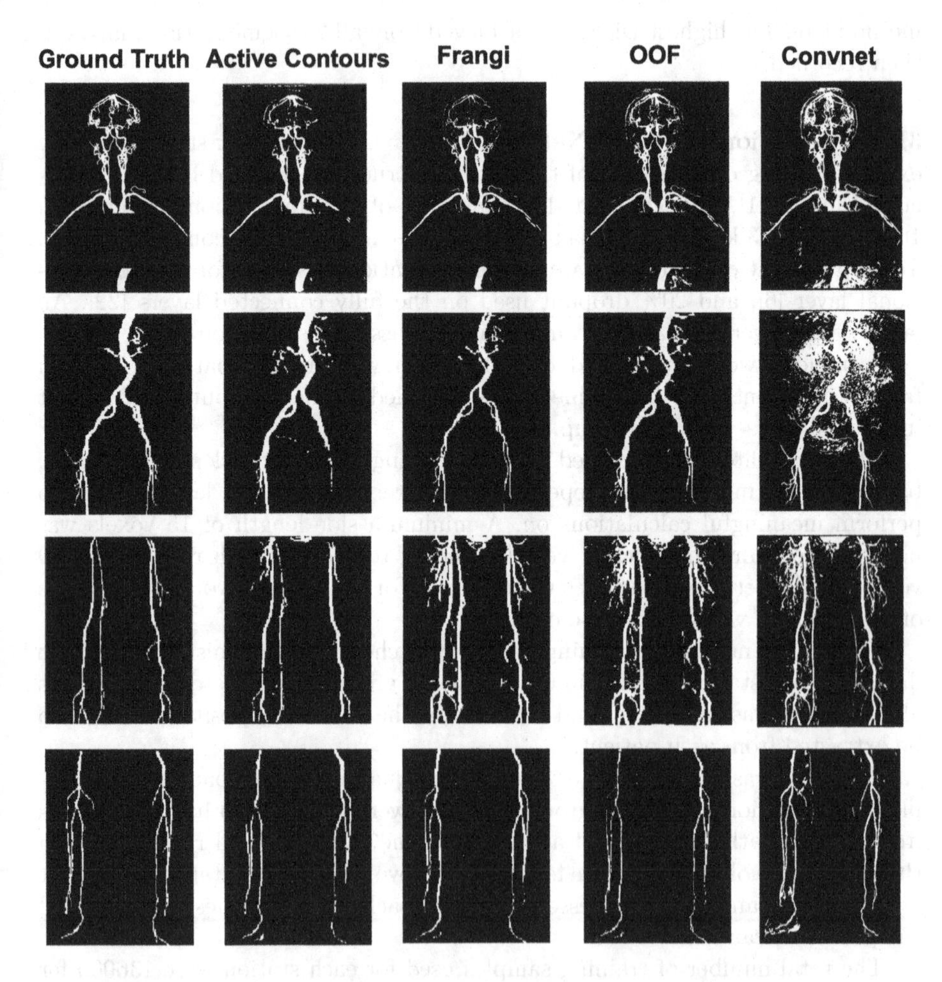

Fig. 4. Segmentation results for patient 1, shown as coronal projections.

It can be seen from Table 2 that for our case of only three patients, the OOF filter achieved the greatest mean DSC of 0.705. The 3D Convnet typically

Table 2. Dice coefficients for each method

Patient (station)	Level set	Frangi	OOF	3D convnet
1 (1)	0.785	0.697	**0.795**	0.727
1 (2)	**0.837**	0.817	0.812	0.618
1 (3)	0.647	**0.702**	0.657	0.605
1 (4)	0.672	0.780	**0.843**	0.806
2 (1)	**0.803**	0.689	0.794	0.752
2 (2)	0.712	**0.722**	0.674	0.552
2 (3)	**0.353**	0.323	0.281	0.303
2 (4)	0.662	0.776	**0.844**	0.692
3 (1)	0.572	**0.665**	0.663	0.567
3 (2)	**0.504**	0.488	0.469	0.422
3 (3)	0.803	0.747	0.835	**0.845**
3 (4)	0.618	**0.843**	0.791	0.697
Mean DSC	0.660	0.690	**0.705**	0.632

outperforms at least one of the other techniques, except for station 2. The main reason for this appears to be because of the additional artefacts left over from the imperfect volume registration and subtraction procedure (particularly the kidneys and bladder). The network approach had the most difficulty distinguishing between these artefacts and the arteries, causing it to over-segment station 2, resulting in a lower Dice score.

Looking at the segmentation results in Fig. 4, a number of observations can be made. The active contour method often produces broken vessels, such as the right branch in station 2, and has the most difficulty segmenting the finest vessels in station 4.

The Frangi and OOF enhancement filters produce visually similar results, though the OOF performs better at rejecting non-vessel artefacts (most noticeably in the brain and abdomen of station 1). Both filters do exhibit difficulties segmenting the finest vessels, such as at the bottom of station 2, and in different cases tend to either underestimate (Frangi in station 2, OOF in station 4) or overestimate (Frangi and OOF in station 3) the true diameter of the vessels as compared to the ground truth.

The Convnet performs poorest at rejecting non-arterial artefacts in station 2, but it also has the highest sensitivity to extracting fine vessels. Indeed, some of the finest vessels in the lower half of stations 2 and 3 were not present in the ground truth, having been either overlooked or rejected due to low contrast.

4 Conclusions

In this paper we have presented a quantitative comparison between four automated vessel segmentation technique for whole-body MRA data, using three manually segmented patient datasets.

In this regime of having limited ground truth data, it has been found that the Optimally Oriented Flux filter provides the best average DSC of 0.705. Visually, the Convnet approach segments vessels most consistently, with the least number of breaks, picking up finer vessels, and having the most consistently accurate diameters when compared with the ground truth. However it performed poorest at rejecting non-arterial artefacts, resulting in a lower DSC overall. It was also noted that some of the fine vessels segmented by the Convnet were not present in our ground truth. Due to having ground truth from a single observer, we are unable to estimate the quality and reliability of the ground truth data, and therefore the impact of this on the DSC results cannot be easily estimated for our data. We are not aware of any publicly available sets of manually annotated vascular networks for WBMRA volumes.

The Convnet approach appears to be mainly limited by the lack of training data. Other deep learning approaches which integrate large amounts of data augmentation, such as U-Net [18], may achieve better results at rejecting non-arterial artefacts, however those techniques have not been explored here. Another approach often used is to fine-tune a previously trained network such as GoogLeNet, a 22 layer network trained on a database of 1 million natural images [23]. Given that currently no pre-trained 3D networks exist for medical data, the training of deeper networks will require a larger database of ground truth segmentations.

References

1. Chan, T.F., Vese, L.A.: Active contours without edges. IEEE Trans. Image Process. **10**(2), 266–277 (2001)
2. de Brbisson, A., Montana, G.: Deep neural networks for anatomical brain segmentation. In: 2015 IEEE Conference on Computer Vision and Pattern Recognition Workshops (CVPRW), pp. 20–28, June 2015
3. Dehkordi, M.T., Sadri, S., Doosthoseini, A.: A review of coronary vessel segmentation algorithms. J. Med. Sig. Sens. **1**(1), 49 (2011)
4. Fedorov, A., Beichel, R., Kalpathy-Cramer, J., Finet, J., Fillion-Robin, J.-C., Pujol, S., Bauer, C., Jennings, D., Fennessy, F., Sonka, M., Buatti, J., Aylward, S., Miller, J., Pieper, S., Kikinis, R.: 3D slicer as an image computing platform for the quantitative imaging network. Magn. Reson. Imaging **30**(9), 1323–1341 (2012)
5. Frangi, A.F., Niessen, W.J., Vincken, K.L., Viergever, M.A.: Multiscale vessel enhancement filtering. In: Wells, W.M., Colchester, A., Delp, S. (eds.) MICCAI 1998. LNCS, vol. 1496, pp. 130–137. Springer, Heidelberg (1998). doi:10.1007/BFb0056195
6. Goodfellow, I., Bengio, Y., Courville, A.: Deep Learning. MIT Press, Cambridge (2016). http://www.deeplearningbook.org
7. Kass, M., Witkin, A., Terzopoulos, D.: Snakes: active contour models. Int. J. Comput. Vis. **1**(4), 321–331 (1988)

8. Kirbas, C., Quek, F.: A review of vessel extraction techniques and algorithms. ACM Comput. Surv. **36**(2), 81–121 (2004)
9. Krizhevsky, A., Sutskever, I., Hinton, G.E.: ImageNet classification with deep convolutional neural networks. In: Pereira, F., Burges, C.J.C., Bottou, L., Weinberger, K.Q. (eds.) Advances in Neural Information Processing Systems, vol. 25, pp. 1097–1105. Curran Associates Inc. (2012)
10. Law, M.W.K., Chung, A.C.S.: Three dimensional curvilinear structure detection using optimally oriented flux. In: Forsyth, D., Torr, P., Zisserman, A. (eds.) ECCV 2008. LNCS, vol. 5305, pp. 368–382. Springer, Heidelberg (2008). doi:10.1007/978-3-540-88693-8_27
11. LeCun, Y., Bengio, Y., Hinton, G.: Deep learning. Nature **521**(7553), 436–444 (2015)
12. Lesage, D., Angelini, E.D., Bloch, I., Funka-Lea, G.: A review of 3D vessel lumen segmentation techniques: models, features and extraction schemes. Med. Image Anal. **13**(6), 819–845 (2009)
13. Lupaşcu, C.A., Tegolo, D., Trucco, E.: Accurate estimation of retinal vessel width using bagged decision trees and an extended multiresolution hermite model. Med. Image Anal. **17**(8), 1164–1180 (2013)
14. McRobbie, D.W., Moore, E.A., Graves, M.J., Prince, M.R.: MRI from Picture to Proton, 2nd edn. Cambridge University Press, Cambridge (2007)
15. Moeskops, P., Viergever, M.A., Mendrik, A.M., de Vries, L.S., Benders, M.J.N.L., Igum, I.: Automatic segmentation of MR brain images with a convolutional neural network. IEEE Trans. Med. Imaging **35**(5), 1252–1261 (2016)
16. Otsu, N.: A threshold selection method from gray-level histograms. IEEE Trans. Syst. Man Cybern. **9**(1), 62–66 (1979)
17. Pereira, S., Pinto, A., Alves, V., Silva, C.A.: Brain tumor segmentation using convolutional neural networks in MRI images. IEEE Trans. Med. Imaging **35**(5), 1240–1251 (2016)
18. Ronneberger, O., Fischer, P., Brox, T.: U-Net: Convolutional Networks for Biomedical Image Segmentation, pp. 234–241. Springer International Publishing, Cham (2015)
19. Ruehm, S.G., Goehde, S.C., Goyen, M.: Whole body MR angiography screening. Int. J. Cardiovasc. Imaging **20**(6), 587–591 (2004)
20. Shelhamer, E., Long, J., Darrell, T.: Fully convolutional networks for semantic segmentation. IEEE Trans. Pattern Anal. Mach. Intell. **39**(4), 640–651 (2017)
21. Simonyan, K., Zisserman, A.: Very deep convolutional networks for large-scale image recognition. CoRR, abs/1409.1556 (2014)
22. Srivastava, N., Hinton, G.E., Krizhevsky, A., Sutskever, I., Salakhutdinov, R.: Dropout: a simple way to prevent neural networks from overfitting. J. Mach. Learn. Res. **15**(1), 1929–1958 (2014)
23. Szegedy, C., Liu, W., Jia, Y., Sermanet, P., Reed, S., Anguelov, D., Erhan, D., Vanhoucke, V., Rabinovich, A.: Going deeper with convolutions. In: 2015 IEEE Conference on Computer Vision and Pattern Recognition (CVPR), pp. 1–9, June 2015
24. Taha, A.A., Hanbury, A.: Metrics for evaluating 3D medical image segmentation: analysis, selection, and tool. BMC Med. Imaging **15**(1), 29 (2015)
25. Zhang, Y., Matuszewski, B.J., Shark, L.K., Moore, C.J.: Medical image segmentation using new hybrid level-set method. In: 2008 5th International Conference BioMedical Visualization: Information Visualization in Medical and Biomedical Informatics, pp. 71–76, July 2008

Evaluating Classifiers for Atherosclerotic Plaque Component Segmentation in MRI

Arna van Engelen[1(✉)], Marleen de Bruijne[2], Torben Schneider[3], Anouk C. van Dijk[4], M. Eline Kooi[5], Jeroen Hendrikse[6], Aart Nederveen[7], Wiro J. Niessen[2], and Rene M. Botnar[1]

[1] Division of Imaging Sciences, Department of Biomedical Engineering, King's College London, London, UK
arna.van_engelen@kcl.ac.uk
[2] Biomedical Imaging Group Rotterdam, Erasmus MC, Rotterdam, The Netherlands
[3] Philips Healthcare, Guildford, UK
[4] Department of Radiology, Erasmus MC, Rotterdam, The Netherlands
[5] Department of Radiology and Nuclear Medicine, CARIM School for Cardiovascular Diseases, Maastricht University Medical Center, Maastricht, The Netherlands
[6] Department of Radiology, University Medical Center Utrecht, Utrecht, The Netherlands
[7] Department of Radiology, Academic Medical Center Amsterdam, Amsterdam, The Netherlands

Abstract. Segmentation of tissue components of atherosclerotic plaques in MRI is promising for improving future treatment strategies of cardiovascular diseases. Several methods have been proposed before with varying results. This study aimed to perform a structured comparison of various classifiers, training set sizes, and MR image sequences to determine the most promising strategy for methodology development. Five different classifiers (linear discriminant classifier (LDC), quadratic discriminant classifier (QDC), random forest (RF), and support vector classifiers with both a linear (SVM_{lin}) and radial basis function kernel (SVM_{rbf})) were evaluated. We used carotid MRI data from 124 symptomatic patients, scanned in 4 centres with 2 different MRI protocols (45 and 79 patients). Firstly, learning curves of accuracy as a function of increasing training data size showed stabilisation of performance after using ~10–15 patients for training. Best results were found for LDC, QDC and RF. Intraplaque haemorrhage was most accurately classified in both protocols, and lowest accuracy was found for the lipid-rich necrotic core. Secondly, for LDC and RF it was shown that leaving out different MRI sequences usually negatively affects results for one or more classes. However, leaving out T2-weighted scans did not have a big impact. In conclusion, several classifiers obtain generally good results for classification of plaque components in MRI. Identification of intraplaque haemorrhage is the most promising, and lipid-rich necrotic core remains the most difficult.

M. Valdés Hernández and V. González-Castro (Eds.): MIUA 2017, CCIS 723, pp. 156–168, 2017.
DOI: 10.1007/978-3-319-60964-5_14

1 Introduction

Cardiovascular diseases are the leading cause of death and disability worldwide [1]. In ischaemic strokes and transient ischaemic attacks (TIA), atherosclerosis plays an important causal role. One of the major hazards of an atherosclerotic plaque is plaque rupture, which can result in a clinical event. One way to prevent such events is by performing carotid endarterectomy in which a high-risk plaque is removed surgically. Early diagnosis and risk stratification is therefore important to accurately select patients for treatment.

Plaque composition is considered an important determinant of plaque rupture [2]. However, due to the lack of techniques to derive high-risk plaque characteristics accurately and reproducibly in daily clinical practice, treatment still mainly relies on the degree of arterial narrowing [3]. The imaging modality that has shown most promising results in imaging of plaque composition is magnetic resonance imaging (MRI) [4,5]. Previous research has shown that MRI can visualise plaque characteristics such as a lipid-rich necrotic core and intraplaque haemorrhage [6,7], and that these MRI-derived parameters are predictive of clinical events [8–10]. However, data analysis is complex and primarily done by visual inspection and manual delineation.

Several studies have developed methods for automatic segmentation of plaque components in MRI [11–16], but these still lack accuracy for use in clinical practice. Moreover, due to differences between studies it remains unclear which segmentation methodology is most promising, how much data is required to develop stable and accurate methods, and which MRI protocol should be used for best identification of the most important plaque components. The most common approach has been to perform voxel classification trained on a ground truth dataset with either manual contours or contours from histology, using a set of imaging-derived features, including normalised intensity and Gaussian filters. Classifiers that have been used are linear discriminant classifiers [13–16], non-linear Bayesian classifiers [11,12], and support vector machines [14] with training sets ranging from 12–22 patients. The MRI-protocols all included T1-weighted and T2-weighted scans, and, in most cases proton-density (PD) weighted and bright-blood time of flight (TOF) scans, and in a number of cases included contrast-enhanced scans or a specific scan to identify intraplaque haemorrhage.

This study therefore aims to compare different classifiers and image sequences for the classification of plaque components, to provide insights to develop improved plaque characterisation methods. We evaluate these techniques on two datasets with a slightly different MRI protocol.

2 Methods

2.1 Data

MRI Study. We used data acquired within the multi-centre Parisk study [17]. In this study patients with a recent (<3 months) ischaemic stroke or TIA and a symptomatic 30–69% carotid artery stenosis as defined on ultrasound or CT, were prospectively recruited. MRI was performed in one of four centres within the Netherlands: Academic Medical Center Amsterdam (centre 1), Erasmus Medical Center, Rotterdam (centre 2), Maastricht University Medical Center (centre 3) and University Medical Center Utrecht (centre 4).

Five MRI sequences were used for plaque characterisation. Patients in centres 1, 3 and 4 were scanned using the same MRI protocol on an Achieva or Ingenia scanner (Philips Healthcare, Best, the Netherlands). We will refer to this protocol as *Protocol 1*. Patients in centre 2 were scanned on a Discovery MR 750 system (GE Healthcare, Milwaukee, MI, USA), (*Protocol 2*).

Both protocols contained a pre-contrast T1-weighted (T1w) and T2-weighted (T2w) scan, a post-contrast T1w scan acquired 6 min after administration of 0.1 mmol/kg body weight of a gadolinium-based contrast medium, and a pre-contrast heavily T1-weighted scan that shows high signal intensity for intraplaque haemorrhage (an 2D inversion-recovery turbo-field echo for Protocol 1, and a 3D spoiled gradient echo for Protocol 2). Protocol 1 contains a 2D bright-blood TOF scan that aims to identify calcifications bordering the lumen. In contrast, in Protocol 2 a 3D fast spoiled gradient echo scan that is specifically aimed at showing all calcifications as hypointense with little or no difference in appearance between other tissues and structures in the image, is used. Due to those differences, in our experiments the data from centres 1, 3 and 4 was taken together, and the data from centre 2 was considered separately. For details on the acquisition parameters, and more details on patient recruitment, we refer to Truijman et al. [17]. For this study we only used patients who completed the five MRI scans above with diagnostic image quality, and whose data was available. This resulted in 9 patients from centre 1, 45 patients from centre 2, 52 patients from centre 3, and 18 patients from centre 4, so in total 79 for Protocol 1 and 45 for Protocol 2.

Manual Annotation. Manual annotation of the symptomatic arteries was performed by a set of six observers, using dedicated software (VesselMASS, Department of Radiology, Leiden University Medical Center, Leiden, The Netherlands). Firstly, the lumen centreline was manually or semi-automatically identified. Then, the other four MRI sequences were registered to the T1w precontrast scan using a region of interest around the centreline [18] and manually adjusted. The vessel wall was subsequently segmented either manually or semi-automatically using a previously published technique [19] and manually adjusted. Plaque components (lipid-rich necrotic core (LRNC), calcification (CA) and intraplaque haemorrhage (IPH)) were fully manually annotated. The remainder of the vessel wall was considered fibrous tissue. All observers received the same training, and annotations were made based on previously published criteria [20–22]. LRNC was defined as a region that shows no contrast-enhancement on the postcontrast T1w scan compared with the precontrast scan, and IPH as hyperintense signal compared to the adjacent sternocleidomastoid muscle in the bulk of the plaque on the IR-TFE or SPGR image, and is considered as part of the LRNC. Calcification was identified as hypointense on at least two sequences, where for Protocol 2 a hypointense signal on the FSPGR scan was the main criterium. For the first 19 patients recruited in centre 3, annotations were made using MRI-Plaque View (VPDiagnostics Inc., Seattle, WA, USA) for a different study and converted for use in VesselMass as described in van Engelen et al. [15].

2.2 Experiments

Classifiers. Five commonly used classifiers were evaluated:

- Linear Discriminant Classifier (LDC): This relatively simple classifier has successfully been used for voxel classification in atherosclerotic plaques [13–16]. It determines the optimum linear boundaries between classes assuming the data is normally distributed with equal covariance matrices for each class, using the class means, class priors, and covariance matrix [23].
- Quadratic Discriminant Classifier (QDC): This classifier is similar to LDC, except that it does not assume equal covariance for all classes, and thereby allows quadratic instead of linear class boundaries.
- Random Forest (RF): Random forests have more recently become popular in medical imaging [24]. They are formed of a set of decision trees where subsets of features are randomly selected at each node. The predictions of all trees are combined for classification. This provides a data-driven way of feature selection and allows for more flexible decision boundaries. The number of trees was optimised in our experiments, with $\sqrt{(\text{nr of features})}$ features selected at each node.
- Linear Support Vector Machine (SVM_{lin}): Support Vector Machines maximise the width of the margin between classes. Therefore, the decision boundary is determined by the samples on the boundary, rather than on the distribution of all data like LDC and QDC. The parameter C that trades-off between maximising the margin and minimising misclassification is optimised in our experiments. Multiclass classification (also for SVM_{rbf}) was performed by combining different 1-vs-1 classifiers.
- Support Vector Machine with radial basis function kernel (SVM_{rbf}): The radial basis function kernel allows for non-linear decision boundaries. The kernel radius, γ, and C are optimised in our experiments.

All experiments were performed in Matlab. LDC and QDC were implemented using the PrTools toolbox [25], SVM_{lin} and SVM_{rbf} using libsvm [26], and RF using a toolbox based on the implementation described in [27].

Data Preparation. Preprocessing and feature computation was carried out in a similar fashion as in [15]. A bias field due to coil inhomogeneity was present in the data from Protocol 2, which was corrected by N4 inhomogeneity correction [28]. Intensity in all MRI sequences for both protocols was normalised by scaling the 5th and 95th percentile in a region of 4×4 cm around the lumen centre between 0 and 1000, on a per-volume basis scaling all slices with the same values.

Image features for classification were based on previous studies [11,13,15]: (1) the normalised intensities for each MRI sequence, (2) the images blurred with a Gaussian filter ($\sigma = 0.3$ mm), (3) First order (gradient magnitude) and second order (Laplacian) derivatives at the same scale, (4) the Euclidean distance to the lumen and to the outer wall, and the product of those two distances.

Learning Curves. To compare the five different classifiers, learning curves were made to determine the performance with increasing size of training data for each classifier. The aim of this was firstly, to compare the accuracy between different classifiers, and secondly, to establish the amount of required training data and to compare the this between the classifiers.

The data of each protocol was randomly split in three groups. For Protocol 1 two groups of 30 patients were used for training and testing, and the remaining 19 were used to optimise classifier-specific parameters. For Protocol 2 two groups of 20 patients were used for training and testing, and 5 patients for optimisation. The distribution of all four classes was kept similar between the three groups (overall, Protocol 1 had 86% F, 3% LRNC, 5% CA and 6% IPH. Protocol 2 had 82% F, 3% LRNC, 8% CA and 7% IPH.). The two non-optimisation groups for each protocol served both as training and testing data in a two-fold cross-validation. For parameter optimisation, classifiers were trained on the full set of 20 or 30 patients, and the average best accuracy on the optimisation data was determined. For RF, the number of decision trees evaluated was 10, 25, 50, 100, 250, 500, 750 and 1000. For SVM, C and (for SVM_{rbf}) γ were evaluated for 0.001, 0.01, 0.1, 1, 10, 100 and 1000.

To create the learning curves, the size of the training set was increased from 1 to 30 (Protocol 1) or 20 (Protocol 2). For training set sizes of 1, 2 and the maximum minus 1, all possibilities were evaluated (so 20 or 30 repeats). For all sizes in between 25 randomly selected combinations of patients were used for training. Those were the same for every classifier. For the maximum training set size only 1 combination, using all training data, was possible. When training on the first fold, patients in the second fold were used for testing and the other way around. Results are presented by averaging over all 60 (Protocol 1) or 40 (Protocol 2) datasets. Overall voxelwise accuracy was determined. Moreover, sensitivity, intraclass correlation coefficients of volumes, and Cohen's kappa (presence/absence in ground truth vs. the result) were determined for all four classes. Since fibrous tissue was present and detected in all cases, its kappa is always 1 and not further presented. Standard deviations were calculated over all repetitions.

MRI Sequences. To determine which sequences are most important in plaque component classification, and to determine whether this is classifier-dependent, classification using the full training sets was repeated after leaving all features of single MRI sequences out. This was performed for two classifiers that were deemed most successful based on the learning curves. Again, sensitivity, ICC and Cohen's kappa were determined and presented for LRNC, CA and IPH. The voxelwise accuracy of classification without each sequence was statistically compared with the full dataset using a Wilcoxon signed ranks test, a non-parametric test for paired differences.

3 Results

3.1 Learning Curves

Parameter optimisation for RF resulted in 250 trees used for Protocol 1, and 100 or 750 trees for Protocol 2 for the two training sets. For SVM_{lin}, for Protocol 1 C was 0.01 for both training sets, and for Protocol 2 C was 0.1 and 0.01. For SVM_{rbf}, for Protocol 1, the parameters for the first training set were $C=1$ and $\gamma=0.01$, and for the second set $C=10$ and $\gamma=0.01$. For Protocol 2 those were C = 10 and $\gamma=0.001$, and $C=1$ and $\gamma=0.01$.

The learning curves for all classifiers are shown in Figs. 1 (Protocol 1) and 2 (Protocol 2). It can be seen that the average overall voxelwise accuracy (bottom-right in the figures) varies little between classifiers, except for a slightly lower performance for QDC, and stabilises after using about ∼10 patients for training. The other curves stabilise after 10–15 datasets (Protocol 1) or 5–10 datasets (Protocol 2), which may be related to all data being acquired in the same centre for Protocol 2. Kappa, which only looks at presence or absence of tissue components, stabilises the quickest. Only for the ICC for LRNC, the smallest class, the plateau might not always be reached. LRNC is better identified in Protocol 1 than in Protocol 2, although the largest differences between classifiers is seen here. LDC, QDC and, for Protocol 1, RF, perform best for LRNC. SVM shows much lower performance, and particularly fails to identify CA in Protocol 1 compared with the other classifiers. For both protocols best results are achieved for IPH, with high ICC and kappa values and reasonable sensitivity. While voxelwise sensitivity for calcification is low in both protocols, ICC is good, with reasonable Kappa values.

3.2 MRI Sequences

LDC and RF were repeated after leaving out each MRI sequence individually. Results are presented in Tables 1 (LDC) and 2 (RF). Leaving out one single sequence has a minor effect on overall classification accuracy for both protocols and classifiers, but more relevant effects on plaque components are seen. For Protocol 1, results for LRNC decrease mostly when the postcontrast (LDC and RF) or precontrast T1w scan (only for LDC) are left out. For Protocol 2, LRNC classification is generally low, but leaving out the SPGR sequence has the biggest effect. Calcification is for both classifiers mostly dependent on the FSPGR scan (Protocol 2) or a combination of several sequences (Protocol 1), for which the TOF scan may be the most important one. Classification of IPH is not strongly affected by leaving out any single sequence, though leaving out the heavility T1-weighted IR-TFE or SPGR has the biggest effect. For both classifiers and protocols, no reduction in performance was seen after leaving out T2w scans.

4 Discussion and Conclusion

We have shown a comparison between five common classifiers on two different MRI datasets. Generally, learning curves show stabilisation of results after

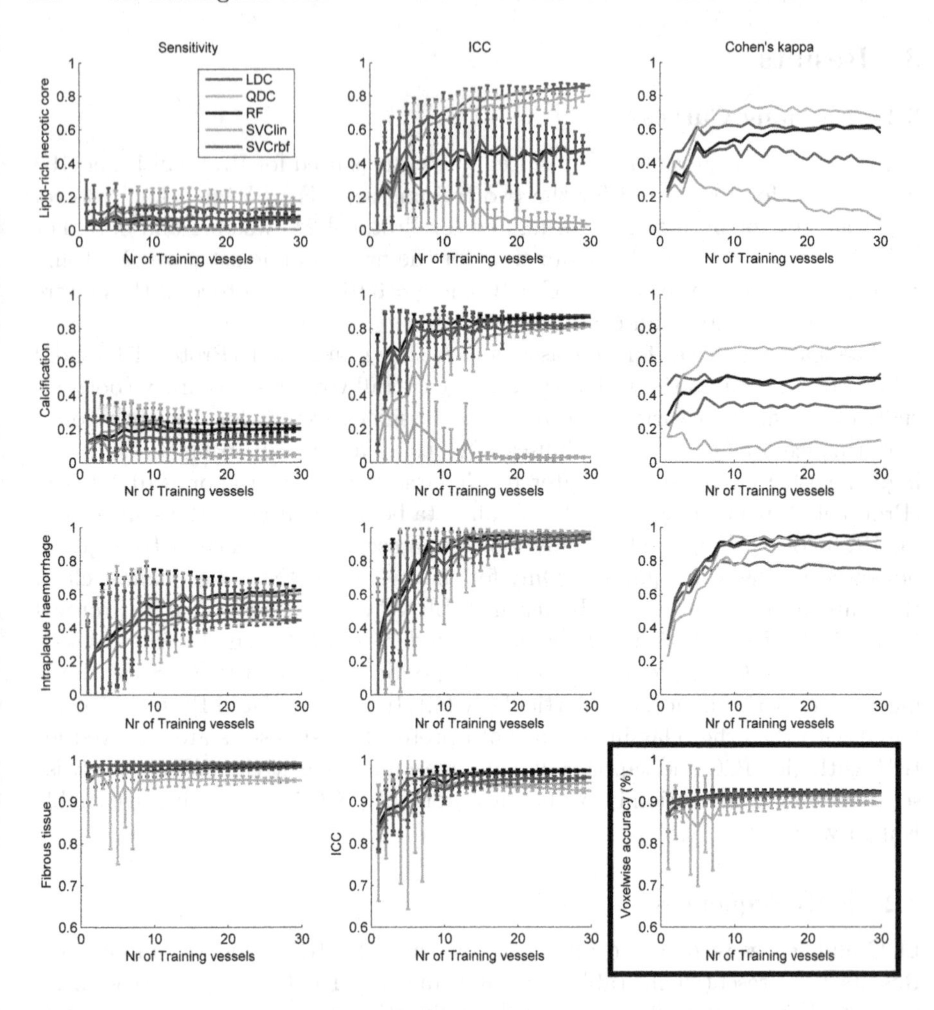

Fig. 1. Learning curves for the data of **Protocol 1**. Cohen's kappa is not shown for fibrous tissue, since it is 1 in all cases. Instead the voxelwise accuracy is shown in the bottom right figure. Note that the figures in the bottom row are scaled from 0.6 to 1 instead of 0 to 1. The error bars indicate the standard deviation for the average of all patients over the number of repetitions.

including 10–20 patients for training. Good results can be obtained for IPH and CA, however, accurate classification of LRNC was shown to be more difficult. The largest differences between classifiers were also seen for LRNC, and, for Protocol 1, CA. LDC, QDC and RF generally showed best performance, while SVM had lower performance.

Lower performance for SVM could be related with more difficult optimisation of the classifier parameters. SVM is likely to suffer more from the considerable class imbalance that was present. Moreover, optimisation on accuracy tends to

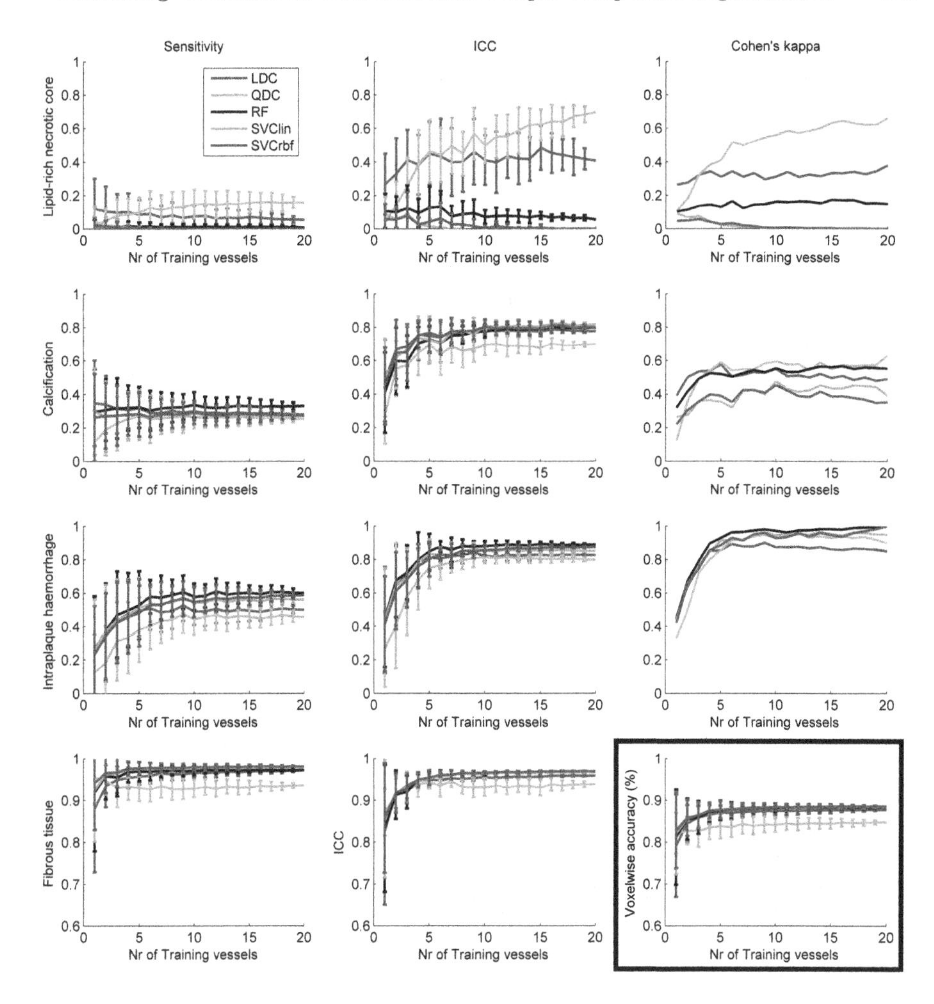

Fig. 2. Learning curves for the data of **Protocol 2**. Cohen's kappa is not shown for fibrous tissue, since it is 1 in all cases. Instead the voxelwise accuracy is shown in the bottom right figure. Note that the figures in the bottom row are scaled from 0.6 to 1 instead of 0 to 1. The error bars indicate the standard deviation for the average of all patients over the number of repetitions.

be biased towards correct classification of fibrous tissue, rather than the other, smaller, classes. A more optimal way for feature selection may be to look at average sensitivity over the four classes, or the F-score, which balances between sensitivity and precision. Furthermore, more differences between the optimised parameters for the two training sets were seen for Protocol 2. This could be due to only five patients being used for optimisation. However, this was chosen to have as much data as possible available to evaluate classification performance.

To improve results for small classes, future research can also investigate the effect of using (more) balanced classes in training, however, measures should be

Table 1. Classification results for varying image protocols for **LDC**. Acc. = accuracy, TOF = time of flight, IR-TFE = inversion-recovery turbo field echo, (F)SPGR = (fast) spoiled gradient echo. *Statistically different from using all features ($p < 0.05$)

	Voxelwise acc. (%)	Sensitivity			ICC			Kappa		
		LRNC	CA	IPH	LRNC	CA	IPH	LRNC	CA	IPH
Protocol 1										
All 5 sequences	92	0.13	0.20	0.45	0.86	0.88	0.95	0.61	0.53	0.75
No postcontrast	91*	0.05	0.16	0.45	0.76	0.84	0.95	0.48	0.40	0.75
No T2w	92	0.12	0.19	0.44	0.86	0.87	0.95	0.61	0.50	0.75
No TOF	92*	0.13	0.12	0.45	0.86	0.79	0.95	0.61	0.30	0.75
No T1w	92*	0.08	0.20	0.45	0.77	0.88	0.95	0.55	0.42	0.75
No IR-TFE	91*	0.10	0.15	0.35	0.81	0.74	0.89	0.58	0.44	0.75
Protocol 2										
All 5 sequences	88	0.05	0.28	0.50	0.41	0.78	0.83	0.38	0.49	0.85
No postcontrast	88	0.05	0.27	0.51	0.35	0.77	0.83	0.23	0.44	0.85
No T2w	88	0.06	0.26	0.51	0.32	0.77	0.83	0.25	0.49	0.90
No FSPGR	87*	0.04	0.12	0.51	0.40	0.77	0.83	0.30	0.16	0.90
No T1w	88	0.04	0.28	0.46	0.30	0.78	0.83	0.25	0.49	0.85
No SPGR	87*	0.03	0.28	0.36	0.25	0.78	0.79	0.17	0.44	0.80

taken to prevent overclassification of small classes in this case. Another reason for suboptimal results for certain classes is that the classes may not be separable with the evaluated features. We have used the ones that have commonly been used for this application in more recent previous studies [13–16]. Other features, such as Gaussian filters at more scales, or texture features, may be interesting to study as well. Instead of evaluating the effect of leaving out individual features, we have evaluated them on a per-MRI-sequence basis. This was chosen because eliminating features on a per-MRI-sequence basis would be advantageous in clinical practice, since it could reduce scan time. In presence of the four other available sequences, leaving out the T2w scan had the smallest effect in our study.

Much more classifiers than the ones evaluated here exist. We have chosen to use the most commonly used ones. Currently, deep learning techniques, using deep neural networks, have gained enormous popularity in image analysis. These techniques need to be considered in future research. Furthermore, both MRI protocols have been considered separately in this study, as previous research has shown that considerable differences exist between them [15]. In future research it could be interesting to see whether some classifiers are better at handling all MRI data combined.

Table 2. Classification results for varying image protocols for **RF**. Acc. = accuracy, TOF = time of flight, IR-TFE = inversion-recovery turbo field echo, (F)SPGR = (fast) spoiled gradient echo. *Statistically different from using all features ($p < 0.05$)

	Voxelwise acc. (%)	Sensitivity			ICC			Kappa		
		LRNC	CA	IPH	LRNC	CA	IPH	LRNC	CA	IPH
Protocol 1										
All 5 sequences	92	0.08	0.20	0.63	0.48	0.87	0.97	0.58	0.50	0.96
No postcontrast	92*	0.02	0.17	0.62	0.25	0.81	0.97	0.61	0.63	0.96
No T2w	92	0.08	0.21	0.62	0.47	0.87	0.97	0.74	0.50	0.96
No TOF	92*	0.09	0.18	0.63	0.53	0.85	0.97	0.61	0.42	0.96
No T1w	92*	0.07	0.20	0.60	0.46	0.87	0.97	0.61	0.50	0.96
No IR-TFE	91*	0.07	0.18	0.33	0.41	0.77	0.78	0.51	0.50	0.79
Protocol 2										
All 5 sequences	88	0.01	0.33	0.60	0.05	0.80	0.89	0.15	0.55	1.00
No postcontrast	88*	0.02	0.34	0.60	0.05	0.79	0.89	0.27	0.55	1.00
No T2w	88*	0.02	0.33	0.60	0.10	0.79	0.89	0.23	0.55	1.00
No FSPGR	87*	0.02	0.15	0.60	0.08	0.73	0.89	0.25	0.39	0.95
No T1w	88*	0.01	0.34	0.60	0.07	0.78	0.89	0.15	0.55	1.00
No SPGR	88*	0.00	0.35	0.36	0.06	0.84	0.83	0.05	0.63	0.95

The most accurate results were obtained for IPH, which also has been considered as one of the most promising imaging characteristics for use in clinical practice due to its high predictive value for future events [9]. This study confirms that also automated techniques can identify IPH well in MRI. Leaving out one single sequence did not have a large effect for IPH. This is probably because two T1-weighted sequences were available in both protocols, so when one if left out the other still provides enough information.

In conclusion, for the evaluated classifiers training set sizes of 15–20 patients are sufficiently large. A simple classifier such as LDC, but also QDC and RF, yields good results. However, improvements can still be made. Especially classification of LRNC remains difficult. Classification of IPH is possible with high accuracy and therefore most promising for implementation into clinical practice.

Acknowledgments. This research has been supported by an EPSRC Technology Strategy Board CR&D Grant (EP/L505304/1). The PARISK study was performed within the framework of the Center for Translational Molecular Medicine (www.ctmm.nl), project PARISK (Plaque At RISK; grant 01C-202) and supported by the Dutch Heart Foundation. This research was also partly funded by the Netherlands Organisation for Scientific Research (NWO).

The Division of Imaging Sciences also receives support from the Centre of Excellence in Medical Engineering (funded by the Welcome Trust and EPSRC; grant number

WT 088641/Z/09/Z) and the Department of Health through the National Institute for Health Research (NIHR) Biomedical Research Centre award to Guy's and St Thomas' NHS Foundation Trust in partnership with King's College London, and by the NIHR Healthcare Technology Co-operative for Cardiovascular Disease at Guys and St Thomas NHS Foundation Trust. The views expressed are those of the author(s) and not necessarily those of the NHS, the NIHR or the Department of Health.

References

1. World Health Organisation: WHO global burden of disease 2000–2015 (2015)
2. Finn, A.V., Nakano, M., Narula, J., Kolodgie, F.D., Virmani, R.: Concept of vulnerable/unstable plaque. Arterioscler. Thromb. Vasc. Biol. **30**(7), 1282–1292 (2010)
3. Paraskevas, K.I., Mikhailidis, D.P., Veith, F.J.: Comparison of the five 2011 guidelines for the treatment of carotid stenosis. J. Vasc. Surg. **55**(5), 1504–1508 (2012)
4. Sanz, J., Fayad, Z.A.: Imaging of atherosclerotic cardiovascular disease. Nature **451**(7181), 953–957 (2008)
5. Huibers, A., de Borst, G., Wan, S., Kennedy, F., Giannopoulos, A., Moll, F., Richards, T.: Non-invasive carotid artery imaging to identify the vulnerable plaque: current status and future goals. Eur. J. Vasc. Endovasc. Surg. **50**(5), 563–572 (2015)
6. Yuan, C., Mitsumori, L.M., Ferguson, M.S., Polissar, N.L., Echelard, D., Ortiz, G., Small, R., Davies, J.W., Kerwin, W.S., Hatsukami, T.S.: In vivo accuracy of multispectral magnetic resonance imaging for identifying lipid-rich necrotic cores and intraplaque hemorrhage in advanced human carotid plaques. Circulation **104**(17), 2051–2056 (2001)
7. Saam, T., Ferguson, M., Yarnykh, V., Takaya, N., Xu, D., Polissar, N., Hatsukami, T., Yuan, C.: Quantitative evaluation of carotid plaque composition by in vivo MRI. Arterioscler. Thromb. Vasc. Biol. **25**(1), 234–239 (2005)
8. Gupta, A., Baradaran, H., Schweitzer, A.D., Kamel, H., Pandya, A., Delgado, D., Dunning, A., Mushlin, A.I., Sanelli, P.C.: Carotid plaque MRI and stroke risk. Stroke **44**(11), 3071–3077 (2013)
9. Saam, T., Hetterich, H., Hoffmann, V., Yuan, C., Dichgans, M., Poppert, H., Koeppel, T., Hoffmann, U., Reiser, M.F., Bamberg, F.: Meta-analysis and systematic review of the predictive value of carotid plaque hemorrhage on cerebrovascular events by magnetic resonance imaging. J. Am. Coll. Cardiol. **62**(12), 1081–1091 (2013)
10. Sun, J., Zhao, X.Q., Balu, N., Neradilek, M.B., Isquith, D.A., Yamada, K., Canton, G., Crouse, J.R., Anderson, T.J., Huston, J., OBrien, K., Hippe, D.S., Polissar, N.L., Yuan, C., Hatsukami, T.S.: Carotid plaque lipid content and fibrous cap status predict systemic CV outcomes: the MRI substudy in AIM-HIGH. JACC: Cardiovasc. Imaging **10**(3), 241–249 (2017)
11. Liu, F., Xu, D., Ferguson, M.S., Chu, B., Saam, T., Takaya, N., Hatsukami, T.S., Yuan, C., Kerwin, W.S.: Automated in vivo segmentation of carotid plaque MRI with morphology-enhanced probability maps. Magn. Reson. Med. **55**(3), 659–668 (2006)
12. Hofman, J., Branderhorst, W., ten Eikelder, H., Cappendijk, V., Heeneman, S., Kooi, M., Hilbers, P., ter Haar Romeny, B.: Quantification of atherosclerotic plaque components using in vivo MRI and supervised classifiers. Magn. Reson. Med. **55**(4), 790–799 (2006)

13. van't Klooster, R., Naggara, O., Marsico, R., Reiber, J., Meder, J.F., van der Geest, R., Touz, E., Oppenheim, C.: Automated versus manual in vivo segmentation of carotid plaque MRI. Am. J. Neuroradiol. **33**, 1621–1627 (2012)
14. van Engelen, A., Niessen, W.J., Klein, S., Groen, H.C., Verhagen, H.J.M., Wentzel, J.J., van der Lugt, A., de Bruijne, M.: Atherosclerotic plaque component segmentation in combined carotid MRI and CTA data incorporating class label uncertainty. PLOS ONE **9**(4), 1–14 (2014)
15. van Engelen, A., van Dijk, A., Truijman, M., van't Klooster, R., van Opbroek, A., van der Lugt, A., Niessen, W., Kooi, M., de Bruijne, M.: Multi-center MRI carotid plaque component segmentation using feature normalization and transfer learning. IEEE Trans. Med. Imaging **34**(6), 1294–1305 (2015)
16. Gao, S., van't Klooster, R., van Wijk, D.F., Nederveen, A.J., Lelieveldt, B.P.F., van der Geest, R.J.: Repeatability of in vivo quantification of atherosclerotic carotid artery plaque components by supervised multispectral classification. Magn. Reson. Mater. Phys., Biol. Med. **28**(6), 535–545 (2015)
17. Truijman, M., Kooi, M., van Dijk, A., de Rotte, A., van der Kolk, A., Liem, M., Schreuder, F., Boersma, E., Mess, W., van Oostenbrugge, R., Koudstaal, P., Kappelle, L., Nederkoorn, P., Nederveen, A., Hendrikse, J., van der Steen, A., Daemen, M., van der Lugt, A.: Plaque At RISK (PARISK): prospective multicenter study to improve diagnosis of high-risk carotid plaques. Int. J. Stroke **9**, 747–754 (2013)
18. van't Klooster, R., Staring, M., Klein, S., Kwee, R.M., Kooi, M.E., Reiber, J.H.C., Lelieveldt, B.P.F., van der Geest, R.J.: Automated registration of multispectral MR vessel wall images of the carotid artery. Med. Phys. **40**(12), 121904 (2013)
19. van't Klooster, R., de Koning, P.J., Dehnavi, R.A., Tamsma, J.T., de Roos, A., Reiber, J.H., van der Geest, R.J.: Automatic lumen and outer wall segmentation of the carotid artery using deformable three-dimensional models in MR angiography and vessel wall images. J. Magn. Reson. Imaging **35**(1), 156–165 (2012)
20. Cai, J., Hatsukami, T.S., Ferguson, M.S., Kerwin, W.S., Saam, T., Chu, B., Takaya, N., Polissar, N.L., Yuan, C.: In vivo quantitative measurement of intact fibrous cap and lipid-rich necrotic core size in atherosclerotic carotid plaque: Comparison of high-resolution, contrast-enhanced magnetic resonance imaging and histology. Circulation **112**(22), 3437–3444 (2005)
21. Cappendijk, V.C., Heeneman, S., Kessels, A.G., Cleutjens, K.B., Schurink, G.W.H., Welten, R.J., Mess, W.H., van Suylen, R.J., Leiner, T., Daemen, M.J., van Engelshoven, J.M., Kooi, M.E.: Comparison of single-sequence T1w TFE MRI with multisequence MRI for the quantification of lipid-rich necrotic core in atherosclerotic plaque. J. Magn. Reson. Imaging **27**(6), 1347–1355 (2008)
22. Kwee, R.M., van Engelshoven, J.M., Mess, W.H., ter Berg, J.W., Schreuder, F.H., Franke, C.L., Korten, A.G., Meems, B.J., van Oostenbrugge, R.J., Wildberger, J.E., Kooi, M.E.: Reproducibility of fibrous cap status assessment of carotid artery plaques by contrast-enhanced MRI. Stroke **40**(9), 3017–3021 (2009)
23. Hastie, T., Tibshirani, R., Friedman, J.H.: The Elements of Statistical Learning, Corrected edn. Springer, New York (2003)
24. Criminisi, A., Shotton, J., Konukoglu, E.: Decision Forests: A Unified Framework for Classification, Regression, Density Estimation, Manifold Learning and Semi-Supervised Learning. NOW Publishers, Breda (2012)
25. Duin, R., Juszczak, P., Paclik, P., Pekalska, E., de Ridder, D., Tax, D., Verzakov, S.: PRTools4.1, A Matlab Toolbox for Pattern Recognition. Delft University of Technology (2007)

26. Chang, C.C., Lin, C.J.: LIBSVM: a library for support vector machines. ACM Trans. Intell. Syst. Technol. **2**, 27:1–27:27 (2011)
27. Liaw, A., Wiener, M.: Classification and regression by randomforest. R News **2**(3), 18–22 (2002)
28. Tustison, N.J., Avants, B.B., Cook, P.A., Zheng, Y., Egan, A., Yushkevich, P.A., Gee, J.C.: N4ITK: improved N3 bias correction. IEEE Trans. Med. Imaging **29**(6), 1310–1320 (2010)

Cardiac Mesh Reconstruction from Sparse, Heterogeneous Contours

Benjamin Villard[1(✉)], Valentina Carapella[2], Rina Ariga[3], Vicente Grau[1], and Ernesto Zacur[1]

[1] Institute of Biomedical Engineering, University of Oxford, Oxford, UK
benjamin.villard@eng.ox.ac.uk
[2] Simula Research Laboratory, Bærum, Norway
[3] Division of Cardiovascular Medicine, Radcliffe Department of Medicine, University of Oxford Centre for Clinical Magnetic Resonance Research, University of Oxford, Oxford, UK

Abstract. We introduce a tool to reconstruct a geometrical surface mesh from sparse, heterogeneous, non coincidental contours and show its application to cardiac data. In recent years much research has looked at creating personalised 3D anatomical models of the heart. These models usually incorporate a geometrical reconstruction of the anatomy in order to understand better cardiovascular functions as well as predict different processes after a clinical event. The ability to accurately reconstruct heart anatomy from MRI in three dimensions commonly comes with fundamental challenges, notably the trade off between data fitting and regularization. Most current techniques requires data to be either parallel, or coincident, and bias the final result due to prior shape models or smoothing terms. Our approach uses a composition of smooth approximations towards the maximization of the data fitting. Assessment of our method was performed on synthetic data obtained from a mean cardiac shape model as well as on clinical data belonging to one normal subject and one affected by hypertrophic cardiomyopathy. Our method is both used on epicardial and endocardial left ventricle surfaces, but as well as on the right ventricle.

Keywords: Computational geometry · Mesh reconstruction · Cardiac modelling

1 Introduction

In recent years much research has looked at creating personalised 3D anatomical models of the heart [1–3]. These models usually incorporate a geometrical reconstruction of the anatomy in order to understand better cardiovascular functions as well as predict different processes after a clinical event. Besides being fundamental to shape analysis, they are the prerequisite to finite element analysis [4]. The ability to accurately reconstruct heart anatomy from magnetic resonance

M. Valdés Hernández and V. González-Castro (Eds.): MIUA 2017, CCIS 723, pp. 169–181, 2017.
DOI: 10.1007/978-3-319-60964-5_15

imaging (MRI) in three dimensions commonly comes with fundamental challenges, notably the trade off between data fitting and regularization. Most works in the literature tend to over regularize, or bias the data fitting process towards a smooth result [5,6]. Other methods work based on templates or shape prior which are usually learned on normal patient anatomies, and tend to fail when presented with pathological data, or any data outside the learned model [7]. Our approach differs in this, as it minimizes the use of a regularization term while maximizes the data fitting. In this paper we discuss a process to reconstruct a geometrical 3D surface mesh from a set of contours available from manual delineations as well as from automatic segmentations. Cardiac contours are typically composed from a stack of short axes (SAX) ranging from 8–12 slices, and a couple long axes (LAX). As such the input data is quite sparse for conventional mesh reconstruction processes such as isosurfacing [8]. In manual contouring, clinicians are seldom able to account for the 3D environment which can lead to spatial discrepancies between long axes and short axes contours. For example, in LAX, papillary muscles can be hard to differentiate from myocardium which can result in their segmentation, in spite of their exclusion in SAX. Furthermore due to the image acquisition occurring at different breath hold, as well as any patient movement inside the scanner, the contours might be misaligned. This would further increase their discrepancies and extra steps might be needed to correct for this alignment [9]. This post-acquisition correction of the pose of the images can lead to loosing the parallelism in the SAX stack. Furthermore, due to the delineations not being able to grasp perfectly the overall shape of the heart, contour to contour distances cannot be minimized to obtain perfect coincidence and therefore spatial 3D consistency [9]. Finally, since many cardiomyopathies are localized in the left ventricle (LV), many previous efforts were focused on methodologies specifically designed or tuned for the LV. The main contributions of this work are:

- Use of our method on non parallel, sparse, heterogeneous contours
- Ability to deal with non coincidental contours
- Use of our technique on the right ventricle as well as the left ventricle.

1.1 State of the Art

Whilst simple heart models can be used for mechanical studies, patient-specific meshes require more complex anatomical models [10]. These models are characterized though 3D meshes, on which shape analysis or finite element methods (FEM) can be performed. From a set of contours or delineated curves, a surface mesh can be rendered and act as an input to generate a volumetric mesh. Depending on the complexity of the mesh needed, many parameters need to be taken into account; notably if the visual appearance of the mesh needs to be smooth [2,10]. In [2], the surface mesh is constructed by optimising over the topology, using a level-sets approach, and having the genus as a control parameter. Whilst this approach allows control over the topology, it can fail for sparse data. In order for the resulting mesh to represent the most accurate anatomical representation, an image volume is necessary. In order to obtain uniformly distributed contour points, the authors in [11,12], first fit a cubic periodic B-spline curve to the contours allowing them to uniformly sample candidate points. The disadvantages of this approach is that it relies on the the number of control points selected which in turn smooths the overall mesh. This is similar to [6], which can cause for the resulting mesh to be highly convoluted when dealing with pathological data. The method in [13] is closest to our own, as it tries to maximize the data fitting, whilst smoothing the interpolated part of the mesh.

2 Materials and Methods

2.1 Data

Both synthetic and clinical datasets were used. Synthetic contours were created from an average MRI mesh obtained from [14,15] and are available online[1]. Using that mean shape, contours where synthesized by slicing the mesh at various spatial positions. Parallel and non parallel contours were produced and small random translations and rotations applied to the contours. We treated the initial mean shape mesh as ground-truth and compared our results against it. Clinical contours belonging to one normal subject and one affected by hypertrophic cardiomyopathy (HCM), were also used, which were delineated by experts on cine images at end-diastole.

2.2 Initial Mesh

Let $\{C_j\}$ be a set of contours, each lying on a plane with normal n_j, where j is the contour index. The normals are clustered and the mean normal of the cluster group containing the most frequent orientation of the planes is calculated. The transformation matrix to align it with the Z axis is determined and is then applied to $\{C_j\}$ such that the contours are in a common frame of reference aligned with respect to the Z axis. The contours with normals belonging to the cluster with the most frequent plane orientations (which are expected to be the ones in SAX view) are then sorted from the most apical to the most basal (see Fig. 1(a)). Two extra points are added to $\{C_j\}$ providing an upper and lower lid, to fully enclose the initial mesh. We then proceed to create a rough initialization of the mesh by constructing a *tubular* surface going through the contours. For each of the contours, distribute K equally spaced contour points. A ruled surface approach is then used to obtain an initial mesh, where the initial triangle making up the mesh are composed of the set $\{(C_j(i), C_j(i+1), C_{j+1}(i)), (C_{j+1}(i), C_{j+1}(i+1), C_j(i+1))\}$, with i being the point index of the contours after resampling (see Fig. 1(b)). Depending on the availability of the LAX information, the algorithm initializes the upper mesh point as the mean position of the topmost contour points lying on the plane that contains such point. Analogously, the lower bound mesh point is initialized from the lowest contour points in that plane.

[1] http://wp.doc.ic.ac.uk/wbai/data/.

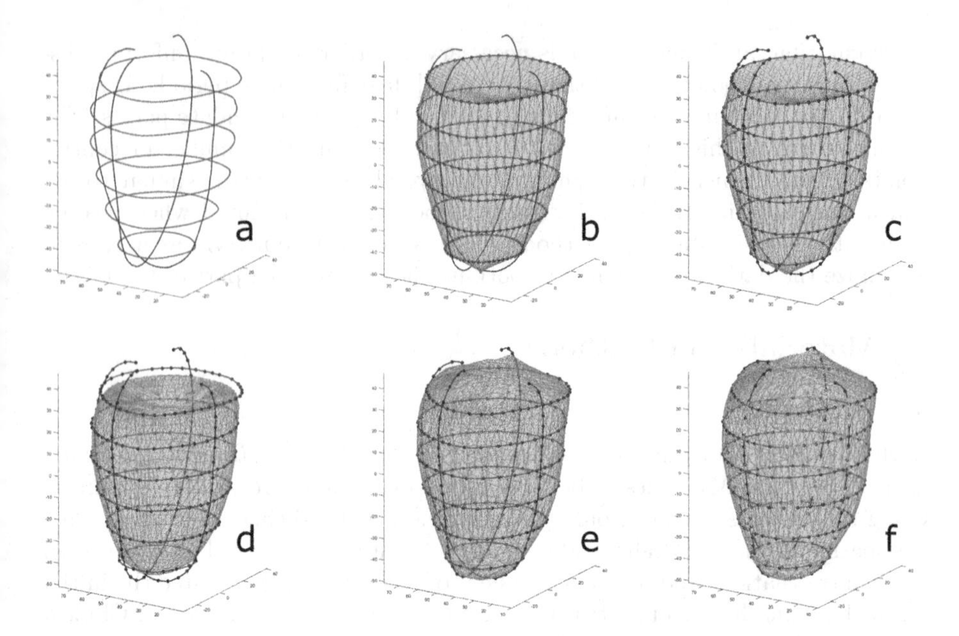

Fig. 1. Overall pipeline for generating a surface mesh from contours. (a) The input contours (red represents the automatically selected cluster group around the mean normal), (b) *Tubular* initial mesh, (c–e) different iterations of the process after deforming, subdividing and decimating the mesh, (f) final mesh. (Color figure online)

2.3 Mesh Deformation

Since the initial selected points make up the mesh, another set of contour points will be selected as the data fitting terms, and will act as attractors for the mesh. N number of points are sampled, where $N \gg K$, using farthest point sampling, allowing for a representation of the global spatial distribution of the contour points. The farthest point sampling method [16] ensures that no contour is sampled more densely than any others, and enforces an almost uniform global distribution of $\{C_j\}$. At this point, the initial mesh $\mathcal{M}$, the contours, and the attractors are all scaled to a common reference frame that is irrespective of the units of the provided contours. Once all the preprocessing is achieved, a deformation field is computed by taking the closest point from each of the attractor points to $\mathcal{M}$, and a dense force field is computed by using approximating thin plate splines [17,18]. The mesh is then deformed by using the *inverse* of the force field. Let $\{P_i\}$ be the set of closest points to $\{Q_i\}$ lying on the surface $\mathcal{M}$. Let be $F : \mathbb{R}^3 \Rightarrow \mathbb{R}^3$ such that:

$$\mathcal{M} = \min \lambda \sum_{i=1}^{N} (||F(P_i) - (Q_i)||)^2 + J_m^d(F) \tag{1}$$

where F is the mapping function, and J is the thin plate spline functional using derivatives of order m, and d the image dimensions, as defined in [17,18]. We perform this deformation iteratively, resulting in a composition of several smooth approximations of $\mathcal{M}$ towards the contours. Once the mesh has been deformed, the mesh undergoes subdivision and quadratic decimation. By subdividing the mesh we increase its resolution allowing for a more appropriate reconstruction of the mesh. However, as this process can lead to extremely high amounts of triangles, the mesh is therefore decimated, which allows for some control over the amount of triangles while preserving its geometric characteristics [19]. Finally, the resulting mesh is smoothed, using Laplacian smoothing. The latter steps of subdivision, decimation and smoothing occurs every S iterations. We chose S empirically, to be 20. The overall pipeline can be seen in Fig. 1. The overall process starting from the farthest point sampling is repeated for a set number of iterations, or until some stopping criterion is achieved, such the contours-to-mesh distance falls bellow a chosen threshold.

3 Results

3.1 Synthetic Data

Short Axis Contours. In order to assess our algorithm, we performed a series of experiments. In a first instance we applied our method to a set of parallel SAX contours and observed our technique's ability to recover the ground truth mesh. We also assessed the effect the amount of SAX had on the quality of the mesh by increasing the amount of SAX contours provided initially from 2 to 15, between the base and the apex. Our validation consisted of calculating the distances between vertices of our generated mesh to the ground truth mesh (mean shape of the statistical model) and can be seen in Fig. 2.

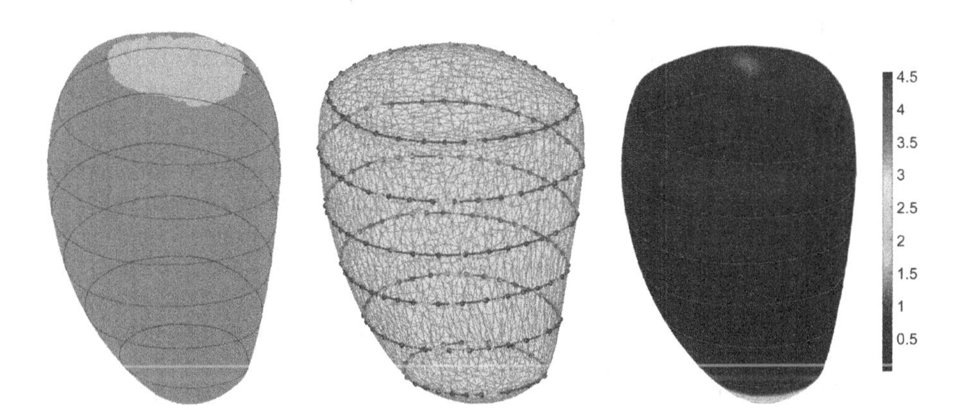

Fig. 2. (a) The ground truth mesh (mean shape of a statistical model of the left epicardium) and the synthesized contours in red, (b) our resulting mesh given the red contours, (c) the mesh-to-mesh distance error. the resulting mesh has been clipped at the level of the most basal contour. (Color figure online)

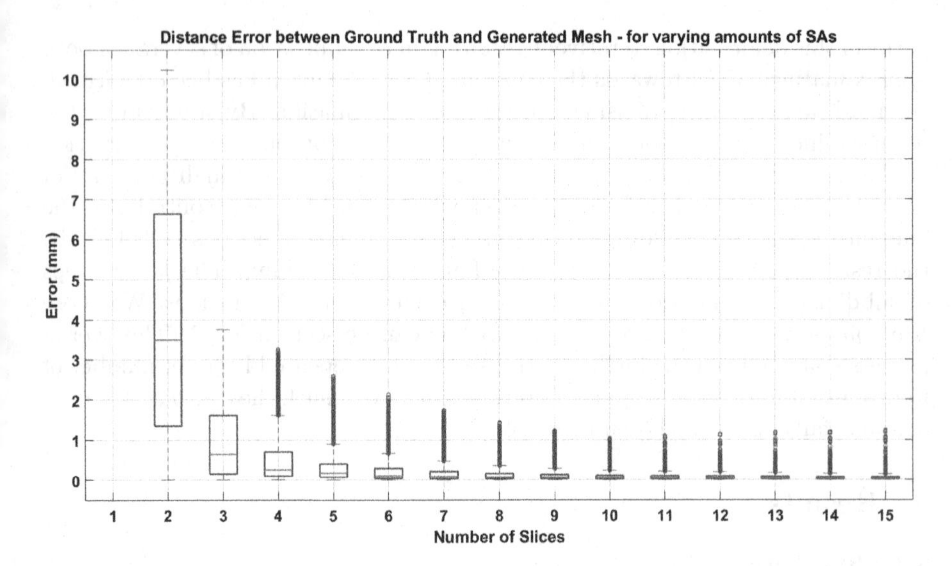

Fig. 3. Impact of increasing the amount of SAX has on the error distance between the generated mesh and the ground truth mesh.

Short Axis and Long Axis Contours. In a second experiment, we assessed the impact providing LAX contour information had. Similarly to the first experiment, we looked at varying the amount of LAX's and the effect it had on the overall mesh result, as well as look at the influence LAX positions had. Four short axes were chosen as the amount of SAX slices, which was chosen empirically due to the previous experiment, and an increasing amount of long axis contours were provided as the initial input by adding a rotated version of the initial LAX contour every iteration. Likewise, the same analysis was performed, but only keeping two LAX contours every time and only changing their position.

Non-parallel, Non-coincidental Contours. As this method is intended to work with clinical data, we simulated a typical set of cardiac contours that is usually obtained. From the mean shape model, 10 SAX contours (resulting in 10 mm inter-slice distance) as well as 2 LAX contours (simulating 2 and 4 chambers view) were generated. This is the equivalent of a typical clinical segmentation set. The contours were then perturbed simulating a breathing misalignment by applying small random translations and rotations (in plane as well as out of plane). The resulting non-parallel, non coincidental contours where given as an input to our algorithm. The resulting mesh can be seen in Fig. 5(a). The distance map overlayed on the mesh is the distance from the resulting mesh to the input contours. The algorithm is also tested on a right ventricle, as seen in Fig. 5(b).

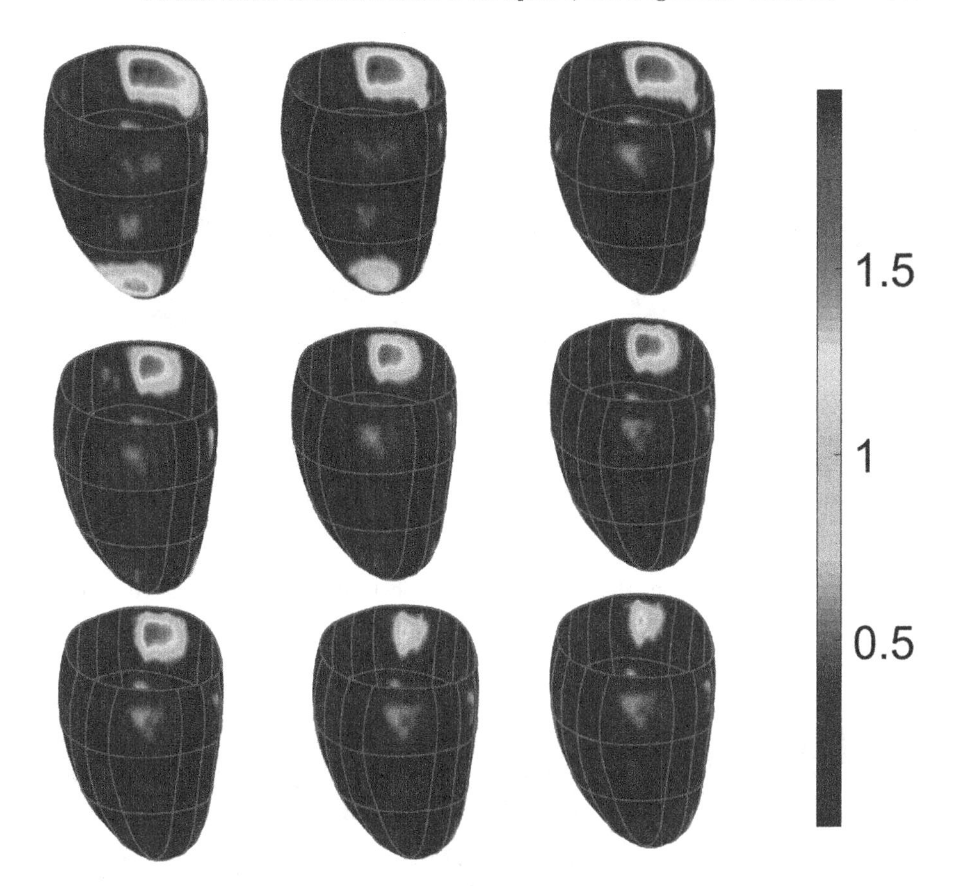

Fig. 4. Effect of increasing the amount of LAX contours have on the error distance between the generated mesh and the ground truth mesh.

3.2 Clinical Data

Our reconstruction was applied to real clinical data to show how it responds to both normal and pathological cases. As we use the provided expert delineated contours to build the mesh, there is no ground truth to assess our method on. However we performed quantitative analysis on the clinical datasets by measuring the discrepancy between individual contours and the resulting mesh. This measurement provides an evaluation of the "goodness of fit" that the contour has with regards to the mesh. This can be seen in Fig. 8, as the blue box-plots (labelled OM). To further evaluate the individual impact each of the contours had on the resulting mesh, we calculated the discrepancy between each of the contours to the mesh, having removed that contour prior to building the mesh. In doing so we assess the ability of our method to interpolate missing information from the dataset. This assessment can be seen in Fig. 8, as the red box-plots (labelled CM). It should be noted that the bottom figure contains more boxplots as our HCM dataset contained more contours (13).

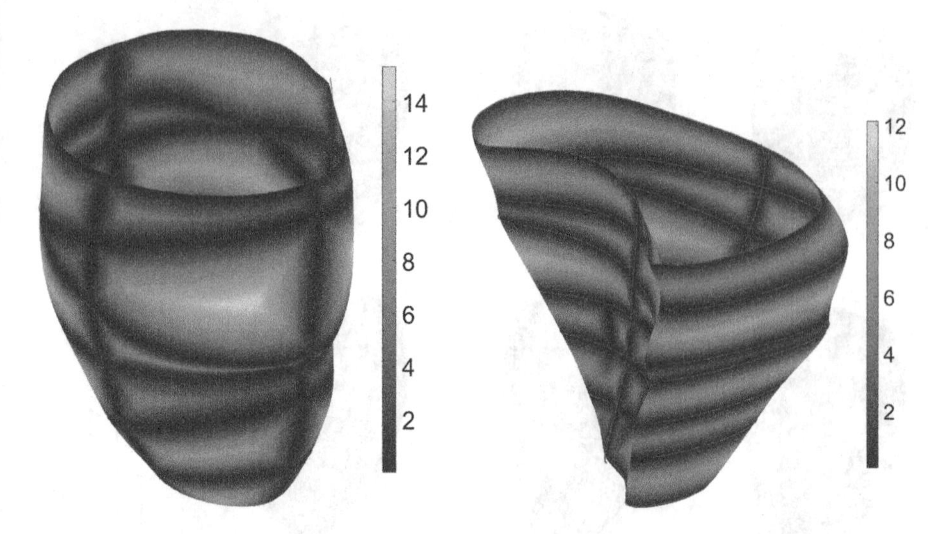

Fig. 5. Distance map from the resulting geometrical mesh to the simulated cardiac input data. The contours were transformed by small out of plane rotations and translations.

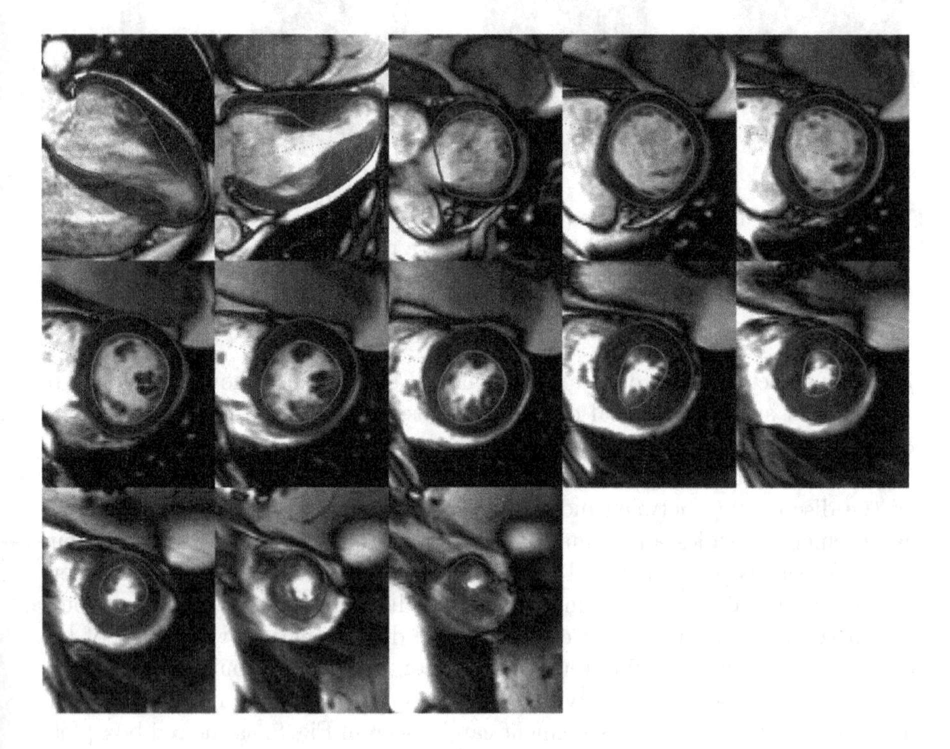

Fig. 6. Different clinical contours, delineated by an expert, on a severe case of HCM.

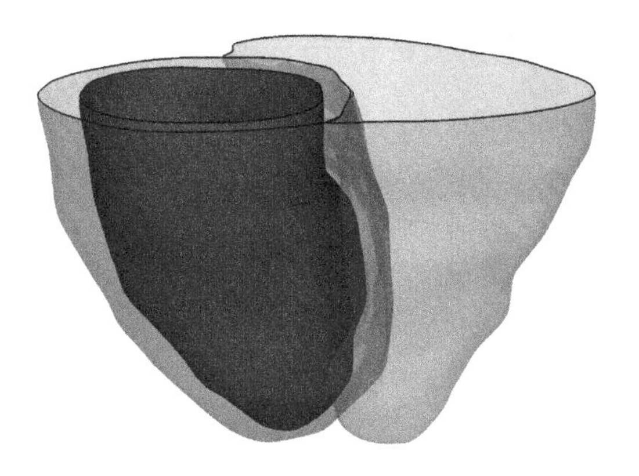
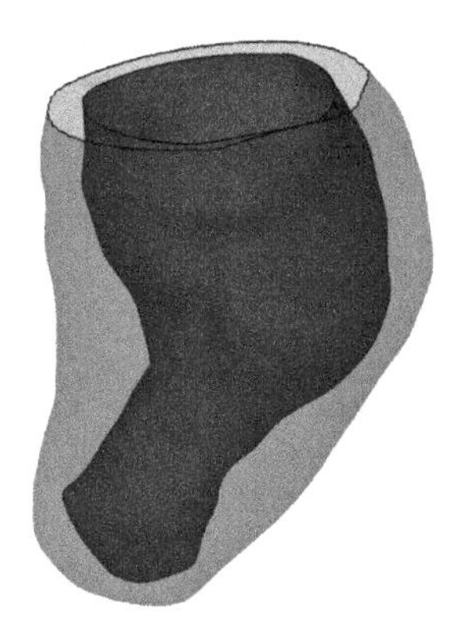

Fig. 7. Resulting surface mesh including the left Endocardial and Epicardial ventricle as well as the right Epicardial ventricle obtained from (a) contours for a normal case, and (b) contours belonging to a severe HCM patient.

4 Discussion

To be able to assess our process, several experiments where run, ranging from synthetic data using a mean statistical shape model, to severe pathological data. The method was applied to a stack of synthesised parallel contours and the result can be seen in Fig. 2. In order to capture the discrepancy between the ground truth and the resulting mesh, vertex-to-mesh distances were calculated. It can be seen in Fig. 2(c) that the maximum discrepancy error is around 4.5 mm. This error is the distance between the two meshes most apical areas. As we do not have information at the apex due to the absence of LAX, it is expected that this is where the discrepancy would be highest. The other distances however are minimal and appear no greater than 2 mm. To be able to assess the impact the amount of SAX contours have on the process, the distance between meshes was calculated by adding an extra short axis to the input data, in an iterative manner, starting with 2 SAX until a stack of 15. The resulting box plot showing the distances can be seen in Fig. 3. Figure 3 shows that there is a considerable reduction of errors up until 8 slices, after which the decrease in the distance error is not as notable. In this experiment, 8–9 slices represented an average clinical dataset as a SAX stack is generally separated by 10 mm in between. The outliers make up the error distances at the base of the mesh, were no information was provided. The impact that long axes have on the provided data was then assessed. A stack of 4 short axis slices were selected at equidistant positions from base to apex, and two long axes were placed at various different positions. Results showed that the position of the LAX contours had no major contribution to minimising the error distance. We then looked at increasing the number of LAX contours and evaluated the errors, which can be seen in Fig. 4. It can be seen that as the amount of LAX increases, the error decreases, as well as the area with no information. We then looked at generating non parallel, non coincidental contours to simulate clinical data, both with the left and right ventricle. It can

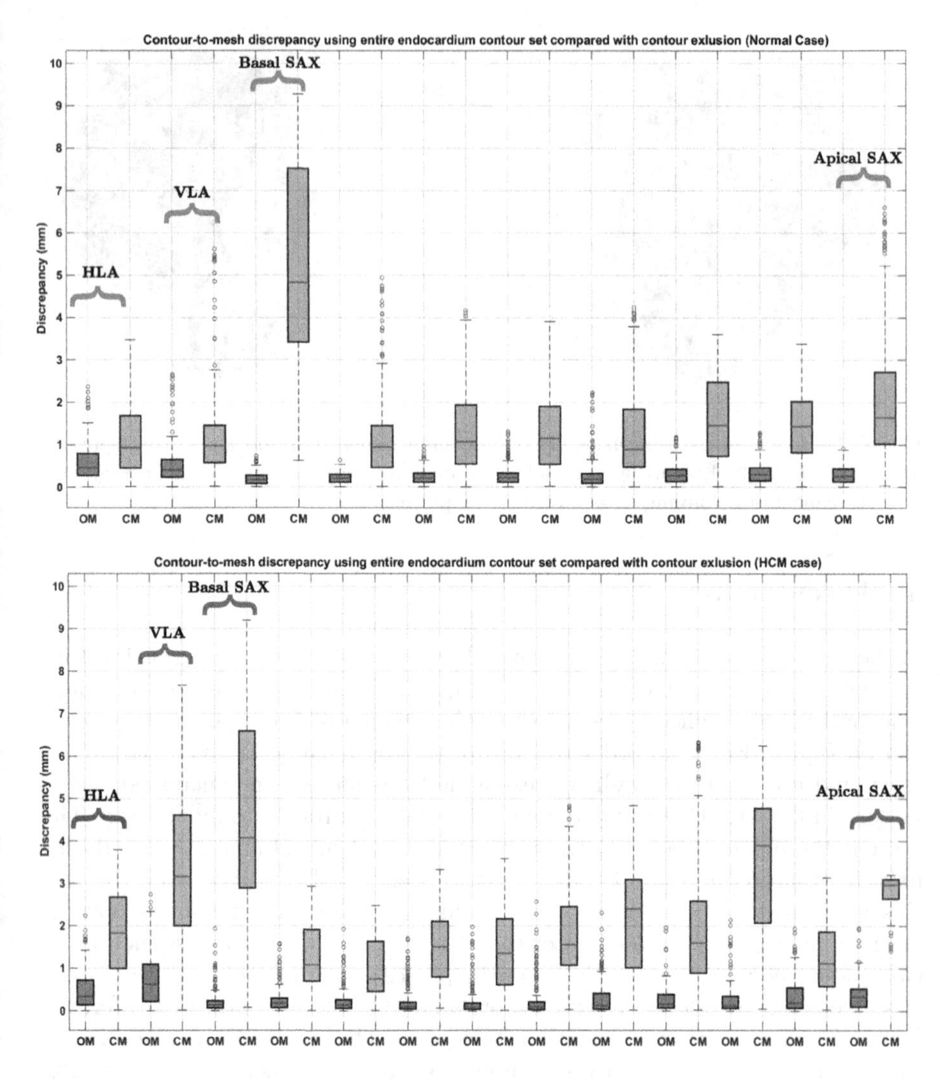

Fig. 8. Contour-to-mesh distance having built the mesh with the entire set of contours (labelled OM, in blue) and excluded contour-to-mesh distance having built the mesh without the excluded contour (labelled CM in red). Top and bottom figures represents the endocardial contours for the normal case, and the HCM case, respectively (Color figure online)

be seen in Fig. 5 that where the contours are, the discrepancy is negligent, and the biggest distance occurs in the areas where there is no information. Despite having non parallel, non coincidental contours, the resulting meshes are visually pleasing. Finally clinical data was used, for both a normal and pathological case. The clinical contours for the pathological case (patient with HCM) can be seen in Fig. 6. The resulting meshes can be observed in Fig. 7. To assess our method's ability to deal with clinical datasets, we measured the individual contour-to-mesh discrepancies, which can be seen in Fig. 8. It can be seen that when all the contours are used (blue box-plots), the discrepancy is within 1 mm for the normal case (Fig. 7(a)), and 2 mm for the HCM case (Fig. 7(b)). As we proceed to iteratively remove a single contour, in a *leave-one-out* fashion, and rebuild the mesh, the distance for that excluded contour to the mesh increases. It can be seen in both the top and bottom figures, that when the most basal SAX is removed, there is a high discrepancy occurring when measuring the distance from that contour to the resulting mesh. As this is the most basal slice, this is expected as the basal slice contains important structural information about the global geometry of the heart. When the basal slice is removed, our process relies on the LAX to fill in the basal information, which can lead to an early closure of the basal part of the mesh, which might be at a lower geometrical position then the contour, leading to high discrepancy. The removal of contours allows us to assess the individual contributions of the contours to the geometrical result of the mesh. As such, it can be noted that in the case of HCM (Fig. 7(b)), at the positions where the thickening is greatest, when the contours containing valuable information are removed, the discrepancy will be high, which is also the case for the apex. However the discrepancy still falls within 4 mm, except for the basal slice, for the normal case and 7 mm for the HCM case.

5 Conclusion

We presented a tool for surface reconstruction applicable to sparse, non parallel, non coincidental, heterogeneous cardiac contours. Our technique works on the left ventricle as well as on the right ventricle. We have empirically shown by means of the exposition of a variety of results, that our technique results in reconstructed interpolating surfaces with very good fitting to the input contours. In the areas to be filled by the interpolated surfaces, our technique presents very pleasant visual appearance. Although the absence of ground truth data precludes fair validations, our technique shows very good agreement against synthetic data.

Acknowledgments. BV acknowledges the support of the RCUK Digital Economy Programme grant number EP/G036861/1 (Oxford Centre for Doctoral Training in Healthcare Innovation). VC was supported by ERACoSysMed through a grant to the project SysAFib - Systems medicine for diagnosis and stratification of atrial fibrillation. RA is supported by a British Heart Foundation Clinical Research Training Fellowship. VG is supported by a BBSRC grant (BB/I012117/1), an EPSRC grant (EP/J013250/1) and by BHF New Horizon Grant NH/13/30238. EZ acknowledges the Marie Sklodowska-Curie Individual Fellowship from the H2020 EU Framework Programme for Research and Innovation [Proposal No: 655020-DTI4micro-MSCA-IF-EF-ST].

References

1. Arevalo, H.J., Vadakkumpadan, F., Guallar, E., Jebb, A., Malamas, P., Wu, K.C., Trayanova, N.A.: Arrhythmia risk stratification of patients after myocardial infarction using personalized heart models. Nature Commun. **7** (2016)
2. Zou, M., Holloway, M., Carr, N., Ju, T.: Topology-constrained surface reconstruction from cross-sections. ACM Trans. Graph. (TOG) **34**, 128 (2015). Proceedings of ACM SIGGRAPH 2015
3. Sunderland, K., Boyeong, W., Pinter, C., Fichtinger, G.: Reconstruction of surfaces from planar contours through contour interpolations. In: Image-Guided Procedures, Robotic Interventions, and Modeling, SPIE Proceedings, vol. 9415 (2015)
4. Deng, D., Zhang, J., Xia, L.: Three-dimensional mesh generation for human heart model. Life Syst. Model. Intell. Comput., Commun. Comput. Inf. Sci. **98**, 157–162 (2010)
5. Young, A.A., Cowan, B.R., Thrupp, S.F., Hedley, W.J., DellItalia, L.J.: Left ventricular mass and volume: fast calculation with guide-point modelling on MR images. Radiology **2**(216), 597–602 (2000)
6. Lamata, P., Niederer, S., Nordsletten, D., Barber, D.C., Roy, I., Hose, D.R., Smith, N.: An accurate, fast and robust method to generate patient-specific cubic Hermite meshes. Med. Image Anal. **15**(6), 801–813 (2011)
7. Albà, X., Pereañez, M., Hoogendoorn, C., Swift, A.J., Wild, J.M., Frangi, A.F., Lekadir, K.: An algorithm for the segmentation of highly abnormal hearts using a generic statistical shape model. IEEE Trans. Med. Imaging **35**(3), 845–859 (2016)
8. Lorensen, W.E., Cline, H.E.: Marching cubes: a high resolution 3D surface construction algorithm. In: ACM SIGGRAPH Computer Graphics, vol. 21, pp. 163–169. ACM (1987)
9. Villard, B., Zacur, E., Dall'Armellina, E., Grau, V.: Correction of slice misalignment in multi-breath-hold cardiac MRI scans. In: Mansi, T., McLeod, K., Pop, M., Rhode, K., Sermesant, M., Young, A. (eds.) STACOM 2016. LNCS, vol. 10124, pp. 30–38. Springer, Cham (2017). doi:10.1007/978-3-319-52718-5_4
10. Lopez-Perez, A., Sebastian, R., Ferrero, J.M.: Three-dimensional cardiac computational modelling: methods, features and applications. Biomed. Eng. Online **14**, 35 (2015)
11. Zhang, Z., Konno, K., Tokuyama, K.: 3D terrain reconstruction based on contours. In: 9th International Conference on Computer Aided Design and Computer Graphics, pp. 325–330 (2005)
12. Wang, Z., Geng, N., Zhang, Z.: Surface mesh reconstruction based on contours. In: International Conference on Computational Intelligence and Software Engineering, pp. 325–330 (2009)
13. Liu, L., Bajaj, C., Deasy, J.O., Ju, T.: Surface reconstruction from non-parallel curve networks. In: Eurographics, vol. 27 (2008)
14. Shi, W., Minas, C., Keenan, N.G., Diamond, T., Durighel, G., Montana, G., Rueckert, D., Cook, S.A., O'Regan, D.P., de Marvao, A., Dawes, T.: Population-based studies of myocardial hypertrophy: high resolution cardiovascular magnetic resonance atlases improve statistical power. J. Cardiovasc. Magn. Reson. **16**, 16 (2015)
15. Bai, W., Shi, A.W., Dawes de Marvao, T.J.W., O'Regan, D.P., Cook, S.A., Rueckert, D.: A bi-ventricular cardiac atlas built from 1000+ high resolution MR images of healthy subjects and an analysis of shape and motion. Med. Image Anal. **26**(1), 133–145 (2015)

16. Moenning, C., Neil, A.D.: Fast marching farthest point sampling for implicit surfaces and point clouds (2003)
17. Rohr, K., Stiehl, H.S., Sprengel, R., Buzug, T.M., Weese, J., Kuhn, M.H.: Landmark-based elastic registration using approximating thin-plate splines. IEEE Trans. Med. Imaging **20**(6), 526–534 (2001)
18. Sprengel, R., Rohr, K., Stiehl, H.S.: Thin-plate spline approximation for image registration. In: Engineering in Medicine and Biology Society, vol. 18 (1996)
19. Garland, M., Heckbert, P.S.: Surface simplification using quadric error metrics. In: Proceedings of 24th Annual Conference on Computer Graphics and Interactive Techniques, pp. 209–216. ACM Press/Addison-Wesley Publishing Co. (1997)

Classification of Cross-sections for Vascular Skeleton Extraction Using Convolutional Neural Networks

Kristína Lidayová[1](✉), Anindya Gupta[2], Hans Frimmel[3], Ida-Maria Sintorn[1], Ewert Bengtsson[1], and Örjan Smedby[4]

[1] Department of IT, Centre for Image Analysis, Uppsala University, Uppsala, Sweden
kristina.lidayova@it.uu.se
[2] T. J. Seebeck Department of Electronics, Tallinn University of Technology, Tallinn, Estonia
[3] Division of Scientific Computing, Department of IT, Uppsala University, Uppsala, Sweden
[4] School of Technology and Health, KTH Royal Institute of Technology, Stockholm, Sweden

Abstract. Recent advances in Computed Tomography Angiography provide high-resolution 3D images of the vessels. However, there is an inevitable requisite for automated and fast methods to process the increased amount of generated data. In this work, we propose a fast method for vascular skeleton extraction which can be combined with a segmentation algorithm to accelerate the vessel delineation. The algorithm detects central voxels - nodes - of potential vessel regions in the orthogonal CT slices and uses a convolutional neural network (CNN) to identify the true vessel nodes. The nodes are gradually linked together to generate an approximate vascular skeleton. The CNN classifier yields a precision of 0.81 and recall of 0.83 for the medium size vessels and produces a qualitatively evaluated enhanced representation of vascular skeletons.

Keywords: Vascular skeleton · CT angiography · Convolutional neural networks · Classification

1 Introduction

Vascular diseases are among the leading causes of death around the world. To diagnose a vascular disease, a detailed description of the state of each major artery in the arterial tree is needed. Such a description can be obtained by non-invasive vascular imaging techniques. The evolutionary success of the Computed Tomography Angiography (CTA), in terms of resolution quality, has benefited the clinicians with enhanced image details but at a cost of huge amount of data. Processing of such amount of data is a monotonous, error-prone and time consuming task, which certainly affects the efficiency of the clinicians. Hence,

M. Valdés Hernández and V. González-Castro (Eds.): MIUA 2017, CCIS 723, pp. 182–194, 2017.
DOI: 10.1007/978-3-319-60964-5_16

automation of the vessels segmentation in CTA is highly desirable to facilitate a quick and accurate diagnosis.

The tubular shape of the blood vessels offers a great possibility to develop a simple and fast method for vessel segmentation. For instance, the vascular skeleton can first be extracted and used as an initialization step for the vessel segmentation in subsequent stages. A method for fast vascular skeleton extraction was presented in our previous work [1]. In that method, a set of knowledge-based filters is applied to the central voxels of potential vessels to distinguish between voxels located inside and outside of the vessel. Voxels that passed through the filters are then connected to generate an approximate vascular skeleton. However, the set of knowledge-based filters also introduces a large number of false positive (FP) nodes. The FPs are further eliminated in the final step of the algorithm - the anatomy-based analysis, which removes most of the spurious branches by examining the shape of their connection.

In this paper, we propose an alternative method for vascular skeleton extraction, where we replace the knowledge-based filters with an efficient convolutional neural networks (CNN) classifier. We evaluate the performance of the CNN classifier using the CTA of the lower limbs in order to compare it with the results obtained by the knowledge-based filters from our previous work. A visual, qualitative, comparison of the resulting skeletons obtained by the two versions of the algorithm is also included.

2 Related Work

Due to the large impact of vascular diseases on public health, many scientists are dedicated to research regarding vascular segmentation or vascular centerline extraction. A detailed overview of other vascular segmentation techniques was presented in papers [2,3]. Here, we briefly review the work that is most relevant for our approach.

Charbonnier et al. [4] recently proposed a method which uses a CNN classifier to improve an airway tree segmentation. In this approach, an initial airway segmentation was provided to classify short airway branch segments into airway or leakage. Each airway candidate was represented by a set of three 2D cross-sectional patches, i.e., the beginning, middle and end of the segment. This set of patches was used as an input for a CNN classifier. Utilizing the CNN classifier significantly improved the quality of a given leaky airway segmentation.

Another method, proposed by Merkow et al. [5] utilized a 3D-CNN to predict the location of the boundary in volumetric data. They demonstrated the performance of their method for the detection of the vascular boundary, but, their approach is not limited to this application. CNN for vessel detection in volumetric images was recently utilized by Gülsün et al. [6]. In their work the blood vessel centerlines were first automatically extracted and then a 1D CNN classifier was used for removing extraneous paths from the detected centerlines. Our proposed method, in comparison, uses the 2D CNN classifier for cross-sectional classification.

3 Dataset

In this work, we utilized 25 CTA volumes of the lower limbs, taken from the clinical routine, to train and validate our proposed method. Initially, four volumes were given to an experienced radiologist for ground-truth labeling. The remaining 21 CTA volumes were kept for independent validation of the method. The radiologist utilized a semi-automatic segmentation tool based on the active contour method, provided by ITK-SNAP [7], to perform the vascular segmentation. The four labeled volumes were used for the detection of the initial nodes and in total provided a set of 352,523 nodes of multi-size vessels and non-vessels.

4 Proposed Method

The proposed method is based on the observation that the vessels, being tubular structures, often appear on orthogonal CT slices as bright elliptical-like regions. The method detects voxels located in the middle of such 2D regions, referred to as vessel nodes, and extracts a 2D patch around each node. A CNN classifier is accommodated to classify the patches into two categories, i.e., vessel or non-vessel nodes. Vessel nodes are connected with straight lines, referred to as edges, which results in tree-graph structures. Given a 3D CTA scan of the lower limbs, the method returns one or more tree-graphs representing an approximate skeleton of the vasculature of the lower limbs.

The proposed method for vascular skeleton extraction is a modified version of our previous algorithm [1]. It comprises of four steps that are shown in Fig. 1. This work focuses on improving the nodes classification step. To do so, we replace the previously reported method based on a set of knowledge-based filters with a CNN classifier. Other steps remain the same.

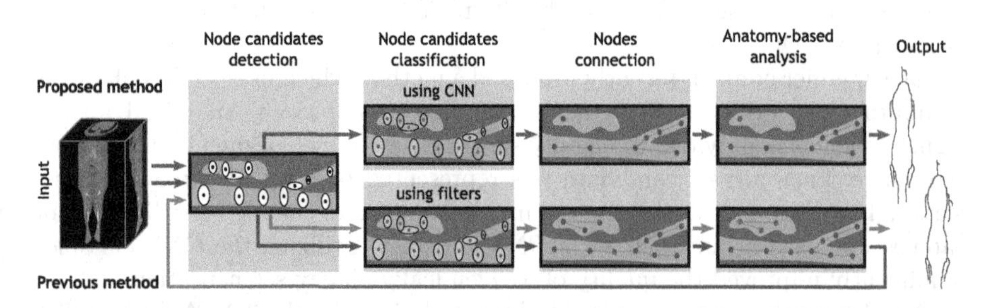

Fig. 1. The pipeline of the proposed method (green, top) produces the final vascular skeleton in one algorithm pass compared to the pipeline of the previous method [1] (blue, bottom) which detect skeletons of larger arteries in the first iteration and adds the skeletons of smaller arteries in the second iteration. (Color figure online)

4.1 Node Candidates Detection

A prerequisite for detecting the bright elliptical regions with the vessel node candidates is the knowledge of the intensity range that corresponds to blood in the input CTA volume. Blood becomes visible on the CTA scan due to the injected contrast medium, which increases its intensity above the intensity of the surrounding muscle tissue. Due to variations in hemodynamics and timing between injection and image acquisition, the blood intensity can differ between different patients and different scans. In [1], we proposed an algorithm based on fitting a sum of Gaussian curves to the image histogram. For each patient input volume, the fitted Gaussian curves automatically define the intensity ranges for three types of tissues: fat, muscle and blood. The intensity range $[\theta^b_{low}, \theta^b_{high}]$ of the blood in each volume is needed for the node candidates detection step and is defined by using this algorithm.

To detect the node candidates, we scan the input volume through all axis oriented planes (axial, sagittal, coronal). Any of the scanned voxels that has an intensity within the range of blood vessel intensities $[\theta^b_{low}, \theta^b_{high}]$ and its position is central within the area of similar intensities, is considered as a node candidate. The central position of the voxel is verified by casting four rays into four main directions starting from the voxel position outwards and confirming that the pair of opposite-pointing rays traversed the same distance until they reached three consecutive voxels with intensities outside of the $[\theta^b_{low}, \theta^b_{high}]$ range. Casting four rays is a sufficient and fast way to verify the central position of the potential nodes. However, not all detected areas are true vessel cross-sections. Due to the partial volume effect, a bone surface, noise, metallic implants or other imaging artifacts, may have intensities similar to blood. Therefore, the detected nodes need to be further classified as either vessel or non-vessel nodes.

4.2 Node Candidates Classification Using CNN

Patch Extraction. For each node candidate detected in the orthogonal slice, we calculated the biggest diameter and added 2 extra pixels around it to ensure the inclusion of the boundary information. We chose a patch size of 31×31 pixels as an input for the CNN classifier. The patch size of 31×31 pixels was chosen to cover sufficient contextual information of the candidates. At the same time, it provides a good trade-off between a detailed view of the smaller vessels and the possibility to include intermediate and large vessels. The patch pixel values were kept in Hounsfield units, in order not to lose the fine-grained details of the candidates. Some examples of extracted patches are shown in Fig. 2.

From all the initially detected candidates, we considered only those having a diameter between 6–27 pixels (ca. 4–20 mm) as a reference set for training the classifier. This resulted in a set of 138,302 potential candidates (24,625 vascular nodes and 113,677 non-vascular nodes). It is not meaningful to train a CNN model on candidates with a diameter smaller than 6 pixels due to insufficient spatial information, leading to inadequate training. Additionally, it is difficult to ensure that the ground truth segmentation of tiny vessels is absolutely correct as it is very tedious if at all possible, to detect all vessels of such small size in lower limb CTA. Candidates having a diameter bigger than 27 pixels were excluded from the reference training set since they were bigger than the chosen patch size (31×31 pixels; including the margin of 2 pixels around the candidate) and

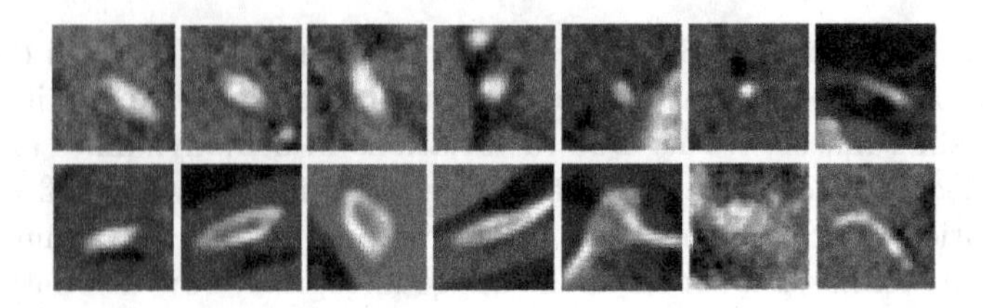

Fig. 2. Some examples of extracted patches of multi-size candidates. The first and the second row shows the patches of vessels and non-vessels, respectively, after intensity normalization.

hence needed to be resampled to this size. However, both, smaller and larger candidates than the reference candidates, were utilized for testing purposes.

Data Partitioning and Augmentation. We randomly split the reference set into two subsets. One subset was utilized for CNN model development (refer as the model-development subset) and the second one was used for its independent evaluation (refer as the model-evaluation subset). The model-development subset consists of 20,000 samples of each class, whereas the model-evaluation subset was kept imbalanced and contained 4,625 vessels and 93,677 non-vessels samples. The reason for such an imbalanced setting is to evaluate the trained model as per the real clinical scenario, where the frequency of false positives (FP) samples is much higher than the true positives (TP) samples. The model-development subset was again randomly split into training, validation, and testing subsets. Training and validation sets are utilized for cross-validation scheme whereas the test set is utilized for the final model selection. The training set contained 12,000 samples of each class (true vessel nodes and false vessel nodes). The validation and the test sets, both consisted of 4,000 samples of each class.

Generally, the vessels have large variability in terms of contextual surrounding, shape, size, and orientation. Lets assume that such variability can be modeled to a CNN by data-driven approaches. In such way, the classifier (CNN) can learn the orientation-invariant features. However, the number of vessel candidates are usually fewer than the number of non-vessel candidates, which can negatively affect the training of the classifier. We applied several transformations to generate a moderate number of new yet correlated training candidates. Each class of candidates was augmented using the image transformations: horizontal and vertical flipping, translation on x and y axes, and six random angular rotations (0–180°). The translation was limited in moving the candidate position 1 pixel from the center, in order to keep the candidate properly in the patch. This augmentation scheme resulted in 10 augmented variations for each candidate. In such way, classifier will learn the orientation-invariant features. This could be important because the candidates identified by the initial stage are not always centered at the local anatomical structures. Each subset (training and validation) is augmented separately to ensure their independency from each other.

False Positive Reduction: CNN Configuration. The false positive reduction stage is constructed by utilizing a CNN classifier. The architectural design of our CNN classifier is empirically determined by modifying the network itself. We modified several parameters (i.e. number of layers, kernel size, and types of pooling layer) in a structured way to obtain a better validation accuracy. Amongst all, we further analyzed the usability of two parameters, namely (1) pooling layer and (2) batch normalization (BN). To do so, we developed two models: (1) CNN model with pooling layers and (2) fully-connected convolutional network (FC-CNN) model without pooling layers.

First, we developed a CNN classifier consisting of four convolutional layers, with a max-pooling layer after every second convolution layer. The first convolutional layer consists of 32 kernels of size 3×3 and padded with a two pixels thick frame of zeros. This is done to keep the spatial sizes of the patches same after the first convolutional layer. The second, third and the last convolutional layers consist of 32, 64 and 64 kernels of size 3×3, respectively. The max-pooling layer reduces the size of feature maps by selecting the maximum feature response in overlapping or non-overlapping windows of size 2×2 (stride of 2). The illustration of the CNN model is shown in Fig. 3.

In the FC-CNN model, the pooling layers (maxpooling) were completely removed from the network which resulted in a network of only four convolutional layers. As reported in [12], the FC-CNN could result in an improved performance if the pooling layers are replaced with convolutional layers. In such setting, the network does not lose the spatial representation of the patch. However, such a network can be computationally expensive due to an increased number of network parameters. On the other hand, the feature-wise ordering of the pooling layers can lead to fast optimization, as well as further improve the translation invariance produced by the convolutional layers [12].

In both architectures, we also implemented the recently published Batch Normalization (BN) method [13], after the non-linear (activation function) layers of the network. It normalizes the activations of a feature map for each mini-batch at every optimization step and improves the overall network performance. For the activation function, we utilized the rectified linear units (ReLU) [14] after every convolutional and dense layer. In both networks, the last layer is followed by a dense layer consisting of 512 neurons, which is further connected to the Softmax layer for the final classification into vessels and non-vessels. A comparative evaluation of both models is reported in the result section.

Training. Before feeding the training data to the network, we normalize the intensity for each patch by subtracting the mean and dividing by the standard deviation. In such a way, the uneven distribution of intensities is scaled into a normalized intensity distribution and lead to a better convergence. The same procedure was applied during testing. The classifier was trained in a 5-fold cross-validation scheme. The candidates were randomly split into five blocks to ensure that each set was utilized as validation set once in each fold.

The RMSProp [8] is used to efficiently optimize the weights of the CNN. It normalizes the gradients by utilizing the magnitude of recent gradients. The

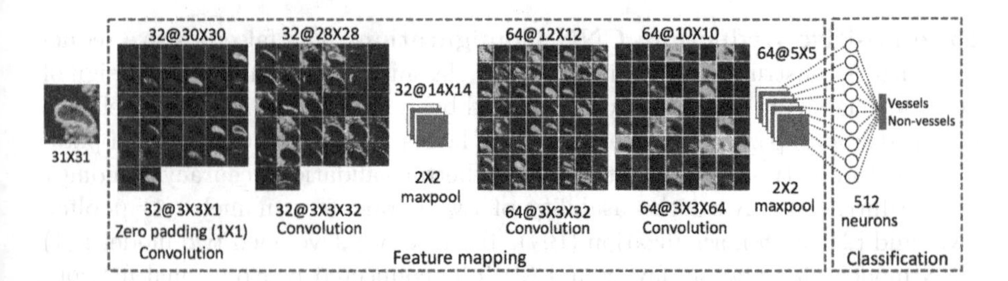

Fig. 3. An overview of the proposed CNN classifier, showing the output of each convolution filter applied to an example patch of a vessel. Here, the greyscale intensities are shown in color for suitable visualization.

weights are initialized as proposed in [9] and updated in a mini-batch scheme of 128 candidates at a rate of 0.001. The biases were initialized with zero. A dropout [10] of 0.25 is implemented on the output of each pooling layer and a dropout of 0.5 is implemented on the output of the dense layer. Softmax loss is utilized to predict the final output. The CNN model is implemented using Theano backend in Keras [11]. The training continued for 20 epochs with an average training time of 103 s/epoch on a GPU GeForce GTX 680.

4.3 Nodes Connection

The nodes that were classified by the CNN classifier as true vessel nodes are, in this step, linked together by using simple connection rules. First, each node is considered a separate graph. A link between two nodes is established, if the two nodes are close neighbors, all voxels on the line connecting these two nodes have an intensity within the range $[\theta^b_{low}, \theta^b_{high}]$ and these two nodes were not yet connected via other nodes. After all possible connections between the nodes have been created the preliminary vascular tree-graph structure is obtained which needs to be cleaned from possible spurious branches.

4.4 Anatomy-Based Analysis

The anatomy-based analysis step cleans the preliminary tree-graph structure from spurious non-vessel graphs or graph segments. The cleaning is based on the observation that vessel nodes are connected into straight or slightly corrugated branches whereas non-vessel nodes, often arising on the bone surface, are linked into unorganized and zigzag branches. Calculating the average angle between the line segments per branch allows distinguishing between true and false graph segments depending on the average angle being greater than 135° or not, respectively. The anatomy-based analysis step also closes small gaps between two segments and removes very short graphs (for details see [1]). Finally, it returns clean tree-graph structure that corresponds to the approximate vascular skeleton.

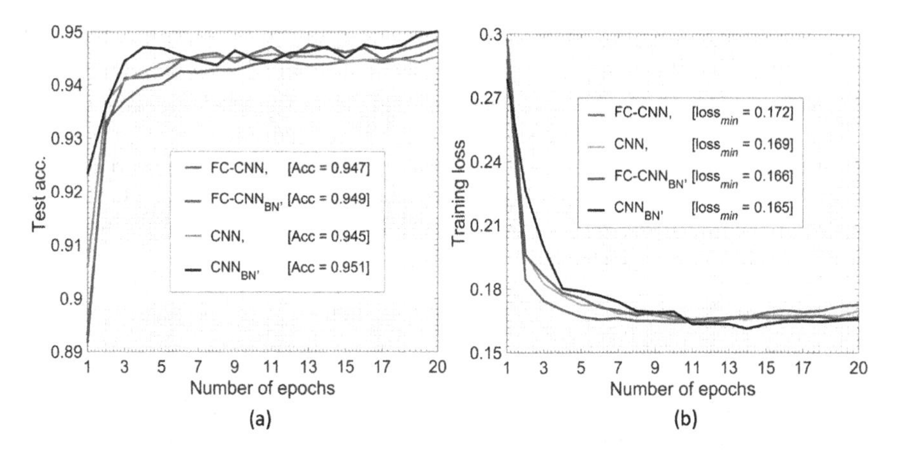

Fig. 4. Performance curves of different configurations at different number of epochs: (a) test accuracy, (b) training loss of both networks with and without batch normalization (BN) method.

5 Evaluation and Results

We compare the performance of two architectures, CNN and FC-CNN. Subsequently, the performance of the whole algorithm for vascular skeleton extraction, using the proposed method is qualitatively evaluated and compared to the previous skeleton extraction method [1].

5.1 CNN Versus FC-CNN

A comparative performance of both networks with and without BN is shown in the Fig. 4(a) and (b). The left figure shows the test accuracy of each configuration at different number of epochs on the test subset of the model development set (consisting 4000 samples of each class). It is noticeable that both networks with BN yield a better accuracy in comparison to the networks without BN. Interestingly, in the case of non-batch normalization configuration, the FC-CNN classifier also achieves a higher accuracy than the CNN classifier. Both networks trained with BN resulted in a better training loss. These results are in line with similar findings on the original BN work [13]. Comparatively, the CNN, trained with BN, resulted in better accuracy as the FC-CNN classifier (validated by paired t-test) with less number of parameters. The FC-CNN and CNN classifiers consist of total ca. 30.1 million and 1.2 million parameters, respectively. Therefore, we decided to utilized the CNN classifier for our application.

5.2 CNN and Knowledge-Based Filters Evaluation

Knowledge-Based Filters. The knowledge-based filters, proposed in [1] are simple filters derived from the characteristic appearance of vessels. They quickly remove the node candidates that do not fulfill the vessel characteristics. These filters examine: (1) if the artery lumen is homogeneously filled with blood, (2) if the inside of the artery is brighter than the outside, (3) if the vascular cross-section has regular elliptical or circular shape and (4) if the close artery neighborhood contain only intensities corresponding to fat, muscles or blood. The reason for the last filter is that a vessel can be adjacent to either fat, muscles or bones, but the partial volume effect will cause the surface of the bone to have decreased intensities that overlap with blood intensities.

Quantitative Evaluation. Quantitative evaluation was performed on the model-evaluation subset (4,625 vessels and 93,677 non-vessels) of patches not used in the model development process. The performance of the proposed CNN model and knowledge-based filters was evaluated on this subset in terms of *Precision*, *Recall*, and *F-score*. Additionally, for the CNN model, the *Area under the Precision-Recall curve (AUC)* is also presented in Fig. 5. The evaluation was performed separately for small vessels (<4 mm), medium-sized vessels (4–20 mm) and large vessels (>20 mm). This division is motivated by the fact that the CNN classifier was trained for the middle-sized vessels, however, it was used to classify small and large vessel candidates as well. Table 1 shows the resulting values for each evaluation measure per classifier and per candidate group.

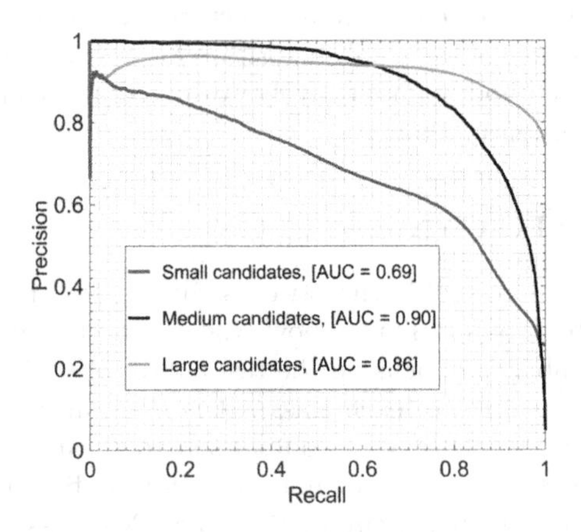

Fig. 5. Precision-Recall curves showing the performance of CNN classifier for the test subset of different sizes of candidates.

Qualitative Evaluation. Qualitative evaluation was performed by visual comparison of the resulting skeletons extracted from 21 CTA volumes of lower limbs by using two pipelines, the proposed algorithm pipeline and the pipeline presented in our previous work [1]. The schematic illustration of the pipelines in Fig. 1, shows that the previous filter-based algorithm needs to run in two iterations. The skeleton of large vessels, extracted in the first iteration serves as a basis for distinguishing between true and spurious graphs of small arteries detected in the second iteration. Small artery candidates are easier to mistake

Table 1. Comparative evaluation of CNN classifier and knowledge-based filters.

	Set	Prec.	Rec.	F-score	AUC
CNN	Small	0.66	0.71	0.65	0.69
	Medium	0.81	0.83	0.82	0.90
	Large	0.70	0.75	0.72	0.86
Filters	Small	0.29	0.78	0.42	–
	Medium	0.28	0.89	0.43	–
	Large	0.06	0.58	0.11	–

for noise or candidates detected on the bone surface and a useful indication about their true belongingness is the connection to the graph established in the first iteration. Since the proposed CNN classifier improves the false positive candidate rate, having two iterations is not relevant anymore and it is possible to simplify the pipeline while still obtaining better results.

It is crucial to determine a suitable threshold level for the decision boundary of the CNN classifier. We experimented with several threshold values ranging from 0.9 (corresponds to high precision) to 0.45 (corresponds to high recall). We observed that the skeleton, resulting from the higher thresholds, rarely contains the spurious branches and does not need the further anatomy-based analysis step. However, it also reduces the number of true vessel candidates and leads to fewer vessel branches. On the other hand, a lower threshold value detects a larger number of false positives and many spurious branches. This resembles the behavior of the simple knowledge-based filters. After this empirical analysis, we finally decided to select 0.65 as the suitable threshold level for our application.

From 21 resulting skeletons, we selected two representative results that demonstrate the merits and demerits of both methods. Figure 6 shows a comparison of the final skeletons along with the partial results obtained after each algorithm step by both algorithms. The results, after the second step, confirmed that the CNN classifier improves the FP rate and keeps a fewer number of candidates from which the majority are true vessel candidates. The final results of volume A (Fig. 6) show that the proposed method detects more vascular branches, and compared to the previous method does not contain spurious graph segments in the pelvis region. On the other hand, we also observed some cases where the CNN classifier discarded a larger amount of true vessel candidates, which led to missing a complete vessel branch. The final results of volume B depict an example of such a case. After an in-depth analysis of these vessels, we noticed that these vessels were either very small or they were very diseased. The CNN classifier was not trained for classifying such small vessels, which explains the lower performance in case of small vessels. In a case of diseased vessels, there exist many different variations between the appearance of diseased vessels depending on the type and the seriousness of the disease. Our dataset is taken from a clinical routine and contains large variations between the patient material. Therefore, using patches of four volumes in the training process was not sufficient enough to cover

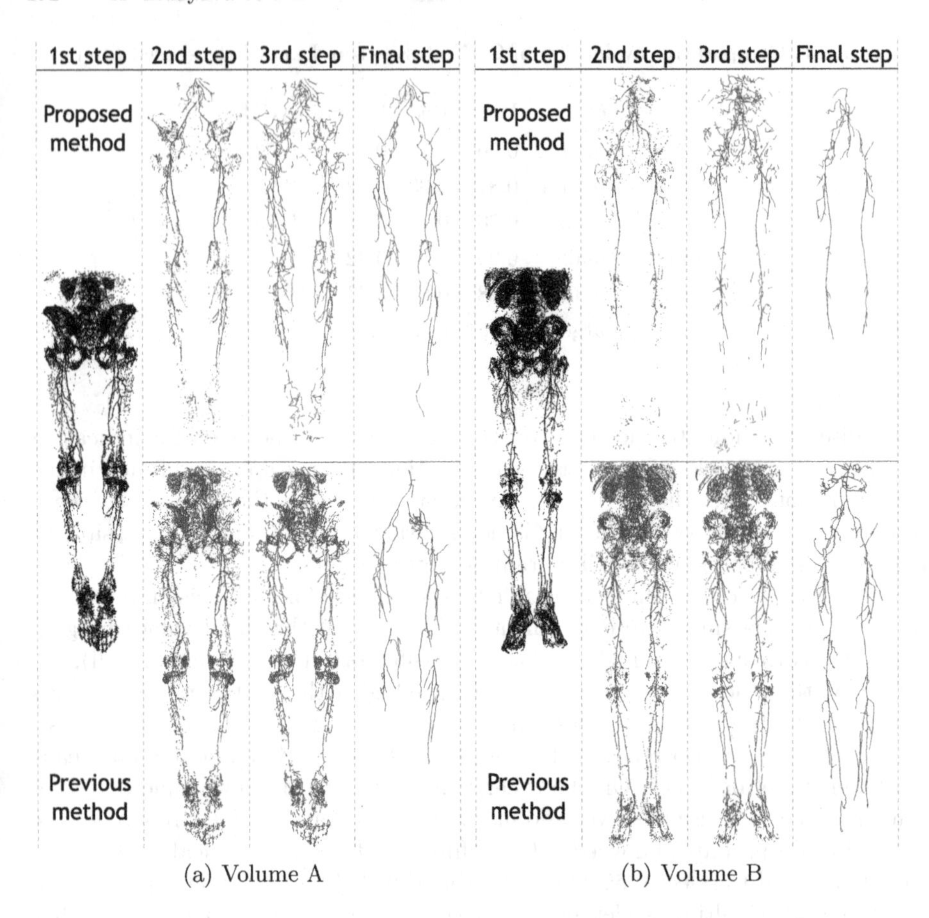

(a) Volume A

(b) Volume B

Fig. 6. Results after each algorithm step for 2 volumes; result after the 1st step is the same for both methods.

all the possible clinical variability. In the case of volume B, the knowledge-based filters were more inclusive in keeping true and false candidates, which resulted in a better skeleton compared to the proposed method.

Computation Time. The computation time needed to process a set of 130,000 patches which is approximately the average number of patches per volume in our dataset, was measured for both classifiers. The knowledge-based filters took ca. 22 s to process the patches, whereas the CNN classifier took ca. 30 s.

6 Discussion and Conclusions

Our main goal to reduce the false positive rate by using a CNN classifier was successfully fulfilled. Both the quantitative and qualitative evaluations support this. In comparison to our previous work, the detection of vessel nodes can now be processed in a single iteration, resulting in a simplified methodological pipeline.

Interestingly, in some cases, our previously proposed simple knowledge-based filters in combination with the anatomy-based analysis step, also performed well in comparison to the proposed method. One reason for the occasional lower performance of the proposed method could be the high variance in the diseased arterial vessels of each patient. To improve on this, the network could be remodeled with a bigger training set of vessels with a wider variation of arterial diseases. To do so, a reliable and consistent labeling of ground truth segmentation is desired, which is a tedious and difficult task; especially for the small and tiny vessels.

In order to further improve the performance, multiple patches from the different axes oriented planes can be utilized to remodeled the CNN model. Alternatively, the 3D clusters of the candidates can be extracted to train a 3D CNN model for classification.

Acknowledgements. Lidayová, Frimmel, Bengtsson, and Smedby have been supported by the Swedish Research Council (VR), grant no. 621-2014-6153. Gupta has been supported by Skype IT Academy Stipend Program, EU institutional grant IUT19-11 of Estonian Research Council.

References

1. Lidayová, K., Frimmel, H., Wang, C., Bengtsson, E., Smedby, Ö.: Fast vascular skeleton extraction algorithm. Pattern Recogn. Lett. **76**, 67–75 (2016)
2. Kirbas, C., Quek, F.: A review of vessel extraction techniques and algorithms. ACM Comput. Surv. (CSUR) **36**(2), 81–121 (2004)
3. Lesage, D., Angelini, E.D., Bloch, I., Funka-Lea, G.: A review of 3D vessel lumen segmentation techniques: models, features and extraction schemes. Med. Image Anal. **13**(6), 819–845 (2009)
4. Charbonnier, J.P., van Rikxoort, E.M., Setio, A.A., Schaefer-Prokop, C.M., van Ginneken, B., Ciompi, F.: Improving airway segmentation in computed tomography using leak detection with convolutional networks. Med. Image Anal. **36**, 52–60 (2017)
5. Merkow, J., Marsden, A., Kriegman, D., Tu, Z.: Dense volume-to-volume vascular boundary detection. In: Ourselin, S., Joskowicz, L., Sabuncu, M.R., Unal, G., Wells, W. (eds.) MICCAI 2016. LNCS, vol. 9902, pp. 371–379. Springer, Cham (2016). doi:10.1007/978-3-319-46726-9_43
6. Gülsün, M.A., Funka-Lea, G., Sharma, P., Rapaka, S., Zheng, Y.: Coronary centerline extraction via optimal flow paths and CNN path pruning. In: Ourselin, S., Joskowicz, L., Sabuncu, M.R., Unal, G., Wells, W. (eds.) MICCAI 2016. LNCS, vol. 9902, pp. 317–325. Springer, Cham (2016). doi:10.1007/978-3-319-46726-9_37
7. Yushkevich, P.A., Piven, J., Hazlett, H.C., Smith, R.G., Ho, S., Gee, J.C., Gerig, G.: User-guided 3D active contour segmentation of anatomical structures: significantly improved efficiency and reliability. Neuroimage **31**(3), 1116–1128 (2006)
8. Tieleman, T., Hinton, G.: Lecture 6.5-RmsProp: divide the gradient by a running average of its recent magnitude. In: COURSERA: Neural Networks for ML (2012)
9. Glorot, X., Bengio, Y.: Understanding the difficulty of training deep feed-forward neural networks. In: AISTATS, vol. 9, pp. 249–256 (2010)

10. Srivastava, N., Hinton, G.E., Krizhevsky, A., Sutskever, I., Salakhutdinov, R.: Dropout: a simple way to prevent neural networks from overfitting. J. Mach. Learn. Res. **15**(1), 1929–1958 (2014)
11. Chollet, F.: Keras (2015). https://github.com/fchollet/keras.
12. Springenberg, J.T., Dosovitskiy, A., Brox, T., Riedmiller, M.: Striving for simplicity: the all convolutional net. In: Proceedings of 3rd International Conference on Learning Representations (ICLR) (2015)
13. Ioffe, S., Szegedy, C.: Batch normalization: accelerating deep network training by reducing internal covariate shift (2015). arXiv preprint arXiv:1502.03167
14. Nair, V., Hinton, G.E.: Rectified linear units improve restricted Boltzmann machines. In: 27th International Conference on Machine Learning, pp. 807–814 (2010)

Segmenting Atrial Fibrosis from Late Gadolinium-Enhanced Cardiac MRI by Deep-Learned Features with Stacked Sparse Auto-Encoders

Guang Yang[1,2(✉)], Xiahai Zhuang[3(✉)], Habib Khan[1,2], Shouvik Haldar[1], Eva Nyktari[1], Xujiong Ye[4], Greg Slabaugh[5], Tom Wong[1], Raad Mohiaddin[1,2], Jennifer Keegan[1,2], and David Firmin[1,2]

[1] Cardiovascular Biomedical Research Unit, Royal Brompton Hospital, London SW3 6NP, UK
g.yang@imperial.ac.uk
[2] National Heart and Lung Institute, Imperial College London, London SW7 2AZ, UK
[3] School of Data Science, Fudan University, Shanghai 201203, China
zxh@fudan.edu.cn
[4] School of Computer Science, University of Lincoln, Lincoln LN6 7TS, UK
[5] Department of Computer Science, City University London, London EC1V 0HB, UK

Abstract. The late gadolinium-enhanced (LGE) MRI technique is a well-validated method for fibrosis detection in the myocardium. With this technique, the altered wash-in and wash-out contrast agent kinetics in fibrotic and healthy myocardium results in scar tissue being seen with high or enhanced signal relative to normal tissue which is 'nulled'. Recently, great progress on LGE MRI has resulted in improved visualization of fibrosis in the left atrium (LA). This provides valuable information for treatment planning, image-based procedure guidance and clinical management in patients with atrial fibrillation (AF). Nevertheless, precise and objective atrial fibrosis segmentation (AFS) is required for accurate assessment of AF patients using LGE MRI. This is a very challenging task, not only because of the limited quality and resolution of the LGE MRI images acquired in AF but also due to the thinner wall and unpredictable morphology of the LA. Accurate and reliable segmentation of the anatomical structure of the LA myocardium is a prerequisite for accurate AFS. Most current studies rely on manual segmentation of the anatomical structures, which is very labor-intensive and subject to inter- and intra-observer variability. The subsequent AFS is normally based on unsupervised learning methods, e.g., using thresholding, histogram analysis, clustering and graph-cut based approaches, which have variable accuracy. In this study, we present a fully-automated multi-atlas propagation based whole heart segmentation method to derive the anatomical structure of the LA myocardium and pulmonary veins. This is followed by a supervised deep learning method for AFS. Twenty clinical LGE MRI scans from longstanding persistent AF patients were entered into this study retrospectively. We have demonstrated that our fully automatic method can achieve accurate and reliable AFS compared to manual delineated ground truth.

M. Valdés Hernández and V. González-Castro (Eds.): MIUA 2017, CCIS 723, pp. 195–206, 2017.
DOI: 10.1007/978-3-319-60964-5_17

1 Introduction

Atrial fibrillation (AF) is the most commonly observed cardiac arrhythmia that occurs in up to 2% of the general population with increased prevalence in the aged population [1]. AF can cause substantial morbidity and mortality; for example, it is associated with a five-fold incidence of stroke, three-fold risk of congestive heart failure and doubles the possibility of dementia that has a major worldwide public health impact [2]. Currently, the pathophysiology of AF is not fully understood; however, previous studies on both animal and human experimental models have shown that multiple disease pathways, e.g., structural, contractile, or electrical alterations, can promote abnormal electrical impulse formation and propagation [3]. Fibrosis in the left atrium (LA) is the hallmark of atrial structural remodeling, and is one of the major risk factors for AF progression [4, 5]. Moreover, studies have shown that ectopic beats from the pulmonary veins (PVs) can frequently trigger the AF [6]. Based on these findings, minimally invasive radio-frequency catheter ablation (CA) using the pulmonary vein antrum isolation method has been developed as a front-line therapy for symptomatic AF patients refractory to drug treatment [7], but suffers a >30% recurrence rate [8].

In order to understand AF and facilitate better management and prognosis, techniques have been developed to evaluate the LA wall composition and assess the circumferential PVs scarring that results from CA. Currently, the electro-anatomical mapping (EAM) system is used as a clinical reference standard technique for the assessment of extent and distribution of native atrial fibrosis and post-ablation scar. This is normally performed during an electrophysiological procedure and has the major drawback of invasiveness and suboptimal accuracy, which has been reported as being up to 10 mm in the localization of atrial fibrosis [9, 10]. Moreover, there are the potentially hazardous effects of ionizing radiation for the patients using EAM.

In contrast, the late gadolinium-enhanced (LGE) MRI, which is noninvasive and without ionizing radiation, allows the detection and quantification of native fibrosis and post-ablation scars by highlighting the slow washout kinetics of the gadolinium in these tissues [11–15]. Firstly, the extent and distribution of native fibrosis identified in the pre-ablation baseline LGE MRI scan has emerged as the strongest independent predictor of AF recurrence after the first ablation [13]. Secondly, LGE MRI can be used as a powerful tool to detect ablation-induced fibrosis formed by radiofrequency energy delivered in the atrial myocardium [11]. This has a potential role in recognizing ablation line gaps, which are the main reason of ablation failure [11, 16]. In addition, LGE MRI can be also useful to guide the ablation procedure [11, 17].

Despite the excellent results of using LGE MRI in the assessment of ventricular fibrosis and its promising potential in the detection of atrial scars, there are still challenges to be addressed when applying LGE MRI for AF patients in clinical practice: (1) frequently the image quality of pre-ablation LGE MRI scans is poor due to residual respiratory motion, heart rate variability, low signal-to-noise ratio (SNR), and gadolinium wash-out during the long acquisition (current scanning time is around 6–10 min per patient); (2) the resolution of LGE MRI images is limited compared to the thinness (about 3 mm) of the LA wall; (3) the various morphologies of the LA wall and PVs anatomy; and (4) confounded enhancement from surrounding heart substructures, e.g., blood, aorta, spine, and esophagus. These can result in poor delineation of the LA myocardium and cause a large number of false positives in the atrial fibrosis delineation.

Essentially, there are two main steps required to analyze fibrotic tissues from LGE MRI images: (1) segmentation of the anatomical structure of the LA and PVs and (2) the atrial fibrosis segmentation (AFS).

For the segmentation of the anatomical structure of the LA and PVs, most previous studies have relied on manual delineation [7, 13, 18, 19], which potentially suffers from large inter- and intra-observer variability and is also very time-consuming. Semi-automatic and automatic methods have been proposed to solve this task, e.g., using thresholding with region growing [20], statistical shape model [21] and atlas propagation [22] based approaches. However, these methods required further operator's manual intervention [20, 21] or used un-gated first-pass MR angiography (MRA) data [22], which may cause difficulties in co-registration with the respiratory and cardiac gated LGE MRI data.

For the AFS, to the best of our knowledge, most studies applied unsupervised learning based methods, e.g., using histogram analysis [13], k-means clustering [18] and graph-cut [21] based approaches. In addition, maximum intensity projection (MIP) can provide intuitive visualization of the atrial fibrosis [11, 12, 20, 22]; however, this is only a visualization technique for hyper-enhancement regions, rather than a segmentation method that can result in volumetric quantification [21]. Recently, a grand challenge was carried out to benchmark different algorithms for solving AFS [19] including 8 submissions for the competition. These benchmarked algorithms were all unsupervised learning based methods [19]. The challenge included data acquired from multiple institutions, and the LA endocardium and cavity for each scan were provided to all the participants beforehand. Promising results were achieved for the best performing algorithms. However, there were large variances in the performances especially for the pre-ablation cases, which may be attributed to the fact that the image quality was generally worse and the native fibrosis is more diffuse. Therefore, the challenge of atrial fibrosis segmentation and assessment remains open. Moreover, the inaccurate AFS could be one of the major reasons for poor reproducibility of the correlation between atrial fibrosis identified by LGE MRI (enhanced regions) and EAM (low voltage regions) [23, 24].

In this study, we present a fully automatic framework for an efficient and objective atrial fibrosis assessment using: (1) a fully-automated multi-atlas based whole heart segmentation (MA-WHS) method to solve the LA and PVs anatomy and (2) a fully automatic supervised deep learning method for the AFS. Compared with the ground truth formed by manual delineation, our fully automatic framework obtains promising segmentation results, which are comparable to other state-of-the-art methods.

2 Method

2.1 Cardiac MRI Data Acquisition

Cardiac MRI acquisitions were performed on a Siemens Magnetom Avanto 1.5T scanner. Transverse navigator-gated 3D LGE MRI [11, 13, 25] was performed using an inversion prepared segmented gradient echo sequence (TE/TR 2.2 ms/5.2 ms) 15 min after gadolinium (Gd) administration when a transient steady-state of Gd wash-in and wash-out of normal myocardium had been reached [26]. LGE MRI images were scanned with a field-of-view 380 × 380 mm^2 and reconstructed to 60–68 slices at 0.75 × 0.75 × 2 mm^3.

In the LGE MRI images, healthy myocardium is 'nulled' and only fibrotic tissues are seen with high signal; therefore, it is hard to extract the anatomical structure of the LA and PVs directly from LGE MRI images. Instead of using un-gated MRA as previous studies described, in our study, a respiratory and cardiac gated 3D Roadmap image, which is acquired using a balanced steady state free precession sequence (TE/TR 1 ms/2.3 ms), has been scanned for each patient to derive the anatomical structure of the LA and PVs. Our roadmap data were acquired with a field-of-view 380 × 380 mm^2 and reconstructed to 160 slices at 0.8 × 0.8 × 1.6 mm^3.

Both 3D LGE MRI and Roadmap data were acquired during free-breathing using a crossed-pairs navigator positioned over the dome of the right hemi-diaphragm with a navigator acceptance window size of 5 mm and continuously adaptive windowing strategy based respiratory motion control [27].

2.2 Patients

Cardiac MRI was performed in longstanding persistent AF patients between 2011–2013 in agreement with the local regional ethics committee. A Likert-type scale was applied to score the image quality of each LGE MRI scan, e.g., 0 (non-diagnostic), 1 (poor), 2 (fair), 3 (good) and 4 (very good) depending on the level of SNR, appropriate inversion time, and the existence of navigator beam and ghost artifacts.

Ten pre-ablation scans with image quality ≥ 2 have been retrospectively entered into this study ($\sim$60% of all the scanned pre-ablation cases). To make a balanced dataset, we randomly selected 10 post-ablation cases from all the 26 post-ablation scans with image quality ≥ 2 ($\sim$92% of all the scanned post-ablation cases).

2.3 Anatomical Structure Delineation for the LA and PVs

In this study, we applied a MA-WHS method to segment the anatomical structure of the LA and PVs [28]. This has been done on the Roadmap images and then mapped to LGE MRI. Our segmentation consists of two major steps: (1) atlas propagation based on image registration algorithms and (2) label fusion from multi-atlas propagated segmentation results as described in [28].

The whole heart atlases were constructed using 30 MRI Roadmap studies retrieved from the Left Atrium Segmentation Grand Challenge organized by King's College London [29]. For each atlas dataset, we have manual labels of all the heart substructures including the right and left ventricles, the right and left atria, the aorta, the pulmonary artery, the pulmonary veins and the appendages. MA-WHS executes the atlas-to-target registration for each atlas dataset $(\boldsymbol{A}_a, \boldsymbol{L}_a | a = 1, \ldots N)$, where $\boldsymbol{A}_a$ and $\boldsymbol{L}_a$ are the intensity image and the corresponding segmentation label image of the a-th atlas ($N = 30$). Then a set of warped atlases can be derived $\{(A_a, L_a)|a = 1, \ldots N\}$ for label fusion, which is achieved using local weighted and multi-scale based label fusion (MSP-LF),

$$L_1(x) = \underset{l \in \{l_{bk}, l_{la}\}}{\arg\max} \sum_a w_a(S(I, A_a, x))\delta(L_a(x), l) \tag{1}$$

where l_{bk} and l_{la} are the labels of the background and LA and PVs, respectively, and the local weight $w_a(\cdot) \propto S(\cdot)$ is determined by the local similarity $S(\cdot)$ between the target image and the atlas. $\delta(p, q)$ is the Kronecker delta function which returns 1 when $p = q$ and returns 0 otherwise.

In MSP-LF, the local similarity of patches using a multi-scale strategy is computed as follows,

$$S_{\text{msp}}(I, A_a, x) = \sum_s S\left(I^{(s)}, A_a^{(s)}, x\right), \tag{2}$$

in which $I^{(s)} = I * \text{Gaussian}(0, \sigma_s)$ is the target image from s scale-space that is computed from the convolution of the target image with Gaussian kernel function with scale s. Here, we computed the local similarity in multi-scale images using the conditional probability of the images, that is

$$S\left(I^{(s)}, A_a^{(s)}, x\right) = p(i_x|j_x) = \frac{p(i_x, j_x)}{p(j_x)} \tag{3}$$

where $i_x = I^{(s)}(x)$ and $j_x = A_a^{(s)}(x)$ and the conditional image probability is obtained from the joint and marginal image probability which can be calculated using the Parzen window estimation [30].

For each patient case, the Roadmap dataset was then registered to the LGE MRI dataset using the DICOM header data, and then refined by affine and nonrigid registration steps. The resulting transformation was applied to the MA-WHS derived cardiac anatomy to segment the anatomical structure of the LA and PVs on the LGE MRI dataset.

In this study, we validated the anatomical structure segmentation against established ground truth from manual segmentations by experienced expert cardiologists specialized in cardiac MRI. We evaluated the segmentation accuracy using Dice score [31], Hausdorff distance (HD) [32] and Average Surface Distance (ASD) [31].

2.4 Training Datasets and Ground Truth Construction for AFS

The fixed-size window or regular pixel-grid is a traditional way to select patches from larger images prior to feature extraction [33], and has been widely used in deep convolutional neural networks based classification. In this study, we used a Simple Linear Iterative Clustering (SLIC) [34] based method to partition pixels into meaningful 'atomic' regions, i.e., super-pixels (SPs), based on certain similarity metrics (Fig. 1(c)). In addition, SLIC based over-segmentation can be more consistent with the object boundaries in the image and is able to avoid producing outputs with zigzag boundaries that can be normally observed using relatively large sliding windows [33].

In order to create training datasets for our further supervised learning based AFS, we asked experienced cardiologists specialized in cardiac MRI to perform manual mouse clicks on the LGE MRI images to label the enhanced atrial fibrotic regions (Fig. 1(b)). These manual mouse clicks were done on the original LGE MRI images without the SPs grid overlaid, as this would reduce the visibility of the enhancement on LGE MRI images. The coordinates of the mouse clicks were used to select the enhanced SPs. Only one mouse click is taken into account if multiple clicks dwell in the same super-pixel.

As aforementioned, the anatomical structure of the LA and PVs was segmented using a MA-WHS method. Then a morphological dilation was applied (assuming the LA wall thickness is 3 mm) to extract the LA wall and PVs. The blood pool regions were derived by another morphological erosion (5 mm) from the endocardial LA boundary and the pixel intensities throughout the image were normalized according to the blood pool intensities [19]. We masked the selected enhanced SPs using the LA wall and PVs segmentation. Only the SPs having a defined overlap ($\geq 20\%$) with the LA wall and PVs segmentation were selected as enhancement (these enhanced SPs were then labeled as 1). Other enhanced SPs were discarded as they were considered to be enhancement from other confounded tissues. As seen in (Fig. 1(e)), the other SPs overlapped with the LA wall and PVs but not selected as enhancement were considered as unenhanced (these unenhanced SPs were then labeled as 0).

Once we extracted the enhanced SPs, they were combined to create a binary image, i.e., 1 for enhanced SPs and 0 for unenhanced. The binary image was overlaid on the original LGE MRI images and our cardiologists performed manual corrections to create the final boundaries (ground truths) for the enhanced atrial fibrosis.

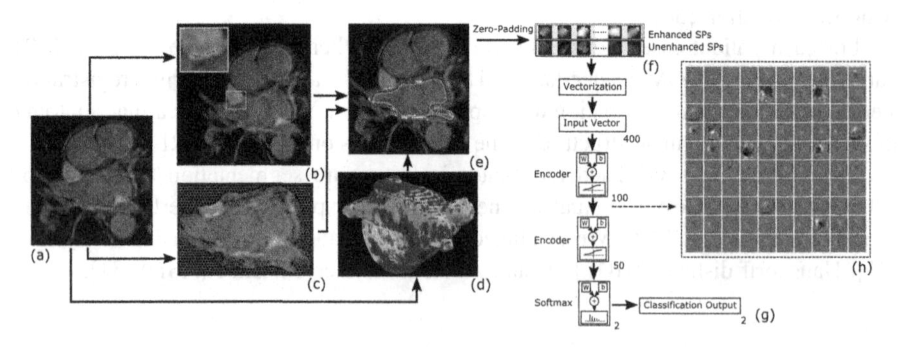

Fig. 1. (a) Original LGE MRI image of a post-ablation example case; (b) Manual mouse clicks performed by cardiologists to identify enhanced fibrotic tissues; (c) SLIC over-segmentation of the original LGE MRI image (only a ROI of the LA region shown); (d) 3D rendering of the MA-WHS results with 3D MIP based visualization of the enhanced fibrotic tissues; (e) Labeled enhanced (yellow) and unenhanced (blue) SPs that will be used to train our classification model; (f) Zero-padded SP patches (yellow box: enhanced SPs and blue box: unenhanced SPs); (g) SSAE based classification; (h) Visualization of the weights derived from the first auto-encoder. (Color figure online)

2.5 Deep Learning via Stacked Sparse Auto-Encoders for AFS

After we obtained the over-segmented SPs, the Stacked Sparse Auto-Encoders (SSAE) [35] were initially pre-trained in an unsupervised manner without using the labels of the SPs. An auto-encoder neural network tries to learn an approximation to the identity function to replicate its input at its output using a back-propagation algorithm, that is $\hat{\mathrm{X}} = h_{\mathrm{W,b}}(\mathrm{X}) \approx \mathrm{X}$, in which $\mathrm{X} = \{x_1, x_2, \ldots x_m\}, \mathrm{X} \in \Re^{n \times m}$ is a matrix storing all the input training vectors $x_i \in \Re^n$. Each input vector x_i was formed by: (1) zero-padding all the SPs into a 20×20 matrix, which is the smallest bounding box for the largest super-pixel dimensions (Fig. 1(f)), and (2) vectorizing the 20×20 matrix into a 400×1 vector. The cost function of this pre-training can be written as

$$\arg\min_{\mathrm{W}^l} J_a(\mathrm{W}^l) = \frac{1}{2m}\sum_{i=1}^{m} \|\hat{x}_i - x_i\|_2^2 + \frac{\lambda}{2}\|\mathrm{W}^l\|_2^2 + \beta\sum_{j=1}^{k} \mathrm{KL}(\rho \parallel \hat{\rho}_j) \tag{4}$$

where m is the number of input training vectors, k is the number of hidden nodes, λ is the coefficient for the L_2 regularization term, β is the weight of sparsity penalty, KL is the Kullback-Leibler divergence function $\mathrm{KL}(\rho \parallel \hat{\rho}) = \rho \log \frac{\rho}{\hat{\rho}_j} + (1-\rho)\log\frac{1-\rho}{1-\hat{\rho}_j}$, ρ is sparsity parameter that specifies the desired level of sparsity, $\hat{\rho}_j$ is probability of firing activity that is $\hat{\rho}_j = \frac{1}{m}\sum_{i=1}^{m} h_j(x_i)$. The unsupervised pre-training is performed one layer at a time by minimizing the error in reconstructing its input and learning an encoder and a decoder, which yields an optimal set of weights W and biases b stored in W^l. If the number of hidden nodes k is less than the number of visible input nodes n, then the network is forced to learn a compressed and sparse representation of the input [35] (Fig. 1(g) and (h)).

Second, a Softmax layer was added as the activity classification model $h_\theta(x_i)$ to accomplish the SPs classification task [35]. In addition, it can be jointly trained with the SSAE during fine-tuning of the parameters with labeled instances in a *supervised fashion*. The weight matrix θ is obtained by solving the convex optimization problem as following.

$$\arg\min_{\theta} J_s(\theta) = -\frac{1}{m}\sum_{i=1}^{m}\sum_{c=1}^{C} \mathbf{1}\{y_i = c\} \times \log P(y_i = c | x_i; \theta) + \frac{\lambda}{2}\|\theta\|_2^2 \tag{5}$$

where $c \in \{1, C = 2\}$ is the class label, $\tilde{\mathrm{X}} = \{(x_1, y_1), (x_2, y_2), \ldots, (x_m, y_m),\}$ represents a set of labeled training instances, and the last term for the L_2 regularization.

Finally, fine-tuning was applied to boost the classification performance, and it treats all layers of the SSAE and the Softmax layer as a single model and improves all the weights of all layers in the network by using the backpropagation technique [35].

Hyper-parameters were not optimized explicitly but were determined via trial and error. Here are the final defined hyper-parameters (values) used in this study: maximum epochs of the SSAE (200), maximum epochs of the Softmax and fine-tuning (500), hidden layers size of the SSAE (100 and 50), sparsity parameter ρ (0.1), sparsity penalty β (5), L_2 regularization term λ for the SSAE and the Softmax (0.0001).

For evaluation of the AFS, we used leave-one-patient-out cross-validation (LOO CV) and reported the cross-validated accuracy, sensitivity, specificity, average area under the receiver operating characteristic (ROC) curve (AUC), and the Dice score.

3 Results and Discussion

In this study, we proposed a fully automatic framework for the AFS, which is based on the segmented anatomical structure of the LA and PVs using MA-WHS.

Figure 1(d) and (e) show the 3D MIP results overlaid on the segmented LA and PVs and LA wall and PVs boundaries overlaid on a LGE MRI slice respectively. The quantitative evaluations show that the MA-WHS based method achieved 0.90 ± 0.12 Dice score, 9.53 ± 6.01 mm HD and 1.47 ± 0.89 mm ASD.

For the SSAE based SPs classification, we obtained LOO CV accuracy of 0.91, sensitivity of 0.95, specificity of 0.75, AUC of 0.95, and the Dice score for the final AFS was found to be 0.82 ± 0.05.

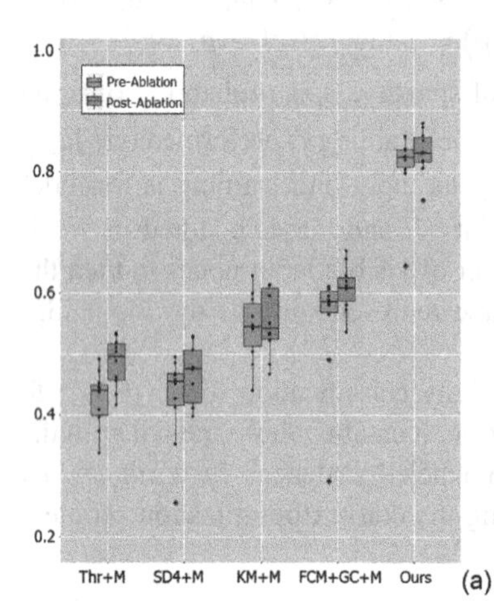

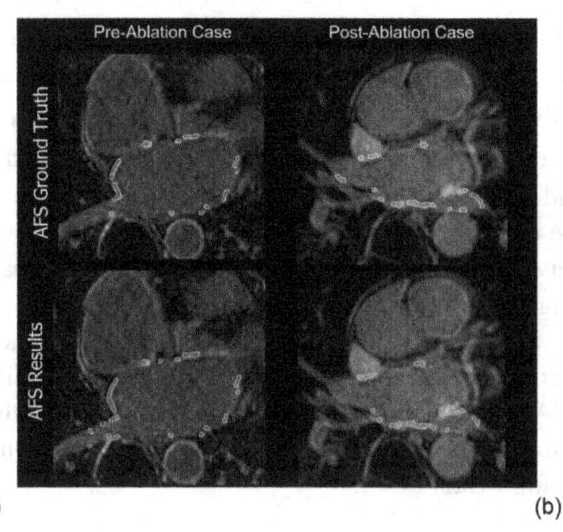

Fig. 2. (a) Boxplot for the comparison results of the Dice scores obtained by our fully automatic framework and other four methods (Thr, SD4, KM and FCM + GC) with manual delineated LA wall and PVs (+M); (b) Final AFS results (cyan regions) for an example pre-ablation (left) and an example post-ablation (right) case compared to the ground truth (yellow regions). (Color figure online)

In addition, for the AFS, we compared our fully automatic framework with existing semi-automatic methods with manually delineated anatomical structure of the LA and PVs. The four methods we compared in this study were described in the benchmarking work [19], namely simple thresholding (Thr), conventional standard deviation (4 SDs were tested, i.e., SD4), k-means clustering (KM) and fuzzy c-means clustering with graph-cuts (FCM + GC). Figure 2(a) shows that our fully automatic framework obtained more accurate and more consistent results across 20 AF patient cases (Fig. 2(a), red dots represent outliers), which can partly be attributed to the fact that our method is based on supervised learning. Of note is that, while in the benchmarking study these algorithms (i.e., Thr, SD4, KM and FCM + GC) were fine-tuned, in our comparison study, we have only implemented standard versions without performing further optimization. This is because details of the fine-tuning implemented for the benchmarking study are not available and in any case, that fine-tuning was done for datasets acquired in a different patient population to ours and may not be ideal. Despite this, similar performances were obtained between our implementation and those reported in [12] especially for the pre-ablation cases. For the post-ablation cases, our AFS results demonstrated similar results to the best-performing method reported in [12] but with smaller variance; however, our method has the advantage of being fully automatic. Of note is that multi-scanner and multi-institution datasets were used in the benchmarking work and this may have resulted in a large variance in the images which could affect the final AFS segmentation. One of our future studies will be applying our fully automatic framework on multi-scanners LGE datasets to validate its robustness.

Figure 2(b) demonstrates that qualitatively our fully automatic AFS is in accordance with the manual segmented atrial fibrosis. However, if there are enhancements from the nearby mitral valve or blood pool regions, our method may mis-classify them as enhanced atrial fibrosis that is the major contribution for the false positives.

Another possible limitation of our study is that the SSAE based classifier has many hyper-parameters, which need to be carefully tuned, e.g., maximum epochs of the SSAE, maximum epochs of the Softmax and fine-tuning, hidden layers size of the SSAE, sparsity parameter ρ, sparsity penalty β, L_2 regularization term λ for the SSAE and the Softmax. Currently these hyper-parameters were tuned via trial and error, which may limit the final classification accuracy.

4 Conclusion

In conclusion, we have developed and validated a fully automatic framework to segment atrial fibrosis from LGE MRI images that is based on accurate anatomical structure delineation via a MA-WHS algorithm. The evaluation has been done on 20 LGE MRI scans for longstanding persistent AF patients that contain both pre-ablation and post-ablation cases. Based on the results, we can envisage a straightforward deployment of our framework for clinical usage. As a future direction, we will develop a more robust parameter tuning method for the applications on multi-scanners datasets.

References

1. Feinberg, W.M., Blackshear, J.L., Laupacis, A., Kronmal, R., Hart, R.G.: Prevalence, age distribution, and gender of patients with atrial fibrillation. Arch. Intern. Med. **155**, 469–473 (1995)
2. January, C.T., Wann, L.S., Alpert, J.S., Calkins, H., Cigarroa, J.E., Cleveland, J.C., Conti, J. B., Ellinor, P.T., Ezekowitz, M.D., Field, M.E., Murray, K.T., Sacco, R.L., Stevenson, W. G., Tchou, P.J., Tracy, C.M., Yancy, C.W.: 2014 AHA/ACC/HRS guideline for the management of patients with atrial fibrillation: executive summary. J. Am. Coll. Cardiol. **64**, 2246–2280 (2014)

3. Pontecorboli, G., i Ventura, F., Rosa, M., Carlosena, A., Benito, E., Prat-Gonzales, S., Padeletti, L., Mont, L.: Use of delayed-enhancement magnetic resonance imaging for fibrosis detection in the atria: a review. Europace **19**(2), 180–189 (2016)
4. Allessie, M.: Electrical, contractile and structural remodeling during atrial fibrillation. Cardiovasc. Res. **54**, 230–246 (2002)
5. Boldt, A., Wetzel, U., Lauschke, J., Weigl, J., Gummert, J., Hindricks, G., Kottkamp, H., Dhein, S.: Fibrosis in left atrial tissue of patients with atrial fibrillation with and without underlying mitral valve disease. Heart **90**, 400–405 (2004)
6. Haïssaguerre, M., Jaïs, P., Shah, D.C., Takahashi, A., Hocini, M., Quiniou, G., Garrigue, S., Le Mouroux, A., Le Métayer, P., Clémenty, J.: Spontaneous initiation of atrial fibrillation by ectopic beats originating in the pulmonary veins. N. Engl. J. Med. **339**, 659–666 (1998)
7. Ravanelli, D., Dal Piaz, E.C., Centonze, M., Casagranda, G., Marini, M., Del Greco, M., Karim, R., Rhode, K., Valentini, A.: A novel skeleton based quantification and 3-D volumetric visualization of left atrium fibrosis using late gadolinium enhancement magnetic resonance imaging. IEEE Trans. Med. Imaging **33**, 566–576 (2014)
8. Fichtner, S., Sparn, K., Reents, T., Ammar, S., Semmler, V., Dillier, R., Buiatti, A., Kathan, S., Hessling, G., Deisenhofer, I.: Recurrence of paroxysmal atrial fibrillation after pulmonary vein isolation: is repeat pulmonary vein isolation enough? A prospective, randomized trial. Europace **17**, 1371–1375 (2015)
9. Schmidt, E.J., Mallozzi, R.P., Thiagalingam, A., Holmvang, G., D'Avila, A., Guhde, R., Darrow, R., Slavin, G.S., Fung, M.M., Dando, J., Foley, L., Dumoulin, C.L., Reddy, V.Y.: Electroanatomic mapping and radiofrequency ablation of porcine left atria and atrioventricular nodes using magnetic resonance catheter tracking. Circ. Arrhythm. Electrophysiol. **2**, 695–704 (2009)
10. Zhong, H., Lacomis, J.M., Schwartzman, D.: On the accuracy of CartoMerge for guiding posterior left atrial ablation in man. Heart Rhythm. **4**, 595–602 (2007)
11. Peters, D.C., Wylie, J.V., Hauser, T.H., Kissinger, K.V., Botnar, R.M., Essebag, V., Josephson, M.E., Manning, W.J.: Detection of pulmonary vein and left atrial scar after catheter ablation with three-dimensional navigator-gated delayed enhancement MR imaging: initial experience. Radiology **243**, 690–695 (2007)
12. McGann, C.J., Kholmovski, E.G., Oakes, R.S., Blauer, J.J.E., Daccarett, M., Segerson, N., Airey, K.J., Akoum, N., Fish, E., Badger, T.J., DiBella, E.V.R., Parker, D., MacLeod, R.S., Marrouche, N.F.: New magnetic resonance imaging-based method for defining the extent of left atrial wall injury after the ablation of atrial fibrillation. J. Am. Coll. Cardiol. **52**, 1263–1271 (2008)
13. Oakes, R.S., Badger, T.J., Kholmovski, E.G., Akoum, N., Burgon, N.S., Fish, E.N., Blauer, J.J.E., Rao, S.N., Dibella, E.V.R., Segerson, N.M., Daccarett, M., Windfelder, J., McGann, C.J., Parker, D., MacLeod, R.S., Marrouche, N.F.: Detection and quantification of left atrial structural remodeling with delayed-enhancement magnetic resonance imaging in patients with atrial fibrillation. Circulation **119**, 1758–1767 (2009)
14. Akkaya, M., Higuchi, K., Koopmann, M., Burgon, N., Erdogan, E., Damal, K., Kholmovski, E., McGann, C., Marrouche, N.F.: Relationship between left atrial tissue structural remodelling detected using late gadolinium enhancement MRI and left ventricular hypertrophy in patients with atrial fibrillation. Europace **15**, 1725–1732 (2013)
15. Bisbal, F., Guiu, E., Cabanas-Grandío, P., Berruezo, A., Prat-Gonzalez, S., Vidal, B., Garrido, C., Andreu, D., Fernandez-Armenta, J., Tolosana, J.M., Arbelo, E., De Caralt, T. M., Perea, R.J., Brugada, J., Mont, L.: CMR-guided approach to localize and ablate gaps in repeat AF ablation procedure. JACC Cardiovasc. Imaging **7**, 653–663 (2014)

16. Badger, T.J., Daccarett, M., Akoum, N.W., Adjei-Poku, Y.A., Burgon, N.S., Haslam, T.S., Kalvaitis, S., Kuppahally, S., Vergara, G., McMullen, L., Anderson, P.A., Kholmovski, E., MacLeod, R.S., Marrouche, N.F.: Evaluation of left atrial lesions after initial and repeat atrial fibrillation ablation; Lessons learned from delayed-enhancement MRI in repeat ablation procedures. Circ. Arrhythmia Electrophysiol. **3**, 249–259 (2010)
17. Arujuna, A., Karim, R., Zarinabad, N., Gill, J., Rhode, K., Schaeffter, T., Wright, M., Rinaldi, C.A., Cooklin, M., Razavi, R., O'Neill, M.D., Gill, J.S.: A randomized prospective mechanistic cardiac magnetic resonance study correlating catheter stability, late gadolinium enhancement and 3 year clinical outcomes in robotically assisted vs. standard catheter ablation. Europace **17**, 1241–1250 (2015)
18. Perry, D., Morris, A., Burgon, N., McGann, C., MacLeod, R., Cates, J.: Automatic classification of scar tissue in late gadolinium enhancement cardiac MRI for the assessment of left-atrial wall injury after radiofrequency ablation. In: van Ginneken, B., Novak, C.L. (eds.) SPIE Medical Imaging, p. 83151D (2012)
19. Karim, R., Housden, R.J., Balasubramaniam, M., Chen, Z., Perry, D., Uddin, A., Al-Beyatti, Y., Palkhi, E., Acheampong, P., Obom, S., Hennemuth, A., Lu, Y., Bai, W., Shi, W., Gao, Y., Peitgen, H.-O., Radau, P., Razavi, R., Tannenbaum, A., Rueckert, D., Cates, J., Schaeffter, T., Peters, D., MacLeod, R., Rhode, K.: Evaluation of current algorithms for segmentation of scar tissue from late gadolinium enhancement cardiovascular magnetic resonance of the left atrium: an open-access grand challenge. J. Cardiovasc. Magn. Reson. **15**, 105–122 (2013)
20. Knowles, B.R., Caulfield, D., Cooklin, M., Rinaldi, C.A., Gill, J., Bostock, J., Razavi, R., Schaeffter, T., Rhode, K.S.: 3-D visualization of acute RF ablation lesions using MRI for the simultaneous determination of the patterns of necrosis and edema. IEEE Trans. Biomed. Eng. **57**, 1467–1475 (2010)
21. Karim, R., Arujuna, A., Housden, R.J., Gill, J., Cliffe, H., Matharu, K., Rinaldi, C.A., O'Neill, M., Rueckert, D., Razavi, R., Schaeffter, T., Rhode, K.: A method to standardize quantification of left atrial scar from delayed-enhancement MR images. IEEE J. Transl. Eng. Heal. Med. **2**, 1–15 (2014)
22. Tao, Q., Ipek, E.G., Shahzad, R., Berendsen, F.F., Nazarian, S., van der Geest, R.J.: Fully automatic segmentation of left atrium and pulmonary veins in late gadolinium-enhanced MRI: towards objective atrial scar assessment. J. Magn. Reson. Imaging **44**, 346–354 (2016)
23. McGann, C., Akoum, N., Patel, A., Kholmovski, E., Revelo, P., Damal, K., Wilson, B., Cates, J., Harrison, A., Ranjan, R., Burgon, N.S., Greene, T., Kim, D., DiBella, E.V.R., Parker, D., MacLeod, R.S., Marrouche, N.F.: Atrial fibrillation ablation outcome is predicted by left atrial remodeling on MRI. Circ. Arrhythmia Electrophysiol. **7**, 23–30 (2014)
24. Harrison, J.L., Sohns, C., Linton, N.W., Karim, R., Williams, S.E., Rhode, K.S., Gill, J., Cooklin, M., Rinaldi, C.A., Wright, M., Schaeffter, T., Razavi, R.S., O'Neill, M.D.: Repeat left atrial catheter ablation: cardiac magnetic resonance prediction of endocardial voltage and gaps in ablation lesion sets. Circ. Arrhythmia Electrophysiol. **8**, 270–278 (2015)
25. Peters, D.C., Wylie, J.V., Hauser, T.H., Nezafat, R., Han, Y., Woo, J.J., Taclas, J., Kissinger, K.V., Goddu, B., Josephson, M.E., Manning, W.J.: Recurrence of atrial fibrillation correlates with the extent of post-procedural late gadolinium enhancement. A pilot study. JACC Cardiovasc. Imaging **2**, 308–316 (2009)
26. Keegan, J., Jhooti, P., Babu-Narayan, S.V., Drivas, P., Ernst, S., Firmin, D.N.: Improved respiratory efficiency of 3D late gadolinium enhancement imaging using the continuously adaptive windowing strategy (CLAWS). Magn. Reson. Med. **71**, 1064–1074 (2014)
27. Keegan, J., Drivas, P., Firmin, D.N.: Navigator artifact reduction in three-dimensional late gadolinium enhancement imaging of the atria. Magn. Reson. Med. **785**, 779–785 (2013)

28. Zhuang, X., Shen, J.: Multi-scale patch and multi-modality atlases for whole heart segmentation of MRI. Med. Image Anal. **31**, 77–87 (2016)
29. Tobon-Gomez, C., Geers, A., Peters, J., Weese, J., Pinto, K., Karim, R., Schaeffter, T., Razavi, R., Rhode, K.: Benchmark for algorithms segmenting the left atrium from 3D CT and MRI datasets. IEEE Trans. Med. Imaging **34**, 1460–1473 (2015)
30. Thévenaz, P., Unser, M.: Optimization of mutual information for multiresolution image registration. IEEE Trans. Image Process. **9**, 2083–2099 (2000)
31. Zhuang, X.: Challenges and methodologies of fully automatic whole heart segmentation: a review. J. Healthc. Eng. **4**, 371–408 (2013)
32. Huttenlocher, D.P., Klanderman, G.A., Rucklidge, W.J.: Comparing images using the Hausdorff distance. IEEE Trans. Pattern Anal. Mach. Intell. **15**, 850–863 (1993)
33. Xu, J., Luo, X., Wang, G., Gilmore, H., Madabhushi, A.: A deep convolutional neural network for segmenting and classifying epithelial and stromal regions in histopathological images. Neurocomputing **191**, 214–223 (2016)
34. Achanta, R., Shaji, A., Smith, K., Lucchi, A.: SLIC superpixels compared to state-of-the-art superpixel methods. IEEE Trans. Pattern Anal. Mach. Intell. **34**, 2274–2281 (2012)
35. Hasan, M., Roy-Chowdhury, A.K.: A continuous learning framework for activity recognition using deep hybrid feature models. IEEE Trans. Multimed. **17**, 1909–1922 (2015)

Improved CTA Coronary Segmentation with a Volume-Specific Intensity Threshold

Muhammad Moazzam Jawaid[1](✉), Ronak Rajani[2], Panos Liatsis[3], Constantino Carlos Reyes-Aldasoro[1], and Greg Slabaugh[1]

[1] City, University of London, London, UK
muhammad.jawaid.2@city.ac.uk
[2] Guys and St. Thomas Hospital, London, UK
[3] The Petroleum Institute, Abu Dhabi, United Arab Emirates

Abstract. State-of-the-art CTA imaging equipment has increased clinician's ability to make non-invasive diagnoses of coronary heart disease; however, an effective interpretation of the cardiac CTA becomes cumbersome due to large amount of imaged data. Intensity based background suppression is often used to enhance the coronary vasculature but setting a fixed threshold to discriminate coronaries from fatty muscles could be misleading due to non-homogeneous response of contrast medium in CTA volumes. In this work, we propose a volume-specific model of the contrast medium in the coronary segmentation process to improve the segmentation accuracy. The influence of the contrast medium in a CTA volume was modelled by approximating the intensity histogram of the descending aorta with Gaussian approximation. It should be noted that a significant variation in Gaussian mean for 12 CTA volumes validates the need of volume-wise exclusive intensity threshold for accurate coronary segmentation. Moreover, the effectiveness of the adaptive intensity threshold is illustrated with the help of qualitative and quantitative results.

Keywords: Computed tomography angiography · Contrast medium · Curve evolution · Coronary segmentation

1 Introduction

Coronary heart disease (CHD) has become a major cause of death worldwide. According to recent statistics [13], CHD is responsible for approximately 73,000 deaths per year (an average of one death every seven minutes). Consequently, clinicians are interested in early detection of CHD to effectively predict and control future cardiac events. The limitations of conventional cardiac-angiography based diagnosis have driven intensive research for non-invasive diagnosis leading to highly sophisticated imaging procedures. The clinical use of computed tomography angiography (CTA) is a prominent example of non-invasive diagnosis, in which blood filled vasculature can be easily discriminated from the background based on high intensity. However, the high volume of imaging data demands

M. Valdés Hernández and V. González-Castro (Eds.): MIUA 2017, CCIS 723, pp. 207–218, 2017.
DOI: 10.1007/978-3-319-60964-5_18

automatic segmentation of coronary vasculature as manual diagnosis becomes cumbersome and prone to inter-observer errors.

Apart from simple threshold and clustering techniques, the sophisticated algorithms employ partial differential equations (PDEs) to detect object boundaries, i.e. an initial guess is evolved under constraints to detect the object boundaries. Commonly used formulations include the *parametric snake model* and the *level set representation*. The parametric snake i.e. active contour model [9] leads to a fast and computationally efficient segmentation but shows greater sensitivity to the topological changes, whereas the level set representation [1,14] provides inherent split and merge mechanisms to accurately detect complex structures at the cost of processing time. It should be noted that for both formulations, the evolution of the initially placed curve is regulated by an image based energy. Methods reported in [1,9] approximate the image-based energy in terms of the intensity gradient strength (edge-map), whereas techniques proposed in [2,19] employs regional intensity statistics for the energy approximation. The region-based methods show robust performance in general as the gradient strength often leads to over segmentation for weak edges. However, the conventional region-based methods fail to address the intensity inhomogeneity problem of medical images due to the underlying piecewise constant assumption. Consequently, Li *et al.* [11] and Lankton and Tannenbaum [10] proposed the use of localized statistics to regulate the curve growth in medical images for minimizing the impact of the intensity inhomogeneity.

In context of blood vessel segmentation in CTA, Harnandez *et al.* [6], Mohr *et al.* [12], Szymczak *et al.* [15], Wang and Liatsis [17] and Yang [18] reported successful segmentations; however, the impact of the externally injected contrast medium has been little employed in the coronary segmentation process. Isgum *et al.* [8] proposed an automated system for the coronary calcification detection, in which all the connected components of intensity value greater than 220 HU were interpreted as potential calcified plaques. Similarly, Hong *et al.* [7] proposed a fixed threshold of 350 HU for the segmentation of coronary calcified plaques in the contrast enhanced CTA.

In this work, we derive the estimate intensity threshold by investigating the impact of the contrast medium in the respective CTA volume to ensure the accuracy of segmentation. Followed with this introduction, we define the proposed coronary segmentation model in Sect. 2. Subsequently, comparative results are presented in Sect. 3, which is followed by the shortcomings and the conclusion.

2 Proposed Model

Based on the fact that externally injected contrast medium enhances the visual brightness of blood filled coronaries in CTA, we propose to adaptively model the contrast medium in the coronary segmentation process. The proposed method is classified as semi-automatic since it requires manual seed points to initialize the segmentation process. We start with the assumption that the coronary segmentation can be improved by suppressing the non-coronary structures using intensity and shape constraints in a pre-processing step. However, the derivation of a generic intensity threshold across the dataset is challenging due to the non-homogeneous diffusion of the contrast medium in different CTA volumes. Consequently, the impact of the contrast agent is mathematically modelled in a first step to derive the volume-specific intensity range in Hounsfield units (HU) for respective CTA volumes. In the following step, we computed the voxel-wise vesselness measure using 3D Hessian matrix of the CTA volume to suppress the non-tubular voxels. In the final step, we applied the localized region based segmentation to extract coronary the tree from the pre-processed CTA volume. For the rest of the paper, let I denote a 3D CTA volume defined on the domain Ω and $\mathbf{x}$, $\mathbf{y}$ denotes two independent spatial variables in the domain Ω. In addition, we employ a mask function $M(\mathbf{x}, \mathbf{y})$, which defines a neighbourhood of radius R_L centred at $\mathbf{x}$. Accordingly, the mask function $M(\mathbf{x}, \mathbf{y})$ will be 1 when a point $\mathbf{y}$ lies within a neighbourhood region of $\mathbf{x}$, and 0 otherwise.

2.1 Contrast Medium Modelling

For enhanced visualization of the coronary vasculature, a contrast medium is often injected intravenously before the cardiac CTA exam. Consequently, the contrast affected blood appears brighter in the CTA volume which allows clinician to distinguish the coronary vasculature from the background as shown in Fig. 1a–b the diffusion of the contrast medium is non-homogeneous across patients as it depends upon several factors including the type and amount of contrast medium, the total scan time and the heart rate. This clinical fact leads to the assumption that despite of similar visual appearance of the blood filled coronaries, there exists a statistically significant difference in the blood intensity values for different CTA volumes. Consequently, the intensity based suppression of the non-coronary structures requires volume specific threshold values for optimal segmentation. Therefore, the use of a fixed threshold from the literature [7,8] may result in erroneous segmentation. Based on the fact that the contrast affected blood flows into coronaries from the descending aorta, we therefore segmented the aorta in the first step to estimate the volume-specific HU intensity range. For aorta segmentation, we started with the background suppression in CTA using an intensity threshold of 100HU as shown in Fig. 1c. In the following step, we applied a circular Hough Transform [4] based shape analysis to segment the aorta from in the blood volume as shown in Fig. 1e. Iteratively, 2D segmentation is performed through axial slices until the circular aorta changes the shape which reflects the origin of coronary vasculature. Next, we computed the intensity histogram of the segmented aorta and the contrast medium response is modelled using Gaussian fitting. Figure 2a shows the Gaussian approximation for four CTA volumes where a significant variation in the mean values emphasize the need of an adaptive intensity threshold for accurate segmentation. It should be noted that the Gaussian mean represents the intensity for blood-filled aorta; however, the concentration of the contrast medium decreases as the blood flows towards distal segments of coronary tree. Moreover, the vessel narrowing towards the distal end points often result in the less diffusion and poor contrast. Thus,

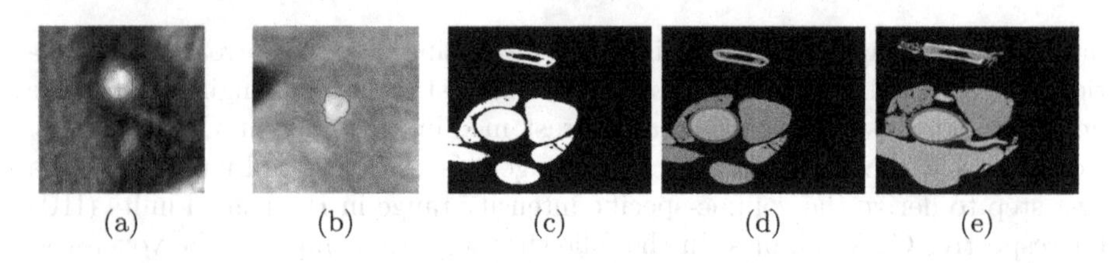

Fig. 1. Coronary appearance and aorta segmentation in axial slices. (a–b) Similar appearance for coronary in two CTA volumes. (c–d) Background suppression mask and the segmented aorta, (e) aorta shape change due to emerging coronary structure.

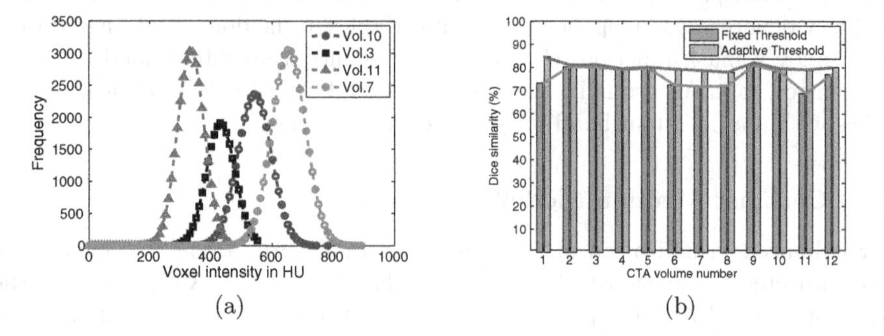

Fig. 2. Intensity approximation and mean accuracy for CTA volumes. (a) Shows the intensity distribution histogram for four CTA volumes in which a significant mean variation demonstrates the need of an adaptive intensity threshold. (b) Represents comparative segmentation accuracy for two intensity thresholds.

to take into account the intensity drop towards distal segments, we estimate the adaptive intensity range R_I for respective CTA volume I as expressed in Eq. 1.

$$R_I = \{\mu_I \pm 3\sigma_I\} \tag{1}$$

where μ_I and σ_I represent the aorta based mean HU and standard deviation for the respective CTA volume. For a quantitative comparison, the Gaussian distribution parameters and the derived intensity range for 12 clinical CTA volumes are presented in Table 1. It should be noted that the lower boundary of adaptive intensity range is meant for suppressing the non-coronary voxels and the upper boundary can be used to segment the calcified plaques (if any) in the arterial tree.

2.2 Enhancement of Tubular Structures

In this step, we employ shape information to effectively suppress the non-coronary voxels. Based on the fact that coronary vessels follow a tubular structure, we enhanced tubular voxels as proposed by Frangi *et al.* [5]. Accordingly, we obtained the 3D Hessian matrix of the CTA volume I in a first step to investigate the structural shape information. Next, we computed the eigenvalues from

Table 1. Volume-specific intensity (HU-range) for 12 CTA volumes.

CTA Vol	Mean HU	Std	Minimum HU	Maximum HU
01	942	62	756	1128
02	495	42	369	621
03	436	45	301	571
04	485	38	371	599
05	542	60	362	722
06	630	50	480	780
07	663	53	504	822
08	463	62	277	650
09	517	53	358	676
10	543	55	378	708
11	335	45	200	470
12	425	53	296	554

the 3D Hessian matrix to identify the geometric patterns and the voxel-wise vesselness is computed as follows:

$$V_o(\mathbf{x}) = \begin{cases} 0 & \text{if } \lambda_2 \; or \; \lambda_3 > 0 \\ \left\{1 - exp\left(-\frac{R_A^2}{2\alpha^2}\right) exp\left(-\frac{R_B^2}{2\eta^2}\right)\left(1 - exp\left(\frac{S^2}{-2\zeta^2}\right)\right)\right\} & \text{otherwise} \end{cases} \quad (2)$$

where $R_A = \frac{|\lambda_2|}{|\lambda_3|}$ discriminates plate-like structures from the cylindrical vessels, $R_B = \frac{|\lambda_1|}{\sqrt{|\lambda_2 \lambda_3|}}$ differentiates blobs from other shapes and S serves as a penalty for the noise suppression. Moreover, the tuning parameters $\alpha = 0.6$, $\eta = 0.5$ and $\zeta = 220$ controls the overall vesselness measure. The response of the vesselness filter for 2D axial slices of CTA volume is shown in Fig. 3b, whereas the vesselness computed for the complete 3D CTA volume is presented in Fig. 3c. Figure 3b–c reflect that the tubular structures have been assigned high vesselness in comparison with the background; however, an inherent limitation of the multi-scale filter is misclassification of the edges, i.e. edges are often assigned comparatively high vesselness as well. This drawback is evident in Fig. 3c where it becomes extremely complex to identify the coronary vasculature. Consequently, the CTA

$$I(\mathbf{x}) = \begin{cases} I(\mathbf{x}) & \text{if } V_o(\mathbf{x}) > T_f \; and \; I(\mathbf{x}) \in R_I \\ 0 & \text{otherwise} \end{cases} \quad (3)$$

volume I is filtered using intensity and vesselness constraints of Eq. 3 with T_f set equal to 10^{-3}.

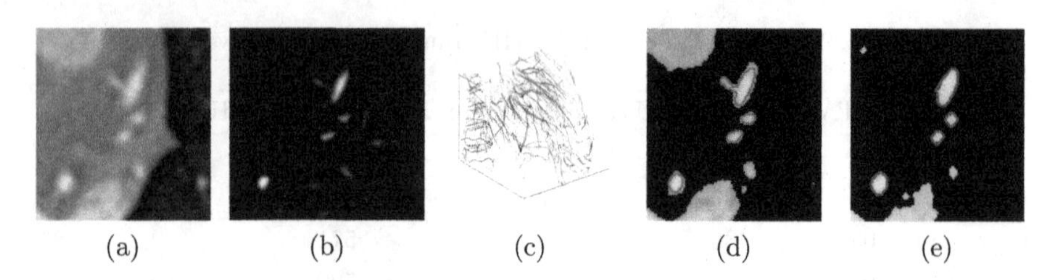

(a) (b) (c) (d) (e)

Fig. 3. Pre-processing for optimal segmentation. (a) Shows a 2D axial slice, (b–c) represent the 2D and 3D vesselness measure for CTA volume with prominent tubular structures and (d–e) show the intensity based background suppression using fixed and adaptive intensity threshold respectively. It can be observed that fixed threshold based segmentation (red) performs over segmentation due to leakage into nearby non-coronary structures, whereas (blue) contour represents the adaptive threshold segmentation. (Color figure online)

2.3 Coronary Tree Segmentation

Once the CTA volume is effectively filtered (as expressed in Eq. 3), the coronary tree is segmented using a 2D level set evolution based on the Chan-Vese [2] localized image energy. The segmentation process starts with the selection of the coronary seed points. To ensure that both coronary structures (left and right arteries) are segmented simultaneously, the coronary seed points are placed on an axial slice at the mid of the caudal-cranial axis. Next, the seed points are used to initialize a localized mask with a radius of 6 mm (i.e. maximum possible coronary diameter on an axial slice [3]). Subsequently, the initial mask evolves under the influence of image-based localized Chan-Vese energy to capture the true boundary of the coronary. Because of the 2D nature of the level set evolution, the evolved mask serves as an initialization to its adjacent axial slice on the caudal-cranial axis to capture the complete coronary tree. To define the mathematical model for level set coronary evolution, we start with the Chan-Vese segmentation in which two regions (the object and the background) are modelled with their mean intensity values as follows:

$$F(c_1, c_2, C) = \int_{in(C)} [I(\mathbf{x}) - c_1]^2 d\mathbf{x} + \int_{out(C)} [I(\mathbf{x}) - c_2]^2 d\mathbf{x} + \gamma length(C) \quad (4)$$

where C is the evolving curve, and c_1, c_2 represents the mean intensity value inside and outside the evolving curve respectively. For the level set formulation, the evolving curve C is embedded into a higher space using a signed distance function ϕ : such that $C = \{\mathbf{x}|\phi(\mathbf{x}) = 0\}$. The internal and external regions of the curve are defined using the Heaviside function (also termed as unit step function) $H\phi$, which is 1 when $\phi(\mathbf{x}) > 0$ and 0 when $\phi(\mathbf{x}) < 0$. Moreover, the evolving curve (zero level set) can be identified using derivative of Heaviside function i.e. the Dirac delta function δ which is 1 when $\phi(\mathbf{x}) = 0$ and 0 far from the interface. Accordingly, we formulate Eq. 4 using level set representation as expressed in Eq. 5. The first term of Eq. 5 is image-based curve driving energy which minimizes the approximation error and the second term is the regularization term added to ensure the curve smoothness. For additional details and complete mathematical derivations, readers are referred to [2,10].

$$\begin{aligned}\frac{\partial \phi}{\partial t}(\mathbf{x}) =& \delta\phi(\mathbf{x}) \int_{\Omega\mathbf{y}} \phi(\mathbf{y}) M(\mathbf{x},\mathbf{y}) \left\{(I(\mathbf{y}) - c_1)^2 - (I(\mathbf{y}) - c_2)^2\right\} d\mathbf{y} \\ &+ \gamma\delta\phi(\mathbf{x}) div \left\{\frac{\nabla\phi(\mathbf{x})}{|\nabla\phi(\mathbf{x})|}\right\}\end{aligned} \tag{5}$$

where γ is the weight assigned to regularization term and c_1, c_2 represents the localized interior and exterior intensity mean values as expressed in 6. It should be noted that localization mask based statistics are used in the segmentation primarily due to the intensity inhomogeneity problem in medical data.

$$c_1 = \frac{\int_{\Omega y} M(\mathbf{x},\mathbf{y}) I(\mathbf{y}) H\phi(\mathbf{y}) d\mathbf{y}}{\int_{\Omega y} M(\mathbf{x},\mathbf{y}) H\phi(\mathbf{y}) d\mathbf{y}}, \quad c_2 = \frac{\int_{\Omega \mathbf{y}} M(\mathbf{x},\mathbf{y}) I(\mathbf{y})(1 - H\phi(\mathbf{y})) d\mathbf{y}}{\int_{\Omega y} M(\mathbf{x},\mathbf{y})(1 - H\phi(\mathbf{y})) d\mathbf{y}} \tag{6}$$

2.4 Auto-correction Feature of the Mask

In general, the coronary tree comes out from the descending aorta and splits into branches along the caudal-cranial axis; hence all the segments are well captured in the level set based active contour evolution. However, due to the wide inter-patient variability and 2D axial slice based data acquisition in CTA, some distal branches emerge away from the main trajectory and become a part of the tree as slices are navigated. To address this issue, one possible solution is the 3D level set segmentation but it increases the computational load. In contrast, we introduced an auto-correction feature in the mask to capture the emerging peripheries during evolution. The proposed method reconstructs the mask in every iteration by scanning the neighbourhood of the trajectory on 2D axial slice. All the individual peripheries that satisfy the constraints (tubular shape and adaptive intensity) are captured as shown in Fig. 4c–d. This self-adjustment feature offers improved accuracy and the computational robustness, whereas the non-connected structures are automatically discarded using connected component analysis.

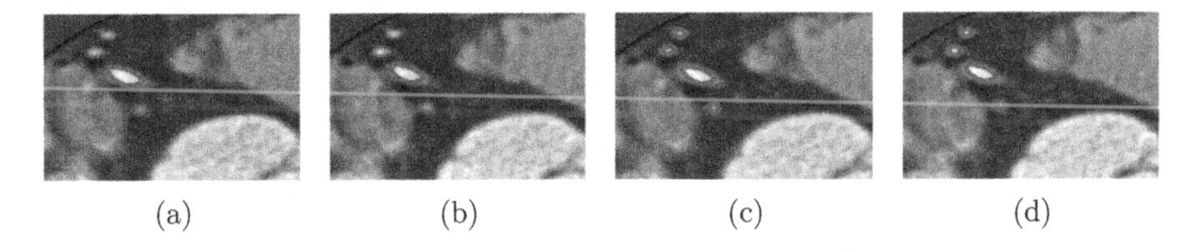

Fig. 4. Auto correction of mask to capture nearby emerging peripheries for CTA Volume (a–b), Emerging peripheries missed during evolution. (c–d), Emerging peripheries are captured for complete tree extraction.

3 Results

To demonstrate the effectiveness of the adaptive intensity modelling, the coronary segmentation was performed using two different intensity thresholds. The comparative results reveal that the use of fixed threshold i.e. 350 HU [7] leads to an erroneous coronary tree in terms of under/over segmentation, whereas the proposed adaptive threshold ensures accurate segmentation by employing the influence of contrast medium in the segmentation process. Moreover, the proposed segmentation shows a greater corroboration with the manual annotations in the cross sectional analysis as illustrated below.

3.1 CPR Based Analysis

Figure 5 shows the segmented right coronary artery (RCA) of CTA volume 1 using two different thresholds. Table 1 indicates that the strong concentration of the contrast medium requires a higher intensity threshold (756 HU) to minimize false positives. It can be observed from the figure that the volume-specific threshold precisely tracks the main progression of the RCA Fig. 5b from aorta to the distal segment with the minimal peripheries, whereas the use of a literature based [7] fixed threshold 350 HU results in numerous side branches for the RCA Fig. 5a. The efficacy of the adaptive intensity threshold is illustrated by constructing the curve planar reformatted (CPR) images along three different axes. CPR visualization from three different views helps to evaluate if there exist any intermediate peripheries for the segmented RCA. The centreline for the right coronary artery is obtained in the first step using sub-voxel skeletonization algorithm of [16]. In the subsequent step, we constructed the 2D CPR images from CTA volume as shown in Fig. 5c–e. It should be noted that distinct views along three different axes substantiate the fact that the right coronary artery is well segmented from aorta to the distal points using the adaptive intensity threshold. Moreover, it can be observed that the peripheries which appear to be a part of the coronary structure in Fig. 5a, are not coronaries indeed but the kissing vasculature in close proximity which were captured mistakenly by active contour during the evolution.

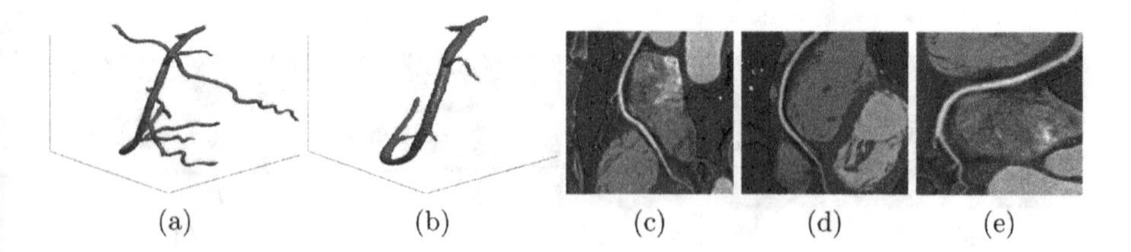

Fig. 5. Visualization of segmented RCA in CTA volume1, (a) RCA obtained using fixed intensity threshold of 350 HU, (b) RCA obtained using adaptive threshold. (c–e) Represent CPR image along three axes to confirm the efficacy of adaptive threshold.

3.2 Cross-Sectional Analysis

The efficacy of the adaptive intensity threshold is further illustrated by comparing the two segmentations in 3D space. Figure 6a–b shows a zoomed version of the segmented Left circumflex artery (LCX) branch of CTA volume 1 obtained using two thresholds. It should be noted that the adaptive threshold (756 HU) results in a smooth segmentation (see Fig. 6a), whereas the fixed threshold (350 HU) leads to over-segmentation in terms of disconnected expansion of the LCX. This is based on the fact that the high concentration of the contrast medium misleads the evolving curve to capture the nearby structures (see Fig. 6b). This over-segmentation is further unfolded using the orthogonal planar analysis as shown in Fig. 6c–d. The impact of the over-segmentation can be clearly observed by viewing the boundary points as the fixed threshold based segmentation shows incorrect expansion of the vessel in cross sectional planes in contrast to the response of adaptive threshold based segmentation.

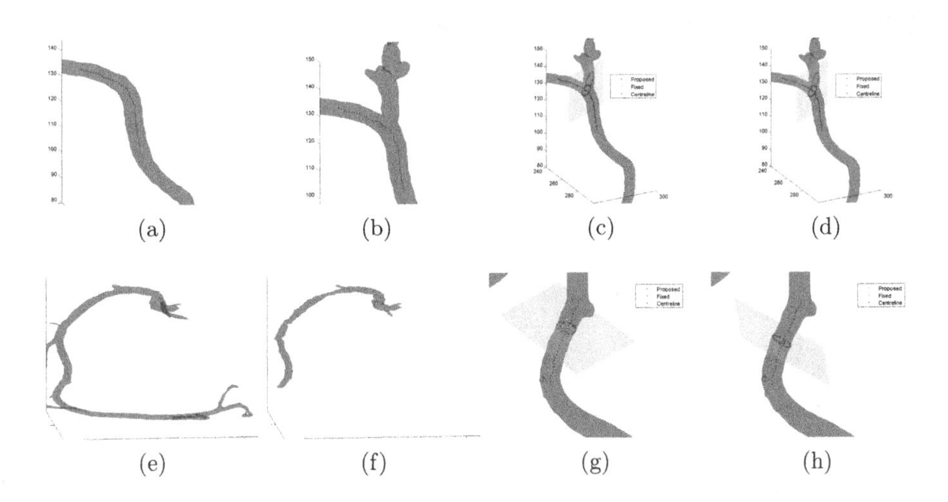

Fig. 6. Top. LCX branch of CTA volume 1. (a–b) LCX segmentation using adaptive (756 HU) and fixed (350 HU) threshold values respectively. (c–d) Illustrates the efficacy of adaptive threshold as planar boundary points show over-segmentation for fixed threshold. (Bottom) RCA branch of CTA volume 11. (e–f) RCA segmentation using adaptive (200 HU) and fixed (350 HU) threshold values respectively. (g–h) illustrates the efficacy of adaptive threshold as planar boundary points show under segmentation for fixed threshold. Red is the boundary for fixed threshold segmentation and green represents the response of adaptive threshold. (Color figure online)

Likewise, Fig. 6e–h present the case where the use of the fixed intensity threshold leads to an under-segmented tree because of low concentration of the contrast medium in CTA. The less concentration of contrast medium results in a lower intensity threshold in coronary segmentation, as Table 1 defines 200 HU for CTA volume 11. Figure 6e shows that adaptive threshold leads to detailed coronary structure, whereas RCA obtained using a fixed intensity threshold of 350 HU shows under-segmentation as a significant portion towards distal RCA is missed (see. Fig. 6f). This under-segmentation becomes more evident in the planar analysis as the segmented lumen shrinks rapidly towards the distal segments. Figure 6g–h shows that the 350 HU based segmentation vanishes through the distal section of RCA in contrast to the response of adaptive threshold based segmentation.

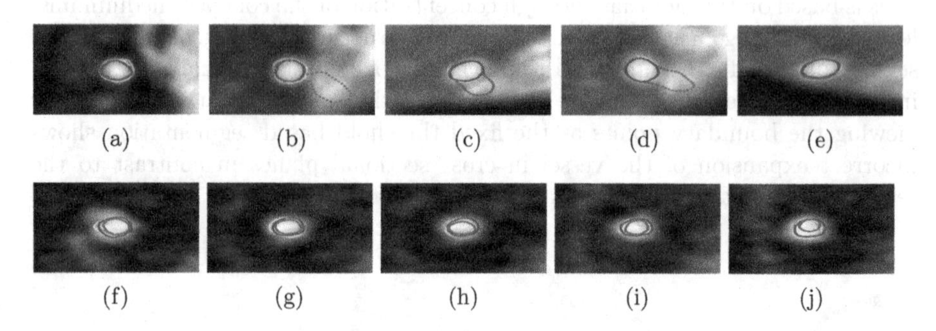

Fig. 7. Fixed and adaptive threshold based segmentation with respect to manual annotations. (Top) analysis for LCX of CTA volume 1 (bottom) analysis for RCA of CTA volume 11. Fixed threshold reflects over segmentation for CTA volume 1 and under segmentation for CTA volume 11. Red is the fixed threshold segmentation and green is the adaptive threshold result. Blue and yellow represents manual annotations. (Color figure online)

3.3 Validation Against Manual Annotations

The effectiveness of the volume-specific intensity threshold is also evaluated with respect to the manual annotations of two independent observers. Two well trained biomedical students were requested to perform the manual annotations of coronary lumen independently in our centre using interactive coronary analysis software. The lumen boundary obtained at the optimal coronary display settings (L/W = 300/800) are recorded for qualitative and quantitative evaluation of two segmentations. It can be observed from Fig. 7a–e) that the adaptive threshold leads to a good agreement with manual observers by suppressing the nearby vasculature, whereas the fixed threshold based segmentation captures the adjacent non-coronary structures that results in increased false positives. Likewise, Fig. 7f–j show the response of two segmentations for RCA of CTA volume 11. It should be noted that the adaptive threshold leads to a true segmentation by allowing expansion towards low intensity voxels, whereas the fixed intensity threshold favours high intensity voxels resulting in under-segmentation. To demonstrate the quantitative efficiency of the adaptive intensity threshold, we computed the segment-wise accuracy with respect to the manual annotation using three metrics i.e. sensitivity, specificity and the Dice similarity coefficient. It can be observed from Fig. 8a that the adaptive threshold in LCX segmentation leads to a reasonable score for all the accuracy metrics, whereas high false positives associated with the fixed threshold leads to low specificity and a decreased Dice coefficient score (see Fig. 8b for the fixed threshold). Similarly, the use of adaptive threshold results in a stable value for all three metrics in RCA segmentation (Fig. 8c), whereas the fixed threshold leads to increased false negatives causing a significant drop in the sensitivity and the corresponding Dice similarity coefficient (Fig. 8d). Moreover, the mean Dice similarity for two segmentation methods is presented in Fig. 2b which clearly demonstrates the efficacy of adaptive threshold over fixed intensity threshold (specifically for volumes with irregular concentration of dye i.e.volume 1, 6, 7, 8 and 11).

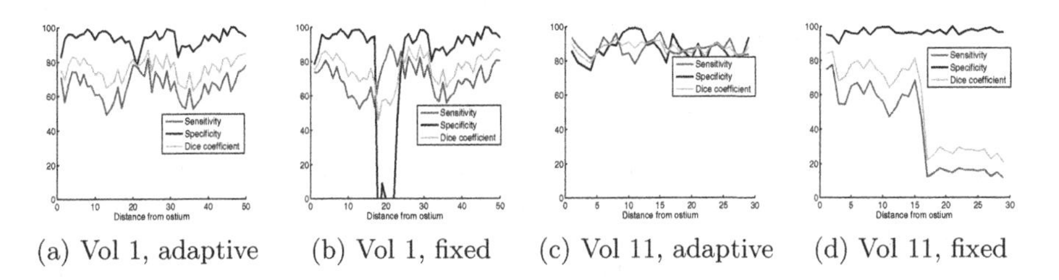

(a) Vol 1, adaptive (b) Vol 1, fixed (c) Vol 11, adaptive (d) Vol 11, fixed

Fig. 8. Segmentation accuracy for fixed and the adaptive threshold based segmentation. (a–b) Results for LCX segment of CTA volume 1 where high false positives lead to low specificity and decreased Dice score, (c–d) results for RCA segment of CTA volume 11 where high false negatives lead to low sensitivity and decreased Dice score.

4 Conclusion

We demonstrated that adaptive modelling of the contrast medium intensity can considerably improve the accuracy of the coronary segmentation. In contrast, the use of a fixed intensity threshold across the dataset may decrease precision by capturing the nearby non-coronary segments or missing the distal parts of coronary tree. After deriving the volume-specific intensity ranges, we employed a bi-directional level set based Chan-Vese evolution to segment the coronary tree from CTA volume. Promising results validating a significant improvement in segmentation quality confirms the need of contrast medium modelling in segmentation process. A limitation of the current method is its failure to detect non-calcified plaques which exhibit an unexpected intensity drop across the lesion regions of the coronary tree, which is being investigated in an ongoing study.

References

1. Caselles, V., Kimmel, R., Sapiro, G.: Geodesic active contours. Int. J. Comput. Vis. **22**(1), 61–79 (1997)

2. Chan, T.F., Vese, L.A.: Active contours without edges. IEEE Trans. Image Process. **10**(2), 266–277 (2001)
3. Dodge, J.T., Brown, B.G., Bolson, E.L., Dodge, H.T.: Lumen diameter of normal human coronary arteries. influence of age, sex, anatomic variation, and left ventricular hypertrophy or dilation. Circulation **86**(1), 232–246 (1992)
4. Duda, R.O., Hart, P.E.: Use of the hough transformation to detect lines and curves in pictures. Commun. ACM **15**(1), 11–15 (1972)
5. Frangi, A.F., Niessen, W.J., Vincken, K.L., Viergever, M.A.: Multiscale vessel enhancement filtering. In: Wells, W.M., Colchester, A., Delp, S. (eds.) MICCAI 1998. LNCS, vol. 1496, pp. 130–137. Springer, Heidelberg (1998). doi:10.1007/BFb0056195
6. Hernandez, M., Frangi, A.F., Sapiro, G.: Three-dimensional segmentation of brain aneurysms in CTA using non-parametric region-based information and implicit deformable models: method and evaluation. In: Ellis, R.E., Peters, T.M. (eds.) MICCAI 2003. LNCS, vol. 2879, pp. 594–602. Springer, Heidelberg (2003). doi:10.1007/978-3-540-39903-2_73
7. Hong, C., Becker, C.R., Schoepf, U.J.: Coronary artery calcium: absolute quantification in nonenhanced and contrast-enhanced multi-detector row ct studies 1. Radiology **223**(2), 474–480 (2002)
8. Isgum, I., van Ginneken, B., Olree, M.: Automatic detection of calcifications in the aorta from CT scans of the abdomen 1: 3D computer-aided diagnosis. Acad. Radiol. **11**(3), 247–257 (2004)
9. Kass, M., Witkin, A., Terzopoulos, D.: Snakes: active contour models. Int. J. Comput. Vis. **1**(4), 321–331 (1988)
10. Lankton, S., Tannenbaum, A.: Localizing region-based active contours. IEEE Trans. Image Process. **17**(11), 2029–2039 (2008)
11. Li, C., Kao, C.Y., Gore, J.C., Ding, Z.: Implicit active contours driven by local binary fitting energy. In: IEEE Conference on Computer Vision and Pattern Recognition, pp. 1–7. IEEE (2007)
12. Mohr, B., Masood, S., Plakas, C.: Accurate lumen segmentation and stenosis detection and quantification in coronary CTA. In: Proceedings of 3D Cardiovascular Imaging: A MICCAI Segmentation Challenge Workshop (2012)
13. NHS, UK: Coronary Heart Disease, Statistics for United Kingdom. http://www.nhs.uk/Conditions/Coronary-heart-disease/Pages/Introduction.aspx. Accessed 11 Nov 2016
14. Sethian, J.A., et al.: Level set methods and fast marching methods. J. Comput. Inf. Technol. **11**(1), 1–2 (2003)
15. Szymczak, A., Stillman, A., Tannenbaum, A., Mischaikow, K.: Coronary vessel trees from 3D imagery: a topological approach. Med. Image Anal. **10**(4), 548–559 (2006)
16. Van Uitert, R., Bitter, I.: Subvoxel precise skeletons of volumetric data based on fast marching methods. Med. Phys. **34**(2), 627–638 (2007)
17. Wang, Y., Liatsis, P.: A fully automated framework for segmentation and stenosis quantification of coronary arteries in 3D CTA imaging. In: DESE 2009, pp. 136–140. IEEE (2009)
18. Yang, Y.: Image segmentation and shape analysis of blood vessels with applications to coronary atherosclerosis. Ph.D. thesis, Georgia Institute of Technology (2007)
19. Yezzi, A., Tsai, A., Willsky, A.: A fully global approach to image segmentation via coupled curve evolution equations. J. Vis. Commun. Image Represent. **13**(1), 195–216 (2002)

Segmentation of Abdominal Aortic Aneurysm (AAA) Based on Topology Prior Model

Safa Salahat[1(✉)], Ahmed Soliman[2], Tim McGloughlin[3], Naoufel Werghi[1], and Ayman El-Baz[2]

[1] Electrical and Computer Engineering Department, Khalifa University, Abu Dhabi, United Arab Emirates
{safa.salahat,naoufel.werghi}@kustar.ac.ae
[2] BioImaging Laboratory, Bioengineering Department, University of Louisville, Louisville, KY, USA
{ahmed.soliman,ayman.elbaz}@louisville.edu
[3] Biomedical Engineering Department, Khalifa University, Abu Dhabi, United Arab Emirates
tim.mcgloughlin@kustar.ac.ae

Abstract. In this paper, we propose a statistical based method using a topology prior model, integrating both intensity and shape information, to segment abdominal aortic aneurysm (AAA) from computed tomography angiography (CTA) scans. The method was tested on a total of 48 slices taken from 6 different patients and has shown competitive performance compared with the best reported results in the literature. Our method has achieved a mean Dice coefficient of $0.9303_{\pm 0.0499}$, and mean Hausdorff distance of $3.5703_{\pm 3.1941}$ mm. This method overcomes the major problem faced by currently existing solutions of similar Hounsfield values of neighboring tissues to that of the AAA thrombus. This is a promising medical tool which can be used to analyze the AAA in order to generate an accurate rupture risk indicator.

Keywords: Abdominal aortic aneurysm · Segmentation · Lumen · Thrombus · Topology · Probability

1 Introduction

Abdominal Aortic Aneurysm (AAA) is a balloon like dilation in abdominal aorta which may lead to patient death if untreated. The expansion is defined as an aneurysm if its maximum diameter reaches 3 cm, and usually requires medical intervention when the maximum diameter reaches 5.5 cm [1–3]. Currently, there are two techniques used for aneurysm repair, open aneurysm repair (OR) and Endovascular Aneurysm Repair (EVAR). In the former, the aneurysm is replaced with a graft, and in the later, a stent graft is inserted inside the aneurysm lumen to avoid the invasive open surgery [4].

Lumen, thrombus, calcification and vessel wall are the components of the aneurysm, as illustrated in Fig. 1. Currently, clinicians' rupture risk assessment is mainly based on the maximum measured diameter of the aneurysm. However, some studies have indicated that AAA maximum diameter doesn't always lead to accurate prediction

M. Valdés Hernández and V. González-Castro (Eds.): MIUA 2017, CCIS 723, pp. 219–228, 2017.
DOI: 10.1007/978-3-319-60964-5_19

of rupture risk [5, 6]. Several contributions from biomechanics have suggested that peak wall stress can predict rupture risk [7–10]. It is also important for surgery planning and accurate patient-specific stent graft fabrication. Segmentation of the aneurysm and extraction of the lumen and the thrombus are required tasks for the aforementioned purposes. Manual segmentation of the aneurysm is a time consuming task for the clinicians. Considering the high number of patients diagnosed with AAA on daily basis worldwide, there is a need to develop automated methods to segment the aneurysm.

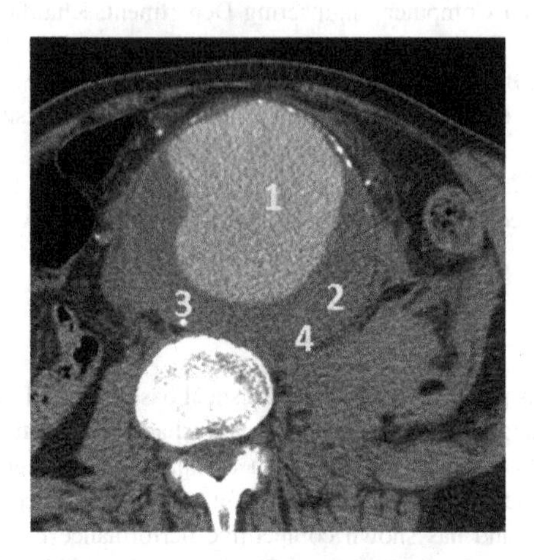

Fig. 1. The four main components of an AAA. The region labeled with 1 is the lumen, while the darker region labeled with 2 is the thrombus. The bright spots around the aneurysm labeled as region 3 are the calcified regions. Label 4 represents the vessel wall.

In response to this need, numerous contributions from the medical imaging community have been proposed providing fully and semi-automated solutions for the AAA segmentation problem. These efforts have mostly concentrated on the lumen and thrombus segmentation, and have relied on deformable models, graph cut, fuzzy c-means clustering, or a combination of other methods to perform the AAA segmentation.

While the problem of lumen segmentation is relatively easy due to the contrast agent injected into the patient which makes the lumen distinguishable from other structures, the segmentation of the thrombus is more challenging since Hounsfield values which represent it are very similar to other neighboring structures which results in weak edges. Further, other organs with strong features, e.g. spinal cord, and presence of calcification may obstruct the segmentation of the thrombus. In addition, the aneurysm doesn't have a specific shape or geometry to aid in the segmentation of the thrombus.

Most of the proposed solutions for aneurysm segmentation are based on deformable models [11–21]. However, this class of solutions has three major weaknesses. Deformable models are more computationally demanding than other methods, the outer edge of the aneurysm is of weak gradient which results in over- or under-segmentation of the aneurysm, and these models can easily get drawn to other structures of high gradient. Therefore, extra stopping criteria, and/or approximate shape constraints are added to the models to improve the results. In addition, some user interaction, other commercial software, or preprocessing step(s) may be needed to remove these structures before the segmentation process starts.

Graph cut based methods also over-segments the aneurysm and includes other structures in addition to it [22–25]. Shape constraint or further analysis is needed to exclude irrelevant regions.

Fuzzy c-means based solutions [26, 27] generate a binary image of all organs in a CT scan, and the result obtained is highly dependent on threshold and other tuned parameters values used. It is also observed that fuzzy c-means cannot distinguish between the thrombus and neighboring tissues of similar Hounsfield values.

Some other contributions utilized low level image processing operations such as thresholding and morphological operation [28], region growing [29, 30], isoperimetric segmentation [31], histogram derived information [32], active learning and random forest classifier [33, 34]. These methods are dependent on user interaction, threshold and parameters tuned, and aneurysm shape fitting into a mathematical shape (e.g. ellipse, circle, or radial function) into the thrombus.

In summary, it is clear that currently existing solutions has one or more of the following three problems: (1) The method cannot fully/partially distinguish the thrombus from other tissues, (2) The segmentation is hindered by other structures of strong features, (3) Computationally demanding, (4) Extra user interaction is needed to guide the segmentation process, or correct it afterwards.

Therefore, it is desirable to develop a robust solution which tackles the above mentioned drawbacks. In this work, we employ a statistical topology prior model based method which utilizes both intensity and shape information to segment thrombus and lumen volumes from CT data. This method is used for the first time to solve the AAA segmentation problem, and have managed to overcome the above mentioned limitations to a great extent while maintaining high accuracy at the same time.

The remaining of the paper is organized as follows. Section 2 describes the methodology used for aneurysm segmentation. Section 3 provides a detailed evaluation and analysis of the results obtained. Section 4, concludes the paper.

2 Methodology

The proposed framework for segmenting the AAA thrombus and the lumen from 3D-CT data (depicted in Fig. 2) utilizes a label propagation, with topology preservation, scheme using both a patient specific shape model and 1st order adaptive intensity model to overcome the problem of intensity homogeneity between thrombus and its adjacent structures. Details of each model component are as follow.

2.1 Patient Specific Shape Model

Accurate segmentation of the Abdominal Aortic Aneurysm (AAA) wall is very difficult since the intensities/grey levels of the AAA wall are very close to the intensities of other abdominal tissues. Thus, inclusion of information on the shape or topology of AAA wall will provide a guiding feature during the segmentation process and

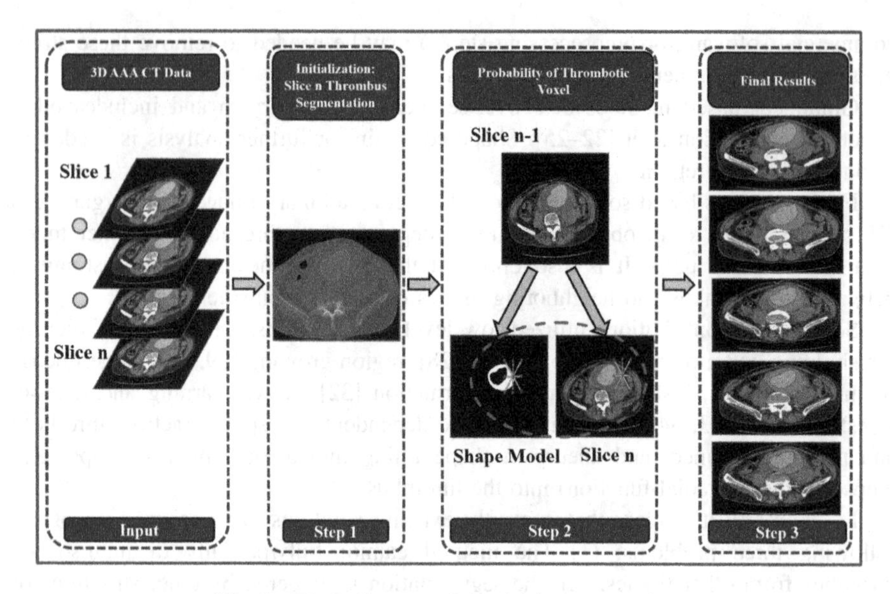

Fig. 2. Block diagram of the AAA segmentation.

potentially enhance the segmentation accuracy. The primary challenge in creating a prior shape model of the AAA wall is the high intra-patient variability, especially due to pathology. To overcome this challenge, in this manuscript, we are introducing adaptive shape-specific model that is based on manual delineation of the inner and outer boarders of AAA. Subsequently, the appearance and the topology of manually segmented AAA wall will be used to guide the segmentation of the adjacent slice. Each slice segmentation will drive its adjacent one which can be viewed as an adaptive label propagation process (see Fig. 3).

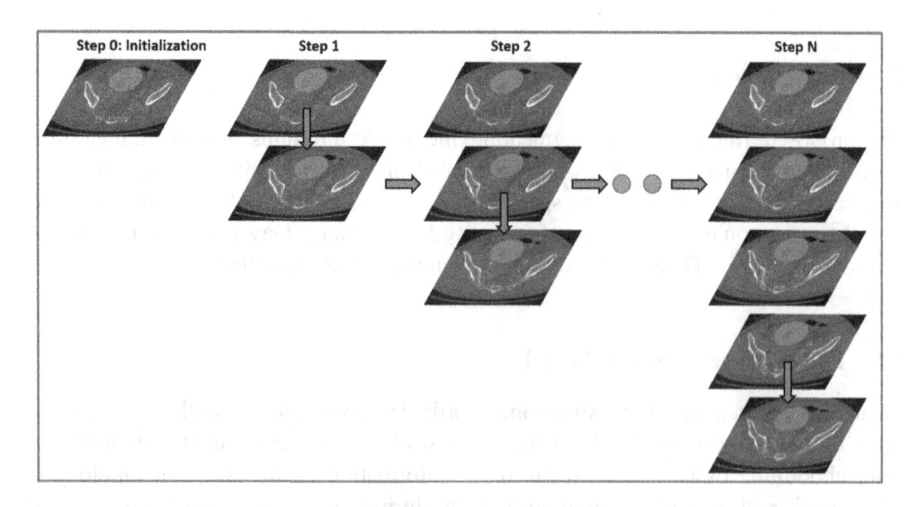

Fig. 3. Step-by-step illustration of the guiding shape model.

2.2 1st Order Adaptive Intensity Model

Unlike traditional shape models that depends only on the mapped voxel location to calculate the probabilistic map, our 1st order adaptive intensity model ensures that only the visually similar voxels will contribute in the probability map calculations for the slice to be segmented to provide an accurate segmentation results.

The complete framework proceeds as follows: Starting from the last slice, the lumen and the thrombus of the last slice are manually segmented by the operator, which results in two binary masks which are subtracted to produce the binary mask representing the thrombotic region in the last slice (see Fig. 4). (2) Then moving backward, each slice i is segmented referring to the previously segmented slice $(i + 1)$. This procedure is performed as follows: at each voxel in the slice i an $N1 \times N2$ window w is generated around its counterpart in slice $(i + 1)$, then voxels in that window whose Hounsfield values fall within a predefined tolerance $\pm\tau$ are select. If no voxels are found, window size is increased until such voxel(s) are found, or maximum window size is reached. (4) Then the probability of each voxel to be part of the thrombus is calculated as the occurrence of positively labeled voxels from the total voxels in slice $i + 1$ which are within the window whose Hounsfield values are close to the voxel in slice i. Therefore, if we have k similar voxels within the window, of which m are labeled as 1, then the probability of this voxel in slice i to belong to the thrombus is simply $p_{TH}(x) = m/k$. If $(p_{TH}(x) > p_{BG}(x))$ then this voxel x belongs to the thrombus, and background otherwise. (5) 2D median filter

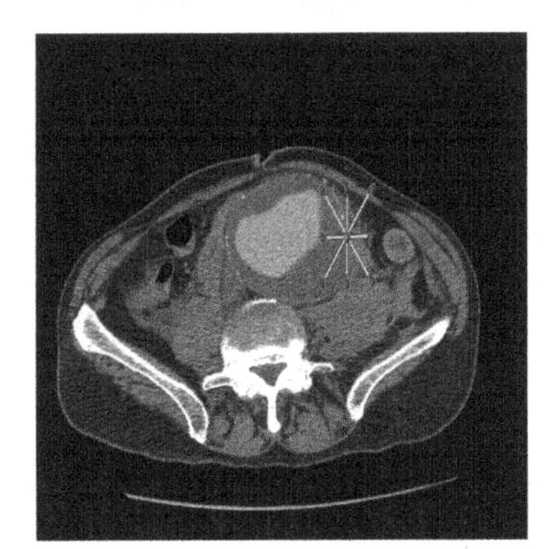

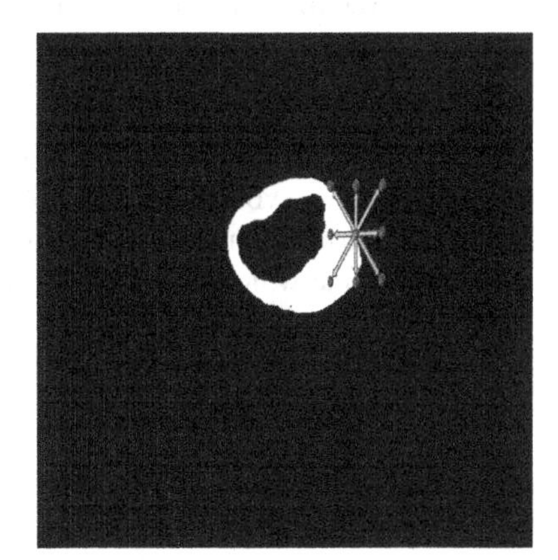

Fig. 4. Slice n and its binary mask resulting from the manual segmentation.

For example, in a 3×3 window, if k voxels in slice n are similar to voxel x (in slice $n{-}1$) Hounsfield value (within the given tolerance), and m out of these k voxels are labeled as 1, then $p_{TH}(x) = m/k$. The related algorithm of the above procedure can be described in more details as follows:

Algorithm 1:- Topology based segmentation

1. Manually label the outer boundaries of the lumen and the thrombus in slice *n*.
2. Derive a masks related to the thrombotic region labeled as 1 for thrombotic voxel and 0 for background.
3. For each slice *i*, i=n-1to1
   I. For each voxel x in slice i
      a) Construct a window w around the counterpart of voxel x in slice $i+1$
      b) Find voxels within with Hounsfield values that fall within a predefined tolerance $\pm\tau$ in w
      c) If no voxels are found to satisfy (e), increase size of w until correspondences are found or the maximum size allowed for w is reached
      d) Calculate the probability $p(x)$ of each voxel belonging to the thrombus based on the occurrences of white voxels from the total corresponding voxels which satisfy (e) in slice $i+1$.
      e) If $(p_{TH}(x) > p_{BG}(x))$, x is part of the thrombus, background otherwise.

   End For

   End For
4. Apply 3D smoothing using 3D median filter on the reconstructed 3D volume

Filter is then applied for each slice independently to improve the 2D segmentation result. (6) The whole volume is finally reconstructed, and 3D-median filter is applied to the volume improve segmentation consistency and surface smoothness. The final result is a binary volume which labels the thrombus across the slices.

3 Method Evaluation

This method has been tested on CT datasets collected from six patients, the region of interest to be segmented appears in eight slices each, who diagnosed with AAA which was provided by Limerick University. The in-plane voxel spacing ranges from 0.7031 × 0.7031 to 0.8984 × 0.8984, while the slice thickness ranges from 1.25 mm to 5 mm. Final results obtained are visualized in Fig. 4. To evaluate the accuracy of our method, we have used the Dice similarity coefficient (DC) and Hausdorff distance, that characterize the spatial overlap and surface-to-surface distances, in addition to other commonly used metrics, for comparison purpose, as illustrated in Table 1. An expert radiologist manually labeled the thrombus and the lumen of the aneurysm to acquire the ground truth data to be utilized to evaluate the accuracy of our proposed method.

Figure 5 shows a 2D axial projection for sample results obtained by our proposed method, from different subjects, where the ground truth edges are plotted in red along our segmentation in addition to false positive and false negative (in green, yellow, and pink respectively). It is clear from the sample results in Fig. 5 that our proposed method accurately segment the thrombus from its neighbors with little false positive segmentation as a result of high homogeneity of neighboring tissues. This proposed method managed to achieve results which are comparable to best results reported in the literature. The mean dice coefficient obtained was $0.9303_{\pm 0.0499}$, and mean thrombus Hausdorff distance was $3.5703_{\pm 3.1941}$. Our results are summarized in Table 1. To provide a comparison with the best results from the literature, we used all the accuracy

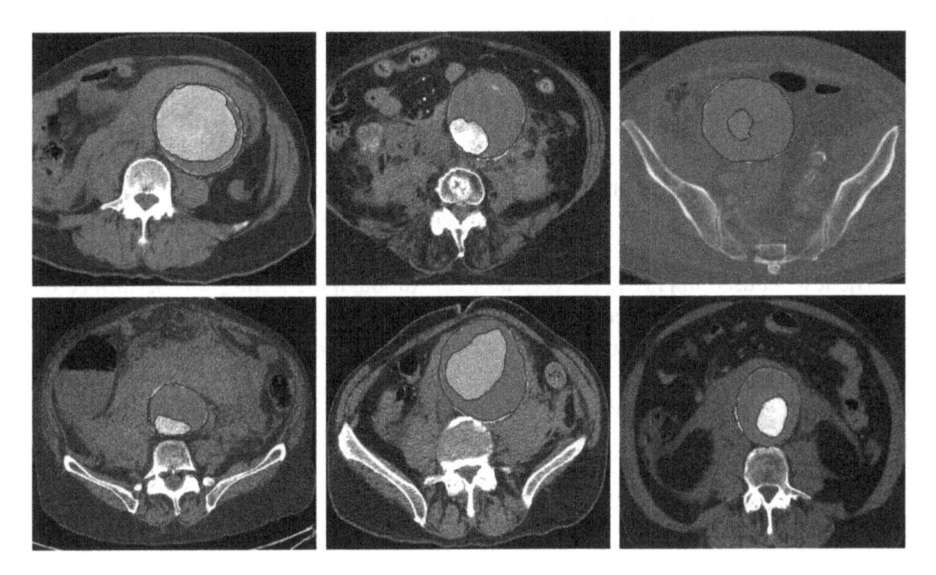

Fig. 5. Example thrombus segmentation result taken from different patients with color-coded ground truth edges, false positive errors, and false negative errors (red, yellow, and pink, respectively). (Color figure online)

Table 1. Evaluation of the proposed thrombus segmentation method tested on the six patients.

Metric	$\text{Mean}_{\pm \text{Std}}$
Dice coefficient	$0.9303_{\pm 0.0499}$
Sensitivity	$0.9138_{\pm 0.0621}$
Specificity	$0.9989_{\pm 0.0005}$
Positive predictive value	$0.9483_{\pm 0.0438}$
Volume overlap %	$87.35_{\pm 8.180}$
Hausdorff distance (mm)	$3.5703_{\pm 3.1941}$
Mean absolute surface distance (mm)	$0.2578_{\pm 0.2274}$
Mean absolute volume difference %	$5.3665_{\pm 2.3786}$
Mean symmetric absolute surface distance	$0.4752602_{\pm 0.4119}$

Table 2. Best literature reported thrombus segmentation results

Reference	Metric	Value	Slices/patients
12	Dice coefficient	0.8508 to 0.9316	7/-
11	Sensitivity	0.9354	5/-
11	Specificity	0.9837	5/-
35a	Volume overlap %	95 ± 3.30	125/17
23	Outer wall Hausdorff distance (mm)	3.09 ± 1.81	24/-
22	Mean unsigned error for thrombotic surface (mm)	1.9 ± 0.72	1300/9
25	Mean absolute volume difference %	8.0 ± 7.00	-/8
25	Average symmetric surface distance	1.46 ± 0.40 mm	-/8

In this reference, the metric is referred as Volume overlap but, in our opinion, it seems rather as Dice coefficient.

metrics used by all the compared work, illustrating total number of slices and/or patients used for testing, are provided in Table 2. The most important parameters which affect the segmentation result are the tolerance τ, which controls the voxels that will contribute in the current voxel probability calculation, and the maximum window size, which determines the search space for these voxels.

The test set used in [11, 12] is very limited. Although the testing was performed on a large dataset in [22], it requires several manual initializations and user guidance throughout the segmentation process, which explains the high performance obtained. It can be observed that our results surpasses the best results reported in the literature for some metrics, and produce comparable results for others, except for the volume overlap metric where our method is approximately 7.65% less than the highest reported value.

4 Conclusion

In conclusion, this paper has suggested a new statistical-based method for Abdominal Aortic Aneurysm segmentation from 3D-CT which utilizes topology (both intensity and shape) information to perform the segmentation. Results obtained are competitive to the best results reported in literature. This makes it a promising robust medical tool to perform aneurysm segmentation to reduce the burden on radiologists. In future, we will perform testing on a larger data set, and aim to detect the calcification which exits within the aneurysm to complete the final objective of building a biomechanical model which works as a rupture risk indicator.

Acknowledgements. This research was funded with a generous grant from Al-Jalila Foundation, Grant no. AJF201551. Ethics approval from University of Limerick, Ireland was acquired for the used data set in this research.

References

1. Moll, F., et al.: Management of abdominal aortic aneurysms clinical practice guidelines of the european society for vascular surgery. Eur. J. Vasc. Endovasc. Surg. **41**(1), 1–58 (2011)
2. Ernst, C.B.: Abdominal aortic aneurysms. N. Engl. J. Med. **328**(16), 1167–1172 (1993)
3. Vardulaki, K.A., et al.: Growth rates and risk of rupture of abdominal aortic aneurysms. Br. J. Surg. **85**(12), 1674–1680 (1998)
4. Zarins, C.K., et al.: AneuRx stent graft versus open surgical repair of abdominal aortic aneurysms: multicenter prospective clinical trial. J. Vasc. Surg. **29**(2), 292–305 (1999)
5. Darling, R.C., Messina, C.R., Brewster, D.C., Ottinger, L.W.: Autopsy study of unoperated abdominal aortic aneurysms. The case for early resection. Circulation **56**(3), 161–164 (1977)
6. Conway, K.P., Byrne, J., Townsend, M., Lane, I.F.: Prognosis of patients turned down for conventional abdominal aortic aneurysm repair in the endovascular and sonographic era: Szilagyi revisited? J. Vasc. Surg. **33**(4), 752–757 (2001)
7. Georgakarakos, E., et al.: The role of geometric parameters in the prediction of abdominal aortic aneurysm wall stress. Eur. J. Vasc. Endovasc. Surg. **39**(1), 42–48 (2010)
8. Venkatasubramaniam, A.K., et al.: A comparative study of aortic wall stress using finite element analysis for ruptured and non-ruptured abdominal aortic aneurysms. Eur. J. Vasc. Endovasc. Surg. **28**(2), 168–176 (2004)
9. O'Leary, S.A., et al.: Determining the influence of calcification on the failure properties of abdominal aortic aneurysm (AAA) tissue. J. Mech. Behav. Biomed. Mater. **42**, 154–167 (2015)
10. Vorp, D.A., Raghavan, M.L., Webster, M.W.: Mechanical wall stress in abdominal aortic aneurysm: influence of diameter and asymmetry. J. Vasc. Surg. **27**(4), 632–639 (1998)
11. Demirci, S., Lejeune, G., Navab, N.: Hybrid deformable model for aneurysm segmentation. Boston, 28 June–1 July 2009
12. Das, B., Mallya, Y., Srikanth, S., Malladi, R.: Aortic thrombus segmentation using narrow band active contour model. New York, August–3 September 2006
13. Zohiosa, C., Kossiorisa, G., Papaharilaou, Y.: Geometrical methods for level set based abdominal aortic aneurysm thrombus and outer wall 2D image segmentation. Comput. Methods Programs Biomed. **107**(2), 202–217 (2012)
14. Subašić, M., Lončarića, S., Sorantin, E.: Model-based quantitative AAA image analysis using a priori knowledge. Comput. Methods Programs Biomed. **80**(2), 103–114 (2005)
15. Loncaric, S., Subasic, M., Sorantin, E.: 3-D deformable model for aortic aneurysm segmentation from CT images, Chicago, IL, 23–28 July 2000
16. Loncaric, S., Subasic, M., Sorantin, E.: 3-D deformable model for abdominal, 23–28 July 2000
17. Subasic, M., Loncaric, S., Sorantin, E.: 3-D Image Analysis of Abdominal Aortic Aneurysm. San Diego, CA (2001)
18. Subasic, M., Loncaric, S., Sorantin, E.: Region-based deformable model for aortic wall segmentation, Rome, 18–20 September 2003
19. Magee, D., Bulpitt, A., Berry, E.: Level set methods for the 3D segmentation of CT images of abdominal aortic aneurysms, pp. 141–144 (2001)
20. Bulpitt, A.J., Berry, E.: Spiral CT of abdominal aortic aneurysms: comparison of segmentation with an automatic 3D deformable model and interactive segmentation, San Diego (1998)
21. Magee, D., Bulpitt, A., Berry, E.: Combining 3D deformable models and level set methods for the segmentation of abdominal aortic aneurysms, Manchester (2001)

22. Lee, K., et al.: Three-dimensional thrombus segmentation in abdominal aortic aneurysms using graph search based on a triangular mesh. Comput. Biol. Med. **40**(3), 271–278 (2010)
23. Duquette, A.A., Jodoin, P.-M., Bouchot, O., Lalande, A.: 3D segmentation of abdominal aorta from CT-scan and MR images. Comput. Med. Imaging Graph. **36**(4), 294–303 (2012)
24. Hraiech, N., Carroll, M., Rochette, M., Coatrieux, J.L.: 3D vascular shape segmentation for fluid-structure modeling, Lyon, 13–15 June 2007
25. Freiman, M., Esse, S.J., Joskowicz, L., Sosna, J.: An iterative model-constrained graph-cut algorithm for abdominal aortic aneurysm thrombus segmentation. Rotterdam, 14–17 April 2010
26. Pham, T.D., Golledge, J.: Geostatistically constrained fuzzy segmentation of abdominal aortic aneurysm CT Images, Hong Kong, 1–6 June 2008
27. Majd, E.M., Sheikh, U.U., Abu-Bakar, S.A.R.: Automatic segmentation of abdominal aortic aneurysm in computed tomography images, Kuala Lumpur, 15–18 December 2010
28. Dehmeshki, J., et al.: Computer aided detection and measurement of abdominal aortic aneurysm using computed tomography digital images, Cancun, February 2009
29. Biasi, H.D., Wangenheim, A.V., Silveira, P.G., Comunello, E.: 3D reconstruction of abdominal aortic aneurysms, Maribor (2002)
30. Macía, I., Legarreta, J.H., Paloc, C., Graña, M., Maiora, J., García, G., Blas, M.: Segmentation of abdominal aortic aneurysms in CT images using a radial model approach. In: Corchado, E., Yin, H. (eds.) IDEAL 2009. LNCS, vol. 5788, pp. 664–671. Springer, Heidelberg (2009). doi:10.1007/978-3-642-04394-9_81
31. Bodur, O., et al.: Semi-automatic aortic aneurysm analysis, San Jose (2007)
32. Hosseini, B., et al.: Automatic segmentation of abdominal aortic aneurysm using logical algorithm. Pisa, 17–19 November 2010
33. Maiora, J., Ayerdi, B., Graña, M.: Random forest active learning for AAA thrombus segmentation in computed tomography angiography images. Neurocomputing **126**, 71–77 (2014)
34. Maiora, J., Graña, M.: Abdominal CTA image analysis through active learning and decision random forests: application to AAA segmentation, Brisbane, 10–15 June 2012

Automated LGE Myocardial Scar Segmentation Using MaskSLIC Supervoxels - Replicating the Clinical Method

Iulia A. Popescu[1(✉)], Alessandra Borlotti[2], Erica Dall'Armellina[3], and Vicente Grau[1]

[1] Department of Engineering and Science, Institute of Biomedical Engineering, University of Oxford, Oxford, UK
iulia.popescu@eng.ox.ac.uk
[2] Department of Cardiovascular Medicine, Acute Vascular Imaging Centre (AVIC), University of Oxford, Oxford, UK
[3] Leeds Cardiovascular Research, University of Leeds, Leeds, UK

Abstract. Cardiovascular diseases (CVD) are one of the major killers in modern society. In the UK alone 70,000 people have died from CVD last year according to the 2016 British Heart Foundation Annual Report [2]. Furthermore, the number of patients suffering from chronic CVDs is likely to rise due to an ageing population, as well as, better survival rates after cardiac events. In the case of a heart attack, accurate quantification of the formed scar is essential for improving and deciding the treatment plan, and therefore improving the patient outcomes. In this work we present an automated method for segmenting the scar from late gadolinium enhancement magnetic resonance images using maskSLIC clustering method and Otsu thresholding to divide the myocardium into regions based on differences in intensity. Our method is fast and simple to use, and is consistent across cases and eliminates the spatial inconsistencies previously reported in the literature. The validation is performed using visual assessment from cmr^{42} clinical software.

1 Introduction

When someone suffers a myocardial infarction - or heart attack, the region where the blood supply is interrupted results in the formation of a scar. It is of critical importance that this region gets assessed as accurately as possible as surrounding the scar is tissue that could potentially be salvaged.

Late gadolinium enhancement (LGE) is a contrast enhancement based technique used to highlight the changes in acute ischaemia using cardiac magnetic resonance (CMR) imaging. These changes are both complex and dynamic [3] and therefore accurately segmenting the enhanced areas is not a trivial task. LGE represents the gold standard for the assessment of scarred tissue in myocardial infarction patients. Automated scar segmentation from LGE images is a highly desired tool in clinical practice.

M. Valdés Hernández and V. González-Castro (Eds.): MIUA 2017, CCIS 723, pp. 229–236, 2017.
DOI: 10.1007/978-3-319-60964-5_20

Some of the earliest proposed methods for scar segmentation from LGE images were using voxelwise intensity based global threshold [7]. However, these methods suffer from spacial discontinuity, resulting in areas that are not part of the scar being classified as scar, when in fact they are just pericardium or artefacts. A common way to compute this threshold is taking three to five standard deviations above the mean value of a region that is considered normal within the myocardium. This approach is still used in clinical practice at present.

Extensive work has been done in developing semi-automated and automated methods for segmenting the scar in LGE images. Previous work in semi-automated LGE scar segmentation include adding an additional clustering step, where in [8] k-means clustering is used to enforce connectivity, however by itself k-means has no spacial connectivity. Another semi-automated method [4] for scar segmentation envolves enforcing spacial constraints, however considers the scar to be generally subendocardial and also significantly large. Other methods proposed in the literature suggest using a completely automated pipeline for segmenting the myocaridium as well as the scar. One of these methods is [10] using Graph-cuts to classify the myocardium into infarcted and non-infarcted. Intensity based thresholding and spacial constrains have been reported previously in the literature [9], however, to authors best knowledge a supervoxel approach was never used for cardiac scar segmentation tasks. Further more, our approach differs from [9] in a way that we first enforce spacial connectivity - to remove the noise, and subsequently apply the thresholding to segment the scar.

In this work we present **a novel method for the automated segmentation of the myocardial scar from LGE MR images using maskSLIC supervoxels**. Specific contributions include: **(a)** The development of a parcellation method for dividing the left ventricle into locally similar regions based on intensity, which has the potential of automate the LGE scar segmentation from MR images. **(b)** Automated scar segmentation that excludes the false-positive enhancement regions, without user interference.

The results are compared with the method used at present in clinical practice. This involves manually segmenting a region considered normal - or remote in respect to the enhancement region, and subsequently calculating the threshold, $\mathrm{t} = \mathrm{mean}\,(\text{normal myocardium region}) + 5\mathrm{std}\,(\text{normal myocardium region})$, in order to obtain the desired threshold for segmenting the scarred area. Where remote represents the healthy myocardium and std represents the standard deviation.

2 Materials

The images used for this publication are part of the OxAMI (Oxford Acute Myocardial Infarction) clinical study. The data was provided by the University of Oxford Centre for Clinical Magnetic Resonance Research at the John Radcliffe Hospital, Oxford. The images were acquired on a 3.0T Siemens TIM-Trio whole-body MRI scanner, where LGE-CMR was performed with a T1-weighted segmented inversion-recovery gradient echo-phase sensitive-inversion recovery

(GRE _ PSIR) sequence 5 to 10 min after the administration of 0.1 mmol/kg contrast agent (Gadodiamide, Omniscan^{TM}, GE Healthcare, Amersham, UK). The segmentation of the left ventricle (LV) and the LGE remote areas segmentation was done manually by an expert cardiologist, where the endocardium and epicardium is segmented using the cmr^{42} software (Circle Cardiovascular Imaging, Calgary, Alberta, Canada), and performed by an expert cardiologist.

3 Methods

This work uses 2D LGE MR images. The aim is to reproduce the clinical method, without the need to manually segment the remote - normal myocardium region. Furthermore using our method there is no need to exclude false-positive areas, as the supervoxels approach provides a good regulariser against such occurrences.

3.1 Left Ventricle Parcellation - MaskSLIC Supervoxels

MaskSLIC [5] is a novel implementation of the Simple Linear Iterative Clustering (SLIC) method, where the supervoxels are generated within a predefined mask, compared with the conventional SLIC method, where supervoxels are generated in a rectangular grid. MaskSLIC algorithm is more suitable for segmentation applications in biomedical image analysis, where is unlikely to have regular regions of interest (ROI) to segment and analyse.

The conventional SLIC algorithm divides the ROI into locally similar regions, called superpixels, using a modified k-means clustering method, where a distance metric is used as a weight between pixel intensity and spatial similarity [1]. In conventional SLIC the cluster centres are generated using a grid, and placed equidistantly in the image, while in maskSLIC the cluster centres are generated within a mask using seed points. The position of the seeds is subsequently recomputed using the lowest gradient position in respect to the mask boundary. This method allows for the supervoxels to be generated only within the mask and therefore eliminates the possibility for voxels belonging only partially to the region of interest. The source code is available at URL: https://github.com/benjaminirving/maskSLIC to encourage reimplementation.

The distance transform, D(x), at location x is given by Eq. 1.

$$D(x) = min_{y \in L}((\sum_{i}^{n}(x_i + y_i))^{\frac{1}{2}} \tag{1}$$

where n represents the number of spatial dimensions of the image.

The furthest distance, p^* is found:

$$p^* = arg_x max D(x) \tag{2}$$

and the cluster centres are recomputed N times. Finally SLIC is applied to the voxels that are defined inside the mask.

A supervoxel based approach was used for a couple of reasons, Firstly, the spatial regularisation removes the occurrence of false-positive enhancement areas - that do not belong to the scar region. This is important because the false-positive enhancement areas have to be manually segmented out by the MR analyst, which means more time spent per case. Secondly, it provides meaningful regions for visual assessment and quantification of the extent of scarred tissue. For this application we choose a number of 60 supervoxels and compactness of 0.01.

3.2 Otsu Thresholding - Dividing the Myocardium into Scar and Background

We use the output of the maskSLIC supervoxels to take the mean value of intensity per supervoxel and subsequently build a histogram of intensities from these mean values per supervoxel. We classify the myocardium into two regions based on this histogram using the Otsu threshold. Otsu method is a binarisation algorithm that takes as an input an image assuming that it has two classes and builds a bimodal histogram of foreground and background pixels, and subsequently calculates the optimum threshold to separate the two classes so that the variability within each class is minimal.

Otsu threshold has been previously used in segmenting the scar from LGE MR images, however to our best knowledge it has never been used in conjunction with a supervoxels based method. By using the maskSLIC [5] supervoxels approach prior to classification, we enforce spacial consistency in therefore overcome the previous limitations reported in the literature [6].

Otsu thresholding is a method used classify the region of interest in two classes: background and foreground pixels. The method computes the threshold that separates the two classes using Eq. 3.

$$\sigma_w^2(t) = \omega_0(t)\sigma_0^2(t) + \omega_1(t)\sigma_1^2(t) \tag{3}$$

where $\omega 0$ and $\omega 1$ are the probabilities of the two classes, t is the threshold, and σ is the variation for each of the classes, with:

$$\omega_0(t) = \sum_{i=0}^{t-1}(p(i)) \tag{4}$$

$$\omega_1(t) = \sum_{i=t}^{L-1}(p(i)) \tag{5}$$

where class propability $\omega_{0,1}$ are computed from the L histograms.

4 Results

The set of experiments was performed on ten myocardial infarction patients using late gadolinium enhancement MR images. The cmr[42] clinical software was used - on the same images, to determine the position and dimension of the scarred tissue in order to evaluate the performance of the proposed method. Worth mentioning that in clinical practice the scars obtained are not directly segmented, but computed using a healthy myocardium area. The LGE images were acquired 24 h after myocardial infarction patient was presented at the emergency care unit - considered the acute stage post infarction. Also the stage at which, if assessed accurately, the extent of the scar can lead to a more accurate treatment plan.

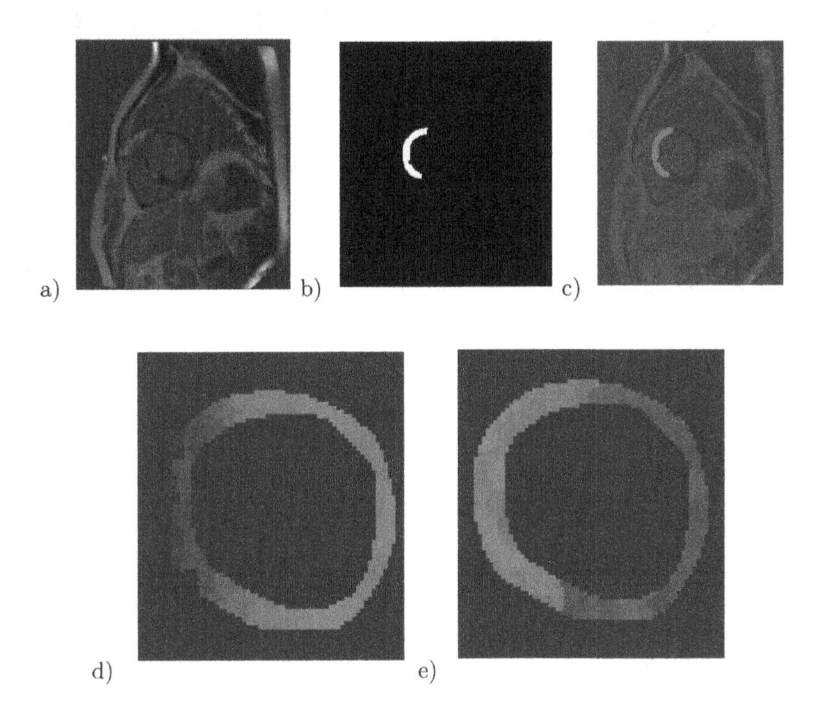

Fig. 1. (a) Late gadolinium enhancement (LGE) for myocardial infarction patient, case1. (b) MaskSLIC and Otsu thresholding generated scar enhancement. (c) MaskSLIC supervoxles and Otsu thresholding over corresponding LGE image. (d) MaskSLIC supervoxels (e) MaskSLIC mean value of intensity per supervoxel. (Color figure online)

As a preprocessing step, the LGE images are normalised to reduce the noise and possible artefacts.

The supervoxels are generated in the 2D left ventricular mask - for each slice, and subsequently the corresponding Otsu threshold is computed. The Otsu threshold is not a global threshold in this case, but a local parameter, which changes with each slice, and it has a local significance in accordance with the 2D distribution of intensities.

In Fig. 1(b) we present the final mask of the enhanced region using maskSLIC supervoxels and Otsu thresholding to separate the background and foreground

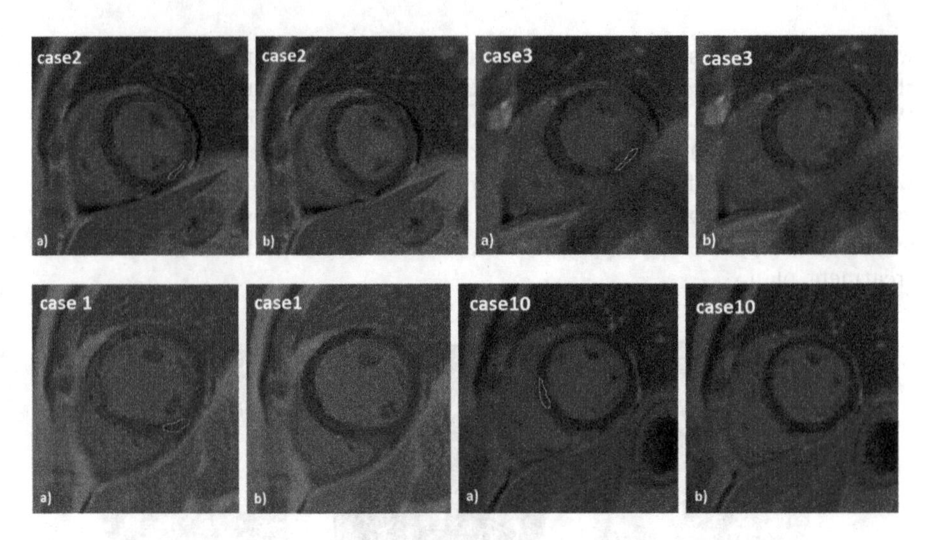

Fig. 2. Four cases with a and b representing the scars generated using mean plus 5 standard deviations of the healthy myocardium, and maskSLIC plus Otsu thresholding on the left ventricular mask respectively.

pixels for one of cases. Figure 1(d) we show how these supervoxels were generated based local similarities in intensity, and subsequently the mean intensity value per supervoxel (Fig. 1(e)). It can be seen that variations in intensity are higher in the enhanced areas (yellow and orange) than in the normal ones (in blue and green).

Using the maskSLIC approach, we eliminate the possibility of supervoxels being generated outside the myocardial mask or across the border of the myocardial mask, therefore increasing the accuracy of the segmentation method. In Fig. 2 we present four cases with the scars segmented using the conventional clinical method (a) and using the proposed maskSLIC and Otsu thresholding method (b).

The supervoxels output is used to compute the mean value of intensity per supervoxel, from which we subsequently build a histogram of intensities.

Otsu thresholding is then used to separate background and foreground pixels and segment the myocardial scar. In Fig. 3(a) we present the Otsu thresholding method on a normalised LGE, where the stem histogram represents the distribution of mean intensities per supervoxel.

We use the Dice similarity coefficient as a validation metric between the proposed method and the scars generated in clinical context. The results are presented in Fig. 3(b) for all the ten cases of the myocardial infarction patients.

The post-processing is done using Matlab and Python. The average processing times using a Intel i5, 32 GB (RAM) and 64-bit Ubuntu operating system. Average running time per case: 2.4 s.

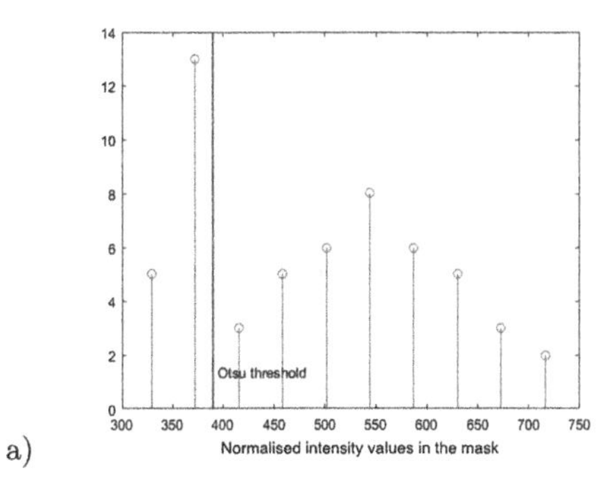

a)

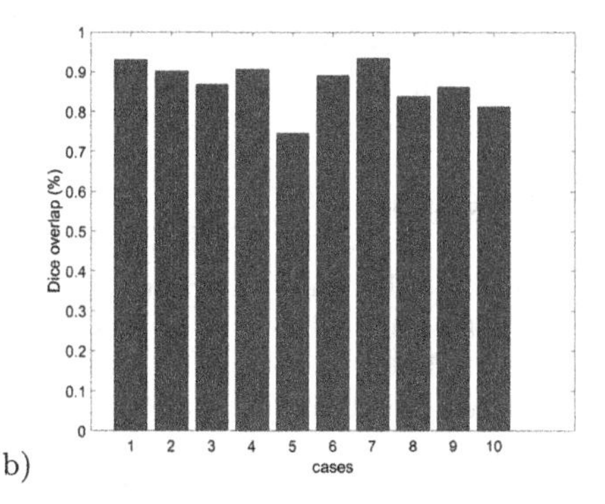

b)

Fig. 3. (a) Late gadolinium enhancement for myocardial infarction patient, case1. (b) MaskSLIC and Otsu thresholding LGE scar enhancement. (c) MaskSLIC supervoxles. (d) MaskSLIC mean value of intensity per supervoxel.

5 Discussion

The proposed method has the potential to automate the LGE scar segmentation process. At present, considerable user input is required in order to correctly estimate the extent of the myocardial scar in a clinical setting. Firstly, because the clinical used approach is not enforcing spacial constraints, meaning that subsequent exclusion areas need to be hand-drawn in order to eliminate the false-positive enhancement areas. Secondly, no global thresholding formula can be applied without having to reposition the remote myocardium area (manually segmented) around, until the best match is achieved. This can be time consuming and obviously subjective to inter-user variability.

Our method does not involve the need of manually segmenting the remote area of the myocardium, as well as having to manually exclude false positive enhancements, as the maskSLIC supervoxels approach represents an excellent regulariser within the region of interest.

It is perhaps worth mentioning that Dice overlap similarity metric can be biased in this case, as we compare against a ground truth that has not been generated manually (but by using a remote area to extract the mean + 5std of), and therefore we argue that the method we propose performs better than the one used at present in clinical practice.

In order to replicate the clinical method as much as possible, we suggest an additional step, where a region can be selected by simply selecting one of supervoxels in the remote area and compute the threshold value above five standard deviations of the average intensity of that supervoxel. However, this is not necessary for the method to perform, as presented in this work the method achieves comparable results with the clinical method.

Acknowledgements. This research is supported by the RCUK Digital Economy Programme grant number EP/G036861/1, Oxford Centre for Doctoral Training in

Healthcare Innovation. VG is supported by a BBSRC grant (BB/I012117/1), an EPSRC grant (EP/J013250/1) and by BHF New Horizon Grant NH/13/30238.

References

1. Achanta, R., Shaji, A., Smith, K., Lucchi, A., Fua, P., Süsstrunk, S.: SLIC superpixels compared to state-of-the-art superpixel methods. IEEE Trans. Pattern Anal. Mach. Intell. **34**(11), 2274–2281 (2012)
2. British Heart Foundation: CVD Statistics - BHF UK Factsheet (CVD) (2016)
3. Dall'Armellina, E., Karia, N., Lindsay, A.C., Karamitsos, T.D., Ferreira, V., Robson, M.D., Kellman, P., Francis, J.M., Fofar, C., Prendergast, B.D., Banning, A.P., Channon, K.M., Kharbanda, R.K., Neubauer, S., Choudhury, R.P.: Dynamic changes of edema and late gadolinium enhancement after acute myocardial infarction and their relationship to functional recovery and salvage index. Circ.: Cardiovasc. Imaging **4**(3), 228–236 (2011)
4. Hsu, L.Y., Ingkanisorn, W.P., Kellman, P., Aletras, A.H., Arai, A.E.: Quantitative myocardial infarction on delayed enhancement MRI. Part II: clinical application of an automated feature analysis and combined thresholding infarct sizing algorithm. J. Magn. Reson. Imaging **23**(3), 309–314 (2006)
5. Irving, B., et al.: maskSLIC: regional superpixel generation with application to local pathology characterisation in medical images, pp. 1–7 (2016). http://arxiv.org/abs/1606.09518
6. Karim, R., Bhagirath, P., Claus, P., James Housden, R., Chen, Z., Karimaghaloo, Z., Sohn, H.M., Lara Rodríguez, L., Vera, S., Albà, X., Hennemuth, A., Peitgen, H.O., Arbel, T., Gonzàlez Ballester, M.A., Frangi, A.F., Götte, M., Razavi, R., Schaeffter, T., Rhode, K.: Evaluation of state-of-the-art segmentation algorithms for left ventricle infarct from late Gadolinium enhancement MR images. Med. Image Anal. **30**, 95–107 (2016)
7. Karim, R., Housden, R.J., Balasubramaniam, M., Chen, Z., Perry, D., Uddin, A., Al-Beyatti, Y., Palkhi, E., Acheampong, P., Obom, S., Hennemuth, A., Lu, Y., Bai, W., Shi, W., Gao, Y., Peitgen, H.O., Radau, P., Razavi, R., Tannenbaum, A., Rueckert, D., Cates, J., Schaeffter, T., Peters, D., MacLeod, R., Rhode, K.: Evaluation of current algorithms for segmentation of scar tissue from late gadolinium enhancement cardiovascular magnetic resonance of the left atrium: an open-access grand challenge. J. Cardiovasc. Magn. Reson.: Off. J. Soc. Cardiovasc. Magn. Reson. **15**, 105 (2013)
8. Metwally, M.K., El-Gayar, N., Osman, N.F.: Improved technique to detect the infarction in delayed enhancement image using K-mean method. In: Campilho, A., Kamel, M. (eds.) ICIAR 2010. LNCS, vol. 6112, pp. 108–119. Springer, Heidelberg (2010). doi:10.1007/978-3-642-13775-4_12
9. Tao, Q., Milles, J., Zeppenfeld, K., Lamb, H.J., Bax, J.J., Reiber, J.H.C., Van Der Geest, R.J.: Automated segmentation of myocardial scar in late enhancement MRI using combined intensity and spatial information. Magn. Reson. Med. **64**(2), 586–594 (2010)
10. Wei, D., Sun, Y., Ong, S.H., Chai, P., Teo, L.L., Low, A.F.: A comprehensive 3-D framework for automatic quantification of late gadolinium enhanced cardiac magnetic resonance images. IEEE Trans. Med. Imaging **60**(6), 1499–1508 (2013)

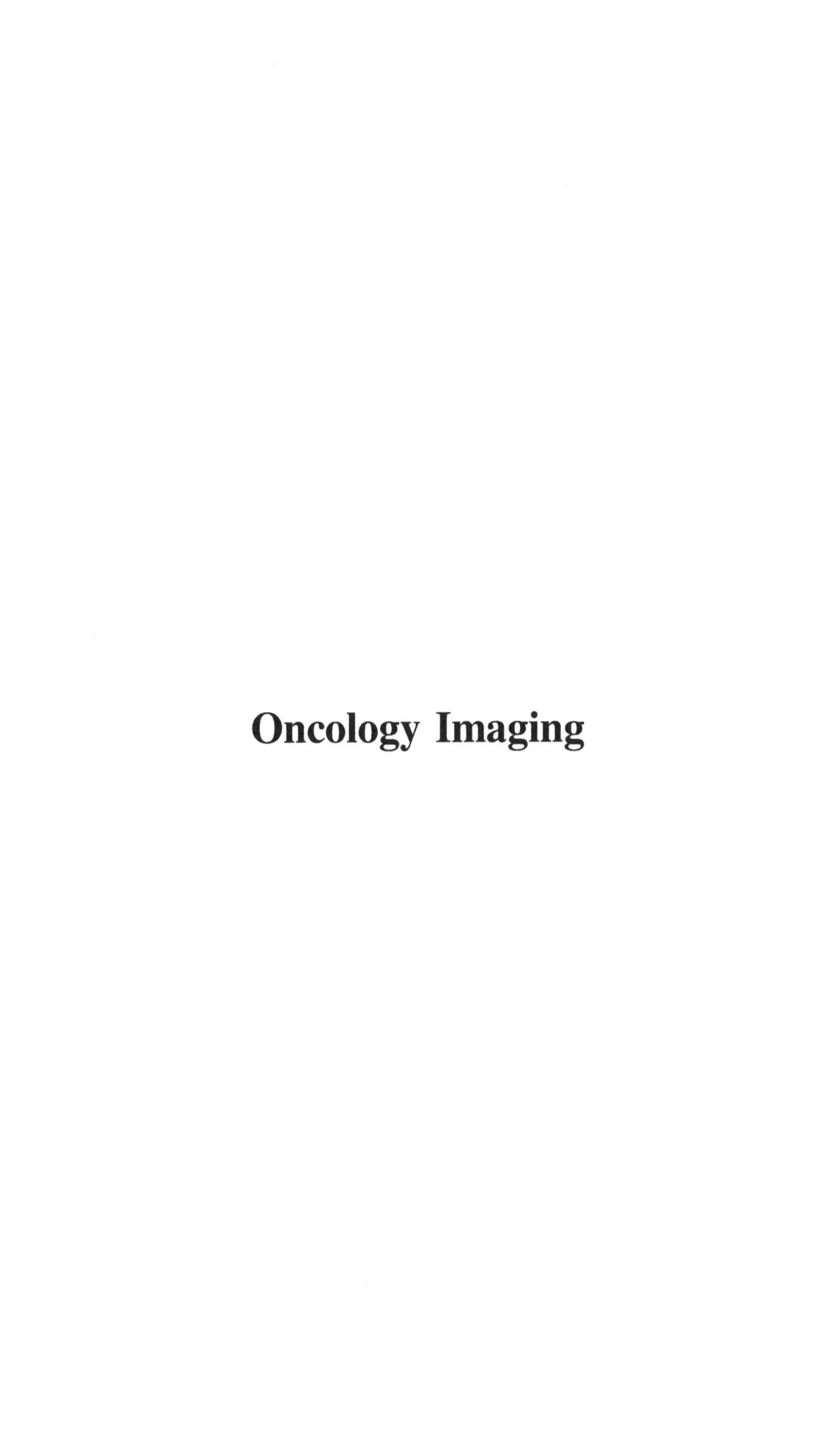

Oncology Imaging

Multi-task Fully Convolutional Network for Brain Tumour Segmentation

Haocheng Shen(✉), Ruixuan Wang, Jianguo Zhang, and Stephen McKenna

Computing, School of Science and Engineering, University of Dundee,
Dundee, UK
hyshen@dundee.ac.uk

Abstract. In this paper, a novel, multi-task fully convolutional network (FCN) architecture is proposed for automatic segmentation of brain tumour. The proposed network builds on the hierarchical relationship between tumour substructures with branch and leaf losses imposed and optimised simultaneously. The network takes multimodal MR images along with their symmetric-difference images as input and extracts multi-level contextual information, firstly by the branch losses which are then fed to the leaf loss in a combination stage. The model was evaluated on BRATS13 and BRATS15 datasets and results show that the proposed multi-task FCN outperforms single-task FCN on all sub-tasks. The method is among the most accurate available and its computational cost is relatively low at test time.

Keywords: Deep learning · Tumour segmentation · Multi-task learning

1 Introduction

Accurate localization of brain tumours in 3D MR images is clinically important for planning treatment, guiding surgery and monitoring the rehabilitation progress of patients. Unreliable segmentation risks potentially irreversible impact from surgery (e.g., difficulty in speaking fluently). Since manually segmenting brain tumour, particularly in 3D images, is a tedious and time-consuming process, computer-aided, automatic and reliable segmentation is desirable and would save clinicians' valuable time.

Among brain tumours, gliomas appear most frequently in adult patients [1] and can be graded as high grade (HG) or low grade (LG) according to aggressiveness. Due to the diversity of size, shape, location and appearance of gliomas, multimodal MRI is often used to enhance the ability to differentiate tumour and tumour substructures. Figure 1(a) shows a representative HG gliomas tumour and its sub-regions whose boundaries have been delineated by experts.

The automatic segmentation of glioma and its substructures is often formulated as a patch-level or voxel-level classification problem in which each (either 2D or 3D) patch or voxel in the 3D MR is classified as one type of substructure and the collection of all patches' or voxels' classifications generates the final,

M. Valdés Hernández and V. González-Castro (Eds.): MIUA 2017, CCIS 723, pp. 239–248, 2017.
DOI: 10.1007/978-3-319-60964-5_21

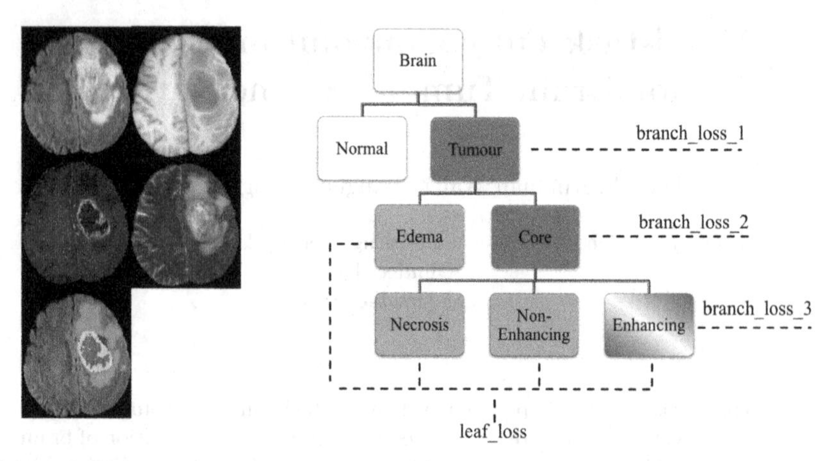

(a)Multimodal brain tumour images (b) Hierarchical labelling tree

Fig. 1. (a) An HG tumour in multimodal MRI. Flair, T1, T1c, T2 modalities (above) and expert delineation (below) showing: edema (green), necrosis (red), non-enhancing (blue), enhancing (yellow). (b) Hierarchical labelling tree of tumour tissues with corresponding *branch loss* and *leaf loss*. Note that the labels at the *leaves* (green blocks) are *mutually exclusive* whereas labels at the *branches* (brown blocks) are not. (Color figure online)

complete segmentation. While hand-crafted features and conditional random field (CRF) incorporating class-label smoothness terms have been adopted for the voxel-level classification [1,2], deep convolutional neural networks (CNNs), which have achieved substantial performance breakthroughs in several natural and medical image analysis benchmarks by automatically learning high-level discriminative feature representations, are not suprissingly achieving state-of-the-art results when applied to MRI brain tumour segmentation [3–5]. Specifically, Pereira et al. [3] trained a traditional 2D CNN as a patch-level classifier, and Havaei et al. [4] trained a 2D CNN to classify larger patches in a cascaded structure in order to capture both small and large-scale contextual information. Very recently, Kamnitsas et al. [5] trained a 3D CNN directly on 3D instead of 2D patches and considered global contextual features via an extra down-sampling path. Note that all these methods are *patch*-level classification.

Fully convolutional networks (FCNs) lack the fully connected layers often used for the last few layers in CNNs. FCNs have achieved promising results for natural image segmentation [9,10] as well as medical image segmentation [11–13]. In FCNs, up-sampling (de-)convolutional layers can be added on top of the traditional down-sampling convolutional layers in order to gain the same spatial size at the network output as at the input. Compared to CNNs applied to a sliding window on the input, FCNs can be applied to the whole input without using a sliding window and generate the classification result for each voxel (or pixel). Therefore, FCNs as voxel-level classifiers are more computationally efficient than traditional CNNs as patch-level classifiers.

In this paper, we propose a tree-structured, multi-task FCN model for brain tumour segmentation. The main contributions of our work are: (1) formulation and application of a tree-structured, multi-task FCN to multimodal brain tumour segmentation that implicitly encodes the hierarchical relationship of tumour sub-structures; (2) experiments providing evidence that the tree-structured, multi-task FCN can improve segmentation performance in all sub-tasks compared to single-task FCN on both BRATS13 and BRATS15 datasets; the proposed method is ranked top on the BRATS 2013 testing set and is more efficient than the closest competing methods.

2 Methodology

2.1 Hierarchical Labeling Tree

A tumour typically contains four sub-structures as shown in Fig. 1(a): edema (green), necrosis (red), non-enhancing (blue) and enhancing (yellow). We observe a hierarchical label relationship of tumour sub-regions, shown as a tree in Fig. 1(b). Specifically, the tree starts from a brain partitioned into non-tumour and tumour. The *complete* tumour normally consists of *edema* and *tumour core*. The *tumour core* can be further divided into *necrosis*, *non-enhancing* and *enhancing* parts. Finally, the leaves of the tree represent the five classes (including background) that are mutually exclusive (Fig. 1(b)). Encoding such a hierarchical relationship into an FCN framework can benefit tumour segmentation. For example, an enhancing part is always labeled as tumour core. We describe an FCN in a multi-task framework designed to implicitly encode the hierarchical relationship. In the following, we first describe a single-task FCN structure, upon which the proposed multi-task FCN is built.

2.2 Single-Task FCN

Our single-task FCN is a variant of FCN [9,12]. It includes a down-sampling path and three up-sampling paths, as shown in Fig. 2. The down-sampling path contains three convolutional blocks separated by max pooling (see yellow arrows in Fig. 2). Each block includes 2–3 convolutional layers similar to the VGG-16 network [6]. This down-sampling path extracts multi-scale features from low-level texture to higher-level context features. The three up-sampling paths are connected to the down-sampling path at different stages, i.e., at the last convolutional layer of each convolutional block in the downsampling path. Such a structure ensures that up-sampled feature maps are from different scales. The final feature maps in each of the three up-sampling paths (purple rectangles in Fig. 2) have the same spatial size as the input to the FCN and are concatenated before being fed to the final classification layer. ReLU activation functions and batch normalization are used after each convolutional layer. Note that the single-task FCN only considers separating the five classes at the leaf level in the hierarchical tree (i.e., a typical multi-class classification task). The efficacy of this single-task FCN was evaluated in [7].

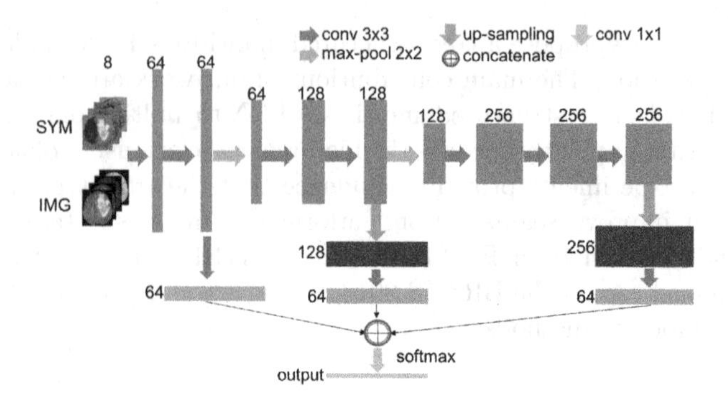

Fig. 2. Single-task FCN. Images and symmetry maps are concatenated as the input to the net [7]. Colored rectangles represent feature maps with numbers nearby being the number of feature maps. Best viewed in color. (Color figure online)

2.3 Multi-task FCN

The single-task FCN predicts the class label for each voxel. Although it can produce good probability maps its architecture ignores any hierarchical relationship shown in Fig. 1(b)). We design a multi-task FCN to implicitly encode such a relationship of tumour tissues labels. Specifically, there are two types of loss in our framework: *branch* loss and *leaf* loss (see Fig. 1(b)). The ground truth labels for branch loss (the brown blocks) are *hierarchical*, e.g., *complete tumour* contains *core* while *core* contains *enhancing* parts. On the other hand, the ground truth labels for leaf loss (the green blocks) are *mutually exclusive.* Note that the *enhancing* parts are involved in both branch loss and leaf loss. When designing a structure to match such a relationship, we also consider that the information flow runs from root to leaves. This implies that the branch loss will be applied earlier whilst leaf loss is the final layer.

The structure of the proposed multi-task FCN is illustrated in Fig. 3. We formulate the segmentation task within a multi-task learning framework, rather than treating it as a single voxel-wise classification problem. Three single-task FCNs with shared down-sampling path and three different up-sampling branches (the blue arrows in Fig. 3) are applied for three separate tasks: *complete tumour*, *tumour core* and *enhancing tumour* classification. Then, the outputs (i.e., probability maps) from the three branches are concatenated and fed to a block of two convolutional layers followed by the final softmax classification layer ('combination stage' in Fig. 3). The 'combination stage' task is a 5-class classification task whereas the others are binary classification tasks. Cross-entropy loss is used for each task. Therefore, the total loss in our proposed multi-task FCN is the sum of branch loss and leaf loss:

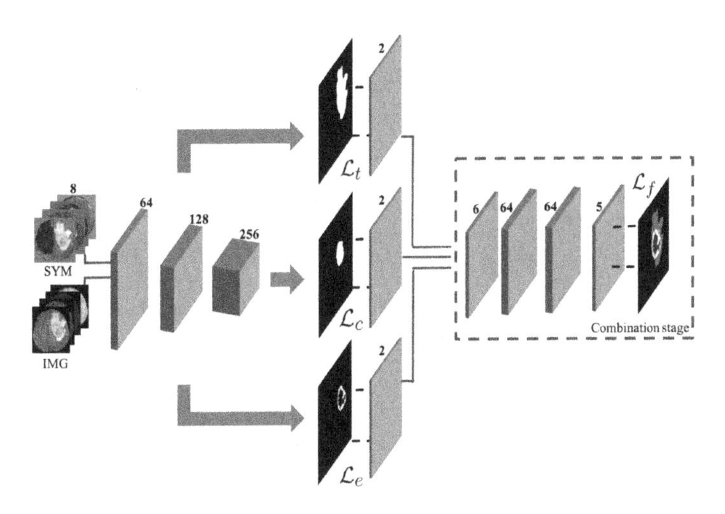

Fig. 3. The structure of multi-task FCN. The three up-sampling branches in the three FCNs are represented by blue arrows. Note that the upsampling paths are connected to the down-sampling path at different stage as described in Sect. 2.2 and Fig. 2. (Color figure online)

$$\begin{aligned}
\mathcal{L}_{total}(w) &= \mathcal{L}_{leaf}(w_{leaf}) + \mathcal{L}_{branch}(w_{branch}) \\
&= \mathcal{L}_{leaf}(w_{leaf}) + \sum_{m \in \{t,c,e\}} \mathcal{L}_m(w_m) \\
&= - \sum_{m \in \{t,c,e,l\}} \sum_n \sum_i \log P_m(l_m(x_{n,i}); x_{n,i}, w_m)
\end{aligned} \tag{1}$$

where $\{t, c, e, l\}$ are the tasks of *complete tumour*, *tumour core*, *enhancing core* and the leaf output by the final combination stage, respectively, and $w = \{w_t, w_c, w_e, w_l\}$ is the set of weight parameters in the multi-task FCN. $\mathcal{L}_m$ refers to the loss function of each task. $x_{n,i}$ is the i-th voxel in the n-th image used for training, and P_m refers to the predicted probability of the voxel $x_{n,i}$ belonging to class l_m.

In the proposed multi-task FCN, 2D slices from 3D MR volumes in axial view are used as part of the input to the network. In addition, since adding brain symmetry information has proved helpful for FCN based tumour segmentation [7], 'symmetric intensity difference' maps are combined with the original slices as input, resulting in 8 input channels to the network (see Figs. 2 and 3).

3 Evaluation

Our model was evaluated on BRATS13 and BRATS15 datasets. Each patient's data in the two datasets includes 4 modalities (T1, T1-contrast or T1c, T2, and

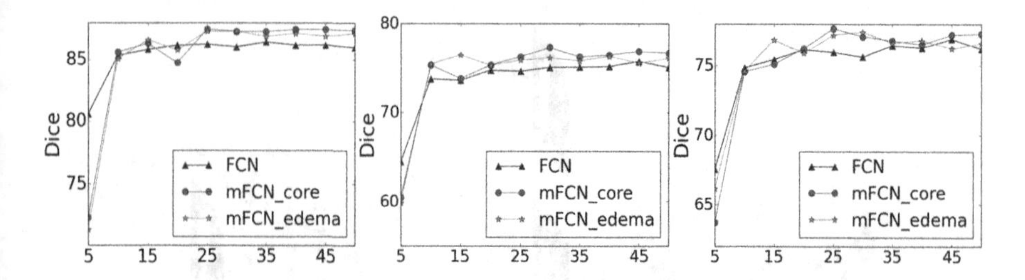

Fig. 4. Validation results of three models on BRATS13. From left to right: *Complete*, *Core* and *Enhancing* tumour task. The vertical axis is Dice while horizontal axis is the number of epochs.

Flair) which were skull-stripped and co-registered. BRATS13 contains 20 high-grade training data with known ground-truth segmentation maps and 10 high-grade testing data with ground-truth segmentation kept only by the BRATS13 organizer. (We do not use the 10 low-grade data; here we focus on high-grade tumour segmentation). For BRATS15, we used 220 released, annotated high-grade patients' images in the original training set for both training and testing. For each MR image, voxel intensities were normalised to have zero mean and unit standard deviation.

Quantitative evaluation was performed on three sub-tasks: (1) the *complete* tumour (including all four tumour sub-structures); (2) the tumour *core* (including all tumour sub-structures except "edema"); (3) the *enhancing* tumour region (including only the "enhancing tumour" sub-structure). For each sub-task, *Dice*, *Sensitivity* and *Positive Predictive Value (PPV)* were computed. Our network model was implemented in Keras with Theano as backend. The network was trained using the Adam optimizer with learning rate 0.001. The down-sampling path was initialized with VGG-16 weights [6] while up-sampling paths were initialized randomly using He's method [14].

3.1 Results on BRATS13 Dataset

A 5-fold cross validation was performed on the 20 high-grade training data in BRATS13. The training folds were augmented by scaling, rotating, and left-right flipping, resulting a dataset which was three times larger than the original one. Besides the proposed multi-task model, a variant of the proposed multi-task model was also evaluated by replacing the loss function of the *core* task with that of the *edema* task whose purpose is to segment edema. The motivation of evaluating such a variant model is from the fact that tumour *core* is a super-structure containing *enhancing*, *non-enhancing* and *necrotic* parts. These sub-structures are different in texture and appearance, e.g., in T1c (see Fig. 1) *enhancing* sub-structure shows hyper-intensity signal whereas *necrosis* has low-intensity signal. This causes large variability of *core* across patients which could be difficult for the network to model. In comparison, the texture and appearance of *edema* are relatively consistent across patients (e.g., hyper-intensity signal in

Flair). As a result, three models were evaluated on both validation set and test set: (1) single-task FCN (Fig. 2), denoted 'FCN' in the following; (2) the multi-task FCN with core task, denoted 'mFCN_core'; (3) the multi-task FCN with edema task, denoted 'mFCN_edema'.

With the validation set, Fig. 4 shows *Dice* values at every 5 epochs for each of the three models and for each of the three tasks. It can be observed that although at the starting points (e.g., the fifth epoch), mFCN_core and mFCN_edema have lower performance due to the extra parameters in the network, the highest *Dice* values achieved by mFCN_core and mFCN_edema are clearly higher than the highest *Dice* value achieved by the FCN in all the three tasks. Also, mFCN_core and mFCN_edema outperform the FCN in all three tasks at most training epochs, especially for mFCN_core. mFCN_edema gives competitive segmentation results in *complete* and *enhancing* tasks while it is slightly worse on the *core* task compared to mFCN_core, which indicates replacing *core* task by *edema* task might be unnecessary in this dataset. This could be partially due to the powerful capability of FCN to handle large appearance variability. However, mFCN_edema still outperforms FCN on all tasks, evidencing the efficacy of the tree-structured, multi-task FCN framework. The validation performances of both mFCN_core and mFCN_edema models were saturated or even decreased around 30 epochs. Therefore, models trained at 30 epochs were used for benchmarking on test data.

Further evaluation was performed on the 10 high-grade testing data (see Table 1). Here, all the 20 high-grade training data were used to train the models. The returned evaluation from the official organizer showed that both mFCN models are ranked higher than the FCN (Table 1). Due to the small size of the testing set, we observe marginal improvements in most tasks in terms of *Dice* while *Sensitivity* and *PPV* changed inversely (e.g., *Sensitivity* of mFCN increased over the FCN while *PPV* decreased a bit). Thus, we conducted a further evaluation by calculating F-scores which is the harmonic mean of *Sensitivity* and *PPV* for each model (see Table 3); mFCN outperformed FCN in all segmentation tasks and mFCN_core was the best on the *core* task. This conclusion is consistent with the results on the validation set.

Table 1 also shows that our proposed models are among the best of the state-of-the-art results on the BRATS13 testing set. Specifically, our models outperformed the best performers (Tustison et al. [2], Meier and Reza) from the BRATS13 challenge [1] as well as a semi-automatic method [8]. For CNN methods, our results are competitive with Pereira's et al. [3] and better than Havaei's et al. [4] while being roughly twice as fast in terms of average computational time (3 min compared to the 8 min reported by Pereira et al. [3]) due to the fast inference property of FCN. A direct comparison with 3D CNN [5] is not applicable as they did not report results on this dataset.

3.2 Results on BRATS15 Dataset

Here we randomly split 220 high-grade data in BRATS15 training set into three subsets at a ratio of 6:2:2, resulting in 132 training data, 44 validation data and 44 test data. No data augmentation was performed on this dataset. The performance curves are shown in Fig. 5.

Table 1. Comparison with the state-of-the-art on the testing set (ranked by VSD evaluation system [1])

Method	Dice			Positive Predictive Value			Sensitivity		
	Complete	Core	Enhancing	Complete	Core	Enhancing	Complete	Core	Enhancing
Pereira [3]	88	83	77	88	87	74	89	83	81
mFCN_core	88	83	75	86	85	70	91	85	83
mFCN_edema	88	82	76	85	82	72	92	85	82
FCN	87	82	75	85	87	72	89	79	80
Kwon [9]	88	83	72	92	90	74	84	78	72
Havaei [4]	88	79	73	89	79	68	87	79	80
Tustison [2]	87	78	74	85	74	69	89	88	83
Meier [1]	82	73	69	76	78	71	92	72	73
Reza [1]	83	72	72	82	81	70	86	69	76

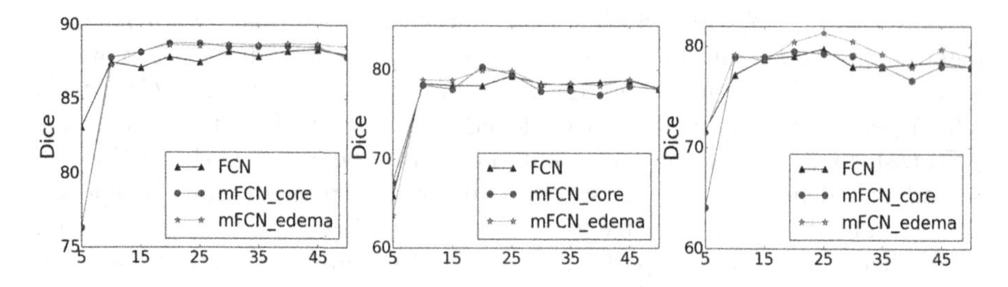

Fig. 5. Validation results of three models on BRATS15. From left to right: *Complete*, *Core* and *Enhancing* tumour task.

For the *Complete* task, both mFCN models outperform the baseline FCN. However, the mFCN_core model becomes overfitted more easily on the other two tasks. This may be due to the more powerful ability of the mFCN_core model to learn the larger appearance variability of the *Core* region in the training data, such that some of the largely varied *Core* region in the testing data may contain some new appearance or texture features which will not be well predicted by the over-trained mFCN_core model. For *Enhancing* task, mFCN_edema performs better than FCN_core and FCN, and its performance peaks at epoch 25.

On the 44 testing data, we trained the model for 25 epochs with the combination of training and validation set. The results of FCN and mFCN_edema are shown in Table 2. We found mFCN_core achieved the best results in all tasks in terms of *Dice* and *Sensitivity* as well as F1 score (see Table 3). This is consistent with the BRATS13 test result while contrary to the BRATS15 validation result where mFCN_edema seems to perform best. We might attribute this to several possible causes such as the relatively noisy ground truth in BRATS15, random initialization, unrepresentative epoch samplings or heterogeneity of data. Overall, from Table 3, we can conclude that both mFCN models appear to be better than the baseline FCN while mFCN_core is perhaps slightly better than mFCN_edema on this dataset.

Table 2. Performance on the BRATS15 44 testing set

Method	Dice			Positive Predictive Value			Sensitivity		
	Complete	Core	Enhancing	Complete	Core	Enhancing	Complete	Core	Enhancing
FCN	88.1	70.9	72.5	**92.2**	**82.7**	**79.7**	86.0	67.5	70.5
mFCN_edema	**88.5**	71.0	73.1	91.2	82.4	78.7	**87.5**	67.9	71.4
mFCN_core	**88.5**	**72.6**	**73.2**	91.1	81.3	78.2	**87.5**	**70.1**	**72.2**

Table 3. F-score on the BRATS13 and BRATS15 testing set

Method	BRATS13			BRATS15		
	Complete	Core	Enhancing	Complete	Core	Enhancing
FCN	87.0	82.8	75.8	89.0	74.3	74.8
mFCN_edema	**88.4**	83.5	**76.7**	**89.3**	74.5	74.9
mFCN_core	**88.4**	**85.0**	75.9	**89.3**	**75.3**	**75.1**

4 Conclusion

In this paper, we introduced a tree-structured, multi-task FCN for brain tumour segmentation. Our approach formulates and jointly learns the *Complete*, *Core* and *Enhancing* tumour segmentation tasks in a multi-task framework that implicitly encodes the hierarchical relationship of tumour subregions. This multi-task FCN achieved state-of-the-art results and improved segmentation in all sub-tasks on BRATS13 and BRATS15 datasets compared to the single-task FCN. Our method is among the top ranked methods and has relatively low computational cost. We would point out that the proposed multi-task network only takes the relationship between branch and leaf, and is one possible implementation of the tree in Fig. 1(b). However, the idea of imposing a loss at branch level is generic. Future work could include designing a structure to encode the hierarchy between branches.

References

1. Menze, B.H., et al.: The multimodal brain tumour image segmentation benchmark (BRATS). Med. Imaging **34**(10), 1993–2024 (2015)
2. Tustison, N.J., et al.: Optimal symmetric multimodal templates and concatenated random forests for supervised brain tumour segmentation (simplified) with ANTsR. Neuroinformatics **13**(2), 209–225 (2015)
3. Pereira, S., et al.: Brain tumour segmentation using convolutional neural networks in MRI images. Med. Imaging **35**(5), 1240–1251 (2016)
4. Havaei, M., et al.: Brain tumour segmentation with deep neural networks. Med. Image Anal. **35**, 18–31 (2017)
5. Kamnitsas, K., et al.: Efficient multi-scale 3D CNN with fully connected CRF for accurate brain lesion segmentation. Med. Image Anal. **36**, 61–78 (2017)

6. Simonyan, K., et al.: Very deep convolutional networks for large-scale image recognition. arXiv preprint arXiv:1409.1556 (2014)
7. Shen, H., et al.: Efficient symmetry-driven fully convolutional network for multimodal brain tumour segmentation (2017). Submitted to ICIP
8. Kwon, D., Shinohara, R.T., Akbari, H., Davatzikos, C.: Combining generative models for multifocal glioma segmentation and registration. In: Golland, P., Hata, N., Barillot, C., Hornegger, J., Howe, R. (eds.) MICCAI 2014. LNCS, vol. 8673, pp. 763–770. Springer, Cham (2014). doi:10.1007/978-3-319-10404-1_95
9. Long, J., et al.: Fully convolutional networks for semantic segmentation. In: CVPR 2015 (2015)
10. Chen, L.-C., et al.: Semantic image segmentation with deep convolutional nets and fully connected CRFs. arXiv preprint arXiv:1412.7062 (2014)
11. Ronneberger, O., Fischer, P., Brox, T.: U-Net: convolutional networks for biomedical image segmentation. In: Navab, N., Hornegger, J., Wells, W.M., Frangi, A.F. (eds.) MICCAI 2015. LNCS, vol. 9351, pp. 234–241. Springer, Cham (2015). doi:10.1007/978-3-319-24574-4_28
12. Chen, H., et al.: Deep contextual networks for neuronal structure segmentation. In: AAAI 2016 (2016)
13. Chen, H., et al.: DCAN: deep contour-aware networks for accurate gland segmentation. In: CVPR 2016 (2016)
14. He, K., et al.: Delving deep into rectifiers: surpassing human-level performance on imagenet classification. In: ICCV 2015 (2015)

FF-CNN: An Efficient Deep Neural Network for Mitosis Detection in Breast Cancer Histological Images

Boqian Wu[1], Tasleem Kausar[1], Qiao Xiao[1], Mingjiang Wang[1(✉)], Wenfeng Wang[2(✉)], Binwen Fan[1], and Dandan Sun[3]

[1] Department of Electronics and Communication Engineering, Harbin Institute of Technology, Shenzhen 518000, China
mjwang@hit.edu.cn

[2] State Key Laboratory of Desert and Oasis Ecology, Xinjiang Institute of Ecology and Geography, Chinese Academy of Sciences, Urumqi 830011, China
wangwenfeng@ms.xjb.ac.cn

[3] The Second Surgical Department, Zouping Hospital of Traditional Chinese Medicine, Zouping 256200, China

Abstract. We utilized deep fused fully convolutional neural network (FF-CNN) based on the knowledge transferred from pre-trained AlexNet model to detect mitoses in hematoxylin and eosin stained histology image of breast cancer. Currently, existing mitosis counting methods are based on either the handcrafted features including morphology, color and texture or the abstract features learning from deep neural network. The handcrafted features are pertinent, corresponding to what learned from lower layers of deep convolutional neural network (CNN), on the other hand, features extracted from higher layers are comprehensive. Both handcrafted and extracted features are significant to mitosis detection. More importantly, this detection suffers from class-imbalance and the inconsistent staining color of H&E images. This paper proposed a modified fully convolutional neural network (FCN) structure combining the rich features from different level layers together and a multi-step fine-tuning strategy to reduce over-fitting. In order to treat class-imbalance, we utilized a cascaded approach to select the most confusing non-mitosis samples which made the training efficiently, and weighted the loss function by the corresponding class frequency. For the inconsistent staining appearance issue, we applied the method of stain-normalization to the augment training samples to improve generalization ability of model and to pre-process testing images to obtain more accurate results. As preliminarily validated on the public 2014 ICPR MITOSIS data, our method achieves a better performance in term of detection accuracy than ever recorded for this dataset with an acceptable detection speed.

Keywords: Breast cancer · Mitosis detection · CNN · Stain-normalization · Training samples · Class-imbalance

M. Valdés Hernández and V. González-Castro (Eds.): MIUA 2017, CCIS 723, pp. 249–260, 2017.
DOI: 10.1007/978-3-319-60964-5_22

1 Introduction

Mitosis counting refers to the number of dividing cell identified in a fixed number of high power fields (40x magnification in this experiment), which is laborious, subjective [1]. In hematoxylin and eosin (H&E) stained breast cancer sections, mitoses are discernible as hyperchromatic objects with dark color that lack clear nuclear membranes and have irregularity shape properties. In fact, mitosis is a complex process during which a cell nucleus undergoes four phase and exhibits highly variable appearance, moreover in most stages a mitotic nucleus looks like a non-mitotic nucleus shown in Fig. 1 [2]. Therefore, the identification of mitosis may often suffer from disagreement between inter-observers. The computer-aid mitosis detector for breast cancer becomes a promising solution for these issues. Because it is a hard task to extract high-efficiency features manually, the performance of early studies based on handcraft features were not so impressive [3–5]. In recent years, deep convolution neural networks (CNN) exhibits outstanding performance in classification [6–8], which can learn high-level feature representation from the raw dataset. Nowadays, more and more researchers begin to apply CNN model to mitosis detection.

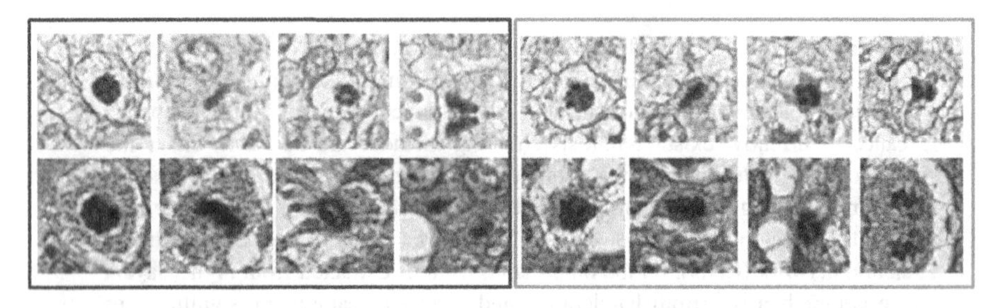

Fig. 1. The example of mitoses and non-mitoses (each column in red rectangle represents prophase, metaphase, anaphase and telophase of mitosis respectively, and non-mitoses with similarity appearance as mitosis were shown in the green rectangle) (Color figure online)

The latest studies show that deep CNN with hierachical feature representation has made breakthroughs in mitosis detection [9]. Ciresan et al. applied a DNN model to classify each pixel of the H&E stained mitosis images, consequently won the ICPR 2012 mitosis detection competition [10]. They proposed an approach to select relatively rare challenging non-mitosis samples based on the output of DNN model, which allowing the model to learn the significant differences between mitotic and non-mitotic nuclei efficiently. However, this training dataset includes 95.4% non-mitosis class samples and only 6.6% mitosis samples, so the detector tend to classify the mitosis as non-mitosis. In addition, the pixel-wise classifier is time-consuming, costing roughly 8 min per image. Then, Hao Chen et al. designed a faster mitosis detector by leveraging the FCN model, moreover, they proposed an innovative cascaded CNN model to detect mitosis [11]. The detection accuracy outperformed other methods by a large margin in 2014 ICPR MITOS-ATYPIA challenge. Haibo Wang et al. proposed an cascaded strategy combining the handcraft features and the features learned from deep CNN model together to generate more comprehensive features for mitosis detection [12]. This method was evaluated on the public ICPR12 mitosis dataset and yielded an F-measure of 0.7345 being the second best performance ever recorded in this dataset.

However, above work didn't discuss the impact of the different level features from CNN on final accuracy, moreover, the problem of class-imbalance was not addressed. While the extraction and combination of features in the work of Haibo Wang are relatively complicated, inspiring by this work, we proposed an end-to-end FF-CNN model to fuse the multiple level features. The establishments of the sub-paths would increase the number of parameters, therefore, the multi-step training strategy is proposed to reduce over-fitting. As for the issue of class-imbalance, we applied the method of Frequency-weighted loss function used in remote sensing images segmentation to this mitosis detection task [13]. Building an efficient training dataset by two stages, the blue-ratio image [14–16] was leveraging for a coarse samples extraction firstly. At the second stage, the final training patches were selected according to the output probability maps of the previous step. We up-sampled the output maps of FF-CNN model by a simpler but effective way, that is slide-widows and nearest neighbor interpolation instead of the de-convolution computation within FCN. In summary, the main jobs of this study include:

- Proposing a fused FCN model combining the features from different level layers together which help to improve the detection accuracy.
- Completing the training of the proposed FF-CNN model by a multi-step fine-tuning strategy.
- Leveraging the cascaded selection strategy to build an effective dataset and adopting the method of stain-normalization to preprocess images and augment the training dataset.
- Up-sampling the output images of FCN model through slide-widows and nearest neighbor interpolating to keep more valuable information of raw data, which contributed to the improvement of accuracy.

Among above jobs, both the FF-CNN model with corresponding training strategy and up-sampling method are first time applied in the field of mitosis detection. The rest of this paper organized as follow. The Sect. 2 will explain the new methodology in detailed. The Sect. 3 will describe the experiment and exhibit the results. The Sect. 4 will present the conclusion for this work.

2 Methodology

2.1 Stain-Normalization Method

The inconsistency of stain condition makes the appearances of H&E stained histology drastically different, so the classify performance was degraded [14, 15]. For instance, many false mitosis may arise when the histopathology slide is over-stained. This experiment performed staining unmixing (separation of the hematoxylin and eosin stains) and normalized each hematoxylin and eosin image separately [17]. This method was based on what was described in [16]. It was utilized not only to preprocess the training and testing image suffering from a poor staining, but also to augment training dataset. The normalized images are shown in Fig. 2.

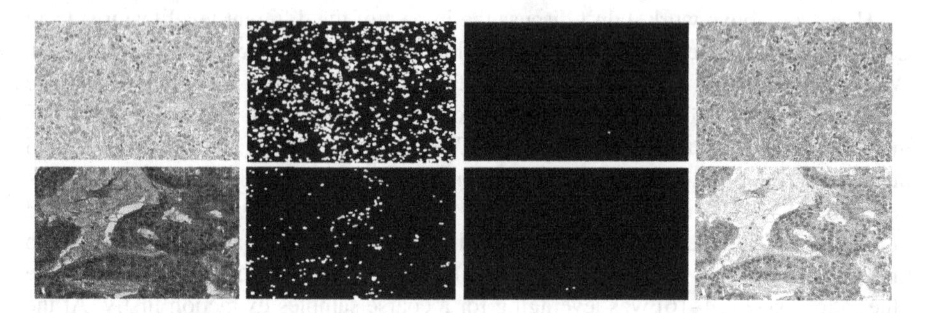

Fig. 2. Original image, blue-ratio image, ground true image and normalized image of A05_01Cd frame and A11_05 Da frame in the first row and the second row respectively (The two leftmost H&E images with different appearances were implemented staining unmixing and normalization, and the results shown in the rightmost images reduce the color inconformity significantly.) (Color figure online)

2.2 Training Samples Selection Strategy

The overall workflow of training samples selection is depicted in Fig. 3. At the first samples selection stage, we took the spots around given mitosis within 30 pixels as the centre of mitosis utilizing to make the ground-truth images shown as Fig. 3. Augmenting the mitosis training samples by rotations of {0;45;90;135;180;215;270} degrees and stain-normalization. Blue-ratio images calculated as the ratio of the blue color channel and the sum of red and green channels were used to select more likely but not mitosis samples, which benefits to remove trivial examples [13, 14]. Compared to the number of mitosis samples, 2.5 times non-mitosis samples were extracted.

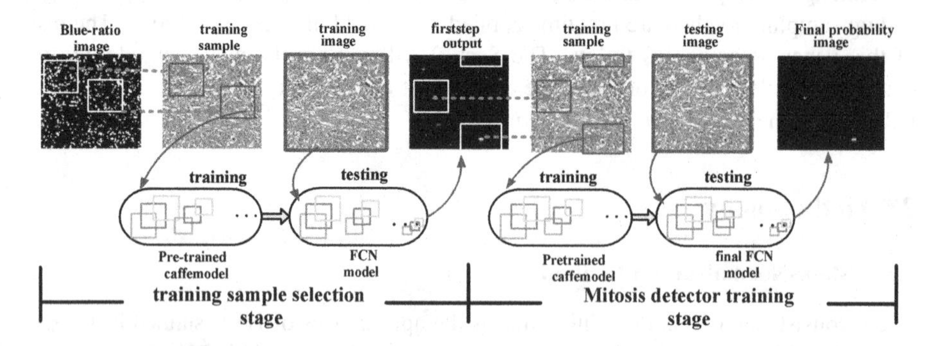

Fig. 3. The overall training process of mitosis detector including samples selection phase and model training phase

At the second samples selection stage, we extracted the non-mitosis patches by leveraging the output probability images of the previous stage. The values of probability images represent the likelihood of mitosis and were used to the select the challenging samples.

2.3 Fused Fully Convolutional Network

The features from lower layers of CNN model responds to the general attributes such as edge, color, texture and so on, while the higher layer features are more class-specific and abstract [18–20]. Inspiring by the previous work which combined the handcrafted features and the features learned from deep CNN [15], this study proposed a sampler and efficient approach in which linked the output of conv3 and conv4 layer to fc6 to combine the extracted different level features. The structure of model is shown as Fig. 4.

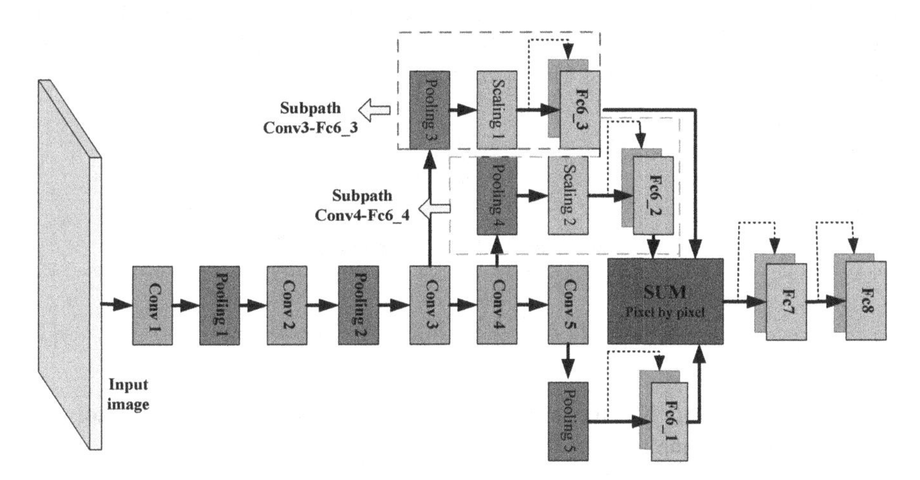

Fig. 4. An overview of proposed FF-CNN model combining different level features together

Fine-Tuning CaffeNet Model. Compared with the data in the other fields of computer vision, the quantity of data in medical domain is limited. The training samples may be still insufficient to train a powerful model even though a few dataset augment approaches were applied. As above mentioned, previous studies have evidenced that the parameters of the lower layers are general even though those in higher layers are more specific to different tasks. Fine tuning a pre-trained model could be a good solution to reduce over-fitting on the small scale dataset and achieve a excellent performance for specific tasks [19]. Therefore, we utilized the off-the-shelf CaffeNet model for initialization and did a series of experiments to determine which layers should be initialized by the pre-trained model. At last, the parameters of all the layers except the last layer in our proposed model is initialized by the CaffeNet model, and the learning rates of the last five layers were set ten times larger than the previous three layers. More importantly, the training process of the whole FF-CNN model including three steps, we trained the plain model without the two sub-paths firstly, later the sub-path between conv4 and fc6_2 and the sub-path between conv3 and fc6_3 were set up successively.

Fully convolutional network (FCN) could speed up the mitosis detection process and operate on an input image of arbitrary size. At the testing phase, we converted the trained model into FCN by replacing all the fully connected layers with the equivalent convolutional layers with the filter size of 1 × 1. In contrast to the previous time consuming pixel-based detector, our result obtained a far faster detection speed.

Frequency-Weighted Loss Function. Mitosis detection is a rare event detection in which one class (i.e. mitotic nuclei) is considerably less prevalent than the other class (i.e. non-mitotic nuclei). Cross-entropy loss function is used in most of image classifier, however, it doesn't deal with the imbalanced classes classification case well, in which the classifier prefers to recognize the smaller scale class as the larger scale class. In this study, we applied the modified loss function by using the mean frequency to weight the loss of the corresponding class. The modified loss function is,

$$L = \sum_{c=1}^{C} L_c \tag{1}$$

Where C is the total number of class in the classification task,

$$L_c = -\frac{1}{N}\sum_{n=1}^{N} \log(p_c^n) w_c \tag{2}$$

L_c is the loss of class c, N is the number of samples in a mini-batch, p_c^n is the softmax probability of n-th sample being class c, w_c is the loss weight of class c,

$$w_c = \frac{1}{C} \frac{\sum_{c=1}^{C} f_c}{f_c} \tag{3}$$

Where f_c is the frequency of class c in training dataset [13].

Up-Sample. From Table 1, we notice that the size of output image is 49 × 44 not 1539 × 1376 as the original images, since the convolutional layer and pooling layers would reduce the data size. The ratio from the larger original image to the smaller output image is defined as "compress-ratio", in this case, $1539/49 \approx 32$ and $1376/44 \approx 32$ are the compress-ratio of the width and height respectively. Actually,

Table 1. The architecture parameter of final mitosis detector based on AlexNet FCN model

Layer	Feature maps	Filter size	Stride	Padding	Learning rate
input	1766 × 1603 × 3	–	–	227 × 227	–
C1	439 × 399 × 96	11	4	–	10^{-5}
P1	219 × 199 × 96	3	2	–	–
C2	219 × 199 × 256	5	1	2 × 2	10^{-5}
P2	109 × 99 × 256	3	2	–	–
C3	109 × 99 × 384	3	1	1 × 1	10^{-3}
P3	54 × 49 × 384	3	2	–	–
C4	109 × 99 × 384	3	1	1 × 1	10^{-3}
P4	54 × 49 × 384	3	2	–	–
C5	109 × 99 × 256	3	1	1 × 1	10^{-3}
P5	54 × 49 × 256	3	2	–	–
C6_1	49 × 44 × 4096	6	1	–	10^{-3}
C6_2	49 × 44 × 4096	6	1	–	10^{-3}
C6_3	49 × 44 × 4096	6	1	–	10^{-3}
C7	49 × 44 × 4096	1	1	–	10^{-3}
C8	49 × 44 × 2	1	1	–	10^{-3}

the "compress-ratio" is roughly equal to the stride S of whole model. According to our compress-ratio, the value of displacement should be chosen as 32. In this up-sample approach, there will be 32 × 32 = 1024 operations to be implemented per image. The computation is redundant and time-consuming, finally, we set the sliding stride as 16, that is sliding number reduced to 4, ensuring a comparative accuracy at the same time.

From the parameters shown in Table 1, the feature maps of p3, p4 and p5 layer are the same size and the size of feature maps did not change after 3-th, 4-th and 5-th convolution operation, therefore, we calculated the corresponding index p_0 in the input image of index $\hat{p}_s$ from the output image was as a plain FCN model:

$$p_0 = \frac{(\hat{p}_s + f_{c_6} - 1 - 1)S_{p_5}S_{p_2}S_{p_1}S_{c1} + (f_{p5} - 1)S_{p_2}S_{p_1}S_{c1} + (f_{p2} - 1)S_{p_1}S_{c1} + (f_{p1} - 1)S_{c1} + f_{c1}}{2} + 1$$

Where f_{c_n} represents the filter size of the n-th convolutional layer, S_{p_n} and S_{c_n} represent the strides of the n-th pooling layer and the n-th convolutional layer, respectively. For instance, position (1,1) from output image is corresponding to the position (1 + 114,1 + 114) in the input image, that is the position (1,1) of 1539 × 1376 original image shown as Fig. 5. The sliding window described as the orange rectangle in the Fig. 5 was slid with the stride of 16. After it implemented right and down sliding operation, the position (1,1) from output image became corresponding to position (1 + 16,1 + 16) in 1539 × 1376 original image. The rest pixels would be filled by the method of nearest interpolation.

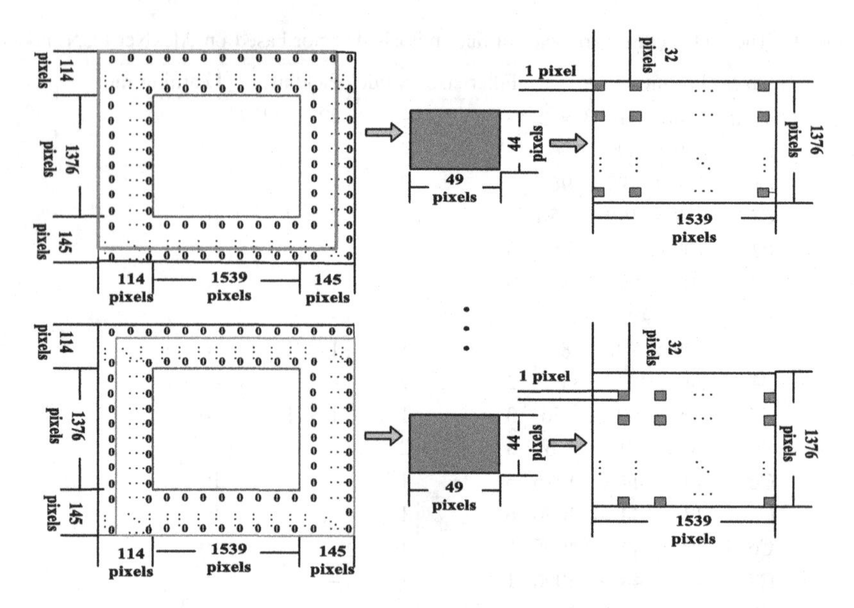

Fig. 5. The detailed implementation of up-sample approach (In the left two images, the red rectangles represent the size of 1539 × 1376 original images, the orange rectangles represent the size of 1766 × 1603 sliding window on the padding images, and the biggest black rectangles represent the images after padding 114 zeros on the direction of up and left and 145 zeros on the direction of down and right. The middle two orange images represent 49 × 44 output images from proposed FF-CNN model) (Color figure online)

3 Experiment and Result

3.1 Dataset

In this work, the performance of model was evaluated on 2014 ICPR MITOSIS dataset including 1,200 training images and 496 testing images. Due to different condition during the tissue acquisition process, the appearance of tissues are various, which makes the detection task more challenging. The spatial resolution of images acquired by the widely-used Apero-XT scanner is 0.25 μm/pixel and the magnification is 40. The stain-normalization method was leveraged to preprocess the images for the slice of A03, A07, A11, A14, A15, A17 and A18, which made the edges of mitosis clearer and color consistent shown as Fig. 6, and to augment the training data for the slice of A04 and A05.

3.2 Evaluation and Performance

Evaluation was performed according to the ICPR 2014 content criteria, if the detected mitoses whose coordinates are closer than 30 pixels (8 μm) to true mitoses are defined as true positives (TP). Those detection locating without 32 pixels of true mitosis are counted as false positive (FP). The true mitoses which were not detected by the model

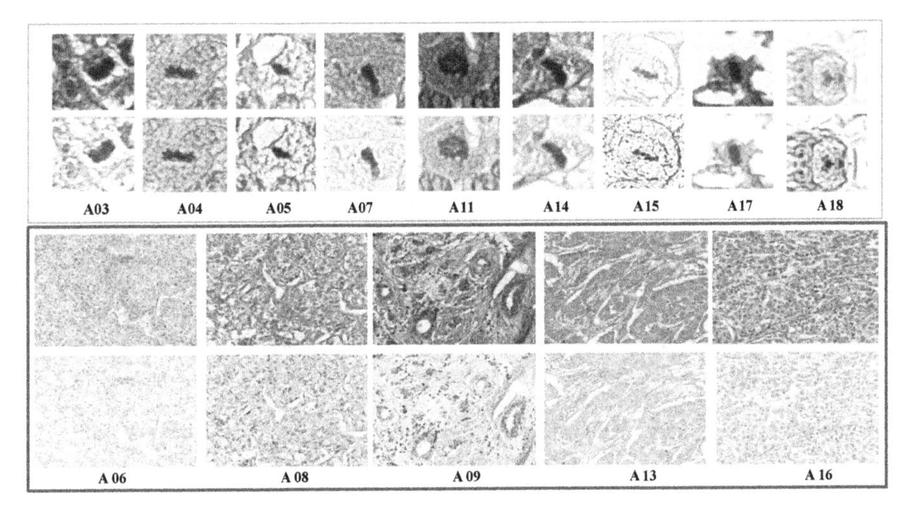

Fig. 6. The train samples and test images from different slices (The raw test images and the normalized test images are show in the first and second row of red rectangle, respectively. The over-staining could be reduced shown as in the slices of A03, A07, A11, A14 and A17, and the staining of A15, A18 slices would be enhanced after stain-normalizing.) (Color figure online)

are defined as false negative (FN). We compute the performance measure including precision: $P = \frac{N_{TP}}{N_{TP}+N_{FP}}$, recall: $R = \frac{N_{TP}}{N_{TP}+N_{FN}}$ and F-measure: $F = \frac{2 \times \mathrm{P} \times \mathrm{R}}{\mathrm{P}+\mathrm{R}}$, Where N_{TP}, N_{FP}, N_{FN} refers to the number of true positive, false positive and false negative, respectively. We evaluated the performance of the proposed FF-CNN model with and without frequency weighted loss function (FF-CNN+FWLF, FF-CNN) and the one-way plain FCN model with the modified loss function (FCN+FWLF).

Detection Accuracy. The results of mitosis detection are showed in the Table 2 ("/" denotes that the results are not released). The FF-CNN+FWLF approach yields the highest F-measure 0.393, which improves by 10.4% comparing with the state-of-art method recorded in 2014 ICPR MITOSIS contest. The total number of detected mitosis (sum of TP and FP) by the method of FCN, FF_CNN (without FC6_3) and FF_CNN increases one by one, from 175 to 183, finally ending to 202, more importantly, the

Table 2. Result of 2014 ICPR MITOSIS Datast

Method	TP	FN	FP	Precision	Recall	F-measure
STRASBOURG	/	/	/	/	/	0.024
YILDIZ	/	/	/	/	/	0.167
MINES-CURIE-INSERM	/	/	/	/	/	0.235
CUHK	/	/	/	/	/	0.356
FCN	60	143	115	0.343	0.296	0.318
FFCN (without FC6_3)	65	138	118	0.355	0.32	0.337
FFCN	76	125	126	0.376	0.374	**0.375**
FFCN+FWLF	81	122	128	0.388	0.399	**0.393**

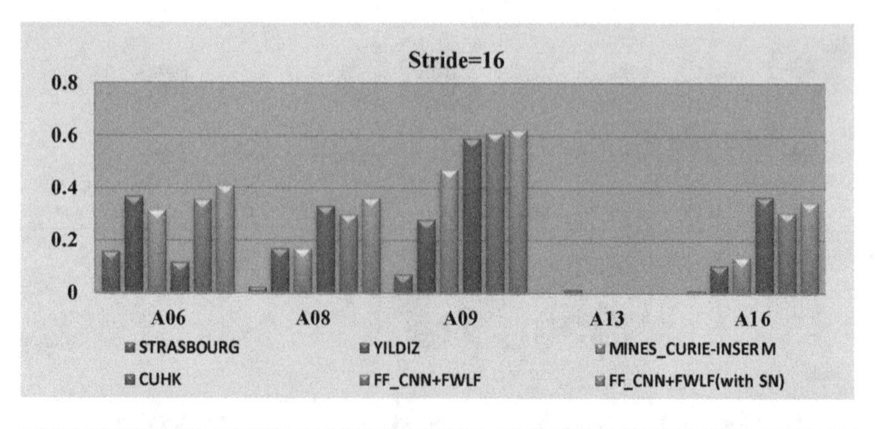

stride	1	2	4	8	16	32
speed (s/per image)	112.5	27.31	7.42	1.55	**0.38**	0.11
F-measure	0.405	0.396	0.4	0.398	**0.395**	0.285

Fig. 7. The performances on different slides of 2014 ICPR MITOSIS test data and the impact of stride on detection speed and accuracy by the method of FF_CNN+FWLF (with SN)

precisions are improved from 0.343 to 0.376 after sub-paths were set up, illustrating that both two sub-paths contribute to model's ability to identify more potential mitoses and obtain a lower error rate. Through weighting the loss function, the detection performances including recall and precision are further enhanced by 6.68% and 3.19% respectively, demonstrating the effectiveness of frequency weighted loss function in this mitosis detection task. The detection results for each slice are shown in the Fig. 7, through comparing the results of FF_CNN + FWLF with and without stained-normalized, the effectiveness of this preprocess method was verified. Even though the overall results is improved, the F-measure of A13 is still 0. Observing the H&E image of slice A13, their appearance are largely various and over-stained, and there are only two mitoses within the 64 images.

Computation Time. In breast cancer diagnosis, a pathologist should look over several slides including thousands of large sized H&E stained images for one patient. So the detection time is crucial in the clinic application. The whole detection of each HPF image including four computations within FF-CNN model and up-sampling process took about 0.375 s at the workstation with a 2.4 GHz Intel(R) Xeon(R) E5-2630 CPU and four NVIDIA Tesla M40 GPUs. This detection speed is acceptable in the practical application.

4 Conclusion

Computer-assisted mitosis detector is meaningful to improve the efficiency and accuracy in cancer grading. In this paper, we proposed a fused fully convolutional network combining the features from different level layers together and fine-tuned by a pre-trained caffe model. In addition, we utilized a cascaded training samples selection strategy and frequency weighted loss function to solve the sparse detection issue. This work achieve a impressive accuracy rate with an accepted detection speed. Future work will include developing an adaptive model for multiple dataset and applying unsupervised learning method in mitosis detection task.

Acknowledgements. This research was financially supported by Basic layout of Shenzhen City (no. JCYJ 20150827165024088), Supporting platform project in Guangdong Province (no. 2014B0909B001), the CAS 'Light of West China' Program (XB BS-2014-16), the "Thousand Talents" plan (Y474161) and the Shenzhen Basic Research Project (JCYJ20150630114942260). The author was grateful to Dr. Ludovic Roux for the data providing and results evaluation.

References

1. Roux, L., Racoceanu, D., Lomenie, N., et al.: Mitosis detection in breast cancer histological images an ICPR 2012 contest. J. Pathol. Inform. **4**(1), 8 (2013)
2. Basavanhally, A., Ganesan, S., Feldman, M., et al.: Multi-field-of-view framework for distinguishing tumor grade in ER+ Breast cancer from entire histopathology slides. IEEE Trans. Biomed. Eng. **60**(8), 2089–2099 (2013)
3. Irshad, H.: Automated mitosis detection in histopathology using morphological and multi-channel statistics features. J. Pathol. Inform. **4**(1), 10 (2013)
4. Irshad, H., Jalali, S., Roux, L., et al.: Automated mitosis detection using texture, SIFT features and HMAX biologically inspired approach. J. Pathol. Inform. **4**(2) (2013)
5. Huang, C., Lee, H.C.: Automated mitosis detection based on eXclusive independent component analysis. In: International Conference on Pattern Recognition, pp. 1856–1859 (2012)
6. Lecun, Y., Bengio, Y., Hinton, G.: Deep learning. Nature **521**(7553), 436–444 (2015)
7. Krizhevsky, A., Sutskever, I., Hinton, G.E., et al.: ImageNet classification with deep convolutional neural networks. In: Neural Information Processing Systems, pp. 1097–1105 (2012)
8. Szegedy, C., Liu, W., Jia, Y., et al.: Going deeper with convolutions. In: Computer Vision Pattern Recognition, pp. 1–9 (2015)
9. Sainath, T.N., Mohamed, A., Kingsbury, B., et al.: Deep convolutional neural networks for LVCSR. In: International Conference on Acoustics, Speech, Signal Processing, pp. 8614–8618 (2013)
10. Cireşan, D.C., Giusti, A., Gambardella, L.M., Schmidhuber, J.: Mitosis detection in breast cancer histology images with deep neural networks. In: Mori, K., Sakuma, I., Sato, Y., Barillot, C., Navab, N. (eds.) MICCAI 2013. LNCS, vol. 8150, pp. 411–418. Springer, Heidelberg (2013). doi:10.1007/978-3-642-40763-5_51
11. Chen, H., Dou, Q., Wang, X., et al.: Mitosis detection in breast cancer histology images via deep cascaded networks. In: Thirtieth AAAI Conference on Artificial Intelligence (2016)
12. Wang, H.: Mitosis detection in breast cancer pathology images by combining handcrafted and convolutional neural network features. J. Med. Imaging **1**(3), 1–8 (2014)
13. Kampffmeyer, M., Salberg, A.B., Jenssen, R.: Semantic segmentation of small objects and modeling of uncertainty in urban remote sensing images using deep convolutional neural networks. In: The IEEE Conference on Computer Vision and Pattern Recognition, pp. 680–688. IEEE (2016)

14. Veta, M., van Diest, P.J., Willems, S.M., et al.: Assessment of algorithms for mitosis detection in breast cancer histopathology images. Med. Image Anal. **20**(1), 237–248 (2014)
15. Chang, H., Loss, L.A., Parvin, B.: Nuclear segmentation in H&E sections via multi-reference graph cut (MRGC). In: Proceedings of the Sixth IEEE International Conference on Symposium on Biomedical Imaging, ISBI 2012 (2012)
16. Veta, M., Van Diest, P.J., Pluim, J.P., et al.: Detecting mitotic figures in breast cancer histopathology images. In: Proceedings of SPIE (2013)
17. Macenko, M., Niethammer, M., Marron, J.S., et al.: A method for normalizing histology slides for quantitative analysis. In: IEEE International Symposium on Biomedical Imaging: From Nano To Macro, pp. 1107–1110. IEEE (2009)
18. Yosinski, J., Clune, J., Bengio, Y., Lipson, H.: How transferable are features in deep neural networks? In: Advances in Neural Information Processing Systems, pp. 3320–3328 (2014)
19. He, K., Zhang, X., Ren, S., et al.: Deep residual learning for image recognition. In: Computer Vision and Pattern Recognition, pp. 770–778 (2015)
20. Zhao, L., Jia, K.: Multiscale CNNs for brain tumor segmentation and diagnosis. Comput. Math. Methods Med. **2016**(7), 1–7 (2016)

Classification of Cervical-Cancer Using Pap-Smear Images: A Convolutional Neural Network Approach

Bilal Taha(✉), Jorge Dias, and Naoufel Werghi

Department of Electrical and Computer Engineering, Khalifa University, Abu Dhabi, UAE
bilal.taha@kustar.ac.ae

Abstract. Cervical cancer is the second most common and the fifth deadliest cancer in women. In this paper, we propose a deep learning approach for detecting cervix cancer from pap-smear images. Rather than designing and training a convolutional neural network (CNN) from the scratch, we show that we can employ a pre-trained CNN architecture as a feature extractor and use the output features as input to train a Support Vector Machine Classifier. We demonstrate the efficacy of such a new employment on the Herlev public database for single cell pap-smear, whereby the experimental results show that our proposed system neatly outperforms other state of the art methods.

Keywords: Pap-smear classification · Deep learning · Convolutional neural network

1 Introduction

For many years, cancer has been one of the biggest threats to human life, and the number of new cases is expected to rise by about 70% over the next 2 decades [2]. Cervical cancer, in particular, is the second most common and the fifth deadliest cancer in women [24]. The low rate of cancer survival is due to the fact that the majority of cancer cases are detected at advanced stages. There is a consensus in the medical community on the vital need of early detection of cancer for effective treatment. Indeed, some studies reported that cervical cancer is the most preventable disease with the incidence rates getting reduced by 80% [12] through early detection, although this sharp increase in figures might have been influenced by lead time bias and over-diagnosis [11]. Accurate and early cancer detection is very important for timely diagnosis and effective treatment. In fact, the inability to detect cancer in its early stages may cause the treatment to be delayed to a more advanced stage with more severe implications for survival rates and resource utilization. On the other hand, false detection of cancer may lead to unnecessary invasive treatments that might be both physically and emotionally traumatic to the patient, in addition to being costly to the health care system

M. Valdés Hernández and V. González-Castro (Eds.): MIUA 2017, CCIS 723, pp. 261–272, 2017.
DOI: 10.1007/978-3-319-60964-5_23

in terms of human and logistic resources. A variety of diagnostic tools are used in screening depending on the type of cancer. These tools include chest x-ray, computerized tomography (CT) scan, bronchoscopy, positron emission tomography (PET) scans and microscopic images. This last modality utilized in the detection of cervical cancer has the advantage of little or no side effects, as the procedure employed for the acquisition of microscopic images is virtually non-invasive. Screening for cervical cancer uses microscopic images of sample of cells collected from the cervix area. The cells undergo Papanicolaou staining method [4] which aims to visualize cells and cell components under the microscope allowing to display the variations of cellular morphology, and to differentiate the main cells from the debris cells.

The Pap smear slides usually contain both of single cells and clusters of cells. Most of cells are found with high degree of overlapping. Similar to other cells in human body, a cervical cell consists of two main components. One is the nucleus located about the center of cell surrounded by the cytoplasm. Normally, nucleus shape is small and almost round. Its intensity is darker than cytoplasm. In dysplastic cells, or abnormal cells, the cell will not grow and divide as it should. This is referred as precancerous cell. A sample of normal and abnormal cells is shown in Fig. 1. The dysplastic cells are categorized into mild, moderate, and severe dysplastic. A high amount of the mild dysplastic cells will disappear without becoming malignant, whereas severe dysplastic cells are likely to turn into malignant cells. The squamous dysplastic cells generally have larger and darker nuclei and tend to cling together in clusters. In severe dysplastic cells, nuclei are large, with dark granules and usually deformed. In Pap Smear image analysis, the cervical cells are divided into 7 classes, categorized by cell appearance, especially related to the nucleus.

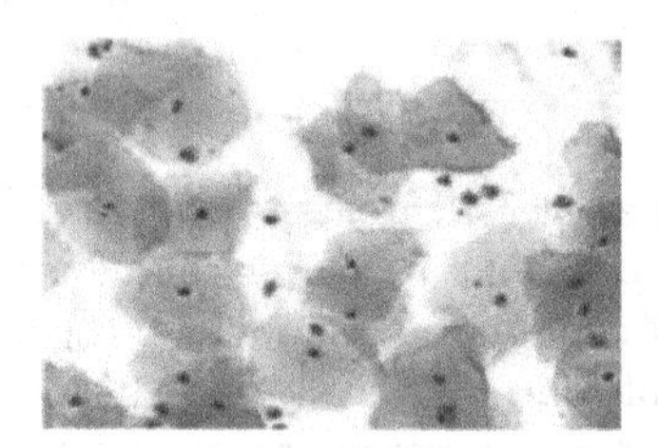
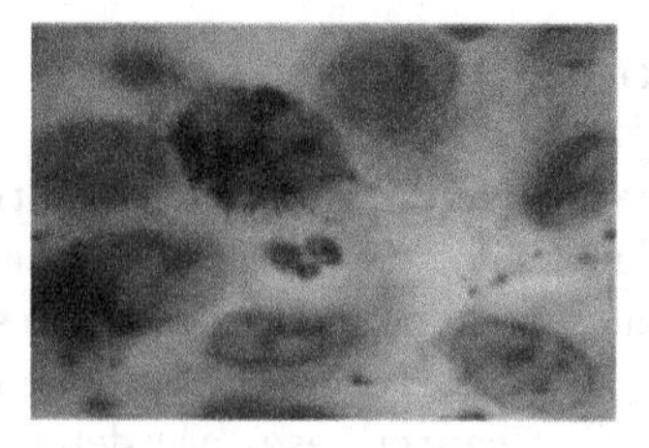

Fig. 1. Sample of cervical cell microscopic images. (a) Normal cells (b) Abnormal cells

Cervix cancer classification using pap-smear images adopts, basically, a three-stage paradigm, namely cell detection, cell segmentation, and cell classification. For cell detection, some methods, as explained in [28] employed contrast limited adaptive histogram equalization and global thresholding applied to the red, green and blue channels of the image. Automatic dell detection reached mature stage, and does not present any particular challenge nowadays.

For the cell segmentation a variety of methods have been proposed. They can be categorized into region-based methods and contour-based methods. In the former, cell pixels are separated into nucleus and cytoplasm based on regional image pixel similarity (or dissimilarity). Methods in this category fall into grey level histogram methods [20] and clustering methods using watersheds [13,19]. Threshold selection is a major issue with the gray-level histogram methods. Clustering methods, though they are threshold-free, result in an over-segmentation scenario as they fix the number of clusters to two (two clusters corresponding to the nucleus and the cytoplasm), whereas the actual number might be larger because of the color variability within the cell region. The watershed methods, though more robust, present an intrinsic limitation arising from the fact that it relies on the principle of touching regions exhibiting a narrow "neck" on the area of contract. Consequently, nucleus exhibiting thick or blurred boundaries with the cytoplasm cannot be detected reliably [26]. Contour-based methods detect the contour in the image by marking the boundary of the nuclei with respect to the cytoplasm. [27] used the concept of active contour, which is a kind of parameterized closed curve that iteratively deform until it fits the boundary edges. However this method requires manual initialization. Other resent work employed edge detector techniques [22] which work fine for clear single cell images, but their performance degrades considerably for cases of non-homogeneous cell regions and overlapping cells. And above all, these methods inherit the sensitivity to image noise and artifacts that characterize edge detection operators.

Once the cell is segmented, features are extracted to be used as input for cell classification. Jantzen et al. [14] proposed several important cell features that are used for Pap smear image analysis, derived from the nucleus and cytoplasm areas, and which include brightness, shorter diameter, longest diameter, elongation, roundness, perimeter, maxima, minima (the number of pixels with the maximum/minimum intensity value in a 3×3 neighborhood of the specific area). In terms of the features that are extracted from both the cytoplasm and the nucleus area, the nucleus position and the ratio *nucleus size)/cytoplasm size* are calculated. Thus, a total of 20 features are considered important for the analysis of Pap smear images. These feature has been used later in [22]. Recently Boral et al. [5] Consolidated the set of features by novel color and texture features extracted using Ripplet Type I transform, Histogram first order statistics and Gray Level Co-occurrence Matrix Ripplet Type I transform, Histogram first order statistics and Gray Level Co-occurrence Matrix.

In the classification stage, most approaches proposed machine learning methods [5,9,21,22,28]. Marinakis and Dounias [21] employed two classifiers in their method namely 1-Nearest Neighbor and the Weighted (w) k-Nearest Neighbor classifier. The wk-Nearest-Neighbor is used to give different weighting for the features according to the distance to the test samples. To accommodate for the huge number of features extracted from the pap-smear images, Ant Colony Optimization method is utilized as a feature reduction mechanism, whereby Plissiti et al. [28] proposed to use an unsupervised learning technique for the classification of pap-smear images. They have focused on the nucleus features only and applied different feature reduction methods to select a subset from the features. The low dimensional features used in Spectral Clustering and fuzzy C-means classifiers for the decision making process. Chen in [9] proposed and integrated system providing tolls for: selecting the cell, automatically detect the nucleus and the cytoplasm regions, extracting 13 morphological and texture descriptors, and using a Support Vector Machine (SVM) for classification. The same classifier has been used in [22]. In the recent work [5], Bora et al. employed a majority-voting fusion method including a Multilayer Perceptron (MLP), Random Forest (RF) and Least Square Support Vector Machine (LSSVM).

In this work, we focus on the classification step where we propose a novel approach for cervix cancer detection. Rather than designing a convolutional neural network from the scratch, we employ a pre-trained CNN architecture coupled with a support vector machine at the back-end, saving thus time and resources. This employment of the pre-trained CNN architecture as a features extractor is deeply instigated with different experiments, and thus comparing the performance of the features across the different architecture layers outputs. The closest work to our approach is [6] whom used a CNN as a feature extractor. However, our method is distinguished by the following aspects:

- CNN was used to transfer different level of features from different layers contradicting to the work in [6] where they assume the best features are obtained from FC7 without any experimentation. Since there is a very few work implemented on the classification of cervix cancer from pap-smear images using deep learning, it turns out essential to perform more experimentation to obtain a deeper understanding for this method.
- Two testing sets was performed in this work by employing the Herlev public database where the other method emphasized more on their own generated database.
- SVM classifier was utilized for the decision making process because of its capability to handle high dimensional features where the other work used least-squares SVM (LSSVM) and softmax regression classifiers.
- Since we are using SVM which is capable to manage huge number of features, feature selection techniques was not empowered because the focus is on the usage of the CNN as a feature extractor and the transferred features from its different layers. However, the work in [6] have implemented Maximal Information Compression Index as a feature selection to reduce the number of features obtained from the CNN.

The evaluation of our paradigm illustrates the superiority of our method when compared with state of the art methods. The remainder of the paper will be organized as follows: Sect. 2 introduces the proposed approach and elaborates on its rational. Section 3 describes the different experiments and the related results. Section 4 concludes the paper.

2 The Proposed Method

In image analysis, designing the appropriate features for a given interpretation task has been a central problem in computer vision and in medical image analysis. Explicit feature design extraction in medical image analysis requires subject-matter expertise. In this process, the visual information on which the physician relies in his assessment is not necessarily reflected into a suitable computational representation. Moreover, the practical considerations in the extraction and the usages of these features make the reproducibility of these related methods often problematic [25]. To overcome these challenges we propose a deep learning approach, whereby Convolution Neural Network (CNN) is employed to replace a handcrafted and customized features, which would be strongly sensitive to multiple parameters.

It is known that through a hierarchical unsupervised or semi-supervised feature design, CNN's can produce effective representation of the visual data [18]. Basically, a CNN deep learning architecture is composed of a sequence of cascading layers performing basic operations such as convolution, subsampling, followed by another sequence of fully connected layers, which act similarly as a classic artificial neural network. These fully connected layers can be replaced by a support vector machine classifier. In another hand, training a CNN network from scratch requires a large data-set, which is a tedious process and often cannot be afforded in medical applications, including database of pap-smear classification. Also, in addition to requiring considerable computational resources, there is no systematic guidelines as for the optimal choice of the architecture in terms of depth (number of layers) and structure.

An economic alternative is to use pre-trained CNN architectures, that are proven to have good performance through training and validation over a huge database, and then tune, via training conducted on a specific application data-set, the pre-trained weights of the architecture. This procedure, known as fine-tuning, can be performed either across the whole CNN or at specific layers. In this paradigm, the lower layers are kept the same since they have learned generic features (e.g. edge, region) that are less dependent on the final application [18], whereas the top layers are removed and the linear classifier is trained to accommodate the new application-specific database.

In our approach, we advocate the hypothesis that a trained CNN architecture embeds sufficiently rich feature representations that can be utilized as input to train a standard classifier, such as the Support Vector Machine, relieving thus the system from laborious training from scratch or fine tuning. Therefore a pre-trained CNN is then deployed as a feature extractor for our specific image interpretation task of pap-smear detection, as depicted in the block diagram in Fig. 2. The database employed for training and testing include different patch sizees of the pap-smear. Therefore, the resolution for the patches less than the standard size 227 $times$227 adopted in AlexNet, was completed by empty regions with a white background.

There are several pretrained CNN architectures that can be investigated, such as GoogleNet [1] and VGGNet [29]. In our method, we explored AlexNet [16]. This CNN architecture was trained with 1.2 million images for 1000 different classes, thus the learned features are expected to span a large spectrum of visual information. The main layers of the AlexNet architecture is briefly described in Table 1.

Fig. 2. Block diagram for the CNN as a feature extractor

Table 1. Summary of AlexNet architecture

Layer	Type	Input	Kernel	Stride	Pad	Output
Data	Input image	$227 \times 227 \times 3$	N/A	N/A	N/A	$227 \times 227 \times 3$
conv1	Conv	$227 \times 227 \times 3$	11×11	4	0	$96 \times 55 \times 55$
pool1	Max pooling	$55 \times 55 \times 96$	3×3	2	0	$96 \times 27 \times 27$
conv2	Conv	$27 \times 27 \times 96$	5×5	1	2	$256 \times 27 \times 27$
pool2	Max pooling	$27 \times 27 \times 256$	3×3	2	0	$256 \times 13 \times 13$
conv3	Conv	$13 \times 13 \times 256$	3×3	1	1	$384 \times 13 \times 13$
conv4	Conv	$13 \times 13 \times 384$	3×3	1	1	$384 \times 13 \times 13$
conv5	Conv	$13 \times 13 \times 384$	3×3	1	1	$256 \times 13 \times 13$
pool5	Max pooling	$13 \times 13 \times 256$	3×3	2	0	$256 \times 6 \times 6$
FC6	Fully connected	$6 \times 6 \times 256$	6×6	1	0	4096×1
FC7	Fully connected	1×4096	1×1	1	0	4096×1
FC8	Fully connected	1×4096	1×1	1	0	1000×1

Usually features from first layers are too generic to be employed as discriminative descriptors (see Fig. 3). As a result, we investigated features from the middle layers and onward, namely, Con4 till FC8. The output from one of these layers will be a sort of feature encoding of the pap smear images. These features will be then fed into the subsequent classifier block.

Here a fully connected neural network (ANN), a SoftMax classifier, or a SVM can be used. We choose the SVM for the following reasons: SVM and has the capacity to deal with a high-dimension input without compromising the computational complexity, and thus, contrary to ANN or SoftMax, can map the huge number of feature vectors across the different layers of the CNN. For instance, in [23], the SVM has been deployed across all six layers from layer 1 to layer 7 except 6 of the ConvNet model, whereas softmax has been used only at layer 5 and 7. Moreover, SVM classifier showed better overall discriminating power on that model where recently it has been observed that coupling SVM as a final layer improves the learning rate [17]. With regard to overfitting, the SVM is assumed to be less sensitive to overfitting, at least in principle, because of aforementioned feature dimensionality, in practice it provides mechanisms to control overfitting through the C parameter. Having said that. Generally, overfitting remains a general problem that practically has to be dealt with for any

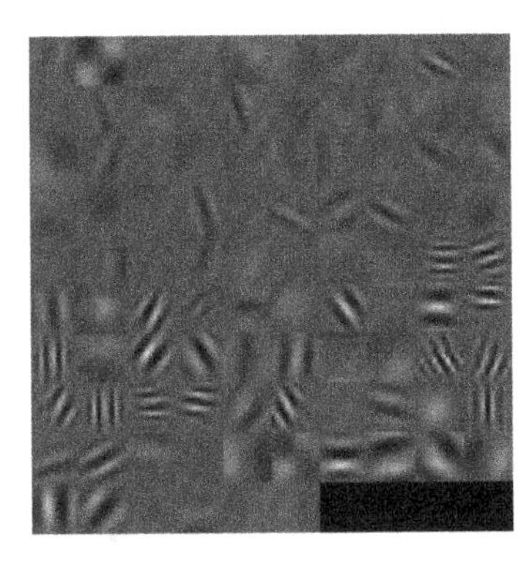

Fig. 3. Visualization for the first convolutional layer weights. Since we have 96 weights in the first layer of the AlexNet, the last 4 weights appeared black.

classifier [8]. In CNN specifically, the problem is more crucial because of the high dimensionality that characterize their architecture. Several mechanism have been proposed to address this issue, which include, dropout [15], data augmentation [15], regularization [3] and stochastic pooling [30].

When training the SVM, each feature vector is given a label either -1 or 1 (normal, abnormal) to create the feature-class pair $\{x, y\}$. Therefore, given L features $\{x_i, y_i\}$ such that $i = 1...L$, and $y_i \in \{1, -1\}$, $x \in \Re^D$, where D is the vector size. A hyper-plane separating the two classes could be written as

$$w^T x + b = 0 \tag{1}$$

the w is known as the weight vector which is normal to the separation hyper-plane, and b is known as the bias. In order to separate the two classes with the hyper-plane the following equation should be optimized

$$min(\frac{1}{2}w^T w + C\sum_{i=1}^{L}\xi_i) \tag{2}$$

subject to the constrain $y_i(w^T x_i + b) >= 1 - \xi_i$, where $\xi_i >= 0$ for $i = 1, ..., L$, and C is known as the penalty parameter. This will lead to the optimal hyper-plane that minimizes the distance between itself and all the training examples. The optimal hyper-plane, allows a classification to be done according to a decision function such as:

$$f(x) = \text{sgn}(w^T x + b) \tag{3}$$

3 Experimentation

To evaluate the performance of our method, Herlev pap smear database was used for training and testing. This database is publicly available and consists of 917 single cell images divided into 7 classes. Four categories are considered as abnormal images with different severity namely light dysplastic, moderate dysplastic,

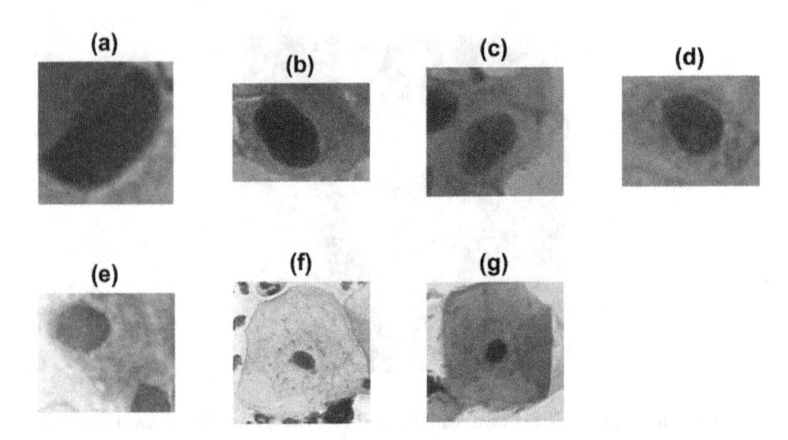

Fig. 4. Sample images from the 7 classes of the Herlev pap smear database (a) carcinoma in situ, (b) severe dysplastic, (c) moderate dysplastic, (d) light dysplastic, (e) normal columnar, (f) normal intermediate, (g) normal superficial.

severe dysplastic, and carcinoma in situ. The other three categories considered as normal are normal columnar, normal intermediate, normal superficial (Fig. 4). However, according to [10] when considering a two class classification problem, the columnar class is not considered as a normal nor an abnormal. As a result, the total number of normal class images is 144 while the total number of abnormal cell images is 675.

We selected and transferred features from different layers of the CNN, from shallow to deep layers, where the aim was to assess the performance of the CNN as a feature extractor and its effectiveness in the detection scenarios on the specified database. The main focus in the experiment is the quality, in terms of classification power, of the extracted features from the CNN. The features were deployed using different layers from the pre-trained CNN. AlexNet was trained using the ImageNet database which consists of non-medical images, therefore there is a need to know the best layer that will provide the best features discriminating normal pap-smear cell images from the abnormal pap-smear ones. As we mentioned earlier, we considered only deep layers, starting from Conv4. In this transfer learning scheme, the layers up to the output features layers are frozen and the output features are used to train the SVM classifier. For example, considering conv5, as the feature output layer, we keep the weights across the layers conv1 to conv5 at their pre-trained values, while training the SVM classifier. While this scheme reduces the number trained entities, the number of features remain large, as an example, the dimension of the obtained feature vector from the fourth convolution layer (C4) is $13 \times 13 \times 384$ which is equal to 64896 features. The two-class classification problem was implemented where the number of normal pap-smear cell images is 144, while the number of the abnormal images is 675. For training protocol we adopted the 70%, 30% for training and testing, respectively. Table 2 reports the best recall and precision performance from each layer. For instance, C5 in Table 2 refers to the features coming out from the layer Conv5. It is interesting to notice that the top performance is obtained with features coming out from a deep layer (fc7), which are more descriptive than their shallow layers counterparts (e.g. Conv4 - see Table 1).

Table 2. The recall and precision values for 2-class classification of pap-smear without the columnar class.

Experiment 1				
		Recall	Precision	Accuracy
CNN feature	Conv4	99.01%	99.02%	98.37%
	Conv5	100%	98.54%	98.78%
	FC6	99.5%	99.01%	98.6%
	FC7	**99.51%**	**99.5%**	**99.19%**
	FC8	98.6%	97.8%	97.9%

In another experiment, we tested our method on the data-set without removing the columnar class which results in having a number of 242 normal cell images and 675 abnormal cell images. The recall, precision, and accuracy are reported in Table 3.

Table 3. The recall and precision values for 2-class classification of pap-smear including the columnar class.

Experiment 2			
		Recall	Precision
CNN feature	Conv4	94.1%	87.6%
	Conv5	97.04%	89.14%
	FC6	99.01%	85.2%
	FC7	64.04%	96.3%
	FC8	59.11%	95.24%

We compared our method with two state of the art methods that used the same database and ignore the columnar class [7,14]. Table 4 reports the performances of the two methods together with the results achieved by our method. We found that our approach outperforms all the existing paradigms in terms of recall, precision, specificity and accuracy with scores of 99.51%, 99.5%, 97.67% 99.19% respectively.

Table 4. Recall, precision, specificity, and accuracy scores in percent by setting the parameters according to the best results in each experiment.

Method	[7]	[14]	Our method
Recall	95.11	-	99.51
Precision	-	-	99.5
Specificity	96.53	-	97.67
Accuracy	95.36	93.75	99.19

4 Conclusion

In this work we presented a deep learning solution for cervix cancer screening using pap-smear images. In this application, we proposed a novel employment whereby a pre-trained architecture, the AlexNet, is used as feature extractor, then coupled with a classic SVM classifier. This approach relives the system form the high computational and resource demanding training from the scratch or even fine-tuning. The evaluation and testing conducted with the Herlev database, confirmed the rational of our hypothesis that the features derived from a CNN architecture (pre-trained by means of colossal datasets), embeds sufficient discriminatory information to be tailored to our specific Herlev pap-smear dataset. In the future work we plan to investigate further our approach on other pap-smear datasets and other standard trained CNN architecture such as VGGNet.

References

1. Szegedy, C., et al.: Going deeper with convolutions. In: Proceedings of Conference on Computer Vision and Pattern Recognition, pp. 1–9 (2015)
2. Ferlay, J., et al.: Cancer incidence and mortality worldwide. In: GLOBOCAN 2012, vol. v1.0 (2010)
3. Srivastava, N., et al.: Dropout: A simple way to prevent neural networks from overfitting. J. Mach. Learn. Res. **15**(1), 1929–1958 (2014)
4. Arbyn, M., Anttila, A., Jordan, J., Ronco, G., Schenck, U., Segnan, N., Wiener, H., Herbert, A., von Karsa, L.: European guidelines for quality assurance in cervical cancer screening. Ann. Oncol. **21**(3), 448 (2010). Second edition summary document
5. Bora, K., Chowdhury, M., Mahanta, L.B., Kundu, M.K., Das, A.K.: Automated classification of Pap smear images to detect cervical dysplasia. Comput. Methods Programs Biomed. **138**, 31–47 (2017)
6. Bora, K., Chowdhury, M., Mahanta, L.B., Kundu, M.K., Das, A.K.: Pap smear image classification using convolutional neural network. In: Proceedings of the Tenth Indian Conference on Computer Vision, Graphics and Image Processing, ICVGIP 2016, NY, USA, pp. 55:1–55:8. ACM, New York (2016)
7. Cawley, G.C., Talbot, N.L.C.: On over-fitting in model selection and subsequent selection bias in performance evaluation. J. Mach. Learn. Res. **113**(2), 2079–2107 (2014)

8. Chankong, T., Theera-Umpon, N., Auephanwiriyakul, S.: Automatic cervical cell segmentation and classification in Pap smears. Comput. Methods Programs Biomed. **113**(2), 539–556 (2010)
9. Chen, Y.F., Huang, P.C., Lin, K.C., Lin, H.H., Wang, L.E., Cheng, C.C., Chen, T.P., Chan, Y.K., Chiang, J.Y.: Semi-automatic segmentation and classification of Pap smear cells. IEEE J. Biomed. Health Inform. **18**(1), 94–108 (2014)
10. Gençtav, A., Aksoy, S., Önder, S.: Unsupervised segmentation and classification of cervical cell images. Pattern Recogn. **45**(12), 4151–4168 (2012)
11. Gigerenzer, G., Wegwarth, O.: Five year survival rates can mislead. BMJ **346**, f548 (2013)
12. Henschke, C.L., et al.: International early lung cancer action program investigators: survival of patients with stage 1 lung cancer detected on CT screening. N. Engl. J. Med. **335**, 1763–1771 (2006)
13. Costa, J.A.F., Mascarenhas, N.D., de Andrade Netto, M.L.: Cell nuclei segmentation in noisy images using morphological watersheds. In: International Society for Optical Engineering, vol. 3164, pp. 314–324 (1997)
14. Jantzen, J., Norup, J., Dounias, G., Bjerregaard, B.: Pap-smear benchmark data for pattern classification. In: Proceedings of NiSIS 2005: Nature Inspired Smart Information Systems, EU Co-ordination, pp. 1–9 (2005)
15. Krizhevsky, A., Sutskever, I., Hinton, E.: Imagenet classification with deep convolutional neural networks (2012)
16. Krizhevsky, A., Sutskever, I., Hinton, G.: Imagenet classification with deep convolutional neural networks. In: Bartlett, P., Pereira, F., Burges, C., Bottou, L., Weinberger, K. (eds.) Advances in Neural Information Processing Systems, vol. 25, pp. 1106–1114 (2012)
17. Berrada, L., Zisserman, A., Kumar, M.P.: Trusting SVM for piecewise linear CNNs. In: Proceedings of International Conference on Learning Representations (2017, to appear)
18. LeCun, Y., Bengio, Y., Hinton, G.: Deep learning. Nature **521**, 436–444 (2015)
19. Lezoray, O., Cardot, H.: Cooperation of color pixel classification schemes and color watershed: a study for microscopic images. IEEE Trans. Image Process. **11**(7), 783–789 (2002)
20. Mahanta, L.B., Nath, D.C., Nath, C.K.: Cervix cancer diagnosis from Pap smear images using structure based segmentation and shape analysis. J. Emerg. Trends Comput. Inf. Serv. **3**(2), 245–249 (2012)
21. Marinakis, Y., Dounias, G.: Nature-inspired intelligent techniques for Pap smear diagnosis: ant colony optimization for cell classification (2006)
22. Mbaga, A., ZhiJun, P.: Pap smear images classification for early detection of cervicel cancer. Int. J. Comput. Appl. **118**(7), 10–16 (2016)
23. Zeiler, M.D., Fergus, R.: Visualizing and understanding convolutional networks. In: Proceedings of European Computer Vision Conference, pp. 818–833 (2014)
24. World Health Organization: Fact Sheet No. 297: Cancer, February 2006
25. Vandewalle, P., Kovacevic, J., Vetterli, M.: Reproducible research in signal processing. IEEE Sig. Process. Mag. **26**(3), 37–47 (2009)
26. Pawley, J.B.: Handbook of Biological Confocal Microscopy. Springer, Heidelberg (2006)
27. Plissiti, M.E., Charchanti, A., Krikoni, O., Fotiadis, D.I.: Automated segmentation of cell nuclei in PAP smear images, October 2006
28. Plissiti, M.E., Nikou, C., Charchanti, A.: Automated detection of cell nuclei in Pap smear images using morphological reconstruction and clustering. IEEE Trans. Inf. Technol. Biomed. **15**(2), 233–241 (2011)

29. Simonyan, K., Zisserman, A.: Very deep convolutional networks for large-scale image recognition. CoRR abs/1409.1556 (2014)
30. Zeiler, M.D., Fergus, R.: Stochastic pooling for regularization of deep convolutional neural networks. CoRR abs/1301.3557 (2013)

New Level Set Model in Follow Up Radiotherapy Image Analysis

Roushanak Rahmat[1(✉)], William Henry Nailon[2], Allan Price[2], David Harris-Birtill[1], and Stephen McLaughlin[3]

[1] School of Computer Science, University of St Andrews, St Andrews, UK
rr77@st-andrews.ac.uk
[2] Edinburgh Cancer Research Centre, Western General Hospital, Edinburgh, UK
[3] Institute of Sensors, Signals and Systems, Heriot-Watt University, Edinburgh, UK

Abstract. In cancer treatment by means of radiation therapy having an accurate estimation of tumour size is vital. At present, the tumour shape and boundaries are defined manually by an oncologist as this cannot be achieved using automatic image segmentation techniques. Manual contouring is tedious and not reproducible, e.g. different oncologists do not identify exactly the same tumour shape for the same patient. Although the tumour changes shape during the treatment due to effect of radiotherapy (RT) or progression of the cancer, follow up treatments are all based on the first gross tumour volume (GTV) shape of the tumour delineated before treatment started. Re-contouring at each stage of RT is more complicated due to less image information being available and less time for re-contouring by the oncologist. The absence of gold standards for these images makes it a particularly challenging problem to find the best parameters for any segmentation model. In this paper a level set model is designed for the follow up RT image segmentation. In this contribution instead of re-initializing the same model for level sets in vector-image or multi-phase applications, a combination of the two best performing models or the same model with different sets of parameters can result in better performance with less reliance on specific parameter settings.

1 Introduction

In 2014 in the United Kingdom (UK) 356,860 new cases of cancer were diagnosed, while 163,444 deaths were directly attributable to it [CS1]. A major part of cancer treatment involves RT, that is the treatment of cancer by ionising radiation. This aims to deliver a dose of radiation to the diseased tissue whilst minimising damage to healthy tissue. In the UK, RT contributes towards 40% of curative treatment for cancer [BR1,BG1]. It is used to treat cancer patients during several treatment sessions that usually last several weeks.

In RT planning, finding the exact shape of a tumour is the most challenging issue therefore delineation of the volume of interest is based on a visual assessment of medical images, such as X-ray computed tomography (CT) and

M. Valdés Hernández and V. González-Castro (Eds.): MIUA 2017, CCIS 723, pp. 273–284, 2017.
DOI: 10.1007/978-3-319-60964-5_24

magnetic resonance imaging (MRI), by an oncologist. The accuracy of the volume of interest is dependent primarily upon the ability to visualise the tumour, interpret anatomy and understand the potential areas of tumour involvement based on tumour biology. Interpretation of these variables is complex, time consuming and requires considerable clinical expertise. Based on the shape of the tumour that the oncologist delineated before RT started, the patient undergoes RT in several sessions. There is an absence of clinically segmented GTV or ground truth data within follow up RT medical images due to the shortcomings in image qualities available in the treatment time and also limitation of time for re-contouring and re-planning by the clinicians. Therefore, a good segmentation model which can converge to the tumour boundaries with the least amount of prior knowledge is the main concern in the selection of model parameters.

Many different segmentation techniques have been tested in tumour segmentation, particularly level set methods due to their shape adjustment capability. Level set is a segmentation method for capturing moving fronts which was designed to overcome the shortcomings of the snake method, by Osher-Sethian in 1988 as a numerical method for tracking interfaces [DB1,OS1]. The level set was initially designed as an Eulerian formulation of a evolving interface, which grows with the speed F perpendicular to the curve. A literature review shows that every approach is robust to a specific application but not for all kinds of images [CM1]. This highlights the importance of parameter setting to tune each model to handle more images, or the importance of not relying on only one technique when the dataset consists of different images. Exploiting the existing models with less computation difficulties could lead to a more efficient segmentation method. In this paper a new level set model is proposed which has the potential to be beneficial in follow-up RT image analysis due to the absence of ground truth in these images. The robustness of this method relies on the ability of combining different models while compensating the error of any single method. The parallel level set model combines Chan-Vese and Li models and averages them in each iteration of operation.

2 Existing Level Set Models

Level set defines the propagating front as the zero level set of a higher dimension on function ϕ, where the height (z axis) corresponds to the minimum distance from each point in a rectangular coordinate (image plane) from the contour C. To start with this evolving function, an initial value (initialization) is needed. Level set can be initialized automatically or semi-automatically in two or more phases depending on the decision of the user on how many different batches of segmentation are expected in an image. The two-phase level set method segments the image into two regions. Wherever three or four-phase level set methods exist, they can divide into three or four categories respectively by applying two separate level set functions at the same time. Re-initialization is repeated during evolution to prevent the occurrence of sharp corners and prevent flatness by calculating new ϕ values depending on the specified speed function.

2.1 Chan-Vese Level Set Model

In the 1990s Chan-Vese developed Osher-Sethian's model by applying an energy minimization model instead of PDE, which allows automatic detection of interior contours [CV1,CV2]. This is performed by using a piecewise constant and piecewise smooth optimal approximations proposed by Mumford-Shah [MS1]. They also proposed a two-phase level set method without edges that could segment the image into two regions and developed their model further to deal with vector-valued images, which performed robustly in the presence of noise. In 2002, they presented a multi-phase level set method that uses the log of the level set function to separate n phases by using piecewise constant [CV3].

Two-Phase Chan-Vese Without Edges. Chan-Vese proposed the two-phase level set method without edges firstly. The main improvement in this version of the Chan-Vese method is the simplification of the energy functional which is based on the mean intensity values in each region of the level set (inside or outside in two-phase) which are c_1 and c_2 in an image I in a closed subset in ω and ϕ is the level set. At each iteration the values of c_1 and c_2 change and must be recalculated based on the level set of a new region to calculate a new speed function as:

$$
\begin{aligned}
F(c_1, c_2, \phi) = & \int_{\Omega} (u_0 - c_1)^2 H(\phi) d_x d_y \\
& + \int_{\Omega} (u_0 - c_2)^2 (1 - H(\phi)) d_x d_y + \int_{\Omega} |\nabla H(\phi)|
\end{aligned}
\tag{1}
$$

c_1 and c_2, which are defined as:

$$
c_1 = \frac{\int_{\Omega} (1 - H(\phi(x,y)))(I(x,y)) d_x d_y}{\int_{\Omega} 1 - H(\phi(x,y)) d_x d_y} \tag{2}
$$

$$
c_2 = \frac{\int_{\Omega} (H(\phi(x,y)))(I(x,y)) d_x d_y}{\int_{\Omega} H(\phi(x,y)) d_x d_y} \tag{3}
$$

H is the Heaviside function,

$$
H(x) = \begin{cases} 1 & \text{if } x \geq 0 \\ 0 & \text{otherwise} \end{cases} \tag{4}
$$

Vector-Valued Image Chan-Vese Method. In this model, Chan-Vese extended the two-phase method to a vector-valued image in 2000 [CV2]. This model is widely used in colour imaging and video imaging for motion of objects and texture images.

$$
\begin{aligned}
F(c^+, c^-, \phi) = & \int_{inside(C)} \frac{1}{N} \sum_{i=1}^{N} \lambda_i^+ |u_{0,i} - c_i^+|^2 dxdy \\
& + \int_{outside(C)} \frac{1}{N} \sum_{i=1}^{N} \lambda_i^- |u_{0,i} - c_i^-|^2 dxdy + \mu.L
\end{aligned}
\tag{5}
$$

Multi-phase Chan-Vese Without Edges. Chan-Vese extended the two-phase method to a multi-phase method by using n level sets to segment 2^n regions in an image [CV3]. The following equations demonstrate this model for four-phase which consist of two level sets. They are initialized separately but the same level set function is applied for both initializations. The mapping plane shows this growth concept in four-phase clearly.

$$\begin{aligned} F(c,\phi) = &\int_{\Omega}(u_0 - c_{11})^2 H(\phi_1)H(\phi_2)d_x d_y \\ &+ \int_{\Omega}(u_0 - c_{10})^2 H(\phi_1)(1 - H(\phi_2))d_x d_y \\ &+ \int_{\Omega}(u_0 - c_{01})^2 (1 - H(\phi_1))H(\phi_2)d_x d_y \\ &+ \int_{\Omega}(u_0 - c_{00})^2 H(\phi_1)(1 - H(\phi_2))d_x d_y \\ &+ \nu\int_{\Omega}|\nabla H(\phi_1)| + \nu\int_{\Omega}|\nabla H(\phi_2)| \end{aligned} \tag{6}$$

$$\begin{aligned} c_{11} &= mean(u_0) \in \{(x,y) : \phi_1(t,x,y) > 0, \phi_2(t,x,y) > 0\} \\ c_{10} &= mean(u_0) \in \{(x,y) : \phi_1(t,x,y) > 0, \phi_2(t,x,y) < 0\} \\ c_{01} &= mean(u_0) \in \{(x,y) : \phi_1(t,x,y) < 0, \phi_2(t,x,y) > 0\} \\ c_{00} &= mean(u_0) \in \{(x,y) : \phi_1(t,x,y) < 0, \phi_2(t,x,y) < 0\} \end{aligned} \tag{7}$$

2.2 Li Level Set Model

Li developed an edge based level set method based on gradient flow for solving the inhomogeneity in intensity, which previous methods had difficulty in solving them [Li1, Li2, Li3]. This model uses the energy minimization technique, similar to the Snake model, by reducing the fitting energy in image segmentation. This energy is:

$$F_{Li}(\phi) = \mu P(\phi) + \lambda L(\phi) + \nu A(\phi) \tag{8}$$

where,

$$P(\phi) = \int_{\Omega}\frac{1}{2}(|\nabla\phi| - 1)d_x d_y \tag{9}$$

$$L(\phi) = \int_{\Omega} g(I)\delta(\phi)|\nabla(\phi)|d_x d_y \tag{10}$$

$$A(\phi) = \int_{\Omega} g(I)H(-\phi)d_x d_y \tag{11}$$

and

$$g(I) = \frac{1}{1 + |\nabla G^{\sigma} * I|^{p}}, p \geq 1 \tag{12}$$

where μ is the penalizing coefficient, λ is the coefficient for length and ν refers to the area. The ratio of λ and ν defines the stopping point of level set evolution because both of these terms contain edge information. The length term keeps the contour tight and the area helps the expansion of the contour. $P(\phi)$ is a penalty term in the energy functional that is used to level set periodical during evolution. The stopping operator, g is based on a Gaussian Kernel that forces the level set to converge to zero when approaching the edges. σ is the standard deviation, L represents the length of the contour with respect to the stopping operator of g as its weight and A is the speed controller of the evolution which makes the contour to shrink if the ν is positive and tends to expand when the ν is negative.

3 Proposed Level Set Models

Poor parameter choice can significantly, and detrimentally, change the performance of a robust model. Therefore, the parameter setting has to take advantage of all existing information. The usage of level set segmentation as the current models are very much dependent on their parameter setting and finding appropriate parameters is particularly challenging in the absence of any ground truth on the data. Therefore a new level set technique is proposed for this matter. First, the parallel level sets in vector-valued image model can choose the best force calculation in each iteration as it has two different level set models competing with each other and takes the average of their force in each iteration. This model benefits from the strength of two models and minimizes any error that could arise with a single model based technique. Second, the parallel level sets is extended in multi-phase technique which can apply different level set models or the same model with different parameters to segment different objects. The multi-phase level set model so far depends on the same level set model applied at the same time with different initialisations.

3.1 Proposed Parallel Level Sets in Vector-Valued Image Model

A new level set configuration is presented based on the concept of vector-valued imaging introduced in Sect. 2. The concept of a vector-valued image in level set is introduced first by applying the two-phase Chan-Vese method at the same time on different images, such as different RGB channels or different texture images, and averaging the force value in each image at each iteration as shown in Fig. 1.

The success of the vector-valued Chan-Vese is in using different features of an image. It needs pre-processing of the image to obtain the features, before applying the same two-phase Chan-Vese on all layers. The proposed vector-valued image model in this paper does not require different features of an image but it applies different level sets on the same image at each iteration. This parallel

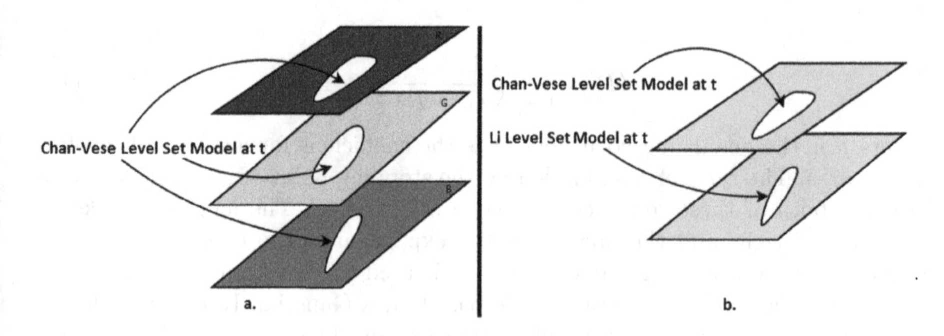

Fig. 1. a. Vector-valued Chan-Vese on RGB channels, b. Proposed parallel level sets in vector-valued image.

implementation of different level sets on the same image takes advantage of the original image while engaging with the other level set performance by averaging their forces in each iteration. The proposed model is illustrated in Fig. 1.b. and Eqs. (13)–(14).

In the proposed approach, the combination of the Chan-Vese and the Li models were used but any other model could be used. The rationale is to exploit the advantages of both techniques at each iteration, for example the Chan-Vese model works well for inhomogeneous images while the Li model for sharp ones. Different level sets can also be used but with different sets of parameters. This implementation is beneficial when the user does not know the best set of parameters to tune the model. The force of the level set is modified at each iteration by calculating the average force of both methods.

$$F = \frac{1}{2}(F_{Li} + F_{CV}) \tag{13}$$

Equation (13) defines the averaging equation for forces which happens in each iteration that new forces are obtained from each model. Finally, to reinitialise level set, new ϕ should be calculated from Eq. (14).

$$\phi(x, y, t+1) = \phi(x, y, t) + \Delta t.F \tag{14}$$

where Δt is the step size.

3.2 Proposed Parallel Level Sets in Multi-phase Method

The proposed parallel level sets model in Sect. 3.1 can be extended to a multi-phase model. The current multi-phase models usually apply the same model twice with a different initialization for the same model. The original multi-phase level set method can fail after a large number of iterations because detected regions from different phases become the same as the level set characteristics and parameters for different phases are the same. Also, as different objects in an image are expected to have different specifications, different level sets can detect them better than the same model. Figure 2, Eqs. (15) and (16) illustrates this model.

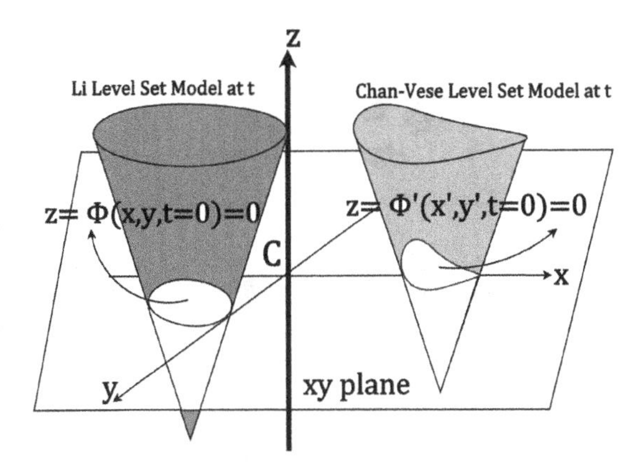

Fig. 2. Proposed parallel level sets in multi-phase shape.

This model also can use the same methods such as the Chan-Vese but with a different set of parameters like the proposed model in Sect. 3.1. It is actually less complex to use the same model with different initialization and different parameters. In this case each level set and the different regions can be calculated in a similar fashion to Eqs. (6) and (7) but different parameters of each level set in PDE can define the new level set differently.

$$\begin{aligned}\frac{\partial \phi_1}{\partial t} &= \delta(\phi_1)[\nu_1 div(\frac{\nabla \phi_1}{|\nabla \phi_1|}) \\ &- \Big((u_0 - c_{11})^2 + (u_0 - c_{01})^2)\Big) H(\phi_2) \\ &+ \Big((u_0 - c_{10})^2 + (u_0 - c_{00})^2))\Big)(1 - H(\phi_2))]\end{aligned} \tag{15}$$

$$\begin{aligned}\frac{\partial \phi_2}{\partial t} &= \delta(\phi_2)[\nu_2 div(\frac{\nabla \phi_2}{|\nabla \phi_2|}) \\ &- \Big((u_0 - c_{11})^2 + (u_0 - c_{01})^2)\Big) H(\phi_1) \\ &+ \Big((u_0 - c_{10})^2 + (u_0 - c_{00})^2))\Big)(1 - Hs(\phi_1))]\end{aligned} \tag{16}$$

where ν_1 and ν_2 are weighting parameters that has been selected differently.

4 Discussion of Proposed Parallel Level Sets

Initially a comparison on non-medical images was carried out, as these had the benefit of ground truth, and this is shown in Fig. 3, where each row shows the performance of one of the Chan-Vese and Li methods on different test images: a. test images, b. two-phase Chan-Vese [CV1], c. multi-phase Chan-Vese [CV3], d. two-phase Li [Li2], e. multi-phase Li [Li1] and the proposed model is shown in row f. as the parallel level sets in vector-valued imaging.

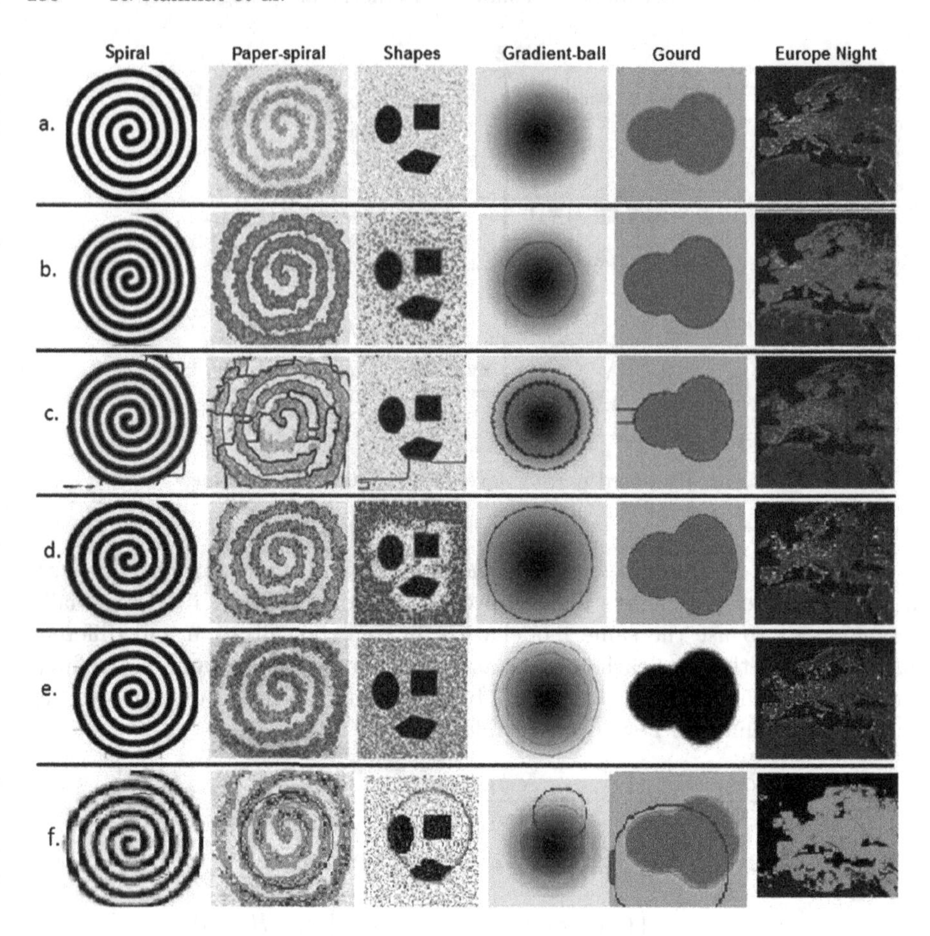

Fig. 3. Non-medical image segmentation with different level set methods: a. test images, b. two-phase Chan-Vese [CV1], c. multi-phase Chan-Vese [CV3], d. two-phase Li [Li2], e. multi-phase Li [Li1]. and f. proposed parallel level sets in vector-valued imaging on test images (red is the initial level set and green is the level set segmentation). (Color figure online)

In Fig. 3, rows b. and c. show the result of applying the Chan-Vese model in two and four-phases for non-medical images. The improvement of the four-phase level set compared to two-phase is obvious and results from considering more regions of interest. Rows d. and e. illustrate Li model in two and four-phase respectively. The proposed configurations of the level set is shown in row f. which can perform better than Chan-Vese and Li models alone. The final segmentation of each method is shown in different colours from rows a. to e. while row f. shows the final segmentation in green colour when the red contour is the initial level set placement before computation starts. Also, Fig. 4 illustrates the performance of proposed model combining Chan-Vese and Li models compared with vector-valued and multiphase Chan-Vese as well as multiphase Li. In this figure, four

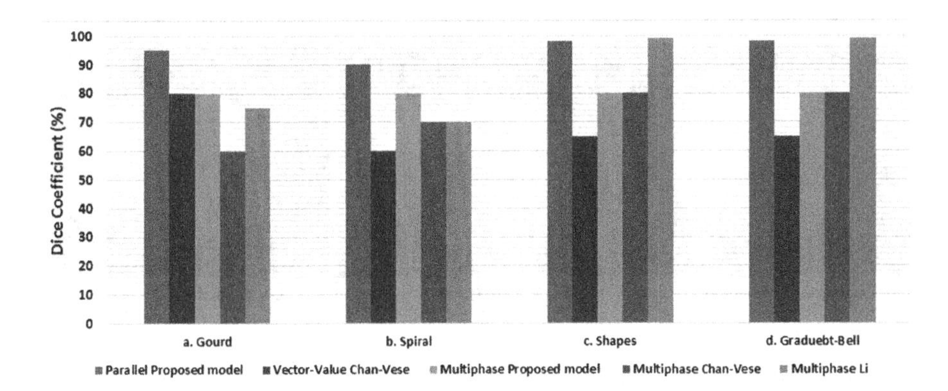

Fig. 4. Testing proposed parallel level sets in vector-valued and multi-phase level set on test images with ground truth: a. gourd, b. spiral, c. shapes and d. gradient-ball.

of the non-medical images which have ground truth are chosen and its similarity with the level set segmenting of them was compared by Dice coefficient. The Dice coefficient can measure the level of similarity between two closed sets, A and B. Equation 17 is used for measuring the Dice coefficient to provide an understanding of the level of overlap between the two closed sets, such as the ground truth boundary and the final computed segmentation.

$$D = \frac{2|A \bigcap B|}{|A| + |B|} \tag{17}$$

The results of Dice coefficient in Fig. 4 illustrate the success of the proposed model over its original forms as parallel level set can perform over 90% accurate for all of these images. The poor performance of other models can be due to low contrast of these images. Other images in Fig. 3 were not assessed qualitatively as their ground truth is not very accurate. The results for these images appear to be subjective assessment through satisfactory compared to the original models.

To consider the suitability of proposed level set model on follow up medical images, two different image modalities were chosen. These images consist of brain MRI and lung CT. Follow-up cone beam CT (CBCT), where CBCT is the only image modality available throughout the whole course of RT. CBCT is generally used in the mechanical procedure of adjusting patient position before starting each session of treatment and is not generally used by clinicians for monitoring the tumour. The brain MRI dataset consists of 20 patients who suffered from Glioma from BRATS 2015 [MB1], where the GTV is known for these images. The Dice coefficient is 84% for 20 patients shown in this figure which is higher than Li and Chan-Vese models. In Fig. 5, the better performance of proposed model is shown for only one selected slice.

Figure 7 shows the Dice coefficient performance of proposed model compared to Li and Chan-Vese models on the follow up CBCT images of lung cancer RT. This dataset includes twenty patients who suffered lung cancer and underwent an

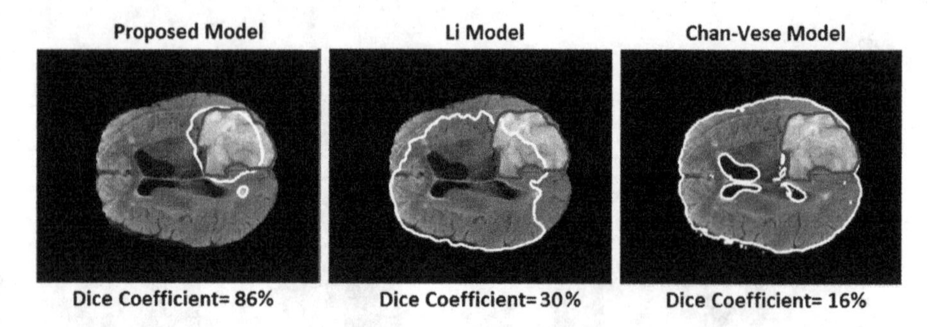

Fig. 5. Performance comparison of the proposed parallel level sets in vector-valued imaging with Li and Chan-Vese models on a slice of brain MR image chosen from 20 patients, GTV is shown in green and the outcome of level set models in blue. (Color figure online)

RT in 2010 and 2011 at the Western General Hospital, Edinburgh, Scotland. All of these patients went through the whole course of treatment over approximately four weeks. CBCT #10 are most commonly used as the oncologists are keen to know the effect of RT while one third of treatment is done (Fig. 6).

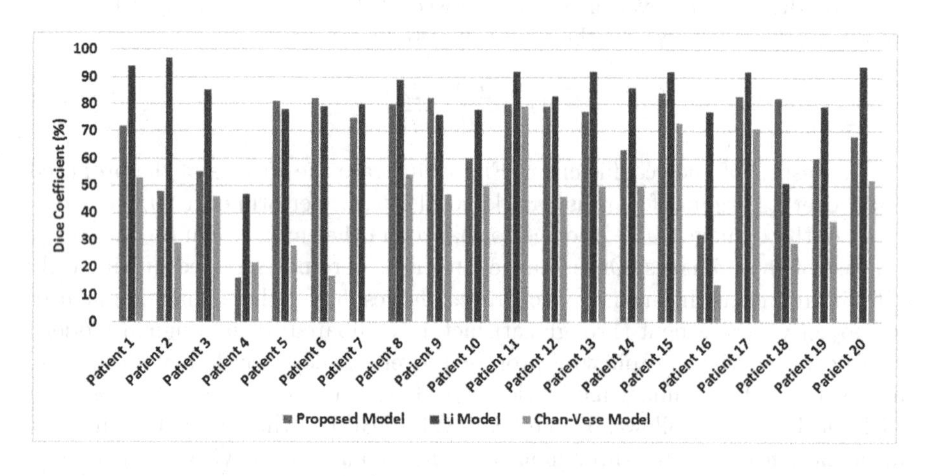

Fig. 6. Dice coefficient analysis of proposed parallel level sets in vector-valued imaging compared to Li and Chan-Vese models on lung CBCT #10 for 20 different patients suffered from non-small cell.

The proposed model can show the expected shrinkage in this period while Li model has very high values for the Dice coefficients and Chan-Vese model was so low. Also, Fig. 7 illustrates some qualitative analysis for a sample slice per patient in the lung datasets for the proposed model.

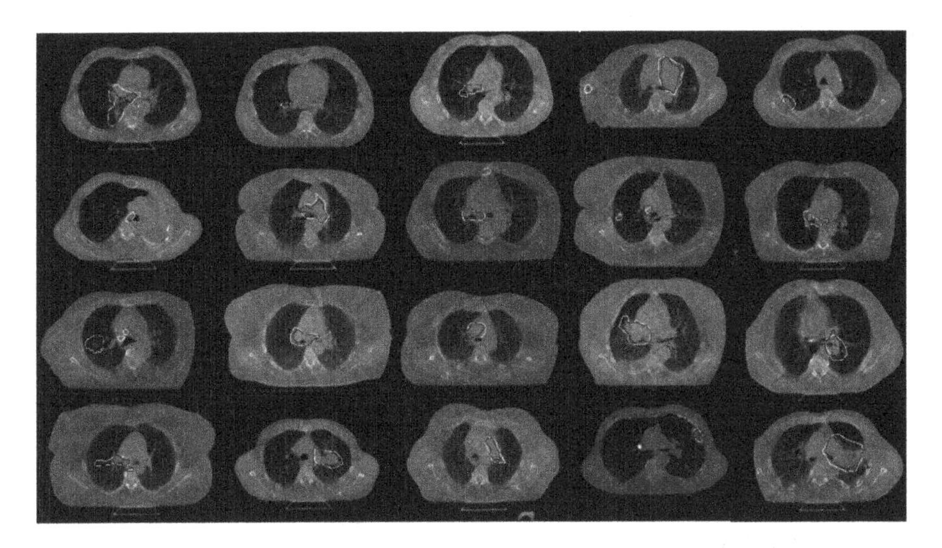

Fig. 7. Performance of proposed parallel level sets in vector-valued imaging compared to Chan-Vese and Li models on lung CBCT #10 for 20 different patients suffered from non-small cell lung cancer (GTV in red colour delineated before starting RT and blue contour represents the result of proposed model). (Color figure online)

5 Conclusions

This paper shows the formation of a new level set model in the application of medical image analysis when the ground truth is not available, specifically on follow up RT medical image segmentation. This model has advantages by combining both Chan-Vese and Li models where both models provide satisfactory results. Using existing models, the parameter settings should be tweaked manually which is both time consuming and challenging, by combining these models single parameter settings can be used by multiple image sets while still maintaining higher accuracies. Different test images in medical and non-medical shapes were used to illustrate the performance of existing Chan-Vese and Li models compared to the combined proposed model. The Dice coefficient of the combined proposed model presents a higher accuracy level compared to each of the previous models. The medical datasets consisted from famous BRATS 2015 in MRI modality of brain and the lung CBCT images for the follow up RT treatment. The Dice coefficient showed 84% of similarity between the proposed model and the GTV in averaging 20 patients form BRATS dataset. For the follow up CBCT images, this similarity was 67% while using the GTV delineated before RT. This new model enables the freedom of two or more level sets to run in parallel on the same image in each iteration to compensate each other's performance and therefore requires less manual parameter tweaking. The concept can be used in vector-valued image level set or multi-phase level sets.

References

[CS1] Cancer Statistics for the UK. https://www.cancerresearchuk.org/health-professional/cancer-statistics. Accessed 17 Feb 2017

[BR1] Baskar, R., et al.: Cancer and radiation therapy: current advances and future directions. Int. J. Med. Sci. **9**(3), 193–199 (2012)

[PR1] Prescribing, recording, and reporting photon-beam intensity-modulated radiation therapy (IMRT): contents. J. ICRU **10**(1) (2010). doi:10.1093/jicru/ndq002

[GM1] Grégoire, V., Mackie, T.R.: State of the art on dose prescription, reporting and recording in intensity-modulated radiation therapy (ICRU report No. 83). Cancer/Radiothérapie **15**(6), 555–559 (2011)

[GR1] Gonzalez, R.C., Woods, R.E.: Digital Image Processing (2007)

[DB1] Dale, A.M., Fischl, B., Sereno, M.I.: Cortical surface-based analysis: I. Segmentation and surface reconstruction. Neuroimage **9**(2), 179–194 (1999)

[OS1] Osher, S., Sethian, J.A.: Fronts propagating with curvature-dependent speed: algorithms based on Hamilton-Jacobi formulations. J. Comput. Phys. **79**(1), 12–49 (1988)

[CV1] Chan, T., Vese, L.: An active contour model without edges. In: Nielsen, M., Johansen, P., Olsen, O.F., Weickert, J. (eds.) Scale-Space 1999. LNCS, vol. 1682, pp. 141–151. Springer, Heidelberg (1999). doi:10.1007/3-540-48236-9_13

[CV2] Chan, T.F., Sandberg, B.Y., Vese, L.A.: Active contours without edges for vector-valued images. J. Vis. Commun. Image Represent. **11**(2), 130–141 (2000)

[MS1] Mumford, D., Shah, J.: Optimal approximations by piecewise smooth functions and associated variational problems. Commun. Pure Appl. Math. **42**(5), 577–685 (1989)

[CV3] Vese, L.A., Chan, T.F.: A multiphase level set framework for image segmentation using the Mumford and Shah model. Int. J. Comput. Vis. **50**(3), 271–293 (2002)

[RM] Rousson, M.: Cue integration and front evolution in image segmentation. Diss. Universit Nice Sophia Antipolis (2004)

[CM1] Cremers, D., Rousson, M., Deriche, R.: A review of statistical approaches to level set segmentation: integrating color, texture, motion and shape. Int. J. Comput. Vis. **72**(2), 195–215 (2007)

[LY1] Lianantonakis, M., Petillot, Y.R.: Sidescan sonar segmentation using active contours and level set methods. In: Oceans 2005-Europe, vol. 1. IEEE (2005)

[Li1] Li, C., et al.: Minimization of region-scalable fitting energy for image segmentation. IEEE Trans. Image Process. **17**(10), 1940–1949 (2008)

[Li2] Zhang, K., Song, H., Zhang, L.: Active contours driven by local image fitting energy. Pattern Recogn. **43**(4), 1199–1206 (2010)

[Li3] Li, C., et al.: A level set method for image segmentation in the presence of intensity inhomogeneities with application to MRI. IEEE Trans. Image Process. **20**(7), 2007–2016 (2011)

[BG1] Barnett, G.C., et al.: Normal tissue reactions to radiotherapy: towards tailoring treatment dose by genotype. Nat. Rev. Cancer **9**(2), 134–142 (2009)

[MB1] Menze, B.H., et al.: The multimodal brain tumor image segmentation benchmark (BRATS). IEEE Trans. Med. Imaging **34**(10), 1993–2024 (2015)

Topological Analysis of the Vasculature of Angiopoietin-Expressing Tumours Through Scale-Space Tracing

Constantino Carlos Reyes-Aldasoro[1(✉)], Meit Bjorndahl[2], Chryso Kanthou[2], and Gillian M. Tozer[2]

[1] City, University of London, London EC1V 0HB, UK
reyes@city.ac.uk
[2] The University of Sheffield, Sheffield S10 2JF, UK

Abstract. This work describes the topological analysis of the vasculature of tumours. The analysis is performed with a scale-space technique, which traces the centrelines of *vessels* as *topological ridges* of the image intensities and then obtains a series of measurements, which are used to compare the vasculatures. Besides the measurements directly associated with the centrelines, the scales obtained allow the estimation of width and thus area covered with vessels.

Tumours of SW1222 human colorectal carcinoma xenografts were observed when growing in dorsal skin-fold window chambers in mice. Three variants of the tumours expressing either endogenous levels of angiopoietins (WT) or over-expressing either angiopoietin-1 (Ang-1) or angiopoietin-2 (Ang-2) were assessed with/without vascular targeted therapy. The scale-space technique was able to discriminate between the vasculatures of the three different tumour types prior to treatment. Results also suggested that over-expression of Ang-2 was associated with susceptibility of the tumour vasculature to the vascular disrupting agent, combretastatin A4 phosphate (CA4P). Substantiation of this finding would point to the potential of tumour Ang-2 expression as a predictive bio-marker for response to CA4P.

Keywords: Scale space analysis · Tracing of vasculature · Vascular disrupting agents · CA4P · Tumour microcirculation · Angiopoietin

1 Introduction

The process of formation of new vessels, in which capillaries sprout from existing vessels is known as *angiogenesis* [1] and, in adults, this process is restricted to relatively few situations such as the female reproductive cycle [2] and pathological processes such as growth of collaterals in response to obstructive arterial disease [3,4], wound healing and tumour growth [5]. In the case of tumours, the new vasculature has become an attractive therapeutic target to prevent nutrients and oxygen reaching the tumour as well as preventing metastasis [6,7].

M. Valdés Hernández and V. González-Castro (Eds.): MIUA 2017, CCIS 723, pp. 285–296, 2017.
DOI: 10.1007/978-3-319-60964-5_25

Angiogenesis is largely driven by a series of growth factors, of which the *angiopoietin* family of proteins play a substantial role. The angiopoietins work in concert with vascular endothelial growth factor A (VEGFA), with VEGFA stimulating formation of the initial vascular plexus and angiopoietin-1 (Ang-1) promoting vascular re-modelling, maturation and stabilisation [8]. During early phases of vascular sprouting, angiopoietin-2 (Ang-2) can act as a functional antagonist of Ang-1 that destabilises blood vessels prior to vascular sprouting or regression, although the precise functions of Ang-2 are highly context dependent [9].

The microtubule depolymerising agent, combretastatin A4 phosphate (CA4P; fosbretabulin) is currently in clinical trial (http://www.mateon.com) as a tumour vascular disrupting agent (VDA) that causes selective tumour blood flow shut-down, necrosis and growth retardation in solid tumours. Susceptibility of the tumour vasculature to CA4P is thought to relate to an immature vascular phenotype [7]. Therefore, we hypothesised that tumours over-expressing Ang-2 would be more susceptible to CA4P than those over-expressing Ang-1 or unmodified tumours, pointing to the potential of tumour angiopoietin expression analysis for predicting response to CA4P.

Murine dorsal skin-fold window chambers [10] provide a means to observe developing tumour blood vessels microscopically in longitudinal studies. Quantitative assessment of the angiogenesis process can be performed via analysis of the topological or chromatic characteristics of the resulting optical images [11].

In this work, we assessed the influence of angiopoietin expression in tumours through the comparison of a series of topological parameters: *number of vessels, length, width* and *relative area*. The mathematical description of the topology was obtained through a fully automatic vessel tracing algorithm [11] based on the scale-space ridge detection algorithm proposed by Lindeberg [12,13].

2 Materials and Methods

2.1 Data Sets

SW1222 human colorectal carcinoma cells, kindly supplied by Professor Barbara Pedley, University College London, UK, were transfected with angiopoietin-1 or angiopoietin-2 cDNA cloned into pcDNA3.1 mammalian expression vector or with empty pcDNA-3.1 vector alone. cDNA was kindly supplied by Professor Pam Jones, University of Leeds, UK. The resulting cells transfected with empty vector were designated wild-type (WT). Stable transfectants were selected in G418-containing medium and translation of appropriate gene products was confirmed by western blotting (data not shown). Cells over-expressing angiopoietin-1 or angiopoietin-2 were designated Ang-1 and Ang-2 respectively.

Methodological details of the animal experimentation have been described previously [14], but are included here briefly for completeness. Animal experiments were conducted in accordance with the United Kingdom Home Office Animals (Scientific Procedures) Act 1986, with local ethical approval and in line with the published Guidelines for the Welfare of Animals in Cancer Research [15].

Surgical procedures on severe-combined immune-deficient (SCID) mice consisted of implanting an aluminium window chamber (total weight ~2 g), designed to hold parallel glass windows 200 μm apart to allow tumour growth, into a dorsal skin-fold under general anaesthesia. SW1222 human colorectal carcinoma cells were grown as hanging drop cultures on the lids of 60 mm plastic petri dishes inverted over dishes containing 4 ml DMEM containing standard antibiotics. Hanging drops were established from 1×10^5 SW1222 cells in 20 μl DMEM supplemented with 10% foetal calf serum and 4 mmol/l glutamine plus standard antibiotics. After 3–4 days of growth, the cell aggregates from the drops were transplanted directly onto the exposed panniculus muscle within the dorsal skin-fold window chamber preparations and the chambers closed with a glass cover-slip. Treatments and image capturing were started approximately 10–12 days after surgery.

Mice were implanted with WT, Ang-1 or Ang-2 tumour cells and, once established, were treated with either 30 mg/kg of CA4P or saline as a control, administered intraperitoneally (i.p.). CA4P was kindly supplied by Professor GR Pettit, Arizona State University, Tempe, AZ, USA.

The distribution of the mice into treatment groups was: Ang-1/CA4P ($n = 5$), Ang-1/control ($n = 6$), Ang-2/CA4P ($n = 6$), Ang-2/control ($n = 6$), WT/CA4P ($n = 4$), WT/control ($n = 2$). All tumours were observed longitudinally before (time = 0) and up to 24 h (time = 2.5, 15, 30, 60, 180, 360, 1440 min) after treatment.

Transmitted light images from a Nikon Eclipse E600FN microscope fitted with a ×2.5 and ×10 objective and incorporating illumination from a 12 V/100 W halogen light source were acquired from 29 restrained window chamber-bearing mice at 2 regions of interest (ROI) at different locations within the tumours, as illustrated in Fig. 1. In total, **four hundred and sixty four** images were acquired and processed.

The variability of the vasculatures obtained is illustrated in Fig. 2. Control tumours experienced remodelling of the vasculature within this time period, as illustrated in Fig. 3.

2.2 Scale Space Analysis

The scale-space approach [12,13] is a multiscale technique in which a progressive low pass filtering is applied to an image with the intention of detecting features of different dimensions. The features can be of different nature, e.g. ridges, edges or blobs. Fine features will be detected at initial scales, with minimal smoothing, whilst large or coarse features will be detected with considerable smoothing of the images. Scale-space algorithms are *intensity-based* as opposed to clustering or deformable models. The details of the implementation of a scale-space vessel tracing implementation have been previously described in [11]. Essentially, the centreline of vessels was detected as a *ridge* with the intensity f of a given pixel at position (x, y) as the height in a topographical analogy. It is important to notice some limitations of the method: (a) the intensity of a physiological vessel in the image will guide the tracing process. Therefore, if a vessel does not show contrast against the

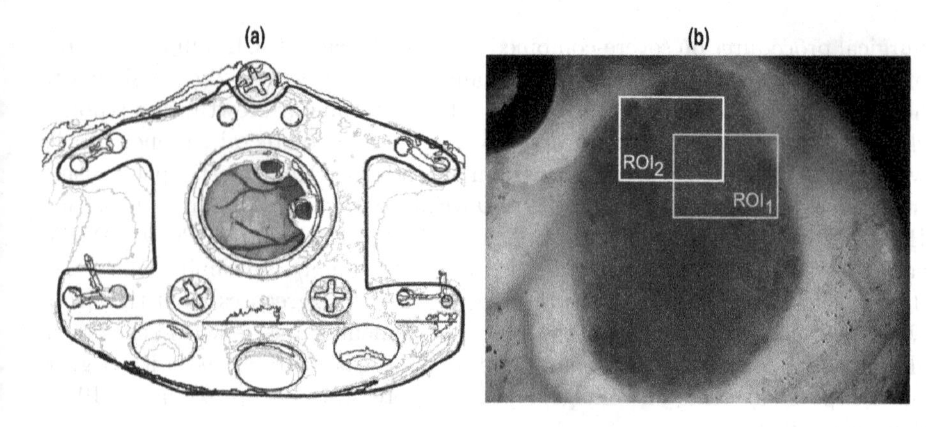

Fig. 1. (a) Illustration of a dorsal skin-fold window chamber, which is implanted in the dorsal skin of a mouse and through which the vasculature of a tumour can be observed over time. (b) One representative image acquired with a $\times 2.5$ microscope objective. The size of the regions of interest to be captured with a $\times 10$ microscope objective is illustrated with white and yellow boxes. (Color figure online)

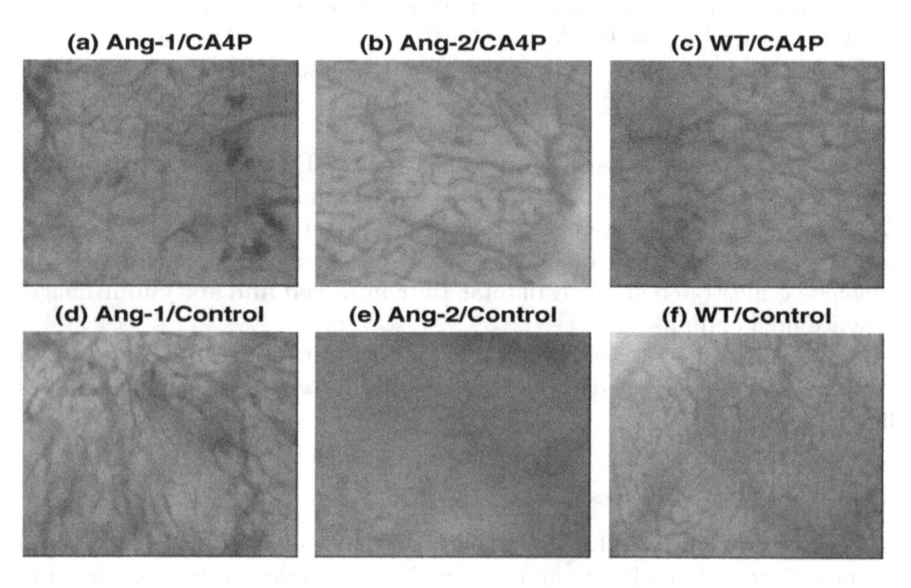

Fig. 2. ROIs from tumours within different treatment groups showing representative vasculatures. Notice the variability of vasculature in terms of size, density, colour, etc. (Color figure online)

tissue it may not be traced; (b) a physiological vessel may experience changes of intensity within its centreline and in some cases a single vessel may appear as two segments or traced vessels; (c) crossings of vessels may imply discontinuity of the tracing. Thus, the measurement of *number of vessels* should be considered as an approximation and may differ from a manually obtained measurement.

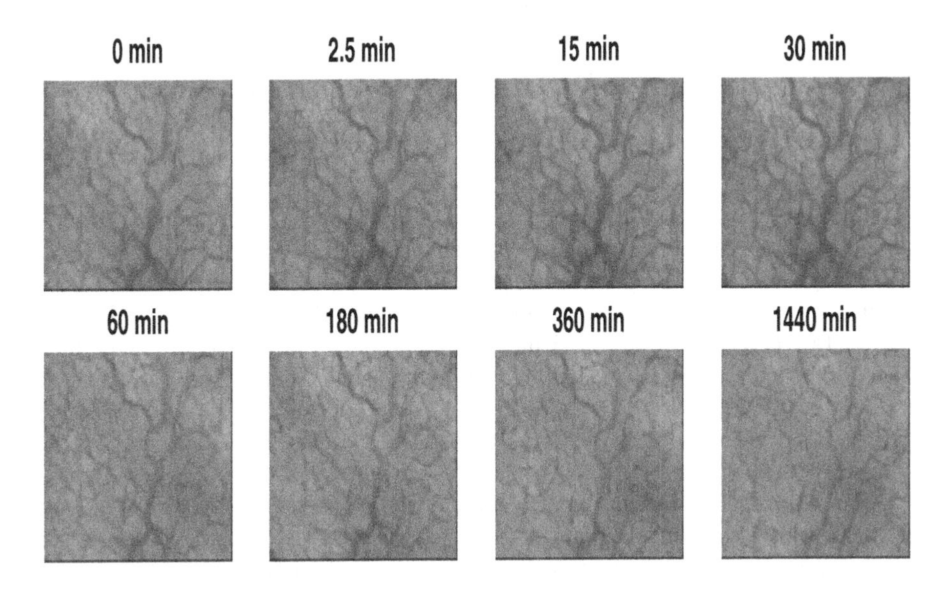

Fig. 3. Longitudinal acquisition of one tumour. Notice the variation of the thickness and position of the vasculature.

The scale-space representation of a function $f(x, y)$ can be defined as the convolution with a Gaussian $g(x, y; t)$ where t corresponds to the width of a Gaussian:

$$L(x, y; t) = g(x, y; t) * f(x, y) = \frac{1}{(2\pi t)} e^{-\frac{(x^2+y2)}{(2t)}} * f(x, y) \tag{1}$$

Then, the Hessian Matrix can be formed with the normalised first and second derivatives in x and y dimensions (Lx, Lxx, Ly, Lyy).

$$H = \begin{bmatrix} \partial_x \partial_x f \ \partial_y \partial_x f \\ \partial_x \partial_y f \ \partial_y \partial_y f \end{bmatrix} = \begin{bmatrix} L_{xx} \ L_{yx} \\ L_{xy} \ L_{yy} \end{bmatrix} \tag{2}$$

Regions of maxima and minima in H were calculated when the derivatives reached zero. To obtain the centrelines of vessels, or ridges in a topological analogy, it was necessary to convert from the (x, y) coordinate system to a local (p, q) system aligned with the eigendirections of H:

$$\begin{aligned} L_p &= \partial pL &&= (sin\beta\partial_x - cos\beta\partial_y)L, \\ L_q &= \partial qL &&= (cos\beta\partial_x + sin\beta\partial_y)L, \\ L_{pq} &= \partial p\partial qL &&= (cos\beta\partial_x + sin\beta\partial_y)(sin\beta\partial_x - cos\beta\partial_y)L, \end{aligned} \tag{3}$$

where β denotes the angle of rotation of the coordinate system and it was defined by:

$$cos\beta\Big|_{(x_0,y_0)} = \sqrt{\frac{1}{2}\left(1+\frac{L_{xx}-L_{yy}}{(L_{xx}-L_{yy})^2+4L_{xy}^2}\right)}\Bigg|_{(x_0,y_0)},$$

$$sin\beta\Big|_{(x_0,y_0)} = (sign(L_{xy})) \times \sqrt{\frac{1}{2}\left(1+\frac{L_{xx}-L_{yy}}{(L_{xx}-L_{yy})^2+4L_{xy}^2}\right)}\Bigg|_{(x_0,y_0)}. \quad (4)$$

The ridges at different scales constituted a scale-space ridge surface and were defined as the points of (Lp, Lpp, Lq, Lqq) that fulfilled the conditions of maxima at every scale. $L, Lx, Ly, Lpp,$ andLqq are illustrated in Fig. 4 for three scales. Finally, a *scale-space ridge* was simplified from the ridge surface by selecting the points where the surface had maximal values by a given norm. Thus, fine ridges were detected at fine scales, whilst coarse ridges had higher norm values at larger scales. A scale-space *vessel* was subsequently defined as an individual ridge detected by the algorithm, noticing that the changes of intensity may break a physical vessel into two or more *vessels*.

A large number of measurements can be obtained from the traced vessels, but for this work we concentrated on the following: number of vessels, average vessel length, average vessel diameter, and relative area of the ROI covered by vessels. The *length* and *number* are direct measurements. The *diameter* is estimated as proportional to the scale where the ridge was selected. The *area* is derived from the combination of diameter and the length of each traced vessel. The *relative area* is the ratio of the area previously calculated over the total area of the

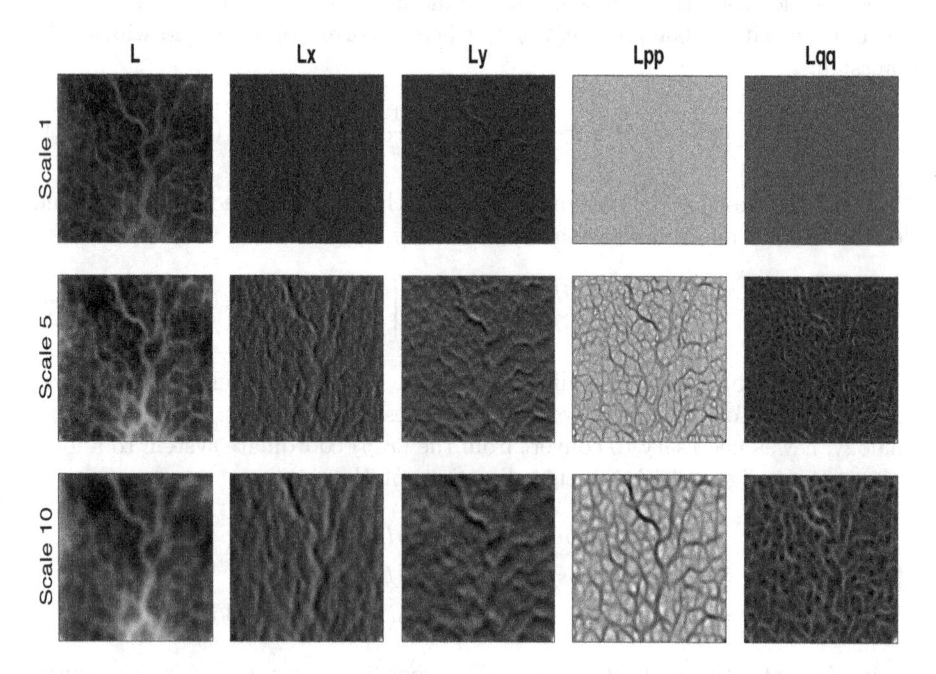

Fig. 4. Illustration of a Scale-Space process where the intensity of an image is filtered at different scales (vertical) and derivatives in different directions highlight features at each scale.

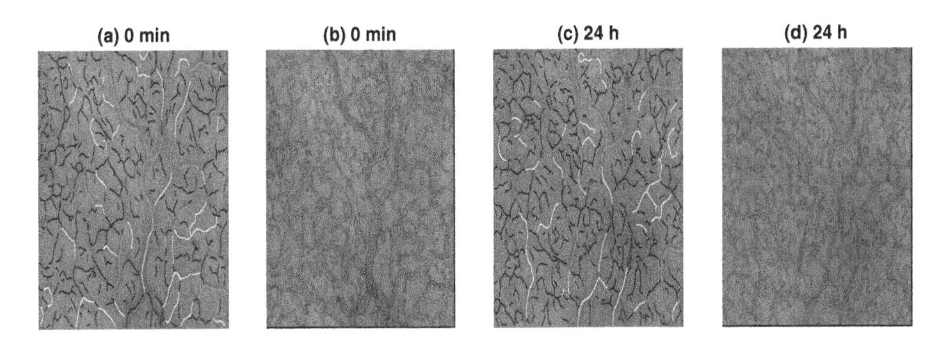

Fig. 5. (a,c) Traced vessels overlaid on two time points from a single tumour. The vessels have been ranked and the top 10 are traced in red, 11–50 are traced in light green and the rest in thinner black lines. Notice the changes of the vascular topology over time. (b,d) Traced vessels and the inferred thickness of each vessel denoted by dark shading. The relative area covered was obtained by the ratio of the shaded pixels over the total area of the image. (Color figure online)

tumour tissue in the ROI. In addition, it is possible to rank the vessels by their *saliency*, i.e. the contrast between the vessel itself and the surrounding regions, in order to distinguish the top 5, 10, 50, etc. vessels. Figure 5 illustrates results of length, diameter, area and saliency for two time points of the tumour of Fig. 3.

All algorithms were implemented in MATLAB® (The Mathworks™, Natick, USA). Statistical tests of one-way Analysis of Variance (ANOVA) followed by a Tukey post-test to compare individual groups were performed in MATLAB and GraphPad Prism v7.0b.

3 Results and Discussion

The 464 images described in Sect. 2.1 were traced with the scale-space algorithm previously described. The following measurements were selected for analysis: *number of vessels, average diameter, average length* and *relative area covered by the vessels*. Each of these measurements was calculated for the three different tumour types *(Ang-1, Ang-2, WT)*, two ROIs *(1, 2)*, treated or control *(CA4P, control)* and eight time points *(1–8)*. Data from the two ROIs were pooled for each tumour.

Results for the three tumour types at time $t = 0$, that is before treatment was administered, are shown in Fig. 6. $p - values$ for a one-way ANOVA test with a Tukey post-test are presented within the figure. Boxplot whiskers indicate minimum and maximum values, and the box is formed by lines at the median, first and third quartiles.

Ang-1 tumours displayed higher average length, diameter and relative area of blood vessels compared with Ang-2 tumours (Fig. 6). Furthermore, the relative area measurement was significantly lower for Ang-2 tumours than either Ang-1 tumours or WT tumours. Ang-1 expression has previously been associated with

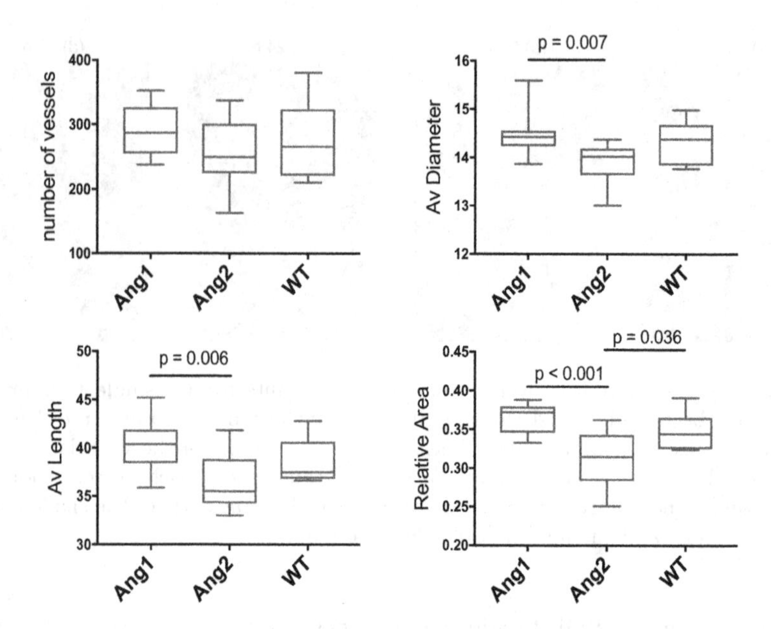

Fig. 6. Boxplots summarising the statistical difference between the three tumour types at time $t = 0$, before the start of treatment. The $p-values$ for 1-way ANOVA followed by a Tukey post-test are included within the figures. There was a statistical difference between Ang-1 and Ang-2 for the diameter, length and relative areas, and between Ang-2 and WT for relative area only.

development of wide vessels, consistent with these findings [9]. Ang-2 can act as an antagonist for Ang-1, causing vessel destabilisation [16–21], which may also have contributed to our results. Interestingly, angiopoietin over-expression

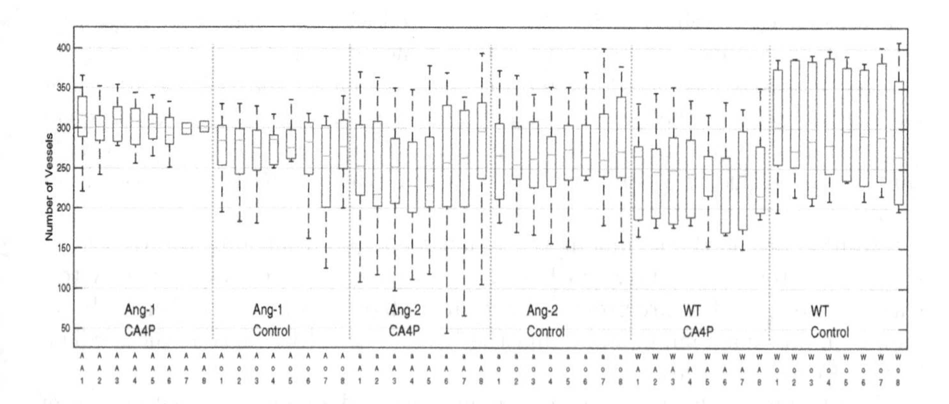

Fig. 7. Boxplots of the number of vessels as determined by the tumour type Ang-1, Ang-2 and WT (**A/a/W**), treatment for CA4P or control (**A/o**) and the time of acquisition (**1–8**).

did not influence the number of vessels identified by the algorithm (Fig. 6(a)), suggesting that effects were confined to the diameter and length and not the number of traced vessels.

Figures 7, 8, 9 and 10 show the effects of CA4P treatment versus control over time on the four vascular parameters shown in Fig. 6, for each of the three tumour types. The number of vessels remained relatively stable in each of the tumour types over time, whether treated with CA4P or not. However, CA4P appeared to reduce average diameter and length of blood vessels in all tumour types within 3 h of treatment (Figs. 7 and 8), leading to a similar effect on vascular area (Fig. 10). This confirms the rapid response of tumour vasculature to CA4P reported in other tumour types (e.g. [22]). In general, Ang-2 tumours were most affected by

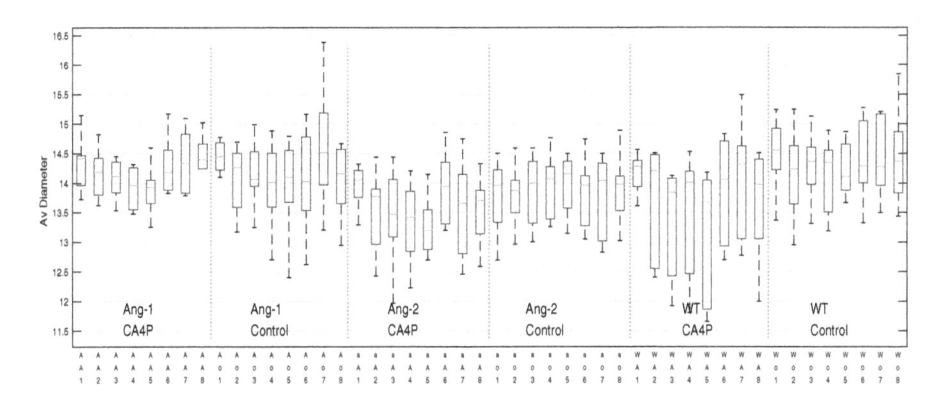

Fig. 8. Boxplots of the diameter [pixels] as determined by the tumour type Ang-1, Ang-2 and WT (**A/a/W**), treatment for CA4P or control (**A/o**) and the time of acquisition (**1–8**).

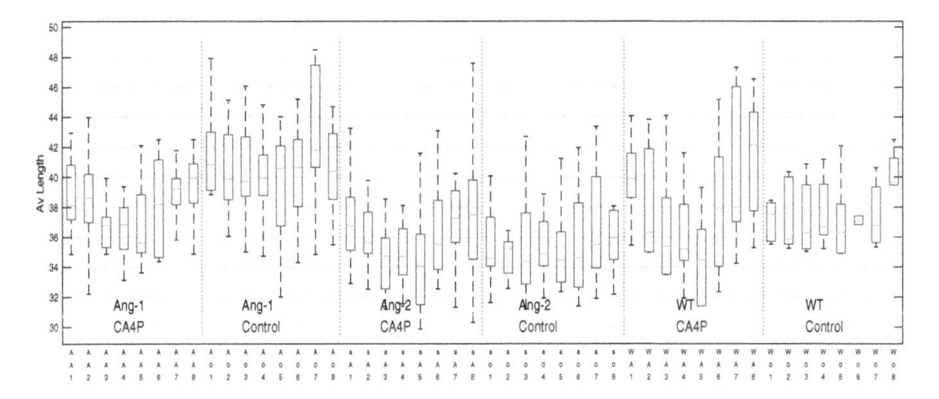

Fig. 9. Boxplots of the length [pixels] as determined by the tumour type Ang-1, Ang-2 and WT (**A/a/W**), treatment for CA4P or control (**A/o**) and the time of acquisition (**1–8**).

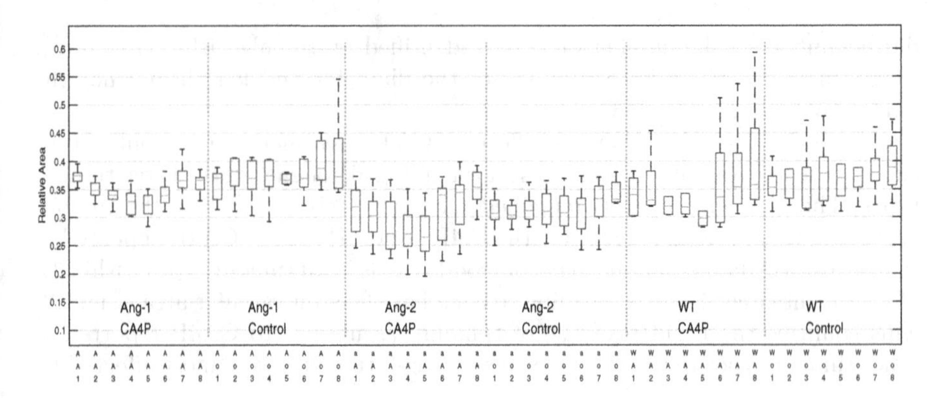

Fig. 10. Boxplots of the relative area as determined by the tumour type Ang-1, Ang-2 and WT (**A/a/W**), treatment for CA4P or control (**A/o**) and the time of acquisition (**1–8**).

CA4P, although further work, including functional assays, would be needed to determine whether differences in response are statistically significant.

4 Conclusions

The scale-space algorithm here described was an effective technique to analyse the topological variation of the vasculature of three different tumour types. Besides the vascular parameters analysed here, it would be possible to extract other parameters such as total length of the vasculature, length of longest vessels and branching patterns or other non-geometrical measurements such as saturation or hue.

The results presented here are consistent with Ang-1 acting as a stabilising influence in angiogenesis and Ang-2 having antagonistic properties. However, further analysis is required as the effects of Ang-2 have been reported to have opposing effects under different circumstances such as in the presence of VEGFA [17,23] or in the lymph versus blood vessel networks [24]. The well known tumour vascular disrupting effect of CA4P was confirmed in the current study and there was a suggestion that tumour cell over-expression of Ang-2 resulted in an increased response to CA4P. If this is substantiated in further studies, it would confirm our hypothesis and point to the potential of tumour angiopoietin expression analysis for aiding the development of predictive biomarkers of response to CA4P.

5 Code

The code and a brief user manual is freely available as Matlab functions from the author's website: http://www.staff.city.ac.uk/~sbbk034/tracingD.php?idMem=software.

Acknowledgements. We thank Professor Barbara Pedley for her gift of the SW1222 tumour cell line and Professor Pam Jones for her gift of Ang-1 and Ang-2 cDNA. We thank Drs Olga Greco and Sheila Harris for their contributions to development of the genetically modified cell lines. We thank staff at the University of Sheffield for their care of the animals used in this study. This work was funded by Programme Grant C1276/A9993 from Cancer Research UK.

References

1. Risau, W.: Differentiation of endothelium. FASEB J. **9**(10), 926–933 (1995)
2. Jiang, Y.F., Hsu, M.C., Cheng, C.H., Tsui, K.H., Chiu, C.H.: Ultrastructural changes of goat corpus luteum during the estrous cycle. Anim. Reprod. Sci. **170**, 38–50 (2016)
3. Zhang, H., van Olden, C., Sweeney, D., Martin-Rendon, E.: Blood vessel repair and regeneration in the ischaemic heart. Open Heart **1**(1), e000016 (2014). http://openheart.bmj.com/content/1/1/e000016
4. Collinson, D., Donnelly, R.: Therapeutic angiogenesis in peripheral arterial disease: can biotechnology produce an effective collateral circulation? Eur. J. Vasc. Endovasc. Surg. **28**(1), 9–23 (2004)
5. Hanahan, D., Folkman, J.: Patterns and emerging mechanisms of the angiogenic switch during tumorigenesis. Cell **86**(3), 353–364 (1996)
6. Tozer, G.M., Bicknell, R.: Therapeutic targeting of the tumor vasculature. Semin. Radiat. Oncol. **14**(3), 222–232 (2004)
7. Tozer, G.M.. Kanthou, C., Baguley, B.C.: Disrupting tumour blood vessels. Nat. Rev. Cancer **5**(6), 423–435 (2005)
8. Augustin, H., Koh, G., Thurston, G., Alitalo, K.: Control of vascular morphogenesis and homeostasis through the angiopoietin-tie system. Nat. Rev. Mol. Cell Biol. **10**(3), 165–177 (2009)
9. Thurston, G., Daly, C.: The complex role of angiopoietin-2 in the angiopoietintie signaling pathway. Cold Spring Harbor Perspect. Med. **2**(9), a006650 (2012). http://perspectivesinmedicine.cshlp.org/content/2/9/a006650
10. Koehl, G.E., Gaumann, A., Geissler, E.K.: Intravital microscopy of tumor angiogenesis and regression in the dorsal skin fold chamber: mechanistic insights and preclinical testing of therapeutic strategies. Clin. Exp. Metastasis **264**, 329–344 (2009)
11. Reyes-Aldasoro, C.C., Bjorndahl, M.A., Akerman, S., Ibrahim, J., Griffiths, M.K., Tozer, G.M.: Online chromatic and scale-space microvessel-tracing analysis for transmitted light optical images. Microvasc. Res. **84**(3), 330–339 (2012)
12. Lindeberg, T.: Feature detection with automatic scale selection. Int. J. Comput. Vis. **30**(2), 79–116 (1998)
13. Lindeberg, T.: Edge detection and ridge detection with automatic scale selection. Int. J. Comput. Vis. **30**(2), 117–154 (1998)
14. Tozer, G.M., Ameer-Beg, S.M., Baker, J., Barber, P.R., Hill, S.A., Hodgkiss, R.J., Locke, R., Prise, V.E., Wilson, I., Vojnovic, B.: Intravital imaging of tumour vascular networks using multi-photon fluorescence microscopy. Adv. Drug Deliv. Rev. **57**(1), 135–152 (2005)
15. Workman, P., Aboagye, E.O., Balkwill, F., Balmain, A., Bruder, G., Chaplin, D.J., Double, J.A., Everitt, J., Farningham, D.A.H., Glennie, M.J., Kelland, L.R., Robinson, V., Stratford, I.J., Tozer, G.M., Watson, S., Wedge, S.R., Eccles, S.A.:

Committee of the national cancer research institute: guidelines for the welfare and use of animals in cancer research. Br. J. Cancer **102**(11), 1555–1577 (2010)
16. Zagzag, D., Hooper, A., Friedlander, D.R., Chan, W., Holash, J., Wiegand, S.J., Yancopoulos, G.D., Grumet, M.: In situ expression of angiopoietins in astrocytomas identifies angiopoietin-2 as an early marker of tumor angiogenesis. Exp. Neurol. **159**(2), 391–400 (1999)
17. Plank, M., Sleeman, B., Jones, P.: The role of the angiopoietins in tumour angiogenesis. Growth Factors **22**(1), 1–11 (2004)
18. Maisonpierre, P.C., Suri, C., Jones, P.F., Bartunkova, S., Wiegand, S.J., Radziejewski, C., Compton, D., McClain, J., Aldrich, T.H., Papadopoulos, N., Daly, T.J., Davis, S., Sato, T.N., Yancopoulos, G.D.: Angiopoietin-2, a natural antagonist for Tie2 that disrupts in vivo angiogenesis. Science **277**, 55–60 (1997)
19. Hanahan, D.: Signaling vascular morphogenesis and maintenance. Science **277**(5322), 48–50 (1997)
20. Holash, J., Maisonpierre, P.C., Compton, D., Boland, P., Alexander, C.R., Zagzag, D., Yancopoulos, G.D., Wiegand, S.J.: Vessel cooption, regression, and growth in tumors mediated by angiopoietins and VEGF. Science **284**(5422), 1994–1998 (1999)
21. Koblizek, T.I., Weiss, C., Yancopoulos, G.D., Deutsch, U., Risau, W.: Angiopoietin-1 induces sprouting angiogenesis in vitro. Curr. Biol. **8**(9), 529–532 (1998)
22. Tozer, G.M., Prise, V.E., Wilson, J., Cemazar, M., Shan, S., Dewhirst, M.W., Barber, P.R., Vojnovic, B., Chaplin, D.J.: Mechanisms associated with tumor vascular shut-down induced by combretastatin A-4 phosphate: intravital microscopy and measurement of vascular permeability. Cancer Res. **61**(17), 6413–6422 (2001)
23. Hata, K., Nakayama, K., Fujiwaki, R., Katabuchi, H., Okamura, H., Miyazaki, K.: Expression of the angopoietin-1, angopoietin-2, Tie2, and vascular endothelial growth factor gene in epithelial ovarian cancer. Gynecol. Oncol. **93**(1), 215–222 (2004)
24. Gale, N.W., Thurston, G., Hackett, S.F., Renard, R., Wang, Q., McClain, J., Martin, C., Witte, C., Witte, M.H., Jackson, D., Suri, C., Campochiaro, P.A., Wiegand, S.J., Yancopoulos, G.D.: Angiopoietin-2 is required for postnatal angiogenesis and lymphatic patterning, and only the latter role is rescued by Angiopoietin-1. Dev. Cell **3**(3), 411–423 (2002)

Quantitative Electron Density CT Imaging for Radiotherapy Planning

Jonathan H. Mason[1](✉), Alessandro Perelli[1], William H. Nailon[2], and Mike E. Davies[1]

[1] Institute for Digital Communications, University of Edinburgh, Edinburgh EH9 3JL, UK
{j.mason,a.perelli,mike.davies}@ed.ac.uk

[2] Oncology Physics Department, Edinburgh Cancer Centre, Western General Hospital, Edinburgh EH4 2XU, UK
bill.nailon@luht.scot.nhs.uk

Abstract. Computed tomography (CT) is the imaging modality used to calculate the deposit of dose in radiotherapy planning, where the physical interactions are modelled based upon the electron density, which can be calculated from CT images. Traditionally this is a three step process: linearising the raw x-ray measurements and correcting for beam-hardening and scatter; inverting the system with analytic or iterative reconstruction algorithms into linear attenuation coefficient; then applying a non-linear calibration into electron density. In this work, we propose a new method for statistically inferring a quantitative image of electron density directly from the raw CT measurements, with no pre- or post-processing necessary, and able to cope with both beam-hardening from a single polyenergetic source and additive scatter. We evaluate this concept with cone-beam CT (CBCT) imaging for bladder cancer, where we demonstrate significantly higher electron density accuracy than other quantitative approaches. We also show through simulated photon and proton beam calculation, that our method may facilitate superior dose estimation, especially with regions containing bony structures.

Keywords: Computed tomography · Quantitative · Imaging · Statistical · Reconstruction · Radiotherapy · Proton therapy

1 Introduction

Computed tomography (CT) imaging facilitates radiation therapy planning by providing a quantitative map of electro-magnetic attenuation from a low x-ray energy source. With this, one can approximate how the higher energy treatment beams will interact with the patient and distribute dose throughout tumours and sensitive organs. For photon and proton therapy, dose calculation methods usually rely upon electron density [1], since it allows the interaction processes at the high energies to be modelled accurately. However, the attenuation with

M. Valdés Hernández and V. González-Castro (Eds.): MIUA 2017, CCIS 723, pp. 297–308, 2017.
DOI: 10.1007/978-3-319-60964-5_26

CT imaging energies is not solely determined by quantity of electrons, but their atomic environment, which induces whether the photon will interact through photoelectric effect, scattering or other [2]. When a specimen contains significantly different materials, such as soft-tissue and bone, the attenuation will be both spatially and spectrally non-linear, so calculating a consistent energy independent electron density is not straight forward.

With traditional approaches to CT imaging, the mapping from raw x-ray measurement to electron density is performed in three distinct steps. Firstly, the measurements are converted to linearised projections, which requires correcting for scatter, taking the logarithm, and calibrating from the polyenergetic to approximate monoenergetic source—known as beam hardening correction. Common assumptions for this conversion are either that the specimen has the spectral properties of water, or a mix of bone and water [3], the latter requiring a preliminary reconstruction and segmentation of the bony structures.

After the preprocessing from measurement to projections comes the 'reconstruction algorithm' [4], that attempts to invert the system into a spatial distribution of linear attenuation coefficient. These are broadly classed into 'analytic methods', where a closed form approximate inverse is applied to the projections, or 'iterative methods', where noise modelling and regularisation can be incorporated to infer a more accurate reconstruction through optimisation. Due to its preprocessing, the resulting images are approximations of a monoenergetic source attenuation.

Finally, the reconstruction is non-linearly converted into electron density through calibration [1]. In general, a three component linear fit is accurate for a biological specimen. At each of the three stages in this reconstruction work-flow, approximations are made that inevitably introduce artefacts that will propagate through the mapping, which will become increasingly pronounced as the dose is lowered. If this processes could be replaced by a direct inference with an accurate measurement model, then it is likely that these errors could be mitigated.

To date, their are several techniques that go some of the way to this direct inference. In [5], the authors propose a statistical algorithm for inferring the mass density from polyenergetic measurements and additive noise. Here, given a prior segmentation of the specimen into distinct material classes, such as water and bone, the attenuation within each region is modelled as a product of mass density and energy dependent mass attenuation coefficient of the corresponding class. They then presented an alternative model in [6], where the material classification is itself a function of mass density, and so no prior segmentation is necessary. A pre-dating method [7] attempts to model the photoelectric and scattering cross-sections, determining the polyenergetic attenuation, as a function of monoenergetic attenuation.

In both of the methods [6,7], there is an explicit assumption that for biological tissues, the polyenergetic attenuation can be modelled as an energy dependent piecewise linear function of either mass density or monoenergetic attenuation coefficient. Since the calibration from either of these into electron density [1] is also a piecewise linear calibration, we suggest it is reasonable to assume that these may be combined into a single mapping, which is what we set out to do in Sect. 3.1.

From the new relation in Sect. 3.1, we then couple this with an accurate CT model incorporating spectral properties of the source, and additive scatter or noise. In Sect. 3.2, we then show how this new model can be used for quantitative reconstruction, and how through properties of the model design, efficient algorithms can be realised that may include many smooth and non-smooth regularisation functions, such as generalized Gaussian Markov random field (GGMRF) [8], total variation (TV) [9] or wavelet sparsity [10].

As a proof of concept, we evaluate the accuracy of our method in the challenging geometry of cone-beam CT (CBCT), with its high scatter [11] and poor sampling [12], for radiation therapy planning of the bladder region. We demonstrate the superior quantitative accuracy of our approach in both image and dose, when compared to standard three-step reconstructions, and the model in [5].

2 Background

2.1 X-ray Attenuation

The mechanism that allows various regions in a heterogeneous specimen to be differentiated is their degree of x-ray attenuation. For biological tissues irradiated with a diagnostic x-ray source, the significant phenomena contributing to the attenuation of incident radiation are photoelectric and scattering effects—consisting of Compton and coherent scatter. The combined attenuation strength of a given element is quantified with the linear attenuation coefficient, defined as

$$\mu = \frac{\rho(\sigma_{\text{p.e.}} + \sigma_{\text{incoh.}} + \sigma_{\text{coh.}})}{uA}, \tag{1}$$

where ρ is the mass density, $\sigma_{\text{p.e.}}$, $\sigma_{\text{incoh.}}$ and $\sigma_{\text{coh.}}$ represent the interaction cross sections of photoelectric, incoherent (Compton) and coherent (Rayleigh and Thompson) effects, A is the atomic mass of the element, and u is a constant—the unified atomic mass unit. For a material, its attenuation coefficient is found by a sum of its constituent elemental coefficients, weighted by their mass fraction.

Since the strength of each attenuation effect varies with energy, we find that μ is dependent on a material's mass density, elemental composition and the incident x-ray source spectrum.

2.2 Probabilistic Measurement Model

In CT, one is able to observe a specimen's attenuation through the radiation intensity after transmission. The magnitude of this is often approximated using the Beer–Lambert law, given for a monoenergetic beam as

$$I_o = I_i \exp\left(-\int_\ell \mu(\ell) d\ell\right),$$

where I_i is the incident intensity, ℓ is the line-of-sight path of the beam through specimen, I_o is the output intensity one is able to measure. Since in practice, μ is energy dependent and we have a polyenergetic source, the output intensity in this case is

$$I_o = \int_{\xi} I_i(\xi) \exp\left(- \int_{\ell} \mu(\ell, \xi)\, d\ell \right) d\xi,$$

where ξ is energy.

For a finite number of photons, the measured intensity will be probabilistic with an approximate Poisson distribution [13]. If we also move the attenuation, measurements and energy spectrum into a discretised setting, we can write the measurement process as

$$y_i \sim \text{Poisson}\left\{ \sum_{j=1}^{N_\xi} b_i(\xi_j) \exp\left(-[\boldsymbol{\Phi}\boldsymbol{\mu}(\xi_j)]_i\right) + s_i \right\} \text{ for } i = 1, ..., N_{\text{ray}}, \quad (2)$$

where N_{ray} is the number of CT measurements, N_ξ is the number of energy bins, $\boldsymbol{b}(\xi) \in \mathbb{R}^{N_{\text{ray}}}$ is a vector of incident intensities, $\boldsymbol{\mu}(\xi) \in \mathbb{R}^{N_{\text{vox.}}}$ is the vector of attenuation coefficients with $N_{\text{vox.}}$ the number of voxels, $\boldsymbol{\Phi} \in \mathbb{R}^{N_{\text{ray}} \times N_{\text{vox.}}}$ is the system matrix describing the paths from source through specimen onto each detector, and $\boldsymbol{s} \in \mathbb{R}^{N_{\text{ray}}}$ is expectation of scatter or other background noise reaching the detector.

2.3 Standard Reconstruction Approaches

Inferring $\boldsymbol{\mu}$ from measurements with the model in (2) is not straightforward due to its the non-linearity and energy dependence. Instead, a standard approach is to assume a monoenergetic model and attempt to linearise the system. In a simple form, this linearised model can be expressed as

$$\boldsymbol{p} = \boldsymbol{\Phi}\boldsymbol{\mu} + \boldsymbol{n}, \quad (3)$$

where $\boldsymbol{n} \in \mathbb{R}^{N_{\text{ray}}}$ is noise, $\boldsymbol{p} \in \mathbb{R}^{N_{\text{ray}}}$ is the linearised projection, calculated by

$$p_i = \log\left(\frac{\bar{b}_i}{y_i - s_i} \right) \text{ for } i = 1, ..., N_{\text{ray}},$$

where $\bar{b}_i = \sum_{j=1}^{N_\xi} b_i(\xi_j)$ is the total incident flux. In practice, $\boldsymbol{p}$ is calibrated to approximate the monoenergetic equivalent before attempting to invert the system in (3).

One class of reconstruction methods—know as 'analytic' [4]—attempt to find a closed form expression for the inverse of (3). These normally are classed as 'filtered back-projection' (FBP) or 'back-projection filtration' (BPF). For the cone-beam geometry, the popular Feldkamp-Davis-Kress (FDK) is FBP. These methods are usually very fast, but perform poorly in high noise or limited measurement scenarios.

Another class of reconstruction algorithm is known as 'iterative' [4], which usually try to find an optimal $\boldsymbol{\mu}$ by some metric. A popular version of iterative reconstruction for (3) is penalised weighted-least-squares (PWLS), where $\boldsymbol{\mu}$ is found as [5]

$$\hat{\boldsymbol{\mu}} = \underset{\boldsymbol{\mu}}{\operatorname{argmin}} \left(\boldsymbol{\Phi}\boldsymbol{\mu} - \boldsymbol{p}\right)^T \boldsymbol{W} \left(\boldsymbol{\Phi}\boldsymbol{\mu} - \boldsymbol{p}\right) + \lambda R(\boldsymbol{\mu}), \tag{4}$$

where $\boldsymbol{W} \in \mathbb{R}^{N_{\text{ray}} \times N_{\text{ray}}}$ is a diagonal weighting matrix with entries $w_{ii} = (y_i - s_i)^2 / y_i$, $R(\boldsymbol{\mu})$ is a regularisation function to promote desirable structure in $\boldsymbol{\mu}$, and λ is usually a scalar constant trade-off between data fit and regularisation. The weights are used here to approximate the Poisson noise model in (2).

In order to calculate dose from reconstructions in $\boldsymbol{\mu}$, whether the linearised system is used within FBP and PWLS, or a more accurate maximum-likelihood polyenergetic model is used [5–7], conversion to electron density is necessary. In [1], this is done through a piecewise linear fit derived from the chemical properties of many biological tissues.

3 Method

3.1 Electron Density to Attenuation

The essence of our method is to use the model in (2) for reconstruction directly into electron density, ρ_e. This involves a single non-linear energy dependent fitting from ρ_e to attenuation, which depends on the existence of a simple relation. To investigate the possibility of such a mapping, we calculated the relative electron density—normalised to that of water—for 52 material classes from the International Commission on Radiological Protection (ICRP) report 89 [14]. This is plotted against the relative attenuation—normalised to maximum attenuation—in Fig. 1.

By observing the data in Fig. 1, it becomes apparent that there is indeed a simple well fitting model that can map from electron density to attenuation. We have superimposed two linear fits around a constant point, and imposed constraints that the first fit should pass through the origin and both fits cross at a consistent 'knee' position, which was calculated to maximise the accuracy of overall fit—1.0083 in this case.

To make the realisation of a practical algorithm easier, we make use of a single continuously differentiable function to approximate the piecewise linear nature from the two fits shown in Fig. 1. For this we use the generalised logistic function

$$f(x) = \frac{1}{1 + \exp\left(-k(x - x_0)\right)},$$

where k is the steepness of its transition and x_0 is the central point, which we set to our knee. The continuous mapping from electron density to attenuation coefficient can then be written as

$$\hat{\mu}(\rho_e, \xi) = [1 - f(\rho_e)]\alpha(\xi)\rho_e + f(\rho_e)[\beta(\xi)\rho_e + \gamma(\xi)], \tag{5}$$

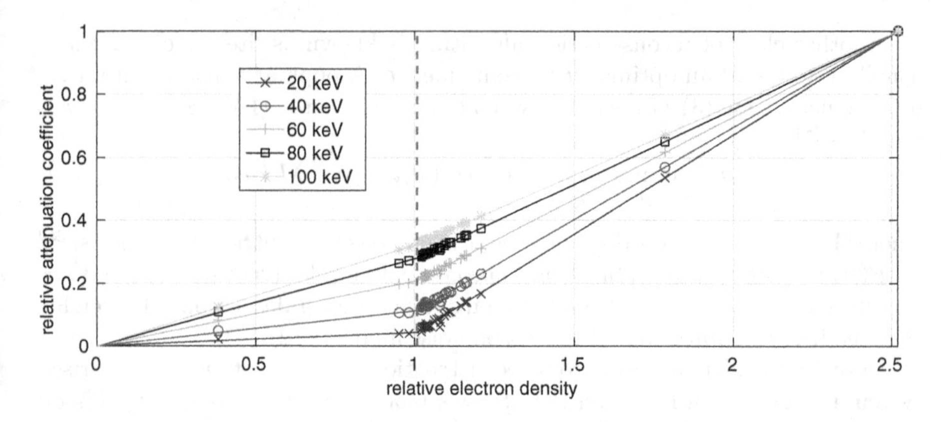

Fig. 1. Plot of relative attenuation coefficient against electron density for ICRP tissues over a range in energy. Two linear fits are made to this data, with the transition between them shown by the dashed line.

where α is the gradient of first fit, and β, γ are equation of the second fit; we use the constraint $\gamma = x_0(\alpha - \beta)$ for continuity. In practice, k is set to be very large to give a sharp transition between fits.

3.2 Direct Quantitative Electron Density Reconstruction

Our quantitative method combines (5) with (2) and infers ρ_e from the model through optimisation. Although this is a complicated task explicitly, we will show how through our choice of fitting functions, simplifications can yield an efficient optimisation object, and we point towards simple algorithms for solving it.

Firstly, combining (5) with (2) results in the relation

$$\sum_{j=1}^{N_\xi} b_i(\xi_j) \exp\left(-[\boldsymbol{\Phi}\boldsymbol{\mu}(\xi_j)]_i\right) \approx \sum_{j=1}^{N_\xi} b_i(\xi_j) \exp\left(-[\boldsymbol{\Phi}\hat{\boldsymbol{\mu}}(\boldsymbol{\rho_e}, \xi_j)]_i\right) \text{ for } i = 1, ..., N_{\text{ray}}.$$

If we introduce a function $\psi(\cdot, \cdot)$ to simplify notation as

$$\psi_i(\boldsymbol{\rho_e}, \xi) \equiv b_i(\xi) \exp\left(-[\boldsymbol{\Phi}\hat{\boldsymbol{\mu}}(\boldsymbol{\rho_e}, \xi)]_i\right) \text{ for } i = 1, ..., N_{\text{ray}},$$

we can write the negative log-likelihood (NLL) for the Poisson model as

$$\text{NLL}(\boldsymbol{\rho_e}; \boldsymbol{y}) = \sum_{i=1}^{N_{\text{ray}}} \sum_{j=1}^{N_\xi} \psi_i(\boldsymbol{\rho_e}, \xi_j) + s_i - y_i \log\left(\sum_{j=1}^{N_\xi} \psi_i(\boldsymbol{\rho_e}, \xi_j) + s_i\right). \tag{6}$$

Reconstruction of the electron density map can be performed by finding a $\boldsymbol{\rho_e}$ that minimises (6). We will look at gradient descent methods, for which we require an expression for the derivative of NLL. If we simplify notation with the following

$$\boldsymbol{d}(\boldsymbol{\rho_e}) = \boldsymbol{y} \oslash \left(\sum_{j=1}^{N_\xi} \boldsymbol{\psi}(\boldsymbol{\rho_e}, \xi_j) + \boldsymbol{s} \right) - \boldsymbol{1},$$

where $\oslash$ represents component-wise division. An expression for the derivative is then

$$\begin{aligned} \frac{\partial \text{NLL}(\boldsymbol{\rho_e}; \boldsymbol{y})}{\partial \boldsymbol{\rho_e}} = (\boldsymbol{1} - \boldsymbol{f}(\boldsymbol{\rho_e})) \odot \boldsymbol{u}_\alpha(\boldsymbol{\rho_e}) \odot \boldsymbol{\Phi}^T & \left[\sum_{j=1}^{N_\xi} \alpha(\xi_j) \boldsymbol{\psi}(\boldsymbol{\rho_e}, \xi_j) \odot \boldsymbol{d}(\boldsymbol{\rho_e}) \right] \\ + \boldsymbol{f}(\boldsymbol{\rho_e}) \odot \boldsymbol{u}_\beta(\boldsymbol{\rho_e}) \odot \boldsymbol{\Phi}^T & \left[\sum_{j=1}^{N_\xi} \beta(\xi_j) \boldsymbol{\psi}(\boldsymbol{\rho_e}, \xi_j) \odot \boldsymbol{d}(\boldsymbol{\rho_e}) \right] \\ + \boldsymbol{u}_\gamma(\boldsymbol{\rho_e}) \odot \boldsymbol{\Phi}^T & \left[\sum_{j=1}^{N_\xi} \gamma(\xi_j) \boldsymbol{\psi}(\boldsymbol{\rho_e}, \xi_j) \odot \boldsymbol{d}(\boldsymbol{\rho_e}) \right], \end{aligned} \tag{7}$$

where $\boldsymbol{\Phi}^T$ represents a transpose of the system matrix or 'back-projection', and $\odot$ is component-wise multiplication. $\boldsymbol{u}_\alpha(\cdot)$, $\boldsymbol{u}_\beta(\cdot)$ and $\boldsymbol{u}_\gamma(\cdot)$ are factors from the derivative of (5) with respect to $\boldsymbol{\rho_e}$. If $\boldsymbol{f}(\cdot)$ were instead independent on $\boldsymbol{\rho_e}$, then we would have $\boldsymbol{u}_\alpha(\cdot) = \boldsymbol{1}$, $\boldsymbol{u}_\beta(\cdot) = \boldsymbol{1}$ and $\boldsymbol{u}_\gamma(\cdot) = \boldsymbol{0}$, which would cancel the third term in (7) for faster computation, and we have noted this to have negligible difference from the exact form for very high k, as is desirable.

With the necessary tools, we can write quantitative reconstruction as

$$\hat{\boldsymbol{\rho}}_e = \underset{\boldsymbol{\rho_e} \in C}{\text{argmin}} \ \text{NLL}(\boldsymbol{\rho_e}; \boldsymbol{y}) + \lambda R(\boldsymbol{\rho_e}), \tag{8}$$

where $R(\cdot)$ is some regularisation function, and C is a set of box constraints on $\boldsymbol{\rho}$ so that $0 \leq \rho_i \leq \zeta$ for $i = 1, ..., N_{\text{vox.}}$, where ζ is the maximum allowable electron density value. In the experimental section, we use the TV regularisation function for $R(\cdot)$ [9] to promote a piece-wise constant image.

We note that the objective function in (8) is both non-linear and nonconvex, so its minimisation should be treated carefully, though there exist several gradient descent algorithms that may be invoked to minimise it, such as VMFB [15] or paraboloid surrogates [16] if only smooth $R(\cdot)$ is used.

4 Experimentation

4.1 Data

The data we used was derived from the 'Adult Reference Computational Phantoms' [17]—a segmented structure with defined biological tissue and elemental composition—from which we isolated the bladder region. A single slice from this data is shown in Fig. 2a, where the voxel intensities simply encode the material type. For example, 30 is muscle tissue, 50 is fat, 10 the femoral head, 52 the rectum, 53 urine and 1 is air. To simulate a planning protocol, we also synthesised a prior image from this data by non-rigidly deforming several of the soft-tissue regions, and this is shown in Fig. 2d. The electron densities of the oracle and planning images are shown in Fig. 2b and e respectively.

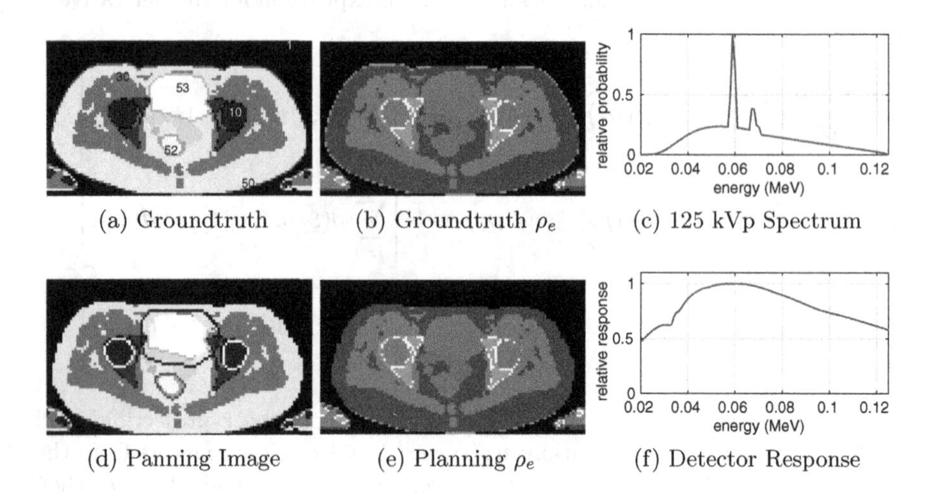

(a) Groundtruth (b) Groundtruth ρ_e (c) 125 kVp Spectrum

(d) Panning Image (e) Planning ρ_e (f) Detector Response

Fig. 2. Experimental data used: (a) is the oracle test image as material index; (b) is the oracle electron density with grey scale [0.8,1.2]; (c) is the 125 kVp source spectrum used; (d) is the simulated planning image with contours shown for the PTV, GI tract and femoral heads; (e) is the planning image electron density with grey scale [0.8,1.2]; and (f) is the detector response function used. (Color figure online)

To generate CBCT measurements, we used the Monte-Carlo engine, Gate [18], with a total of 2×10^{10} photons over 160 projection angles. To try and closely approximate a real clinical CBCT scanner, we derived the filtered energy spectrum and detector response, shown in Fig. 2c and f from a Varian TrueBeam^{TM} system (Varian Medical Systems, Palo Alto CA, USA). We also matched the geometry, source profile from offset detector bow-tie filter to its 'half-fan' mode, and included a model of its trans-axial focused scatter collimation grid. For the scatter term, $\boldsymbol{s}$ in (8), we applied the same scan to eh planning image and isolated the scattered photons as in [19].

We used a number of different methods to map these measurements into electron density, where we utilised CBCT projection and back-projection operators for $\boldsymbol{\Phi}$ and $\boldsymbol{\Phi}^T$ from the Michigan Image Reconstruction Toolbox [20]. From relative electron density (normalised to water), we then calculated dose distributions using the matRad Toolbox (German Cancer Research Centre) for photons and protons. In both cases, we optimised for delivering a dose of 64 Gy (J/kg) into a planning target volume (PTV) for bladder cancer [21], where we added a margin of $\sim$ 1 cm around the planning image bladder—this is shown by the contour (blue) in Fig. 2d, and applied identical beams to each other method under test and the oracle image.

To assess the various methods, we calculated the electron density accuracy as the error euclidean norm to the oracle, and also the mean and maximum absolute deviations in dose estimation within the PTV, femoral heads and GI tract, the latter of which are sensitive to excessive dose and should be carefully monitored.

4.2 Methods Under Test

- Plan: using the planning image directly, with its oracle bone registration and matching.
- FDK: 'analytic' FBP reconstruction for CBCT [22] using prior Monte-Carlo scatter correction [19] and water-based beam hardening compensation [3] as preprocessing, with Hann windowing to suppress noise amplification, weighting for offset detector [23], and non-linear calibration to electron density [1] as postprocessing.
- PWLS: 'statistical iterative' reconstruction with (4), with same pre- and post-processing as FDK, and Total Variation (TV) regularisation.
- PolySIR: polyenergetic reconstruction model from [5] with oracle bone segmentation, prior Monte-Carlo scatter expectation estimation [19], TV regularisation, and electron density calibration [1] as postprocessing.
- ρ_eCT: our proposed quantitative approach with prior Monte-Carlo scatter expectation estimation [19], TV regularisation, and no postprocessing.

In all cases of iterative reconstruction—PWLS, PolySIR and ρ_eCT—the regularisation constant for TV, λ, was individually numerically optimised to give maximum soft tissue accuracy, which ensured fairness throughout the different data-fidelity terms.

4.3 Results

Results are summarised in Table 1, and illustrated visually in Figs. 3 and 4. Since for the application, it may be more critical to control the worst case dose error rather than the average, we have also calculated the maximum absolute errors, which are shown in Table 2.

From the numerical results, it is evident that our proposed method has a significant numerical advantage in calculating accurate electron density, with its error norm 35% lower than the next best tested. This gain appears to translate into average dose calculation, where it the best performing across the board. There are a couple of instances where it is outperformed in the worst case however, in Table 2, which we predict is due to the less smooth soft-tissue structures in the optimal TV regularisation in ρ_eCT than PWLS. Indeed, if we manually increase the regularisation parameter λ, this maximum error does drop below PWLS, but we did not include this result to maintain fairness and consistency. Additionally, from the substantially better performance in the fem. heads and

Table 1. Quantitative results 1: norm ρ_e error and mean dose error into planning target volume (PTV), femoral heads (fem. heads) and gastrointestinal tract (GI tract)

		Mean photon error (Gy)			Mean proton error (Gy)		
Scheme	ρ_e error norm	PTV	fem. heads	GI tract	PTV	fem. heads	GI tract
Plan	56.3	0.587	0.587	0.320	0.321	0.438	0.0976
FDK	135	7.52	3.06	2.79	0.417	3.96	2.90
PWLS	30.7	0.288	0.198	0.229	0.102	0.156	0.141
PolySIR	30.0	0.321	0.209	0.227	0.0862	0.131	0.144
ρ_eCT	**19.5**	**0.275**	**0.182**	**0.209**	**0.0809**	**0.0696**	**0.0683**

Table 2. Quantitative results 2: worst case dose error—maximum absolute difference

	Max. photon error (Gy)			Max. proton error (Gy)		
Scheme	PTV	fem. heads	GI tract	PTV	fem. heads	GI tract
Plan	48.9	47.1	25.0	39.4	45.8	33.5
FDK	10.2	15.6	15.7	39.5	61.9	61.8
PWLS	3.51	7.15	9.87	**9.04**	11.5	12.0
PolySIR	2.78	5.92	**8.48**	24.6	24.4	17.7
ρ_eCT	**2.59**	**5.84**	8.86	11.1	**5.50**	**5.03**

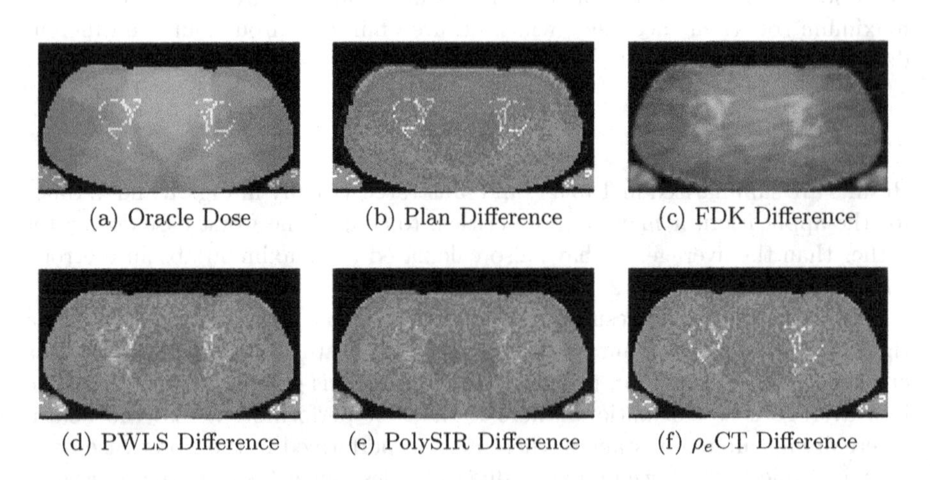

(a) Oracle Dose (b) Plan Difference (c) FDK Difference

(d) PWLS Difference (e) PolySIR Difference (f) ρ_eCT Difference

Fig. 3. Visual illustration of photon dose results: oracle dose maps have colour scale [0.2,64]; dose differences are absolute and have colour scale [0.2,5]; and all electron densities are shown with grey scale [0,1.8]. (Color figure online)

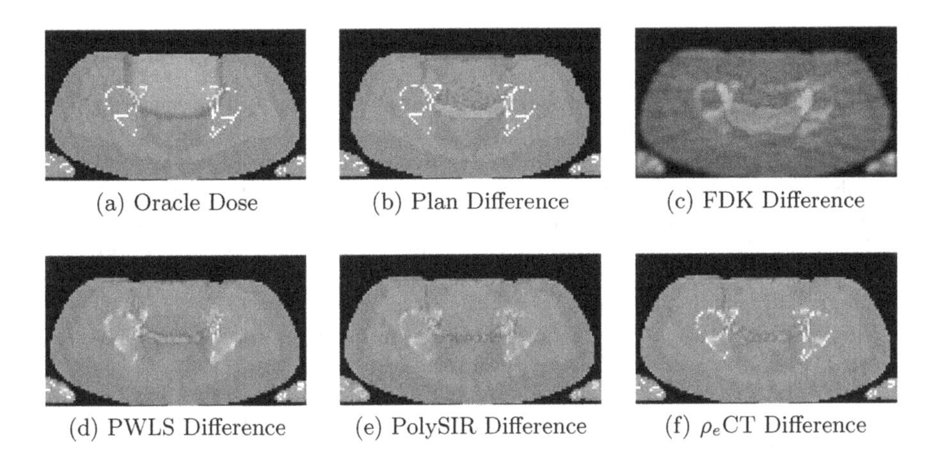

Fig. 4. Visual illustration of proton dose results, with same colour scales as Fig. 3. (Color figure online)

GI tract for proton, we also think this trade-off is acceptable. It may be that invoking varying regularisation for the different tissue classes could be key to pushing our performance even further.

5 Conclusion

In conclusion, we have successfully introduced a method for inferring electron density directly from raw CT measurements, with additive scatter and a single polyenergetic source. From our preliminary experiments, it significantly outperforms existing approaches, and appears to be an accurate quantitative tool, even in the severely low-dose and high scatter setting under test. There are certainly many avenues for further investigation, such as a relation to the underlying physical phenomena and dual energy systems, and more advanced regularisation strategies. We also predict it could be readily extended to model metal objects and implants also.

Acknowledgements. The authors would like to thank the Maxwell Advanced Technology Fund, EPSRC DTP studentship funds and ERC project: C-SENSE (ERC-ADG-2015-694888) for supporting this work.

References

1. Schneider, U., Pedroni, E., Lomax, A.: The calibration of CT Hounsfield units for radiotherapy treatment planning. Phys. Med. Biol. **41**(1), 111–124 (1996)
2. Curry, T.S., Dowdey, J.E., Murry, R.C.: Christensen's Physics of Diagnostic Radiology (1990)
3. Joseph, P.M., Spital, R.D.: A method for correcting bone induced artifacts in computed tomography scanners (1978)

4. Fessler, J.A.: Fundamentals of CT reconstruction in 2D and 3D. Compr. Biomed. Phys. **2**(2.11), 263–295 (2014). Elsevier
5. Elbakri, I.A., Fessler, J.A.: Statistical image reconstruction for polyenergetic X-ray computed tomography. IEEE Trans. Med. Imaging **21**(2), 89–99 (2002)
6. Elbakri, I.A., Fessler, J.A.: Segmentation-free statistical image reconstruction for polyenergetic X-ray computed tomography with experimental validation. Phys. Med. Biol. **48**(15), 2453–2477 (2003)
7. De Man, B., Nuyts, J., Dupont, P., Marchal, G., Suetens, P.: An iterative maximum-likelihood polychromatic algorithm for CT. IEEE Trans. Med. Imaging **20**(10), 999–1008 (2001)
8. Bouman, C., Sauer, K.: A generalized Gaussian image model for edge-preserving MAP estimation. IEEE Trans. Image Process. **2**(3), 296–310 (1993)
9. Rudin, L.I., Osher, S., Fatemi, E.: Nonlinear total variation based noise removal algorithms. Phys. D Nonlinear Phenom. **60**(1–4), 259–268 (1992)
10. Daubechies, I., Fornasier, M., Loris, I.: Accelerated projected gradient method for linear inverse problems with sparsity constraints. J. Fourier Anal. Appl. **14**(5–6), 764–792 (2008)
11. Siewerdsen, J.H., Jaffray, D.A.: Cone-beam computed tomography with a flat-panel imager: magnitude and effects of X-ray scatter. Med. Phys. **28**(2), 220 (2001)
12. Tuy, H.K.: An inversion formula for cone-beam reconstruction. SIAM J. Appl. Math. **43**(3), 546–552 (1983)
13. Chang, Z., Zhang, R., Thibault, J.-B., Sauer, K., Bouman, C.: Statistical X-ray computed tomography imaging from photon-starved measurements. SPIE Comput. Imaging **9020**, 90200G (2014)
14. ICRP Publication 89. Basic anatomical and physiological data for use in radiological protection reference values. Ann. ICRP **32**, 3–4 (2002)
15. Chouzenoux, E., Pesquet, J.C., Repetti, A.: Variable metric forward-backward algorithm for minimizing the sum of a differentiable function and a convex function. J. Optim. Theory Appl. **162**(1), 107–132 (2014)
16. Erdogan, H., Fessler, J.A.: Monotonic algorithms for transmission tomography. IEEE Trans. Med. Imaging **18**(9), 801–814 (1999)
17. ICRP Publication 110. Adult Reference Computational Phantoms. Ann. ICRP **39**(2) (2009)
18. Jan, S., Benoit, D., Becheva, E., Carlier, T., Cassol, F., Descourt, P., Frisson, T., Grevillot, L., Guigues, L., Maigne, L., Morel, C., Perrot, Y., Rehfeld, N., Sarrut, D., Schaart, D.R., Stute, S., Pietrzyk, U., Visvikis, D., Zahra, N., Buvat, I.: GATE V6: a major enhancement of the GATE simulation platform enabling modelling of CT and radiotherapy. Phys. Med. Biol. **56**(4), 881–901 (2011)
19. Xu, Y., Bai, T., Yan, H., Ouyang, L., Pompos, A., Wang, J., Zhou, L., Jiang, S.B., Jia, X.: A practical cone-beam CT scatter correction method with optimized Monte Carlo simulations for image-guided radiation therapy. Phys. Med. Biol. **60**(9), 3567–3587 (2015)
20. Fessler, J.A.: Image Reconstruction Toobox. https://web.eecs.umich.edu/~fessler/code/. Accessed 14 Nov 2014
21. Barrett, A., Morris, S., Dobbs, J., Roques, T.: Practical Radiotherapy Planning. 4th edn (2009)
22. Feldkamp, L.A., Davis, L.C., Kress, J.W.: Practical cone-beam algorithm. J. Opt. Soc. Am. A **1**(6), 612 (1984)
23. Wang, G.: X-ray micro-CT with a displaced detector array. Med. Phys. **29**(7), 1634 (2002)

3D Texton Based Prostate Cancer Detection Using Multiparametric Magnetic Resonance Imaging

Liping Wang(✉) and Reyer Zwiggelaar

Department of Computer Science,
Aberystwyth University, Aberystwyth SY23 3DB, UK
{liw20,rrz}@aber.ac.uk

Abstract. Multiparametric magnetic resonance imaging (mp-MRI) has shown its potential in prostate cancer detection. In this study, we investigate the application of 3D texton based prostate cancer detection using T2-weighted (T2W) MRI, dynamic contrast-enhanced (DCE) MRI and apparent diffusion coefficient (ADC) maps. For the T2W and ADC modalities, the traditional texton based approach is adopted, i.e., for each voxel, a texton histogram is extracted as the feature to perform the classification. For the DCE data, we present a new method, where the textons are extracted from each series and for each voxel, the corresponding textons across all series are used as features. A random forest classifier is applied for classifying all voxels into benign or malignant. The evaluation is conducted by performing a receiver operating characteristics (ROC) analysis and computing the area under the curve (AUC). The experiments on the Initiative for Collaborative Computer Vision Benchmarking (I2CVB) database demonstrate that the texton based approach using mp-MRI data obtains excellent performance in prostate cancer detection and produces 88.3% accuracy, whereas the accuracy produced by an intensity based approach is 79.8%.

Keywords: Texton · Multiparametric MRI · Prostate cancer

1 Introduction

Prostate cancer has been reported as the second most frequently diagnosed cancer of men [1] and the second leading cause of death from cancer among men [2]. Prostate-specific antigen (PSA) test and the transrectal ultrasound (TRUS) has been widely used since the mid-1980s [3,4]. However, this technique is invasive and suffers from low specificity, which causes over-treatment [5]. It is also difficult to evaluate the aggressiveness and progression of the prostate cancer [6]. In order to improve the current stage of prostate cancer detection and diagnosis, a computer-aided system using multiparametric magnetic resonance imaging (mp-MRI) can be incorporated into the traditional PSA screening based approach.

M. Valdés Hernández and V. González-Castro (Eds.): MIUA 2017, CCIS 723, pp. 309–319, 2017.
DOI: 10.1007/978-3-319-60964-5_27

A variety of studies using mp-MRI for prostate cancer detection and diagnosis have been conducted [7–11]. The mp-MRI data used include T2-weighted (T2W), dynamic contrast-enhanced (DCE) and diffusion-weighted (DW) MRI images and magnetic resonance spectroscopy imaging (MRSI) data. As reviewed in [12], image based features are usually extracted from T2W and DW MRI images for classification; for DCE images, either the whole spectra can be used directly or the features/parameters can be extracted by applying quantitative/semi-quantitative approaches; to model the MRSI signal, the whole spectra, quantification and wavelet-based approaches are often performed.

Regarding image based features, the texton based approach produced promising results in retinal vessel segmentation [13] and lung cancer detection [14]. Textons are considered as the fundamental micro-structures in natural images [15]. Due to the lack of a mathematical model, the concept of 'texton' remains vague. This approach has also been applied for prostate cancer detection [16] and produced results comparable to alternative state-of-the-art techniques. However, only T2W MRI images were used and the cancer detection was performed only in the peripheral zone (PZ). The textons were extracted from 2D slices hence the information from the neighbouring slices was not taken into account. In this study, we would like to further investigate the application of a 3D texton based approach for cancer detection in the whole prostate gland. Mp-MRI images are used, including T2W and DCE MRI images and the apparent diffusion coefficient (ADC) maps, which are generated from the DW MRI images. For T2W and ADC images, the traditional texton based approach is extended to 3D space and applied to extract the texton histograms as features. For DCE MRI images, 3D textons are extracted from each series of images; then the textons across all series rather than histograms are used as features to perform the classification. Apart from utilising the whole spectra of DCE data, i.e., the intensity across all series, this is the first study that extracts texture-based features from all series of DCE images for prostate cancer detection. The experiments are conducted on a publicly available database: the Initiative for Collaborative Computer Vision Benchmarking (I2CVB) [12]. We compare the results produced by our texton based approach and the intensity based approach using either mono-modality or mp-MRI data quantitatively.

2 I2CVB Dataset

We have used 17 cases from the I2CVB dataset with biopsy proven prostate cancer in the central gland (CG) or PZ or in both regions. The T2W, DCE and ADC MRI images were taken from a 3.0 Tesla Siemens scanner. The size of the T2W MRI image ranges from $308 \times 384 \times 64$ to $368 \times 448 \times 64$ with the voxel resolution ranging from $0.68\,\text{mm} \times 0.68\,\text{mm} \times 1.25\,\text{mm}$ to $0.79\,\text{mm} \times 0.79\,\text{mm} \times 1.25\,\text{mm}$. ADC maps of different patients have image sizes ranging from $96 \times 96 \times 13$ to $150 \times 150 \times 16$ with the slice thickness of 4 mm. The DCE MRI image of each patient includes 40 series and the image size of each series is $192 \times 256 \times 16$ or $200 \times 256 \times 16$ with 3.5 mm slice thickness. The prostate gland is segmented by an experienced radiologist for all three modalities. The segmentation of the regions of CG, PZ and the prostate cancer is only performed on T2W MRI images. Example slices of T2W, DCE and ADC MRI images are shown in Fig. 1.

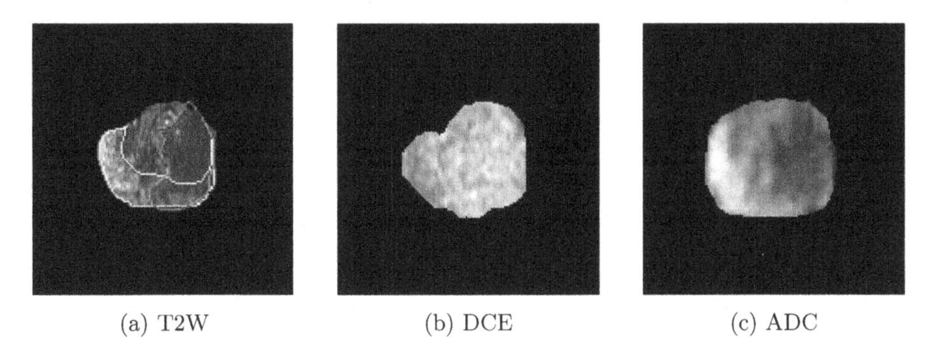

(a) T2W (b) DCE (c) ADC

Fig. 1. Example slices of the prostate gland segmented by the radiologist from the T2W, DCE and ADC images of the same patient in I2CVB. DCE and ADC images were resampled using the spatial information of the T2W images. In the prostate region of T2W image, the PZ is annotated by white contour and the rest part is the CG. The prostate cancer is annotated by red contour. (Color figure online)

3 Method

The methodology constitutes preprocessing, feature extraction and classification. The workflow is illustrated in Fig. 2.

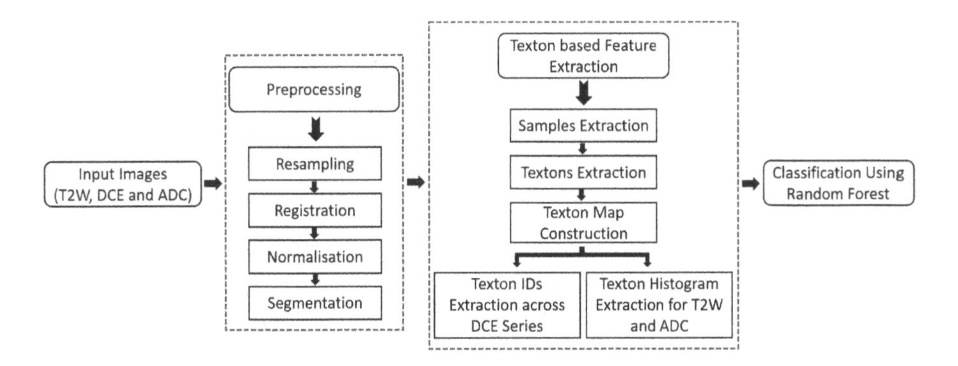

Fig. 2. Flowchart of our texton based classification approach.

3.1 Preprocessing

In order to propagate the ground truth annotated on T2W MRI images to the other two modalities, for each patient, each DCE series and ADC map are resampled using the spatial information of the T2W MRI images. Rigid registration is applied to the resampled DCE series to correct the intra-patient motion; each series of DCE data is registered to the first series by minimising a metric defined by mutual information. Subsequently, a non-rigid registration approach is applied to align the intra-patient motion corrected DCE series to T2W MRI images by successively estimating the rigid transform, the coarse elastic transform and the fine elastic transform. In this process, only the masks of the prostate gland segmented by the radiologist are registered by minimising the mean squared error metric. B-splines transformation is used as the elastic transform. For the ADC maps, only the non-rigid registration is applied to achieve the inter-modality correspondence between the ADC maps and T2W MRI. The optimisation is performed using a regular step gradient descent.

In order to reduce the inter-patient intensity variations, normalisation is performed for each modality. A parametric normalisation method using Rician *a priori* [17], which assumes that the MRI data follows a Rician distribution, is applied to T2W MRI images; a graph-based approach is used in conjunction with a model-based approach to correct the intensity offset, the time offset and the intensity scaling of DCE series [18]; for the ADC maps, a non-parametric piecewise-linear normalisation method is applied [19]. Finally, the segmentation of prostate, CG, PZ and prostate cancer regions of T2W images is propagated to the DCE series and ADC maps. Further processing will only focus on the region of the prostate gland. For all these preprocessing steps, we use the approach developed by Lemaître [18]. The code is publicly available on GitHub.[1]

3.2 Extended Traditional Texton Based Approach for Processing T2W and ADC MRI

The traditional texton based approach is extended to 3D and performed to extract the features from T2W and ADC MRI images. The processing steps are depicted in Fig. 2 and elaborated in this section.

Extracting Samples. For the image of each patient, the prostate gland is divided into cancer (CAP) and non-cancer (NCAP) regions. We randomly select the same number (e.g. 500) of voxels from the CAP and NCAP regions, respectively. Then for each selected CAP or NCAP voxel, a sample within a window (we used $7 \times 7 \times 3$) is extracted from the image, with the selected voxel as its centre. The size of the window in the 3rd dimension is smaller since the slice thickness is almost twice of the in-plane pixel spacing for most cases. Each sample is composed of 147 ($7 \times 7 \times 3$) intensity values. The CAP and NCAP samples extracted from all patients are used to extract the textons.

Extracting Textons. k-means is applied to the extracted CAP and NCAP samples, respectively, and the k clusters are considered as CAP or NCAP textons. Each texton has the same size with the sample, i.e., 147 dimensions. A

[1] https://github.com/I2Cvb/mp-mri-prostate.

unique ID is assigned to each texton. In our experiments, extracting 20 CAP and 20 NCAP textons for each modality empirically produces good results. Hence the texton IDs range from 1 to 20 for CAP textons and 21 to 40 for NCAP textons, respectively. All CAP and NCAP textons with their IDs compose the texton dictionary. Figure 3 illustrate examples of CAP and NCAP textons extracted from T2W MRI images.

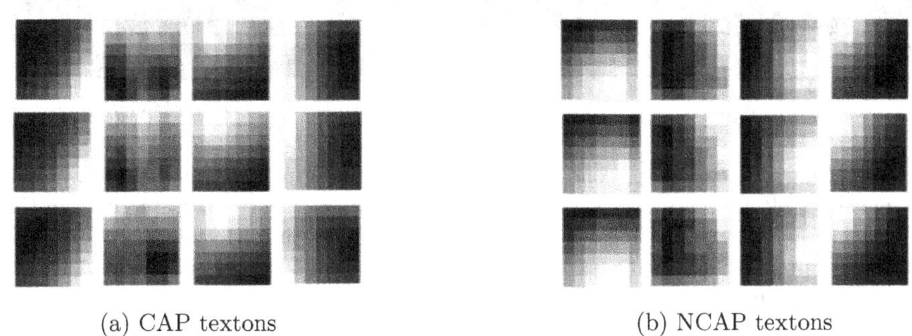

(a) CAP textons (b) NCAP textons

Fig. 3. Examples of CAP and NCAP textons extracted from T2W MRI images. The three patches in each column represent a 3D texton which has the size of $7 \times 7 \times 3$.

Constructing Texton Maps. For each voxel within the prostate gland, a sample with size $7 \times 7 \times 3$ is extracted from the image. The nearest texton can be found from the texton dictionary by calculating the Euclidean distance between the sample and each texton. Other metrics such as the mutual information can be used. The Euclidean distance is chosen because of its simplicity and efficiency. The corresponding texton ID is assigned to the voxel. A texton map is constructed by retrieving the nearest textons for all voxels within the prostate gland. The texton map has the same size with the image and the value of each voxel within the prostate gland represents its corresponding texton ID. Figure 4(a) and (c) illustrate the texton maps of one patient constructed for T2W and ADC MRI images.

Extracting Texton Histograms. In the constructed texton map, for each voxel within the prostate gland, a patch (we used the same size as before $7 \times 7 \times 3$) is extracted around the central voxel. Then a histogram is generated from the values (i.e. textons IDs) of the voxels within the patch (the central voxel is included). The histogram is considered as the feature vector with the dimensionality equals to the total number of CAP and NCAP textons (i.e. 40) and the value of each dimension equals to the frequency of each texton. The feature vectors extracted for all voxels within the prostate gland of the images taken for all patients are used for classification.

3.3 Texton Based Approach for Processing DCE MRI

DCE MRI data has more than one series hence a slightly different approach is used. As depicted in Fig. 2, for each series of images, the first three steps of

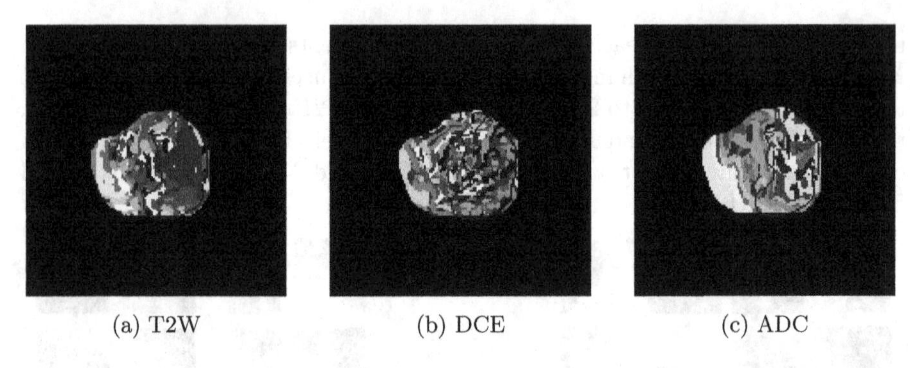

(a) T2W (b) DCE (c) ADC

Fig. 4. Slices of texton maps constructed for T2W, DCE and ADC images. The original image slices after resampling are shown in Fig. 1. The regions of prostate gland in the three slices are the same because the registration has been performed to align DCE and ADC prostate gland masks with T2W prostate gland masks. Prostate cancer is annotated by red contours. (Color figure online)

texton based feature extraction described in the previous section are applied to extract samples, extract textons and construct texton maps. The same window size is used as for T2W and ADC regarding samples extraction. For each series of images, 20 CAP and 20 NCAP textons are extracted, respectively. Different ranges of texton IDs are assigned to textons extracted from different series of images. For example, the texton IDs are 1–40 for the first series and 41–80 for the second series, etc. Thus a total number of 1, 600 textons are extracted from all 40 series of DCE MRI images. Figure 4(b) shows the texton map constructed for the first series of DCE images taken for one patient. Then instead of extracting texton histograms, for each voxel within the prostate gland, the corresponding texton IDs across all series are considered as the feature vector. Hence the dimensionality of the feature vector equals the number of series (i.e. 40) and the values in each feature vector are in the range $[1, 1600]$. The feature vectors extracted from the DCE images of all patients are used for classification.

3.4 Classification

Leave-one-patient-out cross validation is performed in our work. A random forest classifier is applied to classify all voxels into benign or malignant. The classification results are evaluated by conducting a receiver operating characteristics (ROC) analysis and computing the area under the curve (AUC).

4 Experimental Results

We compared the performance of prostate cancer detection of the intensity based and the 2D and 3D texton based approaches using images of each individual modality and all three modalities. In the intensity based approach, for each voxel, its corresponding intensity value was used as the feature. In the 2D texton based

Table 1. Dimensionality of the feature vector

Modality	Intensity based	Texton based
T2W	1	40
DCE	40	40
ADC	1	40
Mp-MRI	43	121

approach, only the neighbourhood within the corresponding slice was taken into account. Hence the window size used to extract the samples and the texton histograms was 7×7 rather than $7 \times 7 \times 3$ used in the 3D approach. In the approach using mp-MRI images, a spatial feature with a boolean value, which indicates the location of the voxel (in the PZ or not), was also concatenated with the aggregated intensity or texton based feature vectors. In Table 1, we list the dimensionality of feature vector extracted from each modality and all three modalities used in each approach.

4.1 Comparisons of Classification Performance Using Mono-Modality Images

Figure 5(a–c) illustrates the classification results produced by intensity based, 2D and 3D texton based approaches using T2W, DCE and ADC MRI images, respectively. It can be seen that both texton based approaches outperform the intensity based approach for each modality. The 3D texton based approach significantly outperforms the 2D approach using the T2W and ADC MRI images ($p < 0.05$). The two texton based approaches produce nonsignificantly different results using DCE MRI images. Compared with the intensity based approach, applying the 3D texton based approach improves the cancer detection accuracy by 19.1%, 9.2% and 12.2% using T2W, DCE and ADC MRI, respectively. The intensity based approach obtains the best performance using DCE MRI while the highest accuracy is obtained by the 3D texton based approach using T2W MRI. Using ADC maps generates the worst classification by applying either intensity based or texton based approach. We tried combining intensity and 3D texton based features extracted from each modality to perform classification. However, it does not significantly increase the accuracy (less than 1%) or even slightly deteriorates the classification performance. It indicates that the texton based features could better model the intensity information; combing the original intensity is redundant and might introduce noise which could bias the classification. Overall, in the experiments using mono-modality images, the best classification is achieved by the 3D texton based approach using T2W MRI. It can also be observed from Fig. 4 that the appearance of the texton map seems more distinct between CAP and NCAP regions in T2W MRI than in the other two modalities.

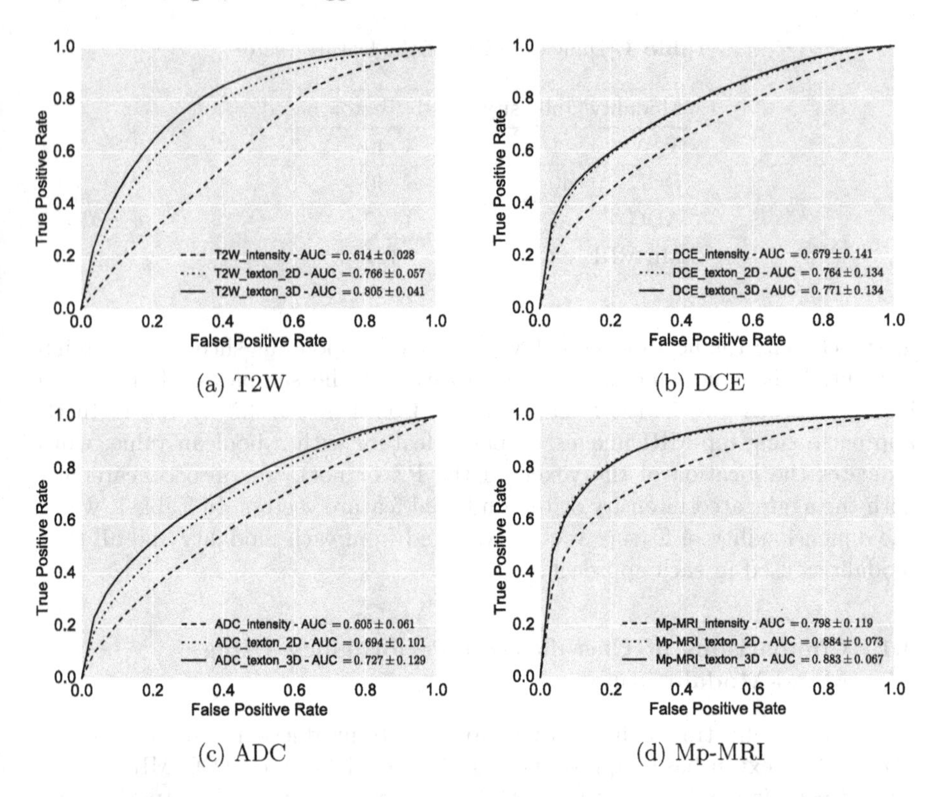

Fig. 5. Classification results produced by the intensity based, 2D and 3D texton based approaches using images of each modality and mp-MRI.

4.2 Comparisons of Classification Performance Using Mp-MRI Images

We concatenated the intensity or texton based features extracted from all three modalities to perform the classification. The results are demonstrated in Fig. 5(d). It shows that for intensity based and both texton based approaches, the classification accuracy obtained by using mp-MRI is better than that using each individual modality. Again, compared with intensity based approach, the 2D and 3D texton based approaches improve the classification accuracy by 8.6% and 8.5%, respectively, using mp-MRI images. 2D and 3D texton based approaches perform equally using mp-MRI. According to the clinical setting [20], by contrast to the intensity based approach using mp-MRI or both approaches using mono-modality, the texton based approach using mp-MRI improves the performance of prostate cancer detection from 'acceptable' (with AUC ranging from 0.7 to 0.8) to 'excellent' (with AUC ranging from 0.8 to 0.9). We also tried combining intensity and texton based features extracted from mp-MRI images for classification; however, the performance is slightly deteriorated.

Figure 6 shows the probability maps of prostate cancer detection for some patients with cancer in the PZ (a, b), CG (c) and both regions (d, e, f). The

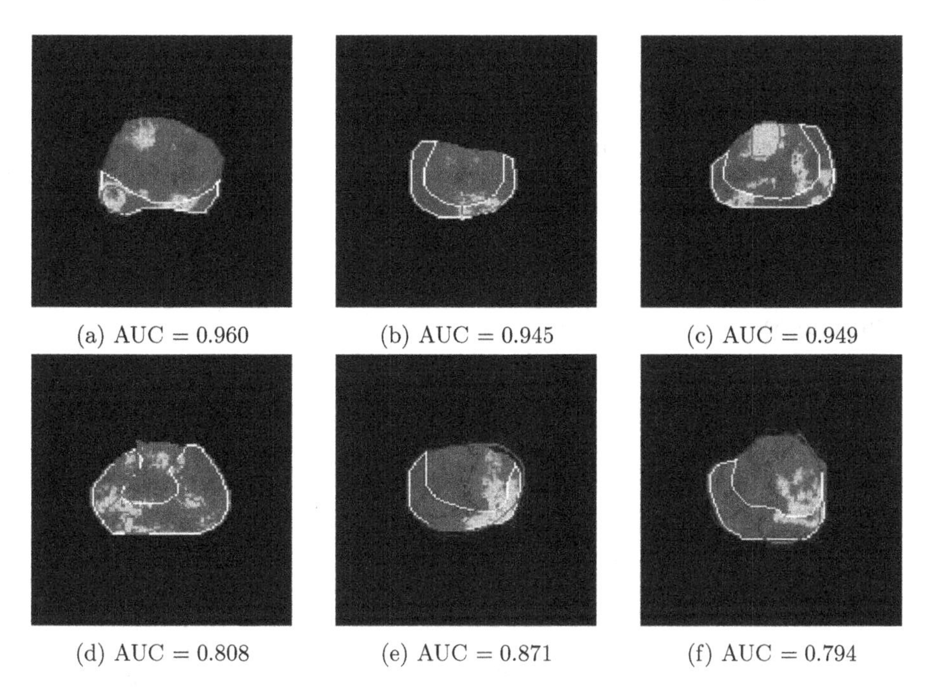

Fig. 6. Results of prostate cancer detection produced by the 3D texton based approach using mp-MRI. The PZ region is depicted by white contour while the cancer region is depicted by red contours. The jet overlap represents the probability map of cancer detection. (Color figure online)

results were produced by the 3D texton based approach using mp-MRI. We can see that for most cases, the cancer region is accurately identified. Cancer in the PZ can be detected with high accuracy. The performance of prostate cancer detection is poor for some cases with cancer in the CG or in both the CG and PZ, as shown in Fig. 6(d, f). Among all 17 cases, the accuracy measured by AUC is below 85% for 5 cases and 4 of them have cancer in the CG, which manifests the difficulty of prostate cancer detection in the CG.

5 Conclusions

This paper presents a 3D texton based approach for prostate cancer detection using mp-MRI. The texton histograms were extracted as features from T2W and ADC MRI images by applying the extended traditional texton based approach. For DCE MRI, a new approach has been proposed to extract texton IDs across all series as features to perform classification. The experimental results have demonstrated that the texton based approach outperforms the intensity based approach using either mono-modality or mp-MRI data. The 3D texton based approach outperforms the 2D approach using T2W and ADC MRI images; the two texton based approaches perform equally using DCE and mp-MRI data.

Using mp-MRI produces higher classification accuracy by applying both intensity based and texton based approaches, compared with using mono-modality. And the texton based approach using mp-MRI obtains excellent discrimination between malignant and benign prostate tissues. The qualitative and quantitative results also illustrate the difficulty of prostate cancer detection in the CG. In the future, we will incorporate more image-based and signal-based features into the framework.

References

1. Ferlay, J., Shin, H.R., Bray, F., Forman, D., Mathers, C., Parkin, D.M.: Estimates of worldwide burden of cancer in 2008: GLOBOCAN 2008. Int. J. Cancer **127**(12), 2893–2917 (2008)
2. Siegel, R., Naishadham, D., Jemal, A.: Cancer statistics. CA Cancer J. Clin. **63**(1), 11–30 (2013)
3. Etzioni, R., Penson, D.F., Legler, J.M., Di Tommaso, D., Boer, R., Gann, P.H., Feuer, E.J.: Overdiagnosis due to prostate-specific antigen screening: lessons from US prostate cancer incidence trends. J. Natl Cancer Inst. **94**(13), 981–990 (2002)
4. Chou, R., Croswell, J.M., Dana, T., Bougatsos, C., Blazina, I., Fu, R., Gleitsmann, K., Koenig, H.C., Lam, C., Maltz, A., Rugge, J.B.: Screening for prostate cancer: a review of the evidence for the US preventive services task force. Ann. Intern. Med. **155**(11), 762–771 (2011)
5. Schröder, F.H., Carter, H.B., Wolters, T., van den Bergh, R.C., Gosselaar, C., Bangma, C.H., Roobol, M.J.: Early detection of prostate cancer in 2007: part 1: PSA and PSA kinetics. Eur. Urol. **53**(3), 468–477 (2007)
6. Delpierre, C., Lamy, S., Kelly-Irving, M., Molini, F., Velten, M., Tretarre, B., Woronoff, A.S., Buemi, A., Laptre-Ledoux, B., Bara, S., Guizard, A.V.: Life expectancy estimates as a key factor in over-treatment: the case of prostate cancer. Cancer Epidemiol. **37**(4), 462–468 (2013)
7. Delongchamps, N.B., Peyromaure, M., Schull, A., Beuvon, F., Bouazza, N., Flam, T., Zerbib, M., Muradyan, N., Legman, P., Cornud, F.: Prebiopsy magnetic resonance imaging and prostate cancer detection: comparison of random and targeted biopsies. J. Urol. **189**(2), 493–499 (2013)
8. Vos, P.C., Barentsz, J.O., Karssemeijer, N., Huisman, H.J.: Automatic computer-aided detection of prostate cancer based on multiparametric magnetic resonance image analysis. Phys. Med. Biol. **57**(6), 1527 (2012)
9. Viswanath, S., Bloch, B.N., Chappelow, J., Patel, P., Rofsky, N., Lenkinski, R., Genega, E., Madabhushi, A.: Enhanced multi-protocol analysis via intelligent supervised embedding (EMPrAvISE): detecting prostate cancer on multiparametric MRI. In: Proceedings of SPIE 7963, Medical Imaging 2011: Computer-Aided Diagnosis, p. 79630U (2011)
10. Tiwari, P., Kurhanewicz, J., Madabhushi, A.: Multi-kernel graph embedding for detection, Gleason grading of prostate cancer via MRI/MRS. Med. Image Anal. **17**(2), 219–235 (2013)
11. Trigui, R., Mitran, J., Walker, P.M., Sellami, L., Hamida, A.B.: Automatic classification and localization of prostate cancer using multi-parametric MRI/MRS. Biomed. Sig. Process. Control **31**, 189–198 (2017)

12. Lemaître, G., Mart, R., Freixenet, J., Vilanova, J.C., Walker, P.M., Meriaudeau, F.: Computer-aided detection and diagnosis for prostate cancer based on mono and multi-parametric MRI: a review. Comput. Biol. Med. **60**, 8–31 (2015)
13. Zhang, L., Fisher, M., Wang, W.: Retinal vessel segmentation using Gabor filter and textons. In: Medical Image Understanding and Analysis, pp. 155–160 (2014)
14. Gangeh, M.J., Srensen, L., Shaker, S.B., Kamel, M.S., De Bruijne, M., Loog, M.: A texton-based approach for the classification of lung parenchyma in CT images. In: International Conference on Medical Image Computing and Computer-Assisted Intervention, pp. 595–602 (2010)
15. Julesz, B.: A theory of preattentive texture discrimination based on first-order statistics of textons. Biol. Cybern. **41**(2), 131–138 (1981)
16. Rampun, A., Tiddeman, B., Zwiggelaar, R., Malcolm, P.: Computer aided diagnosis of prostate cancer: a texton based approach. Med. Phys. **43**(10), 5412–5425 (2016)
17. Lemaître, G., Dastjerdi, M.R., Massich, J., Vilanova, J.C., Walker, P.M., Freixenet, J., Meyer-Baese, A., Mriaudeau, F., Mart, R.: Normalization of T2W-MRI prostate images using Rician a priori. In: SPIE Medical Imaging 2016: Computer-Aided Diagnosis, p. 9785 (2016)
18. Lemaître, G.: Computer-aided diagnosis for prostate cancer using multi-parametric magnetic resonance imaging. Doctoral dissertation, Universite de Bourgogne; Universitat de Girona (2016)
19. Nyúl, L.G., Udupa, J.K., Zhang, X.: New variants of a method of MRI scale standardization. IEEE Trans. Med. Imaging **19**(2), 143–150 (2000)
20. Hosmer Jr., D.W., Lemeshow, S., Sturdivant, R.X.: Applied logistic regression, vol. 398. Wiley, Hoboken (2013)

Tumor Segmentation in Whole Slide Images Using Persistent Homology and Deep Convolutional Features

Talha Qaiser[1], Yee-Wah Tsang[2], David Epstein[3], and Nasir Rajpoot[1](✉)

[1] Department of Computer Science, University of Warwick, Coventry CV4 7AL, UK
N.M.Rajpoot@warwick.ac.uk
[2] University Hospital of Coventry and Warwickshire, Coventry CV2 2DX, UK
[3] Mathematics Institute, University of Warwick, Coventry CV4 7AL, UK

Abstract. This paper presents a novel automated tumor segmentation approach for Hematoxylin & Eosin stained histology images. The proposed method enhances the segmentation performance by combining the topological and convolution neural network (CNN) features. Our approach is based on 3 steps: (1) construct enhanced persistent homology profiles by using topological features; (2) train a CNN to extract convolutional features; (3) employ a multi-stage ensemble strategy to combine Random Forest regression models. The experimental results demonstrate that proposed method outperforms the conventional CNN.

Keywords: Tumor segmentation · Persistent homology · Histology image analysis · Deep learning · Digital pathology

1 Introduction

Automated identification of tumor areas in standard whole slide images (WSIs) of Hematoxylin & Eosin (H&E) stained tissue slides is one of the primary task for a computer-aided diagnosis system. Accurate segmentation of tumor-rich areas may assist the experts in selection of high power fields for tumor proliferation grading and scoring of immunohistochemical (IHC) stained slides by restricting the analysis to tumor regions only. Hence, automated tumor segmentation methods could speed up the diagnostic process and reduce the inter-observer variability of conventional methods [1].

Recently proposed tumor segmentation methods are mostly based on hand crafted features like intensity [2] and morphological appearance [3,4] extracted from different tissue components. The performance of such approaches may vary due to heterogeneity found within tumour and normal regions. In recent years deep convolution neural network (CNN) have significantly improved the performance for a wide range of computer vision [5,6] and medical image analysis [7] [8] problems. Here we propose a combination of handcrafted topological features and CNN features to attain high accuracy for tumor segmentation.

M. Valdés Hernández and V. González-Castro (Eds.): MIUA 2017, CCIS 723, pp. 320–329, 2017.
DOI: 10.1007/978-3-319-60964-5_28

The pivotal aspect of the presented approach is to explore the topological features by employing the concept of persistent homology. The topological features provides a local description of the connectedness of a given space. The degree of connectivity in tumor and normal regions is significantly different due to arrangement of nuclei structures as shown in Fig. 1. In normal regions nuclei retain their morphological appearance and structure whereas in tumor regions nuclei display atypical characteristics by filling the inter-cellular space. We quantify these features by employing persistent homology.

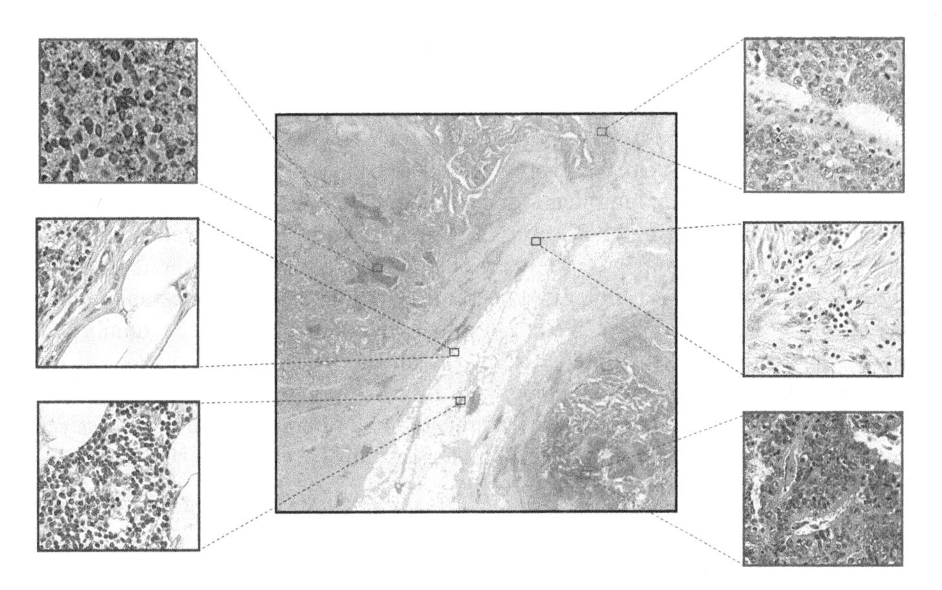

Fig. 1. Regions of interest (ROIs) extracted at 1.25× magnification from a whole slide image (center) depict the heterogeneity lies among tumor and normal regions. The six sub-windows shows the zoomed-in regions at 20× magnification. Sub-regions with green boundaries shows non-tumor region whereas sub-region with red boundaries shows tumor areas. (Color figure online)

This paper is an extension of our previous work [9]. In this paper, we propose a novel approach to tumor segmentation for whole-slide images (WSIs) by combining topological and deep convolutional features. Our contributions are summarized as follows: a modified approach for generating persistent homology profiles (PHPs) by reducing the parameters and handling the unwanted noise components. We also present a multi-stage ensemble strategy to combine the topological and deep convolutional based regression models. The proposed approach has been validated on significantly large dataset than that in [9].

2 Methods

The proposed approach takes an H&E stained WSI (containing gigapixel data) as input and outputs the segmentation of image into tumor regions. At first we split the WSI into patches and predict the probability of each patch being tumor or normal. For each patch, we compute persistent homology features by exploring the degree of connectivity among nuclei and deep convolutional features by training a CNN. The extracted features were separately fed into Random Forest regression model and finally the output label was predicted by multi-stage ensemble strategy.

2.1 Persistent Homology

Overview: In homology theories, persistent homology is a way of computing topological features that persist over certain range of parameters and more likely to constitute true features of a given space. For a more comprehensive and intuitive explanation of homology concepts, refer to [10–12].

Persistent features can be readily described by observing the forming and merging of connected components of homological classes. In this study we are concerned with exploring the degree of connectivity inside 2D planes with $0th$ and $1st$ homology groups. Fortunately this can be explained without in-depth knowledge of algebraic topology. The rank of i-dimensional persistent homology group is known by its ith *Betti* number, denoted by β_i. *Betti* numbers is a measure to distinguish topological spaces on the basis of their degree of connectivity. The rank of $0th$ homology group corresponds to $0th$ *Betti* number $\beta 0$, which is counting the number of connected components. Similarly, $1st$ *Betti* number $\beta 1$ defines the rank of $1st$ homology group by counting the number of one-dimensional "circular" holes or voids.

Persistent Homology Profiles: To calculate the persistent homology features the given space should be represented as a filtered space. Suppose, I represents a single channel (2D) image of size $m \times n$ extracted from a WSI. Let X represents a filtered space contains a sequence of subspaces $X(t)$. For each value of threshold (t), we attained a binary image referring as a subspace $X(t) \subseteq X$ that contains the union of pixels of I. Similarly, for certain range of threshold (t) $(0 \leq t_l \leq t_h \leq 255)$ over 2D image plane I, we get a filtered space X as shown in (1).

$$X(t_0) \subseteq X(t_1) \subseteq \cdots \subseteq X(t_{255}) = \mathbf{X} \tag{1}$$

The corresponding Betti numbers ($\beta 0$, $\beta 1$) have been recorded for entire threshold range. By normalizing so that areas under the curves are equal to one, we can think of these as statistical distributions. We named those distributions as *persistent homology profiles* (PHP) associated with image I. The algorithm for deriving the persistent homology profiles is further explained in [9].

2.2 Enhanced Persistent Homology Profiles Features

First we performed stain deconvolution [13] and extracted the hematoxylin channel in order to overcome the stain variations. In enhanced persistent homology profiles (*e*PHP) we have improved the PHPs [9] by proposing following advancements: (a) reduce the range of threshold parameters by retaining the PHP characterstics (b) apply edge-preserving smoothing [14] before thresholding.

In *e*PHP, the topological features ($\beta 0$, $\beta 1$) were only recorded for limited threshold values by alternatively selecting one and leaving two values from the entire thresholds range. The modified filtration space is shown in (2). The *e*PHPs retains the discriminative characteristics of PHPs by using three times less parameters. The existing PHPs and *e*PHPs are shown in Fig. 2. In tumor regions, nuclie fills the inter-cellular regions and lie relatively close to each other as can be seen in Fig. 1. Hence, when the threshold increase the inter-cellular regions filled gradually whereas in normal regions nuclei retain their structure and homology classes show high rate of change while merging and forming into new classes. This advancement reduce the number of parameters required to construct a PHP and makes the them less flexible and more robust to noise.

$$X'(t_0) \subseteq X'(t_3) \subseteq X'(t_6) \subseteq \cdots \subseteq X'(t_{255}) = \mathbf{X}' \tag{2}$$

We also observed while varying the threshold from lower limit (t_l) to its upper limit (t_h) that some unwanted objects also contributed to the *e*PHP. In order to overcome the participation of artifacts and significantly small noisy regions, we employed a multi-scale edge-preserving smoothing approach [14]. The edge-preserving method is based on weighted least squares optimization framework that progressively capture the details at multiple scales. It can be also be seen in Fig. 3 that Gaussian smoothing is limited in their ability and distort the morphological appearance of tissue components, in contrast to edge-preserving smoothing method.

2.3 Deep Convolutional Features

We train a deep convolutional neural network (CNN) to extract the deep convolutional features for tumor and normal patches. The detailed architectures of implemented CNN is shown in Table 1 and is inspired by [15].

The CNN architecture contains 4 convolution layers followed by max-pooling operation, 2 fully connected and a softmax classification layer to assign probabilities to the normal and tumor class. After each convolution layer an exponential linear unit (ELU) [16] was placed to overcome the vanishing gradient problem during back propagation. We also used dropout layer [17] at the end of the FC2 layer in order to overcome the over fitting problem. The CNN was trained to minimize the cross entropy loss L as in (3).

$$L(g, y) = -\sum_{x} g(x) log(y(x)) \tag{3}$$

where, for input x, g represents the ground-truth label and y is the predicted probability by CNN.

We extracted deep convolutional features after the FC1 layer, immediately after the convolution layers. The motivation behind acquiring the features after

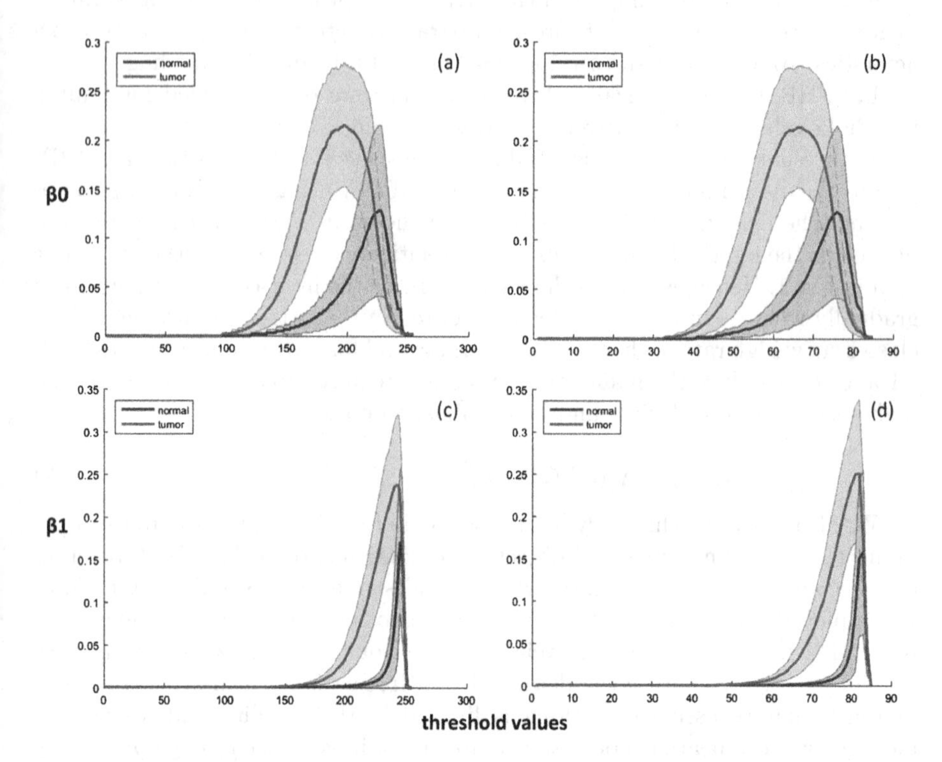

Fig. 2. Persistent homology profiles of training dataset with median values, first and third quartile, for tumor (red) and non-tumor (blue) (a) PHPs for $\beta 0$ with complete range of threshold (b) ePHP for $\beta 0$ with limited threshold values (c) PHPs for $\beta 1$ with complete range of threshold (b) *e*PHP $\beta 1$ with limited threshold values. (Color figure online)

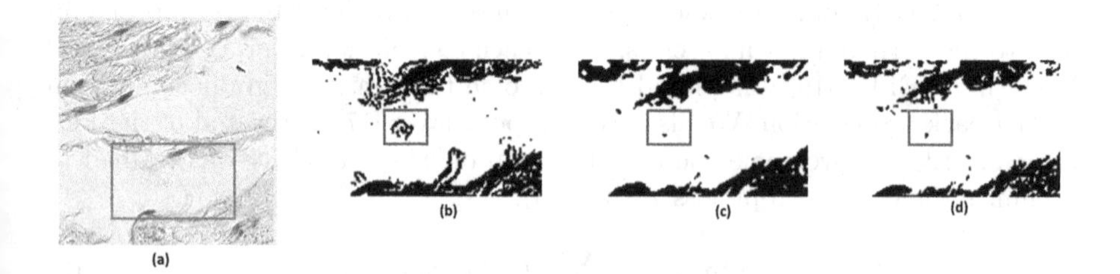

Fig. 3. (a) a patch from training dataset with selected region of interest (ROI). (b) thresholded ROI region without smoothing (c) Gaussian smoothing results for ROI region. (d) edge-preserving smoothing on ROI region.

Table 1. Detailed architecture of the convolution neural network for extracting the deep convolutional features.

Type	Patch size/stride	Input size
conv-1	$11 \times 11/1$	$256 \times 256 \times 3$
max-pool-1	$1 \times 1/2$	$123 \times 123 \times 32$
conv-2	$8 \times 8/1$	$116 \times 116 \times 64$
max-pool-2	$1 \times 1/2$	$58 \times 58 \times 64$
conv-3	$5 \times 5/1$	$54 \times 54 \times 128$
max-pool-3	$1 \times 1/2$	$27 \times 27 \times 128$
conv-4	$4 \times 4/1$	$24 \times 24 \times 256$
max-pool-4	$1 \times 1/2$	$12 \times 12 \times 256$
FC1	$12 \times 12/1$	$1 \times 1 \times 512$
FC2	$1 \times 1/1$	$1 \times 1 \times 1024$
Softmax	classifier	$1 \times 1 \times 2$

FC1 is that convolution layer contains spatially learned features with combination of extremely high and low frequency information. The convolutional features may also refer as the set of local features describing particular image regions [18]. For a given patch we obtained an activation matrix of $(1 \times 1 \times 512)$.

2.4 Ensemble Strategy

After obtaining the convolutional and *e*PHP features, we first converted the *e*PHPs into discrete probability distributions and then concatenate $(\beta 0, \beta 1)$ them to form a combined feature vector. Furthermore, the Random Forest (RF) regression model was separately trained for both features. The RF model was further optimized by selecting an ensemble of 200 bagged trees, randomly selecting one third of variables for each decision split and setting the minimum leaf size as 5. The multi-stage ensemble strategy is shown in Fig. 4.

We combined the output from both regression models (O_1, O_2) as in (4). The multi-stage ensemble strategy is based on two steps (a) average the O_1, O_2 probabilities $(p(x))$ and assign the output label for those patches where both regression models shows good agreement (b) for the remaining few patches ($\approx$1% from testing data) where the average probabilities lie in critical range $(0.49 < p(x) < 0.51)$ we assign the output label by rounding the probabilities from O_1 as our experiments shows comparatively good F1-score with this setting. In (4), 0 represents the output label for tumor patches and 1 represents the output for normal patches.

$$\hat{O}(x) = \begin{cases} 0, & \text{if } \frac{(O_1 + O_2)}{2} < 0.49. \\ 1, & \text{else if } \frac{(O_1 + O_2)}{2} > 0.51. \\ \lfloor O_1(x) \rceil & \text{otherwise} \end{cases} \quad (4)$$

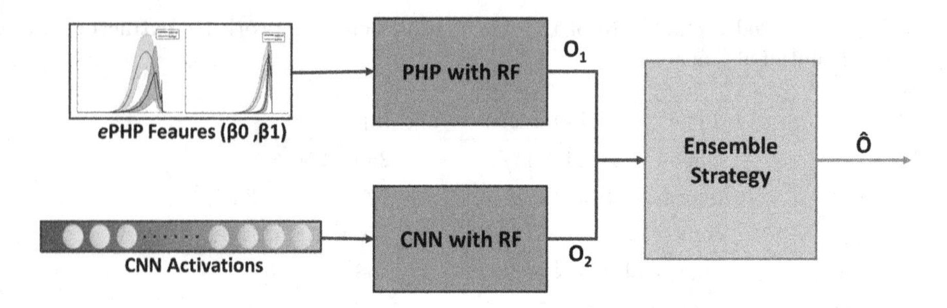

Fig. 4. A multi-stage ensemble strategy ($\hat{O}$) where O_1 shows the regression output from topological features and O_2 show the regression output for CNN features.

3 Experimental Results

3.1 Dataset and Implementation Details

We collected 50 H&E stained colorectal adenocarcinoma tumor WSIs. The tumor areas for this dataset were marked by a expert pathologist. The dataset was randomly split by selcting 35 WSIs for training and 15 WSIs for testing. For training, we randomly selected 1,000 patches from normal and 1,000 patches from tumor region of each WSI. In total, we have 50,000 (35,000 training; 15,000 test) patches each of size 256 × 256 pixels at 20× magnification. The selected dataset for this study was approximately 10 times more as compare to [9]. For testing, we split the WSI into patches and then fed in to proposed method and some of the patches can be seen in Fig. 5.

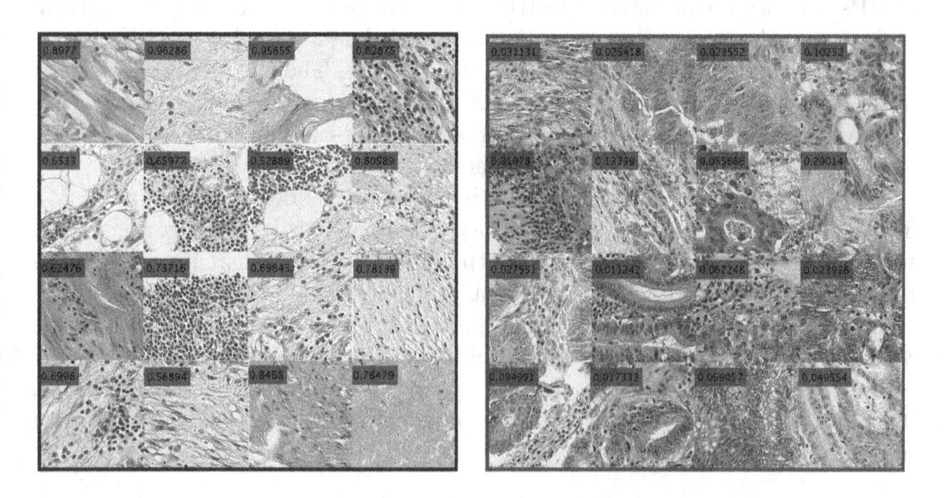

Fig. 5. Results of the proposed method for randomly selected patches (*left*) normal patches (*right*) tumor patches. Text in yellow box shows the confidence value for each patch. The confidence value range for tumor class is ≤0.50 and >0.50 for normal class. (Color figure online)

The CNN was optimized by using AdamOptimizer [19] with mini-batch stochastic gradient descent (learning rate = 10^{-4} & momentum = 0.9) with batch size of 120. The presented method is implemented in Python and MATLAB by using TensorFlow library [20]. For evaluating the segmentation accuracy we used F1 score as defined in (5)

$$F1 = 2 \times \frac{Precision \times Recall}{(Precision + Recall)}. \quad (5)$$

3.2 Results and Discussion

In order to evaluate the performance of the proposed algorithm, we first compare our method with conventional CNN. In addition, we separately trained RF regression models on *e*PHP and CNN features to highlight the significance of multi-stage ensemble strategy. The architecture for conventional CNN is shown in Table 1. In Table 2, RF (CNN) represents the regression model for CNN based features, similarly RF (*e*PHP) represents the regression model for *e*PHP based features and RF (CNN + *e*PHP) refers the combined results after using ensemble strategy.

Table 2. Comparative results for proposed approach and conventional neural network features.

Method	Precision	Recall	F1 score
Conventional CNN	0.8691	0.9146	0.8913
RF (CNN)	0.8760	0.9058	0.8906
RF (*e*PHP)	0.8768	0.9059	0.8911
RF (CNN + *e*PHP)	**0.8825**	**0.9157**	**0.9004**

A CNN learns a stack of feature maps at each convolution layer that mainly contains the data driven local features (first layer learn localized, oriented edges, second layer learn contour, arcs and boundary edges etc.). In additional to CNN features, if we combine the local description information of a surface by using *e*PHP, then we can increase the performance of tumor segmentation as shown in Table 2.

4 Conclusions

In this paper, we presented a patch based tumor segmentation method for histology whole slide images. Here we combined the enhanced persistent homology profiles (*e*PHP) and convolution neural network (CNN) features for tumor segmentation. We also proposed a multi-stage ensemble strategy to combine the output of Random Forest regression models. Experimental results demonstrated that the combination of topological and CNN features outperformed the conventional CNN. Our future work will look into large-scale validation and exploring this approach for tumor segmentation in histology images of other cancers.

Acknowledgments. The first author (Qaiser) acknowledges the financial support provided by the University Hospital Coventry Warwickshire (UHCW) and the Department of Computer Science at the University of Warwick.

References

1. Litjens, G., Sanchez, C.I., Timofeeva, N., Hermsen, M., Nagtegaal, I., Kovacs, I., Hulsbergen-Van De Kaa, C., Bult, P., Van Ginneken, B., Van Der Laak, J.: Deep learning as a tool for increased accuracy and efficiency of histopathological diagnosis. Scientific reports 6 (2016)
2. Sieren, J.C., Weydert, J., Bell, A., De Young, B., Smith, A.R., Thiesse, J., Namati, E., McLennan, G.: An automated segmentation approach for highlighting the histological complexity of human lung cancer. Ann. Biomed. Eng. **38**(12), 3581–3591 (2010)
3. Khan, A.M., El-Daly, H., Rajpoot, N.: RanPEC: random projections with ensemble clustering for segmentation of tumor areas in breast histology images. In: Medical Image Understanding and Analysis, pp. 17–23 (2012)
4. Khan, A.M., El-Daly, H., Simmons, E., Rajpoot, N.M.: HyMaP: a hybrid magnitude-phase approach to unsupervised segmentation of tumor areas in breast cancer histology images. J. Pathol. Inform. **4**(2), 1 (2013)
5. Long, J., Shelhamer, E., Darrell, T.: Fully convolutional networks for semantic segmentation. In: Proceedings of the IEEE Conference on Computer Vision and Pattern Recognition, pp. 3431–3440 (2015)
6. Russakovsky, O., Deng, J., Hao, S., Krause, J., Satheesh, S., Ma, S., Huang, Z., et al.: Imagenet large scale visual recognition challenge. Int. J. Comput. Vis. **115**(3), 211–252 (2015)
7. Sirinukunwattana, K., Pluim, J.P.W., Chen, H., Qi, X., Heng, P.-A., Guo, Y.B., Wang, L.Y., et al.: Gland segmentation in colon histology images: the glas challenge contest. Med. Image Anal. **35**, 489–502 (2017)
8. Camelyon 2016. https://camelyon16.grand-challenge.org/. Accessed 10 Mar 2017
9. Qaiser, T., Sirinukunwattana, K., Nakane, K., Tsang, Y.-W., Epstein, D., Rajpoot, N.: Persistent homology for fast tumor segmentation in whole slide histology images. Procedia Comput. Sci. **90**, 119–124 (2016)
10. Carlsson, G.: Topology and data. Bull. Am. Math. Soc. **46**(2), 255–308 (2009)
11. Zomorodian, A.J.: Topology for Computing, vol. 16. Cambridge University Press, Cambridge (2005)
12. Cerri, A., Fabio, B.D., Ferri, M., Frosini, P., Landi, C.: Betti numbers in multidimensional persistent homology are stable functions. Math. Methods Appl. Sci. **36**(12), 1543–1557 (2013)
13. Khan, A.M., Rajpoot, N., Treanor, D., Magee, D.: A nonlinear mapping approach to stain normalization in digital histopathology images using image-specific color deconvolution. IEEE Trans. Biomed. Eng. **61**(6), 1729–1738 (2014)
14. Farbman, Z., Fattal, R., Lischinski, D., Szeliski, R.: Edge-preserving decompositions for multi-scale tone and detail manipulation. In: ACM Transactions on Graphics (TOG), vol. 27, no. 3, p. 67. ACM (2008)

15. Krizhevsky, A., Sutskever, I., Hinton, G.E.: Imagenet classification with deep convolutional neural networks. In: Advances in neural information processing systems, pp. 1097–1105 (2012)
16. Clevert, D.-A., Unterthiner, T., Hochreiter, S.: Fast and accurate deep network learning by exponential linear units (elus). arXiv preprint arXiv:1511.07289 (2015)
17. Srivastava, N., Hinton, G.E., Krizhevsky, A., Sutskever, I., Salakhutdinov, R.: Dropout: a simple way to prevent neural networks from overfitting. J. Mach. Learn. Res. **15**(1), 1929–1958 (2014)
18. Babenko, A., Lempitsky, V.: Aggregating local deep features for image retrieval. In: Proceedings of the IEEE International Conference on Computer Vision, pp. 1269–1277 (2015)
19. Kingma, D., Ba, J.: Adam: a method for stochastic optimization. arXiv preprint arXiv:1412.6980 (2014)
20. Abadi, M., Agarwal, A., Barham, P., Brevdo, E., Chen, Z., Citro, C., Corrado, G.S., et al.: Tensorflow: large-scale machine learning on heterogeneous distributed systems. arXiv preprint arXiv:1603.04467 (2016)

Multispectral Biopsy Image Based Colorectal Tumor Grader

Suchithra Kunhoth(✉) and Somaya Al Maadeed

Department of Computer Science and Engineering,
Qatar University, Doha, Qatar
{suchithra, s_alali}@qu.edu.qa

Abstract. Automated tumor cell grading systems have an immense potential in improving the speed and accuracy of cancer diagnostic procedures. It can boost the confidence level of pathologists who perform the manual assessment of tumor cells. The application of image processing and machine learning techniques on the digitized biopsy slides enables the discrimination between various cell types. Deployment of multispectral imaging technique for biopsy slide digitization serves to provide spectral information along with the spatial information. Multispectral imaging allows to acquire several images of the sample in multiple wavelengths including the infrared ranges. This paper presents a multispectral image based colorectal tumor grading system. The algorithm validation is performed on our biopsy image database comprising 200 samples from 4 classes, viz. normal, hyperplastic polyp, tubular adenoma low grade as well as carcinoma cells. In addition to the visible bands, we have incorporated the spectral bands in near infrared ranges. Rotation invariant Local phase quantization (LPQ) feature extraction on our multispectral images have yielded a classification accuracy of 86.05% with an SVM classifier. Moreover, the experiments were carried out on another small multispectral image dataset which had 3 categories of cells.

Keywords: Multispectral image · Colorectal tumor · LPQ · LBP · Infrared

1 Introduction

Cancer, a commonly used term nowadays have become the deadliest disease worldwide. Despite all the latest technologies and treatments available, it seems to be a devastating disease when fails to be detected at an early stage. The American Cancer society records show that Prostate, lung and colon cancers are those leading to highest rates of mortality [1]. Even though the cancer burden is lower in middle east compared to US and Europe, it is predicted to be doubled by 2030 [2]. In terms of the mortality rate, colorectal cancer holds the third position among all other types [3]. It is the colon or rectum, parts of the large intestine that are affected by this disease. Most colon cancers appear initially as colorectal polyps, which are actually abnormal growths inside the colon or rectum. Among the various diagnostic methodologies available, the

M. Valdés Hernández and V. González-Castro (Eds.): MIUA 2017, CCIS 723, pp. 330–341, 2017.
DOI: 10.1007/978-3-319-60964-5_29

major types are Colonoscopy, Molecular, Histopathologic and Spectroscopic diagnosis [4]. Histopathologic diagnosis, or in other terms biopsy is considered to be the ultimate screening aid for majority of cancers.

When it comes to the case of a huge mass of population, the analysis of biopsy slide by an experienced pathologist can be a time consuming task. Moreover, the expertise of the concerned pathologist can be a main contributing factor that effect the result of histopathologic analysis [5]. The tedioustask of looking under the microscope may result in several false positives and misinterpretations. If we have a computer aided diagnostic system which can detect and classify the various tumor grades, the reliability and rapidity of the screening procedure can be guaranteed. Until now, a wide range of image processing based approaches are introduced as part of automated colorectal diagnosis [6]. It is the digitization of biopsy slides, that comes as the foremost step in such approaches. Commonly utilized techniques include RGB, HSV, grayscale etc. Multispectral imaging involves capturing images in specific wavelength bands and can image even in the invisible infrared range, which was unable to be visualized by a normal human eye. Classification of cells in tissue imagery which includes cancer detection have already been adopted multispectral imaging methods [7].

The novelty of our work include (1) Using asufficiently large dataset of multi-spectral colorectal images for our experiments (2) multispectral images incorporating the infrared bands as well (3) application of LPQ, rotation invariant LPQ features which have not been attempted for cancer cell grading. The remainder of paper is organized as follows. Section 2 gives a brief review of related literature, and system overview in Sect. 3. The database aspects are detailed in Sect. 4. Section 5 gives the overview of our approach, followed by experimental results in Sect. 6. A discussion is then provided in Sect. 7. Finally concludes the paper with Sect. 8.

2 Related Work

Zhang and Liu have developed an automatic cervical cancer detection system from multispectral microscopic thin Pap smear images using micro-interferometric spectral imaging setup [8]. Both the local multispectral and textural properties in the 52 spectral bands within 400–690 nm is utilized. An SVM based screening algorithm served to remove certain unwanted features from the detection task. Candidate cancer regions were then detected based on pixel level classification.

An earlier work which has recognized the significance of multispectral imaging for the grading of prostate neoplasia is [9]. Features extracted from the 16 spectral bands led to the discrimination between normal and three different grades of cancerous tissue. They have overcome the drawbacks of PCA based feature selection adopted here, with the tabu search in [10] and a round robin tabu search in [11]. In the later paper [12], their different approaches towards detection and classification of prostatic tissues using multispectral imagery was analyzed. A Round-Robin (RR) classification algorithm using a sequential forward selection/nearest neighbor (SFS/1NN) classifier was proposed in [13] with an intention to improve the existing classification accuracy. And they were able to achieve a best classification accuracy of 99.9%. In [14], the concept of gray level co-occurrence matrix is extended for the texture characterization in multiband images and hence to discriminate the healthy and pathological prostatic tissues. In order to reduce the multispectral band representation, a band selection technique was employed which selects the best relevant bands of multispectral prostate cancer database.

Rajpoot and Rajpoot [15] have accomplished the classification of normal and malignant colorectal tissue cells using biopsy images captured in 20 spectral bands within the range 450–640 nm. Multiscale morphological features were used to train the classifier which performed well to distinguish the colorectal cells. Masood et al. tried using morphological as well as grey level co-occurrence features from the hyperspectral data for the tissue cell classification in [16]. Experiments were carried out using the LDA and SVM for classification. Later they have investigated and found that a single spectral band within the range of 400–800 nm have the sufficient textural features for the biopsy classification [17]. Although PCA, 2DPCA were tried out, the textural features with an SVM proved to be very efficient. A spatial analysis of the same single spectral band with the circular local binary pattern features was done in their further work [18]. Maggioni et al. classified the colon biopsy tissue samples into normal, precancerous or cancerous in [19]. They analyzed to discriminate among tissue features: gland nuclei, gland cytoplasm and lamina propria/lumens.

Chaddad et al. [20] proposed an improved version of the snake algorithm for segmentation along with extraction of Haralick texture features in the multispectral segmented colorectal images. 16 spectral bands in the range of 500–650 nm was utilized to achieve an efficient classification of cancer cells of type Carcinoma (Ca), Intraepithelial Neoplasia (IN) and Benign Hyperplasia (BH). Apart from the 5 haralick features, 9 morphological parameters were also included in their similar work [21]. But only three morphologic parameters (Area, Xor convex and Solidity) and 3 Haralick's features (Correlation, Entropy and Contrast) was found to be having sufficient discriminatory capability. Reference [22] have proposed a method for colorectal cancer cell detection using shape features along with a nearest neighbor classifier technique. The three dominant shape features, Area, Xor Cell-Convex and Solidity, was proved to be effective in detecting the carcinoma cells from the other grades of cancer cells, BH and IN. In their later work [23] a method for the characterization of continuum of colorectal cancer was proposed using multiple texture features extracted from multispectral optical microscopy images. Active contour segmentation approach was performed for the initial step of region of interest extraction. It was followed by the computation of texture features based on the Laplacian-of-Gaussian (LoG) filter, discrete wavelets (DW) and gray level co-occurrence matrices (GLCM). The combination of all texture features had resulted in a classification accuracy of 98.92%.

The system [24] was intended to analyze the significance of multispectral imaging in order to discriminate cancer and non-cancer nuclei in liver tissues. Textural features were extracted using Gabor descriptors, which include 30 Gabor patterns at different scales and orientations. The results have indicated approximately 99% classification accuracy. It was the images taken in wavelength bands 418–467 nm, 481–513 nm and 548–641 nm that proved efficient to classify normal and hepatocellular carcinoma in high-magnification. Meaningful features offered for classification by multispectral imaging technology is well exhibited in this work. Qi et al. [25] have conducted a performance study to compare the classification accuracy using both multispectral imaging (MSI) and standard bright-field imaging (RGB) to characterize breast tissue microarrays. Feature extraction comprised texton library training and histogram construction. In conclusion, their experimental result showed multispectral images achieved higher classification rate than RGB images. Reference [26] aims to compare the performances of various texture features including local binary patterns, Haralick features and local intensity order patterns for multispectral image based colorectal tissue classification. Using SVM and random forest classifier, the superiority of multispectral imaging technology over the classical panchromatic approach was well proved. It was the local binary patterns when combined with SVM classifier that gave the best classification accuracy of 91.3%. The method in [27] is based on the analysis of spectral information of the cells as well as their morphological properties. A large number of descriptors is extracted for each cell location, which are used to train a supervised classifier which discriminates between normal and cancer cells. The detection rate (recall) turned out to be 81–100% for the cancer class.

Most of the works described here have conducted experiments on a small number of images except for [13, 14]. This is not sufficient to quantify the success rate of specific algorithms which may need extensive training for learning purpose. Although multispectral imaging offers image acquisition capability in infrared bands, all the methods till date have utilized only the spectral bands in visible range. The majority of works have utilized texture features except the shape based features in few papers like [15, 21]. Obviously, it was the texture features who scored better in terms of classification accuracy. Among the texture descriptors, GLCM based method seems to be a commonly adopted approach [16, 20, 21] etc. The comparative studies conducted in [25, 26] indicates that the multispectral images can provide additional features that serve to achieve a greater classification accuracy.

3 Tumor Grading System Overview

The automatic tumor grader system comprises a multispectral colorectal image database for training and testing the algorithm. A preprocessing step is executed in order to enhance the raw images. Next stage performs the feature extraction and the classification of tumor types. The user interface of our system can be visualized at the center portion of Fig. 1. The interface is built to be operated with rotation invariant LPQ algorithm and SVM classifier, which gave the maximum classification accuracy for our dataset.

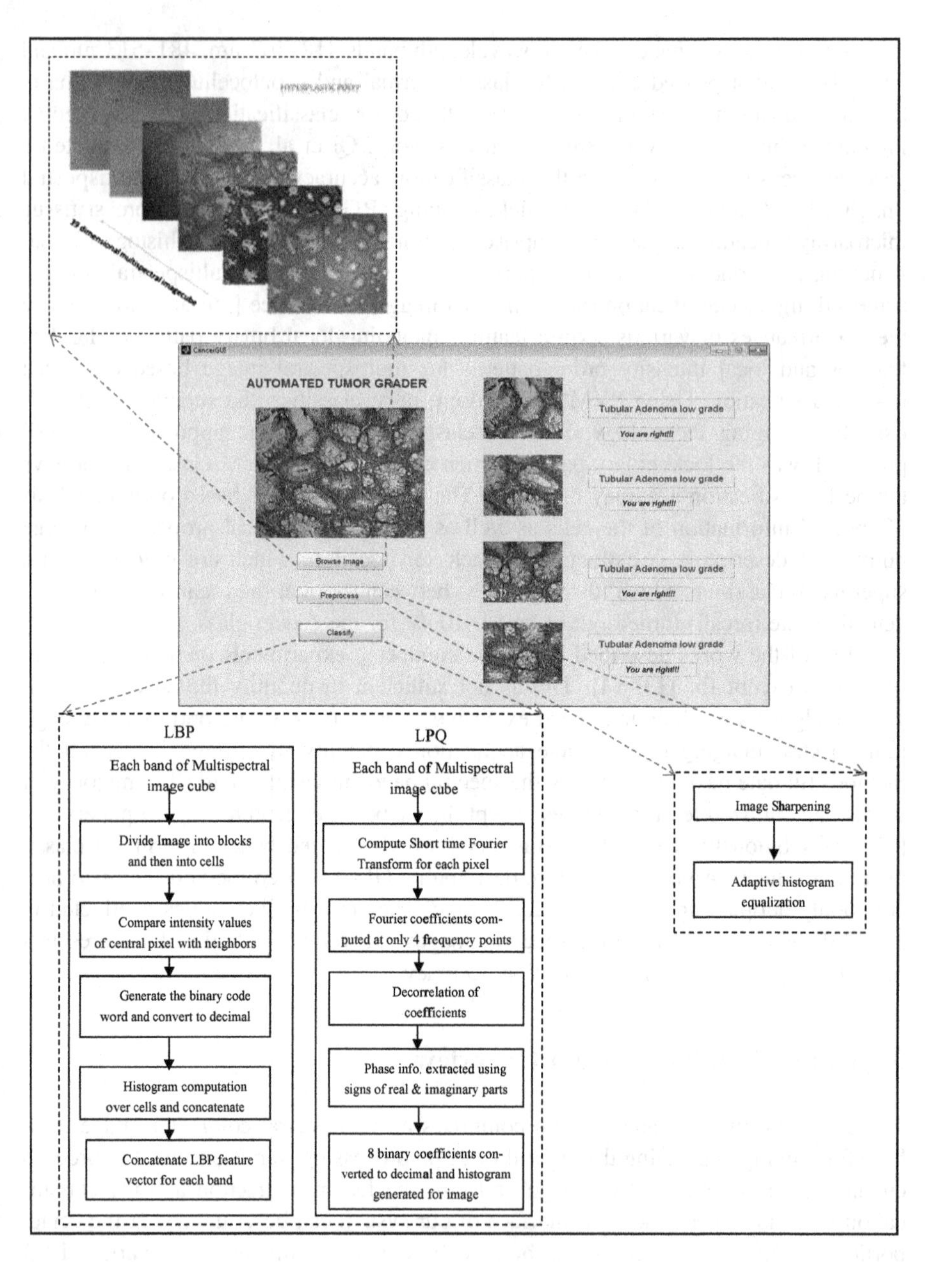

Fig. 1. Tumor grading system overview

4 Datasets

4.1 Dataset I

Our colorectal tissue collection comprises the biopsy specimens from Al Ahli Hospital, Qatar. The slide set consists of 50 samples each from the 4 different type of colorectal cells, viz normal, hyperplastic polyps, tubular adenoma with low grade dysplasia and carcinoma. Hence this set encompass the non-pathologic tissues, the two intermediate stages of colorectal tumor as well as the carcinoma, which is the final and deadliest stage of disease. In our acquisition, the images are collected following the magnification of samples at 10x. We have captured 13 multispectral visible images for each specimen at wavelength intervals of 20 nm starting from 470 nm. And in the near infrared range, 26 bands from 1150 nm to 1650 nm are chosen for acquisition. All the captured images are of grayscale nature. The resolution of the entire set of images is 320×256[1].

4.2 Dataset II

It consists of multispectral biopsy imagery with a magnification power of 40x. The database [26], which is obtained from the University of Texas, consists of 29 three-dimensional images with 512*512 pixels in the spatial dimensions and 16 along spectral dimension. These spectral dimensions correspond to 16 individual wavelengths within 500–650 nm of the visible spectrum. The images belong to 3 classes of tumor: carcinoma (Ca), benign hyperplasia (BH), intraepithelial neoplasia (IN).

5 Method

5.1 Preprocessing

The images in our dataset are undergone through the preprocessing steps which include an image sharpening followed by an adaptive histogram equalization. All the multispectral images are sharpened using the unsharp masking techniquewhich basically involves the application of a low pass filter, subtraction of low pass filtered image from the original to get the image details and then performing an addition of this obtained imagewith the original image.

5.2 Feature Extraction

Local Phase Quantization (LPQ)

It is a feature descriptor proposed recently by Ojansivu [28]. The method is based on quantized phase of the discrete Fourier transform (DFT) which is computed in local

[1] http://imaging.qu.edu.qa/datasets/.

image windows. The LPQ operator is applied to texture identification by a local computation at every pixel location and concatenating the resulting codes as a histogram similar to local binary pattern (LBP) texture algorithm [29]. It has already been deployed in applications including recognition of blurred faces, fingerprint liveness detection etc. and proved its purpose. The codes produced by the LPQ operator are insensitive to centrally symmetric blur. The utilization of mere phase information makes it invariant to uniform illumination changes.

The local phase information is extracted using the 2-D DFT (short-term Fourier transform or STFT) computed over a rectangular M-by-M neighborhood N_x at each pixel position x of the image f(x) defined by

$$F(u,x) = \sum\nolimits_{y \in N_x} f(x-y) e^{-j2\pi u^T y} = w_u^T f_x \tag{1}$$

Let G(u), F(u) and H(u) are the discrete Fourier transforms (DFT) of the blurred image g(x), the original image f(x), and the Point spread function of the blur h(x). Then the phase of observed image ∠G(u) at the frequencies, where H(u) is positive, is invariant to centrally symmetric blur. This results in the blur invariance property of LPQ. The local Fourier coefficients are then computed at only four frequency points (satisfying H(u) > 0). These four complex coefficients corresponds to the 2D frequencies as:

$$u1 = [a,0]^T, u2 = [0,a]^T, u3 = [a,a]^T, u4 = [a,-a]^T, \tag{2}$$

where a is a sufficiently small scalar to satisfy H(u) > 0

For each pixel, there will be a

$$F(x) = [F(u_1,x), F(u_2,x), F(u_3,x), F(u_4,x)] \tag{3}$$

Maximum information can be preserved in scalar quantization if the samples to be quantized are statistically independent. These independence can be achieved by the decorrelation of coefficients. Assuming Gaussian distribution, it is based on the whitening transform

$$G_x = V^T F_x, \tag{4}$$

where V is an orthonormal matrix derived from the singular value decomposition of the matrix D that is

$$D = U \sum V^T \tag{5}$$

G_x is computed for all image positions, i.e., x ∈ {x1, x2, …, xN}, and then the resulting vectors are quantized. The phase information in the Fourier coefficients is hence recorded by observing the signs of the real and imaginary parts of each component in G(x)

$$q_j = 1, g_j \geq 0 \\ = 0, \text{ otherwise} \tag{6}$$

g_j is the j-th component of the vector Gx.

The resulting eight binary coefficients q_j are represented as integer values between 0–255 using binary coding. Histogram of these values from all positions is composed, and used as a 256-dimensional feature vector in classification.

A rotation invariant extension to the blur insensitive local phase quantization texture descriptor was later proposed in [30]. It consists of a first stage which estimates the local characteristic orientation, and the second one that extracts a binary descriptor vector. Both stages apply the phase of the locally computed Fourier transform coefficients. These coefficientsare insensitive to centrally symmetric image blurring.

Local Binary Pattern (LBP)
It is a texture descriptor which is a specific case of texture spectrum model [29], proposed by Ojala et al. The computation of LBP involves an initial stage that divide the entire image into blocks and further into cells. Based on the difference in intensity values of each pixel with its neighbors, a binary code word is derived. The decimal conversion of these binary values followed by a histogram computation over each cell will produce a feature vector. Concatenating feature vectors over all the cells will finally yield the LBP feature descriptor. There are several variants for the LBP including uniform LBP, uniform rotation invariant LBP [31]. All these will reduce the original LBP feature size by a representation that involve same bin for several similar patterns.

5.3 Classification

All types of features are extracted from the multispectral images, and the classification performance was compared for two supervised learning techniques: SVM and RF. Support vector machine(SVM) was introduced in 1963 by Vladimir. It is a discriminative classifier which is defined by a separating hyperplane. Given a labeled training data input, the algorithm outputs an optimal hyperplane which categorizes new data, test dataset [32]. Random forests(RF) which was introduced later in 1995, operates by constructing a multitude of decision trees during training and outputs the class that is the mode of the classes for classification purpose [33].

6 Results

All the experiments were carried out in MATLAB along with libsvm toolbox for the SVM implementation [34], and statistics toolbox for RF. The SVM is built on radial basis function kernel with the parameter estimation(cost and gamma) using grid search method. RF uses a 100 tree structure for the classification. An 8 neighborhood schema is used for both the basic LBP and the uniform rotation invariant LBP implementations. For the LPQ, alocal window size of 3 is used with frequency estimation based on STFT with Gaussian window. A precomputed filter of window size 9 is applied with the rotation invariant LPQ descriptor.

6.1 Dataset I

In order to enlarge our database, each image was split into 4 patches which can be treated as individual images with all sufficient features encompassed. This resulted in 800 images among which 70% were used for training and remaining for testing. The features are extracted from each band of the multispectral image cube and concatenated together to create the final feature vector. The results of LPQ, rotation invariant LPQ are then compared to the LBP and uniform rotation invariant LBP algorithm results. Preprocessing stage is not used with the LBP implementation since it was already seen to reduce the accuracy with the LPQ descriptors. The experimental results are shown in Table 1. All the classification accuracies are based on a 50 fold data shuffle method of holdout validation [35].

Table 1. Results on Dataset I (Accuracy values in %)

Method	SVM	RF
LBP	77.86	62.10
Uniform rotation invariant LBP	83.61	72.18
LPQ with preprocessing	63.56	57.63
LPQ	67.52	60.14
Rotation invariant LPQ with preprocessing	82.73	69.55
Rotation invariant LPQ	**86.05**	72.04

6.2 Dataset II

The 29 images of database are split into 16 multispectral patches of dimensions128*128*16. This size allows us to have an enlarged database. These patches are thenlabeled with the same label as the image from which theywere extracted. Therefore, the database on which the tests areconducted consists of 160 CA, 160 BH and 144 IN imagesof size 128*128*16. LPQ, Rotation invariant LPQ as well as LBP results on the dataset are shown in Table 2.

Table 2. Results on dataset II (Accuracy values in %)

Method	SVM	RF
Uniform rotation invariant LBP	**91.49**	88.81
LPQ	80.92	79.91
Rotation invariant LPQ	90.39	84.53

7 Discussion

The results on our dataset indicate the superior performance of rotation invariant LPQ for the grading of cells into 4 specific classes. A comparison with the simple LBP feature extractor shows a drastic increase in classification accuracy for the rotation invariant LPQ descriptor. Also shows a measurable increase in accuracy with respect to the uniform rotation invariant LBP feature. Moreover, the preprocessing stage is not giving any substantial accuracy improvement. It is the SVM classifier which proved out to be better for the classification.

As far as the results of the dataset II is concerned, it is the uniform rotation invariant LBP that gave the highest classification accuracy. This results demonstrate the effectiveness of the uniform rotation invariant LBP for a 3 class tumor grading scenario. But the rotation invariant LPQ has also yielded a comparable accuracy. Reference [26] have carried out LBP feature extraction on the same dataset and has resulted in 88.27% accuracy with simple LBP and 91.28% with a multiscale LBP. Compared to the simple LBP, rotation invariant feature descriptor has yielded higher accuracy. But the multiscale LBP has given only a small accuracy improvement compromising the very huge size of feature vector. Even though the dataset I uses infrared range bands, its accuracy is less than that of dataset II results. This is accounted to the 4 class classification problem which is obviously more complex than the 3 class classification as with dataset II.

8 Conclusion

We have developed an automatic tumor grading system from multispectral colorectal biopsy images. The system was validated with two separate datasets: one with a 4 class grading and the other with 3 class grading problem. The former one has the multispectral images from both visible and near infrared bands, apart from the visible bands alone for the latter dataset. The performance of LBP and LPQ feature descriptors were evaluated with the help of SVM and Random Forest Classifier. The four class grading was very well performed with a rotation invariant LPQ texture descriptor, yielding satisfactory results of 86.05%. For the three class grading case, it was the uniform rotation invariant feature descriptor that delivered the highest accuracy of 91.49%.

Acknowledgment. This publication was made possible using a grant from the Qatar National Research Fund through National Priority Research Program (NPRP) No. 6-249-1-053. The contents of this publication are solely the responsibility of the authors and do not necessarily represent the official views of the Qatar National Research Fund or Qatar University.

References

1. American Cancer Society. http://old.cancer.org/cancer/colonandrectumcancer/index
2. The Lancet Oncology: Addressing the burden of cancer in the Gulf. Lancet Oncol. **15**(13), 1407, **2045**(14), 71141–71146 (2014). Epub 24 November 2014
3. Siegel, R.L., Miller, K.D., Jemal, A.: Cancer statistics. CA Cancer J. Clin. **66**(1), 7–30 (2016)
4. Kunhoth, S., Al Maadeed, S., Bouridane, A., Al, S.R.: Medical and computing insights into colorectal tumors. Int. J. Life Sci. Biotechnol. Pharma Res. **4**(2), 122 (2015)

5. Turner, J.K., Williams, G.T., Morgan, M., Wright, M., Dolwani, S.: Interobserver agreement in the reporting of colorectal polyp pathology among bowel cancer screening pathologists in Wales. Histopathology **62**(6), 916–924 (2013)
6. Rathore, S., Hussain, M., Ali, A., Khan, A.: A recent survey on colon cancer detection techniques. IEEE/ACM Trans. Comput. Biol. Bioinform. **10**(3), 545–563 (2013)
7. Boucheron, L.E., Bi, Z., Harvey, N.R., Manjunath, B., Rimm, D.L.: Utility of multispectral imaging for nuclear classification of routine clinical histopathology imagery. BMC Cell Biol. **8**(1), 1 (2007)
8. Zhang, J., Liu, Y.: Cervical cancer detection using SVM based feature screening. In: Barillot, C., Haynor, David R., Hellier, P. (eds.) MICCAI 2004. LNCS, vol. 3217, pp. 873–880. Springer, Heidelberg (2004). doi:10.1007/978-3-540-30136-3_106
9. Roula, M., Diamond, J., Bouridane, A., Miller, P., Amira, A.: A multispectral computer vision system for automatic grading of prostatic neoplasia. In: Proceedings 2002 IEEE International Symposium on Biomedical Imaging. IEEE (2002)
10. Tahir, M.A., Bouridane, A., Kurugollu, F., Amira, A.: A novel prostate cancer classification technique using intermediate memory tabu search. EURASIP J. Adv. Sig. Process. **2005**(14), 1–9 (2005)
11. Tahir, M.A., Bouridane, A.: Novel round-robin tabu search algorithm for prostate cancer classification and diagnosis using multispectral imagery. IEEE Trans. Inf Technol. Biomed. **10**(4), 782–793 (2006)
12. Tahir, M.A., Bouridane, A., Roula, M.A.: Prostate cancer classification using multispectral imagery and metaheuristics. Comput. Intell. Med. Imaging: Tech. Appl., 139 (2009)
13. Bouatmane, S., Roula, M.A., Bouridane, A., Al-Maadeed, S.: Round-Robin sequential forward selection algorithm for prostate cancer classification and diagnosis using multispectral imagery. Mach. Vis. Appl. **22**(5), 865–878 (2011)
14. Khelifi, R., Adel, M., Bourennane, S.: Multispectral texture characterization: application to computer aided diagnosis on prostatic tissue images. EURASIP J. Adv. Sig. Process. **2012** (1), 1 (2012)
15. Rajpoot, K., Rajpoot, N.: SVM optimization for hyperspectral colon tissue cell classification. In: Barillot, C., Haynor, David R., Hellier, P. (eds.) MICCAI 2004. LNCS, vol. 3217, pp. 829–837. Springer, Heidelberg (2004). doi:10.1007/978-3-540-30136-3_101
16. Masood, K., Rajpoot, N.M., Qureshi, H.A., Rajpoot, K.: Co-occurrence and morphological analysis for colon tissue biopsy classification (2006)
17. Masood, K., Rajpoot, N.M.: Classification of colon biopsy samples by spatial analysis of a single spectral band from its hyperspectral cube (2007)
18. Masood, K., Rajpoot, N.: Texture based classification of hyperspectral colon biopsy samples using CLBP. In: 2009 IEEE International Symposium on Biomedical Imaging: From Nano to Macro. IEEE (2009)
19. Maggioni, M., Davis, G.L., Warner, F.J., Geshwind, F.B., Coppi, A.C., DeVerse, R.A., et al.: Hyperspectral microscopic analysis of normal, benign and carcinoma microarray tissue sections. Biomedical Optics 2006; International Society for Optics and Photonics (2006)
20. Chaddad, A., Tanougast, C., Dandache, A., Al Houseini, A., Bouridane, A.: Improving of colon cancer cells detection based on Haralick's features on segmented histopathological images. In: 2011 IEEE International Conference on Computer Applications and Industrial Electronics (ICCAIE). IEEE (2011)
21. Chaddad, A., Tanougast, C., Dandache, A., Bouridane, A.: Extracted haralick's texture features and morphological parameters from segmented multispectrale texture bio-images for classification of colon cancer cells. WSEAS Trans. Biol/Biomed. **8**(2), 39–50 (2011)
22. Chaddad, A., Tanougast, C., Golato, A., Dandache, A.: Carcinoma cell identification via optical microscopy and shape feature analysis. J. Biomed. Sci. Eng. **6**(11), 1029 (2013)

23. Chaddad, A., Desrosiers, C., Bouridane, A., Toews, M., Hassan, L., Tanougast, C.: Multi texture analysis of colorectal cancer continuum using multispectral imagery. PLoS ONE **11** (2), e0149893 (2016)
24. Oranit, B., Chamidu, A., Hiroshi, N., Kota, A., Fumikazu, K., Masahiro, Y.: Multispectral band analysis: application on the classification of hepatocellular carcinoma cells in high-magnification histopathological images. J. Cytol. Histol. **3**(3), 1 (2015)
25. Qi, X., Xing, F., Foran, D.J., Yang, L.: Comparative performance analysis of stained histopathology specimens using RGB and multispectral imaging. In: SPIE Medical Imaging. International Society for Optics and Photonics (2011)
26. Peyret, R., Bouridane, A., Al-Maadeed, S.A., Kunhoth, S., Khelifi, F.: Texture analysis for colorectal tumour biopsies using multispectral imagery. In: 2015 37th Annual International Conference of the IEEE Engineering in Medicine and Biology Society (EMBC). IEEE (2015)
27. Zimmerman-Moreno, G., Marin, I., Lindner, M., Barshack, I., Garini, Y., Konen, E., et al.: Automatic classification of cancer cells in multispectral microscopic images of lymph node samples. In: 2016 IEEE 38th Annual International Conference of the Engineering in Medicine and Biology Society (EMBC). IEEE (2016)
28. Ojansivu, V., Heikkilä, J.: Blur insensitive texture classification using local phase quantization. In: Elmoataz, A., Lezoray, O., Nouboud, F., Mammass, D. (eds.) ICISP 2008. LNCS, vol. 5099, pp. 236–243. Springer, Heidelberg (2008). doi:10.1007/978-3-540-69905-7_27
29. Ojala, T., Pietikäinen, M., Harwood, D.: A comparative study of texture measures with classification based on featured distributions. Pattern Recogn. **29**(1), 51–59 (1996)
30. Ojansivu, V., Rahtu, E., Heikkila, J.: Rotation invariant local phase quantization for blur insensitive texture analysis. In: 19th International Conference on Pattern Recognition, ICPR 2008. IEEE (2008)
31. Ojala, T., Pietikainen, M., Maenpaa, T.: Multiresolution gray-scale and rotation invariant texture classification with local binary patterns. IEEE Trans. Pattern Anal. Mach. Intell. **24** (7), 971–987 (2002)
32. Scholkopf, B., Smola, A.J.: Learning with Kernels: Support Vector Machines, Regularization, Optimization, and Beyond. MIT press, Cambridge (2001)
33. Genuer, R., Poggi, J., Tuleau-Malot, C.: Variable selection using random forests. Pattern Recog. Lett. **31**(14), 2225–2236 (2010)
34. Chang, C., Lin, C.: LIBSVM: a library for support vector machines. ACM Trans. Intell. Syst. Technol. (TIST) **2**(3), 27 (2011)
35. Kuncheva LI. Combining pattern classifiers: methods and algorithms. John Wiley & Sons; 2004

Semi-automatic Bone Marrow Evaluation in PETCT for Multiple Myeloma

Patrick Leydon(✉), Martin O'Connell, Derek Greene, and Kathleen Curran

UCD Institute for Discovery, O'Brien Centre for Science, University College Dublin, Dublin 4, Ireland
patrick.leydon@ucdconnect.ie

Abstract. Approximately 10% of all haematologic cancers are related to Multiple Myeloma (MM). Whole-body 18F-FDG PETCT is an extremely useful imaging tool for the assessment of patients with MM. The software developed in this research performs a pixel thresholding based segmentation and a semi-automatic placement of regions of interest at key anatomical sites for the assessment of bone marrow metabolism. The proposed method offers an automated and objective approach for evaluation of PET scans of patients with MM. This method has also allowed for quantitative statistical comparisons between patients with normal bone marrow metabolism and those with myeloma to be performed, establishing a baseline against which future scans may be referenced. In cases where the suspicion of myeloma exists, the tools developed in this research may be used to support the diagnosis of the disease, and could potentially be useful in staging of the disease in cases positive for myeloma, monoclonal gammopathy of undetermined significance or smouldering myeloma.

Keywords: Multiple myeloma · Bone marrow · SUV · PETCT · Segmentation

1 Introduction

1.1 Multiple Myeloma

Multiple myeloma (MM) is a malignant haematologic disorder characterized by bone marrow infiltration with neoplastic plasma cells. Approximately 10% of all haematologic cancers are related to MM, with an incidence of approximately 4/100,000 per year [2]. The clinical symptoms of MM are a combination of anaemia, hypercalcaemia, renal impairment, and bone lesions, which are present in approximately 80% of symptomatic patients. The extent of bone lesions will impact later choices of therapy and it is therefore vital to diagnose myelomatous lesions accurately and as non-invasively as possible. In order to standardize treatment approaches it is essential that the disease be clearly characterised at the time of diagnosis, this is achieved through the Durie-Salmon staging system for MM which was introduced in 1975 [3]. Advances in diagnostic imaging techniques led to the introduction of the Durie-Salmon PLUS staging system in

M. Valdés Hernández and V. González-Castro (Eds.): MIUA 2017, CCIS 723, pp. 342–351, 2017.
DOI: 10.1007/978-3-319-60964-5_30

2003 [5]. The Durie-Salmon PLUS system highlights the benefits of 18F-FDG PETCT in accurate staging of MM, with some of the main advantages being the identification of monoclonal gammopathy of undetermined significance (MGUS) and low level smouldering myeloma, which is consistently negative on the standard planar x-ray skeletal surveys preformed as part of the standard diagnostic work-up [4].

1.2 18F-FDG PETCT and Standard Uptake Value

Positron Emission Tomography Computed Tomography (PETCT) imaging utilizes 18F-fluorodeoxyglucose (18F-FDG), which is an analogue of glucose that has positron-emitting radionuclide 18 F attached to the molecule. Metabolically active cells take up the FDG which remains trapped within the cell as there is no further metabolization. The resulting intracellular accumulation of 18F-FDG is then imaged using PET [2]. The index of activity most commonly applied in 18F-FDG PETCT and other studies to determine a regions level of metabolism is the Standard Uptake Value (SUV). It is used as a diagnostic tool to semi-quantitatively differentiate between benign and malignant anatomic regions, such as the lesions caused by MM [6]. It is of clinical importance to be aware that 18F-FDG is not a cancer specific agent and uptake may be observed in benign tumours, inflammatory foci, or the normal physiological activity in healthy organs which may result in false positives if not taken into consideration by the reporting clinician [8]. The semi-quantitative nature of SUV measurements allows for a more objective assessment of PET images. 18F-FDG PETCT imaging is useful to clarify disease classification and can substitute for MRI for staging, restaging, and/or serial monitoring. At present, there is an ever growing body of evidence to suggest PETCT's increasing importance in the future of MM diagnosis [16]. This study applies pixel thresholding methods to create a binary mask and semi-automatically define bone marrow regions of interest that are used to extract SUVs that correspond to only the bone marrow portions of the PET images.

2 Methods

2.1 PETCT Datasets

Inclusion criteria for the MM group, was any patient referred for a 18F-FDG PETCT scan that had a clinical diagnosis of MM, and did not meet any of the exclusion criteria. The inclusion criteria for the normal bone marrow metabolism group was considered to be any patient referred for a whole body 18F-FDG PETCT scan that did not did not meet any of the exclusion criteria. Exclusion criteria was considered to be any factor, which could produce misleading regions of FDG uptake/activity or otherwise influence the normal metabolic activity of the bone marrow under examination. These factors included, but were not limited to, (i) Hyperglycaemia, (ii) Plasmacytoma, (iii) Patients under 18 years of age, (iv) Chemotherapy within the previous 8 weeks, (v) Radiotherapy within the previous 8 weeks, (vi) Any therapy which has an immunosuppressant effect, (vii) Presence of metastasises to any site, (vii) Any haematological disorder, (viii) Unusual haemoglobin levels, (ix) Any active infections or inflammations, (x) Any image artifacts on PETCT (motion etc.). These exclusion criteria were assessed by reviewing the clinical reports and patient referrals available through the local hospital and radiology information systems. This yielded a main cohort of 33 patient datasets, consisting of 18 patients with MM and 15 patients meeting the criteria for normal bone marrow metabolism. A further 12 patients with MM that were receiving, or had received, chemotherapy within 8 weeks of their PET scan was also included in order to assess the impact of treatment on bone marrow metabolism. Scans were acquired under the local protocols, consisting of low dose whole body CT component and attenuation corrected PET images that used the ordered subset expectation maximization (OSEM) method of iterative reconstruction. All patient data was anonymized prior to analysis, and this study was granted full ethical approval in advance of data compilation (Table 1).

Table 1. Patient groups summary

Group	n	Male	Female	Age $\pm$ s.d	Weight $\pm$ s.d. (kg)	Dose $\pm$ s.d. (MBq)
MM*	18	10	8	68.6 $\pm$10.1	70.0 $\pm$16.3	411.5 $\pm$53.8
MMc†	12	6	6	61.9 $\pm$8.1	75.8 $\pm$12.7	423.4 $\pm$62.4
NBM‡	15	9	6	50.1 $\pm$12.7	73.9 $\pm$11.9	407.2 $\pm$60.8

* MM is multiple myeloma, † MMc is multiple myeloma with chemotherapy, and ‡NBM is normal bone marrow.

2.2 Semi-automatic Region of Interest (ROI) Definition

The in-house software was developed using Matlab to define the required ROIs and to subsequently extract a series of SUVs from the corresponding PET images. The software first segments cortical bone from the CT component of the PETCT data. Segmentation was based upon thresholding of pixel values followed by a series of region filling and region growing steps, to ensure both the cortical bone and red marrow bone are part of the segmentation [7,12]. The red marrow was then further isolated from the original CT by subtracting the initial cortical bone segmentation from the segmentation with region filling. This results in a bone marrow binary mask that can be applied to the co-registered PET data through a pixel-wise matrix multiplication operation [14]. The initial stages in the creation of a bone marrow binary mask are shown in Fig. 1.

Next, the software prompts the user to place a series of large cuboid volumes of interest (VOI) over the various anatomical sites for evaluation [17]. This VOI placement was carried out in both the Sagittal and Coronal planes in order to localise the volume in 3-D space. The large VOI was of a fixed size and was used

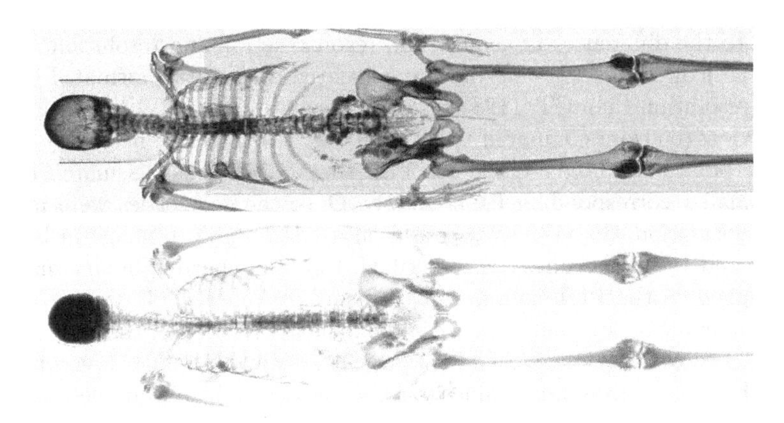

Fig. 1. There are several steps involved in the bone marrow segmentation process. The original image is used to create a binary mask based on the pixel values of cortical bone, this can be applied to the original image resulting in the segmentation of cortical bone (top). Through the application of a series of image processing steps the regions relating to bone marrow can be isolated resulting in a bone marrow binary mask (bottom) which can then be applied to the PET component.

as an initial positioning tool, the actual SUV measurement was obtained by a subsequent automatic placement of a standard spherical VOI (15 mm radius) over the max pixel within the larger volume. The SUVmax is frequently used clinically as it less susceptible to intra-observer variability.

However, SUVmax values have been found to be sensitive to image noise and have demonstrated poor reproducibility [15]. In order to account for this the SUVmedian was then calculated from the standard spherical VOI centred around the max pixel value which represents the SUVmax (Fig. 2).

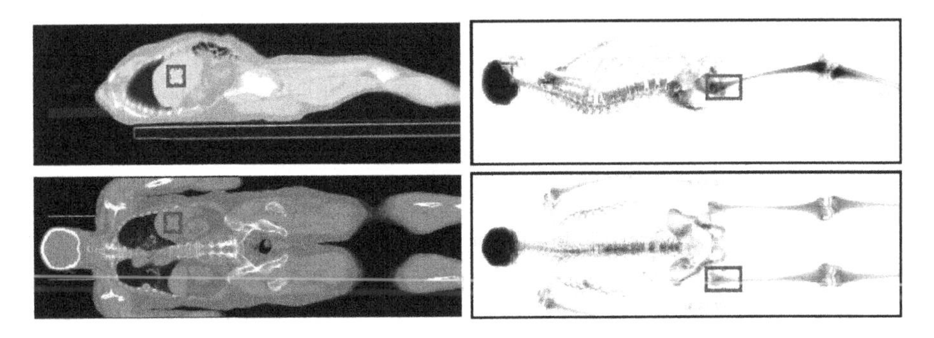

Fig. 2. The user must define the regions on images in two planes to establish the 3-D volume to be evaluated. The images to the left show volume placement for evaluation of the liver. The images to the right show volume placement for evaluation of the right proximal femur.

Due to the differences in acquisition, reconstruction and resolution, the CT and PET dicom dataset geometries do not match in their raw format. PET is of a lower resolution than CT, (128×128 compared to 512×512) but has a wider field of view (600 mm compared to 500 mm). It was therefore necessary to crop and interpolate the larger CT dataset so that the dimensions matched those of the smaller corresponding PET exactly. Once the geometries were matched image registration was visually assessed on a fused PETCT image in both the Sagittal and Coronal planes. Once verified, the bone marrow binary mask was then applied to the PET component resulting in a new PET dataset relating to bone marrow regions only. The results of the application of the bone marrow mask applied to PET images of a MM patient is shown in Fig. 3. The conversion of the PET images from pixel values to SUV was determined from metadata contained within the dicom header to account for the injected activity, radioactive isotope decay, and the patients weight. This study evaluated the bone marrow contained in the medullary cavities of the proximal femura, left and right iliac crests, the lumbar portion of the vertebral column, and the sternum. An internal control measurement of the uptake in the liver was also performed. Data from contra lateral sites were combined resulting in one dataset each for right/left femura and iliac crests.

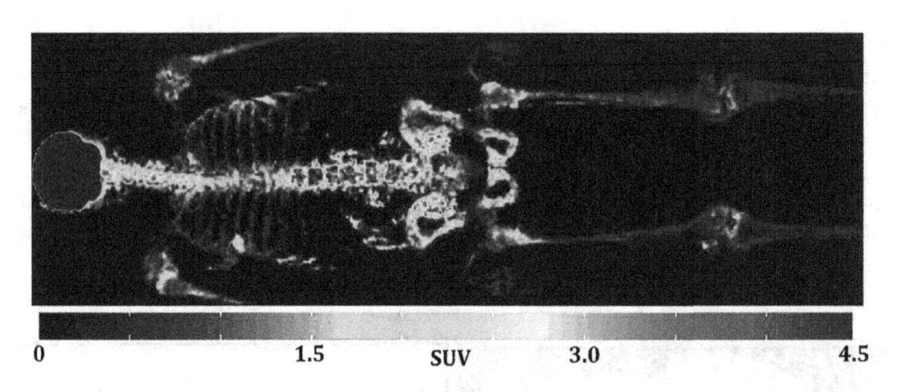

Fig. 3. The application of the bone marrow mask to the PET images allows for assessment of metabolism in bone marrow regions only. This is an example of metabolism for a patient with myeloma showing several regions of focal uptake.

3 Results

Analysis of the SUVmedian data was performed using Minitab 17 statistical software. Figure 4 shows boxplots of the results of the analysis for the various anatomical assessment sites for Myeloma, Myeloma with chemotherapy and Normal Bone Marrow groups. The control check of liver uptake across all groups was also performed but is not included as a boxplot, however the liver SUVs were within expected limits for all three groups - with the Myeloma with chemotherapy group demonstrating the highest mean of 2.9 ± 1.2, which was in close agreement for similar studies [9]. Both the Myeloma and Myeloma with chemotherapy groups showed consistently higher uptake across all bone marrow assessment regions. The highest differences were observed in the lumbar region, with the Myeloma and Myeloma with chemotherapy measuring a SUVmedian of 2.8, and 2.9 respectively compared to an SUVmedian of 1.9 observed in the normal bone marrow group. The Myeloma groups also demonstrated greater variation across all regions, reflecting the presence of more extreme values than the healthy bone marrow group. The Myeloma with chemotherapy group demonstrated slightly higher uptake values in comparison to the Myeloma group in all sites except the left iliac region.

Table 2. One-way ANOVA p-values (95% C.I.)

	Liver	Femura	Iliac crest	Lumbar	Sternum
MM & NBM	0.962	0.000	0.000	0.039	0.002
MMc & NBM	0.745	0.000	0.000	0.026	0.008
MM & MMc	0.727	0.225	0.863	0.376	0.810

A p-value <0.05 indicates a statistically significant difference with an assumption of equal variance between groups.

In order to determine if there were any significant differences between means at the assessment sites an analysis of variance (ANOVA) between each of the three groups was performed. The p-values for this are presented in Table 2. The liver did not show any significant differences between the groups which is to be expected as this is an internal control check which is included to confirm the validity of the SUV measurement itself. There were also no significant differences observed between the Multiple Myeloma and Multiple Myeloma with chemotherapy groups at any of the bone marrow assessment sites. The Normal Bone Marrow group did show significant differences across all bone marrow assessment sites when compared with both the Multiple Myeloma and Multiple Myeloma with chemotherapy groups.

4 Conclusions

A total of 45 patient datasets have been compiled, consisting of 18 Myeloma, 15 Myeloma with chemotherapy and 12 Non-myeloma and although the initial dataset is quite small, there are significant differences when a comparison of the SUVs at the various assessment sites of each group is made. Both the Myeloma groups demonstrated higher uptake in the bone marrow regions in comparison to the healthy bone marrow group. The more subtle differences observed between the Myeloma and Myeloma with chemotherapy groups may be an indicator of the impact of the chemotherapy drug on a patients immune response or normal bone marrow physiology.

The proposed method offers a more automated and objective approach for the segmentation of anatomical regions relating to bone marrow in comparison to manual contouring and segmentation. The inclusion of a fixed, systematic placement of large VOIs at the key anatomical sites ensures a highly structured approach to bone marrow assessment and ensures a thorough, yet efficient, evaluation of image data. The subsequent automatic placement of a standard spherical VOI over the maximum pixel value within the large VOI reduces the intra-observer variability that may result from manual placement, but also accounts for image noise errors that are associated with the sole reliance on a single maximum pixel value.

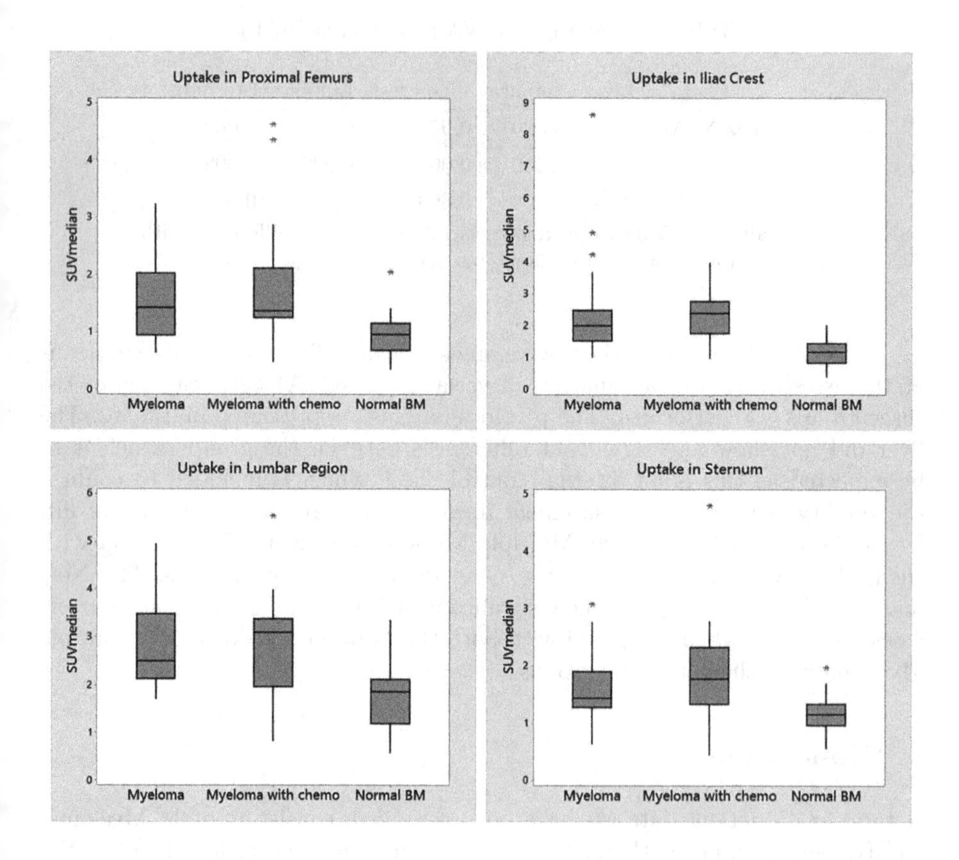

Fig. 4. Boxplots of the SUVs for anatomical sites assessed for both groups.

5 Discussion

This study focused on segmentation across the left and right proximal femura, left and right iliac crests, the lumbar portion of the vertebral column, and the sternum. Future work may also include other typical anatomical sites for assessment of MM patients such as, humeri, ribs, clavicles, and skull base. Accurate segmentation of bone marrow in these regions may prove more challenging due to the small size of the medullary cavities. The skull may be especially challenging as any lesions located in the skull are often quite small and the normal physiological activity observed in the brain tends to impair detectability [9]. The low image resolution of the PET component means that partial volume effects may lead to an underestimation of FDG uptake for small regions. Partial volume effects increase with decreasing image resolution, and need to be considered for lesions smaller than three times the full width at half max resolution of the PET scanner [1].

The simple, yet effective, pixel value thresholding approach to bone marrow segmentation had the advantages of being computationally inexpensive and easy to implement. Other studies have also demonstrated excellent segmentation of bone marrow using other algorithms such as Graph Cuts [10] and Active Contours [14], however due to the extra computational requirements may not be feasible time-wise for routine clinical evaluation.

Prior to progression, MM patients often go through an asymptomatic stage of monoclonal gammopathy of undetermined significance (MGUS), which is estimated to be prevalent in approximately 3–4% of those over the age of 50 [13]. There is also an intermediate stage between MGUS and MM termed as smouldering MM. It is in the accurate staging of these lower risk patients that the proposed system of bone marrow evaluation from PETCT will be most clinically beneficial, as these are the more difficult cases to diagnose accurately.

The outlined PET bone marrow segmentation method using a CT derived bone marrow mask also provides a useful means of visually assessing bone marrow activity in a particular scan. Although the application of this method to MM was the primary focus of this study, it may also prove useful when applied to other pathologies which cause changes in bone marrow metabolism when compared to normal physiological activity. The method could be used to develop a database of PETCT images against which quantitative statistical comparisons between patients with normal bone marrow metabolism and those with MM can be made and from a much larger dataset, machine learning classification software could be created to further assist in diagnosis by flagging cases with suspicious, or elevated, bone marrow activity [7].

Applying such techniques will remove the confounding variability introduced due to the subjective nature of ROI placement and, considering the large number of ROIs required per patient scan, this approach would greatly expedite the diagnosis of MM from PETCT. This would reduce the risk of patients developing bone disease complications, such as hypercalcemia and fracture, that often result from delays in treatment [11].

References

1. Boellaard, R., Oyen, W.J., Hoekstra, C.J., Hoekstra, O.S., Visser, E.P., Willemsen, A.T., Arends, B., Verzijlbergen, F.J., Zijlstra, J., Paans, A.M., et al.: The netherlands protocol for standardisation and quantification of FDG whole body pet studies in multi-centre trials. Eur. J. Nucl. Med. Mol. Imaging **35**(12), 2320–2333 (2008)
2. Bredella, M.A., Steinbach, L., Caputo, G., Segall, G., Hawkins, R.: Value of FDG pet in the assessment of patients with multiple myeloma. Am. J. Roentgenol. **184**(4), 1199–1204 (2005)
3. Durie, B., Salmon, S.E.: A clinical staging system for multiple myeloma correlation of measured myeloma cell mass with presenting. Cancer **36**(3), 842–854 (1975)
4. Durie, B.G.: The role of anatomic and functional staging in myeloma: description of Durie/Salmon plus staging system. Eur. J. Cancer **42**(11), 1539–1543 (2006)
5. Fechtner, K., Hillengass, J., Delorme, S., Heiss, C., Neben, K., Goldschmidt, H., Kauczor, H.U., Weber, M.A.: Staging monoclonal plasma cell disease: comparison of the Durie-Salmon and the Durie-Salmon plus staging systems 1. Radiology **257**(1), 195–204 (2010)
6. Hallett, W.A., Marsden, P.K., Cronin, B.F., O'doherty, M.J.: Effect of corrections for blood glucose and body size on [18f] FDG pet standardised uptake values in lung cancer. Eur. J. Nucl. Med. **28**(7), 919–922 (2001)
7. Martínez-Martínez, F., Kybic, J., Lambert, L., Mecková, Z.: Fully automated classification of bone marrow infiltration in low-dose CT of patients with multiple myeloma based on probabilistic density model and supervised learning. Comput. Biol. Med. **71**, 57–66 (2016)
8. Nakamoto, Y.: Clinical contribution of PET/CT in myeloma: from the perspective of a radiologist. Clin. Lymphoma Myeloma Leuk. **14**(1), 10–11 (2014)
9. Nanni, C., Zamagni, E., Versari, A., Chauvie, S., Bianchi, A., Rensi, M., Bellò, M., Rambaldi, I., Gallamini, A., Patriarca, F., et al.: Image interpretation criteria for FDG PET/CT in multiple myeloma: a new proposal from an italian expert panel. IMPeTUs (Italian myeloma criteria for pet use). Eur. J. Nucl. Med. Mol. Imaging **43**(3), 414–421 (2016)
10. Nguyen, C., Havlicek, J., Duong, Q., Vesely, S., Gress, R., Lindenberg, L., Choyke, P., Chakrabarty, J.H., Williams, K.: An automatic 3D CT/PET segmentation framework for bone marrow proliferation assessment. In: 2016 IEEE International Conference on Image Processing (ICIP), pp. 4126–4130. IEEE (2016)
11. Pianko, M.J., Terpos, E., Roodman, G.D., Divgi, C.R., Zweegman, S., Hillengass, J., Lentzsch, S.: Whole-body low-dose computed tomography and advanced imaging techniques for multiple myeloma bone disease. Clin. Cancer Res. **20**(23), 5888–5897 (2014)
12. Puri, T., Blake, G.M., Curran, K.M., Carr, H., Moore, A.E., Colgan, N., O'Connell, M.J., Marsden, P.K., Fogelman, I., Frost, M.L.: Semiautomatic region-of-interest validation at the femur in 18f-fluoride pet/ct. J. Nucl. Med. Technol. **40**(3), 168–174 (2012)
13. Rajkumar, S.V., Dimopoulos, M.A., Palumbo, A., Blade, J., Merlini, G., Mateos, M.V., Kumar, S., Hillengass, J., Kastritis, E., Richardson, P., et al.: International myeloma working group updated criteria for the diagnosis of multiple myeloma. Lancet Oncol. **15**(12), e538–e548 (2014)

14. Sambuceti, G., Brignone, M., Marini, C., Massollo, M., Fiz, F., Morbelli, S., Buschiazzo, A., Campi, C., Piva, R., Massone, A.M., et al.: Estimating the whole bone-marrow asset in humans by a computational approach to integrated PET/CT imaging. Eur. J. Nucl. Med. Mol. Imaging **39**(8), 1326–1338 (2012)
15. Sattarivand, M., Caldwell, C., Poon, I., Soliman, H., Mah, K.: Effects of ROI placement on pet-based assessment of tumor response to therapy. International J. Mol. Imaging **2013** (2013)
16. Touzeau, C., Moreau, P.: Multiple myeloma imaging. Diagn. Intervent. Imaging **94**(2), 190–192 (2013)
17. Valadares, A.A., Duarte, P.S., Carvalho, G., Ono, C.R., Coura-Filho, G.B., Sado, H.N., Sapienza, M.T., Buchpiguel, C.A.: Receiver Operating Characteristic (ROC) curve for classification of 18F-NAF uptake on PET/CT. Radiol. Bras. **49**(1), 12–16 (2016)

Mammography Image Analysis

A Texton-Based Approach for the Classification of Benign and Malignant Masses in Mammograms

Zobia Suhail[1], Azam Hamidinekoo[1], Erika R.E. Denton[2], and Reyer Zwiggelaar[1(✉)]

[1] Department of Computer Science, Aberystwyth University, Aberystwyth, UK
rrz@aber.ac.uk
[2] Department of Radiology, Norfolk and Norwich University Hospitals NHS Foundation Trust, Norwich, UK

Abstract. Classification of benign and malignant masses in mammograms is a complex task due to the appearance similarities in both classes. Thus, classification of masses in mammograms is considered an important step in the development of current Computer Aided Diagnosis (CAD) systems. In this paper, we present a way to classify masses without the need for segmentation. A supervised texton-based approach is developed using filter bank responses. Subsequently, a Support Vector Machine (SVM) classifier is used to classify the images. We evaluated the results on a subset of publicly available dataset (DDSM) and obtained classification accuracy of 96% which is comparable to the state-of-the-art techniques developed for the task of mammographic mass classification.

Keywords: Classification · Mammogram · Masses · Computer Aided Diagnosis

1 Introduction

Breast cancer is considered to be one of the most common types of cancer and the second leading cause of death after lung cancer [10]. Unfortunately, the actual reason of breast cancer is not known although there are some known risk factors associated with breast cancer [10,15] in order to identify women at increased risk, but there is still a question-mark on the actual cause of breast cancer. Mammography is considered to be one of the most reliable modalities used to screen the human breast [9,14]. Mammography scans (mammograms) are used to identify the abnormalities at an early stage when it is more treatable. Calcifications and masses are two of the most commonly found abnormalities. Unlike calcifications that appear as more bright pixels compared to their surroundings [18,21], mammographic masses are difficult to detect due to their variation in shape and structural properties [6].

M. Valdés Hernández and V. González-Castro (Eds.): MIUA 2017, CCIS 723, pp. 355–364, 2017.
DOI: 10.1007/978-3-319-60964-5_31

Research has focused on classification of both micro-calcification [4,5,20] and characteristics masses in mammograms [12,19,22]. Huo et al. [12] proposed a method for the classification of masses by automating the segmentation, feature-extraction and characterisation of mass types. Similar work in [19] follows a three steps procedure for the classification of benign and malignant masses. Segmentation of the Region of Interest (RoI) is used with a group of 32 Zernike moments that were extracted to train a neural network. In [22] Vaidehi and Subashini proposed a method using Gray Level Co-occurrence Matrix (GLCM) features including contrast, correlation, energy and homogeneity to classify benign and malignant masses. After RoI extraction, a fuzzy C-mean technique is used to extract GLCM features at a distance equal to 1 and four orientations ($0°$, $45°$, $90°$ and $135°$). They reported the results using three different classifiers: AdaBoost, Back Propagation Neural Network (BPNN) and Sparse Representation based Classifier (SRC).

According to Breast Imaging Reporting and Data Systems (BI-RADS) [1], the shape of the mass can be used as a way of defining the mass as benign or malignant (See Table 1). Accordingly, if a mass appears as an irregular shape, it can be considered liekly to be malignant. This has been used to explore the relationship between the shape of the mammographic mass and the related category in order to achieve good classification results [7,17,23]. Valarmathie et al. [23] used shape, margin and textural features from the segmented mass area and report the best results using mass and shape features. In similar work, Rangayyan et al. [17] proposed a region-based measure of image edge profile acutance and use it to differentiate between benign and malignant masses. In addition, they investigated several other shape features like compactness, Fourier descriptors, moments, and chord-length in order to classify between circumscribed and spiculated tumours. Ertas et al. [7] computed the geometric parameters (area, perimeter, circularity, normalised circularity, Fourier descriptors, etc.) for the basic shape of the segmented mass area and reported that normalised circulatory area and Fourier coefficients can be used more effectively in order to classify benign and malignant masses.

Table 1. BIRADS mass characterisation based on shape

Shape	Likely benign	Suspicious	Highly suspicious of malignancy
Round	X		
Oval	X		
Lobular	X	X	
Irregular			X

Instead of selecting features from the segmented mass area, it is possible to extract eatures from RoIs in order to classify the abnormality as benign or malignant [2,8]. In [8], a CAD system for three different types of mammogram patches i.e. normal, benign and malignant is proposed. The Local Configuration

Pattern (LCP) algorithm was used for feature extraction which are then concatenated with statistical and frequency domain features. They reported the results of classifying new ensemble features using four different classifiers. Another three class classification method has been proposed in [2], in which the directional features are extracted by filtering image patches using Gabor wavelets. After reducing the data dimension (for both filtered and original mammogram patches) using Principal Component Analysis (PCA), they reported better classification results when Gabor features were used instead of using original mammogram images. Recently, texton-based approaches have been used effectively for the texture classification using single images [25], where the textures are represented as histograms of texton contained in the texton dictionary. In the original work [25] for textons, Varma and Zisserman used filter responses in order to build texton dictionary. Whereas in later work [24], they demonstrate that textures can be classified using joint distribution of intensity values which can outperform the classification results achieved using filter responses. In medical images, the application of a texton-based approach has been found in prostate images for the task of characterising benign and malignant tissue [16], modelling mammographic tissue [3], retinal vessel segmentation [26] and classifying masses in mammograms as benign/malignant [13].

We propose a method for the classification of benign and malignant masses in mammograms using a texton-based approach. Unlike the texton-based work proposed in the literature [13], where the modified version of the texton-based approach is used, we use the original version of the work proposed in [25]. Filter banks are used to construct the texton dictionary. Initial study reveals a good classification results and is comparable to state of the art.

2 Dataset

The dataset used in this paper is the publicly available Digital Database for Screening Mammography (DDSM) [11], provided by the University of South Florida, which is a wide ranging annotated film mammography repository. It provides approximately 2620 studies of both mediolateral oblique (MLO) and cranio caudal (CC) views of each breast provided with a lesion mask if a lesion is present. In our experiment, we considered mass lesions and used a subset of 400 cases, comprising 200 benign and 200 malignant image lesions. DDSM images are grey-level images with the bit depth of 16 bits per pixel. The cases were randomly selected from the whole dataset. Figure 1 represents benign and malignant examples from this database accompanied by their respective annotations. Using the annotated RoI, a bounding box of the mass was extracted and subsequently all the lesion patches were resized into 256×256 pixels as shown in Fig. 1. Additionally, Fig. 2 shows various typical patches used in our experiment from both benign and malignant classes.

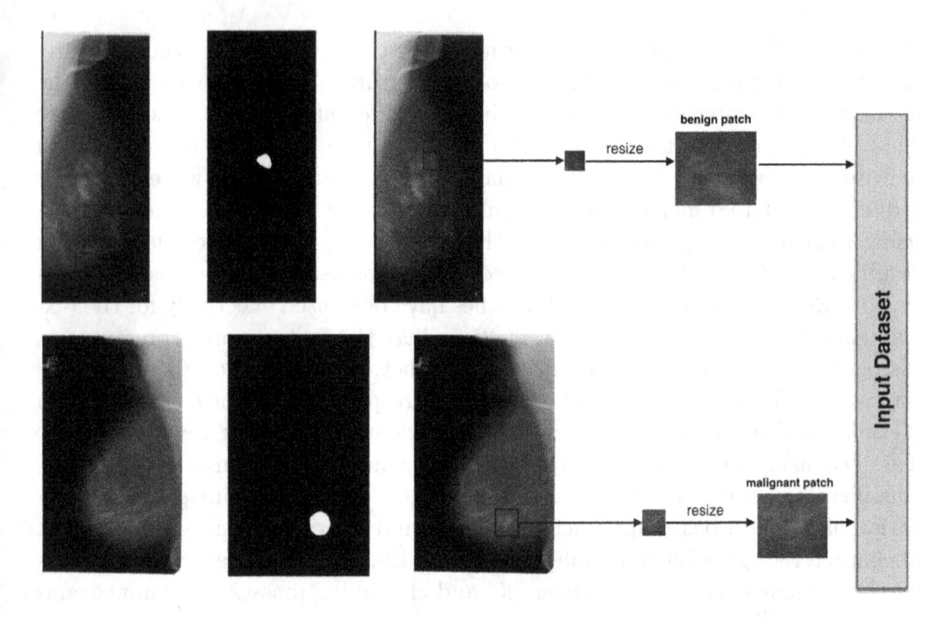

Fig. 1. Process of extracting patches from mammographic images containing abnormality, top row shows an example of a mammogram containing a benign abnormality, the annotation, bounding box around the mass, the extracted lesion and the resized RoI. The bottom row represents the same process of RoI selection for malignant mass.

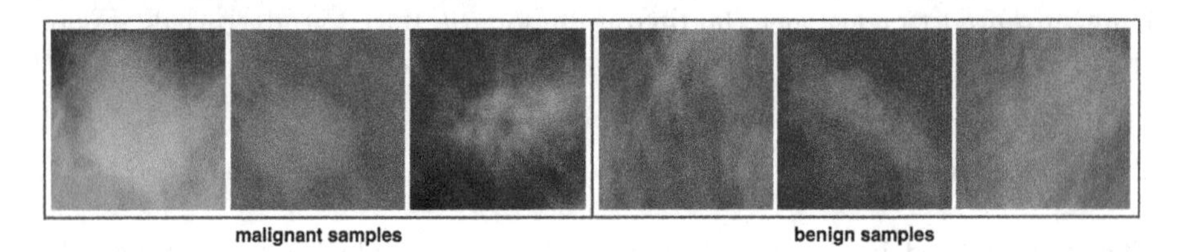

Fig. 2. Samples of malignant and benign mammogram patches from the reference database.

3 Current Approach

We used a texton-based approach for the classification of benign and malignant masses in mammograms using statistical distribution filter responses. We convolve 53 different filters with the training images to get textons. The detail of the proposed method can be found in subsequent sections. After getting the filter responses and applying some preprocessing, we proceed to the learning phase which is composed of two steps; Step I is generating the texton dictionary and Step II is model generation. The last stage we report are the classification results for benign and malignant mammographic masses by using a Support Vector Machine (SVM) as classifier.

3.1 Getting Filter Response

As the first step, we pick the images from both classes and convolve them with filters. In this study, we have selected a combination of 53 filters from the proposed Maximum Response (MR) and Schmid (S) filter banks [25][1] which are depicted in Fig. 3. The first 38 filters are obtained from the MR8 filter bank and is a mixture of an edge filter at six orientations and three scales $((\sigma_x, \sigma_y) = (1,3), (2,6), (4,12))$, a bar filter at the same orientations and scales, a Gaussian and a Laplacian of Gaussian at $\sigma = 10$. The other 13 filters are obtained from the S filter bank which consists of various isotropic Gabor-like filters of the form:

$$F(r,\sigma,\tau) = F_0(\sigma,\tau) + cos(\frac{\pi\tau r}{\sigma})e^{-\frac{r^2}{2\sigma}} \quad (1)$$

with: $(\sigma,\tau) = \{(2,1),(4,1),(4,2),(6,1),(6,2),(6,3),(8,1),(8,2),(8,3),(10,1),(10,2),(10,3),(10,4)\}$. The characteristics of these selected filters are: rotationally invariant, isotropic and anisotropic, multi-scale and multiple orientations which make the filters capable of generating good features for various types of mass textures. The applied results of the mentioned filters on one of the patches from the input dataset is shown in Fig. 4.

Before going to the subsequent steps (training and classification), we normalized all the filter responses to zero mean and unit variance. This standardisation will take all the filter responses to the same range. Experiment showed that this preprocessing improves the overall classification results.

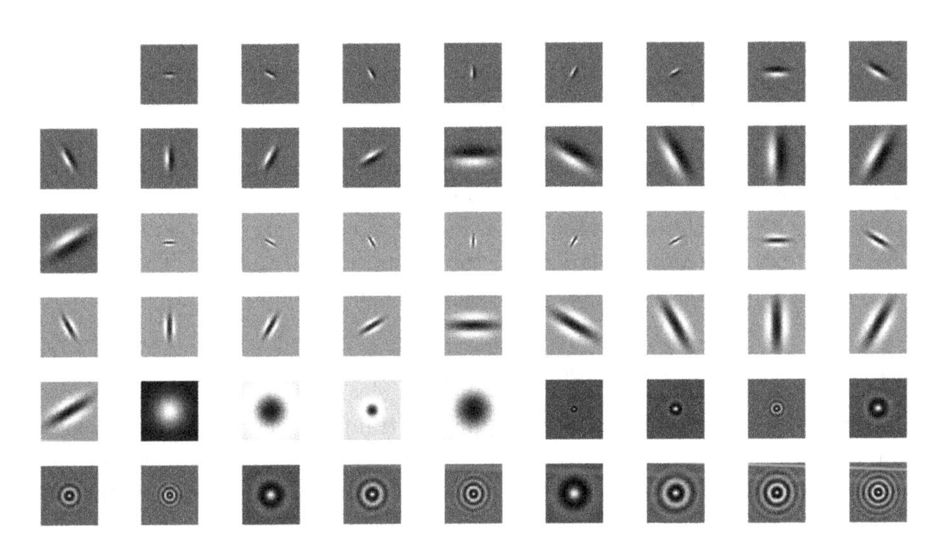

Fig. 3. Combination of MR and S filter banks.

[1] http://www.robots.ox.ac.uk/~vgg/research/texclass/filters.html.

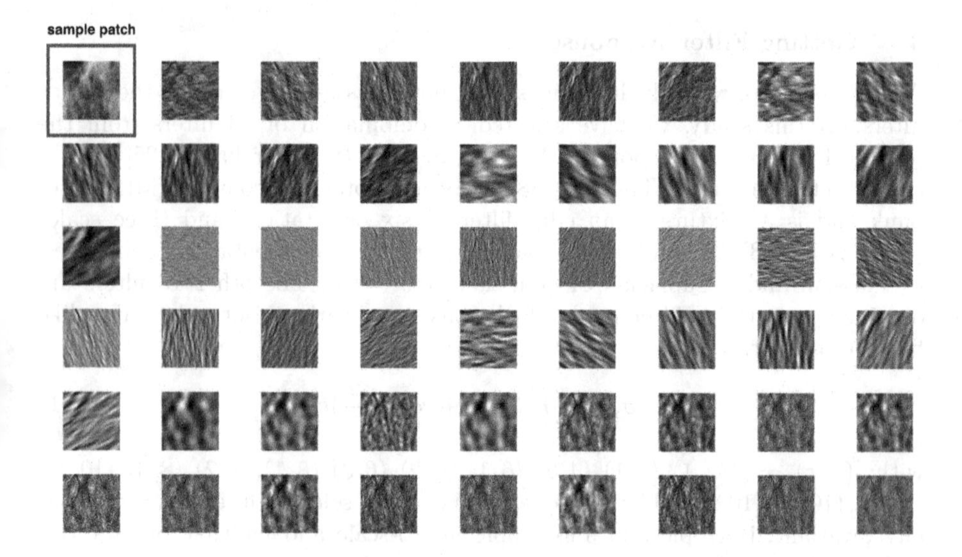

Fig. 4. Filter responses on a sample image from the input dataset.

3.2 Learning: Step I [Generating Texton Dictionary]

Training stage starts with making a texton dictionary. After getting filter responses for all the input RoIs (53 response images per RoI) for each class, we apply K-means clustering on the filter responses to define the clusters for each class separately. At this stage, we define 15 clusters per class i.e. for benign and malignant. Experiments shows that 15 clusters is an appropriate cluster representation for 200 samples/class (some more results on the different setting of number of clusters could be found in Sect. 3.4). Therefore, 15 clusters for benign and 15 for malignant class has been defined based on the filter responses. The cluster centroids are referred to as textons. In this way, we get 15 textons for benign and 15 for the malignant class and aggregating the textons from both classes, a texton dictionary is formed that composes 30 textons. The detailed steps involved in creating texton for each class can be found in Fig. 5.

3.3 Learning: Step II [Model Generation]

Step II of the learning process is also a part of training, where a model is generated for each training image using the texton dictionary. For each image (both for benign and malignant class) a texton histogram will be computed. As explained in Sect. 3.2 each image is convolved with 53 filters and the filter responses for all the images in the dataset have been generated. Each filter response has been assigned to the closest texton from the texton dictionary. In order to measure the distance between two textons, the Euclidean distance between each image response and the texton is calculated. After assigning all image responses to textons, the frequency histogram for all the textons occurrences in a ROI is

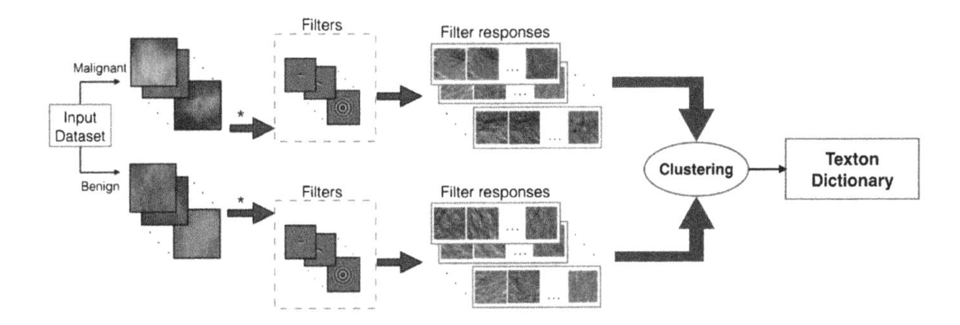

Fig. 5. Process of generating texton dictionary.

computed. The final model consists of normalised frequency histograms for filter responses which represents the probabilistic distribution of textons. Figure 6 is showing the steps required to generate the model for the texton-based approach.

3.4 Classification

The normalised texton frequency histograms are utilised as features for classifying a ROI as benign or malignant. A Support Vector Machine (SVM) is used for classifying the images. As explained in Sect. 2, 200 images for the benign class and 200 for the malignant class are used for this experiment. In addition, we took 1 lesion/patient to make sure no sample image for the same patient is used for training and testing phase (as in DDSM database more than one mass annotations are available for the same patient). Using SVM (10 Fold Cross-Validation) as a classier, the accuracy of 96% is achieved by using histogram of textons frequency map as the feature set. From 200 benign images, 192 are correctly classified as benign. Whereas, for malignant image, 193 samples are classified as malignant from total of 200 images. Applying 10 runs of 10 Fold Cross-Validation (10FCV), again the accuracy of 96% (with standard deviation 2.98) maintained that shows the stability of the developed model.

The results of texton-based classification depends on the number of clusters or the size of the texton dictionary used during the learning phase of texton dictionary. We evaluated the proposed method on cluster size 10 and 5 for each of benign and malignant class in order to show the effect of different cluster sizes on the overall classification accuracy (in addition to cluster size 15 that is used for the experiments). The classification accuracy is 92% when 10 clusters per class were used and 89% for 5 clusters per class. Classification accuracy increases to 96% by increasing the size of texton dictionary to 15 textons per class. This also verifies the argument of Varma and Zisserman [25] that the performance of the classification using texton methods improves by using largest texton dictionaries. On the contrary, the algorithm takes more time to execute by increasing the number of clusters, therefore keeping in view the time and memory requirements we keep the cluster size to 15 without compromising the classification accuracy.

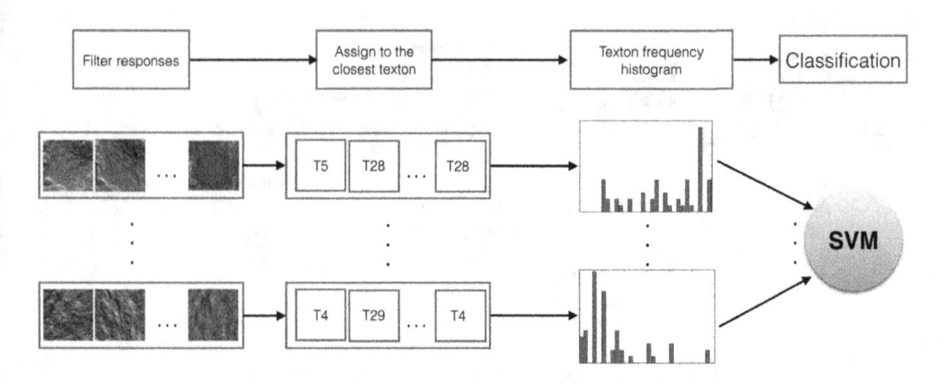

Fig. 6. Process of generating a texton-based model.

4 Future Work

According to texton-based work proposed in the past [24] for texture classification, image intensity values can be used for generating textons instead of using filter responses. Such intensity based textons have been investigated in the past for medical images [3,13,16] in order to achieve good classification results. In the future, we will try to implement the approach proposed by Varma and Zisserman [24] in which they used the intensities to generate the texton dictionary. In addition, in the current method, we used the whole mass patch to extract the filter responses, in future we will try to investigate the texton-based approach for the classification of benign and malignant masses in mammograms by using mass patches (patches of size 3×3, 5×5.... from the mass area) as input and then compute the joint probability map for the whole image. Previous study revealed a considerable effect of the size of the patch for using image sub-patches for texton-based approach [16]. We will also try to investigate the effect of using different patch sizes on the overall classification performance.

5 Conclusion

In this study, a supervised texton-based approach is utilised to classify pre-detected mass abnormalities in mammograms as benign or malignant. Accordingly, filter band responses on a specific input patch size are extracted to construct the texton dictionary. Subsequently, normalised texton frequency histograms are used as features for a SVM classifier. We evaluated the results on a subset of DDSM dataset and obtained the accuracy of 96% which is comparable to the state-of-the-art results from the techniques developed for this task so far.

References

1. Breast Imaging Reporting and Data System atlas (BI-RADS atlas). Reston, VA: American College of Radiology 98 (2003)
2. Buciu, I., Gacsadi, A.: Directional features for automatic tumor classification of mammogram images. Biomed. Sig. Process. Control **6**(4), 370–378 (2011)
3. Chen, Z., Denton, E., Zwiggelaar, R.: Modelling breast tissue in mammograms for mammographic risk assessment. In: MIUA, pp. 37–42 (2011)
4. Chen, Z., Oliver, A., Denton, E., Boggis, C., Zwiggelaar, R.: Classification of microcalcification clusters using topological structure features. In: Medical Image Understanding and Analysis, pp. 37–42 (2012)
5. Chen, Z., Strange, H., Oliver, A., Denton, E., Boggis, C., Zwiggelaar, R.: Topological modeling and classification of mammographic microcalcification clusters. IEEE Trans. Biomed. Eng. **62**(4), 1203–1214 (2015)
6. Djaroudib, K., Ahmed, A., Zidani, A.: Textural approach for mass abnormality segmentation in mammographic images. arXiv preprint arXiv:1412.1506 (2014)
7. Ertas, G., Gulcur, H., Aribal, E., Semiz, A.: Feature extraction from mammographic mass shapes and development of a mammogram database. In: Proceedings of the 23rd Annual International Conference of the IEEE Engineering in Medicine and Biology Society, vol. 3, pp. 2752–2755. IEEE (2001)
8. Esener, İ.I., Ergin, S., Yüksel, T.: A new ensemble of features for breast cancer diagnosis. In: 2015 38th International Convention on Information and Communication Technology, Electronics and Microelectronics (MIPRO), pp. 1168–1173. IEEE (2015)
9. Feig, S., Shaber, G., Schwartz, G., Patchefsky, A., Libshitz, H., Edeiken, J., Nerlinger, R., Curley, R., Wallace, J.: Thermography, mammography, and clinical examination in breast cancer screening: review of 16,000 studies 1. Radiology **122**(1), 123–127 (1977)
10. Ferlay, J., Héry, A., Autier, P., Sankaranarayanan, R.: Global burden of breast cancer. In: Li, C. (ed.) Breast Cancer Epidemiol., pp. 1–19. Springer, Heidelberg (2010). doi:10.1007/978-1-4419-0685-4_1
11. Heath, M., Bowyer, K., Kopans, D., Moore, R., Kegelmeyer, W.P.: The digital database for screening mammography. In: Proceedings of the 5th International Workshop on Digital Mammography, pp. 212–218. Medical Physics Publishing (2001)
12. Huo, Z., Giger, M., Vyborny, C., Wolverton, D., Metz, C.: Computerized classification of benign and malignant masses on digitized mammograms: a study of robustness. Acad. Radiol. **7**(12), 1077–1084 (2000)
13. Li, Y., Chen, H., Rohde, G., Yao, C., Cheng, L.: Texton analysis for mass classification in mammograms. Pattern Recogn. Lett. **52**, 87–93 (2015)
14. Lissner, J., Kessler, M., Anhalt, G., Hahn, D., Wendt, T., Seiderer, M.: Developments in methods for early detection of breast cancer. In: Zander, J., Baltzer, J. (eds.) Early Breast Cancer, pp. 93–112. Springer, Heidelberg (1985). doi:10.1007/978-3-642-70192-4_11
15. Maass, H.: Identification of a high-risk population. In: Zander, J., Baltzer, J. (eds.) Early Breast Cancer, pp. 87–92. Springer, Heidelberg (1985). doi:10.1007/978-3-642-70192-4_10
16. Rampun, A., Tiddeman, B., Zwiggelaar, R., Malcolm, P.: Computer aided diagnosis of prostate cancer: a texton based approach. Med. Phys. **43**(10), 5412–5425 (2016)

17. Rangayyan, R., El-Faramawy, N., Desautels, J., Alim, O.: Measures of acutance and shape for classification of breast tumors. IEEE Trans. Med. Imaging **16**(6), 799–810 (1997)
18. Sahakyan, A., Sarukhanyan, H.: Segmentation of the breast region in digital mammograms and detection of masses. Int. J. Adv. Comput. Sci. Appl. **3**(2) (2012)
19. Serifovic-Trbalic, A., Trbalic, A., Demirovic, D., Prljaca, N., Cattin, P.: Classification of benign and malignant masses in breast mammograms. In: 2014 37th International Convention on Information and Communication Technology, Electronics and Microelectronics (MIPRO), pp. 228–233. IEEE (2014)
20. Strange, H., Chen, Z., Denton, E.E., Zwiggelaar, R.: Modelling mammographic microcalcification clusters using persistent mereotopology. Pattern Recogn. Lett. **47**, 157–163 (2014)
21. Suhail, Z., Sarwar, M., Murtaza, K.: Automatic detection of abnormalities in mammograms. BMC Med. Imaging **15**(1), 53 (2015)
22. Vaidehi, K., Subashini, T.: Automatic characterization of benign and malignant masses in mammography. Procedia Comput. Sci. **46**, 1762–1769 (2015)
23. Valarmathie, P., Sivakrithika, V., Dinakaran, K.: Classification of mammogram masses using selected texture, shape and margin features with multilayer perceptron classifier. Biomed. Res. (2016)
24. Varma, M., Zisserman, A.: Texture classification: are filter banks necessary? In: Proceedings of 2003 IEEE computer society conference on Computer vision and pattern recognition, vol. 2, pp. II-691. IEEE (2003)
25. Varma, M., Zisserman, A.: A statistical approach to texture classification from single images. Int. J. Comput. Vis. **62**(1–2), 61–81 (2005)
26. Zhang, L., Fisher, M., Wang, W.: Retinal vessel segmentation using gabor filter and textons. Med. Image Underst. Anal. (MIUA 2014), 155–160 (2014)

Breast Density Classification Using Multiresolution Local Quinary Patterns in Mammograms

Andrik Rampun[1(✉)], Philip Morrow[1], Bryan Scotney[1], and John Winder[2]

[1] School of Computing and Information Engineering, Ulster University, Coleraine BT52 1SA, Northern Ireland, UK
{y.rampun,pj.morrow,bw.scotney}@ulster.ac.uk
[2] School of Health Sciences, Institute of Nursing and Health, Ulster University, Newtownabbey BT37 0QB, Northern Ireland, UK
rj.winder@ulster.ac.uk

Abstract. This paper presents a method for breast density classification using local quinary patterns (LQP) in mammograms. LQP operators are used to capture the texture characteristics of the fibroglandular disk region (FGD_{roi}) instead of the whole breast region as the majority of current studies have done. To maximise the local information, a multiresolution approach is employed followed by dimensionality reduction by selecting dominant patterns only. Subsequently, the Support Vector Machine classifier is used to perform the classification and a stratified ten-fold cross-validation scheme is employed to evaluate the performance of the method. The proposed method produced competitive results up to 85.6% accuracy which is comparable with the state-of-the-art in the literature. Our contributions are two fold: firstly, we show the role of the fibroglandular disk area in representing the whole breast region as an important region for more accurate density classification and secondly we show that the LQP operators can extract discriminative features comparable with the other popular techniques such as local binary patterns, textons and local ternary patterns (LTP).

1 Introduction

In 2014 there were more than 55,000 malignant breast cancer cases diagnosed in the United Kingdom (UK) with more than 11,000 mortality [1]. In the United States (US), an estimate of more than 246,000 malignant breast cancer were expected to be diagnosed in 2016 with approximately 16% women expected to die [2]. Many studies have indicated that breast density is a strong risk factor for developing breast cancer [3–5, 7–13] because dense tissues are very similar in appearance to breast cancer this making it more difficult to detect in mammograms. Therefore, women with dense breasts can be six times more likely to develop breast cancer, which means an accurate breast density estimation is an important step during the screening procedure. Although most

M. Valdés Hernández and V. González-Castro (Eds.): MIUA 2017, CCIS 723, pp. 365–376, 2017.
DOI: 10.1007/978-3-319-60964-5_32

experienced radiologists can do this task, manual classification is impractical, tiring, time consuming and often suffers from results variability among radiologists. Computer Aided Diagnosis (CAD) systems can reduce these problems providing robust, reliable, fast and consistent diagnosis results. Based on the Breast Imaging Reporting and Data System (BI-RADS), there are four major categories used for classifying breast density: (a) predominantly fat, (b) fat with some fibroglandular tissue, (c) heterogeneously dense and (d) extremely dense. These representations are illustrated in Fig. 1.

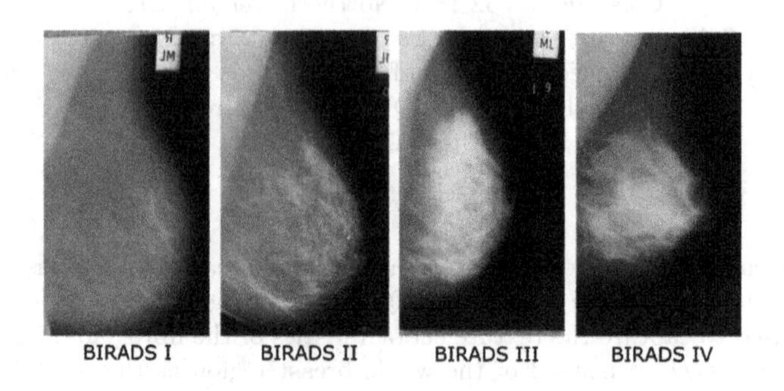

Fig. 1. Mammograms with different breast densities

One of the earliest approaches to breast density assessment was a study by Boyd *et al.* using interactive thresholding known as Cumulus, where regions with dense tissue were segmented by manually tuning the greylevel threshold value. The most popular approaches are based on the first and second-order (e.g. Grey Level Co-occurrence Matrix) statistical features as used by Oliver *et al.* [3], Bovis and Singh [4], Muštra *et al.* [6] and Parthaláin *et al.* [7]. Texture descriptors such as local binary patterns (LBP) were employed in the study of Chen *et al.* [8] and Bosch *et al.* [9] and textons were used by Chen *et al.* [8], Bosch *et al.* [9] and Petroudi *et al.* [13]. Other texture descriptors also have been evaluated such as fractal-based [5,11], topography-based [10], morphology-based [3] and transform-based features (e.g. Fourier, Discrete Wavelet and Scale-invariant feature) [4,9].

Many breast density classification methods in mammograms have been proposed in the literature but only very small number of studies have achieved accuracies above 80%. The methods of Oliver *et al.* [3] and Parthaláin *et al.* [7] extract a set of features from dense and fatty tissue regions segmented using a fuzzy c-means clustering technique followed by feature selection before feeding them into the classifier. Oliver *et al.* [3] achieved 86% accuracy and Parthaláin *et al.* [7] who used a sophisticated feature selection framework achieved 91.4% accuracy. Bovis and Singh [4] achieved 71.4% accuracy based on a combined classifier paradigm in conjunction with a combination of features extracted using the Fourier and Discrete Wavelet transforms, and first and second-order statistical features. Chen *et al.* [8] made a comparative study on the performance of local binary patterns (LBP), local greylevel appearance (LGA), textons and basic image features (BIF) and reported accuracies of 59%, 72%, 75% and 70%, respectively. Later, they proposed a method by modelling the distribution of the dense region in topographic representations and reported a slightly higher accuracy of 76%. Petroudi *et al.* [13] implemented the textons approach based on the Maximum Response 8 (MR8) filter bank. The χ^2 distribution was used to compare each of the resulting histograms from the training set to all the learned histogram models from the training set and reported 75.5% accuracy. He *et al.* [12] achieved an accuracy of 70% using the relative proportions of the four Tabár's building blocks. Muštra *et al.* [6] captured the characteristics of the breast region using multi-resolution of first and second-order statistical features and reported 79.3% accuracy.

Based on the results reported in the literature, the majority of the proposed methods achieved below 80% which indicate that breast density classification is a difficult task due to the complexity of tissue appearance in the mammograms such as a wide variation and obscure texture patterns within the breast region. In this paper, texture features were extracted from the FGD_{roi} only (see Fig. 4 later) to obtain more descriptive information instead of from the whole breast region. The motivation behind this approach is that because in most cases the non FGD_{roi} contains mostly fatty tissues regardless its BI-RADS class and most dense tissues are located and start to develop within the FGD_{roi}. Therefore, extracting features from the whole breast region means extracting overlapping texture information which makes the extracted features less discriminant in corresponding to the BIRADS classes. The reminder of the paper is organised as follows: We present the technical aspects of our proposed method in Sect. 2 and discuss experimental results in Sect. 3 which covers the quantitative evaluation and comparisons, Sect. 4 presents conclusions and future work.

2 Methodology

Figure 2 shows an overview of the proposed methodology.

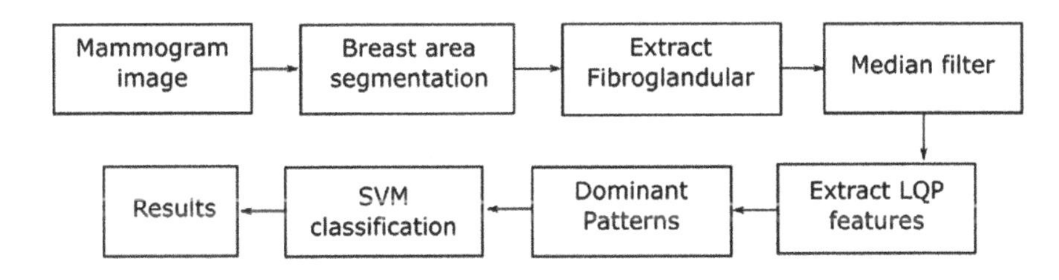

Fig. 2. An overview of the proposed breast density methodology

Firstly, we segment the breast area and estimate the FGD_{roi} from this breast region. Subsequently, we use a simple median filter using a 3×3 window size for noise reduction and employ the multiresolution LQP operators to capture the micro-structure information within the FGD_{roi}. Finally, we train the SVM classifier to build a predictive model and use it to test each unseen case.

2.1 Pre-processing

To segment the breast and pectoral muscle region, we used the method in [18] which is based on Active Contours without edges for the breast boundary estimation and contour growing with edge information for the pectoral muscle boundary estimation. The left most image in Fig. 4 shows the estimated FGD_{roi} area. To extract FGD_{roi}, we find B_w which is the longest perpendicular distance between the y-axis and the breast boundary (magenta line). The width and the height of the square area of the FGD_{roi} (amber line Fig. 4) can be computed as $B_w \times B_w$ with the center located at the intersection point between B_h and B_w lines. B_h is the height of the breast which is the longest perpendicular distance between the x-axis and the breast boundary. B_h is then relocated to the middle of B_w to get the intersection point. The size of the FGD_{roi} varies depending on the width of the breast.

2.2 Feature Extraction

The Local Binary Pattern (LBP) operators were first proposed by Ojala *et al.* [14] to encode pixel-wise information. Tan and Triggs [16] modified it by introducing LTP operators which thresholds the neighbouring pixels using a three-value encoding system based on a threshold constant set the user. Later, Nanni *et al.* [15] introduced a five-value encoding system called LQP. The LBP, LTP and LQP are similar in terms of architecture as each are defined using a circle centered on each pixel and a number of neighbours but the main difference is that the LBP, LTP and LQP threshold the neighbouring pixels into two (1 and 0), three (−1, 0 and 1) and five (2, 1, 0, −1 and −2) values, respectively. This means the difference between the grey level value of the center pixel (g_c) and a neighbour's grey level (g_p) can assume five values. The value of LQP code of the pixel (i, j) is given by:

$$LQP_{(P,R)}^{pattern}(i,j) = \sum_{p=0}^{(P-1)} s_{pattern}(g_p)2^p \tag{1}$$

$$s(x) = \begin{cases} 2, & x \geq g_c + \tau_2 \\ 1, & g_c + \tau_1 \leq x < g_c + \tau_2 \\ 0, & g_c - \tau_1 \leq x < g_c + \tau_1 \\ -1, & g_c - \tau_2 \leq x < g_c - \tau_1 \\ -2, & \text{otherwise} \end{cases} \tag{2}$$

where R is the circle radius, P is the number of pixels in the neighbourhood, g_c is the grey level value of the center pixel, g_p is the grey level value of the

p^{th} neighbour, and *pattern* $\in \{1, 2, 3, 4\}$. Once the LQP code is generated, it is split into four binary patterns by considering its positive, zero and negative components, as illustrated in Fig. 3 using the following conditions

$$s_1(x) = \begin{cases} 1, & \text{if } s(x) = 2 \\ 0, & \text{otherwise} \end{cases} \tag{3}$$

$$s_2(x) = \begin{cases} 1, & \text{if } s(x) = 1 \\ 0, & \text{otherwise} \end{cases} \tag{4}$$

$$s_3(x) = \begin{cases} 1, & \text{if } s(x) = -1 \\ 0, & \text{otherwise} \end{cases} \tag{5}$$

$$s_4(x) = \begin{cases} 1, & \text{if } s(x) = -2 \\ 0, & \text{otherwise} \end{cases} \tag{6}$$

In this section, feature extraction is the process of computing the frequencies of all binary patterns and present the occurrences in a histogram which represents the number of appearances of edges, corners, spots, lines, etc. within the FGD_{roi}. This means, the size of histogram is depending on the value of P. To enrich texture information, we extract feature histograms at different resolutions which can be achieved by concatenating histograms using different values of R and P. In this paper resolution means the use of different radii of circle (e.g. different window sizes). Figure 3 shows an example of converting neighbouring pixels to a LQP code and binary code, resulting to four binary patterns.

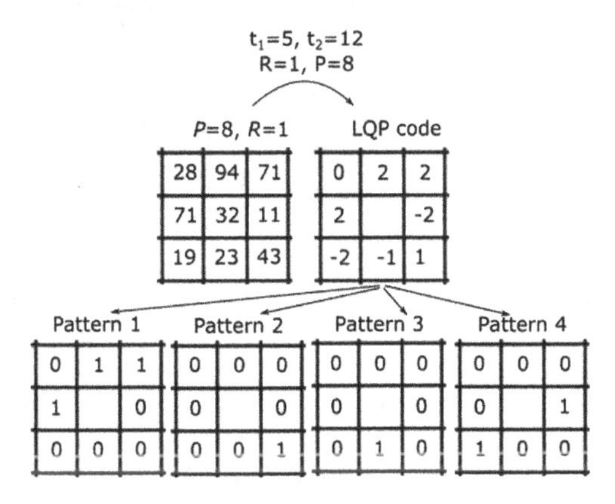

Fig. 3. An illustration of computing the LQP code using $P = 8$ and $R = 1$, resulting to four binary patterns.

Figure 4 shows an example of the feature extraction process using multiresolution LQP operators. Note that each resolution produces four binary patterns

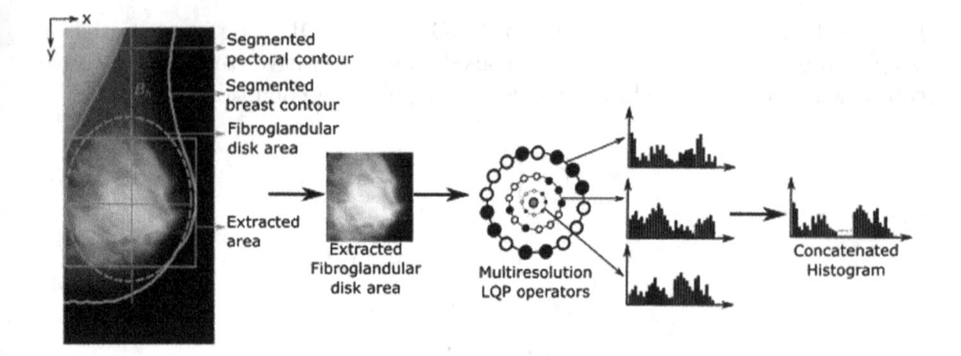

Fig. 4. An overview of the feature extraction using multiresolution LQP operators. Black dots in multiresolution LQP operators means neighbours with less value than the central pixel (red dot). (Color figure online)

(see Fig. 3) resulting to four histograms. Subsequently these four histograms are concatenated into a single histogram which represents the feature occurrences of a single resolution. The illustration in Fig. 4 uses three resolutions producing three histograms and finally concatenated to be a single histogram as the final representation of the feature occurrences in the multiresolution approach.

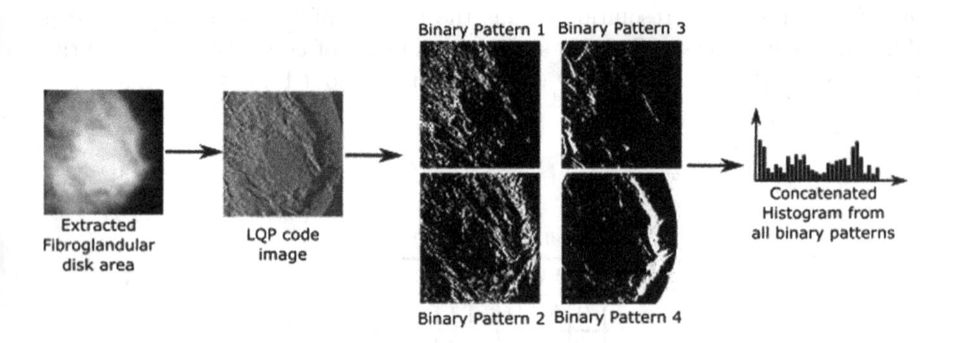

Fig. 5. Four images of binary patterns generated from the LQP code image.

Figure 5 shows four images of binary patterns generated based on the conditions in Eqs. (2), (3), (4) and (5). The LQP code image is generated using the condition in Eq. (6). All histograms from binary patterns are concatenated to produce a single histogram. Similar to LBP and LTP approaches, the LQP approach achieves rotation invariant by rotating the orientation (θ) of the circle's neighbourhood in a clockwise direction as shown in Fig. 6.

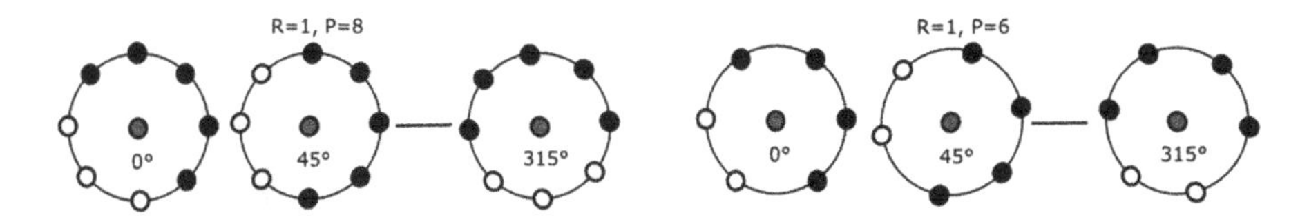

Fig. 6. An illustration of neighbourhood rotation in a clockwise direction. The black and white dots are determined based on one of the conditions in (2), (3), (4), (5) derived from (6).

2.3 Dominant Patterns

Selecting dominant patterns is an important process for dimensionality reduction purposes due to the large number of features (e.g. concatenation of several histograms). According to Guo *et al.* [19] a dominant patterns set of an image is the minimum set of pattern types which can cover $n(0 < n < 100)$ of all patterns of an image. In other words, dominant patterns are patterns that frequently occurred (or have high occurrence) in training images. Therefore, to find the dominant patterns we apply the following procedure. Let $I_1, I_2...I_j$ be images in the training set. Firstly, we compute the multiresolution histogram feature ($H_{I_j}^{LQP}$) for each training image. Secondly, we perform a bin-wise summation for all the histograms to find the pattern's distribution from the training set. Subsequently, the resulting histogram (H^{LQP}) is sorted in descending order and the patterns corresponding to the first D bins are selected. Where D can be calculated using the following equation:

$$D = \arg\min_{N} \frac{\sum_{i=1}^{N-1} H^{LQP}(i)}{\sum_{i=1}^{2^P} H^{LQP}(i)} > 0.01 \times n \tag{7}$$

where N is the total number of patterns and n is the threshold value set by the user. For example $n = 97$ means removing patterns which have less than 3% occurrence in H^{LQP}. This means only the most frequently occurring patterns will be retained for training and testing.

2.4 Classification

Once the feature extraction was completed, the Support Vector Machine (SVM) was employed as our classification approach using a polynomial kernel. The GridSearch technique was used to explore the best two parameters (complexity (C) and exponent (e)) by testing all possible values of C and e ($C = 1, 2, 3...10$ and $e = 1, 2, 3...5$ with 1.0 interval) and selecting the best combination based on the highest accuracy in the training phase. The SVM classifier was trained and in the testing phase, each unseen FGD_{roi} from the testing set is classified as BIRADS I, II, III or IV.

3 Experimental Results

To test the performance of the method, we used the Mammographic Image Analysis Society (MIAS) database [17] which consists of 322 mammograms of 161 women. Each image contains BIRADS information (e.g. BIRADS class I, II, III or IV) provided by an expert radiologist. A stratified ten runs 10-fold cross validation scheme was employed, where the patients are randomly split into 90% for training and 10% for testing and repeated 100 times. The metric accuracy (*Acc*) is used to measure the performance of the method which represents the total number of correctly classified images compared to the total number of images. We evaluate the method using three different parameters combinations ($LQP_{(R,P)}$): (a) Small multiresolution ($LQP^{small}_{((1,8)+(2,12)+(3,16))}$) (b) Medium multiresolution ($LQP^{medium}_{((5,10)+(7,14)+(9,18))}$) and (c) Large multiresolution ($LQP^{large}_{((11,16)+(13,20)+(15,24))}$). For the threshold values we have investigated several combinations of $\tau_{\in\{1,2\}}$ and found $\tau_1 = 5$ and $\tau_2 = 12$ produced the best accuracy for all experiments. However, for the sake of comparison we will report the performance of the method using $[\tau_1 = 4, \tau_2 = 9]$, $[\tau_1 = 3, \tau_2 = 13]$ and $[\tau_1 = 5, \tau_2 = 15]$. We will also show the effect on performance when varying n from 90 to 99.9 with 0.1 interval.

We present the quantitative results of the proposed method in Fig. 7 using small, medium and large multiresolution approaches which show that LQP^{medium} outperformed LQP^{small} and LQP^{large} regardless of the number of dominant patterns selected with the best accuracy of 84.91% ($n = 99.2$) followed by the small multiresolution approach of 81.81% (n = 98.9).

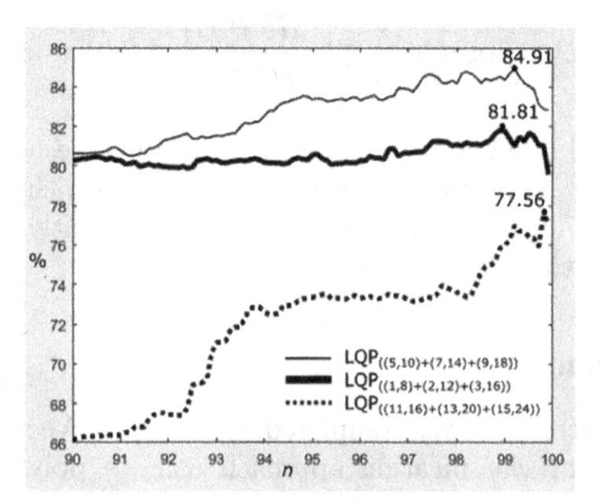

Fig. 7. Quantitative results using different multiresolutions $LQP_{((1,8)+(2,12)+(3,16))}$, $LQP_{((5,10)+(7,14)+(9,18))}$ and $LQP_{((11,16)+(13,20)+(15,24))}$.

Figure 8, shows quantitative results of $LQP_{((5,10)+(7,14)+(9,18))}$ using the following thresholds ($[\tau_1, \tau_2]$): [5, 12], [4, 9], [3, 13] and [5, 15] (note that these parameters are determined empirically). It can be observed that threshold values of

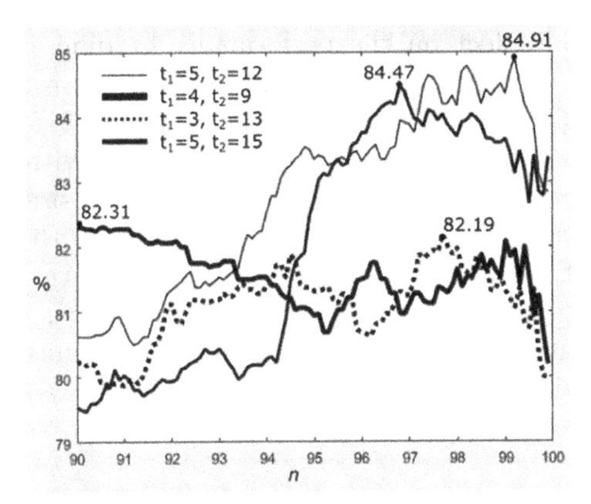

Fig. 8. Quantitative results using different values of τ_1 and τ_2.

$[\tau_1 = 5, \tau_2 = 12]$ produced the best accuracy of 84.91% with $n = 99.2$ followed by $[\tau_1 = 5, \tau_2 = 15]$ with $n = 96.8$. The other threshold values still produced good results in comparison to most of the proposed methods in the literature. For performance comparison when extracting features from the whole breast (wb), as majority of the current studies have done, compared to extracting features from FGD_{roi} only we conducted two experiments both using $LQP_{((5,10)+(7,14)+(9,18))}$ with $[\tau_1 = 5, \tau_2 = 12]$. Results can be seen in Fig. 9 which suggests that textures from the fibroglandular disk region are sufficient to differentiate breast density. In fact, it produced classification results of between 5–8% better depending on the value of n. Extracting features from the whole breast produced up to 77.88% accuracy with $n = 97.2$ which is 7% less than when feature extracted from the FGD_{roi} only.

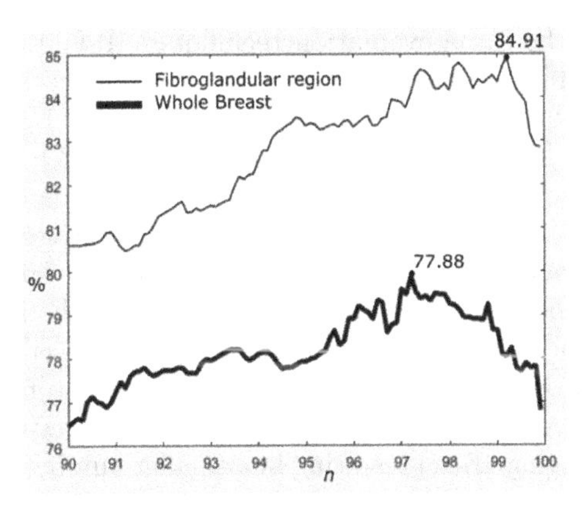

Fig. 9. Quantitative results of $LQP_{((5,10)+(7,14)+(9,18))}$ based on features extracted from the whole breast versus fibroglandular disk region.

To investigate the effect on the performance at different orientations (θ), we conducted eight experiments by varying θ clockwise rotation with the following values: 0°, 45°, 90°, 135°, 180°, 225°, 270° and 315°. The multiresolution approach for $LQP_{((5,10)+(7,14)+(9,18))}$ was applied with $[\tau_1 = 5, \tau_2 = 12]$. Figure 10 shows experimental results on eight different orientations chosen in this study which revealed that $LQP_{((5,10)+(7,14)+(9,18))}$ with $\theta = 270°$ produced the best accuracy of 85.6% which indicates density patterns are more visible at this orientation followed by $\theta = 180°$ with accuracy 85%. Overall results suggest that multiresolution LQP can produce consistent results (>83%) regardless ofthe parameter θ with $97 \leq n \leq 99.5$.

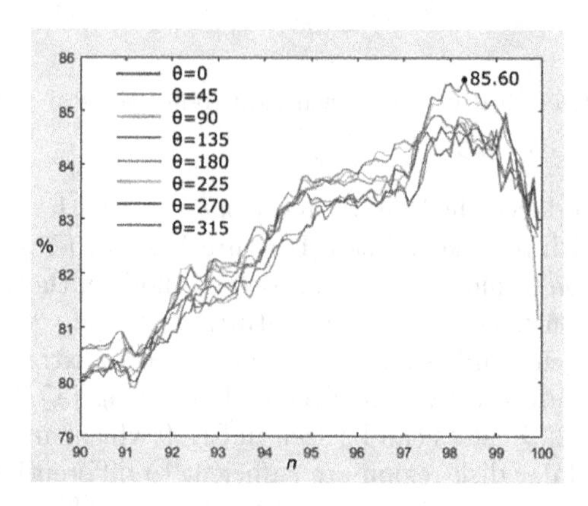

Fig. 10. Quantitative results using different orientation values.

For quantitative comparison with the other methods in the literature we selected those studies that have used the MIAS database [17], four-class classification, and using the same evaluation technique (10-fold cross validation) as in this study to minimise bias. The proposed method achieved up to 85.60% accuracy which is better than the methods proposed by Muštra *et al.* [6] (79.3%), Chen *et al.* [8,10] (59%, 70%, 72%, 75% and 76%), Bovis and Singh [4] (71.4%), and He *et al.* [12] (70%). However, the methods of Parthaláin *et al.* [7] and Oliver *et al.* [3] achieved 91.4% and 86%, respectively. In comparison to the other popular local statistical features such as LBP and textons, Chen *et al.* [8] reported the best accuracy achieved by these methods as 59% and 75%, respectively using the same evaluation approach and dataset. Recently, Rampun *et al.* [20] obtained over 82% accuracy using LTP operators, whereas the proposed method achieved up to 85.6%. In this study features were extracted only from one orientation (e.g. 270°) resulting in a smaller number of features whereas the method in [20] extracted features from eight different orientations and concatenated them, resulting in a large number of features. Furthermore, this study conducted feature selection by taking of account dominant patterns only which reduced the number of features significantly compared to the study in [20].

4 Conclusion

In conclusion, we have presented and developed a breast density classification method using multireslution LQP operators applied only within the fibroglandular disk area which is the most prominent region of the breast instead of the whole breast region as suggested in current studies [3–5, 7–13]. The multiresolution LQP features are robust in comparison to the other methods such as LBP, texton based approaches and LTP due to the five encoding system which generates more texture patterns. Moreover the multiresolution approach provides complementary information from different parameters which cannot be captured in a single resolution. The proposed method produced competitive results compared to some of the best accuracies reported in the litearture. For future work, we plan to develop a method that can automatically estimate τ_1 and τ_2 as well as combining multiresolution LQP features with features from the texton approach.

Acknowledgments. This research was undertaken as part of the Decision Support and Information Management System for Breast Cancer (DESIREE) project. The project has received funding from the European Union's Horizon 2020 research and innovation programme under grant agreement No. 690238.

References

1. Cancer Research UK: Breast cancer statistics (2014). http://www.cancerresearchuk.org/health-professional/cancer-statistics/statistics-by-cancer-type/breast-cancer. Accessed 6 Jan 2017
2. Breast Cancer: U.S. Breast Cancer Statistics (2016). http://www.breastcancer.org/symptoms/understand_bc/statistics. Accessed 6 Jan 2017
3. Oliver, A., Freixenet, J., Martí, R., Pont, J., Perez, E., Denton, E.R.E., Zwiggelaar, R.: A novel breast tissue density classification methodology. IEEE Trans. Inf Technol. Biomed. **12**(1), 55–65 (2008)
4. Bovis, K., Singh, S.: Classification of mammographic breast density using a combined classifier paradigm. In: 4th International Workshop on Digital Mammography, pp. 177–180 (2002)
5. Oliver, A., Tortajada, M., Lladó, X., Freixenet, J., Ganau, S., Tortajada, L., Vilagran, M., Sentś, M., Martí, R.: Breast density analysis using an automatic density segmentation algorithm. J. Digit. Imaging **28**(5), 604–612 (2015)
6. Muštra, M., Grgić, M., Delać, K.: A Novel breast tissue density classification methodology. Breast density classification using multiple feature selection. Automatika **53**(4), 362–372 (2012)
7. Parthaláin, N.M., Jensen, R., Shen, Q., Zwiggelaar, R.: Fuzzy-rough approaches for mammographic risk analysis. Intell. Data Anal. **14**(2), 225–244 (2010)
8. Chen, Z., Denton, E., Zwiggelaar, R.: Local feature based mamographic tissue pattern modelling and breast density classification. In: The 4th International Conference on Biomedical Engineering and Informatics, pp. 351–355 (2011)
9. Bosch, A., Munoz, X., Oliver, A., Martí, J.: Modeling and classifying breast tissue density in mammograms. In: Computer Vision and Pattern Recognition (CVPR 2006), pp. 1552–1558 (2006)

10. Chen, Z., Oliver, A., Denton, E., Zwiggelaar, R.: Automated mammographic risk classification based on breast density estimation. In: Sanches, J.M., Micó, L., Cardoso, J.S. (eds.) IbPRIA 2013. LNCS, vol. 7887, pp. 237–244. Springer, Heidelberg (2013). doi:10.1007/978-3-642-38628-2_28
11. Byng, J.W., Boyd, N.F., Fishell, E., Jong, R.A., Yaffe, M.J.: Automated analysis of mammographic densities. Phys. Med. Biol. **41**(5), 909–923 (1996)
12. He, W., Denton, E., Stafford, K., Zwiggelaar, R.: Mammographic image segmentation and risk classification based on mammographic parenchymal patterns and geometric moments. Biomed. Sig. Process. Control **6**(3), 321–329 (2011)
13. Petroudi, S., Kadir, T., Brady, M.: Automatic classification of mammographic parenchymal patterns: a statistical approach. In: Proceedings IEEE Conference Engineering in Medicine Biology Society, vol. 1, pp. 798–801 (2003)
14. Ojala, T., Pietikainen, M., Maenpaa, T.: Multiresolution gray-scale and rotation invariant texture classification with local binary patterns. IEEE Trans. Pattern Anal. Mach. Intell. **24**(7), 971–987 (2002)
15. Nanni, L., Luminia, A., Brahnam, S.: Local binary patterns variants as texture descriptors for medical image analysis. Artif. Intell. Med. **49**(2), 117–125 (2010)
16. Tan, X., Triggs, B.: Enhanced local texture feature sets for face recognition under difficult lighting conditions. In: Analysis and Modelling of Faces and Gestures, pp. 168–182 (2007)
17. Suckling, J., et al.: The mammographic image analysis society digital mammogram database. In: Proceedings of Exerpta Medica. International Congress Series, pp. 375–378 (1994)
18. Rampun, A., Morrow, P.J., Scotney, B.W., Winder, R.J.: Fully automated breast boundary and pectoral muscle segmentation in mammograms. Artif. Intell. Med. (2017, under review)
19. Gio, Y., Zhao, G., Pietikäinen, M.: Discriminative features for feature description. Pattern Recogn. **45**, 3834–3843 (2012)
20. Rampun, A., Winder, R.J., Morrow, P.J., Scotney, B.W.: Breast density classification in mammograms using local ternary patterns. In: International Conference on Image Analysis and Recognition (2017)

Rich Interaction and Feedback Supported Mammographic Training: A Trial of an Augmented Reality Approach

Qiang Tang(✉), Yan Chen, and Alastair G. Gale

Applied Vision Research Centre, Loughborough University, Loughborough, UK
{Q.Tang,Y.Chen}@lboro.ac.uk

Abstract. The conventional 'keyboard and workstation' approach allows complex medical image presentation and manipulation during mammographic interpretation. Nevertheless, providing rich interaction and feedback in real time for navigational training or computer assisted detection of disease remains a challenge. Through computer vision and state of the art AR (Augmented Reality) technique, this study proposes an 'AR mammographic workstation' approach which could support workstation-independent rich interaction and real-time feedback. This flexible AR approach explores the feasibility of facilitating various mammographic training scenes via AR as well as its limitations.

Keywords: Augmented Reality · Mammographic training · Interaction · Feedback

1 Introduction

To maintain quality in the UK Breast Screening Programme, a nationwide mandatory self-assessment scheme has been developed and deployed across the UK for the past 30 years. The current work presented here builds on this and addresses the particular need for supplying richer interaction and real time feedback in medical training or assessment by incorporating AR based interaction and feedback.

Training plays a significant role in maintaining mammographic interpretation skills as demonstrated by the UK self-assessment scheme (Personal Performance in Mammographic Screening PERFORMS®) [1]. According to [2], effective and sufficient training keeps being challenged by the shortage of high cost DICOM mammographic workstations. Early research [3,4] suggests appropriate interaction approaches could highly facilitate mammographic training. [6,7] indicate that feedback, as a critical training characteristic, provides a positive influence on trainee behaviour in medical training. The fast development of AR devices opens new opportunities to incorporating rich interaction and feedback with medical training without considering time or location [8]. Also, [5] suggests appropriate novel interaction approaches are important components of employing various

M. Valdés Hernández and V. González-Castro (Eds.): MIUA 2017, CCIS 723, pp. 377–385, 2017.
DOI: 10.1007/978-3-319-60964-5_33

display size in different mammographic interpretation training situations. The current study describes a device-independent AR approach which can provide interaction and feedback to support medical interpretation training in real time.

2 Method

The initial AR approach is conceptually constructed as the following steps:

(1) A series of mammographic cases which are either normal or have malignant or benign radiologic features are first examined by an expert radiologist on a workstation who annotates the cases. If any malignant or benign features are present on a case, then there are also feedback marks (areas of interest identified by radiologists) associated with the case which can be presented to the trainee radiologists by the AR system.
(2) A trainee radiologist wears the AR system as they inspect the mammographic images. Once the AR system appropriately identifies the particular case being examined then appropriate feedback information can be displayed on the AR system. The AR system has a forward facing camera and a display which enables the scene (i.e. here the mammographic images) to be captured and feedback to be overlaid on the viewed scene.
(3) Computer vision (object recognition) is implemented for this AR approach to effectively detect the working overall scene in front of the radiologist and extract the actual screen areas of the workstation into the AR system as time-sequenced images. These time-sequenced images are captured in real time so that a perspective transformation has to be applied to enable accurate overlaying feedback marks for each image. This approach then allows for the radiologist to move their head whilst constantly permitting accurate overlay of feedback information on the viewed mammographic cases. The perspective transformation [11] is a general 3*3 matrix expression for spatial transformations. Then an inverse transformation of the above enables pathology feedback information to overlay the real world scene appropriately.
(4) As a result, feedback marks are co-registered with the transformed images and presented to the radiologist via the AR display. This feedback information is only visible to the radiologist wearing the AR device.

The initial setup (Fig. 1) comprises a GE mammographic workstation, Google Glass and an Android virtual machine which is used for image processing. Considering that Google Glass has a critical performance limitation of image processing capability, a subsequent setup (Fig. 2) employs a simulated AR approach which uses a fixed camera (numbered as 3) and a laptop (numbered as 2) instead of Google Glass and the Android virtual machine to appropriately simulate an AR environment. Both setups are unobtrusive and allow radiologists quickly to access response feedback information. The only difference is that augmented views can be presented to the user via the laptop screen rather than directly presented by the see-through display of a wearable AR, whereas the laptop-based AR approach maximally provides a simulated AR training scene.

A further OCR approach recognises radiological case identity. Additionally, a natural writing method is implemented which allows a recognised stylus (numbered as 1 in Fig. 2) held by the radiologist logging its position on the workstation images. These both allow rich interaction and synchronising of the cases on the workstation and the AR device so that the same mammographic case displayed on the dedicated workstation is loaded from the PERFORMS® database and is also presented on the AR device at the same time.

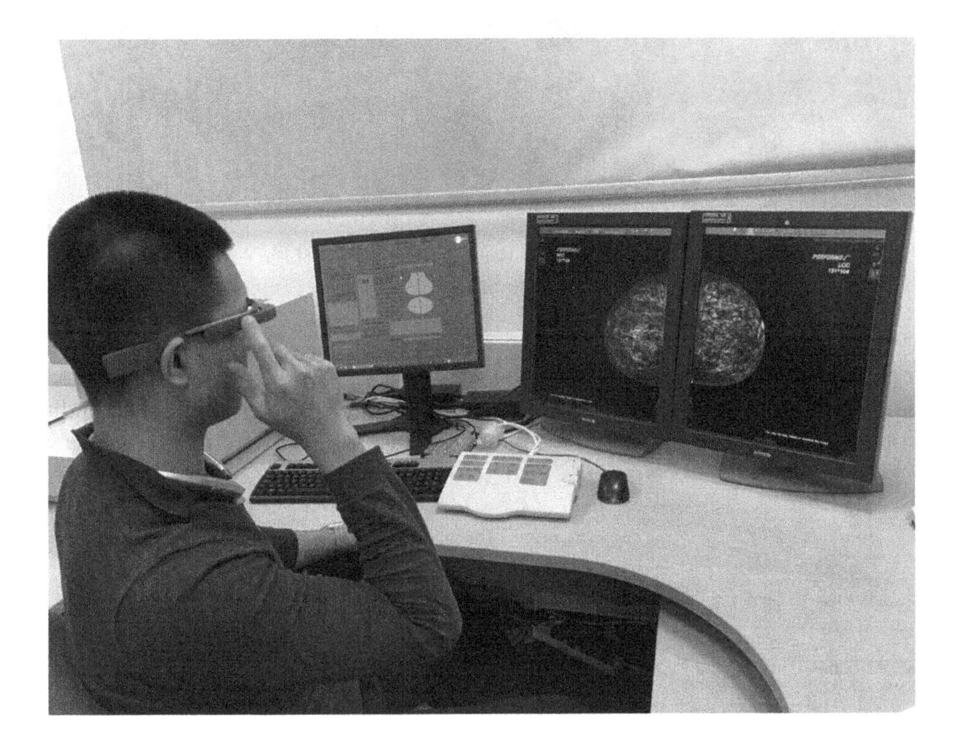

Fig. 1. The setup of the Google Glass AR environment

An effectively designed mammographic training process is able to incorporate virtual objects with human vision based reality so that augmented vision is obtained. Simultaneously, it allows seamless interaction between the human, the workstation and the AR device so that behaviour-dependent feedback information is supplied in real time. Such an augmented interaction is achieved via computer vision. Figure 3 shows the overall concept of incorporating augmented interaction and vision with conventional mammographic training.

An experimental AR prototype is implemented to support the delivery of mammographic training. The detailed workflow in Fig. 4 shows the mechanism of incorporating AR with conventional mammographic training. In this Figure, blue labels represent the general workflow of existing training approaches. Labels E and F respectively represent existing approaches and the AR approach. The AR

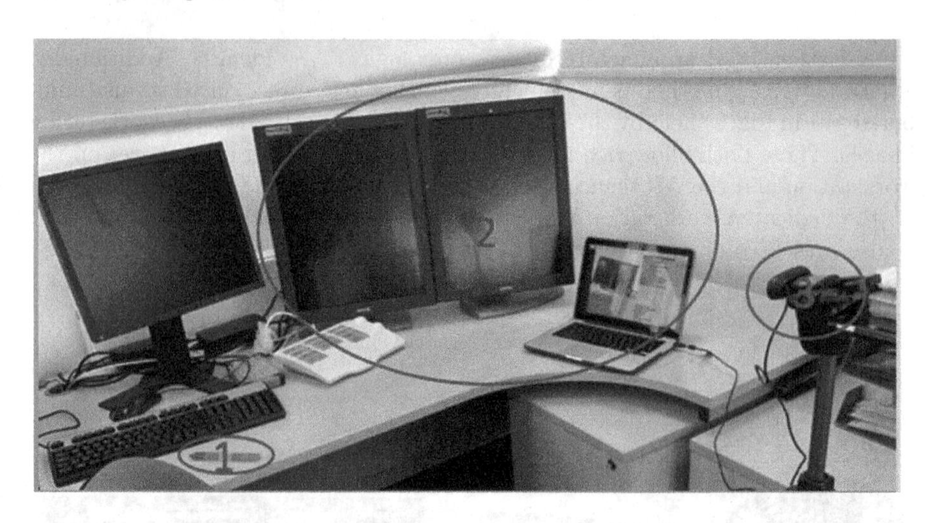

Fig. 2. The setup of the laptop-based AR environment has three essential elements: a GE workstation, a single camera and a laptop.

approach workflow consists of augmented vision sub-workflow and augmented interaction sub-workflow, which are represented by green and grey labels individually. Augmented vision and augmented interaction are described separately below in numbered steps.

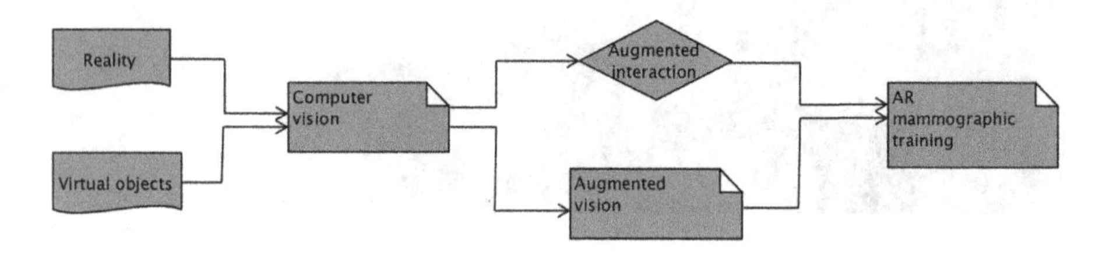

Fig. 3. The concept of AR mammographic training.

(i) Augmented vision (green labels in Fig. 4)
  (a) (1 in Fig. 4) The radiologist's view is time-sequentially captured by a high resolution AR camera. These time-sequenced images comprise the training scene (A in Fig. 4).
  (b) (2 in Fig. 4) Through computer vision, the AR prototype extracts the dedicated mammographic workstation screens and their contents from the overall scene.
  (c) (3 in Fig. 4) Through performing further image processing, including hough transformation [10] and perspective transformation [11], the irregular appearing mammographic image captured by the single camera is then

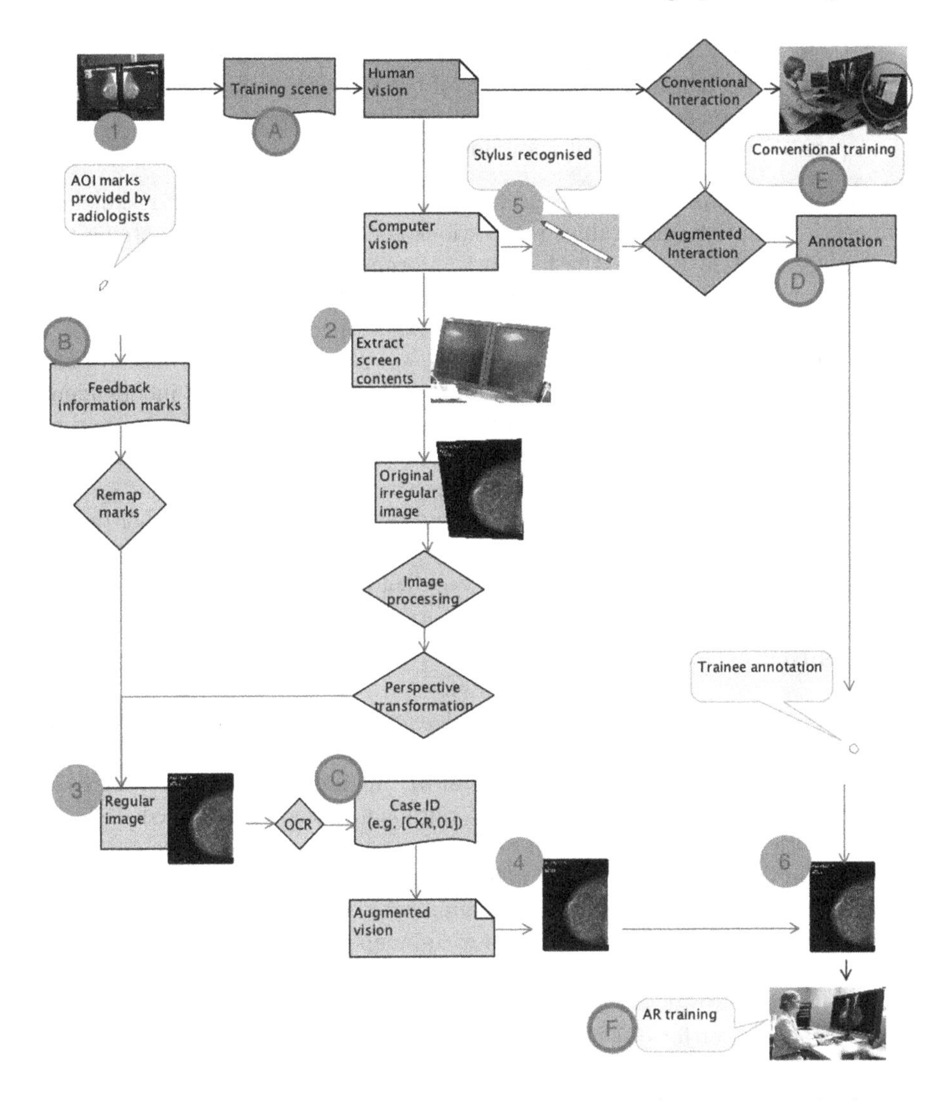

Fig. 4. The workflow of AR mammographic training. (Color figure online)

reconstructed as a corrected rectangular medical image. To be adapted for dedicated workstations or standard lightweight devices, screen distribution information is calculated after the hough transformation so that the AR approach can support complex training scenes with more than two devices which are different in their display size.

(d) (4 in Fig. 4) Feedback information marks are provided by radiologists (B in Fig. 4) and case identity is recognised via computer vision (C in Fig. 4) are both synchronised with the reconstructed regular mammographic image.

(ii) Augmented interaction (grey labels in Fig. 4)

(a) (5 in Fig. 4) The prototype is designed to recognise a stylus and track the natural handwriting behaviour of the radiologist via the stylus. This allows virtual annotation feedback (D in Fig. 4) to be generated and incorporated with the original mammographic images.

Finally, the enhanced mammographic image overlaying the expert radiologist feedback information and trainee annotations is constructed by assembling the training scene (A in Fig. 4), the feedback information marks (B in Fig. 4), the recognised mammographic case identity (C in Fig. 4) and the trainee specified annotations (D in Fig. 4). A sequential inverse perspective transformation of 3 in Fig. 4 is applied to accurately project the overlaid virtual information onto the real training scene. Therefore, AR training (F in Fig. 4) can be independent of any dedicated DICOM viewing workstation or software and enables superimposition of rich feedback information and interaction in real time via assembling reality and accurate virtual information registration as needed. Compared to conventional training (E in Fig. 4) which requires transcribing on another device [9], the AR approach can be vendor neutral and adapted for various viewing platforms (e.g. dedicated workstation or handheld device).

3 Results

The proposal demonstrates that AR can deliver rich interaction interpretation training via dynamically remapping radiological feedback into the real world using perspective transformation. A natural writing method (a stylus recognised through computer vision) enables rich interaction and feedback as needed. Feedback presentation and interaction both suggest a pixel level experimental accuracy for hardware. Synchronising of mammographic cases, both on a workstation and an AR system, is enabled, although complicated image manipulations are not yet encompassed. The AR approach is device-independent so that it can be used for different training scenarios (for example, a mixted environment of a tablet and a conventional workstation). This flexible generic approach works without any artificial marker for object registration and no calibration is required.

The registration accuracy of the overlaid feedback marks is gauged to be at a pixel level after correct detection of the training scene and workstation monitor screens via computer vision. Two metrics of hardware, camera resolution and display resolution, limit the accuracy of AR registration.

Hand input via the stylus is estimated by a series of carefully designed vision-guided experiments. Visual cues are used to identify the positions of the stylus components which are distinguished by different colours (Fig. 5: the drawing path of the stylus is represented by pink colour and a coloured dot indicates the centre of the stylus). To establish accuracy the AR prototype generates random points in the human visual field and requires a participant to identify each point by the stylus. The distance for a computer generated point to each participant identifying point is then calculated automatically and the mean distance derived. Repeated experiments show that the hand input accuracy (mean distance between computer generated points and identified points) is within ± 5

pixels. Nevertheless, hand input is significantly affected by handheld stability and the accuracy of the stylus detection via computer vision. High resolution and large size images are also factors in improving hand input accuracy. [12] claim consistent results while transcribing mammographic image feature positions.

4 Limitation

Conventional complicated medial image manipulations, for instance, zooming or panning are not currently mirrored in the proposed AR approach. Due to the image processing requirements and complex calculation procedures high performance hardware is essential for AR. Appropriate parallel computing for computer vision by GPU programming could significantly accelerate the process of AR detection and registration. However, the scarcity of high performance AR devices limits the wide adoption of AR in radiology training. A compromised solution is to employ an individual computing unit (e.g. a high performance laptop) for the complex image processing while presenting augmented views to radiologists via a see-through display.

5 Discussion

Following the perceived need of situated medical education, particular complicated medical training regarding difficulties in responding and gaining quick feedback information associated with self performance [13], immersive and high realistic situated training show the immense potential of enabling complicated perception of the training process and direct interaction by superimposing appropriate virtual information. Especially for mammographic training, there is a critical limitation in synchronising supplementary radiological case-associated information and allowing direction interaction or complex feedback. An important reason is the commercial edges maintained either by hardware or software vendors so that key technical details are not divulged to allow flexible training. AR provides a vendor neutral method which can superimpose device-independent interaction and feedback as needed via computer vision. It allows mammographic training to be independent of time, location, and device type because each realistic object is wrapped and complemented with a virtual property. For instance, a dedicated mammographic workstation and a tablet can form a flexible training scene together (Fig. 5). However, input accuracy can be affected by display size and resolution.

It is also acknowledged that existing wearable AR devices have no capability for high performance computing which is essentially required by image processing, in particular with breast screening. Incorporating an individual computing unit (e.g. a laptop with a dedicated graphic card) with a low cost AR display can be a superior compromise.

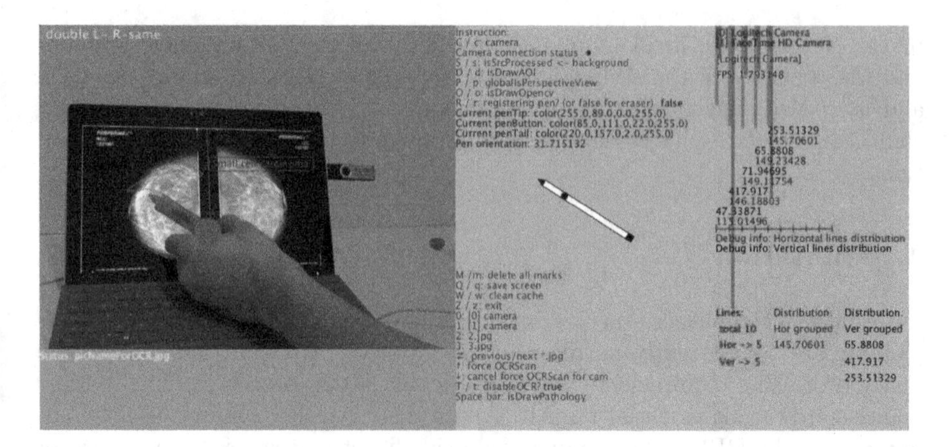

Fig. 5. AR mammographic training with a tablet. (Color figure online)

Although this study is concerned with mammographic images, the application is not only limited to mammographic screening training. Other similar training scenarios could also adopt this method to facilitate interaction and provide rich feedback. Moreover, incorporating contextual and accurate location-related CAD prompts can be achieved in interpretation training by recognising virtual contents (mammographic images) through computer vision.

In the future, appropriate experiments will be conducted to decide whether the AR training approach could effectively facilitate medical training quality by comparing AR and existing training approaches.

6 Conclusion

It is concluded that it is feasible to employ Augmented Reality to deliver situated learning and facilitate perception of complex situations in mammographic training. AR is a superior interface which seamlessly connects reality with rich virtual interaction and overlaid vision. However, the current development platform could not realise the AR approach's full potential. Appropriate hardware performance is the critical important factor in the wide adoption of AR in medical imaging education. The AR approach described here provides a compromised solution to build a maximally simulated AR training environment. A future series of experiments will investigate possible mammographic image manipulations via AR.

References

1. Chen, Y., Gale, A., Scott, H.: Mammographic interpretation training: what exactly do film-readers want? Breast Cancer Res. **10**(Suppl 3), P85 (2008). doi:10.1186/bcr2083

2. Chen, Y., Gale, A.: Intelligent computing applications based on eye gaze: their role in medical image interpretation. In: Huang, D.-S., McGinnity, M., Heutte, L., Zhang, X.-P. (eds.) ICIC 2010. CCIS, vol. 93, pp. 320–325. Springer, Heidelberg (2010). doi:10.1007/978-3-642-14831-6_43
3. Preim, B.: HCI in medical visualization. In: Dagstuhl Follow-Ups, vol. 2, p. 310 (2011). doi:10.4230/DFU.Vol2.SciViz.2011.292
4. Weiss, D.L., Siddiqui, K.M., Scopelliti, J.: Radiologist assessment of PACS user interface devices. J. Am. Coll. Radiol.: JACR **3**(4), 265–273 (2006). doi:10.1016/j.jacr.2005.10.016
5. Chen, Y., Gale, A.G., Scott, H.: Mammographic interpretation training in the UK: current difficulties and future outlook. In: Sahiner, B., Manning, D.J. (eds.) SPIE Medical Imaging, SPIE, p. 72631C-12 (2009). doi:10.1117/12.810730
6. Wood, B.P.: Feedback: a key feature of medical training. Radiology **215**(1), 17–19 (2000). doi:10.1148/radiology.215.1.r00ap5917
7. Karl, K.A., O'Leary-Kelly, A.M., Martocchio, J.J.: The impact of feedback and self-efficacy on performance in training. J. Organ. Behav. **14**(4), 379394 (1993). doi:10.2307/2488290?ref=search-gateway:c3129bc069220854d7dbb24db580bf8f
8. Chen, Y., Gale, A., Scott, H.: Anytime, anywhere mammographic interpretation training. Contemp. Ergon. 375–380 (2008). doi:10.1201/9780203883259.ch59
9. Gale, A.: A self-assessment scheme for radiologists in breast screening. Semin. Breast Dis. **6**(3), 148–152 (2003). doi:10.1053/j.sembd.2004.03.006
10. Mukhopadhyay, P., Chaudhuri, B.B.: A survey of hough transform. Pattern Recogn. **48**(3), 993–1010 (2015). doi:10.1016/j.patcog.2014.08.027
11. Wolberg, G.: Geometric Transformation Techniques for Digital Images: A Survey. Department of Computer Science, Columbia University, New York (1988)
12. Hatton, J.W, et al.: Accuracy of transcribing locations on mammograms: implications for the user interface of a system to record and assess breast screening decisions. In: Chakraborty, D.P., Krupinski, E.A. (eds.) Proceedings of the SPIE, SPIE, pp. 32–41 (2003). doi:10.1117/12.480086
13. Kamphuis, C., et al.: Augmented reality in medical education? Perspect. Med. Educ. **3**(4), 300–311 (2014). doi:10.1007/s40037-013-0107-7

A Robust Algorithm for Automated HER2 Scoring in Breast Cancer Histology Slides Using Characteristic Curves

Ramakrishnan Mukundan(✉)

Department of Computer Science and Software Engineering,
University of Canterbury, Christchurch, New Zealand
mukundan@canterbury.ac.nz

Abstract. This paper presents a novel feature descriptor and classification algorithms for automated scoring of HER2 in Whole Slide Images (WSI). Since a large amount of processing is involved in analyzing WSI images, the primary design goal has been to keep the computational complexity to the minimum possible level. We propose an efficient method based on characteristic curves which encode all relevant information in a smooth polynomial curve with the percentage of stained membranes plotted against variations in intensity/saturation of the colour thresholds used for segmentation. Our algorithm performed exceedingly well at a recent online contest held by the University of Warwick [1], obtaining the second best points score of 390 out of 420 and the overall seventh position in the combined leaderboard [2]. The paper describes three classification algorithms with features extracted from characteristic curves and provides experimental results and comparative analysis.

Keywords: Whole Slide Image Processing · Automated HER2 scoring · Medical image classification · Characteristic curves · Digital pathology

1 Introduction

The most commonly used method for breast cancer grading is the Immuno HistoChemistry (IHC) test which is a staining process performed on biopsy samples of breast cancer tissues [3]. The IHC stained slides are normally observed under a microscope by pathologists to determine the level of over-expression of Human Epidermal Growth factor Receptor 2 (HER2) protein in cancer cells. The tissue sample is then assigned a HER2 score of 0 to 3+ representing the grade of cancer present in the sample [4]. Manual grading and annotations of breast cancer slides are time consuming, and there are huge maintenance costs associated with collecting, archiving, and transporting tissue specimens. It is also well documented that manual grading can have significant variability in pathologist assessments due to the subjective process of determining the intensity and uniformity of staining in the presence of variable staining patterns and heterogeneity of tumor grade [5]. Automated methods can also suffer from errors due to inaccuracies in the training algorithm and its inability to segment faint and complex tissue structures [6].

M. Valdés Hernández and V. González-Castro (Eds.): MIUA 2017, CCIS 723, pp. 386–397, 2017.
DOI: 10.1007/978-3-319-60964-5_34

In the rapidly growing field of digital pathology, several Whole Slide Image (WSI) processing algorithms are currently being developed as diagnostic tools to help pathologists in the assessment of disease patterns [7]. WSIs have a pyramidal structure to enable optimized viewing across multiple magnification levels, and they provide a high resolution overview of the entire slide [7, 8]. Typically, at 40× magnification, the images have a resolution of approximately 0.25 microns per pixel. At this resolution, a slide region of size 15 mm × 15 mm could correspond to 60,000 × 60,000 pixels. WSIs were originally used as a computer aided digital microscopy tool, where pathologists could view different parts of a sample at different magnifications to improve the accuracy of their scores [5]. Powerful computational algorithms are being developed to automatically extract features related to cytological and protein structures in the image for accurately quantifying biomarkers like HER2 [9]. In the past, similar studies for quantitative IHC were performed using images of lower resolution [10].

Recently, an online contest was organized by the University of Warwick in conjunction with the UK/Ireland Pathology Society annual meeting 2016, with the aim of advancing research in the field of WSI-based automated HER2 scoring algorithms [1]. This contest was the primary motivation for our research work presented in this paper. Our algorithm (registered with team name UC-CSSE-CGIP) performed exceedingly well in the contest, obtaining the second best points score of 390 out of 420 and the overall seventh position in the combined leader board [2]. The teams that were on the top of the leader board, including our team, were invited to submit a very brief (one paragraph) summary of the algorithms used for inclusion in a journal paper prepared by the contest organizers [12].

WSIs contain voluminous amounts of data. One of the primary design goals has been to keep the computational complexity to the minimum possible level and to develop an efficient method that can process relevant tiles of an input WSI image quickly and classify the image into one of the four classes corresponding to the four HER2 scores. The second design goal was to have a feature set whose correlation to the percentage of membrane staining in the given sample could be easily visualized and interpreted by pathologists. The third design goal was to reduce the amount of information redundancy in the feature set by extracting a minimal set of characteristic features that would adequately represent the staining pattern. This paper presents classification algorithms using characteristic curves, providing detailed descriptions of the processing stages, development and selection of features, and the experimental analysis performed. We hope that the methods presented in this paper will contribute significantly to the development of faster and accurate automatic HER2 scoring techniques in the area of breast cancer histopathology.

The paper is organized as follows: The next section gives a description of the dataset used, an outline of HER2 assessment scheme and an overview of the stages of the processing pipeline. Section 3 provides an introduction to a novel set of features called characteristic curves. Section 4 gives a description of the classification algorithms using characteristic curves, and Sect. 5 presents experimental results and comparative analysis. Section 6 gives a summary of the work reported in this paper and outlines future directions.

2 Materials and Methods

2.1 HER2 Assessment

The assessment of HER2 protein over-expression is done based on the percentage of membrane staining observed in tumor cells as well as the intensity of staining [4]. The mapping between the level of membrane staining and the reported HER2 score is shown in Table 1.

Table 1. Correspondence between HER2 scores and membrane staining [4].

HER2 score	Assessment	Staining pattern
0	Negative	No staining is observed, or membrane staining is observed in less than 10% of tumor cells
1+	Negative	A faint/barely perceptible membrane staining is detected in greater than 10% of tumor cells. The cells exhibit incomplete membrane staining.
2+	Weakly Positive	A weak to moderate membrane staining is observed in greater than 10% of tumor cells.
3+	Positive	A strong complete membrane staining is observed in greater than 10% of tumor cells.

The WSI image segments of Immunohistochemical (IHC) stained slides given in Fig. 1 correspond to different HER2 scores and show the variations in the level of membrane staining.

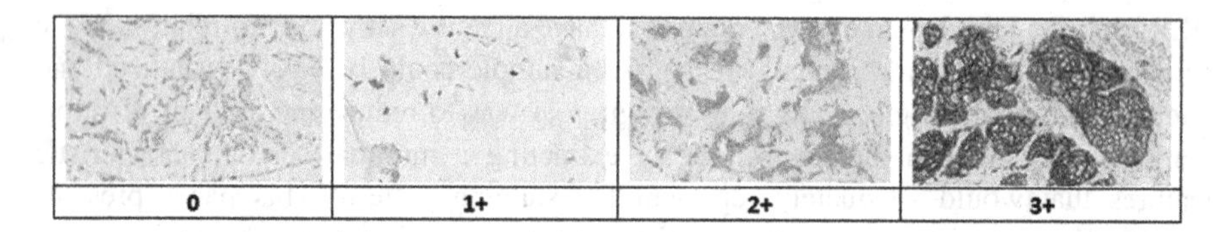

Fig. 1. WSI tiles showing different levels of staining and corresponding HER2 scores.

2.2 Dataset

The dataset used in this research work was provided by the University of Warwick as part of the online HER2 scoring contest [1]. Permission was granted by the contest organizers to participating teams for the use of the dataset for research and academic purposes. The dataset consisted of a total of 172 whole slide images in Nano-zoomer Digital Pathology (NDPI) format. These WSIs were extracted from 86 cases of patients with invasive breast carcinomas [12]. For each case, WSIs of both Hematoxylin and Eosin (H&E) stained and Immunohistochemical (IHC) stained slides were provided. There were two HER2 scoring contests, and the number of WSIs provided for training and testing the classification algorithm is given in Table 2. The training data included ground truth provided by expert pathologists and consisted of the HER2 score assigned for each case and also the percentage of membrane staining in the tissue sample.

Table 2. Number of WSIs provided for training and testing the classification algorithm.

Training set		Test set	
HER2 score (ground truth)	Number of WSIs	Contest-1 no. of WSIs	Contest-2 no. of WSIs
0	13	**28**	**6**
1+	13		
2+	13		
3+	13		
Total	**52**		

2.3 Processing Stages

Various stages of the processing pipeline are shown in Fig. 2. We used the OpenSlide API [11] to read WSIs of IHC stained slides, and a region of interest (ROI) containing a significant portion of the imaged tissue is extracted from the middle segment of the image. Rectangular tiles of size 1800 × 1200 pixels at 20× magnification that contain at most 20% background pixels are then created and used as inputs for the method that computes characteristic curves. At least six tiles at randomly selected locations within the ROI are generated for each WSI. The remaining part of the pipeline computes the percentage of staining in the tissue sample to obtain the characteristic curve as detailed in the next section.

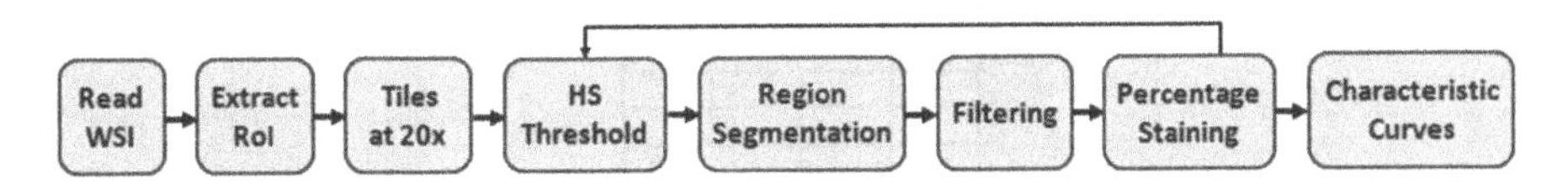

Fig. 2. Processing stages in the extraction of characteristic curves.

3 Characteristic Curves

In this section, we introduce a novel feature vector called a characteristic curve. An important parameter in HER2 assessment is the percentage of membrane staining perceived in an image segment. Assuming that we can compute the percentage of membranes stained in a particular colour range (this computation will be discussed in detail below), we can analyse the variations in this percentage value with respect to changes in the colour saturation threshold. Specifically, if $[h, s, v]$ represent the stain colour components in HSV space, and if $p(s_{low})$ denotes the percentage of staining with colour in the range given by the following inequalities:

$$\begin{aligned} h_1 \leq h < h_2 \\ s > s_{low} \\ v_1 \leq v < v_2 \end{aligned} \tag{1}$$

then, the variation of $p(s_{low})$ plotted against s_{low} gives the characteristic curve (or the percentage-saturation curve) of the image. In Eq. (1), $[h_1, h_2]$ denote fixed hue thresholds specifying allowable variations in the hue value, and similarly $[v_1, v_2]$ denote value thresholds. Since we specify only the lower bound for saturation, progressively increasing s_{low}, typically from 0.1 to 0.5, produces a non-increasing characteristic curve (Fig. 3). In our experiments, we used the following threshold values: $h_1 = 0$, $h_2 = 0.1$, $s_{low} = 0.1$, $v_1 = 0$, $v_2 = 1$.

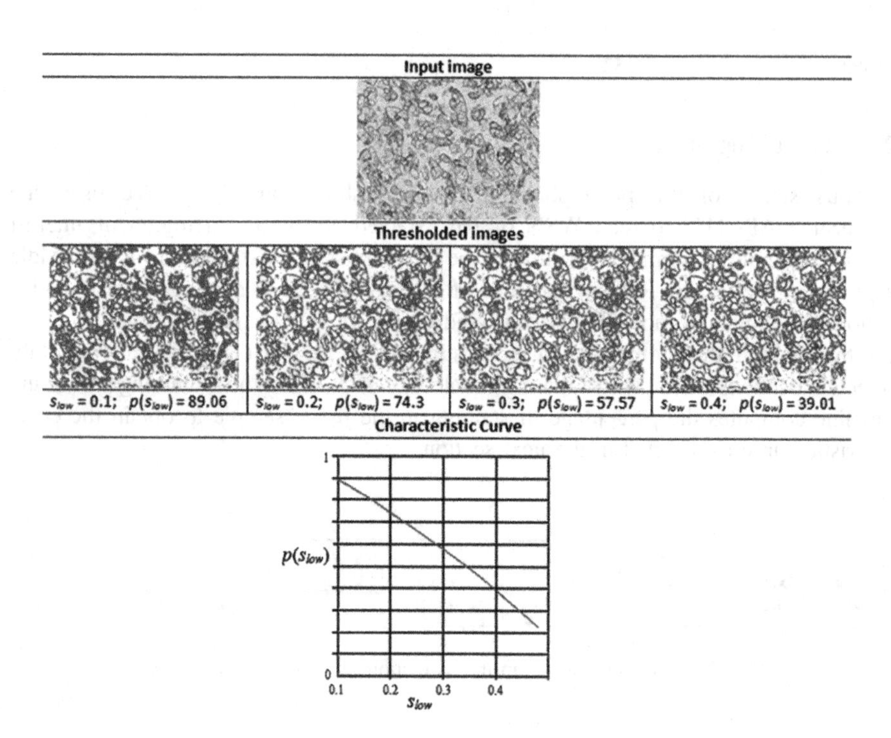

Fig. 3. Intermediate stages in the generation of a characteristic curve. (Color figure online)

The base components of the stain colour $[h, s, v]$ are computed using the training set where the given percentage of staining is above 80%. While computing the percentage of staining for the test (or cross-validation) sets, it is important to eliminate not only the background region but also other segments that are not part of the membrane region such as connective tissues, lobules and nuclei. These regions can be segmented using colour (nuclei are stained in a distinctly different colour) or using a distance measure evaluated in colour space over a neighbourhood mask around each pixel (for identifying regions of nearly constant colour value). Figure 3 shows thresholded images with stained regions in red colour as the value of s_{low} is increased from 0.1 to 0.4. The resulting characteristic curve is also shown. The characteristics curves have the property that they are always monotonically decreasing smooth curves. They allow accurate polynomial approximations using cubic curves. The shape of the curve can be directly matched with the staining patterns given in the HER2 assessment guidelines (Table 1) for a straightforward interpretation of the derived score (Fig. 4). For example, the characteristic curve always lies below the 10% threshold when the score is 0, and only a small initial segment of the curve lies above the 10% mark when the score is 1. If the score is 3+, the curve lies completely above the 30% mark showing a strong and complete membrane staining. As seen in Fig. 4, the curve passes through a much wider range of values of percentage staining when the score is 2+.

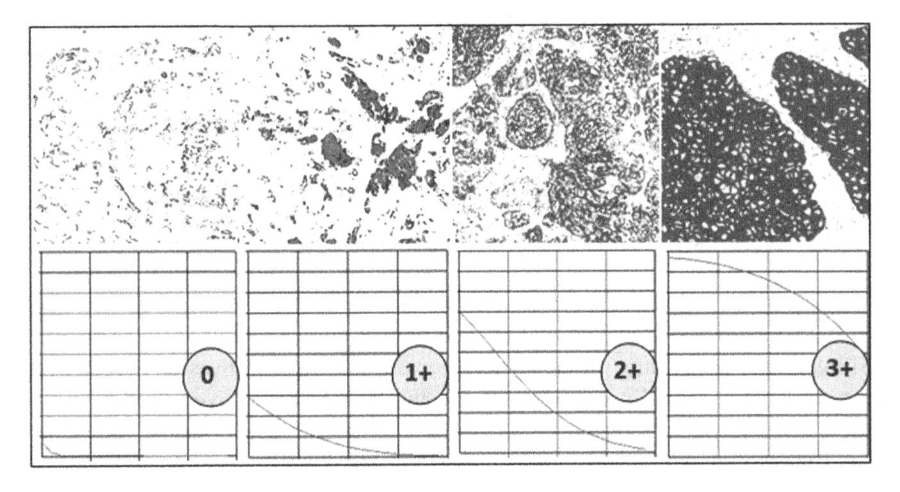

Fig. 4. Variations in the shapes of the characteristic curves with different levels of staining.

4 Classification

The properties of the characteristic curve outlined in the previous section, particularly the fact that the curve is non-increasing, can be effectively used for developing a rule-based classification algorithm as follows.

- if z_0 ($= p(0.1)$) < 10%, then the whole curve lies below 10%, and the score is 0 (rule 1)
- else if z_{n-1} ($= p(0.5)$) > 30%, then the whole curve lies above 30%, and the score is 3+ (rule 2)
- else if 10% $\leq$ z_0 ($= p(0.1)$) < 40% and $p(0.2)$ < 15%, the score is 1+ (rule 3)
- else if $p(0.4)$ < 15%, then the score is 2+ (rule 4)
- else, the score is 3+ (rule 5).

The rules were formed by analyzing the shapes of characteristic curves for several image tiles with ground truth values of HER2 scores assigned by pathologists. Note that for the above simple classification algorithm, we sample the curve at only four key points $p(0.1)$, $p(0.2)$, $p(0.4)$, and $p(0.5)$. As discussed in the next section on experimental results, the rule based algorithm is primarily used to assess the feature representation capability of the characteristic curves.

For more accurate classification, we use the 'one-vs-all' multi-class classification algorithm using logistic regression [13]. For a given training example with index j, the points sampled along its characteristic curve $x_i^{(j)} = p(s_i), i = 1..n, j = 1..m$ are used as features. The class labels are denoted by $y_j \in [0, 3]$, $j = 1..m$. We denote the feature matrix by $X \in \Re^{m\times(n+1)}$, the output vector of labels by $Y \in \Re^{m\times1}$, and the classifier parameter vector for each class by $\theta_k \in \Re^{(n+1)\times1}$, $k = 1..4$. Here, class-1 corresponds to the set of training examples with HER2 score 1+ , class-2 with HER2 score 2+ , class-3 with HER2 score 3+ and class-4 with HER2 score 0. The hypothesis function vector $H \in \Re^{m\times1}$ is given by $H = g(X\theta_k)$, where $g()$ denotes the sigmoid function. For prediction, the points x_i on the characteristic curve of given sample are combined with the trained values of class parameters θ_k for each class $k = 1..4$, and the class that gives the maximum value for $g(x_i^{\prime}\, \theta_k)$ is chosen. In the next section, we provide the result of classification experiments using the above methods.

5 Experimental Results and Analysis

First, we provide the results for the rule-based classification algorithm. The percentage of staining values $p()$ obtained from the characteristic curves computed for some of the WSI images in the training set are given below. The table also gives the ground truth values and the predicted scores computed using the five rules given in the previous section. For each case, three segments of the WSI (tiles) at 20× magnification were used. The incorrect predictions are highlighted in red colour.

The overall performance of the rule-based classification algorithm can be seen in the confusion matrix below. 52 WSIs with 3 tiles at 20× from each image (comprising of 156 images) were used in this experiment as the training data. Another set of 3tiles from each of the 52 cases formed the cross-validation set. Out of the total of 156 image tiles in the cross-validation set, 39 belonged to each of the four classes corresponding to four HER2 scores. As seen in Table 4, a few images for cases with score 1+ were wrongly classified as having either 0 or 2+ scores, while all images with score 3+ were correctly classified. These results of the rule based algorithm are presented here only to show that one could roughtly estimate the HER2 scores directly from the shapes of the characteristic curves (Table 3).

The results given above show that even a minimal set of four points derived from the characteristic curves can have a good discriminating power. The accuracy can be further improved by including the slope information at the key points also in the classification rules (Table 5). The slope at the point $p(s_i)$ is computed as

$$p'(s_i) = \frac{1}{0.02}\left(p(s_i) - p(s_i + 0.02)\right) \tag{2}$$

Table 3. Sampled values of the characteristics curves and the predicted scores obtained a set of WSI images in the training set.

CaseNo (tileNo)	$p(0.1)$	$p(0.2)$	$p(0.4)$	$p(0.5)$	Ground truth	Predicted	Rule
1 (1)	0.72	0	0	0	0	0	Rule-1
1 (2)	7.16	0.01	0	0	0	0	Rule-1
1 (3)	7.21	0.01	0	0	0	0	Rule-1
12 (1)	14.31	4.09	0.10	0	1	1	Rule-3
12 (2)	35.81	13.02	0.61	0.04	1	1	Rule-3
12 (3)	28.76	13.44	1.25	0.19	1	1	Rule-3
15 (1)	76.16	22.07	0.02	0.00	1	2	Rule-4
15 (2)	74.68	22.6	0.18	0	1	2	Rule-4
15 (3)	17.97	0.33	0	0	1	1	Rule-3
16 (1)	8.09	0.91	0	0	1	0	Rule-1
16 (2)	11.44	0.63	0	0	1	1	Rule-3
16 (3)	1.98	0.09	0	0	1	0	Rule-1
25 (1)	75.79	36.06	1.21	0.08	2	2	Rule-4
25 (2)	48.12	15.33	0.52	0.03	2	2	Rule-4
25 (3)	61.18	24.24	0.56	0.02	2	2	Rule-4
33 (1)	88.64	83.33	67.65	46.96	3	3	Rule-2
33 (2)	90.72	87.27	70.15	50.13	3	3	Rule-2
33 (3)	86.22	82.62	68.48	50.39	3	3	Rule-2
84 (1)	77.53	66.29	38.37	16.4	3	3	Rule-5
84 (2)	75.20	63.42	39.64	20.35	3	3	Rule-5
84 (3)	57.81	44.88	21.42	6.65	3	3	Rule-5

Table 4. Confusion matrix showing the performance of the rule based classification method.

		Predicted				Accuracy = 80.12%	
		0	1+	2+	3+	Precision	Recall
Actual	0	30	7	2	0	0.81	0.77
	1+	7	23	9	0	0.74	0.59
	2+	0	1	33	5	0.75	0.84
	3+	0	0	0	39	0.88	1.0

For generating feature vectors for classification using logistic regression, it was found that a step size of 0.02 for the saturation threshold would provide an adequate number of 20 points (features) within the saturation range [0.1, 0.5]. The feature matrix X in Eq. (2) therefore had the dimension 156×20. The gradient descent algorithm used 100 iterations to converge to the solution with a learning rate of 0.001 (Fig. 5).

The confusion matrix showing the improvement of accuracy on the rule-based method is given in Table 6. Note that this method had 20 features for each sample, while the rule-based method used only four points from the characteristic curve.

Table 5. Confusion matrix for the rule based classification method augmented by the slope information.

		Predicted				Accuracy = 85.25%	
		0	1+	2+	3+	Precision	Recall
Actual	0	35	4	0	0	0.83	0.89
	1+	7	25	7	0	0.83	0.64
	2+	0	1	34	4	0.82	0.87
	3+	0	0	0	39	0.9	1.0

The smoothness and monotonically decreasing properties of the characteristic curve can be effectively made use of in reducing the dimensionality of the features in the logistic regression algorithm. As in the case of the rule based classification method, we can sample the curve at only four key points $p(0.1)$, $p(0.2)$, $p(0.4)$, and $p(0.5)$, and also use the slope information at those points $p'(0.1)$, $p'(0.2)$, $p'(0.4)$, and $p'(0.5)$ to get a feature vector of size 8 instead of 20. The cost functions converge to almost similar values with only a slight increase in the magnitudes (Fig. 6).

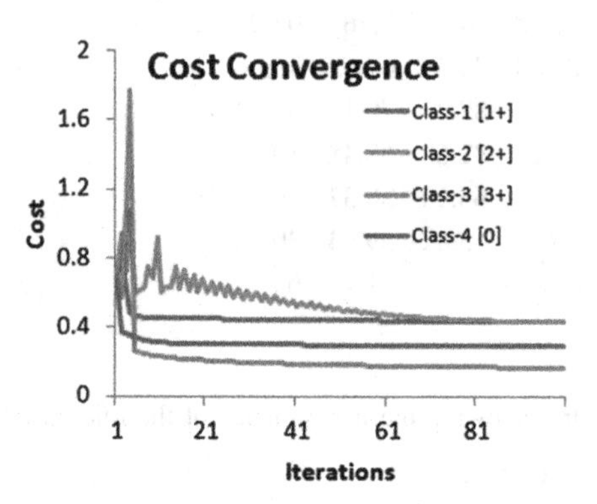

Fig. 5. Convergence of the cost functions of the multi-class logistic regression algorithm.

The confusion matrix obtained by running the algorithm with the reduced set of features of the characteristic curve is shown in Table 7.

The rule-based classification algorithm with augmented slope information is computationally fast, and provides very good results for all classes except class-1

Table 6. Confusion matrix for the multi-class logistic regression algorithm.

		Predicted				Accuracy = 88.46%	
		0	1+	2+	3+	Precision	Recall
Actual	0	37	2	0	0	0.86	0.95
	1+	6	29	4	0	0.83	0.74
	2+	0	4	34	1	0.87	0.87
	3+	0	0	1	38	0.97	0.97

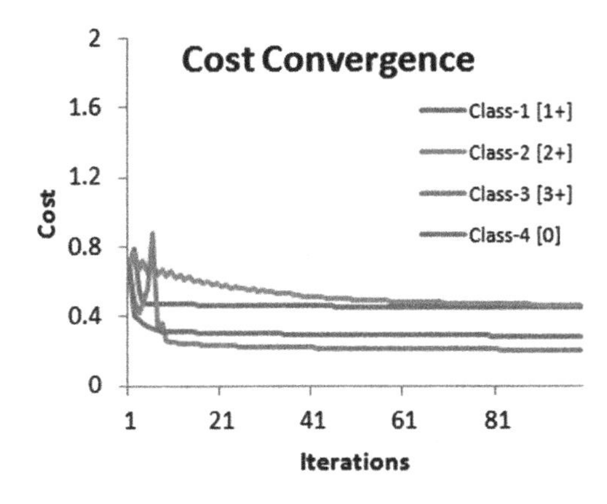

Fig. 6. Convergence of the cost functions with reduced feature set.

(corresponding to HER2 score 1+) where the recall rate is 0.64 (Table 5). All methods gave the lowest recall rate for this class. This is because the characteristic curves for several tiles with HER2 score 1+ grossly resembled the shape of curves with score 0 or 2+. However, the performance of the rule-based method for this class is even better than logistic regression with reduced feature set (Table 7). Multi-class logistic regression gave better results in all remaining classes. Reducing the dimensionality of the feature set from 20 to 8 only affected the recall rates of classes 1 and 2. Overall, logistic regression with 20 feature points gave the highest accuracy of 88.5%.

Analysing the staining patterns in tiles that were wrongly classified revealed a common problem in the automatic extraction of tiles from WSIs (see Fig. 2). Some of the samples with scores 1+ and 2+ had large tissue regions without any staining. The example shown in Fig. 7 contains a tissue sample at 10× magnification with an assigned score of 2+.

In Fig. 7, the tile on the top didn't contain any stained membrane regions and was assigned a ground truth value of 2+ at the training stage, and a predicted value of 0 at the cross-validation stage. This tile could have been a valid part of any WSI with a score 0, and therefore there is no way by which such tiles can be identified and discarded by the automatic tile extraction method. Manually identifying such tiles from the training and cross-validation sets significantly improved the scores of the classification algorithms. The tile on the bottom half of Fig. 7 was assigned the correct score of 2+.

Table 7. Confusion matrix for the multi-class logistic regression algorithm with reduced feature set.

		Predicted				Accuracy = 83.3%	
		0	1+	2+	3+	Precision	Recall
Actual	0	37	2	0	0	0.80	0.95
	1+	8	24	7	0	0.75	0.61
	2+	1	6	31	1	0.79	0.79
	3+	0	0	1	38	0.97	0.97

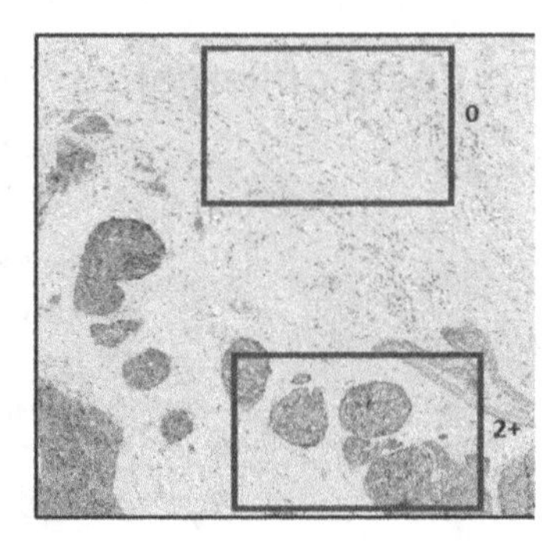

Fig. 7. An example showing two tile positions with varying image characteristics within the same WSI.

6 Conclusions

This paper has introduced a novel feature descriptor called a characteristic curve that could be effectively used in classification algorithms for automated scoring of HER2 in breast cancer histology slides. The computational aspects of characteristic curves and their shape features that embed information on the staining patterns for different HER2 scores have been discussed in detail. The usefulness of features based on characteristic curves and their applications in classification algorithms have been demonstrated through experimental results obtained using a comprehensive WSI dataset provided by the University of Warwick [1]. The results show that the features used with a multi-class classification algorithm such as logistic regression can provide very good levels of accuracy. The paper also outlined computational stages in the overall processing pipeline for automatic HER2 scoring using WSI files as inputs.

Experimental results showed the need for further improving the discriminating power of the characteristic curves by developing methods for accurate identification of membrane morphology and region segmentation, particularly for samples with an assigned HER2 score 1+. It is also necessary to assess the reproducibility of results, specifically inter-scanner variability [14] of the rule-based classification algorithm as the rules were formed using data produced by a single scanner.

References

1. Department of Computer Science, University of Warwick: Her2 Scoring Contest. http://www2.warwick.ac.uk/fac/sci/dcs/research/combi/research/bic/her2contest/
2. Department of Computer Science, University of Warwick: Her2 Contest Results. http://www2.warwick.ac.uk/fac/sci/dcs/research/combi/research/bic/her2contest/outcome
3. Hicks, D.G., Schiffhauer, L.: Standardized assessment of the HER2 status in breast cancer by immunohistochemistry. Lab. Med. **42**(8), 459–467 (2015). doi:10.1309/LMGZZ58CTS0D BGTW

4. Rakha, E.A., et al.: Updated UK recommendations for HER2 assessment in breast cancer. J. Clin. Pathol. **68**, 93–99 (2015). doi:10.1136/jclinpath-2014-202571
5. Gavrielides, M.A., Gallas, B.D., Lenz, P., Badano, A., Hewitt, S.M.: Observer variability in the interpretation of HER2 immunohistochemical expression with unaided and computer aided digital microscopy. Arch. Pathol. Lab. Med. **135**(2), 233–242 (2011). doi:10.1043/1543-2165-135.2.233
6. Akbar, S., Jordan, L.B., Purdie, C.A., Thompson, A.M., McKenna, S.J.: Comparing computer-generated and pathologist-generated tumor segmentations for immunohistochemical scoring of breast tissue microarrays. Br. J. Cancer **113**(7), 1075–1080 (2015). doi:10.1038/bjc.2015.309
7. Hamilton, P.W., et al.: Digital pathology and image analysis in tissue biomarker research. Methods **70**(1), 59–73 (2014). doi:10.1016/j.ymeth.2014.06.015
8. Farahani, N., Parwani, A.V., Pantanowitz, L.: Whole slide imaging in pathology: advantages, limitations and emerging perspectives. Pathol. Lab. Med. Int. **7**, 23–33 (2015). doi:10.2147/PLMI.S59826
9. Ghaznavi, F., Evan, A., Madabhushi, A., Feldman, M.: Digital imaging in pathology: whole-slide imaging and beyond. Ann. Rev. Pathol. Mech. Dis. **8**, 31–59 (2013). doi:10.1146/annurev-pathol-011811-120902
10. Matkowskyj, K.A., Cox, R., Jensen, R.T., Benya, R.V.: Quantitative immunohistochemistry by measuring cumulative signal strength accurately measures receptor number. J. Histochem. Cytochem. **51**(2), 205–214 (2003). doi:10.1177/002215540305100209
11. Goode, A., Gilbert, B., Harkes, J., Jukie, D., Satyanarayanan, M.: OpenSlide: a vendor-neutral software foundation for digital pathology. J. Pathol. Inf. **4**(27) (2013). doi:10.4103/2153-3539.119005
12. Qaiser, T., et.al.: HER2 challenge contest: a detailed assessment of HER2 scoring algorithms and man vs machine in whole slide images of breast cancer tissues. Submitted to Histopathology. Wiley (2017)
13. Watt, J., Borhani, R., Katsaggelos, A.K.: Machine Learning Refined: Foundations, Algorithms and Applications. Cambridge University Press, Cambridge (2016)
14. Keay, T., et.al.: Reproducibility in the automated quantitative assessment of HER2/neu for breast cancer. J. Pathol. Inf. **4**(19) (2013). doi:10.4103/2153-3539.115879

Investigating the Effect of Various Augmentations on the Input Data Fed to a Convolutional Neural Network for the Task of Mammographic Mass Classification

Azam Hamidinekoo[1], Zobia Suhail[1], Talha Qaiser[2], and Reyer Zwiggelaar[1](✉)

[1] Department of Computer Science, Aberystwyth University, Aberystwyth, UK
rrz@aber.ac.uk

[2] Department of Computer Science, Warwick University, Coventry, UK

Abstract. Along with the recent improvement in medical image analysis, exploring deep learning based approaches in the context of mammography image processing has become more realistic. In this paper, we concatenate on both conventional machine learning and deep learning approaches to classify mass abnormalities in mammographic images. Using a deep convolutional neural network (CNN) architecture, the effect of performing various augmentation approaches on the raw pre-detected masses fed to the network is investigated. We propose an extended augmentation method, specific filter bank responses and also a texton-based approach to generate characteristic filtered features for various types of mass textures and eventually use the resulting image data as input for training the CNN. Evaluating our proposed techniques on the DDSM dataset, we show that mammographic mass classification can be tackled effectively by employing an extended augmentation scheme. We obtained 87% accuracy which is comparable to the currently reported results for this task.

Keywords: Breast cancer · Mammographic mass classification · Data augmentation · Convolutional neural network (CNN)

1 Introduction

Breast cancer is the most frequently diagnosed type of cancer worldwide [2] which accounts for 25.2% of the total cancer related death among women followed by colorectum (9.2%), lung (8.7%), cervix (7.9%), and stomach cancers (4.8%) according to the report by International Agency for Research on Cancer, WHO in 2014 [26]. The assessment process for breast screening is based on a imaging for finding early changes in breast tissue, plus clinical assessment and needle biopsy if required [1]. Mammography scans (mammograms) are used as a primary imaging modality to identify the abnormalities at early stages. Moreover, two main appearances shown in mammograms are masses and microcalcifications [6]; and Computer Aided Diagnosis (CAD) systems have been developed for each.

M. Valdés Hernández and V. González-Castro (Eds.): MIUA 2017, CCIS 723, pp. 398–409, 2017.
DOI: 10.1007/978-3-319-60964-5_35

However, the variability among different masses with respect to shape and boundary causes misdiagnosis of masses in mammographic images.

Up to now, various conventional machine learning methods and deep learning approaches have been explored for the task of breast mass classification and diagnosis which will be covered in the next section. In this paper, we try to concatenate both conventional and deep learning approaches to classify benign and malignant breast masses. So the main aim of this paper is to classify mammographic masses using a deep convolutional neural network (CNN) architecture and investigate the effect of performing data augmentation on the raw pre-detected masses fed to the network. For this aim, 3 different new augmentation schemes are proposed and the results are evaluated on a public dataset.

2 Related Work

Various conventional machine learning methods for the task of mammographic mass classification are reviewed in [20]. Moreover, a three steps procedure for the classification of benign and malignant masses is proposed in [25] in which the RoI segmentation is done with a group of 32 Zernike moments that were extracted to train a neural network. Vaidehi and Subashini [28] proposed a method using Gray Level Co-occurrence Matrix (GLCM) features including contrast, correlation, energy and homogeneity accompanied with the Fuzzy C-Mean technique to extract GLCM features for mass classification. Buciu and Gacsadi [4] proposed directional features extracted by filtering image patches using Gabor wavelets. Reducing the data dimension by Principal Component Analysis (PCA), they improved classification results when Gabor features are used instead of using original mammogram images. Additionally, texton-based approaches have been used effectively for the texture classification using single images [30], where the textures are represented as histograms of texton contained in texton dictionary. In a similar work, Varma and Zisserman used filter responses in order to build texton dictionary. They demonstrated that such textures could be classified using joint distribution of intensity values [29]. However, these methods - considered as the conventional methods - rely on hand designed feature extractors that require a considerable amount of engineering skill and domain expertise. In recent years, the hand-crafted feature extraction aspect is proposed to be replaced by trainable deep networks which are able to extract discriminant features automatically from the images.

Considering the layer aspect of deep networks along with parallelisable algorithms and the vast acceleration property of GPUs, exploring deep learning based approaches in the context of mammography image processing has also become more realistic. Accordingly, Petersen *et al.* [21] presented a generic multi-scale denoising autoencoder for contextual breast density segmentation. Kallenberg *et al.* [16] tested a convolutional sparse autoencoder network with a sparsity regularisation on their own unlabelled sets of mammograms expanding the idea of [23] but for pixel-wise labelling and for large scale images. Jamieson *et al.* [13] explored the use of Adaptive Deconvolutional Networks for learning high-level features of mass lesions. Arevalo *et al.* [3] presented various depth CNN frameworks for automatic supervised feature learning of film mammograms. Fonseca *et al.* [9] evaluated the performance of an architecture search procedure [22] and Dhungel *et al.* [8] presented a multi-scale deep belief network combined with a Gaussian Mixture Model classifier for mass candidate generation. Kooi *et al.* [17] and Huynh *et al.* [12] took advantage of transfer learning to extract tumour information from medical images via CNNs that were originally pre-trained with non-medical data. In a similar way, Jiao *et al.* [15] used a pre-trained CNN on LSVRC images [7] and fine-tuned on a subset of breast mass images.

Deep learning is becoming a powerful tool for image classification. In medical imaging the lack of images/examples has been described as the main drawback for their development [10]. Considering the use of deep learning methods to address breast mass classification, deep networks are also difficult to train from scratch for breast mass images because of the small number of samples and the existing variance in various mammographic appearances which makes it difficult to combine all available images. Transfer learning and data augmentation have been proposed as promising solutions for overcoming such lack of data [18,27]. As new solutions for this issue in the mammographic mass classification problem, in this study, firstly, data augmentation as was suggested in [18] is extended conceptually to acquire more data to train a deep CNN architecture. In another experiments, two texture discovery techniques are proposed to generate informative mass pattern representations to alleviate the value of train data.

3 Dataset

The benchmark dataset used in our experiments is publicly available Digital Database for Screening Mammography (DDSM) [11] which is a wide ranging annotated film mammography based repository provided by the University of South Florida. DDSM provides nearly 2620 studies of both mediolateral oblique (MLO) and cranio caudal (CC) views of each breast accompanied by a mask if an abnormality is present. The images are grey-level mammograms with the bit depth of 16 bits per pixel.

Using the annotated mass lesions, first of all, 947 cases were selected after manually removing distorted images for our designed experiments. Eventually a subset of 990 and 943 images containing annotated regions of interests (RoIs) were used identifying benign and malignant lesions respectively. The women cases were randomly split into train, validation and test sets by 80%, 10% and 10% of the whole database respectively. Using the approach proposed in [19], we extracted two patch forms (bounding box and double bounding box) from the annotated images and used them as input patches for further procedures described in Sect. 4.1.

In the first and the second experiment, we performed augmentation on the selected train dataset which contained 367 benign cases and 352 malignant cases. In the third experiment, we considered a subset of 400 cases, comprising 200 benign and 200 malignant cases. These cases were randomly selected from the whole women in dataset dedicated for training. Using the annotated RoI, a bounding box of the mass was extracted and subsequently resized into 224×224 pixels to be used as an input patch for further processing that will be explained more in details in Sect. 4.3.

4 Train Data Augmentation Scheme

One of the aims of this study is to compare the effect of performing image preprocessing on the raw extracted patches. Thus, we performed three separate experiments. In the first experiment we used raw image patches with augmentation as the input training data for the deep network. In the second experiments, we used specific filter bank responses from the extracted RoIs. In the third experiment, we used a texton-based approach to generate characteristic features for various types of mass textures and eventually used the resulted textons as the input data for training a convolutional neural network.

4.1 Experiment I

Inspired by [18,19], one approach to alleviate the constraint size of training is to consider the area surrounding the bounding box of an annotated mass. Therefore, two different patch contexts from the training images were selected without implementing any variations in the image intensity. However, data augmentation was performed to compensate for the small amount of data. So, as shown explicitly in Fig. 1, original images, selected for training, are randomly rotated by angles in the $0 < \theta < 360$ range and the following patch contexts are extracted from each rotated image:

1. Bounding box of the annotated mass: (BB).
2. Two times the bounding box (double bounding box) of the annotated mass: (D-BB).

The resulting black corners from the rotation were filled with the mean pixel value of that image in order to avoid significant distortion for the abnormalities that are located near the image borders. In the next step, each patch is resized to 224 along the shorter side and then five random crops are performed on each rotated and extracted context. Finally, mirroring was done resulting in totally 50 augmented images for each training example beside its original sample. The idea behind this augmentation is that not only masses can have different orientations but also the mass neighbourhood may have relevant informative values. Therefore, by these transformations, we were able to generate new labelled samples for a supervised mass lesion classification. This data augmentation scheme is depicted in Fig. 1.

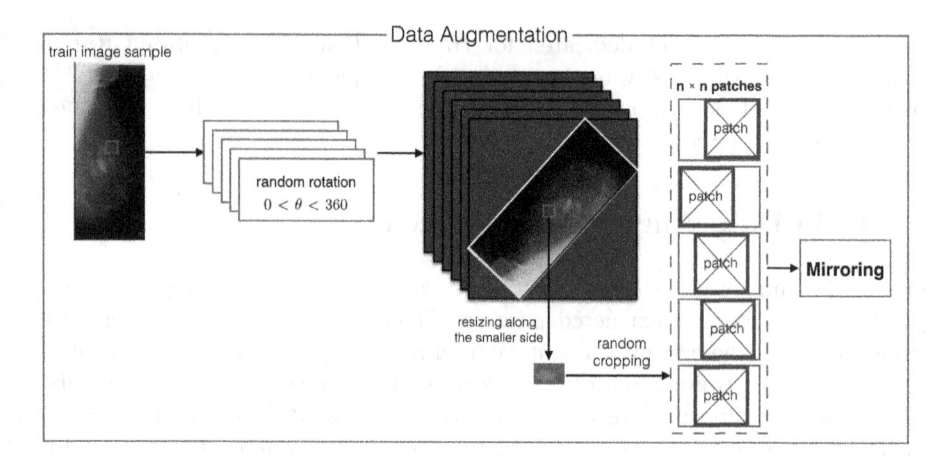

Fig. 1. Data augmentation scheme in experiment I resulting in $5 \times 5 \times 2$ augmented samples.

4.2 Experiment II

In the second experiment, the bounding box of the annotated masses were extracted and resized to 224×224. Samples of malignant and benign mammogram patches used in this experiment are shown in Fig. 2. Inspired by the texton-based methodology proposed for the texture classification [29,30], a texture-based and a texton-based approach using statistical distribution filter responses are used to prepare the input data for the deep network. To achieve this, firstly, we have selected a combination of 53 filters from the proposed Maximum Response (MR) and Schmid (S) filter banks [30] which are depicted in Fig. 3. The first 38 filters are obtained from the MR8 filter bank and are a mixture of edge and bar filters at six orientations and three scales, a Gaussian and a Laplacian of Gaussian filters. The other 13 filters are obtained from the S filter bank which consists of various isotropic Gabor-like filters.

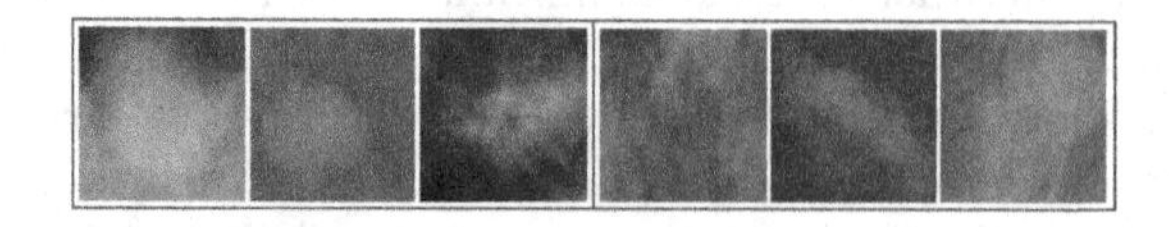

Fig. 2. Samples of malignant (3 left examples) and benign (3 right examples) mammogram patches from DDSM database.

Using the selected filter bank, in the second experiment, the training database was convolved with the mentioned filters to generate filter responses. These responses actually extract various textural information from the RoI images and are possible candidates to be used in the augmented train set. The resulted responses for one RoI example from the train set is presented in Fig. 4.

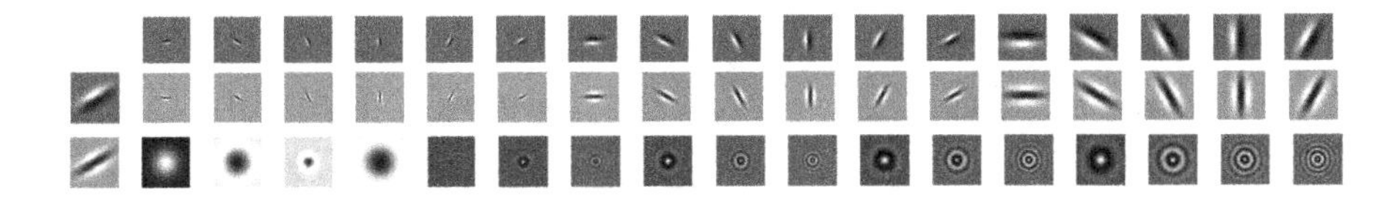

Fig. 3. Combination of MR and S filter banks used to make filter responses for generating a RoI-texton-model in experiment II.

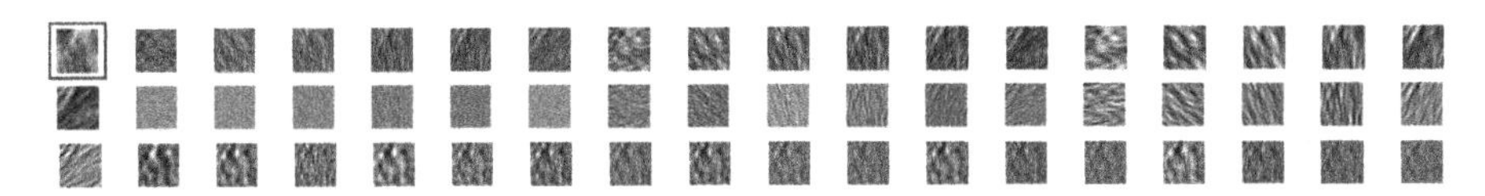

Fig. 4. Filter bank responses on one sample extracted RoI image that is shown in the red box. (Color figure online)

4.3 Experiment III

Subsequently, all the filter responses from the second experiment are normalised to zero mean and unit variance to put them in the same range. After getting filter responses for the selected train dataset, K-mean clustering is performed to define 15 clusters (textons) for each class. The reason to select 15 textons is related to the balance between accuracy and memory requirements. The higher the texton numbers, the higher the precision but at the same time memory requirements will increase.

Aggregating the textons from both classes, a texton dictionary is formed that composes 30 textons. The detailed steps involved in creating textons for each class can be found explicitly in Fig. 5. Afterwards, each filter response from a RoI is assigned to the closest texton from the texton dictionary. Summing up the assigned textons, a new image is generated which we call a RoI-texton-model. Eventually, these created RoI-texton-models for each image are fed into the deep network as augmented samples. No cropping, rotation or mirroring is performed on these texton maps since it can change the inherent information stored in each RoI-texton-model related to a raw sample from a specific class.

The characteristics of the selected filters are: rotationally invariant, isotropic and anisotropic, multi-scale and multiple orientations which make the filters capable of generating good features for various types of mass textures. Thus, the idea behind the last two methodologies is that each filter in the filter bank reveals specific statistical information from the extracted RoIs which is addressed in the filter responses. Subsequently, created clusters of filter response distributions are expected to be mutual representative of a specific class. Additionally, due to the inherent characteristics of benign and malignant abnormalities, the generated clusters or textons in the texton dictionary for each class can be sufficiently discriminative to express the patterns hidden in a benign or a malignant lesion.

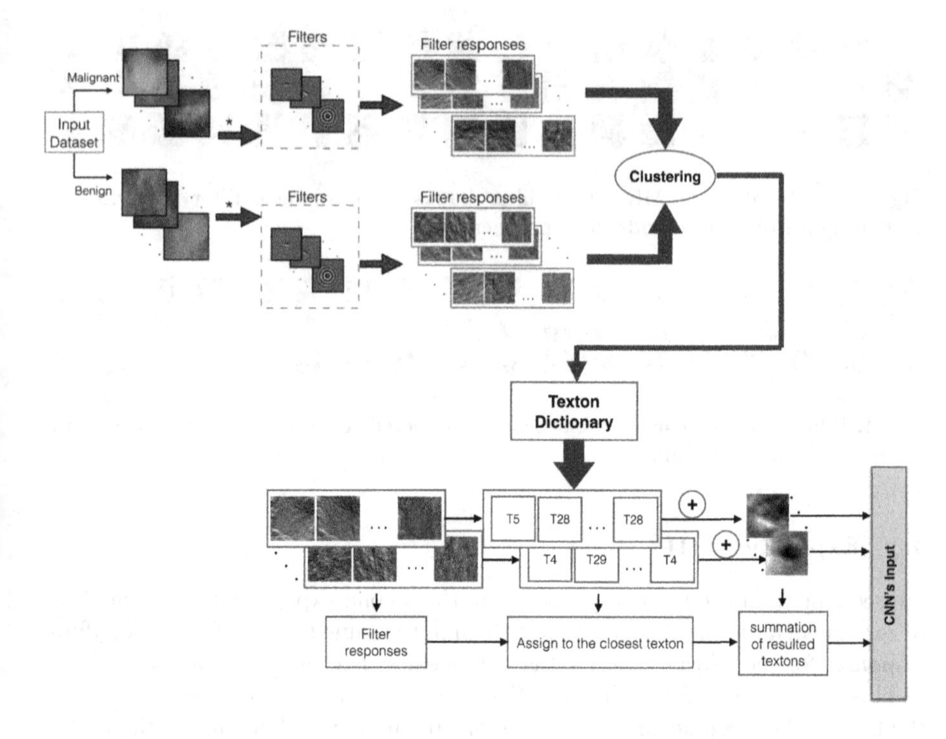

Fig. 5. Process of generating a texton dictionary and RoI-texton-models to augment the training data for CNN.

5 CNN Architecture

The first work that popularised Convolutional Networks in Computer Vision was AlexNet, developed by Krizhevsky *et al.* [18] which was submitted to the ImageNet ILSVRC challenge in 2012 and significantly outperformed the runner-up. The Network has a deep architecture with five featured Convolutional Layers stacked on top of each other followed by pooling and finally three fully connected layers. In this work, we have used the same original architecture in all our experiments. Moreover, the last fully connected layer was changed into 2 outputs representing benign and malignant classes.

On the other hand, transfer learning is a machine learning technique, where knowledge gained during training on one type of problem is used to train on another similar type of problem. Tajbakhsh *et al.* [27] investigated the performance of deep CNNs trained from scratch compared with the pre-trained CNNs fine-tuned in a layer-wise manner. They tested various medical imaging applications from different imaging modalities and consistently showed that the use of a pre-trained CNN with adequate fine-tuning outperformed or, in the worst case, performed as well as a CNN trained from scratch. Some other studies [7,12,17,19] have also confirmed the effectiveness of transfer learning. Accordingly, in this study, we initialised the AlexNet with pre-trained weights on the

ImageNet dataset [24] (which consists of 1000 classes of images and 1.2 million training images) and fine-tuned it on our specific augmented training dataset. Using the same knowledge of ImageNet dataset features, the network is expected to identify mass abnormalities with less samples and training time. For this aim, convolutional layers are initialised with the pre-trained weights to identify features of the problem while the fully connected layers are removed from the trained network and retrained with fresh layers by a Gaussian distribution for our target classification task. Table 1 gives the specific parameter values we used during training AlexNet in our experiments. We also performed mean image subtraction during training the network. All classification implementations are performed within the Caffe framework [14] and the pre-trained model weights from the Caffe library and the visualisation facility of NVIDIA-DIGITS [1] are also utilised. Moreover, the computations were carried out using a NVIDIA GeForce GTX 1080 GPU on Intel Core i7-4790 Processor with Ubuntu 16.04 operating system.

Table 1. AlexNet parameter information.

Optimisation scheme	SGD
Base learning rate	0.001
Learning policy	step-down
Weight decay	0.005
Momentum	0.9
Dropout rate	0.5
Epochs	30
Train batch size	128
Validation batch size	64

6 Results and Discussion

Using the pre-trained AlexNet model with the mentioned parameter values and fine-tuning on augmented training data from experiment I (AUG-1) and using the unseen test dataset, classification results of a deep network trained on two extracted context sizes are reported in Table 2. First of all, selecting a larger bounding box (double size) around the annotated mass to achieve higher classification accuracy (Acc: 81%) is confirmed since neighbouring context of the abnormality contains useful information which adds to the value of the samples considered in training data.

Moreover, adding the filter responses to the augmented data from experiment I, increased the accuracy result of the classification (85.19%) which we show with AUG-2. To understand why such accuracy is achieved, we should understand how

[1] https://developer.nvidia.com/digits.

the network manages to achieve mass classification ability in both experiments. In AlexNet, the most interpretable convolution layer is filters in the first layer which are shown in Fig. 6. These filters can capture horizontal, vertical and parallel edges. Moreover, some of them have diagonal orientations and some look for high frequencies. This filter bank is applied to the input image and the maximum response from each 11×11 block is taken by pooling, giving these filter responses to the next layer. These are the low level features that are extracted from the input images. However, the filter responses from our suggested filter bank, in experiment II, are applied on the whole extracted RoI. These filters are proved to extract the most significant features from texture patterns [30]. Furthermore, some of these filters are not included in the first layer of the CNN which trys to captures low level features. So, the idea was to impose such filter responses to the network and we did it one time with bounding box contexts and another time by double bounding box contexts. The results showed better results while using filter responses from bounding box around the RoIs (85.8% using the bounding box compared to 85.1% using responses from double the bounding box around the RoIs). The reason can be interpreted as a possible zooming behaviour that can happen while using a bounding box context among double the bounding box extracted samples which eventually helped to the learning procedure.

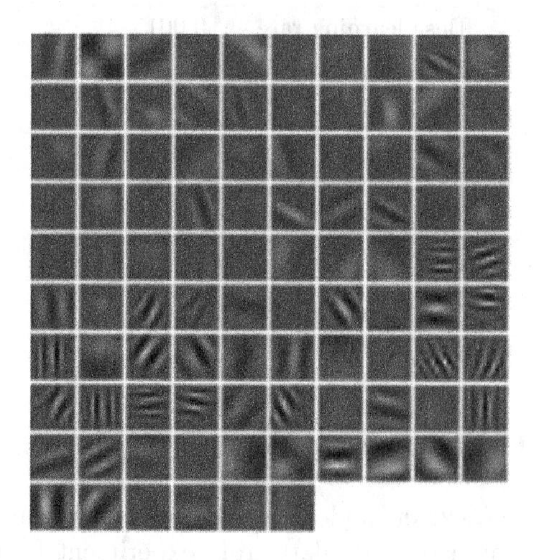

Fig. 6. Generated filters of the first convolutional layer in AlexNet.

On the other side, while making a RoI-texton-model of some random train samples, clusters of salient textural patterns for each class capture the relevant information related to one specific class and presents such textural characteristics in the resulted RoI-texton-model. The more textons selected from one class, the more common behaviour of that class is recorded in the RoI-texton-model. Considering this effect and adding this augmentation on a subset of training data to the previously augmented data, eventually the trained network can extracts more discriminative features for classification and obtaining 87% accuracy.

Comparing the results reported in the confusion matrix, TP rate representing the number of true classified malignant masses are significantly improved and FN rate representing the number of misclassified malignant cases as benign cases are decreased which is very crucial in developing CAD systems. Beside these improvements, by increasing the number of TN which represents the number of true classified benign cases the need for further biopsy as an invasive procedure can be avoided.

Table 2. Classification results using various augmentations. (BB: bounding box of the training RoI image, D-BB: double bounding box of the training RoI image; AUG-1/2/3 represent augmentation method explained in experiments I, II and III respectively)

Type of training data	TN	FP	FN	TP	Acc
AUG-1 on BB	53	12	48	49	62.96%
UG-1 on D-BB	73	13	17	59	81.48%
AUG-1 & AUG-2 on D-BB	71	15	9	67	85.19%
AUG-1 on D-BB & AUG-2 on BB	72	14	9	67	85.8%
AUG-1 on D-BB, AUG-2 & AUG-3 on BB	76	10	11	65	87.04%

Comparing classification results with the competing methods, it is comparable to the current approach outcomes [5,19]. However, using different number of images in DDSM dataset and various evaluation metrics does not make a fair comparison. Nevertheless, the augmentation effect in improving the final result was significant that was not covered in the previous studies.

7 Conclusion

In this study, a supervised deep learning based approach is employed to address mass lesion classification from mammographic data. For this aim, we have proposed and compared various ways of train data augmentation for training the AlexNet architecture. In the first experiment, an extended augmentation scheme is applied on the raw extracted annotated mass and the network learns directly from this training data. In the second experiment, responses of a suggested filter bank are also fed to the network. Finally, in the third experiment, a texton-based approach is utilised to model pre-detected mass abnormalities in mammograms in the form of particular RoI-texton-models to explain salient features contained in each input image. Subsequently, these processed RoI-texton-models are fed to the CNN architecture along with the other augmented train samples. We evaluated the results on the DDSM dataset and obtained the accuracy of 87% which is comparable to the current approach outcomes and can be considered in real world settings.

Acknowledgments. The authors would like to gratefully acknowledge Sandy Spence and Alun Jones for their support and maintenance of the GPU and the systems used for this research.

References

1. Breasr Cancer Biopsy (2015). http://www.breastcancer.org/symptoms/testing/types/biopsy
2. National Health Service-Breast screening: professional guidance, 31 August 2016. https://www.gov.uk/government/collections/breast-screening-professional-guidance
3. Arevalo, J., González, F.A., Ramos-Pollán, R., Oliveira, J.L., Lopez, M.A.G.: Representation learning for mammography mass lesion classification with convolutional neural networks. Comput. Methods Programs Biomed. **127**, 248–257 (2016)
4. Buciu, I., Gacsadi, A.: Directional features for automatic tumor classification of mammogram images. Biomed. Sig. Process. Control **6**(4), 370–378 (2011)
5. Carneiro, G., Nascimento, J., Bradley, A.P.: Unregistered multiview mammogram analysis with pre-trained deep learning models. In: Navab, N., Hornegger, J., Wells, W., Frangi, A. (eds.) MICCAI 2015. LNCS, vol. 9351, pp. 652–660. Springer, Cham (2015). doi:10.1007/978-3-319-24574-4_78
6. Cheng, S.C., Huang, Y.M.: A novel approach to diagnose diabetes based on the fractal characteristics of retinal images. IEEE Trans. Inf. Technol. Biomed. **7**(3), 163–170 (2003)
7. Deng, J., Dong, W., Socher, R., Li, L.J., Li, K., Fei-Fei, L.: ImageNet: a large-scale hierarchical image database. In: IEEE Conference on Computer Vision and Pattern Recognition, 2009, CVPR 2009, pp. 248–255 (2009)
8. Dhungel, N., Carneiro, G., Bradley, A.P.: Automated mass detection in mammograms using cascaded deep learning and random forests. In: IEEE International Conference on Digital Image Computing: Techniques and Applications (DICTA), pp. 1–8 (2015)
9. Fonseca, P., Mendoza, J., Wainer, J., Ferrer, J., Pinto, J., Guerrero, J., Castaneda, B.: Automatic breast density classification using a convolutional neural network architecture search procedure. In: SPIE Medical Imaging, vol. 9414 (2015)
10. Greenspan, H., van Ginneken, B., Summers, R.M.: Guest editorial deep learning in medical imaging: overview and future promise of an exciting new technique. IEEE Trans. Med. Imaging **35**(5), 1153–1159 (2016)
11. Heath, M., Bowyer, K., Kopans, D., Moore, R., Kegelmeyer, W.P.: The digital database for screening mammography. In. In Proceedings of the 5th International Workshop on Digital Mammography, pp. 212–218. Medical Physics Publishing (2001)
12. Huynh, B.Q., Li, H., Giger, M.L.: Digital mammographic tumor classification using transfer learning from deep convolutional neural networks. J. Med. Imaging **3**(3), 034501 (2016)
13. Jamieson, A.R., Drukker, K., Giger, M.L.: Breast image feature learning with adaptive deconvolutional networks. In: SPIE Medical Imaging, vol. 8315 (2012)
14. Jia, Y., Shelhamer, E., Donahue, J., Karayev, S., Long, J., Girshick, R., Guadarrama, S., Darrell, T.: Caffe: convolutional architecture for fast feature embedding. arXiv preprint arXiv:1408.5093 (2014)
15. Jiao, Z., Gao, X., Wang, Y., Li, J.: A deep feature based framework for breast masses classification. Neurocomputing **197**, 221–231 (2016)

16. Kallenberg, M., Petersen, K., Nielsen, M., Ng, A.Y., Diao, P., Igel, C., Vachon, C.M., Holland, K., Winkel, R.R., Karssemeijer, N.: Unsupervised deep learning applied to breast density segmentation and mammographic risk scoring. IEEE Trans. Med. Imaging **35**(5), 1322–1331 (2016)
17. Kooi, T., et al.: A comparison between a deep convolutional neural network and radiologists for classifying regions of interest in mammography. In: Tingberg, A., Lång, K., Timberg, P. (eds.) IWDM 2016. LNCS, vol. 9699, pp. 51–56. Springer, Cham (2016). doi:10.1007/978-3-319-41546-8_7
18. Krizhevsky, A., Sutskever, I., Hinton, G.E.: ImageNet classification with deep convolutional neural networks. In: Advances in Neural Information Processing Systems, pp. 1097–1105 (2012)
19. Lévy, D., Jain, A.: Breast mass classification from mammograms using deep convolutional neural networks. 30th Conference on Neural Information Processing Systems (NIPS), Barcelona, Spain, arXiv preprint arXiv:1612.00542 (2016)
20. Oliver, A., Freixenet, J., Marti, J., Pérez, E., Pont, J., Denton, E.R., Zwiggelaar, R.: A review of automatic mass detection and segmentation in mammographic images. Med. Image Anal. **14**(2), 87–110 (2010)
21. Petersen, K., Chernoff, K., Nielsen, M., Ng, A.Y.: Breast density scoring with multiscale denoising autoencoders. In: STMI workshop at MICCAI, vol. 2012 (2012)
22. Pinto, N., Doukhan, D., DiCarlo, J.J., Cox, D.D.: A high-throughput screening approach to discovering good forms of biologically inspired visual representation. PLoS Comput. Biol. **5**(11), e1000579 (2009)
23. Ranzato, M., Poultney, C., Chopra, S., Cun, Y.L.: Efficient learning of sparse representations with an energy-based model. In: Advances in Neural Information Processing Systems, pp. 1137–1144 (2006)
24. Russakovsky, O., Deng, J., Su, H., Krause, J., Satheesh, S., Ma, S., Huang, Z., Karpathy, A., Khosla, A., Bernstein, M., BergLi, A.C., Fei-Fei, L.: Imagenet large scale visual recognition challenge. Int. J. Comput. Vis. **115**(3), 211–252 (2015)
25. Serifovic-Trbalic, A., Trbalic, A., Demirovic, D., Prljaca, N., Cattin, P.: Classification of benign and malignant masses in breast mammograms. In: 37th International Convention on Information and Communication Technology, Electronics and Microelectronics (MIPRO), pp. 228–233. IEEE (2014)
26. Stewart, B., Wild, C.P.: International Agency for Research on Cancer, W: World Cancer Report (2014)
27. Tajbakhsh, N., Shin, J.Y., Gurudu, S.R., Hurst, R.T., Kendall, C.B., Gotway, M.B., Liang, J.: Convolutional neural networks for medical image analysis: full training or fine tuning? IEEE Trans. Med. Imaging **35**(5), 1299–1312 (2016)
28. Vaidehi, K., Subashini, T.: Automatic characterization of benign and malignant masses in mammography. Procedia Comput. Sci. **46**, 1762–1769 (2015)
29. Varma, M., Zisserman, A.: Texture classification: are filter banks necessary? In: IEEE Conference on Computer Vision and Pattern Recognition, vol. 2, pp. II-691. IEEE (2003)
30. Varma, M., Zisserman, A.: A statistical approach to texture classification from single images. Int. J. Comput. Vis. **62**(1–2), 61–81 (2005)

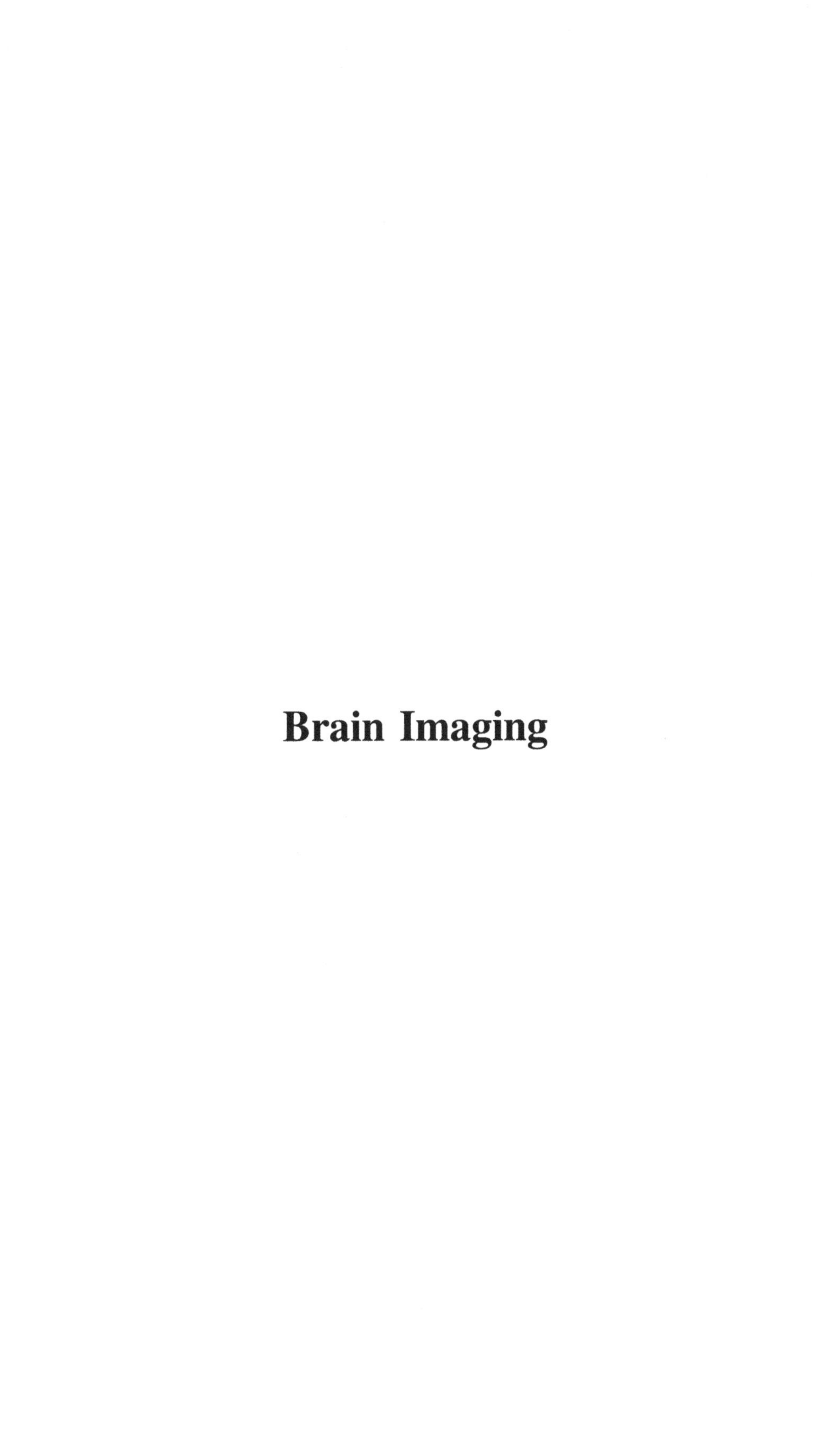

Brain Imaging

Learning Longitudinal MRI Patterns by SICE and Deep Learning: Assessing the Alzheimer's Disease Progression

Andrés Ortiz[1(✉)], Jorge Munilla[1], Francisco J. Martínez-Murcia[2], Juan M. Górriz[2], Javier Ramírez[2], and for the Alzheimer's Disease Neuroimaging Initiative

[1] Communications Engineering Department, University of Málaga, 29004 Málaga, Spain
aortiz@ic.uma.es

[2] Department of Signal Theory, Communications and Networking, University of Granada, 18060 Granada, Spain

Abstract. Automatic knowledge extraction from medical images constitutes a key point in the construction of computer aided diagnosis tools (CAD). This takes a special relevance in the case of neurodegenerative diseases such as the Alzheimer's disease (AD), where an early diagnosis makes the treatments easier and more effective. Moreover, the study of the evolution of the illness results crucial to differentiate the neurodegenerative process associated to the disease from the natural degeneration due to the ageing process. In this paper we present a method to construct longitudinal models from subjects using a series of MRI images. Specifically, the method presented here aims to model Gray matter (GM) variation at different brain areas of a subject across subsequent examinations, being possible to relate those regions which degenerate jointly. Hence, it allows determining variation patterns that differentiate controls from AD patients. Additionally, White matter (WM) density is also incorporated to the longitudinal model to complement the information provided by GM. The results obtained demonstrated the effectiveness of the method in the extraction of these patterns, that can be used to classify between Controls (CN) and AD subjects with 94% of accuracy, outperforming other previous methods.

1 Introduction

Alzheimer's Diasese (AD) is the most common cause of dementia among older people. Nowadays, 44 million people worldwide and it is expected AD affects

Alzheimer's Disease Neuroimaging Initiative—Data used in preparation of this article were obtained from the Alzheimer's Disease Neuroimaging Initiative (ADNI) database (adni.loni.ucla.edu). As such, the investigators within the ADNI contributed to the design and implementation of ADNI and/or provided data but did not participate in analysis or writing of this report. A complete listing of ADNI investigators can be found at: http://adni.loni.ucla.edu/wpcontent/uploads/how_to_apply/ADNI_Acknowledgement_List.pdf.

M. Valdés Hernández and V. González-Castro (Eds.): MIUA 2017, CCIS 723, pp. 413–424, 2017.
DOI: 10.1007/978-3-319-60964-5_36

60 million people worldwide over the next 50 years. AD progressively causes the loss of nerve cells, whose symptoms usually start with mild memory problems, turning into severe brain damage in several years. The early diagnosis of AD results crucial, since there is no cure and currently developed drugs can only help to temporarily slow down the progression of the disease [3]. This way, the use of current functional or structural image systems have provided a way to explore new insights of the disease and to improve the diagnosis accuracy. In fact, many previous works used functional images such as Single Emission Computerized Tomography (SPECT) [8,14,22] or Positron Emission Tomography (PET) [1,23] to reveal functional differences between controls and AD patients and then to evaluate the loss of brain functions as the disease progress. On the other hand, since AD also causes structural changes in the brain, these can be figured out by means of Magnetic Resonance Image (MRI) analysis [5,6,16,17]. These works use GM or WM images on whole brain volume to classify controls and AD patients [5,6] or to compute Regions of Interest (ROI), searching for common patterns in Controls (CN) and AD subjects. On the other hand, the construction of neurodegeneration models to study the progression of the disease plays an important role for a better understanding of the neurodegeneration process. The main goal of this work, unlike previous works, is to model the degeneration process in different brain areas for individual subjects from a series of MRI images, rather than searching for static patterns in CN subjects or AD patients. This is addressed by modelling the regional grey matter density by means of the Sparse Inverse Covariance (SIC) and building a per-subject model that shows the relationship between areas in which the GM density covaries. This has important implications from the functional point of view, as positive covariance in GM evidences the existence of a fiber tract connection between these two areas [4,21]. These models, as explained later, can be further used to extract patterns that can be used to discriminate between CN and AD. Although most part of the information regarding the neurodegeneration process associated to AD is included in GM patterns, we also included WM density covariation to the model, as it provides complementary information that leverages the classification accuracy. On the other hand, we opted to use a deep-learning (DL) classifier instead of statistical classifiers such as SVM, since DL has demonstrated its effectiveness in extracting relevant information from noisy patterns [18]. In particular, we use a stacked denoising autoencoder to train a deep neural network which is fine-tuned by backpropagation.

The rest of the paper is organized as follows. Section 2 describes the database used in this work. Section 3 introduces the methods used in this work, including image preprocessing, density computation and the methods used for longitudinal modelling (and feature extraction) and classification in Subsects. 3.1, 3.2, 3.3, and 3.4, respectively. Section 4 shows details on the conducted experiments and the results obtained using longitudinal MRI data from the ADNI database. Finally, the main conclusions are drawn in Sect. 5.

2 Database

The database used in this work contains longitudinal MRI image data from 138 subjects, comprising 68 Controls (CN), and 70 AD patients from the ADNI database [2], comprising one evaluation each 6 months per subject in a period of three years. Thus, 6 evaluations are usually available for each subject, although in a few subjects, only 4 or 5 images were available in the database. This repository, which was created to study AD and provide a means for its early diagnosis, collects a vast amount of MRI and PET images as well as blood biomarkers and cerebrospinal fluid analyses. In this work, however, only MRI data have been used. Patients' demographics are shown in Table 1.

Table 1. Patient demographics. Mini-Mental State Examination Scores (MMSE) are indicated for diagnosis reference

Evaluation	Sex (M/F)	Mean age $\pm$ std (1st visit)	Mean MMSE $\pm$ std (1st visit)
NC	61/61	75.74 ± 5.12	29.22 ± 0.02
AD	47/49	75.19 ± 7.49	23.14 ± 1.94

3 Methods

3.1 Image Preprocessing

MRI images from the ADNI database have been spatially normalized according to the Voxel-based morphometry (VBM) T1 template and segmented into White Matter (WM) and Grey Matter (GM) tissues using the VBM toolbox for Statistical Parametric Mapping (SPM) software [7,24]. This ensures each image voxel corresponds to the same anatomical position. After image registration, all the images from the ADNI database were resized to $121 \times 145 \times 121$ voxels with voxel-sizes of 1.5 mm (sagittal) $\times$ 1.5 mm (coronal) $\times$ 1.5 mm (axial). MRIs are further segmented to obtain information about GM and WM tissue distributions, which can be used to differentiate AD from CN patients [13,16,17]. This process is guided by means of tissue probability maps of grey matter, white matter or cerebro-spinal fluid. A nonlinear deformation field is estimated that best overlays the tissue probability maps on the individual images. The tissue probability maps provided by the International Consortium for Brain Mapping (ICBM) are derived from 452 T1-weighted scans, which were aligned with an atlas space, corrected for scan inhomogeneities, and classified into grey matter, white matter and cerebro-spinal fluid. Segmentation through SPM/VBM provides values in the range $[0, 1]$ which denote the membership probability to a specific tissue.

3.2 Density Computation

Features used in this work are based on brain volumes at specific regions delineated by the Automatic Anatomical Labelling atlas (AAL). Although this brain atlas delimitates 116 regions, we just used 42 of them which are considered the most closely related to AD [11,19,25], discarding cerebellum and vermis regions. The volume corresponding to the i-region can be computed using the following expression,

$$\mathrm{Vol}_i = \frac{\#voxels_i > thr}{1000} * \text{voxel size} \tag{1}$$

where thr denotes the probability threshold that indicates whether a voxel belongs to a specific tissue and voxel size is indicated in mm^3. Similarly, tissue density for each region can be computed as follows,

$$D_i = \frac{\#voxels_i > thr}{Vol_i} \tag{2}$$

Since SPM segmentation provides the membership probability values to a specific tissue for each voxel, the threshold thr, in Eqs. 1 and 2, indicates how the partial volume effect is taken into account. Thus, the lower the thr the less the relative importance of one tissue over the other, and for thr=1 no partial volume effect is taken into account. In this work, we selected $thr \geq 0.3$, meaning that voxels with a value ≥ 0.3 in the GM probability map are considered belonging to GM.

3.3 Sparse Inverse Covariance Estimation (SICE)

The characterization of the tissue covariation between brain areas is usually addressed by correlation analysis. Correlation between two variables is, however, a necessary but not sufficient condition for a causal relationship. In other words, a correlation between two variables captures pairwise information but does not guarantee that the occurrence of one variable causes the other, as the correlation could be originated by third-party effects. Partial correlation can be used then to effectively characterize the interaction of two brain areas varying together while factoring out the influence of the rest of the regions [20]. When applied to AD, this mathematical tool can be used to identify patterns associated to cerebral neurodegeneration by identifying conditional independence between regions given the rest as constant. Partial correlations, in turn, coincide with the off-diagonal entries of the inverse covariance matrix, also known as precision matrix. Partial correlations are thus usually computed using the Maximum Likelihood Estimation (MLE) of the inverse covariance matrix. MLE, however, is not recommended when the sample size is not considerably higher than the number of variables; e.g. the number of patients is not higher than the number of regions of interest. In those cases, and taking into account the inherent sparseness of the brain network [27], sparse computation can be employed [9]. Sparse Inverse Covariance Estimation (SICE), also known as known as Gaussian graphical model or graphical LASSO (Least Absolute Shrinkage and Selection Operator), uses a regularization parameter that controls the number of zero entries. Next, we explain this with more detail.

Let $\mathbf{x_1}, \mathbf{x_2}, ..., \mathbf{x_n} \sim \mathcal{N}(\boldsymbol{\mu}, \Sigma)$ denote p-dimension vectors, corresponding to n samples measured at p selected ROIs, which follow a multivariate Gaussian distribution where $\boldsymbol{\mu} \in \mathbb{R}^p$ is the mean and $\Sigma \in \mathbb{R}^{p \times p}$ is the covariance, then the empirical covariance is:

$$S = \frac{1}{n} \sum_{i=1}^{n} (\mathbf{x_i} - \mu)(\mathbf{x_i} - \mu)^T. \tag{3}$$

Let also $\Theta = \Sigma^{-1}$ be the inverse covariance (or precision) matrix, the maximum log likelihood estimation (or MLE) of Θ under a multivariate Gaussian model can be obtained as follows:

$$\widehat{\Theta} = \underset{\Theta \succ 0}{argmax} \left(\log(\det \Theta) - tr(S\Theta)\right), \tag{4}$$

where $tr(S\Theta)$ is the trace of $(S\Theta)$. If S is not singular, by deriving with regards to Θ and setting it to zero, we would get, as expected, that the estimate of the inverse covariance is $\widehat{\Theta} = S^{-1}$. However, because $p > n$ the empirical estimate of S becomes singular and a regularization must be applied so that a shrunken estimate of Θ can be obtained through a maximization of the penalized log likelihood function. In particular, an estimate of the inverse covariance matrix $\widehat{\boldsymbol{\Theta}}$ of the brain regions is computed by solving the following optimization problem using the algorithm proposed in [12]:

$$\widehat{\boldsymbol{\Theta}} = \underset{\Theta \succ 0}{argmax} \left(\log(\det \Theta) - tr(S\Theta) - \lambda ||\Theta||_1\right), \tag{5}$$

where $|| \cdot ||_1$ denotes the "entrywise" l_1-norm regularization [11], which corresponds to the sum of absolute values of all the entries in a matrix, and $\lambda > 0$ is the pre-selected regularization parameter. The larger the value of λ the more sparse are the estimates for Θ provided by SICE. Conversely, when λ is small the constraint has little effect and SICE becomes the conventional MLE.

SICE reports conditional independence between two variables (given the other variables in the multivariate Gaussian distribution) and therefore, once it is computed, neurodegenerative patterns can be developed. Two brain regions are connected if and only if they are not conditionally independent.

3.4 Deep Learning-Based Classifier

Deep learning (DL) classifiers have demonstrated their effectiveness under noisy patterns. Moreover, they discover the underlying structure of the data and model it, allowing the computation of representative and/or discriminative features from the original patterns [10,18]. Thus, we used a deep-learning based classifier to learn patterns from the covariance values computed using the SICE method described above. Specifically, we used a stacked denoising autoencoder (SDA) to train a deep neural network in a stepwise manner. Stacked autoencoders compute representative features in their most inner layer while trying to reconstruct the original sample. Note also that, as this is the layer with fewer neurons, it usually becomes the bottleneck in the neural structure. Subsequently, these most inner layer from each autoencoders are used as hidden layers of a deep neural network. Positive saturating linear transfer functions are used at the encoder part and Linear transfer function at the decoder part of the autoencoders. The learning rate for the scaled conjugate gradient descent training algorithm was selected to 0.005. These learning parameters make the network to converge in less than 500 iterations. Additionally, the training phase is stopped when no improvements in the loss function are obtained during 100 iterations to avoid overfitting.

At the top most, a single unit softmax layer is included to implement the classifier. This method provides the unsupervised pre-training stage of the deep neural structure, which is further fine-tuned by backpropagation using not the samples (as in the autoencoder phase) but the labels and a learning rate of 0.001. This way, the first stage using autoencoders is focused in minimizing the representation error provided by the features computed in the hidden layer, while the second stage tries to minimize the classification error. Figure 1 shows the neural structure used in this work, where the first stage consists of three autoencoders (with 1000, 500 and 30 neurons in the hidden layer) and the deep neural network contains 5 layers (1 visible layer, 3 hidden layers and 1 softmax layer).

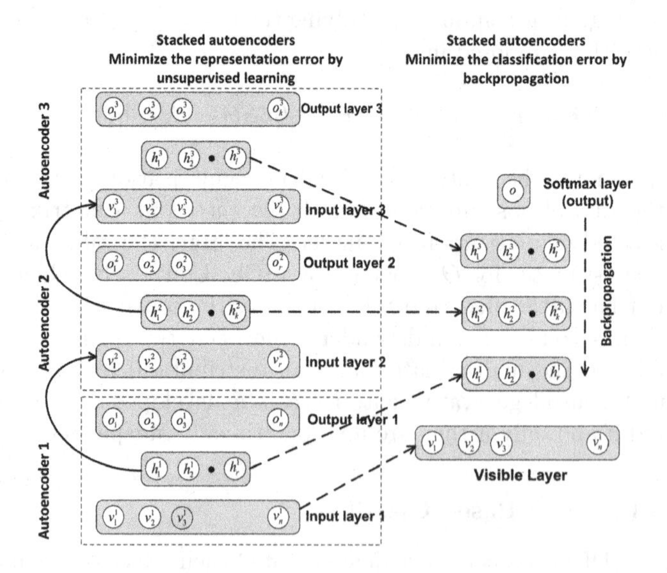

Fig. 1. Deep learning architecture used for classification. Unsupervised pre-training is addressed by three autoencoders using 1000, 500 and 30 neurons, respectively in the hidden layer. The multilayer perceptron is composed of autoencoder hidden layers and a softmax layer. Fine-tuning is performed by backpropagation.

In order to improve the robustness of the classifier and its noise immunity, inputs at the first autoencoder are corrupted by adding white gaussian noise [26] with a certain power level to keep a specific Signal-to-Noise ratio (SNR). This way, the autoencoder will be trained to reconstruct each input from noisy versions. In this work, we experimentally determined a SNR = 30 dB as it provides the best results, which means that the input units at the first autoencoder are corrupted with gaussian noise while keeping a SNR of 30 dB for each sample.

4 Experimental Results

For experimental results the deep learning architecture described in the previous section was applied to GM and WM data in the database as shown in Fig. 2. GM and WM voxels from 42 regions closely related to AD according to the literature [11] were used to compute the regional GM and WM densities for all the available examinations (visits) of each subject in the database. This results in a $N \times 42$ matrix for each subject, where N is the number of available examinations. These matrices are then used to compute the corresponding SICE; i.e. an estimate of the corresponding inverse covariance between regions.

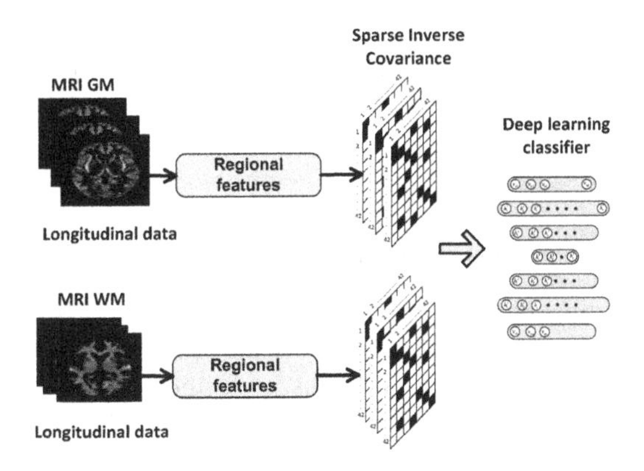

Fig. 2. Deep learning architecture applied to MRI-GM and MRI-WM data through SICE. Regional features refer to regional densities computed according to Eq. 2 and as indicated in [19]

Besides the discriminative information contained in SICE data (as we will see later), these also allows us to carry out certain exploratory analysis by identifying covarying regions. Figure 3 shows these data graphically for $\lambda = 0.05$, when $thr = 0.3$ is applied (see Subsect. 3.2). As shown in Subsect. 3.3, larger λ values provide sparser SICE matrices (i.e. more zero entries) and only stronger relationships between regions are kept. On the other hand, small λ values provide less sparse SICE matrices, capturing weak between-regions relationships that can introduce non-relevant information to the classifier. Hence, value $\lambda = 0.05$ was selected by experimentation as it is a good trade-off between sparsity and computational burden, at the time it provided the best classification results. In the same way, the probability threshold to determine whether a voxel is considered to belong to GM or WM has also been determined by experimentation to obtain the best classification results. Thus, these values of the parameters have been used for the rest of this work. Finally, we also note that to feed the deep learning architecture only the lower triangle elements are employed, since SICE matrix is symmetric (diagonal elements are also discarded).

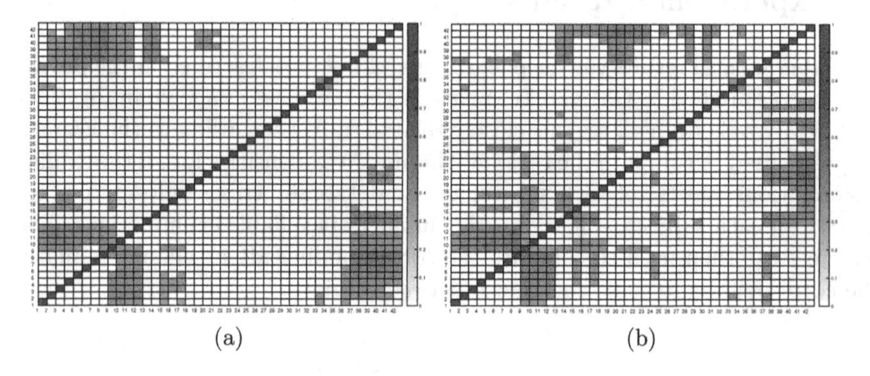

Fig. 3. Representation of the mean SICE ($\lambda = 0.05$) for CN (a) and AD (b) subjects thresholded by 0.3

4.1 Disease Patterns and Diagnostic Relevance of Brain Regions

Information regarding regions which covary without any third-party influence can be extracted from the SICE model. This shows regions whose GM/WM densities vary jointly during the examinations performed in the three-years period. This way, variations due to the ageing process as well as variations due to a possible neurodegenerative process are captured by the inverse covariance. Hence, analysing statistically significant differences between CN and AD SICEs by means of the Wilcoxon test, we found different between-region relationships. In Fig. 4, we show these relationships using BrainNet Viewer [15] where edges connect the 20 most significant covarying regions according to the rank provided by the Wilcoxon test. As shown in this figure, the parahipoccampal and hippocampal regions are related to the rectus and frontal orbital regions, indicating that the GM density in these regions varies jointly in AD.

4.2 Classification

Covariance values computed by SICE determine a per-subject, compact longitudinal model. In fact, we used these values as features to train the deep neural structure described above to classify subjects. The overall method here presented

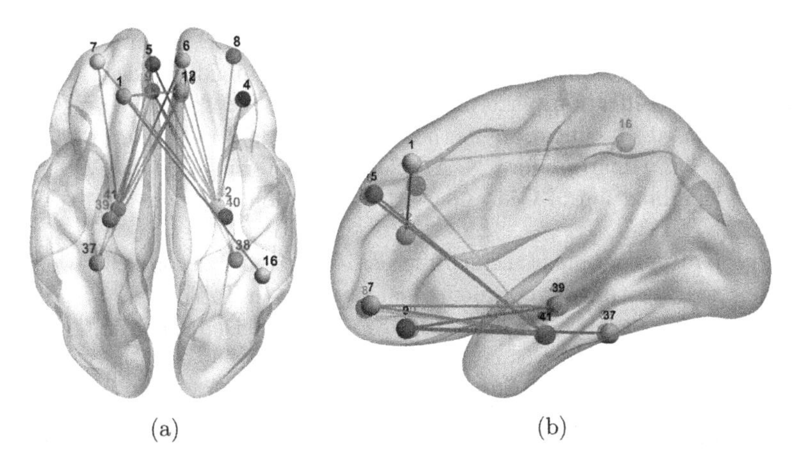

Fig. 4. Axial and sagittal views of the surface model of the WM of a representative brain, generated using brainviewer [15] with the 42 ROIs selected and their association (indicated with a line between points). An edge indicates covariation between regions. 1 = Left Precentral Gyrus; 4 = Right superior frontal gyrus (dorsolateral); 5 = Left superior frontal gyrus (dorsolateral); 6 = Right superior frontal gyrus (orbital); 7 = Left middle frontal gyris (lateral); 8 = Right middle frontal gyris (lateral); 9 = Left middle frontal gyrus (orbital), 10 = Right middle frontal gyrus (orbital); 12 = Right opercular part of inferior frontal gyrus; 16 = Right Orbital part of inferior frontal gyrus; 37 = Left Hippocampus; 38 = Right Hippocampus; 39 = Left parahippocampal gyrus; 40 = Right parahippocampal gyrus; 41 = Left amygdala; 42 = Right amygdala

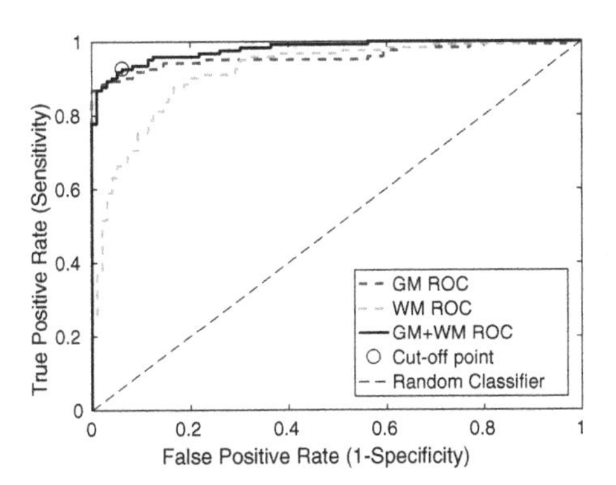

Fig. 5. ROC curve for SICE features computed from GM and WM. Classification performance using GM and WM features at the same time is also shown.

was assessed by k-fold cross-validation (specifically, k = 10), which ensures that training and testing subsets do no share any sample and estimates the generalization error. Hence, testing samples are never used during the training stage and double-dipping is avoided.

The ROC curve in Fig. 5 summarizes the classification performance using GM, WM and information from both tissues simultaneously. As expected, this figure shows that most part of information linked to AD is contained in GM, while WM provides some extra knowledge not contained in GM that slightly leverages the classification performance.

The proposed classification method provides an Area under ROC curves (AUC) of 0.94 and 0.90 and for CN/AD classification using GM and WM, respectively. The combination of GM and WM information improves the AUC up to 0.98.

Additionally, experiments using different classification approaches were carry out. The results of these experiments, summarized in Table 2, show the superiority of the deep learning approach based on Stacked Denoising Autoencoders.

Table 2. Results obtained for different classification approaches using the same descriptors. Standard deviations obtained through k-fold (k = 10) cross-validation is indicated.

Method	Accuracy	Sensitivity	Specificity	AUC
Decision tree	0.90 ± 0.11	0.86 ± 0.10	0.97 ± 0.04	0.90
Linear SVM	0.90 ± 0.09	0.88 ± 0.09	0.93 ± 0.07	0.92
Linear LP-SVM	0.85 ± 0.09	0.84 ± 0.10	0.87 ± 0.12	0.85
DL-SDA	$\mathbf{0.94 \pm 0.07}$	$\mathbf{0.90 \pm 0.08}$	$\mathbf{0.96 \pm 0.07}$	**0.98**

5 Conclusions and Future Work

This paper proposes a method to compute longitudinal models from tissue densities at different brain regions defined by the AAL atlas. In the conducted experiments these densities were used as features to assess their capability to distinguish between CN and AD subjects from longitudinal MRI data. Thus, the exploratory analysis performed using the SICE model on density data obtains discriminant regions corresponding to those found in medical literature, such as the hippocampus in both hemispheres. In addition to these, relationships between hippocampal areas and other regions such as the inferior temporal gyrus have also been revealed. Finally, classification experiments (using k-fold cross-validation) were performed using SICE features computed from GM and WM density values, showing an accuracy of up to 94%, and AUC of up to 0.98.

As future work we plan to include other biomarkers jointly with GM and WM densities in order to study possible interdependencies in CN and AD subjects. The development of new methods to corrupt the inputs, specially for MRI data, becomes also particularly relevant to improve the robustness of the DL classifiers. Lastly, the proposed method can be extended to MCI subjects, which is an interesting task that could reveal neurodegeneration patterns moving towards the early diagnosis of AD.

Acknowledgements. This work was partly supported by the MINECO/FEDER under TEC2015-64718-R and PSI2015-65848-R projects and the Consejería de Innovación, Ciencia y Empresa (Junta de Andalucía, Spain) under the Excellence Project P11-TIC-7103.

References

1. Alvarez, I., Górriz, J., Ramírez, J., Salas-González, D., Lopez, M., Segovia, F., Chaves, R., Gomez-Rio, M., García-Puntonet, C.: 18F-FDG PET imaging analysis for computer aided Alzheimer's diagnosis. Inf. Sci. **184**(4), 903–916 (2011)
2. Alzheimer's Disease Neuroimaging Initiative. http://adni.loni.ucla.edu/. Accessed May 2017
3. Alzheimer's Disease Society: Factsheet: Drug Treatments for Alzheimer's Disease, February 2017. https://www.alzheimers.org.uk. Accessed May 2017
4. Chen, Z.J., He, Y., Rosa-Neto, P., Germann, J., Evans, A.C.: Revealing modular architecture of human brain structural networks by using cortical thickness from MRI. Cereb. Cortex **18**(10), 2374–2381 (2008)
5. Chyzhyk, D., Graña, M., Savio, A., Maiora, J.: Hybrid dendritic computing with kernel-lica applied to Alzheimer's disease detection in MRI. Neurocomputing **75**(1), 72–77 (2012)
6. Cuingnet, R., Gerardin, E., Tessieras, J., Auzias, G., Lehéricy, S., Habert, M., Chupin, M., Benali, H., Colliot, O., Alzheimer's Disease Neuroimaging Initiative: Automatic classification of patients with Alzheimer's disease from structural MRI: a comparison of ten methods using the ADNI database. Neuroimage **56**(2), 766–781 (2010)
7. Friston, K., Ashburner, J., Kiebel, S., Nichols, T., Penny, W.: Statistical Parametric Mapping: The Analysis of Functional Brain Images. Academic Press, Cambridge (2007)
8. Górriz, J., Segovia, F., Ramírez, J., Lassl, A., Salas-González, D.: Gmm based spect image classification for the diagnosis of Alzheimer's disease. Appl. Soft Comput. **11**, 2313–2325 (2011)
9. Hilgetag, C., Kötter, R., Stephan, K., Sporns, O.: Computational methods for the analysis of brain connectivity. In: Ascoli, G.A. (ed.) Computational Neuroanatomy, pp. 295–335. Humana Press, New York (2002)
10. Hinton, G.: Where do features come from? Cogn. Sci. **38**(6), 1078–1101 (2014)
11. Huang, S., Li, J., Sun, L., Jun, L., Wu, T., Chen, K., Fleisher, A., Reiman, E., Jieping, Y.: Learning brain connectivity of Alzheimer's disease from neuroimaging data. In: Bengio, Y., Schuurmans, D., Lafferty, J., Williams, C., Culotta, A. (eds.) Advances in Neural Information Processing Systems 22, pp. 808–816. Curran Associates Inc., Red Hook (2009)
12. Liu, J., Ji, S., Ye, J.: SLEP: Sparse Learning with Efficient Projections. Arizona State University (2009). http://www.public.asu.edu/jye02/Software/SLEP
13. Liu, M., Zhang, D., Shen, D.: Ensemble sparse classification of Alzheimer's disease. NeuroImage **60**(2), 1106–1116 (2012)
14. López, M., Ramírez, J., Górriz, J., Álvarez, I., Salas-Gonzalez, D., Segovia, F., Chaves, R., Padilla, P., Gómez-Río, M.: Principal component analysis-based techniques and supervised classification schemes for the early detection of Alzheimer's disease. Neurocomputing **74**(8), 1260–1271 (2011). Selected Papers from the 3rd International Work-Conference on the Interplay between Natural and Artificial Computation (IWINAC 2009)

15. Mingrui, X., Jinhui, W., Yong, H.: BrainNet viewer: a network visualization tool for human brain connectomics. PLoS ONE **8**(7), e68910 (2013)
16. Ortiz, A., Górriz, J.M., Ramírez, J., Martinez-Murcia, F.J., Alzheimer's Disease Neuroimaging Initiative: Automatic roi selection in structural brain MRI using som 3D projection. PLOS ONE **9**(4), e93851 (2014)
17. Ortiz, A., Górriz, J.M., Ramírez, J., Martínez-Murcia, F.J.: Lvq-SVM based CAD tool applied to structural MRI for the diagnosis of the Alzheimer's disease. Pattern Recogn. Lett. **34**(14), 1725–1733 (2013)
18. Ortiz, A., Munilla, J., Górriz, J.M., Ramírez, J.: Ensembles of deep learning architectures for the early diagnosis of the Alzheimer's disease. Int. J. Neural Syst. **26**(07), 1650025 (2016)
19. Ortiz, A., Munilla, J., Illán, I.Á., Górriz, J.M., Ramírez, J., Alzheimer's Disease Neuroimaging Initiative: Exploratory graphical models of functional and structural connectivity patterns for Alzheimer's disease diagnosis. Front. Comput. Neurosci. **9**, 132 (2015)
20. Pourahmadi, M.: High-Dimensional Covariance Estimation, 1st edn. Wiley, Hoboken (2013)
21. Raamana, P.R., Weiner, M.W., Wang, L., Beg, M.F.: Thickness network features for prognostic applications in dementia. Neurobiol. Aging **36**(1), S91–S102 (2015)
22. Ramirez, J., Chaves, R., Gorriz, J.M., Lopez, M., Alvarez, I.A., Salas-Gonzalez, D., Segovia, F., Padilla, P.: Computer aided diagnosis of the Alzheimer's disease combining spect-based feature selection and random forest classifiers. In: Proceedings of IEEE Nuclear Science Symposium Conference Record (NSS/MIC), pp. 2738–2742 (2009)
23. Segovia, F., Górriz, J.M., Ramírez, J., Álvarez, I., Jiménez-Hoyuela, J.M., Ortega, S.J.: Improved Parkinsonism diagnosis using a partial least squares based approach. Med. Phys. **39**(7), 4395–4403 (2012)
24. Structural Brain Mapping Group: Department of Psychiatry. http://dbm.neuro.uni-jena.de/vbm8/VBM8-Manual.pdf. Accessed Oct 2014
25. Sun, L., Patel, R., Liu, J., Chen, K., Wu, T., Li, J., Reiman, E., Ye, J.: Mining brain region connectivity for Alzheimer's disease study via sparse inverse covariance estimation. In: Proceedings of the 15th ACM SIGKDD International Conference on Knowledge Discovery and Data Mining, pp. 1335–1344. ACM, New York (2009)
26. Vincent, P., Larochelle, H., Lajoie, I., Bengio, Y., Manzagol, P.A.: Stacked denoising autoencoders: learning useful representations in a deep network with a local denoising criterion. J. Mach. Learn. Res. **11**, 3371–3408 (2010)
27. Zalesky, A., Fornito, A., Bullmore, E.: Network-based statistic: identifying differences in brain networks. NeuroImage **53**(4), 1197–1207 (2010)

Improved Reference Tracts for Unsupervised Brain White Matter Tractography

Susana Muñoz Maniega[1,2(✉)], Mark E. Bastin[1,2], Ian J. Deary[2,3], Joanna M. Wardlaw[1,2], and Jonathan D. Clayden[4]

[1] Department of Neuroimaging Sciences, University of Edinburgh, Edinburgh, UK
s.m.maniega@ed.ac.uk
[2] Centre for Cognitive Ageing and Cognitive Epidemiology, University of Edinburgh, Edinburgh, UK
[3] Department of Psychology, University of Edinburgh, Edinburgh, UK
[4] UCL Great Ormond Street Institute of Child Health, University College London, London, UK

Abstract. Neighbourhood tractography aims to automatically segment equivalent brain white matter tracts from diffusion magnetic resonance imaging (dMRI) data in different subjects by using a "reference tract" as a prior for the shape and length of each tract of interest. In the current work we present a means of improving the technique by using references tracts derived from dMRI data acquired from 80 healthy volunteers aged 25–64 years. The reference tracts were tested on the segmentation of 16 major white matter tracts in 50 healthy older people, aged 71.8 (±0.4) years. We found that data-generated reference tracts improved the automatic white matter tract segmentations compared to results from atlas-generated reference tracts. We also obtained higher percentages of visually acceptable segmented tracts and lower variation in water diffusion parameters using this approach.

Keywords: MRI · Brain · White matter · Unsupervised segmentation · Tractography

1 Introduction

Tractography uses dMRI data to reconstruct in vivo the white matter connections within the brain [1]. Clinical applications of tractography typically involve group analysis, where tract characteristics are examined across a patient group of interest, or compared to a matched control group. In these instances, sources of nuisance variance within and between groups—and in particular any variability introduced by the tract segmentation method—need to be kept to a minimum to facilitate detection of true biological differences and avoid spurious findings. Probabilistic neighbourhood tractography (PNT) aims to reduce operator interaction, and therefore any potential variability induced by it, during the tract segmentation process. PNT automatically segments the same white matter fasciculus in different subjects by scoring the similarity between a predefined "reference tract" and a group of candidate tracts generated with

M. Valdés Hernández and V. González-Castro (Eds.): MIUA 2017, CCIS 723, pp. 425–435, 2017.
DOI: 10.1007/978-3-319-60964-5_37

different initial seed points within a neighbourhood [2, 3]. Other automated tract segmentation tools informed by prior information have also been developed [4].

Reference tracts can be generated directly from dMRI data, or from an atlas or similar reference point. In either case, the underlying reference dataset should be representative of the population. A suitably large and diverse "training" dataset is subsequently used to capture the variability typically observed around each reference tract. However, this set of training data should generally be kept separate from the data that will be used for hypothesis testing, to prevent any potential bias during analysis. To avoid use of valuable testing data in the creation of reference tracts, and for consistency across studies, a set of reference tracts has been previously derived from a white matter atlas, which is independent of all new subject data acquired [5, 6]. These atlas-based reference tracts improved significantly the results from PNT, however a small a proportion of the segmented tracts still needed excluding after visual inspection [7].

In the current work, we are proposing a new set of reference tracts directly derived from dMRI data acquired from a large group of healthy volunteers with a wide age range, so as to capture the variability due to age. We then test these new reference tracts on a different set of healthy older volunteers.

2 Methods

2.1 Participants

Training Data. The reference and training data consisted of brain dMRI from 80 clinically normal, right-handed, healthy volunteers (40 males, 40 females) aged 25–64 years. All participants gave written informed consent. Health status was assessed using medical questionnaires and all structural MRI scans were reported by a fully qualified neuroradiologist. More details can be found in previous publications [8].

Testing Data. The testing data consisted of brain dMRI data from 50 healthy, community-dwelling older participants from the Lothian Birth Cohort 1936 (LBC1936), all born in the same year, with average age 71.8 $\pm$ 0.4 years at the time of scanning. All participants gave written informed consent. More details of this cohort have been published previously [9].

2.2 MRI

All brain MRI data were acquired using the same GE Signa Horizon HDxt 1.5T clinical scanner (General Electric, Milwaukee, WI, USA) equipped with a self-shielding gradient set (33 mT/m maximum gradient strength) and manufacturer supplied eight-channel phased-array head coil. The same dMRI protocol was used for both training and testing data. The acquisition consisted of seven T_2-weighted (T2W; b = 0 s/mm^2) and sets of diffusion-weighted (b = 1000 s/mm^2) single-shot, spin-echo, echo-planar (EP) imaging volumes, acquired with diffusion gradients applied in 64 non-collinear directions [10] and 2 mm isotropic spatial resolution.

2.3 Image Analysis

dMRI volumes were preprocessed using FSL tools (http://www.fmrib.ox.ac.uk/fsl) to extract the brain [11], remove bulk motion and correct eddy current induced distortions by registering all subsequent volumes to the first T2W EP volume [12]. The water self-diffusion tensor was calculated, and parametric maps of fractional anisotropy (FA) and mean diffusivity (MD) derived from its eigenvalues using DTIFIT.

2.4 Creation of Reference Tracts

We followed the standard reference tract construction steps for PNT in the TractoR software package v.2.1 for all reference datasets (http://www.tractor-mri.org.uk/reference-tracts#creating-custom-reference-tracts; [13]). Briefly, for each tract of interest, a seed point was chosen in standard space and registered linearly to each of the 80 training datasets. A cuboidal region of interest (ROI) was created in the $7 \times 7 \times 7$ voxel neighbourhood around each of these original seeds in native space. A probabilistic tract was then created for each voxel in the neighbourhood with FA > 0.2, using BEDPOSTx/PROBTRACKx as the underlying tractography algorithm [14], with 2000 streamlines and a two-fibre model. All the tracts generated were reviewed visually, and for each dataset we manually chose the seed that produced the tract most closely representing the expected shape and length of the fasciculus of interest. In the cases where there was more than one potential candidate available, we chose the one generated from the seed closest to the centre of the neighbourhood, i.e. closest to the seed point selected originally in standard space.

We therefore obtained 80 representative training tracts for each tract of interest. Each of them was reduced to a single streamline by obtaining the spatial median [3], and then mapped into the standard MNI brain (with its corresponding seed point) by applying the reverse linear transformation. A reference tract was then created by obtaining the median seed point and median streamline from the 80 training tracts, and fitting a B-spline to it, with a distance between knots of approximately 6 mm. A maximum bending angle restriction of 90° was also applied to avoid unrealistic 'twists' at the ends of the tracts, where uncertainty is larger.

2.5 Creation of Matching Models

The "matching model" describes typical deviations in shape and length that matching tract pathways make from the reference tract, using maximum likelihood estimation. The model for a tract of interest may be fitted in a supervised fashion by manually choosing a set of training tracts representing good matches to the reference [3], or following an unsupervised approach using an expectation-maximisation (EM) algorithm which will train the model and select at the same time the best segmentations from each dataset [2].

With the centroid reference tract created from the training data as explained above, the whole set of 80 training tracts were used to fit a matching model in a supervised fashion [3].

We then used an unsupervised approach in the 50 testing datasets (LBC1936), based on an EM algorithm, whereby the model was trained and applied iteratively using the same data [2]. Using this approach, a matching model was obtained from the testing data as well as the best candidate tract for each dataset. We therefore obtained two matching models for each tract of interest, one created from the 80 training datasets (ages 25–64 years) and one created from the 50 testing datasets (age 71.8 ± 0.4 years).

2.6 Testing of Reference Tracts and Matching Models

The new reference tracts were used to segment the fasciculi of interest in the LBC1936 testing data with PNT by evaluating novel candidate tracts for plausibility against the model fitted to the training data and against the model created with the unsupervised approach in the own testing data. This allows us to test the influence of the matching model on the selection of candidate tracts in the testing data.

The unsupervised fitting process was also repeated using the reference tracts previously created from an atlas [5, 6], which are currently provided with the TractoR package. This allows the new data-based reference tracts to be compared with the previous atlas-based reference tracts.

We therefore obtained three segmentations for each fasciculus of interest for each testing dataset (LBC1936): (a) using a supervised matching model from the training dataset and the data-based reference tract, (b) using an unsupervised matching model from the testing LBC1936 dataset and the data-based reference tract, and (c) using an unsupervised matching model from the testing LBC1936 dataset and the atlas-based reference tract.

For all methods, an additional shape modelling-based approach was used to reject false positive streamlines from the final tracts [15]. The resulting segmented tracts were then visually assessed, blinded to the method used, and tracts were considered unacceptable if any significant portion of the tract (i.e. with high visitation count) ran in a direction different from that expected from anatomy, or if they were severely truncated or bent in an unrealistic angle.

Tract-averaged FA and MD values were then calculated in tracts that passed this visual quality check, weighting the values in each voxel by the streamline visitation count. To compare the three segmentations, the proportions of visually plausible tracts were recorded and the coefficients of variation (CV) of the mean FA and MD values extracted from the resulting tracts calculated and compared.

To obtain an impression of the relative importance of the reference tracts and the fitted model, the degree of agreement on the best-matching candidate tract was assessed across the 50 LBC1936 testing datasets between the three methods

3 Results

3.1 Reference Tracts

The data-based reference tracts were created for 16 main brain white matter fasciculi: the genu and splenium of the corpus callosum, the anterior thalamic radiations (ATR),

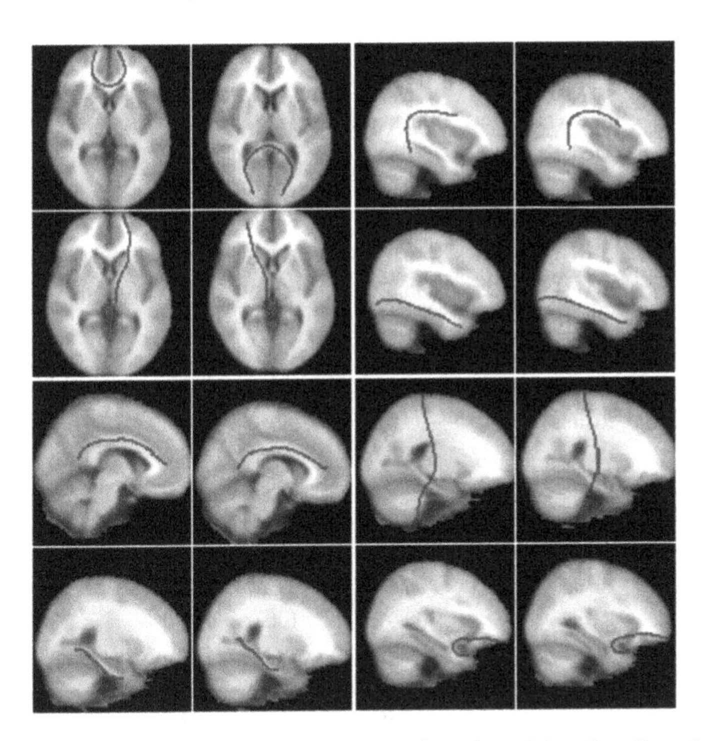

Fig. 1. 2D projections of the reference tracts created form the training data. From left to right and top to bottom: genu and splenium of the corpus callosum, left (L) and right (R) arcuate fasciculi, L and R anterior thalamic radiations, L and R inferior longitudinal fasciculi, L and R frontal cingula, L and R corticospinal tracts, L and R ventral cingula, and L and R uncinate fasciculi. Note: images are in radiological convention (left is shown on the right)

the arcuate (Arc), uncinate (Unc), and inferior longitudinal fasciculi (ILF), the frontal and ventral cingula (Cing), and the corticospinal tract (CST), bilaterally. Figure 1 shows a representation of all the reference tracts created from the training data as a projection in a plane.

3.2 Testing of Reference Tracts and Matching Models

Visual Assessments. The use of the data-based reference tracts improved the number of visually acceptable tracts when compared with the same segmentations created from the previous atlas-based reference tracts. Table 1 shows the percentage of successful segmentations for each white matter tract using each method.

When comparing tracts created with the same testing data model, the data-based reference tracts improved the consistency of the segmentations, with >92% of successful segmentations for all tracts. By contrast, atlas-based reference tracts had a lower average performance, particularly due to the poor performance segmenting the ATR, bilaterally, where only 32 and 76% of the cases could be segmented successfully.

Table 1. Proportion of segmented tracts visually acceptable when using two different matching models and each set of reference tracts as priors.

Reference tracts	Data-based		Atlas-based
Model trained on	Training data	Testing data	
Genu	100.0%	100.0%	96.0%
Splenium	98.0%	96.0%	98.0%
LArc	100.0%	100.0%	98.0%
RArc	96.0%	96.0%	94.0%
LATR	100.0%	100.0%	32.0%
RATR	96.0%	100.0%	76.0%
LCing	98.0%	98.0%	100.0%
RCing	98.0%	92.0%	98.0%
LCing_ventral	98.0%	100.0%	98.0%
RCing_ventral	94.0%	98.0%	100.0%
LILF	100.0%	100.0%	100.0%
RILF	100.0%	100.0%	100.0%
LUnc	96.0%	92.0%	88.0%
RUnc	100.0%	100.0%	100.0%
LCST	100.0%	98.0%	100.0%
RCST	100.0%	100.0%	100.0%
Mean	*98.3%*	*98.1%*	*92.4%*

When comparing the two models, both perform well, with an average of >98% visually plausible tracts, suggesting that a model can be trained in a separate dataset and still successfully segment the tracts in the testing (LBC1936) data.

Figure 2 shows the group maps created by overlaying the segmented tracts from the 50 older age volunteer LBC1936 testing data set into the standard brain as maximum intensity projections. These images show that the segmentations obtained from the two sets of reference tracts are similar, except for the left and right ATR, where many of the segmentations using the atlas-based reference followed the wrong path, thereby failing the visual check. Some small differences are, however, obvious in other tracts, specifically regarding their lengths. In particular, the segmentations of the corpus callosum genu, the arcuate fasciculi and the ventral cingula were longer when using the new data-based reference tracts, with more of the tract included in the segmentation.

The group maps from tracts generated with each training model showed that the choice of training model had a modest effect on the segmented tracts.

FA and MD Variability. Table 2 shows the mean values and CV of FA and MD, measured along the tracts extracted by the three methods. One-way analysis of variance (ANOVA) tests, corrected for multiple comparisons, showed that the parameters measured in tracts generated by each method were generally not significantly different. Only the corpus callosum splenium, the RATR and RCST produced significantly different mean parameters. Without multiple comparison correction, genu (FA), LCing (FA) and LCST (FA and MD) also became significant. However, for both the FA and

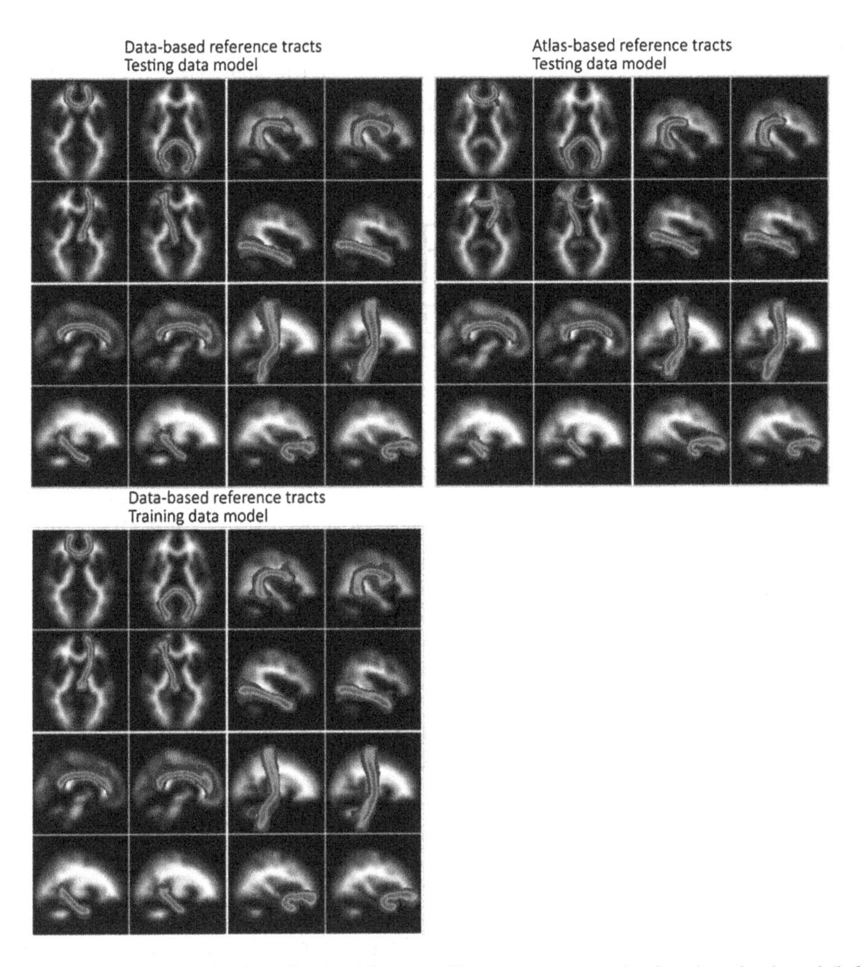

Fig. 2. Group maps projections for the 16 tracts of interest segmented using the atlas-based (left) and data-based (right) reference tracts. Top row used a matching model trained in the testing data, and the bottom row used a model trained in the training data. Colour scale represents the voxel visitation frequency, from 1 (dark blue) to 50 (yellow). Maps are projected into the plane of the voxel with maximum visitation value. (Color figure online)

MD, the variation across the 50 LBC1936 datasets is lower for most tracts when generated with the data-based reference tracts.

Comparison Between Fitted Models. The source of training data used to fit the model appeared to be less influential than the choice of reference tract. Models trained with the separate training data or with the testing data (in the unsupervised framework), but with the reference tracts in common, resulted in agreement on the best candidate tract in an average of 39% of subjects. By contrast, the two models fitted in an unsupervised fashion on the same testing data (LBC1936), but with different reference tracts, agreed only 9% of the time.

Table 2. Averaged values of fractional anisotropy (FA) and mean diffusivity (MD) measured along the tracts segmented with two different matching models, and atlas-based or data-based reference tracts as priors in 50 older age volunteers. The coefficients of variation (CV) for each parameter are shown in the shaded columns. Bold type indicates that the mean parameters were significantly different ($p < 0.05$) between the tracts created with each method (One-way ANOVA after Bonferroni-Holm adjustment for multiple comparisons across tracts).

	FA						*MD ($10^{-6}mm^2/s$)*					
Reference	*Atlas-based*		*Data-based*				*Atlas-based*		*Data-based*			
Model training	*Testing data*				*Training data*		*Testing data*				*Training data*	
	Mean (sd)	*CV*	*Mean (sd)*	*CV*	*Mean (sd)*	*CV*	*Mean (sd)*	*CV*	*Mean (sd)*	*CV*	*Mean (sd)*	*CV*
Genu	0.41 (0.05)	0.11	0.39 (0.05)	0.12	0.39 (0.05)	0.12	776.91 (65.59)	0.08	799.20 (75.46)	0.09	799.85 (74.59)	0.09
Splenium	**0.45 (0.09)**	0.20	**0.52 (0.06)**	0.12	**0.51 (0.08)**	0.15	**1117.26 (220.22)**	0.20	**807.61 (108.59)**	0.13	**837.77 (162.71)**	0.19
LArc	0.46 (0.05)	0.10	0.45 (0.04)	0.09	0.45 (0.04)	0.10	663.30 (49.21)	0.07	661.30 (49.26)	0.07	659.82 (49.73)	0.08
RArc	0.43 (0.05)	0.12	0.42 (0.04)	0.10	0.43 (0.04)	0.09	646.56 (55.00)	0.09	645.36 (48.93)	0.08	644.13 (45.30)	0.07
LATR	0.34 (0.05)	0.14	0.34 (0.03)	0.10	0.34 (0.03)	0.10	757.89 (81.23)	0.11	755.39 (60.94)	0.08	746.41 (60.30)	0.08
RATR	**0.35 (0.04)**	0.10	**0.36 (0.03)**	0.08	**0.33 (0.04)**	0.12	**747.07 (54.08)**	0.07	**704.05 (50.40)**	0.07	**766.81 (74.85)**	0.10
LCing	0.45 (0.05)	0.12	0.46 (0.06)	0.12	0.46 (0.06)	0.12	647.29 (51.00)	0.08	638.39 (45.15)	0.07	640.95 (47.46)	0.07
RCing	0.42 (0.06)	0.13	0.43 (0.04)	0.10	0.42 (0.05)	0.11	619.92 (36.16)	0.06	626.56 (36.03)	0.06	630.97 (33.82)	0.05
LCing_ventral	0.32 (0.06)	0.19	0.29 (0.04)	0.12	0.29 (0.04)	0.12	752.54 (155.54)	0.21	728.86 (62.50)	0.09	733.07 (69.52)	0.09
RCing_ventral	0.30 (0.06)	0.20	0.30 (0.05)	0.15	0.29 (0.04)	0.14	760.68 (95.07)	0.12	748.37 (79.00)	0.11	748.73 (88.67)	0.12
LILF	0.42 (0.05)	0.12	0.41 (0.05)	0.12	0.40 (0.05)	0.12	740.50 (75.45)	0.10	752.41 (67.06)	0.09	745.86 (61.13)	0.08
RILF	0.39 (0.05)	0.14	0.40 (0.04)	0.11	0.38 (0.05)	0.12	788.00 (142.54)	0.18	750.31 (83.70)	0.11	755.39 (87.47)	0.12
LUnc	0.34 (0.03)	0.10	0.33 (0.03)	0.10	0.34 (0.04)	0.11	767.04 (53.54)	0.07	767.63 (60.41)	0.08	764.88 (60.65)	0.08
RUnc	0.33 (0.03)	0.10	0.33 (0.03)	0.10	0.33 (0.04)	0.11	756.22 (41.27)	0.05	758.75 (41.27)	0.05	754.75 (41.77)	0.06
LCST	0.48 (0.03)	0.07	0.46 (0.04)	0.08	0.46 (0.04)	0.08	655.47 (36.72)	0.06	672.26 (37.18)	0.06	675.52 (38.65)	0.06
RCST	0.49 (0.03)	0.07	0.49 (0.03)	0.07	0.50 (0.04)	0.07	**653.82 (32.72)**	0.05	**676.03 (32.36)**	0.05	**676.37 (31.99)**	0.05
Mean	*0.40 (0.06)*	*0.13*	*0.40 (0.07)*	*0.10*	*0.40 (0.07)*	*0.11*	*740.65 (115.51)*	*0.10*	*718.28 (58.64)*	*0.08*	*723.83 (61.36)*	*0.09*

4 Discussion

The reference tract represents the "matching" target for PNT automatic segmentation, and it is therefore crucial that this prior epitomises the topological characteristics of the fasciculus of interest correctly. Using a large group of healthy volunteers, with a wide age range, we were able to capture the variability in tract topology better. Our results showed that the results from PNT can be improved, even when the testing data corresponds to an age group outside the age range used during training to generate the reference tracts or the matching models (72 vs 25–64 years old). We also demonstrated that the source of training data used to fit the model was less influential than the choice of reference tract, and that matching models previously fitted in training data can be used to apply PNT in separate testing datasets. This enables the possibility of using PNT in small samples of testing data, where the number of datasets might not be large enough for fitting the matching model in an unsupervised fashion.

The large percentage of successful segmentations obtained in the older population (>98%) when using the new reference tracts suggests that these can be used as priors in different populations, and not just in a population matching the training data characteristics. Although the improvement is significant, is it still not sufficient to make manual checking of the segmented tracts entirely unnecessary, but this is true for most automated methods. Further tests would also be required to investigate whether these reference tracts would still be good priors to perform PNT segmentation in diseased populations with potentially large changes in brain topology, such as in the presence of tumours or stroke, but preliminary work suggests that the general approach is robust to even quite substantial mass effects [16].

The most obvious improvement with the new reference tracts is the high success rate obtained for the ATR, indicating that the prior for this tract generated from real data is a much better representation of the ATR topology. Another improvement is the extraction of longer segments of some of the tracts of interest, such as the genu of the corpus callosum, the arcuate and the ventral cingulum, which arises due to the greater difficulty of inferring accurate pathways near the ends of tracts when using an atlas as the reference, leading to a shorter reference tract. The segmentation of a larger section of the genu projections into the frontal cortex (where FA tends to be lower than in the centre of the tract) could explain the slightly lower mean values of FA obtained for this tract when using the new reference tracts. There was also a very subtle shift in the overall position of the splenium of the corpus callosum, with the segmentations for this tract obtained with the atlas-based reference tract being generally closer to the boundary with the ventricles, while the data-based reference producing segmentations within the middle of this fasciculus. This is also reflected in the higher MD and lower FA of the atlas-based splenium, suggesting more partial volume averaging with cerebrospinal fluid from the ventricles.

There could be two main reasons for the differences in parameters measured with each method. Firstly, the atlas used to generate the previous references tracts was obtained using data from subjects with an average age of 29 $\pm$ 7.9 years [6], while the training data for the new priors had a wider age range of 25–65 years. The new reference tracts will therefore represent better the characteristics of the white matter in older age, and particularly the changes due to ageing such as atrophy and enlarged ventricles. This is reflected in the better segmentations, and in the change in the parameters measured, in the tracts running closer to the ventricles, such as the ATR, the CST, and the genu and splenium of the corpus callosum. Secondly, the native-space tractography data used for generating the reference tracts here is a much richer dataset than the subject-averaged tract probability maps that constitute the atlas.

The CVs in the parameters measured in the segmentations created from the new set of reference tracts are lower than those created from the atlas-based reference tracts, particularly for the splenium and the ventral Cing. This suggests a lower variability introduced by the tract segmentation method, which should facilitate detection of true biological differences and avoid spurious findings.

In summary, we have created a new set of data-based reference tracts to be used as priors for PNT, which improved the segmentations of 16 tracts of interest. We have also demonstrated that the matching model could be fitted in separate training data, which will make the use of PNT in small testing datasets newly practicable.

Acknowledgements. LBC1936 was supported by the Age UK-funded Disconnected Mind project, with additional funding from the UK Medical Research Council (MR/M013111/1). MRI scanning for the training dataset was funded under NIH grant R01 EB004155-03. The scanning was performed at the Brain Research Imaging Centre, Edinburgh, part of Edinburgh Imaging (www.ed.ac.uk/clinical-sciences/edinburgh-imaging) and the SINAPSE Collaboration (Scottish Imaging Network, A Platform for Scientific Excellence, www.sinapse.ac.uk).

References

1. Tournier, J.D., Mori, S., Leemans, A.: Diffusion tensor imaging and beyond. Magn. Reson. Med. **65**, 1532–1556 (2011). doi:10.1002/mrm.22924
2. Clayden, J.D., Storkey, A.J., Muñoz Maniega, S., Bastin, M.E.: Reproducibility of tract segmentation between sessions using an unsupervised modelling-based approach. Neuroimage **45**, 377–385 (2009). doi:10.1016/j.neuroimage.2008.12.010
3. Clayden, J.D., Storkey, A.J., Bastin, M.E.: A probabilistic model-based approach to consistent white matter tract segmentation. IEEE Trans. Med. Imaging **26**, 1555–1561 (2007). doi:10.1109/TMI.2007.905826
4. Yendiki, A., Panneck, P., Srinivasan, P., Stevens, A., Zöllei, L., Augustinack, J., Wang, R., Salat, D., Ehrlich, S., Behrens, T., Jbabdi, S., Gollub, R., Fischl, B.: Automated probabilistic reconstruction of white-matter pathways in health and disease using an atlas of the underlying anatomy. Front. Neuroinform. **5**, 23 (2011). doi:10.3389/fninf.2011.00023
5. Muñoz Maniega, S., Bastin, M.E., McIntosh, A.M., Lawrie, S.M., Clayden, J.D.: Atlas-based reference tracts improve automatic white matter segmentation with neighbourhood tractography. In: ISMRM (ed.) Proceedings of ISMRM 16th Scientific Meeting and Exhibition, p. 3318 (2008)
6. Hua, K., Zhang, J., Wakana, S., Jiang, H., Li, X., Reich, D.S., Calabresi, P.A., Pekar, J.J., van Zijl, P.C., Mori, S.: Tract probability maps in stereotaxic spaces: analyses of white matter anatomy and tract-specific quantification. Neuroimage **39**, 336–347 (2008). doi:10.1016/j.neuroimage.2007.07.053
7. Penke, L., Muñoz Maniega, S., Houlihan, L.M., Murray, C., Gow, A.J., Clayden, J.D., Bastin, M.E., Wardlaw, J.M., Deary, I.J.: White matter integrity in the splenium of the corpus callosum is related to successful cognitive aging and partly mediates the protective effect of an ancestral polymorphism in ADRB2. Behav. Genet. **40**, 146–156 (2010). doi:10.1007/s10519-009-9318-4
8. Dickie, D.A., Mikhael, S., Job, D.E., Wardlaw, J.M., Laidlaw, D.H., Bastin, M.E.: Permutation and parametric tests for effect sizes in voxel-based morphometry of gray matter volume in brain structural MRI. Magn. Reson. Imaging **33**, 1299–1305 (2015). doi:10.1016/j.mri.2015.07.014
9. Deary, I.J., Gow, A.J., Taylor, M.D., Corley, J., Brett, C., Wilson, V., Campbell, H., Whalley, L.J., Visscher, P.M., Porteous, D.J., Starr, J.M.: The Lothian Birth Cohort 1936: a study to examine influences on cognitive ageing from age 11 to age 70 and beyond. BMC Geriatr. **7**, 28 (2007). doi:10.1186/1471-2318-7-28
10. Jones, D.K., Williams, S.C.R., Gasston, D., Horsfield, M.A., Simmons, A., Howard, R.: Isotropic resolution diffusion tensor imaging with whole brain acquisition in a clinically acceptable time. Hum. Brain Mapp. **15**, 216–230 (2002). doi:10.1002/hbm.10018
11. Smith, S.M.: Fast robust automated brain extraction. Hum. Brain Mapp. **17**, 143–155 (2002). doi:10.1002/hbm.10062
12. Jenkinson, M., Smith, S.: A global optimisation method for robust affine registration of brain images. Med. Image Anal. **5**, 143–156 (2001). doi:10.1016/S1361-8415(01)00036-6
13. Clayden, J.D., Muñoz Maniega, S., Storkey, A.J., King, M.D., Bastin, M.E., Clark, C.A.: TractoR: magnetic resonance imaging and tractography with R. J. Stat. Softw. **44**, 1–18 (2011). doi:10.18637/jss.v044.i08

14. Behrens, T.E.J., Berg, H.J., Jbabdi, S., Rushworth, M.F.S., Woolrich, M.W.: Probabilistic diffusion tractography with multiple fibre orientations: what can we gain? Neuroimage **34**, 144–155 (2007). doi:10.1016/j.neuroimage.2006.09.018
15. Clayden, J.D., King, M.D., Clark, C.A.: Shape modelling for tract selection. In: Yang, G.Z., Hawkes, D., Rueckert, D., Noble, A., Taylor, C. (eds.) MICCAI 2009. LNCS, vol. 5762, pp. 150–157. Springer, Heidelberg (2009). doi:10.1007/978-3-642-04271-3_19
16. Hill, C.S., Clayden, J.D., Kitchen, N., Bull, J., Harkness, W., Clark, C.A.: A feasibility study of neighbourhood tractography in the presence of paediatric brain tumours. In: Proceedings of Autumn Meeting of the Society of British Neurological Surgeons (2009)

Review of Fast Density-Peaks Clustering and Its Application to Pediatric White Matter Tracts

Shichao Cheng[1], Yuzhuo Duan[1], Xin Fan[1], Dongyu Zhang[1], and Hua Cheng[2](✉)

[1] School of Software, Dalian University of Technology, Dalian, China
xin.fan@ieee.org

[2] Medical Imaging Center, Beijing Childrens Hospital, Beijing, China
chhuaer@hotmail.com

Abstract. Clustering white matter (WM) tracts from diffusion tensor imaging (DTI) is primarily important for quantitative analysis on pediatric brain development. A recently developed algorithm, density peaks (DP) clustering, demonstrates great robustness to the complex structural variations of WM tracts without any prior templates. Nevertheless, the calculation of densities, the core step of DP, is time consuming especially when the number of WM fibers is huge. In this paper, we propose a fast algorithm that accelerates the density computation about 50 times over the original one. We convert the *global* calculation for the density as well as critical parameter in the process into *local* computations, and develop a binary tree structure to orderly store the neighbors for these local computations. Hence, the density computation turns out to direct access of the structure, rendering significantly computational saving. Experiments on synthetic point data and the JHU-DTI data set validate the efficiency and effectiveness our fast DP algorithm compared with existing clustering methods. Finally, we demonstrate the application of the proposed algorithm on the analysis of pediatric WM tract development.

Keywords: Fast clustering · White matter tracts · DTI

1 Introduction

Diffusion-tensor imaging (DTI), capturing the diffusion of water molecules in human brain, provides a non-invasive assessment of microstructural organization of white matter (WM) [1]. Fiber bundles detected by DTI typically form a set of tract bunddles that share common neurophysiological connectivity and neuropsychiatric function. For example, the inferior longitudinal fasciculus (ILF) connects ipsilateral temporal and occipital lobes, and integrates auditory and speech nuclei [2]. There exist numerous studies investigating the integrity evolution of WM pathways in human brain via these clustered WM tracts. One interesting aspect for these studies is to discover the development process of WM tracts [3,4]. However, classical methods typically analyze spatially averaged parameters transformed into a common template space by registration

M. Valdés Hernández and V. González-Castro (Eds.): MIUA 2017, CCIS 723, pp. 436–447, 2017.
DOI: 10.1007/978-3-319-60964-5_38

algorithms that smooth local alterations along tracts. In this paper, we develop a fast algorithm to automatically cluster fibers into WM bundles in the original space for tract-wise development analysis [5,6].

In the past decade, researchers resorted to classical clustering algorithms in the machine learning community, including K-means [7,8], Spectral Clustering (SC) [9], Gaussian mixture models (GMM) [10], Affinity Propagation (AP) [11,12], and DBSCAN [13], to address the issue of tract clustering. These clustering algorithms gained some success as they grouped fiber tracts having similar intrinsic characteristics. However, one major drawback of these methods lies in that they need to specify the number of clusters. A wrong configuration of this parameter can lead to inaccurate segmentation. Recently, prior knowledge of brain anatomy is incorporated into clustering to guide the tract segmentation process. Li *et al.* adopted a hybrid approach for automatic clustering white fibers, using a top-down atlas [14], while Jin *et al.* fuse labels from multi-atlas for tract clustering [15]. Considering variations among individuals especially pediatric ones undertaking rapid developments, it is an open issue how largely the bundle clustering specific to a subject can rely on a universal atlas. Additionally, little attention has been paid to fiber outliers, greatly affecting the classical algorithms.

Those problems are solved by a newly developed algorithm, density peaks (DP) clustering, is able to tackle the challenges for clustering WM tracts [16]. The algorithm discovers cluster centers and assigns cluster labels based on the density per data point. Users can intuitively determine the number of clusters from a graph characterizing the density distribution of a subject's fibers, so that need no prior atlas. The density graph also removes the outliers by specifying the border density for each cluster. The outputs of the algorithm show high consistency with manual parcellation [17,18]. Unfortunately, the generation of density graph has to numerate all fiber pairs (the square of the fiber count) in order to generate the decision graph, resulting in expensive computational load for over 100,000 fibers.

In this paper, we propose a fast algorithm for the density-peaks clustering, and apply the algorithm to the analysis of WM tract development. We observe that the density of a data point largely relies on its neighboring points, and the farther points including outliers have few effect on density calculation. This observation enables us to use *local* points to compute the *global* density of a point over all the other points. The neighboring points (fibers) of any individual fiber are orderly stored in a binary tree. Hence, a one-pass traverse on the tree generates the densities for all fibers, and thus circumvents the repeated calculations on all point pairs. This computation process on the tree significantly reduces the expensive computational expenses on the density calculation in the original DP algorithm, yet yields an almost equivalent density graph. We validate our fast DP on the synthetic data and DTI images, and compare the performance with the classical K-means, GMM, Mean-shift, SC, AP, and DBSCAN algorithms on the JHU-DTI data set (http://lbam.med.jhmi.edu/). Finally, we apply the fast DP to automatically cluster several WM tracts, and investigate their development across 18 to 36 months. This automatic process produces coincide findings with the latest results discovered by an atlas-based approach [4].

2 DTI Preprocessing

We downloaded DTI images from JHU brain MRI laboratory[1], and randomly selected 12 subjects for algorithm validation. Clinical pediatric images in our brain development experiments were obtained from 3 Tesla(T) Philips scanner with 30 or 31 diffusion gradient directions, manually inspected by radiologists, and no structural abnormality was found. DtiStudio[2] [19] and FSL[3] [20] were mainly used for the preprocessing.

DTI data contains 35 MRI images and the first 5 images are minimally diffusion-weighted images. All the raw diffusion-weighted images (DWIs) were coregistered to the first minimally diffusion-weighted images to correct participant's motion effect and the linear portion of eddy current distortions. A 12-parameter affine transformation of automated image registration AIR[4] accompanied in Dtistudio was used for the registration [21]. Coregistered DWIs were sent to brain extraction tool (BET[5]) to strip brain's skull and outer tissues [22]. Option "Apply to 4D FMRI data" was chose and parameter "Fractional intensity threshold" was set as 0.5. In order to keep the conforming characteristics of all subjects' image, sampling function of DiffeoMap[6] was used to resample images to a specified resolution, image resolution $181 * 217 * 181\,\text{mm}^3$ and physical resolution $1.00 * 1.00 * 1.00\,\text{mm}^3$ was adopted. DTI mapping of DtiStudio then was used on the resampled DWIs, tensor data was generated. Images of DTI derived parameters including FA and ADC were also preserved in this step for further analysis.

For each subject, after its tensor data was generated, which was sent into DtiStudio to do fiber tracking based on Fiber Assignment by Continuous Tracking (FACT) algorithm and a brute-force reconstruction approach. We set the fiber tracking starting from all seed points with FA value larger than 0.4, and stopping when reaching one voxel with FA value smaller than 0.25 or direction change larger than 42°. These values were chosen according to practical experiment in which we can tract some immature fibers better. Then we draw a crossing area containing ILF, IFO and Fmajor named ROI fibers according to the protocol in [2] which can be seen in Fig. 1 **A**. The complete fiber tracts generated by fiber tracking include more than 100,000 fibers. We removed those fibers shorter than 35 mm as considered as "artifact" [23].

3 Fast Density-Peaks Algorithm

In this section, we briefly introduce the density peaks clustering [16] with the analysis on its bottleneck for efficiency, and then provide our new algorithm illustrated in Fig. 1 as well as its complexity analysis.

[1] http://lbam.med.jhmi.edu.
[2] https://www.mristudio.org.
[3] http://fsl.fmrib.ox.ac.uk.
[4] http://bishopw.loni.ucla.edu/air5.
[5] http://fsl.fmrib.ox.ac.uk/fsl/fslwiki/BET.
[6] https://www.mristudio.org.

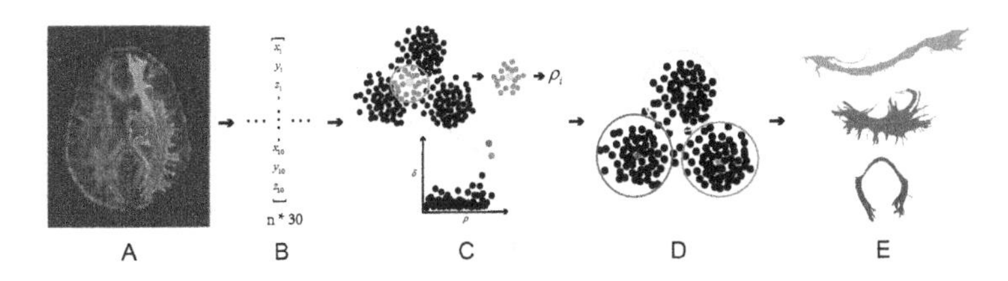

Fig. 1. Pipeline of fiber clustering. The crossing area named ROI fibers can be segmented into 3 clusters correctly including Fmajor (red), IFO (green) and ILF (red) by FDP algorithm, this figure shows procedures including **A**. tractography (to draw a crossing area contains Fmajor, ILF and IFO), **B**. fiber representation, **C**. calculate local density of each fiber to choose clustering center, **D**. fiber clustering and **E**. visualization results, respectively. (Color figure online)

3.1 Density Peaks Clustering

The DP clustering algorithm finds density peaks as cluster centers, and assigns labels for every point (fiber in our context) based on the similarity and density with the neighboring points [16]. Cluster centers are those points as following:

- the density of a cluster center should be higher than its neighbors;
- a cluster center ought to be far away from other points with higher density than itself.

Therefore, the calculation for the density of per point is the core of DP.

Considering a set of fibers $\{f_1, f_2, ..., f_n\}$, the distance between f_i and f_j is denoted as d_{ij} whose definition can be found in [17]. The density $\bar{\rho_i}$ of f_i is defined as:

$$\bar{\rho_i} = \sum_j \exp(-\frac{d_{ij}^2}{d_c^2}), \tag{1}$$

where d_c is the cutoff distance. Another critical attribute parameter δ_i is defined as the minimum distance from f_i to all other fibers whose density is higher than ρ_i:

$$\delta_i = \min_{j:\rho_j > \rho_i} (d_{ij}). \tag{2}$$

The clustering process using these two attributes is given as follows: (1) determine d_c; (2) compute $\bar{\rho}$ for each fiber according to Eq. 2; (2) compute δ for each fiber and keep the nearest fiber with higher density as its neighbor; (3) select cluster centers according to the decision graph, given by the values of $\bar{\rho}$ and δ; (4) assign each fiber to the same cluster of its neighbor according to fiber density descending order; (5) calculate border density for each cluster, those fibers with smaller density than border density are set as outliers.

Complexity analysis:

- The choice of d_c is so critical that we cannot specify its value in an *Ad Hoc* way but have to derive from the data set itself. Typically, the value is the tail in the top 5% of all distances d_{ij} in the ascending order. We have to sort the sequence of all distance pairs at the complexity of $O(n^2 \log(n^2))$, where n is the number of fibers obtained by tractography.

- For the calculation of $\bar{\rho}$ for a fiber, say f_i, we need to sum all the distances between f_i and other fibers, $d_{ij}(j = 1, ..., n, i \neq j)$. Consequently, the total complexity for $\bar{\rho}$ turns out to be $O(n^2)$.

3.2 Fast Algorithm

The bottleneck to efficiently compute the density for one point (fiber) lies in the enumeration of all the others. Actually, the nearer a fiber is, the more contribution it makes as the contributions of other points to the density exponentially decay with the distances to the point of interest. As shown in the above subsection, $\exp(-\frac{d_{ij}^2}{d_c^2})$ is smaller than 0.02 even though $d_{ij} = 2d_c$, and furthermore, $\exp(-\frac{d_{ij}^2}{d_c^2})$ is so subtle as 10^{-11}. We only need to consider a small fraction of neighboring points (the K nearest neighbors) as the approximation of the density $\bar{\rho}$ for the fiber:

$$\rho_i = \sum_{j \in A} \exp(-\frac{d_{ij}^2}{d_c^2}), A = \{j | j \in \text{K nearest neighbors of fiber } i\}. \quad (3)$$

The approximation enables us to use *local* points to compute the *global* density of a fiber. Distances between every two fibers have been obtained and the neighboring fibers of any individual fiber are orderly stored in leaf nodes of a binary tree. Hence, a one-pass traverse on the tree searching nearer neighbors for all fibers without sorting all distances data, and circumvents the repeated calculations on all fiber pairs. This computation process on the tree significantly reduces the expensive computational cost on the density calculation in the original DP algorithm, but yields an almost equivalent decision graph.

We take a point i as an example and Fig. 2 to illustrate the process. Firstly, we calculate the mean distance over all distances between the fiber i and another, and keep the data points with the distances less than the mean shown as the orange dots in Fig. 2). Then, we update the mean value for current set(orange ones), and finds those points closer to fiber i than the updated mean. These mean updating and neighbor searching processes runs iteratively until K neighbors of fiber i are found as colored in green. The upper row of Fig. 2 demonstrates this iterative process. All the distances between fiber i and the others can be stored in a binary tree where the left subtree always store distances smaller than the updated mean. So we only transverse on left subtree reaching the most left leaf to get K nearest neighbors of any fiber within time complexity of $O(\text{logn})$. Then the total computation is significantly reduced by this strategy. Both the storage and computational complexity reduce from n to K for each fiber's density.

Subsequently, we use these K neighbors to calculate ρ. The value of ρ is smaller than $\bar{\rho}$ but make few effects on the distribution of density. It can not only preserve the density information, but also, actually more importantly, lowers

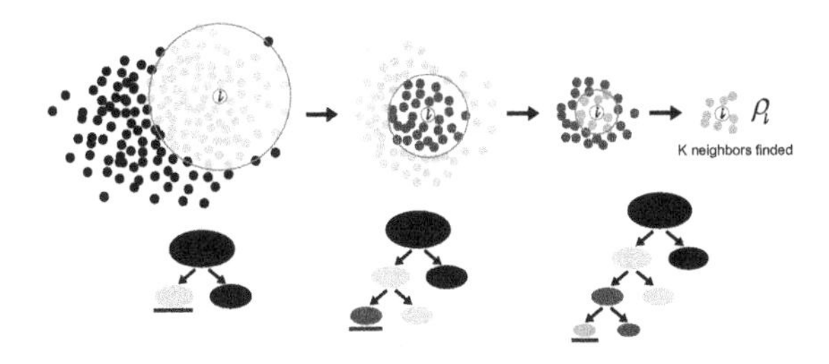

Fig. 2. Neighbor-searching process. Find the K nearest neighbors of one data point by iteratively updating mean distances via a binary tree. The red circle denotes the mean distance for the current set. The bottom row gives the binary tree for the searching process.

down the complexity for ρ calculation of n points from $O(n^2)$ to $O(Kn)$. We validate its effectiveness and efficiency in Sect. 4

The cutoff distance d_c can be calculated by this strategy instead of using all the other distances. We can obtain d_c as the top 5% of the K nearest neighbors given the tree instead of sorting all distances. We generate an equivalent d_c value, but the time cost for sorting can be reduced to $O(log(n^2))$.

Table 1 compares the Algorithm complexity of DP and our improved FDP in d_c and ρ calculation.

Table 1. Algorithm complexity comparison between DP and FDP. n represents the number of data point

	DP	FDP
ρ	$O(n^2)$	$O(Kn)$
d_c	$O(n^2\ log(n^2))$	$O(log(n^2))$

4 Clustering Results on Synthetic and Benchmark Sets

We validate the proposed FDP algorithm on several synthetic data sets widely used in the machine learning community to demonstrate the accuracy of FDP equivalent to DP, but extremely faster implementation. We also compared FDP with other classical clustering algorithms on clustering fibers on a benchmark set.

4.1 Comparisons Between FDP and DP

We apply FDP and DP on the sets including Aggregation, R15, Compound, and Jain, showing complex distributions[7]. Figure 3 shows the original data, and their corresponding decision graphs derived from DP and FDP, respectively. The decision graphs present quite close with unnoticeable differences, and accordingly their accuracies are almost the same as given in Table 2.

[7] http://cs.joensuu.fi/sipu/datasets.

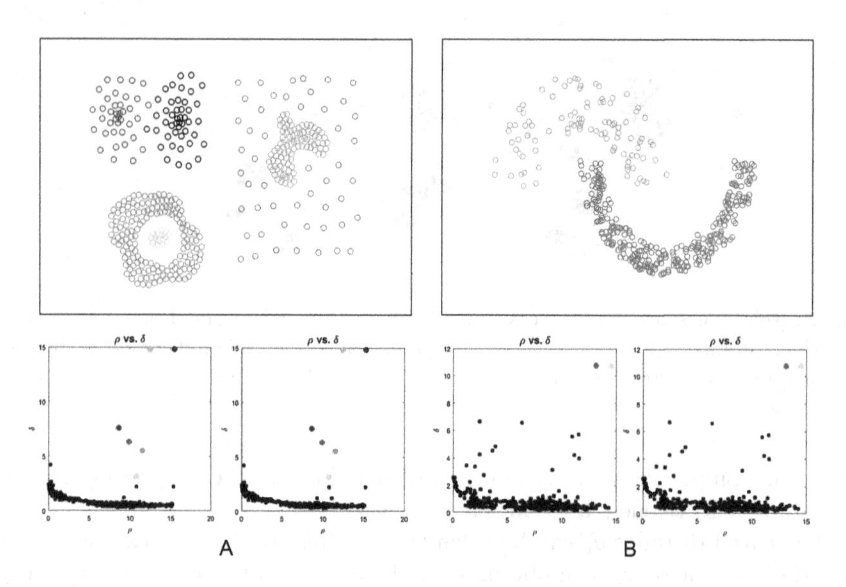

Fig. 3. Decision graphs comparison between DP and FDP in the 2 datasets. Part A and B upper panel are Compound, jain data, lower panel displays their decision graphs (DP in the left and FDP in the right side) respectively.

Time costs of FDP considerably decrease compared with DP. We apply DP and FDP on 3 subjects from the JHU dataset, and calculate the averaged time expenses listed in Table 3. All the time expenses for distance sorting, d_c and ρ calculating significantly decrease. For more than 10^3 fibers, DP takes 10 times computations to FDP on the distance sorting and d_c calculating. More significantly, the time for calculating ρ of FDP is only 1/100 of DP.

4.2 Comparisons with Classical Cluster Algorithms

We compare FDP with six classical clustering algorithms, including SC, GMM, K-means, Mean-shift, Affinity AP, and DBSCAN on the JHU-DTI dataset. For three fiber tracts of interest, we calculate the Dice ratio (DR) of the resultant bundles to manually labeled ones (ground truth) following the protocol in [2]. Figure 4 shows the box plots of the Dice ratios for clustering algorithms, where

Table 2. Accuracy comparison between DP and FDP. The 4th and 5th columns shows the results corresponding to A, B part of Fig. 3.

Accuracy(%) Data / Method	Aggregation	R15	Compound	Jain
DP	100	99.83	87.47	85.52
FDP	100	99.67	87.72	85.52

Table 3. Time expenses for DP and FDP.

Subject	Distance sorting (s)		d_c (s)		ρ (s)	
	DP	FDP	DP	FDP	DP	FDP
1	0.2040	0.0069	0.2068	0.0090	0.2118	0.0167
2	0.2972	0.0312	0.3070	0.0393	0.5344	0.0723
3	0.1259	0.0175	0.1312	0.0210	0.3141	0.0369

the dotted line, blue rectangle and red short line denote the data range, the top 25% to 75% bounds, and the mean of DR, respectively. The mean value close to 1 as well as narrow bounds indicate high consistency with the ground truth. For the tracts of IFO and ILF, FDP has obvious more consistent overlapping with the ground truth than the other four algorithms listed in Fig. 4. For Fmajor, FDP outputs higher mean DR value than the others, but works a bit less stable than the mean-shift. The visualization results in Fig. 5 also demonstrate the clustering outputs of FDP are the most consistent with the manually labeled ones. Detailed discussions are given below.

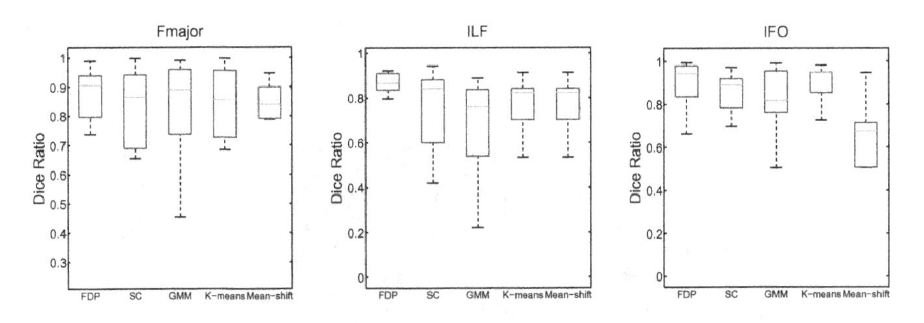

Fig. 4. Fiber's DR of different clustering algorithms. FDP stands for Fast density-peaks clustering.

SC applies the eigen decomposition to the similarity matrix of data points, and then clusters the eigen vectors derived from the decomposition. In Fig. 4, SC results have close mean values with those of k-means, but work less stable than k-means and ours. This can be explained by the crossing regions exist in the right part of Fmajor highlighted by yellow in the SC results of Fig. 5.

The solution to GMM is found by the expectation maximization (EM) algorithm, which may fall into local extremes. Also, correct initialization is quite critical to the solution. Thus, the mean values of the algorithm fall as the median of the compared algorithms, but it works quite unstable as shown by larger bounds in Fig. 5.

K-means outputs close performance to ours for IFO, but worse on Fmajor and ILF. The computational load of the classical k-means are quite high. There exists accelerated version of k-means in the context of fiber clustering [8]. However, the number of clusters cannot be set by an intuitive way as FDP does. Additionally, FDP is able to locate the 'core' fiber that is important for statistical analysis on WM tracts, but k-means only gives a mean of the clusters.

The Mean-shift algorithm finds cluster centers having the largest density. It applies to the case when probability density function has only one extreme in a local region. As we can see from Fig. 4, the mean-shift is not suitable for fiber clustering especially for IFO and Fmajor in Fig. 5. Also, the crossing area (right side of Fmajor) cannot be well separated by the mean shift. Fibers in the crossing region also have relatively larger densities so that the algorithm may converge to the local optima giving wrong clustering centers.

AP is famous for its automatic clustering without the need of setting the cluster number *a priori*. Unfortunately, AP cannot automatically find a correct cluster number in our experiments. AP is quite sensitive to outliers so that identify much more clusters than three in our experimental setup. Also, over one hundred times of iterations are necessary for clustering one tract, spending extremely long time to generate a plausible solution.

DBSCAN shares the similar philosophy with FDP in the sense that both focus on density of data points. The parameter setting has a great influence on the output. These parameters include the number of neighbors and neighboring radius. We found in our experiments that slight changes on these parameters output significant variations on the clustering. Thus, we can hardly find a stable and plausible solution on these three tracts of interest.

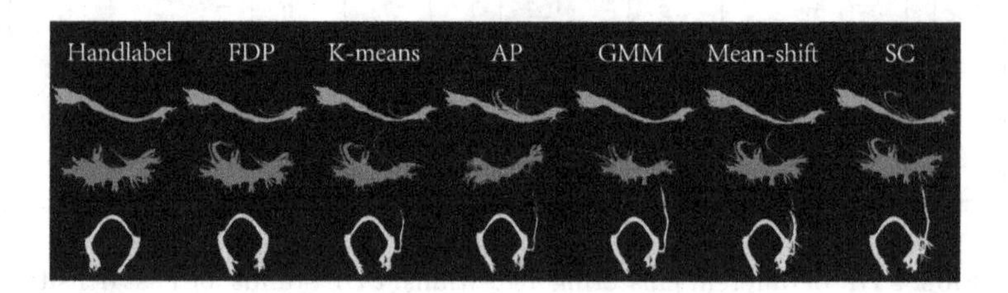

Fig. 5. Fiber clustering results visualization. It illustrates fiber clustering results using FDP, K-means, Affinity Propagation (AP), Gaussian Mixture Model (Gmm), Mean-shift and Spectral Clustering (SC) algorithms respectively. Take handlabel tracts as ground truth.

5 Tract-Based Analysis on Pediatric Brain Development

We analyze the development on the clustered tracts, Fmajor, ILF and IFO, given by FDP, and provide the preliminary results below.

5.1 Fiber Core Properties

Using FDP clustering method, there will be a core fiber tract for each fiber bundles shown in Fig. 6. We take this core fiber's characteristic to describe the property of the clustered fiber bundles [24]. We sampled each fiber tracts into ten 3D points. As the points in the boundary is not very stable compared with points in the middle of fiber tracts. Diffusion properties are then calculated for the inner six points of each fiber tract. Every point's property was taken by interpolating the image value (FA, MD).

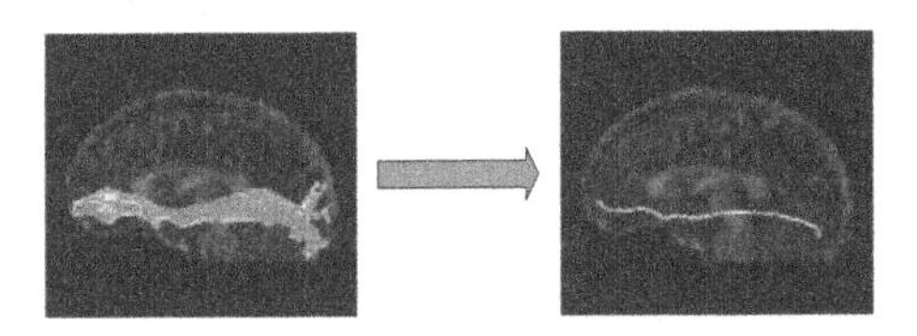

Fig. 6. Fiber bundel core. Inferior fronto-occipital fasciculus (IFO)'s core fiber tract. Statistics is performed along this fiber.

5.2 Tract-Wise Development Analysis

About 11 healthy subjects aged 18–36 months participate in this investigation. We calculate FA value of the core fiber tract for each fiber bundles. The concrete FA value and the correlation coefficient are different, but there is no doubt that they both have the similar increasing trend. P values are small than 0.05 in all groups, it indicates that there are significant correlations between ages and bundles' FA through center fiber. The results suggest that the progressive increase of FA values in each fiber bundles (including ILF, IFO and Fmajor) are observed for healthy Pediatrics with significant correlations ($P < 0.05$ for all scatter plots in Fig. 7) between FA and age. It maintains highly consistent with the conclusion of clinic and other researchers' study [4].

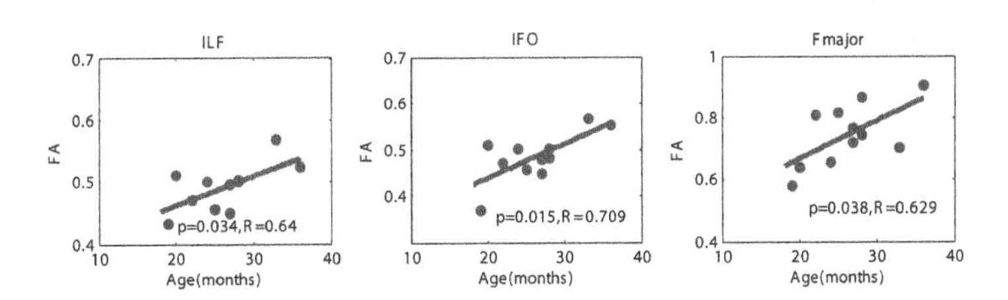

Fig. 7. Tract-based Analysis results according to core fibers.

6 Conclusion

In this paper, we propose a fast algorithm for clustering based on density peaks, applying the algorithm to segment fibers and analyzing the development of WM tract. We also propose using K nearest neighbors to calculate density. Experimental results show that Fast density-peaks clustering has a prefect performance in accuracy and algorithm accelerating. In the future, we will evaluate our method using more DTI data, and explore functions of different fiber bundles in Pediatric's brain.

Acknowledgments. We would like to thank Dr. Hua Cheng from Beijing Childrens Hospital for providing research data. It's acknowledged that paediatric data used for the experiment has been granted by the children's legal guardians.

References

1. Basser, P.J., Mattiello, J., Lebihan, D.: Mr diffusion tensor spectroscopy and imaging. Biophys. J. **66**(1), 259–267 (1994)
2. Wakana, S., Caprihan, A., Panzenboeck, M.M., Fallon, J.H., Perry, M., Gollub, R.L., Hua, K., Zhang, J., Jiang, H., Dubey, P., Blitz, A., van Zijl, P., Mori, S.: Reproducibility of quantitative tractography methods applied to cerebral white matter. NeuroImage **36**, 630–644 (2007)
3. Lebel, C., Gee, M., Camicioli, R., Wieler, M., Martin, W., Beaulieu, C.: Diffusion tensor imaging of white matter tract evolution over the lifespan. Neuroimage **60**, 340–352 (2012)
4. Ouyang, M., Cheng, H., Mishra, V., Gong, G., Mosconi, M.W., Sweeney, J., Peng, Y., Huang, H.: Atypical age-dependent effects of autism on white matter microstructure in children of 2–7 years. Hum. Brain Mapp. **37**, 819–832 (2015)
5. Jin, Y., Huang, C., Daianu, M., Zhan, L., Dennis, E.L., Reid, R.I., Jack, C.R., Zhu, H., Thompson, P.M.: 3D tract-specific local and global analysis of white matter integrity in Alzheimer's disease. Human Brain Mapp. **38**(3), 1191–1207 (2016)
6. O'Donnell, L.J., Golby, A.J., Westin, C.-F.: Fiber clustering versus the parcellation-based connectome. NeuroImage **80**, 283–289 (2013)
7. O'Donnell, L., Westin, C.-F.: White matter tract clustering and correspondence in populations. In: Duncan, J.S., Gerig, G. (eds.) MICCAI 2005. LNCS, vol. 3749, pp. 140–147. Springer, Heidelberg (2005). doi:10.1007/11566465_18
8. Reichenbach, A., Goldau, M., Heine, C., Hlawitschka, M.: V–bundles: clustering fiber trajectories from diffusion MRI in linear time. In: Navab, N., Hornegger, J., Wells, W.M., Frangi, A. (eds.) MICCAI 2015. LNCS, vol. 9349, pp. 191–198. Springer, Cham (2015). doi:10.1007/978-3-319-24553-9_24
9. O'Donnell, L.J., Kubicki, M., Shenton, M.E., Dreusicke, M.H., Grimson, W.E.L., Westin, C.-F.: A method for clustering white matter fiber tracts. Am. J. Neuroradiol. **27**(5), 1032–1036 (2006)
10. Wang, Q., Yap, P.T., Wu, G., Shen, D.: Application of neuroanatomical features to tractography clustering. Hum. Brain Mapp. **34**, 2089–2302 (2013)
11. Zhang, T., Chen, H., Guo, L., Li, K., Li, L., Zhang, S., Shen, D., Xiaoping, H., Liu, T.: Characterization of u-shape streamline fibers: methods and applications. Med. Image Anal. **18**(5), 795–807 (2014)
12. Jin, Y., Cetingül, H.E.: Tractography-embedded white matter stream clustering. In: 2015 IEEE 12th International Symposium on Biomedical Imaging, pp. 432–435 (2015)
13. Mai, S.T., Goebl, S., Plant, C.: A similarity model and segmentation algorithm for white matter fiber tracts. In: 2012 IEEE 12th International Conference on Data Mining, pp. 1014–1019 (2012)

14. Li, H., Xue, Z., Guo, L., Liu, T., Hunter, J., Wong, S.T.C.: A hybrid approach to automatic clustering of white matter fibers. NeuroImage **49**(2), 1249–1258 (2010)
15. Jin, Y., Shi, Y., Zhan, L., Gutman, B.A., de Zubicaray, G.I., McMahon, K.L., Wright, M.J., Toga, A.W., Thompson, P.M.: Automatic clustering of white matter fibers in brain diffusion MRI with an application to genetics. NeuroImage **100**, 75–90 (2014)
16. Rodriguez, A., Laio, A.: Clustering by fast search and find of density peaks. Science **344**, 1492–1296 (2014)
17. Chen, P., Fan, X., Liu, R., Tang, X.: Fiber segmentation using a density-peaks clustering algorithm. Biomed. Imaging **20**, 633–637 (2015)
18. Kamali, T., Stashuk, D.: Automated segmentation of white matter fiber bundles using diffusion tensor imaging data and a new density based clustering algorithm. Artif. Intell. Med. **73**, 14–22 (2016)
19. Jiang, H., van Zijl, P.C.M., Kim, J., Pearlson, G.D., Mori, S.: Dtistudio: resource program for diffusion tensor computation and fiber bundle tracking. Comput. Methods Programs Biomed. **81**, 106–116 (2006)
20. Jenkinson, M., Beckmann, C.F., Behrens, T.E.J., Woolrich, M.W., Smith, S.M.: Fsl. NeuroImage **62**, 782–790 (2012)
21. Woods, R.P., Grafton, S.T., Watson, J.D.G., Sicotte, N.L., Mazziotta, J.C.: Automated image registration: Ii. intersubject validation of linear and nonlinear models. J. Comput. Assist. Tomogr. **22**, 153–165 (1998)
22. Smith, S.M.: Fast robust automated brain extraction. Hum. Brain Mapp. **17**, 143–155 (2002)
23. Mayer, A., Zimmerman-Moreno, G., Shadmi, R., Batikoff, A., Greenspan, H.: A supervised framework for the registration and segmentation of white matter fiber tracts. IEEE Trans. Med. Imaging **30**, 131–145 (2011)
24. Johnson, R.T., Yeatman, J.D., Wandell, B.A., Buonocore, M.H., Amaral, D.G., Nordahl, C.W.: Diffusion properties of major white matter tracts in young, typically developing children. NeuroImage **88**, 143–154 (2014)

A Deep Learning Pipeline to Delineate Proliferative Areas of Intracranial Tumors in Digital Slides

Zaneta Swiderska-Chadaj[1](✉), Tomasz Markiewicz[1,2], Bartlomiej Grala[2], Malgorzata Lorent[2], and Arkadiusz Gertych[3]

[1] Faculty of Electrical Engineering, Warsaw University of Technology, Warsaw, Poland
zaneta.swiderska@gmail.com

[2] Department of Pathomorphology, Military Institute of Medicine, Warsaw, Poland

[3] Bioimage Informatics Laboratory, Department of Surgery, Cedars-Sinai Medical Center, Los Angeles, CA, USA

Abstract. Separating tumor cells from other tissues such as meninges and blood is one of the vital steps towards automated quantification of the proliferative index in digital slides of brain tumors. In this paper, we present a deep learning based pipeline to delineate areas of tumor in meningioma and oligodendroglioma specimens stained with Ki-67 marker. A pre-trained convolutional neural network (CNN) was fine-tuned with 7057 image tiles to classify whole slide images (n = 15) in a tile-by-tile mode. The performance of the model was evaluated on slides manually annotated by the pathologist. The CNN model detected tumor areas with 89.4% accuracy. Areas with blood and meninges were respectively classified with 98.2% and 89.8% accuracy. The overall classification accuracy was 88.7%, and the Cohen's kappa coefficient reached 0.748, indicating a very good concordance with the manual ground truth. Our pipeline can process digital slides at full resolution, and has the potential to objectively pre-process slides for proliferative index quantification.

Keywords: Deep learning · Whole slide quantitative image analysis · Brain tumor · Proliferation index

1 Introduction

Whole slide image (WSI) analysis in digital pathology is an actively growing area of research. New methods of image processing, texture analysis, and pattern recognition are developed to aid pathologists in detecting specific areas on the digital slide for in-depth evaluation. One of the ongoing challenges in this domain is to reduce inter- and intra-observer biases attributed to the manual slide assessment by pathologists [1, 2]. Assessments that suffer from reproducibility problems often relate to the underlying subjectivity of the manual evaluation, and support is mostly needed in counting of cells, evaluation of tumor burden, and tumor grading [1, 2].

M. Valds Hernndez and V. Gonzlez-Castro (Eds.): MIUA 2017, CCIS 723, pp. 448–458, 2017.
DOI: 10.1007/978-3-319-60964-5_39

In clinical setting, the assessment of tumor proliferative index requires counting of tumor cells expressing the Ki-67 antigen and cells that do not express it. The expression of Ki-67 is visualized by immunostaining. Nuclei of immunopositive cells are brown, whereas immunonegative nuclei are blue. The cells can be counted manually using regular microscope, or automatically by a computerized technique if a digital - image of the stained tissue is available [3]. In meningiomas and oligodendrogliomas which are common types of intracranial neoplasms, the Ki-67 proliferative index is helpful in predicting clinical outcome, tumor recurrence, overall survival, and serves as an ancillary technique in tumor grading.

The proliferative index is measured in hot-spots [4–7]. We recently developed an automated system for hot-spot selection in whole digital slides of brain tumors [3, 8]. The automated hot-spot selection and the proliferative index assessment were highly correlated to their manual counterparts. However, in some slides, blood cells reacted with the antigen and stained brown, and thus compromised the performance of hot-spot detection.

Most of the currently available state-of the-art machine learning approaches employ feature extraction and segmentation techniques to detect and exclude such artifacts. Yet, when applied to processing of whole slide images, these techniques are slow and computing resources demanding [2, 9–11]. Recently convolutional neural networks (CNNs) have emerged as efficient tools for image classification in histopathology. Conceptually, CNNs belong to the family of Deep Learning (DL) tools as they can learn generally useful feature detectors. CNNs can model the image data on low and high abstraction levels that are organized in interconnected processing blocks [12–14]. The most common DL architectures include convolutional neural networks, deep belief networks and recurrent neural networks. Of these, the CNNs are most commonly employed. The CNNs have demonstrated superior time performance to classical feature-based approaches, achieve high image classification rates, and have the potential to outperform humans in certain image recognition tasks [15]. In digital pathology, CNNs have for instance been applied to localize metastatic tumor areas, tumor grading, or detection of mitotic figures [11, 14]. Recent advances include developments of CNN-bases tools to delineate cancer in digital slides of hematoxylin and eosin (H&E) stained tissues. In [16], a custom designed CNN model achieved an accuracy of 67.4% in detecting cancer and non-cancer areas. This model was however inferior to a random forest classifier (76.4% accuracy) that learned RGB histograms. On the other hand, the same CNN model outperformed the random forest in detecting necrotic areas. In [17] two VGG-net models were applied to detect tumor associated stroma in H&E stained breast biopsies. The area under the ROC curve of the best model was 0.92. In [18] CNN models previously trained on H&E images were combined with a genetic algorithm to detect invasive tumor regions in whole digital slides. According to the recent review of DL applications in medical image analysis [19], CNNshave been predominantly developed for the analysis of tissue images stained with H&E, and only a handful number of solutions exists for analyzing images if tissues stained with immunohistochemical markers. In this domain [20, 21] describe approaches to Ki-67 cell detection and counting. Yet, no research on the detection of whole Ki-67 positive areas has so far been undertaken. Example application from other medical imaging domains include automated brain tumor detection, or lung nodule segmentation in computer tomography scans [22, 23].

The goal of our work is to develop a pre-processing pipeline that would delineate tumor area and exclude the artifacts in whole slide images of intracranial tumors stained with Ki-67 marker. In this work, we apply a CNN-based technique for image classification to detect three types of areas: (1) tumor including proliferating and non-proliferating cells, (2) areas of blood, and (3) other tissue such as meninges in whole slide images. We employed a previously trained CNN model, and fine-tuned it with a handful set of image tiles from digital slides. The performance of the model in whole digital slides was assessed in slides manually annotated by a pathologist.

2 Materials

20 glass slides with brain tumors (13 meningiomas and 7 oligodendrogliomas) previously stained with Ki-67 (3,3'-diaminobenzidine tetrahydrochloride, DAB marker, brown color) and counterstained with hematoxylin(blue) were obtained from the archives of the Department of Pathology at the Military Institute of Medicine in Warsaw, Poland. Slides were digitized using the Pannoramic 250 Flash II (3DHISTECH, Budapest, Hungary) whole slide scanner equipped with a 20× objective. Whole slide images outputted from the scanner were encoded as 24 bit RGB matrices with pixel size 0.38895 μm × 0.38895 μm, and saved in MRXS file format. Depending on the size of tissue fixed on slide, our digital slides contained 30,000 × 80,000 pixels that occupy 1.5–5.0 GB of hard drive space.

Each WSI was reviewed by a pathologist who manually generated ground truth annotations for areas of tumor, blood, and meninges directly on the WSIs. Since annotating the entire WSI was very time consuming, the pathologist selected the areas randomly without giving any preferences. Example WSIs with markings by the pathologist are shown in Fig. 1. Annotations from 5 randomly selected WSIs were used to obtain 256 × 256 pixel image tiles for CNN training. The remaining 15 WSIs (10 meningiomas and 5 oligodendrogliomas) were left to assess the CNN concordance with expert. All patient's data were removed from glass slides. This research was approved by the Research Ethics Board of the Military Institute of Medicine (Number: 30/WIM/2016).

3 Methods

The proposed workflow of WSI analysis includes three steps: CNN model training, WSI pre-processing to mark tissue area, and patch-based WSI classification to find areas of tumor, blood and meninges.

3.1 CNN Model Training

For our study, we chose AlexNet that was previously trained with images from LSVRC-2010 ImageNet database that contains 1000 different classes of natural objects [24]. AlexNethas eight learned layers of neurons; five of them are convolutional and

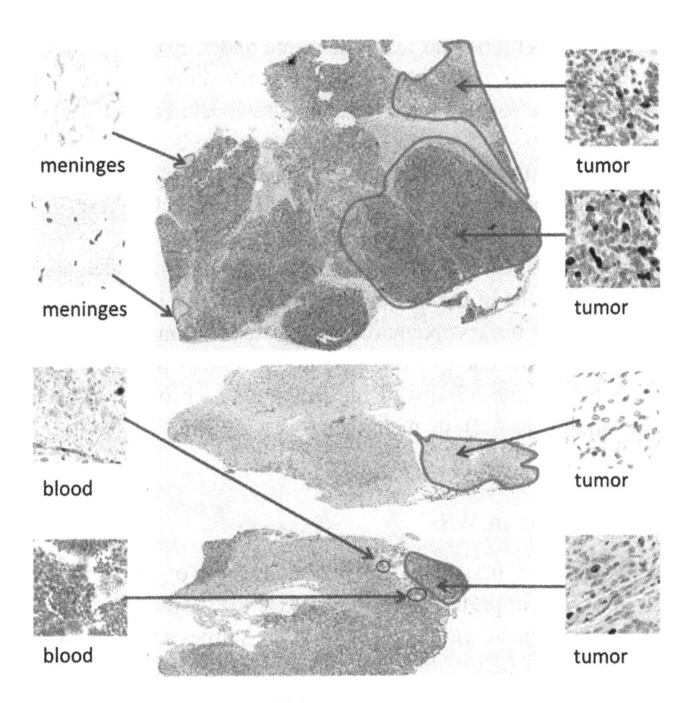

Fig. 1. Example WSI (middle column) with areas of tumor (blue), meninges (yellow) and blood (red) marked by a pathologist. Example image tiles extracted for trained are shown in the left and right columns. Image tiles in the right and left columns are 256 × 256 pixel large. (Color figure online)

three are fully-connected. The top-5 error rate of the pretrained AlexNetis 15.3%. The model was downloaded from [25] and installed in Caffe environment [26] for fine-tuning and adaptation to our project.

Fine-tuning (FT) of a CNN is one of the transfer learning techniques that has been shown effective in many image classification tasks [27, 28]. This strategy is to not to replace and retrain the classifier on top of the CNN, but to fine-tune the weights of the pretrained network by continuing the backpropagation. The fine-tuning is based on transferring the features and the parameters of the network from the broad domain to the specific one. The re-training operation allows for optimization of the network to minimize the error in the other, more specific domain. The FT may also be a solution when only a small cohort of training images is available, or when other transfer learning approaches such as the feature extraction provided at the output of the second to last fully connected layer of neurons are not suitable [23].

Areas annotated by the pathologist were divided into 256 × 256 non-overlapping pixel image tiles. After discarding poor quality tiles and those with mixed content, 2530 tumor, 718 blood, and 855 meningeal image tiles were retained. Selection of image tiles with tumor was guided by the pathologist who identified areas of very low, low, moderate, high and very-high proliferative tumor activity (Fig. 1, right column). These areas were then sampled with about 500 image tiles per category. In this collection, tiles labeled as blood and meninges were underrepresented due to the lack of sufficiently large areas with these tissue types in the 5 slides that we selected. A data augmentation step was performed to address this drawback. Each image tile displaying blood or meningeal tissue was respectively rotated by 90, 180, and 270° and then saved. Ultimately, the number of images with blood and meninges was increased 4 times, and the final number of all tiles was leveraged to 8822. Of these, 7057 were used for training (fine-tuning), and 1765 for testing of the CNN.

The CNN model was fine-tuned by modifying the network weights and by trimming the number of outputs in the last layer to three. The weights in the convolutional layers were fixed and the backpropagation was carried out only in the last three layers using the training set. The fine-tuned CNN model returns a three-value vector with class scores representing the affinity of the image to each of the predefined classes: blood (class 1), tumor (class 2), or meninges (class 3).

3.2 Detection of Tissue in WSI

Tissue localization is the first step in WSI analysis. To obtain a tissue mask, a down-sampled WSI was imported from the MRSX file, and then converted to a gray level image. The size of the down-sampled WSI was 8 times smaller than the size of the full-resolution WSI. Otsu image thresholding applied to this image yielded a binary tissue mask. The mask was refined with morphological operations (hole filling, small object removal), and then up-sampled to match the size of the full-resolution WSI (Fig. 2). WSIs encoded in MRXS format were read through open source libraries OpenSlide [29] and Matlab (Mathworks, Natick, MA) wrappers [30] that we included in the pipeline.

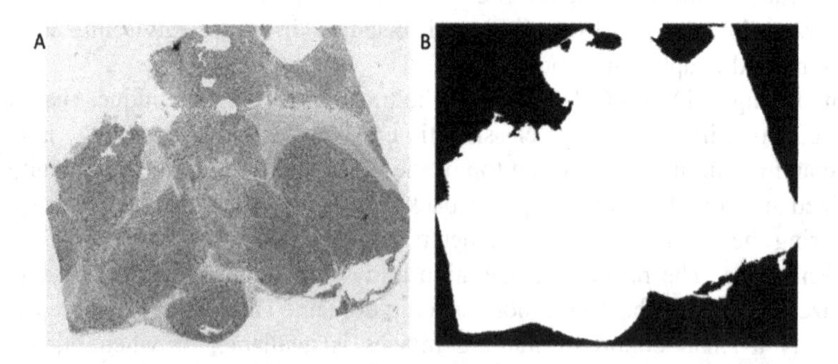

Fig. 2. Example of WSI (A) with a corresponding tissue mask (B).

3.3 WSI Classification

The up-sampled tissue mask was divided into non-overlapping 256 $\times$ 256 pixel tiles. The X and Y coordinates of tiles from the tissue mask were transferred to the original WSI image file to sequentially read corresponding image tiles at full resolution. Classification of image tiles by the fine-tuned CNNwas carried by finding the maximum class score for each tile. The tissue class corresponding to that score was then assigned to the tile as a classification result. In other words, image tiles in the original image were replaced with class score tiles of the same size. Tiles at the WSI border that are narrower than 256 pixels were augmented to perform classification. Finally, classification results from all tiles were saved as an image that had the same size as the full resolution WSI (Fig. 3).

3.4 Performance Evaluation

WSI classification performance was evaluated by comparing computer predicted and human (true) generated class labels at the pixel level. True positive (TP), true negative (TN), false positive (FP) and false negative (FN) classifications were collected in a 3×3 confusion matrix, and the Accuracy ACC = (TP + TN)/(TP + TN + FP + FN), Sensitivity (Recall) = TP/(TP + FN), Precision = TP/(TP + FP), Specificity = TN/(FP + TN), and F1score = 2TP/(2TP + FP + FN) were found for each tissue class. The overall classification accuracy and the Cohen's kappa (κ) concordance coefficient were calculated as well.

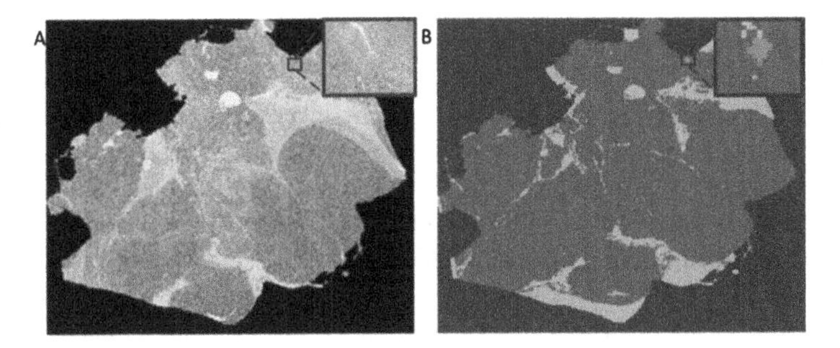

Fig. 3. An example of whole slide brain tumor image classification: (A) isolated tissue area, (B) classification results: tumor (blue), meninges (yellow), and blood (khaki). Insert in (B) shows tissue example area with all detected components. One point in the insert is equivalent to the tissue region with 256×256 pixels. (Color figure online)

4 Results

During the fine-tuning phase the fine-tuned CNN model achieved the 94% accuracy (n = 1765 images used for testing). The model was then embedded into the pipeline for processing of the 15 WSIs with brain tumors. The performance evaluation was carried out in regions annotated by the pathologist. The classification accuracy for slides with meningiomas and oligodendrogliomas was respectively 90.9% (n = 11) and 86.3% (n = 4). The overall system accuracy regardless the tumor type was ACC = 88.7%, and the Cohen's kappa was κ = 0.748. Figure 4 shows the confusion matrix and Table 1 contains performance metrics calculated for individual tissue classes. Example results of image classification with superimposed ground truth are illustrated in Fig. 5. The average processing time of a single WSI did not exceed 35 min.

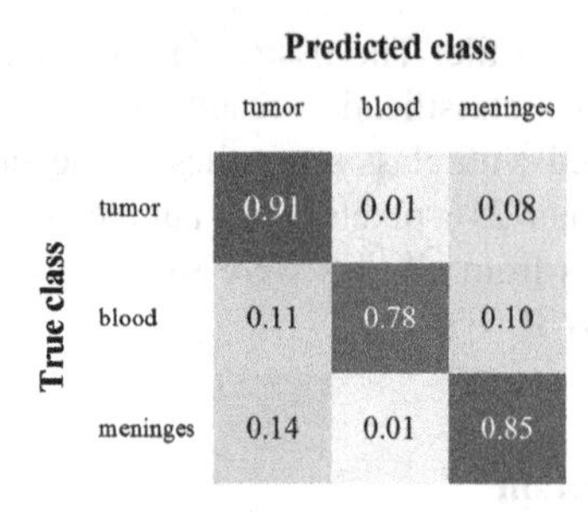

Fig. 4. Normalized confusion matrix.

Table 1. Performance metrics of brain tumor tissue classification by the fine-tuned AlexNet.

	Sensitivity (Recall)	Specificity	Precision	Accuracy	F-score
Tumor	0.906	0.867	0.942	0.894	0.924
Blood	0.784	0.99	0.745	0.982	0.764
Meninges	0.850	0.914	0.774	0.898	0.810

5 Discussion

Expression of nuclear antigen Ki-67 has been linked to proliferative activity and prognosis of many tumor types. Contemporary digital image analysis algorithms can reduce the subjectivity and improve the reproducibility of the manual slide evaluation and automatically assess the proliferative index in hot-spots [3, 8]. However, they are prone to errors arising from artifactual Ki-67 staining that is present in many specimens

The main goal of the study was to establish the quantitative method to delineate areas of tumor in brain cancers stained with Ki-67 and evaluate its performance. As previously reported, areas with blood constituted the major source of discrepancies in detecting tumor hot-spots and measuring the proliferative index between the pathologist and a SVM classifier [8]. The previous method utilized hand-crafted color and texture features that were extracted from image tiles down-sampled 8 times. The down-sampling was applied to reduce the computational burden of feature extraction. However, the down-sampling compromised the discriminatory power of the classifier which frequently assigned small areas with blood to the tumor category. As a result, regions with false positive tumor detection contributed to the overestimation of the proliferative index in the entire slide.

In this approach, image tiles extracted from a full-resolution WSI are presented to the CNN model for classification. Based on the F-score, the model performed best in identifying areas with tumor cells. However, about 9% of tumor, 15% of meningeal, and 21% of blood pixels were erroneously assigned to a different class. These errors can possibly arise from two sources: the insufficient accuracy of manual delineations, and the size of the image tile. It has been reported that manual delineation of fragmented, heterogenous or arbitrary shaped tissues is inherently less accurate and reproducible thana tracing around fine and homogenous tissue areas [1, 10, 31]. In our study, this

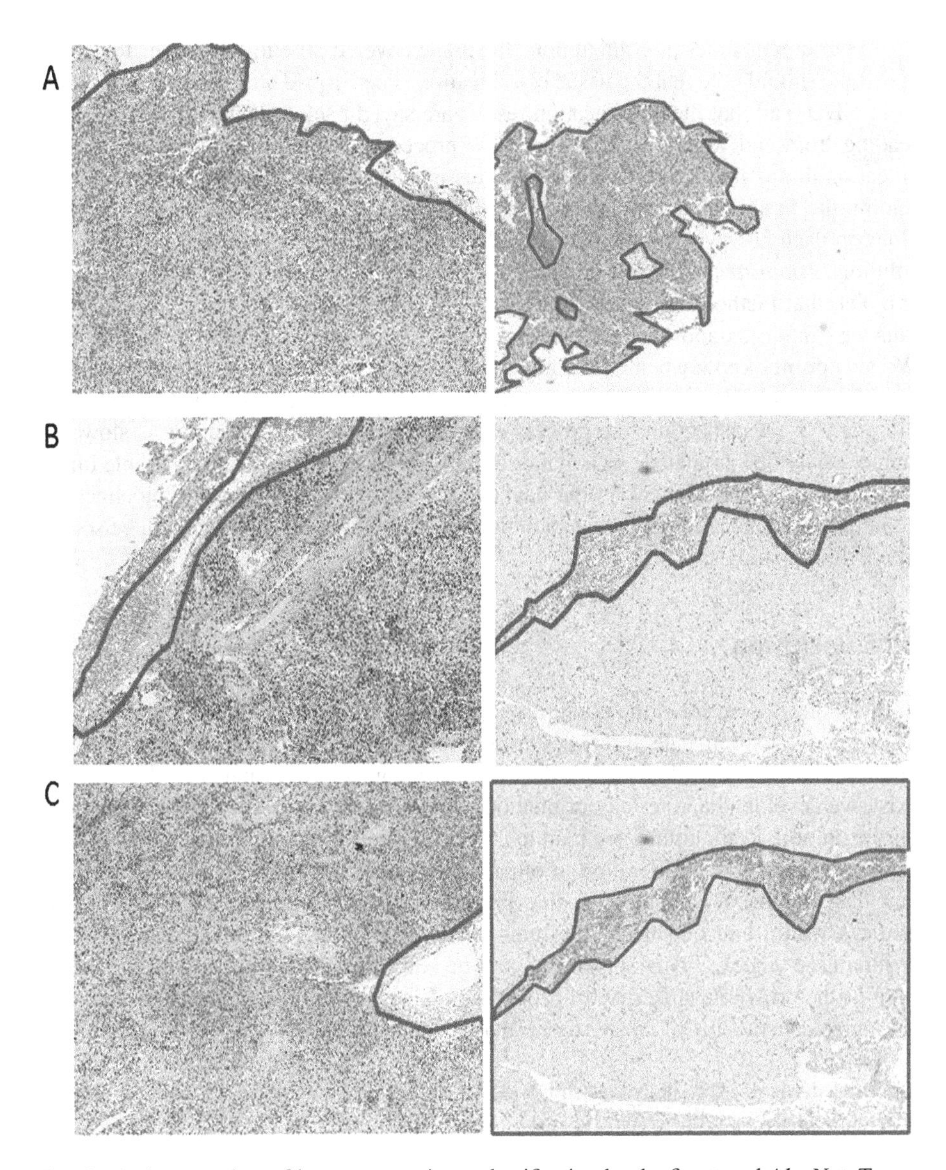

Fig. 5. A close-up view of image tumor tissue classification by the fine-tuned AlexNet. Tumor (A), areas with blood (B), and meninges (C) are outlined. Yellow outlines - regions marked by expert, pink area - regions detected by the fine-tuned AlexNet. (Color figure online)

problem is attributed to delineating small areas with blood cells that are scattered across the specimen. We noticed that the smallest areas with blood are about 500×500 pixels large. Hence, we chose to analyze WSI tile-by-tile at full image resolution. Our tile size was fixed at 256×256 and was smaller than the smallest blood areas. Nevertheless, processing of WSIs with 256×256 tile size lead to a very good concordance with the ground truth ($\kappa = 0.748$), and the overall accuracy reached 88.7%.

In our digital slides of brain tumors, the tissue covered 30% to 70% of the total WSI area. One should note that the tissue classification is performed after reading image data from HDD, and that the classification results are saved back to HDD afterwards. Both reading from and writing to HDD are slow procedures in contrast to the actual classification that is performed on GPU and thus much faster. Removing background and finding the tissue helps limit interactions with HDD and thus speeds up the analysis. Our approach however, turned out to be slower comparing to our previously published solution (2 min of processing time) which involved the 8-fold image down-sampling [8]. Yet, that method did not have the capacity to process the entire digital slide, and thus we cannot juxtapose its performance to the performance of the proposed pipeline. We also do not know whether the amount of training images was optimal to fine-tune the CNN model. However, since each image tile for CNN training needs to undergo a thorough visual inspection, the process of collecting tiles for the training is slow and cannot easily be automated. As a matter of fact, the process of collecting reliable image data constitutes a bottleneck in the development of many deep learning pipelines for quantitative image analysis in histopathology. A solution to this problem will be sought in our future studies.

6 Conclusion

In this work, we explored the possibility of implementing a pipeline with a pre-trained CNN model to delineate three tissue components in whole-slide images of brain tumors. Since standard benchmark data for this image analysis challenge does not yet exist, we developed and tested our methodology on brain tumor specimens acquired at our institution. In the future, we plan to expand our research and adjust our pipeline to capture and analyze proliferation in other tumor types. Our preliminary data suggests that this pipeline is suitable for slide pre-processing to delineate tumor and isolate artifacts within and outside of the tumor that can compromise quantification of the proliferative index. This pipeline can be an attractive alternative for many state-of-the-art whole slide quantification approaches that utilize handcrafted feature to recognize constituents of cancer tissue that is stained with a nuclear biomarker.

Acknowledgment. This study was supported by the National Centre for Research and Development, Poland (grant PBS2/A9/21/2013).

References

1. Stålhammar, G., Martinez, N.F., Lippert, M., Tobin, N.P., Mølholm, I., Kis, L., Rosin, G., Rantalainen, M., Pedersen, L., Bergh, J., Grunkin, M., Hartman, J.: Digital image analysis outperforms manual biomarker assessment in breast cancer. Mod. Pathol. **29**, 318–329 (2016)
2. Gurcan, M., Boucheron, L., Can, A., Madabhushi, A., Rajpoot, N., Yener, B.: Histopathological image analysis: a review. IEEE Rev. Biomed. Eng. **2**, 147–171 (2009)

3. Swiderska, Z., Korzynska, A., Markiewicz, T., Lorent, M., Zak, J., Wesolowska, A., Roszkowiak, L., Slodkowska, J., Grala, B.: Comparison of the manual, semiautomatic, and automatic selection and leveling of hot spots in whole slide images for Ki-67 quantification in meningiomas. Anal. Cell. Pathol. (Amst), **2015**, Article no. 498746 (2015)
4. Bruna, J., Brell, M., Ferrer, I., Gimenez-Bonafe, P., Tortosa, A.: Ki-67 proliferative index predicts clinical outcome in patients with atypical or anaplastic meningioma. Neuropathology **27**(2), 114–120 (2007)
5. Torp, S.H., Lindboe, C.F., Grønberg, B.H., Lydersen, S., Sundstrøm, S.: Prognostic significance of Ki-67/MIB-1 proliferation index in meningiomas. Clin. Neuropathol. **24**(4), 170–174 (2005)
6. Coleman, K.E., Brat, D.J., Cotsonis, G.A., Lawson, D., Cohen, C.: Proliferation (MIB-1 expression) in oligodendrogliomas: assessment of quantitative methods and prognostic significance. Appl. Immunohistochem. Mol. Morphol. **14**(1), 109–114 (2006)
7. Kros, J.M., Hop, W.C., Godschalk, J.J., Krishnadath, K.K.: Prognostic value of the proliferation-related antigen Ki-67 in oligodendrogliomas. Cancer **78**(5), 1107–1113 (1996)
8. Swiderska-Chadaj, Z., Markiewicz, T., Grala, B., Lorent, M.: Content-based analysis of Ki-67 stained meningioma specimens for automatic hot-spot selection. Diagn. Pathol. **11**(1), 93 (2016)
9. Janowczyk, A., Madabhushi, A.: Deep learning for digital pathology image analysis: a comprehensive tutorial with selected use cases. J. Pathol. Inform. **7**, 29 (2016)
10. Gertych, A., Ing, N., Ma, Z., Fuchs, T.J., Salman, S., Mohanty, S., Bhele, S., Velásquez-Vacca, A., Amin, M.B., Knudsen, B.S.: Machine learning approaches to analyze histological images of tissues from radical prostatectomies. Comput. Med. Imaging Graph. **46**(Pt 2), 197–208 (2015)
11. Litjens, G., Sánchez, C.I., Timofeeva, N., Hermsen, M., Nagtegaal, I., Kovacs, I., Hulsbergen - van de Kaa C., Bult, P., van Ginneken, B., van der Laak, J.: Deep learning as a tool for increased accuracy and efficiency of histopathological diagnosis. Sci. Rep. **6**, Article no. 26286 (2016)
12. LeCun, Y., Bengio, Y., Hinton, G.: Deep learning. Nature **521**(7553), 436–444 (2015)
13. Schmidhuber, J.: Deep learning in neural networks: an overview. Neural Netw. **61**, 85–117 (2015)
14. Cruz-Roa, A., Basavanhally, A., González, F., Gilmore, H., Feldman, M., Ganesan, S., Shih, N., Tomaszewski, J., Madabhushi, A.: Automatic detection of invasive ductal carcinoma in whole slide images with convolutional neural networks. In: Proceedings of SPIE 9041, Medical Imaging 2014: Digital Pathology, p. 904103 (2014)
15. Ciresan, D.C., Meier, U., Schmidhuber, J.: Multi-column deep neural networks for image classification. In: IEEE Conference on Computer Vision and Pattern Recognition CVPR 2012, pp. 3642–3649. Arxiv preprint arXiv:1202.2745 (2012)
16. Sharma, H., Zerbe, N., Klempert, I., Hellwich, O., Hufnagl, P.: Deep convolutional neural networks for histological image analysis in gastric carcinoma whole slide images. Diagn. Pathol. **1**(8), 1–3 (2016)
17. Bejnordi, B.E., Linz, J., Glass, B., Mullooly, M., Gierach, G.L., Sherman, M.E., Beck, A.H.: Deep learning-based assessment of tumor-associated stroma for diagnosing breast cancer in histopathology images. arXiv preprint arXiv:1702.05803 (2017)
18. Puerto, M., Vargas, T., Cruz-Roa, A.: A digital pathology application for whole-slide histopathology image analysis based on genetic algorithm and Convolutional Networks. In: 2016 IEEE Latin American Conference on Computational Intelligence (LA-CCI), pp. 1–7. IEEE (2016)

19. Litjens, G., Kooi, T., Bejnordi, B.E., Setio, A.A.A., Ciompi, F., Ghafoorian, M., Sánchez, C. I.: A survey on deep learning in medical image analysis. arXiv preprint arXiv:1702.05747 (2017)
20. Xie, W., Noble, J.A., Zisserman, A.: Microscopy cell counting and detection with fully convolutional regression networks. Comput. Methods Biomech. Biomed. Eng.: Imaging Vis. **2016**, 1–10 (2016)
21. Xie, Y., Kong, X., Xing, F., Liu, F., Su, H., Yang, L.: Deep voting: a robust approach toward nucleus localization in microscopy images. In: Navab, N., Hornegger, J., Wells, W.M., Frangi, A.F. (eds.) MICCAI 2015. LNCS, vol. 9351, pp. 374–382. Springer, Cham (2015). doi:10.1007/978-3-319-24574-4_45
22. Havaei, M., Davy, A., Warde-Farley, D., Biard, A., Courville, A., Bengio, Y., Larochelle, H.: Brain tumor segmentation with deep neural networks. Med. Image Anal. **35**, 18–31 (2017)
23. Anthimopoulos, M., Christodoulidis, S., Ebner, L., Christe, A., Mougiakakou, S.: Lung pattern classification for interstitial lung diseases using a deep, convolutional neural network. IEEE Trans. Med. Imaging **35**(5), 1207–1216 (2016)
24. Krizhevsky, A., Sutskever, I., Hinton, G.E.: Imagenet classification with deep convolutional neural networks. In: Advances in Neural Information Processing Systems, pp. 1097–1105 (2012)
25. https://github.com/BVLC/caffe/wiki/Model-Zoo
26. Jia, Y., Shelhamer, E., Donahue, J., Karayev, S., Long, J., Girshick, R., Guadarrama, S., Darrell, T.: Caffe: convolutional architecture for fast feature embedding. arXiv preprint arXiv:1408.5093 (2014)
27. Tajbakhsh, N., Shin, J.Y., Gurudu, S.R., Hurst, R.T., Kendall, C.B., Gotway, M.B., Liang, J.: Convolutional neural networks for medical image analysis: full training or fine tuning? IEEE Trans. Med. Imaging **35**(5), 1299–1312 (2016)
28. CS231n: Convolutional Neural Networks for Visual Recognition. http://cs231n.github.io/transfer-learning/
29. Goode, A., Benjamin, G., Harkes, J., Jukic, D., Satyanarayanan, M.: OpenSlide: a vendor-neutral software foundation for digital pathology. J. Pathol. Inform. **4**, 27 (2013)
30. https://github.com/fordanic/openslide-matlab
31. Rizzardi, A.E., Zhang, X., Vogel, R.I., Kolb, S., Geybels, M.S., Leung, Y.K., Henriksen, J. C., Ho, S.M., Kwak, J., Stanford, J.L., Schmechel, S.C.: Quantitative comparison and reproducibility of pathologist scoring and digital image analysis of estrogen receptor β2 immunohistochemistry in prostate cancer. Diagn. Pathol. **11**(1), 63 (2016)

Tree-Based Ensemble Learning Techniques in the Analysis of Parkinsonian Syndromes

J.M. Górriz[1(✉)], J. Ramírez[1], M. Moreno-Caballero[2], F.J. Martinez-Murcia[1], A. Ortiz[4], I.A. Illán[3], F. Segovia[1], D. Salas-González[1], and M. Gomez-Rio[2]

[1] Department of Signal Theory and Communications, University of Granada, Granada, Spain
gorriz@ugr.es
[2] Hospital Virgen de Las Nieves, Granada, Spain
[3] Department of Scientific Computing, The Florida State University, Tallahassee, USA
[4] Department Communication Engineering, University of Malaga, Malaga, Spain

Abstract. ^{123}I-ioflupane single photon emission computed tomography (SPECT) is a standard and well-known imaging modality in the medical practice for the diagnosis of Parkinson's disease (PD). That said, atypical parkinsonian syndrome (APS), a symptom-related disease to PD, detection is yet considered inconsistent at least based on visual inspection on region of interests (ROIs). Although some machine learning approaches have been proposed in this regard, in this paper we take up this matter again by applying advanced image processing techniques based on ensemble learning in order to discriminate PD from the various APS, included in the group denominated as P plus. This study enrolled 168 subjects including followed-up patients with degenerative parkinsonism and normal controls undergoing ^{123}I-ioflupane SPECT at the "Virgen de las Nieves" Hospital, Spain in the last years: 45 Normal, 75 PD, 31 APs (including multiple system atrophy (MSA), progressive supranuclear palsy (PSP) and corticobasal degeneration (CBD) patients) and 17 controls. Several advanced ensemble techniques and feature extraction methods were applied voxel-wise to the analysis of the SPECT images using robust classifiers based on decision trees. The system is trained by means of boosting and bagging algorithms and their performance control is specified in terms of the classification error and the received operating characteristics curve (ROC) using 10-fold cross validation. By the use of these statistical validation methods it was possible to confirm that this modality may be useful for discriminating the abnormal patterns under study.

Keywords: Atypical parkinsonism · Single photon emission computed tomography (SPECT) · Ensemble pattern recognition · Boosting · Bagging · Decision tress

M. Valdés Hernández and V. González-Castro (Eds.): MIUA 2017, CCIS 723, pp. 459–469, 2017.
DOI: 10.1007/978-3-319-60964-5_40

1 Introduction

The parkinsonian syndromes (PS) include a set of neurodegeneration diseases with similar clinical symptoms such as bradykinesia, extrapyramidal rigidity, tremor and gait disturbance [26]. The forms of degenerative parkinsonism include idiopathic Parkinson disease (PD) and the collectively referred to as atypical parkinsonian syndromes (APs), such as progressive supranuclear palsy (PSP), multiple system atrophy (MSA), corticobasal degeneration (CBD), among other rarer causes of parkinsonism. The prevalence of these diseases is up to 2% in the population above 65 years of age, although with the increasing life expectancy they will become much more prevalent in the next decades.

The discrimination among all these pathologies, mainly APs vs PD, is nowadays a challenge. As an example, the clinical diagnosis of PD using ^{123}I-ioflupane-SPECT can be achieved by an expert with a high accuracy [14] (around 90%), nevertheless it is true that the detection of APs with that sensibility is unlikely to be reached even by using post-synaptical modalities [25]. Moreover, before differential diagnosis among them becomes an issue, parkinsonism and tremor conditions that are not associated with neurodegeneration but may mimic "true" parkinsonism, i.e. drug-induced and psychogenic parkinsonism; essential, drug-induced, psychogenic tremor, have to be reliably ruled out [15]. The importance of an accurate diagnosis is related to the fact that it enables follow-up investigations and may also be valuable for potentially monitoring disease progression and therapeutic effects, i.e. PD patients typically respond well to dopaminergic medications unlike APS.

In the last decade, there has been increasing interest in pattern recognition techniques, such as support vector machines (SVMs) and linear discriminant classifiers (LDC) to assist diagnosis of PD in brain imaging data, such as magnetic resonance imaging (MRI) [2,11], SPECT [7,10,14,24]. Structural imaging approaches (MRI) have limited value in the early diagnosis of Parkinsonism since structural abnormalities can be "seen" only when the disease is advanced in most cases. Conversely, the approaches based on SPECT imaging has a high sensitivity to detect PD versus controls even at an early stage, yet the alterations on the postsynaptic level may be quite similar between PD and APS. Finally, recent advances have tackled the comparison between PD and APS in two different ways. On SPECT imaging [1] used ensemble of LDC and leave-one-out (LOO) cross-validation in a large cohort of consecutive patients with a clinical diagnosis of PD, MSA, PSP and CBD. They reached the highest specificity of 84–90% for PD versus APS and AUCs around 0.7 in distinguishing among APS. Unfortunately, it is widely appreciated that LOO is a suboptimal method for cross-validation, as it gives estimates of the prediction error with higher variance than other method for CV such as K-fold or bootstrap [12]. The LOO is only suitable to be used with small datasets ($N < 100$) and may lead to overly optimistic cross-validation results. On the other hand, using other ligands to assess the post-synaptic function with ^{18}F-DMFP positron emission tomography (PET) in a cohort study ($N = 87$) [25], an accuracy up to 78% was reached using the same validation methodology as the latter approach (LOO-CV).

Ensemble learning based pattern recognition techniques are multivariate, supervised approaches that can take into account characteristics of different distributed populations encoded in complex high dimensional features and use them to categorize the data. Boosting and bagging are popular ensemble methods for improving the accuracy of classification algorithms [4]. They are based on a combination of the outputs of many "weak" classifiers to produce a powerful "committee" [12], i.e. AdaBoostM1 and all its later versions [3,5,23], etc. They have been successfully applied to neuroimaging not only in the classification problem [19–21] but for MRI subcortical surface-based analysis [13], automated hippocampal segmentation [18], automated mapping of hippocampal atrophy [17], etc. The aim of the paper is to propose these techniques in combination with feature extraction methods, such as principal component analysis (PCA) [9] or partial least squares (PLS) [8], that provide a significant reduction in computational time effectively retaining the latent structure in the data. In this sense, a novel method based on RobustBoost tree ensemble technique [3] in combination with PLS feature extraction is proposed that yields a significant reduction in the K-fold cross-validation error when compared with other standard tree ensemble techniques such as LogitBoost or Adaboost or Bagging, i.e. an accuracy up to 90% in the more complex classifications tasks.

2 Materials and Methods

2.1 Database

The present retrospective study of parkinsonism includes 151 patients (Male: 83, age: 37–85, ave: 71.04; Female: 68, age: 31–89, ave: 70.3) and a group of 17 healthy people for research purposes. In a clinical setting, we try to combine the use of semi-quantitative information approaches to achieve a better diagnostic performance in the diagnosis of neurodegeneration of the dopaminergic nigrostriatal pathway. Beyond the aim of this paper, overall experimental results shall be compared taking into account the visual interpretation, the analysis of computer-aided diagnosis systems (CAD) and the final clinical diagnoses, that are established by monitoring the progress of the disease (gold standard). The data has been collected in a tertiary hospital ("Virgen de las Nieves", Granada) between January 1, 2007 and December 31, 2012. Patients were evaluated in the Movement Disorders Unit of this hospital by a group of trained neurologists, showing among their symptoms tremor and other presynaptic impairments for uncertain clinical Parkinsonism. The 17 controls (CNTRL) were healthy people, relatives of patients who agreed to undergo the procedure for research purposes.

After obtaining a thorough medical history, all subjects were examined by ^{123}I-ioflupane-SPECT imaging following the available international guidelines. The scans were acquired at our center four hours after the intravenous injection of 5 mCi of radiotracer using a triple-head gamma camera (Philips-Picker Prism 3000; low-energy, high-resolution NeuroFAN collimation; 128×128 matrix; radius: 12.9–13.2 cm; step-and-shoot mode; 35 s/step; 15–20 kcts/step; 3°/step; circular orbit). Raw data was reconstructed by iterative algorithm (OS MLEM; 1

subset; 4 iterations) and attenuation correction was performed using the Chang method (linear correction coefficient: $0.12\,\mathrm{cm}^{-1}$). The processed images were exposed according to the intercommissural line and transaxial orthogonal planes and were interpreted and labeled, with masking of the clinical orientation, under the criterium of nuclear medicine specialists grouping them into three classes: Normal (NOR 45 + CNTRL 17), PD (75) and APs (31), with several levels of disease.

The image preprocessing steps [16] consist of spatial-intensity normalization and binary masking on the striatum area, which reduces the number of voxels used in the subsequent analysis. In I-123-ioflupane-SPECT imaging the non-specific region is not considered for diagnostics. All the images were spatially normalized using the SPM 8 software [6] and yielding, a $79 \times 95 \times 69$ functional activity map for each patient. This method assumes a general affine model with 12 parameters and a Bayesian framework that maximizes the product of the prior function (which is based on the probability of obtaining a particular set of zooms and shears) and the likelihood function, derived from the residual squared difference between the template and the processed image. The template was obtained as a result in [22] by registering all control images to a randomly chosen one by affine transformations. Finally, a simple segmentation method (voxels whose intensity is higher than a specific intensity threshold I_{th}) is applied to discriminate the striatum area from the background with clear advantages in terms of computational load ($D \sim 10^3$).

2.2 Feature Extraction Methods

Two classical feature extraction methods, such as PCA and PLS, are considered at this step. They have been widely used in neuroimaging for performing image analysis and classification [24]. PCA orthogonally decomposes the dataset into a linear combination of principal components or loadings, in such a way that they maximize the explained variance. The main idea in this dimension reduction is to truncate the number of components to the first K components ($K < D$), thus the $D \times K$ mixing matrix $\mathbf{W}$ contains the transformation of the D original features to the new K-dimensional space, that is, $\tilde{\mathbf{X}}_{N\times K} = \mathbf{X}_{N\times D}\mathbf{W}_{D\times K}$. On the other hand, PLS is a statistical method which models relationships among sets of observed variables by means of latent variables [24]. It includes regression analysis and classification tasks, and is intended as a dimension reduction technique as well. The starting point for PLS is the very simple assumption that the observed data is generated by a system or process which is driven by a smaller number of latent (not directly observed or measured) variables. PLS finds the relationship between the input data $\mathbf{X}$ and the set of labels $\mathbf{Y}$ as linear combinations of the score matrices via the matrices of loadings assuming an error matrix ($\mathbf{X} = \mathbf{X}_s\mathbf{X}_l^T + \mathbf{E}$). Again, a number of K components can be truncated to obtain a novel set of reduced dimension features.

2.3 Tree-Based Ensemble Learning Techniques

Boosting and Bagging are the most powerful methods for statistical validation in machine learning. In a nutshell, ensemble learning is based on building a prediction model by combining the strengths of a collection of simpler base models. Bagging is an ensemble method for classification, where a committee of trees each cast a vote for the predicted class [12], whereas in Boosting the committee of weak learners evolves over time [4]. Consider a two-class problem $Y \in \{1, -1\}$ and a N-sample realization of predictor variables $\mathbf{X}$, given the set of weak classifiers $G_m(x)$, the predictions from all of them are combined through a weighted majority vote to compose the final classifier as:

$$G(\mathbf{x}) = sign\left(\sum_{m=1}^{M} \alpha_m G_m(\mathbf{x})\right) \tag{1}$$

where $\alpha_m = \log\left((1 - e_m)/e_m\right)$ is a weighting error-dependent parameter of each weak learner:

$$e_m = \frac{\sum_{i=1}^{N} w_i I(y_i \neq G_m(\mathbf{x}_i))}{\sum_{i=1}^{N} w_i} \tag{2}$$

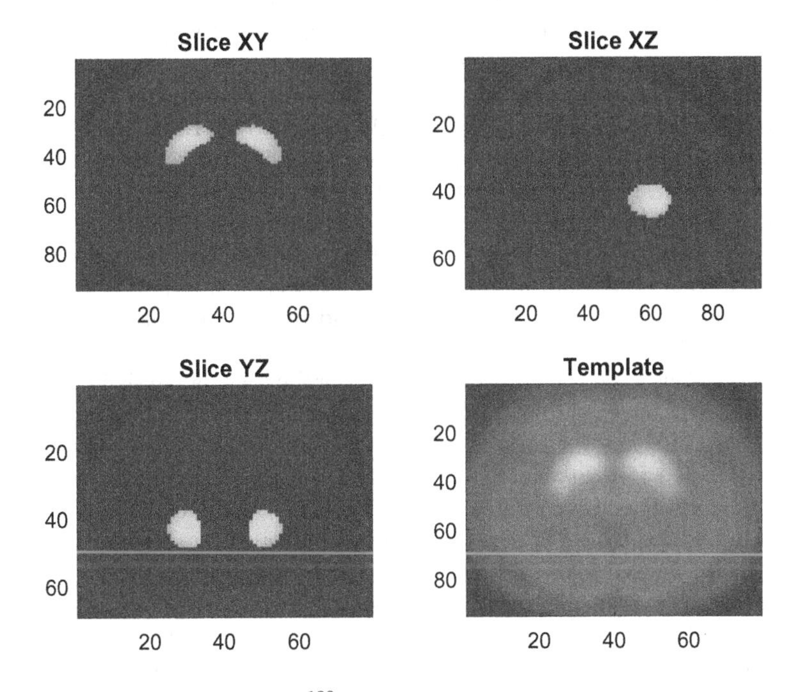

Fig. 1. Preprocessing result of a ^{123}I-ioflupane-SPECT image from a healthy subject and the template used in spatial normalization. The image is represented by its slices on the specific region.

Recently, several alternatives have been proposed following these learning ideas, such as LogitBoost, GentleBoost or RobustBoost [3,5]. The LogitBoost algorithm uses Newton steps for fitting an adaptive symmetric logistic model by maximum likelihood [5] where the weights of the training sample are updated in terms of probability $w_i = p(\mathbf{x}_i)(1-p(\mathbf{x}_i))$, where $p(x) = e^{G(\mathbf{x})}/(e^{G(\mathbf{x})}+e^{-G(\mathbf{x})})$. GentleBoost is an hybrid procedure between AdaBoost and LogitBoost that minimizes the exponential loss by using adaptive steps similar to the ones in LogitBoost. The main benefit of the last approach is that RobustBoost is more robust against label noise than either LogitBoost or Adaboost [3]. Thus, this alternative is clearly fitted to be applied on neuroimaging datasets, whose labeling is not directly linked to the contents of the images or is subjected to visual error, i.e. cognitive test or clinical-based labeling. In this paper we demonstrate this claim by using a real-world image dataset as detailed in Sect. 2.1. The effectiveness of Robustboost is based on the fact that, after an approximate classifier has been learned, it can be beneficial to down-weight examples that are far from the decision boundary regardless of their label [3] (Fig. 1).

3 Results

The system was implemented using the Matlab Software and the Signal Processing Toolbox. Several experiments were carried out in order to analyze the performance of the feature extraction methods in combination with several boosting and bagging ensemble tree learning approaches. In particular, ensembles of decision trees with an increasing number of trees were trained by the use of the most representative methods (AdaBoostM1, LogitBoost, Bagging and RobustBoost algorithms). The loss of the resulting predictors was estimated through a 10-fold CV process, where the feature extraction methods were applied in the CV loop to avoid overfitting. Three group comparisons with increasing task complexity were conducted in order to highlight the differences among the disease patterns, in particular the differences of dopaminergic uptakes between Norm and PD + APs (G1), Norm and APs (G2) and finally, the differences between PD and APs (G3).

In the current experimental framework, a preliminary analysis is firstly performed in terms of explained variance for the selection of experimental parameters. The PLS analysis of the current datasets, including all the patterns and class labels PD, APs, Norm and Control, and the 3D representation of the PL components reveal an interesting finding that can be exploited in further classification stages (see Fig. 2). As shown in this figure the selected functional modality may be used to discriminate between the four classes since the patterns are clearly drawn from different distributions. Note the normal-labeled images by visual inspection clearly represent an overlapped class between controls and PS, and the differences between PD and APs pattern distribution.

In the second experiment we analyze in Fig. 3 the influence of the number of PLS components K by the ROC curves of the proposed system, where the RobustBoost algorithm was selected following the motivations in Sect. 2.3. The curves are shifted

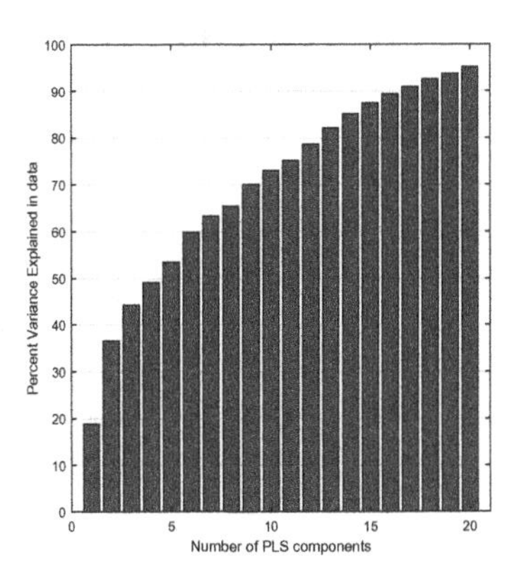

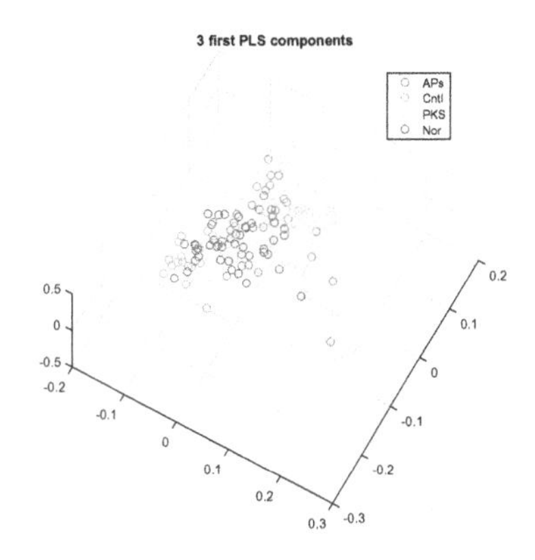

Fig. 2. Multi-class PLS analysis of the complete dataset. Left: the explained variance (accumulated). Right: 3D representation of the PLS components

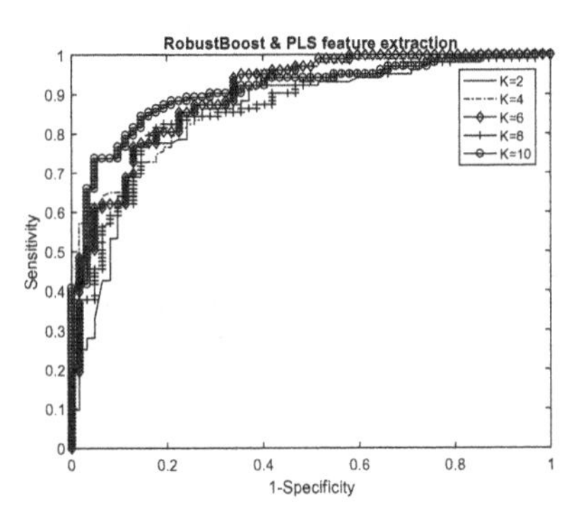

Fig. 3. ROC curves of the RobustBoost-based system for several values of the K PLS parameter on the G1 task.

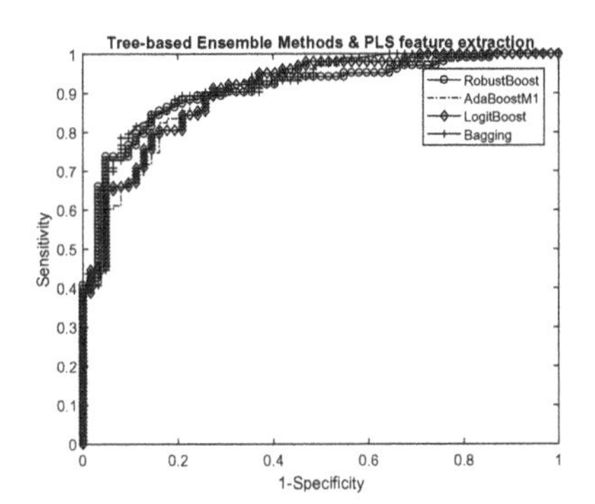

Fig. 4. ROC curves of selected tree-ensemble methods for $K = 10$ (PLS) on the G1 task.

up to the left when increasing the component dimension from $K = 2$ up to a value of $K = 10$ with $AUCs = \{0.8603, 0.8922, 0.8984, 0.8644, 0.9075\}$, respectively. This dimension value is then used to perform a comparison with the rest of ensemble methods (Fig. 4) for the same comparison group, G1: Normal vs PS. As shown in the figure the performance of all methods are quite similar in the operating points of the curves on this simple task with $AUCs$: Bagging 0.9153; AdaBoostM1 0.9001; LogitBoost 0.9040.

Finally, by selecting a number of components of $K = 10$ corresponding to a 70% explained variance in the PLS analysis, we compare, in Fig. 5, the AdabostM1, LogitBoost and Bagging ensemble methods with our RobustBoost-based system. All of them were applied to the defined classification tasks (G1, G2 and G3) in combination with the analyzed feature extraction methods (PLS and PCA). For space reasons, we only show the results using PLS for the complete

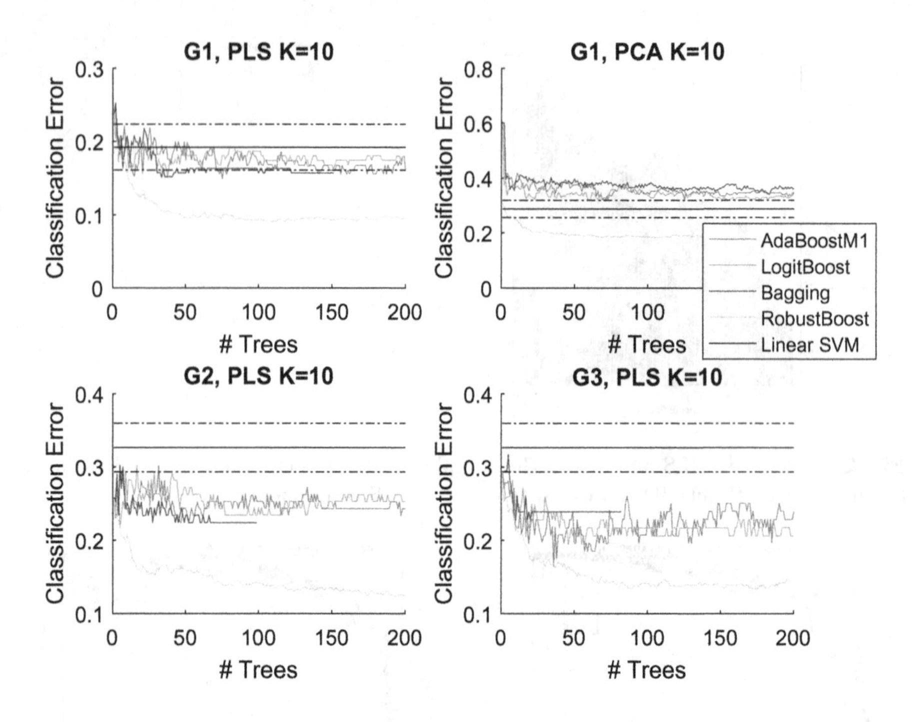

Fig. 5. Ensemble loss for $K = 10$ and 10-fold validation as a function of the number of trees in the ensemble in different complexity image classification tasks. Note: Normal group (No: Normal, Co: Control). The confidence interval for the SVM approach is also provided.

study groups (G1, G2 and G3). Moreover, We show the resulting loss-function representations for a different number of trees (T). In these figures a full comparison with another baseline method, consisting of a PLS(PCA)-based SVM linear classifier with the confidence interval computed in the K-folds, is also provided. The residual error after convergence of the RobustBoost ensemble is significantly smaller under the most complex tasks for all the selected values of the K parameter, whereas the performance in terms of convergence rate is similar for the analyzed ensemble methods. For K = 10-fold RobustBoost provides AUC values of $G1_{PLS}$: 0.9075; $G1_{PCA}$:0.6948; $G2_{PLS}$:0.7833; $G3_{PLS}$:0.7833.

As shown from all the set of experiments, none of the standard methods achieved an accuracy rate over than 90% for the most complex task, i.e. APs vs PD, unlike the RobustBoost-based approach that clearly outperformed all of them. This accuracy is motivated by the fact that the gold standard used in this paper refers to the visual labeling of the database by a group of experts with masking of the clinical orientation, thus all the available information, derived from the follow-up or the clinical suspicion of PS, is not used to establish the "ground truth". This is the main reason for obtaining a better performance using the RobustBoost algorithm whose main feature is the robustness against label noise unlike the rest of the analyzed ensemble methods. Surely, a step forward may be to validate the outcome of the system with the final clinical diagnosis of the proposed database.

4 Conclusions

This paper proposed an improved method for SPECT classification using feature extraction methods, such as PCA and PLS, to enable a significant dimensionality reduction without loss of information, and tree-based ensemble learning techniques. The dimension of the feature space K was empirically selected by 10-fold cross-validation of the proposed methods. It was shown that the residual error after convergence of the RobustBoost ensemble is significantly smaller than the ones obtained by other ensemble methods, such as AdaBoostM1, LogitBoost, Bagging, etc. and the linear SVM classifier-based paradigm for a wide range of dimension K in the feature space. Thus, the proposed FE methods, especially the PLS approach, were found to be suitable for reducing the dimensionality of the input data without a penalty on the performance of the ensembles. The performance in terms of convergence rate is similar for the analyzed ensemble methods, and the minimum residual loss obtained by the RobustBoost algorithm for high complexity classification tasks (i.e. Normal vs. Aps and Aps vs. PD) is crucially lower. In future works we will compare the outcome of the proposed ensemble methods with the final clinical diagnoses in order to assess the ability of the proposed methods in the early detection of typical and atypical PS.

Acknowledgement. This work was partly supported by the MINECO/FEDER under the TEC2015-64718-R project and the Consejería de Innovación, Ciencia y Empresa (Junta de Andalucía, Spain) under the Excellence Project P11-TIC-7103.

References

1. Badoud, S., Ville, D.V.D., Nicastro, N., Garibotto, V., Burkhard, P.R., Haller, S.: Discriminating among degenerative parkinsonisms using advanced ^{123}I-ioflupane SPECT analyses. NeuroImage: Clin. **12**, 234–240 (2016)
2. Focke, N.K., Helms, G., Scheewe, S., Pantel, P.M., Bachmann, C.G., Dechent, P., Ebentheuer, J., Mohr, A., Paulus, W., Trenkwalder, C.: Individual voxel-based subtype prediction can differentiate progressive supranuclear palsy from idiopathic Parkinson syndrome and healthy controls. Hum. Brain Mapp. **32**(11), 1905–1915 (2011)
3. Freund, Y.: A more robust boosting algorithm. arXiv preprint arXiv:0905.2138 (2009)
4. Freund, Y., Schapire, R.E.: A decision-theoretic generalization of on-line learning and an application to boosting. J. Comput. Syst. Sci. **55**(1), 119–139 (1997)
5. Friedman, J., Hastie, T., Tibshirani, R.: Additive logistic regression: a statistical view of boosting (with discussion and a rejoinder by the authors). Ann. Statist. **28**(2), 337–407 (2000)
6. Friston, K., et al. (eds.): Statistical Parametric Mapping: The Analysis of Functional Brain Images. Academic Press, Cambridge (2007)

7. Górriz, J.M., Lassl, A., Ramírez, J., Salas-Gonzalez, D., Puntonet, C., Lang, E.: Automatic selection of ROIs in functional imaging using gaussian mixture models. Neurosci. Lett. **460**(2), 108–111 (2009)
8. Gorriz, J.M., Ramirez, J., Illan, I.A., Martinez-Murcia, F.J., Segovia, F., Salas-Gonzalez, D.: Case-based statistical learning applied to SPECT image classification. In: SPIE, Medical Imaging, Computer-Aided Diagnosis, vol. 78, pp. 1–4 (2017)
9. Górriz, J.M., Ramírez, J., Lassl, A., Salas-González, D., Lang, E.W., Puntonet, C.G., Álvarez, I., López, M., Gómez-Río, M.: Automatic computer aided diagnosis tool using component-based SVM. In: 2008 IEEE Nuclear Science Symposium Conference Record, pp. 4392–4395 (2008)
10. Górriz, J.M., Segovia, F., Ramírez, J., Lassl, A., Salas-Gonzalez, D.: GMM based SPECT image classification for the diagnosis of Alzheimer's disease. Appl. Soft Comput. **11**(2), 2313–2325 (2011). http://dx.doi.org/10.1016/j.asoc.2010.08.012
11. Haller, S., Badoud, S., Nguyen, D., Barnaure, I., Montandon, M.L., Lovblad, K.O., Burkhard, P.: Differentiation between Parkinson disease and other forms of Parkinsonism using support vector machine analysis of susceptibility-weighted imaging (SWI): initial results. Eur. Radiol. **23**(1), 12–19 (2013)
12. Hastie, T., Tibshirani, R., Friedman, J.: The Elements of Statistical Learning. Springer, New York (2001)
13. Hu, Z., Pan, Z., Lu, H., Li, W.: Classification of Alzheimer's disease based on cortical thickness using AdaBoost and combination feature selection method. In: Wu, Y. (ed.) ICCIC 2011. CCIS, vol. 234, pp. 392–401. Springer, Heidelberg (2011). doi:10.1007/978-3-642-24091-1_51
14. Illán, I., Górriz, J., Ramírez, J., Segovia, F., Jiménez-Hoyuela, J., Ortega Lozano, S.: Automatic assistance to Parkinson's disease diagnosis in datscan spect imaging. Med. Phys. **39**(10), 5971–5980 (2012)
15. Tatsch, K.: Extrapyramidal syndromes: PET and SPECT. In: Hodler, J.J., von Schulthess, G.K., Zollikofer, C.L. (eds.) Diseases of the Brain, Head & Neck, Spine, pp. 234–239. Springer, Milan (2008). doi:10.1007/978-88-470-0840-3_36
16. Martínez-Murcia, F.J., Górriz, J.M., Ramírez, J., Illán, I., Ortiz, A., Initiative, P.P.M., et al.: Automatic detection of Parkinsonism using significance measures and component analysis in datscan imaging. Neurocomputing **126**, 58–70 (2014)
17. Morra, J.H., Tu, Z., Apostolova, L.G., Green, A.E., Avedissian, C., Madsen, S.K., Parikshak, N., Toga, A.W., Jack, C.R., Schuff, N., et al.: Automated mapping of hippocampal atrophy in 1-year repeat MRI data from 490 subjects with Alzheimer's disease, mild cognitive impairment, and elderly controls. Neuroimage **45**(1), S3–S15 (2009)
18. Morra, J.H., Tu, Z., Apostolova, L.G., Green, A.E., Toga, A.W., Thompson, P.M.: Comparison of AdaBoost and support vector machines for detecting Alzheimer's disease through automated hippocampal segmentation. IEEE Trans. Med. Imaging **29**(1), 30 (2010)
19. Ramírez, J., Górriz, J., Segovia, F., Chaves, R., Salas-Gonzalez, D., López, M., Álvarez, I., Padilla, P.: Computer aided diagnosis system for the Alzheimer's disease based on partial least squares and random forest spect image classification. Neurosci. Lett. **472**(2), 99–103 (2010)
20. Ramírez, J., Górriz, J.M., Ortiz, A., Padilla, P., Martínez-Murcia, F.J.: Ensemble tree learning techniques for magnetic resonance image analysis. In: Chen, Y.-W., Toro, C., Tanaka, S., Howlett, R.J., Jain, L.C. (eds.) Innovation in Medicine and Healthcare 2015. SIST, vol. 45, pp. 395–404. Springer, Cham (2016). doi:10.1007/978-3-319-23024-5_36

21. Ramírez, J., Górriz, J.M., Martínez-Murcia, F.J., Segovia, F., Salas-Gonzalez, D.: Magnetic resonance image classification using nonnegative matrix factorization and ensemble tree learning techniques. In: 2016 IEEE 18th International Workshop on Multimedia Signal Processing (MMSP). pp. 1–5, September 2016
22. Salas-González, D., Górriz, J.M., Ramírez, J., Lassl, A., Puntonet, C.G.: Improved Gauss-Newton optimization methods in affine registration of SPECT brain images. IET Electron. Lett. **44**(22), 1291–1292 (2008)
23. Schapire, R.E., Freund, Y., Bartlett, P., Lee, W.S.: Boosting the margin: a new explanation for the effectiveness of voting methods. Ann. Statist. **26**(5), 1651–1686 (1998)
24. Segovia, F., Górriz, J., Ramírez, J., Alvarez, I., Jiménez-Hoyuela, J., Ortega, S.: Improved Parkinsonism diagnosis using a partial least squares based approach. Med. Phys. **39**(7), 4395–4403 (2012)
25. Segovia, F., Illán, I.A., Górriz, J.M., Ramírez, J., Rominger, A., Levin, J.: Distinguishing Parkinson's disease from atypical Parkinsonian syndromes using PET data and a computer system based on support vector machines and bayesian networks. Front. Comput. Neurosci. **9**, 137 (2014)
26. Williams, D., Litvan, I.: Parkinsonian syndromes. Continuum?: Lifelong Learn. Neurol. **5**, 1189–1212 (2013)

Evaluating Alzheimer's Disease Diagnosis Using Texture Analysis

Francisco Jesús Martinez-Murcia[1(✉)], Juan Manuel Górriz[1], Javier Ramírez[1], Fermin Segovia[1], Diego Salas-Gonzalez[1], Diego Castillo-Barnes[1], Ignacio A. Illán[2], Andres Ortiz[3], and for the Alzheimer's Disease Neuroimaging Initiative

[1] Department of Signal Theory, Networking and Communications, University of Granada, 18071 Granada, Spain
fjesusmartinez@ugr.es

[2] Department of Scientific Computing, The Florida State University, Tallahassee, FL 32306-4120, USA

[3] Department of Communications Engineering, University of Málaga, 29071 Málaga, Spain

Abstract. Many advanced automated systems have been proposed for the diagnosis of Alzheimer's Disease (AD). Most of them use Magnetic Resonance Imaging (MRI) as input data, since it provides high resolution images of the structure of the brain. Usually, Computer Aided Diagnosis (CAD) systems are based on massive univariate test and classification, although many strategies based on signal decomposition have been proposed for feature extraction in MRI images. In this work, we propose a novel analysis technique comprising the texture analysis of different cortical and subcortical structures in the brain. The procedure shows promising results, achieving up to 81.3% accuracy in the diagnosis task, and up to 79.6% accuracy using only one texture measure at the most discriminant region. These results prove the ability of textural analysis in the characterization of structural neurodegeneration of the brain, and paves the way to future longitudinal and conversion analyses.

1 Introduction

Alzheimer's Disease (AD) is the most common neurodegenerative disorder in the world, with more than 46 million people affected [2]. With the current population ageing in developed countries, this number is expected to increase up to 131.1 millions by 2050 [2]. Therefore, an early diagnosis is needed for an early

Data used in preparation of this article were obtained from the Alzheimer's Disease Neuroimaging Initiative (ADNI) database (https://adni.loni.usc.edu). As such, the investigators within the ADNI contributed to the design and implementation of ADNI and/or provided data but did not participate in analysis or writing of this report. A complete listing of ADNI investigators can be found at: http://adni.loni.usc.edu/wp-content/uploads/how_to_apply/ADNI_Acknowledgement_List.pdf.

M. Valdés Hernández and V. González-Castro (Eds.): MIUA 2017, CCIS 723, pp. 470–481, 2017.
DOI: 10.1007/978-3-319-60964-5_41

intervention and improving the life expectancy and quality of life of the affected subjects and their families.

Currently, one of the most extended techniques to explore neurodegeneration in AD is Magnetic Resonance Imaging (MRI). It provides us a non-invasive tool to explore the internal structure of the brain and the distribution of Gray Matter (GM) and White Matter (WM), which usually has a correlation with neurodegeneration [3,11], in contrast to other neuroimaging modalities, such as Single Photon Emission Computed Tomography (SPECT) or Positron Emission Tomography (PET), which require the injection of a radiopharmaceutical. Analysis of these images is usually performed visually or using semiquantitative tools to assess the degree of neurodegeneration.

In contrast to traditional visual and semiquantitative analysis of images, many fully automated systems for analysing MRI images have been proposed. Apart from the widely extended Voxel Based Morphometry (VBM) [3], numerous feature extraction algorithms have been proposed. Some recent approaches include decomposition via Principal Component Analysis (PCA) [8,25], Independent Component Analysis (ICA) [17], Partial Least Squares (PLS) [22], or projecting information of the brain to a bidimensional plane using Spherical Brain Mapping (SBM) [14,18]. In a recent review [10], a wide variety of algorithms using shape, volume and texture analysis were reported for the diagnosis of AD. Of these, texture analysis has already been used in the analysis of neuroimaging of different modalities with great success [13,16,28].

In this work, we propose a system that combines brain region segmentation of T1-weighted images using a strategy based on atlas masking, and a posterior texture analysis of each region. These features extracted at each region are used to quantify which combination of measures and regions are useful to characterize neurodegeneration in AD.

This article is organized as follows. First, the methodology used to analyse MRI images, and evaluate our system is detailed in Sect. 2. Later, in Sect. 3, the results are presented. Finally, we draw some conclusions about our system at Sect. 4.

2 Methodology

2.1 Atlas Segmentation

In our work we wanted to test the hypothesis that structural changes in different areas of the brain can predict changes in neurodegeneration, and therefore, be related to Alzheimer's Disease (AD). We will compute the structural changes at different regions using a naive atlas segmentation.

Atlas segmentation is a technique that uses a brain atlas, such as the Automated Anatomical Labelling (AAL) [26], the Montreal Neurological Institute (MNI) or IBASPM [1], to mask out different regions in images registered to the same space. Since in our case, T1-weighted images are registered to the MNI space, we have applied the IBASPM atlas to extract regions to which a subsequent texture analysis has been applied.

The IBASPM atlas consist of 90 cortical and subcortical regions divided by hemisphere, that have been set using three elements: gray matter segmentation, normalization transform matrix (the matrix used to map voxels from individual to the MNI space) and the MaxPro MNI atlas (Fig. 1).

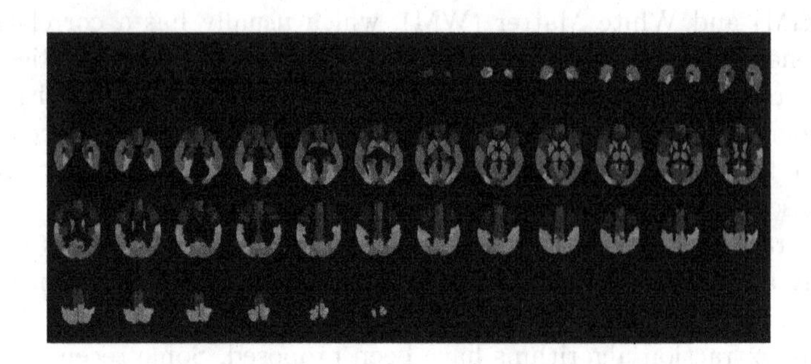

Fig. 1. Some axial cuts of the IBASPM atlas.

2.2 Haralick Texture Analysis

Texture analysis is usually based on the computation of a Gray-Level Co-occurrence (GLC) matrix. This matrix is defined over an image to be the distribution of co-occurring values at a given offset. Mathematically, we can define the co-occurrence matrix over a $n \times m$ bidimensional image $\mathbf{I}$ as:

$$\mathbf{C}_{\Delta x,\Delta y}(i,j) = \sum_{p=1}^{n}\sum_{q=1}^{m}\begin{cases}1, & \text{if } \mathbf{I}(p,q) = i \text{ and } \mathbf{I}(p+\Delta x, q+\Delta y) = j\\ 0, & \text{otherwise}\end{cases} \tag{1}$$

where i and j are the different gray levels. For simplicity, the image is usually quantized to N_g gray levels. In this work we have used $N_g = 16$.

The parametrization of the GLC matrix using the offsets $(\Delta x, \Delta y)$ can make it sensitive to rotation. Therefore, we will use different offsets at different angles to get to some degree of rotational invariance. For simplicity, we will use the same distance d in all directions, and therefore, we can rewrite the offset vector $\Delta_\mathbf{p}$ as:

$$\Delta_\mathbf{p} = d(\Delta_x, \Delta_y) \quad \text{so that} \quad \Delta_x, \Delta_y \in \{-1, 0, 1\} \tag{2}$$

Using this parametrization, it is easy to expand the GLC matrix to a three-dimensional image $\mathbf{I}$ of size $n \times m \times k$, parametrized this time by a 3D offset [20]:

$$\mathbf{C}_{\Delta_p}(i,j) = \sum_{\mathbf{p}=(1,1,1)}^{(n,m,k)}\begin{cases}1, & \text{if } \mathbf{I}(\mathbf{p}) = i \text{ and } \mathbf{I}(\mathbf{p}+\Delta_\mathbf{p}) = j\\ 0, & \text{otherwise}\end{cases} \tag{3}$$

We use thirteen spatial directions to compute every GLCM in the 3D space [20]. In this work we will use $d = \{1, 2, 3\}$, and therefore, $3 \times 13 = 39$ GLC matrices will be computed for each region. To extract features from these GLC matrices, let us define the probability matrix $\mathbf{P}$ as:

$$\mathbf{P}(i,j) = \frac{\mathbf{C}_{\Delta_p}(i,j)}{\sum_{i,j}\mathbf{C}_{\Delta_p}(i,j)} \tag{4}$$

With this new matrix of probabilities, we can compute the thirteen Haralick Texture measures that were defined in the original Haralick paper [7], with the following expressions:

$$f_1 = \sum_i \sum_j \mathbf{P}(i,j)^2 \tag{5}$$

$$f_2 = \sum_{n=0}^{N_g-1} n^2 \left\{ \sum_{|i-j|=n} \mathbf{P}(i,j) \right\} \tag{6}$$

$$f_3 = \frac{\sum_i \sum_j ij\mathbf{P}(i,j) - \mu_x \mu_y}{\sigma_x \sigma_y} \tag{7}$$

$$f_4 = \sum_i \sum_j (i-\mu)^2 \mathbf{P}(i,j) \tag{8}$$

$$f_5 = \sum_i \sum_j \frac{\mathbf{P}(i,j)}{1+(i-j)^2} \tag{9}$$

$$f_6 = \sum_{k=2}^{2N_g} k \sum_{i+j=k} \mathbf{P}(i,j) \tag{10}$$

$$f_7 = \sum_{k=2}^{2N_g} (k-f_6)^2 \sum_{i+j=k} \mathbf{P}(i,j) \tag{11}$$

$$f_8 = -\sum_{k=2}^{2N_g} \sum_{i+j=k} \mathbf{P}(i,j) \log \left\{ \sum_{i+j=k} \mathbf{P}(i,j) \right\} \tag{12}$$

$$f_9 = -\sum_i \sum_j \mathbf{P}(i,j) \log(\mathbf{P}(i,j)) \tag{13}$$

$$f_{10} = \mathrm{VAR} \left\{ \sum_{|i-j|=k} \mathbf{P}(i,j) \right\} \tag{14}$$

$$f_{11} = -\sum_{k=0}^{N_g-1} \sum_{|i-j|=k} \mathbf{P}(i,j) \log \left\{ \sum_{|i-j|=k} \mathbf{P}(i,j) \right\} \tag{15}$$

where f_1 is the Angular Second Moment (ASM), f_2 is Contrast, f_3 is Correlation, f_4 is Sum of Squares: Variance, f_5 is the Inverse Difference Moment, f_6 is the Sum Average, f_7 is the Sum Variance, f_8 is Sum Entropy, f_9 is Entropy, f_{10} is Difference Variance and f_{11} is Difference Entropy.

For the last two texture measures, f_{12} and f_{13}, let us define:

$$p_x(i) = \sum_{j=1}^{N_g} \mathbf{P}(i,j) \tag{16}$$

$$p_y(j) = \sum_{i=1}^{N_g} \mathbf{P}(i,j) \tag{17}$$

Let us also note H_X and H_Y the entropies of p_x and p_y respectively, and:

$$H_{XY} = -\sum_i \sum_j \mathbf{P}(i,j) \log \mathbf{P}(i,j) \tag{18}$$

$$H_{XY1} = -\sum_i \sum_j \mathbf{P}(i,j) \log \{p_x(i)p_y(i)\} \tag{19}$$

$$H_{XY2} = -\sum_i \sum_j p_x(i)p_y(i) \log \{p_x(i)p_y(i)\} \tag{20}$$

where H_{XY} is their joint entropy, H_{XY1} and H_{XY2} would be the join entropy of X and Y assuming independent distributions.

With all these notations, the last two measures, known as Information Measures of Correlation (IMC-1 and IMC-2) can be defined as:

$$f_{12} = \frac{H_{XY} - H_{XY1}}{\max\{H_X, H_Y\}} \tag{21}$$

$$f_{13} = (1 - \exp[-2(H_{XY2} - H_{XY})])^{1/2} \tag{22}$$

and f_{12} and f_{13} are Information Measures of Correlation (IMC-1 and IMC-2).

Having 39 GLC matrices per region from which 13 measures are computed, we will obtain 507 measures per region, and having 90 region, this makes a total of 45630 measures per patient. In one of the experiments, we have used all these measures in the classification task, therefore a strategy to perform feature selection is desired.

2.3 Feature Selection

Different feature selection methods have been proposed throughout the literature [15]. These methods use statistical measures to assess significance of the Haralick measures. To do so, either a parametrical or empirical approach can be used. In this work, we have used both approaches.

For the parametrical approach, the widely known independent two-sample t-test has been used. This test computes the t-statistic for each element in the feature vector, and then, its statistical significance can be estimated by using the t-distribution. The t-statistic is computed as:

$$t = \frac{\bar{X}_1 - \bar{X}_2}{\sqrt{\frac{\sigma^2_{X_2} + \sigma^2_{X_1}}{n}}} \tag{23}$$

where $\sigma^2_{X_i}$ is the variance and $\bar{X}_i$ is the average within class i.

On the other hand, we can use the Kullback-Leibler (KL) divergence to assess statistical significance, by computing the KL measure as in [24]:

$$KL = \frac{1}{2}\left(\frac{\sigma^2_{X_2}}{\sigma^2_{X_1}} + \frac{\sigma^2_{X_1}}{\sigma^2_{X_2}} - 2\right) + \frac{1}{2}\left(\bar{X}_1 - \bar{X}_2\right)^2\left(\frac{1}{\sigma^2_{X_1}} + \frac{1}{\sigma^2_{X_2}}\right) \tag{24}$$

and then perform a battery of permutation tests so that we can obtain the empirical distribution of the KL values for each feature, from which the p-values can be obtained.

2.4 Database and Preprocessing

Data used in the preparation of this article were obtained from the Alzheimer's Disease Neuroimaging Initiative (ADNI) database (https://adni.loni.usc.edu). The ADNI was launched in 2003 as a public-private partnership, led by Principal Investigator Michael W. Weiner, MD. The primary goal of ADNI has been to test whether serial magnetic resonance imaging (MRI), positron emission tomography (PET), other biological markers, and clinical and neuropsychological assessment can be combined to measure the progression of mild cognitive impairment (MCI) and early Alzheimers disease (AD). For up-to-date information, see www.adni-info.org.

The database used in this article was extracted from the ADNI1: Screening 1.5T (subjects who have a screening data) and contains 1075 T1-weighted MRI images, comprising 229 NOR, 401 MCI and 188 AD images. In this work, only the first session of 188 AD and 229 subjects was used. The images were spatially normalized using the SPM software [6], after a skull removing procedure.

2.5 Evaluation

We have evaluated the texture measures by means of a classification analysis. In this analysis, different sets of measures from the train set are used to train a Support Vector Classifier (SVC) [27], and performance results are estimated by testing the SVC with the test set. In this work, we have used a 10-Fold cross validation strategy [9]. In this strategy, we divide the whole dataset in 10 parts (folds), and each of them are used as a test set to the SVC when training with the remaining 9. The procedure is repeated 10 times and performance values of accuracy, sensitivity and specificity (and their corresponding standard deviations) are obtained.

We have evaluated our system in two different experiments:

- **Experiment 1:** We test our system with only one feature at a time, and evaluate the performance obtained with that feature. Each feature is one texture measure at a certain distance and offset for a given region. From these performance values, we can obtain an estimation of which texture measures provide a higher discrimination.

- **Experiment 2:** We pool together all values measures computed in all regions, in all directions and offset vectors. Then, we apply hypothesis testing to select the most significant measures at different p thresholds.

3 Results and Discussion

3.1 Experiment 1

Multiple performance results (90 regions, 13 texture measures, 13 offset and 3 distances) have been computed for this experiment. We cannot detail all 45630 accuracy, sensitivity or specificity values, but we examine the distribution of those using a boxplot.

In Fig. 2, we show a boxplot of the distribution of the accuracy values obtained using all texture measures in each region. We can see that higher performance is obtained at the Hippocampus and surrounding regions, especially at the right Hippocampus. The parahippocampal lobes and amygdalas (especially the left amygdala) get notable results. The hippocampal and parahippocampal lobes have received high interest in the literature, since it plays a relevant role in memory. Particularly, Gray Matter loss due to neurodegeneration has been consistently reported in many works [4,5]. This neurodegeneration probably leads to a change in some of the texture measures used here, therefore, it reveals the usefulness of texture measures in the parametrization of AD.

We will focus then on the analysis of which texture measures perform better within the Hippocampus, our area of interest. For this purpose, we pool again the performance values obtained at the right Hippocampus and plot them at Fig. 3 after grouping them by texture measure.

In Fig. 3, one measure clearly stands out: Angular Second Moment. Other measures such as Inverse Difference Moment and Entropy obtain good, although more variable, performance.

3.2 Experiment 2

In this experiment, we test the performance that can be achieved using all measures and selecting the most significant ones by means of a hypothesis test. We have used two different strategies to assess significance: the Student's t-test and the Kullback-Leibler (KL) divergence. Table 1 displays the values obtained for the different strategies, compared to the performance of using all measures computed at the right Hippocampus.

We can se that the selection improves the system's ability to detect changes related to AD, when compared to using the measures computed at each region, or even using the best measure at the best scoring region.

When compared to a commonly used voxel-wise baseline, Voxels As Features (VAF) [23], we can see that it clearly outperforms the typical approach where segmented GM and WM maps are used. When using the whole segmented T1-weighted image, the difference is smaller, although both the system using feature

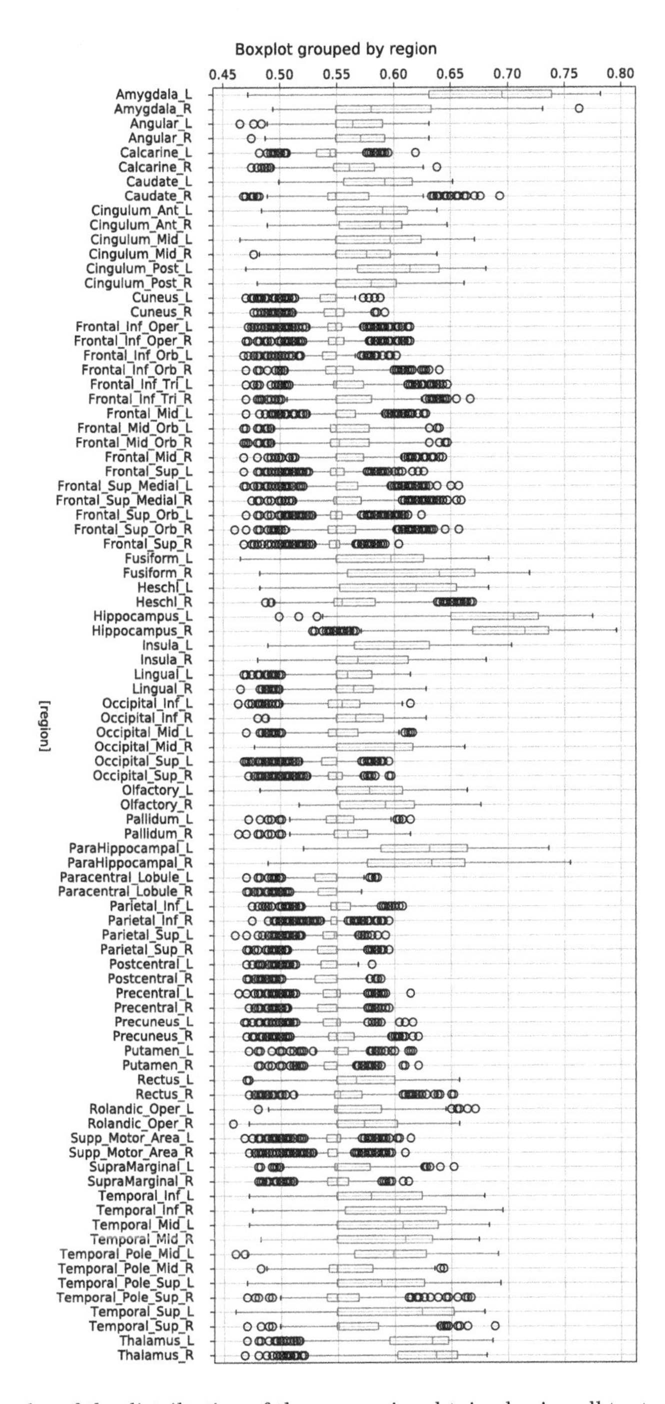

Fig. 2. Boxplot of the distribution of the accuracies obtained using all texture measures in each region.

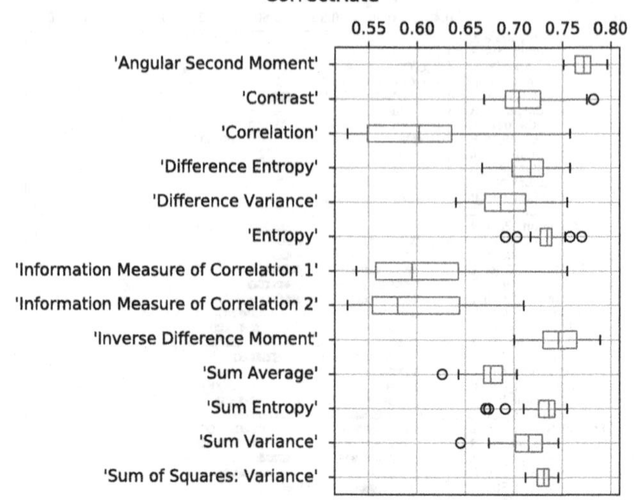

Fig. 3. Boxplot of the distribution of the accuracies obtained using different texture measures at the Hippocampus R.

Table 1. Results of our system using all measures with selection, compared to performance in some regions and the Voxels As Features (VAF) approach [23].

Strategy	Accuracy	Sensitivity	Specificity
Selection via t-Test ($p < 0.05$)	0.813 ± 0.083	0.775 ± 0.122	0.843 ± 0.129
Selection via t-Test ($p < 0.01$)	0.796 ± 0.079	0.732 ± 0.150	0.847 ± 0.150
Selection via t-Test ($p < 0.005$)	0.810 ± 0.081	0.769 ± 0.136	0.843 ± 0.156
Selection via KL ($p < 0.05$)	0.803 ± 0.081	0.743 ± 0.151	0.852 ± 0.154
Selection via KL ($p < 0.01$)	0.791 ± 0.078	0.732 ± 0.132	0.838 ± 0.143
Selection via KL ($p < 0.005$)	0.803 ± 0.086	0.822 ± 0.155	0.786 ± 0.150
All Features at Right Hippocampus	0.751 ± 0.063	0.664 ± 0.232	0.821 ± 0.232
Hippocampus_R, ASM, $\Delta_{\mathbf{p}} = (3, 0, 0)$	0.796 ± 0.051	0.766 ± 0.083	0.821 ± 0.075
VAF (T1)	0.791 ± 0.056	0.707 ± 0.098	0.860 ± 0.117
VAF (GM)	0.768 ± 0.011	0.752 ± 0.016	0.785 ± 0.016
VAF (WM)	0.642 ± 0.009	0.668 ± 0.012	0.617 ± 0.013

selection and the one using only one texture measure at the hippocampus still achieve better performance, but this time, providing a significant feature reduction of two magnitude orders (from more than half a million voxels to thousands of measures).

These results highlights the prospective of using texture measures to characterize per-region structural changes due to neurodegeneration in MRI images.

This is a preliminary analysis of the utility of these texture measures, using the more traditional Haralick texture analysis. Other more advanced techniques have been developed in the late 90s and the 2000s, for example the Local Binary Patterns (LBP) [19], the watershed transform [12] or the orientation pyramid (OP) [21]. These algorithms have previously shown to outperform the Haralick texture features, and are valid candidates to improve AD diagnosis and its possible application to model the progression of AD, which is the real challenge.

4 Conclusions and Future Work

In this work, we have proposed a texture analysis framework for characterizing the structural changes in Alzheimer's Disease (AD). The preliminary results that we show prove that the texture measures are an excellent descriptor of structural changes in different regions of the brain. The most discriminant regions are located in and surrounding the Hippocampus, and the discrimination accuracy obtained at these regions is close to 80%. On the other hand, we have demonstrated that pooling all the texture measures and selecting the most significant achieves higher performance than any other region by itself, which proves that there exist other regions which could play a significant role in neurodegeneration and can be characterized by texture measures. The system using texture measures provides a significant feature reduction and obtains similar and even higher performance that the common baseline Voxels As Features (VAF). In future works, we will extend this texture analysis with other texture features and apply those to Mild Cognitive Impairment (MCI) affected patients, and see whether these measures can be applied to the prediction of MCI conversion to AD and its progression.

Acknowledgements. This work was partly supported by the MINECO/ FEDER under the TEC2015-64718-R project and the Consejería de Economía, Innovación, Ciencia y Empleo (Junta de Andalucía, Spain) under the Excellence Project P11-TIC-7103.

Data collection and sharing for this project was funded by the Alzheimer's Disease Neuroimaging Initiative (ADNI) (National Institutes of Health Grant U01 AG024904) and DOD ADNI (Department of Defense award number W81XWH-12-2-0012). ADNI is funded by the National Institute on Aging, the National Institute of Biomedical Imaging and Bioengineering, and through generous contributions from the following: AbbVie, Alzheimer's Association; Alzheimer's Drug Discovery Foundation; Araclon Biotech; BioClinica, Inc.; Biogen; Bristol-Myers Squibb Company; CereSpir, Inc.; Cogstate; Eisai Inc.; Elan Pharmaceuticals, Inc.; Eli Lilly and Company; EuroImmun; F. Hoffmann-La Roche Ltd and its affiliated company Genentech, Inc.; Fujirebio; GE Healthcare; IXICO Ltd.; Janssen Alzheimer Immunotherapy Research & Development, LLC.; Johnson & Johnson Pharmaceutical Research & Development LLC.; Lumosity; Lundbeck; Merck & Co., Inc.; Meso Scale Diagnostics, LLC.; NeuroRx Research; Neurotrack Technologies; Novartis Pharmaceuticals Corporation; Pfizer Inc.; Piramal Imaging; Servier; Takeda Pharmaceutical Company; and Transition Therapeutics. The Canadian Institutes of Health Research is providing funds to support ADNI clinical sites in Canada. Private sector contributions are facilitated by the Foundation for the

National Institutes of Health (www.fnih.org). The grantee organization is the Northern California Institute for Research and Education, and the study is coordinated by the Alzheimers Therapeutic Research Institute at the University of Southern California. ADNI data are disseminated by the Laboratory for Neuro Imaging at the University of Southern California.

References

1. Alemán, Y., Melie, L., Valdés, P.: Ibaspm: toolbox for automatic parcellation of brain structures. In: 12th Annual Meeting of the Organization for Human Brain Mapping, pp. 11–15, June 2006
2. Alzheimer's Association: 2016 Alzheimer's disease facts and figures. Alzheimer's Dement. **12**(4), 459–509 (2016)
3. Ashburner, J., Friston, K.J.: Voxel-based morphometry–the methods. Neuroimage **11**(6), 805–821 (2000)
4. Baron, J.C., Chételat, G., Desgranges, B., Perchey, G., Landeau, B., de la Sayette, V., Eustache, F.: In vivo mapping of gray matter loss with voxel-based morphometry in mild Alzheimer's disease. Neuroimage **14**(2), 298–309 (2001)
5. Dubois, B., Feldman, H.H., Jacova, C., DeKosky, S.T., Barberger-Gateau, P., Cummings, J., Delacourte, A., Galasko, D., Gauthier, S., Jicha, G., et al.: Research criteria for the diagnosis of Alzheimer's disease: revising the NINCDS-ADRDA criteria. Lancet Neurol. **6**(8), 734–746 (2007)
6. Friston, K., Ashburner, J., Kiebel, S., Nichols, T., Penny, W.: Statistical Parametric Mapping: The Analysis of Functional Brain Images. Academic Press, Cambridge (2007)
7. Haralick, R., Shanmugam, K., Dinstein, I.: Textural features for image classification. IEEE Trans. Syst. Man Cybern. **3**(6), 610–621 (1973)
8. Khedher, L., Ramírez, J., Górriz, J., Brahim, A., Segovia, F.: Early diagnosis of Alzheimers disease based on partial least squares, principal component analysis and support vector machine using segmented MRI images. Neurocomputing **151**, 139–150 (2015)
9. Kohavi, R.: A study of cross-validation and bootstrap for accuracy estimation and model selection. In: Proceedings of International Joint Conference on AI, pp. 1137–1145 (1995)
10. Leandrou, S., Petroudi, S., Kyriacou, P.A., Reyes-Aldasoro, C.C., Pattichis, C.S.: An overview of quantitative magnetic resonance imaging analysis studies in the assessment of Alzheimer's disease. In: Kyriacou, E., Christofides, S., Pattichis, C.S. (eds.) XIV Mediterranean Conference on Medical and Biological Engineering and Computing 2016. IP, vol. 57, pp. 281–286. Springer, Cham (2016). doi:10.1007/978-3-319-32703-7_56
11. Lerch, J.P., Pruessner, J.C., Zijdenbos, A., Hampel, H., Teipel, S.J., Evans, A.C.: Focal decline of cortical thickness in Alzheimer's disease identified by computational neuroanatomy. Cereb. Cortex **15**(7), 995–1001 (2005)
12. Malpica, N., Ortuño, J.E., Santos, A.: A multichannel watershed-based algorithm for supervised texture segmentation. Pattern Recogn. Lett. **24**(9), 1545–1554 (2003)

13. Martinez-Murcia, F., Górriz, J., Ramírez, J., Moreno-Caballero, M., Gómez-Río, M., Initiative, P.P.M., et al.: Parametrization of textural patterns in 123i-ioflupane imaging for the automatic detection of Parkinsonism. Med. Phys. **41**(1), 012502 (2014)
14. Martinez-Murcia, F., Górriz, J., Ramírez, J., Ortiz, A., The Alzheimers Disease Neuroimaging Initiative: A spherical brain mapping of MR images for the detection of Alzheimers disease. Curr. Alzheimer Res. **13**(5), 575–588 (2016)
15. Martínez-Murcia, F., Górriz, J., Ramírez, J., Puntonet, C., Salas-González, D.: Computer aided diagnosis tool for Alzheimer's disease based on Mann-Whitney-Wilcoxon U-test. Expert Syst. Appl. **39**(10), 9676–9685 (2012)
16. Martinez-Murcia, F.J., Ortiz, A., Górriz, J.M., Ramírez, J., Illán, I.A.: A volumetric radial LBP projection of MRI brain images for the diagnosis of Alzheimer's disease. In: Ferrández Vicente, J.M., Álvarez-Sánchez, J.R., de la Paz López, F., Toledo-Moreo, F.J., Adeli, H. (eds.) IWINAC 2015. LNCS, vol. 9107, pp. 19–28. Springer, Cham (2015). doi:10.1007/978-3-319-18914-7_3
17. Martínez-Murcia, F.J., Górriz, J., Ramírez, J., Puntonet, C.G., Illán, I.: Functional activity maps based on significance measures and independent component analysis. Comput. Methods Programs Biomed. **111**(1), 255–268 (2013)
18. Martínez-Murcia, F.J., Górriz, J.M., Ramírez, J., Alvarez Illán, I., Salas-González, D., Segovia, F., A.D.N.I.: Projecting MRI brain images for the detection of Alzheimer's disease. Stud. Health Technol. Inform. **207**, 225–233 (2015)
19. Ojala, T., Pietikäinen, M., Mäenpää, T.: Multiresolution gray-scale and rotation invariant texture classification with local binary patterns. IEEE Trans. Pattern Anal. Mach. Intell. **24**(7), 971–987 (2002)
20. Philips, C., Li, D., Raicu, D., Furst, J.: Directional Invariance of Co-occurrence Matrices within the Liver. In: International Conference on Biocomputation, Bioinformatics, and Biomedical Technologies, pp. 29–34 (2008)
21. Reyes-Aldasoro, C.C., Bhalerao, A.: The Bhattacharyya space for feature selection and its application to texture segmentation. Pattern Recogn. **39**(5), 812–826 (2006)
22. Segovia, F., Górriz, J., Ramírez, J., Salas-Gonzalez, D., Álvarez, I.: Early diagnosis of Alzheimers disease based on partial least squares and support vector machine. Expert Syst. Appl. **40**(2), 677–683 (2013)
23. Stoeckel, J., Ayache, N., Malandain, G., Koulibaly, P.M., Ebmeier, K.P., Darcourt, J.: Automatic classification of SPECT images of Alzheimer's disease patients and control subjects. In: Barillot, C., Haynor, D.R., Hellier, P. (eds.) MICCAI 2004. LNCS, vol. 3217, pp. 654–662. Springer, Heidelberg (2004). doi:10.1007/978-3-540-30136-3_80
24. Theodoridis, S., Pikrakis, A., Koutroumbas, K., Cavouras, D.: Introduction to Pattern Recognition: A Matlab Approach. Academic Press, Cambridge (2010)
25. Towey, D.J., Bain, P.G., Nijran, K.S.: Automatic classification of 123I-FP-CIT (DaTSCAN) SPECT images. Nucl. Med. Commun. **32**(8), 699–707 (2011)
26. Tzourio-Mazoyer, N., Landeau, B., Papathanassiou, D., Crivello, F., Etard, O., Delcroix, N., Mazoyer, B., Joliot, M.: Automated anatomical labeling of activations in SPM using a macroscopic anatomical parcellation of the mni MRI single-subject brain. Neuroimage **15**(1), 273–289 (2002)
27. Vapnik, V.N.: Statistical Learning Theory. Wiley, New York (1998)
28. Zhang, J., Yu, C., Jiang, G., Liu, W., Tong, L.: 3D texture analysis on MRI images of Alzheimers disease. Brain Imaging Behav. **6**(1), 61 (2012)

Evaluation of Four Supervised Learning Schemes in White Matter Hyperintensities Segmentation in Absence or Mild Presence of Vascular Pathology

Muhammad Febrian Rachmadi[1,2](✉), Maria del C. Valdés-Hernández[2], Maria Leonora Fatimah Agan[2], Taku Komura[1], and The Alzheimer's Disease Neuroimaging Initiative

[1] School of Informatics, University of Edinburgh, Edinburgh, UK
m.f.rachmadi@sms.ed.ac.uk
[2] Centre for Clinical Brain Sciences, University of Edinburgh, Edinburgh, UK

Abstract. We investigated the performance of four popular supervised learning algorithms in medical image analysis for white matter hyperintensities segmentation in brain MRI with mild or no vascular pathology. The algorithms evaluated in this study are support vector machine (SVM), random forest (RF), deep Boltzmann machine (DBM) and convolution encoder network (CEN). We compared these algorithms with two methods in the Lesion Segmentation Tool (LST) public toolbox which are lesion growth algorithm (LGA) and lesion prediction algorithm (LPA). We used a dataset comprised of 60 MRI data from 20 subjects from the ADNI database, each scanned once in three consecutive years. In this study, CEN produced the best Dice similarity coefficient (DSC): mean value 0.44. All algorithms struggled to produce good DSC due to the very small WMH burden (*i.e.*, smaller than 1,500 mm^3). LST-LGA, LST-LPA, SVM, RF and DBM produced mean DSC scores ranging from 0.17 to 0.34.

Keywords: Brain MRI · White matter hyperintensities · Segmentation · Supervised learning · Deep neural network

1 Introduction

White hyperintensities (WMH) segmentation is an important problem in medical image analysis because it is believed that WMH are associated with the

Data used in preparation of this article were obtained from the Alzheimers Disease Neuroimaging Initiative (ADNI) database (http://adni.loni.usc.edu). As such, the investigators within the ADNI contributed to the design and implementation of ADNI and/or provided data but did not participate in analysis or writing of this report. A complete listing of ADNI investigators can be found at: http://adni.loni.usc.edu/wp-content/uploads/how_to_apply/ADNI_Acknowledgement_List.pdf.

M. Valdés Hernández and V. González-Castro (Eds.): MIUA 2017, CCIS 723, pp. 482–493, 2017.
DOI: 10.1007/978-3-319-60964-5_42

progression of dementia [1,21]. WMH are brain regions that have higher gray-scale intensities than normal tissues in T2-Fluid Attenuation Inversion Recovery (FLAIR) magnetic resonance images (MRI).

There have been many attempts to automatically segment WMH in the past few years. Most of the works used support vector machine (SVM) and random forest (RF) for which some image features need to be extracted first. Some notable works were done in [10,11] where several feature extraction methods and learning algorithms were evaluated to find the best possible combination for this purpose. Both studies concluded that SVM was the best performer in the experiments. In another study, RF was compared with SVM and the former performed better [8]. However, these studies cannot be compared to each other directly because they use different feature extraction methods. Feature extraction and selection are as important as the learning algorithm itself for WMH segmentation. Fortunately, machine learning algorithms have developed into more sophisticated approaches of deep learning, which are now commonly used in image analysis. In these approaches, the algorithm extracts the features automatically from the data to get the best results possible. Some algorithms of this type like deep Boltzmann machine (DBM) [12] and convolutional encoder network (CEN) [2,3] have been successfully tested to work well with medical image data, including brain MRI.

In this study, we investigate performances of supervised learning algorithms of SVM, RF, DBM and CEN for WMH segmentation in brain MRI with mild or no vascular pathology. We choose brain MRI with mild or no vascular pathology because it is important to detect the presence of WMH as early as possible. It is also notably more challenging to do WMH segmentation in this type of data because the WMH burden for each patient is smaller. We also compare the results of these algorithms with those from with a publicly available toolbox for WMH segmentation named Lesion Segmentation Tool (LST) [19].

2 Data, Processing Flow and Experiment Setup

Data used in this study are obtained from the Alzheimers Disease Neuroimaging Initiative (ADNI) public database [14,22]. Our dataset contains MRI data from 20 ADNI participants, randomly selected and blind from any clinical, imaging or demographic information at the time of selection. MRI data were acquired on three consecutive years, resulting in data from a total of 60 MRI scans. Three of them were cognitively normal (CN), 12 had early mild cognitive impairment (EMCI) and 5 had late mild cognitive impairment (LMCI). Ground truth segmentation of the respective MRI data is produced by an experienced image analyst, semi-automatically by thresholding the T2-FLAIR images using the region-growing algorithm in the Object Extractor tool of Analyze^{TM} software, simultaneously guided by the co-registered T1- and T2-weighted sequences. A subset of manually delineated WMH masks from another observer is also used for validation purposes. Each brain scan was processed independently, blind to any clinical, cognitive or demographic information and to the results of the WMH segmentations from the same individual at different time points. For more details and to access these segmentations, please refer to http://hdl.handle.net/10283/2186.

The preprocessing steps of the data comprise of co-registration of the MRI sequences on each scanning session, skull stripping and intracranial volume mask generation, cortical grey matter, cerebrospinal fluid and brain ventricle extraction and intensity value normalisation. FSL-FLIRT [9] is used for rigid-body linear registration of the T1-W to the T2-FLAIR. Whereas, optiBET [13] and morphological fill holes operation are used for skull stripping and generation of the intracranial volume mask. On the other hand, two steps intensity value normalisation, which are adjustment of maximum grey scale value of the brain without skull to 10 percent of the maximum T2-FLAIR intensity value and histogram matching algorithm for MR images [16], are done. Furthermore, zero-mean and unit-variance grey scale value normalisation is also used for CEN to ensure a smooth gradient in the back propagation. In addition, scaling the features to [0...1] is used for DBM and SVM training processes as it is needed for the binary type of DBM and for easing the SVM training process. After the normalisation is finished, patch-wise data of WMH and non-WMH from MRI with ratio of 1:1 (*i.e.*, the same number of patches from WMH and non-WMH regions) are extracted for SVM and RF training processes while ratio of 1:4 is used for DBM training process (*i.e.*, there are four times more patches from non-WMH regions than patches extracted from WMH regions in the data used for training the DBM). On the other hand, in CEN, one slice of MRI is treated as one training data.

Two different tests are done, which are 5-fold cross validation test and longitudinal test. Cross validation is done with 16 individuals for training and 4 individuals for testing in each fold. Whereas, longitudinal test is done using MRI data from the first year of acquisition for training and the second and the third years of acquisition for testing. Dice similarity coefficient (DSC) [6], sensitivity (TPR), specificity (TNR), precision (PPV) and volume difference (VD) and its ratio (VDR) are used as performance metrics. VDR is computed using Eq. 1 where $Volume(Seg.)$ is the WMH volume resulting from segmentation and *Volume(GT)* is the WMH volume from ground truth. VD is computed using the same Eq. 1 without normalisation of the ground truth volume. All evaluation metrics are computed after probability map values of WMH, resulting from automatic segmentation method, are cut-off using threshold value $t \geq 0.7$. This value was chosen after the results were reviewed by a neuro-radiologist.

$$VDR = \frac{Volume(Seg.) - Volume(GT)}{Volume(GT)} \tag{1}$$

3 Methods

In this section, all methods used in this study for WMH segmentation are discussed. The methods are Lesion Segmentation Tool (LST) toolbox, support vector machine (SVM), random forest (RF), deep Boltzmann machine (DBM) and convolutional encoder network (CEN).

3.1 Lesion Segmentation Tool, Support Vector Machine and Random Forest

Lesion Segmentation Tool (LST) is a public toolbox developed for segmenting multiple sclerosis (MS) lesions in MRI [19]. It also claims to be useful in other brain diseases including WMH in normal aging. In this study, we use both algorithms available on LST version 2.0.15[1] toolbox, which are lesion growth algorithm (LGA), an unsupervised algorithm, and lesion prediction algorithm (LPA), a supervised algorithm pre-trained with data from 53 MS patients.

Support vector machine (SVM) is a supervised machine learning algorithm that separates data points by a hyperplane [5]. Whereas, random forest (RF) is a collection of decision trees trained individually to produce outputs that are collected and combined together [17]. These two algorithms are commonly used in segmentation and classification tasks. For reproducibility and repeatability reasons, and also to make comparison easier, we modified a public toolbox: W2MHS[2] [8], which uses RF for WMH segmentation. We retrained the RF model using the following parameters: 300 trees, 2 minimum samples in a leaf and 4 minimum samples before splitting. The feature extraction of the toolbox is used without any change. The features extracted and used in the training process comprise of 125 MR image grey scale values and 1875 response values from a filter bank of low pass filter, high pass filter, band pass filter and edge filter (see [8] for full explanation), all of them extracted from 3D ROIs with the size of $5 \times 5 \times 5$. In total, for training the SVM and RF classifiers we used 200,000 samples: 100,000 patches from WMH regions and 100,000 from non-WMH regions. We also modified the toolbox so that we can now choose from which MRI modality, T2-FLAIR or both T2-FLAIR and T1W, these features are extracted from. These extracted features are also used to train the SVM classifier after the feature's dimensionality is reduced to 10 using PCA and then whitened before training. In this study, radial basis (RBF) kernel is used for SVM classifier.

3.2 Deep Boltzmann Machine

Deep Boltzmann Machine (DBM) is a variant of restricted Boltzmann machine (RBM), a generative neural network that works by minimizing its energy function, where multiple layers of RBM are used instead of only one layer. Each hidden layer captures more complex high-order correlations between activities of hidden units than the layer below [18]. Each layer can be independently trained first (pre-trained) to get better initialization of the weight matrix. A DBM with two hidden layers is used in this study (depicted by Fig. 1a). It has the energy function defined by Eq. 2 where $\mathbf{v}$ is the visible layer, $\mathbf{h}^1$ and $\mathbf{h}^2$ are the first and second hidden layers and $\Theta = \{\mathbf{W}^1, \mathbf{W}^2\}$ is the model's parameters where $\mathbf{W}^1$ and $\mathbf{W}^2$ are weight matrices for symmetric relation of visible-hidden and

[1] http://www.statisticalmodelling.de/lst.html.
[2] https://www.nitrc.org/projects/w2mhs/.

hidden-hidden layers. The objective function is the probability of the model that generates back visible variables of $\mathbf{v}$ using the model's parameter Θ, as per Eq. 3. Given a restricted structure where each layer units are conditionally independent of each other, the conditional distribution of the probability for a unit in a layer given other layers can be computed as in Eqs. 4, 5 and 6 where σ is a sigmoid function. Full mathematical derivation of RBM and its learning algorithm can be read in [7]. Whereas, derivation of DBM and its learning algorithm can be read in [18].

$$E\left(\mathbf{v}, \mathbf{h}^1, \mathbf{h}^2; \Theta\right) = -\mathbf{v}^\top \mathbf{W}^1 \mathbf{h}^1 - (\mathbf{h}^{1\top})\mathbf{W}^2 \mathbf{h}^2 \tag{2}$$

$$p(\mathbf{v}; \Theta) = \frac{1}{Z(\Theta)} \sum_{\mathbf{h}^1, \mathbf{h}^2} \exp\left[-E\left(\mathbf{v}, \mathbf{h}^1, \mathbf{h}^2; \Theta\right)\right] \tag{3}$$

$$p\left(h_k^2 = 1 | \mathbf{h}^1\right) = \sigma\left(\sum_j W_{jk}^2 h_j^1\right) \tag{4}$$

$$p\left(h_j^1 = 1 | \mathbf{v}, \mathbf{h}^2\right) = \sigma\left(\sum_i W_{ij}^1 v_i + \sum_k W_{jk}^2 h_k^2\right) \tag{5}$$

$$p\left(v_i = 1 | \mathbf{h}^1\right) = \sigma\left(\sum_j W_{ij}^1 h_j\right) \tag{6}$$

We use $5 \times 5 \times 5$ 3D ROIs to get grayscale intensity values from the MRI's T2-FLAIR modality for the DBM's training process. The intensity values are feed-forwarded into a 2-layer DBM with 125-50-50 structure where 125 is the number of units of the input layer and 50 is the number of units of both hidden layers. Each RBM layer is pre-trained for 200 epochs, and the whole DBM is trained for 500 epochs. After the DBM training process is finished, a label layer is added on top of the DBM's structure and *fine-tuning* is done using gradient descent for supervised learning of WMH segmentation. We modified and used Salakhutdinov's public code for DBM implementation[3].

3.3 Convolutional Encoder Network

Convolutional encoder network (CEN) is one of deep learning models which is usually used to generate a negative data (*i.e.*, synthesised data) learned from a dataset. In this study, CEN is used to generate a WMH segmentation of an MRI data learned from the dataset. CEN is trained using a whole image of MRI, just like in natural images where a whole image is feed-forwarded into the network, rather than using a patch-wise approach (*i.e.*, uses image segments) of MRI like in other medical image analysis studies that use deep learning algorithms. This approach has been applied before in [2,3] for MS lesions segmentation and the

[3] http://www.cs.toronto.edu/~rsalakhu/DBM.html.

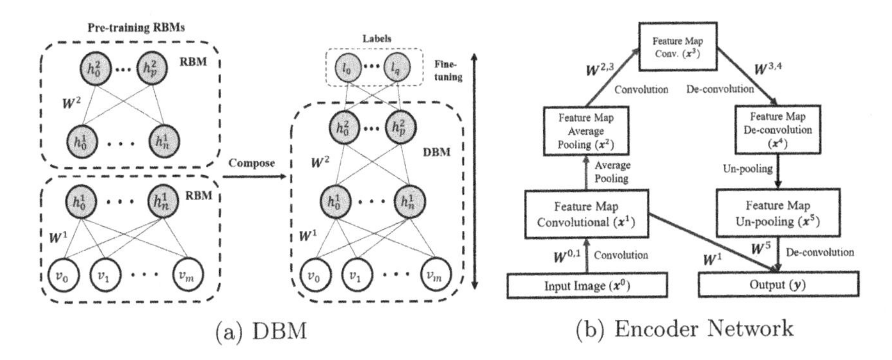

(a) DBM (b) Encoder Network

Fig. 1. Illustrations of (a) DBM and (b) convolutional encoder network (CEN) used in this study. In (a), two RBMs are stacked together for pre-training (left) to form a DBM (right). In (b), input image is encoded by using two convolutional layers and an average pooling layer and decoded to WMH segmentation using two de-convolutional layers and an un-pooling layer. This architecture is inspired from [2,3].

results were reported as promising. However, we used a 2D CEN instead of a 3D CEN like in the previous studies due to the anisotropy of the MR images used in this study (*i.e.*, the T2-FLAIR MRI from ADNI database have dimensions of $256 \times 256 \times 35$ and voxel size of $0.86 \times 0.86 \times 5$ mm^3).

In this study, we use a simple CEN composed of 1 input layer, 5 hidden layers (*i.e.*, feature maps or FM in deep learning study) and 1 output layer. The input layer is made of the MRI slices with size 256×256 and 2 channels (*i.e.*, T2-FLAIR MRI and brain mask). Whereas, the output layer is a simple binary mask of WMH labels for the corresponding T2-FLAIR MRI. The first feature map (FM) is produced by convolving a 9×9 kernel to the input layer. The second FM is produced by doing average pooling operation to the first FM. The third FM is produced by convolving a 5×5 kernel to the second FM. All together, they are called an encoding path. All convolution operations in the encoder path use the following Eq. 7 where $\mathbf{x}$ is input/output vector, l is index layer, $\mathbf{W}^{l-1,l}$ is weight matrix from layer $l-1$ to layer l, $*$ is convolution operation, $\mathbf{b}$ is bias vector and σ is non-linear ReLU activation function [15].

$$\mathbf{x}^l = \sigma(\mathbf{W}^{l-1,l} * \mathbf{x}^{l-1} + \mathbf{b}^l) \quad (7)$$

On the other hand, the fourth and the fifth FMs are produced by using deconvolution (with 5×5 kernel) and un-pooling operations respectively to the previous FMs. Output layer is produced by a deconvolution operation (with 9×9 kernel) to a merged FM composed by the fifth and the first FMs. This merger is called a *skip connection* which provides richer information before pooling and un-pooling operations. All together, these operations formed a decoding path. Also, please note that the same size of kernel is used at the same level of encoding-decoding. Deconvolution at the fourth layer follows the same Eq. 7 except that $*$ is now a deconvolution operation. On the other hand, the output layer follows Eq. 8 where $\mathbf{y}$ is output vector of output layer, $\mathbf{W}^1$ and $\mathbf{W}^5$ are weight matrices connecting FM #1 and FM #5 to output layer respectively, $\mathbf{x}^1$ and $\mathbf{x}^5$ are FM #1 and FM #5, $\mathbf{b}^y$ is bias vector of output layer and σ is non-linear sigmoid activation function.

$$\mathbf{y} = \sigma(\mathbf{W}^5 * \mathbf{x}^5 + \mathbf{W}^1 * \mathbf{x}^1 + \mathbf{b}^y) \tag{8}$$

For optimising the CEN, we use Dice similarity coefficient (DSC) [6] as objective function of CEN as we want to get the best DSC metric as possible in the evaluation. This is different from [2] where they use a combination of specificity and sensitivity as objective function. CEN is implemented by using Keras [4], with its default values of layer's hyper-parameter are used. The CEN itself is trained for 2500 epochs without early stopping (*i.e.*, the same epoch and approach suggested in a previous study [3] for limited number of training dataset), learning rate of 1E-5 and batch size of 5 in each epoch. The number of FM in all layers is 32 feature maps.

4 Results and Discussion

Table 1 shows the overall results for all methods tested in this study. This table is interesting because the highest overall DSC score is produced by CEN whereas the highest scores of sensitivity, specificity and precision are all produced by different methods which are LST-LPA and RF-FLAIR respectively. If we look closely, all methods have high scores of sensitivity and precision, but all of them have different scores of specificity. The highest specificity score, 0.8133, is produced by RF-FLAIR which also has a high sensitivity score of 0.9705. However, RF-FLAIR produce a low DSC score, 0.2215. If we compare to CEN, which has the highest DSC score of 0.4400, it has 0.9985 and 0.4287 for sensitivity and specificity scores respectively. From this observation, we can conclude that DSC score is highly related to sensitivity. The relationship between DSC and

Table 1. Experiment results based on several metrics which are dice similarity coefficient (DSC), sensitivity (Sen.), specificity (Spe.), precision (Pre.), volume difference ratio (VDR) and DSC for longitudinal test (DSC-Long.).

No.	Method	DSC	Sen.	Spe.	Pre.	VDR	DSC-Long.
1	LST-LGA [19]	0.2894	0.9964	0.3051	0.9964	0.5458	-
2	LST-LPA [19]	0.1938	0.9990	0.1330	0.9957	−0.7227	-
3	SVM_FLAIR	0.1919	0.9697	0.7336	0.9987	15.5927	0.1587
4	SVM_FLAIR_T1W	0.1736	0.9881	0.3474	0.9966	4.3564	0.1800
5	RF_FLAIR	0.2215	0.9705	0.8133	0.9991	13.5706	0.1977
6	RF_FLAIR_T1W	0.2252	0.9752	0.7132	0.9985	12.2179	0.2178
7	DBM	0.3405	0.9975	0.3517	0.9964	0.1434	0.3326
8	CEN	0.4400	0.9985	0.4287	0.9967	0.2070	0.4713

sensitivity is stronger than between DSC and specificity. A small drop in the sensitivity score (*e.g.*, 2.85% drop from CEN to RF-FLAIR) changes the DSC score considerably (*i.e.*, 22.15% lower) independently from the specificity score (*i.e.*, RF-FLAIR is 38.46% higher than CEN). This means that there should be a balance between the DSC, sensitivity and specificity, to get the best result possible.

To see the distribution of segmentation performance based on WMH burden for each subject, we grouped our data into 5 groups based on WMH volume of each patient. The groups are: (1) Very Small (VS) where WMH volume range is $(0, 1500]$ mm^3, (2) Small (S) where WMH volume range is $(1500, 4500]$ mm^3, (3) Medium (M) where WMH volume range is $(4500, 13000]$ mm^3, (4) Large (L) where WMH volume range is $(13000, 24000]$ mm^3 and (5) Very Large (VL)

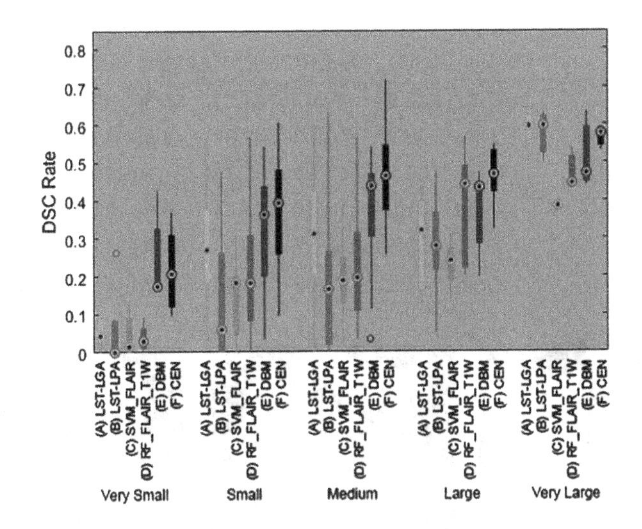

Fig. 2. Distribution of DSC scores for each group based on WMH volume burden. A, B, C, D, E and F represent methods listed in Table 2, which are (A) LST-LGA, (B) LST-LPA, (C) SVM_FLAIR, (D) RF_FLAIR_T1W, (E) DBM and (F) CEN.

Table 2. Average values of dice similarity coefficient (DSC) and volume difference ratio (VDR) for grouped MRI data based on its WMH burden. VS, S, M, L and VL stand for 'Very Small', 'Small', 'Medium', 'Large' and 'Very Large' which are names of the groups.

	Method	DSC					VDR				
		VS	S	M	L	VL	VS	S	M	L	VL
A	LST-LGA	0.0687	0.2800	0.3076	0.2901	0.5905	3.1403	0.3258	0.1116	0.3874	−0.3832
B	LST-LPA	0.0581	0.1215	0.1707	0.2805	0.5761	−0.9084	−0.9085	−0.6457	−0.8094	−0.5651
C	SVM_FLAIR	0.0466	0.1498	0.1855	0.2304	0.3882	25.6947	7.4457	3.0759	1.1032	1.1821
D	RF_FLAIR_T1W	0.0384	0.2063	0.2252	0.3801	0.4743	71.9682	20.1032	7.9772	2.5138	1.9135
E	DBM	0.2451	0.3372	0.3806	0.3706	0.5152	1.3297	0.2583	−0.0976	−0.6816	−0.2448
F	CEN	0.2179	0.3736	0.4649	0.4636	0.5670	4.2237	0.6528	−0.0581	−0.0443	−0.5444

where WMH volume is bigger than 24000 mm^3. We then plotted and listed DSC scores based on the group in Fig. 2 and Table 2. From both the figure and the table, we can see that all methods do not have any problems in segmenting very large WMH burden from a subject, but their performances are decreasing greatly in smaller WMH burdens, except for DBM and CEN where the decrease in performance with WMH load is not much.

Some visual examples of WMH segmentation can be seen in Fig. 3 where visualisations from ground truth, LST-LGA, SVM-FLAIR, RF-FLAIR-T1W, DBM and CEN in a subject with small WMH burden are shown. We can see

Table 3. Volumetric disagreement (D) with observers' measurements for LST-LGA, LST-LPA, SVM_FLAIR, RF_FLAIR_T1W, DBM and CEN.

	Method	Intra-D (%)				Inter-D (%)			
		Label.1	SD	Label.2	SD	Obs.1	SD	Obs.2	SD
A	LST-LGA	67.78	32.37	77.07	44.94	55.36	40.61	52.84	35.22
B	LST-LPA	155.89	48.85	157.03	47.28	146.45	44.95	146.27	52.13
C	SVM_FLAIR	148.80	34.57	154.99	35.34	161.52	29.32	158.37	27.39
D	RF_FLAIR_T1W	145.38	41.04	152.42	40.44	159.99	34.65	157.72	28.81
E	DBM	129.71	50.50	138.67	48.53	153.03	44.56	150.53	36.87
F	CEN	62.28	44.42	78.60	50.68	74.29	41.18	63.33	48.97

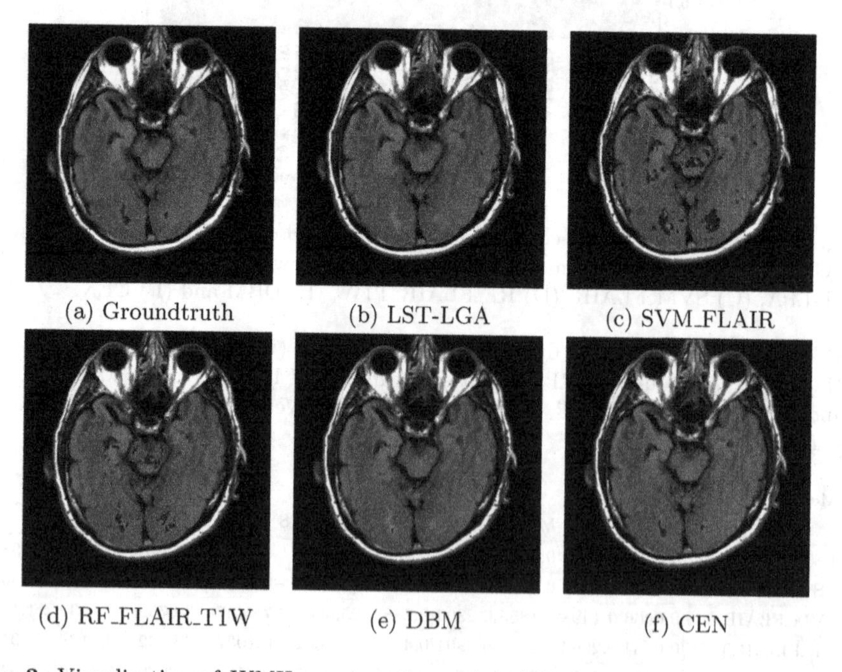

(a) Groundtruth (b) LST-LGA (c) SVM_FLAIR

(d) RF_FLAIR_T1W (e) DBM (f) CEN

Fig. 3. Visualisation of WMH segmentation using different method from Subject 3 which has WMH burden of 3537.74 mm^3.

that CEN produced much better results than the other methods (Fig. 3 and Table 3).

In addition to the evaluations that have been mentioned in all figures, tables and previous paragraphs, we also keep records on the time training and testing processes take in the experiments. SVM, RF and DBM took roughly 26, 37 and 1341 min respectively for the training process. Whereas, it took 83, 41 and 17 s for SVM, RF and DBM to complete one MRI data in the testing process from a workstation in a Linux server with 32 Intel(R) Xeon(R) CPU E5-2665 @ 2.40 GHz processors. On the other hand, Linux Ubuntu desktop with Intel(R) Core(TM) i7-6700 CPU @ 3.40 GHz and EVGA NVIDIA GeForce GTX 1080 8 GB GAMING ACX 3.0 was used to train and test the CEN; and the training and testing processes took 152 min and 5 s respectively. An image analyst can take from 15 to 60 min to segment WMH on a single dataset depending on the level of experience [20].

5 Conclusion and Future Work

In this study, we have seen performances from different supervised learning methods for WMH segmentation in brain MRI with mild or no vascular pathology. We tested SVM, RF, DBM and CEN and compared them with a public toolbox LST which provides two different algorithms, LGA and LPA. From the experiments, we can see that WMH volume is the most challenging problem in this study because WMH segmentation results using all methods on subjects with low and very low WMH produce low DSC scores. Furthermore, we also find that there are strong dependencies between DSC, sensitivity and specificity, especially between DSC and sensitivity. To produce a high score of DSC, we need to find a good balance between these three metrics. In this study, we use DSC as objective function on CEN. If DSC, sensitivity and specificity are used all together on CEN on objective function, better results of WMH segmentation may be obtained. Furthermore, the MRI could be re-sampled to isotropic images so that a 3D CEN can be tested and compared with the 2D CEN evaluated in this study.

Acknowledgement. The first author wants to thank Indonesia Endowment Fund for Education (LPDP) of Ministry of Finance, Republic of Indonesia, for funding his study at School of Informatics, the University of Edinburgh. Funds from Row Fogo Charitable Trust (MCVH) are also gratefully acknowledged. Data collection and sharing for this project was funded by the Alzheimer's Disease Neuroimaging Initiative (ADNI) (National Institutes of Health Grant U01 AG024904) and DOD ADNI (Department of Defense award number W81XWH-12-2-0012). ADNI is funded by the National Institute on Aging, the National Institute of Biomedical Imaging and Bioengineering, and through generous contributions from the following: AbbVie, Alzheimer's Association; Alzheimers Drug Discovery Foundation; Araclon Biotech; BioClinica, Inc.; Biogen; Bristol-Myers Squibb Company; CereSpir, Inc.; Cogstate; Eisai Inc.; Elan Pharmaceuticals, Inc.; Eli Lilly and Company; EuroImmun; F. Hoffmann-La Roche Ltd. and its affiliated company Genentech, Inc.; Fujirebio; GE Healthcare; IXICO Ltd.; Janssen

Alzheimer Immunotherapy Research and Development, LLC.; Johnson and Johnson Pharmaceutical Research and Development LLC.; Lumosity; Lundbeck; Merck and Co., Inc.; Meso Scale Diagnostics, LLC.; NeuroRx Research; Neurotrack Technologies; Novartis Pharmaceuticals Corporation; Pfizer Inc.; Piramal Imaging; Servier; Takeda Pharmaceutical Company; and Transition Therapeutics. The Canadian Institutes of Health Research is providing funds to support ADNI clinical sites in Canada. Private sector contributions are facilitated by the Foundation for the National Institutes of Health (www.fnih.org). The grantee organization is the Northern California Institute for Research and Education, and the study is coordinated by the Alzheimers Therapeutic Research Institute at the University of Southern California. ADNI data are disseminated by the Laboratory for Neuro Imaging at the University of Southern California.

References

1. Birdsill, A.C., Koscik, R.L., Jonaitis, E.M., Johnson, S.C., Okonkwo, O.C., Hermann, B.P., LaRue, A., Sager, M.A., Bendlin, B.B.: Regional white matter hyperintensities: aging, Alzheimer's disease risk, and cognitive function. Neurobiol. Aging **35**(4), 769–776 (2014)
2. Brosch, T., Tang, L.Y.W., Yoo, Y., Li, D.K.B., Traboulsee, A., Tam, R.: Deep 3D convolutional encoder networks with shortcuts for multiscale feature integration applied to multiple sclerosis lesion segmentation. IEEE Trans. Med. Imaging **35**(5), 1229–1239 (2016)
3. Brosch, T., Yoo, Y., Tang, L.Y.W., Li, D.K.B., Traboulsee, A., Tam, R.: Deep convolutional encoder networks for multiple sclerosis lesion segmentation. In: Navab, N., Hornegger, J., Wells, W.M., Frangi, A.F. (eds.) MICCAI 2015. LNCS, vol. 9351, pp. 3–11. Springer, Cham (2015). doi:10.1007/978-3-319-24574-4_1
4. Chollet, F.: Keras (2015). https://github.com/fchollet/keras
5. Cortes, C., Vapnik, V.: Support-vector networks. Mach. Learn. **20**(3), 273–297 (1995)
6. Dice, L.R.: Measures of the amount of ecologic association between species. Ecology **26**(3), 297–302 (1945)
7. Hinton, G.: A practical guide to training restricted boltzmann machines. Momentum **9**(1), 926 (2010)
8. Ithapu, V., Singh, V., Lindner, C., Austin, B.P., Hinrichs, C., Carlsson, C.M., Bendlin, B.B., Johnson, S.C.: Extracting and summarizing white matter hyperintensities using supervised segmentation methods in Alzheimer's disease risk and aging studies. Hum. Brain Mapp. **35**(8), 4219–4235 (2014)
9. Jenkinson, M., Bannister, P., Brady, M., Smith, S.: Improved optimization for the robust and accurate linear registration and motion correction of brain images. Neuroimage **17**(2), 825–841 (2002)
10. Klöppel, S., Abdulkadir, A., Hadjidemetriou, S., Issleib, S., Frings, L., Thanh, T.N., Mader, I., Teipel, S.J., Hüll, M., Ronneberger, O.: A comparison of different automated methods for the detection of white matter lesions in MRI data. NeuroImage **57**(2), 416–422 (2011)
11. Leite, M., Rittner, L., Appenzeller, S., Ruocco, H.H., Lotufo, R.: Etiology-based classification of brain white matter hyperintensity on magnetic resonance imaging. J. Med. Imaging **2**(1), 014002 (2015)

12. Liu, M., Zhang, D., Yap, P.-T., Shen, D.: Hierarchical ensemble of multi-level classifiers for diagnosis of Alzheimer's disease. In: Wang, F., Shen, D., Yan, P., Suzuki, K. (eds.) MLMI 2012. LNCS, vol. 7588, pp. 27–35. Springer, Heidelberg (2012). doi:10.1007/978-3-642-35428-1_4
13. Lutkenhoff, E.S., Rosenberg, M., Chiang, J., Zhang, K., Pickard, J.D., Owen, A.M., Monti, M.M.: Optimized brain extraction for pathological brains (optibet). PLoS ONE **9**(12), e115551 (2014)
14. Mueller, S.G., Weiner, M.W., Thal, L.J., Petersen, R.C., Jack, C., Jagust, W., Trojanowski, J.Q., Toga, A.W., Beckett, L.: The Alzheimer's disease neuroimaging initiative. Neuroimaging Clin. N. Am. **15**(4), 869–877 (2005)
15. Nair, V., Hinton, G.E.: Rectified linear units improve restricted boltzmann machines. In: Proceedings of the 27th International Conference on Machine Learning (ICML-2010), pp. 807–814 (2010)
16. Nyúl, L.G., Udupa, J.K., Zhang, X.: New variants of a method of MRI scale standardization. IEEE Trans. Med. Imaging **19**(2), 143–150 (2000)
17. Opitz, D., Maclin, R.: Popular ensemble methods: an empirical study. J. Artif. Intell. Res. **11**, 169–198 (1999)
18. Salakhutdinov, R., Hinton, G.E.: Deep boltzmann machines. In: International Conference on Artificial Intelligence and Statistics, pp. 448–455 (2009)
19. Schmidt, P., Gaser, C., Arsic, M., Buck, D., Förschler, A., Berthele, A., Hoshi, M., Ilg, R., Schmid, V.J., Zimmer, C., et al.: An automated tool for detection of flair-hyperintense white-matter lesions in multiple sclerosis. Neuroimage **59**(4), 3774–3783 (2012)
20. Valéds Hernández, M.D.C., Armitage, P.A., Thrippleton, M.J., Chappell, F., Sandeman, E., Muoz Maniega, S., Shuler, K., Wardlaw, J.M.: Rationale, design, methodology of the image analysis protocol for studies of patients with cerebral small vessel disease, mild stroke. Brain Behav. **5**(12), e00415 (2015)
21. Wardlaw, J.M., Smith, E.E., Biessels, G.J., Cordonnier, C., Fazekas, F., Frayne, R., Lindley, R.I., O'Brien, J.T., Barkhof, F., Benavente, O.R., et al.: Neuroimaging standards for research into small vessel disease and its contribution to ageing and neurodegeneration. Lancet Neurol. **12**(8), 822–838 (2013)
22. Weiner, M.W., Veitch, D.P., Aisen, P.S., Beckett, L.A., Cairns, N.J., Green, R.C., Harvey, D., Jack, C.R., Jagust, W., Liu, E., et al.: The Alzheimers disease neuroimaging initiative: a review of papers published since its inception. Alzheimer's Dement. **8**(1), S1–S68 (2012)

Context-Aware Convolutional Neural Networks for Stroke Sign Detection in Non-contrast CT Scans

Aneta Lisowska[1,2(✉)], Alison O'Neil[1], Vismantas Dilys[1], Matthew Daykin[1,2], Erin Beveridge[1], Keith Muir[3], Stephen Mclaughlin[2], and Ian Poole[1]

[1] Toshiba Medical Visualization Systems Europe Ltd., Edinburgh, UK
alisowska@tmvse.com
[2] School of Engineering and Physical Sciences, Heriot-Watt University, Edinburgh, UK
[3] Queen Elizabeth University Hospital, Glasgow, UK

Abstract. Detection of acute stroke signs in non-contrast CT images is a challenging task. The intensity and texture variations in pathological regions are subtle and can be confounded by normal physiological changes or by old lesions. In this paper we investigate the use of contextual information for stroke sign detection. In particular, the appearance of the contralateral anatomy and the atlas-encoded spatial location are incorporated into a Convolutional Neural Network (CNN) architecture. CNNs are trained separately for the detection of dense vessels and of ischaemia. The network performance is evaluated on 170 datasets by cross-validation. We find that atlas location is important for dense vessel detection, but is less useful for ischaemia, whereas bilateral comparison is crucial for detection of ischaemia.

1 Introduction

In acute ischaemic stroke the ultimate goal of treatment is tissue reperfusion via recanalisation of an occluded vessel. This can be successfully achieved, with subsequent improved clinical outcome, by reperfusion therapies or mechanical thrombectomy. However, the offsetting of risks versus benefit in the selection of patients for treatment options is currently suboptimal, and thus there is a recognised need for imaging biomarkers [1]. Whilst the debate of a gold standard imaging profile and optimal modalities continues, the role of non-contrast computed tomography (NCCT) in the emergency setting is undisputed. Beyond its primary role of intracerebral haemorrhage detection, signs of early ischaemic change can be identified. These subtle imaging features include the parenchymal changes as well as the dense vessel sign, which represents causative vessel occlusion. Objective, automated methods to detect and quantify these signs of early ischaemic change hold value in a time critical clinical setting for treatment triage, especially out-of-hours in the absence of a specialist.

Automatic detection of acute stroke signs in NCCT images is challenging. Dense vessels and areas of ischaemia are difficult to detect due to the proximity

M. Valdés Hernández and V. González-Castro (Eds.): MIUA 2017, CCIS 723, pp. 494–505, 2017.
DOI: 10.1007/978-3-319-60964-5_43

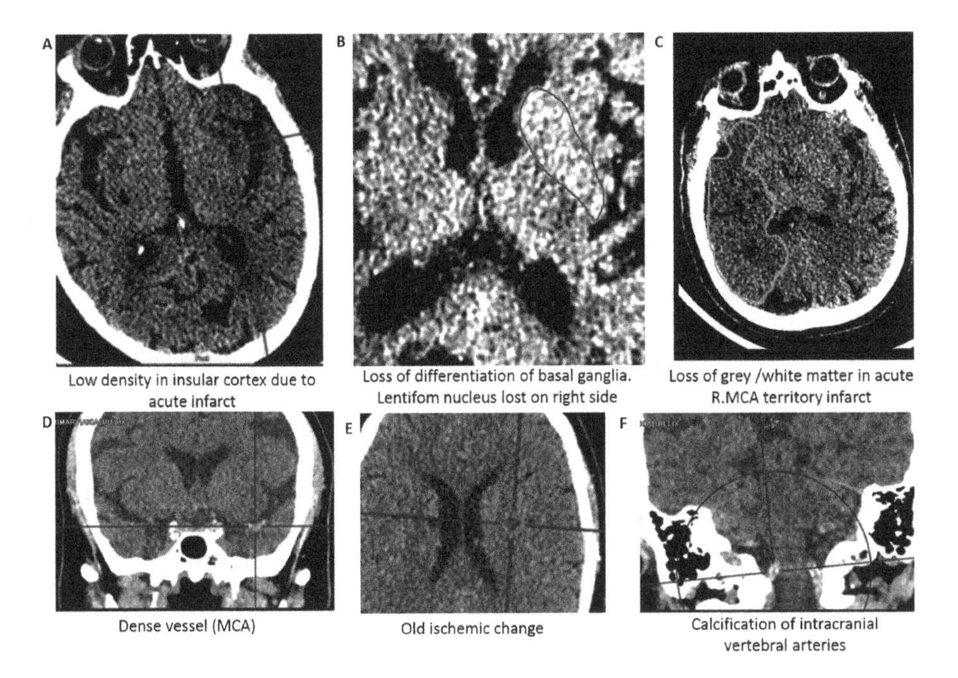

Fig. 1. A-E show examples of various subtle stroke signs. F shows normal calcification of arteries, which could be confused with dense vessel signs by a naïve classifier.

of bone in the former case, and to the subtlety of intensity and texture changes in the latter case (see Fig. 1).

For the easier problem of haemorrhagic tissue detection in NCCT, common solutions include histogram-based thresholding [2] or clustering [3], followed by morphological operations. More sophisticated methods of stroke sign detection use classifiers with handcrafted features [4]. Chawala *et al.* developed a system for the classification of both ischaemic and haemorrhagic stroke signs [5]. Their system computes slicewise histograms for each hemisphere and classification of the image slice is based on a comparison between the left and right side histograms (see Sect. 5). This bilateral comparison is an example of where the provision of contextual information might be helpful to a system for stroke sign identification.

There have been CNN-based methods suggested for ischaemia lesion segmentation in MR images [6, 7] (see Sect. 5), nevertheless the architecture for CT may require different design choices. CT brain images show less structural detail in regions of soft tissue and take on a textured appearance, therefore there are fewer higher-level concepts for a CNN to learn than for MR brain images.

To inform the network design, we observed an experienced neuroradiologist during reading of the NCCT scan. They routinely compared the appearance and Hounsfield Unit intensities of the left and right hemispheres when searching for stroke signs and used their knowledge of brain anatomy to navigate straight to the regions most commonly affected in stroke episodes. Therefore, we hypothesise that incorporation of the bilateral comparison and atlas information into the

CNN architecture might be helpful for the detection of the dense vessels and ischaemia (see Fig. 2). These insights have been previously incorporated in a CNN applied to a thrombus detection problem [8]. In this study we examine the impact of each of the architectural choices on the detection performance of different type of stroke signs, specifically we:

- Demonstrate a CNN-based solution applied to the detection of dense vessel and ischaemic regions in non-contrast CT scans.
- Evaluate the impact of including contralateral features in a CNN architecture for dense vessel and ischaemia detection.
- Evaluate the impact of including atlas location features in a CNN architecture for dense vessel and ischaemia detection.

2 Proposed Solution: A CNN for Stroke Sign Detection

Bilateral Comparison: The exploitation of anatomical symmetry has previously been incorporated in unsupervised approaches to pathology detection.

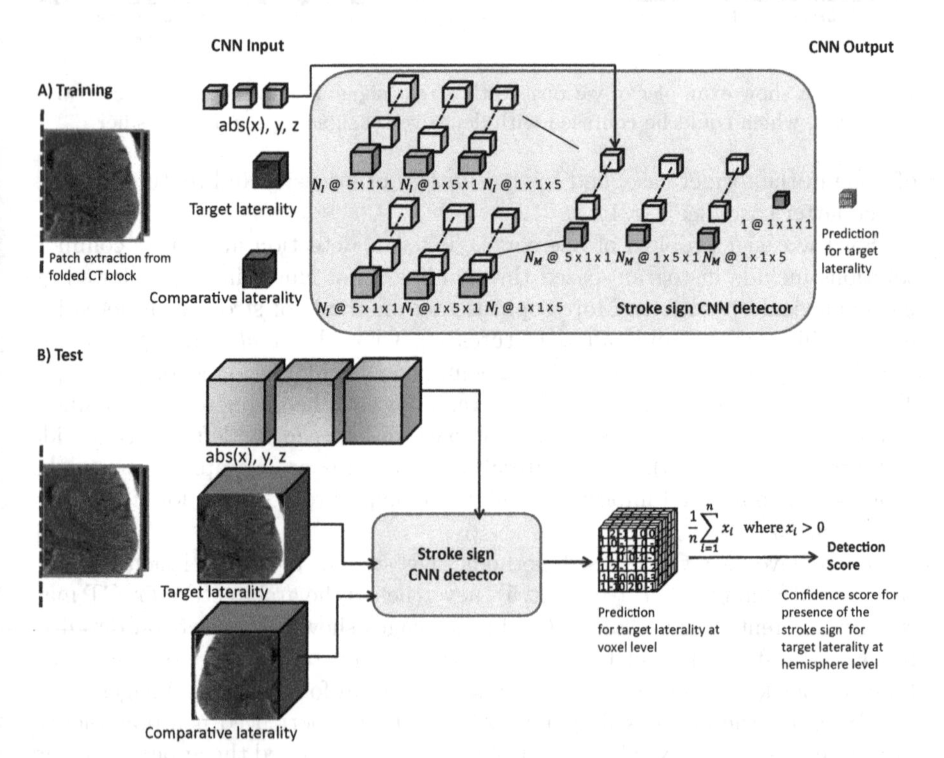

Fig. 2. (A) Schematic of the CNN, including filter sizes and number of layers. Pairs of contralateral 3D image intensity patches are input to the network at training time. Atlas coordinate inputs are fused at the merge point of the intensity channels. (B) Application of the detector at test time. The whole folded block is input to the network, and predictions are generated separately for each side.

Researchers utilised within-organ symmetry for the detection of brain tumours [9] and examined symmetry between a pair of organs for the detection of breast tumours [10]. In these approaches, the pathology is found by searching for the most dissimilar regions between the left and right sides of the organ [9], or by identification of asymmetry between paired organs [10]. We incorporate right and left hemisphere comparison in the CNN architecture by inputting image patches extracted from both hemispheres to parallel CNN channels and allow left/right comparison functionality to emerge as a part of supervised training. To ensure extraction of *corresponding* patches from contralateral regions, we first align the CT volume to the reference dataset, then we extract symmetrical blocks of interest (Fig. 4) about the sagittal midline, and finally we fold the CT block along the midline (Fig. 2A).

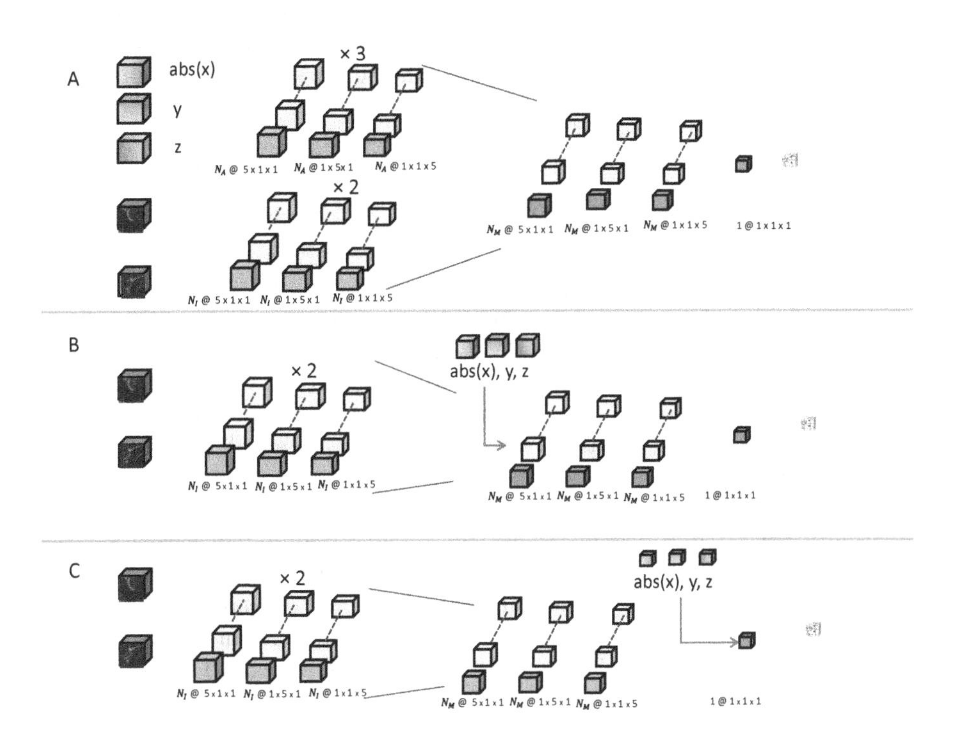

Fig. 3. Adding atlas information at different network stages: (A) Adding the atlas location by creating an additional 3 input channels (×3) alongside two intensity channels (×2). (B) Adding the atlas location midway, at the merge point of the bilateral intensity channels. (C) Adding the atlas location to the pre-classification layer.

Anatomical Context: One way of discovering abnormalities in the image is to compare the patient image with a normative atlas created from healthy examples of anatomy. In this approach, the patient image is registered to the normative atlas and the pathology is identified by the differences between the reference atlas and the examined image [11]. Anatomical atlases have also been used in supervised medical imaging applications as they provide useful anatomical context information. Researchers have previously employed explicit anatomical context mechanisms when training random forest classifiers [12]. We propose to supply the CNN with this information by adding three channels encoding the x, y and z atlas locations. Furthermore, we investigate the level at which this information should be provided to the network (see Fig. 3).

3D Architecture: NCCT scans are three-dimensional (3D), therefore architectures designed for 2D images cannot be directly applied. When moving to 3D data, 3D convolutions [6,13] may be used in place of 2D convolutions. A 3D CNN could potentially lead to better results than a 2D CNN applied slicewise [13] since information within a 3D neighbourhood may be leveraged at a deep level. In this paper we adopt a 3D CNN, but we apply spatial decomposition of the kernels such that convolutions are applied one dimension at the time. This allows us to reduce the risk of overfitting as the network has smaller number of parameters.

3 Data Sets

We use data from the following studies: ATTEST [14], POSH and WYETH [15]. Ground truth was collected on the acute NCCT scans for 170 patients with suspected acute ischaemic stroke within 6 hours of onset.

Capturing ground-truth for subtle stroke signs is only achievable by manual segmentation, given the diversity of shape, size and location of these signs. 3D Slicer 4.5.0 was used for this task which generated a label map for each of the acute NCCT datasets. Manual segmentations of ischaemia and dense vessels were generated by a clinical researcher under the supervision of an experienced neuroradiologist. Annotations were blind to additional scans (e.g. CT angiography, CT perfusion, follow-up scans) and clinical information except for the radiology report which included laterality of symptoms.

Thromboembolism is most frequently seen in the middle cerebral artery, thus we currently focus our detection on dense vessels within the anterior circulation and on ischaemic regions which lie downstream in the associated vascular territory (see Fig. 4).

4 Experiments

Data Preparation: Volumetric datasets are pre-aligned to a reference dataset, designated as an atlas. The transformation between a given dataset and the reference atlas is discovered via landmarks, which are detected in the novel volume by a random forest as proposed in [16]. The datasets are isotropically resampled to 1 mm per voxel for dense vessels and 2 mm per voxel for ischaemia. Blocks of interest are extracted symmetrically about the sagittal midline and folded. Volume intensities are clamped in the range $\{-125, 225\}$ HU for dense vessels

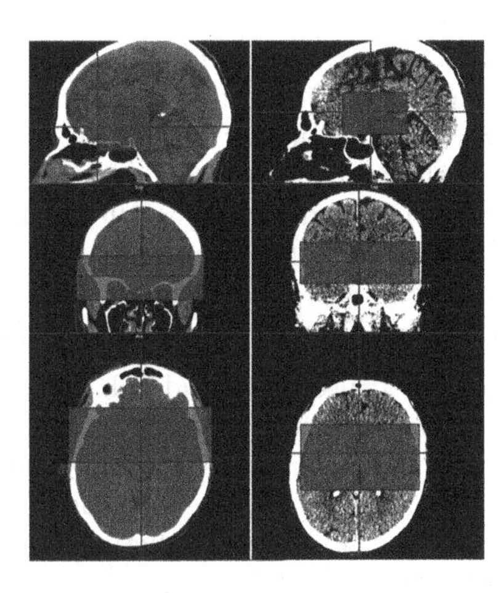

Fig. 4. Left: anterior circulation block used in dense vessel detection experiments. Right: subcortex block used in ischaemia detection experiments.

and {0, 80} HU for ischaemia to imitate the typical window levels chosen by a radiologist when searching for these signs.

CNN Model: There are two data input channels for 3D patches selected symmetrically from the left and right hemispheres. Weights are shared between intensity channels for ischaemia, but not for dense vessels. We also insert x, y and z atlas coordinates into the architecture. Each full convolution operation comprises a series of orthogonal one-dimensional convolutions, with kernels of $5 \times 1 \times 1$, $1 \times 5 \times 1$ and $1 \times 1 \times 5$. $N_I = 32$ kernels are used for the data channels and $N_A = 4$ kernels are used for the atlas channels. Channels are then merged and another convolution operation is applied, with $N_M = 32$ kernels in the case of dense vessels and $N_M = 2$ in the case of ischaemia. The number of kernels and the filter size were chosen empirically. ReLU activation functions are used. The output is fully convolutional allowing for the efficient prediction of all voxels of the dataset in a single pass (see Fig. 2). The models were implemented in Python using the Keras [17] library built on top of Theano [18].

CNN Training: We use 71 datasets for training, 48 for validation and 51 for testing. To compensate for the large imbalance between abnormal voxels and normal voxels, we adopt a biased patch selection process and use a weighting factor w, which is defined as the ratio of normal to abnormal voxels in the training set. The patches for training of ischemia are of size $18 \times 18 \times 18$ voxels and for dense vessels are $24 \times 24 \times 24$ voxels. There is inevitably some uncertainty around precise ground truth segmentation of stroke signs in NCCT, due to their diffuse appearance, especially in case of ischaemia. Therefore we are not interested in refining the segmentation boundary, but in the detection of the presence or absence of a stroke sign. We mark the regions around the pathology as "do not care". We adjust the loss function to ignore the voxels with this label. For implementation, we use labels of -1 and $+1$ for the normal and abnormal classes respectively and use the label 0 to represent the "do not care" voxels for which loss is not computed. The border is created by a dilation operation and it is 1 mm thick for dense vessel and 6 mm for ischaemia. Training is performed at the voxel level, meaning that each voxel in the patch is a training example. Training is performed using the Adam optimiser [19] on normalised data samples, to optimise the squared *hinge loss* function, with a learning rate of 0.001, a momentum of 0.9 and L2 regularisation of 0.002. For training and testing of the CNN classifier we used a Titan X GPU.

Post-processing: We train the dense vessel and ischaemia detectors at the voxel level as this gives us a larger number of training samples. Nevertheless, we are interested in determining whether the stroke sign is present or absent at the level of the brain hemisphere. To arrive at the detection confidence score at half block level we compute the mean of all voxel level predictions with a score above 0. This threshold was necessarily as the majority of voxels are negative (normal tissue) and had more influence over the mean than the confident positive detection of smaller stroke signs.

Evaluation: We evaluate the performance of the ischaemia (Sect. 4.1) and dense vessel (Sect. 4.2) detectors at the brain hemisphere level (half-patient) in terms of the area under the curve for the Receiver Operating Characteristic (ROC AUC) and the Precision-Recall (PR AUC) curve. It is suggested that PR curves should be used when the positive class samples are rare compared to the negative class samples [20], because precision is more sensitive to any change in the number of false positives, while specificity is not due to the large number of negative samples. To determine whether the inclusion of bilateral comparison or the atlas are helpful in stroke sign detection we compare the detection performance of four different CNN architectures.

4.1 Ischaemia Detection

Table 1 presents the detection results obtained for ischaemia detection. Incorporation of bilateral features in the CNN architecture improves detection performance compared to a single intensity channel CNN, which is not trained with pairs of contralateral patches, but follows standard single intensity patch training. Incorporation of atlas information is not helpful for the ischaemia detection task as there is no significant different in performance between CNN architectures with and without atlas coordinates provided to the network.

4.2 Dense Vessel Detection

Table 2 presents the dense vessel detection results. The addition of atlas coordinates to the network has a large impact on the performance of the dense vessel detector. It gives rise to larger improvements than inclusion of bilateral channels, which has little impact on dense vessel detection. We incorporated atlas information in the network at three different levels (see Fig. 3) to investigate if the point at which this information is injected to the network affects the detection results. The earliest incorporation of the atlas led to the best performance.

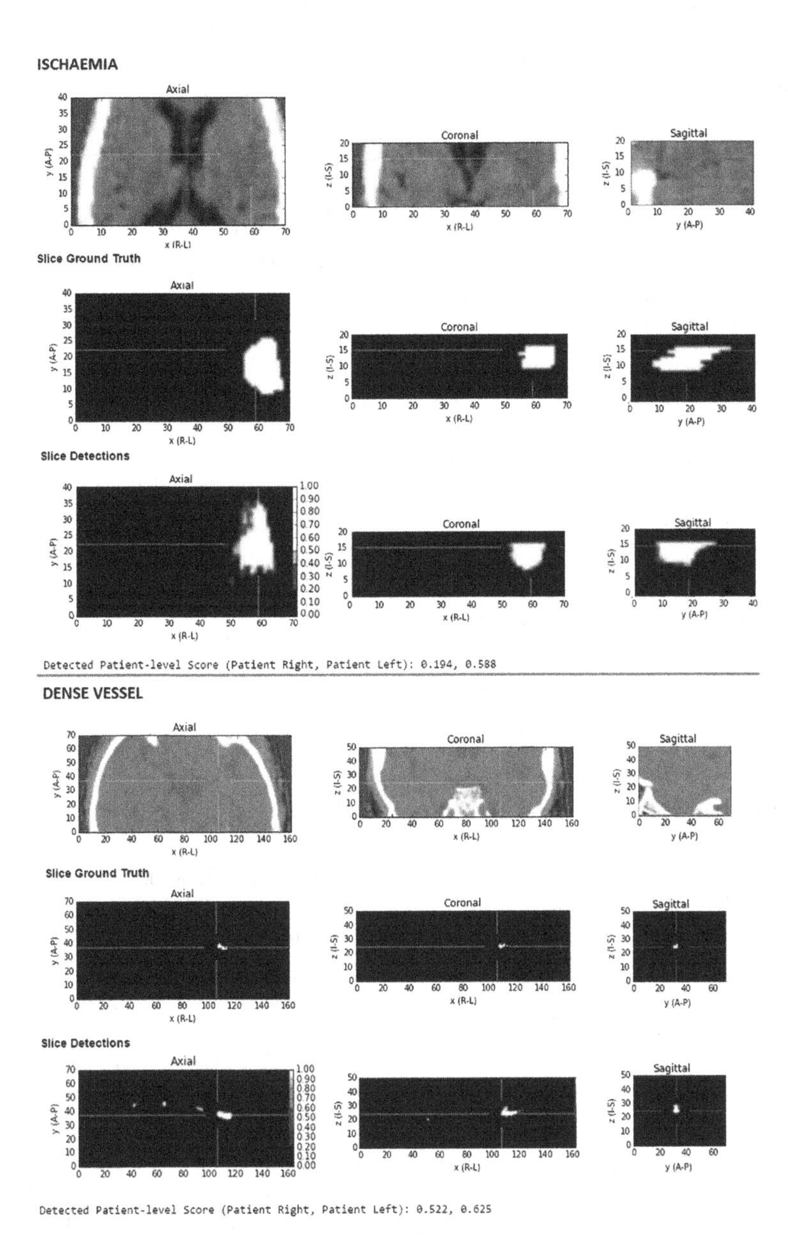

Fig. 5. Examples of ischaemia detection (top) and dense vessel detection (bottom). For each example, we display the volume slice with the highest number of abnormal voxels. Nevertheless the predictions are computed for all voxels in the volume. Image brightness corresponds to confidence level. For dense vessels we selected an example with both true and false positive detections. The false positives are detected with lower confidence, as indicated by the brightness. The *hemisphere*-level scores for the right and left hemisphere are printed below each example.

Table 1. Ischaemia detection results. Detection of the presence/absence of a sign was performed at the hemisphere level. Each experiment was run 3 times with different random seeds, and we report the mean and standard deviation

Model	ROC AUC [std]	PR AUC [std]
Bilateral CNN + atlas	0.915 [0.006]	0.783 [0.014]
Bilateral CNN	0.912 [0.007]	0.782 [0.006]
Single intensity channel CNN + atlas	0.738 [0.003]	0.483 [0.021]
Single intensity channel CNN	0.743 [0.012]	0.461 [0.022]

Table 2. Dense vessel detection results. Detection of the presence/absence of a sign was performed at the hemisphere level. Each experiment was run 3 times with different random seeds, and we report the mean and standard deviation.

Model	ROC AUC [std]	PR AUC [std]
Bilateral CNN + atlas (A)	0.964 [0.005]	0.898 [0.029]
Bilateral CNN + atlas (B)	0.950 [0.011]	0.817 [0.062]
Single intensity channel CNN + atlas (A)	0.936 [0.026]	0.790 [0.063]
Bilateral CNN + atlas (C)	0.927 [0.019]	0.718 [0.072]
Bilateral CNN	0.891 [0.011]	0.691 [0.036]
Single intensity channel CNN	0.876 [0.013]	0.514 [0.060]

5 Related Work and Discussion

In this paper we were inspired by the workflow of a radiologist, to investigate the role of contextual information when interpreting NCCT scans to identify acute stroke signs. Frequently in medical images, local and global spatial context is informative. Pathology might have a characteristic distribution relative to different anatomical structures and tissues. This is the case with dense vessel signs which are expected to appear along the vasculature. Precise anatomical location is less useful for ischaemic changes, which are diffuse in appearance and more texture-like than dense vessels, therefore the type of context required to detect those signs may differ.

Some researchers have tried the use of foveation [21] or the similar method of non-uniform sampling [22] in order to capture anatomical contextual information. Foveation refers to under-sampling of pixels closed to a window boundary, whilst keeping the central pixels at the original sampling level. Ciresan *et al.* [21] propose that this method is well suited to the training of networks whose task is to classify the central pixel of the window, as the network is then forced to ignore fine details at the periphery of the window, whilst still having access to general context information. The challenge with this approach is that it assumes classification of one voxel at a time using sliding window approach, which leads to long detection times. Although these methods may offer efficient training, it is not obvious how to incorporate them in a fully convolutional network, and application to stroke sign detection might not be feasible due to constraints on detection time in the clinical setting.

In tackling ischaemia lesion detection in MR, a couple of authors have pursued a dual-pathway approach, in which two pathways are devoted to local and global context respectively, before combining at the pre-classification stage. Kamnitsas *et al.* [6] achieves this through the use of patches at two different image scales, whereas Dutil *et al.* [7] uses kernels of different sizes. As the pathways are disjoint the network can process them in parallel, which is convenient for run time speed up. We adopted a similar architectural design, but designated the two input pathways for bilateral inputs, rather than two scales.

We are not the first authors to notice that comparison of the appearance of the left and right hemisphere might be helpful in ischaemia detection. Chawla *et al.* proposed a two stage system for differentiation between chronic infarcts, haemorrhagic and ischaemic stroke [5]. For each image slice, histograms of intensity values are computed for each hemisphere and hemisphere similarity is assessed using the correlation coefficient as a measure. In the first stage they classify the datasets containing chronic infarcts or haemorrhage by histogram thresholding. In the second stage they employ a wavelet decomposition of the histograms for slices assigned to neither of the categories to further discriminate between acute infarcts and normal slices. The authors report average recall of 90% for acute stoke categorisation at slice level [5]. By contrast, our system first produces a prediction for all voxel in the block so we are able to highlight the region suspected of ischaemia alongside the hemisphere detection score computed at half patient level (see Fig. 5).

Since we have a strong prior belief about the spatial location of dense vessels, we also injected explicit atlas location information. In this we follow [23] who, for coronary calcium scoring in cardiac CT angiography, registered each image to an atlas image and fed the resulting atlas coordinates $\{x, y, z\}$ of each voxel as three additional inputs to the network at the dense fully connected layer. We build upon their work by showing that atlas coordinates may be effectively utilised even if the fully connected layers are implemented as convolutions. Furthermore, we investigated at which level to inject atlas information to the architecture and we found that providing it at the input layer alongside the intensity inputs led to best performance. This agrees with the finding of Havaei et al., who investigated at which level the information from the output of an initial CNN should be incorporated to a second CNN in a cascade of classifiers. They also tried providing this additional information at three different levels and found that concatenation of the prediction of the first classifier with intensity patches at the input to the second classifier gave best brain tumour segmentation in MR [24].

6 Conclusion and Future Work

The design of the CNN classifier depends on the task at hand. We have investigated the type of contextual information required for two different types of stroke sign and suggested how this information might be incorporated in the CNN architecture. We found that providing atlas information is helpful for dense vessel detection, where the signs appear in typical locations. Furthermore, the earlier this information is provided to the network, the sooner the detector is able to focus in on the critical spatial region. Atlas coordinates are less useful for detecting ischaemic regions since they vary in location and size. However, the incorporation of contralateral features in our CNN design enables bilateral comparison, and leads to more successful ischaemia detection.

In future work it would be interesting to compare the CNN-based solutions designed for ischaemia lesion segmentation in MR, such as [6], with the proposed CNN architecture for stoke sign detection in NCCT. This compares solutions designed for similar tasks but for different imaging modalities and so would enable evaluation of the extent to which the CNN architectures designs should be modality specific.

References

1. Wintermark, M., Albers, G.W., Broderick, J.P., Demchuk, A.M., Fiebach, J.B., Fiehler, J., Grotta, J.C., Houser, G., Jovin, T.G., Lees, K.R., et al.: Acute stroke imaging research roadmap II. Stroke **44**(9), 2628–2639 (2013)
2. Chan, T.: Computer aided detection of small acute intracranial hemorrhage on computer tomography of brain. Comput. Med. Imaging Graph. **31**(4), 285–298 (2007)
3. Dhawan, A.P., Loncaric, S., Hitt, K., Broderick, J., Brott, T.: Image analysis and 3-D visualization of intracerebral brain hemorrhage. In: Proceedings of Sixth Annual IEEE Symposium on Computer-Based Medical Systems, pp. 140–145. IEEE (1993)
4. Usinskas, A., Dobrovolskis, R.A., Tomandl, B.F.: Ischemic stroke segmentation on CT images using joint features. Informatica **15**(2), 283–290 (2004)
5. Chawla, M., Sharma, S., Sivaswamy, J., Kishore, L.: A method for automatic detection and classification of stroke from brain CT images. Eng. Med. Biol. Soc. **2009**, 3581–3584 (2009)
6. Kamnitsas, K., Ledig, C., Newcombe, V.F., Simpson, J.P., Kane, A.D., Menon, D.K., Rueckert, D., Glocker, B.: Efficient multi-scale 3D CNN with fully connected CRF for accurate brain lesion segmentation. Med. Image Anal. **36**, 61–78 (2017)
7. Dutil, F., Havaei, M., Pal, C., Larochelle, H., Jodoin, P.-M.: A convolutional neural network approach to brain segmentation. In: Ischemic Stroke Lesion Segmentation, p. 53 (2015)
8. Lisowska, A., Bereridge, E., Muir, K., Poole, I.: Thrombus detection in ct brain scans using a convolutional neural network. In: Proceedings of the 10th International Joint Conference on Biomedical Engineering Systems and Technologies (BIOSTEC 2017), Bioimaging, vol. 2, pp. 24–33. SCITEPRESS (2017)
9. Hasan, A., Meziane, F., Khadim, M.: Automated segmentation of tumours in MRI brain scans. In: Proceedings of the 9th International Joint Conference on Biomedical Engineering Systems and Technologies (BIOSTEC 2016), pp. 55–62. SCITEPRESS (2016)
10. Erihov, M., Alpert, S., Kisilev, P., Hashoul, S.: A cross saliency approach to asymmetry-based tumor detection. In: Navab, N., Hornegger, J., Wells, W.M., Frangi, A.F. (eds.) MICCAI 2015. LNCS, vol. 9351, pp. 636–643. Springer, Cham (2015). doi:10.1007/978-3-319-24574-4_76

11. Doyle, S., Vasseur, F., Dojat, M., Forbes, F.: Fully automatic brain tumor segmentation from multiple MR sequences using hidden Markov fields and variational EM. In: Proceedings of NCI-MICCAI BraTS, pp. 18–22 (2013)
12. O'Neil, A., Murphy, S., Poole, I.: Anatomical landmark detection in CT data by learned atlas location autocontext. In: Medical Image Understanding and Analysis (MIUA), pp. 189–194 (2015)
13. Payan, A., Montana, G.: Predicting Alzheimers disease: a neuroimaging study with 3D convolutional neural networks. arXiv:1502.02506 (2015)
14. Huang, X., Cheripelli, B.K., Lloyd, S.M., Kalladka, D., Moreton, F.C., Siddiqui, A., Ford, I., Muir, K.W.: Alteplase versus tenecteplase for thrombolysis after ischaemic stroke (ATTEST): a phase 2, randomised, open-label, blinded endpoint study. Lancet Neurol. **14**(4), 368–376 (2015)
15. Wardlaw, J.M., Muir, K.W., Macleod, M.J., Weir, C., McVerry, F., Carpenter, T., Shuler, K., Thomas, R., Acheampong, P., Dani, K., Murray, A.: Clinical relevance and practical implications of trials of perfusion and angiographic imaging in patients with acute ischaemic stroke: a multicentre cohort imaging study. J. Neurol. Neurosurg. Psychiatry **84**(9), 1001–1007 (2013). http://jnnp.bmj.com/content/84/9/1001
16. Dabbah, M.A., Murphy, S., Pello, H., Courbon, R., Beveridge, E., Wiseman, S., Wyeth, D., Poole, I.: Detection and location of 127 anatomical landmarks in diverse CT datasets. In: SPIE Medical Imaging, pp. 903415–903415. International Society for Optics and Photonics (2014)
17. Chollet, F.: Keras (2015). https://github.com/fchollet/keras
18. Theano Development Team. Theano: a Python framework for fast computation of mathematical expressions. arXiv e-prints, vol. abs/1605.02688, May 2016
19. Kingma, D., Ba, J.: Adam: a method for stochastic optimization. arXiv preprint arXiv:1412.6980 (2014)
20. Davis, J., Goadrich, M.: The relationship between precision-recall and ROC curves. In: Proceedings of the 23rd International Conference on Machine Learning, ICML 2006, pp. 233–240 (2006)
21. Ciresan, D., Giusti, A., Gambardella, L.M., Schmidhuber, J.: Deep neural networks segment neuronal membranes in electron microscopy images. In: Advances in Neural Information Processing Systems, pp. 2843–2851 (2012)
22. Ghafoorian, M., Karssemeijer, N., Heskes, T., van Uder, I., de Leeuw, F., Marchiori, E., van Ginneken, B., Platel, B.: Non-uniform patch sampling with deep convolutional neural networks for white matter hyperintensity segmentation. In: 2016 IEEE 13th International Symposium on Biomedical Imaging (ISBI), pp. 1414–1417. IEEE (2016)
23. Wolterink, J.M., Leiner, T., Viergever, M.A., Išgum, I.: Automatic coronary calcium scoring in cardiac CT angiography using convolutional neural networks. In: Navab, N., Hornegger, J., Wells, W.M., Frangi, A.F. (eds.) MICCAI 2015. LNCS, vol. 9349, pp. 589–596. Springer, Cham (2015). doi:10.1007/978-3-319-24553-9_72
24. Havaei, M., Davy, A., Warde-Farley, D., Biard, A., Courville, A., Bengio, Y., Pal, C., Jodoin, P.-M., Larochelle, H.: Brain tumor segmentation with deep neural networks. Med. Image Anal. **35**, 18–31 (2017)

Automatic Brain Tumor Detection and Segmentation Using U-Net Based Fully Convolutional Networks

Hao Dong[1], Guang Yang[2,3], Fangde Liu[1], Yuanhan Mo[1], and Yike Guo[1](✉)

[1] Data Science Institute, Imperial College London, London SW7 2AZ, UK
{hao.dong11, fangde.liu, y.mo16, y.guo}@imperial.ac.uk
[2] Neurosciences Research Centre, Molecular and Clinical Sciences Institute, St. George's, University of London, London SW17 0RE, UK
g.yang@imperial.ac.uk
[3] National Heart and Lung Institute, Imperial College London, London SW7 2AZ, UK

Abstract. A major challenge in brain tumor treatment planning and quantitative evaluation is determination of the tumor extent. The noninvasive magnetic resonance imaging (MRI) technique has emerged as a front-line diagnostic tool for brain tumors without ionizing radiation. Manual segmentation of brain tumor extent from 3D MRI volumes is a very time-consuming task and the performance is highly relied on operator's experience. In this context, a reliable fully automatic segmentation method for the brain tumor segmentation is necessary for an efficient measurement of the tumor extent. In this study, we propose a fully automatic method for brain tumor segmentation, which is developed using U-Net based deep convolutional networks. Our method was evaluated on Multimodal Brain Tumor Image Segmentation (BRATS 2015) datasets, which contain 220 high-grade brain tumor and 54 low-grade tumor cases. Cross-validation has shown that our method can obtain promising segmentation efficiently.

1 Introduction

Primary malignant brain tumors are among the most dreadful types of cancer, not only because of the dismal prognosis, but also due to the direct consequences on decreased cognitive function and poor quality of life. The most frequent primary brain tumors in adults are primary central nervous system lymphomas and gliomas, which the latter account for almost 80% of malignant cases [1]. The term glioma encompasses many subtypes of the primary brain tumor, which range from slower-growing 'low-grade' tumors to heterogeneous, highly infiltrative malignant tumors. Despite significant advances in imaging, radiotherapy, chemotherapy and surgical procedure, certain cases

H. Dong, G. Yang and F. Liu—contributed equally to this study.

M. Valdés Hernández and V. González-Castro (Eds.): MIUA 2017, CCIS 723, pp. 506–517, 2017.
DOI: 10.1007/978-3-319-60964-5_44

of malignant brain tumors, e.g., high-grade glioblastoma and metastasis, are still considered untreatable with a 2.5-year cumulative relative survival rate of 8% and 2% at 10 years [2]. Moreover, there are variable prognosis results for patients with low-grade gliomas (LGG) with an overall 10-year survival rate about 57% [3].

Previous studies have demonstrated that the magnetic resonance imaging (MRI) characteristics of newly identified brain tumors can be used to indicate the likely diagnosis and treatment strategy [4–6]. In addition, multimodal MRI protocols are normally used to evaluate brain tumor cellularity, vascularity, and blood-brain barrier (BBB) integrity. This is because different image contrasts produced by multimodal MRI protocols can provide crucial complementary information. Typical brain tumor MRI protocols, which are used routinely, include T_1-weighted, T_2-weighted (including Fluid-Attenuated Inversion Recovery, i.e., FLAIR), and gadolinium enhanced T_1-weighted imaging sequences. These structuralMRI images yield a valuable diagnosis in the majority of cases [7].

Image segmentation is a critical step for the MRI images to be used in brain tumor studies: (1) the segmented brain tumor extent can eliminate confounding structures from other brain tissues and therefore provide a more accurate classification for the subtypes of brain tumors and inform the subsequent diagnosis; (2) the accurate delineation is crucial in radiotherapy or surgical planning, from which not only brain tumor extend has been outlined and surrounding healthy tissues has been excluded carefully in order to avoid injury to the sites of language, motor, and sensory function during the therapy; and (3) segmentation of longitudinal MRI scans can efficiently monitor brain tumor recurrence, growth or shrinkage. In current clinical practice, the segmentation is still relied on manual delineation by human operators. The manual segmentation is a very labor-intensive task, which normally involves slice-by-slice procedures, and the results are greatly dependent on operators' experience and their subjective decision making. Moreover, reproducible results are difficult to achieve even by the same operator. For a multimodal, multi-institutional and longitudinal clinical trial, a fully automatic, objective and reproducible segmentation method is highly in demand.

Despite recent developing in semi-automatic and fully automatic algorithms for brain tumor segmentation, there are still several opening challenges for this task mainly due to the high variation of brain tumors in size, shape, regularity, location and their heterogeneous appearance (e.g., contrast uptake, image uniformity and texture) [6, 8]. Other potential issues that may complicate the brain tumor segmentation include: (1) the BBB normally remains intact in LGG cases and the tumor regions are usually not contrast enhanced; therefore, the boundaries of LGG can be invisible or blurry despite FLAIR sequence may provide differentiation between normal brain and brain tumor or edema to delineate the full extent of the lesion; (2) in contrast, for high-grade gliomas (HGG) cases, the contrast agent, e.g., gadolinium, leaks across the disrupted BBB and enters extracellular space of the brain tumor causing hyper-intensity on T_1-weighted images. Therefore, the necrosis and active tumor regions can be easily delineated. However, HGG usually exhibit unclear and irregular boundaries that might also involve discontinuities due to aggressive tumor infiltration. This can cause problems and result in poor tumor segmentation; (3) varies tumor sub-regions and tumor types can only be visible by considering multimodal MRI data. However, the co-registration across multiple MRI sequences can be difficult especially when these sequences are acquired in different spatial resolutions; and (4) typical clinical MRI images are normally acquired with higher in-plane resolution and much lower inter-slice resolution in order to balance between adequate image slices to cover the whole tumor volume with good quality cross-sectional views and the restricted scanning time. This can cause inadequate signal to noise ratio and asymmetrical partial volume effects may also affect the final segmentation accuracy.

Previous studies on brain tumor segmentation can be roughly categorized into unsupervised learning based [9–12] and supervised learning based [13–18] methods. A more detailed topical review on various brain tumor segmentation methods can be found elsewhere, e.g., in [6]. In addition, a dedicated annual workshop and challenge, namely Multimodal Brain Tumor Image Segmentation (BRATS), is held to benchmark different algorithms that developed for the brain tumor segmentation [19]. Here we only reviewed some most recent and closely relevant studies for this topic.

Unsupervised learning based clustering has been successfully used for brain tumor segmentation by grouping data based on certain similarity criteria. Hsieh et al. [20] combined fuzzy clustering with region-growing for brain tumor cases scanned by T_1-weighted and T_2-weighted sequences and achieved a segmentation accuracy of 73%. In [9], a multi-stage fuzzy c-means framework was proposed to segment brain tumors scanned by multimodal MRI and obtained promising results, but the proposed framework was tested on a very limited number of datasets. Recently, a study [11] has been carried out to evaluate different clustering algorithms for glioblastoma segmentation, and results showed that Gaussian hidden Markov random field outperformed k-means, fuzzy k-means and Gaussian mixture model for this task. However, the best performing algorithm described in this study still only achieved 77% accuracy.

On the other hand, supervised learning based methods require training data-and-label pairs to learn a classification model, based on which new instances can be classified and then segmented. Wu et al. [13] employed superpixel features in a conditional random fields framework to segment brain tumors, but the results varied significantly among different patient cases and especially underperformed in LGG images. A study was proposed in which extremely randomized forest was used for classifying both appearance and context based features and 83% Dice score was achieved [14]. More recently, Soltaninejad et al. [16] combined extremely randomized trees classification with superpixel based over-segmentation for a single FLAIR sequence based MRI scan that obtained 88% overall Dice score of the complete tumor segmentation for both LGG and HGG tumor cases. Nevertheless, the tuning of superpixel size and compactness could be tricky and influence the final delineation.

Recently, supervised deep convolutional neural networks (CNN) have attracted lots of interests. Compared to conventional supervised machine learning methods, these deep learning based methods are not dependent on hand-crafted features, but automatically learn a hierarchy of increasingly complex features directly from data [21]. Currently, using BRATS datasets and their benchmarking system, deep learning based methods have been ranked on top of the contest [21–23]. This can be attributed to the fact that deep CNN is constructed by stacking several convolutional layers, which involve convolving a signal or an image with kernels to form a hierarchy of features that are more robust and adaptive for the discriminative models. Despite recent advances in these deep learning based methods, there are still several challenges:

(1) essentially tumor segmentation is an abnormal detection problem, it is more challenging than other pattern recognition based tasks; (2) while most methods provided satisfied segmentation for HGG cases, in general the performance of the LGG segmentation is still poor; (3) compared to complete tumor segmentation, the delineation of core tumor regions and enhanced infiltrative regions is still underperformed; (4) a more computing-efficient and memory-efficient development is still in demand because existing CNN based methods require considerable amount of computing resources.

In this study, we developed a novel 2D fully convoluted segmentation network that is based on the U-Net architecture [24]. In order to boost the segmentation accuracy, a comprehensive data augmentation technique has been used in this work. In addition, we applied a 'Soft' Dice based loss function introduced in [25]. The Soft Dice based loss function has a unique advantage that is adaptive to unbalanced samples, which is very important for brain tumor segmentation because some sub-tumoral regions may only count for a small portion of the whole tumoral volume. The proposed method has been validated using datasets acquired for both LGG and HGG patients. Compared with manual delineated ground truth, our fully automatic method has obtained promising results. Also compared to other state-of-the-art methods, we have achieved comparable results for delineating the complete tumor regions, and superior segmentation for the core tumor regions.

2 Method

2.1 Brain MRI Data Acquisitions and Patients

The proposed method was tested and evaluated on the BRATS 2015 datasets [19], which contain 220 high-grade glioma (HGG) and 54 low-grade glioma (LGG) patient scans. Multimodal MRI data is available for every patient in the BRATS 2015 datasets and four MRI scanning sequences were performed for each patient using T_1-weighted (T1), T_1-weighted imaging with gadolinium enhancing contrast (T1c), T_2-weighted (T2) and FLAIR. For each patient, the T1, T2 and FLAIR images were co-registered into the T1c data, which had the finest spatial resolution, and then resampled and interpolated into $1 \times 1 \times 1$ mm^3 with an image size of $240 \times 240 \times 155$. We have applied data normalization for each sequence of the multimodal MRI by subtracting the mean of each sequence and dividing by its standard deviation.

In addition, manual segmentations with four intra-tumoral classes (labels) are available for each case: necrosis (1), edema (2), non-enhancing (3), and enhancing tumor (4). The manual segmentations have been used as the ground truth in both segmentation model training and final segmentation performance evaluation. In previous studies, multimodal data were stacked like the multichannel RGB images [21–23]. In this study, we used FLAIR images to segment the complete tumor regions and tumor regions except edema that has been proved to be effective [16]. Additionally, T1c data were used to delineate the enhancing tumor. In so doing, our framework is not only more efficient, but also requires less clinical inputs because frequently multimodal MRI data are not available due to patient symptoms and limited acquisition time.

2.2 Data Augmentation

The purpose of data augmentation is to improve the network performance by intentionally producing more training data from the original one.In this study, we applied a set of data augmentation methods summarized in Table 1. Simple transformation such as flipping, rotation, shift and zoom can result in displacement fields to images but will not create training samples with very different shapes. Shear operation can slightly distort the global shape of tumor in the horizontal direction, but is still not powerful to gain sufficient variable training data, as tumors have no definite shape. To cope with this problem, we further applied elastic distortion [26] that can generate more training data with arbitrary but reasonable shapes.

Table 1. Summary of the applied data augmentation methods (γ controls the brightness of the outputs; α and σ control the degree of the elastic distortion).

Methods	Range
Flip horizontally	50% probability
Flip vertically	50% probability
Rotation	$\pm 20^{o}$
Shift	10% on both horizontal and vertical direction
Shear	20% on horizontal direction
Zoom	$\pm 10\%$
Brightness	$\gamma = 0.8$–1.2
Elastic distortion	$\alpha = 720$, $\sigma = 24$,

2.3 U-Net Based Deep Convolutional Networks

Biomedical images usually contain detailed patterns of the imaged object (e.g., brain tumor), and the edge of the object is variable. To cope with the segmentation for the objects with detailed patterns, Long et al. [27] proposed to use the skip-architecture that combined the high-level representation from deep decoding layers with the appearance representation from shallow encoding layers to produce detailed segmentation. This method has demonstrated promising results on natural images [27] and is also applicable to biomedical images [28]. Ronneberger et al. [24] introduced the U-Net, which employed the skip-architecture, to solve the cell tracking problem.

Our network architecture, which is based on the U-Net, consists of a down-sampling (encoding) path and an up-sampling (decoding) path as shown in Fig. 1. The down-sampling path has 5 convolutional blocks. Every block has two convolutional layers with a filter size of 3×3, stride of 1 in both directions and rectifier activation, which increase the number of feature maps from 1 to 1024. For the down-sampling, max pooling with stride 2×2 is applied to the end of every blocks except the last block, so the size of feature maps decrease from 240×240 to 15×15. In the up-sampling path, every block starts with a deconvolutional layer with filter size of 3×3 and stride of 2×2, which doubles the size of feature maps in both directions but decreases the number of feature maps by two, so the size of feature maps increases from 15×15 to 240×240. In every up-sampling block, two convolutional layers reduce the number of feature maps of concatenation of deconvolutional feature maps and the feature maps from encoding path. Different from the original U-Net architecture [24], we use zero padding to keep the output dimension for all the convolutional layers of both down-sampling and up-sampling path. Finally, a 1×1 convolutional layer is used to reduce the number of feature maps to two that refect the foreground and background segmentation respectively. No fully connected layer is invoked in the network. Other parameters of the network are tabulated in Table 2.

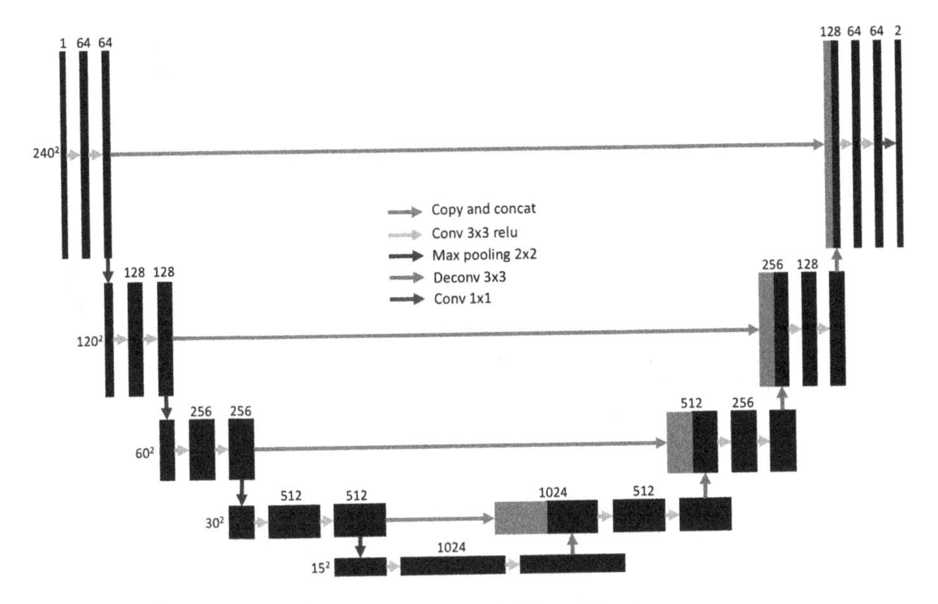

Fig. 1. Our developed U-Net architecture.

Table 2. Parameters setting for the developed U-Net.

Parameters	Value
Number of convolutional blocks	[4–6]
Number of deconvolutional blocks	[4–6]
Regularization	L1, L2, dropout

2.4 Training and Optimization

During the training process, the Soft Dice metric described in [25] was used as the cost function of the network rather than the cross-entropy based or the quadratic cost function. Soft Dice can be considered as a differentiable form of the original Dice Similarity Coefficient (DSC) [25].

Training deep neural networks requires stochastic gradient-based optimization to minimize the cost function with respect to its parameters. We adopted the adaptive moment estimator (Adam) [29] to estimate the parameters. In general, Adam utilizes the first and second moments of gradients for updating and correcting moving average of the current gradients. The parameters of our Adam optimizer were set as: learning rate = 0.0001 and the maximum number of epochs = 100. All weights were initialized by normal distribution with mean of 0 and standard deviationof 0.01, and all biases were initialized as 0.

2.5 Experiments and Performance Evaluation

The evaluation has been done using a five-fold cross-validation method for the HGG and LGG data, respectively. For each patient, we have validated on three sub-tumoral regions as described by

a. The *complete* tumor region (including all four intra-tumoral classes, labels 1, 2 3, and 4).
b. The *core* tumor region (as above but excluded "edema" regions, labels 1, 3, and 4).
c. The *enhancing* tumor region (only label 4).

For each tumoral region, the segmentations have been evaluated using the DSC, and the Sensitivity was also calculated. The DSC provides the overlap measurement between the manual delineated brain tumoral regions and the segmentation results of our fully automatic method that is

$$\mathrm{DSC} = \frac{2\mathrm{TP}}{\mathrm{FP} + 2\mathrm{TP} + \mathrm{FN}}, \tag{1}$$

in which TP, FP and FN denote the true positive, false positive and false negative measurements, respectively.

In addition, Sensitivity is used to evaluate the number of TP and FN that is

$$\mathrm{Sensitivity} = \frac{\mathrm{TP}}{\mathrm{TP} + \mathrm{FN}}. \tag{2}$$

We reported the mean DSC results of the five-fold cross-validation and also showed boxplots of the corresponding sensitivities. In this study, the HGG and LGG cases are trained and cross-validated separately.

3 Results and Discussion

In this study, we proposed and developed U-Net based fully convolutional networks for solving the brain tumor segmentation problem. Essentially, tumor detection and segmentation belongs to the task of semantic segmentation. Compared to previous deep learning based studies on this topic, we employed a comprehensive data augmentation scheme that not only contains rigid or affine based deformation, but also includes brightness and elastic distortion based transformation, and this has then been coupled with the U-Net that incorporates the skip-architecture.

Table 3 tabulates the DSC results of our cross-validated segmentation results for the HGG and LGG cases, respectively. In our current study, we only compared with three different deep learning based studies that published recently. All these three studies currently ranked on the top of the BRATS challenge. To the best of our knowledge, most published full papers were still focused on the BRATS 2013 datasets, which contains much less patient cases than the BRATS 2015 datasets. For example, in [21, 22] the model building has been done on the BRATS 2013 training datasets and then tested on the BRATS 2015 challenge datasets. Compared to the results on the BRATS 2015 challenge datasets [21, 22], the cross validation demonstrated that our method obtained superior results for the complete and core tumor segmentations. By using our method, the enhancing tumor segmentation for the LGG cases by using the T1c images only is not successful. This may be attributed to three reasons: (1) the BBB remains intact in most of these LGG cases and the tumor regions are rarely contrast enhanced; (2) the LGG cohort contains only 54 cases and the training datasets might be insufficient and (3) in these LGG cases, the borders between enhanced tumor and non-enhanced regions are more diffused and less visible that causes problems for both manual delineated ground truth and our fully automated segmentation model. Nevertheless, our method achieved 0.81 DSC for the enhancing tumor segmentation in the HGG cohort. Compared to Kamnitsas et al.'s work on the BRATS 2015 datasets [23], our method obtained comparable complete tumor segmentation results while achieving higher DSC for the core tumor delineation. Figure 2 displays the boxplots of the calculated Sensitivities and Fig. 3 shows some exemplar qualitative overlaid segmentation results compared to the ground truth.

Table 3. Quantitative results of our proposed fully automatic brain tumor segmentation method compared to the results from other recently published deep learning based methods. Here we tabulated the Dice Similarity Coefficient (DSC) for HGG, LGG and combined cases, respectively. Grey background highlighted the experiments on the BRATS 2015 datasets. Bold numbers highlighted the results of the best performing algorithm.

DSC					
Method	Data	Grade	Complete	Core	Enhancing
Proposed	Cross-validation on BRATS 2015 training datasets	HGG	0.88	0.87	0.81
		LGG	0.84	0.85	0.00
		Combined	0.86	**0.86**	0.65
Pereira16	BRATS 2013 leaderboard	HGG	0.88	0.76	0.73
		LGG	0.65	0.53	0.00
		Combined	0.84	0.72	0.62
	BRATS 2013 challenge	HGG	0.88	0.83	0.77
	BRATS 2015 challenge	Combined	0.79	0.65	0.75
Havaei16	BRATS 2013 training	Combined	0.88	0.79	0.73
	BRATS 2013 challenge	Combined	0.88	0.79	0.73
	BRATS 2013 leaderboard	Combined	0.84	0.71	0.57
	BRATS 2015 challenge	Combined	0.79	0.58	0.69
Kamnitsas17	BRATS 2015 training	Combined	**0.90**	0.76	**0.73**
	BRATS 2015 challenge	Combined	0.85	0.67	0.63

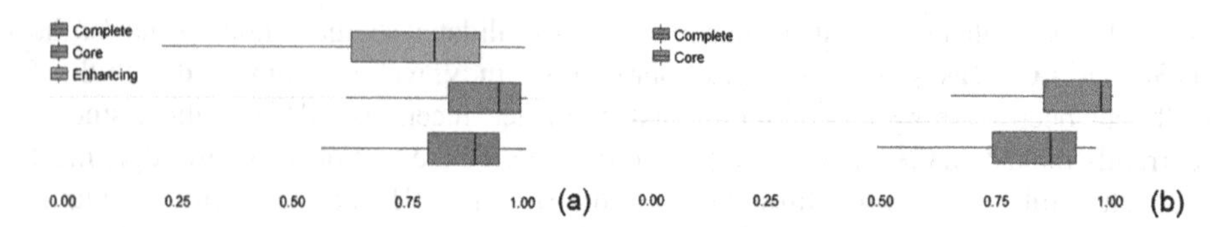

Fig. 2. Boxplots of the sensitivity; (a) for the HGG cases and (b) for the LGG cases.

Our network has been implemented using the TensorFlow and the TensorLayer libraries. For the HGG cohort, each cross validation training session requires approximately 18 h to finish on an NVIDIA Titan X (Pascal) graphics processing unit (GPU) with 12G memory, while the LGG cohort takes about ¼ of the training timeof the HGG cohort. There is a trade-off in choosing between 2D and 3D models. Due to the memory limits of the GPU, a 2D U-Net model can process a full slice in one go while a 3D convolution system can only process a small patch cover a small portion of the 3D volume. Therefore, in this study we used a 2D based network.

For the prediction once we fixed the model, the computation time is approximately 2 to 3 s per case regardless a HGG or a LGG study. Compared to our computational time, previous studies were less computational efficient: ~30 s [23], 25 s to 3 min [22], and about 8 min [21] to predict one tumor study.

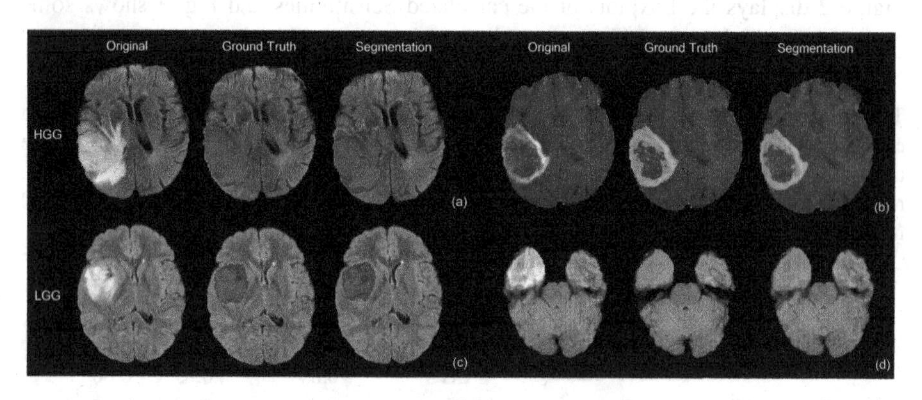

Fig. 3. Segmentations results for the exemplar HGG and LGG cases compared with manual delineated ground truth; (a) segmented complete tumor (red) of a HGG case overlaid on the FLAIR image; (b) segmented enhancing tumor (cyan) of a HGG case overlaid on the T1c image; (c) segmented complete tumor (red) of a LGG case overlaid on the FLAIR image; (a) segmented core tumor (green) of a LGG case overlaid on the FLAIR image. (Color figure online)

There are still some limitations of the current work. First, our segmentation method has been evaluated using a cross-validation scheme, which can provide an unbiased predictor, but running our model on a separate and independent testing dataset may produce a more objective evaluation. Secondly, there are several parameters need to be carefully tuned in our network. Currently all the parameters were determined via empirical study. In particular, for the regularization, we did not find a significant performance improvement after applying L1, L2 or dropout to the network. This may be attributed to the fact that an effective image distortion has been applied during our model training, and it is difficult to overfit a network with large amount of training data. Moreover, our current framework is less successful in segmenting enhancing tumor regions of the LGG cohort. By stacking all the multimodal MRI channels and performing joint training with HGG datasets may solve the problem. Despite these limitations, the developed model has still demonstrated promising segmentation results with efficiency. We can certainly envisage its application and effectiveness on an independent testing dataset. In addition, validating our method on multi-institutional and longitudinal datasets, especially for clinical datasets with anisotropic resolutions, will be one of the future directions.

4 Conclusion

In this paper, we presented a fully automatic brain tumor detection and segmentation method using the U-Net based deep convolution networks. Based on the experiments on a well-established benchmarking (BRATS 2015) datasets, which contain both HGG and LGG patients, we have demonstrated that our method can provide both efficient and robust segmentation compared to the manual delineated ground truth. In addition, compared to other state-of-the-art methods, our U-Net based deep convolution networks can also achieve comparable results for the complete tumor regions, and superior results for the core tumor regions. In our current study, the validation has been carried out using a five-fold cross-validation scheme; however, we can envisage a straightforward application on an independent testing datasets and further applications for multi-institutional and longitudinal datasets. The proposed method makes it possible to generate a patient-specific brain tumor segmentation model without manual interference, and this potentially enables objective lesion assessment for clinical tasks such as diagnosis, treatment planning and patient monitoring.

References

1. Schwartzbaum, J.A., Fisher, J.L., Aldape, K.D., Wrensch, M.: Epidemiology and molecular pathology of glioma. Nat. Clin. Pract. Neurol. **2**, 494–503 (2006)
2. Smoll, N.R., Schaller, K., Gautschi, O.P.: Long-term survival of patients with glioblastoma multiforme (GBM). J. Clin. Neurosci. **20**, 670–675 (2013)
3. Ramakrishna, R., Hebb, A., Barber, J., Rostomily, R., Silbergeld, D.: Outcomes in reoperated low-grade gliomas. Neurosurgery **77**, 175–184 (2015)
4. Mazzara, G.P., Velthuizen, R.P., Pearlman, J.L., Greenberg, H.M., Wagner, H.: Brain tumor target volume determination for radiation treatment planning through automated MRI segmentation. Int. J. Radiat. Oncol. Biol. Phys. **59**, 300–312 (2004)
5. Yamahara, T., Numa, Y., Oishi, T., Kawaguchi, T., Seno, T., Asai, A., Kawamoto, K.: Morphological and flow cytometric analysis of cell infiltration in glioblastoma: a comparison of autopsy brain and neuroimaging. Brain Tumor Pathol. **27**, 81–87 (2010)
6. Bauer, S., Wiest, R., Nolte, L.-P., Reyes, M.: A survey of MRI-based medical image analysis for brain tumor studies. Phys. Med. Biol. **58**, R97–R129 (2013)

7. Jones, T.L., Byrnes, T.J., Yang, G., Howe, F.A., Bell, B.A., Barrick, T.R.: Brain tumor classification using the diffusion tensor image segmentation (D-SEG) technique. Neuro. Oncol. **17**, 466–476 (2014)
8. Soltaninejad, M., Yang, G., Lambrou, T., Allinson, N., Jones, T.L., Barrick, T.R., Howe, F. A., Ye, X.: Automated brain tumour detection and segmentation using superpixel-based extremely randomized trees in FLAIR MRI. Int. J. Comput. Assist. Radiol. Surg. **12**(2), 183–203 (2016)
9. Szilágyi, L., Lefkovits, L., Benyó, B.: Automatic brain tumor segmentation in multispectral MRI volumes using a fuzzy c-means cascade algorithm. In: 2015 12th International Conference on Fuzzy Systems and Knowledge Discovery (FSKD), pp. 285–291 (2015)
10. Mei, P.A., de Carvalho Carneiro, C., Fraser, S.J., Min, L.L., Reis, F.: Analysis of neoplastic lesions in magnetic resonance imaging using self-organizing maps. J. Neurol. Sci. **359**, 78–83 (2015)
11. Juan-Albarracín, J., Fuster-Garcia, E., Manjón, J.V., Robles, M., Aparici, F., Martí-Bonmatí, L., García-Gómez, J.M.: Automated glioblastoma segmentation based on a multiparametric structured unsupervised classification. PLoS ONE **10**, e0125143 (2015)
12. Dhanasekaran, R.: Fuzzy clustering and deformable model for tumor segmentation on MRI brain image: a combined approach. Procedia Eng. **30**, 327–333 (2012)
13. Wu, W., Chen, A.Y.C., Zhao, L., Corso, J.J.: Brain tumor detection and segmentation in a CRF (conditional random fields) framework with pixel-pairwise affinity and superpixel-level features. Int. J. Comput. Assist. Radiol. Surg. **9**(2), 241–253 (2013)
14. Pinto, A., Pereira, S., Correia, H., Oliveira, J., Rasteiro, D.M.L.D., Silva, C.A.: Brain tumour segmentation based on extremely randomized forest with high-level features. In: 2015 37th Annual International Conference of the IEEE Engineering in Medicine and Biology Society (EMBC), pp. 3037–3040 (2015)
15. Gotz, M., Weber, C., Blocher, J., Stieltjes, B., Meinzer, H., Maier-Hein, K.: Extremely randomized trees based brain tumor segmentation. In: Proceeding of BRATS Challenge-MICCAI (2014)
16. Soltaninejad, M., Yang, G., Lambrou, T., Allinson, N., Jones, T.L., Barrick, T.R., Howe, F. A., Ye, X.: Automated brain tumour detection and segmentation using superpixel-based extremely randomized trees in FLAIR MRI. Int. J. Comput. Assist. Radiol. Surg. **12**(2), 183–203 (2016)
17. Jafari, M., Kasaei, S.: Automatic brain tissue detection in MRI images using seeded region growing segmentation and neural network classification. Aust. J. Basic Appl. Sci. **5**, 1066–1079 (2011)
18. Subbanna, N., Precup, D., Arbel, T.: Iterative multilevel MRF leveraging context and voxel information for brain tumour segmentation in MRI. In: Proceedings of the IEEE Computer Society Conference on Computer Vision and Pattern Recognition, pp. 400–405 (2014)
19. Menze, B.H., Jakab, A., Bauer, S., Kalpathy-Cramer, J., Farahani, K., Kirby, J., Burren, Y., Porz, N., Slotboom, J., Wiest, R., Lanczi, L., Gerstner, E., Weber, M.-A., Arbel, T., Avants, B. B., Ayache, N., Buendia, P., Collins, D.L., Cordier, N., Corso, J.J., Criminisi, A., Das, T., Delingette, H., Demiralp, Ç., Durst, C.R., Dojat, M., Doyle, S., Festa, J., Forbes, F., Geremia, E., Glocker, B., Golland, P., Guo, X., Hamamci, A., Iftekharuddin, K.M., Jena, R., John, N.M., Konukoglu, E., Lashkari, D., Mariz, J.A., Meier, R., Pereira, S., Precup, D., Price, S.J., Raviv, T.R., Reza, S.M.S., Ryan, M., Sarikaya, D., Schwartz, L., Shin, H.-C., Shotton, J., Silva, C.A., Sousa, N., Subbanna, N.K., Szekely, G., Taylor, T.J., Thomas, O.M., Tustison, N.J., Unal, G., Vasseur, F., Wintermark, M., Ye, D.H., Zhao, L., Zhao, B., Zikic, D., Prastawa, M., Reyes, M., Van Leemput, K.: The multimodal brain tumor image segmentation benchmark (BRATS). IEEE Trans. Med. Imaging **34**, 1993–2024 (2015)

20. Hsieh, T.M., Liu, Y.-M., Liao, C.-C., Xiao, F., Chiang, I.-J., Wong, J.-M.: Automatic segmentation of meningioma from non-contrasted brain MRI integrating fuzzy clustering and region growing. BMC Med. Inform. Decis. Mak. **11**, 54 (2011)
21. Pereira, S., Pinto, A., Alves, V., Silva, C.A.: Brain tumor segmentation using convolutional neural networks in MRI images. IEEE Trans. Med. Imaging **35**, 1240–1251 (2016)
22. Havaei, M., Davy, A., Warde-Farley, D., Biard, A., Courville, A., Bengio, Y., Pal, C., Jodoin, P.-M., Larochelle, H.: Brain tumor segmentation with deep neural networks. Med. Image Anal. **35**, 18–31 (2016)
23. Kamnitsas, K., Ledig, C., Newcombe, V.F.J., Simpson, J.P., Kane, A.D., Menon, D.K., Rueckert, D., Glocker, B.: Efficient multi-scale 3D CNN with fully connected CRF for accurate brain lesion segmentation. Med. Image Anal. **36**, 61–78 (2017)
24. Ronneberger, O., Fischer, P., Brox, T.: U-Net: convolutional networks for biomedical image segmentation. In: Navab, N., Hornegger, J., Wells, W.M., Frangi, A.F. (eds.) MICCAI 2015. LNCS, vol. 9351, pp. 234–241. Springer, Cham (2015). doi:10.1007/978-3-319-24574-4_28
25. Milletari, F., Navab, N., Ahmadi, S.-A.: V-Net: Fully Convolutional Neural Networks for Volumetric Medical Image Segmentation. arXiv, pp. 1–11 (2016)
26. Simard, P.Y., Steinkraus, D., Platt, J.C.: Best practices for convolutional neural networks applied to visual document analysis. In: Proceedings of Seventh International Conference on Document Analysis and Recognition, pp. 958–963. IEEE Computer Society (2003)
27. Long, J., Shelhamer, E., Darrell, T.: Fully convolutional networks for semantic segmentation. In: 2015 IEEE Conference on Computer Vision and Pattern Recognition (CVPR), pp. 3431–3440. IEEE (2015)
28. Drozdzal, M., Vorontsov, E., Chartrand, G., Kadoury, S., Pal, C.: The importance of skip connections in biomedical image segmentation. In: Carneiro, G., et al. (eds.) LABELS/DLMIA -2016. LNCS, vol. 10008, pp. 179–187. Springer, Cham (2016). doi:10.1007/978-3-319-46976-8_19
29. Kingma, D., Ba, J.: Adam: A Method for Stochastic Optimization (2014)

Modeling Diffusion Directions of Corpus Callosum

Safa Elsheikh[1], Andrew Fish[1], Roma Chakrabarti[1], Diwei Zhou[2](✉), and Mara Cercignani[3,4]

[1] School of Computing, Engineering and Mathematics, University of Brighton, Brighton, UK
[2] Department of Mathematical Sciences, Loughborough University, Loughborough, UK
D.Zhou2@lboro.ac.uk
[3] Clinical Imaging Sciences Centre, Department of Neuroscience, Brighton and Sussex Medical School, University of Sussex, Brighton, UK
[4] Neuroimaging Laboratory, Santa Lucia Foundation, Rome, Italy

Abstract. Diffusion Tensor Imaging (DTI) has been used to study the characteristics of Multiple Sclerosis (MS) in the brain. The von Mises-Fisher distribution (vmf) is a probability distribution for modeling directional data on the unit hypersphere. In this paper we modeled the diffusion directions of the Corpus Callosum (CC) as a mixture of vmf distributions for both MS subjects and healthy controls. Higher diffusion concentration around the mean directions and smaller sum of angles between the mean directions are observed on the normal-appearing CC of the MS subjects as compared to the healthy controls.

Keywords: Diffusion tensor imaging · Multiple Sclerosis · von Mises-Fisher · Corpus Callosum · Concentration · Mean directions

1 Introduction

The von Mises-Fisher distribution (vmf) is one of the most basic probability distributions for modeling directional data on the unit hypersphere [1]. The use of a finite mixture of vmf distributions to cluster directional data on the unit sphere has been proposed by [2]. They used an Expectation Maximization (EM) algorithm to estimate the parameters of the mixture model. Much research has been performed to model the diffusion imaging of the brain as a vmf distribution. For example, [3] model the orientation distribution function (ODF) of High Angular Resolution Diffusion Imaging (HARDI) as a mixture of vmf distributions. They also use this model for segmentation using synthetic and real HARDI data. In [4], they proposed a $5D$ hyper spherical model for HARDI data using a mixture of vmf distributions. A method for reconstructing Diffusion Weighted Magnetic Resonance (DW-MR) signal using a continuous mixture of vmf distributions has been introduced in [5]. They validate the method using synthetic

M. Valdés Hernández and V. González-Castro (Eds.): MIUA 2017, CCIS 723, pp. 518–526, 2017.
DOI: 10.1007/978-3-319-60964-5_45

and real brain data. A probabilistic fiber tracking algorithm using a particle filtering technique and vmf sampling has been proposed in [6]. In [7], they build a model of glioma growth using diffusion tensor imaging (DTI) and a bimodal vmf distribution. The vmf distribution, spherical harmonic (SH) expansion and Fractional anisotropy (FA) have been used in [8] for the classification of Alzheimer's disease. Comparing the results, they found that vmf and SH outperform the FA in the classification.

Multiple Sclerosis (MS) is an immune-mediated neurological disease, characterized by recurring inflammatory events (relapsing-remitting MS), associated with demyelinating lesions within the white matter of the brain and spinal cord. The majority of patients after a variable period of relapsing-remitting course develop the so-called secondary-progressive form of MS, characterised by a chronic accumulation of disability, with less and less acute events. Microscopic damage is known to occur also outside of macroscopic lesions, and DTI was found to be sensitive to such abnormalities e.g. [9–14]. The Corpus Callosum (CC) is the largest bundle of commissural fibers in the brain, and damage to it has been associated with increased risk of developing disability in MS. The FA of the CC has been reported to be decreased and the Mean Diffusivity to be increased for MS subjects.

In this paper, we modeled the diffusion directions in the CC as a mixture of vmf distributions. This allows us to determine the distinct mean diffusion directions in the CC. Also it helps to cluster the CC depending on the diffusion directionality. We study the diffusion directions in the whole normal-appearing CC and the lesions are not included. We refer to normal-appearing CC briefly as CC. We briefly review single and finite mixtures of vmf distributions in Sect. 2. In Sect. 3, we directly apply the mixture of vmf to model the CC directions for healthy controls and MS patients. We find that there are at least three distinct mean directions in the three dimensional CC. The results are presented in Sect. 4. The diffusion concentrations around the mean directions are higher for MS subjects. Moreover, the sum of angles between the three mean directions are smaller for MS subjects. We conclude with a brief summary and future work in Sect. 5.

2 Von Mises-Fisher Distribution

The vmf is a probability distribution on the $(d-1)$ dimensional sphere in $\mathbb{R}^d$. The vmf density function of the unit vector $\mathbf{x}$, given the mean direction $\boldsymbol{\mu}$ and the concentration around the mean K, is given by:

$$f(\mathbf{x}|\boldsymbol{\mu}, K) = c_d(K) e^{K \boldsymbol{\mu}^T \mathbf{x}}, \tag{1}$$

where $\|\boldsymbol{\mu}\| = 1$ and $K > 0$. The normalizing constant $c_d(K)$ is given by:

$$c_d(K) = \frac{K^{d/2-1}}{(2\pi)^{d/2} I_{d/2-1}(K)}, \tag{2}$$

where $I_v(.)$ is the modified Bessel function of the first kind at order v. If $d = 2$ this distribution is von Mises distribution on the circle, which is the circular analogue of the normal distribution. If $d = 3$ the normalization constant can be written as

$$c_3(K) = \frac{K}{4\pi \sinh(K)}. \tag{3}$$

The density of finite mixtures H of vmf distributions is given by

$$p(\mathbf{x}|\Theta) = \sum_{h=1}^{H} w_h f(\mathbf{x}|\theta_h), \tag{4}$$

where $\Theta = \{w_1, w_2, \ldots, w_H, \theta_1, \theta_2, \ldots, \theta_H\}$ are the parameters of the mixture density and $f(\mathbf{x}|\theta_h)$ is the density of the vmf distribution with parameter $\theta_h = \{\mu_h, K_h\}$. The water diffusion in the white matter in the brain is parallel to the principal eigenvector of the tensor. In the following Sect., we modeled the principal eigenvectors of the tensors in the CC as a mixture of vmf distributions to study the means and the concentrations of water diffusion in the CC.

3 Modeling the Diffusion Direction of the Corpus Callosum

The data was collected at the Neuroimaging Laboratory, Santa Lucia Foundation in Rome (Italy). The study was approved by Ethics Committee of Santa Lucia Foundation. Written informed consent was obtained from all subjects before entering the study. The data consists of four healthy and nine secondary progressive MS subjects. Diffusion-weighted imaging was obtained using a head-only 3.0 T scanner (Siemens Magnetom Allegra, Siemens Medical Solutions, Erlangen, Germany), using a twice-refocused spin echo echo-planar imaging (SE EPI)

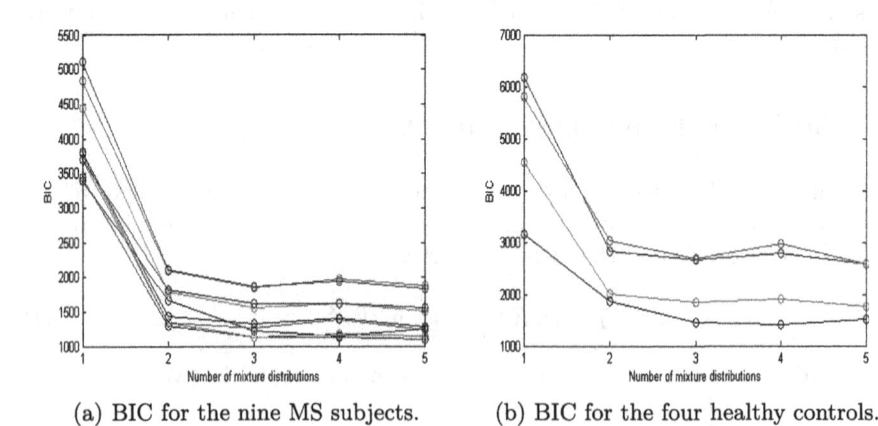

(a) BIC for the nine MS subjects. (b) BIC for the four healthy controls.

Fig. 1. Bayesian information criterion (BIC) for MS subjects and healthy controls.

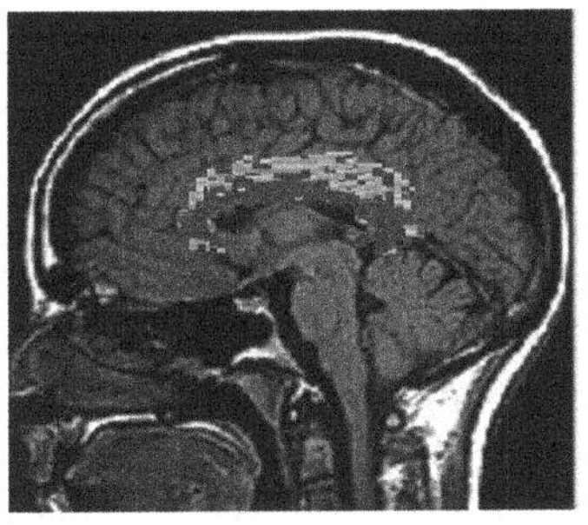

(a) The mixture directions for a MS subject.

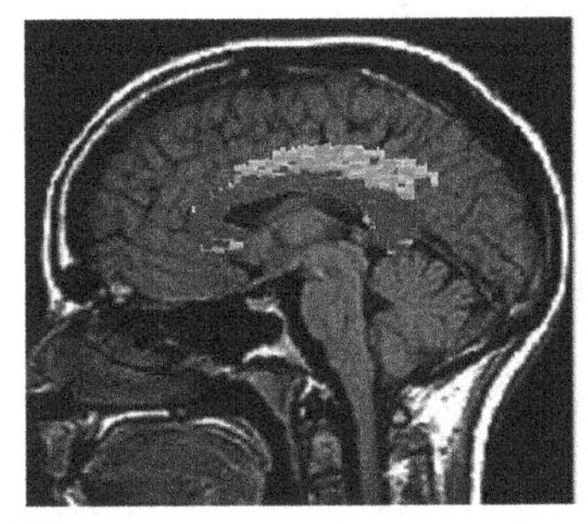

(b) The mixture directions for a healthy control.

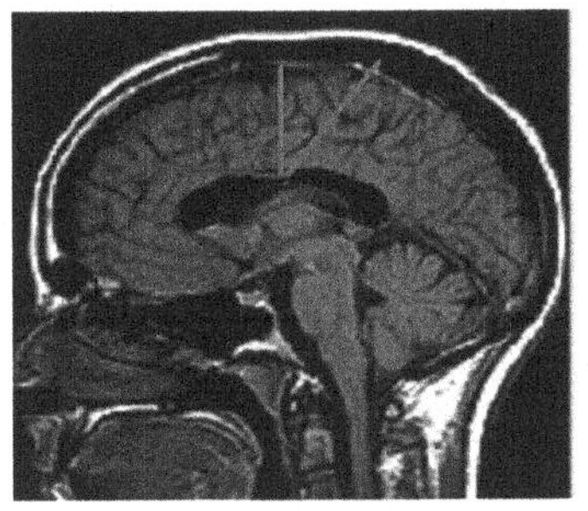

(c) The mean directions for the three mixture distributions in (a). The sum of angles between each two of the mean directions is 77 .99.

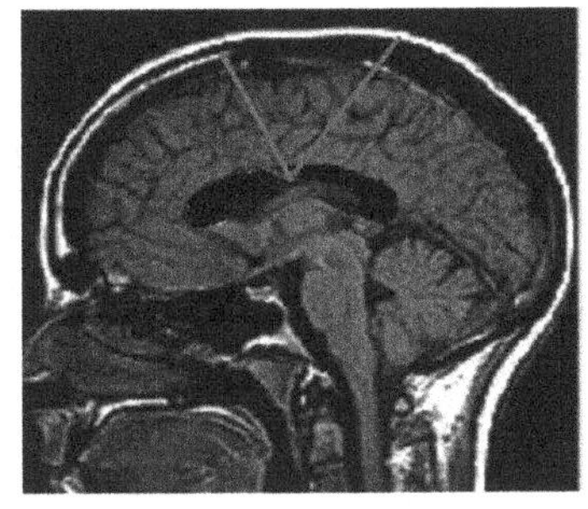

(d) The mean vectors for the three mixture distributions in (b). The sum of angles between each two of the mean directions is 83 .97.

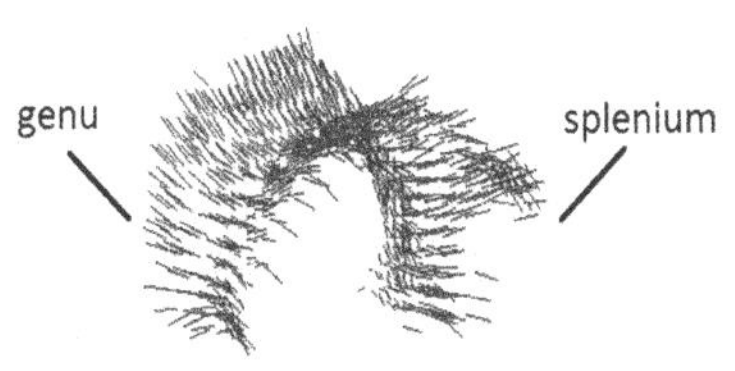

(e) The $3D$ view of the mixture directions in (a). The concentration values for the regions (blue, green, red) are (23.36, 13.48, 20).

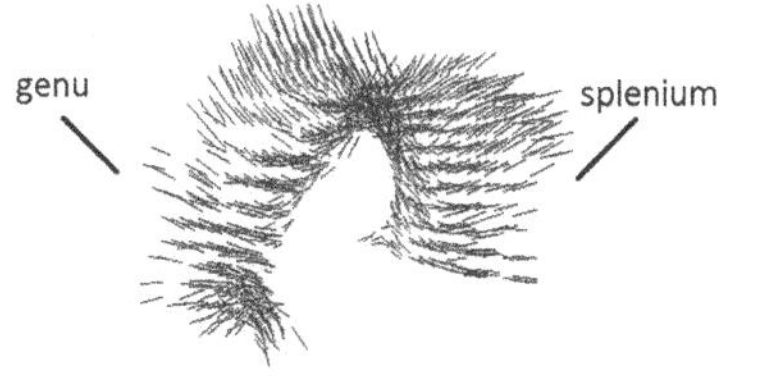

(f) The $3D$ view of the mixture directions in (b). The concentration values for the regions (blue, green, red) are (18.21, 12.02, 14.29).

Fig. 2. The CC is clustered into three regions (blue, green and red) using the mixture of vmf distributions for a MS subject and a healthy control. The sum of angles between each of the three mean directions are smaller for the MS subject. For both the MS subject and healthy control, the diffusion directions on the blue and red regions are more concentrated than on the green region. The diffusion concentrations around the mean directions for MS subject are higher than for the healthy control. (Color figure online)

sequence (TR = 7000 ms, TE = 85 ms, maximum b factor = 1000 s/mm^2, isotropic resolution 2.3 mm^3; matrix = 96 × 96; 60 slices), accomplished by collecting 7 images with no diffusion weighting ($b = 0$) and 61 images with diffusion gradients, applied in 61 non-collinear directions (scan time: 11 min).

The three dimensional CC is segmented using the Euclidean method discussed in [15]. The principal eigenvectors of the diffusion tensors in the CC are parallel to the diffusion directions in the CC. Hence, the unit vector $\mathbf{x}$ represents the principal eigenvector, $\boldsymbol{\mu}$ is the mean diffusion direction and K is the concentration of the diffusion directions around $\boldsymbol{\mu}$. Intuitively, the concentration is the opposite of the variation and spread. As the diffusion tensor is a 3×3 matrix, the principal eigenvector $\mathbf{x}$ of the tensor is a 3 × 1 vector so $d = 3$. Substituting (1) and (3) into (4) gives:

$$p(\mathbf{x}|\Theta) = \frac{1}{4\pi}\sum_{h=1}^{H}\left[\frac{w_h K_h}{\sinh(K_h)} e^{K_h \boldsymbol{\mu}_h^T \mathbf{x}}\right], \tag{5}$$

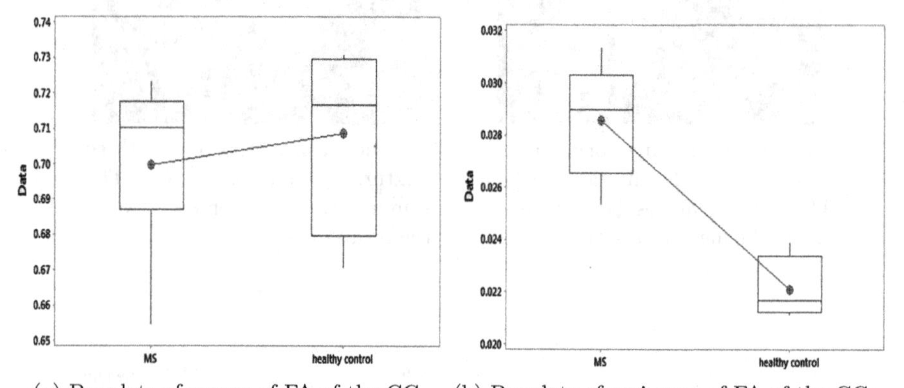

(a) Boxplots of means of FA of the CC. (b) Boxplots of variances of FA of the CC.

Fig. 3. Boxplots of means and variances of FA for MS subjects and healthy controls.

Fitting the mixture of vmf distributions has been performed using the Expectation Maximization (EM) [2]. For implementation of the EM algorithm using R language see [16]. The Bayesian information criterion (BIC) can be used for model selection to choose the most suitable model. The model with the smallest BIC is the preferred model. The BIC is defined as:

$$BIC = h\log(n) - 2\log(L), \tag{6}$$

where h is the number of parameters to be estimated, n is the sample size and L is the maximized likelihood function of the model. The BIC values showed that a mixture of vmf distributions for the CC direction is preferred over a single vmf distribution for both healthy and MS patients (see Fig. 1(b) and (a)).

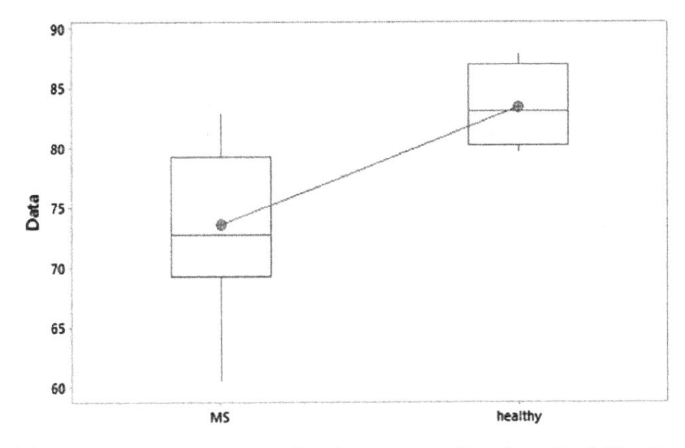

(a) Sum of angles between the three mean directions for MS subjects and healthy controls.

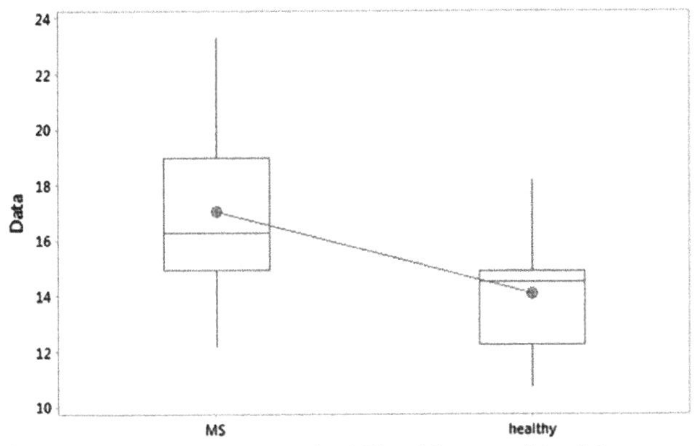

(b) Diffusion concentrations for MS subjects and healthy controls.

Fig. 4. Boxplots of concentrations and sums of angles between the three mean directions.

From the figures, it is clear that the BIC values for three mixture distributions are lower than one and two mixture distributions in all cases for both the MS and healthy controls. Thus, at least three mixtures of vmf distributions are preferred, to model the diffusion in the whole CC. The BIC values for four and five mixture distributions are either bigger or slightly smaller than BIC values for three mixture distributions. Hence, to be able to compare the models for all the subjects with the same number of mixture components, we choose to model the data using three mixtures of vmf distributions (i.e. $H = 3$ in (5)). The mixture distributions of the CC for one healthy control and one MS subject are shown in Fig. 2.

4 Results

Although the average of the FA over the whole CC is higher for the healthy controls than for the MS subjects (Fig. 3(a)), the difference is not significant (p-value = 0.604). However, there is a significant difference in the FA variation (p-value = 0.000) in the CC between the MS and healthy control (Fig. 3(b)). To compare the diffusion direction concentration between MS subjects and healthy controls, we need to model the data (the principal eigenvectors of the tensors in the CC) as vmf distributions. The BIC values show that there are at least three different diffusion directions, so we modeled the data as a mixture of three vmf distributions. We calculate the angles between each two of the three mean directions. As the diffusion is a symmetric process (we cannot distinguish between left to right diffusion and right to left, this means the angle 0 is equivalent to the angle π), the angle is calculated as the minimum of θ and $(\pi - \theta)$ where θ is the angle between two mean vectors and thus we get angles between 0 and $\pi/2$. We find that the sum of the angles between each two of the three mean directions are significantly higher for healthy controls than MS subjects (p-value = 0.007). This result shown in Fig. 4(a). Then we compare the concentration values between MS subjects and the healthy controls (Fig. 4(b)). The diffusion concentrations of MS subjects are significantly higher than the diffusion concentrations of the healthy controls (p-value = 0.001). These might be a result of the atrophy in the CC of MS subjects which leads to decrease the spread of the diffusion directions over the CC.

5 Conclusion

We have used a mixture of vmf distributions to model the diffusion in the three dimensional CC. There are at least three different mean diffusion directions in the CC. On the normal-appearing CC of MS subjects the sum of angles between the mean diffusion directions are smaller with higher concentration around the mean directions compared with the healthy controls. This due to the atrophy in the CC of the MS subjects. Future work will address how these results will be affected when the regions with MS lesions in the CC are included in the analysis. The result in this paper used the Euclidean method for segmenting the CC as it is faster than the segmentation methods using non-Euclidean measures but it is less accurate than other non-Euclidean methods [15]. Future work will also include using the segmentation results obtained by non-Euclidean methods and by other segmentation methods to investigate more about the consistency of the results in this paper when using different segmentation methods. Using the vmf to obtain similar results as Witelson subdivisions of the CC [17] is also of interest. Furthermore, a larger data set will be considered.

References

1. Mardia, K.V., Jupp, P.E.: Directional Statistics, vol. 494. Wiley, Hoboken (2009)
2. Banerjee, A., Dhillon, I.S., Ghosh, J., Sra, S.: Clustering on the unit hypersphere using von Mises-Fisher distributions. J. Mach. Learn. Res. **6**, 1345–1382 (2005)
3. McGraw, T., Vemuri, B., Yezierski, R., Mareci, T.: Segmentation of high angular resolution diffusion MRI modeled as a field of von Mises-Fisher mixtures. In: Leonardis, A., Bischof, H., Pinz, A. (eds.) ECCV 2006. LNCS, vol. 3953, pp. 463–475. Springer, Heidelberg (2006). doi:10.1007/11744078_36
4. Bhalerao, A., Westin, C.-F.: Hyperspherical von Mises-Fisher mixture (HvMF) modelling of high angular resolution diffusion MRI. In: Ayache, N., Ourselin, S., Maeder, A. (eds.) MICCAI 2007. LNCS, vol. 4791, pp. 236–243. Springer, Heidelberg (2007). doi:10.1007/978-3-540-75757-3_29
5. Kumar, R., Barmpoutis, A ., Vemuri, B.C., Carney, P.R., Mareci, T.H.: Multi-fiber reconstruction from DW-MRI using a continuous mixture of von Mises-Fisher distributions. In: IEEE Computer Society Conference on Computer Vision and Pattern Recognition Workshops, CVPRW 2008, pp. 1–8. IEEE (2008)
6. Zhang, F., Hancock, E.D., Goodlett, C., Gerig, G.: Probabilistic white matter fiber tracking using particle filtering and von Mises-Fisher sampling. Med. Image Anal. **13**(1), 5–18 (2009)
7. Painter, K.J., Hillen, T.: Mathematical modelling of glioma growth: the use of diffusion tensor imaging (DTI) data to predict the anisotropic pathways of cancer invasion. J. Theor. Biol. **323**, 25–39 (2013)
8. Reynolds, G.K., Nir, T.M., Jahanshad, N., Prasad, G., Thompson, P.M.: Using the raw diffusion MRI signal and the von Mises-Fisher distribution for classification of Alzheimer's disease. In: 2014 IEEE 11th International Symposium on Biomedical Imaging (ISBI), pp. 1027–1030. IEEE (2014)
9. Werring, D.J., Clark, C.A., Barker, G.J., Thompson, A.J., Miller, D.H.: Diffusion tensor imaging of lesions and normal-appearing white matter in multiple sclerosis. Neurology **52**(8), 1626–1626 (1999)
10. Guo, A.C., MacFall, J.R., Provenzale, J.M.: Multiple sclerosis: diffusion tensor MR imaging for evaluation of normal-appearing white matter 1. Radiology **222**(3), 729–736 (2002)
11. Ge, Y., Law, M., Johnson, G., Herbert, J., Babb, J.S., Mannon, L.J., Grossman, R.I.: Preferential occult injury of corpus callosum in multiple sclerosis measured by diffusion tensor imaging. J. Magn. Reson. Imaging **20**(1), 1–7 (2004)
12. Hesseltine, S.M., Law, M., Babb, J., Rad, M., Lopez, S., Ge, Y., Johnson, G., Grossman, R.I.: Diffusion tensor imaging in multiple sclerosis: assessment of regional differences in the axial plane within normal-appearing cervical spinal cord. Am. J. Neuroradiol. **27**(6), 1189–1193 (2006)
13. Shu, N., Liu, Y., Li, K., Duan, Y., Wang, J., Chunshui, Y., Dong, H., Ye, J., He, Y.: Diffusion tensor tractography reveals disrupted topological efficiency in white matter structural networks in multiple sclerosis. Cereb. Cortex **21**(11), 2565–2577 (2011)
14. Yaldizli, Ö., Pardini, M., Sethi, V., Muhlert, N., Liu, Z., Tozer, D.J., Samson, R.S., Wheeler-Kingshott, C.A.M., Yousry, T.A., Miller, D.H., et al.: Characteristics of lesional and extra-lesional cortical grey matter in relapsing-remitting and secondary progressive multiple sclerosis: a magnetisation transfer and diffusion tensor imaging study. Mult. Scler. J. **22**(2), 150–159 (2016)

15. Elsheikh, S., Fish, A., Chakrabarti, R., Zhou, D.: Cluster analysis of diffusion tensor fields with application to the segmentation of the corpus callosum. Procedia Comput. Sci. **90**, 15–21 (2016)
16. Hornik, K., Grün, B.: movMF: an R package for fitting mixtures of von Mises-Fisher distributions. J. Stat. Softw. **58**(10), 1–31 (2014)
17. Witelson, S.F.: Hand and sex differences in the isthmus and genu of the human corpus callosum. Brain **112**(3), 799–835 (1989)

Feature Extraction and Classification to Diagnose Hypoxic-Ischemic Encephalopathy Patients by Using Susceptibility-Weighted MRI Images

Sisi Wu[1](✉), Sasan Mahmoodi[1], Angela Darekar[2], Brigitte Vollmer[3], Emma Lewis[2], and Maria Liljeroth[2]

[1] Electronics and Computer Science, University of Southampton, University Road, Southampton SO17 1BJ, UK
sw5e12@soton.ac.uk, sm3@ecs.soton.ac.uk
[2] Department of Medical Physics, University Hospital Southampton NHS Foundation Trust, Tremona Road, Southampton SO16 6YD, UK
[3] Clinical Neurosciences, Clinical and Experimental Sciences, University of Southampton, Southampton SO17 1BJ, UK

Abstract. In this paper a method is presented to enable automatic classification of the degree of abnormality of susceptibility-weighted images (SWI) acquired from babies with hypoxic-ischemic encephalopathy (HIE), in order to more accurately predict eventual cognitive and motor outcomes in these infants. SWI images highlight the cerebral venous vasculature and can reflect abnormalities in blood flow and oxygenation, which may be linked to adverse outcomes. A qualitative score based on magnetic resonance imaging (MRI) analyses is assigned to SWIs by specialists to determine the severity of abnormality in an HIE patient. The method allows the detection of image ridges, representing the vessels in SWIs, and the histogram of the ridges grey scales. A curve with only four parameters is fitted to the histograms. These parameters are then used to estimate the SWI abnormality score. The images are classified by using a *k*NN- and multiple SVM classifiers based on the parameters of the fitting curves. The algorithm is tested on an SWI-MRI dataset consisting of 10 healthy infants and 48 infants with HIE with a range of SWI abnormality scores between 1 and 7. The accuracy of classifying babies with HIE vs. those without (i.e.: healthy controls) using our algorithm with a leave-one-out strategy is measured as 91.38%. Our method is fast and could increase the prognostic value of these scans, thereby improving management of the condition, as well as elucidating the disease mechanisms of HIE.

Keywords: HIE · SWI · Soft ridge detection · Histogram · Curve fitting · Automatic diagnosis

1 Introduction

Hypoxic-ischemic encephalopathy (HIE) is a type of neurological impairment or injury in infants caused by low oxygen during the perinatal period, and it is a long-term, frequent, severe disease [1, 2]. In the developed world, HIE is a common illness affecting

M. Valdés Hernández and V. González-Castro (Eds.): MIUA 2017, CCIS 723, pp. 527–536, 2017.
DOI: 10.1007/978-3-319-60964-5_46

approximately 1.5–2 patients per 1,000 new-born infants per year [3]. Infants with moderate encephalopathy have a 10% risk of death, and those who survive have a 30% risk of disability. Sixty percent of infants with severe encephalopathy die, and many, if not all, survivors are handicapped [4, 5]. Hence, specialists need to diagnose the infant in good time and look at the severity of abnormality. So far, MRI plays an important role in HIE diagnosis. An MR image can confirm a normally developed brain, assess the severity and pattern of any neonatal injury, or even potentially predict the outcome from the pattern of injury and clinical details. Susceptibility weighted imaging (SWI) is a type of MR image; an SWI image highlights the cerebral venous vasculature, reflecting abnormalities in blood flow and oxygenation which may be linked to adverse outcomes. A qualitative score based on MRI analyses can be assigned to SWIs by specialists to provide an assessment of the degree of severity of an abnormality in an HIE patient [19].

In previous research, numerous studies have focused on the cerebral vessel structure and vessel detection. In the medical research, a number of studies on HIE diagnosis have focused on Diffusion Weighted Imaging (DWI) [6, 7, 21] and Electroencephalogram (EEG) diagnosis [8, 9]. In [21], DWIs are acquired for the apparent diffusion coefficient (ADC) map to evaluate the quantification of ischemic injury by employing the watershed and superpixel methods. This is considered the first step of diagnosis in which the 2D slices of DWIs are inspected. The main issue in the work [21] is that the number of abnormal cases is limited to three patients. However the score predicted by the method proposed in [3] has a 0.84 correlation coefficient. The other type of MRIs, which is T1W1 and proton MR spectroscopy is useful to distinguish between HIE and normal infants, However the study presented in [20] is performed manually on MRIs. In my opinion, these two areas have been not evaluated on large datasets or fail to give a good result for the largest number of images in screening processing. To the best of our knowledge, no attempt has been made to automate the HIE diagnosis so far in the medical image processing community. Some approaches are focused on ridge detection of the vessels' centreline and rebuilt the vessel 3D structure for the detection of the sickle cell disease [10]. Some methods are used to develop the application of Hessian matrix eigenvalues and eigenvectorsto find the ridge-valley inflection points and the application of eigenvalues [11]. The aforementioned methods are not automated to provide advice to medical personnel.

In this paper, a method is proposed to aid diagnosis in infants who have HIE automatically by processing their SW images. In our work, we extract features from SW images to classify babies into healthy and HIE infants. Our feature extraction technique proposed here is based on the fact that the appearance of anoxic vessels is different from the healthy vessels in that the anoxic vessels look darker than the healthy ones. Here a ridge detection method is applied to detect the vessels. A pixel on the detected ridges (vessels) is then assigned a value proportional to the eigenvalue of Hessian matrix in that pixel to obtain a soft ridge map of the SWI. The histogram of such a soft ridge map for each infant's SWIs is then calculated. A curve characterized by four parameters is also fit to the soft ridge histograms to avoid overfitting issues. In our method, SW images are classified based on the parameters of this curve. Our method provides a reliable diagnosis without the need for specialists to spend a great deal of time to examine SWIs. The diagnosis provided by our system is also free from human error.

2 Methodology

2.1 Ridge Detection

Ridge detection in this research is used to define the centreline of deoxygenated vessels. We apply a local median filter and Gaussian filter to remove the noise first. A median filter removes the noise and preserves most of the original features of SWI. As the SWIs are three dimensional images consisting of two dimensional slices, we, therefore, start our analysis of ridge detection for a two dimensional grey scale image. This ridge detection method is based on the retinal vessel segmentation method [12–14, 18].

Let us denote $I(\mathbf{x})$ with $\mathbf{x} = (x, y)^T$ to be the SW image, where the first derivative of the image $I(\mathbf{x})$ in the direction perpendicular to the ridge tangent has a zero-crossing. Because imagescontain noisea Gaussian filter with a standard deviation of σ is convolved with the image derivative, i.e.: [12].

$$I(x_i) = \frac{\partial I(\mathbf{x}, \sigma)}{(\partial x_j)} = \frac{1}{2\pi\sigma^2} \int_{\mathbf{x}' \in \mathbb{R}^2} \frac{\partial e^{-} \parallel \mathbf{x} - \mathbf{x}' \parallel^2 / 2\sigma^2}{\partial x_j} L(\mathbf{x}') d\mathbf{x}' \tag{1}$$

where x_j is the image coordinate with respect to which the derivative is taken. The higher order derivatives could be computed by taking a higher order derivative of the Gaussian kernel.

The ridge point is detected by setting pixels corresponding to local minima (valley ridges) to −1 pixels corresponding to local maxima (hill ridges)to +1 and 0 otherwise by using the following equation in a simple ridge detector [12]:

$$R(\mathbf{x}, \sigma) = -\frac{1}{2} sign(\lambda) |sign(\nabla I(\mathbf{x} + \hat{\mathbf{v}}, \sigma) \cdot \hat{\mathbf{v}}) - sign(\nabla I(\mathbf{x} - \hat{\mathbf{v}}, \sigma) \cdot \hat{\mathbf{v}})| \tag{2}$$

where the gradient operator ∇ is defined as $(\partial/\partial x, \partial/\partial y)^T$ and $\lambda(\mathbf{x}, \sigma)$ is the largest absolute value of the eigenvalue for Hessian matrix H, in each pixel as calculated in Eq. (3). Eigenvalue λ and $\hat{\mathbf{v}}$ are then evaluated at $(\mathbf{x}, \sigma)$. Ridge map R calculated in Eq. (2) contain both valley and hill ridges. Such a ridge map is known as hard ridge since the ridge location is specified in the map however the ridge magnitudes are ignored in this ridge map.

$$H = \begin{bmatrix} I_{xx} & I_{xy} \\ I_{xy} & I_{yy} \end{bmatrix} \tag{3}$$

where $I_{ab} = \frac{\partial^2 I}{\partial a \partial b}$.

Since we are interested in only vessels corresponding to valley ridges, we aim to construct a ridge map containing only valley ridges. SWIs of the deoxygenated infants demonstrate darker and wider vessels corresponding to deeper valleys. It is, therefore, interesting to know how deep or shallow a valley (vessel) is for the diagnosis of HIE. To this end, the absolute maximum eigenvalue of the Hessian matrix associated with each pixel on the ridge is calculated and then the corresponding pixel value in the ridge map is set to the absolute maximum eigenvalue. Otherwise, the pixel values of the ridge map in non-ridge locations are considered zero. The eigenvalues of a Hessian matrix on a ridge point is proportional to how deep the valley ridge (vessel) is, and the direction of the eigenvector on that ridge point is perpendicular to the ridge tangent [13]. The ridge map produced with the aforementioned process is known as soft ridge map as calculated by Eq. (4).

$$R_{Soft\,Ridge}[\mathbf{x}] = \begin{cases} \lambda_i(\mathbf{x}) & \max_i(|\lambda_i(\mathbf{x})|)\; i = 1,2 \\ 0 & otherwise \end{cases} \tag{4}$$

where λ_1, λ_2 are the eigenvalues of the Hessian matrix.

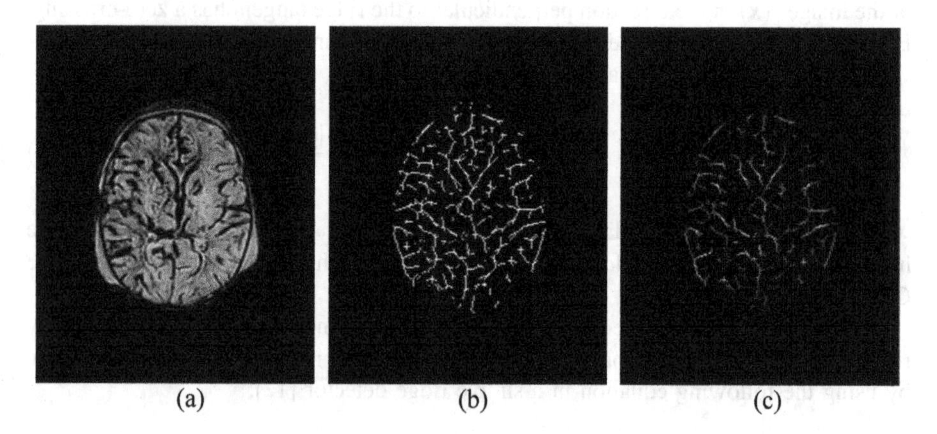

Fig. 1. Ridge maps for an SW image of an infant with HIE. (a) Original SW image smoothed with a Gaussian kernel, $\sigma = 2.0$ pixels. (b) The hard ridge corresponding to local minima ridges. (c) The soft ridge map.

By comparing with abnormal (Fig. 1-c) and normal (Fig. 2-c) soft ridge maps, the soft ridge maps of an SWI indicates the presence and severity of the HIE. Therefore, the soft ridge detection is an effective method to detect and determine the presence and severity of deoxygenated brain in SWIs.

2.2 Feature Extraction

The deoxygenated vessels in the cerebrum of SWIs are used to produce the soft ridge maps as explained in the previous section. The histograms of soft ridge maps are initially calculated. In these histograms, we ignore the pixels with grey scale zero, because they correspond to the background and are of no importance. The soft ridge maps are calculated for all slices of SW images of an infant. Then the histogram of the soft ridge maps for all slices is computed for an infant. In fact, this histogram is considered a feature representing all slices of SWIs of an infant. Such a histogram is shown in Fig. (3-a) for an infant suffering from HIE. The histogram for a healthy baby is also depicted in Fig. (3-b). In these figures, the bins start from 0.00001 with the bin

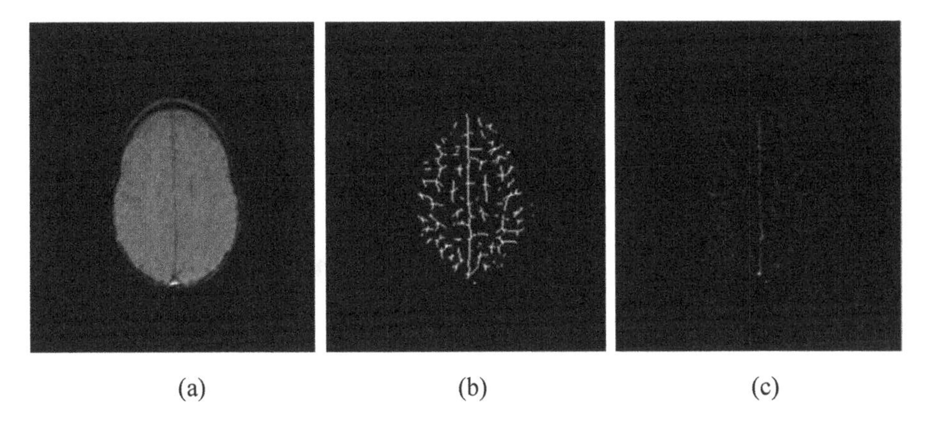

Fig. 2. Ridge maps for an SW image of a healthy infant. (a) Original SW image smoothed with Gaussian kernel, $\sigma = 2.0$ pixels. (b) The hard ridge corresponding to local minima ridges. (c) The soft ridge map.

size of 0.001. In our numerical experiments, the maximum value of these histograms in our dataset is around 30. In our dataset, there are 10 healthy infants and 48 infants suffering from HIE with various scores of severity. In order to avoid overfitting problems, we fit a curve whose function is presented in Eq. (5). This function is characterised with four parameters a, b, c, n, i.e.:

$$D = \frac{a}{\sqrt{x^n}} e^{-(bx + c/x)} \quad (5)$$

where x is the grey scale of the soft ridge map (the horizontal axis in Fig. 3).

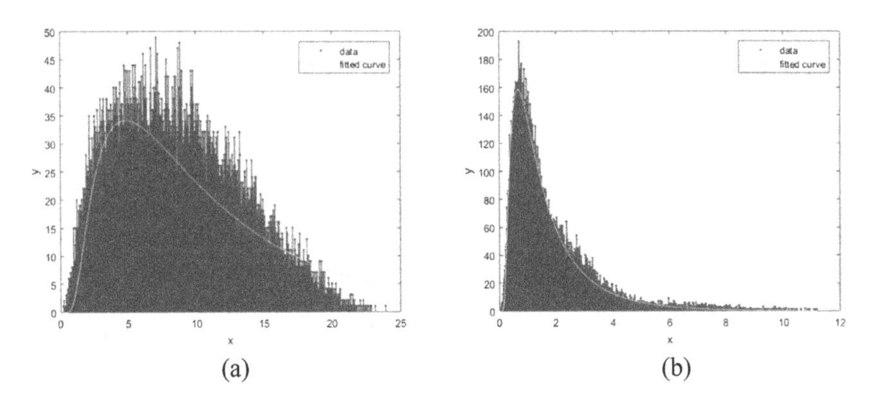

Fig. 3. (a) Histogram with a fitting curve of the infant of Fig. 1. (b) Histogram with a fitting curve of the infant of Fig. 2. The yellow curves are the fitting curving using Eq. (5). (Color figure online)

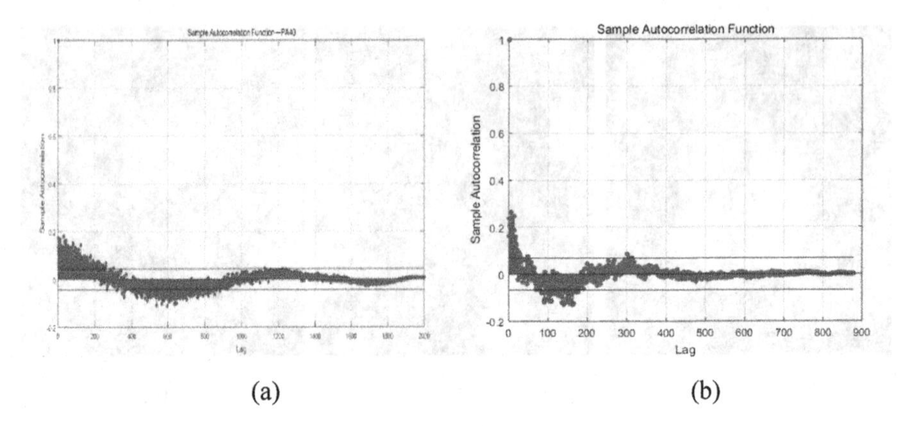

Fig. 4. The autocorrelation plot of the errors between the histogram and fitting curve. (a) The autocorrelation plot for a patient with HIE. (b) The autocorrelation plot for a healthy infant. The red points illustrate the correlation error between the histogram and the fitting curve. Two parallel lines on the both side of the coordinate axis are the error bounds, which have a default value of 2 standard deviations. (Color figure online)

Equation (5) is fit to the histograms by using a nonlinear least square technique with 95% confidence bounds. The results of the fitting process are also presented in Fig. 3 for both histograms. The autocorrelation of the errors between the actual histogram data and the fitted curve is also depicted in Fig. (4-a) can be observed from Fig. 4, most of the autocorrelation data is random, however, there is a pattern inside the autocorrelation function. This is due to that fact that function proposed in Eq. (5) can be modified for a better fit until the autocorrelation function contains only random data. Since the modification of Eq. (5) is an intuitive process which is based on trial and error, we leave this for our future work. These autocorrelations are calculated with the estimation error bounds 2 standard deviations away from zero which corresponds to 95% confidence bounds.

3 Results

The deoxygenated vessels in SW images are detected with the ridge detection method discussed in Sect. 2. It is, therefore, important to evaluate the performance of our ridge detection algorithm by plotting its receiver operating characteristic (ROC) curve [15, 16], before the feature extraction process. In our data set, there are over 2400 SW 45images, which come from 58 infants and each infant dataset contains more than 40 slides. We manually detect the ridges for 200 slices of these SW images and also apply our ridge detection algorithm to these 200 slices of SW images for comparison purposes. The ridges are manually annotated to compare the annotated ridges with the ridges detected by our algorithm to measure the performance of our algorithm using ROC curves. We have applied Gaussian and local mean filters on the SW images prior to the ridge detection. The performance evaluations of our method are plotted as ROC curves shown in Fig. 5.

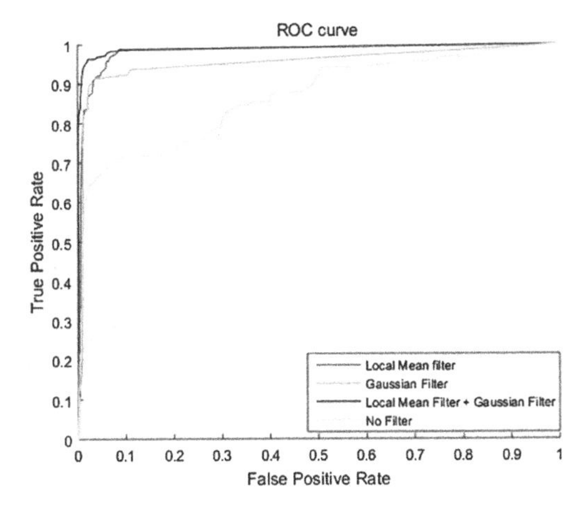

Fig. 5. ROC curves to evaluate the performance of our ridge detection algorithm when various filtering schemes are used before the ridge detection. The cyan line is the ROC curve corresponding to the case no filter is used. The green ROC curve is when the original SW images are filtered with a Gaussian kernel with the standard deviation of 2. The red ROC curve is associated with the case where a local 3×3 mean filter is employed prior to ridge detection. The black ROC curve represents the performance of our ridge detection when both filters (Gaussian and mean filters) are applied to SW images. (Color figure online)

The ROC curve is created by plotting the rate of the manually detected vessel points (true positive rate) against the rate of the auto-detected vessel points which are not detected by manual work (false positive rate). The cyan ROC curve corresponds to the case where no filter is applied to the original SW images. The green ROC curve is associated with the case where the Gaussian filter with standard deviations of 2 is applied to the SW images prior to the ridge detection. The red ROC curve is obtained by employing a local 3×3 mean filter. Finally, the black ROC curve corresponds to the case where both Gaussian and the local mean filters are employed. As can be seen from Fig. 5, the best performance is achieved, when the Gaussian filter is applied before the local mean filter on SWI slices prior to ridge detection process.

In the next experiment, we would like to classify the infants whose SWIs are in our dataset, into two groups: healthy infants and infants with HIE. Initially, the histograms of the soft ridge maps are recalculated with 24 bins. Therefore there are 24 features (parameters) for each histogram. These features (parameters) are then fed into various kinds of classifiers presented in Table 1.

As can be seen from this table, the SVM classifier with a polynomial kernel with order 2 produces the best classification accuracy of 94.83% and the cross-validation rate of approximately 5%. It is important to note that every infant in our dataset is already diagnosed by specialists to be either healthy or with HIE illness. To calculate, the classification accuracies presented in Table 1, a leave-one-out strategy in which the data of an infant is taken out as a test sample without any knowledge of the specialists' diagnosis, is employed. Leave-one-out strategy is a type of cross validated, which the predicted infant's data is picked out of the validation dataset (soft ridge map dataset of whole infants).

Table 1. Classification accuracy with 24-bin histograms to represent SWIs

Classification method	Branch method and the coefficient		Accuracy result
*k*NN	$k = 1$		89.66%
	$k = 3$		91.83%
	$k = 5$		91.83%
	$k = 7$		89.66%
SVM	Gaussian kernel linear		87.93%
			86.21%
	Polynomial (Kernel order: p)	$p = 2$	94.83%
		$p = 3$	93.10%
		$p = 4$	91.83%

Having fit the curve (Fig. 5) to the histograms of the SWIs in our dataset, every infant is represented by a feature vector containing the values of the four parameters a, b, c and n. These vector features are then fed to *k*NN and SVM classifiers with various classification parameters and various kernels to produce the accuracy classification results presented in Table 2. As explained before for Table 1, a leave-one-out strategy is also employed here to calculate the classification accuracies. The best classification result achieved with the curve fitting method is 91.38% which is with an SVM classifier with a polynomial kernel of order 3. It is important to note that the fitting curve method may be characterised with a slightly lower accuracy result but it avoids the problems related to overfitting issues from which the first method associated with 24 bin histograms, may suffer.

To evaluate the classifiers' performance, the precision (specificity) and recall (sensitivity) values of the HIE infants is presented in Table 3. Table 3 is clearly and directly to show the advantages of our classification features and classifier, which shows the sensitivity and specificity of this automatic algorithm. The method presented in [21], has a sensitivity and specificity of 0.72 and 0.99 respectively. Therefore the performance of our algorithm is better than the method presented in [21] in terms of sensitivity and specificity values.

Table 2. Classification accuracy with parameters of Eq. (5) chosen as features to represent SWIs

Classification method	Branch method and the coefficient		Accuracy result
*k*NN	$k = 1$		84.48%
	$k = 3$		89.66%
	$k = 5$		86.20%
	$k = 7$		84.48%
SVM	Gaussian kernel linear		87.93%
			81.03%
	Polynomial (Kernel order: p)	$p = 2$	87.93%
		$p = 3$	91.38%
		$p = 4$	84.48%

Table 3. The precision and recall values for the classifiers employed in Table 2.

Classification method	Branch method and the coefficient		Precision value	Recall value
kNN	$k = 1$		92%	93.88%
	$k = 3$		91.84%	91.84%
	$k = 5$		90.57%	100%
	$k = 7$		90.38%	100%
SVM	Gaussian kernel linear		87.27%	100%
			85.71%	100%
	Polynomial (Kernel order: p)	$p = 2$	94.83%	95.92%
		$p = 3$	93.10%	92.31%
		$p = 4$	91.83%	92%

4 Conclusion and Future Work

The method in this paper presents an automatic diagnosis of HIE based on SWI-MRIs. Unlike the previous ridge detection methods, the eigenvalues of ridge pixels intuitively determine if an infant suffers from HIE or not. The image containing the eigenvalues of the Hessian matrix for all ridge pixels is known as the soft ridge map. The soft ridge maps consist of the vessels detected as ridges with eigenvalues of the Hessian matrix calculated for each ridge pixel. In the method proposed here, we find the features as histograms of soft ridge maps to represent each infant for classification. To reduce the dimensionality of our feature space, we have used two techniques: (1) simply reducing the number is bins into 24 and (2) exploiting the parameters of the curve fitted to the histograms. In the second technique, after the soft ridge detection, the histogram of each patient's soft map is represented by a curve. This curve has been optimally fitted to that histogram. Such a curve is described by four parameters which are used as features for classification stage. Then these features (each consisting of four parameters) are used to classify infants as healthy or a baby suffering from HIE, with kNN and SVM classifiers. The SVM classifier with polynomial kernel of order 2 presents the best accuracy rate of 91.38%. A leave-one-out strategy is employed for the measurements of classification accuracies. Our automatic classification method can, therefore, produce reliable diagnosis without the specialists' interventions and inspections. As a result, our method is faster than a specialist for diagnosis, is not prone to human errors and would save the specialists' time taken for diagnosis. Our method could increase the prognostic value of these SWIs, thereby it improves the management of the illness as well as elucidates the disease mechanisms. This method can also be extended to estimate different MR scores [17, 19] assigned by specialists to HIE infants. Such an extension is an interesting topic for our future work.

Acknowledgment. The research database is provided by Dr. Brigitte Vollmer and Dr. Angela Darekar, Southampton General Hospital.

References

1. Perlman, J.M.: Summary proceedings from the neurology group on hypoxic-ischemic encephalopathy. Pediatrics **117**(Suppl. 1), S28–S33 (2006)

2. Fatemi, A., Wilson, M.A., Johnston, M.V.: Hypoxic-ischemic encephalopathy in the term infant. Clin. Perinatol. **36**(4), 835–858 (2009)
3. James, A., Patel, V.: Hypoxic ischaemic encephalopathy. Paediatr. Child Health **24**(9), 385–389 (2014)
4. Shankaran, S., et al.: Effect of depth and duration of cooling on deaths in the NICU among neonates with hypoxic ischemic encephalopathy: a randomized clinical trial. Jama **312**(24), 2629–2639 (2014)
5. Friberg, H., Cronberg, T.: Hypoxic–ischemic encephalopathy. In: Seminars in Neurology, vol. 37. no. 01. Thieme Medical Publishers (2017)
6. Ou, C.-X., Xiao, F.-Y., Sun, D.-C.: The value of diffusion weighted imaging in the early diagnosis and prognostic evaluation of neonatal HIE. Chin. J. CT MRI **1**, 011 (2013)
7. Bozzao, A., et al.: Diffusion-weighted MR imaging in the early diagnosis of periventricular leukomalacia. Eur. Radiol. **13**(7), 1571–1576 (2003)
8. Massaro, A.N., et al.: Short-term outcomes after perinatal hypoxic ischemic encephalopathy: a report from the children's hospitals neonatal consortium HIE focus group. J. Perinatol. **35** (4), 290–296 (2015)
9. Vergales, B.D., et al.: Depressed heart rate variability is associated with abnormal EEG, MRI, and death in neonates with hypoxic ischemic encephalopathy. Am. J. Perinatol. **31**(10), 855–862 (2014)
10. Winchell, A.M., et al.: Evaluation of SWI in children with sickle cell disease. Am. J. Neuroradiol. **35**(5), 1016–1021 (2014). Hladůvka, J., König, A., Gröller, E.: Exploiting eigenvalues of the Hessian matrix for volume decimation (2001)
11. Bofill, J.M., Quapp, W.: Analysis of the valley-ridge inflection points through the partitioning technique of the Hessian eigenvalue equation. J. Math. Chem. **51**(3), 1099–1115 (2013)
12. Staal, J., et al.: Ridge-based vessel segmentation in color images of the retina. IEEE trans. Med. Imaging **23**(4), 501–509 (2004)
13. Annunziata, R., et al.: Leveraging multiscale hessian-based enhancement with a novel exudate inpainting technique for retinal vessel segmentation. IEEE J. Biomed. Health Inform. **20**(4), 1129–1138 (2016)
14. Wang, Y., et al.: Retinal vessel segmentation using multiwavelet kernels and multiscale hierarchical decomposition. Pattern Recogn. **46**(8), 2117–2133 (2013)
15. Hajian-Tilaki, K.: Receiver operating characteristic (ROC) curve analysis for medical diagnostic test evaluation. Caspian J. Intern. Med. **4**(2), 627 (2013)
16. Hand, D.J.: Measuring classifier performance: a coherent alternative to the area under the ROC curve. Mach. Learn. **77**(1), 103–123 (2009)
17. Barkovich, A.J., et al.: Prediction of neuromotor outcome in perinatal asphyxia: evaluation of MR scoring systems. Am. J. Neuroradiol. **19**(1), 143–149 (1998)
18. Chen, D., Cohen, L.D.: Automatic tracking of retinal vessel segments using radius-lifted minimal path method. In: MIUA 2015 (2015)
19. Kitamura, G., et al.: Hypoxic-ischemic injury: utility of susceptibility-weighted imaging. Pediatr. Neurol. **45**(4), 220–224 (2011)
20. Guo, L., Wang, D., Bo, G., Zhang, H., Tao, W., Shi, Y.: Early identification of hypoxic-ischemic encephalopathy by combination of magnetic resonance (MR) imaging and proton MR spectroscopy. Exp. Ther. Med. **12**(5), 2835–2842 (2016). doi:10.3892/etm.2016.3740
21. Murphy, K., van der Aa, N.E., Negro, S., Groenendaal, F., de Vries, L.S., Viergever, M.A., Boylan, G.B., Benders, M.J.N.L., Išgum, I.: Automatic quantification of ischemic injury on diffusion-weighted MRI of neonatal hypoxic ischemic encephalopathy. NeuroImage: Clin. **14**, 222–232 (2017). doi:10.1016/j.nicl.2017.01.005

Evaluation of an Automatic ASPECT Scoring System for Acute Stroke in Non-Contrast CT

Matt Daykin[1,2](✉), Erin Beveridge[1], Vismantas Dilys[1], Aneta Lisowska[1,2], Keith Muir[3], Mathini Sellathurai[2], and Ian Poole[1]

[1] Toshiba Medical Visualization Systems Europe Ltd., Edinburgh, UK
MDaykin@tmvse.com
[2] School of Engineering and Physical Sciences, Heriot-Watt University, Edinburgh, UK
[3] Institute of Neuroscience and Psychology, University of Glasgow, Glasgow, UK

Abstract. Determining the severity of ischemic stroke in non-contrast CT is a difficult problem due to a low signal to noise ratio. This leads to variable interpretation of ischemic stroke severity. We investigate the level of agreement between four methods including the use of an automated system with the aim of identifying early ischemic changes within the brain. For the evaluation we divide the middle cerebral artery territory of each hemisphere into ten regions defined according to the Alberta Stroke Programme Early CT Score (ASPECTS). The automatic system uses a specialised Convolutional Neural Network (CNN) based regressor to produce voxel-level confidence masks of which voxels are suspected as showing early ischemic change and from this we compute the score. Additionally, we obtain the score from three other methods that involved trained human graders. We compare the level of agreement between these methods at both a patient level and a territory level through Simultaneous Truth and Performance Level Estimation (STAPLE) and Cohen's kappa coefficient. We analyse possible causes of disagreement between the methods and statistically validate the performance of the CNN model against the performance of clinical staff. We find that the CNN produces scores that correlate the greatest with its training data at the patient level, but the training data could be improved to strengthen the correlation with the professional standard.

1 Introduction

During an ischemic stroke, blood supply is lost to a region of the brain. This requires immediate medical attention as the affected tissue will die if blood flow is not promptly restored. To select the appropriate treatment for a patient exhibiting stroke symptoms, the cause of the symptoms must be identified - ischemic stroke, hemorrhagic stroke, or stroke mimic - and the severity determined [11]. Non-Contrast CT (NCCT) is a widely available and rapid means of imaging the brain that is frequently used as the first step in identifying the stroke cause and classifying its severity [10]. However, early ischemic changes are challenging to detect and the boundary between affected and normal brain

M. Valdés Hernández and V. González-Castro (Eds.): MIUA 2017, CCIS 723, pp. 537–547, 2017.
DOI: 10.1007/978-3-319-60964-5_47

tissue is poorly defined [18]. A range of features with different pathophysiological bases may present including hyperdense vessels, loss of the insular ribbon, obscuration of the lentiform nucleus, loss of grey-white matter differentiation, sulcal effacement, and hypoattenuation [6]. Conflation of these features contributes to variable interpretation of stroke severity [5,16]. This makes it difficult to compare the performance of an external automated system for detecting early ischemic changes to clinical diagnosis.

The Alberta Stroke Programme Early CT Score (ASPECTS) is a system of scoring stroke severity that concentrates on hypoattenuation as this is the most relevant indication of the severity of ischemia [15]. ASPECTS determines the presence or absence of acute hypoattenuation in ten regions within the middle cerebral artery territory of each hemisphere and can influence treatment choice [1]. The score starts at 10 for a normal brain volume and reduces by one point for each affected territory in the symptomatic hemisphere. Patients with a score of 7 and higher are more likely to respond positively to treatment than patients with a lower score [7].

There are many ways to arrive at the ASPECTS ranging from visual observation to precise segmentation of the lesion to automated systems. This paper presents a statistical comparison and discussion of four methods of determining ASPECTS for ischemic stroke using Simultaneous Truth and Performance Level Estimation (STAPLE) [19] and Cohen's kappa [2]. These methods consist of three based on manual reading, and one using an automatic method. The performance is judged at a dichotomised patient level and an ASPECTS territory level.

2 Materials

We used data from three previous studies on patients with suspected stroke [8,14,17] provided by the Institute of Neuroscience and Psychology, University of Glasgow, Southern General Hospital. Specifically, we made use of the NCCT volumes, radiology reports, and ASPECTS for 156 patients. The NCCT scan was taken within 6 h of symptom onset in all cases. The ASPECTS provided by these studies represents the professional standard.

Further to this ASPECTS we have calculated the score via three other means, leading to four methods of determining ASPECTS for each patient:

1. Clinical studies.
2. Manual segmentation of ischemic and non-ischemic regions.
3. Visual inspection of the NCCT volumes.
4. Automatic segmentation from a Convolutional Neural Network (CNN) algorithm.

These scores are referred to as the Clinical score, Ground Truth score, Observed score, and CNN score respectively throughout. Details of each method follows.

Clinical Score: These scores were generated as part of the aforementioned studies. ASPECTS was calculated by two clinicians reviewing the data independently. A third person resolved discrepancies. The experience and background of the reviewers varied.

Ground Truth Score: An in-house member of staff, who we will refer to as the clinical expert throughout, with 16 h of training in detecting stroke signs segmented regions of acute ischemia in the NCCT volumes under the supervision of an experienced neuroradiologist. The clinical expert additionally created an atlas volume of the twenty ASPECTS territories. These two tasks were completed in 3D Slicer 4.5.0 [4]. The atlas was rigidly aligned to each ground truth volume via landmark registration using automatic landmarks [3] to identify the territories for that patient.

To convert the ground truth to ASPECTS two rules were applied. The first rule - the 5% rule - stated that any territory with ground truth ischemia present in more than 5% of its volume is classified as ischemic. The second rule - the small volume rule - was applied if no territories met the 5% rule but there was ground truth ischemia present in at least one territory. This rule stated that the territory with the highest amount of ischemia present in it, and only this territory, is classified as ischemic, resulting in a score of 9. These rules were based on work by Kosior et al. [9] in the absence of an accepted standard for calculating ASPECTS.

Observed Score: Several months after completing the ground truth segmentations, the clinical expert visually examined each of the data volumes by using 5 mm slabbed slices and noted which territories presented acute ischemia. The ASPECTS came directly from the number of affected territories.

CNN Score: A CNN architecture based on previous work by Lisowska et al. [12, 13] was used to produce a voxel-level confidence mask of early ischemic change in each volume. The CNN featured two identical intensity channels consisting of a series of three orthogonal one-dimensional convolutions with kernal size of $9 \times 1 \times 1$, $1 \times 9 \times 1$, and $1 \times 1 \times 9$ and 16 kernals. The two channels then merged, and a final repeat of the convolutions was completed with two kernals. The performance of the CNN improved with the number of samples used. We used samples of around 20 million voxels. Voxels were given positive confidences if ischemia was suspected, and negative confidences if no ischemia was suspected. There was no upper or lower bound on the confidence values.

The ground truth segmentations were used to train and validate the CNN with five-fold validation. For each fold the model was trained with 80% of the total data as training data and 20% as testing data, with 40% of the training data used as a validation set. The data volumes were randomly divided into these groups at the start of the five-fold validation, and the volumes in the train and test sets were cycled after each fold such that each volume appeared in the test set once. The ASPECTS atlas was aligned via landmarks to the voxel confidence masks to provide the ASPECTS territories; Fig. 1 demonstrates this with an example detection and ground truth.

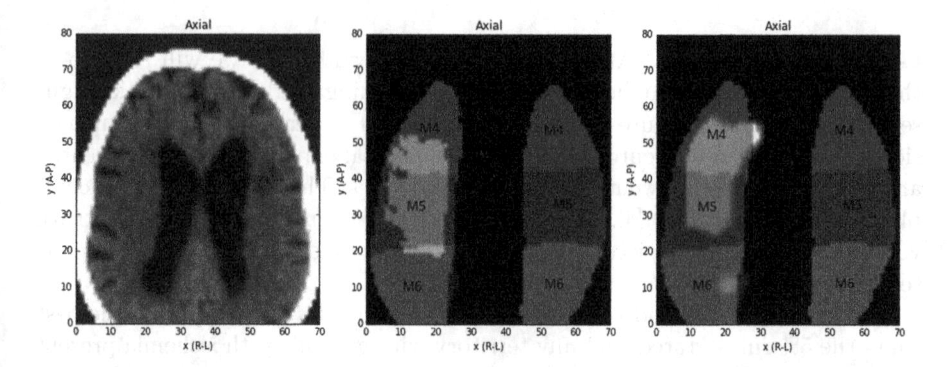

Fig. 1. Left: NCCT volume slice of one of the datasets. Middle and right: the ASPECTS atlas for the shown NCCT slice with the M4–M6 territories marked for each hemisphere. The brighter region in the left of the images shows segmented ground truth ischemia by the clinical expert (middle) or suspected ischemic voxels from the CNN (right) for that slice.

The voxel confidences were converted into ischemic territories via Eq. 1, which averages across the suspected ischemic (positive) voxels in a territory and applies a threshold to determine if this average is sufficient for the region to be classed as ischemic. The threshold was selected by ROC analysis to balance false positives with false negatives relative to the ground truth in the training set, resulting in a score that is unbiased across a large number of datasets.

$$\text{Territory Ischemia} = \text{Present if } \left(\frac{1}{|D|} \sum_{(i \in D \cap c_i > 0)} c_i \right) \geq \text{t} \tag{1}$$

where D represents the set of voxels comprising the region, c_i is the confidence score of voxel i, and t is the threshold.

We have dichotomised the four scores at ≥7 for all the methods. A dichotomised ASPECTS can be clinically useful to weight treatment decisions and the use of a dichotomised score is increasingly common in clinical trials [15]. For all the methods except the Clinical score we also have a territory breakdown of where the acute ischemia is located. This level of detail was not provided with the Clinical score. We refer to this as the territory data.

3 Evaluation Procedure

We cannot take any of these methods as a definitive gold standard. Hence, different statistical measures are required to compare the overall accuracy of each method. We applied two measures: Cohen's kappa [2] and STAPLE [19]. Cohen's kappa was applied to pairs of methods at a time for all possible pairings. STAPLE was applied to the four dichotomised ASPECTS methods and to the ASPECTS territory data.

Cohen's kappa is a statistic to measure agreement between two sets of data while taking into account the probability of random chance agreement. It allows direct comparison with other studies in the field. Cohen's kappa is defined as

$$\mathcal{K} = \frac{p_o - p_e}{1 - p_e} \tag{2}$$

where p_o is the observed agreement among methods, and p_e is the probability of chance agreement.

$$p_o = \frac{TP + TN}{N} \tag{3}$$

$$p_e = \frac{(TP + FN) * (TP + FP) + (TN + FN) * (TN + FP)}{N^2} \tag{4}$$

where N is the number of patients or regions in the data and *TP*, *FN*, *FP*, and *TN* are the number of True Positives, False Negatives, False Positives, and True Negatives respectively.

STAPLE is an algorithm that compares the performance of two or more *experts*, in this case the four methods, with each other. It does this by estimating the *true score* for each patient based on the score from each expert and the perceived ability of each expert. The algorithm then calculates the performance of each expert against the true score, which gives the sensitivity and specificity of each expert. The sensitivity is the fraction of positive cases correctly identified as being positive. The term *positive* here refers to an ASPECTS on the "more severe" side of the dichotomised scale (i.e. ASPECTS < 7) in the case of the patient-level evaluation, or to an ischemic territory in the case of the territory-level evaluation. The specificity is the fraction of negative cases correctly identified as being negative.

$$\text{Sensitivity} = \frac{TP}{TP + FN} \tag{5}$$

$$\text{Specificity} = \frac{TN}{TN + FP} \tag{6}$$

4 Results

We used data for all available patients (156 patients) from the studies and for each patient the ASPECTS was collected via four methods at the patient level, and three methods at the territory level. At the patient level these methods were the Clinical score, the Ground Truth score, the Observed score, and the CNN score. At the territory level the Clinical score was unavailable. Out of the 156 patients, 52 of these reported a Clinical score of 10, however the clinical exert recorded an Observed score of 10 in 62 cases, and there were 66 cases of a score of 10 in the Ground Truth score. Figure 2 illustrates the distribution of the scores for each method. Each method generally tails off towards the low scores, but there are notable exceptions. The Clinical score features a collection of patients around the central scores that the other methods do not exhibit. The Ground Truth score has a peak at score 9 that is 33% larger than the method with the next highest bar of that score (CNN Score) and 2.9 times that of the professional standard (Clinical Score).

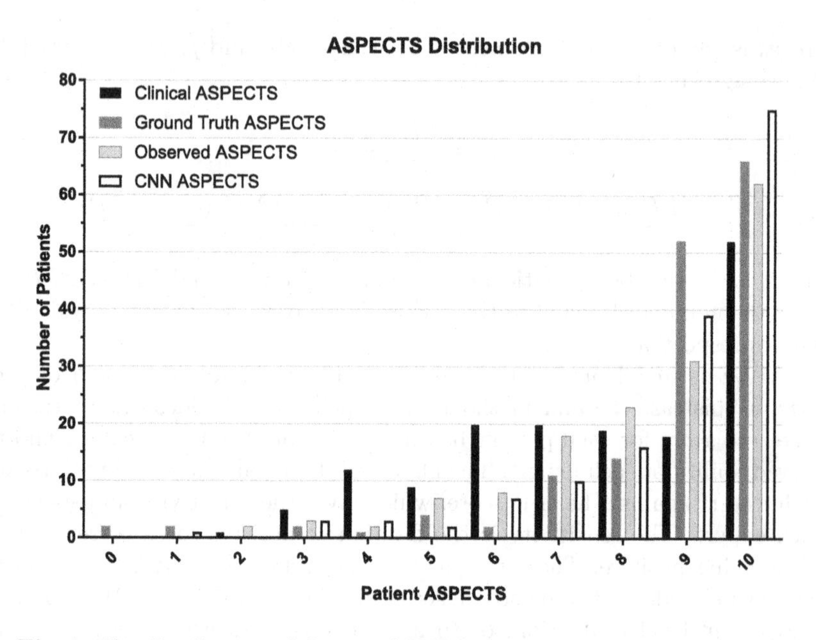

Fig. 2. The distribution of the raw ASPECTS for each of the four methods.

At the patient level, the fraction of patients above this threshold varied between the methods: 70% of the Clinical Scores, 92% of the Ground Truth Scores, 86% of the Observed Scores, and 90% of the CNN Scores were above the threshold (≥7).

A total of 3120 (156 × 20) ASPECTS territories were scored by each of the three methods (Ground Truth, Observed, and CNN scores). Each territory was binary scored as presenting ischemic signs or not. Cohen's kappa coefficient was calculated between each pair of dichotomised scores and each pair of territory scores. Figure 3 displays these values. The highest correlation is seen between the Ground Truth and CNN dichotomised scores, while the lowest set of correlations at the dichotomised level are the three comparisons with the Clinical Score. At the territory level there is a moderate correlation between the territories noted as ischemic between the Ground Truth and the Observed methods, while a weaker kappa is present between the CNN and both of these methods.

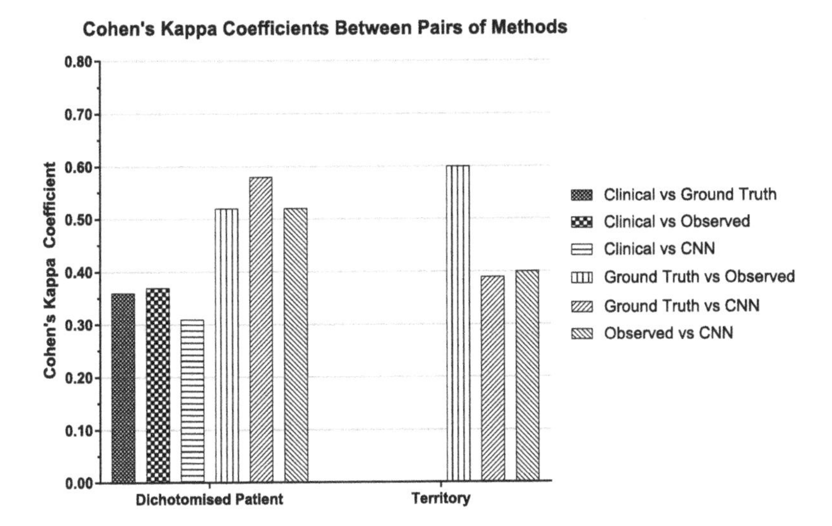

Fig. 3. Cohen's kappa coefficient between pairs of ASPECTS methods for the dichotomised data and territory data.

The voxel-level masks from the CNN were evaluated against the ground truth segmentations in terms of Area Under Curve (AUC) for the Receiver Operating Characteristic (ROC) curve and the Precision-Recall (PR) curve: ROC AUC = 0.94, PR AUC = 0.41.

The STAPLE algorithm determined the sensitivity and specificity of each of the methods at the dichotomised level and territory level. Table 1 shows these values. Further to the evaluation against the STAPLE true score, each of the methods were evaluated against each other to produce sensitivity and specificity for each. Table 2 shows the evaluation at the dichotomised level, while Table 3 displays it at the territory level. In both cases the assumed true class is on the left and the assumed predicted class is along the top. Sensitivity for methods evaluated against the Clinical score is low in all cases, but specificity is high. Specificity is expected to be high due to the large number of true negatives in the data relative to true positives. This means any false positives only have a minimal effect on the specificity each. On the other hand, the sensitivity is more prone to false negatives due to the low number of true positives in the data relative to true negatives.

5 Discussion

We have presented results from two statistical measures applied to methods of scoring ASPECTS to determine if a CNN algorithm is producing results comparable to the professional standard and to explore the methods against each other. We used Cohen's kappa as a method resilient to dataset imbalance to compare the correlation between pairs of methods, and we used STAPLE analysis to determine the sensitivity and specificity of each method.

Table 1. The sensitivity and specificity from binary STAPLE analysis on the dichotomised and territory data.

Evaluation	Clinical	Observed	Ground truth	CNN
Dichotomised score sensitivity	0.87	0.72	0.48	0.50
Dichotomised score specificity	0.82	0.98	1.00	0.97
Region sensitivity	N/A	0.84	0.72	0.49
Region specificity	N/A	0.98	0.99	0.97

Table 2. The sensitivity and specificity between pairs of scores at the dichotomised patient level.

		Clinical	Observed	Ground Truth	CNN
Clinical	Sensitivity:		0.37	0.28	0.28
	Specificity:		0.96	1.00	0.97
Observed	Sensitivity:			0.46	0.50
	Specificity:			0.98	0.96
Ground Truth	Sensitivity:				0.69
	Specificity:				0.95
CNN	Sensitivity:				
	Specificity:				

Table 3. The sensitivity and specificity between pairs of scores at the territory level.

		Observed	Ground Truth	CNN
Observed	Sensitivity:		0.57	0.40
	Specificity:		0.98	0.97
Ground Truth	Sensitivity:			0.43
	Specificity:			0.96
CNN	Sensitivity:			
	Specificity:			

5.1 Comparison of Score Methods

CNN Score: At the patient level, the CNN score is correlated the strongest with the Ground Truth score, which is expected as the CNN was trained on the ground truth. The similarity between the CNN and Ground Truth scores is influenced by the choice of the CNN score threshold, which was chosen to balance false positives and false negatives, and thus minimise the bias between the two scores. The quality of the ground truth provides an upper bound on the ability of the CNN. At the territory level the performance of these methods diverge. A STAPLE sensitivity of 0.5 for the CNN score indicates that half of the positive cases are missed.

Ground Truth Score and Observed Score: The Kappa coefficient of 0.6 between the Ground Truth territory scores and the Observed territory scores was lower than expected given that both were marked by the same clinical expert. It could be reasonably assumed that the clinical expert will have chosen similar territories for both methods, so the Kappa coefficient should be high. However, the Observed territories were identified directly the clinical expert, the Ground Truth territories came from an indirect route through atlas alignment and thresholding rules. There are associated errors with each of these processing steps such as ground truth segmentation accuracy, atlas accuracy and misalignment, threshold rules inconsistently applied due to human judgement, and different tools being available at the time of the method - namely the availability of a slabbing tool. The high frequency for the Ground Truth score at an ASPECTS of 9 in Fig. 2 is an indication that the small volume rule is being applied too regularly or inconsistently with the clinical expert's manual observations.

Clinical Score: The Clinical score returned a much lower STAPLE specificity than all of the other methods. Figure 2 indicates that this method is substantially biased towards lower scores, which leads to a lower specificity. Cohen's kappa between the Clinical score and each of the other methods was poor. A reason why the Clinical score was markedly lower than the other methods could be that the hospital clinicians assigning the score may have been able to see subtle signs due to their expertise level that were missed in the other evaluations.

6 Conclusions

We use two statistical methods: STAPLE and Cohen's kappa to compare four methods of scoring ASPECTS on NCCT datasets at the dichotomised patient level and three methods at the ASPECTS territory level for 156 patients. We explore how an automated method using a CNN compares to the professional standard. Initial results are encouraging, but we find that our ground truth does not capture the level of detail that is visible to expert clinicians, which leads to an overall higher ASPECTS for methods based on the ground truth. The CNN performs similar to the ground truth on which it was trained when viewed at the patient level, but the performance diverges at the territory level.

We are thus examining ways to improve the ground truth for the CNN to reduce the difference between the CNN predictions and the professional standard. Routes for this may involve using slabbed NCCT or follow-up scan data to improve the accuracy of ground truth marking, or generating multiple annotations for each volume by different human markers and using averaging to generate the ground truth. We believe an automatic method can provide benefits such as improved diagnosis consistency and reduced analysis time.

References

1. Barber, P.A., Demchuk, A.M., Zhang, J., Buchan, A.M., ASPECTS Study Group, et al.: Validity and reliability of a quantitative computed tomography score in predicting outcome of hyperacute stroke before thrombolytic therapy. Lancet **355**(9216), 1670–1674 (2000)
2. Cohen, J.: A coefficient of agreement for nominal scales. Educ. Psychol. Meas. **20**(1), 37–46 (1960)
3. Dabbah, M.A., Murphy, S., Pello, H., Courbon, R., Beveridge, E., Wiseman, S., Wyeth, D., Poole, I.: Detection and location of 127 anatomical landmarks in diverse CT datasets. In: SPIE Medical Imaging, p. 903415. International Society for Optics and Photonics (2014)
4. Fedorov, A., Beichel, R., Kalpathy-Cramer, J., Finet, J., Fillion-Robin, J.C., Pujol, S., Bauer, C., Jennings, D., Fennessy, F., Sonka, M., et al.: 3D slicer as an image computing platform for the quantitative imaging network. Magn. Reson. Imaging **30**(9), 1323–1341 (2012)
5. Grotta, J.C., Chiu, D., Lu, M., Patel, S., Levine, S.R., Tilley, B.C., Brott, T.G., Haley, E.C., Lyden, P.D., Kothari, R., et al.: Agreement and variability in the interpretation of early CT changes in stroke patients qualifying for intravenous rtPA therapy. Stroke **30**(8), 1528–1533 (1999)
6. Grunwald, I., Reith, W.: Non-traumatic neurological emergencies: imaging of cerebral ischemia. Eur. Radiol. **12**(7), 1632–1647 (2002)
7. Hill, M.D., Buchan, A.M., The Canadian Alteplase for Stroke Effectiveness Study (CASES) Investigators: Thrombolysis for acute ischemic stroke: results of the Canadian Alteplase for stroke effectiveness study. Can. Med. Assoc. J. **172**(10), 1307–1312 (2005)
8. Huang, X., Cheripelli, B.K., Lloyd, S.M., Kalladka, D., Moreton, F.C., Siddiqui, A., Ford, I., Muir, K.W.: Alteplase versus tenecteplase for thrombolysis after ischaemic stroke (ATTEST): a phase 2, randomised, open-label, blinded endpoint study. Lancet Neurol. **14**(4), 368–376 (2015)
9. Kosior, R.K., Lauzon, M.L., Steffenhagen, N., Kosior, J.C., Demchuk, A., Frayne, R.: Atlas-based topographical scoring for magnetic resonance imaging of acute stroke. Stroke **41**, 455–460 (2010)
10. Kunst, M.M., Schaefer, P.W.: Ischemic stroke. Radiol. Clin. N. Am. **49**(1), 1–26 (2011)
11. Larrue, V., von Kummer, R., Müller, A., Bluhmki, E.: Risk factors for severe hemorrhagic transformation in ischemic stroke patients treated with recombinant tissue plasminogen activator. Stroke **32**(2), 438–441 (2001)
12. Lisowska, A., Beveridge, E., Muir, K., Poole, I.: Thrombus detection in CT brain scans using a convolutional neural network. In: Proceedings of the 10th International Joint Conference on Biomedical Engineering Systems and Technologies (BIOSTEC), BIOIMAGING, vol. 2, pp. 24–33 (2017). ISBN 978-989-758-215-8
13. Lisowska, A., O'Neil, A., Dilys, V., Daykin, M., Beveridge, E., Muir, K., Mclaughlin, S., Poole, I.: Context-aware convolutional neural networks for stroke sign detection in non-contrast CT scans. J. Imaging (2017, in press)
14. MacDougall, N., McVerry, F., Huang, X., Welch, A., Fulton, R., Muir, K.: Post-stroke hyperglycaemia is associated with adverse evolution of acute ischaemic injury. In: Cerebrovascular Diseases, vol. 37, pp. 267–267. Karger, Basel (2014)

15. Pexman, J.W., Barber, P.A., Hill, M.D., Sevick, R.J., Demchuk, A.M., Hudon, M.E., Hu, W.Y., Buchan, A.M.: Use of the alberta stroke program early CT score (ASPECTS) for assessing CT scans in patients with acute stroke. Am. J. Neuroradiol. **22**(8), 1534–1542 (2001)
16. Wardlaw, J., Dorman, P., Lewis, S., Sandercock, P.: Can stroke physicians and neuroradiologists identify signs of early cerebral infarction on CT? J. Neurol. Neurosurg. Psychiatry **67**(5), 651–653 (1999)
17. Wardlaw, J.M., Muir, K.W., Macleod, M.J., Weir, C., McVerry, F., Carpenter, T., Shuler, K., Thomas, R., Acheampong, P., Dani, K., Murray, A.: Clinical relevance and practical implications of trials of perfusion and angiographic imaging in patients with acute ischaemic stroke: a multicentre cohort imaging study. J. Neurol. Neurosurg. Psychiatry **84**(9), 1001–1007 (2013). http://jnnp.bmj.com/content/84/9/1001
18. Wardlaw, J.M., Von Kummer, R., Farrall, A.J., Chappell, F.M., Hill, M., Perry, D.: A large web-based observer reliability study of early ischaemic signs on computed tomography. the acute cerebral CT evaluation of stroke study (ACCESS). PLoS ONE **5**(12), e15757 (2010)
19. Warfield, S.K., Zou, K.H., Wells, W.M.: Simultaneous truth and performance level estimation (STAPLE): an algorithm for the validation of image segmentation. IEEE Trans. Med. Imaging **23**(7), 903–921 (2004)

Image Enhancement and Alignment

Pre-processing Techniques for Colour Digital Pathology Image Analysis

Wael Saafin(✉) and Gerald Schaefer

Department of Computer Science, Loughborough University, Loughborough, UK
W.Saafin@lboro.ac.uk, gerald.schaefer@ieee.org

Abstract. Digital pathology (DP) can provide extensive information from captured tissue samples and support accurate and efficient diagnosis, while image analysis techniques can offer standardisation, automation, and improved productivity of DP. Since DP images are intrinsically captured as colour images, such analysis should appropriately exploit the colour or spectral information residing in the obtained data. However, before analysing colour DP images, some pre-processing is typically necessary to both support and enable effective analysis. Colour calibration aims to ensure accurate colour information is recorded, while colour enhancement is useful to be able to obtain robust performance of image analysis algorithms which are otherwise sensitive to imaging conditions and scanner variations. Employed methods range from calibrating the cameras and scanners over correcting the displayed colours to transferring the image to another colour representations that in turn can improve e.g. segmentation or other subsequent tasks. Colour deconvolution allows to effectively separate the contributions of stains localised in the same area and thus allows analysis of stain specific images, while variations in colour appearance of histopathology images due to e.g. scanner characteristics, chemical colouring concentrations, or different protocols, can be reduced through application of colour normalisation algorithms. In this paper, we give an overview of commonly employed colour image pre-processing techniques for digital pathology, summarising important work in colour calibration, colour enhancement, colour deconvolution and colour normalisation.

Keywords: Medical imaging · Digital pathology · Whole slide images · Image analysis · Colour imaging · Image pre-processing

1 Introduction

Digital pathology (DP) captures tissue samples under a microscope in form of digital images. Its application can provide extensive information from the captured samples and can hence help in quality control and assurance, archiving, automated image analysis, automated decision making, diagnosis, telepathology, etc.

DP imaging starts with sample preparation, and then proceeds with scanning the sample using a virtual slide scanner with the scanned view being sampled

M. Valdés Hernández and V. González-Castro (Eds.): MIUA 2017, CCIS 723, pp. 551–560, 2017.
DOI: 10.1007/978-3-319-60964-5_48

using a digital camera. DP also typically includes image compression, transmission and display of the obtained images [46].

Pathology imaging has been applied to histopathology to study the signs of diseases by examining biopsy samples or surgical specimen fixed on a glass slide. Histopathological tissue analysis is devoted to detect a disease, to determine its grading or progression, and to distinguish between different subtypes of a disease. DP has been also applied to cytology to study the cell structure, function, and chemistry.

During slide preparation, staining is used to highlight microscopic components of the tissue under study. This staining is dependent on many factors like the manufacturer, storage conditions before use, and method of application [11,24,25]. For example, haematoxylin-eosin (H&E) stainings are routinely used to stain cell nuclei in blue/purple and cytoplasm and connective tissue in pink. Instead of dying the tissue with stains, immunohistochemistry (IHC) highlights specific antigens in the tissue by injecting some antibodies in the slide. In a tissue microarray (TMA), small cylindrical biopsies of tissue from histological sections are extracted and arranged in matrix form in a recipient paraffin block, then sections are cut and mounted on glass slides.

The desire to obtain a digital version of classical glass slides led to the technology of whole slide imaging (WSI) to obtain high resolution digital images of complete glass histopathology or cytopathology slides. Effectively, WSI is based on two successive procedures: converting a whole glass slide into a digital image, and viewing this image using a virtual slide viewer. Digitised WSI is useful for quality assurance, image analysis, and tracking how an image was viewed, and ultimately to aid in diagnosis.

Since WSI (or wide-field microscopy) is based on digitising histopathology slides to store them in digital form, it is amenable to the application of computerised image analysis and machine learning techniques, and consequently computer-aided diagnosis (CAD) algorithms have been developed to complement and support the expertise and opinion of the pathologist [12,13,31]. Commonly employed techniques in this context include detection, segmentation, classification, and feature extraction methods. Typically, prior to such an analysis some image pre-processing is performed in order to improve the reliability and quality of the following analysis.

In contrast to monochromatic images, a standard three channel red-green-blue (RGB) camera allows for colour detection and processing, and DP images are typically captured in colour. In the RGB representation, obtained usually by a charge coupled device (CCD) camera, every pixel in the image plane is represented by three values corresponding to the red, green and blue components together constituting the overall colour of a pixel. While colour imaging supports improved analysis for DP [40], it also leads to several challenges which need to be addressed, including during the pre-processing stage.

In this paper, we provide an overview of colour pre-processing techniques that are useful in the context of DP image analysis. In particular we review commonly employed techniques for colour calibration and colour enhancement in Sect. 2, colour deconvolution in Sect. 3, and colour normalisation in Sect. 4. Section 5 concludes the paper.

2 Colour Calibration and Colour Enhancement for Digital Pathology

Since the transfer function of a WSI system including all subsequent procedures is typically not ideal, the colour information in the viewed WSI is expected to deviate from the original information, especially if, as is often the case, the WSI is not standardised. Colour enhancement aims to obtain colour output images that are closer to the "original" ones. They allow simpler and more accurate visualisation of the WSI, ameliorate degradations, and improve the image contrast. Colour enhancement is useful to obtain a stable performance of image analysis algorithms and to reduce sensitivity to imaging conditions and scanner variations.

The enhancement procedure might start before capturing the images by appropriately calibrating the scanners and viewers to be able to correct the captured and displayed colours which can be affected due to many factors like thickness of specimen, staining, scanner, viewer and display. For this purpose, colour chart slides can be employed. A calibration slide was proposed in [47], where the slides are scanned and the displayed images are then compared to the defined standard which has been produced by the calibration images. A calibration procedure for scanners was proposed in [39] using a film of 246 colour batches and 24 skin tones. While reference colour values are provided by the manufacturer, these are correlated with the scanned colours of the film to obtain the colour calibration information. A similar approach has been followed in [5].

In [2], colour correction is performed in the linear RGB colour space using a colour correction matrix followed by gamma correction to transfer the input RGB image to a corrected image for display. This correction matrix is estimated from nine reference colour patches, and the required colour correction matrix is calculated so as to identify the minimum deviation between the scanned colour batches and the target spectral colours.

Image standardisation has been used to obtain images suitable to the performance of an individual pathologist including image colourisation, white colour balance, or individual adjusted brightness [17,27,29]. Colour and illumination correction and contrast enhancement have been applied in automated virtual microscopy in [29], where the colours of RGB test WSIs were corrected by a white balancing adjustment to match the colours of a reference WSI.

Image artefacts including tissue folds, blurred regions, pen marks, shadows, and chromatic aberration can be eliminated using colour saturation and intensity methods. A folded tissue usually appears as a dark region while blur can affect the visualisation of nuclei [16,30]. Chromatic aberration might be the result of a failure of the lens to focus all colours to the same convergence point [20,41] and can be corrected by quantifying the amount of colour dispersion at the edges to realign colour components.

Sometimes, there will be differences between batches of data which is referred to as batch effects. TMAs might not be well prepared, for example not geometrically ordered, which can lead to artefacts and aberration effects that may show a TMA slide with a disordered array and some missing tissue discs [9]. These effects can be addressed by normalising the colours of an image to a reference image or by converting the image to a different colour space for subsequent processing [19,30]. These effects can also be reduced by artefact compensation and colour decomposition in order to correct for aberration effects based on the Hough transform to automatically locate, delineate and index each individual disc [9]. Colour decomposition performing polar transformations and peak detection in a multi-dimensional colour space has also been applied in [32].

Colour WSIs can also be enhanced using different colour spaces. In [3], the HSV (hue-saturation-value) colour space is employed to be able to modify the saturation and luminance components independently. This was shown to improve visualisation and detection of tissue folds based on simulated RGB images. The authors of [8] converted RGB TMA images to HSV to support improved segmentation of brown colour which corresponds to vessel membranes. In [6], images are mapped to the L*u*v* colour space to improve the segmentation and further analysis.

3 Colour Deconvolution for Digital Pathology

When staining a section with multiple stains, the stains will unavoidably mix, so that the captured colours in the image indicate mixtures of stains. Figure 1 (on the left) shows examples of slide images stained with haematoxylin and eosion (top), and diaminobenzidine (DAB) and haematoxylin (bottom) respectively.

Since different stains are associated with different features in the images, being able to separate information with respect to stains is of central importance in DP. When two or more stains are used and thus appear mixed in the captured images, this will still be the case after converting the images from the source RGB space to other colour spaces such as HSV. Simple colour transformations are hence not sufficient to be able to reliably separate the contributions of the stains. Rather what is required is a method that effectively unmixes the dyes, that is, separates the image into channels that correspond to actual stain contributions. Not surprisingly, this is a well researched topic [10,35,43].

Colour deconvolution (CD) is an orthonormal transformation of the image data that effectively separates the contributions of (for RGB input images) up to three stains localised in the same area [14,18,36]. The transformation is defined by the constituent stains to correspond to the actual colours of the stains used depending on their concentrations in the slide. CD is routinely used as a preprocessing step for visualising DP slides and for automated image analysis procedures such as detection, classification, object segmentation, stain normalisation, quantification of IHC TMAs, and to score cell death processes. CD is able to separate two or three stains provided that their colours are sufficiently different in their red, green or blue absorption characteristics [35]. After CD, the input

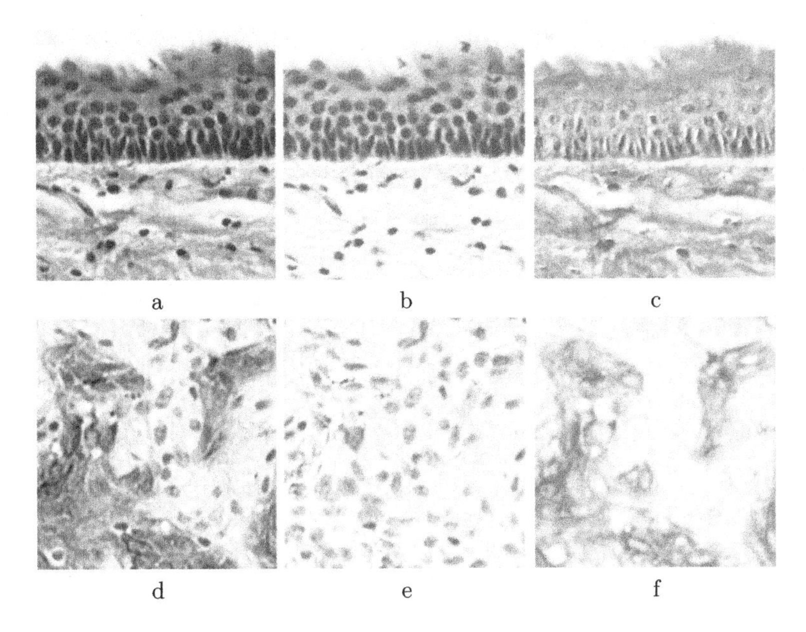

Fig. 1. Examples of slides stained with haematoxylin and eosion (a) and diaminobenzidine (DAB) and haematoxylin (d). Also shown are the results of colour deconvolution giving the separated haematoxylin (b) and eosin stain images (c) for (a) and separated haematoxylin (e) and DAB images (f) for (d) [21]. (Color figure online)

RGB images are separated into channels corresponding to the stain components as shown in Fig. 1.

Stain decomposition is usually applied before spectral normalisation [22]. Stain decomposition/separation is a process which estimates stain spectra and corresponding stain proportions/depths in pathology images as a pre-processing step to improve subsequent diagnosis. CD is one of the most widely used methods utilising the Lambert-Beer law which relates stain amount to light absorption to separate light absorbing stains and is often employed for stain normalisation [25,26,33,35,43].

CD has been applied to isolate individual IHC stains in [15,23,34] and in [37] to isolate colour information from histological multiple stained RGB images to isolate DAB (which represents the K19 positive area) from haematoxylin (which represents the epithelial area).

In [4,14], CD and its errors were assessed as a pre-processing step which affects later processing and image analysis. Based on the experiments conducted, it was shown that sequential narrow band illumination in combination with a wide band sensor leads to lower errors and it was proposed to integrate microscopic RGB LED illumination into microscopic instruments to potentially reduce the CD-step error and improve the image quality.

4 Colour Normalisation for Digital Pathology

Colour variations in histopathology images can be the result of changes in slide scanner, chemical colouring, concentration, timing, raw materials, stain vendors or protocols, and light transmission due to change in thickness [18]. Such variations in colour appearance between samples produced in the same lab or different labs can be reduced by application of colour normalisation or standardising image colours before further analysis as in [1,44]. Colour normalisation typically also enhances the image contrast and can address image artefacts [19,38]. Colour normalisation has also been addressed in [7,16].

Colour batch effects can be limited by normalising the colour image to a reference image or transferring the image into a different colour space such as the L*a*b* colour space as in [19]. Normalisation of colour channels to a target image has been addressed in [26,28,33,45] where the mean and standard deviation of each channel are matched to that of the target after performing linear transforms. The approach in [33] converts the image into the $l\alpha\beta$ colour space defined as

$$\begin{aligned} l_m &= \frac{l_o - M_{l_o}}{\sigma_{l_o}} \sigma_{l_t} + M_{l_t} \\ \alpha_m &= \frac{\alpha_o - M_{\alpha_o}}{\sigma_{\alpha_o}} \sigma_{\alpha_t} + M_{\alpha_t} \\ \beta_m &= \frac{\beta_o - M_{\beta_o}}{\sigma_{\beta_o}} \sigma_{\beta_t} + M_{\beta_t} \end{aligned} \tag{1}$$

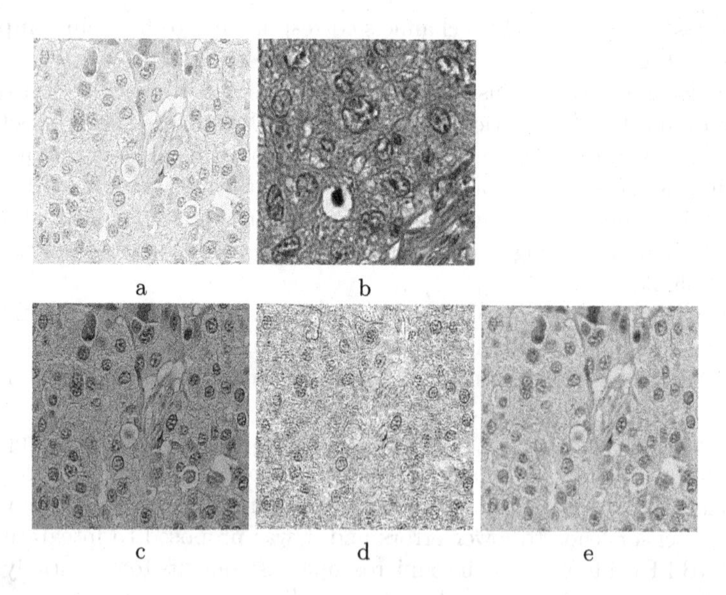

Fig. 2. Examples of colour normalisation techniques (from the Stain Normalisation Toolbox [42]): (a) source DP image; (b) reference/target DP image; (c) normalisation result using [33]; (d) normalisation result using [25]; (e) normalisation result using [18]. (Color figure online)

where M represents the mean, σ the standard deviation, and subscripts m, o, t denote the mapped, original, and target image respectively. The result of the normalisation algorithm applied to a sample image is shown in Fig. 2, together with the outputs of two other approaches, namely [18,25].

Normalisation has been used before pixel-based colour segmentation of H&E stained histology images in [26], although a colour deconvolution-based approach was shown to outperform it. Normalisation can be applied at pixel level using one model for the complete image, as in [18,26,45], or at stain level using multiple models for each stain as in [43]. In [45], normalisation to a target image was applied before performing colour-based segmentation, while in [43], the colour image appearance of a source image was changed to match a target image without losing structural information.

5 Conclusions

Digital pathology presents many opportunities but at the same time also faces various obstacles. While a wealth of information is recorded in DP virtual slides, accurately analysing this information is challenging. DP images are typically captured in colour and appropriately analysing and exploiting this colour information is crucial. Unfortunately, the obtained image colours are affected by many factors including scanner characteristics, stain properties and concentration, and protocols. In this paper, we have thus provided an overview of colour-based image processing techniques that are employed to pre-process digital pathology images. In particular, we have summarised important work in colour calibration, colour enhancement, colour deconvolution and colour normalisation in the context of digital pathology. Application of these techniques can lead to improved and more robust performance of subsequent analysis stages such as segmentation or classification. While significant progress has been made within a fairly short timeframe, digital pathology is still a relatively new field and thus undoubtedly a wealth of further research is still to come in this exiting area.

Acknowledgements. This work was supported by the EC under Marie Curie grant actions, grant No. 612471, Academia and Industry Collaboration for Digital Pathology (AIDPATH) project (http://aidpath.eu/).

References

1. Basavanhally, A., Madabhushi, A.: EM-based segmentation-driven color standardization of digitized histopathology. In: SPIE Medical Imaging 2013: Digital Pathology (2013)
2. Bautista, P.A., Hashimoto, N., Yagi, Y.: Color standardization in whole slide imaging using a color calibration slide. J. Pathol. Inform. **5**(1), 4 (2014)
3. Bautista, P.A., Yagi, Y.: Improving the visualization and detection of tissue folds in whole slide images through color enhancement. J. Pathol. Inform. **1**(1), 25 (2010)

4. Bueno, G., Déniz, O., Fernández-Carrobles, M.D.M., Vállez, N., Salido, J.: An automated system for whole microscopic image acquisition and analysis. Microsc. Res. Tech. **77**(9), 697–713 (2014)
5. Cheng, W., Keay, T., O'Flaherty, N., Wang, J., Ivansky, A., Gavrielides, M.A., Gallas, B.D., Badano, A.: Assessing color reproducibility of whole-slide imaging scanners. In: SPIE Medical Imaging 2013: Digital Pathology (2013)
6. Comaniciu, D., Meer, P., Foran, D., Medl, A.: Bimodal system for interactive indexing and retrieval of pathology images. In: 4th IEEE Workshop on Applications of Computer Vision, pp. 76–81 (1998)
7. Demir, C., Yener, B.: Automated cancer diagnosis based on histopathological images: a systematic survey. Rensselaer Polytechnic Institute, Technical report (2005)
8. Fernandez-Carrobles, M., Tadeo, I., Bueno, G., Noguera, R., Déniz, O., Salido, J., García-Rojo, M.: TMA vessel segmentation based on color and morphological features: application to angiogenesis research. Sci. World J. **2013**, 1–11 (2013)
9. Foran, D.J., Chen, W., Yang, L.: Automated image interpretation and computer-assisted diagnostics. Anal. Cell. Pathol. **34**(6), 279 (2011)
10. Ghosh, B., Karri, S.P.K., Sheet, D., Garud, H., Ghosh, A., Ray, A.K., Chatterjee, J.: A generalized framework for stain separation in digital pathology applications. In: IEEE Annual India Conference, pp. 1–4 (2016)
11. Glatz-Krieger, K., Spornitz, U., Spatz, A., Mihatsch, M.J., Glatz, D.: Factors to keep in mind when introducing virtual microscopy. Virchows Arch. **448**(3), 248–255 (2006)
12. Gurcan, M.N., Boucheron, L.E., Can, A., Madabhushi, A., Rajpoot, N.M., Yener, B.: Histopathological image analysis: a review. IEEE Rev. Biomed. Eng. **2**, 147–171 (2009)
13. Hamilton, P.W., Bankhead, P., Wang, Y., Hutchinson, R., Kieran, D., McArt, D.G., James, J., Salto-Tellez, M.: Digital pathology and image analysis in tissue biomarker research. Methods **70**(1), 59–73 (2014)
14. Haub, P., Meckel, T.: A model based survey of colour deconvolution in diagnostic brightfield microscopy: error estimation and spectral consideration. Sci. Rep. **5**, 12096 (2015)
15. Helps, S.C., Thornton, E., Kleinig, T.J., Manavis, J., Vink, R.: Automatic non-subjective estimation of antigen content visualized by immunohistochemistry using color deconvolution. Appl. Immunohistochem. Mol. Morphol. **20**(1), 82–90 (2012)
16. Irshad, H., Veillard, A., Roux, L., Racoceanu, D.: Methods for nuclei detection, segmentation, and classification in digital histopathology: a review. IEEE Rev. Biomed. Eng. **7**, 97–114 (2014)
17. Kayser, K., Görtler, J., Borkenfeld, S., Kayser, G.: Interactive and automated application of virtual microscopy. Diagn. Pathol. **6**(1), 1 (2011)
18. Khan, A.M., Rajpoot, N., Treanor, D., Magee, D.: A nonlinear mapping approach to stain normalization in digital histopathology images using image-specific color deconvolution. IEEE Trans. Biomed. Eng. **61**(6), 1729–1738 (2014)
19. Kothari, S., Phan, J.H., Stokes, T.H., Wang, M.D.: Pathology imaging informatics for quantitative analysis of whole-slide images. J. Am. Med. Inform. Assoc. **20**(6), 1099–1108 (2013)
20. Kruger, P.B., Mathews, S., Aggarwala, K.R., Sanchez, N.: Chromatic aberration and ocular focus: Fincham revisited. Vis. Res. **33**(10), 1397–1411 (1993)
21. Landini, G.: Colour deconvolution. http://www.mecourse.com/landinig/software/cdeconv/cdeconv.html

22. Li, X., Plataniotis, K.N.: Circular mixture modeling of color distribution for blind stain separation in pathology images. IEEE J. Biomed. Health Inform. **21**(1), 150–161 (2017)
23. Liang, Y., Wang, F., Treanor, D., Magee, D., Teodoro, G., Zhu, Y., Kong, J.: Liver whole slide image analysis for 3D vessel reconstruction. In: 12th IEEE International Symposium on Biomedical Imaging, pp. 182–185 (2015)
24. Ljungberg, A., Johansson, O.: Methodological aspects on immunohistochemistry in dermatology with special reference to neuronal markers. Histochem. J. **25**(10), 735–745 (1993)
25. Macenko, M., Niethammer, M., Marron, J.S., Borland, D., Woosley, J.T., Guan, X., Schmitt, C., Thomas, N.E.: A method for normalizing histology slides for quantitative analysis. In: IEEE International Symposium on Biomedical Imaging, pp. 1107–1110 (2009)
26. Magee, D., Treanor, D., Crellin, D., Shires, M., Smith, K., Mohee, K., Quirke, P.: Colour normalisation in digital histopathology images. In: MICCAI Workshop on Optical Tissue Image analysis in Microscopy, Histopathology and Endoscopy, vol. 100 (2009)
27. Marchevsky, A.M., Khurana, R., Thomas, P., Scharre, K., Farias, P., Bose, S.: The use of virtual microscopy for proficiency testing in gynecologic cytopathology: a feasibility study using ScanScope. Arch. Pathol. Lab. Med. **130**(3), 349–355 (2006)
28. Mosquera-Lopez, C., Agaian, S.: Iterative local color normalization using fuzzy image clustering. In: SPIE Mobile Multimedia/Image Processing, Security, and Applications 2013 (2013)
29. Murakami, Y., Abe, T., Hashiguchi, A., Yamaguchi, M., Saito, A., Sakamoto, M.: Color correction for automatic fibrosis quantification in liver biopsy specimens. J. Pathol. Inform. **4**(1), 36 (2013)
30. Palokangas, S., Selinummi, J., Yli-Harja, O.: Segmentation of folds in tissue section images. In: 29th Annual International Conference of the IEEE Engineering in Medicine and Biology Society, pp. 5641–5644 (2007)
31. Pantanowitz, L.: Digital images and the future of digital pathology. J. Pathol. Inform. **1**(1), 15 (2010)
32. Rabinovich, A., Agarwal, S., Laris, C., Price, J.H., Belongie, S.J.: Unsupervised color decomposition of histologically stained tissue samples. In: Advances in Neural Information Processing Systems, pp. 667–674 (2003)
33. Reinhard, E., Adhikhmin, M., Gooch, B., Shirley, P.: Color transfer between images. IEEE Comput. Graph. Appl. **21**(5), 34–41 (2001)
34. Rizzardi, A.E., Johnson, A.T., Vogel, R.I., Pambuccian, S.E., Henriksen, J., Skubitz, A.P., Metzger, G.J., Schmechel, S.C.: Quantitative comparison of immunohistochemical staining measured by digital image analysis versus pathologist visual scoring. Diagn. Pathol. **7**(1), 1 (2012)
35. Ruifrok, A.C., Johnston, D.A.: Quantification of histochemical staining by color deconvolution. Anal. Quant. Cytol. Histol. **23**(4), 291–299 (2001)
36. Ruifrok, A.C., Katz, R.L., Johnston, D.A.: Comparison of quantification of histochemical staining by hue-saturation-intensity (HSI) transformation and color-deconvolution. Appl. Immunohistochem. Mol. Morphol. **11**(1), 85–91 (2003)
37. Safadi, R.A., Musleh, A.S., Al-Khateeb, T.H., Hamasha, A.A.: Analysis of immunohistochemical expression of K19 in oral epithelial dysplasia and oral squamous cell carcinoma using color deconvolution - image analysis method. Head Neck Pathol. **4**(4), 282–289 (2010)

38. Schaefer, G., Rajab, M.I., Celebi, M.E., Iyatomi, H.: Colour and contrast enhancement for improved skin lesion segmentation. Comput. Med. Imaging Graph. **35**(2), 99–104 (2011)
39. Shrestha, P., Hulsken, B.: Color accuracy and reproducibility in whole slide imaging scanners. J. Med. Imaging **1**(2), 027501 (2014)
40. Stack, E.C., Wang, C., Roman, K.A., Hoyt, C.C.: Multiplexed immunohistochemistry, imaging, and quantitation: a review, with an assessment of tyramide signal amplification, multispectral imaging and multiplex analysis. Methods **70**(1), 46–58 (2014)
41. Thibos, L.N., Bradley, A., Still, D.L., Zhang, X., Howarth, P.A.: Theory and measurement of ocular chromatic aberration. Vis. Res. **30**(1), 33–49 (1990)
42. Trahearn, N., Khan, A.: Stain normalisation toolbox for matlab. http://www2.warwick.ac.uk/fac/sci/dcs/research/tia/software/sntoolbox/
43. Vahadane, A., Peng, T., Sethi, A., Albarqouni, S., Wang, L., Baust, M., Steiger, K., Schlitter, A.M., Esposito, I., Navab, N.: Structure-preserving color normalization and sparse stain separation for histological images. IEEE Trans. Med. Imaging **35**(8), 1962–1971 (2016)
44. Veta, M., Pluim, J.P., van Diest, P.J., Viergever, M.A.: Breast cancer histopathology image analysis: a review. IEEE Trans. Biomed. Eng. **61**(5), 1400–1411 (2014)
45. Wang, Y., Chang, S., Wu, L., Tsai, S., Sun, Y.: A color-based approach for automated segmentation in tumor tissue classification. In: 29th Annual International Conference of the IEEE Engineering in Medicine and Biology Society, pp. 6576–6579 (2007)
46. Weinstein, R.S., Graham, A.R., Richter, L.C., Barker, G.P., Krupinski, E.A., Lopez, A.M., Erps, K.A., Bhattacharyya, A.K., Yagi, Y., Gilbertson, J.R.: Overview of telepathology, virtual microscopy, and whole slide imaging: prospects for the future. Hum. Pathol. **40**(8), 1057–1069 (2009)
47. Yagi, Y.: Color standardization and optimization in whole slide imaging. Diagn. Pathol. **6**(1), 1 (2011)

Motion Compensation Using Range Imaging in C-Arm Cone-Beam CT

Bastian Bier[1(✉)], Mathias Unberath[1,2], Tobias Geimer[1,2], Jennifer Maier[1], Garry Gold[3], Marc Levenston[3], Rebecca Fahrig[3,4], and Andreas Maier[1,2]

[1] Pattern Recognition Lab, Friedrich-Alexander-University Erlangen-Nuremberg, Erlangen, Germany
bastian.bier@fau.de

[2] Erlangen Graduate School in Advanced Optical Technologies, Friedrich-Alexander-University Erlangen-Nuremberg, Erlangen, Germany

[3] Department of Radiology, School of Medicine, Stanford University, Stanford, USA

[4] Siemens Healthcare GmbH, Erlangen, Germany

Abstract. Cone-beam C-arm CT systems allow to scan patients in weight-bearing positions to assess knee cartilage health under more realistic conditions. Involuntary patient motion during the acquisition results in motion artifacts in the reconstructions. The current motion estimation method is based on fiducial markers. They can be tracked with a high spatial accuracy in the projection images, but only deliver sparse information. Further, placement of the markers on the patient's leg is time consuming and tedious. Instead of relying on a few well defined points, we seek to establish correspondences on dense surface data to estimate 3D displacements.

In this feasibility study, motion corrupted X-ray projections and surface data are simulated. We investigate motion estimation by registration of the surface information. The proposed approach is compared to a motion free, an uncompensated, and a state-of-the-art marker-based reconstruction using the SSIM.

The proposed approach yields motion estimation accuracy and image quality close to the current state-of-the-art, reducing the motion artifacts in the reconstructions remarkably. Using the proposed method, Structural Similarity improved from 0.887 to 0.975 compared to uncorrected images. The results are promising and encourage future work aiming at facilitating its practical applicability.

1 Introduction

C-arm cone-beam CT (CBCT) offers a great variety of applications in interventional and diagnostic medicine. One of the reasons for that is their high degree of flexibility, which allows the C-arm to scan objects on arbitrary trajectories, even enabling scanning of patients in upright, weight-bearing positions [6,7]. Weight-bearing imaging was found to be particularly beneficial for the assessment of knee cartilage health, since the knee joint has different mechanical properties

M. Valdés Hernández and V. González-Castro (Eds.): MIUA 2017, CCIS 723, pp. 561–570, 2017.
DOI: 10.1007/978-3-319-60964-5_49

under load [19]. To this end, the C-arm rotates around the standing patient on a horizontal trajectory and acquires images from different views [13]. A schematic scheme of the imaging geometry is shown in Fig. 1. One of the major problems is that patients tend to show motion during the scan time of approximately ten seconds. This results in motion artifacts in the reconstructed volumes, which manifest as streaks, double edges, and blurring, as can be seen in Fig. 3(b). In order to increase the image quality and the diagnostic value of the reconstructions, motion has to be estimated and compensated for.

Recently, motion correction in extremity imaging under weight-bearing conditions gained increasing attention. Choi et al. and Berger et al. use fiducial markers placed on the patient's knees. These markers can be detected in the 2D projection images and their position can be used for alignment with their respective 3D reference position. Although this method has been used in various clinical studies [4,7,16,17], several limitations exist. Marker placement is time consuming, associated with patient discomfort, and tedious since overlapping markers in the projections would result in inaccurate motion estimates. Another category of approaches do not rely on markers and are image-based only. Unberath et al. estimate motion by aligning the projection images with the maximum intensity projections of a motion corrupted reconstruction [22]. However, only 2D detector shifts are evaluated. Berger et al. uses 2D/3D registration of a segmented bone from a motion free supine acquisition with the acquired projection images [4,18]. The reconstructions show a much sharper bone outline, however, a previously acquired supine scan might not always be available. Sisniega et al. proposed a method to estimate 6D rigid motion by optimizing the image sharpness and entropy in a region of interest in the reconstruction [20,21]. Further, epipolar consistency conditions were investigated to estimate motion and showed promising results [1,5]. However, motion estimation proved to be robust only parallel to the detector.

Range imaging has been used in different applications in medicine [2] such as augmented reality [10], motion estimation in PET/SPECT imaging [15], and patient positioning and motion estimation in radiotherapy [12]. Recently, Fotouhi et al. investigated the feasibility on improving iterative reconstruction quality by incorporating information of a RGBD camera [11]. In this work, a feasibility study is conducted to investigate the motion estimation capability of range imaging in the scenario of acquisitions under weight-bearing conditions. Therefore, surface point cloud data and X-ray projections are simulated under the influence of motion. Using registration of the point clouds, a motion estimate is obtained. We compare the reconstructed images of the proposed method with the results of the uncompensated reconstruction and the result of a marker-based method [7]. Further, the estimated motion signals are compared to the marker-based method using the correlation coefficient. To investigate the effects of motion compensation on the image quality, we compare the reconstructed images of the uncompensated case, the marker-based method and the proposed approach to the ground truth using Structural Similarity (SSIM).

2 Materials and Methods

The method relies on 3D point clouds and X-ray projection images generated for each time point in the same motion state. Exemplary input data are shown in Fig. 2. The section is structured as follows: First, we describe the range camera simulation; second, we present the point cloud registration, and finally, the X-ray simulation and reconstruction is described.

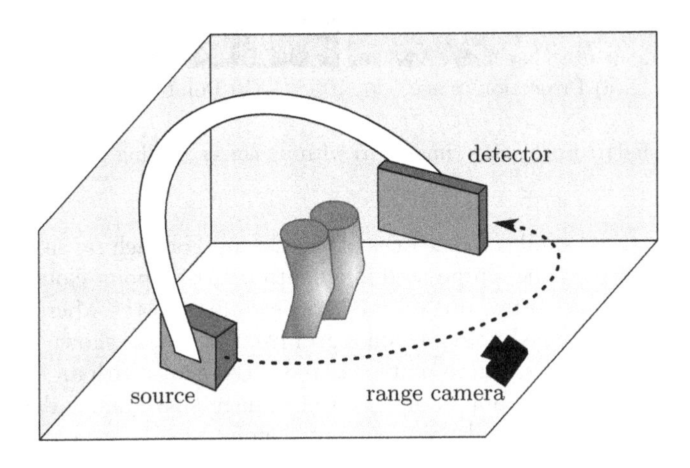

Fig. 1. Schematic scheme of the imaging setup. The X-ray source and the detector rotate around the standing patient. A range imaging camera is placed statically on the ground in front of the object.

2.1 Depth Camera Simulation

Camera Position. Different camera positions are feasible in combination with a C-arm CT system. Either the camera is mounted on the detector or statically on the ground. If mounted on the C-arm, only one calibration between the two modalities is necessary. However, in the weight-bearing scenario occlusion from the second leg occurs. Further, the registration task is more difficult due to partial overlap. Especially for imaging cylindrical, smooth objects such as the knees, ambiguities in the surface will make the registration task difficult. For a static camera position, this is not the case. Therefore, in this study, the camera is placed in front of the standing patient, such that the object of interest is always completely in the field of view, as shown in Fig. 1. Further, the registration problem simplifies since partial overlap at the surface border is smaller and some anatomical structures, e.g. the patella, are visible.

Creation of the Point Clouds. Depth images are simulated by casting rays through a mask volume, which is obtained from a motion free supine reconstruction, where one leg is segmented. The geometry of the camera is described by

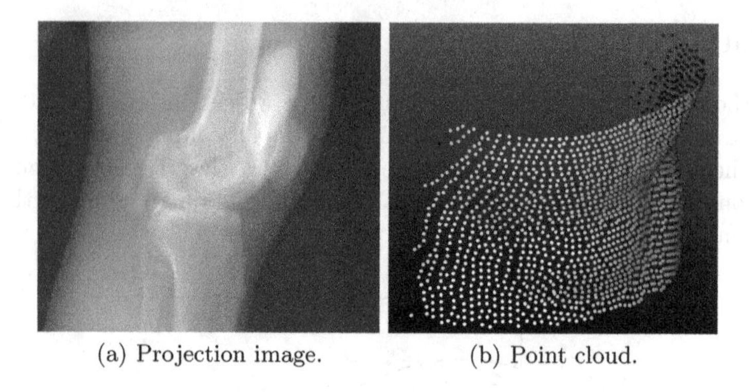

(a) Projection image. (b) Point cloud.

Fig. 2. Simulated projection image (a) with its corresponding point cloud (b).

projection matrices similar to the ones of the C-arm. For each ray, a depth value is obtained, which can be represented in a depth map or a point cloud. The camera and its noise properties are selected to be similar to the Microsoft Kinect One v2 [24]: The sampled points lie on a grid like pattern, as shown in Fig. 2(b). In the area of the object, the resolution is 160×120 pixels with an approximate distance of the sampling points of 1.4 mm in image space and a depth resolution of 1 mm. Noise free and noise corrupted point clouds are created for our experiments. For this purpose, Gaussian noise with a zero mean and a standard deviation of $\sigma = 1$ and $\sigma = 2$ is added on the depth coordinates. The camera to object distance is 75 cm.

2.2 Motion Simulation

In order to simulate rigid 3D translation on the depth camera, the volume is shifted for each time point by a reference motion vector in the 3D volume, resulting in a different point cloud for each time point. The applied reference motion is 3D translation derived from real patient motion captured with a motion capture system [6].

The same reference motion is used to create motion corrupted X-ray projections. To this end, the projection matrices $\mathbf{P} \in \mathbb{R}^{4\times3}$, which describe the imaging geometry of the C-arm system, are modified. Rigid motion can be directly incorporated in the projection matrices using the following formula:

$$\hat{\mathbf{P}}_j = \mathbf{P}_j \cdot \begin{pmatrix} \mathbf{R}_j & \mathbf{t}_j \\ 0 & 1 \end{pmatrix} \tag{1}$$

$\mathbf{P}_j$ is the calibrated projection matrix from the system and $\mathbf{R}_j \in \mathbb{R}^{3\times3}$ and $\mathbf{t}_j \in \mathbb{R}^{1\times3}$ are the rotation and translation for each time step j, respectively. These motion corrupted projection matrices are used to create motion corrupted projection images, shown in Fig. 2(a).

2.3 Point Cloud Registration

The registration is initialized by aligning the centroids of all point clouds. In order to speed up the registration, all point clouds are registered to the point cloud corresponding to the first frame that is assigned the translation vector $\mathbf{t}_0 = (0, 0, 0)$. Afterwards, a point-to-surface Iterative Closest Point (ICP) algorithm [3] is used to refine the estimation result. In contrast to the common ICP algorithm that optimizes for point-to-point distance, point-to-surface ICP seeks to minimize the distance of points to the other point cloud's surface. This strategy proved more accurate in this application. Further, the common ICP is heavily influenced by the grid pattern of the point clouds, which results in an estimate that is a multiple of the spacing between the points. The optimal motion estimate is found by solving the following objective function for a pair of point clouds:

$$\hat{\boldsymbol{t}} = \arg\min_{\boldsymbol{t}} \left(\sum_{i=1}^{N} ||((\boldsymbol{p}_i + \boldsymbol{t}) - \boldsymbol{q}_i)\boldsymbol{n}_i|| \right), \quad (2)$$

where $\hat{\boldsymbol{t}}$ is the translation to be estimated, N is the number of points in the source point cloud with $\boldsymbol{p}_i$ and $\boldsymbol{q}_i$ being a point on the source point cloud and its corresponding point on the target point cloud, respectively. The corresponding point is defined by the projection of $\boldsymbol{p}_i$ along the normal vector $\boldsymbol{n}_i$ of the target point cloud. This is implemented by triangulating the point cloud and computing the distance of the point to the respective triangles with a defined normal vector. The normal vector is obtained from the cross product of two vectors of the triangle. The function is solved with a gradient descent algorithm[1].

2.4 Reconstruction

Images are reconstructed using the Feldkamp-David-Kress backprojection algorithm [9]: the projection data is preprocessed using cosine weighting, Parker redundancy weighting, and row-wise ramp filtering. In a last step the projections are backprojected. The projection matrices used are real calibrated projection matrices from a clinical C-arm system, which has been operated on a horizontal trajectory. In order to incorporate the motion estimate into the reconstruction, Eq. 1 can be used to modify the projection matrices before the backprojection step.

2.5 Experiment

The simulation is performed on a segmented knee, extracted from a clinical high quality supine reconstruction. The data was acquired on a clinical C-arm CT system (Artis Zeego, Siemens Healthcare GmbH, Erlangen, Germany). We use the real scanner geometry with 248 projection images on a horizontal trajectory

[1] https://www5.cs.fau.de/research/software/java-parallel-optimization-package/.

over a rotation of 200°. Detector resolution is 1240×960 pixels with an isotropic pixel spacing of 0.308 mm. Volumes with a size of 512^3 and an isotropic voxelsize of 0.5 mm are reconstructed. The entire processing, simulation, and reconstruction pipeline is implemented in the open-source framework CONRAD, dedicated to the simulation and reconstruction of CBCT data [14]. In order to compare the image quality of the results, the SSIM [23] is computed. To this end, all reconstructions are registered to the motion free reference reconstruction using the open-source software 3D slicer [8].

3 Results

In Fig. 3, axial and sagittal slices of the reconstruction results of the motion free scan, the motion corrupted acquisition, the marker-based method, and the proposed method (on noise free data) are shown. Severe streaking and blurring artifacts are present if no correction is applied. Both, the marker-based method as well as the new proposed method substantially reduce the corruption by artifacts. In the marker-based result (Fig. 3(c)) streaks from one direction are visible. They appear due to marker overlap in these views resulting in an inaccurate estimation. In contrast, images obtained using the proposed approach have more but smaller streak artifacts, which appear especially at the bone outline.

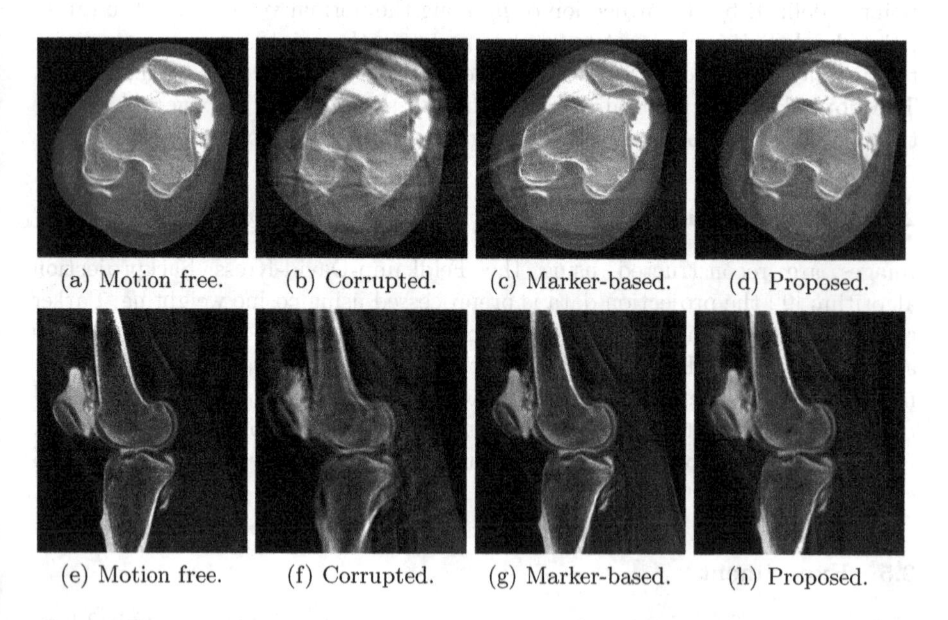

(a) Motion free. (b) Corrupted. (c) Marker-based. (d) Proposed.

(e) Motion free. (f) Corrupted. (g) Marker-based. (h) Proposed.

Fig. 3. Axial and sagittal slices of the reconstruction results of the motion free reference volume (a), the motion corrupted (b), the marker-based corrected (c), and the proposed method (d).

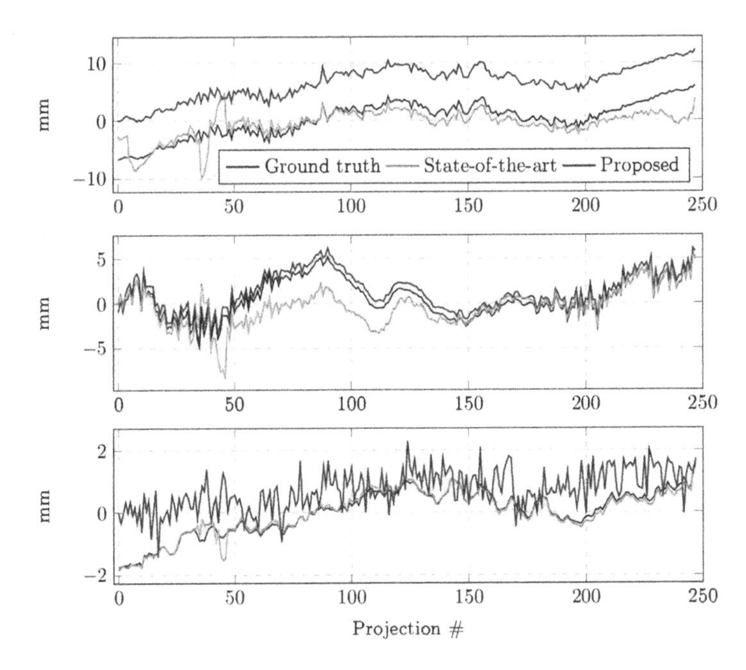

Fig. 4. Ground truth motion parameters compared to the estimation of the marker-based and the proposed method for the x (top), y (middle), and z (bottom) direction.

In Fig. 4, the estimated motion parameters of the marker-based and proposed method are compared with the ground truth motion: x is the object motion parallel to the detector, y is the depth and z corresponds to the direction parallel to the rotation axis. Considering the estimation results for x and y direction, both methods were able to recover the motion well, especially the high frequencies. Note, however, that due to the current optimization scheme of the proposed method, all projections are aligned to the first frame resulting in a constant offset in the estimated signal. This offset will cause a shift of the reconstruction in the volume. The views in which the markers overlap are visible in the motion estimation around projection number 40, where sudden jumps are visible in the marker-based estimation. They correspond to the heavy streaks in the reconstruction images. In contrast, the z direction is estimated poorly with the proposed method. One reason for that might be that the cylindrical shape does not contain enough information to nicely align the point clouds. Further, the object is also truncated in the z direction, which might lead to these errors.

In Table 1 the SSIM between the shown reconstructions compared to the motion free volume are shown. The quantitative values are in agreement with the visual impression of the reconstructions: the marker-based and the proposed method show comparably good results that are far superior to the uncompensated reconstruction. The SSIM decreases with the level of noise applied to the point clouds. However, achieved results are superior towards the uncorrected reconstruction.

Table 1. Structural Similarity (SSIM) results of the reconstructed images.

Method	SSIM
Uncorrected	0.887
Marker-based [7]	0.971
Proposed	0.975
Proposed with noise $\sigma = 1$	0.961
Proposed with noise $\sigma = 2$	0.952

Table 2. Correlation coefficients (CC) between the estimated and the ground truth motion signals.

Method	x-direction	y-direction	z-direction
Marker-based [7]	0.77	0.72	0.98
Proposed	0.98	0.98	0.59

In order to quantify the accuracy of the motion estimation, we calculated the correlation coefficient between the estimated and the ground truth signal. The results are shown in Table 2, and show that the proposed method outperforms the marker-based method in the motion estimate of the x and y direction, but achieved only poor results in the z direction with a correlation coefficient of 0.58 only.

4 Discussion and Conclusion

In this feasibility study, a novel method to estimate patient motion in acquisitions under weight-bearing conditions using range imaging is investigated. Our method registers point clouds to a reference frame to estimate motion, without the need to place markers on the patient's skin. We validated our approach on simulated projection and point cloud data of a high quality supine reconstruction. The method reduces the motion artifacts notably and suggests high potential of using depth imaging for motion correction in CBCT reconstruction.

Surface and projection data are simulated with the same motion pattern for a full acquisition. All point clouds are registered using a point-to-surface ICP algorithm resulting in a motion estimate used to obtain the motion corrected reconstruction. In the performed experiments, the proposed method showed an improvement of the image quality comparable to the marker-based approach.

Future work will address the current limitations of the method. Compared to the 6D rigid motion estimation of the marker-based method, we tested translational motion only in our experiments that does not accurately reflect real knee motion. However, the method might have the potential to estimate more complex non-rigid motion of the knee as the dense surface information could be used to estimate more sophisticated displacements. Currently, motion along z direction is estimated insufficiently. One possibility to address this issue would be to combine the presented approach with image-based consistency conditions, which usually estimate this direction robustly. Further, a more comprehensive study on the method's robustness against sensor noise has to be conducted. This is especially important when dealing with real depth data containing a high amount of noise and physical artifacts. Finally, for practical applicability in the dynamic clinical environment, the camera has to be placed such that the field of view is not temporarily obstructed and in a distance, which allows to acquire data in good quality. Synchronization and cross calibration of the depth camera with the C-arm system have to be considered.

Despite the mentioned limitations, this work is yet another step towards motion estimation in imaging under weight-bearing conditions without the need of prior knowledge or the placement of fiducial markers.

References

1. Aichert, A., Wang, J., Schaffert, R., Dörfler, A., Hornegger, J., Maier, A.: Epipolar consistency in fluoroscopy for image-based tracking. In: Proceedings of BMVC, pp. 82.1–82.10 (2015)
2. Bauer, S., et al.: Real-time range imaging in health care: a survey. In: Grzegorzek, M., Theobalt, C., Koch, R., Kolb, A. (eds.) Time-of-Flight and Depth Imaging. LNCS, vol. 8200, pp. 228–254. Springer, Heidelberg (2013)
3. Bellekens, B., Spruyt, V., Weyn, M.: A survey of rigid 3D pointcloud registration algorithms. In: The Fourth International Conference on Ambient Computing, Applications, Services and Technologies, AMBIENT 2014, pp. 8–13 (2014)
4. Berger, M., Müller, K., Aichert, A., Unberath, M., Thies, J., Choi, J.H., Fahrig, R., Maier, A.: Marker-free motion correction in weight-bearing cone-beam CT of the knee joint. Med. Phys. **43**(3), 1235–1248 (2016)
5. Bier, B., Aichert, A., Felsner, L., Unberath, M., Levenston, M., Gold, G., Fahrig, R., Maier, A.: Epipolar consistency conditions for motion correction in weight-bearing imaging. In: Maier-Hein, K.H., et al. (eds.) Bildverarbeitung für die Medizin 2017. Springer, Heidelberg (2017). doi:10.1007/978-3-662-54345-0_47
6. Choi, J.H., Fahrig, R., Keil, A., Besier, T.F., Pal, S., McWalter, E.J., Beaupré, G.S., Maier, A.: Fiducial marker-based correction for involuntary motion in weight-bearing C-arm CT scanning of knees. Part I. Numerical model-based optimization. Med. Phys. **41**(6), 061902 (2014)
7. Choi, J.H., Maier, A., Keil, A., Pal, S., McWalter, E.J., Beaupré, G.S., Gold, G.E., Fahrig, R.: Fiducial marker-based correction for involuntary motion in weight-bearing C-arm CT scanning of knees. II. Experiment. Med. Phys. **41**(6), 061902 (2014)
8. Fedorov, A., Beichel, R., Kalpathy-Cramer, J., Finet, J., Fillion-Robin, J.C., Pujol, S., Bauer, C., Jennings, D., Fennessy, F., Sonka, M., Buatti, J., Aylward, S., Miller, J.V., Pieper, S., Kikinis, R.: 3D Slicer as an image computing platform for the Quantitative Imaging Network. Magn. Reson. Imaging **30**(9), 1323–1341 (2012)
9. Feldkamp, L.A., Davis, L.C., Kress, J.W.: Practical cone-beam algorithm. J. Opt. Soc. Am. A **1**(6), 612 (1984)

10. Fischer, M., Fuerst, B., Lee, S.C., Fotouhi, J., Habert, S., Weidert, S., Euler, E., Osgood, G., Navab, N.: Preclinical usability study of multiple augmented reality concepts for K-wire placement. Int. J. Comput. Assist. Radiol. Surg. **11**(6), 1007–1014 (2016)
11. Fotouhi, J., Fuerst, B., Wein, W., Navab, N.: Can real-time RGBD enhance intraoperative Cone-Beam CT? Int. J. Comput. Assist. Radiol. Surg. (2017). https://link.springer.com/journal/11548/onlineFirst/page/3
12. Geimer, T., Unberath, M., Taubmann, O., Bert, C., Maier, A.: Combination of markerless surrogates for motion estimation in radiation therapy. In: Computer Assisted Radiology and Surgery (CARS) 2016, pp. 59–60 (2016)
13. Maier, A., Choi, J.H., Keil, A., Niebler, C., Sarmiento, M., Fieselmann, A., Gold, G., Delp, S., Fahrig, R.: Analysis of vertical and horizontal circular C-arm trajectories. In: SPIE Medical Imaging, vol. 7961, pp. 796123-1–796123-8 (2011)
14. Maier, A., Hofmann, H.G., Berger, M., Fischer, P., Schwemmer, C., Wu, H., Müller, K., Hornegger, J., Choi, J.H., Riess, C., Keil, A., Fahrig, R.: CONRAD - a software framework for cone-beam imaging in radiology. Med. Phys. **40**(11), 111914 (2013)
15. McNamara, J.E., Pretorius, P.H., Johnson, K., Mukherjee, J.M., Dey, J., Gennert, M.A., King, M.A.: A flexible multicamera visual-tracking system for detecting and correcting motion-induced artifacts in cardiac SPECT slices. Med. Phys. **36**(5), 1913–1923 (2009)
16. Müller, K., Berger, M., Choi, J., Maier, A., Fahrig, R.: Automatic motion estimation and compensation framework for weight-bearing C-arm CT scans using fiducial markers. In: IFMBE Proceedings, pp. 58–61 (2015)
17. Müller, K., Berger, M., Choi, J.H., Datta, S., Gehrisch, S., Moore, T., Marks, M.P., Maier, A.K., Fahrig, R.: Fully automatic head motion correction for interventional C-arm systems using fiducial markers. In: Proceedings of the 13th Fully Three-Dimensional Image Reconstruction in Radiology and Nuclear Medicine, pp. 534–537 (2015)
18. Ouadah, S., Stayman, J.W., Gang, G.J., Ehtiati, T., Siewerdsen, J.H.: Self-calibration of cone-beam CT geometry using 3D2D image registration. Phys. Med. Biol. **61**(7), 2613–2632 (2016)
19. Powers, C.M., Ward, S.R., Fredericson, M.: Knee extension in persons with lateral subluxation of the patella: a preliminary study. J. Orthop. Sports Phys. Ther. **33**(11), 677–685 (2013)
20. Sisniega, A., Stayman, J.W., Cao, Q., Yorkston, J., Siewerdsen, J.H., Zbijewski, W.: Image-based motion compensation for high-resolution extremities cone-beam CT. In: SPIE Medical Imaging, vol. 9783, p. 97830K (2016)
21. Sisniega, A., Stayman, J., Yorkston, J., Siewerdsen, J., Zbijewski, W.: Motion compensation in extremity cone-beam CT using a penalized image sharpness criterion. Phys. Med. Biol. **62**(9), 3712 (2017)
22. Unberath, M., Choi, J.H., Berger, M., Maier, A., Fahrig, R.: Image-based compensation for involuntary motion in weight-bearing C-arm cone-beam CT scanning of knees. In: SPIE Medical Imaging, vol. 9413, p. 94130D (2015)
23. Wang, Z., Bovik, A.C., Sheikh, H.R., Simoncelli, E.P.: Image quality assessment: from error visibility to structural similarity. IEEE Trans. Image Process. **13**(4), 600–612 (2004)
24. Wasenmüller, O., Stricker, D.: Comparison of Kinect v1 and v2 depth images in terms of accuracy and precision. In: Chen, C.S., Lu, J., Ma, K.K. (eds.) ACCV 2016. LNCS, vol. 10117, pp. 34–45. Springer, Cham (2016). doi:10.1007/978-3-319-54427-4_3

Medical Image Colorization for Better Visualization and Segmentation

Muhammad Usman Ghani Khan[1(✉)], Yoshihiko Gotoh[2], and Nudrat Nida[1]

[1] AlKhawarizmi Institute of Computer Science, UET Lahore, Lahore, Pakistan
usman.ghani@kics.edu.pk
[2] University of Sheffield, Sheffield, UK

Abstract. Medical images contain precious anatomical information for clinical procedures. Improved understanding of medical modality may contribute significantly in arena of medical image analysis. This paper investigates enhancement of monochromatic medical modality into colorized images. Improving the contrast of anatomical structures facilitates precise segmentation. The proposed framework starts with preprocessing to remove noise and improve edge information. Then colour information is embedded to each pixel of a subject image. A resulting image has a potential to portray better anatomical information than a conventional monochromatic image. To evaluate the performance of colorized medical modality, the structural similarity index and the peak signal to noise ratio are computed. Supremacy of proposed colorization is validated by segmentation experiments and compared with greyscale monochromatic images.

Keywords: Medical image enhancement · Colorization · Visualization

1 Introduction

Digital image and signal processing encompasses a vast arena of technologies for analysis of internal biological structure, function and treatment [8]. Medical imaging utilizes fundamental physical phenomena, stretching from acoustic wave dissemination to X-ray propagation, to understand the patient health parameters. Previously medical images represent structural appearance information only, however today they are capable of examining complex and sophisticated internal biological processes such as mutation, metabolism, blood circulation, chemical reactions and many others. Medical imaging is not only contributing in disease diagnoses, but also playing its role in understanding the human anatomy along with evaluation of drugs chemical reaction.

Another versatile contribution of medical imaging is guided surgical assistance during the procedure [3,26]. There are various studies validating the robotic assisted surgery using medical images [1,26]. The entire foundation of clinical processes from diagnostic, treatment, surgical procedures and case studies are incomplete without medical images. Machine learning algorithms can provide a strong foundation to build clinical decision support systems using medical

M. Valdés Hernández and V. González-Castro (Eds.): MIUA 2017, CCIS 723, pp. 571–580, 2017.
DOI: 10.1007/978-3-319-60964-5_50

imaging [25,29]. Since past few decades, computer-aided detection and diagnosis has emerged as a vibrant research area. Lesions and organs may be diagnosed by examining the biological pattern, thus supporting accurate prognosis and treatment. Medical images hold a precious contribution in visual representation of biological structures and diagnosis of diseases. To embed colour information in traditional medical images play a significant role in clinical decision making. Colourization empowers visual discrimination of biological structures and supports diagnostic decision by surgeons [25].

Computer vision and image processing experts have been contributing in various technologies for decades, whereas colorization is rather new in the medical field. Potentially colorization can play an effective role in medical applications. A number of colorization methodologies, with varying computational costs, have been utilized to colorize greyscale image. A seed of color is used to disseminate color information to similar texture pixels [4]. Colorization technique was initially proposed by Welsh; the technique fell in the category of semi-automatic coloring of natural images. Unfortunately the same methodology generates poor quality colorization for medical images [28]. In the medical domain, technique of false coloring demonstrates induction of coloring in CT (computed tomography) modality images. The short fall of this technique is a consumption of higher system resources during execution [15,17,27]. Comparing the luminance intensity among images and colorized MRI (magnetic resonance imaging) images [2].

Image fusion is another approach that has been explored in order to introduce colorized medical images [7,14]. Using statistical parameters estimation using pervious information with the maximum likelihood criteria the most suitable color may be predicted [19]. In [13] colorized images were generated by cyphering false color code within the image. Texture information was also utilized to predict the similar pattern, potentially propagating colors across the similar texture [16]. Further, video frames were colorized through seeding color in keyframes [9,12].

2 Literature Survey

Colorization of medical images has been a popular research area for a decade. Identification of effected body cells form grey scale images is challenging as well as time consuming assignment for medical professionals. For assistance of medical experts, colorization of medical images is proposed by image processing specialists for automatic detection of effected parts and better understanding of bio-medical substances. Colorized medical images assist in prevention and treatment of diseases. Colorization techniques can be classified into three types—automatic, semi-automatic and user defined coloring techniques [21].

A seeded colorization scheme is one of recent techniques that propagates color to pixels based on neighbouring pixels. This technique maps colorized scribbles on original gary scale images. Scribbles act as seed and are responsible for describing color of pixels. This technique applies a seeded cellular automaton based on the scribbled method. It was demonstrated that a seeded cellular automaton was able to generates chromatic images from gray scale images with acceptable visual quality [4]. Some studies applied coloring techniques based on threshold. The drawbacks of the approach is that processing time and memory consumption are very high [15].

In another technique brain MRI classification was achieved by applying two independent methods, (i) highlighting the variability of input images, (ii) segmentation for outlining gray images [2]. It was then followed by a colorization technique to create chromatic brain MRI image. It was reported that there was luminance distance of input image and target image. In [12] a colorization technique was applied on videos using a color seed, populating chromatic information to remaining pixels. Colorized frames were then referred to when colorizing the remaining frames.

To measure performances by various algorithms some standardized quality parameters have been tested, such as a measure of enhancement, a similarity index of the structure, peak signals and entropy [20]. A feature set consisting of mean gray values, mode gray values, area fraction, aspect ratio and standard deviation was used to classify medical images for identification of a tumor, where a number of classification algorithms were tested including Naive Bayes, Tree J48, artificial neural network and Lazy-Ibk. It was found that a neural network classifier produced the most accurate results [11].

Furthermore, image segmentation based on a threshold leading to watershed segmentation and morphological operators was proposed to locate tumorous areas [18]. Similarly, a tumor size and a location were detected by segmentation based on texture information through morphological operators and the dual tree wavelet decomposition [22]. Another work designed an algorithm for detecting regions affected by a breast cancer [6]. Initially the image was enhanced through a Gaussian smoothing filter, which was then followed by morphological operations. The resultant image was disintegrated into two scales using the DWT (discrete wavelet transform) and reconstructed to form a binary image. Classification was made using the artificial neural network and the SVM (support vector machine). It was found that SVM using RBF (radial basis function) and linear kernel exhibited the highest accuracy rate.

3 Pre-processing

The proposed framework requires pre-processing where images are enhanced to get finer details by removing noise and applying normalization techniques. Pre-processing magnifies contrast, brightness and structural details of images. It adds in significant visual illustration, generating colorized medical images with the increased overall quality.

Series of steps are performed to pre-process a grey scale medical image. The standard resolution for images is set to 256×256 pixels. Textual content is removed from the image. Noise removal is achieved by implementing a weighted averaging filter. The source image is convolved with a weighted kernel of the size 3×3. A sliding kernel of the size $m \times n$ is centered at (x, y) of the source $Image(x, y)$ where $m, n = 3$.

Suppose that z_j is the pixel under process, where j represents the number of pixels less than 9. R_{xy} stand for the area to calculate the average of the source $Image(x, y)$. Smoothing of $Image(x, y)$ is achieved by

$$\text{Weighted Kernel} = \begin{bmatrix} -1 & -1 & -1 \\ -1 & +2 & -1 \\ -1 & -1 & -1 \end{bmatrix} \quad (1)$$

Noise is removed from the source and further processed to fix the contrast ratio accentuating the details of image. After completing the noise removal step the contrast is enhanced to highlight the minor details of the image.

The transform intensity $I(x, y)$ is the sum of fraction of the pixel under process z_i and the total number of pixels:

$$I(x, y) = \sum_{i=0}^{n} \frac{z_i}{\text{Total Pixels}} \quad (2)$$

The image is passed to the edge enhancement phase after the contrast improvement. Edge enhancement is done using the Sobel edge detector. The Sobel kernel find the gradient in the vertical and the horizontal direction of the image. The gradients are used to find the magnitude and the direction of edges. Mathematical expressions of the Sobel kernel and operations are presented below. The horizontal and the vertical derivatives $Kernel_x$, $Kernel_y$ are given by

$$Kernel_x = \begin{vmatrix} -1 & -2 & -1 \\ 0 & 0 & 0 \\ +1 & +2 & +1 \end{vmatrix}, \quad Kernel_y = \begin{vmatrix} -1 & 0 & +1 \\ -2 & 0 & +2 \\ -1 & 0 & +1 \end{vmatrix} \quad (3)$$

Information is enhanced across edges by adding and subtracting these derivatives near the edges. Convolution $*$ is calculated with the source image:

$$Image_x = Kernel_x * Image(x, y), \quad Image_y = Kernel_y * Image(x, y) \quad (4)$$

resulting in two images with the horizontal and vertical derivative approximations. They are used to calculate the gradient magnitude:

$$Image'(x, y) = \sqrt{Image_x^2 + Image_y^2} \quad (5)$$

and the gradient's direction:

$$\theta = tan^{-1} \frac{Image_y}{Image_x} \quad (6)$$

The resultant image is more informative than the input because the edge intensity has been increased noticeably. The image is processed further using the negative transformation. Suppose that the maximum intensity of the image is M. The negative of the image is

$$Image''(x, y) = M - Image'(x, y) \quad (7)$$

After the enhancement steps the image is used as an input for the colorization phase. The outcomes of each phase are presented in Fig. 1.

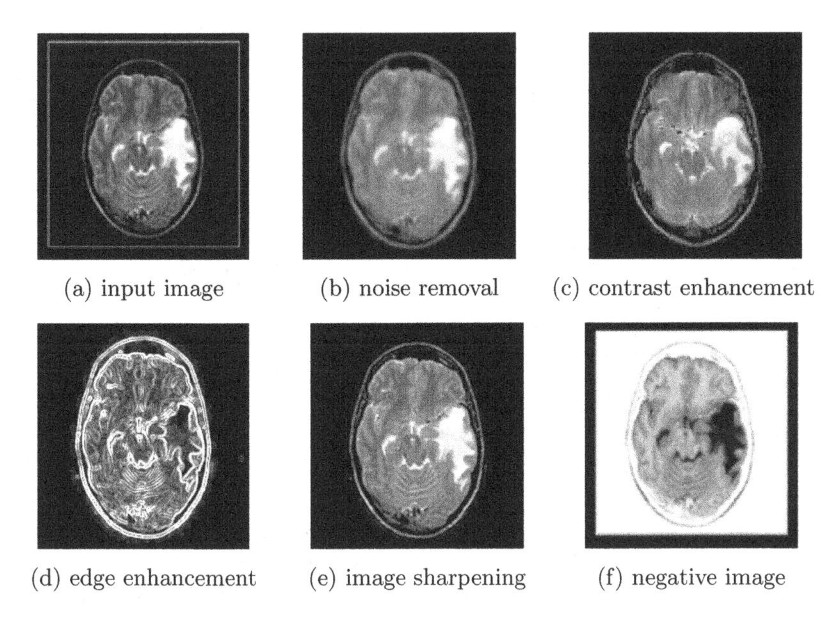
(a) input image (b) noise removal (c) contrast enhancement

(d) edge enhancement (e) image sharpening (f) negative image

Fig. 1. Outputs at pre-processing phases.

4 Colorization

Colorization phase requires a pre-processed image and a chromatic image of to generate a colorized target image of the size 256×256 pixels. The output needs to be three dimensional to hold chromatic information, hence the second and the third channels are populated to the input image.

Normalization is implemented with the input image, leading to the pixel intensity in accordance with the target image. This allows the intensity of the input image to be confined within the defined range. A normalized pixel intensity value is expressed by

$$\text{Normalized value of pixel} = \frac{I(x, y) \times 255}{255 - (L - P)} \tag{8}$$

where L and P are the maximum and minimum values of the intensity.

The source and the target images are both converted to YCbCr color. A comparison of pixel values between two images is computed at Y channel of YCbCr. All pixels from the source image are mapped to chromatic values using chromatic information of the target image. After successful assignment of chromatic information, the source image is converted to RGB colors for better visualization and understanding. Figure 2 presents the full flow chart of the algorithm.

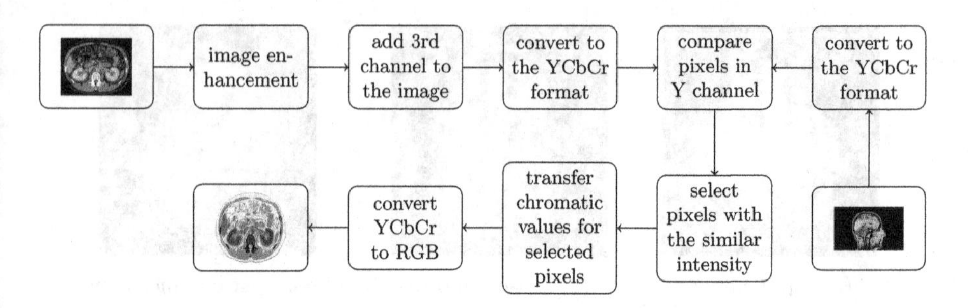

Fig. 2. Flow chart of the proposed algorithm.

5 Segmentation

The colorized medical image is segmented to regions to find the area of interest. Segmentation is an important step to identify images that are more relevant. After segmentation the image becomes a collection of segments that canvas the entire image. Neighboring pixels are similar in terms of the color, the texture and the intensity.

In the experiment, a PCNN (pulse coupled neural network) and a thresholding are used for segmentation of colorized medical images. A PCNN is used for elicitation of edge information, noise removal and segmentation by analyzing the region of interest. Raw estimation of locating the interested region and background is achieved by key point distribution. A PCNN is able to separate the front end from background. Pixels are processed line by line, starting from left to right or from top to bottom.

6 Result and Evaluation

The approach was tested using several medical image datasets acquired from open source online repositories. Table 1 summarizes their modality and the source.

Table 1. Datasets for the experiment

Imaging modality	Source
1. CT	Images of normal heart, brain and kidney [23]
2. Mammogram	Mammography images of above 60 year old women [5]
3. MRI	Normal brain, spine and knee images [?]
4. Nuclear medicine	Full body, spine and knee images [10]
5. PET	PET images of abdomen, heart and brain [23]
6. Ultrasound	Liver images of healthy people [23]
7. X-Ray	X-ray images [24]

6.1 Result

The outcome of the experiment presents better visual description of the image. The proposed algorithm has resulted in meaningful perception of images, where different parts such as tissues, muscles and bones were visualized separately. Table 2 compares the input grey scale images and their colorized images.

Table 2. Input images and their colarization.

imaging modality	input image	output image
1. CT	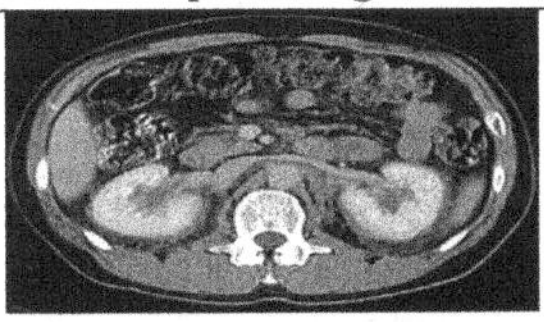	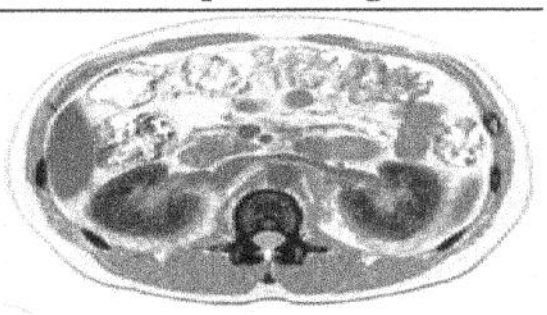
2. Mammogram	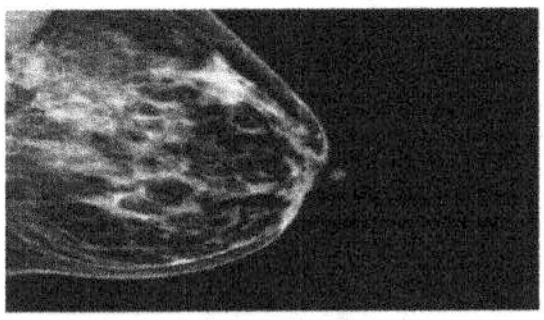	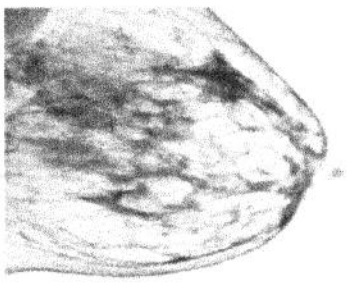
3. MRI	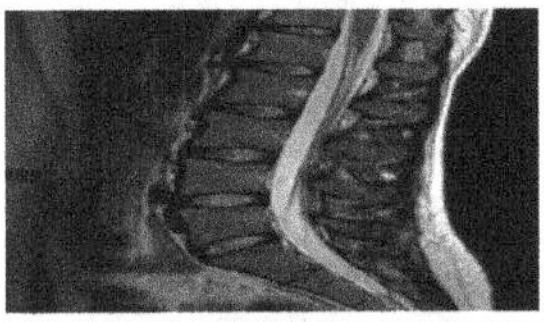	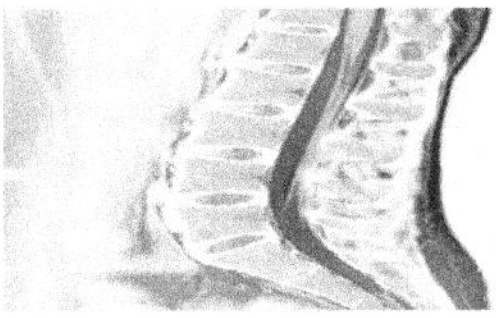
4. Nuclear Medcine	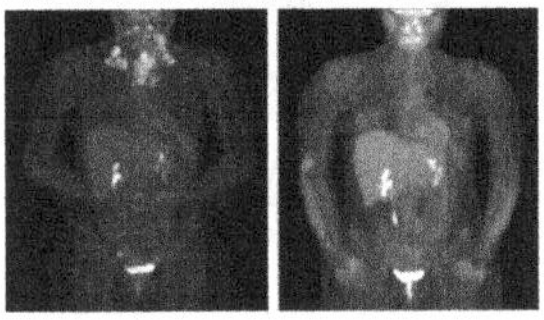	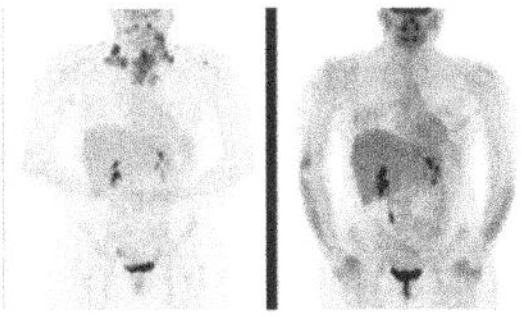
5. PET	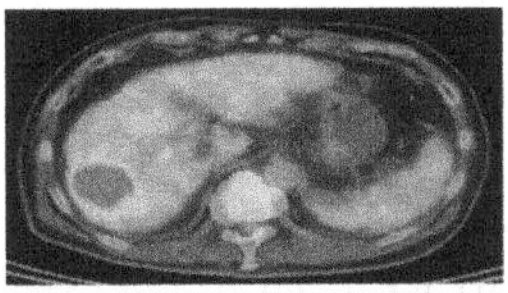	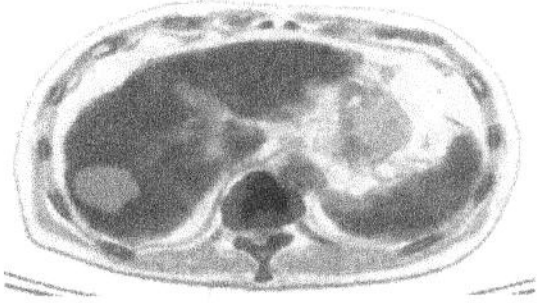
6. Ultrasound	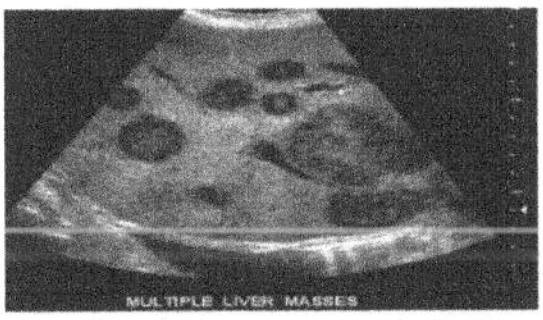	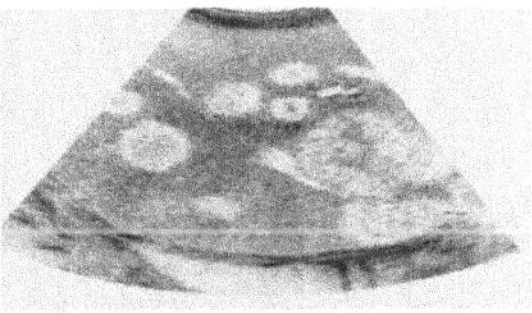
7. X-Ray	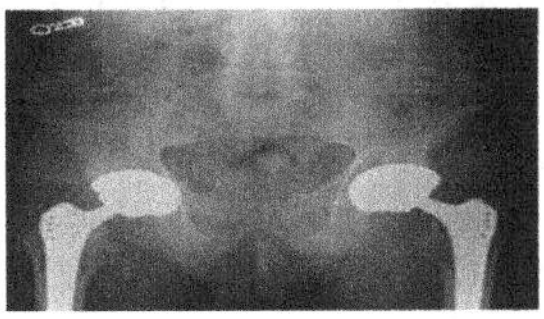	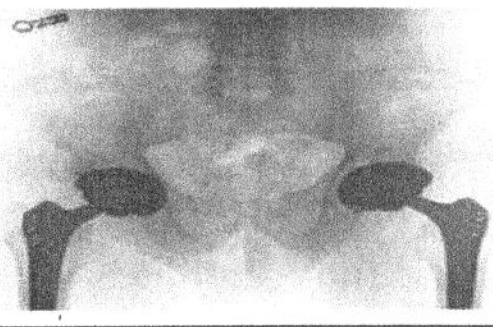

6.2 Comprehensive Comparison with the State-of-the-Art Techniques

Various statistical performance metrics such as peak signal to noise ratio PSNR; structural similarity index, SSIM; entropy and measure of enhancement EME; are available to compare proposed algorithm against the other state of the art work for image colorization. PSNR value is calculated between input grey scale image and resultant colorized image where higher value of PSNR shows a good conversion and vice versa. The approach is compared with the recent state-of-the-art techniques [9,13,16,19,28]. The parameters of all the algorithms were kept optimally same, to achieve highest computable PSNR. Figure 3 presents the comparison chart using the PSNR (peak signal-to-noise ratio). It is clearly observed that approach performed far better than other algorithms. The major factor for the significant improvement is caused by image enhancement prior to colorization.

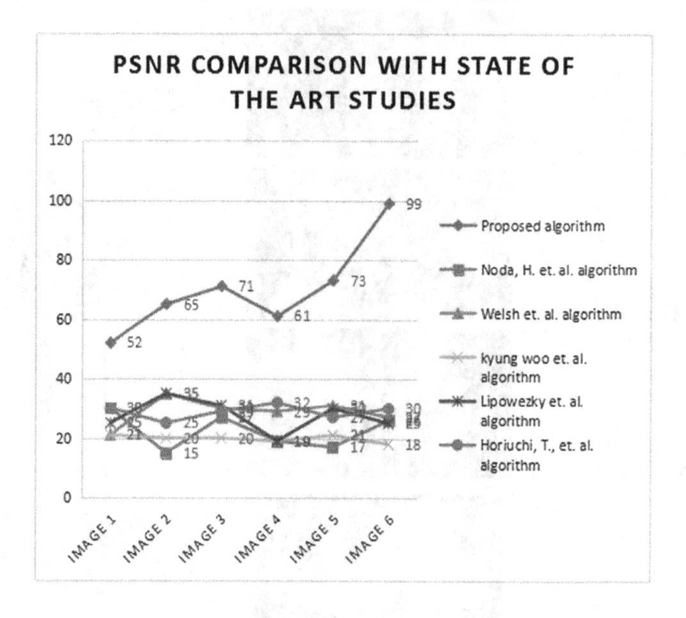

Fig. 3. Comparison chart with the state-of-the-art techniques.

7 Conclusion

Visual information is an important factor for doctors to analyze and diagnose diseases. The proposed algorithm can offer better visualization by clearly differentiating muscle, tissue and the bones area. The advantage of the algorithm is that the structure of an image remain the same during the entire process. Future work may include automatic detection of the disease area based on colored images to support professionals.

Acknowledgment. This work was carried on during the research project entitle "Automatic Surveillance System for Video Streams" funded by ICT Rnd, Pakistan.

References

1. Ahmad, A., Ahmad, Z.F., Carleton, J.D., Agarwala, A.: Robotic surgery: current perceptions and the clinical evidence. Surg. Endosc. **31**(1), 255–263 (2017)
2. Attique, M., Gilanie, G., Mehmood, M.S., Naweed, M.S., Ikram, M., Kamran, J.A., Vitkin, A., et al.: Colorization and automated segmentation of human T2 MR brain images for characterization of soft tissues. PLoS ONE **7**(3), e33616 (2012)
3. Barbash, G.I., Glied, S.A.: New technology and health care costs—the case of robot-assisted surgery. N. Engl. J. Med. **363**(8), 701–704 (2010)
4. Celeste, N.L.U., Yusiong, J.P.T.: Grayscale image colorization using seeded cellular automaton. Int. J. Adv. Res. Comput. Sci. **6**(1) (2015)
5. Clark, K., Vendt, B., Smith, K., Freymann, J., Kirby, J., Koppel, P., Moore, S., Phillips, S., Maffitt, D., Pringle, M., et al.: The cancer imaging archive (TCIA): maintaining and operating a public information repository. J. Digit. Imaging **26**(6), 1045–1057 (2013)
6. Dinsha, D., Manikandaprabu, N.: Breast tumor segmentation and classification using SVM and Bayesian from thermogram images. Unique J. Eng. Adv. Sci. **2**(2), 147–151 (2014)
7. Giesel, F.L., Mehndiratta, A., Locklin, J., McAuliffe, M.J., White, S., Choyke, P.L., Knopp, M.V., Wood, B.J., Haberkorn, U., von Tengg-Kobligk, H.: Image fusion using CT, MRI and PET for treatment planning, navigation and follow up in percutaneous RFA. Exp. Oncol. **31**(2), 106 (2009)
8. Gonzalez, R.C., Woods, R.E.: Image processing. Digit. Image Process. 2 (2007)
9. Horiuchi, T.: Colorization algorithm using probabilistic relaxation. Image Vis. Comput. **22**(3), 197–202 (2004)
10. Hutton, C., Bork, A., Josephs, O., Deichmann, R., Ashburner, J., Turner, R.: Image distortion correction in FMRI: a quantitative evaluation. Neuroimage **16**(1), 217–240 (2002)
11. Kar, A.K.: Bio inspired computing-a review of algorithms and scope of applications. Expert Syst. Appl. **59**, 20–32 (2016)
12. Khan, T.H., Mohammed, S.K., Imtiaz, M.S., Wahid, K.A.: Efficient color reproduction algorithm for endoscopic images based on dynamic color map. J. Med. Biol. Eng. **36**(2), 226–235 (2016)
13. Ko, K.-W., Jang, I.-S., Kyung, W.-J., Ha, Y.-H.: Saturation compensating method by embedding pseudo-random code in wavelet packet based colorization. J. Inst. Electron. Eng. Korea SP **47**(4), 20–27 (2010)
14. Kumar, Y.K.: Comparison of fusion techniques applied to preclinical images: fast discrete curvelet transform using wrapping technique & wavelet transform. J. Theor. Appl. Inf. Technol. **5**(6), 668–673 (2009)
15. Li, F., Zhu, L., Zhang, L., Liu, Y., Wang, A.: Pseudo-colorization of medical images based on two-stage transfer model. Chin. J. Stereol. Image Anal. **2**, 008 (2013)
16. Lipowezky, U.: Grayscale aerial and space image colorization using texture classification. Pattern Recogn. Lett. **27**(4), 275–286 (2006)
17. Martinez-Escobar, M., Foo, J.L., Winer, E.: Colorization of CT images to improve tissue contrast for tumor segmentation. Comput. Biol. Med. **42**(12), 1170–1178 (2012)

18. Mustaqeem, A., Javed, A., Fatima, T.: An efficient brain tumor detection algorithm using watershed & thresholding based segmentation. Int. J. Image Graph. Sig. Process. **4**(10), 34 (2012)
19. Noda, H., Korekuni, J., Niimi, M.: A colorization algorithm based on local map estimation. Pattern Recogn. **39**(11), 2212–2217 (2006)
20. Peruzzo, D., Arrigoni, F., Triulzi, F., Righini, A., Parazzini, C., Castellani, U.: A framework for the automatic detection and characterization of brain malformations: validation on the corpus callosum. Med. Image Anal. **32**, 233–242 (2016)
21. Popowicz, A., Smolka, B.: Overview of grayscale image colorization techniques. In: Celebi, E., Lecca, M., Smolka, B. (eds.) Color Image and Video Enhancement, pp. 345–370. Springer, Cham (2015). doi:10.1007/978-3-319-09363-5_12
22. Prema, C., Vinothini, G.A., Nivetha, P., Suji, A.S.: Dual tree wavelet based brain segmentation and tumor extraction using morphological operation. Int. J. Eng. Res. Technol. **2**. ESRSA Publications (2013)
23. Rosset, A., Spadola, L., Ratib, O.: Osirix: an open-source software for navigating in multidimensional DICOM images. J. Digit. Imaging **17**(3), 205–216 (2004)
24. Shiraishi, J., Katsuragawa, S., Ikezoe, J., Matsumoto, T., Kobayashi, T., Komatsu, K., Matsui, M., Fujita, H., Kodera, Y., Doi, K.: Development of a digital image database for chest radiographs with and without a lung nodule: receiver operating characteristic analysis of radiologists' detection of pulmonary nodules. Am. J. Roentgenol. **174**(1), 71–74 (2000)
25. Suzuki, K., Zhou, L., Wang, Q.: Machine learning in medical imaging. Pattern Recogn. **63**, 465–467 (2017)
26. Talamini, M.A., Chapman, S., Horgan, S., Melvin, W.S.: A prospective analysis of 211 robotic-assisted surgical procedures. Surg. Endosc. Interv. Tech. **17**(10), 1521–1524 (2003)
27. Tofangchiha, M., Bakhshi, M., Shariati, M., Valizadeh, S., Adel, M., Sobouti, F.: Detection of vertical root fractures using digitally enhanced images: reverse-contrast and colorization. Dent. Traumatol. **28**(6), 478–482 (2012)
28. Welsh, T., Ashikhmin, M., Mueller, K.: Transferring color to greyscale images. In: ACM Transactions on Graphics (TOG), vol. 21, pp. 277–280. ACM (2002)
29. Wernick, M.N., Yang, Y., Brankov, J.G., Yourganov, G., Strother, S.C.: Machine learning in medical imaging. IEEE Sig. Process. Mag. **27**(4), 25–38 (2010)

Fetoscopic Panorama Reconstruction: Moving from Ex-vivo to In-vivo

Floris Gaisser[1(✉)], Suzanne H.P. Peeters[2], Boris Lenseigne[1], Pieter P. Jonker[1], and Dick Oepkes[2]

[1] BioMechanical Engineering, Delft University of Technology, 2628CD Delft, The Netherlands
f.gaisser@tudelft.nl
[2] Department of Obstetrics, Leiden University Medical Center, 2333ZA Leiden, The Netherlands

Abstract. Twin-to-Twin Transfusion Syndrome (TTTS) is a condition that occurs in about 10% of pregnancies involving monochorionic twins. This complication can be treated with fetoscopic laser coagulation. The procedure could greatly benefit from panorama reconstruction to gain an overview of the placenta. Current state-of-the-art methods focus on panorama reconstruction in an ex-vivo setting. However, these methods fail in the in-vivo surgical setting. This paper describes the panorama reconstruction approach, the challenges posed by the in-vivo setting and the influence of these challenges on the panorama reconstruction. With experiments we show that the viewing quality is greatly reduced and that the limited motion of the fetoscope complicates and limits the precision of the image registration. We also identify the aspect necessary to shift from ex-vivo to in-vivo panorama reconstruction. Following our recommendations it should be possible to develop an approach that can be applied to TTTS surgery.

Keywords: TTTS · Fetoscopy · Panorama reconstruction · In-vivo

1 Introduction

Twin-to-Twin Transfusion Syndrome (TTTS) is a condition that occurs in about 10% of pregnancies involving monochorionic twins (twins with a shared placenta). In TTTS, an unbalanced exchange of blood caused by vascular anastomoses (shunts) in the placenta can lead to fatal complications for both twins [10]. Fetoscopic surgery is a common technique used to separate the fetal circulations by coagulating the connecting vessels with a laser beam. This technique increases the survival rate over other treatments [6]; even though far from optimal, it is currently a widely applied procedure [14]. To successfully perform this procedure, all vascular anastomoses have to be found. The surgeon moves the fetoscope around and scans the placenta for shunts, which have to be coagulated. This scanning procedure is complicated by the very limited view of the fetoscope, lack of landmarks and bad visibility conditions. A map of the relevant locations

M. Valdés Hernández and V. González-Castro (Eds.): MIUA 2017, CCIS 723, pp. 581–593, 2017.
DOI: 10.1007/978-3-319-60964-5_51

has to be made either on paper by an assistant, which is often inaccurate or in the mind of the surgeon. A rare spatial memory skill is required to do this. Therefore, only a few surgeons can actually perform this procedure successfully. Simplifying the procedure will result in a lower mental load and surgeons can perform this surgery with less complications. Key is to find all shunts on the placenta which is much easier with an overview of the whole placenta. This can be obtained by creating a large view panorama from fetoscopic video while the surgeon scans the placenta.

Reconstructing such views of the internal anatomical structures has been a large field of research and found many applications, such as retina [18], bladder [19] and oesophagus [5] reconstruction. However, for fetoscopic panorama reconstruction [8,12,16,20] this has mostly been done ex-vivo. Obtaining usable fetoscopic videos is challenging. Such data can only be obtained from living subjects and extending the length of the procedure increases the risks significantly. Therefore, a placenta after birth (ex-vivo) is generally used to obtain fetoscopic video. However, the ultimate goal is applying fetocsopic panorama reconstruction to TTTS surgery. Currently, surgeons are trained using simulators removing the risk to human life [14]. Such a simulator, if visually realistic, can also be used to obtain fetoscopic video close to in-vivo surgery settings. Furthermore, parameters influencing visibility, illumination and movement can be controlled. Since realistic fetoscopic data is available, it is possible to make a move towards in-vivo panorama reconstruction. Applying the state-of-the-art panorama reconstruction methods revealed some previously unknown issues [8]. Identifying the underlying problems and arising awareness, is crucial before panorama reconstruction can be applied in TTTS surgery. Therefore, this paper will focus on the differences between ex- and in-vivo fetoscopic settings and the resulting challenges for panorama reconstruction.

The aspects that are necessary to shift from ex- to in-vivo panorama reconstruction are identified in this paper. First, the general approach to panorama reconstruction is introduced (Sect. 2). Next, the surgical settings and its effects will be described and compared to the state-of-the-art in endoscopic panorama reconstruction (Sect. 3). Following, the resulting differences for image processing and the challenges posed are discussed (in Sect. 4). Further, the applicability of several image processing methods are evaluated on in-vivo Fetoscopic video (in Sect. 5). Finally, a conclusion is given with the steps (research topics) towards successful in-vivo fetoscopic panorama reconstruction.

2 Panorama Reconstruction

Panorama Reconstruction combines multiple images into one larger image [21]. These images individually contain only a part of the panorama. But every image has at least one other partly overlapping image. A chain of images can be created, so that each successive image is a pair with overlapping areas. From this chain the whole panorama can be created. In creating a panorama, it is generally assumed that the visual information of the images is on the same surface.

Thus, the panorama is a reconstruction of a plane (placenta) [8,16], cylinder (oesophagus) [5] or sphere (bladder) [2,19]. Furthermore, this means that all points in the overlapping area can be transformed with a rigid transformation to the other image. Key to panorama reconstruction is finding this rigid transformation. This process is generally referred to as image registration. There are two general approaches to this: dense [11] and interest-point based [3,21]. Dense methods use the whole image. The difference between the pixels of both images is minimized. This approach is accurate but also computational intensive. Furthermore, as there are many local minima in the optimization space, finding the correct transform cannot be guaranteed. Therefore, this method is generally used for stereo vision [11] or fine-tuning the panorama. Point-based methods find points in both images that describe the same location or area in both images. These pairs of points are then used to estimate the transformation [3]. This approach has many applications and is generally used in endoscopic panorama reconstruction [2,16,19]. These different steps are described in more detail in the next sections.

For point-based panorama reconstruction methods, only a select set of points are used. These are generally described as *keypoints*, because they describe key locations of the viewed scene [13]. As these keypoints have to be found reliably in multiple images of different conditions, they should be unique, easy to find repeatedly and accurately describe their location. The Harris corner detection method is an example of such a method [9]. Corners are considered to accurately describe their location and easy to find, however not unique. This method uses difference between neighboring pixels to detect a change along the x or y axis. If in both directions the change is large, a corner is found. Similarly edges are found with a change in only one direction. However, this method is not robust to scale and rotation changes. Therefore, methods such as the Difference of Gaussian add scale and rotation invariance to the detection of keypoints [13].

For every two overlapping images, pairs of keypoints have to be obtained from the set of keypoints created for each image. This process is called *matching*. Since there is no prior information on what keypoints are matching, each keypoint is described by its visual information. These descriptions are compared between images and the best matching pairs are chosen. A keypoint description should accurately describe the visual appearance as well as handle changes in orientation, scale, etc. Therefore local methods such as the Histogram of Gradients are generally used as in the SIFT method [13]. Other methods include SURF [1], BRIEF [4] and ORB [17]. Even though, assuming keypoints can be described perfectly, the existence of multiple visually similar keypoints is not taken into account. Therefore, not all matches can be considered correct matches (*mismatches*). To improve the ratio of correct matches, a common practice is to also review the second best match. If the difference between the best and second best match is small, then it is probable that there are multiple visually similar keypoints. Furthermore, the matches can be validated after transform estimation based on the rigid transform assumption.

The final step in image registration is estimating the transformation between the two images. In panorama reconstruction, this transformation is assumed to be a *rigid transformation*. This means that all points in the image keep their spatial relation and the surface is not deforming due to the transformation. Therefore, a point in the first image $[x_1, y_1]$ can be transformed by matrix multiplication to a point in the second image $[x_2, y_2]$. There exist two types of rigid transformations: affine and perspective. In affine transformations the camera has no out-of-plane rotations and the viewing direction is generally considered to be perpendicular to the surface. The camera can move in the image plane and viewing direction. Therefore, there is rotation (θ), translation (t_x, t_y) and scale (s) as described in Eq. 1. As a side note; Skew is considered to be part of affine transformation as well. However, skew changes the spatial relation and therefore not part of rigid transformations.

$$\begin{bmatrix} x_2 \\ y_2 \\ 1 \end{bmatrix} = \begin{bmatrix} s + \cos\theta & -\sin\theta & t_x \\ \sin\theta & s + \cos\theta & t_y \\ 0 & 0 & 1 \end{bmatrix} \cdot \begin{bmatrix} x_1 \\ y_1 \\ 1 \end{bmatrix} \tag{1}$$

For perspective transformations also the out-of-plane rotations are considered. This introduces four extra parameters in the transformation matrix (2), describing the rotation around the x and y axis, (p_x, p_y, s_x, s_y) and makes use of homogeneous coordinates.

$$\begin{bmatrix} x'_2 \\ y'_2 \\ w'_2 \end{bmatrix} = \begin{bmatrix} s + \cos\theta & s_y - \sin\theta & t_x \\ s_x + \sin\theta & s + \cos\theta & t_y \\ p_x & p_y & 1 \end{bmatrix} \cdot \begin{bmatrix} x_1 \\ y_1 \\ 1 \end{bmatrix} \tag{2}$$

$$x_2 = x'_2 / w'_2$$
$$y_2 = y'_2 / w'_2$$

The parameters of the rigid transform can be estimated from the matches. A common approach is least-squares, which minimizes the squared error. However, this method cannot handle large number of mismatches, as they have a large influence on the error. Therefore, another common method is *RANSAC* [7]. This method randomly takes samples and solves for the parameters. Then a confidence is created based on the projection error of the remaining probable pairs. After a given amount of iterations, the transformation with the highest confidence is used. However, small inaccuracies in the keypoint locations have a large influence on the accuracy of the transform. Therefore this method can handle mismatches, but has a sub-optimal accuracy. To both account for mismatches and inaccurate keypoint locations, the method *LmedS* can be used [15]. It uses the median of the error to reject mismatches. Moreover, it can find an optimal solution by minimizing the error. However, with more than 50% of mismatches, an optimal solution can not be guaranteed, due to local minima in the optimization space.

A panorama is reconstructed from a chain of image pairs, based on the assumption that all images view the same surface. However, with perspective transformations it does not mean that the image planes are parallel with the viewed surface. Therefore, a suitable common projection surface has to be chosen, similarly to the physical surface. However, these image registrations describe the relation between images and not to the projection surface. Therefore, a reference image is generally chosen and its image plane as the base for the projection surface. Every following image is transformed to the reference image using the image registrations. However, image registrations generally have small inaccuracies due to mismatches and localization errors of the keypoints. As images are transformed down the chain to the reference image, an error is build up in the overall projection. This error can become large, which results in visual inconsistencies in the panorama.

3 Differences in Setting

The differences between an ex-vivo research setting and our in-vivo simulator (Fig. 1a) will be described in this section. Each paragraph describes one specific aspect and examples of the visibility conditions are shown in Fig. 3.

Ex-vivo research uses a placenta after birth, which is larger compared to an earlier stage of the pregnancy. Furthermore, the bloodvessels of the placenta are dye injected for extra contrast with the underlying tissue. Figure 1 compares the ex-vivo dye-injected with our simulator placenta. To record data, the placenta is placed on a flat working space, making the placenta also flat. Therefore, in contrast to in-vivo settings the observed surface is a plane. Furthermore, compared to in-vivo settings, the placenta does not move and gaining a good view is possible from every angle.

With ex-vivo research, the impact of the amniotic fluid is overlooked. This fluid is generally far from clear, as the fetus normally pees in the amniotic fluid, giving it a yellow-brownish color. However, if the fetus is in distress, which is often the case with TTTS, the fetus might release some bowel movement, giving the amniotic fluid a green turbid color, reducing the range of visibility.

In the ex-vivo setting, the illumination condition can be completely controlled. There can be ambient lights as well as high intensities of light. With this controllable setting, near optimal illumination conditions can be created at all times. However, with in-vivo settings, there is no ambient light. The only

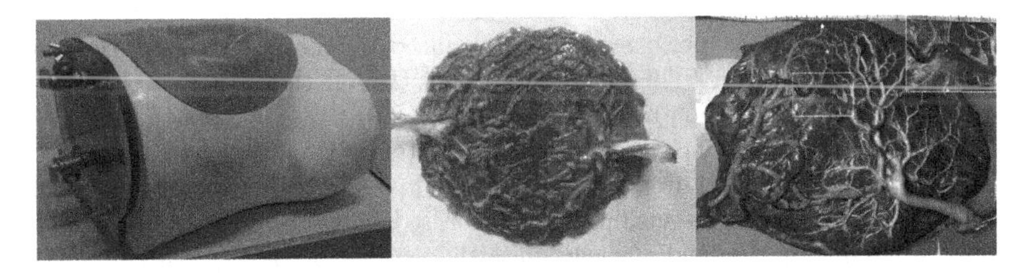

Fig. 1. (a) Our simulator and (b) placenta, (c) compared with dye-injected placenta

light source is from the endoscope itself creating an uneven distribution of light. There is too much light in the center of the image and it gets too dark towards the edges. Moreover, too strong light might blind the fetus, so that only less ideal illumination conditions can be created.

The motion of the fetoscope is the last difference between an ex- and in-vivo setting, and often not considered. In the work of Reeff et al. [16] and Tella-Amo et al. [20] the motion of the endoscope is described as an outwards spiralling motion. This mostly in-plane translation motion is shown in Fig. 2a. However, surgeons cannot make an outwards spiralling motion. Instead, they start at the umbilical cord of the recipient and follow veins one by one to the end and back. Furthermore, the motion of the fetoscope is restricted at the point of entry in the in-vivo setting. This restriction only allows a combination of in-plane translation and out-of-plane rotations or forward motion as shown in Fig. 2c.

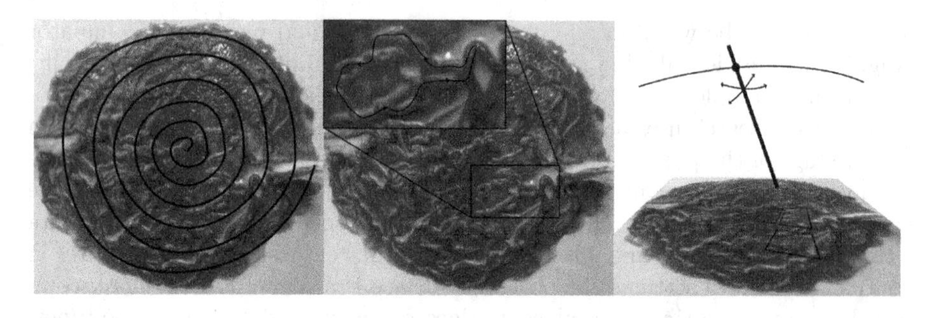

Fig. 2. Fetoscopes movement (a) ex-vivo: spiral motion (b) in-vivo: followed path (c) perspective motion due to rotation around entry point

The combination of bad visibility and illumination conditions force the surgeon to move the endoscope closer to the placenta. However, this reduces the imaged surface significantly. The sample views of related work contain multiple bloodvessels [12,16]. Whereas for in-vivo images, one or two bloodvessels, sometimes a crossing or split can be seen.

4 Influence on Panorama Reconstruction

The image registration performance is influenced by the previously described differences in the setting. This section describes this effect for each of the steps of panorama reconstruction as described in Sect. 2.

Keypoint methods internally create an intensity map of the difference between pixels in x and y direction. Dye injected placentas have better contrast, while illumination and the visibility condition reduce the amount of observable contrast. Reducing the field of view limits the amount of visible structure even more. Therefore the more complicated viewing conditions reduce the number of detected keypoints.

Generally it is assumed that the strongest points in the intensity map represent corners. However, this assumption is not valid when the image contains only a few or no corners. Because of image noise a point along an edge is seen as stronger than other points along the edge and thus selected as a keypoint. Since noise is random, this keypoint is not reproducible and as shown in Fig. 3 occurs quite strongly when there is not enough light. Moreover, keypoints have similar appearance along an edge, thus the location is not unique. These aspects make the selected keypoints unreliable, as they are not unique and not reproducible.

Matches are obtained from the visual appearance around each keypoint. However, the underlying tissue of the placenta does not contain a lot of structure, creating weak descriptions. Moreover, most keypoints are on an edge of a bloodvessel. Since bloodvessels have very similar appearance, their description is also very similar. The weak and similar description of keypoints complicate the matching process, resulting in less reliable matches.

The motion of the fetoscope has a direct relation to the transformation between two successive images. The motion of the fetoscope is not restricted in ex-vivo experiments. Therefore, the fetoscope is generally moved sideways, creating translation transformations. However, because of rotation around the entry point, perspective transformations as shown in Fig. 2c are created. In previous work we showed that estimating translational motion results in a two times lower pixel error compared to other types of motion [8]. Furthermore, rotation around the entry point changes the distance to the placenta. Resulting in a change of light and unfavourable illumination conditions is the result as shown in the two most outer columns of Fig. 3.

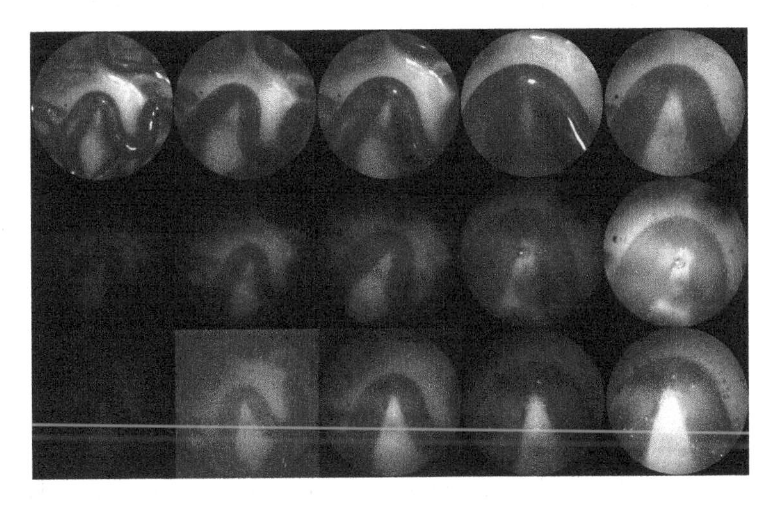

Fig. 3. Variations in viewing conditions. Top row left to right: ex-vivo - far, ex-vivo with water - far, ex-vivo - nominal, ex-vivo - close, ex-vivo with water - close. Middle row: yellow liquid, bottom row: green turbid liquid; left to right: far - dark, far - nominal, nominal for both, close - nominal, close - bright (Color figure online)

With an outward spiralling motion of the fetoscope, a continuous set of loops can be created between two images between the inner and outer spiral allowing for bundle adjustment or loop-closure. The difference between an incorrect and a feasible panorama reconstruction lies with the refinement step as can be seen in Fig. 2 of the work of Reeff et al. [16]. However, this spiralling motion is not possible with in-vivo fetoscopy and combined with all previous mentioned aspects, panorama reconstruction is much more complicated compared to ex-vivo settings.

5 Experiments

To investigate the influence of the previously described viewing conditions on image registration experiments are devised. Images are captured for every condition using our placenta (Fig. 1b) and for in-vivo settings with our similator (Fig. 1a). A 5 mm solid core endoscope with a medical xeon light source and a GigVision BlackFly camera have been used. The effective area of the endoscope is 1030×1030 pixels and after undistortion the inner square has an area of 850×850 pixels.

The visibility condition caused by the color and turbidity of the amniotic fluid is varied in three settings; One simulating the ex-vivo setting and 2 in-vivo settings simulating the two different types of amniotic fluid that can be encountered; normal *yellow* fluid and the distressed case of *green turbid* fluid. These settings are created by adding dye to the water and shown in Fig. 3 middle and bottom row.

The illumination condition is varied in ex-vivo and in-vivo settings. The *ex-vivo* setting is created by placing the placenta on a table with enough abient light. For the in-vivo settings, the placenta is placed inside the simulator and the illumination is manually adjusted to create a *dark*, *nominal* and *bright* setting.

The field of view is varied by changing the distance to the placenta. Three settings are created: *close*, *nominal* and *far*, respectively at ± 1, 2 and 3 cm from the placenta. The far setting is choosen such that the camera is at it limitations for the green turbid fluid and the close setting always gives a clear view.

The different visibility and illumination conditions combined with different viewing distances can be seen in Fig. 3. A set of sequential images is recorded for all possible variation combinations along the path described in Fig. 2b. Every following image is taken such that the movement is about $\frac{1}{3}$ of the visible range, resulting in more images for decreasing quality and viewing area.

Experiment 1: Number of Keypoints. With decreasing quality of the viewing condition also the number of keypoints is expected to decrease. In this experiment three different keypoint methods are evaluated; SIFT, SURF and ORB. Figure 4 shows the number of obtained keypoints for changing illumination, visibility condition and field of view. As expected the results show that with decreasing distance the number of keypoints decrease. For the illumination condition more light gives better contrast and thus more keypoints. However, for the far

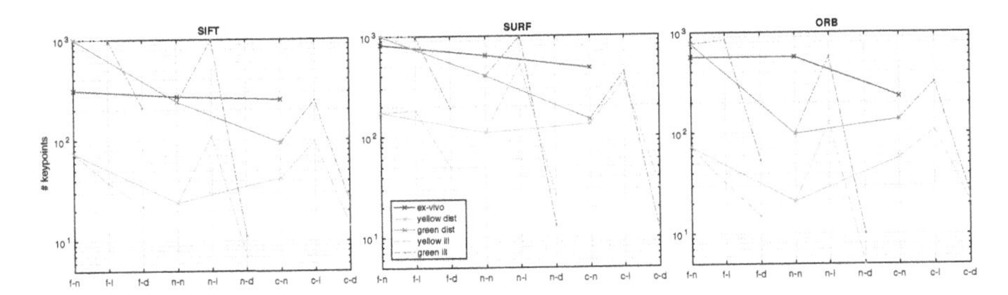

Fig. 4. Comparison of number of keypoints; x-axis labels are far, nominal and close distance and nominal, light and dark illumination. Example: f-l is far and light

distance, increase in contrast is lost due to the increased distance. Furthermore, for the green turbid liquid, there is not enough light, resulting in much noise and many keypoints are obtained on the noise itself.

Experiment 2: Reproducability of Keypoints. Lack of structure in the form of corners reduces the number of reproducible keypoints. By manually establishing the transformation between two successive images, the ratio of reproducable keypoints can be obtained. Figure 5 shows for the different keypoint methods the number of retained keypoints for varying viewing conditions. The number of reproducable keypoints is about 15–25% of the number of detected keypoints. However, looking at the results of individual images pairs the variation is large, sometimes many and sometimes no reproducable keypoints at all are detected.

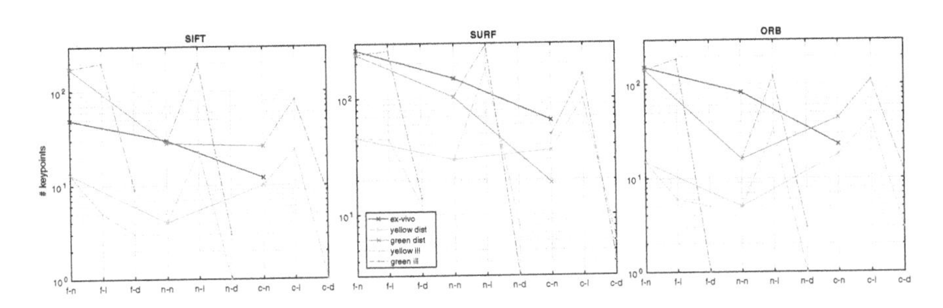

Fig. 5. Comparison of reproducable keypoints; x-axis labels are far, nominal and close distance and nominal, light and dark illumination. Example: f-l is far and light

Experiment 3: Matchability of Keypoints. The matching performance is limited due to similar appearance of the bloodvessel and weak description of the keypoints. The previously obtained keypoints and manually established transformations are used to evaluate the matching performance. For the ex-vivo situation the ratio of correctly matched keypoints is about 10%. However, for the in-vivo situation three groups can be created; First, good visibility condition, with close to ex-vivo matching. Second, low illumination condition, resulting in too much noise and many unreliable keypoints. Third, low structure situation, with too few keypoints to do image registration. The latter two groups have close to zero correctly matchable keypoints and no image registration can be obtained.

6 Discussion

With the previous experiments we have shown, that the visibility conditions encountered in in-vivo TTTS fetoscopic surgery complicate the image registration process. The number and the reproducability of keypoints is reduced to the point that no valid matches can be found. This reduction in performance can be explained by three key aspects; detecting reliable keypoints, matching of keypoints and motion of the fetoscope.

Corners are the source of reliable keypoints, even though they can be found on a placenta, the number of corners in images from ex-vivo settings is limited. Changing to an in-vivo setting reduces the contrast and the field of view. Less corners and with less constrast appear in the image. Furthermore, with reduced illumination the image contains more noise. Since keypoint detection methods adjust themselves to the structure present in the image. Points along the edge of bloodvessel are obtained as keypoints, though they cannot be considered reliable keypoints.

Keypoints are matched based on their visual appearance. For corners, where two bloodvessels cross or split, there is clear and unique visual information. However, for a keypoint on the edge of a bloodvessel, the visual appearance is very similar to any point along the edge of a bloodvessel. This is even true for a curved vein, but the feature is rotated. This results in many mismatches and only a few correct matches.

The fetoscope can only change the view in lateral position by rotating around the point of entry. This motion also changes the distance to the placenta, thus creating unfavourable viewing conditions. Both the distance, but also the illumination condition that has to be manually adjusted, create visibility conditions that are unfavourable for image registration. Moreover, this type of motion also complicates the image registration as more parameters have to be estimated, thus requiring more and better matches.

7 Conclusion

The goal of this paper is to identify and rise awareness to the challenges encountered with in-vivo fetoscopic panorama reconstruction. In the previous section we have discussed three key aspects that complicate image registration. In this section we recommend four key areas to improve the image registration approach and create a panorama reconstruction system applicable to TTTS.

The first area of improvement would be the keypoint detection and description method, so that it can better handle unfavourable viewing conditions. In previous work a step in that direction has already been taken [8]. With deep-learning better keypoint matching is achieved. However, keypoints are selected in a grid, decreasing the transformation estimation accuracy. Possibly this approach can be extended to also detect points or areas of interest. With better keypoint matches higher image registration performance and better panoramas can be obtained.

Second, the panorama reconstruction process could be optimized. Generally a chain of images is used to reconstruct a panorama. When an image registration in the chain is inaccurate or unavailable, as noted in experiment 3, this has a large influence on all following image pairs. However, if this image registration can be detected or even taken out, the quality of the panorama would improve greatly. Therefore, we propose not to create chains, but register the new image to an area of the panorama. This area is obtained from a motion model and the location of the previous image in the panorama. If the variation in the matches is large, the image registration can be considered as inaccurate, and therefore not used in the panorama reconstruction.

Third, the motion of the fetoscope can be improved by providing extra information to the surgeon. The distance to the placenta is changed by moving the fetoscope around. By calculating the fetoscopes position relative to the placenta, the change in distance can be predicted. Unfavourable viewing conditions can be predicted and by giving feedback to the surgeon to move the fetoscope fore- or backward, more favourable conditions can be maintained. Moreover, when the image registration performance is low, the system can ask the surgeon to move back to a previous position and obtain new and better images. Overall, the surgeons should be considered an integral part of the panorama reconstruction process.

Lastly, the equipment also plays a role in the image registration performance. The field of view is depended on the viewing angle of the fetoscope, thus choosing a good fetoscope is crucial. The changing illumination condition can be managed using a high dynamic range camera and higher quality images in a larger range of illumination conditions can be obtained. Improving the viewing conditions will result in better keypoint matches and more accurate image registration.

A system applicable to TTTS surgery can be created following the previous recommendations. Most important is to obtain good keypoints and matches, despite the much more difficult viewing conditions. Better image registration can be achieved by improving keypoint methods, as well as limiting bad viewing conditions. Including the surgeon in the process and improving the equipment will improve the panorama reconstruction. Better image registration combined with a better panorama reconstruction process will allow to create the required overview. This supports the surgeon in better performing the TTTS procedure by reducing the risk of missing connecting bloodvessel and the length of the procedure, thus increasing the chance of survival.

References

1. Bay, H., Ess, A., Tuytelaars, T., Van Gool, L.: Speeded-up robust features (surf). Comput. Vis. Image Underst. **110**(3), 346–359 (2008)
2. Behrens, A., Stehle, T., Gross, S., Aach, T.: Local and global panoramic imaging for fluorescence bladder endoscopy. In: 2009 Annual International Conference of the IEEE Engineering in Medicine and Biology Society, pp. 6990–6993. IEEE (2009)
3. Brown, M., Lowe, D.G.: Automatic panoramic image stitching using invariant features. Int. J. Comput. Vis. **74**(1), 59–73 (2007)
4. Calonder, M., Lepetit, V., Strecha, C., Fua, P.: BRIEF: binary robust independent elementary features. In: Daniilidis, K., Maragos, P., Paragios, N. (eds.) ECCV 2010. LNCS, vol. 6314, pp. 778–792. Springer, Heidelberg (2010). doi:10.1007/978-3-642-15561-1_56
5. Carroll, R.E., Seitz, S.M.: Rectified surface mosaics. Int. J. Comput. Vis. **85**(3), 307–315 (2009)
6. Chmait, R.H., Kontopoulos, E.V., Korst, L.M., Llanes, A., Petisco, I., Quintero, R.A.: Stage-based outcomes of 682 consecutive cases of twin-twin transfusion syndrome treated with laser surgery: the usfetus experience. Am. J. Obstet. Gynecol. **204**(5), 393-e1 (2011)
7. Fischler, M.A., Bolles, R.C.: Random sample consensus: a paradigm for model fitting with applications to image analysis and automated cartography. Commun. ACM **24**(6), 381–395 (1981)
8. Gaisser, F., Jonker, P.P., Chiba, T.: Image registration for placenta reconstruction. In: Proceedings of the IEEE Conference on Computer Vision and Pattern Recognition Workshops, pp. 33–40 (2016)
9. Harris, C., Stephens, M.: A combined corner and edge detector. In: Alvey Vision Conference, vol. 15, p. 50 (1988). Citeseer
10. Lewi, L., Deprest, J., Hecher, K.: The vascular anastomoses in monochorionic twin pregnancies and their clinical consequences. Am. J. Obstet. Gynecol. **208**(1), 19–30 (2013)
11. Li, Y., Shum, H.-Y., Tang, C.-K., Szeliski, R.: Stereo reconstruction from multiperspective panoramas. IEEE Trans. Pattern Anal. Mach. Intell. **26**(1), 45–62 (2004)
12. Liao, H., Tsuzuki, M., Kobayashi, E., Dohi, T., Chiba, T., Mochizuki, T., Sakuma, I.: Fast image mapping of endoscopic image mosaics with three-dimensional ultrasound image for intrauterine treatment of twin-to-twin transfusion syndrome. In: Dohi, T., Sakuma, I., Liao, H. (eds.) MIAR 2008. LNCS, vol. 5128, pp. 329–338. Springer, Heidelberg (2008). doi:10.1007/978-3-540-79982-5_36
13. Lowe, D.G.: Distinctive image features from scale-invariant keypoints. Int. J. Comput. Vis. **60**(2), 91–110 (2004)
14. Peeters, S.: Training and teaching fetoscopic laser therapy: assessment of a high fidelity simulator based curriculum. Ph.D. thesis, Leiden University Medical Center, January 2015
15. Rousseeuw, P.J.: Least median of squares regression. J. Am. Stat. Assoc. **79**(388), 871–880 (1984)
16. Reeff, M., Gerhard, F., Cattin, P.C., Székely, G.: Mosaicing of endoscopic placenta images. In: Informatik fr Menschen, vol. 1. Hartung-Gorre Verlag (2006)
17. Rublee, E., Rabaud, V., Konolige, K., Bradski, G.: ORB: an efficient alternative to SIFT or SURF. In: 2011 IEEE International Conference on Computer Vision (ICCV), pp. 2564–2571, November 2011

18. Seshamani, S., Lau, W., Hager, G.: Real-time endoscopic mosaicking. In: Larsen, R., Nielsen, M., Sporring, J. (eds.) MICCAI 2006. LNCS, vol. 4190, pp. 355–363. Springer, Heidelberg (2006). doi:10.1007/11866565_44
19. Soper, T.D., Porter, M.P., Seibel, E.J.: Surface mosaics of the bladder reconstructed from endoscopic video for automated surveillance. IEEE Trans. Biomed. Eng. **59**(6), 1670–1680 (2012)
20. Tella-Amo, M., Daga, P., Chadebecq, F., Thompson, S., Shakir, D.I., Dwyer, G., Wimalasundera, R., Deprest, J., Stoyanov, D., Vercauteren, T., et al.: A combined EM and visual tracking probabilistic model for robust mosaicking: application to fetoscopy. In: Proceedings of the IEEE Conference on Computer Vision and Pattern Recognition Workshops, pp. 84–92 (2016)
21. Zitova, B., Flusser, J.: Image registration methods: a survey. Image Vis. Comput. **21**(11), 977–1000 (2003)

Finite Element Based Interactive Elastic Image Registration

Yechiel Lamash[1(✉)], Anath Fischer[1], and Jonathan Lessick[2]

[1] Technion – Israel Institute of Technology, 32000 Haifa, Israel
shiliksterQgmail.com
[2] Rambam Health Care Campus, Haifa, Israel

Abstract. Image registration is a central field in biomedical image analysis. Despite the large body of work on automatic image registration algorithms, in practice, results are often imperfect and can benefit from manual editing. In the current study, we propose a novel interactive registration tool that can be used for editing and refinement of automatic image registration results. The method is based on displacements to control points applied by the user to a finite element mesh representation of an organ of interest. The user-applied-displacements are translated to local forces using the organ's elastic properties. The local user-derived forces are applied on the mesh and deform the image accordingly. We test the current method on 2D x-ray hand simulated data with non-continuous motion field and on 3D cardiac CT image data.

1 Introduction

Image registration and segmentation are central fields in medical image analysis [7,8]. However, while for segmentation problems, a large set of interactive editing tools exist [9], for elastic image registration problems only very few methods were proposed [14–16,18,20]. Imperfect results of automatic algorithms often lead to sub-optimal diagnosis and less accurate outcomes of many biomedical image analyses such as organ/tumor perfusion, tumor follow up, motion extraction etc. Several algorithms for interactive image editing were proposed for regular camera 2D images [2]. These, however, are not suitable for objects with elastic properties and higher dimensional data. The main challenge in a registration editing tool here is how to warp images of 3D objects using interactive manipulation in 2D slices. Several common image registration algorithms are based on forces applied to control points with associated basis functions using differences in image intensities. It is trivial, therefore, that such forces could also be applied interactively by the user. The B-spline basis functions [17] are very common in deformation modeling of many biomedical image registration problems. However, when used interactively, the relation between the displacement of a control point on a regular grid, to the deformation applied to a nearby organ's boundary is not intuitive. Other common basis functions are radial or thin plate spline basis functions [3,4,16,20], wherein the control points can be positioned on the organ edges and landmarks. These, however, affect the entire image, and do not

M. Valdés Hernández and V. González-Castro (Eds.): MIUA 2017, CCIS 723, pp. 594–604, 2017.
DOI: 10.1007/978-3-319-60964-5_52

allow for efficient, local control of an object's shape, independent of surrounding background. In addition, in such frameworks, the elastic properties of the organ are not accounted for.

In the current study we developed an interactive image registration algorithm that is based on finite element (FE) analysis. We assume that a segmentation of the organ in one frame is available, so that a mesh of the organ can be generated. Our assumption is straightened by the availability of many organ segmentation algorithms [8]. We use the elastic properties of the segmented organ to generate a mesh representation and calculate the organ's stiffness matrix. The user can then interactively drag the mesh control points and deform the image while keeping the object's elasticity constraints. Finite element approach to warping images interactively presents several advantages: First, unlike methods that work on a regular grid of control points, differentiation between regions with different material properties can be obtained. Second, elastic properties of the deformed region are used, rather than an arbitrary smooth interpolation. Third, when used on a sub-volume of interest, less control points and less computation time are needed. Finally, the method can be connected with an automatic registration method [12,13].

2 Method

2.1 Mesh Generation

We begin our analysis with generating a mesh of the organ of interest. Selection of control point locations is important for efficient interactive editing. Control points should be positioned on a set of image slices to allow the user to modify them on the image background. Setting control points uniformly in the organ's boundary and volume is recommended for compact number of control points with efficient mesh representation.

For 2D images, a multi-stage approach of the Farthest point sampling (FPS) algorithm [6] can be used: The first stage is initiated with points of the organ's boundary, and FPS is used for sub-sampling. In the second stage, points are added iteratively from a dense sampling of the organ volume to boundary set of points (determined in the first stage).

For 3D deformations, it is beneficial to distribute control points equally on a regular grid inside the organ of interest or on a cylindrical axis system (Fig. 2). Such an arrangement allows to view and modify the image in two perpendicular planes simultaneously. After creating a relatively uniform sampling of the domain, we used the Delaunay algorithm [5] to mesh the object domain. We use segmentation again to eliminate elements that are external to the organ of interest. To ensure that the global orientation is consistent (i.e. the Jacobians have equal signs), the ordering of the vertices in 2D meshes should be either clockwise or counterclockwise for all triangles. For 3D meshes, the ordering of the three vertices i, j, k when looking from vertex v should be either clockwise or counterclockwise for all tetrahedra.

Figure 1 demonstrates 2D mesh generation using the multi-stage FPS algorithm. Figure 2 demonstrates 3D mesh generation of the left ventricle where the control points where selected with equal angle sampling in a cylindrical axis system representation.

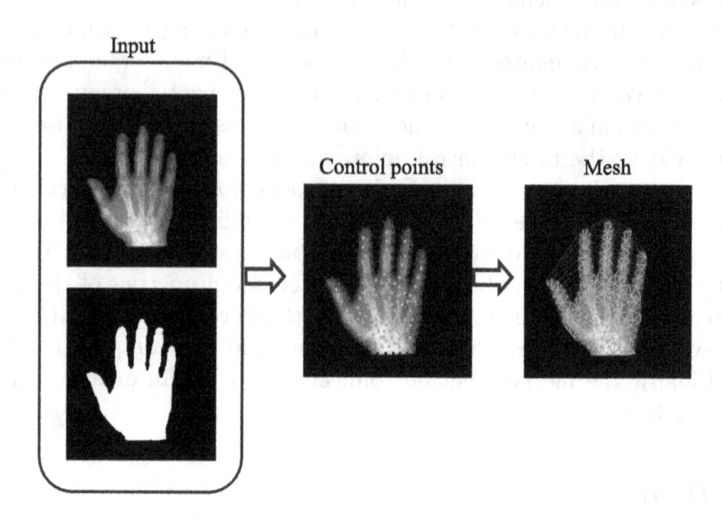

Fig. 1. The proposed generic meshing algorithm. On the left are the image and the segmented object inputs. The middle image describes the input (green dots) and the final result (magenta dots) of the FPS algorithm. The right images show the Delaunay triangulation. (Color figure online)

2.2 Biomechanical Model for Tissue Deformation

Tissue deformation can be characterized as a minimum variation of total energy described as

$$E = \frac{1}{2}\int_{\Omega} \sigma^T \varepsilon d\Omega + \int_{\Omega} f u d\Omega \tag{1}$$

where, Ω represents the continuous domain of an elastic body, f is the external force and u is the displacements in Ω. σ and ε are the stress and strain vectors, respectively.

In linear three dimensional continuum mechanics, a strain tensor is written as $[\varepsilon_{xx}, \varepsilon_{yy}, \varepsilon_{zz}, \varepsilon_{xy}, \varepsilon_{yz}, \varepsilon_{xz}]^T$ and stress vector is written as $[\sigma_{xx}, \sigma_{yy}, \sigma_{zz}, \sigma_{xy}, \sigma_{yz}, \sigma_{xz}]^T$. The strain-stress relation is given by Hooke's law

$$\sigma = C \cdot \varepsilon \tag{2}$$

where $C[6 \times 6]$ is the material stiffness tensor.

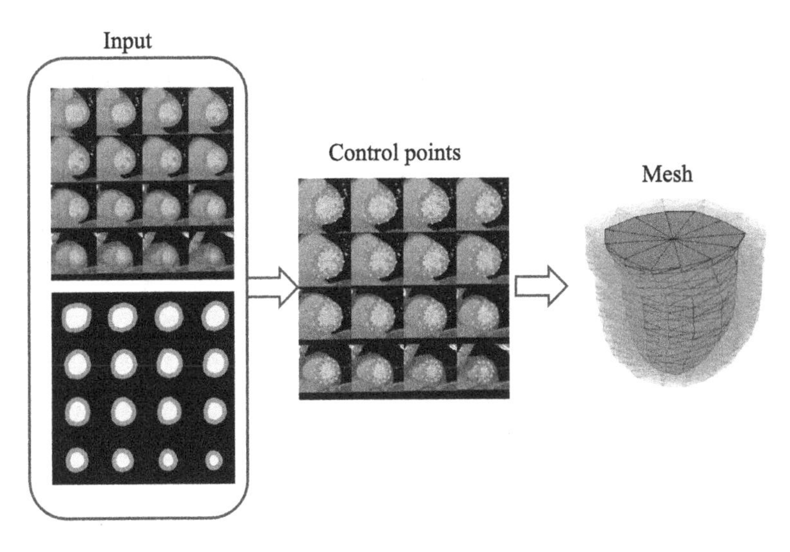

Fig. 2. Meshing of 3D cardiac CT data. Left images show the short axis and segmentation slices of the left ventricle. Middle image shows the control points using cylindrical coordinates. Right images show the resulting 3D mesh.

By assuming isotropic material, the tensor can be expressed using two constants,

$$C = \frac{E}{(1+\nu)(1-2\nu)} \begin{bmatrix} 1-\nu & \nu & \nu & 0 & 0 & 0 \\ \nu & 1-\nu & \nu & 0 & 0 & 0 \\ \nu & \nu & 1-\nu & 0 & 0 & 0 \\ 0 & 0 & 0 & 1/2-\nu & 0 & 0 \\ 0 & 0 & 0 & 0 & 1/2-\nu & 0 \\ 0 & 0 & 0 & 0 & 0 & 1/2-\nu \end{bmatrix} \tag{3}$$

where E, ν are the material Young's modulus and Poisson's ratio respectively.

Using an FE approach, the problem domain Ω can be divided into discrete elements, with each element consisting of several nodes. The continuous displacement field is obtained through the interpolation of nodal displacements using the shape functions N_i^e.

$$u_e(x) = \sum_{i=1}^{4} N_i^e(x)\hat{u}_i^e \tag{4}$$

where,

$$N_i^e(x) = \frac{1}{6V^e}(a_i^e + b_i^e x + c_i^e y + d_i^e z) \tag{5}$$

$\hat{u}_i^e$ are the element's nodal displacements.

The minimization of total energy in (1) using the variation principle simplifies to the following system of linear algebraic equations [21]:

$$K \cdot \hat{u} = f \tag{6}$$

where f is vector of applied forces, $\hat{u}$ is a vector of nodal displacements. K is the global stiffness matrix given by applying the assembly operator on the elements

$$K = \sum_e \left(\int_{V^e} (B^T C B) \right) \tag{7}$$

where B is strain-displacement matrix that relates nodal strain, ε, to nodal displacement $\hat{x}$, as $\varepsilon = B\hat{x}$. For tetrahedral element, $B[6 \times 12]$ is as follows:

$$B = \begin{bmatrix} B_1 \; B_2 \; B_3 \; B_4 \end{bmatrix}$$

where each sub-matrix B_i is as follows:

$$B_i = \begin{bmatrix} N_{i,x} & 0 & 0 \\ 0 & N_{i,y} & 0 \\ 0 & 0 & N_{i,z} \\ N_{i,y} & N_{i,x} & 0 \\ 0 & N_{i,z} & N_{i,y} \\ N_{i,z} & 0 & N_{i,x} \end{bmatrix}$$

where $N_i^e (i = 1, .., 4)$ is the ith shape function.

B_i can be obtained using algebraic manipulations using the element's matrix of coordinates [21].

2.3 Interactive User-Derived and Regularization Forces

The solution for (6) requires boundary conditions that add constraints to the FE system of linear equations. Here instead, we add the identity matrix to the stiffness matrix with a weighting parameter α to inhibit the movement of the organ. In other words, the user-derived local force deforms the organ which has separate forces that keep it in its place. Such forces can be seen as motion resistors that work against the user-derived force.

The force to the control point is extracted from the control point's displacement applied by the user. The user displaces control points one at a time, the local displacement $\hat{u}_{local}$ and the local force f_{local} are related by:

$$f_{local} = k\hat{u}_{local} \tag{8}$$

where, k is a predefined elastic scalar.

After calculating the local force on the displaced control point, it is inserted at the appropriate indices of the global force vector f, with zero values at the rest of the indices. Global mesh displacements are now obtained by:

$$\hat{u} = (K + \alpha I)^{-1} \cdot f \tag{9}$$

where, K is the elastic global stiffness matrix. α is a predefined parameter that controls mesh inhibitory forces.

2.4 3D Image Warping Using Barycentric Coordinates

The current section describes the implementation of the image warping model. (Formulation in this section is for 3D representation, but can be easily adapted to 2D objects as well). Rasterization is an old problem, well known in computer graphics, that find the inner voxels of each tetrahedron. Sermesant et al. [19] used a multi-stage approach for this problem by first considering the tetrahedron intersection with each the relevant horizontal planes. Then, finding the bounding edges and inner voxels located in horizontal lines of each such plane. We propose here a more convenient and less complicated solution to the rasterization problem compared to [19].

Given a tetrahedron T,any point $p \in T$ divides it into four sub tetrahedrons. The vector e of the point p with respect to vertex v can be expressed by

$$e = \alpha e_i + \beta e_j + \gamma e_k \tag{10}$$

where the barycentric coordinates $(\alpha, \beta, \gamma) \in (0, 1)$ are the volume ratios between each sub-tetrahedron and tetrahedron T.

$$\alpha = \frac{\det(e, e_j, e_k)}{\det(e_i, e_j, e_k)} \beta = \frac{\det(e_i, e, e_k)}{\det(e_i, e_j, e_k)} \gamma = \frac{\det(e_i, e_j, e)}{\det(e_i, e_j, e_k)} \tag{11}$$

e_i, e_j, e_k are the tetrahedron edge vectors with respect to vertex v.

The warping within each tetrahedral element is given by:

$$W_e(\alpha, \beta, \gamma) = v + \alpha e_i + \beta e_j + \gamma e_k \tag{12}$$

We implement the image warping as follows: For each tetrahedron, we find the barycentric coordinates (α, β, γ) of its inner voxels that reside on the image grid. We then find the respective value in the reference image and insert it in the warped image. To find the grid of voxels located inside the tetrahedron, we take the grid pixels of the minimal box that bounds the tetrahedron and look up the voxels whose barycentric coordinates apply:

$$0 \leq \alpha \leq 1, 0 \leq \beta \leq 1, 0 \leq \gamma \leq 1, \alpha + \beta + \gamma \leq 1.$$

2.5 Experiment

To quantify the performance of the current tool, we measured reduction in the sum of square difference (SSD) with respect to the initial difference as a function of the users' time and number of mouse (click and drag) operations. We used two types of datasets: 2D hand articulation and 3D cardiac CT data.

The 2D hand articulation dataset includes six articulations of a hand [1] that were obtained using FE simulation. As only part of the fingers were displaced, the motion field is only piecewise continuous, and the background is partly occluded or contains holes. In such cases, we would like to use the elasticity constraint to operate on the inner pixels of the deforming object with rapid motion from its background. To fill holes due to rapid movements, we initialize the warped image by the static image and then update the pixels inside the deformed mesh. The cardiac CT dataset includes six pairs of 3D images of the diastolic and systolic phases of the cardiac cycle from six different patients. The left ventricle was segmented semi-automatically. The diastolic phases were warped to match the diastolic ones by the user. To reduce warping runtime, the image was deformed in the LV mesh region (the LV myocardial and blood pool area). To create continuality with the background, the warped images were initialized by the static images.

Parameter values for the elastic stiffness matrix applied to the myocardial elements were 75,000 and 0.45 for Young's modulus and Poisson's ratio, respectively. The α parameter was 15,000 and k was 200,000. For the blood pool elements we used a smoothness rather than elastic stiffness constraint. The smoothness constraint was implemented using a weighted Laplacian matrix. In each matrix row i refers to the nodes of the blood pool elements, the mesh neighbor indices j get the normalized Euclidean distances $\frac{d(i,j)}{\sum_{j \in Ne(i)}(d(i,j))}$ and the diagonal value gets -1.

3 Results

Figures 3 and 4 demonstrate the current tool on the 2D simulated hand articulation data and for the 3D cardiac CT image data, respectively. The fused images show the misaligned images on different color channels, with the control points, to allow the user to warp one image on the other. Figure 5 shows the reduction

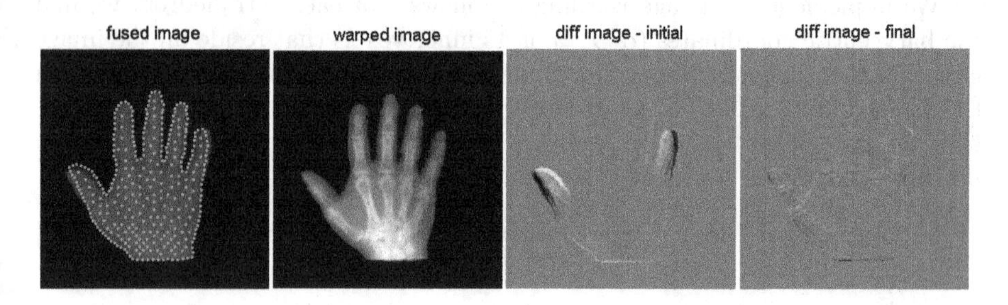

Fig. 3. Demonstration of the current tool on a 2D simulated hand articulation images. The left image shows fusion of the moving and static images with control points of the segmented moving image. In this view, the user can efficiently drag the control points to the right location. The warped image is second image from left. The right images show the initial and final difference images, respectively. (Color figure online)

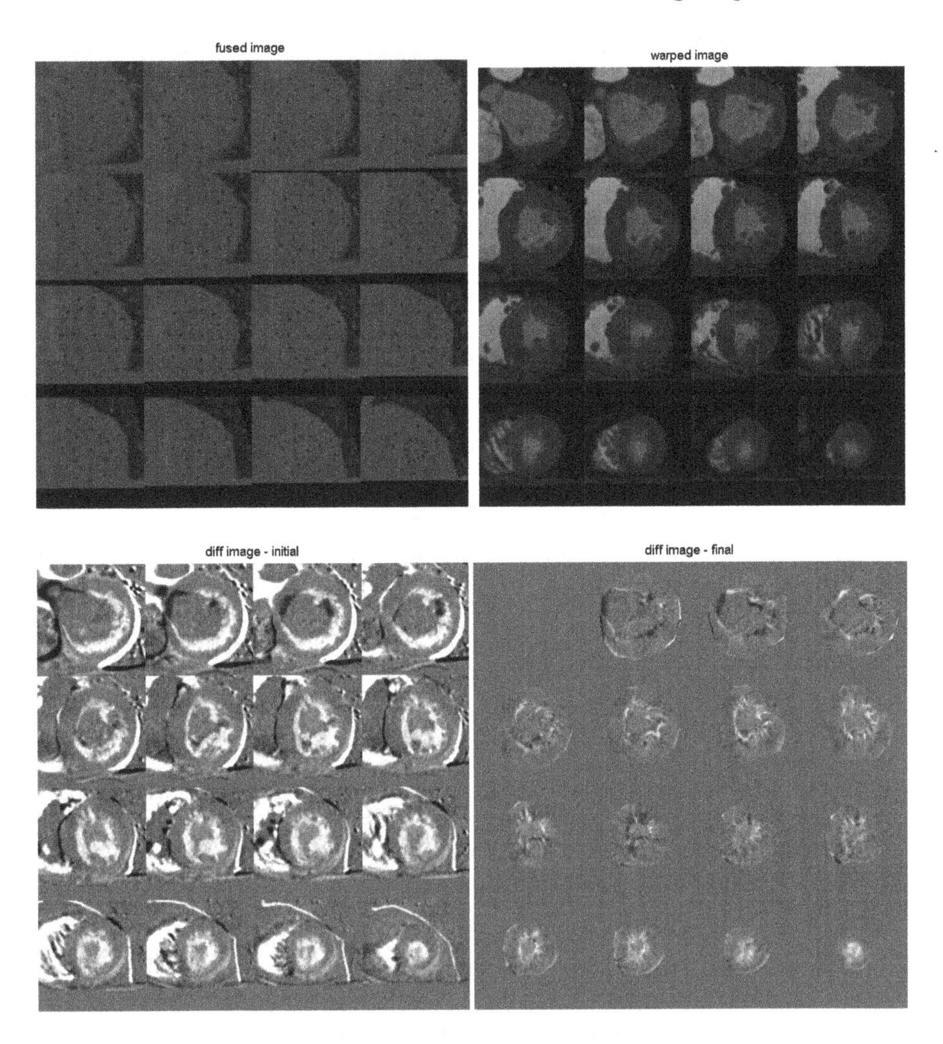

Fig. 4. Demonstration of the current tool on a 3D cardiac CT image data. The upper-left image shows fusion of the moving and static images with control points of the segmented moving image. In this view, the user can efficiently drag the control points to the right location. The neighboring control points (also in different slices) are simultaneously displaced to match the local force applied to the elastic tissue. The warped image is on the upper left. The lower left and right images show the initial and final difference images, respectively. (Color figure online)

of the relative SSD as a function of editing time and as a function of number of user mouse operations for the 2D simulated hand articulation data and for the 3D cardiac CT image data, respectively.

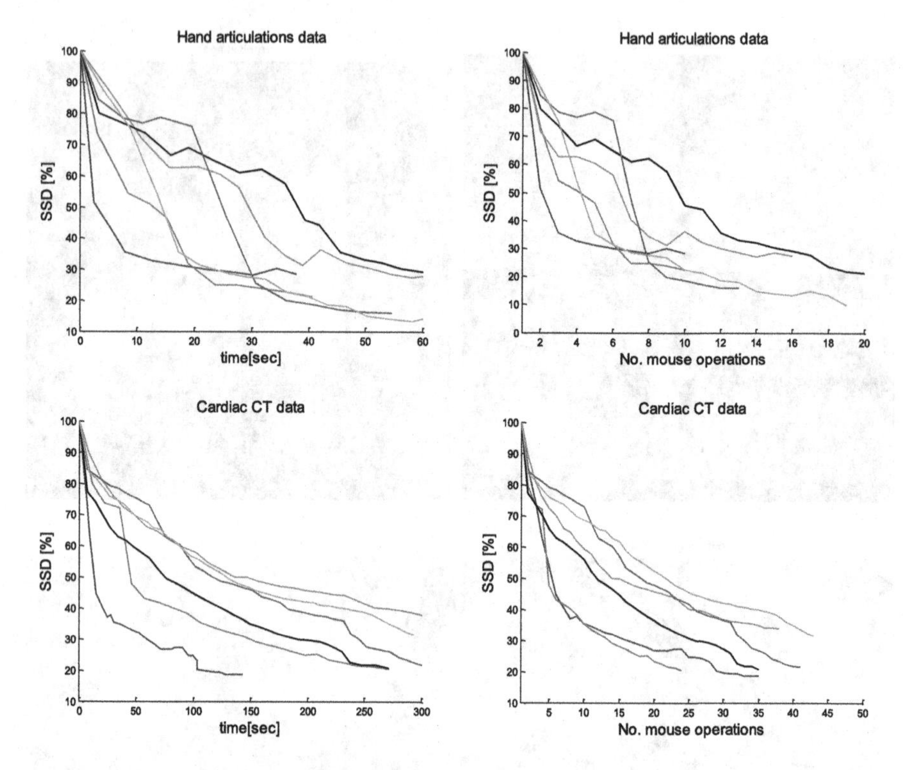

Fig. 5. The reduction of the relative SSD as a function of editing time (left images) and as a function of number of user mouse operations (right images) on the simulated hand articulation dataset (upper images) and on 3D cardiac CT dataset (lower images). Each plot represents different articulation.

4 Discussion and Conclusions

We propose here an interactive tool for elastic image registration. The current tool has an important practical impact, as automatic algorithms for image registration are often imperfect and might benefit from manual editing. Such manual editing may improve the diagnosis and analysis of tissue perfusion, tumor follow up, motion extraction, etc. Our concept was demonstrated on 2D simulated hand articulation data and on 3D cardiac CT image data. Yet, the method is generic and can be applied to many other biomedical image applications. The reduction in SSD as a function of the user operation with reasonable editing time demonstrates the concept of the proposed method. The current method was implemented in a MATLAB workspace, and therefore took up to 5 s to calculate 3D warpings (which require a loop on the elements in each user manual operation). Hence, we expect that, by compiling the warping process, user run time will be much faster. The deformations can also be applied using B-spline finite elements [10,11]. The proposed method can serve other useful applications: For example, image segmentation can be performed interactively by warping a labeled atlas image on a patient's specific statistic image. Another useful application might be the generation of datasets for training and validation of automatic image registration algorithms, as the displacements are known. These applications were beyond the scope of the current paper.

References

1. Amit, Y.: A nonlinear variational problem for image matching. SIAM J. Sci. Comput. **15**(1), 207–224 (1994)
2. Barrett, W.A., Cheney, A.S.: Object-based image editing. In: ACM Transactions on Graphics (TOG), vol. 21, pp. 777–784. ACM (2002)
3. Beier, T., Neely, S.: Feature-based image metamorphosis. In: ACM SIGGRAPH Computer Graphics, vol. 26, pp. 35–42. ACM (1992)
4. Bookstein, F.L.: Principal warps: thin-plate splines and the decomposition of deformations. IEEE Trans. Pattern Anal. Mach. Intell. **11**(6), 567–585 (1989)
5. De Berg, M., Cheong, O., Van Kreveld, M.: Computational Geometry: Algorithms and Applications. Springer, Heidelberg (2008)
6. Eldar, Y., Lindenbaum, M., Porat, M., Zeevi, Y.Y.: The farthest point strategy for progressive image sampling. IEEE Trans. Image Process. **6**(9), 1305–1315 (1997)
7. Hajnal, J.V., Hill, D.L.G.: Medical Image Registration. CRC Press, Boca Raton (2010)
8. Heimann, T., Meinzer, H.-P.: Statistical shape models for 3D medical image segmentation: a review. Med. Image Anal. **13**(4), 543–563 (2009)
9. Heimann, T., van Ginneken, B., Styner, M.A., Arzhaeva, Y., Aurich, V., Bauer, C., Beck, A., Becker, C., Beichel, R., Bekes, G., et al.: Comparison and evaluation of methods for liver segmentation from CT datasets. IEEE Trans. Med. Imaging **28**(8), 1251–1265 (2009)
10. Kagan, P., Fischer, A., Bar-Yoseph, P.Z.: New B-spline finite element approach for geometrical design and mechanical analysis. Int. J. Numer. Methods Eng. **41**(3), 435–458 (1998)
11. Kagan, P., Fischer, A., Bar-Yoseph, P.Z.: Mechanically based models: adaptive refinement for B-spline finite element. Int. J. Numer. Methods Eng. **57**(8), 1145–1175 (2003)
12. Lamash, Y., Fischer, A., Carasso, S., Lessick, J.: Strain analysis from 4-D cardiac CT image data. IEEE Trans. Biomed. Eng. **62**(2), 511–521 (2015)
13. Lamash, Y., Fischer, A., Lessick, J.: Elastic image registration using linear basis functions for non-continuously deforming objects. In: 2014 IEEE 11th International Symposium on Biomedical Imaging (ISBI), pp. 572–575. IEEE (2014)
14. Malmberg, F., Strand, R., Kullberg, J.: Interactive deformation of volume images for image registration. In: Interactive Medical Image Computing (IMIC) Workshop at MICCAI 2015, Munich, Germany, 5–9 October 2015
15. Pieper, S., Halle, M., Kikinis, R.: 3D slicer. In: IEEE International Symposium on Biomedical Imaging: Nano to Macro, pp. 632–635. IEEE (2004)
16. Rohr, K., Stiehl, H.S., Sprengel, R., Buzug, T.M., Weese, J., Kuhn, M.H.: Landmark-based elastic registration using approximating thin-plate splines. IEEE Trans. Med. Imaging **20**(6), 526–534 (2001)
17. Rueckert, D., Sonoda, L.I., Hayes, C., Hill, D.L.G., Leach, M.O., Hawkes, D.J.: Nonrigid registration using free-form deformations: application to breast MR images. IEEE Trans. Med. Imaging **18**(8), 712–721 (1999)

18. Schiwietz, T., Georgii, J., Westermann, R.: Interactive model-based image registration. In: VMV, pp. 213–222 (2007)
19. Sermesant, M., Forest, C., Pennec, X., Delingette, H., Ayache, N.: Deformable biomechanical models: application to 4D cardiac image analysis. Med. Image Anal. **7**(4), 475–488 (2003)
20. Yoshizawa, S., Takemoto, S., Takahashi, M., Muroi, M., Kazami, S., Miyoshi, H., Yokota, H.: Interactive registration of intracellular volumes with radial basis functions. Int. J. Comput. Intell. Appl. **9**(03), 207–224 (2010)
21. Zienkiewicz, O.C., Taylor, R.L.: The Finite Element Method: Solid Mechanics, vol. 2. Butterworth-heinemann, Oxford (2000)

Significance of Magnetic Resonance Image Details in Sparse Representation Based Super Resolution

Prabhjot Kaur(✉), Srimanta Mandal, and Anil K. Sao

Indian Institute of Technology Mandi, Kamand, India
{prabhjot_kaur,srimanta_mandal}@students.iitmandi.ac.in,
anil@iitmandi.ac.in

Abstract. Diverse constraints on image acquisition environment often limit the resolution in cross-slice direction of Magnetic Resonance (MR) image volume, which does not meet the requirement of isotropic 3D MR images in accurate medical diagnosis. This paper proposes an algorithm to restore isotropic 3D MR images from anisotropic 2D multi-slice volumes, by preserving the MR details that play significant role in medical diagnosis. The MR image details are preserved using dictionaries, which are learned using fine to coarse patch details, extracted from different scales of MR image. Learned dictionaries provide detail information for restoring MR patch details. Furthermore, a constraint is used to preserve edges within the restored MR image by minimizing an energy cost. Here, the constraint is weighted adaptively according to the dominant edge orientation of the image, to preserve the details along different orientations effectively. Experimental results demonstrate the ability of our approach to preserve MR image details.

Keywords: MRI super resolution · Sparse representation · Edge preservation · Slice thickness · Multiple dictionaries

1 Introduction

Magnetic Resonance Imaging (MRI) is an invaluable non-invasive imaging technique which is being widely used for early medical diagnosis of internal structures and functionality, thus requires High Resolution (HR) isotropic 3D MR images of organ [1,2]. However, due to various limitations like hardware constraints on strength and homogeneity of magnetic field, linearity of gradient fields, along with acquisition time and relaxation time considerations, it is common practice to acquire thick 2D multi-slices. For example, in Baltimore Longitudinal Study on Aging (BLSA) project [3], resolution of T1 weighted images is $1 \times 1 \times 1.6$ mm^3. Also, voxel size obtained by 1.5T and 3T is approximately $1.5 \times 1.5 \times 4$ mm^3 and $1 \times 1 \times 6$ mm^3, respectively with optimum Signal-to-Noise Ratio (SNR) [1,4]. Thus the resolution along slice-select (z) direction is generally poorer than resolution along in-plane (x, y) direction in 2D MR multi-slice volume, which makes

M. Valdés Hernández and V. González-Castro (Eds.): MIUA 2017, CCIS 723, pp. 605–615, 2017.
DOI: 10.1007/978-3-319-60964-5_53

the resolution of resultant 3D MR volume anisotropic. However, for many medical applications like 3D data visualization for diagnosis, 3D volume registration and accurate post-processing, voxel size in all directions needs to be same [5].

Basic image interpolation methods can be employed to achieve isotropic 3D MR images. However, these methods smear the edges which are very crucial for diagnosis of internal structures. Hence, various Super Resolution (SR) approaches have been reported in the literature to improve the resolution along slice-select direction (through-plane) [1,6]. Multiple 2D Low Resolution (LR) MR slices from same patient with sub-pixel shift have been used to extract high frequency information to super resolve MR volume [6–8]. However, acquiring sub-pixel shifted 2D slices increases acquisition time which leads to increase in motion artifacts due to the patient's discomfort. Hence, it is not always possible to acquire multiple 2D LR images. Moreover, the acquired LR images may not meet the criteria of sub-pixel shift.

Some of the SR approaches use the framework of sparse representation, which is widely employed for several signal processing applications, to generate HR images using dictionaries learned from either HR-LR image pair or input LR image [4,5,9,10]. HR-LR patch pairs of different modalities (e.g. T2WI LR - T1WI HR or T1WI LR - T2W1 HR) are also used for SR [11]. As availability of example HR images is not viable in all the cases, single image super resolution approach is preferred [4,5]. Non-local patch similarity is also explored without any HR volume to construct HR 3D MR cube with linear combination of other 3D cubes [12]. However, the computation of non-local mean, which involves weighted averaging, may smear the finer details in super resolved MR volume.

In this paper, we have explored the significance of MR image details in the case of single image super resolution using framework of sparse representation without any HR example images, to construct isotropic 3D MR images (MR volume) from anisotropic 2D multi-slice volumes. Dictionaries are learned from coarse to fine patch details, extracted from image pyramid, formed by up/down-sampling the LR image. Coarse to fine information from different scales can restore target patch details. Thus, it helps in restoring local image details. In order to preserve edges, edge maps for LR image and the downscaled version of estimated HR image are computed for two orthogonal directions and difference between the edge maps is minimized by adding a constraint to the cost function. Further, the constraint is adaptively weighted as per dominant edge orientation of the image to preserve the details along different directions effectively. Experimental results demonstrate that the proposed approach can preserve the edges better than the existing approaches.

Work done in [13,14], address the problem of edge-preserving single image super resolution, which is closely related with our approach. The weights for edge preserving constraint are fixed in [13] irrespective of edge content of the image, which may not be suitable for different edge orientations. Ref. [14] preserves edges along different orientations separately. Moreover, it requires example HR images for learning dictionaries and it has demonstrated the results for only intensity images.

The rest of the paper is organized as follows: Sect. 2 briefly discusses the super resolution of MR images in the sparse representation framework. Section 3 describes the method of image detail restoration using effective dictionary learning and its adaptive selection depending on the test image patches. The edge preserving constraint is adaptively weighted for global edge consistency in Sect. 4. Qualitative and quantitative comparisons are demonstrated in Sect. 5, and the paper is concluded in Sect. 6.

2 Super Resolution of MR Image in Sparse Representation Framework

In 2D MR multi-slice acquisition, spatially encoded frequencies are received for in-plane data and thus is inherently band limited. However for through-plane, slice is excited by Radio Frequency (RF) excitation pulse which generally has lesser sharp cut off in k-space domain [6]. Hence, sampling in cross-slice direction leads to aliasing and super resolution approach can be used to increase through-plane resolution.

Super resolution for 3D MR images can be framed either in 3D cubes or 2D patches [4,5,10,12]. Here, we extract 2D patches of each sagittal plane to super resolve along direction of slice using sparse representation framework, which is quite effective in image restoration techniques [1].

Super resolution framework assumes degradation model for obtaining LR images from HR images. Among such models, one commonly used model is

$$\mathbf{y} = \mathbf{DHx} + \boldsymbol{\eta}, \tag{1}$$

where $\mathbf{y} \in \mathbb{R}^m$ is the LR image, obtained by blurring ($\mathbf{H} \in \mathbb{R}^{n \times n}$), followed by downsampling ($\mathbf{D} \in \mathbb{R}^{m \times n}$) the HR image $\mathbf{x} \in \mathbb{R}^n$. $\boldsymbol{\eta} \in \mathbb{R}^m$ is generally considered as additive noise and our aim is to find $\mathbf{x}$ from $\mathbf{y}$. Since $n >> m$, Eq. (1) will have many solutions and appropriate constraints need to be added to regularize the problem. Sparsity inducing norm (l_0 norm) is one of the popular constraint (regularizer) in such image restoration problems, and can be well approximated using l_1 norm [15–18].

In sparse domain an image $\mathbf{x}$ is assumed to be linear combination of few columns of overcomplete matrix $\mathbf{A}$, named as dictionary, such that $\mathbf{x} = \mathbf{Ac}$, where $\mathbf{c}$ is sparse in nature. Thus Eq. (1) can be rewritten as:

$$\mathbf{y} = \mathbf{DHAc} + \boldsymbol{\eta}. \tag{2}$$

In general, HR image is assumed to have same sparse coefficient vector ($\mathbf{c}$) as of LR image [16] and can be estimated as following [19]:

$$\hat{\mathbf{c}} = \arg\min_{\mathbf{c}} \left\{ ||\mathbf{y} - \mathbf{DHAc}||_2^2 + \lambda \, ||\mathbf{c}||_1 \right\}. \tag{3}$$

Once, $\hat{\mathbf{c}}$ is computed, one can achieve the HR image. However, computing $\hat{\mathbf{c}}$ requires $\mathbf{y}$, $\mathbf{D}$, $\mathbf{H}$ and $\mathbf{A}$. Among these, $\mathbf{y}$ is LR image which is already available and $\mathbf{A}$ is learned using certain criteria, discussed in following section. Here, $\mathbf{H}$ and $\mathbf{D}$ are assumed based on MR image acquisition process, in which through-plane resolution of 3D MR volume is degraded due to RF pulse characteristics. Thus downsampling and blur operator in MR image acquisition depends on RF pulse. In the proposed algorithm, Point Spread Function (PSF) in slice direction is assumed to be rectangular kernel as in [21], with width equal to the upsampling factor. Rectangular kernel can be defined as [20]

$$h_z(z) = rect\left(\frac{z}{\Delta z}\right). \tag{4}$$

Here, $rect(.)$ is the rectangular-window-function, Δz indicates the slice thickness i.e. width of the excitation pulse and z denotes the axis of slice direction.

3 Restoring Image Details Using Dictionary

MR image details play significant role in medical diagnosis. Hence, we restore image details using dictionaries, learned from coarse to fine patch details, extracted from different scales of the input image. Here dictionaries act as source of information in SR, as they contain fine information related to coarser details.

In order to capture such information, we construct MR image pyramid by up and downsampling the input image by factors $s^k, k = 1$ to 3 and by extracting patches from the same as $\mathbf{x}_i = \mathbf{P}_i\mathbf{x}$, where $\mathbf{P}_i$ is the operator to extract i^{th} patch. It can be observed in Fig. 1 that similar patches exist in the same resolution as well as in different resolution of image, highlighted by red and yellow colors. Patches highlighted with red color contains vertical edge, and can be found in different locations across different scales. It helps in enrichment of collection of similar patches with different resolutions to approximate the target patch efficiently. From the extracted patches, we compute the detail component $\mathbf{d}_{\mathbf{x}_i}$ of ith patch as

$$\mathbf{d}_{\mathbf{x}_i} = \left| \mathbf{x}_i - \sum_{m \in \boldsymbol{\zeta}} \frac{1}{z} e^{\frac{-||\mathbf{x}_i - \mathbf{x}_{i,m}||^2}{h}} \mathbf{x}_i \right|. \tag{5}$$

Here $\boldsymbol{\zeta}$ denotes the set of similar patches, $\mathbf{x}_{i,m}$ denotes the mth similar patch to $\mathbf{x}_i$ and h controls the decay rate in proportion to the dis-similarity measured by Euclidean distance between $\mathbf{x}_{i,m}$ and $\mathbf{x}_i$. These details are further clustered using K-means clustering. As a consequence, we will have K clusters and their associated centroids $\boldsymbol{\mu}_k$. In order to design compact dictionary, Principal Component Analysis (PCA) is performed on each cluster and significant principal components are arranged column-wise to obtain a dictionary, denoted by $\mathbf{A}_k$ for k^{th} cluster.

For a target patch $\mathbf{x}_i$ with patch detail $\mathbf{d}_{\mathbf{x}_i}$, the restored detail information $\hat{\mathbf{d}}_{\mathbf{x}_i}$ is estimated as $\mathbf{A}_k\hat{\mathbf{c}}_{\mathbf{d}_i}$ and $\hat{\mathbf{c}}_{\mathbf{d}_i}$ is computed by solving the following cost function:

$$\hat{\mathbf{c}}_{\mathbf{d}_i} = \arg\min_{\mathbf{c}_{\mathbf{d}_i}} \left\{ ||\mathbf{d}_{\mathbf{x}_i} - \mathbf{A}_k \mathbf{c}_{\mathbf{d}_i}||_2^2 + \lambda \, ||\mathbf{c}_{\mathbf{d}_i}||_1 \right\}. \tag{6}$$

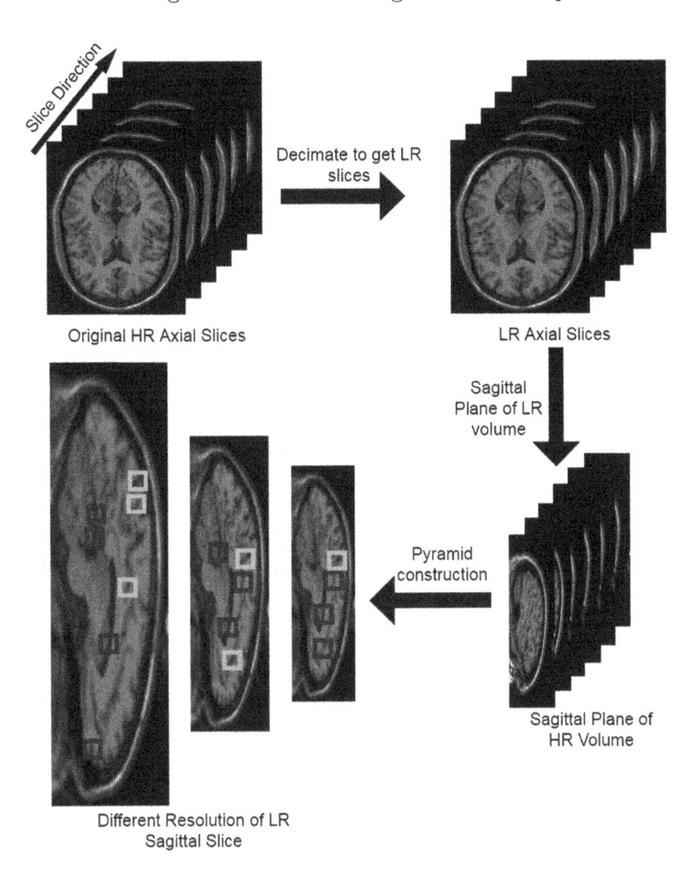

Fig. 1. Illustration of synthesizing LR volume from HR volume and non-local patch similarity. The similarity is demonstrated by boxes highlighted with red and yellow colors which represents vertical and diagonal edges, respectively. Similar vertical and diagonal edge can be found within scale as well as different scales. (Color figure online)

Here $\mathbf{A}_k$ is selected adaptively based on minimizing $||\mathbf{d}_{\mathbf{x}_i} - \boldsymbol{\mu}_k||_2$, and the cost function is solved by iterative shrinkage algorithm [22]. HR patch is estimated as follows:

$$\hat{\mathbf{x}}_i = \mathbf{A}_k \hat{\mathbf{c}}_{\mathbf{d}_i} + \sum_{m \in \zeta} \frac{1}{z} e^{\frac{-||\mathbf{x}_i - \mathbf{x}_{i,m}||^2}{h}}. \tag{7}$$

Restoring patch details helps in restoring local image details, and is useful for diagnosis. Hence, all the patches can be restored as mentioned in (7) and are further used to restore the entire image, as discussed in following section.

4 Preserving Edges Using Constraint

Edges represent change in grey level values in MR images, hence provide information about tissue contrast, which are crucial in medical diagnosis and post processing of the images like segmentation. Hence, edges present in LR image should be preserved. Here, we have computed the difference between edge maps of LR image and downscaled version of estimated HR image. The difference is weighted adaptively based on dominant edge orientation of the image. It helps in better preservation of the edges in estimated HR image.

The edges are computed using 1-D processing of image as done in the edge preserving SR work [14], where smoothening filter (e.g. Gaussian filter) is applied along one direction and the derivative filter (e.g. derivative of Gaussian) is applied along the orthogonal direction.

In this paper, we have computed edge maps along 0° and 90°, denoted as $\mathbf{e}_0$ and $\mathbf{e}_{90}$ respectively. Magnitude of edge map is computed as

$$\mathbf{e}_g = \sqrt{(\mathbf{e}_0^2 + \mathbf{e}_{90}^2)}. \tag{8}$$

$\mathbf{E}_0$ and $\mathbf{E}_{90}$ can be assumed to be the operators that extract vertical edge evidence ($\mathbf{e}_0$) and horizontal edge evidence ($\mathbf{e}_{90}$) respectively. The variance of edge maps along the two orientations can be used as a measure to find the dominant edge information in a given image. Thus,

$$\mathbf{e}_{\max} = \begin{Bmatrix} \mathbf{e}_0, & \text{if } \sigma^2_{\mathbf{e}_0} > \sigma^2_{\mathbf{e}_{90}} \\ \mathbf{e}_{90}, & \text{if } \sigma^2_{\mathbf{e}_{90}} > \sigma^2_{\mathbf{e}_0} \end{Bmatrix}$$

can be used to compute the weight adaptively as

$$\rho = \frac{\sigma^2_{\mathbf{e}_{\max}}}{\sigma^2_{\mathbf{e}_{\max}} + \sigma^2_{\mathbf{e}_{\min}}}. \tag{9}$$

Here, $\sigma^2_{\mathbf{e}_0}$ and $\sigma^2_{\mathbf{e}_{90}}$ are variances of edge map along 0° and 90°, respectively. A similar interpretation can be realized for $\sigma^2_{\mathbf{e}_{\max}}$ and $\sigma^2_{\mathbf{e}_{\min}}$. ρ is obtained using Eq. (9) and contains the information about relatively dominant edge evidence orientation and is always less than one. ρ is used to assign weight to the edge preserving constraint $||\mathbf{E}\{\mathbf{y}\} - \mathbf{E}\{\mathbf{DH\hat{x}}\}||_2^2 < \epsilon$, $\mathbf{E} = \{\mathbf{E}_{\min}, \mathbf{E}_{\max}\}$ which is used to preserve edges along different orientations. Here, $\mathbf{E}_{\max}$ extracts edge evidence along the orientation with maximum edge map variance. For example, $\mathbf{E}_{\max} = \mathbf{E}_0$ and $\mathbf{E}_{\min} = \mathbf{E}_{90}$ if $\sigma^2_{\mathbf{e}_0} > \sigma^2_{\mathbf{e}_{90}}$. With the edge preserving constraint, the final equation which we have to solve becomes:

$$\hat{\hat{\mathbf{x}}} = \arg\min_{\hat{\mathbf{x}}} \Big\{ \sum_i ||\mathbf{P}_i\mathbf{x} - \hat{\mathbf{x}}_i||_2^2 + ||\mathbf{y} - \mathbf{DH\hat{x}}||_2^2 + \rho\, ||\mathbf{E}_{\max}\{\mathbf{y}\} - \mathbf{E}_{\max}\{\mathbf{DH\hat{x}}\}||_2^2 + (1-\rho)\, ||\mathbf{E}_{\min}\{\mathbf{y}\} - \mathbf{E}_{\min}\{\mathbf{DH\hat{x}}\}||_2^2 \Big\}. \tag{10}$$

First term in Eq. (10) represents the construction of HR from patches and second term is the data constraint, which assures the data continuity. Third term is introduced to minimize the difference between edge maps along dominant edge orientation for LR image and degraded version of estimated HR image. Similarly,

Algorithm 1

Input data: LR image $\mathbf{y}$
Output: HR image $\mathbf{x}$

Initialization
(a) Set the stopping criteria (a) Threshold Error Thr (b) Maximum iteration number = M
(b) Set parameter λ as 1, s=0.8
(c) Reshape LR volume in sagittal plane and interpolate using spline interpolation to get initial estimate of HR volume $\mathbf{x}'$.

For each LR sagittal slice
Step 1 - Restoring Image Details using Dictionary
Extract patches $\mathbf{x}_i$ by $\mathbf{x}_i = \mathbf{P}_i\mathbf{x}'$.
Construct image pyramid and cluster the detail components.
Construct multiple PCA dictionaries and compute estimate ($\hat{\mathbf{x}}$) of HR image by solving eq. (5)(6)(7)

Step 2 - Preserving Image Details using Constraint
Compute ratio between variance of $\mathbf{e}_0$ and $\mathbf{e}_{90}$ edge maps
Solve the optimization in eq. (10) and eq. (11) to get regularized output ($\hat{\hat{\mathbf{x}}}$)

Step 3 - Repeat step 1 and step 2
Put $\hat{\mathbf{x}} = \hat{\hat{\mathbf{x}}}$, and repeat for M iterations or till error convergence to Thr.

fourth term is for preservation of edges orthogonal to dominant edge orientation. Third term is weighted with ρ so as to provide more weightage to preservation of dominant edge evidence as compared to evidence along the orthogonal direction (fourth term). This helps in better preservation of finer details along dominant edge orientation as well as in reducing the smearing of finer details in other orientations.

If the weight ρ is less than a threshold for some MR slices, we do not use $\mathbf{E}_{\text{max}}$ or $\mathbf{E}_{\text{min}}$, instead we use $\mathbf{E}_g$ to compute magnitude of the edge evidence of the slice: $\mathbf{e}_g = \mathbf{E}_g\{\mathbf{x}\}$. Hence the equation for those slices becomes [14]:

$$\hat{\hat{\mathbf{x}}} = \arg\min_{\hat{\mathbf{x}}} \left\{ \sum_i ||\mathbf{P}_i\mathbf{x} - \hat{\mathbf{x}}_i||_2^2 + ||\mathbf{y} - \mathbf{DH}\hat{\mathbf{x}}||_2^2 + ||\mathbf{E_g}\{\mathbf{y}\} - \mathbf{E_g}\{\mathbf{DH}\hat{\mathbf{x}}\}||_2^2 \right\}. \quad (11)$$

Hence in this paper, instead of fixed weightage for edge preserving constraints, adaptation of weights is done as per dominant edge orientation present in the image. Thus, preservation of global image details is expected and is demonstrated in experimental results section. The pseudo code of the proposed algorithm is summarized in Algorithm 1.

5 Experimental Results

The experimental results are demonstrated using simulated T1-weighted brain MR image dataset obtained from BrainWeb database [23] with voxel size $1 \times 1 \times 1$ mm^3 and spatial resolution $217 \times 180 \times 181$. We have considered patches

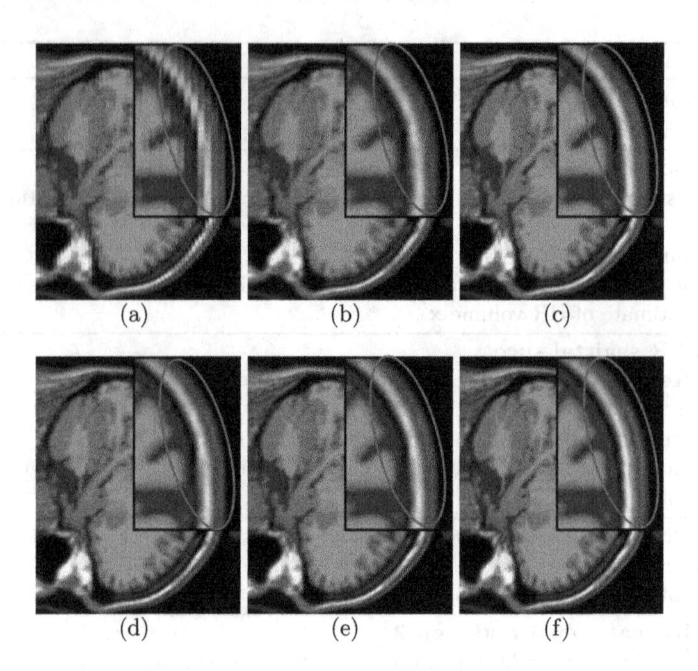

Fig. 2. Demonstration of edge preservation using different approaches: (a) zoomed-in LR image, (b) spline interpolated image, (c) results of Ref. [12], (d) results of edge preserving SR [14], (e) results with proposed algorithm, (f) original HR image. (Color figure online)

of size 7×7 pixels, and number of clusters in K-means clustering is 64. The LR image is obtained by averaging HR images in the direction of slice. LR image is interpolated by spline interpolation to get an initial estimate of HR image. We consider $s = 0.8$ and $k = 1, 2, 3$ in s^k used for obtaining different scales of image. The value of λ is 1 and ρ is computed using Eq. (9).

The performance of approaches is compared with spline interpolation, approach using NLM strategy without using sparse-representation framework [12] and edge preserving SR in sparse-representation framework [14]. The code for [12] is available online[1], and results for same environment are calculated to compare with the performance of proposed algorithm.

HR images obtained from different approaches with zoomed-in windows are shown and differences are highlighted by red color in Figs. 2 and 3. In Fig. 2, it can be observed that most of the information is blurred in spline interpolated image (b) as compared to resultant image of the proposed algorithm (e). If we carefully examine the zoomed-in window, one can point out that the edges of image (e) are sharper than the edges of image obtained by NLM strategy [12]. Same way, in comparison with edge preserving SR [14], proposed algorithm (e) preserves edges better than in image (d). This is because the approach in ref. [14] preserves the edges along different orientations separately. However in

[1] Available at http://personales.upv.es/jmanjon/demo2.zip.

case of vertical edge as shown in zoomed-in parts of Fig. 2, preservation of edge in [14] along 90° will tend to smear the edge along 0°. In our algorithm, as weights for edge preserving constraints are adaptively chosen based on dominant edge orientation of image, there is lesser smearing in (e) as compared to (d). Similarly in Fig. 3, which represents the inner structural information of organ in zoomed-in window, it can be observed that details in case of Fig. 3(e) are preserved better as compared to other approaches.

Table 1. PSNR evaluated for different algorithms

Slice thickness	Approaches				
	Cubic	Spline	SR-Edge [14]	Algorithm in Ref. [12]	Proposed algorithm
2 mm	36.91	37.80	37.26	41.26	**41.43**
3 mm	31.61	32.40	33.56	35.41	**35.43**
5 mm	25.95	26.06	27.06	**28.44**	27.80

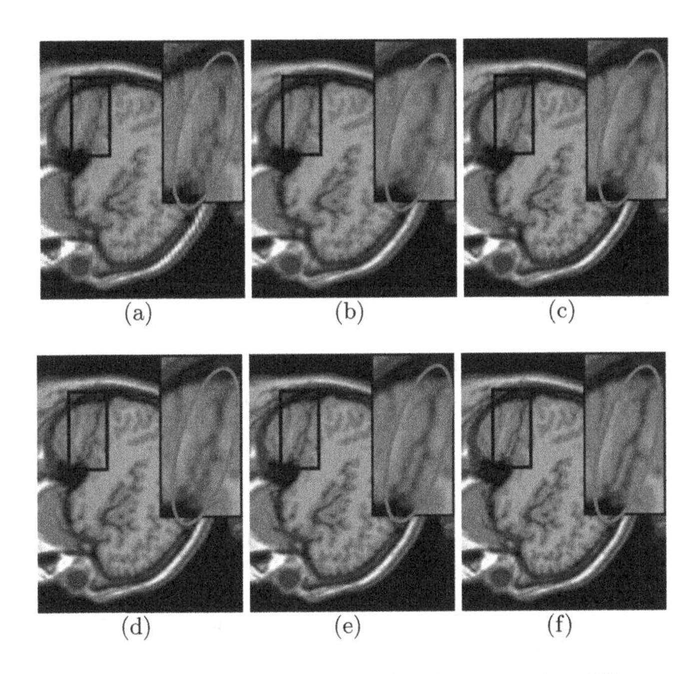

Fig. 3. Preservation of inner structural details of organ using different approaches: (a) zoomed-in LR image, (b) spline interpolated image, (c) results of Ref. [12], (d) results of edge preserving SR [14], (e) results with proposed algorithm, (f) original HR image. (Color figure online)

We have also performed quantitative comparison using Peak Signal-to-Noise Ratio (PSNR) [24]. 2D multi-slice volumes with 1mm $\times$ 1mm in-plane resolution and different slice thickness (2 mm, 3 mm and 5 mm), are super resolved to isotropic resolution of $1 \times 1 \times 1$ mm^3. Table 1 describes PSNR values for obtained MR image volumes, super resolved using proposed approach along with other existing approaches [12,14]. One can observe the comparable performance of the proposed approach in quantitative analysis. Further, our algorithm outperforms the results of ref. [5] for reconstruction of isotropic 3D HR images with $1 \times 1 \times 1$ mm^3 from $1 \times 1 \times 3$ mm^3 with quoted PSNR as 35.07dB. It shows proposed algorithm performs more or less better than existing algorithms quantitatively. Moreover, it can be seen from Figs. 2 and 3 that our algorithm gives better results in terms of edges as well as restoration of MR image details which are very crucial for medical diagnosis of internal structures and also for localization of details.

6 Conclusion

This paper proposed a super resolution approach in slice-select direction of 2D MR multi-slices in the framework of sparse representation. The main focus of the approach is to restore MR image details as it contains diagnostically significant information. The details have been restored using dictionaries, learned from coarse to fine patch details that are extracted from image pyramid. Further, an edge preserving constraint is added in the cost function with adaptive weight, based on dominant edge orientation of the MR image. The constraint helps in preserving edges suitably in the estimated MR image. The experimental results on Brainweb simulated data set demonstrate that the proposed approach can preserve image details better as compared to the existing approaches.

References

1. Van Reeth, E., Tham, I.W.K., Tan, C.H., Poh, C.L.: Super-resolution in magnetic resonance imaging: a review. Concepts Magn. Resona.-Part A **40**(6), 306–325 (2012)
2. Dagia, C., Ditchfield, M.: 3T MRI in paediatrics: challenges and clinical applications. Eur. J. Radiol. **68**(2), 309–319 (2008)
3. Resnick, S.M., Goldszal, A.F., Davatzikos, C., Golski, S., Kraut, M.A., Metter, E.J., Zonderman, A.B.: One-year age changes in MRI brain volumes in older adults. Cereb. Cortex **10**(5), 464 (2000)
4. Zhang, M., Nie, H., Pei, Y., Tao, L.: Volume reconstruction for MRI. In: Proceedings of International Conference on Pattern Recognition, pp. 3351–3356, August 2014
5. Iwamoto, Y., Han, X.H., Sasatani, S., Taniguchi, K., Xiong, W., Chen, Y.W.: Super-resolution of MR volumetric images using sparse representation and self-similarity. In: Proceedings of the 21st International Conference on Pattern Recognition, pp. 3758–3761, November 2012
6. Greenspan, H., Oz, G., Kiryati, N., Peled, S.: MRI inter-slice reconstruction using super-resolution. Magn. Reson. Imaging **20**(5), 437–446 (2002)
7. Greenspan, H., Oz, G., Kiryati, N., Peled, S.: Super-resolution in MRI. In: Proceedings IEEE International Symposium on Biomedical Imaging, pp. 943–946 (2002)

8. Hefnawy, A.A.: An efficient super-resolution approach for obtaining isotropic 3-D imaging using 2-D multi-slice MRI. Egypt. Inform. J. **14**(2), 117–123 (2013)
9. Rueda, A., Malpica, N., Romero, E.: Single-image super-resolution of brain MR images using overcomplete dictionaries. Med. Image Anal. **17**(1), 113–132 (2013)
10. Bahrami, K., Shi, F., Zong, X., Shin, H.W., An, H., Shen, D.: Reconstruction of 7T-like images from 3T MRI. IEEE Trans. Med. Imaging **35**(9), 2085–2097 (2016)
11. Rousseau, F.: Brain hallucination. In: Forsyth, D., Torr, P., Zisserman, A. (eds.) ECCV 2008. LNCS, vol. 5302, pp. 497–508. Springer, Heidelberg (2008). doi:10.1007/978-3-540-88682-2_38
12. Manjn, J.V., Coup, P., Buades, A., Fonov, V., Collins, D.L., Robles, M.: Non-local MRI upsampling. Med. Image Anal. **14**(6), 784–792 (2010)
13. Mandal, S., Bhavsar, A., Sao, A.K.: Super-resolving a single intensity/range image via non-local means and sparse representation. In: Proceedings of the Indian Conference on Computer Vision Graphics and Image Processing, (ICVGIP), pp. 1–8, December 2014
14. Mandal, S., Sao, A.K.: Edge preserving single image super resolution in sparse environment. In: IEEE International Conference on Image Processing, pp. 967–971, September 2013
15. Yang, J., Wang, Z., Lin, Z., Cohen, S., Huang, T.: Coupled dictionary training for image super-resolution. IEEE Trans. Image Process. **21**(8), 3467–3478 (2012)
16. Yang, J., Wright, J., Huang, T.S., Ma, Y.: Image super-resolution via sparse representation. IEEE Trans. Image Process. **19**(11), 2861–2873 (2010)
17. Elad, M., Figueiredo, M.A., Ma, Y.: On the role of sparse and redundant representations in image processing. Proc. IEEE **98**(6), 972–982 (2010)
18. Dong, W., Li, X., Zhang, L., Shi, G.: Sparsity-based image denoising via dictionary learning and structural clustering. In: Proceedings of IEEE Computer Society Conference on Computer Vision and Pattern Recognition, pp. 457–464 (2011)
19. Elad, M., Figueiredo, M.A.T., Ma, Y.: On the role of sparse and redundant representations in image processing. Proc. IEEE **98**(6), 972–982 (2010)
20. Bai, Y., Han, X., Prince, J.L.: Super-resolution reconstruction of MR brain images. In: Proceedings of the 38th Annual Conference on Information Sciences and Systems (CISS 2004) (2004)
21. Kwan, R.K.S., Evans, A.C., Pike, G.B.: MRI simulation-based evaluation of image-processing and classification methods. IEEE Trans. Med. Imaging **18**(11), 1085–1097 (1999)
22. Daubechies, I., Defrise, M., De Mol, C.: An iterative thresholding algorithm for linear inverse problems with a sparsity constraint. Commun. Pure Appl. Math. **57**(11), 1413–1457 (2004)
23. Cocosco, C.A., Kollokian, V., Kwan, R.K.S., Pike, G.B., Evans, A.C.: Brainweb: online interface to a 3D MRI simulated brain database. NeuroImage **5**, 425 (1997)
24. Hore, A., Ziou, D.: Image quality metrics: PSNR vs. SSIM. In: 20th International Conference on Pattern Recognition, pp. 2366–2369, August 2010

Restoration of Intensity Uniformity of Bi-contrast MRI Data with Bayesian Co-occurrence Coring

Stathis Hadjidemetriou[1](✉), Marios Nikos Psychogios[2], Paul Lingor[3], Kajetan von Eckardstein[4], and Ismini Papageorgiou[2]

[1] Department of Electrical Engineering and Informatics, Cyprus University of Technology, Pavlou Mela, 3036 Limassol, Cyprus
s.hadjidemetriou@cut.ac.cy

[2] Department of Diagnostic and Interventional Neuroradiology, University Medical Center Goettingen, University of Goettingen, Robert Koch str. 40, 37075 Goettingen, Germany

[3] Department of Neurology, University Medical Center Goettingen, University of Goettingen, Robert Koch str. 40, 37075 Goettingen, Germany

[4] Department of Neurosurgery, University Medical Center Goettingen, University of Goettingen, Robert Koch str. 40, 37075 Goettingen, Germany

Abstract. The reconstruction in MRI assumes a uniform radio-frequency field. However, this is violated, which leads to anatomically inconsequential intensity non-uniformities. An anatomic region can be imaged with multiple contrasts that result in different non-uniformities. A method is presented for the joint intensity uniformity restoration of two such images. The effect of the intensity distortion on the auto-co-occurrence statistics of each image as well as on the joint-co-occurrence statistics of the two images is modeled. Their non-stationary deconvolution gives Bayesian coring estimates of the images. Further constraints for smoothness, stability, and validity of the non-uniformity estimates are also imposed. The effectiveness and accuracy of the method has been demonstrated extensively with both BrainWeb phantom images as well as with real brain anatomic data of 29 Parkinson's disease patients.

Keywords: Bi-contrast MRI reconstruction · Co-occurrence statistics · Bayesian coring estimate

1 Introduction

The quantification of 3D MRI data is hampered by spatial intensity non-uniformities. They stem from the inhomogeneity of the Radio-Frequency (RF) field within the subject. There have been attempts to calibrate for these non-uniformities with physical phantoms and parameterized acquisitions [7,10]. However, additional physical acquisitions are time-consuming and valid only for particular MRI sequences and anatomies.

M. Valdés Hernández and V. González-Castro (Eds.): MIUA 2017, CCIS 723, pp. 616–628, 2017.
DOI: 10.1007/978-3-319-60964-5_54

Post-acquisition restoration methods have also been proposed. An approach assumes that the image is the union of piecewise constant intensity regions with uniform intensity transitions magnitudes using the total variation [18]. Piecewise smoothness has also been used with local fuzzy C-means intensity clustering [5]. The latter method has been combined with level sets for spatial segmentation [6,17]. The image histogram has been analyzed with coring to remove noise [12,15]. The coring has been combined with spatial smoothness for the non-uniformity [13]. The histogram entropy has also been used [8]. However, the histogram based corrections can lead to an unstable dynamic range [13].

A clinical imaging protocol has multiple sequences giving images that suffer from different non-uniformities. Post-acquisition methods have been developed for their joint restoration. A data dependent method fuses MRI with PET data [11]. A variational method preserves the differential structure of two images [2] and enforces a smooth non-uniformity. A statistical method minimizes the entropy of the joint histogram of two images [16]. Another method restores multiple co-registered images by minimizing the sum of the entropies of voxel-wise stack vectors throughout an image [9]. The bi-contrast and multi-contrast restoration methods assume all images considered have identical signal regions.

Post-acquisition methods benefit from regularity properties of the anatomy and of the physical RF field. They are applicable to a range of MRI contrasts and anatomies. The post-acquisition method developed performs a joint restoration of two images of the same anatomic region with different contrasts. The method uses a non-parametric Bayesian coring formulation for intensity restoration. It is applied to the auto-co-occurrence statistics of each image as well as to the joint-co-occurrence statistics between the two images [3,4]. The effect of the intensity distortions on both types of co-occurrence statistics is modeled and restored. Additional constraints have also been imposed for stability. The restoration also considers the inevitable difference between the signal regions of the two images. It has been tested extensively with BrainWeb phantom images [1] and with real brain anatomic data sets of 29 Parkinson's patients. The method improves both accuracy and efficiency.

2 General Bayesian Formulation

2.1 Spatial and Statistical Image Representation

An anatomic image $v(\mathbf{x})$, with $\mathbf{x} = (x, y, z)$, is assumed to be the product of a latent anatomic image $u(\mathbf{x})$ multiplied with spatial intensity non-uniformity $b(\mathbf{x})$ due to MRI radio-frequency inhomogeneities. There is also an additive and independent noise, n, to give $v = b.u + n$, where . is the voxelwise product. The voxelwise probability distribution of $u(\mathbf{x})$ and $b(\mathbf{x})$, $P_u(u(\mathbf{x}))$ and $P_b(b(\mathbf{x}); 1, \sigma_b^2)$, respectively, are assumed independent. The distortion of the joint intensity statistics is modeled by the non-stationary distribution $P_{(v|u)}(v|u) = P_b\left(v - u|u\right) = P_b(v - u; 0, (\sigma_b u)^2)$. Thus, the intensity statistics of v, $P_v(v)$, result from the convolution, $*$, $P_v(v) = P_{(v|u)}(v|u) * P_u(u) = P_u(u) * P_b\left(v - u; 0, (\sigma_b u)^2\right)$. The probability of the distortion is assumed to be bimodal with a distribution for

$b(\mathbf{x}) > 1$ and another for $b(\mathbf{x}) < 1$ and given by $P_b(b;0,\sigma_b^2) = \frac{G(b;0,\sigma_b^2)}{G^2(b;0,\sigma_b^2)+\epsilon^2}$, where ϵ^2 is for regularization.

2.2 Bayesian Coring Estimates of the Images

An overview of the method is in Fig. 1. The posterior expectation of the latent intensity, $\hat{u} = E(u|v)$, is the minimum mean squared error. The Bayesian expansion for $P_{(u|v)}(u|v)$ gives:

$$\hat{u} = E(u|v) = \int P_{(u|v)}(u|v)udu = \frac{\int P_{(v|u)}(v|u)P_u(u)udu}{\int P_{(v|u)}(v|u)P_u(u)du}. \tag{1}$$

This expression involves the distribution of u, $P_u(u)$, that is estimated with the deconvolution of the distortion $P_b(v-u|u)$ from the intensity distribution $P_v(v)$ to provide the non-parametric prior $\tilde{P}_u(u)$.

The Point Spread Function (PSF) for the distortion of the intensity co-occurrences increases linearly with intensity. Thus, the deconvolution is non-stationary and in direct domain. The Van Cittert algorithm is iterative with

$$P_u^{n+1} = P_u^n + \beta(P_v^0 - P_b * P_u^n), \tag{2}$$

where β is for regularization, P_v^0 are the statistics of the image, and P_b is the distortion PSF. The last iteration provides the estimate $\tilde{P}_u(u)$.

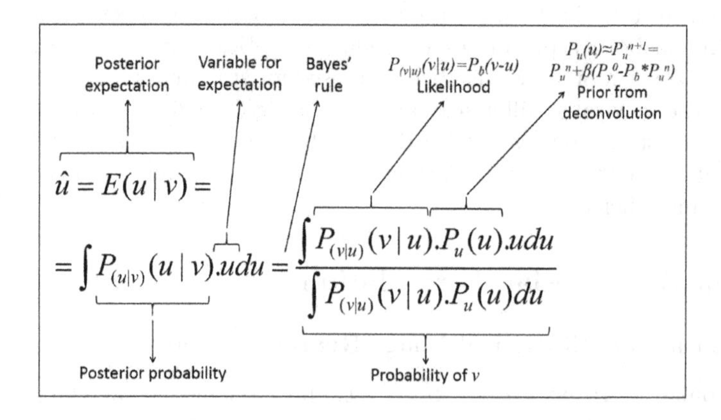

Fig. 1. Non-parametric Bayesian coring for image restoration.

The likelihood $P_{(v|u)}(v|u)$ in Eq. (1) is the non-stationary distortion as given in Subsect. (2.1) for the joint intensity statistics. The prior $P_u(u)$ in Eq. (1) is $\tilde{P}_u(u) = \tilde{C}_u$ as estimated from Eq. (2) for the co-occurrence statistics of image u. These two terms are substituted into Eq. (1) to give $\hat{u} = E(u|v) = \frac{\int P_b(v-u;0,(\sigma_b u)^2).\tilde{C}_u(u).udu}{\int P_b(v-u;0,(\sigma_b u)^2).\tilde{C}_u(u)du}$. The $P_b(v-u;(\sigma_b u)^2)$ is significant for neighborhood

$\Delta u \in \mathcal{N}_v$ and for intensities u far from v, it tends to zero. Thus, the discretization of $\hat{u} = E(u|v)$ becomes:

$$\hat{u} = E(u|v) = \frac{\sum_{\Delta u \in \mathcal{N}_v} P_b(\Delta u; 0, (\sigma_b u)^2).\tilde{C}_u(v + \Delta u).(v + \Delta u)}{\sum_{\Delta u \in \mathcal{N}_v} P_b(\Delta u; 0, (\sigma_b u)^2).\tilde{C}_u(v + \Delta u)}. \tag{3}$$

The extent of P_b and $\mathcal{N}_u$ increase linearly with intensity u. The restoration field, b^{-1}, is computed using $\hat{u}$ given from Eq. (3) with:

$$\frac{1}{b(\mathbf{x})} = E\left(\frac{u(\mathbf{x})}{v(\mathbf{x})}\middle| v(\mathbf{x})\right) = E\left(\frac{u(\mathbf{x})|v(\mathbf{x})}{v(\mathbf{x})}\right) = \frac{\hat{u}}{v}. \tag{4}$$

This provides a voxelwise restoration independent of $\mathbf{x}$. Thus, it is precomputed and stored in a 2D matrix with sizes equal to the dynamic ranges.

3 Methods

3.1 Spatial and Statistical Image Representation

The images of the two different contrasts, $v_i(\mathbf{x})$, where $i = 0, 1$, have the same sampling grid. They can be affected by different multiplicative spatial b_i over the latent anatomic images u_i. Each image is also corrupted with additive and independent noise, n_i, and the images model is

$$v_i = b_i u_i + n_i, \qquad i = 0, 1. \tag{5}$$

The first order term of $b_i(\mathbf{x})$ provides a linear approximation within a spherical neighborhood $\mathcal{N}_\rho$ of radius ρ. The statistical representation of images v_i are based on intensities η_i and their counts within neighborhood $\mathcal{N}_\rho$ to give the co-occurrence statistics as [4]:

$$C_{v_i v_j}(v_i, v_j, \eta_i, \eta_j) = C_{v_i v_j}(\eta_i, \eta_j) = \int_{v_i^{-1}(\eta_i)} \left(\int_{v_j^{-1}(\eta_j)} (\|\mathbf{x} - \mathbf{x}'\|_2 \leq \rho)\, d\mathbf{x}' \right) d\mathbf{x}. \tag{6}$$

The auto-co-occurrences result from $i = j$ and the joint-co-occurrences are C_{01}. The auto-co-occurrences are dominated by their diagonal and thus they are weighted down with the sigmoid: $1/(1 + e^{-(k_1|\eta_0 - \eta_1| + k_2)})$. Examples of the co-occurrences of pairs of T_1 and T_2 images are in Figs. (2) and (3). Median filtering applied to v_i removes high frequency noise n_i.

3.2 Intensity Non-uniformities in the Co-occurrence Statistics

The statistics of the products $b_i.u_i$ are modeled as the convolutions of $C_{u_i u_i}$ and $C_{u_0 u_1}$ with the respective PSF of the intensity distortions that are non-stationary to account for the multiplication. The effect of b_i in $\mathcal{N}_\rho$ around $\mathbf{x}_0$ are approximated by the zero order term $b(\mathbf{x}_0)$ that scales the auto-co-occurrences of $C_{u_iu_i}$ radially and the first order term $\nabla b(\mathbf{x})|_{\mathbf{x}_0}$ that rotates them around the origin [4]. The PSF affecting $C_{u_iu_i}$ is represented in polar coordinates (r_i, ϕ_i). The σ_{r_i} of the PSF is scaled linearly with r_i, $\sigma_{r_i} \propto r_i$. The σ_ϕ of the PSF increases with ρ, is largest along the diagonal and is zero along the axes. The application of the PSFs of the distortions to $P_{u_i}(u_i) = C_{u_iu_i}$ give $P_{v_i}(v_i) = C_{v_iv_i}$ with convolution

$$C_{v_iv_i}(r_i, \phi_i) = C_{u_iu_i}(r_i, \phi_i) * P_b(r_i; 0, \sigma_{r_i}^2) * P_b(\phi_i; 0, \sigma_{\phi_i}^2), i = 0, 1, \quad (7)$$

that represent the auto-co-occurrences of the distorted images.

The effects of the zero order terms of b_i on $C_{u_0u_1}$ of u_i are in Cartesian coordinates. The σ_{η_i} of the PSFs are linear with η_i, $\sigma_{\eta_i} \propto \eta_i$ [3]. The η_i is related to r_i with $\eta_i = r_i/\sqrt{2}$. The application of the PSFs of the distortions to $P_{u_0u_1}(u_0, u_1) = C_{u_0u_1}$ give $P_{v_0v_1}(v_0, v_1) = C_{v_0v_1}$ with convolution

$$C_{v_0v_1}(\eta_0, \eta_1) = C_{u_0u_1}(\eta_0, \eta_1) * P_b(\eta_0; \sigma_{\eta_0}^2) * P_b(\eta_1; \sigma_{\eta_1}^2) \quad (8)$$

that represents the joint-co-occurrences of the distorted images. The PSFs of the distortions are assumed separable. The Van Cittert deconvolutions are non-stationary for both the radial and the angular dimensions of the auto-co-occurrences. The restorations of $C_{v_iv_i}$ provide $\tilde{C}_{u_iu_i}$, for $i = 0, 1$. The restoration of $C_{v_0v_1}$ gives $\tilde{C}_{u_0u_1}$.

3.3 Bayesian Coring Estimates of the Images

The general Eq. (3) for the auto-co-occurrence statistics gives the posterior expectation of the intensities in polar coordinates $(\hat{r_i}, \hat{\phi_i})^T = E((r_i, \phi_i)^T | (r_i', \phi_i')^T)$ in neighborhoods $\Delta r_i \in \mathcal{N}_{r_i}$ and $\Delta\phi_i \in \mathcal{N}_{\phi_i}$ as

$$\begin{pmatrix} \hat{r_i} \\ \hat{\phi_i} \end{pmatrix} = \frac{\sum_{\mathcal{N}_{r_i}} \sum_{\mathcal{N}_{\phi_i}} P_b\left(\begin{pmatrix} \Delta r_i \\ \phi_i \end{pmatrix}; \begin{pmatrix} (\sigma_{b,r_i} r_i)^2 \\ \sigma_{b,\phi_i}^2 \end{pmatrix}\right) . \tilde{C}_{u_iu_i}\begin{pmatrix} r_i + \Delta r_i \\ \phi_i + \Delta\phi_i \end{pmatrix} \cdot \begin{pmatrix} r_i + \Delta r_i \\ \phi_i + \Delta\phi_i \end{pmatrix}}{\sum_{\mathcal{N}_{r_i}} \sum_{\mathcal{N}_{\phi_i}} P_b\left(\begin{pmatrix} \Delta r_i \\ \phi_i \end{pmatrix}; \begin{pmatrix} (\sigma_{b,r_i} r_i)^2 \\ \sigma_{b,\phi_i}^2 \end{pmatrix}\right) . \tilde{C}_{u_iu_i}\begin{pmatrix} r_i + \Delta r_i \\ \phi_i + \Delta\phi_i \end{pmatrix}}, \quad (9)$$

where $P_b(\Delta r_i, \phi_i)$ is as in Eq. (7). Equation (3) for the joint-co-occurrence statistics provides the posterior expectation of the intensities in Cartesian coordinates $(\hat{\eta_0}, \hat{\eta_1})^T = E((\eta_0, \eta_1)^T | (\eta_0', \eta_1')^T)$ in neighborhoods $\Delta\eta_0 \in \mathcal{N}_{\eta_0}$ and $\Delta\eta_1 \in \mathcal{N}_{\eta_1}$ with

$$\begin{pmatrix} \hat{\eta_0} \\ \hat{\eta_1} \end{pmatrix} = \frac{\sum_{\mathcal{N}_{\eta_0}} \sum_{\mathcal{N}_{\eta_1}} P_b\left(\begin{pmatrix} \Delta\eta_0 \\ \Delta\eta_1 \end{pmatrix}; \begin{pmatrix} (\sigma_{b,\eta_0}\eta_0)^2 \\ (\sigma_{b,\eta_1}\eta_1)^2 \end{pmatrix}\right) . \tilde{C}_{u_0u_1}\begin{pmatrix} \eta_0 + \Delta\eta_0 \\ \eta_1 + \Delta\eta_1 \end{pmatrix} \cdot \begin{pmatrix} \eta_0 + \Delta\eta_0 \\ \eta_1 + \Delta\eta_1 \end{pmatrix}}{\sum_{\mathcal{N}_{\eta_0}} \sum_{\mathcal{N}_{\eta_1}} P_b\left(\begin{pmatrix} \Delta\eta_0 \\ \Delta\eta_1 \end{pmatrix}; \begin{pmatrix} (\sigma_{b,\eta_0}\eta_0)^2 \\ (\sigma_{b,\eta_1}\eta_1)^2 \end{pmatrix}\right) . \tilde{C}_{u_0u_1}\begin{pmatrix} \eta_0 + \Delta\eta_0 \\ \eta_1 + \Delta\eta_1 \end{pmatrix}}, \quad (10)$$

where $P_b(\Delta\eta_0, \Delta\eta_1)$ is as given in Eq. (8). The gain in Eq. (4) for $C_{u_iu_i}$ in Eq. (9) becomes $R_{i,t}^s(r, \phi) = \frac{\hat{r_i}}{r_i}$, $i = 0, 1$. The gain for $C_{u_0u_1}$ in Eq. (10) becomes $R_{i,t}^b = \frac{\hat{\eta_i}}{\eta_i}$, $i = 0, 1$. They give 2D restoration matrices of dimensions equal to the corresponding dynamic ranges.

3.4 Iterative Estimation of Cumulative Intensity Restoration

At $t = 0$, the restoration fields are initialized to $b^{-1}_{i,t=0}(\mathbf{x}) = 1, \forall \mathbf{x}$. At $t > 0$ the intensity co-occurrences index the restoration matrices to provide an initial incremental estimate of the restoration with

$$b^{-1}_{i,inc,t}(\mathbf{x}) = \frac{1}{2} E_{\Delta\mathbf{x}\in N}\left(R^{s}_{i,t}\left(v_i(\mathbf{x}), v_i(\mathbf{x}+\Delta\mathbf{x})\right) + R^{b}_{i,t}\left(v_0(\mathbf{x}), v_1(\mathbf{x}+\Delta\mathbf{x})\right)\right). \quad (11)$$

The estimates $b^{-1}_{i,inc,t}$ at iteration $t > 0$ multiply the cumulative estimate of the restoration, $W_{i,t-1}$ of iteration $t-1$ to give $W'_{i,t} = W_{i,t-1} \times b^{-1}_{i,inc,t}$. The estimates $W'(\mathbf{x})$ are smoothed with a spatial Gaussian filter $G(\mathbf{x}; \sigma^2_{s,i})$ to give the cumulative restorations

$$W_{i,t}(\mathbf{x}) = W'_{i,t}(\mathbf{x}) * G(\mathbf{x}; \sigma^2_{s,i}), i = 0, 1. \quad (12)$$

The $W_{i,t}$ are applied to $v_{i,t-1}$ to provide $v_{i,t} = u_{i,t-1}$ that are the updated estimates for the latent images. The end condition of the iterations involves the standard deviations of $W_{i,t}$, $\sigma(W_{i,t}) = \frac{\|W_{i,t}-1\|_2}{\|W_{i,t}-1\|_0}$. The iterations terminate at t_{min} when $\sigma_{i,t}$ reaches a minimum for at least one of the two images i. A maximum number of iterations t_{max} is also imposed.

3.5 Constraints for Stability and Valid Image Domains

The sequences are designed to give the Regions of Interest (ROI) of the images, $I_{ROI,i}$, valid signal, tissues contrasts, and dynamic ranges. The non-uniformity characterizes the signals regions of the images. The upper parts of the images dynamic ranges may correspond to artifacts and are compressed linearly. Reference intensities $\eta_{i,ref}$ for high cumulative percentages, 90%, of the dynamic ranges are set. The dynamic range of the noise is delimited by the minimum signal intensity, $\eta_{i,min} = 0.1 \times \eta_{i,ref}$. The dynamic ranges are preserved up to $\eta_{i,upp} = 1.5 \times \eta_{i,ref}$. Beyond this values the intensity ranges are compressed linearly up to $\eta_{i,max} = 3.0 \times \eta_{i,ref}$ for ranges $[1.5 \times \eta_i^{0.9}, 3.0 \times \eta_i^{0.9}]$. The intensity ranges of $[\eta_i^{min}, \eta_i^{max}] = [0.1\eta_{i,ref}, 3\eta_{i,ref}]$ correspond to the valid signal regions of the ROIs, $I_{ROI,i} = 1$. The regions in the images ROIs with intensities in ranges $[0, \eta_i^{min}]$ and intensities beyond η_i^{max} are invalid, $I_{ROI,i} = 0$. The C_{ii} are computed over $([\eta_{i,min}, \eta_{i,max}]^2)$. The same ranges are used for the deconvolution of the C_{ii} to give the R^s_i. The deconvolution of C_{ii} for R^b_i considers $[\eta_{i,min}, \eta_{i,max}]$ for i and the complete range $[0, \eta_{j,max}]$ for image j. Beyond the valid ranges, the entries of R are set to unity, $R^s_i = R^b_i = 1$.

The estimates of the initial restoration fields $b^{-1}_{i,inc}$ in Eq. (11) involve window $\mathcal{N}_{\mathbf{x}}$. The resulting estimates are smoothed to give W_i with a window depending on σ_s in Eq. (12). The windows consider a multiplicative field with unit weight in the corresponding valid region $I_{ROI,i} = 1$ and with value much less than unity in the remaining ROI. The $b^{-1}_{i,inc}$ is back-projected from Eq. (11) over the valid part of the ROI of image i and is set to unity outside. The smoothing of the non-uniformity with Eq. (12) gives smoothly unity farther from the valid region and towards the invalid region of the image with $I_{ROI,i}(\mathbf{x}) = 0$. The first constraint for the stability of the dynamic range is applied to the output of Eq. (11) to set the pixel-wise average of the restoration to unity with normalization $b^{-1}_{i,inc,t}(\mathbf{x}) \leftarrow b^{-1}_{i,inc,t}(\mathbf{x})/\|b^{-1}_{i,inc,t}(\mathbf{x})\|_1$. A second constraint preserves the reference intensity so that $\eta_{i,ref,t} = \eta_{i,ref,t=0}$, by rescaling the restorations from Eq. (12) with $W_{i,t} \leftarrow W_{i,t} \times (\eta_{i,ref,t=0}/\eta_{i,ref,t})$.

4 Experimental Results

4.1 Image Sets and Implementation

The images were from two brain datasets of pairs of T_1 and T_2 weighted contrasts. The first set was from the BrainWeb MR simulator. It consists of seven representative pairs of T_1w and T_2w brain images from the 1.5 *Tesla* BrainWeb phantom [1]. Both images have a resolution of $(1.0\,\text{mm})^3$ and a matrix size of $181 \times 181 \times 217$. They are corrupted with simulated non-uniformities of levels $B = 0\% - 20\% - 40\% - 60\% - 80\% - 100\%$ and noise of $N = 5\%$ as well as an image with $B = 40\%$ and noise of $N = 3\%$. The brain region available from the BrainWeb simulator was used.

The second brain dataset was of patients with Parkinson's disease. The study was approved by the local ethical committee. The images were acquired under anaesthesia to reduce motion. The quality of the images was verified. The data were from 29 patients, 8 women of average age 66 years and 21 men of average age 60 years. The patients were imaged at 3.0 *Tesla* (SIEMENS, TrioTim). A T_1w image was acquired with resolution $1.0 \times 1.0 \times 1.1\,\text{mm}^3$ and matrix size of $240 \times 256 \times 160$. A T_2w image was acquired with resolution $0.63 \times 0.63 \times 1.80\,\text{mm}^3$ and size of $308 \times 384 \times 80$. The brain regions were extracted with BET [14]. The real brain datasets were placed in the same sampling grid with reference the T_1w image. The images were smoothed with a median filter of size $3 \times 3 \times 3\,\text{mm}^3$ along the axes.

The implementation is in *C++*. The parameters were identical for all datasets. The co-occurrence parameters were $\rho = 6$ *pixels* and $\sigma_{\phi_i} = 4^o$. The variable parameters are $\sigma_{r_0} = \sigma_{r_1}$. The parameters σ_{r_i}, $i = 0, 1$ are expressed as a fraction of the dynamic range and set to 0.02 with $\sigma_{u_i} = \left(1/\sqrt{2}\right)\sigma_{r_i}$. The parameters for spatial smoothing are $\sigma_{s_0} = \sigma_{s_1} = 140$ *pixels* and is performed separably. The iteration parameter is $t_{max} = 20$. The parameters of the Van Cittert deconvolution are $\beta = 0.3$, $\epsilon^2 = 0.01$ with a total of 4 iterations.

4.2 Validation and Experiments for the BrainWeb Images

The validation used the Coefficient of Joint Variation between the White Matter (WM) and the Gray Matter (GM) regions,

$$CJV(GM, WM) = \frac{\sigma_{GM} + \sigma_{WM}}{\|\mu_{GM} - \mu_{WM}\|_2}, \tag{13}$$

where μ_{GM} and μ_{WM} are the means of the two tissues intensities, and σ_{GM} and σ_{WM} are their standard deviations. This measure represents the contrast between the tissues regions. The T_1w and the T_2w images give CJV_{T_1} and CJV_{T_2}, respectively. The ratio of the CJV of the restored image $CJV_{v_{t,i}}$ to the one of the initial image $CJV_{v_{0,i}}$ gives $CJV^{ratio}_{v_{t,i}} = \frac{CJV_{v_{t,i}}}{CJV_{v_{0,i}}}$, for $i = T_1, T_2$. An effective restoration would result in a low value for CJV_{v_t} and on a ratio $CJV^{ratio}_{v_t}$ below unity. The BrainWeb phantom makes available the undistorted latent templates u_i [1]. Their tissues regions are used to compute the CJV. The validation is also performed by considering the absolute value of the difference between the restored images and the corresponding u_i, $diff_{v_{t,i}} = \|v_{t,i} - u_i\|$. This difference is desired to be low. The ratio of the difference from the restored $\|v_{t,i} - u_i\|$ to the difference from the initial $v_{0,i}$, $\|v_{0,i} - u_i\|$, gives $diff^{ratio}_{v_{t,i}} = \frac{\|v_{t,i} - u_i\|}{\|v_{0,i} - u_i\|}$. It is low and less than unity for an effective restoration.

An example of the analysis of images of a BrainWeb phantom with $N = 5\%$ and $B = 40\%$ is in Fig. 2. This figure contains slices from the initial and from the restored images as well as their co-occurrence statistics. In this example the cerebellum in both the T_1w and the T_2w images becomes brighter and thus its statistics become closer to the mean tissues statistics over the remaining images. Table 1 shows the values for the initial corrupted image pairs and the values of the intensity restored images. In parentheses is the ratio $CJV^{ratio}_{v_t,T_i}$, $i = 1, 2$. The improvements resulting from the restoration are particularly apparent for higher non-uniformities. The differences between the restored images and the corresponding phantom u_i images, $diff_{t,i}$, are also computed and given in Table 2. In parentheses is the ratio $diff^{ratio}_{t,i}$. It reduces the ratios to much less than unity, which shows the large improvements particularly for higher levels of non-uniformities. This is at the level of the noise.

Table 1. Validation for phantom T_1 and T_2 data with CJV_i of GM and WM tissues. In parentheses is the CJV^{ratio}_i. Low values indicate improved performance.

BrainWeb\method	Original		Joint Co-occurrences	
	T_1	T_2	T_1	T_2
N = 0, RF = 0	0.581369	0.770175	0.581369 (1)	0.770175 (1)
N = 3, RF = 40	0.765254	1.1697	0.660506 (0.86312)	1.02978 (0.880374)
N = 5, RF = 0	0.720008	1.13181	0.720008 (1)	1.13181 (1)
N = 5, RF = 20	0.735217	1.21963	0.762611 (1.03726)	1.24902 (1.02409)
N = 5, RF = 40	0.815382	1.37113	0.758738 (0.930531)	1.26971 (0.926038)
N = 5, RF = 60	1.29703	2.33456	0.843404 (0.650259)	1.39622 (0.598069)
N = 5, RF = 80	1.29703	2.33456	0.847001 (0.653031)	1.39361 (0.59695)
N = 5, RF = 100	1.29703	2.33456	0.861492 (0.664204)	1.39666 (0.598253)

Table 2. Validation for phantom T_1 and T_2 data with $diff_{t,i}$. In parentheses is the $diff_i^{ratio}$. Low values indicate improved performance.

BrainWeb\method	Original		Joint Co-occurrences	
	T_1	T_2	T_1	T_2
N = 0, RF = 0	0.0203729	0.0397313	0.0203729 (1)	0.0397313 (1)
N = 3, RF = 40	0.0453967	0.0604082	0.0253967 (0.55944)	0.0502752 (0.832258)
N = 5, RF = 0	0.0331167	0.0591677	0.0331167 (1)	0.0591677 (1)
N = 5, RF = 20	0.0385987	0.0610285	0.0325415 (0.843074)	0.0582324 (0.954184)
N = 5, RF = 40	0.0494876	0.0681559	0.0326034 (0.658818)	0.0584461 (0.857535)
N = 5, RF = 60	0.116259	0.105483	0.0503919 (0.433445)	0.0655376 (0.621312)
N = 5, RF = 80	0.116255	0.105479	0.0506253 (0.435466)	0.0651502 (0.617661)
N = 5, RF = 100	0.116252	0.105476	0.0523669 (0.450459)	0.0655749 (0.621706)

4.3 Validation and Experiments with the Parkinson's Brain Images

The entropy of the original image, H_{orig}, and the entropy of the restored image, H_{rest} give the validation measure

$$H_{ratio} = \frac{e^{H_{rest}} - e^{H_{orig}}}{e^{H_{orig}}}. \tag{14}$$

A successful restoration decreases the histogram entropy and the measure H_{ratio} becomes negative. Table 3 gives the statistics of H_{ratio} for all 29 patients. The table shows that it is indeed negative for all pairs of T_1w image and T_2w image restorations. Thus, the restorations are successful for all pairs and for all images.

Table 3. Statistics of H_{ratio} for the T_1w and T_2w images of the 29 Parkinson's patients. They are significantly negative for all images and hence the restorations are successful.

	Mean	Stand. dev	Median	Minimum	Maximum
H_{ratio} for T_1	−0.1328237	0.01646287	−0.134414	−0.159326	−0.0961817
H_{ratio} for T_2	−0.2375128	0.05627061	−0.247903	−0.32561	−0.113616

The restoration of the T_1w and the T_2w images of a representative patient is in Fig. 3. The tissues in the cortex, subcortical regions, and cerebellum become of uniform intensity in both the T_1w and the T_2w images. The intensities of the WM also become more uniform. The statistics of the T_1w image in Fig. 3 show three different distributions corresponding to the three brain tissues. The statistics of the T_2w images also show distinct distributions for the WM and the GM even though they have a low contrast. They also show a sharper distribution corresponding to the border between the WM and the GM as well as sharper statistics for the CSF. The successful restoration for the T_2w image gives a higher

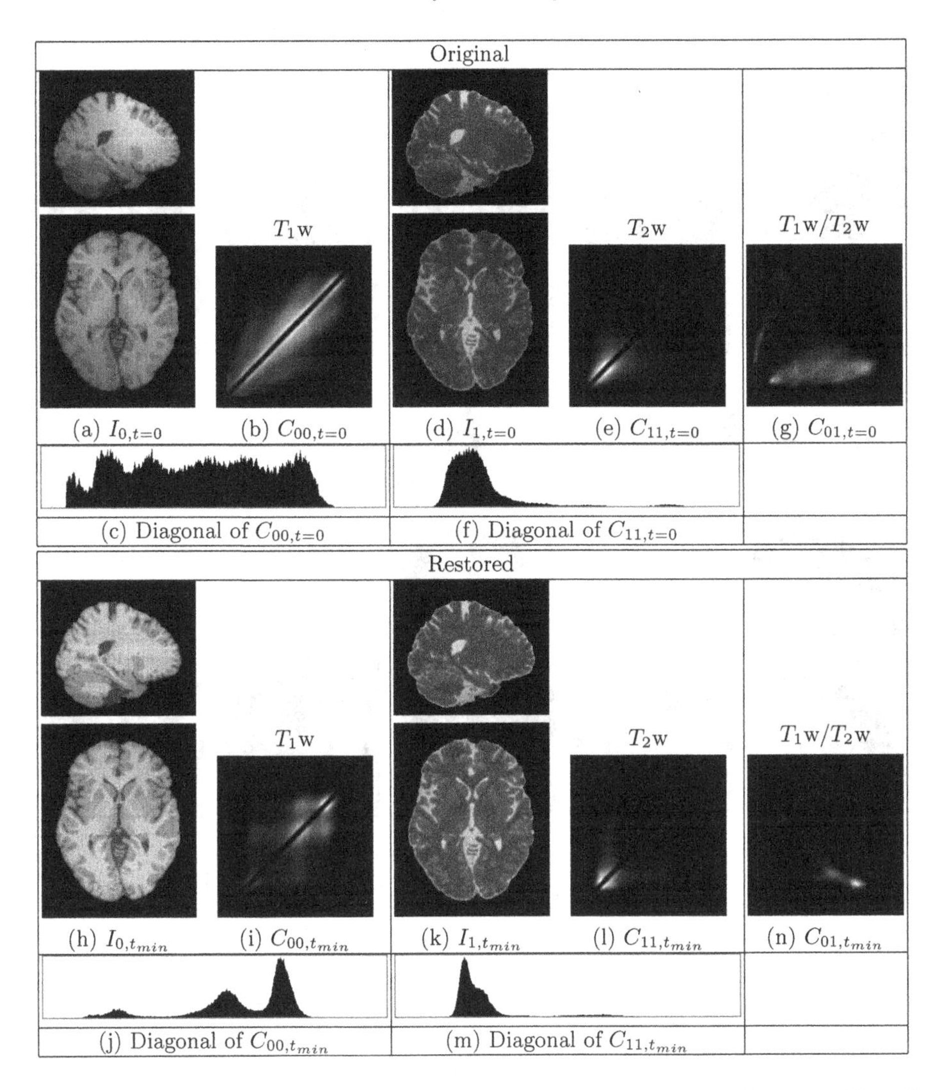

Fig. 2. The restoration of pairs of T_1 and T_2 BrainWeb images with $B = 40\%$ and $N = 5\%$. The cerebellum becomes brighter and the statistics sharper. The images are displayed contrast enhanced and the 2D co-occurrences are shown logarithmic.

contrast. The presence of separate tissue distributions in the statistics of the restored images gives the entropy based measure a semantic meaning for tissue intensity uniformity. The correction for a pair of images lasts approximately 2 h with an Intel processor of 2.60 GHz.

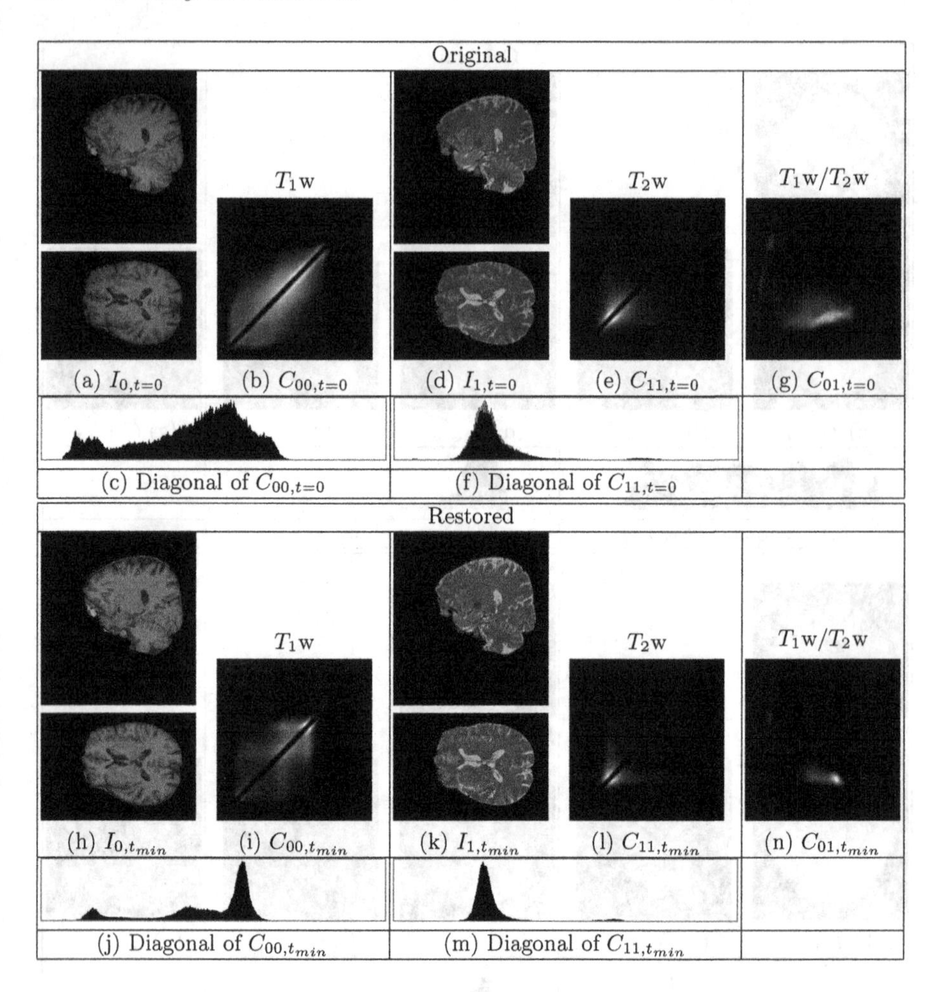

Fig. 3. Restoration of a pair of real T_1w and T_2w images. The intensities of the WM in both images become more uniform. The images are displayed contrast enhanced and the 2D co-occurrences are shown logarithmic.

5 Discussion and Conclusion

The method restores images despite the high intensity distortions that are present in most of them and the lower tissue quality of Parkinson's patients data. It is also efficient since it computes the histogram only once in an iteration. The co-occurrences decrease the spread of the dominant distributions in the data and improve their contrast. The sigmoid in the auto-co-occurrences removes the dominance of their diagonal and increases the sensitivity to the statistics across regions borders in 2D statistical space. In brain images this is from the extensive interface between the GM and the WM tissues. The distortion filter is non-stationary and is used for the non-stationary restoration and for the conditional expectation. The method is stable and accommodates the inevitable difference between the signal regions of the images.

References

1. Cocosco, C., Kollokian, V., Kwan, R.S., Evans, A.: BrainWeb: online interface to a 3D MRI simulated brain database. NeuroImage **5**(4–2/4), S425 (1997)
2. Fan, A., Wells III, W.M., Fisher III, J.W., Çetin, M., Haker, S., Mulkern, R., Tempany, C., Willsky, A.S.: A unified variational approach to denoising and bias correction in MR. In: Taylor, C., Noble, J.A. (eds.) IPMI 2003. LNCS, vol. 2732, pp. 148–159. Springer, Heidelberg (2003). doi:10.1007/978-3-540-45087-0_13
3. Hadjidemetriou, S., Buechert, M., Ludwig, U., Hennig, J.: Joint restoration of bi-contrast MRI data for spatial intensity non-uniformities. In: Székely, G., Hahn, H.K. (eds.) IPMI 2011. LNCS, vol. 6801, pp. 346–358. Springer, Heidelberg (2011). doi:10.1007/978-3-642-22092-0_29
4. Hadjidemetriou, S., Studholme, C., Mueller, S., Weiner, M., Schuff, N.: Restoration of MRI data for intensity non-uniformities using local high order intensity statistics. Med. Image Anal. **13**(1), 36–48 (2009)
5. Li, C., Gore, J., Davatzikos, C.: Multiplicative intrinsic component optimization (MICO) for MRI bias field estimation and tissue segmentation. Magn. Reson. Imaging **32**, 913–923 (2014)
6. Li, C., Huang, R., Ding, Z., Gatenby, J., Metaxas, D., Gore, J.: A level set method for image segmentation in the presence of intensity inhomogeneities with application to MRI. IEEE Trans. Image Process. **20**(7), 2007–2016 (2011)
7. Lui, D., Modhafar, A., Haider, M., Wong, A.: Monte Carlo-based noise compensation in coil intensity corrected endorectal MRI. BMC Med. Imaging **15**, 43 (2015)
8. Mangin, J.: Entropy minimization for automatic correction of intensity nonuniformity. In: Proceedings of IEEE Workshop on MMBIA, pp. 162–169 (2000)
9. Learned-Miller, E.G., Jain, V.: Many heads are better than one: jointly removing bias from multiple MRIs using nonparametric maximum likelihood. In: Christensen, G.E., Sonka, M. (eds.) IPMI 2005. LNCS, vol. 3565, pp. 615–626. Springer, Heidelberg (2005). doi:10.1007/11505730_51
10. Noterdaeme, O., Brady, M.: A fast method for computing and correcting intensity inhomogeneities in MRI. In: Proceedings of ISBI, pp. 1525–1528 (2008)
11. Renugadevi, M., Varghese, D., Vaithiyanathan, V., Raju, N.: Variational level set segmentation and bias correction of fused medical images. Asian J. Med. Sci. **4**(2), 66–74 (2012)
12. Simoncelli, E., Adelson, E.: Noise removal via Bayesian wavelet coring. In: Proceedings of 3rd IEEE ICIP, vol. I, pp. 379–382 (1996)
13. Sled, J., Zijdenbos, A., Evans, A.: A nonparametric method for automatic correction of intensity nonuniformity in MRI data. IEEE Trans. Med. Imaging **17**(1), 87–97 (1998)
14. Smith, S.: Fast robust automated brain extraction. Proc. Hum. Brain Mapp. **17**, 143–155 (2002)
15. Vidal-Pantaleoni, A., Mart, D.: Comparison of different speckle reduction techniques in SAR images using wavelet transform. Int. J. Remote Sens. **25**(22), 4915–4932 (2004)
16. Vovk, U., Pernus, F., Likar, B.: Intensity inhomogeneity correction of multispectral MR images. NeuroImage **32**, 54–61 (2006)

17. Zhang, H., Ye, X., Chen, Y.: An efficient algorithm for multiphase image segmentation with intensity bias correction. IEEE Trans. Image Process. **22**(10), 3842–3851 (2013)
18. Zheng, Y., Gee, J.: Estimation of image bias field with sparsity constraints. In: Proceedings of ISBI, pp. 255–262 (2010)

Can Planning Images Reduce Scatter in Follow-Up Cone-Beam CT?

Jonathan H. Mason[1(✉)], Alessandro Perelli[1], William H. Nailon[2], and Mike E. Davies[1]

[1] Institute for Digital Communications, University of Edinburgh, Edinburgh EH9 3JL, UK
{j.mason,a.perelli,mike.davies}@ed.ac.uk
[2] Oncology Physics Department, Edinburgh Cancer Centre, Western General Hospital, Edinburgh EH4 2XU, UK
bill.nailon@luht.scot.nhs.uk

Abstract. Due to its wide field of view, cone-beam computed tomography (CBCT) is plagued by large amounts of scatter, where attenuated photons hit the detector, and corrupt the linear models used for reconstruction. Given that one can generate a good estimate of scatter however, then image accuracy can be retained. In the context of adaptive radiotherapy, one usually has a low-scatter planning CT image of the same patient at an earlier time. Correcting for scatter in the subsequent CBCT scan can either be self consistent with the new measurements or exploit the prior image, and there are several recent methods that report high accuracy with the latter. In this study, we will look at the accuracy of various scatter estimation methods, how they can be effectively incorporated into a statistical reconstruction algorithm, along with introducing a method for matching off-line Monte-Carlo (MC) prior estimates to the new measurements. Conclusions we draw from testing on a neck cancer patient are: statistical reconstruction that incorporates the scatter estimate significantly outperforms analytic and iterative methods with pre-correction; and although the most accurate scatter estimates can be made from the MC on planning image, they only offer a slight advantage over the measurement based scatter kernel superposition (SKS) in reconstruction error.

Keywords: Computed tomography · Cone-beam · Scatter estimation · Prior information · Statistical · Reconstruction

1 Introduction

Cone-beam computed tomography (CBCT) is an imaging modality that is seeing increased use for image guidance procedures, such as radiation therapy [1]. A key challenge of this geometry is the vast quantities of scattered photons that reach the detector [2], and contaminate other line-of-sight measurements. Usually in this context however, one has a planning scan of the same patient from

M. Valdés Hernández and V. González-Castro (Eds.): MIUA 2017, CCIS 723, pp. 629–640, 2017.
DOI: 10.1007/978-3-319-60964-5_55

a more accurate CT acquisition, such as a helical fan-beam, which has significantly lower scatter due to better collimation and narrower field-of-view. Typical approaches to CBCT scatter correction either form a self-consistent model based solely on the new measurements [3–6], or exploit the prior image as a basis for its estimation [7–9].

In this study, we will adopt the perspective that various scatter models allow one to make an estimate of its expectation, given a set of new CBCT projections, and potential access to a low-scatter prior CT image. We will then investigate two key aspects: how accurate can scatter be estimated in a moderate and low dose setting; and how should these estimates be incorporated into reconstruction.

In the first case, the scatter estimation methods we will look at fall into two distinct classes—methods that are blind to the planning image, and methods that exploit it. In our 'prior blind' category are a view dependent uniform estimation [5], the scatter kernel superposition (SKS) [4] and fast asymmetric SKS (fASKS) [6], along with simulating the scatter through a Monte-Carlo (MC) engine on a preliminary reconstruction with the fast Feldkamp-Davis-Kress (FDK) [10] algorithm. Conversely, based upon a rigid registration of the planning image, we will look at the effectiveness of taking scatter as a smooth projection difference [7,11], and from using the MC engine on this registered plan. Here, we will look at both calculating the planning MC estimate on-line, after registration [9], along with the notion of matching an off-line pre-calculated estimate to the measurements.

In the second case, in the subsequent reconstruction with each of the estimation methods, we make a distinction between 'scatter correction', where measurements are pre-processed to remove its effect, and 'scatter-aware inference', where the imaging operates based on the raw uncorrected measurements and knowledge of the scatter estimate. Most popular 'analytic' and 'iterative' techniques, such as FDK and PWLS [12], fall into the former category. Taking the second approach, although more challenging due to its non-linearity and nonconvexity, represents a more accurate data model that may mitigate reconstruction artefacts and errors.

We begin this article with relevant background material, where we explain the system model in Sect. 2.1, an overview of scatter estimation methods in Sect. 2.2, and standard reconstruction based on pre-corrected measurements in Sect. 2.3. Next, we give details of matching an off-line planning MC scatter given a rigid translation of the specimen in Sect. 3.1, along with the model for statistical scatter aware reconstruction in Sect. 3.2.

From a dataset derived from repeat CT images of a neck caner patient, we then evaluate both the scatter estimation accuracy and reconstruction error with the range the methods under test. The results are then presented in Sect. 4.3.

2 Background

2.1 System Description

The system we will directly study in this work is a circular scan CBCT. This consists of a point source and flat panel detector, which rotate throughout 360°

around the specimen where photons are emitted and measured after a fixed angular increment. Assuming that recorded x-rays are drawn from independent Poisson distributions [13,14], then we can write the distribution using monoenergetic Beer-Lambert law with additive scatter component as

$$y_k \sim \text{Poisson}\left\{b_k \exp(-\left[\boldsymbol{\Phi}\boldsymbol{\mu}\right]_i) + s_k\right\} \text{ for } k = 1, \ldots, N, \tag{1}$$

where $\hat{\boldsymbol{y}} \in \mathbb{R}^N$ is a column vector of measurements, $\boldsymbol{b} \in \mathbb{R}^N$ is a vector of input source fluxes, $\boldsymbol{\mu} \in \mathbb{R}^M$ is a vector of linear attenuation coefficients, and $\boldsymbol{\Phi} \in \mathbb{R}^{N \times M}$ is a system matrix describing the path of each ray through the specimen and onto the detector.

2.2 Estimating Scatter

In this section, we give brief overviews of the various scatter estimation techniques that we evaluate in this study.

Uniform: A simple method that calculates a constant scatter at each projection angle, we denote as 'uniform' [5,15]. Here, using the assumption that a scatter-to-primary ratio (SPR) is known a-priori, along with a distinction between air and object containing projections, we can write

$$\boldsymbol{s}\{i\} = \frac{\text{SPR}}{N_{\text{air}}(i)} \sum_{k \in \mathcal{C}_{\text{air}}(i)} y(k) \text{ for } i = 1, \ldots, P, \tag{2}$$

where $\mathcal{C}_{\text{air}}(i)$ is the set of air containing measurements at the ith angle, where the set satisfies $y(k) \geq t_{\text{air}}\, \forall k \in \mathcal{C}_{\text{air}}$ with some scalar threshold t_{air}. This essentially calculates the mean scatter given a constant ratio. In practice, this SPR can be found by observing the magnitude of signal in the air region with and without a specimen present, and assume the difference is scatter. To ensure the scatter is less than the minimum value in a given projection, a non-negativity constraint can be added [15].

SKS/fASKS: The scatter kernel superposition (SKS) [4] and fast asymmetric SKS (fASKS) [6], perform estimation as a convolution of the scatter free incident beam with an appropriate kernel. Since the incident beam is itself unknown, the methods iteratively estimate this as the difference between raw measurements and the updated scatter estimate from the previous iteration. Due to the ability calculate convolution rapidly through the FFT, both of these methods are relatively fast, especially when the projections are sub-sampled. Although they implicitly model the scatter media as homogeneous, the estimates are accurate in practice [6].

Diff. filt.: A simple concept for predicting the scatter contribution based upon a planning image is though the smooth difference—'diff. filt.'—between CBCT measurements and projections of a registered plan [7]. This model can be expressed as

$$\boldsymbol{s} = \boldsymbol{\mathcal{F}}(\boldsymbol{y} - \boldsymbol{b} \odot \mathbf{exp}(-\boldsymbol{\Phi}\boldsymbol{\mu}_{\text{reg}})), \tag{3}$$

where $\boldsymbol{\mathcal{F}}$ is a projection-wise filter, or set of filters—sequential median and Gaussian filters are used in [7]—and $\boldsymbol{\mu}_{\text{reg}}$ is the registered planning CT onto a preliminary reconstruction of the CBCT measurements.

Monte-Carlo (MC): Monte-Carlo scatter estimation techniques essentially draw a number of samples from an accurate probabilistic model of physical interactions. Given that the model is a faithful representation of reality, then the true expectation of scatter can be found with infinite samples.

We denote this process as

$$\tilde{\boldsymbol{s}} \sim \mathcal{MC}(\boldsymbol{b}, \boldsymbol{\Phi}, \boldsymbol{\mu}_{\text{est.}}, N_{\mathcal{MC}}),$$

where $\tilde{\boldsymbol{s}}$ is the estimate after $N_{\mathcal{MC}}$ photon simulations distributed throughout several projection angles and $\boldsymbol{\mu}_{\text{est.}}$ is the image onto which the estimation is based. We will test the ability of estimating MC scatter onto both a preliminary FDK with appropriate prior blind estimation, such as SKS, and onto the planning image.

If the MC simulation is made after the measurements are taken, then it may be appropriate to sub-sample both the image $\tilde{\boldsymbol{s}}$ and the number of photons, in order to complete the calculation on-line [9]. In the off-line setting, there is no immediate limitation on computational time, as it can be performed days or weeks ahead of the follow-up CBCT. Eventual matching of this off-line estimate is detailed in Sect. 3.1.

2.3 Reconstruction from Scatter Correction

In most cases, reconstruction is performed by inferring the attenuation coefficient given the model in (1), which follows an effective 'correction' of scatter. In the crudest form, this involves simply subtracting the scatter estimate from the measurements. An advantage of pre-correcting for scatter in this manner, allows a linear system to be exposed and solved, of the form

$$\boldsymbol{p} = \boldsymbol{\Phi}\boldsymbol{\mu} + \boldsymbol{n}, \tag{4}$$

where $\boldsymbol{n} \in \mathbb{R}^N$ is noise, $\boldsymbol{p} \in \mathbb{R}^N$ is the linearised projection, calculated by

$$p_i = \log\left(\frac{b_i}{y_i - s_i}\right) \text{ for } i = 1, ..., N, \tag{5}$$

where it can be solved by analytic filtered back-projection methods, such as FDK for CBCT [10], or with iterative methods, that approximate the noise model in (1) and incorporate regularisation, such as penalised weighted-least-squares (PWLS). Reconstruction through PWLS involves solving the problem

$$\hat{\boldsymbol{\mu}} = \underset{\boldsymbol{\mu}}{\operatorname{argmin}} \left(\boldsymbol{\Phi}\boldsymbol{\mu} - \boldsymbol{p}\right)^T \boldsymbol{W} \left(\boldsymbol{\Phi}\boldsymbol{\mu} - \boldsymbol{p}\right) + \lambda R(\boldsymbol{\mu}), \tag{6}$$

where $\boldsymbol{W} \in \mathbb{R}^{N \times N}$ is a diagonal weighting matrix with entries $w_{ii} = (y_i - s_i)^2 / y_i$, $R(\boldsymbol{\mu})$ is a regularisation function to promote desirable structure in $\boldsymbol{\mu}$, and λ is usually a scalar constant trade-off between data fit and regularisation.

3 Method

3.1 Off-line Scatter Matching

We propose that the expectation of scatter may be calculated off-line to a high accuracy based upon a prior image, then matched to the measurements during replanning. Conceptually, this is very similar to the notion of SKS/fASKS [6], where the scatter point spread function of a scanner are measured through blocks of material, and combined with convolution. Instead here, the entire global scatter profile is estimated, and simply shifted to fit the current pose of the patient. Our framework for this off-line scheme is illustrated in Fig. 1.

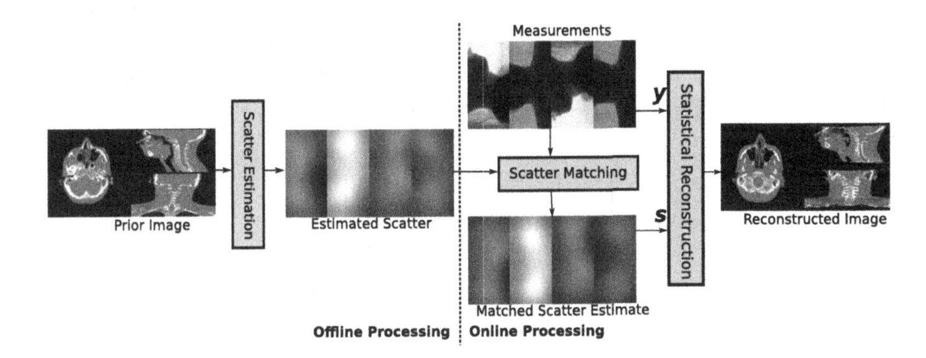

Fig. 1. Flow diagram for off-line prior MC scatter estimation and statistical reconstruction.

For the match, we seek a transformation in the coordinates of detecting elements at each projection angle, for which we adopt the notation

$$s_M\{i\} = \mathcal{I}\left(s\{i\}; (u_M, v_M)_i\right) \text{ for } i = 1, \ldots, P, \tag{7}$$

where P is the number of projections, $\mathcal{I}(\cdot)$ is the 2D linear interpolation of the image $s\{i\}$ corresponding to the ith projection angle of scatter estimate, $s_M\{i\}$

is the matched estimate, and $(u_M, v_M)_i$ are the transformed 2D coordinates according to

$$\begin{bmatrix} u' \\ v' \\ 1 \end{bmatrix} = \begin{bmatrix} w_{1,1} & w_{1,2} & w_{1,3} \\ w_{2,1} & w_{2,2} & w_{2,3} \\ 0 & 0 & 1 \end{bmatrix} \begin{bmatrix} u \\ v \\ 1 \end{bmatrix}, \tag{8}$$

where u and v are the original vector of coordinates for the detector, and w are parameters we wish to calculate through the matching process.

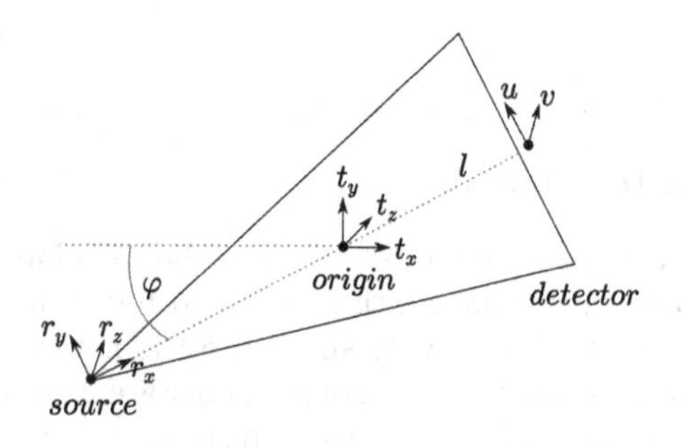

Fig. 2. Geometry of scatter shifting.

With reference to Fig. 2, each scatter projection image with coordinates u, v is updated based on a rigid translation of an object at the centre of rotation by $[t_x, t_y, t_z]^T$ onto a FDK reconstruction of the measurements. We define a new set of coordinates $[r_x, r_y, r_z]^T$, due to rotation of the source around by the ith angle φ_i by

$$\begin{bmatrix} r_x \\ r_y \\ r_z \end{bmatrix} = \begin{bmatrix} \cos\varphi_i & -\sin\varphi_i & 0 \\ \sin\varphi_i & \cos\varphi_i & 0 \\ 0 & 0 & 1 \end{bmatrix} \begin{bmatrix} t_x \\ t_y \\ t_z \end{bmatrix}.$$

We note that due to the projection, a translation along r_x leads to a change in scaling, and translations in r_y, r_z lead to shifting. If we define the distances $l_{\text{SO}}, l_{\text{SD}}$ as the lengths from source to origin and the detector respectively, the transformation to adjust the projection is

$$\begin{bmatrix} u_M \\ v_M \\ 1 \end{bmatrix} = \begin{bmatrix} \frac{l_{\text{SO}}}{l_{\text{SO}}+r_x} & 0 & \frac{l_{\text{SD}}}{l_{\text{SO}}+r_x} r_y \\ 0 & \frac{l_{\text{SO}}}{l_{\text{SO}}+r_x} & \frac{l_{\text{SD}}}{l_{\text{SO}}+r_x} r_z \\ 0 & 0 & 1 \end{bmatrix} \begin{bmatrix} u \\ v \\ 1 \end{bmatrix},$$

3.2 Reconstruction with Scatter Estimate

Instead of using the approximate linearisation of the model in (1), one could use it exactly. This is done for general additive Poisson noise in [16], and we will repeat it here explicitly for the incorporation of a scatter expectation in CBCT. In this case, reconstruction is taken as a maximum a posteriori (MAP) estimate, given (1) and a regularisation function to impose desirable properties in the image, as in PWLS (6).

As in [16], we pursue finding the maximum likelihood, by minimising the negative log-likelihood of (1), which is denoted

$$\mathrm{NLL}(\boldsymbol{\mu};\boldsymbol{y},\boldsymbol{s}) = \sum_{i=1}^{N} b_i \exp(-\left[\boldsymbol{\Phi}\boldsymbol{\mu}\right]_i) + s_i - y_i \log\left(b_i \exp(-\left[\boldsymbol{\Phi}\boldsymbol{\mu}\right]_i) + s_i\right). \quad (9)$$

It should be noted that for $s_i > 0$, (9) is nonconvex, so it may not be minimised with the same ease of the PWLS. Nevertheless, it is continuously differentiable with respect to $\boldsymbol{\mu}$, and can therefore be treated with an appropriate first order method. We note that reconstruction is then solution of

$$\hat{\boldsymbol{\mu}} = \underset{\boldsymbol{\mu}\in\mathcal{C}}{\operatorname{argmin}}\, \mathrm{NLL}(\boldsymbol{\mu};\boldsymbol{y},\boldsymbol{s}) + \lambda R(\boldsymbol{\mu}), \quad (10)$$

where $\mathcal{C}$ is a set of box constraints on $\boldsymbol{\mu}$ so that $0 \leq \mu_i \leq \zeta$ for $i = 1,...,N$, where ζ is the maximum allowable attenuation coefficient.

Although some may consider the difference between our notions of 'correction' with PWLS and 'estimation' to be trivial, there is a compelling distinction. Whilst in the corrective case, one must carefully design the process to well approximate the model used, in estimation, the expectation of scatter can be used directly, and reconstruction may be considered as the direct inference from the raw measurements. How this translates into practical reconstruction accuracy will be studied in the experimental section.

4 Experimentation

4.1 Data

The data set we are using is derived from repeat CT scans of a neck cancer patient from the Cancer Image Archive [17,18]. With these, we will use the first CT scan as the planning image, then synthesise CBCT measurements on the follow up after 5 months—these are shown in Figs. 3a and b respectively. A strong advantage of using this approach is that one has access to a ground truth, against which one can perform valid quantitative assessments.

To generate the CBCT data, we used the Monte-Carlo simulation tool Gate [19] with a 60 keV monoenergetic source on the oracle image, where we did runs with 5×10^{10} and 1×10^{10} photons over 160 projection angles to represent two levels of dose.

4.2 Methods Under Test

Scatter Method Implementation

- Oracle: this is using the true scatter signal from the measurement synthesis, to represent the ultimate conceivable estimate, and ground-truth for assessment.

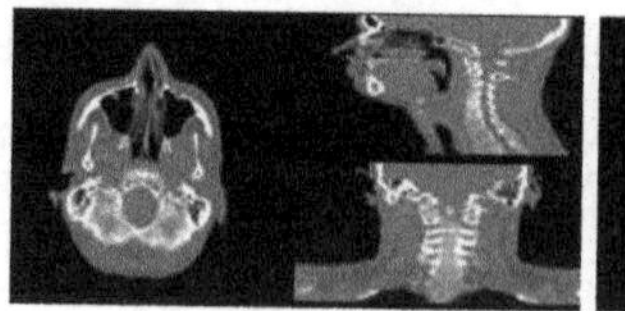

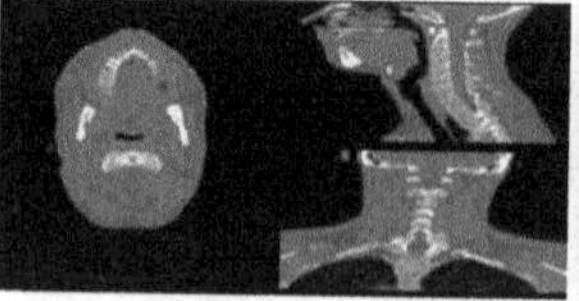

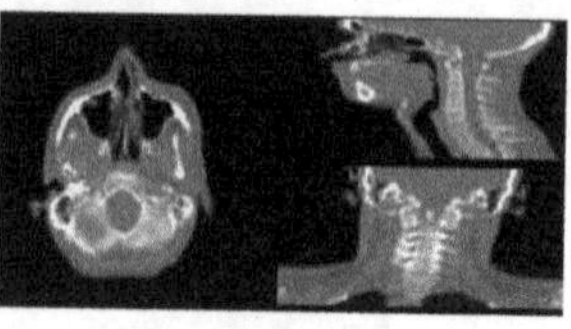

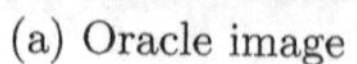

(a) Oracle image (b) Planning image (c) Registered plan

Fig. 3. Experimental data used: (a) is the oracle follow-up CT image; (b) is an unregistered initial planning image; and (c) is the plan registered rigidly onto an FDK (with SKS correction) reconstruction of the raw data—shown is the moderate dose, but a separate registration was used in low dose.

- None: scatter signal is not estimated at all.
- Uniform: calculated using (2) [5] with SPR $= 0.04$ and $t_{\text{air}} = 4000, 800$ for moderate and low doses.
- SKS/fASKS: implemented with same parameters as [6] for 'full-fan' acquisition and 20 iterations each.
- Diff. filt.: using (3) based upon the registered plan in Fig. 3c.
- FDK-MC: a sub-sampled MC estimate based upon preliminary FDK (with SKS correction) reconstruction.
- Online-prior-MC: the work-flow in [9], with sub-sampled MC applied to the registered planning image in Fig. 3c.
- Offline-prior-MC: using a detailed MC of the unregistered planning image in Fig. 3b, and matching as in Sect. 3.1 from same registration parameters as other planning methods.

For illustrative purposes, selected scatter estimates, along with the noisy ground-truth are shown in Fig. 4.

Reconstruction Implementation. The reconstruction methods under test were FDK, PWLS according to (6), and NLL according to (10). All iterative methods were run for 200 iterations, which was deemed ample for convergence, and all λ in the case of PWLS and NLL was set to 2×10^5, which was numerically tuned for good performance in both cases.

4.3 Results

Results are summarised in Tables 3 and 2, and selected reconstruction images are shown in Table 1.

The first observation that can be made from the scatter accuracy in both moderate and low dose cases, is that the prior-MC methods are the best performing on this data. This is interestingly opposed to the FDK-MC, especially since it uses the FDK with SKS, over which it has a significantly worse scatter estimate and only a slight decrease of analytic reconstruction accuracy. Another significant result is the very poor performance of the diff. filt. method, giving the

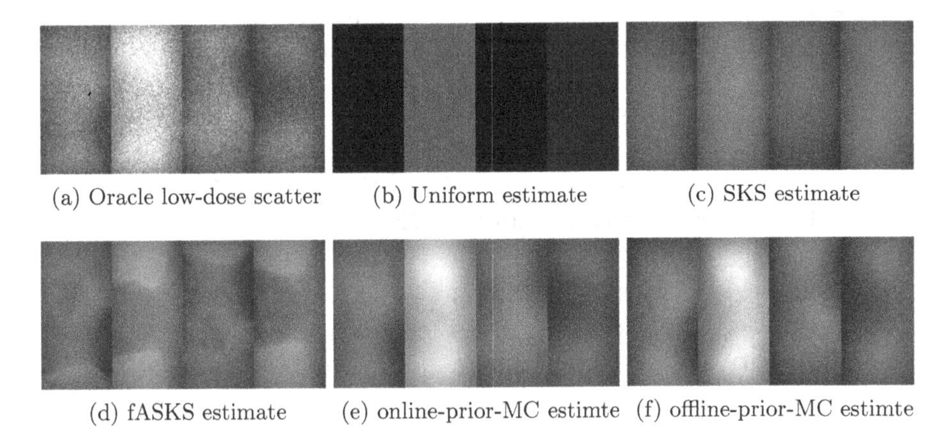
(a) Oracle low-dose scatter (b) Uniform estimate (c) SKS estimate

(d) fASKS estimate (e) online-prior-MC estimte (f) offline-prior-MC estimte

Fig. 4. Examples of low-dose scatter estimates and ground-truth noisy signal shown with grey scale [10, 70]: (a) is the oracle noisy scatter from the measurement synthesis; (b) is a uniform estimate; (c) and (d) are SKS and fASKS estimates respectively; (e) and (f) are on-line and off-line planning MC estimates respectively.

Table 1. Visual experimental results from low-dose measurements

Experiment	FDK	NLL
oracle		
none		
SKS		
offline-prior-MC		

Table 2. Quantitative results for moderate dose (5×10^5 photons). All errors are given as root-mean-squared (RMS), and reconstruction errors are in Hounsfield units.

Scheme	Scatter error	FDK error	PWLS error	NLL error
Oracle	0	51.8	19.8	19.3
None	218	74.35	78.3	76.3
Measurement based online scatter calculation				
Uniform	77.7	53.9	44.1	36.4
SKS	38.7	60.3	24.2	21.7
fASKS	33.9	50.8	27.1	23.1
FDK-MC	102	56.0	46.0	40.1
Prior based online scatter calculation				
Filt. diff.	135	63.3	66.7	62.5
Online-prior-MC	18.5	54.1	22.5	21.2
Prior based offline scatter calculation				
Offline-prior-MC	24.6	58.1	22.7	21.8

Table 3. Quantitative results for low dose (1×10^5 photons). All errors are given as root-mean-squared (RMS), and reconstruction errors are in Hounsfield units

Scheme	Scatter error	FDK error	PWLS error	NLL error
Oracle	0	51.6	23.7	22.5
None	43.9	74.1	78.5	77.4
Measurement based online scatter calculation				
Uniform	27.3	71.9	65.0	60.6
SKS	9.69	56.9	29.3	24.2
fASKS	8.95	51.5	34.1	26.0
FDK-MC	21.2	55.7	49.1	41.4
Prior based online scatter calculation				
Filt. diff.	27.8	63.5	44.7	43.2
Online-prior-MC	6.90	53.0	29.6	24.1
Prior based offline scatter calculation				
Offline-prior-MC	7.65	55.6	29.1	24.6

worst estimates of scatter, which is likely due to large errors from mismatches between the registered plan. Perhaps this would decrease with a non-rigid registration as in [7], though this will inevitably be increasingly difficult and unstable in the lower dose settings.

In general, the relationship between relative errors in SKS/fASKS and the prior MC methods is enlightening. Although fASKS is the best performer in FDK, this does not hold true for the iterative reconstructions. Apart from this however, the relative performance of these methods is very similar within the iterative results, all of which become rather close to the oracle scatter reconstruction in NLL. Of these, SKS may be the most appealing due to its fast computation and no reliance to planning registration.

One global trend in both the moderate and low dose results in Tables 2 and 3 is that NLL is more accurate than PWLS on every count. This is unsurprising, since PWLS may be considered an approximation to NLL, but is motivating for avoiding pre-correction in the high scatter setting of CBCT.

5 Conclusions

In this study, we have provided evidence for the differences between various scatter estimation strategies, and how these may best be incorporated into reconstruction. The most conclusive message is that opting for the NLL is more accurate than pre-correcting for scatter and using the PWLS, and in a lower dose setting this difference becomes significant. In terms of scatter estimation, several of the methods aided rather accurate reconstruction—SKS, fASKS, and both on-line/off-line planning MC estimating all performed similarly. At least in the case of the head and neck we look at in this study, it is perhaps not worth the extra computation of MC estimation from a planning scan, though this may prove to differ as larger regions such as the pelvis are imaged with an offset detector. As a second conclusion, we therefore offer that planning images can help reduce scatter in follow-up CBCT, but one can do almost as well without this information.

Acknowledgements. The authors would like to thank the Maxwell Advanced Technology Fund, EPSRC DTP studentship funds and ERC project: C-SENSE (ERC-ADG-2015-694888) for supporting this work.

References

1. Button, M.R., Staffurth, J.N.: Clinical application of image-guided radiotherapy in bladder and prostate cancer. Clin. Oncol. **22**(8), 698–706 (2010)
2. Siewerdsen, J.H., Jaffray, D.A.: Cone-beam computed tomography with a flat-panel imager: magnitude and effects of x-ray scatter. Med. Phys. **28**(2), 220 (2001)
3. Poludniowski, G., Evans, P.M., Hansen, V.N., Webb, S.: An efficient monte carlo-based algorithm for scatter correction in keV cone-beam CT. Phys. Med. Biol. **54**(12), 3847–3864 (2009)
4. Love, L.A., Kruger, R.A.: Scatter estimation for a digital radiographic system using convolution filtering. Med. Phys. **14**(2), 178–185 (1987)
5. Boellaard, R., van Herk, M., Mijnheer, B.J.: A convolution model to convert transmission dose images to exit dose distributions. Med. Phys. **24**(2), 189–199 (1997)
6. Sun, M., Star-Lack, J.M.: Improved scatter correction using adaptive scatter kernel superposition. Phys. Med. Biol. **55**(22), 6695–6720 (2010)
7. Niu, T., Sun, M., Star-Lack, J., Gao, H., Fan, Q., Zhu, L.: Shading correction for on-board cone-beam CT in radiation therapy using planning MDCT images. Med. Phys. **37**(10), 5395–5406 (2010)

8. Marchant, T.E., Moore, C.J., Rowbottom, C.G., MacKay, R.I., Williams, P.C.: Shading correction algorithm for improvement of cone-beam CT images in radiotherapy. Phys. Med. Biol. **53**(20), 5719–5733 (2008)
9. Xu, Y., Bai, T., Yan, H., Ouyang, L., Pompos, A., Wang, J., Zhou, L., Jiang, S.B., Jia, X.: A practical cone-beam CT scatter correction method with optimized Monte Carlo simulations for image-guided radiation therapy. Phys. Med. Biol. **60**(9), 3567–3587 (2015)
10. Feldkamp, L.A., Davis, L.C., Kress, J.W.: Practical cone-beam algorithm. J. Opt. Soc. Am. A **1**(6), 612 (1984)
11. Park, Y.-K., Sharp, G.C., Phillips, J., Winey, B.A.: Proton dose calculation on scatter-corrected CBCT image: feasibility study for adaptive proton therapy. Med. Phys. **42**(8), 4449–4459 (2015)
12. Fessler, J.A.: Fundamentals of CT reconstruction in 2D and 3D. In: Comprehensive Biomedical Physics, pp. 263–295. Elsevier (2014)
13. Elbakri, I.A., Fessler, J.A.: Statistical image reconstruction for polyenergetic X-ray computed tomography. IEEE Trans. Med. Imaging **21**(2), 89–99 (2002)
14. Chang, Z., Zhang, R., Thibault, J.-B., Sauer, K., Bouman, C.: Statistical X-ray computed tomography imaging from photon-starved measurements. In: SPIE Computational Imaging, vol. 9020, p. 90200G (2014)
15. Rit, S., Vila Oliva, M., Brousmiche, S., Labarbe, R., Sarrut, D., Sharp, G.C.: The reconstruction toolkit (RTK), an open-source cone-beam CT reconstruction toolkit based on the insight toolkit (ITK). J. Phys. Conf. Ser. **489**, 012079 (2014)
16. Erdogan, H., Fessler, J.A.: Accelerated monotonic algorithms for transmission tomography. In: Proceedings 1998 International Conference Image Processing ICIP 1998 (Cat. No. 98CB36269), vol. 2, pp. 680–684. IEEE Computer Society (1998)
17. Clark, K., Vendt, B., Smith, K., Freymann, J., Kirby, J., Koppel, P., Moore, S., Phillips, S., Maffitt, D., Pringle, M., Tarbox, L., Prior, F.: The cancer imaging archive (TCIA): maintaining and operating a public information repository. J. Digit. Imaging **26**(6), 1045–1057 (2013)
18. Fedorov, A., Clunie, D., Ulrich, E., Bauer, C., Wahle, A., Brown, B., Onken, M., Riesmeier, J., Pieper, S., Kikinis, R., Buatti, J., Beichel, R.R.: DICOM for quantitative imaging biomarker development: a standards based approach to sharing clinical data and structured PET/CT analysis results in head and neck cancer research. PeerJ **4**, e2057 (2016)
19. Jan, S., Benoit, D., Becheva, E., Carlier, T., Cassol, F., Descourt, P., Frisson, T., Grevillot, L., Guigues, L., Maigne, L., Morel, C., Perrot, Y., Rehfeld, N., Sarrut, D., Schaart, D.R., Stute, S., Pietrzyk, U., Visvikis, D., Zahra, N., Buvat, I.: GATE V6: a major enhancement of the GATE simulation platform enabling modelling of CT and radiotherapy. Phys. Med. Biol. **56**(4), 881–901 (2011)

Super Resolution Convolutional Neural Networks for Increasing Spatial Resolution of ^{1}H Magnetic Resonance Spectroscopic Imaging

Sevim Cengiz[1], Maria del C. Valdes-Hernandez[2], and Esin Ozturk-Isik[1](✉)

[1] Institute of Biomedical Engineering, Bogazici University, Istanbul, Turkey
{sevim.cengiz,esin.ozturk}@boun.edu.tr
[2] Department of Neuroimaging Sciences, Centre for Clinical Brain Sciences, University of Edinburgh, Edinburgh, UK
m.valdes-hernan@ed.ac.uk
https://cil.boun.edu.tr
http://www.ed.ac.uk/clinical-brain-sciences

Abstract. Proton magnetic resonance spectroscopic imaging (^{1}H-MRSI) provides noninvasive information regarding metabolic activity within the tissues. One of the main problems of ^{1}H-MRSI is low spatial resolution due to clinical scan time limitations. Advanced post-processsing algorithms, like convolutional neural networks (CNN) might help with generation of super resolution ^{1}H-MRSI. In this study, the application of super resolution convolutional neural networks (SRCNN) for increasing the spatial resolution of ^{1}H-MRSI is presented. Fluid Attenuated Inversion Recovery (FLAIR), T1-weighted, T2-weighted magnetic resonance imaging (MRI) data and a fused MRI, which contained the three different structural MR images in each RGB channel, were used in training the SRCNN scheme. The spatial resolution of ^{1}H-MRSI images were increased by a factor of three using the models trained with the anatomical MR images. The results of the proposed technique were compared with bicubic resampling in terms of peak signal to noise ratio and root mean square error. Our results indicated that SRCNN would contribute to reconstructing higher resolution ^{1}H-MRSI.

Keywords: Convolutional neural network · Super resolution · Proton magnetic resonance spectroscopic imaging

1 Introduction

Proton magnetic resonance spectroscopic imaging (^{1}H-MRSI) is commonly used in clinical settings for obtaining information about brain tissue metabolism. Acquisition of ^{1}H-MRSI in addition to standard anotomical MR images, like T1 weighted MRI (T1w MRI), T2 weighted MRI (T2w MRI), and fluid attenuated inversion recovery (FLAIR) MRI, has been proposed to aid in better

M. Valdés Hernández and V. González-Castro (Eds.): MIUA 2017, CCIS 723, pp. 641–650, 2017.
DOI: 10.1007/978-3-319-60964-5_56

defining disease characteristics, including multiple sclerosis, brain tumors and Parkinson's disease [1–5]. For instance, studies reported that there was lower N-acetyl aspartate to creatine ratio (NAA/Cr) in occipital lobe of patients diagnosed with Parkinson's disease with mild cognitive impairment [4,5], and brain tumors with a mutation in isocitrate dehyrogenase (IDH) have been observed to have higher 2-hydroxyglutarate (2HG) [6–8]. ^{1}H-MRSI detects a number of metabolites present in the tissue in relatively much lower concentrations than water. As a result, higher voxel sizes are employed for ^{1}H-MRSI to increase the signal to noise ratio (SNR). ^{1}H-MRS images typically have a spatial resolution that is 10 times lower than anatomical MR images. It is possible to increase the spatial resolution of ^{1}H-MRSI, but it would require a long scan time unless data undersampling strategies are employed [9]. An alternative approach that would result in higher spatial resolution ^{1}H-MRSI without a cost of scan time is advanced post-processing methods.

One of such post-processing techniques is convolutional neural network (CNN), which dates back to the 1980s [10]. CNN has been applied in many fields including handwriting [11] or face recognition [12], and pattern recognition [13] and classification [14]. Additionally, super-resolution CNN (SRCNN) has more recently been proposed to generate higher resolution images out of low resolution versions [15–17]. SRCNN has not been previously applied to increase the spatial resolution of anatomical MRI or ^{1}H-MRSI. In this study, we propose to increase the spatial resolution of ^{1}H-MRSI using SRCNN. For this purpose, we present an SRCNN pipeline for post-processing ^{1}H-MRSI images using the anatomical information present in T1w, T2w and FLAIR MRI.

2 Materials and Methods

2.1 MR Data Acquisition and Preprocessing

Three healthy subjects (age = 51.33 ± 5.03 years; 1F/2M), who provided written informed consent before the data acquisition, were included in this study. The imaging experiments were performed on a 3 T clinical MR scanner (Philips Medical Systems, Best, Holland) with a 32-channel head coil. For each subject, MRI data acquisition frames were aligned parallel to the anterior commissure (AC) - posterior commissure (PC) line. For anatomical scans, a T1w MRI (TR/TE = 8.3/3.8 ms, FOV = 250 × 250 × 180 mm, voxel size = 1 × 1 × 1 mm), T2w MRI (TR/TE = 10243/80 ms, 90° flip angle, FOV = 240 × 240 × 180 mm, voxel size = 2 × 2 × 2 mm), and FLAIR MRI (TR/TE = 4800/1650 ms, FOV = 250 × 250 × 180 mm, voxel size = 1 × 1 × 3 mm) were obtained. Afterwards, three dimensional ^{1}H-MRSI data was acquired by using Point-RESolved Spectroscopy (PRESS) sequence (TR/TE = 1000/52 ms, FOV = 140 × 140 × 36 mm, voxel size = 10 × 10 × 12 mm, 14 × 14 × 3 voxels, scan time = 8 min). T2w MRI was used as the reference image for defining ^{1}H-MRSI region of interest (ROI), which covered a 110 × 110 × 36 mm region.

Raw ^{1}H-MRSI data were exported out and the spectra were quantified by using the LCModel program [18]. Metabolite concentrations including total N-acetyl aspartate (tNAA) were quantified for each voxel. An in-house software written in MATLAB (The Mathworks Inc., Natick, MA) was used to combine the metabolite concentrations of each voxel into a single tNAA map for each slice.

T1w and FLAIR MRI were rigidly registered to reference T2w MRI using FSL-FLIRT [19] to align all anatomical scans (Fig. 1). Additionally, a fused MR image (Fused MRI) was formed by placing T1w, T2w, and FLAIR MRI into three distinct channels of an RGB image using the MCMxxxVI-RGBExplorer tool[1].

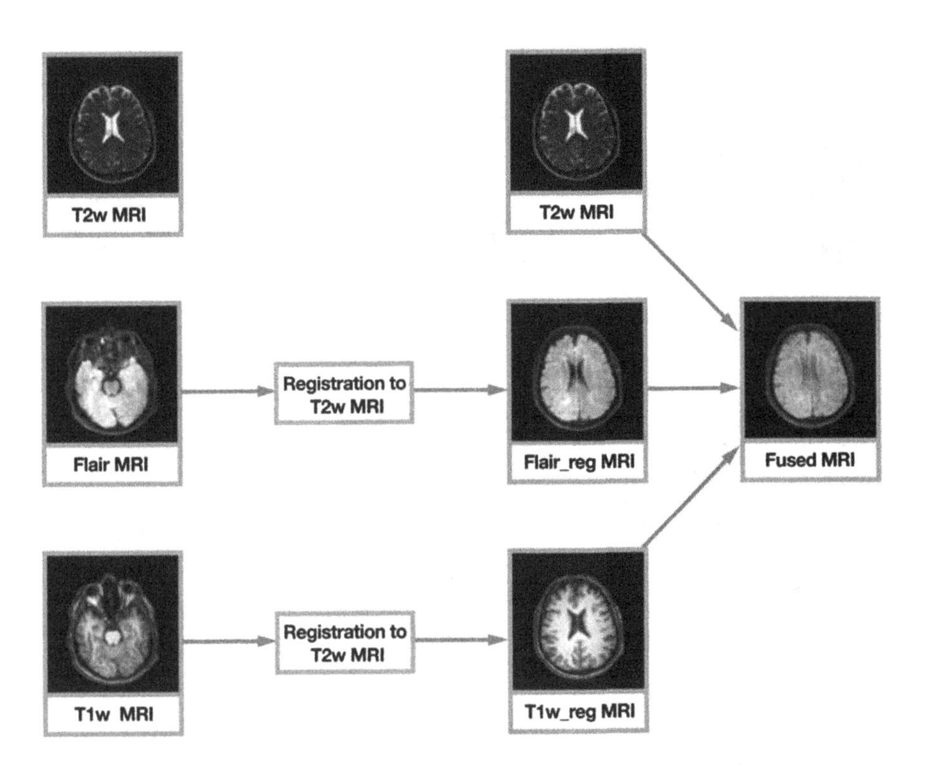

Fig. 1. A schematic of MR image registration and fusion of T1w, T2w, and FLAIR MRI

The spatial resolution of tNAA maps were upscaled by a factor of 5 along the in-plane dimensions using nearest neighbor interpolation to match the T2w MR image resolution. T1w, T2w, FLAIR, and Fused MR image regions that corresponded to the region of interest of tNAA maps were extracted (Fig. 2).

[1] https://sourceforge.net/projects/bric1936/files/MATLAB/.

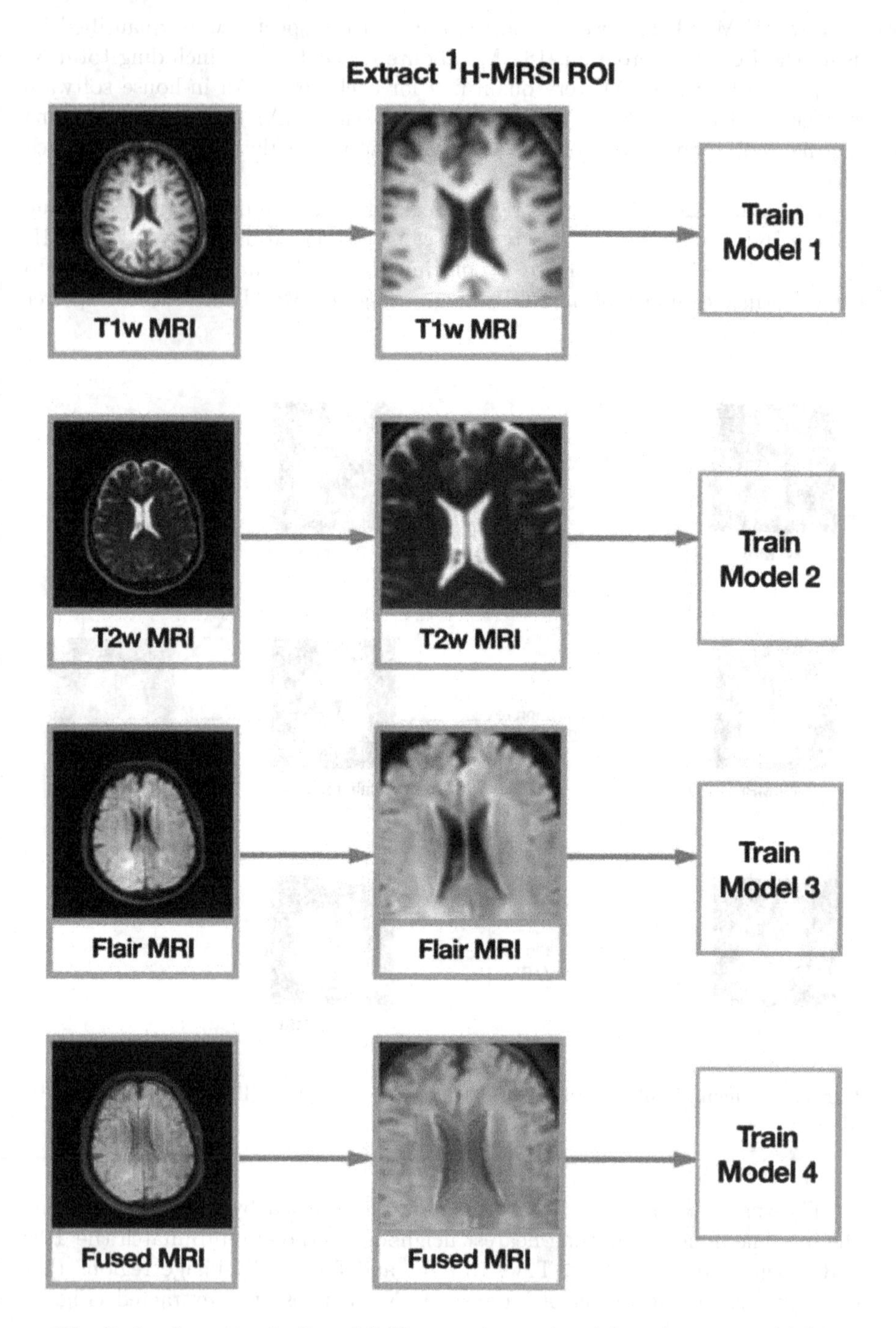

Fig. 2. A schematic pipeline of ROI extraction and training of anatomical MRI

Each spectral slice was 12 mm thick, and it contained six 2 mm thick anatomical MR images. Our ^{1}H-MRSI data was composed of three slices. As a result, we obtained 18 images for each anatomical MR imaging modality for each subject, resulting in a total of 54 slices for all of the three subjects.

2.2 SRCNN Post-processing

Caffe [20] was installed as a deep learning framework for SRCNN to train super-resolution models. SRCNN structure included three convolutional layers, which performed patch extraction and representation, non-linear mapping and reconstruction [16]. At the first layer, the image was convolved with a set of filters followed by an application of Rectified Linear Unit (ReLU, max(0, x)) on filter responses [21]. For the first layer, a kernel size of $3 \times 9 \times 9$ was used for each convolution, and 64 feature maps were produced as the output. The second layer mapped the 64-dimensional features of each patch onto a 32-dimensional feature space of the high resolution image. The final layer served the purpose of combining the overlapping high-resolution patches to produce the final high resolution image. The kernel sizes of the second and the third layers were 1×1 and 5×5, respectively. The weight filler type was set as Gaussian, base learning rate was set as 0.0001, and the learning policy was fixed. A Euclidean loss function was employed. As per the training/testing strategy from [22], the extracted regions of the structural MR images and Fused MRI were downsampled and fed into the SRCNN to train four separate models (Fig. 2). Thirty-nine MR images were used for training the SRCNN framework, and fifteen MR images were used for testing purposes for each model. Afterwards, tNAA maps were used as the testing dataset, and the four distinct models trained on different structural MR images or Fused MRI were employed in SRCNN to upscale the spatial resolution of tNAA maps by a factor of three. SRCNN was run three times with 10,000, 100,000, or 1,000,000 iterations for each model to determine the number of necessary iterations for reconstructing a high quality super resolution tNAA map. The results of the SRCNN were compared with bicubic interpolation.

2.3 Image Quality Evaluation Metrics

Peak signal to noise ratio (PSNR), and root mean square error (RMSE) were used as evaluation metrics of accuracy on all experiments in our study. The tNAA map that was upsampled by nearest neighbor interpolation was used as the reference image for comparison purposes.

3 Results

SRCNN was first applied to increase the spatial resolution of anatomical and fused MRI by using the corresponding MRI for both training and test datasets. Figure 3 displays our SRCNN results for increasing the spatial resolution of

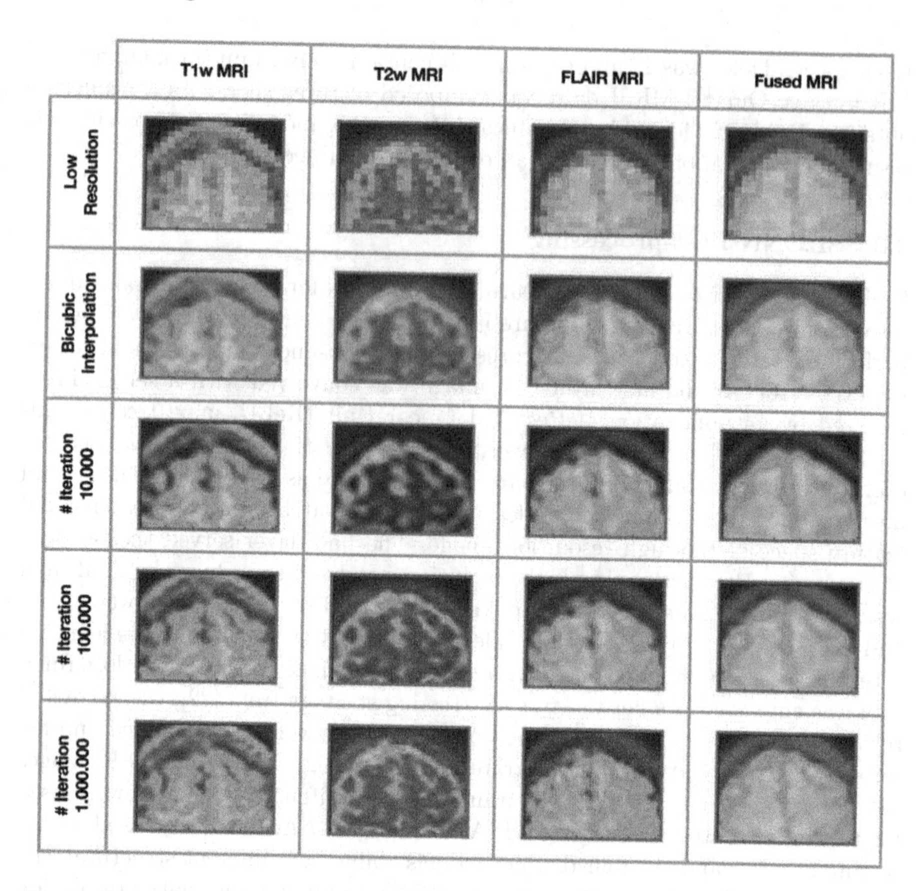

Fig. 3. Example SRCNN results for increasing the spatial resolution of anatomical MRI using 10,000, 100,000, and 1,000,000 iterations.

anatomical MRI. SRCNN resulted in less blurry and more detailed MRI than bicubic interpolation for all anatomical and Fused MRI. Gyri and sulci were better resolved in high resolution images obtained by SRCNN for all anatomical MRI. SRCNN resulted in a higher mean PSNR than bicubic interpolation for all anatomical and fused MRI after 10,000 iterations (Table 1). When T2w or Fused MRI were used as SRCNN training datasets, 10,000 iterations was not sufficient to outperform bicubic interpolation in terms of RMSE. Highest mean PSNR and lowest RMSE values were obtained when SRCNN was trained with 100,000 iterations for T1w and FLAIR MRI, and 1,000,000 iterations for T2w and Fused MRI.

Four distinct training models obtained by using SRCNN algorithm on different anatomical or Fused MRI were applied to increase the spatial resolution of tNAA maps. Table 2 displays the PSNR and RMSE values when bicubic interpolation or SRCNN with varying number of iterations were employed for super-resolution ^{1}H-MRSI. T1w MRI model did not result in a higher PSNR or

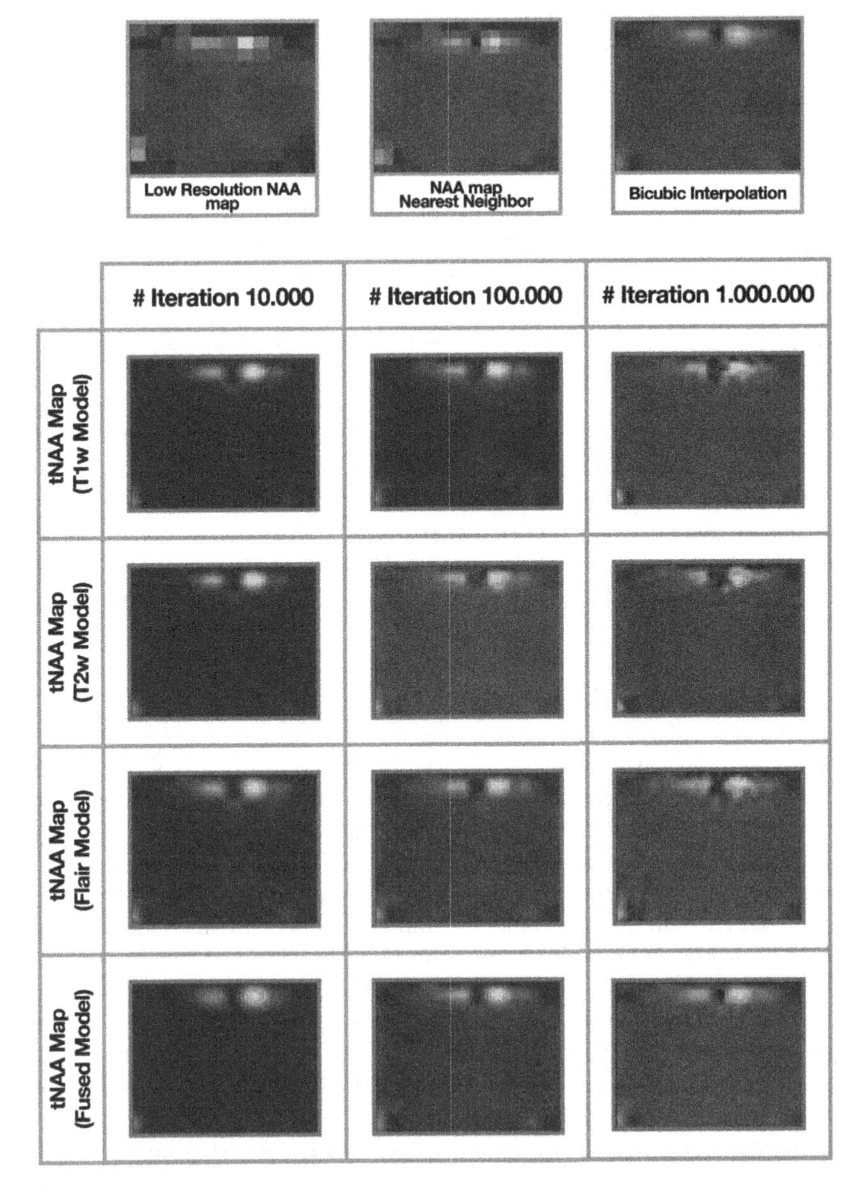

Fig. 4. SRCNN results of an example tNAA map upscaled by using T1w, T2w, FLAIR, and Fused MRI filter models with 10,000, 100,000, and 1,000,000 iterations.

lower RMSE than bicubic interpolation for any of the iteration levels. Fused MRI model with 1,000,000 iterations resulted in the highest PSNR and lowest RMSE. Figure 4 shows our SRCNN results for increasing tNAA map spatial resolution.

Table 1. The mean PSNR and RMSE obtained by using SRCNN or bicubic interpolation for increasing the spatial resolution of anatomical and Fused MRI.

Method	# Iteration	T1w MRI		T2w MRI		FLAIR MRI		Fused MRI	
		PSNR	RMSE	PSNR	RMSE	PSNR	RMSE	PSNR	RMSE
Bicubic	-	25.11	14.14	25.11	14.14	27.23	11.09	31.56	6.73
SRCNN	10,000	25.21	13.98	24.81	14.64	27.52	10.71	30.67	7.46
SRCNN	100,000	**25.86**	**12.98**	25.92	12.89	**28.13**	**9.99**	32.11	6.32
SRCNN	1,000,000	25.85	13	**26.1**	**12.63**	27.77	10.42	**32.36**	**6.2**

Table 2. The mean PSNR and RMSE results of SRCNN for super-resolution MRSI based on different anatomical MRI training models.

Method	# Iteration	T1w MRI		T2w MRI		FLAIR MRI		Fused MRI	
		PSNR	RMSE	PSNR	RMSE	PSNR	RMSE	PSNR	RMSE
Bicubic	-	27.01	11.37	27.01	11.37	27.01	11.37	27.01	11.37
SRCNN	10,000	26	12.77	23.47	17.09	27.05	11.32	24.015	16.06
SRCNN	100,000	26.69	11.79	**27.88**	**10.28**	**27.58**	**10.64**	27.77	10.41
SRCNN	1,000,000	25.94	12.86	26.08	12.65	26.29	12.35	**28.17**	**9.95**

4 Conclusion and Discussion

In this paper, we have presented a novel application of SRCNN deep learning method for increasing the spatial resolution of ^{1}H-MRSI based on the anatomical image definition of T1w, T2w, FLAIR, and Fused MRI. One of the main limitations of this study is the use of a single MR spectral frequency point that corresponds to a tNAA map as an example spectral image instead of all the MR spectral points. Our results could be similarly applied to increase the spatial resolution of other metabolite maps that could be obtained by ^{1}H-MRSI, or a better approach would be the application of the SRCNN models for increasing the spatial resolution of the whole MR spectral array. The proposed approach may contribute to the clinical utility of ^{1}H-MRSI, e.g. better radiotherapy treatment planning based on higher resolution ^{1}H-MRSI. Further work could increase the number of input channels (e.g. adding other structural MR sequences or diffusion parametric maps to the 3-channel [T1w, T2w, FLAIR] input array that was evaluated here), to investigate the optimal input configuration for increasing the spatial resolution of ^{1}H-MRSI. Future studies will be conducted to investigate the use of other deep learning methods, like fast SRCNN and patch-based super-resolution, to increase the spatial resolution of ^{1}H-MRSI.

Acknowledgements. This study was supported by the Royal Society through the Newton Mobility Grant NI150340. MCVH is funded by Row Fogo Charitable Trust, grant no. BRO-D.FID3668413.

References

1. Nelson, S.J.: Multivoxel magnetic resonance spectroscopy of brain tumors. Mol. Cancer Ther. **2**(5), 497–507 (2003)
2. Filippi, M., Agosta, F.: Imaging biomarkers in multiple sclerosis. J. Magn. Reson. Imaging **31**, 770–788 (2010). doi:10.1002/jmri.22102
3. Rovira, A., Auger, C., Alonso, J.: Magnetic resonance monitoring of lesion evolution in multiple sclerosis. Ther. Adv. Neurol. Disord. **6**(5), 298–310 (2013). doi:10.1177/1756285613484079
4. Camicioli, R.M., Korzan, J.R., Foster, S.L., Fisher, N.J., Emery, D.J., Bastos, A.C., Hanstock, C.C.: Posterior cingulate metabolic changes occur in Parkinsons disease patients without dementia. Neurosci. Lett. **354**(3), 177–180 (2004). https://doi.org/10.1016/j.neulet.2003.09.076
5. Griffith, H.R., Hollander, J.A., Okonkwo, O.C., O'Brien, T., Watts, R.L., Marson, D.C.: Brain N-acetylaspartate is reduced in Parkinson disease with dementia. Alzheimer Dis. Assoc. Disord. **22**(1), 54–60 (2008). doi:10.1097/WAD.0b013e3181611011
6. Andronesi, O.C., Kim, G.S., Gerstner, E., Batchelor, T., Tzika, A.A., Fantin, V.R., Vander Heiden, M.G., Sorensen, A.G.: Detection of 2-hydroxyglutarate in IDH-mutated glioma patients by in vivo spectral-editing and 2D correlation magnetic resonance spectroscopy. Sci. Transl. Med. **4**(116), 116ra4 (2012). doi:10.1126/scitranslmed.3002693
7. Elkhaled, A., Jalbert, L.E., Phillips, J.J., Yoshihara, H.A., Parvataneni, R., Srinivasan, R., Bourne, G., Berger, M.S., Chang, S.M., Cha, S., Nelson, S.J.: Magnetic resonance of 2-hydroxyglutarate in IDH1-mutated low-grade gliomas. Sci. Transl. Med. **4**(116), 116ra5 (2012). doi:10.1126/scitranslmed.3002796
8. Choi, C., Ganji, S.K., DeBerardinis, R.J., Hatanpaa, K.J., Rakheja, D., Kovacs, Z., Yang, X.L., Mashimo, T., Raisanen, J.M., Marin-Valencia, I., Pascual, J.M., Madden, C.J., Mickey, B.E., Malloy, C.R., Bachoo, R.M., Maher, E.A.: 2-hydroxyglutarate detection by magnetic resonance spectroscopy in IDH-mutated patients with gliomas. Nat. Med. **18**(4), 624–629 (2012). doi:10.1038/nm.2682
9. Nelson, S.J., Ozhinsky, E., Li, Y., Park, I., Crane, J.: Strategies for rapid in vivo 1H and hyperpolarized 13C MR spectroscopic imaging. J. Magn. Reson. **229**, 187–197 (2013). doi:10.1016/j.jmr.2013.02.003
10. LeCun, Y., Boser, B., Denker, J.S., Henderson, D., Howard, R.E., Hubbard, W., Jackel, L.D.: Backpropagation applied to handwritten zip code recognition. Neural Comput. **1**(4), 541–551 (1989). doi:10.1162/neco.1989.1.4.541
11. Pang, S., Yang, X.: Deep convolutional extreme learning machine and its application in handwritten digit classification. Comput. Intell. Neurosci. **2016**, 10 (2016). doi:10.1155/2016/3049632. Article ID 3049632
12. Lawrence, S., Giles, C.L., Tsoi, A.C., Back, A.D.: Face recognition: a convolutional neural network approach. IEEE Trans. Neural Netw. **8**(1), 98–113 (1997). doi:10.1109/72.554195
13. Xu, Z., Cheng, X.E.: Zebrafish tracking using convolutional neural networks. Sci. Rep. **7**, 42815 (2017). doi:10.1038/srep42815
14. Pang, S., Yu, Z., Orgun, M.A.: A novel end-to-end classifier using domain transferred deep convolutional neural networks for biomedical images. Comput. Methods Programs Biomed. **140**, 283–293 (2017). doi:10.1016/j.cmpb.2016.12.019

15. Saurabh, J., Diana, M.S., Faezeh, S.N., Gilbert, H., Wolfgang, B., Williams, S., Van Huffel, S., Maes, F., Smeets, D.: Patch-based super-resolution of MR spectroscopic images: application to multiple sclerosis. Front. Neurosci. **11**(13) (2017). doi:10.3389/fnins.2017.00013
16. Dong, C., Loy, C.C., He, K., Tang, X.: Image super-resolution using deep convolutional networks. IEEE Trans. Pattern Anal. Mach. Intell. **38**(2), 295–307 (2016). doi:10.1109/TPAMI.2015.2439281
17. Kim, J., Lee, J.K., Lee, K.M.: Accurate image super-resolution using very deep convolutional networks. In: Proceedings of the IEEE Conference on Computer Vision and Pattern Recognition (2016)
18. Provencher, S.W.: Automatic quantitation of localized in vivo 1H spectra with LCModel. NMR Biomed. **14**(4), 260–264 (2001). doi:10.1002/nbm.698
19. Smith, S.M., Jenkinson, M., Woolrich, M.W., Beckmann, C.F., Behrens, T.E., Johansen-Berg, H., Bannister, P.R., De Luca, M., Drobnjak, I., Flitney, D.E., Niazy, R.K., Saunders, J., Vickers, J., Zhang, Y., De Stefano, N., Brady, J.M., Matthews, P.M.: Advances in functional and structural MR image analysis and implementation as FSL. NeuroImage **23**(1), S208–S219 (2004). doi:10.1016/j.neuroimage.2004.07.051
20. Jia, Y., Shelhamer, E., Donahue, J., Karayev, S., Long, J., Girshick, R., Guadarrama, S., Darrell, T.: Caffe: convolutional architecture for fast feature embedding (2014). arXiv preprint: arXiv:1408.5093
21. Nair, V., Hinton, G.E.: Rectified linear units improve restricted Boltzmann machines. In: Proceedings of the 27th International Conference on Machine Learning (ICML 2010), pp. 807–814 (2010)
22. Valdes, M.DelC., Inamura, M.: Improvement of remotely sensed low spatial resolution images by back-propagated neural networks using data fusion techniques. Int. J. Remote Sens. **22**(4), 629–642 (2001)

Radial Basis Function Interpolation for Rapid Interactive Segmentation of 3-D Medical Images

Negar Mirshahzadeh[1](✉), Tanja Kurzendorfer[1], Peter Fischer[2], Thomas Pohl[2], Alexander Brost[2], Stefan Steidl[1], and Andreas Maier[1]

[1] Pattern Recognition Lab, FAU Erlangen-Nuremberg, Erlangen, Germany
n.mirshahzadeh@gmail.com,
{tanja.kurzendorfer,stefan.steidl,andreas.maier}@fau.de
[2] Siemens Healthcare GmbH, Forchheim, Germany
{peterfischer,thomas.tp.pohl,alexander.brost}@siemens-healthineers.com

Abstract. Segmentation is one of the most important parts of medical image processing. Manual segmentation is very cumbersome and time-consuming. Fully automatic segmentation approaches require a large amount of labeled training data and may fail in difficult cases. In this paper, we propose a new method for 2-D segmentation and 3-D interpolation. The Smart Brush functionality quickly segments the ROI in a few 2-D slices. Given these annotated slices, our adapted formulation of Hermite Radial Basis Functions reconstructs the 3-D surface. Effective interactions with less number of equations accelerate the performance and therefore, a real-time and an intuitive, interactive segmentation can be supported effectively. The proposed method was evaluated on 12 clinical 3-D MRI data sets from individual patients and were compared to gold standard annotations of the left ventricle from a clinical expert. The 2-D Smart Brush resulted in an average Dice coefficient of 0.88 ± 0.09 for individual slices. For the 3-D interpolation using Hermite Radial Basis Functions an average Dice coefficient of 0.94 ± 0.02 was achieved.

Keywords: Smart Brush · Segmentation · 3-D interpolation · HRBF

1 Introduction

A great deal of effort has gone into interactive segmentation. Many segmentation techniques have been developed such as Intelligent Scissors, Graph Cuts, and Random Walker [3,5,9]. There are two important applications that these techniques can speed up. First, manual segmentation is still widespread in clinical routine and which is arduous. Second, the training of machine learning methods for segmentations needs ground truth annotations that have to be generated manually. In particular, deep learning is known to require huge amounts of annotated data. Therefore, the challenge is to design a fast, generic and easy segmentation tool that allows to generate clinical segmentations as well as fast ground truth annotations. The most related 2-D segmentation technique is a

M. Valdés Hernández and V. González-Castro (Eds.): MIUA 2017, CCIS 723, pp. 651–660, 2017.
DOI: 10.1007/978-3-319-60964-5_57

Smart Brush tool [7,10]. However, the drawback of this method is that it does not control the boundary smoothness [2].

In surface reconstruction, there is a vast literature which is mainly grouped into direct meshing and implicit approaches. Nowadays, those methods based on implicit surface reconstruction have gained more and more attention. In this approach, first a signed scalar field $f(\cdot)$ is obtained. The scalar value of this scalar field is zero at all scattered points (here control points $\boldsymbol{p}$), $f(\boldsymbol{p}) = 0$ and negative/positive for inside/outside of the surface [8]. Then, the desired surface is reconstructed by extracting the zero-level set of the mentioned field. In previous related work [6], this filed $f(\cdot)$ is computed in a bilateral domain where the spatial and intensity range domain are joined. The interpolation is done using the Radial Basis Function (RBF) with a Hermite data type which incorporates normals and gradients of the scalar field directly, $\nabla f(\boldsymbol{p}) = \boldsymbol{n}$.

In this work, we propose a new formulation of surface reconstruction which is independent of the 3-D intensity gradient information. The interpolation is mainly based on 2-D normal vectors obtained from the segmented slices in 2-D. Hence, a Smart Bush formulation is introduced which can handle medical acquisitions with higher noise level and ambiguous boundaries using a Gaussian Mixture Model (GMM). Furthermore, 3-D normal vectors are estimated for the intersections of annotated planes from different orientations. From this it follows, that the surface is reconstructed using both 2-D and 3-D normal vectors. In contrast to previous implicit methods, this combination can be applied to images with a high noise level, as it is not dependent on any intensity information or well defined borders.

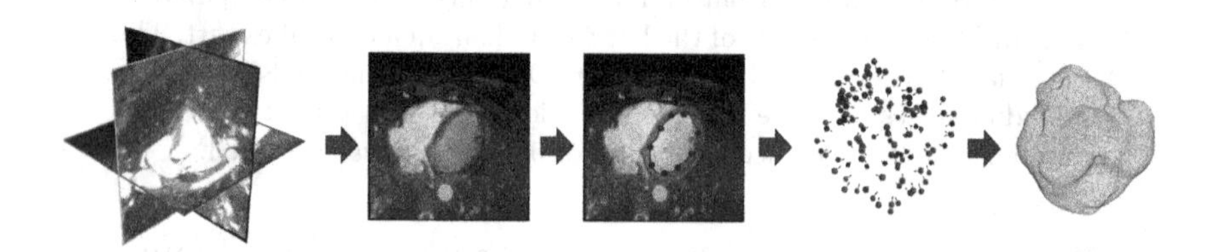

Fig. 1. Segmentation pipeline. The first image from left to right shows the 3-D volume as input. In the next step, single slices are segmented using the Smart Brush functionality. Third, the control points of the contours are extracted. Fourth, the 2-D and 3-D normal vectors are computed for the Hermite Radial Basis Function (HRBF) interpolation. In the final image, the interpolated surface is visualized.

2 Methods

Our approach combines advantages of semi-automatic segmentation methods as well as the user's high-level anatomical knowledge to generate segmentations quickly and accurately with fewer interactions. Using our method, the user first segments a few slices with the Smart Brush, then the scattered data points are extracted by computing the 2-D gradient information of the annotated slices.

Applying our new formulation of Hermite Radial Basis Function (HRBF), the desired surface is reconstructed. In Fig. 1 the segmentation pipeline is illustrated.

2.1 Smart Brush

The 2-D segmentation functionality classifies pixels into foreground and background based on intensity. Initially, a small initial area in the foreground has to be segmented manually by the user. The mean intensity of the initial area is required for the smart brush functionality.

When the user selects a new ROI with the brush, an unsupervised GMM with two components is fitted for the ROI. A threshold for pixel-wise classification is derived as the mean of the two mixture component means. The pixels of the component whose mean is closer to the mean intensity of the initial area are classified as foreground. Finally, to reduce false positives, the morphological connectivity of each pixel in the ROI to the initial ROI is checked using a 4-connected structuring element. This way, pixels that has the same intensity value but are not connected to the previous segmentation are removed.

2.2 Control Point Extraction

We assume that multiple slices are segmented in axial, sagittal, and coronal orientation using the Smart Brush functionality. First, the contours are extracted from the segmentations. Then, control points (CPs) are computed from the contours adaptively according to the shape of the object.

The contour is sampled equidistantly with a predefined sampling size $\delta \in \mathbb{Z}$. The number of control points $n_e \in \mathbb{Z}$ is based on the contour length $l_c \in \mathbb{Z}$ and computed as $n_p = \lfloor \frac{l_c}{\delta} \rfloor$. Furthermore, $n_c \in \mathbb{Z}$ convexity defect points, where the contour has the maximum distance to its convex hull, are added. To increase the accuracy of the 3-D interpolation for complex objects, the number of CPs is increased at rough areas. Therefore, the local curvature $\kappa \in \mathbb{R}$ is checked for all CPs and additional points are added in case of roughness. To compare curvature values, a reference quantity $r \in \mathbb{R}$ (global roughness) is defined which is the ratio of the convex hull area A_h and the entire curve area A_c, $r = A_c/A_h$ [1]. New CPs are added at a certain distance to the investigated CP, if the criterion

$$\frac{\kappa}{r} \geqslant \theta_r \tag{1}$$

is fulfilled, where the threshold $\theta_r \in \mathbb{R}$ is obtained heuristically. The number of additional CPs due to curvature is denoted as n_κ. The total number of CPs is $N = n_e + n_c + n_\kappa$. Figure 2 depicts two methods of control point extraction.

The subsequent interpolation requires Hermite data, i.e., function values and their derivatives. In this case, we need the normal vector for each control point. The first derivative of the contour approximates the tangent vector of the curve. Having the 2-D tangent vector $\boldsymbol{t} = (d_x, d_y)^T$, the orthogonal normal vector is obtained by $\boldsymbol{n} = (-d_y, d_x)^T$.

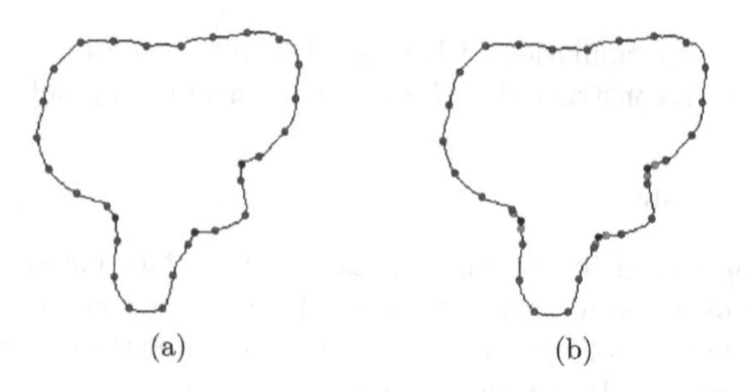

Fig. 2. (a) A rough surface with initial equidistant points in red and convexity defect points in blue. (b) A rough surface with increased number of points in green. (Color figure online)

2.3 3-D Interpolation

A new formulation of HRBF is introduced that allows to reconstruct the 3-D surface based on scattered control points and their associated 2-D normal vectors only. Assume that N Hermite data points $\{(\boldsymbol{p}_i, \boldsymbol{n}_i) | \boldsymbol{p}_i \in \mathbb{R}^3, \boldsymbol{n}_i \in \mathbb{R}^2, i = 1, ..., N\}$ are generated from Sect. 2.2. In RBF interpolation, the final segmentation is given as the zero level set of a scalar field. The scalar field f is formulated as

$$f(\boldsymbol{x}) = \sum_{i=1}^{N} \alpha_i \varphi(\|\boldsymbol{x} - \boldsymbol{p}_i\|) - \boldsymbol{\beta}_i^T \cdot s_i^{2\mathrm{D}}(\nabla \varphi(\|\boldsymbol{x} - \boldsymbol{p}_i\|)) + g(\boldsymbol{x}), \tag{2}$$

where $g(\boldsymbol{x})$ is a low-degree polynomial, $s_i^{2\mathrm{D}}(\boldsymbol{x})$ is a function that selects the 2-D gradient direction that is available for control point i, and the RBF coefficients $\alpha_i \in \mathbb{R}$, $\boldsymbol{\beta}_i \in \mathbb{R}^2$. According to previous work [6], the commonly used tri-harmonic kernel $\varphi(t) = t^3$ with a linear polynomial $g(\boldsymbol{x}) = \boldsymbol{a}^T \boldsymbol{x} + b$ yields adequate results in terms of shape aesthetics. To determine the coefficients α_i and $\boldsymbol{\beta}_i$, constraints are derived from the CPs [6]

$$f(\boldsymbol{p}_i) = 0 \tag{3}$$

$$s_i^{2\mathrm{D}}(\nabla f(\boldsymbol{p}_i)) = \boldsymbol{n}_i \ . \tag{4}$$

In addition, the orthogonality conditions $\sum_{i=1}^{N} \alpha_i = 0$ and $\sum_{i=1}^{N} \alpha_i s_i^{2\mathrm{D}}(\boldsymbol{p}_i) + \boldsymbol{\beta}_i = \boldsymbol{0}$ have to be fulfilled. This yields a linear system of equations of $3(N+1) \cdot 3(N+1)$ order which can be denoted with a block form as

$$\begin{pmatrix} \boldsymbol{0} & \boldsymbol{S}_1^T & \cdots & \boldsymbol{S}_N^T \\ \boldsymbol{S}_1 & \boldsymbol{K}_{1,1} & \cdots & \boldsymbol{K}_{1,N} \\ \vdots & \vdots & \ddots & \vdots \\ \boldsymbol{S}_N & \boldsymbol{K}_{N,1} & \cdots & \boldsymbol{K}_{N,N} \end{pmatrix} \begin{pmatrix} \boldsymbol{s} \\ \boldsymbol{w}_1 \\ \vdots \\ \boldsymbol{w}_N \end{pmatrix} = \begin{pmatrix} \boldsymbol{0} \\ \boldsymbol{c}_1 \\ \vdots \\ \boldsymbol{c}_N \end{pmatrix}, \tag{5}$$

where for CPs with Hermite data based on 2-D normals, the blocks $\boldsymbol{K}_{i,j}$, $\boldsymbol{S}_i \in \mathbb{R}^{3\times 3}$ and vectors $\boldsymbol{s}$, $\boldsymbol{w}_i$ and $\boldsymbol{c}_i$ are defined as:

$$\boldsymbol{K}_{i,j} = \begin{pmatrix} \varphi(\|\boldsymbol{p}_i - \boldsymbol{p}_j\|) & -s_i^{2D}\left(\nabla\varphi(\|\boldsymbol{p}_i - \boldsymbol{p}_j\|)\right)^T \\ s_i^{2D}\left(\nabla\varphi(\|\boldsymbol{p}_i - \boldsymbol{p}_j\|)\right) & -\nabla^T s_i^{2D}\left(\nabla\varphi(\|\boldsymbol{p}_i - \boldsymbol{p}_j\|)\right) \end{pmatrix},$$
$$\boldsymbol{S}_i = \begin{pmatrix} s_i^{2D}(\boldsymbol{p}_i)^T & 1 \\ \boldsymbol{I} & 0 \end{pmatrix}, \quad \boldsymbol{s} = \begin{pmatrix} \boldsymbol{a} \\ b \end{pmatrix}, \quad \boldsymbol{w}_i = \begin{pmatrix} \alpha_i \\ \boldsymbol{\beta}_i \end{pmatrix}, \quad \boldsymbol{c}_i = \begin{pmatrix} 0 \\ \boldsymbol{n}_i \end{pmatrix}, \tag{6}$$

where $\boldsymbol{I} \in \mathbb{R}^{2\times 2}$ is the identity matrix and $\nabla^T s_i^{2D}(\nabla \boldsymbol{x}) \in \mathbb{R}^{2\times 2}$ is the Hessian of the available 2-D dimensions. There is always a unique solution to the system of equations if the points $\boldsymbol{p}_i$ are pairwise distinct [4,6]. The unknown parameters $\alpha_i, \boldsymbol{\beta}_i, \boldsymbol{a}$ and b can be obtained directly as the matrix is square and non-singular.

However, through the segmentation of several slices from different orientations, control points can be close together at the intersection of the planes. The normal vectors of these points, point in a different direction according to the segmented contour. To make the interpolation result even more robust, these points are combined and a 3-D normal vector is estimated. The merging is performed with in a certain user defined radius around each intersection area. Hence, the HRBF interpolation is a combination of 2-D and 3-D normal vectors. Assuming N CPs with 2-D normals and M CPs with 3-D normal, the extended system of equations is

$$\begin{pmatrix} \boldsymbol{0} & \boldsymbol{S}_1^T & \cdots & \boldsymbol{S}_M^T & \boldsymbol{S}_{M+1}^T & \cdots & \boldsymbol{S}_{M+N}^T \\ \boldsymbol{S}_1 & \boldsymbol{K}_{1,1} & \cdots & \boldsymbol{K}_{1,M} & \boldsymbol{K}_{1,M+1} & \cdots & \boldsymbol{K}_{1,M+N} \\ \vdots & \vdots & \ddots & \vdots & \vdots & \ddots & \vdots \\ \boldsymbol{S}_M & \boldsymbol{K}_{M,1} & \cdots & \boldsymbol{K}_{M,M} & \boldsymbol{K}_{M,M+1} & \cdots & \boldsymbol{K}_{M,M+N} \\ \boldsymbol{S}_{M+1} & \boldsymbol{K}_{M+1,1} & \cdots & \boldsymbol{K}_{M+1,M} & \boldsymbol{K}_{M+1,M+1} & \cdots & \boldsymbol{K}_{M+1,M+N} \\ \vdots & \vdots & \ddots & \vdots & \vdots & \ddots & \vdots \\ \boldsymbol{S}_{M+N} & \boldsymbol{K}_{M+N,1} & \cdots & \boldsymbol{K}_{M+N,M} & \boldsymbol{K}_{M+N,M+1} & \cdots & \boldsymbol{K}_{M+N,M+N} \end{pmatrix} \begin{pmatrix} \boldsymbol{s} \\ \boldsymbol{w}_1 \\ \vdots \\ \boldsymbol{w}_M \\ \boldsymbol{w}_{M+1} \\ \vdots \\ \boldsymbol{w}_{M+N} \end{pmatrix} = \begin{pmatrix} \boldsymbol{0} \\ \boldsymbol{c}_1 \\ \vdots \\ \boldsymbol{c}_M \\ \boldsymbol{c}_{M+1} \\ \vdots \\ \boldsymbol{c}_{M+N} \end{pmatrix}, \tag{7}$$

where different color in the matrix implies the points and the corresponding normal vectors with different dimensionality (3-D blue, 2-D green, mixed purple).

3 Evaluation and Results

The evaluation was performed on 12 MRI data sets. Gold standard annotations of the left ventricle were provided by a clinical expert. The Dice coefficient was evaluated as a quantitative score for the segmentation overlap. The 2-D ground truth annotation was used to assess the 2-D segmentation and the complete 3-D ground truth for the 3-D interpolation scheme. The main problem with evaluating the Smart Brush is that it inherently involves human interaction. Therefore, objective testing without human interaction is difficult. To address this, we mimicked user interactions such as slice selection, mouse movement, brush size, etc. Iteratively, a 2-D slice was selected and one patch of the ground

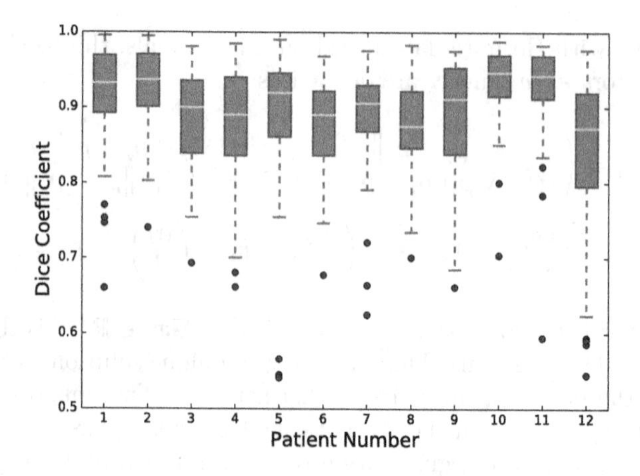

Fig. 3. The evaluation results of the 2-D segmentation result using the Smart Brush.

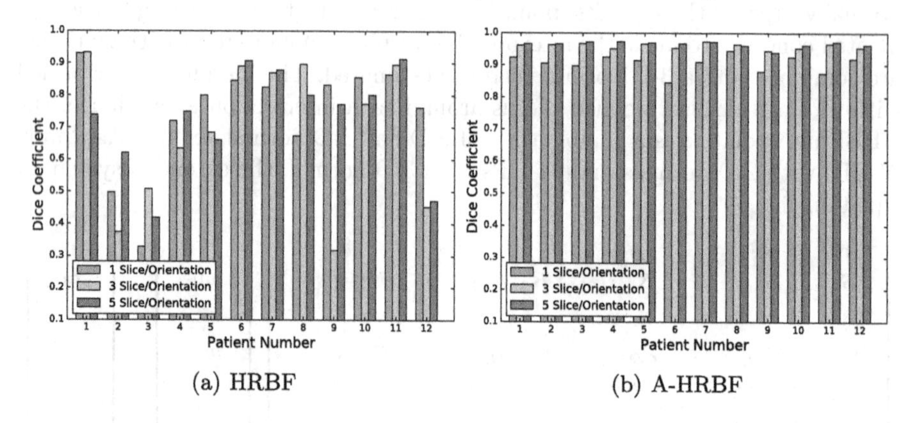

Fig. 4. The 3-D interpolation evaluation results: (a) The HRBF result with average Dice coefficient of 0.69, 0.63 and 0.69 for 1, 3, and 5 slice per orientation, respectively. (b) The A-HRBF result with average Dice coefficient of 0.91, 0.95 and 0.96 for 1, 3, and 5 slice per orientation, respectively.

truth was used for training. The evaluation of the Smart Brush was performed on a different patch by computing the Dice coefficient per patch. For evaluation of the 3-D interpolation, we compare our method (A-HRBF) to a reference method that extracts 3D gradients on the control points based on intensity (HRBF) [6].

The results of the 2-D evaluation of our Smart Brush are depicted in Fig. 3. For most patients, an average Dice coefficient of around 0.9 is achieved. The outliers will be studied in next section. The results of the 3-D segmentation are depicted in Fig. 4. For each data set, the evaluation was performed with a different number of segmented slices per orientation. We evaluated 1, 3, and 5 slices per orientation which means to have a total number of 3, 9, and 15 slices, respectively. The slice selection was randomly. The same method of control point extraction was used for control point computation. It can be seen that by increasing the number of slices the Dice coefficient usually increases slightly. Comparing the different methods, the Dice coefficient for the proposed A-HRBF is consistently higher than for HRBF.

Our experiments showed that three slices per orientation is sufficient to get a good segmentation result. Furthermore, in order to achieve more accurate interpolation results, the user has to segment those slices which have the maximum mismatch with the actual ground truth. In fact, for 3-D interpolation, the user selects those slices which are a good representation of the complete volume. Hence, the actual result of the interpolation is even better than the evaluation result shows. Figure 5 depicts the qualitative results of the A-HRBF 3-D interpolation scheme for one data set.

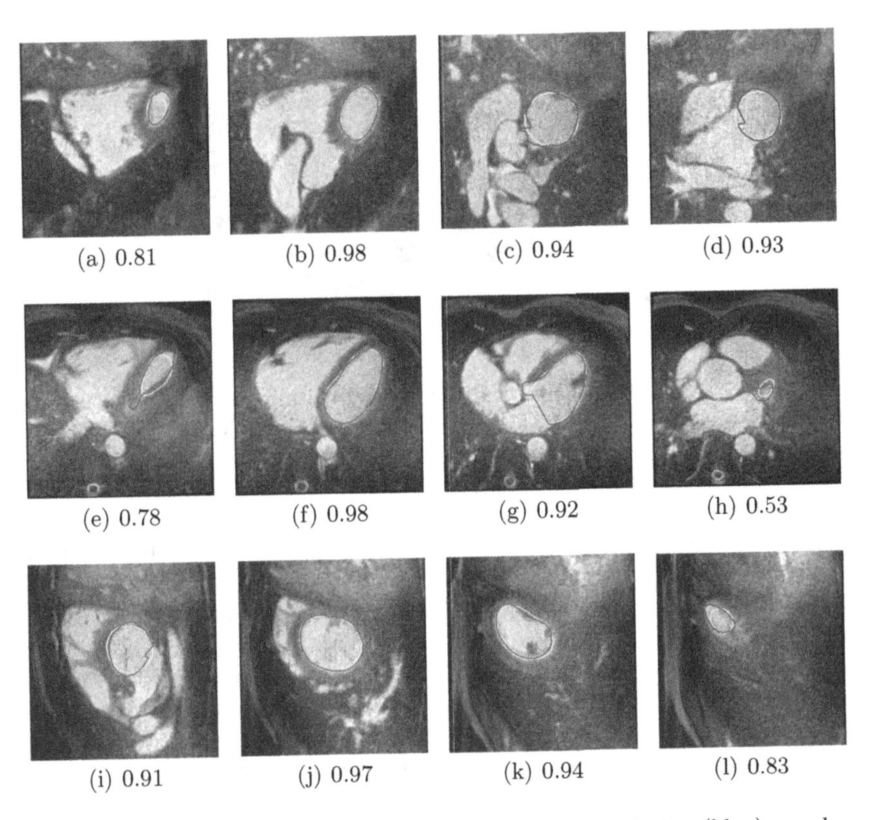

Fig. 5. The ground truth (red) and the result of 3-D interpolation (blue) are shown. The interpolation is obtained based on only one reference slice per orientation. Each row depicts a different orientation (axial, sagittal, and coronal), where the caption is the 2-D Dice coefficient of the respective slice. It is expected that the closer to the reference slice, the higher Dice coefficient is obtained. (Color figure online)

4 Discussion and Conclusion

As it is shown in Fig. 3, there are some outliers in the evaluation of the smart brush. The smart brush functionality involves human interactions inherently. Therefore, in the automatic smart brush evaluation, the main challenge was to simulate and mimic the human interactions. Note, that the process of evaluation was done fully automatic. In this case, the low Dice coefficients occur when different anatomical areas have the same intensity. For example, for heart chamber segmentation, the intensities of the left ventricle and left atrium are the same, see Fig. 6. Without any prior knowledge, the smart brush accuracy decreases as it is difficult to simulate human intelligence. However, these outliers rarely appear in a real segmentation as the user can use the normal brush and paint over the desired area manually.

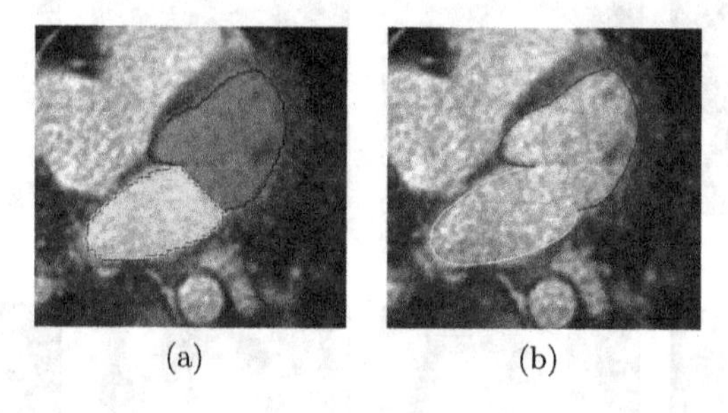

Fig. 6. (a) The left ventricle and left atrium ground truth segmentation shown in (red) and (yellow) areas, respectively. (b) The boundaries of two different labeled areas and the constant intensity while they are two different tissues. (Color figure online)

In contrast to previous implicit methods for 3-D interpolation [6], this method can not only be used for high-contrast images, but also for images with high noise level or other confounding factors due to the independence of intensity information. The gradient information is used in the smart brush functionality and therefore, there is an indirect influence of the gradient in the 3-D interpolation. If the boundary between the organ of interest and the surrounding tissue is sharp and has a high intensity gradient, other fully or semi-automatic methods will be faster. The main advantage of our method happens when there is an ambiguous boundary which only an expert can recognize (e.g. between left ventricle and left atrium). This is show-cased in our evaluation for the normal HRBF interpolation, where the results get worse if the intensities are considered for the 3-D interpolation. In this case, normal vector computation fails based on the previous method [6], while using our method, the normal vectors are orienting properly, see Fig. 7.

We showed that the 3-D interpolation is already quite good with one slice per orientation. However, this was only evaluated for the left ventricle, which is a convex object. Considering more complex objects more annotations would be

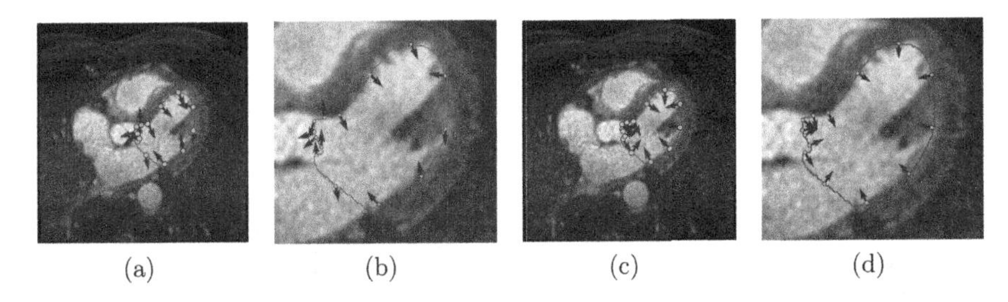

Fig. 7. Normal vector orientation for left ventricle segmentation with an ambiguous boundary: (a,b) Control points (yellow) and associated normal vector (blue) based on intensity gradients for the HRBF method. (c,d) Control points (yellow) and associated normal vector (blue) based on the drawn contour (red) for our proposed method. (Color figure online)

necessary. In general, the slices with the maximum information about the shape should be selected. In most cases, these are the center slices. Furthermore, the more slices are annotated, the better will be the interpolation accuracy. However, this will also increase the computation complexity and therefore the runtime.

The benefit of the method is that the user can correct the segmentation result easily by segmenting an additional slice with the maximum mismatch. Furthermore, no prior knowledge is involved which leads to the ability to generate any arbitrary segmentation of any 3-D data set, irrespective of image modality, displayed organ, or clinical application.

Disclaimer: The methods and information presented in this paper are based on research and are not commercially available.

References

1. Abbena, E., Salamon, S., Gray, A.: Modern Differential Geometry of Curves and Surfaces with Mathematica. CRC Press, Boca Raton (2006)
2. Boykov, Y., Kolmogorov, V.: Computing geodesics and minimal surfaces via graph cuts. In: ICCV, vol. 3, pp. 26–33 (2003)
3. Boykov, Y., Veksler, O., Zabih, R.: Fast approximate energy minimization via graph cuts. IEEE Trans. Pattern Anal. Mach. Intell. **23**(11), 1222–1239 (2001)
4. Brazil, E.V., Macedo, I., Sousa, M.C., de Figueiredo, L.H., Velho, L.: Sketching variational hermite-RBF implicits. In: Proceedings of the Seventh Sketch-Based Interfaces and Modeling Symposium, pp. 1–8. Eurographics Association (2010)
5. Grady, L.: Random walks for image segmentation. IEEE Trans. Pattern Anal. Mach. Intell. **28**(11), 1768–1783 (2006)
6. Ijiri, T., Yoshizawa, S., Sato, Y., Ito, M., Yokota, H.: Bilateral hermite radial basis functions for contour-based volume segmentation. Comput. Graph. Forum **32**, 123–132 (2013). Wiley Online Library
7. Malmberg, F., Strand, R., Kullberg, J., Nordenskjöld, R., Bengtsson, E.: Smart paint a new interactive segmentation method applied to MR prostate segmentation. In: MICCAI Grand Challenge: Prostate MR Image Segmentation 2012 (2012)

8. Morse, B.S., Yoo, T.S., Rheingans, P., Chen, D.T., Subramanian, K.R.: Interpolating implicit surfaces from scattered surface data using compactly supported radial basis functions. In: ACM SIGGRAPH 2005 Courses, p. 78. ACM (2005)
9. Mortensen, E.N., Barrett, W.A.: Interactive segmentation with intelligent scissors. Graph. Models Image Process. **60**(5), 349–384 (1998)
10. Parascandolo, P., Cesario, L., Vosilla, L., Pitikakis, M., Viano, G.: Smart brush: a real time segmentation tool for 3D medical images. In: 2013 8th International Symposium on Image and Signal Processing and Analysis (ISPA), pp. 689–694. IEEE (2013)

Modeling and Ssegmentation of Preclinical, Body and Histological Imaging

Deep Quantitative Liver Segmentation and Vessel Exclusion to Assist in Liver Assessment

Benjamin Irving[1(✉)], Chloe Hutton[1], Andrea Dennis[1], Sid Vikal[1], Marija Mavar[1], Matt Kelly[1], and Sir J. Michael Brady[1,2]

[1] Perspectum Diagnostics, Oxford, UK
ben.irving@perspectum-diagnostics.com
[2] Department of Oncology, University of Oxford, Oxford, UK

Abstract. Liver disease, especially Non-Alcoholic Fatty Liver Disease has reached high levels, and there is a need for non-invasive tests based on quantitative MRI to replace biopsy in order to better assess liver health. An automated quantitative liver segmentation approach is required to automate these tests and in this work we propose a fully convolutional framework with a novel objective function for quantitative liver segmentation. The method has (to date) been tested on quantitative T1 maps generated from the UK Biobank study. We obtained extremely encouraging results on an unseen test set with a Dice score of 0.95, and Sensitivity 0.98 and Specificity 0.99.

Keywords: Segmentation · MRI · Liver · Convolutional neural networks · Deep learning

1 Introduction

Liver disease has already reached high levels worldwide [12]. In some developed countries, up to one third of all adults have some form of liver disease, increasingly Non-Alcoholic Fatty Liver Disease (NAFLD). Up to 12% of people with NAFLD go on to develop the more severe Non-Alcoholic Steatohepatitis (NASH) [2]. The current reference standard for the diagnosis and grading of liver disease is biopsy, but this is limited by its invasiveness, frequent complications, and sampling: a liver biopsy typically represents just 1/50,000th of the whole liver volume and is unable to characterise heterogeneous liver tissue. Furthermore, there can be considerable variability in histological interpretation of liver biopsy samples [3]. These factors highlight the need for a non-invasive method to assess quantitatively a greater volume of the liver.

Quantitative MRI has been shown to be one of the key modalities for non-invasive assessment of NAFLD [10]. It enables quantitative and repeatable assessment of the whole liver region. In practice, quantitative measurements are typically performed manually by a trained human operator who places regions of

M. Valdés Hernández and V. González-Castro (Eds.): MIUA 2017, CCIS 723, pp. 663–673, 2017.
DOI: 10.1007/978-3-319-60964-5_58

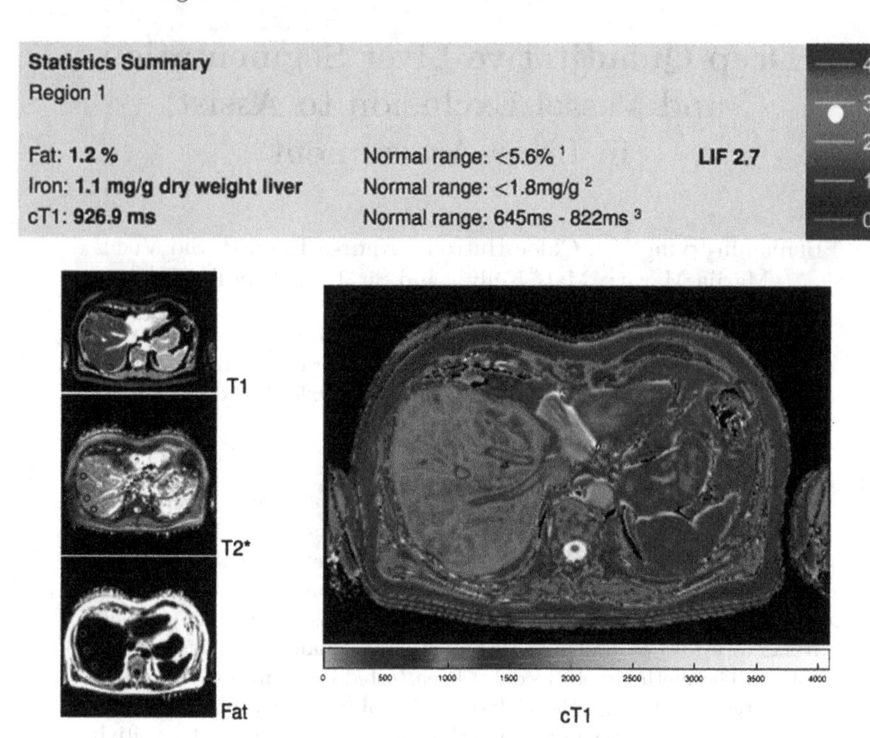

Fig. 1. Perspectum Diagnostics's LIF liver score calculation and corrected T1 based on multi-parametric quantitative analysis

interest within the liver image, avoiding vessels and image artefacts. Such manual analysis inevitably leads to inter-rater variability and possibly bias, not least when the liver tissue is particularly heterogeneous. This highlights the need to automate this part of the analysis, and a critical step is automated liver segmentation.

Perspectum Diagnostics (www.perspectum-diagnostics.com) provides a cloud-based service that enables quantification of liver health based on cT1 (quantitative T1 corrected for iron), T2* (to measure iron burden), and Proton Density Fat Fraction (PDFF), whose fusion as LiverMultiscan is effective for detecting NAFLD [1]. The cT1 is measured in milliseconds, typically in the range 500–1500 ms, but is often mapped onto a scale of 0–4, called the Liver Inflammation and Fibrosis (LIF) Score [10] (see Fig. 1). This enables a hepatologist to relate the LIF score to histology grades such as the Ishak score. The LIF score is based on the distribution of cT1s in patients whose disease status is confirmed by biopsy. Perspectum is analysing many thousands of cases per annum, including those from the ongoing UK Biobank study (http://www.ukbiobank.ac.uk) which currently has over 10,000 cases [13]. Studies of this scale further emphasise the need for automated and objective, quantitative analyses of data in order to provide population based biomarkers for use in prospective studies of liver disease. Semi-automatic liver segmentation methods have already been built into Perspectum's analysis workflow; however, this paper reports a fully automatic segmentation method to reduce user interaction and inter-rater variability.

Liver segmentation from MRI is challenging, not just because of the intrinsic variability of (diseased) liver tissue, but because of the variability in acquisitions and protocols, as well as motion artefacts arising from potentially longer acquisition times compared to CT. However, quantitative imaging sequences enable calculation of the underlying tissue parameters such as T1, T2* and PDFF, which are robust to variation in acquisition (T1 and T2* are still related to field strength). These techniques require the acquisition of a series of images to construct the quantitative maps and so currently just one (or a limited number of) 2D slices are acquired to minimise acquisition time and, therefore, liver motion effects. This makes the segmentation more challenging because we cannot use volumetric shape information as a prior and regions of the liver may appear disconnected in the axial slice. In addition, for quantitative assessment of liver parenchyma, it is also important to exclude ducts and larger blood vessels from the segmentation – making this a unique challenge. In NAFLD, there is also considerable variation in liver health due to fibrosis, steatosis, and any segmentation method must be robust to such variations.

Deep convolutional networks have shown considerable potential in image analysis for detection and segmentation [8,11]. In this paper, we demonstrate that a deep network is effective for the delineation of the liver and exclusion of vessels in quantitative images, even where there is considerable variation of liver health. This work makes a number of novel contributions: we are not aware of any studies that have used quantitative MRI scans to perform automated liver segmentation even though this provides a reliable quantification of the underlying tissue, and deep learning in quantitative MRI also appears to be novel.

Our method builds on fully convolutional neural network research and applies it to quantitative liver segmentation. We make a number of improvements including modification of the loss function, to make it more appropriate to segmentation in biological images. In Sect. 2 we discuss previous methods for liver segmentation. We introduce our approach in Sect. 3. Our method has been tested on a cohort from the Biobank trial (Sect. 4) and results from an independent test set are shown in Sect. 5.

2 Background

Few methods have been developed to segment the liver region from MRI because of the challenges in scanner and sequence variability; rather, most reported methods have been developed for CT [6]. Cheng et al. propose a MRI liver segmentation method that uses a 2D liver shape model as a prior [4]. However, it assumes that the liver appears as a single connected region in the acquisition slice, which is not always the case for certain slices through the liver. We also wish to exclude liver vessels as part of the segmentation and so a shape model is not appropriate.

Masoumi et al. [9] combine the watershed transform with a neural network to optimise the segmentation; the network is used to iteratively optimise the parameters of the watershed transform. This approach is limited by the definition of the watershed transform and would not be effective for cases with a large variation in liver health and pathology.

Instead, we aim to take advantage of modern approaches to segmentation – in particular, fully convolutional neural networks such as U-Nets [8,11]. Such methods use stacked convolutional, ReLu, and pooling layers to automatically learn features (low level features such as edges in the first layers to high level features such as textures and objects in the later layers). Fully convolutional networks do not have any fully connected layers and have been shown to be effective for pixelwise labelling of an image [8]. In the U-Net formulation, the first half of the network combines convolutional and max pooling layers to learn higher order representations within the image, while the second part is based on upsampling and convolution, and translates the representation back into a pixelwise labelling. Merge layers combine low level features into the final segmentation [11].

3 Method

A fully convolutional deep neural network was developed for this analysis based on U-Nets [11]. The input is a 2D T1 quantitative map whose dimensions are 288×384. The network architecture is shown in Fig. 2 and uses 15 stacked convolutional layers with 3×3 kernels. Between every second layer of the first 6 there are 2×2 pooling layers to produce a high level representation and 2×2 upsampling layers in the second half to convert that representation into a pixelwise segmentation. Convolutional weights were initialised with a Glorat uniform initialiser and biases were set to 0. The upsampling followed by a merge layer is used to combine the low level features with higher level features into the final segmentation (upsampling is used to create images of the same dimension for merging). The network is somewhat shallower than U-Nets because of the currently limited training data. We also modify the network to avoid representionational bottlenecks by extending the number of filters in the layer before applying maxpooling rather than the layer after maxpooling as shown in Fig. 2. This minimises the loss of the representation during the pooling stage. The number of filters per layer are shown in the figure. The final layer translates the features into a fuzzy image segmentation. A threshold of 0.5 is used to convert the segmentation into a binary labelling.

Data Augmentation. To date, we have worked with a relatively small dataset and there is an inevitable risk of overfitting to the training set. For this reason, data augmentation is required. We applied random affine transforms to each batch during training in a range of 4° rotation, 10% translation, and 10% scaling. This transformation is applied to each case at training time, so at every epoch the same case will have a different transformation.

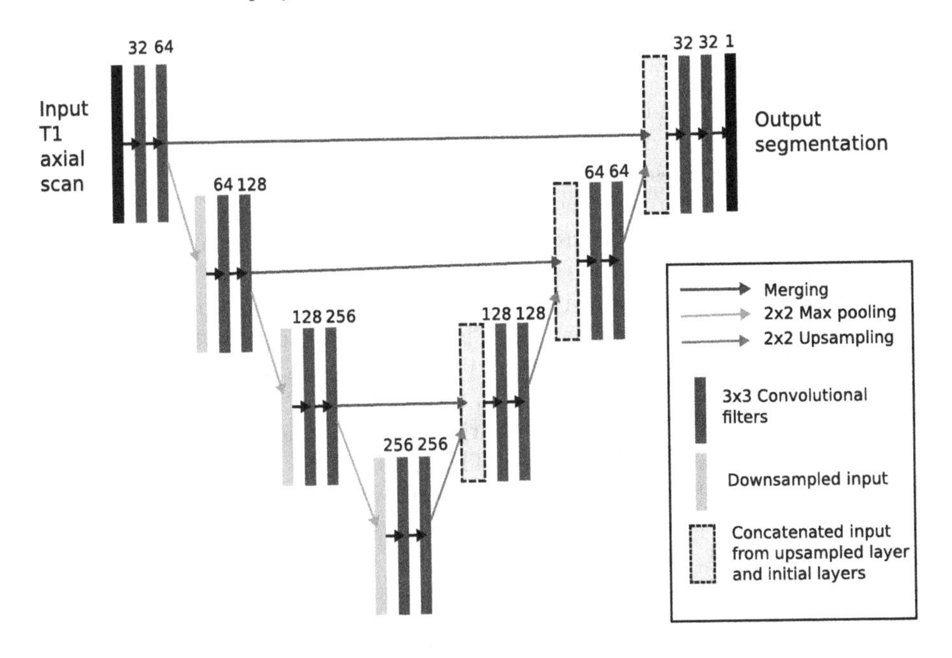

Fig. 2. Fully convolutional segmentation framework for T1 liver segmentation

***SensSpec* Objective Function.** A novel objective function was used to train the method. This score aims to jointly optimise sensitivity and specificity of the detection compared to the ground truth, which we found to produce a stable optimisation.

We chose this objective function because in medical images the labelled region often occupies only a small proportion of the entire image – unlike scene labelling problems used in computer vision. We found that this makes commonly used objective functions such as cross-entropy and mean squared error less effective because there was a bias towards the background label on problems such as this where the training set is unbalanced. The SensSpec objective that we propose is as follows:

$$L(\theta, \hat{\theta}) = -\ln(Se\ Sp) \tag{1}$$

where

$$Se = \sum \theta_i \hat{\theta}_i / \sum \theta_i,\ \forall \theta_i = 1, \quad \text{and} \quad Sp = \sum (1-\theta_i)(1-\hat{\theta}_i) / \sum (1-\theta_i),\ \forall \theta_i = 0 \tag{2}$$

where Se and Sp are Sensitivity and Specificity respectively. θ is a vector of ground truth image pixel labels and $\hat{\theta}$ are the fuzzy predicted labels. The $\forall$ operator means that only ground truth foreground values are used to calculate Sensitivity and only ground truth background values are used to calculate specificity.

4 Data

UK Biobank is a groundbreaking trial for assessing risk factors in an apparently healthy population aged between 40–69 years. The trial is ongoing and aims to collect biomarkers, lifestyle factors, and medical images from the UK population. As part of this study, Perspectum Diagnostics performs Liver Inflammation and Fibrosis (LIF) analysis using quantitative MRI acquired in the study participants [13].

MR imaging was performed on a Siemens 1.5T MAGNETOM Aera at the dedicated Biobank Imaging Centre at Cheadle (UK). The shMOLLI acquisition protocol, which samples the T1 recovery curve using a single-shot steady state free precession acquisition, was used to acquire single slice T1 relaxation time maps in a transverse plane through the right lobe of the liver and spleen. Acquisition parameters were TR = 4.94 ms, TE = 1.93 ms, flip angle = 35° and voxel size = $1.15 \times 1.15 \times 8$ mm. Quantitative T1 parameter maps were then calculated by fitting T1 recovery curves to the acquired data.

For this initial study, we used 170 cases that had been segmented by an expert. Since the UK Biobank participant population mainly represents a disease-free population, this study included 100 participants with LIF ≥ 2 and 70 with LIF < 2. This is because LIF ≥ 2 has been shown to correlate closely with subsequent adverse liver events [10]. The dataset ensures a "sufficient" number of unhealthy cases in the cohort. The cohort was then randomly split into 80% training and 20% test (unseen by the algorithm). A second cohort of 100 Biobank cases (also unseen by the algorithm) was acquired after the initial submission as a general test with no specific requirement for pathology.

As noted above, "ground truth" liver segmentations had been delineated by an expert using an in-house semi-automatic liver segmentation tool for all cases. The tool used level sets based on user-defined landmark points to segment the

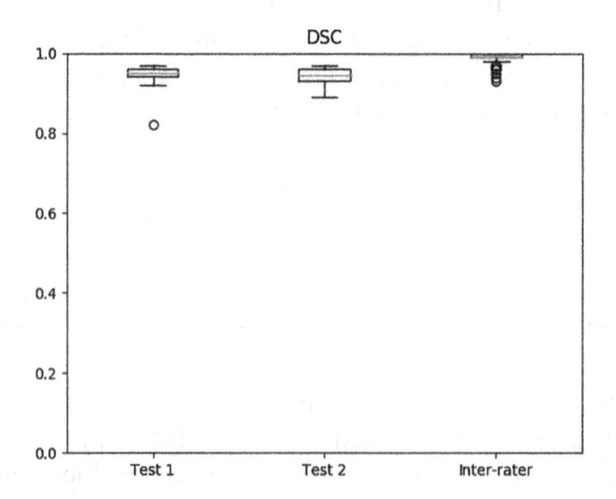

Fig. 3. A boxplot showing the DSC of the original test set (Test 1 with n = 34), an extended test set (Test 2 with n = 100) and the DSC inter-rater agreement using a semi-automatic method.

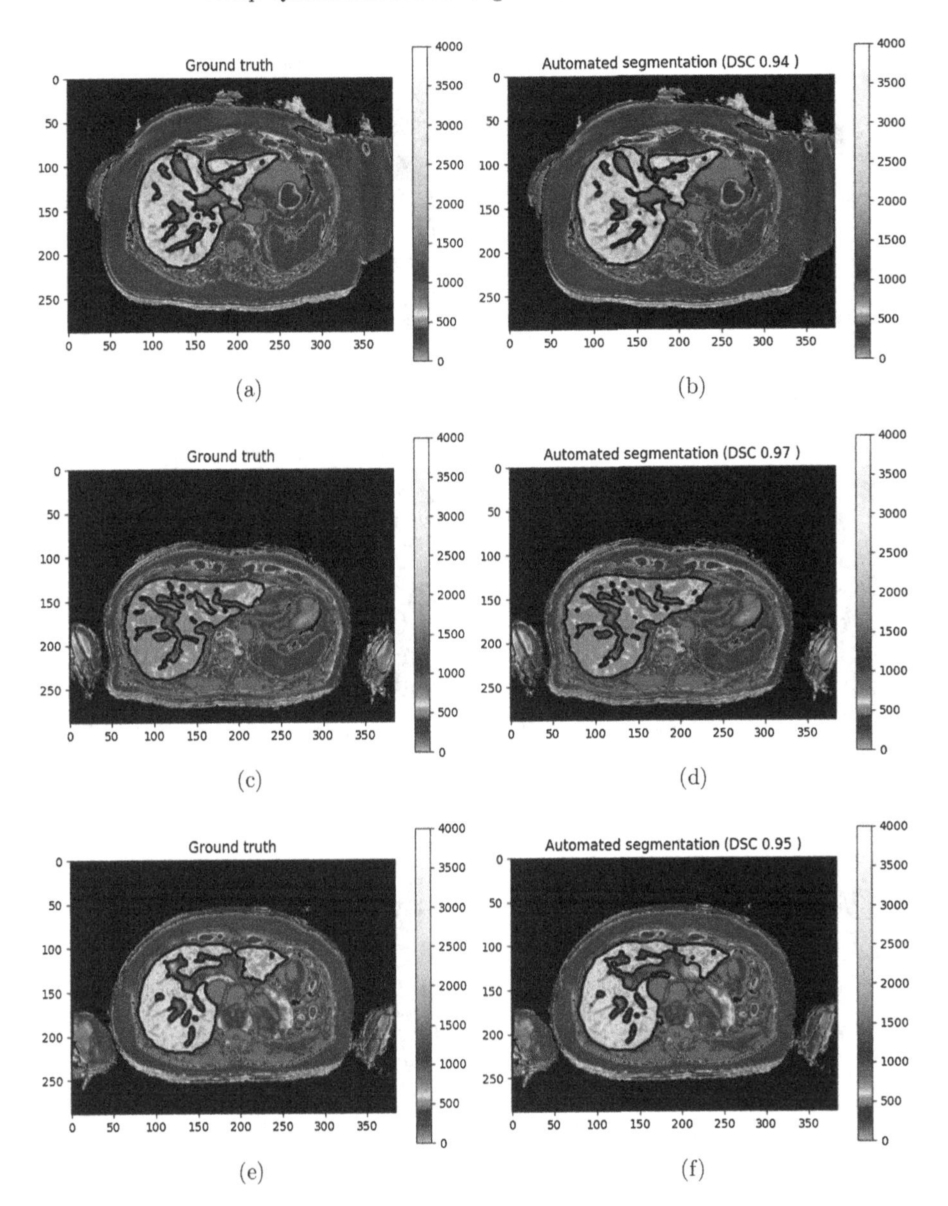

Fig. 4. Semi-automatic ground truth segmentation vs the proposed automated segmentation scheme on quantitative T1 maps. The colormap visualises normal and abnormal T1 ranges (in ms, green = normal, red = high) (Color figure online)

liver and remove vessels. The user can then manually refine the segmentation if required. Finally, a third cohort of 166 cases was semi-automatically segmented by two readers, and was used to assess the DSC inter-rater agreement. This third cohort partly overlaps the first two. It is not exactly the same because we selected data that had already been annotated by two readers for other purposes.

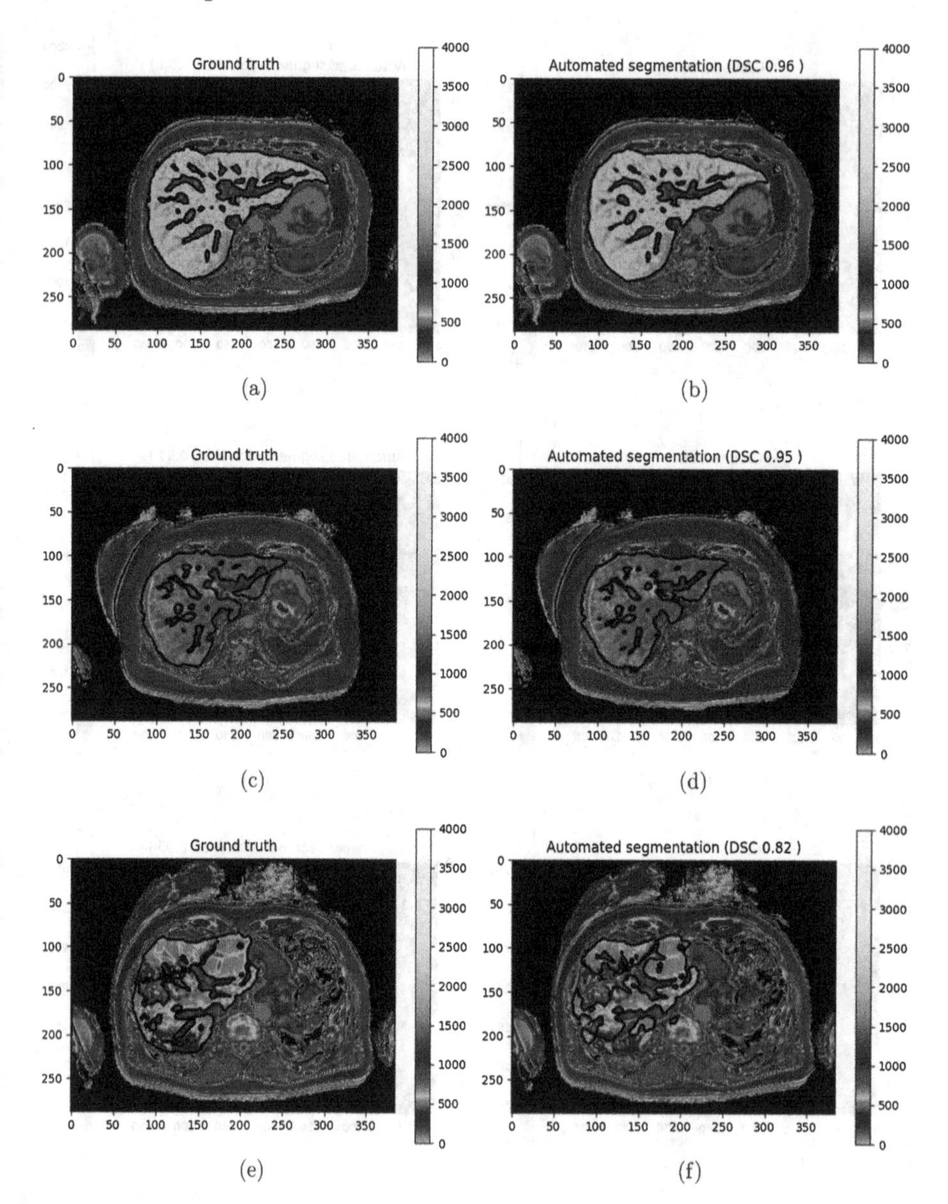

Fig. 5. Semi-automatic ground truth segmentation vs the proposed automated segmentation scheme on quantitative T1 maps (in ms)

5 Experimentation and Results

The fully convolutional network was implemented in Python using Keras [5] with Tensorflow as the backend. The method was trained on the 136 training set for 1500 epochs in batches of 20 cases (with augmentation). The Adam optimiser

[7] with a learning rate of 5E−5 was used with our proposed SensSpec objective function. Training time took 290 min using an Ubuntu system with an Nvidia Titan X GPU. Once trained, the mean time was 2.81 s per case on a Macbook with a dual core i5 CPU (The TitanX machine is only used during training).

Dice Similarity Coefficient (DSC) was used to evaluate the similarity between the semi-automatic ground truth and the automatic segmentation defined as follows:

$$\frac{2|\theta \cap \hat{\theta}|}{|\theta| + |\hat{\theta}|} \tag{3}$$

where θ and $\hat{\theta}$ are the automated segmentation and ground truth. A DSC of 1 is a perfect match between the regions and a DSC of 0 means that there was no overlap. On the unseen test sets, the method achieved a DSC 0.95 and 0.94, respectively for the two test sets. The inter-rater variability using the semi-automatic method was 0.99. A boxplot of the DSC scores for the entire test set is shown in Fig. 3. Mean sensitivity and specificity for the test set were 0.98 and 0.99, respectively. Figures 4 and 5 show the ground truth and automated segmentation for example cases in the test sets.

6 Discussion and Conclusion

This method provides a highly accurate and completely automated approach to segmenting the liver (and excluding vessels and ducts) from quantitative T1 images, and is the first step in complete automation of T1 liver tissue assessment from MRI. The method achieved a DSC of 0.95 and 0.94 for the two test sets compared to the semi-automatic ground truth. The inter-reader variability was 0.99. However, this was performed using the same in-house semi-automatic software so it is likely that there is a tendency towards similarity. Next, we plan to have the automatic and semi-automatic ground truth blindly assessed to determine which appears because we have had qualitative feedback that the automated method often appears to be better than the ground truth.

Once trained, the segmentation method is fast with a mean time of 2.81 s just on the CPU, and therefore a GPU is only required for training the network. This is useful for deployment in production systems.

The poorest performing case (outlier in Test 1 of Fig. 3) had a DSC of 0.82 and is shown on the bottom row of Fig. 5. This is a particularly complex case due to the presence of large vessels and ducts. There also appears to be breathing artefacts in this case. It could be argued that the automated segmentation method is more sensitive to liver parenchyma and excludes more vessels than the ground truth.

To understand if the method is overfitting we also calculated the DSC for the training set. The mean DSC for the training data was 0.96 which suggests that data augmentation has been effective in making the training robust and avoiding overfitting. We still expect that using a larger training set will make the method more robust.

The loss function that we propose provides a stable optimisation and is robust to training against unbalanced dataset where the region of interest is much smaller than the background. The loss function could be adapted to maximise a different classification operating point of the Receiver Operating Characteristics, if desired. For comparison, we attempted to use the mean squared error but, due to the unbalanced labelling, this loss function quickly fell into a local minima where the entire image was classified as non-liver.

UK Biobank has over 10,000 images, and is scheduled to grow to 100,000, which highlights the need for an automated approach. In this initial study we used 136 cases for training, an initial test set of 34 and later an additional test set of 100 because of the time taken to create a ground truth. We have shown that with data augmentation the fully convolutional network is capable of learning from relatively small datasets. In future we intend to train the approach on a much larger cohort once we have performed manual labelling. This would also allow testing of a deeper implementation of the network.

Acknowledgements. This research has been conducted using the UK Biobank Resource under application 9914.

References

1. Banerjee, R., Pavlides, M., Tunnicliffe, E.M., Piechnik, S.K., Sarania, N., Philips, R., Collier, J.D., Booth, J.C., Schneider, J.E., Wang, L.M., Delaney, D.W., Fleming, K.A., Robson, M.D., Barnes, E., Neubauer, S.: Multiparametric magnetic resonance for the non-invasive diagnosis of liver disease. J. Hepatol. **60**(1), 69–77 (2014)
2. Blachier, M., Leleu, H., Peck-Radosavljevic, M., Valla, D.C., Roudot-Thoraval, F.: The burden of liver disease in europe: a review of available epidemiological data. J. Hepatol. **58**(3), 593–608 (2013)
3. Castera, L., Pinzani, M.: Non-invasive assessment of liver fibrosis: are we ready? Lancet **375**(9724), 1419 (2010)
4. Cheng, K., Gu, L., Wu, J., Li, W., Xu, J.: A novel level set based shape prior method for liver segmentation from MRI images. In: Dohi, T., Sakuma, I., Liao, H. (eds.) MIAR 2008. LNCS, vol. 5128, pp. 150–159. Springer, Heidelberg (2008). doi:10.1007/978-3-540-79982-5_17
5. Chollet, F.: Keras (2015). https://github.com/fchollet/keras
6. Heimann, T., van Ginneken, B., Styner, M.A., Arzhaeva, Y., Aurich, V., Bauer, C., Beck, A., Becker, C., Beichel, R., Bekes, G., Bello, F., Binnig, G., Bischof, H., Bornik, A., Cashman, P.M.M., Chi, Y., Cordova, A., Dawant, B.M., Fidrich, M., Furst, J.D., Furukawa, D., Grenacher, L., Hornegger, J., Kainmüller, D., Kitney, R.I., Kobatake, H., Lamecker, H., Lange, T., Lee, J., Lennon, B., Li, R., Li, S., Meinzer, H.P., Nemeth, G., Raicu, D.S., Rau, A.M., van Rikxoort, E.M., Rousson, M., Rusko, L., Saddi, K.A., Schmidt, G., Seghers, D., Shimizu, A., Slagmolen, P., Sorantin, E., Soza, G., Susomboon, R., Waite, J.M., Wimmer, A., Wolf, I.: Comparison and evaluation of methods for liver segmentation from CT datasets. IEEE Trans. Med. Imaging **28**(8), 1251–1265 (2009)
7. Kingma, D.P., Ba, J.: Adam: a method for stochastic optimization. CoRR abs/1412.6980 (2014). http://arxiv.org/abs/1412.6980

8. Long, J., Shelhamer, E., Darrell, T.: Fully convolutional networks for semantic segmentation. In: The IEEE Conference on Computer Vision and Pattern Recognition (CVPR), June 2015
9. Masoumi, H., Behrad, A., Pourmina, M.A., Roosta, A.: Automatic liver segmentation in mri images using an iterative watershed algorithm and artificial neural network. Biomed. Signal Process. Control **7**(5), 429–437 (2012)
10. Pavlides, M., Banerjee, R., Sellwood, J., Kelly, C.J., Robson, M.D., Booth, J.C., Collier, J., Neubauer, S., Barnes, E.: Multiparametric magnetic resonance imaging predicts clinical outcomes in patients with chronic liver disease. J. Hepatol. **64**(2), 308–315 (2016)
11. Ronneberger, O., Fischer, P., Brox, T.: U-Net: convolutional networks for biomedical image segmentation. CoRR abs/1505.04597 (2015). http://arxiv.org/abs/1505.04597
12. Wang, F.S., Fan, J.G., Zhang, Z., Gao, B., Wang, H.Y.: The global burden of liver disease: the major impact of China. Hepatology **60**(6), 2099–2108 (2014)
13. Wilman, H.R., Kelly, M., Garratt, S., Matthews, P.M., Milanesi, M., Herlihy, A., Gyngell, M., Neubauer, S., Bell, J.D., Banerjee, R., et al.: Characterisation of liver fat in the UK Biobank cohort. PLoS One **12**(2), e0172921 (2017)

Initial Results of Multilevel Principal Components Analysis of Facial Shape

D.J.J. Farnell[1(✉)], J. Galloway[1], A. Zhurov[1], S. Richmond[1], P. Perttiniemi[2], and V. Katic[3]

[1] School of Dentistry, Cardiff University, Heath Park, Cardiff CF14 4XY, UK
{FarnellD, GallowayJL, ZhurovAI, RichmondS}@cardiff.ac.uk

[2] Faculty of Medicine, University of Oulu, P.O.Box 5000, 90014 Oulu, Finland
pertti.pirttiniemi@oulu.fi

[3] Department of Orthodontics, School of Medicine, University of Rijeka, Kresimirova 40, HR-51000 Rijeka, Croatia
visnja.katic@gmail.com

Abstract. Traditionally, active shape models (ASMs) do not make a distinction between groups in the subject population and they rely on methods such as (single-level) principal components analysis (PCA). Multilevel principal components analysis (mPCA) allows one to model between-group effects and within-group effects explicitly. Three dimensional (3D) laser scans were taken from 250 subjects (38 Croatian female, 35 Croatian male, 40 English female, 40 English male, 23 Welsh female, 27 Welsh male, 23 Finnish female, and 24 Finnish male) and 21 landmark points were created subsequently for each scan. After Procrustes transformation, eigenvalues from mPCA and from single-level PCA based on these points were examined. mPCA indicated that the first two eigenvalues of largest magnitude related to within-groups components, but that the next eigenvalue of largest magnitude related to between-groups components. Eigenvalues from single-level PCA always had a larger magnitude than either within-group or between-group eigenvectors at equivalent eigenvalue number. An examination of the first mode of variation indicated possible mixing of between-group and within-group effects in single-level PCA. Component scores for mPCA indicated clustering with country and gender for the between-groups components (as expected), but not for the within-group terms (also as expected). Clustering of component scores for single-level PCA was harder to resolve. In conclusion, mPCA is viable method of forming shape models that offers distinct advantages over single-level PCA when groups occur naturally in the subject population.

Keywords: Multilevel principal components analysis · Active shape models · Facial shape

1 Introduction

Active shape models (ASMs) and active appearance models (AAMs) [1–8] are common techniques in image processing that are used to search for specific features or shapes in images. However, if clustering or multilevel data structures exist naturally in

M. Valdés Hernández and V. González-Castro (Eds.): MIUA 2017, CCIS 723, pp. 674–685, 2017.
DOI: 10.1007/978-3-319-60964-5_59

the data set, e.g., as illustrated by the flowchart in Fig. 1, the eigenvectors and eigenvalues from principal components analysis (PCA) will only be partially reflective of the true variation in the set of images/shapes. Multilevel principal components analysis (mPCA) provides a convenient method of modelling both the underlying structures within the images and also any groupings between images. mPCA carries out PCA at both within-group and between-group levels independently. Note that the within-group level might be thought of as being "nested" within the broader between-group level, e.g., as shown in Fig. 1 for human facial expression. This approach also retains the desirable feature that any segmentation can still be constrained so that a fit of the model never "strays too far" from the training set used in forming the model (described in the methods section below).

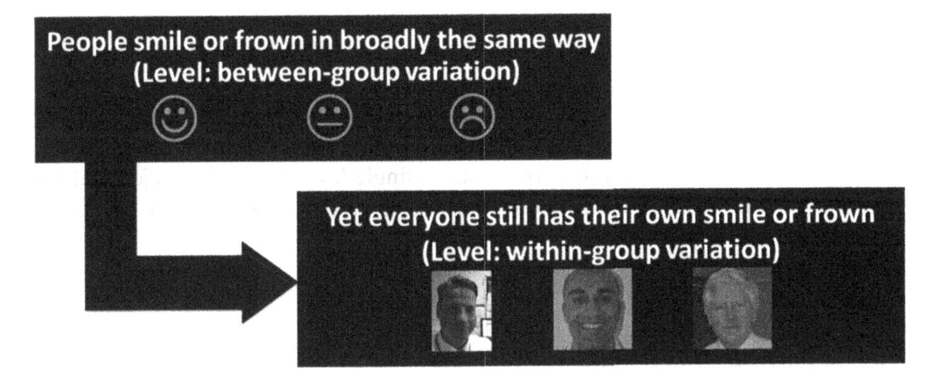

Fig. 1. Flowchart illustrating the "nested" nature of multilevel data

A previous application of mPCA to form ASMs related to the segmentation of the human spine [9]. The results of this study showed that mPCA offers more flexibility and allows deformations that classical statistical models cannot generate. Another recent application of using mPCA to form ASMs related to the field of dental imaging [10]. Proof-of-principle was tested by applying mPCA to model basic peri-oral expressions that were approximated to the junction between the mouth/lips. Monte Carlo simulation was used to create the data set, where a simple quadratic function $y = cx^2$ was used to represent the centreline of the lips and the value of controlled "expression." Different expressions (i.e., −ve c = sad; $c \approx 0$ = neutral; +ve c = happy) were modelled correctly at the between-group level of the model and changes in lip width were modelled correctly at the within-group level. Some evidence was seen that those cases that were extreme (yet still possible) in the training set in terms of both the within-group variation (width of lips) and also the between-group variation (expression) were modelled adequately by mPCA but not by standard (single-level) PCA. mPCA was also used to analyse a dataset that had landmark points placed on panoramic mandibular radiographs by two different clinicians (see also Ref. [8]), thus leading to two sets of such landmark points for the set of images. Variations in the shape of the cortical bone were modelled by one level of mPCA (within-group) and variations between the experts at another (between-group). Not surprisingly, eigenvalues indicated that variation due to changes the shape of the cortical bone were much more important than variation due to any disagreements in placement between the clinicians. Indeed, these clinicians had reported anecdotally [8] that placement of the point along the boundaries was difficult and it was observed the first mode of variation for the between-group level correctly reflected this type of variation. The authors concluded [10] that mPCA was found to provide more control and flexibility than standard "single-level" PCA when multiple levels occurred naturally in the dataset.

Here we apply mPCA to study landmark points of three-dimensional (3D) laser scans of the heads of English, Croatian, Finnish, and Welsh subjects who were of both genders. The improved understanding of the relative contributions to dental and facial development will lead to more informative descriptions of population samples. This improved understanding is useful in multiple medical fields including orthodontics and anthropology. The effect of traditional or novel treatment regimes on detailed homogenous aetiologies and morphologies will yield invaluable treatment outcome information. Details of the mathematics that underpins mPCA for ASMs and also of the 3D laser scanning procedure are presented in the methods section. Results are then presented for the eigenvectors and eigenvalues, and component scores are found by fitting the mPCA model to each set of points for each subject in the dataset. Results of mPCA are compared to those results of standard (single-level) PCA. The major modes of variation are explored. The implications of our research are presented in the discussion.

2 Methods

2.1 Mathematical Formalism

The ASM method has been extensively documented in the literature (see, e.g., Refs. [1–8]), and therefore this topic is not discussed here. One carries out PCA for the covariance matrix as discussed in Ref. [8], and the eigenvalues and eigenshapes (i.e., eigenvectors) are found readily enough for this matrix using standard software. Landmark points (i.e., mark-up points) are represented by a vector, z_i, and the kth element of this vector is given by z_{ik}. The total number of such points is n, and the mean shape vector (averaged over all N subjects) is given by $\bar{z}$. The covariance matrix is found by evaluating

$$C_{k_1,k_2} = \frac{1}{N-1}\sum_{i=1}^{N}(z_{ik_1} - \bar{z}_{ik_1})(z_{ik_2} - \bar{z}_{ik_2}) \tag{1}$$

where k_1 and k_2 indicate elements of the covariance matrix. We find the eigenvalues λ_l and eigenvectors u_l of this matrix. Note that all of the eigenvalues are non-negative, real numbers because covariance matrices are symmetric and (indeed) positive semi-definite. We rank all of the eigenvalues λ_l into descending order and we choose the m eigenvalues of largest magnitude to be retained in the model. Any new shape is given by

$$z = \bar{z} + \sum_{l=1}^{m} a_l u_l. \tag{2}$$

The eigenvectors u_l are orthonormal and so we can determine the coefficients, a_l, for a fit of the model to a new shape vector, z, readily by using

$$a_l = u_l \cdot (z - \bar{z}). \tag{3}$$

Constraints may be placed on these a-coefficients, such as $|a_l| \leq 3\sqrt{\lambda_l}$, which ensures that subsequent model fits to a new shape vector never "strays too far" from the cases in the training set.

The formalism is slightly more complicated for mPCA and details are presented in Ref. [10]. However, we remark here that we form two covariance matrices for a two-level model, namely: a *within-group* covariance matrix which is the covariance matrix evaluated over all subjects with a group and with respect to their local group means or centroids, and this matrix is then averaged all groups; and, a *between-group* covariance matrix that is covariance matrix of the centroids of the groups with respect to an "grand" mean shape $\bar{\bar{z}}$ of the average of these centroids. The rank of this matrix is limited by the number of groups.

We carry out PCA for the (positive semi-definite) *within-group* covariance matrix of the above equation and the eigenvalues are non-negative, real numbers. The lth eigenvalue is denoted λ_l^w and its eigenvector is denoted by u_l^w. Independently, we carry out PCA also for the (positive semi-definite) *between-group* covariance matrix given above and the eigenvalues are non-negative, real numbers. The lth eigenvalue is denoted λ_l^b and its eigenvector is denoted by u_l^b. We rank all of the eigenvalues λ^b and λ^w into descending order for the between- and within-group levels separately, and then we retain the m_b and m_w eigenvectors of largest magnitude, respectively. Any new shape is now given by

$$z = \bar{\bar{z}} + \sum_{l_1=1}^{m_w} a_{l_1}^w u_{l_1}^w + \sum_{l_2=1}^{m_b} a_{l_2}^b u_{l_2}^b. \tag{4}$$

Constraints may again be placed on these a-coefficients, such as $|a_l^b| \leq 3\sqrt{\lambda_l^b}$ and $|a_l^w| \leq 3\sqrt{\lambda_l^w}$, which ensures that subsequent model fits to a new shape vector again never "stray too far" form the cases in the training set with respect to both within-group variation *and* between-group variation.

The covariance matrices are symmetrical and so all "within" eigenvectors u_l^w are orthogonal to all other "within" eigenvectors (and similarly for the "between" eigenvectors). However, the eigenvectors u_l^w and u_l^b do not necessarily have to orthogonal with respect to each other, and so an equivalent projection to Eq. (3) for mPCA becomes problematic. A fit of the model given by Eq. (4) to a set of candidate pointsis achieved by minimising the overall (squared) error with respect to the coefficients a_l^w and a_l^b. A gradient descent method (see Ref. [10] for details) may be implemented straightforwardly to solve this problem iteratively. Importantly, note that the within and between components of variation are fitted to the set of candidate points *simultaneously*. All analyses were carried out using MATLAB R2014a.

2.2 3D Laser Scanning

Two Konica Minolta Vivid laser cameras were used to capture the images of the subjects and this has been reported extensively in the literature [11–13]. Twenty-one reliable facial landmarks were manually identified for each subject. Each landmark point vector z was of size 63 (= 21 $\times$ 3). These facial landmarks are shown in Fig. 2. The numbers of subjects in 8 groups were: Croatian female (n = 38), Croatian male (n = 35), English female (n = 40), English male (n = 40), Welsh female (n = 23), Welsh male (n = 27), Finnish female (n = 23), and Finnish male (n = 24). The age range of the subjects is from 15 to 22 years of age. The sample is not of sufficient size to model age. Landmark points were scaled by Procrustes transformation so that all sets of points were on broadly the same scale.

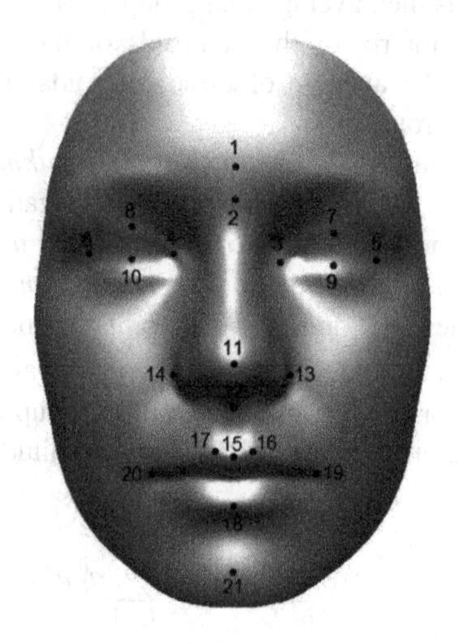

Fig. 2. Twenty-one anthropometric landmarks which were identified on facial laser scans of participants (shown in the coronal plane). (1) Glabella (g); (2) Nasion (n); (3) Endocanthion left (enl); (4) Endocanthion right (enr); (5) Exocanthion left (exl); (6) Exocanthion right (exr); (7) Palpebrale superius left (psl); (8) Palpebrale superius right (psr); (9) Palpebrale inferius left (pil); (10) Palpebrale inferius right (pir); (11) Pronasale (prn); (12) Subnasale (sn); (13) Alare left (all); (14) Alare right (alr); (15) Labiale superius (ls); (16) Crista philtri left (cphl); (17) Crista philtri right (cphr); (18) Labiale inferius (li); (19) Cheilion left (chl); (20) Cheilion right (chr); (21) Pogonion (pg). Definitions by Farkas [14] were used. Reprinted from the author's previous publication with permission from 'John Wiley and Sons'.

3 Results

Results for the 3D facial scans averaged over all subjects in each group (country and gender) are shown in Fig. 3. As one might expect [15], strong differences in facial shapes can be seen qualitatively by both country and gender. Hence, we might reasonably expect to see commensurate differences in landmark points between groups.

Eigenvalues from standard (single-level) PCA and also within-group and between-group eigenvalues from mPCA are presented in Fig. 4. Results of between-groups

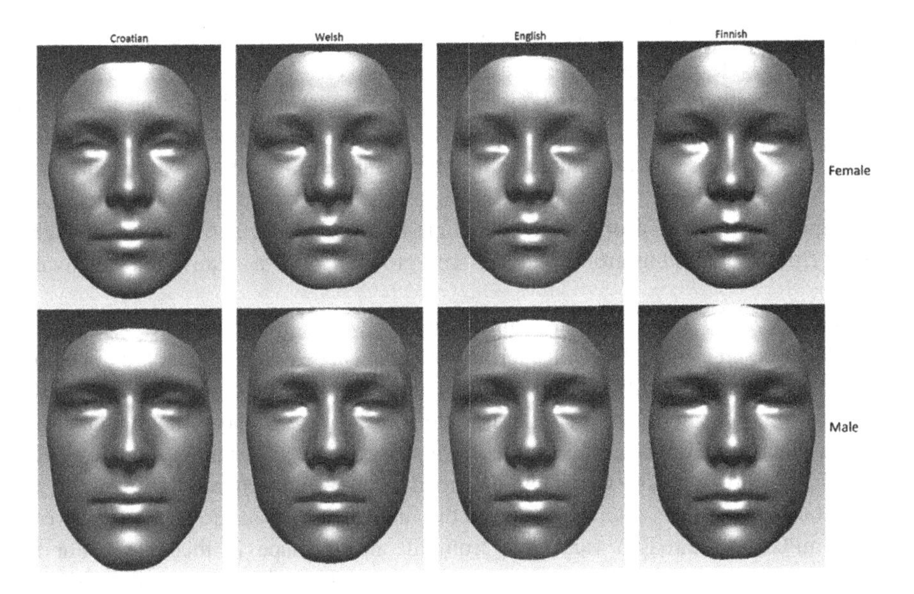

Fig. 3. 3D facial scans averaged over all subjects in each group (country and gender)

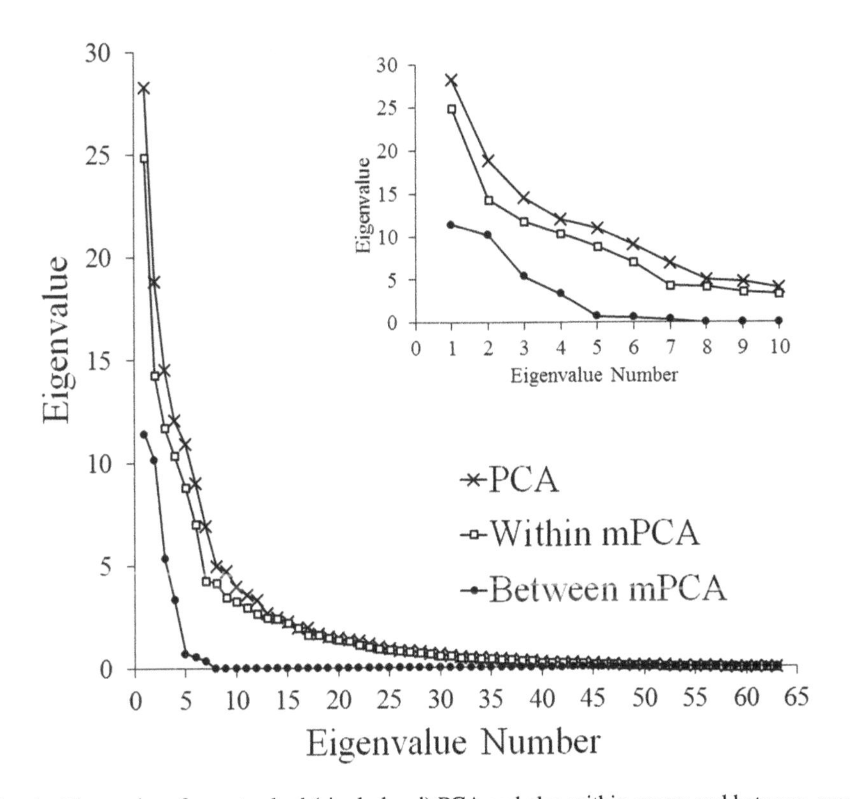

Fig. 4. Eigenvalues from standard (single-level) PCA and also within-group and between-group eigenvalues from mPCA. (Inset: the first ten eigenvalues in more detail.)

mPCA demonstrate that between-group eigenvalues are non-zero for the first seven eigenvalues. Note that within-group eigenvalues are clearly non-zero to much higher eigenvalue numbers. The first two eigenvalues of largest magnitude from mPCA are due to within-groups effects and then the first and second between-groups eigenvalues are broadly of the same magnitude as the third and fourth within-group eigenvalues. These results indicates that both within-groups and between-groups effects are important. Eigenvalues from standard (single-level) PCA lie above those of eigenvalues from both within-group and between-group mPCA at equivalent eigenvalue number.

The effects of the principal components of within-group and between-groups mPCA can be investigated by considering the mean shapes $\pm$ 2 $\times$ standard deviations (i.e., $\lambda^{0.5}$) multiplied by its corresponding eigenvector. The first mode for single-level PCA and the first mode for the within-group mPCA were found to be similar in the coronal plane, e.g., deviations from the mean shape are large at the exocanthion (right and left) positions, as shown in Fig. 5. Broadly, one might equate this mode to the aspect ratio of the face. Variations due to the first mode of variation for the between-group mPCA appeared to vary little from the mean shape in this plane. For the transverse and sagittal planes (note shown here), deviations from the mean shape were large at the exocanthion (right and left) positions; again, the first mode for single-level PCA and the first mode for the within-group mPCA are similar here. By contrast, the first mode for between-groups mPCA had strong deviations from the mean at many points for the transverse and sagittal planes, and again it is quite different to the first mode for within-groups mPCA. Particularly, strong deviations from the mean were

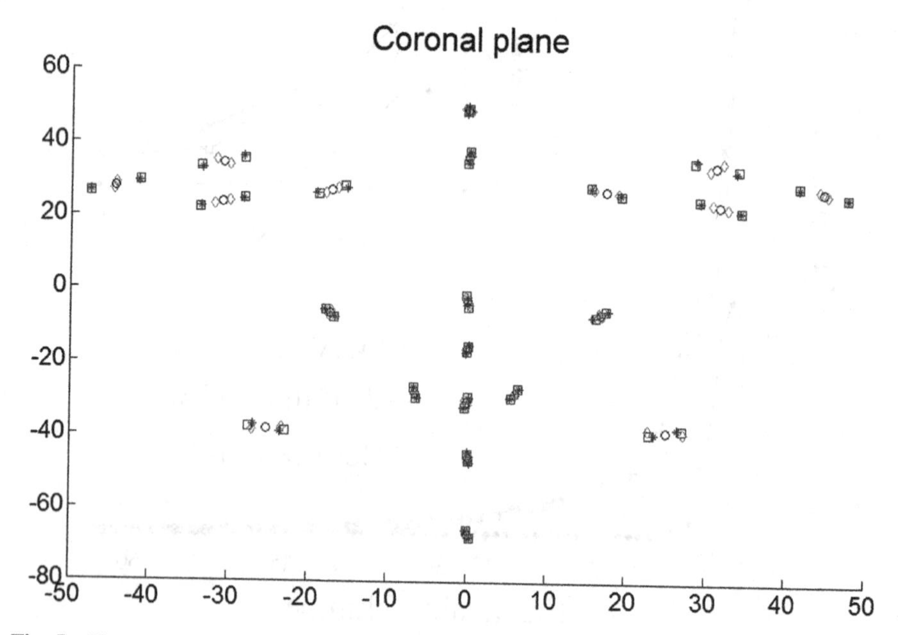

Fig. 5. First modes of variation in the coronal plane. (○ = mean; □ = mode 1 for PCA; ◇ = mode 1 for between-group mPCA; * = mode 1 for within-group mPCA.)

seen for the first mode for between-groups mPCA for the pronasale, and this was also seen at this point for the first mode for PCA. This hints that this mode might govern the length and shape of the nose and/or face in this plane. In any case, it is clear though that the first modes for within-group mPCA and between-group mPCA are quite different. We might also speculate that the first mode of single-level PCA might mix the effects of modes from within-group and between-group mPCA, although this is difficult to judge in 2D. Indeed, many such subtle effects occur even in the first major modes and visualising such subtle changes in 2D plots is difficult. However, the 3D visualisation of these modes, and their subsequent interpretation, lies beyond the scope of this initial analysis.

Component "scores" (i.e., coefficients a in Eqs. (2) and (4)) may be found by fitting the single-level PCA and mPCA models to each set of landmark points for each subject in the data set. Note that no constraints are placed on these coefficients in this case. Results for single-level PCA, shown in Fig. 6, indicate some evidence of clustering for the different groups. Centroids for males are on the right-had side of the figure for a_1

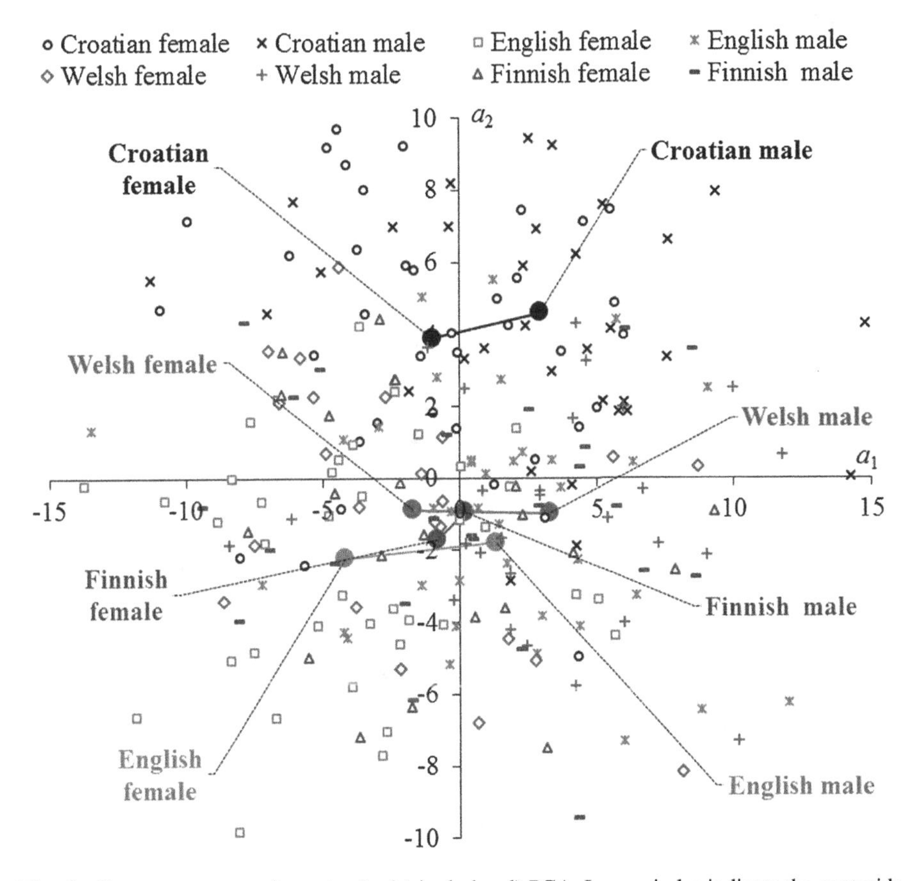

Fig. 6. Component scores from standard (single-level) PCA. Large circles indicate the centroids and results for females and males from the same country are linked by a line.

versus a_2 for PCA and those for females are on the left-hand side. Furthermore, centroids by country tend to be quite close to each other for a_1 versus a_2 and this is shown by the solid lines connecting centroids of the different genders for the same country. However, there is considerable overlap between the groups with respect to the scores for individual subjects for single-level PCA for these component scores (i.e., a_1 versus a_2).

Results for between-groups components of mPCA are shown in Fig. 7 for $m_b = 7$ and $m_w = 40$ and distinct indications of clustering for the different groups is seen. (These results were typical of mPCA results generally for $m_b \geq 3$ and $m_w \geq 3$). It is remarkable that males and females are connected by a vector that is of similar direction and magnitude for all countries. Note that this result is not imposed by assumptions of the model (as far as we are aware) and that it seems to emerge naturally from the data. This result is shown by the solid lines connecting centroids of the different genders for the same country. Furthermore, we see that centroids are being separated quite strongly by country more clearly for the between-group mPCA. The centroids of each group are certainly much easier to resolve for single-level PCA. Although there is overlap in the individual scores between the groups, this overlap appears to be less than for single-level PCA.

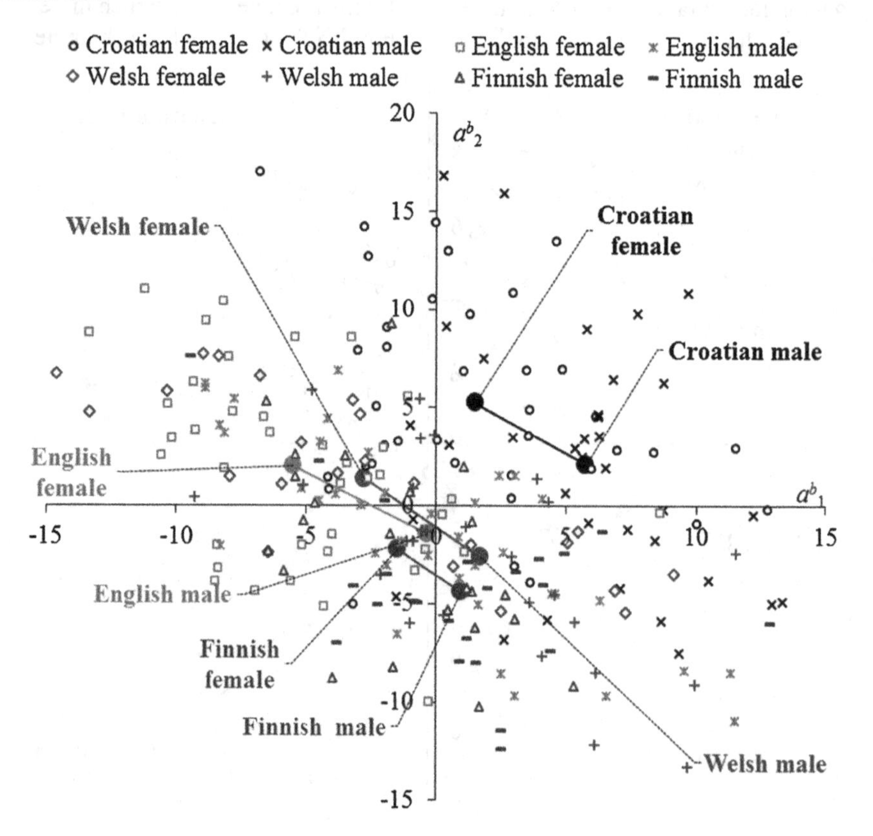

Fig. 7. Component scores from between-group mPCA. Large circles indicate the centroids and results for females and males from the same country are linked by a line.

Results for within-groups mPCA (not shown here) indicate no evidence of clustering for the different groups and all of the centroids were found to lie very close to each other (i.e., at the origin); this is exactly what one would expect as differences between groups ought to be accounted for by between-group components alone. The scatter of points for individual scores about their centroids was found be broadly uniform and of similar magnitude for all of the groups. Indeed, one would also expect the centroids to lie near to the origins because subjective variations ought to be equally spaced both "above" and "below" the group averages.

Results for component scores for the first versus the third components or the second versus third components also showed strong clustering for the between mPCA scores. The centroids for the within mPCA component scores for the first versus the third components or the second versus third components were near to the origin. Some evidence of clustering was seen for the first versus the third components and the second versus third components for single-level PCA.

Results for the point-to-point errors are shown in Table 1. Note that this was not an image search as such, rather a model fit using Eq. (3) for PCA and the iterative procedure for mPCA. Indeed, "test" cases were taken to be the same as those used the training set. As expected, point-to-point errors are found (trivially) to reduce for both PCA and mPCA as we increase the number of eigenvectors retained. However, this is still a reasonable check of mPCA because it demonstrates that the correct iterative solution is probably being identified. Point-to-point errors for mPCA are of the same magnitude (or slightly lower) than single-level PCA at broadly "equivalent" numbers of eigenvectors retained, which is also reasonable. If we include all possible eigenvectors then point-to-point errors appear to go to zero for both PCA and mPCA.

Table 1. Point-to-point errors with respect toall 21 points and all subjects

		Mean	Standard deviation	Maximum
PCA	$m = 3$	1.851	0.333	2.993
	$m = 5$	1.617	0.271	2.452
	$m = 7$	1.421	0.242	2.299
	$m = 20$	0.781	0.127	1.199
	$m = 40$	0.262	0.059	0.446
mPCA	$m_b = 3$ and $m_w = 3$	1.580	0.290	2.637
	$m_b = 5$ and $m_w = 5$	1.289	0.227	2.084
	$m_b = 7$ and $m_w = 7$	1.106	0.189	1.876
	$m_b = 7$ and $m_w = 20$	0.636	0.121	1.049
	$m_b = 7$ and $m_w = 40$	0.167	0.048	0.340

4 Conclusions

The effect of naturally occurring groups in the subject population for facial shape data has been explored in this article. The formalism for mPCA has been described briefly and we have shown that mPCA allows us to model variations at different levels of structure in the data (i.e., between-group and within-group levels). Examination of eigenvalues showed that both between and within-group sources of variation were important. Modes of variation appeared to make sense: the first mode of the within-group mPCA seemed to govern width of the face in the coronal plane and the first mode of the between-group mPCA seemed to govern length and shape of the nose and/or face in the transverse and sagittal planes. We have also demonstrated that initial results of mPCA for facial shape data appear to show evidence clustering in the between-group component scores. Indeed, a consistent relationship between genders was observed for each country. Evidence of clustering was also observed for single-level PCA, although the nature of this clustering was less clear. Point-to-point errors of model fits for both mPCA and PCA reduced with the number of eigenvectors retained, as expected. This research is an excellent first step in evaluating the usefulness and feasibility of mPCA in analysing facial shape.

References

1. Cootes, T.F., Hill, A., Taylor, C.J., Haslam, J.: Use of active shape models for locating structure in medical images. Image Vis. Comput. **12**, 355–365 (1994)
2. Cootes, T.F., Taylor, C.J., Cooper, D.H., Graham, J.: Active shape models - their training and application. Comput. Vis. Image Underst. **61**, 38–59 (1995)
3. Hill, A., Cootes, T.F., Taylor, C.J.: Active shape models and the shape approximation problem. Image Vis. Comput. **14**, 601–607 (1996)
4. Taylor, C.J., Cootes, T.F., Lanitis, A., Edwards, G., Smyth, P., Kotcheff, A.C.W.: Model-based interpretation of complex and variable images. Philos. Trans. R. Soc. Lond. Ser. B-Biol. Sci. **352**, 1267–1274 (1997)
5. Cootes, T.F., Taylor, C.J.: A mixture model for representing shape variation. Image Vis. Comput. **17**, 567–573 (1999)
6. Cootes, T.F., Edwards, G.J., Taylor, C.J.: Active appearance models. IEEE Trans. Pattern Anal. Mach. Intell. **23**, 681–685 (2001)
7. Cootes, T.F., Taylor, C.J.: Anatomical statistical models and their role in feature extraction. Br. J. Radiol. **77**, S133–S139 (2004)
8. Allen, P.D., Graham, J., Farnell, D.J.J., Harrison, E.J., Jacobs, R., Nicopolou-Karayianni, K., Lindh, C., van der Stelt, P.F., Horner, K., Devlin, H.: Detecting reduced bone mineral density from dental radiographs using statistical shape models. IEEE Trans. Inf Technol. Biomed. **11**, 601–610 (2007)
9. Lecron, F., Boisvert, J., Benjelloun, M., Labelle, H., Mahmoudi, S.: Multilevel statistical shape models: a new framework for modeling hierarchical structures. In: 9th IEEE International Symposium on Biomedical Imaging (ISBI), pp. 1284–1287 (2012)
10. Farnell, D.J.J., Popat, H., Richmond, S.: Multilevel principal component analysis (mPCA) in shape analysis: a feasibility study in medical and dental imaging. Comput. Methods Programs Biomed. **129**, 149–159 (2016)
11. Kau, C.H., Cronin, A., Durning, P., Zhurov, A.I., Sandham, A., Richmond, S.: A new method for the 3D measurement of postoperative swelling following orthognathic surgery. Orthod. Craniofac. Res. **9**, 31–37 (2006)

12. Kau, C.H., Hartles, F.R., Knox, J., Zhurov, A.I., Richmond, S.: Natural head posture for measuring three-dimensional soft tissue morphology. In: Middleton, J., Shrive, M.G., Jones, M.L. (eds.) Computer Methods in Biomechanics and Biomedical Engineering – 5 First Numerics Ltd., Cardiff University (2005)
13. Kau, C.H., Richmond, S.: Three-dimensional analysis of facial morphology surface changes in untreated children from 12 to 14 years of age. Am. J. Orthod. Dentofac. Orthop. **134**, 751–760 (2008)
14. Farkas, L.G.: Anthropometry of the Head and Face. Raven Press, New York (1994). pp 21–25
15. Hopman, S.M., Merks, J.H., Suttie, M., Hennekam, R.C., Hammond, P.: Face shape differs in phylogenetically related populations. Eur. J. Hum. Genet. **22**, 1268–1271 (2014)

Estimating Rodent Brain Volume by a Deformable Contour Model

Julio Camacho-Cañamón[1(✉)], María J. Carreira[1], Pedro Antonio Gutiérrez[2], and Ramón Iglesias-Rey[3]

[1] Centro de Investigación en Tecnoloxías da Información (CiTIUS), Universidade de Santiago de Compostela, Santiago de Compostela, Spain
julio.camacho@uco.es, mariajose.carreira@usc.es

[2] Department of Computer Science and Numerical Analysis, Universidad de Córdoba, Córdoba, Spain
pagutierrez@uco.es

[3] Clinical Neurosciences Research Laboratory, Clinical University Hospital, Health Research Institute of Santiago de Compostela (IDIS), Universidade de Santiago de Compostela, Santiago de Compostela, Spain
Ramon.Iglesias.Rey@sergas.es

Abstract. Cerebral stroke is a cerebrovascular disease caused by an alteration of blood flow to the brain. Rodents are used to experiment with drugs provoking a stroke and studying the effects of different drugs as a measure of the relation of lesion volume to brain volume. Nowadays, clinicians are performing these experiments manually, leading to interhuman errors and not repeatability, of results, as well as being time-consuming tasks. This paper presents a methodology to automate this task, performing an automatic computation of the brain volume from the brain area for each slice of the rodent brain. Although in its initial state, results are very promising, and so work will follow in this way with the computation of lesion volume.

Keywords: Brain volume · Contour model · Image segmentation

1 Introduction

Cerebral stroke is a cerebrovascular disease caused by an alteration of blood flow to the brain. Stroke is the third main cause of mortality and one of the leading causes of disability in industrialised countries [1].

The most common types of stroke are cerebral ischemia and brain bleeding. The former is caused by an occlusion of a cerebral artery with the consequent focal or global circulatory disorder. It approximately represents 88% of the strokes, therefore, its non-invasive detection represents a priority in the neurological field. Brain bleeding, despite being less frequent than ischaemia, has a much higher mortality rate. However, the survivors of these strokes usually present less severe sequelae in the medium term.

M. Valdés Hernández and V. González-Castro (Eds.): MIUA 2017, CCIS 723, pp. 686–697, 2017.
DOI: 10.1007/978-3-319-60964-5_60

Magnetic resonance imaging (MRI) and magnetic resonance spectroscopy (MRS) can be used to detect cerebral infarcts produced by ischemia. However, the first step to perform this automatic detection is to correctly segment the brain. This paper presents an automatic method to segment the brain area of rodents from structural T_2 MRI. The goal is to avoid a manual analysis of the different slices to locate a brain injury caused by ischemia, as manual selection of the lesion contour can easily lead to errors. Moreover, a contrast enhancement is needed to make injury manual detection easier, saturating the images and resulting in an important loss of data. Finally, the manual process is considerably time consuming.

The method proposed to avoid a manual segmentation is based on an active contour model and other morphological processes. The set of images considered is obtained from a Bruker Biospec 9.4 T magnetic resonance scanner. An ischemia is induced in the rodent by the Mouse Model of Middle Cerebral Artery Occlusion (MCAo) method [2]. Transient focal ischemia (60 min) was induced by intraluminal occlusion of the middle cerebral artery (MCAo), following the method described previously. Male Sprague-Dawley rats weighing between 280 and 330 g were used [3].

Figure 1 represents the rodent brain and the 14 slices obtained. It can be seen that the shape of the brain change in the upper slices. The proposed automatic system aims to resolve and streamline the procedure, offering support to researchers in order to obtain an objective, repeatable and reliable measurement of the volume of the rodent brain. In other paper, the brain volume has been determined by measuring the slice area in a brain rodent [9].

The rest of the paper is organized as follows. Section 2 describes the images considered and the methodology proposed to perform the segmentation. Section 3 discusses the experiments performed and the associated results. Finally, Sect. 4 concludes the paper.

2 Materials and Methods

If we look for the characteristics of the sequence of images in Fig. 1, and take into account the high contrast existent between brain and skull bone, we could think about using a deformable contour model. As we will show later in this section, this technique will have some problems as the shape and homogeneity of a brain is different as we go through the slices and so the skull. These differences can be clearly seen in Fig. 1, wherein the first slices the skull is finer, in the intermediate slices the brain region is more heterogeneous and in the latter slices the size is considerably smaller.

2.1 Material

Magnetic resonance imaging (MRI) and magnetic resonance spectroscopy (MRS) were considered, both conducted on a 9.4 T horizontal bore magnet (Bruker BioSpin, Ettligen, Germany) with a 20 cm wide actively shielded gradient coils

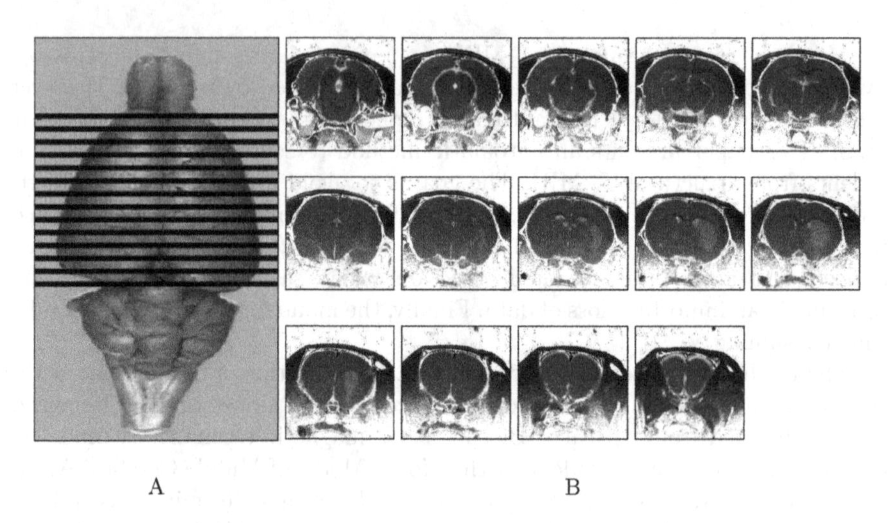

Fig. 1. (A) Representation of the rodent brain with 14 lines where slices were captured from bottom to top; (B) Slices 0 to 13 from left to right and up to down.

(440 mT/m). Radio frequency transmission was achieved with a birdcage volume resonator, and the signal was detected using a four-elements surface coil positioned over the head of the animals. In this way, infarct sizes were assessed from MRI images.

To evaluate the status of MCAo in a non-invasive manner diffusion-weighted images were used. The progression of ischemic lesions and infarct volumes were determined from T2-maps calculated from T2 images at 24 h, 7 and 14 days after occlusion. Apparent diffusion coefficient (ADC) maps were captured before removing the filament from the animal (between 45 and 90 min after the occlusion), using a spin-echo echo-planar imaging sequence with the following acquisition parameters: image matrix 128 × 128 (in-plane resolution, 0.15 mm/pixel), field-of-view 19.2 × 19.2 mm^2, 14 consecutive slices of 1 mm thickness, repetition time = 4 s, echo time = 30 ms, and diffusion b values: 0, 100, 300, 600, 800, 1000, and 1400 s/mm^2.

T2-weighted images were acquired 7 days after the induction of ischemia using a multi-slice multi-spin-echo (MSME) sequence with the following acquisition parameters: image matrix 192 × 192 (isotropic in-plane resolution of 0.1 mm/pixel), field-of-view 19.2 × 19.2 mm^2, 14 consecutive slices of 1 mm thickness, repetition time = 3 s, and 16 echoes with echo time = 9 ms.

After this, 70 images from 5 rodents (each one with 14 slices) were analised and processed by an expert using Bruker's Paravision 5.1 software and Image-J [5]. Once the images are stored appropriately, the process of segmenting the brain begins with the definition of properties that will be used in a deformable contour evolution. As the characteristics of each slice, specially first and last, are different, some additional corrections must be introduced in order to have a good segmentation in all slices, as the objetive is to measure all slices areas to sum them as brain volume.

2.2 Methods

Geodesic active contours (GAC), also known as *snakes*, may be the most well-known curve evolution algorithms. By iteratively solving a partial differential equation, the snake curve is deformed in order to minimise internal and external energy along its edge. The internal forces keep the curve smooth, while the external forces are used to represent the edges, lines and protrusions of the image. This model has very desirable properties such as its natural parametrization, an intuitive mathematical formulation and the ease of adaptation to unknown topology forms. These properties make snakes one of the algorithms most widely used for segmentation and search of objects in images [4].

In the context of this paper, the snake has to adapt its form to the brain. In order to do so, the function to be minimised should be able to tackle different situations, resulting in three different terms with different weighting parameters:

$$E_{\text{Total}} = E_{\text{int}} + E_{\text{ext}} = \alpha \cdot E_{\text{Contour}} + \beta \cdot E_{\text{Curvature}} + \gamma \cdot E_{\text{Image}}, \tag{1}$$

where E_{Total}, E_{int}, E_{ext}, E_{Contour}, $E_{\text{Curvature}}$ and E_{Image} are the total, internal, external, contour, curvature and image energies, respectively, and α, β and γ are the weighting parameters. The internal energy, composed by the energy of contour and curvature, directly affects the flexibility of the snake curve while the image energy is related to the properties of the image, their edges already calculated previously in this case. Given the edge of the brain, $C(v) = (\rho(v), \theta(v))$, which represents a snake with arc length v, the energy function to be minimized is:

$$E_{\text{Total}} = \int_0^1 E_{\text{int}}(C(v)) + E_{\text{ext}}(C(v))dv, \tag{2}$$

where E_{int} is expressed as:

$$E_{\text{int}} = \alpha \cdot E_{\text{Contour}} + \beta \cdot E_{\text{Curvature}} = \alpha\left(\frac{\partial C(v)}{\partial v}\right)^2 + \beta\left(\frac{\partial^2 C(v)}{\partial v}\right)^2. \tag{3}$$

Once the internal energy is defined, we must define the external energy. According to a preliminary analysis, the energy has to be specifically adapted to each slice. An edge detector is applied to all slices from the third to the penultimate one, while the two first slices, which tend to present gaps in the area representing cranium bone, are not computed. Canny edge detector is used, which is a multi-stage algorithm able to detect a wide range of edges in images [6]. The parameter σ represents the width of the Gaussian filter used in Canny detector in order to smooth the image and filter out the noise to prevent false detection caused by noise.

Mathematical morphology methods must be applied to the result of edge detection to achieve a number of regions specifically adapted to image itself.

Erosion and dilation of edges are frequently considered [7] in order to remove the false edges and connect regions that, by their very nature, are heterogeneous.

The threshold parameter of Canny, σ, needs to be large enough to avoid overdetection of edges, so that dilation and erosion (closing) can successfully detect the regions of the image. From this analysis, we selected $\sigma = 2.0$ for all the experiments in this paper. Figure 2 shows the results of applying closing algorithm after edges detection.

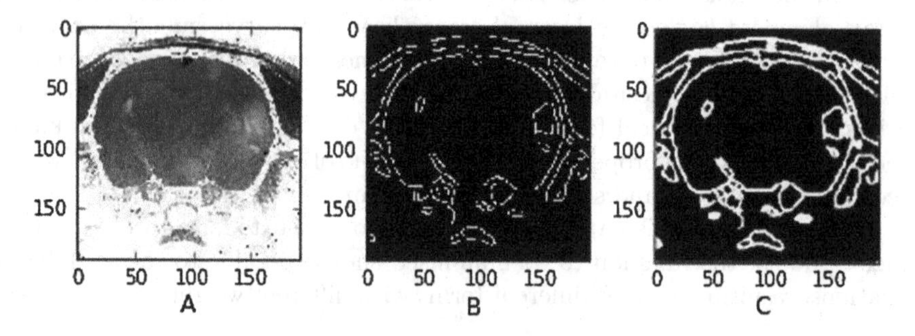

Fig. 2. Results of applying closing algorithms after edges detection with Canny algorithm (A) Original image; (B) Canny edges with $\sigma = 2.0$; (C) Edges after closing procedure with a circle structural element of size 6.

After the processes of edge detection and closing, it will be possible to retain the active contour inside the brain, as we will show in what follows.

The snake algorithm has a stop criterion based on the image gradient, that is, the contour stops expanding when it detects an edge. However, detection of an inner edge within a region can prevent the algorithm from further expand into the region [8]. In this work, given that the brain lesion is better distinguished in slices from 4 to 9, this kind of inner edges are found in these slices, and the snake tends to detect only "healthy" regions as brain (see Fig. 3A). This is because the external energy avoid the snake to include lesions inside the brain region.

To avoid this problem, Chan and Vese [8] proposed to use segmentation algorithms with active contour without edges (ACWE). This variant of the algorithm explores regions beyond edges by comparing the properties of external and internal regions. The snake returns to an earlier state if the pixels found beyond the edges do not have properties similar to those located within the boundary. Figure 3 shows an example of this algorithm, that returns after finding a border too thick (the skull).

The mathematical model is the following one [8]:

$$\frac{\partial u}{\partial t} = |\nabla u| \left(\mu div \left(\frac{\nabla u}{|\nabla u|} \right) - \nu - \lambda_1 (I - c_1)^2 + \lambda_2 (I - c_2)^2 \right), \tag{4}$$

where the behaviour of the algorithm is defined by parameters μ, λ_1, and λ_2 and other variables are time t, curve u and parameters c_1 and c_2 related to gray level intensity (see [8] for details).

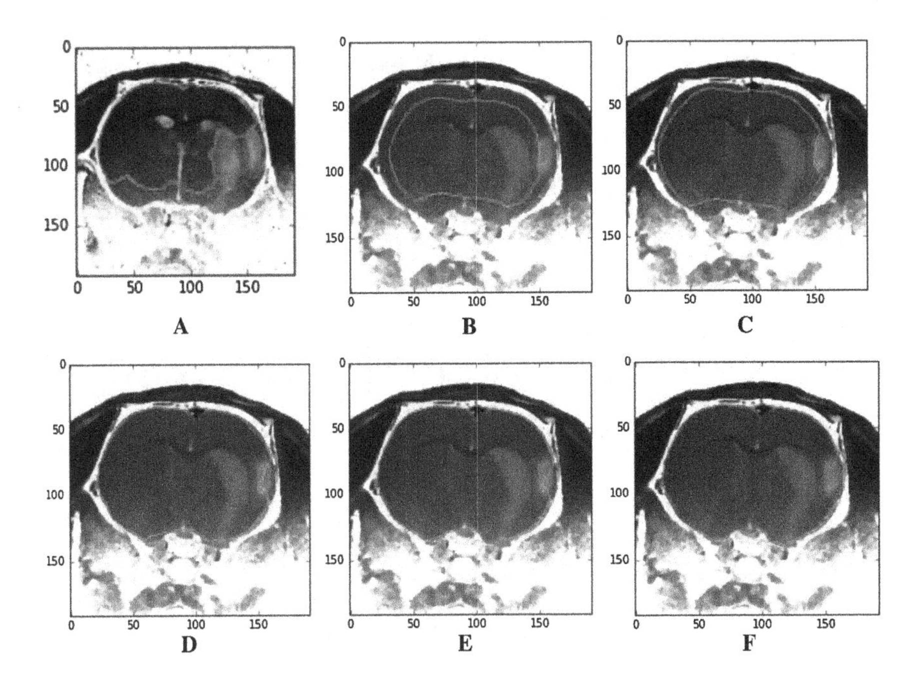

Fig. 3. (A) Final deformable contour without using ACWE in which only healthy region was extracted as "brain"; (B–F) Deformable contour evolution using the ACWE algorithm.

Each parameter has a specific role, and their values should be adjusted according to the characteristics of the image:

Parameter μ: it is related to the smoothness of the algorithm, since it is applied on $div\left(\frac{\nabla u}{|\nabla u|}\right)$, the curvature operator. In this way, μ (smoothing force) is the number of times that the smoothing function is run in each iteration.

Parameters λ_1 and λ_2: they are related to the relative importance given to the properties of the pixels lying inside or outside the snake boundary, respectively. A parametric combination where $\lambda_1 >> \lambda_2$ causes the snake not to explore new regions even when the edges are very weak. By contrast, a combination where $\lambda_1 << \lambda_2$ causes the snake to look in new regions beyond the boundaries that contain it.

It is experimentally verified that the parameters can be fixed to $\mu = 2$, $\lambda_1 = 1$ and $\lambda_2 = 1$.

Once defined the set of internal and external energies influencing the evolution of the snake, it must be defined the initial contour from which the snake will evolve. As we have seen in Fig. 1, the brain in first and last slices has different shape, so we have adapted the initial contour to this situation in the following way (see Fig. 4):

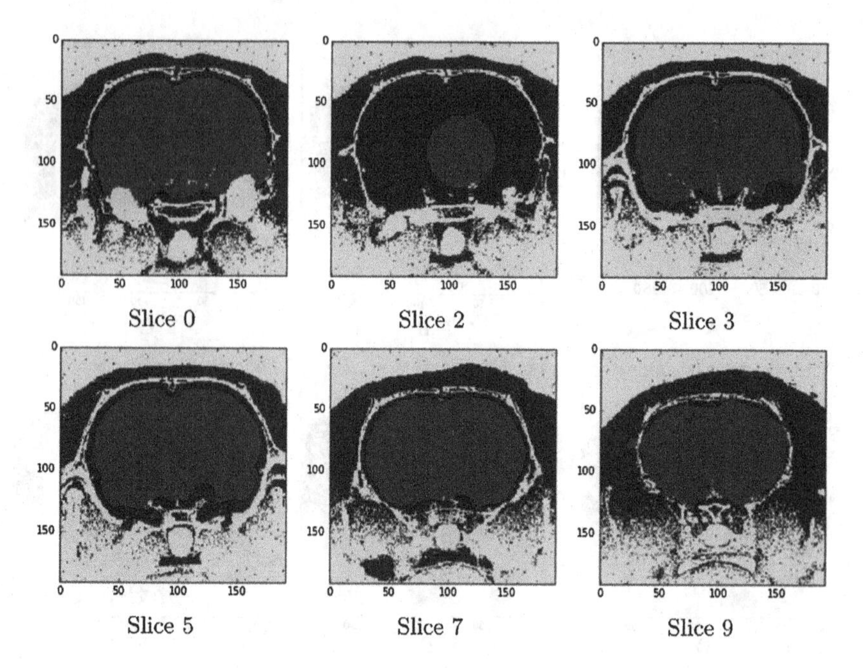

Fig. 4. Initial contour for some slices: slice 2 is selected as to perform the initial circle contour because it is usually the most standard slice, with no lesion and homogeneous brain region. Its final contour will be used in a reduced size for previous and following slices.

- Slices 0 to 1: an initial contour of the same shape but smaller than the best fit of slice 2. The size reduction is carried out by applying a morphological erosion operation with a circular disk of 6 pixels. In these slices, the initial contour can not be a circle due to the heterogeneous nature of the intensity of the pixels within the brain region.
- Slice 2: An initial circular contour of radius 30 pixels centred on a point with an offset of 10 pixels to the left and up of the geometric centre of the image.
- Slices 3 to 13: an initial contour of the same shape but smaller than the best fit of the previous slice. The size reduction is performed by applying a morphological erosion operation with an 8-pixel circular disc. The number of pixels has been set up to prevent the initial contour from being outside the skull.

As shown in Fig. 1, last slices are very different, so the rules used to obtain the segmentation of the brain region in all slices are not valid for them. Therefore, algorithm restart criteria are proposed to solve these issues.

An additional problem is related to the fact that the snake can evolve more than desired, considering the skull as part of the brain and moving beyond its limits (see Fig. 5). As shown, for those slices the snake leaves the current region looking for new zones and does not return back to the brain. This is because the light intensity characteristics of the pixels are very similar inside and outside the

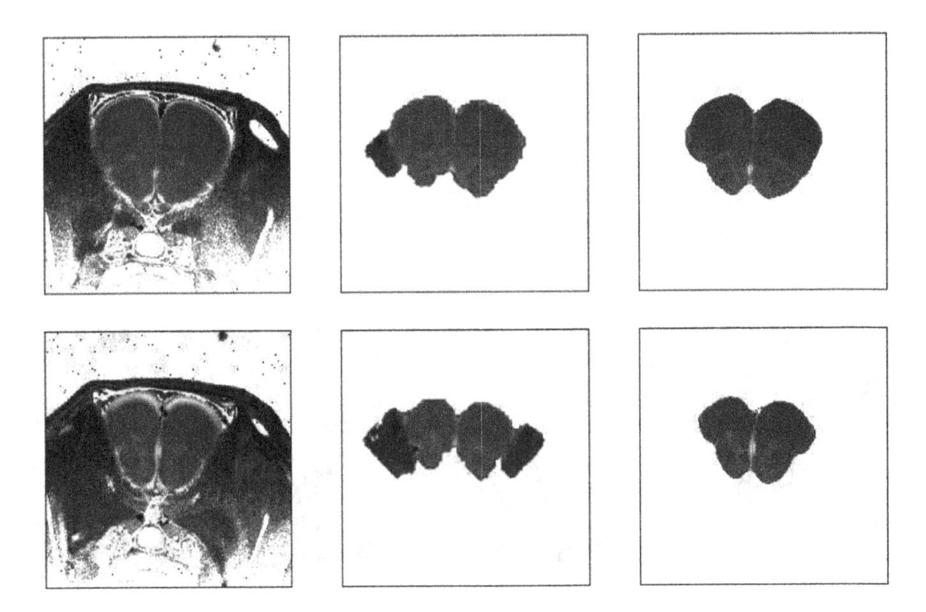

Fig. 5. Final brain segmentation. First row: slice 12 original image, brain region after standard ACWE and brain region after ACWE combined with restart criteria. Second line: the same for slice 13.

brain region, due to the fact that this is an axial slice in the external proximities of the skull. All of the above, combined with the high difference in light intensity between the edges of the upper skull and the sides, makes the snake to ignore the lateral edges, assuming that it may be an inner border of the brain. This is avoided by using a double chained restart criterion:

Ratio of successive sizes: The ratio between the cerebral area of one slice and the next slice is calculated. When this ratio exceeds a certain threshold, the size of the initial contour is reduced. The threshold is set to 120 % in the Sections 1 through 5 and 105 % in the remaining slices. These values are based on the experimental observation of the enlargement of the brain region in the first slices and the decrease in the latter slices.

The eccentricity values of an ellipse: The brain takes the form of an irregular ellipse, and, according to the slice, its eccentricity varies. Therefore it is possible to control when the snake has exceeded the limits of the brain, depending on the value of its eccentricity. The values of the eccentricity are forced to be between 0.7 and 0.75, so when the snake exceeds these values the algorithm is restarted, given that it has left the limits of the head bone.

Figure 6 shows the final result of snake in all slices, where the conditions of ratio and eccentricity help to make a good segmentation for all slices.

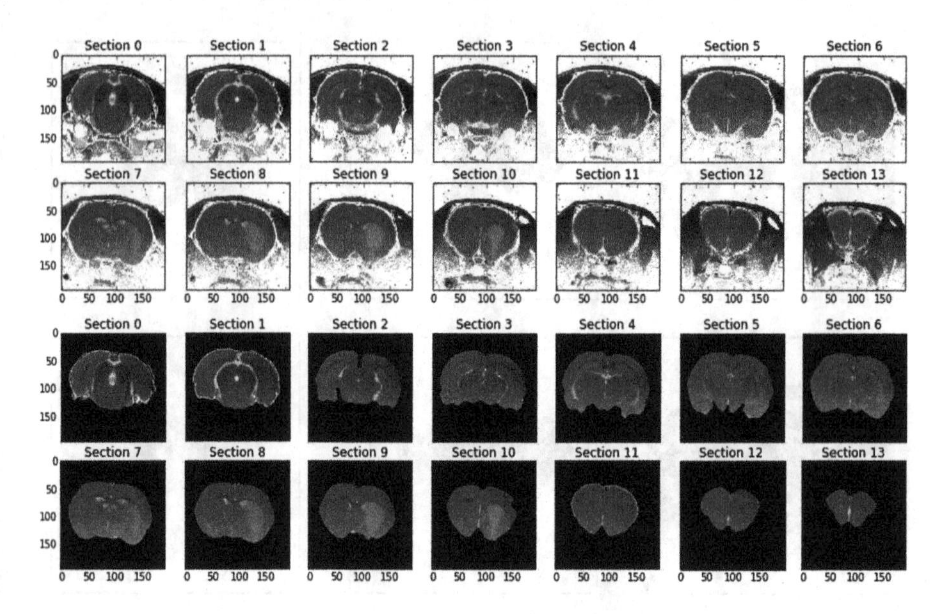

Fig. 6. Complete segmentation of the brain of a rodent, applying the restarting criteria.

3 Results

We have analysed the behaviour of the algorithm when segmenting a brain (see Fig. 6). It can be seen that the results are visually correct and the brain is correctly isolated from the rest of the image.

Moreover, two different LINC [10] experts have manually segmented the brain region for each slice. The objective is double-fold: checking the consistency of the different experts in the segmentation task and evaluating the performance of the automatic segmentation algorithm.

Three widely used performance measures, namely Sensitivity (SEN), Specificity (SPC) and Accuracy (ACC) are used for evaluation purposes. They are expressed as follows:

$$SEN = \frac{TP}{TP + TN}, \tag{5}$$

$$SPC = \frac{TN}{TN + FP}, \tag{6}$$

$$ACC = \frac{TP + TN}{TP + TN + FP + FN}, \tag{7}$$

where TP is the number of pixels classified as brain, FP is the number of pixels incorrectly classified as brain, TN is the number of pixels correctly classified as no brain and FN is the number of pixels incorrectly classified as no brain.

Table 1 shows the results of the comparison of the three segmentations (two given by experts and one given by the proposed algorithm). Two comparisons are performed: the specialist comparison refers to the performance measures obtained from the labels given by both experts, while the automatic one is comparing the labels given by the automatic method with the labels given by the first expert, considered the gold standard.

In the table, the identifier of each image (MCAo) and the results of 3 performance metrics are shown. Sets of performance parameters have their opposites. To obtain these results, for each MCAo and for each slice, two confusion matrices has been derived. One compares the results obtained between the two specialists and the other compares the results of the first specialist with the results obtained automatically by the system.

For each MCAo, 28 sets with these rates are obtained, one for each slice and each comparison. Subsequently, the average of all sets for each procedure is calculated, thus obtaining two sets for each MCAo. As can be checked in Table 1, the results from both specialists are very close and the results from the automatic segmentation algorithm are also very similar to those of the first expert.

Table 1. Performance metrics for the different comparisons, averaged across the different slices. Specialist: comparison between expert 1 and expert 2. Automatic: comparison between expert 1 and proposed method.

MCAo	Comparison	SEN	ACC	SPC
1354001	Specialist	0.7004	0.9795	0.9893
	Automatic	0.7027	0.9566	0.9711
1357814	Specialist	0.7378	0.9654	0.9889
	Automatic	0.7368	0.9258	0.9791
1360314	Specialist	0.7525	0.9611	0.9896
	Automatic	0.7578	0.8608	0.9739
1360307	Specialist	0.7345	0.9741	0.9909
	Automatic	0.7345	0.9309	0.9777
1361814	Specialist	0.7480	0.9611	0.9897
	Automatic	0.7459	0.9373	0.9788

The brain area for each slice is calculated as the sum of pixels classified as brain (inside the deformable contour) after applying the proposed method. Moreover, we can also compare the average of the brain area measurements obtained for each slice in each MCAo. Figure 7 shows the divergence in area between the segmentation algorithm and the first specialist, across the different image slices.

As can be checked, both areas are very similar for almost all slices, with slight differences in first and last slices, as it would be expected. This is because the initial contours of the snake for these slices have a different treatment to the rest.

The time required to perform the automatic segmentation of the cerebral area of one entire MCAo set (14 slices) is always less than 10 s. This greatly

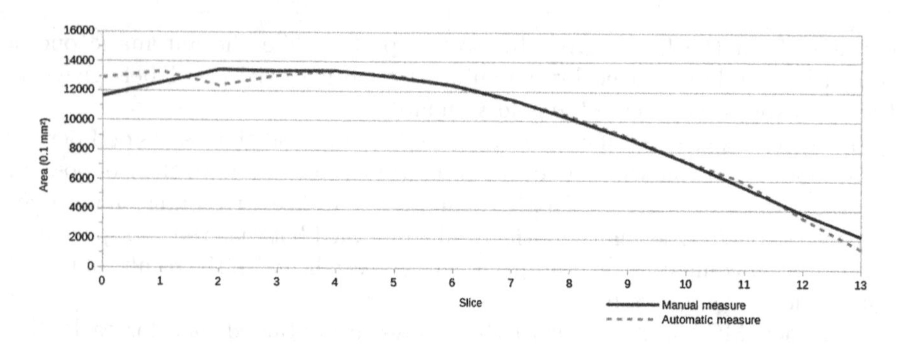

Fig. 7. Mean areas estimated for each slice summarizing all images considered in this experiment.

improves the time consumed by a specialist for segmentation by the manual method (approximately 20 min).

From these areas, two different error metrics have been also obtained: the Standard Error of Prediction (SEP) and the Root-Mean-Square Error (RMSE). Table 2 shows the errors for all the rodents, where, as it can be observed, the SEP is below 10% in all cases and the RMSE is below 103 mm^2.

Table 2. Total standard error of prediction (SEP) and root mean square error (RMSE) computed as a mean of the slices for each rodent.

MCAo	SEP	RMSE
1354001	0.07778	86.6391
1357814	0.04912	48.5153
1360307	0.04943	49.0863
1360314	0.08223	76.0193
1361814	0.07415	70.2949

4 Conclusions

This paper presents an automatic method to segment brain area from structural T_2 MRI, where an active contour model and other morphological processings are applied. The active contour model is coupled with different initial contours according to the slice analysed, and a specific restarting criteria (based on the successive sizes detected and the eccentricity of the brain region) is used to control the evolution of the algorithm.

The results obtained using this method are promising, where the automatic segmentation method is able to increase the accuracy and objectivity of the segmentation, avoiding a time consuming manual process.

As the main objective of this work is to build an automatic system to help the specialist in performing an accurate, repeatable and fast method to evaluate the influence of stroke under the effect of some drugs, next step will be to separate the healthy brain from lesion.

As a future research line, a specifically adapted version of this algorithm could be used to perform the detection of the stroke.

Acknowledgments. This work has been subsidized by the Centro de Investigación en Tecnoloxías da Información (CiTIUS) from Universidade de Santiago de Compostela with financial support from the Consellería de Cultura, Educación e Ordenación Universitaria (accreditation 2016–2019, ED431G/08) and the European Regional Development Fund (ERDF) and partially supported by projects TIN2014-54583-C2-1-R and TIN2015-70308-REDT of the Spanish Ministerial Commission of Science and Technology (MINECO, Spain) and FEDER funds (EU).

References

1. Bramlett, H.M., Dietrich, W.D.: Pathophysiology of cerebral ischemia and brain trauma: similarities and differences. J. Cereb. Blood Flow Metab. **24**(2), 133–150 (2004)
2. Chiang, T., Messing, R.O., Chou, W.-H.: Mouse model of middle cerebral artery occlusion. JoVE (J. Vis. Exp.) **48**, e2761 (2011)
3. Longa, E.Z., et al.: Reversible middle cerebral artery occlusion without craniectomy in rats. Stroke **20**(1), 84–91 (1989)
4. Álvarez, L., et al.: Morphological snakes. In: 2010 IEEE Conference on Computer Vision and Pattern Recognition (CVPR). IEEE (2010)
5. Rasband, W.S.: ImageJ, U. S. National Institutes of Health, Bethesda, MD, USA (1997–2011). http://imagej.nih.gov/ij/
6. Canny, J.: A computational approach to edge detection. IEEE Trans. Pattern Anal. Mach. Intell. **6**, 679–698 (1986)
7. Haralick, R.M., Sternberg, S.R., Zhuang, X.: Image analysis using mathematical morphology. IEEE Trans. Pattern Anal. Mach. Intell. **4**, 532–550 (1987)
8. Chan, T.F., Vese, L.A.: Active contours without edges. IEEE Trans. Image Process. **10**(2), 266–277 (2001)
9. Mayhew, T.M., Olsen, D.R.: Magnetic resonance imaging (MRI) and model-free estimates of brain volume determined using the Cavalieri principle. J. Anat. **178**, 133–144 (1991)
10. Clinical Neurosciences Research Laboratory. http://www.linc-stg.eu/

MIMONet: Gland Segmentation Using Multi-Input-Multi-Output Convolutional Neural Network

Shan E Ahmed Raza[1], Linda Cheung[2], David Epstein[3], Stella Pelengaris[2], Michael Khan[2], and Nasir M. Rajpoot[1](✉)

[1] Department of Computer Science, University of Warwick, Coventry CV4 7AL, UK
n.m.rajpoot@warwick.ac.uk

[2] School of Life Sciences, University of Warwick, Coventry CV4 7AL, UK

[3] Mathematics Institute, University of Warwick, Coventry CV4 7AL, UK

Abstract. Morphological assessment of glands in histopathology images is very important in cancer grading. However, this is labour intensive, requires highly trained pathologists and has limited reproducibility. Digitisation of tissue slides provides us with the opportunity to employ computers, which are very efficient in repetitive tasks, allowing us to automate the morphological assessment with input from the pathologist. The first step in automated morphological assessment is the segmentation of these glandular regions. In this paper, we present a multi-input multi-output convolutional neural network for segmentation of glands in histopathology images. We test our algorithm on the publicly available GLaS data set and show that our algorithm produces competitive results compared to the state-of-the-art algorithms in terms of various quantitative measures.

Keywords: Gland segmentation · Digital pathology · Deep convolutional neural networks

1 Introduction

One of the key criteria in colon cancer grading is histological assessment of glands [1]. This histological assessment requires a highly trained pathologist, is labour intensive, suffers inter and intra-observer variability, and has limited reproducibility. The introduction of whole slide scanners has made it possible to digitise and allow computerised assessment of tissue slides [2], thus increasing reproducibility and decreasing the pathologist workload. Automatic segmentation of glands is challenging due to high variation in texture, size and structure of glands especially in malignant tissue as shown in Fig. 1 which shows sample images taken from the GLaS data set [3]. The top row shows sample images from benign cases, whereas the bottom row shows sample images from malignant cases. Figure 1(c) has been taken from a moderately differentiated colon cancer tissue and (d) has been taken from a poorly differentiated colon cancer

M. Valdés Hernández and V. González-Castro (Eds.): MIUA 2017, CCIS 723, pp. 698–706, 2017.
DOI: 10.1007/978-3-319-60964-5_61

tissue section. It is evident from these images that there is a large variation in the size, texture and structure of glands in both malignant and benign cases although the variation is greater in malignant cases. In this paper, we present a deep learning framework that focuses on segmentation of glands in colon histology images. We compare our results on the publicly available GLaS data set with the state-of-the-art algorithms [3].

Previous studies on gland segmentation can be broadly classified into two main categories: Hand crafted feature based methods and deep learning based methods. Most of the early attempts on gland segmentation work have been using hand crafted features. Wu *et al.* [4] presented a region growing algorithm for segmentation of intestinal glands. They first identified initial seed regions based on large vacant lumen regions and expanded the seed to a surrounding chain of epithelial nuclei. The surrounding epithelial nuclei were identified by thresholding the image. Farjam *et al.* [5] proposed segmentation by clustering texture features calculated using a variance filter. However, robust segmentation requires more domain knowledge and texture features calculated using just the variance filter might not provide enough information for the local structure of the tissue. Naik *et al.* [6] employed a Bayesian classifier to detect lumen regions and then refined using level set curve stopping the curve based on likelihood of nuclei. This method might work well on benign cases, but can fail on malignant cases where morphology of glands is quite complex. Nguyen *et al.* [7] grouped the nuclei, cytoplasm and lumen using colour space analysis and grew the lumen region with constraints to achieve segmentation. Gunduz-Demir *et al.* [8] represented each tissue component as a circular disc and constructed a graph with nearby discs joined by an edge. They performed region growing on lumen discs that were constrained by lines joining the nuclear discs. Nosrati and Hamarneh [9] and Cohen *et al.* [10] first classify tissue regions into different constituents and then employ a constrained level set algorithm to segment the glands. Sirinukunwattana *et al.* [11] identified epithelial superpixels and used epithelial regions as vertices of a polygon approximating boundaries of glands. Most of the methods discussed above first distinguish tissue regions and then employ region growing or level sets to segment glandular regions. Recently, Li *et al.* [12] proposed a slightly different approach where they first determine potential epithelial regions using lumen/background information and then identify connected epithelial cells to segment the glands using a multi-resolution cell orientation descriptor.

Methods based on hand crafted features require careful tuning of features, which is difficult to achieve across a wide range of tissue types especially in the presence of different variations. Deep learning algorithms have recently become popular and produced promising results on solving various image recognition tasks including the analysis of digital pathology images [3]. The fully convolutional network (FCN) for segmentation is considered to be a benchmark for segmentation tasks using convolutional neural networks (CNN) [13]. The network performs pixel-wise classification to get the segmentation mask and consists of downsampling and upsampling paths. The downsampling path consists of convolution and max-pooling whereas the upsampling path consists of convolution and deconvolution (convolution transpose) layers. U-Net [14], inspired by FCN, connects intermediate downsampling and upsampling paths to conserve the context information. DCAN [15] employs a modified FCN and trains the network for both object and contour features to perform segmentation. Another recently proposed multi-scale convolutional neural network [16] trains the network at different scales of the Laplacian pyramid and merges the network in the upsampling path to perform segmentation. Recently, Xu *et al.* [17] proposed a network that performs side supervision of boundary maps in addition to the foreground.

In this paper, we propose a CNN that adds extra layers in the downsampling path, bypassing max pooling operation in order to learn the parameters for segmentation ignored during the max pooling operation. The network retains context information, interprets the output at multiple resolutions and trains the parameters at multiple input image resolutions in downsampling path to learn features for variable gland sizes and shapes in the presence of variable texture.

2 Materials and Methods

2.1 Data Set

In this paper, we used the publicly available Warwick-QU data set published as part of the GLand Segmentation (GLaS) challenge [3]. The data set consists of 165 images with the associated ground truth marked by expert pathologists. The composition of the data set is detailed in Table 1. Sample images from the data set are shown in Fig. 1.

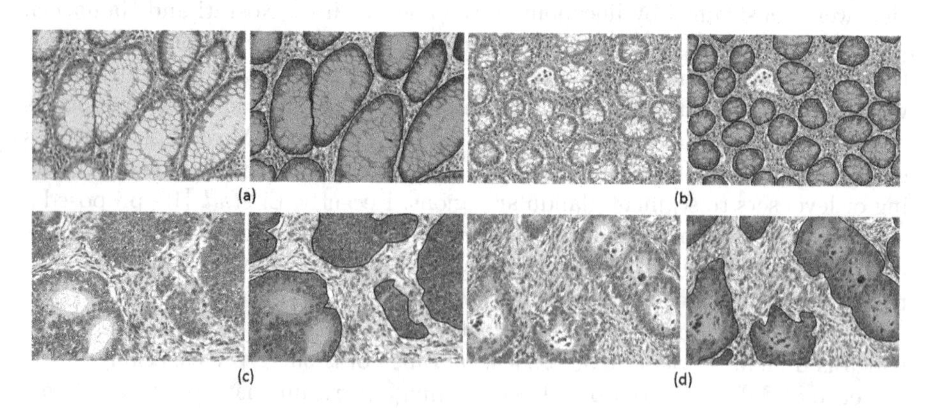

Fig. 1. Sample images from the GLaS data set [3]. The images are shown in pairs, where the sample image on the left is overlaid on the right with the ground truth. The top row shows sample images from benign cases and the bottom row shows sample images from malignant cases. (a) & (b) show variation in size and structure of glands in benign cases, whereas (c) & (d) show variation in malignant colon cancer, where (c) is taken from a moderately differentiated sample and (d) is taken from a poorly differentiated (higher grade) cancerous sample.

Table 1. Composition of Warwick-QU data set.

Histologic grade	Number of images		
	Training	Test A	Test B
Benign	37	33	4
Malignant	48	27	16

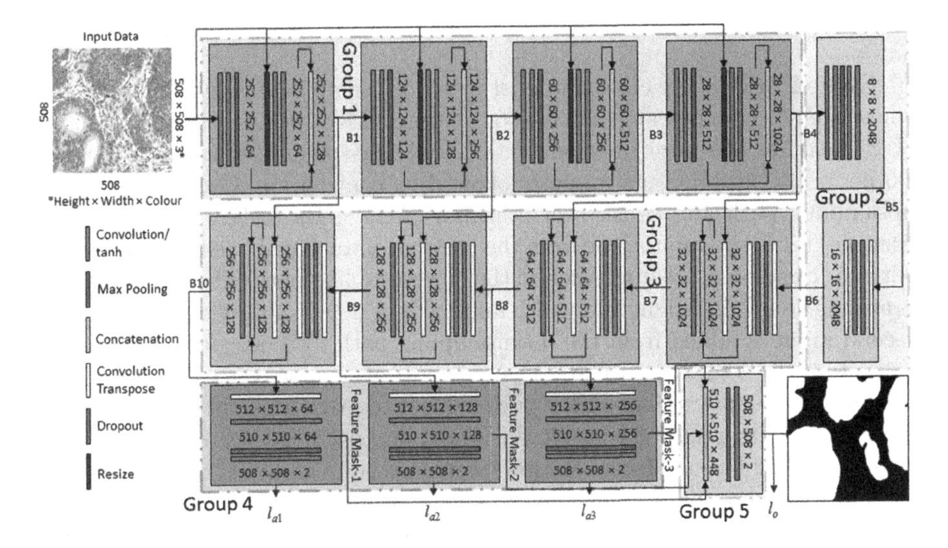

Fig. 2. Architecture of the proposed algorithm.

2.2 The Proposed Method

To reduce the effect of stain variation from different labs and staining conditions, as a preprocessing step we perform stain normalisation using the method proposed by [18][1]. The architecture of the proposed MIMO-Net is shown in Fig. 2. The input to the network is the stain normalised RGB image. The network is divided into five groups and fourteen branches, the division depending on their function and the set of layers/filters. The first group, which consists of four branches with output B1–B4, constructs the downsampling path. Each branch in Group 1 consists of convolution, max-pooling, resize and concatenation layers. The convolution and max-pooling layers perform standard operations as in conventional CNNs. We use *tanh* activation after each convolution layer as our experiments showed that the network converges faster with *tanh* activation compared to ReLU. The resize layer resizes the image using bicubic interpolation so that the resized image dimension matches the corresponding dimension of the max-pooling output. We add the lower resolution input to retain the information from pixels that do not have the maximum response, because they are in the

[1] http://www2.warwick.ac.uk/fac/sci/dcs/research/tia/software/sntoolbox.

vicinity of a noisy neighbourhood. This is particularly useful when we are trying to retain tiny feature details ignored during the max-pooling operation. Another aspect of the resizing operation is to train the network on different sized glands as explained in Sect. 1. The output of branch 1 (B1) has feature depth of size 128 where the first half (64) of the features are the result of the max-pooling operation and the next half (64) are obtained by performing convolutions only on the resized image. The following branches in Group 1 double the feature depth of the previous branch but follow the same protocol in generating the branch output.

Group 2 bridges the connection between downsampling and upsampling path whose architecture is very similar to conventional CNN architectures and consists of B5 and B6. B5 performs convolution and max-pooling operations whereas B6 performs convolution transpose followed by convolution. Group 3 forms the upsampling path and consists of branches B7, B8, B9 & B10. Each of these branches takes two inputs, one from the previous branch and one from the branch with the closest feature dimension in the downsampling path. The output of each branch is double in height and width and half the depth of previous branch. The second input is added from the downsampling path for better localization and to capture the context information as in [14]. It also passes the convolution-only features to the upsampling path, which helps it to learn from those features that do not have maximum response in downsampling path. Compared to the U-Net [14], we add additional deconvolution layers instead of cropping the feature from the downsampling path. This allows us to produce a segmentation map of the same size as the input image and an overlap-tile strategy is not required. It also reduces the number of patches required to produce the desired segmentation output thus removing computational steps.

Groups 4 & 5 generate auxiliary and main output and calculate the loss function. Group 4 consists of three branches where each branch takes the output from one of B8-B10 and generates three auxiliary feature masks, which are fed into the main output branch. The output branch concatenates feature masks and performs convolution followed by softmax classification to get the segmentation output map $p_o(x)$ where x represents a pixel location. The outputs of branches B8-B10 are of different resolutions and so the deconvolution layer in each of the auxiliary branches is set to generate the output of the same size [15]. The deconvolution is followed by a convolution layer which produces the auxiliary feature mask. Each of the auxiliary feature masks is followed by a dropout layer (set to 50%) and the convolution layer followed by softmax classification to get the auxiliary outputs ($p_{a1}(x)$, $p_{a2}(x)$, $p_{a3}(x)$).

For training, we calculate weighted cross entropy loss for the main output (l_o) and the auxiliary outputs (l_{a1}, l_{a2}, l_{a3}) as

$$l_k = \sum_{x \in \Omega} w(x) \log(p_{k(x)}(x)) \tag{1}$$

where $k \in \{o, a1, a2, a3\}$ and Ω is the set of pixel locations in the input image. The weight function $w(x)$ gives higher weights to pixels which are at the merging gland boundaries, leading to a higher penalty [14]. The total loss(l) is calculated by combining auxiliary and main loss using $l = l_o + (l_{a1} + l_{a2} + l_{a3})/epoch$ where $epoch > 0$ represents the number of training passes through the data. This strategy reduces the contribution of auxiliary losses for a higher *epoch* exponentially instead of large steps [15].

3 Results and Discussion

As deep learning algorithms require large amounts of data for training, we augment the data using barrel, pincushion, mustache distortion and introduce Gaussian blur with a Gaussian filter of size 12×12, with σ ranging from 0.2 to 2. While adjusting parameters for barrel, pincushion and mustache distortion we made sure by visual examination that the distortions created by these parameters were realistic and are not too strong to generate unrealistic images. The value of σ is randomly selected for each patch. In addition we rotate, and flip the images left, right, up and down. To train the network, we first extract 600×600 patches from the training data. If the size of image is smaller than 600 in height or width, we symmetrically pad the image to increase its size. During training the network picks these patches in a random order for each *epoch* and crops them to a size of 508×508 patch at random locations in the image as centre before inputting it to network. The proposed network was implemented using TensorFlow v0.12 [19]. We start with a learning rate ($lr = 0.001$) and reduce it according to $lr = 0.001/(10^{(epoch/5)})$, which reduces the learning rate by 10 times for every fifth *epoch*.

The description of the data set used in this paper is given in Sect. 2.1. We used 85 images for training and 80 for testing, where 60 of the test images correspond to Test set A and the remaining 20 to Test set B. For quantitative analysis, we used measures which include Dice coefficient, F1 score, object Dice, pixel accuracy and object Hausdorff [3]. In the case of Hausdorff distance lower values are better; for other measures higher are better. The quantitative results are given in Table 2 which shows that our method produces competitive compared to the state-of-the-art algorithms from the contest and ranks first according the rank sum criteria set by the organisers. On test A, the proposed algorithm performed best in terms of F1 and object Dice but ranked third and fifth on test B. In terms of object Hausdorff, which measures the shape similarity, it ranked second after CUMedVision2 [15] on test A and Xu *et al.* [17] on test B. Lower F1 and object dice on test B suggests that our method missed more glands in malignant cases whereas lower Hausdorff suggests higher shape similarity to the ground truth extracted by our method. Qualitative results of the algorithm for sample images in Fig. 1 are shown in Fig. 3, where for each pair the image on left shows ground truth in green, output of algorithm in red and overlap of ground truth and output of algorithm in yellow. The image on right shows the output of the algorithm overlaid on the sample image. These results show that the algorithm clearly misses a few glands on the boundary of the image for which there is insufficient information. In Fig. 3(b) it merges the glands at the

Table 2. Quantitative comparison with the state-of-the-art methods. S and R in the table correspond to score and rank.

Method	F1 score				Obj Dice				Obj Hausdorff				Rank sum
	Test A		Test B		Test A		Test B		Test A		Test B		
	S	R	S	R	S	R	S	R	S	R	S	R	
Proposed	**0.913**	1	0.724	3	**0.906**	1	0.785	5	49.15	2	133.98	2	**14**
Xu et al.	0.858	8	**0.771**	1	0.888	3	**0.815**	1	54.202	3	**129.93**	1	17
CUMedVision2	0.912	2	0.716	5	0.897	2	0.781	7	**45.418**	1	160.347	8	25
ExB1	0.891	5	0.703	6	0.882	6	0.786	3	57.413	8	145.575	3	31
ExB3	0.896	3	0.719	4	0.886	4	0.765	8	57.350	7	159.873	7	33
Freiburg2	0.870	6	0.695	7	0.876	7	0.786	4	57.093	5	148.463	5	34
CUMedVision1	0.868	7	0.769	2	0.867	9	0.800	2	74.596	9	153.646	6	35
ExB2	0.892	4	0.686	8	0.884	5	0.754	9	54.785	4	187.442	10	40
Freiburg1	0.834	9	0.605	9	0.875	8	0.783	6	57.194	6	146.607	4	42
CVML	0.652	11	0.541	10	0.644	12	0.654	10	155.433	12	176.244	9	64
LIB	0.777	10	0.306	12	0.781	10	0.617	11	112.706	11	190.447	11	65
vision4GlaS	0.635	12	0.527	11	0.737	11	0.610	12	107.491	10	210.105	12	68

bottom of the image and misses one gland. In malignant cases the algorithm seems to be rather 'conservative' in its approach when marking the boundary of glands. It can be observed in Fig. 3(c) for the smaller gland in the middle and in Fig. 3(d) for the two large glands at the bottom. All these glands show significant green inside the ground truth boundary, which at first suggests that the algorithm segmented the gland well inside the ground truth marking for the gland. However, when carefully observed in the overlay with the sample images the algorithm is faithfully following the boundary with tumor cells. For the large gland on the top right in Fig. 3(d) the algorithm 'oversegments' the gland

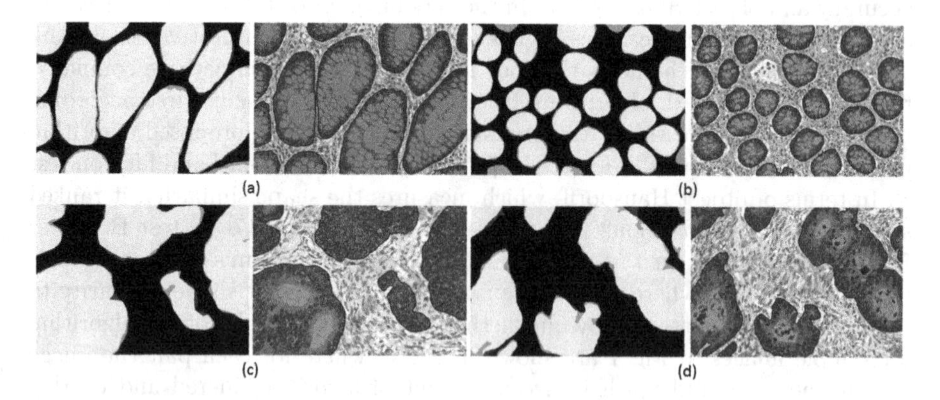

Fig. 3. Results for sample images in Fig. 1. For each pair the image on the left shows the output of algorithm in red, ground truth in green and overlap of ground truth and output in yellow. The image on right shows the output overlaid on the sample image.(Color figure online)

compared to the ground truth but again, looking at the overlay with the sample image, the algorithm has included tumor cells in the segmentation. Overall the algorithm performs a good job in segmenting the glands but needs to improve on the glands at the boundary of a patch. But this limitation could be overcome by using overlapping patches from the whole slide and then merging the results.

Conclusions

In this paper, we presented a multi-input multi-output convolutional neural network for segmentation of glands in colon histopathology images. The proposed architecture allows the network to visualize input and output at multiple resolutions, thus allowing the network to better train its parameters by incorporating the context. We used a publicly available data set for our analysis and showed that the proposed algorithm produces competitive results compared to the state-of-the-art and ranks first according to the criteria set by the publishers of the data. The algorithm lacks accuracy while dealing with glands on the boundary of a patch but this discrepancy can be overcome by extracting overlapping patches from the whole slide image.

Acknowledgements. We are grateful to the BBSRC UK for supporting this study through project grant BB/K018868/1.

References

1. Fleming, M., Ravula, S., Tatishchev, S.F., et al.: Colorectal carcinoma: pathologic aspects. J. Gastrointest. oncol. **3**(3), 153–173 (2012)
2. Gurcan, M.N., Boucheron, L.E., Can, A., et al.: Histopathological image analysis: a review. IEEE Rev. Biomed. Eng. **2**, 147–171 (2009)
3. Sirinukunwattana, K., et al.: Gland segmentation in colon histology images: the glas challenge contest. Med. Image Anal. **35**, 489–502 (2016)
4. Wu, H.S., Xu, R., Harpaz, N., et al.: Segmentation of intestinal gland images with iterative region growing. J. Microsc. **220**(3), 190–204 (2005)
5. Farjam, R., Soltanian-Zadeh, H., et al.: An image analysis approach for automatic malignancy determination of prostate pathological images. Cytom. Part B: Clin. Cytom. **72B**(4), 227–240 (2007)
6. Naik, S., Doyle, S., Agner, S., Madabhushi, A., Feldman, M., Tomaszewski, J.: Automated gland and nuclei segmentation for grading of prostate and breast cancer histopathology. In: ISBI 2008 5th IEEE International Symposium on Biomedical Imaging: From Nano to Macro, pp. 284–287. IEEE (2008)
7. Nguyen, K., Sarkar, A., Jain, A.K.: Structure and context in prostatic gland segmentation and classification. In: Ayache, N., Delingette, H., Golland, P., Mori, K. (eds.) MICCAI 2012. LNCS, vol. 7510, pp. 115–123. Springer, Heidelberg (2012). doi:10.1007/978-3-642-33415-3_15
8. Gunduz-Demir, C., Kandemir, M., Tosun, A.B., Sokmensuer, C.: Automatic segmentation of colon glands using object-graphs. Med. Image Anal. **14**(1), 1–12 (2010)

9. Nosrati, M.S., Hamarneh, G.: Local optimization based segmentation of spatially-recurring, multi-region objects with part configuration constraints. IEEE Trans. Med. Imaging **33**(9), 1845–1859 (2014)
10. Cohen, A., Rivlin, E., Shimshoni, I., Sabo, E.: Memory based active contour algorithm using pixel-level classified images for colon crypt segmentation. Comput. Med. Imaging Graph. **43**, 150–164 (2015)
11. Sirinukunwattana, K., Snead, D., Rajpoot, N.: A stochastic polygons model for glandular structures in colon histology images. IEEE Trans. Med. Imaging **34**, 2366–2378 (2015)
12. Li, G., Raza, S.E.A., Rajpoot, N.M.: Multi-resolution cell orientation congruence descriptors for epithelium segmentation in endometrial histology images. Med. Image Anal. **37**, 91–100 (2017)
13. Shelhamer, E., et al.: Fully convolutional networks for semantic segmentation. IEEE Trans. Pattern Anal. Mach. Intell. **99**, 1–1 (2016)
14. Ronneberger, O., et al.: U-net: convolutional networks for biomedical image segmentation. In: Navab, N., Hornegger, J., Wells, W., Frangi, A. (eds.) Medical Image Computing and Computer-Assisted Intervention – MICCAI 2015, vol. 9351, pp. 234–241. Springer, Cham (2015)
15. Chen, H., Qi, X., Yu, L., et al.: Dcan: deep contour-aware networks for object instance segmentation from histology images. Med. Image Anal. **36**, 135–146 (2017)
16. Song, Y., et al.: Accurate cervical cell segmentation from overlapping clumps in pap smear images. IEEE Trans. Med. Imaging **99**, 1 (2016)
17. Xu, Y., Li, Y., Liu, M., Wang, Y., Lai, M., Eric, I., Chang, C.: Gland instance segmentation by deep multichannel side supervision. In: Ourselin, S., Joskowicz, L., Sabuncu, M., Unal, G., Wells, W. (eds.) Medical Image Computing and Computer-Assisted Intervention – MICCAI 2016, vol. 9901, pp. 496–504. Springer, Cham (2016)
18. Reinhard, E., Adhikhmin, M., Gooch, B., Shirley, P.: Color transfer between images. IEEE Comput. Graph. Appl. **21**(5), 34–41 (2001)
19. Abadi, M., Agarwal, A., et al.: TensorFlow: large-scale machine learning on heterogeneous systems (2015). tensorflow.org

Automated Polyp Segmentation in Colonoscopy Frames Using Fully Convolutional Neural Network and Textons

Lei Zhang[1(✉)], Sunil Dolwani[2], and Xujiong Ye[1]

[1] Laboratory of Vision Engineering, Computer Science, University of Lincoln, Lincoln, UK
{lzhang,xye}@lincoln.ac.uk

[2] Division of Population Medicine, School of Medicine, Cardiff University, Cardiff, UK
dolwanis@cardiff.ac.uk

Abstract. In this paper, we presented a novel hybrid classification based method for fully automated polyp segmentation in colonoscopy video frames. It contains two main steps: initial region proposals generation and regions refinement. Both machine learned features and hand crafted features are taken into account for polyp segmentation. More specifically, the hierarchical features of polyps are learned by fully convolutional neural network (FCN), while the context information related to the polyp boundaries is modeled by texton patch representation. The FCN provides pixel-wise prediction and initial polyp region candidates. Those candidates are further refined by patch-wise classification using texton based spatial features and a random forest classifier. The segmentation results are evaluated on a publicly available CVC-ColonDB database. On average, our method achieves 97.54% of accuracy, 75.66% of sensitivity, 98.81% of specificity and DICE of 0.70%. The fast execution time (0.16 s/frame) demonstrates the promise of our method to be used in real-time clinical colonoscopic examination.

Keywords: Optical colonoscopy · Polyp segmentation · Fully convolutional neural network (FCN) · Textons · Random forest classifier

1 Introduction

Colonoscopy is currently considered as the gold standard diagnostic method for colorectal cancer diagnosis as well as contributing to early detection of colonic polyps before they develop into colorectal cancer, which in turn reduces mortality from colorectal cancer [1, 2]. However, colonoscopy is subject to operator skill and experience and may result in the missed detection of polyps and or cancer during the test [3]. Computer-aided polyp detection (CAD) may help endoscopists reduce the miss-detection rates. It may also help in the differentiation of benign and malignant polyps or benign polyps with malignant potential. Many methods [4–11] have been developed to automatically detect colonic polyps. However, automated polyp detection still remains an open challenge, due to its complex nature in colonoscopy frames. For instance, there are intrinsic frame formation artifacts such as light reflections, similar appearances of different structures

M. Valdés Hernández and V. González-Castro (Eds.): MIUA 2017, CCIS 723, pp. 707–717, 2017.
DOI: 10.1007/978-3-319-60964-5_62

such as fluid bubbles, folds and polyps etc. Moreover, the diversity of polyp type remains a major barrier to train a general model for accurate detection. In addition, even for the same polyp, its appearance may vary from one viewpoint to another, and occlusions are frequently encountered in frames.

The existing automated polyp detection methods can be roughly categorized into two primary groups: segmentation based detection [4–6] and patch classification based detection [7–11]. Hwang et al. [4] proposed an ellipse fitting method based on watershed segmentations. Bernal et al. [5] presented a polyp detection method using the valley information, in which the potential regions are segmented using watersheds followed by region merging and classification. In Li et al. [6], the region detection problem was treated as an image segmentation problem and a patch-wise segmentation method was proposed. Park et al. [7] employed the spatio-temporal features to detect polyps, the dependences between features in adjacent frames are taken into account and modeled by conditional random fields (CRF). In [8], the imbalanced datasets problem was overcome by adopting a data sampling-based boosting framework, and the partial least square analysis was used to enhance the ability to discriminate between polyps and non-polyps. Tajbakhsh et al. [9] proposed an automatic polyp detection method using global geometric constraints and local intensity patterns, a two-stage polyp boundary classification framework was used to refine the polyp edges which was then followed by a voting scheme for localizing polyps. This work was extended by integrating color, temporal and shape features using convolutional neural networks (CNN) to reduce the false positives [10] and further evaluations were reported in [11].

In this paper, a fully automated polyp segmentation framework is presented which combines both data driven and hand-crafted features for accurate differentiation of polyp candidates and non-polyp candidates in colonoscopy frames. The polyp segmentation is obtained using pixel-wise prediction based on fully convolutional neural networks (FCN) [12] rather than edge detection [4–6], further a novel texton based patch descriptor is proposed to refine the initial region of interests (ROIs) obtained from the FCN. The accurate boundaries of ROIs can be used for further analysis (e.g. characterization of polyp into benign and malignant).

The rest of this paper is organized as follows. We describe our method in Sect. 2. Our experimental results are presented in Sect. 3. Finally, the paper is concluded in Sect. 4.

2 Method

Our segmentation method is composed of two main stages: region proposal stage using the FCN [12] and region refinement stage using texton based patch representation followed by a random forest (RF) classifier [13]. The Fig. 1 illustrates the overall framework of our method, where the FCN provides initial polyp region candidates, and the texton based patch representation further discriminates polyp regions from non-polyp regions. Both data driven features and hand designed features are taken into account for the segmentation. More specifically, the hierarchical features of polyps are learned by FCN, while the context information related to the polyp boundaries is modeled by texton patch representation.

In our method, the intrinsic features of polyp in color space and hand-crafted feature in gray-level space are learned through two complementary frameworks. In terms of filter based features, the principle of the convolutional neural network (CNN) is similar to the texton representation that is able to decompose high order visual features to elementary visual features or so-called primitive elements (e.g. edges, corners, and endpoints). However, the CNN adopts the deep architecture while the texton uses shallow architecture. Moreover, the CNN aims to learn filter kernels automatically from raw image data, while the texton learns primitive elements by manual designing filter kernels based on the context information. In our method, these two frameworks are integrated to enable more accurate polyp segmentations.

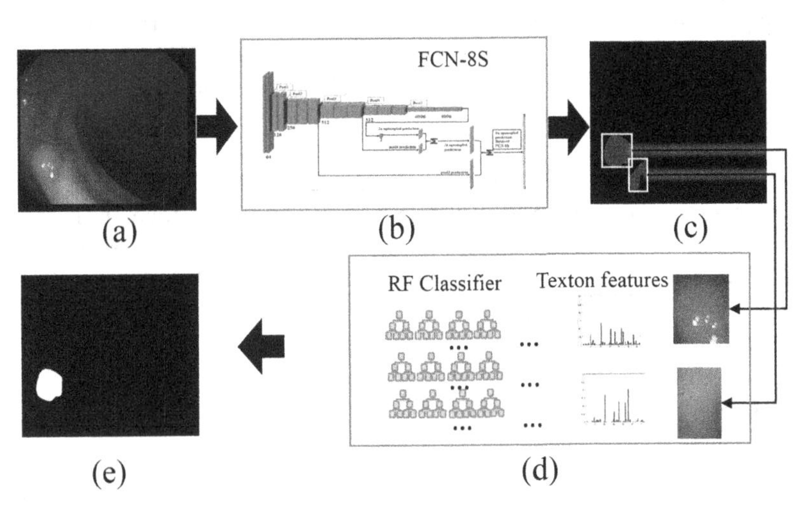

Fig. 1. The framework of our approach. (a) an example of one polyp frame; (b) region proposal network using FCN-8S; (c) initial polyp candidates from FCN-8S; (d) texton based patch representation and classification; (e) the final segmentation result.

2.1 Initial Region Proposals Using FCN

Recently, the CNNs have achieved superior performance in many computer vision tasks such as image classification, segmentation, object detection, localization and tracing etc. [14, 15]. For a regular CNN architecture, it is composed of two main functional parts: multilayer feature extractor and trainable classifier, where the former learns hierarchies of relevant features via trainable filters, activation functions and pooling operations. The activation operation could be considered as a feature selection procedure while the pooling operation is crucial to obtain spatial invariant features. The FCN [12] is one of the CNN variants that has been proven to be of state of art for semantic segmentation in natural images. It is an end-to-end training with supervised learning manner and the image segmentation is obtained by a pixel-wise prediction/ classification. A FCN is implemented by transforming the last fully connected layer of a regular CNN net (e.g. VGG16 [16]) into convolutional layers, and then adding the upsampling or de-convolutional layer to the converted CNN net. The replacement of the fully connected layer enables the net to accept image inputs with arbitrary sizes. The upsampling layer allows the output activation map to be consistent with the size level of input image that enables pixel-wise prediction.

In this paper, we adopted FCN-8s architecture in [12] as a region proposal net. Figure 2 shows FCN architecture with VGG16 (CNN classification net) [16]. The VGG16 net is composed of 5 stacks followed by 3 transformed convolutional layers, where each block contains several convolutional layers, ReLU layers and max pooling layers. The VGG16 net was employed due to the following reasons: (1) a stack of two or three convolutional layers with smaller filters (3 × 3) contains non-linearities that allow the net to learn more discriminative features. (2) The problem of the limited number of training data in polyp segmentation task can be addressed via transferring learning [17–19], namely using the learned VGG16 model which is pre-trained using abundant natural images. The FCN-8s is improved from FCN-16s skip net and FCN-32s coarse net. The skip layer is used to combine coarse predictions at deep layers and fine scale predictions at shallow layers and to improve segmentation detail. More specifically, the FCN-16s is a skip net which fuses 2 × upsampled predictions computed on the last layer at stride 32 with predictions from Pool4 at stride 16. The sum of the two predictions is then upsampled back to the image with stride 16. In the same way, in Fig. 2, the FCN-8s is implemented by fusing predictions of shallower layer (Pool3) with 2 × upsampling of the sum of two predictions derived from Pool4 and last layer. Then the stride 8 predictions are upsampled back to the image.

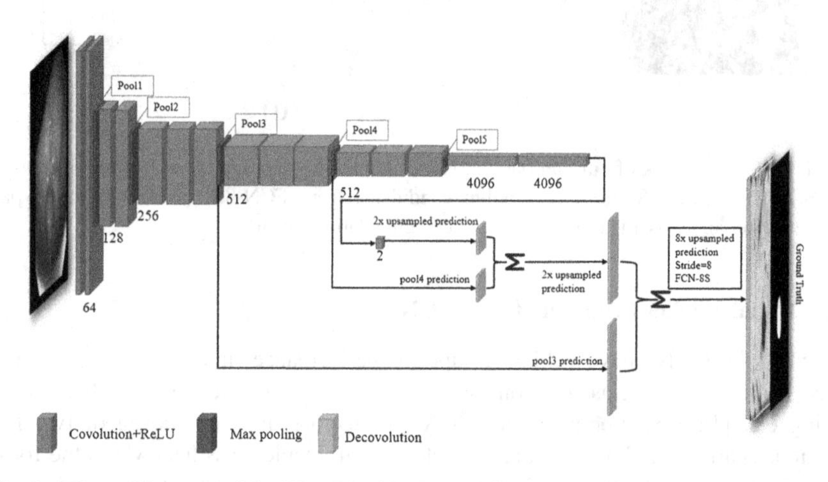

Fig. 2. The architecture of the FCN-8s with VGG16 for polyp region proposals generation.

The FCN-8s produces more detailed segmentations compared to the FCN-16s. However, some false positives (or false negatives) may present due to the lack of the spatial regularization for FCN. In the next section, we introduce a novel region refinement framework using texton based patch-wise predictions to further refine the regions obtained from FCN-8s.

2.2 Patch Based Classification Using Textons and RF

Previous studies [5, 11] have shown the importance of polyp boundary features in patch discrimination. In our method, the context information related to polyp boundaries is modeled using texton patch representation [20]. The region proposals produced by FCN- 8s (described in Sect. 2.1) are then refined by patch based classifications. The overall framework contains three stages: textons generation, patch description and random forest classification.

Texture is represented by its responses to a set of filter kernels (f_1,f_2,…f_n) [21]:

$$R = [f_1 * \mathrm{I(x,y)}, f_2 * \mathrm{I(x,y)}, \ldots, f_n * \mathrm{I(x,y)}] \quad (1)$$

Where * indicates the convolutional operation, n is the number of filter kernels. A set of feature vectors is generated by clustering filter responses in R. Those feature vectors are then used to calculate textons [21]. As a typical filter based feature extraction, it is crucial to design appropriate filter kernels to extract specific features according to given context information. Considering the prior knowledge of polyp boundaries, a two dimensional Gabor filter [22] is employed, which is defined as

$$\begin{aligned} G_{\theta,\varphi,\sigma,\gamma}(x,y) &= e^{\left(-\frac{x'^2+\gamma^2+y'^2}{2\sigma^2}\right)}\cos(2\pi f x' + \varphi) \\ \mathrm{x}' &= \mathrm{x}\cos\theta + \mathrm{y}\sin\theta \\ \mathrm{y}' &= -\mathrm{x}\sin\theta + \mathrm{y}\cos\theta \end{aligned} \quad (2)$$

where γ indicates the spatial aspect ratio of the kernel e.g. if $\gamma = 1$ the kernel is circular, it works together with the orientation parameter θ which determines orientations of the rotated kernel. σ is the standard deviation of Gaussian envelope, φ is the phase of the sinusoidal wave which controls the symmetry of filter kernel and f is center frequency. Given the characteristics of polyp boundaries, we design the Gabor filter kernel which is more likely to be an edge detector, with $\sigma = 2$, $\gamma = 0.5$, $\varphi = \pi$ and $\theta = [0,15,30,45,60,75,90,105,120,135,150,175]$ considering polyp may rotate in any orientations. The size of kernel is 21 by 21 that enables the local spatial information to be considered. The 12 filter kernels are shown in the first row (second column) in the Fig. 3. In addition, we employ standard Gaussian at scale $\sigma = 1$ to exstract non-boundary structures, this also imposes a simple smoothness constrain for feature representations. The textons are generated by employing a *k-means* clustering algorithm on the filter responses, which are then aggregated based on the distances calculated between membership and clustering centers. The *k-means* clustering enables those pixels with similar filter responses to cluster into the same group. Depending on the number of k (i.e. $k = 32$), we are able to not only discriminate boundaries from non-boundaries pixels, but also be able to distinguish strong boundary pixels and weak boundary pixels. The textons generation is only implemented in the training stage. Two sets of patches (polyp and non-polyp) were prepared using ground truths. In the training stage, a 2-dimensional Gabor filter bank with 12 orientations defined in Eq. 2 are applied to every patch in each set, the filter responses derived from all patches in the same set are clustered to generate textons. As a result, there are $k * c$ number of textons, where the k is the number of cluster centroids in the *k-means* and c is the number of classes (i.e. $c = 2$, polyp and non-polyp). All trained textons are stored into a dictionary (D) which will be used for patch description.

Once the textons have been generated, image patch can be represented using texton histogram. More specifically, given a patch, it first convolves with the Gabor filter bank to produce filter responses, and then each pixel in the patch is assigned to one of the texton labels l_i ($l_i \in D, \forall i = (1,2,3\ldots,k * c)$) from the texton dictionary (D) based on the minimum distance between the texton and the filter responses at the pixel. Through this process, for the given patch, a texton label map is generated. A histogram calculated from the texton label map is used to represent the patch. Examples of texton histograms of two patches (polyp and non-polyp) are shown in the top right corner in the Fig. 3. In the training stage, each patch in the training set is represented by texon histogram, and then all these histograms are fed into a RF classifier for patch classification. As a result, the training stage produces a textons dictionary and a trained RF classifier model which are then used in the refinement stage of polyp segmentation.

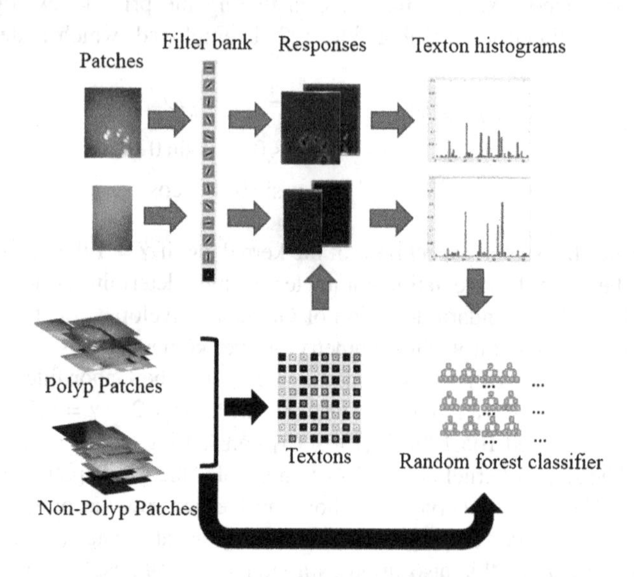

Fig. 3. The framework of patch based classification using textons in the refinement stage of polyp segmentation.

During the refinement stage, image patches are generated from initial region (polyp) candidate obtained from the FCN -8s, Each patch is represented by a texton histogram from the texton label map using the same fashion described above (see top flowchart in Fig. 3). The texton histogram is used to classify the patch into polyp or non-polyp patch using the trained random forest model. The region candidates which are classified as non-polyp candidates are removed from the initial region candidates. Finally, the refined segmentations are further improved by morphological operations.

3 Experiments

3.1 Materials and Evaluation Metrics

A publicly available CVC-ColonDB database [5] is used in our experiment which contains 300 colonoscopy images with size of 500 × 574 pixels. Each image contains one polyp. Manual delineations of polyp boundary (ground truth images) are also provided which polyp and non-polyp regions are labeled by 1 and 0 in the ground truth images. We split the dataset into training and testing sets. 200 images are randomly selected as training samples, and the rest of 100 images are used to evaluate our method. In order to train region refinement framework (i.e. texton dictionary generation and the trained RF classifier as discussed in Sect. 2.2), two patch sets (polyp and non-polyp) are produced from the training set guided by the ground truth images.

A set of standard segmentation evaluation metrics is employed to evaluate our method. These region based evaluation metrics include accuracy, sensitivity, specificity, and Dice Similarity Coefficient [23]. The results per image are averaged to obtain the overall automatic segmentation performance compared with ground truth images.

3.2 Training Protocol

In our experiment, the FCN-8s was trained with two classes of polyp and non-polyp (background) given by the ground truth images. We trained the FCN-8s using MatConvNet [24], which is commonly used in deep learning framework. It was trained by stochastic gradient descent with momentum of 0.9, a batch size of 20 images and leaning rate with 0.0001. The FCN-8s model was initialized by the pre-trained VGG16 model.

For region refinement framework, we empirically chose the number of textons in each class as k = 32. Therefore, there are a total of 64 textons in the dictionary. The number of trees in random forest is 500. Given the balanced patch training sets, we set the cutoff parameter to be 0.5.

3.3 Evaluation Results

The proposed automatic polyp segmentation method is evaluated on the testing set. All individual image evaluations are averaged and presented in Table 1. We can observe that on average, our method reaches 97.54% accuracy, 75.66% sensitivity, 98.81% specificity and DICE of 0.70. As a comparison, the polyp segmentation using FCN-8s only was also implemented and evaluated on the same dataset which are reported in the second row of Table 1. We can see our method (FCN-8s+Texton representation) has better segmentation performances in all metrics than that using the FCN-8s only. We ascribe this to the significant reduction of the false positive pixels of segmentations using our proposed method. We can visually observe these differences in Fig. 4. The false positive reduction is an important step in polyp detection scenario.

We also compare our final segmentation results to the other segmentation method (i.e. Bernal's method [5]). As we can see in Table 1, our method significantly outperforms to the Bernal's method in terms of sensitivity and DICE, with 75.66% of sensitivity and 70.14% of DICE against 61.91% and 55.33%, respectively, which shows our segmentations can detect more true positive regions with less false positive regions compared to Bernal's segmentation results.

In addition, we summarized the execution time of each method in the last row of Table 1. It can be seen that both our method (FCN-8s+Texton representation) and FCN-8s only are very fast (e.g. less than 0.2s/frame). Adding the second step of texton based refinement to FCN-8s, the total computation time only adds 0.06 s/frame. The fast and accurate segmentation of colonic polyp demonstrates the promise of our method to be used in real-time clinical colonoscopic examination.

Table 1 also shows that our method is much faster than Bernal's method. This is because the watershed segmentation used in Bernal's method needs to calculate the gradients in the whole image which is relatively time consuming. Meanwhile, the method applied pre-processing step to remove the specular reflection. While, in our method, we consider all information in the image including the light reflections as cues for polyp segmentation, without pre-processing step.

Table 1. Evaluation results of our method for polyp segmentation in colonoscopy frames

Metrics	FCN-8s+Texton	FCN-8s	Bernal et al. [5]
Acc.(%)	97.54	97.27	–
Sen.(%)	75.66	73.89	61.91%
Spe.(%)	98.81	98.75	–
DICE(%)	70.14	69.39	55.33%
Speed (sec/frame)	0.16	0.10	19

Figure 4 provides comparative examples of automatic segmentations, where column (b) illustrates the ground truth images, and (c) and (d) are the segmentation results of FCN-8s and FCN-8s+texton. Visually, we can observe the segmentations of FCN-8s+texton have less false positive regions compared to the segmentations of FCN-8s only. This indicates the regions refinement framework has good performance in discriminating between polyp and non-polyp patches. For example, the case in the second row of Fig. 4 has several structures such as fold, light reflection and vessel which can influence the segmentation. Especially, the mixture of light reflection and concave region with shadow may have similar appearances to the polyp that leads to the prediction errors in FCN-8s. We can see false positive region in the segmentation result of FCN-8s (column c in second row). This is due to two similar structures presented in the image (column a in second row), both of them look like polyps in terms of shapes. While, using our FCN-8s+texton method, more accurate polyp-specific information can be extracted through texton based representation in the refinement step, the false positive region can then be correctly removed, leading to more accurate segmentation of colonic polyp.

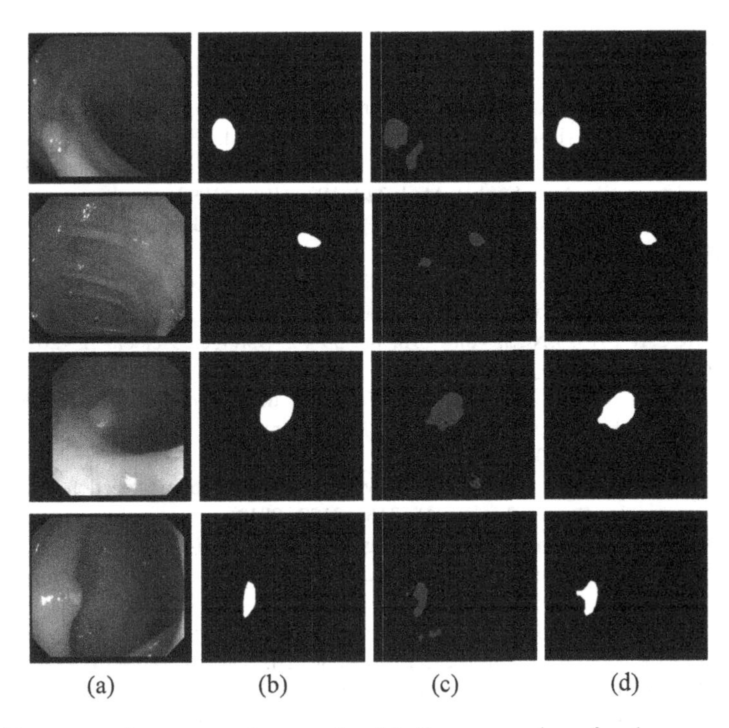

Fig. 4. The automatic segmentation results. (a) three examples of colonoscopy frames. (b) Corresponding ground truth images. The segmentation results of FCN-8s are shown in the (c), and (d) are the segmentations of our method (FCN-8s+textons).

4 Conclusion

In this paper, we present a novel fully automated polyp segmentation method from colonoscopy frames. The method contains two main stages: initial region proposals generation and regions refinement. The intrinsic features of polyp in color space and hand-crafted feature in gray-level space are learned through two complementary frameworks. Both machine learned features and hand crafted texton features are taken into account for accurate polyp segmentation. More specifically, the hierarchical features of polyps are learned by FCN-8s, while the context information related to the polyp boundaries is modeled by texton patch representation. The segmentation evaluation results show that on average, there are 97.54% of accuracy, 75.66% of sensitivity, 98.81% of specificity and DICE of 0.70%. The fast execution time (0.16s/frame) demonstrates the promise of our method to be used in real-time clinical colonoscopic examination.

Acknowledgment. This research was supported by Cancer Research UK (CRUK) funded project "Bowels - inside out" (A22873).

References

1. Winawer, S.J., Zauber, A.G., Ho, M.N., Obrien, M.J., Gottlieb, L.S., Sternberg, S.S., Waye, J.D., Schapiro, M., Bond, J.H., Panish, J.F., Ackroyd, F., Shike, M., Kurtz, R.C., Hornsbylewis, L., Gerdes, H., Stewart, E.T.: Prevention of colorectal-cancer by colonoscopic polypectomy. New Engl. J. Med. **329**, 1977–1981 (1993)
2. Lieberman, D.: Quality and colonoscopy: a new imperative. Gastrointest. Endosc. **61**, 392–394 (2005)
3. Tajbakhsh, N., Gurudu, S.R., Liang, J.: A classification-enhanced vote accumulation scheme for detecting colonic polyps. In: Yoshida, H., Warfield, S., Vannier, M.W. (eds.) ABD-MICCAI 2013. LNCS, vol. 8198, pp. 53–62. Springer, Heidelberg (2013). doi:10.1007/978-3-642-41083-3_7
4. Hwang, S., Oh, J., Tavanapong, W., Wong, J., de Groen, P.C.: Polyp detection in colonoscopy video using elliptical shape feature. In: IEEE International Conference on Image Processing, pp. 1029–1032 (2007)
5. Bernal, J., Sanchez, J., Vilarino, F.: Towards automatic polyp detection with a polyp appearance model. Pattern Recogn. **45**, 3166–3182 (2012)
6. Li, P., Chan, K.L., Krishnan, S.M.: Learning a multi-size patch-based hybrid kernel machine ensemble for abnormal region detection in colonoscopic images. In: Proceedings of CVPR, pp. 670–675. IEEE (2005)
7. Park, S.Y., Sargent, D., Spofford, I., Vosburgh, K.G., A-Rahim, Y.: A colon video analysis framework for polyp detection. IEEE Trans. Bio-Med. Eng. **59**, 1408–1418 (2012)
8. Bae, S.H., Yoon, K.J.: Polyp detection via imbalanced learning and discriminative feature learning. IEEE Trans. Bio-Med. Imaging **34**, 2379–2393 (2015)
9. Tajbakhsh, N., Gurudu, S.R., Liang, J.: Automatic polyp detection using global geometric constraints and local intensity variation patterns. In: Golland, P., Hata, N., Barillot, C., Hornegger, J., Howe, R. (eds.) MICCAI 2014. LNCS, vol. 8674, pp. 179–187. Springer, Cham (2014). doi:10.1007/978-3-319-10470-6_23
10. Tajbakhsh, N., Gurudu, S.R., Liang, J.M.: Automatic polyp detection in colonoscopy videos using an ensemble of convolutional neural networks. In: International symposium on Biomed Imaging, pp. 79–83 (2015)
11. Tajbakhsh, N., Gurudu, S.R., Liang, J.M.: Automated polyp detection in colonoscopy videos using shape and context information. IEEE Trans. Med. Imaging **35**, 630–644 (2016)
12. Long, J., Shelhamer, E., Darrell, T.: Fully convolutional networks for semantic segmentation. In: 2015 IEEE Conference on Computer Vision and Pattern Recognition (CVPR), pp. 3431–3440 (2015)
13. Breiman, L.: Random forests. Mach. Learn. **45**, 5–32 (2001)
14. LeCun, Y., Bengio, Y., Hinton, G.: Deep learning. Nature **521**, 436–444 (2015)
15. LeCun, Y., Kavukcuoglu, K., Farabet, C.: Convolutional networks and applications in vision. In: Ieee International Symposium on Circuits and systems, pp. 253–256 (2010)
16. Simonyan, K., Zisserman, A.: Very deep convolutional networks for large-scale image recognition. In: CORR abs/1409.1556 (2014)
17. Tajbakhsh, N., Shin, J.Y., Gurudu, S.R., Hurst, R.T., Kendall, C.B., Gotway, M.B., Jianming, L.: Convolutional neural networks for medical image analysis: full training or fine tuning? IEEE Trans. Med. Imaging **35**, 1299–1312 (2016)
18. Zeiler, M.D., Fergus, R.: Visualizing and understanding convolutional networks. In: Fleet, D., Pajdla, T., Schiele, B., Tuytelaars, T. (eds.) ECCV 2014. LNCS, vol. 8689, pp. 818–833. Springer, Cham (2014). doi:10.1007/978-3-319-10590-1_53

19. Girshick, R., Donahue, J., Darrell, T., Malik, J.: Rich feature hierarchies for accurate object detection and semantic segmentation. In: 2014 IEEE Conference on Computer Vision and Pattern Recognition (CVPR), pp. 580–587 (2014)
20. Varma, M., Zisserman, A.: A statistical approach to texture classification from single images. Int. J. Comput. Vis. **62**, 61–81 (2005)
21. Leung, T., Malik, J.: Representing and recognizing the visual appearance of materials using three-dimensional textons. Int. J. Comput. Vis. **43**, 29–44 (2001)
22. Daugman, J.G.: Uncertainty relation for resolution in space, spatial-frequency, and orientation optimized by two-dimensional visual cortical filters. J. Opt. Soc. Am. **2**, 1160–1169 (1985)
23. Dice, L.R.: Measures of the amount of ecologic association between species. Ecology **26**, 297–302 (1945)
24. Vedaldi, A., Lenc, K.: MatConvNet convolutional neural networks for MATLAB. In: MM 2015: Proceedings of the 2015 Acm Multimedia Conference, pp. 689–692 (2015)

Model-Based Correction of Segmentation Errors in Digitised Histological Images

David A. Randell[1(✉)], Antony Galton[2], Shereen Fouad[1], Hisham Mehanna[3], and Gabriel Landini[1]

[1] School of Dentistry, College of Medical and Dental Sciences, University of Birmingham, Birmingham, UK
{d.a.randell,s.a.fouad,g.landini}@bham.ac.uk
[2] Department of Computer Science, University of Exeter, Exeter, UK
apgalton@ex.ac.uk
[3] Institute of Cancer and Genomic Sciences, University of Birmingham, Birmingham, UK
h.mehanna@bham.ac.uk

Abstract. This paper describes an application of topological, model-based methods for the algorithmic correction of segmentation errors in digitised histological images. The topological analysis is provided by the spatial logic Discrete Mereotopology and integrates qualitative spatial reasoning and constraint satisfaction methods with classical image processing methods. A set of eight topological relations defined on binary segmented regions are factored out and reworked as nodes of a set of directed graphs. The graphs encode and constrain a set of set-theoretic and topological segmentation operations on regions, so that the interpreted images and any proposed changes made to the regions can be made to conform to a valid histological model. Worked examples are given using images of H&E stained H400 cell line cultures.

Keywords: Mereotopology · Graph theory · Histological image processing

1 Introduction

This paper describes an application of model-based methods for the algorithmic correction of segmentation errors in digitised histological images. This is a real-world application where qualitative spatial reasoning (QSR) and constraint-satisfaction programming methods have been integrated with classical image processing methods to develop context-based histological imaging algorithms.[1] The context here arises from: (i) making an ontological stand whereby regions rather than pixels in digitised images are deemed to be the main carriers of

[1] See our project page at http://www.mecourse.com/landinig/software/intellimic.html.

M. Valdés Hernández and V. González-Castro (Eds.): MIUA 2017, CCIS 723, pp. 718–730, 2017.
DOI: 10.1007/978-3-319-60964-5_63

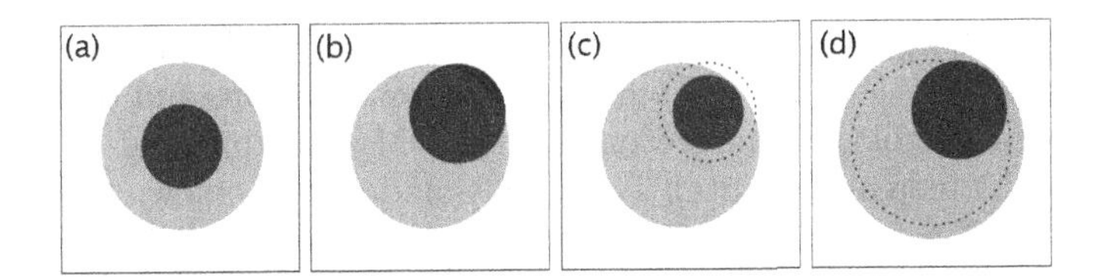

Fig. 1. Idealised image of an H&E-stained nucleated cell, with pink-stained cytoplasm and blue-stained nucleus; overlap between cytoplasm and nucleus is shown in magenta. In (a) the nucleus forms a proper part of the cytoplasm, as expected; (b) shows an anomalous image in which the nucleus partially overlaps the cytoplasm; (c) and (d) show possible morphological corrections of (b), with dotted lines marking the original extent of nucleus and cytoplasm respectively. (Colour figure online)

histological content, and (ii) highlighting the importance of and explicitly representing topological (and in particular relational) information encoded in digitised histological images. The topological analysis and constraints are provided by the spatial logic Discrete Meterotopology (DM) [1,2], which we use to augment the operations of classical Mathematical Morphology (MM) by explicitly encoding a set of binary relations such as contact, overlap and the part-whole relation on pairs of regions. These mereotopological relations are used both to model the domain and to guide algorithms for correcting segmentation results so that they conform to the requirements for a valid histological model.

In three-dimensional reality a cell consists of a nucleus surrounded by a body of cytoplasm; the two are in contact but do not overlap, the nucleus exactly filling a cavity within the cytoplasm. In a correctly-formed H&E (haematoxylin and eosin)-stained two-dimensional image, on the other hand, based on the differential staining of cellular components, the cytoplasm region appears as a simply connected whole, lacking a cavity, and the nucleus forms a proper part of this (see the idealised image in Fig. 1(a)). In practice, however, cell and nuclear segmentations are most often achieved independently of each other, which can result in imperfect relations, where, e.g., the nucleus partially overlaps its cytoplasm, as in Fig. 1(b). Such errors can often be corrected by a process of *resegmentation*, whereby one or both of the cell component images are manipulated using morphological or other operators so that the expected spatial relation between them holds. In Fig. 1(c) this is accomplished by eroding the nucleus; in (d), by dilating the cytoplasm. If the changes required to achieve this are too extreme, then the presence of an imaging artefact or other anomaly may be indicated.

Our approach to resegmentation uses a set of directed graphs defined on region-region relations. Each graph details transformations to the relations arising from application of specified changes to their relata. These changes use discrete topological and set-theoretic operators defined on regions and mapped to their MM equivalents. The graphs enable transitions between relations (e.g., from partial overlap to proper part) to be implemented as sequences of operations on the regions themselves. This correspondence between relations and operations underpins the model-based correction of segmentation described in this paper.

This paper builds on a series of papers that apply DM to the interpretation and segmentation of histological images [3–6]. While the mereotopological interpretation of digitised images is not new (see e.g., [7]), what is novel is its role in enabling systematic context-based manipulation of regions and their relations in quantitative histological image processing. DM first appeared in [1], which arose out of the mereotopological theory RCC and its well-known eight-element relation set RCC8 [8]. The GRCC (Generalised Region Connection Calculus) [9] offers another way to model discrete domains using mereotopology. As a formalism for modelling cellular processes, mereotopology was used to model phagocytosis and exocytosis in [10]. A motivating example of DM used in a constraint-based graph traversal problem is given in [11]. In general, though, while RCC8 has been used extensively in the development of efficient constraint-based algorithms (for deciding the consistency of a set of constraints) the algorithms are typically restricted to operations on symbolic state-state models and not grounded in interpreted segmented digitised images as done here. For a good introduction to QSR literature and its methods, see [12].

This paper extends the framework that was originally reported in [6]. Specifically, the four directed graphs originally described in [6] have been extended to include a set of eight set-theoretical as well as topological operations on regions. The method is generic and does not favour any one particular image segmentation method over another. All that is assumed is that the segmentation of images into regions forming relations with other regions can be ultimately represented as binary regions, e.g. as masks. These relations can be used to test whether the segmented regions conform to a histological model, or not. Where they fail to conform to the segmentation model they may be either be rejected completely, or corrected by applying one of several operations on the regions in question. For this reason, we do not necessarily require a set of gold-standard segmented images with which to compare and analyse the results (though of course this could be done). We discuss this further below. In its place we consider the set of topological segmentation solutions satisfying our assumed histological model and work with those. Reasons of space and the scope of this paper mean we cannot adequately compare and contrast other segmentation methods and the various validation methods used in the literature here, though we acknowledge there is a large body of literature on the subject, e.g. [13,14]. However, none of these frameworks adopt the topological framework assumed here, nor do they address systematic methods of resegmentation to bring the results of initial segmentation into line with the expectations of histological theory, which is the main focus of the present paper.

2 Discrete Mereotopology (DM)

Of the different variants of DM that have been proposed, the version we adopt is the one described in [1,6]. The domain is a set of possibly empty regions which are defined as sets of pixels. We denote regions by lower-case letters $(x, y, \ldots)$; individual pixels are denoted $\hat{x}, \hat{y}, \ldots$. Set inclusion is defined in the standard way, the part-whole relation (P) defined as inclusion restricted to non-null regions, and overlap (O) between regions restricted to regions sharing a part in common.

Adjacency between pixels is axiomatised as a reflexive, symmetric relation. This is used to define the *contact* relation (C) between regions. The set of all pixels is defined as the universal region u and the null set as $\emptyset$. Singleton pixel-sized regions are treated as atoms. The *neighbourhood* $N(\hat{x})$ of pixel $\hat{x}$ is defined as the set of pixels in u that are adjacent to $\hat{x}$. In our application domain u is a rectangular pixel array, and $N(\hat{x})$ a 3×3 pixel array centred on $\hat{x}$. Implemented in MM, the function $N(\hat{x})$ maps to a morphological 3×3-pixel, 8-connected *structuring element* which we now assume by default.

As is common practice in QSR when developing these spatial logics and algebras, subsets of dyadic relations forming *jointly exhaustive and pairwise disjoint* (JEPD) sets are singled out. In DM two JEPD relation sets are used: first, the eight-element set RCC8 comprising DC, EC, PO, TPP, NTPP, TPPi, NTPPi and EQ, respectively disconnected, externally connected, partial overlap, tangential proper part, non-tangential proper part, with their inverses, and equals, and second, the five-element relation set RCC5 comprising the relations DR (= DC|EC), PO, PP (= TPP|NTPP), PPi (= TPPi|NTPPi), and EQ, respectively disjoint, partial overlap, proper part, and inverse proper part. Here the symbol '|' signifies disjunction, e.g., $\mathsf{DR}(x, y) \leftrightarrow \mathsf{DC}(x, y) \vee \mathsf{EC}(x, y)$. These respectively form the eight and five base relations of two relational subsumption lattices with top and bottom elements interpreted as the universal and null dyadic relations. Given that RCC8 is predicated on a continuous embedding space and DM on a discrete one, the relations sharing the same name are strictly not identical. For this reason, DM's JEPD relation sets are identified in the text by the suffix 'D' as in 'RCC8D'.

In [6] the discrete *interior* (int_D) and *closure* (cl_D) are pseudo-topological operators defined on regions; they share some but not all of the usual properties of the interior and closure operators in standard treatments of topology. A map between these operators and the MM operations *erosion* and *dilation* [5] enables one to define a notion of approximate equality that underpins transitions encoded in DM's conceptual neighbourhood graphs, and it also enables the RCC8D relation set to be easily implemented in any image-processing programs featuring standard MM libraries. Other properties of regions can be defined in DM, e.g., regions with or without an interior, regular regions (i.e., those without pixel-wide spikes and fissures), self-connected regions, connected components.

In the histological domain, cells and their parts, groups of cells forming tissues and compartments, and the background of a digitised histological preparation can all be modelled using DM. Simple regions and arbitrary sets of pixels forming regions which may or may not be spatially contiguous all yield potential models. If a histological preparation is thresholded as a single binary image, then segmented regions of interest will form connected components, so that in any one image, the only possible relations between pairs of regions are DC and EQ. The methods presented here, however, assume that two or more independent imaging or segmentation modalities are used, e.g., separating out the contribution of the different dyes in stained sections (as in Fig. 1) or confocal microscopy channels. In this case, pairs of regions, segmented from each channel, can be compared, all RCC5D and RCC8D relations being now possible.

3 Conceptual Neighbourhood Graphs, Continuity and Change, and Composition Tables

In [1,6] a set of *conceptual neighbourhood diagrams* (CNDs), or graphs, were defined on dyadic relations. In RCC8, relation R' is a conceptual neighbour of relation R if some pair of regions related by R can be continuously deformed so that R changes to R' with no other relation holding during that deformation. In the discrete setting of DM, continuous deformation is recast in terms of *minimal change*. In [1], this was defined using the discrete interior (int_D) and closure (cl_D) operators; here we extend this to include changes to regions produced using the set-theoretic operators union (sum), intersection (prod) and difference (diff) on region pairs. The universal region u, representing the image, is assumed to be self-connected, i.e. $\mathsf{SC(u)}$.

In general, an *RCC8D conceptual neighbourhood* of a binary relation R can be defined as

$$\mathsf{nbhd}_{\langle\alpha,\beta\rangle}(R) = \{R' \in \mathrm{RCC8D} \mid \exists x, y(R(x,y) \wedge R'(\alpha,\beta))\}, \tag{1}$$

where α and β are designated functions of the region variables x, y. The elements of $\mathsf{nbhd}_{\langle\alpha,\beta\rangle}(R)$ are called the $\langle\alpha,\beta\rangle$-neighbours of R. They are all the possible relations that can hold after regions x and y are modified in accordance with α, β. Given a segmented image, by a *resegmentation* we understand the replacement of a set S of regions in the image by a new set S' defined from S using some sequence of conceptual neighbourhood transitions. Such a resegmentation is chosen in order to correct anomalous relations in the original segmentation so that it satisfies the constraints of the domain being modelled.

Figure 2 shows a set of graphs that encode the conceptual neighbourhood relations defined in terms of the operators $\mathsf{sum}, \mathsf{prod}, \mathsf{diff}, \mathsf{int}_D$, and cl_D. In the first graph, for example, showing

$$\mathsf{nbhd}_{\langle x,\mathsf{sum}(x,y)\rangle}(R) = \{R' \in \mathrm{RCC8D} \mid \exists x, y(R(x,y) \wedge R'(x,\mathsf{sum}(x,y))\}, \tag{2}$$

the arrow from DC to TPP represents the case $\mathsf{nbhd}_{\langle x,\mathsf{sum}(x,y)\rangle}\mathsf{DC} = \{\mathsf{TPP}\}$, meaning that if $\mathsf{DC}(x,y)$ holds then we must have $\mathsf{TPP}(x,\mathsf{sum}(x,y))$. In this case only one resegmentation exists; other cases, such as $\mathsf{nbhd}_{\langle x,\mathsf{sum}(x,y)\rangle}\mathsf{PO} = \{\mathsf{TPP}, \mathsf{NTPP}\}$, may allow more than one. Loops in the CND indicate where a change to a region of a designated pair does not necessarily result in a corresponding change of relation. Isolated nodes or nodes without outgoing edges arise where an operator returns null (e.g., $\mathsf{diff}(x,y)$ where $\mathsf{PP}(x,y)$).

For graph n, the outgoing edges from the vertex labelled with relation R are designated with the mnemonic n_R; for example, in the case of the pair of outgoing edges from PO to TPP and NTPP in graph 1, this is notated as 1_{PO}. Note that four of the RCC8D relations are *self-inverse*, i.e., $R(x,y)$ implies $R(y,x)$; these are DC, EC, PO, and EQ. The other four relations form two mutually inverse pairs: $\mathsf{TPPi}(x,y)$ if and only if $\mathsf{TPP}(y,x)$, and $\mathsf{NTPPi}(x,y)$ if and only if $\mathsf{NTPP}(y,x)$. These inverse relations will sometimes be exploited in our reasoning. In graph 4, for example, we see that $\mathsf{PO}(x,y)$ implies $\mathsf{TPPi}(y,\mathsf{prod}(x,y))$;

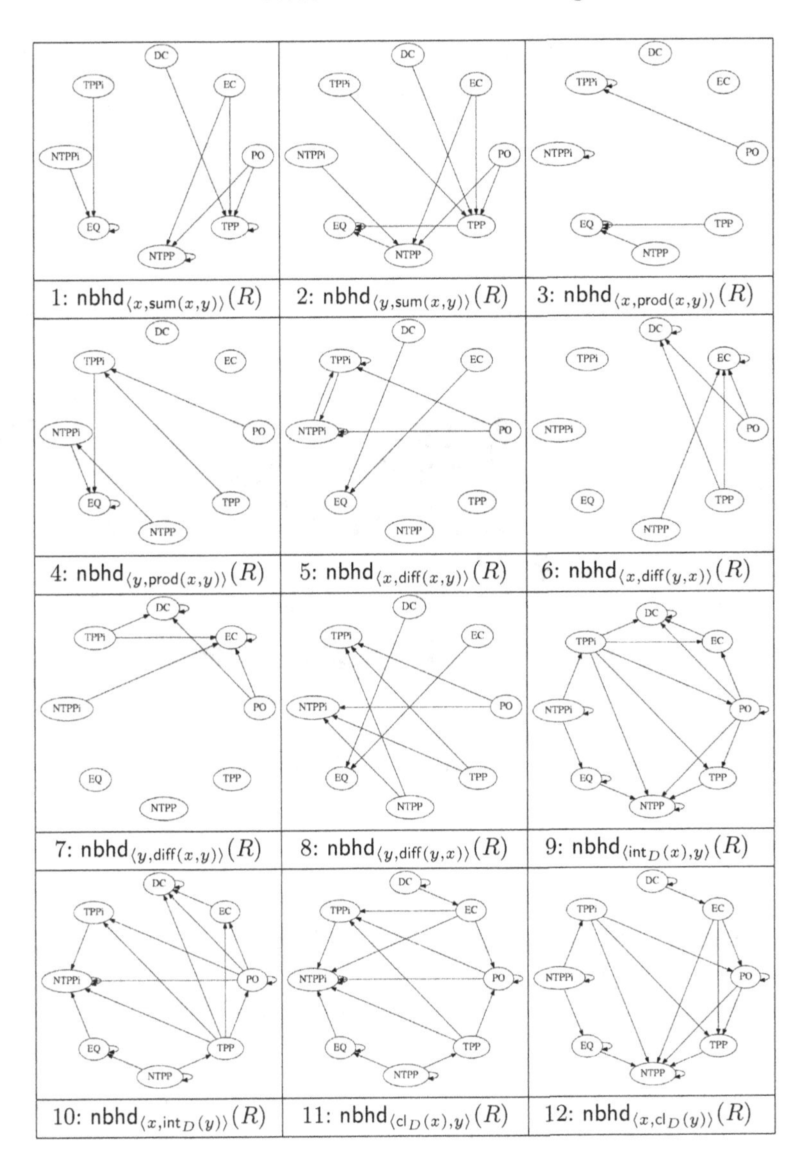

Fig. 2. Directed graphs encoding set theoretic and discrete topological operators. In each case the regions and the resulting operation on them are non-null.

sometimes we will find it more convenient to rewrite this as $\mathsf{TPP}(\mathsf{prod}(x,y),y)$, in which case we would cite the graph operation as $4'_{\mathsf{PO}}$. An example of this is seen later in Table 1.

We also use RCC8D's composition table (RCC8D-CT). The notion of composition is well-known in AI as it provides an efficient inference mechanism for many QSR constraint satisfaction programs, where it is typically implemented as a simple look-up table. Following [15], weak composition of DM's JEPD relation sets is defined as follows. Given relation set Σ, the weak composition RCC8D-CT(R,S), where $R, S \in \Sigma$, is defined to be the smallest subset $\{T_i\} \subseteq \Sigma$ such that DM $\models \forall x, y, z((R(x,y) \wedge S(y,z)) \rightarrow T_1(x,z) \vee \cdots \vee T_n(x,z))$. The elements of RCC8D-CT defined on non-null regions is identical to RCC8's composition table entailed by RCC. This was mechanically proved using the sorted theorem prover SPASS [16] to verify that all entailments of the above form were included in the composition table; and constructing a set of graphical models satisfying each $T_i(x,z)$ disjunct. The same method was used to verify the sets of directed edges of the graphs depicted in Fig. 2.

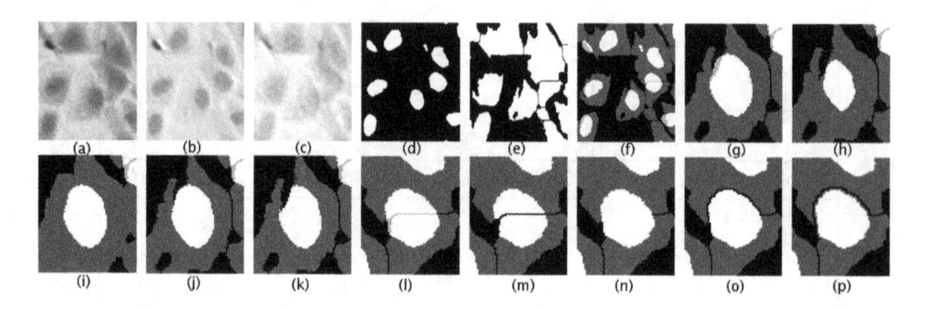

Fig. 3. (a) Original RGB image; (b) Haematoxylin and (c) Eosin channels; binary segmented (d) nuclei and (e) cytoplasm; (f) RGB colour merge: magenta = cytoplasm, green = nucleus, white = cytoplasm/nucleus overlap; (g,l) cropped details of (f); (h–k, m–p) resegmentations of g,l respectively, satisfying PP(nucleus,cytoplasm). See text for explanation. (Colour figure online)

4 Example: Segmenting Cells in Culture

In Fig. 3, image (a) depicts an H&E-stained culture of H400 cells grown on glass. Various image pre-processing operations are done. First a Gaussian filter (kernel radius 2) is applied to the original image to remove noise and reduce the fragmentation artefacts near the region boundaries. Next, colour deconvolution [17] is used to unmix the dye contributions and identify cell nuclei (H-stain, image (b)) from the rest of the cell bodies (E-stain, image (c)). Several standard image processing operations then follow (k-means clustering on the H,E stain images using 3 clusters, Boolean compositions of thresholded clusters, binary watershed separation), which are used to generate the two binary images of cell nuclei and their associated cells (images (d) and (e) respectively). The colour composite merge depicted in (f) illustrates the extent of conformity to the assumed histological constraints; binary segmented nuclei (d) are mapped to green, cytoplasm (e) to magenta, and overlap between the two to white. Where a nucleus forms a proper part (PP) of its cytoplasm, the latter appears as magenta surrounding a white nucleus; in less common cases where EQ holds, the whole cell appears white. First we test the RCC8D relation between cell nuclei and cytoplasm, where each typed set of spatially disjoint regions is treated as a mereological whole (regions n and c respectively).[2] In the case illustrated, we obtain $\mathsf{PO}(n, c)$, as indicated by the presence of green regions in image (f). As this fails the test of conformity (which require nuclei to fall within their associated cytoplasm) the task is to repair the segmentation. In (f) seven candidate nuclei partially overlap (PO) cytoplasm regions and six form proper parts (PP). But these include several small 'slivers' adhering closely to the image boundary. Typically, regions bordering the edge of the image frame (so their true extent is not known) are removed from the analysis.

Using the directed graphs, we look for resegmentation operations on candidate nucleus/cytoplasm pairs that take us from PO to PP. Consider the enlarged detail shown in Fig. 3(g), where candidate nucleus n is PO to cytoplasm component c. One possible solution (image (h)) successively erodes n (i.e., replaces it by its discrete interior) until it becomes PP to c; this requires three successive erosions. Another (image (i)) replaces c by its discrete closure (dilation) until the same result is achieved. In (j) we extend c to cover all of n so that once again the nucleus is PP to its cytoplasm. In (k) we achieve the same result by subtracting from n the part that lies outside c.

Image (l) shows a second enlarged detail of (f), in which another candidate nucleus n' partially overlaps *two* cytoplasm components c_1 and c_2. One correction (image (m)) splits the nucleus into two by taking its intersections with c_1 and c_2. Each nuclear component is now PP to one cytoplasm component. In (n), the lower cytoplasm component (c_2) is extended to cover the whole of n'. This, though, has the effect of merging two cytoplasm components; to compensate for this, the upper component (c_1) is reduced by subtracting from it the closure of n' (image (o)). Finally, in (p), the nucleus is completely surrounded by cytoplasm; this is achieved by extending c_2 to cover not just n' but its closure; the compensatory reduction of c_1 must now subtract the closure of the closure of n' in order to ensure complete separation from the extended c_2.

These resegmentations, associated graphs and inferences used to generate them are summarised in Table 1.

In detail, the steps for c_1 in (p) are as follows:

1. Start with $\mathsf{PO}(n', c_1)$.
2. By 11_{PO} this gives $\mathsf{PO|NTPPi|TPPi}(\mathsf{cl}_D(n'), c_1)$.
3. By $1_{\mathsf{PO|NTPPI|TPPI}}$ this gives $\mathsf{EQ|NTPP|TPP}(\mathsf{cl}_D(n'), \mathsf{sum}(\mathsf{cl}_D(n'), c_1))$.
4. Next, from $\mathsf{EQ}(n', n')$ and 12_{EQ} we have $\mathsf{EQ|NTPP}(n', \mathsf{cl}_D(n'))$.
5. The RCC8D weak composition $\mathsf{EQ|NTPP} \circ \mathsf{EQ|NTPP|TPP}$ is $\mathsf{EQ|NTPP|TPP}$.
6. Hence from 3 and 4 using 5 we have $\mathsf{EQ|NTPP|TPP}(n', \mathsf{sum}(\mathsf{cl}_D(n'), c_1))$.

In step 2, we apply graph 1 to each disjunct of PO|TPPi|NTPPi separately to generate the new disjunction EQ|TPP|NTPP in step 3: here PO and TPPi (NTPPi)

[2] The justification for this step is a potential speed-up of computation in certain cases: if $\mathsf{DC}(n, c)$ is returned, then all pairs of connected components of n and c are DC, if $\mathsf{PP}(n, c)$ then every connected component in n is part of some component of c (and similarly for PPi), and if $\mathsf{EQ}(n, c)$ then every connected component in n maps to an identical component in c. These are theorems of DM.

Table 1. Resegmentation details for Fig. 3. Here CT refers to the RCC8D composition table.

Figure	*Initial relation*	*Relation after resegmentation*	*Graph operation*
(h)	$\mathsf{PO}(n,c)$	$\mathsf{TPP}(\mathsf{int}^3_D(n),c)$	9_{PO}
(i)	$\mathsf{PO}(n,c)$	$\mathsf{TPP}(n,\mathsf{cl}^3_D(c))$	12_{PO}
(j)	$\mathsf{PO}(n,c)$	$\mathsf{TPP}(n,\mathsf{sum}(n,c))$	1_{PO}
(k)	$\mathsf{PO}(n,c)$	$\mathsf{TPP}(\mathsf{prod}(n,c),c)$	$4'_{\mathsf{PO}}$
(m)	$\mathsf{PO}(n',c_1)$	$\mathsf{TPP}(\mathsf{prod}(n',c_1),c_1)$	$4'_{\mathsf{PO}}$
	$\mathsf{PO}(n',c_2)$	$\mathsf{TPP}(\mathsf{prod}(n',c_2),c_2)$	
(n)	$\mathsf{PO}(n',c_1)$	$\mathsf{TPP}(n',\mathsf{sum}(n',c_1))$	1_{PO}
(o)	$\mathsf{PO}(n',c_1)$	$\mathsf{TPP}(n',\mathsf{sum}(n',c_1))$	1_{PO}
	$\mathsf{PO}(n',c_2)$	$\mathsf{DC}(n',\mathsf{diff}(c_2,\mathsf{cl}_D(n')))$	$11_{\mathsf{PO}}, 6_{\mathsf{PO}}, 12_{\mathsf{EQ}}, \mathrm{CT}$
(p)	$\mathsf{PO}(n',c_1)$	$\mathsf{NTPP}(n',\mathsf{sum}(\mathsf{cl}_D(n'),c_1))$	$11_{\mathsf{PO}}, 1_{\mathsf{PO}}, 12_{\mathsf{EQ}}, \mathrm{CT}$
	$\mathsf{PO}(n',c_2)$	$\mathsf{DC}(n',\mathsf{diff}(c_2,\mathsf{cl}^2_D(n'))$	$11_{\mathsf{PO}}, 6_{\mathsf{PO}}, 12_{\mathsf{EQ}}, 12_{\mathsf{NTPP}}, \mathrm{CT}$

are mapped to TPP|NTPP and EQ by 1_{PO} and 1_{TPPi} (1_{NTPPi}) respectively; the combined operation is notated $1_{\mathsf{PO|TPPi|NTPPi}}$.[3] The soundness and completeness of the inference procedures ensure not only that steps 1–4 encode the DM theorem $\mathsf{PO}(x,y) \rightarrow \mathsf{EQ|TPP|NTPP}(x,\mathsf{sum}(\mathsf{cl}_D(x),y))$, but also that the disjunctive relation EQ|TPP|NTPP is the strongest obtainable. Space limitations preclude similarly detailed analysis of the resegmentation of c_2.

5 Discussion

The 12 graphs reveal six resegmentation operations that take us *directly* from PO to PP (i.e., TPP|NTPP), hence guaranteeing the transition to PP,[4] and another four that merely *allow* that possibility.[5] The number (and complexity) of potential resegmentations increases when several graphs are combined and node-node paths through these networks of length $n > 2$ are considered, as in the segmentation operations used to generate the cell depicted in Fig. 3(n).

Strategies for selecting optimal resegmentations would be relatively straightforward if the segmented cells were widely separated from each other, but in Fig. 3 this is not so, since several cells are separated by a pixel-width distance and in (f) several nuclei overlap more than one cytoplasm component. Hence when applying 12_{PO} to (g) to generate (i), or applying 1_{PO} to (l) to generate (o), hitherto separated cytoplasm components are merged. One way to avoid this is to restrict the discrete closure operation so as to prevent merging with a neighbouring component: $\mathsf{cl}^-_D(x,y) =_{\mathrm{def}} \{\hat{x} \mid \mathsf{O}(N(\hat{x}),x) \wedge \neg\mathsf{O}(N(\hat{x}),y)\}$. This picks out those pixels whose immediate neighbourhoods overlap x but are disjoint from y, giving the largest subset of $\mathsf{cl}_D(x)$ not connected to y. It cannot be

[3] Other disjunctions are handled similarly.

[4] Namely 1_{PO}, 2_{PO}, $3'_{\mathsf{PO}}$, $4'_{\mathsf{PO}}$, $5'_{\mathsf{PO}}$, and $8'_{\mathsf{PO}}$.

[5] Namely, 9_{PO}, $10'_{\mathsf{PO}}$, $11'_{\mathsf{PO}}$, an 12_{PO}. In these cases the predicted models include PP, but PP is not entailed.

applied *carte blanche*, however, since histological domain constraints may *require* some regions (e.g., fragmented nuclei) to be merged.

Within DM proper, different classes of regions operate as filters, e.g., atoms, regions without interiors, regular regions (lacking spikes or fissures) or connected components; and all these give rise to varying constraints on the mereotopological relations defined on them. At the image scales typically used in digital microscopy, most segmented regions mapping to histological objects have interiors, and very few are atomic, but restricting these topological properties will constrain the possible segmentation models and relations that can be defined on them. Constraints other than those defined directly within DM can be also used to reduce the number of segmentation models. A simple example is where the range of sizes of histological objects can be used as a filter, either using MM granulometry methods or filtering by morphological thickness [1], enabling us to rule out the segmentation models depicted in Fig. 3(i), (n) (cell bodies too large), and (h) (cell nucleus too small).

Another extra-logical constraint that can be used exploits *empirical* information about the histological stains and their known selectivity in dye take-up with respect to targeted tissues and their parts. Given that the H-stain offers better segmentations for nuclei than the eosin counter-stain does for cytoplasm, resegmentations can be ranked so as to favour those that minimise the changes to *nuclei*. Other assumed empirical and ontological dependencies can also be exploited. For example, depending on microscope resolution, each cell nucleus should fall wholly within some cytoplasm component, whether this is segmentable from the original image or not; in DM this can be captured by adding the histological domain axiom $Nuc(x) \rightarrow \exists y(Cell(y) \wedge P(x,y))$. This constraint also justifies the assumption underlying the transition 1_{PO}, where a cytoplasm component partially overlapping a nucleus is extended to cover the nucleus so that it forms part of that cell. We can also add the correspondence between histological features and stain selectivity: given a PO relation defined on a poorly segmented nucleus and its host cytoplasm, we favour 12_{PO} (dilating the cytoplasm) over 9_{PO} (eroding the nucleus). Also worthy of note is that the boundaries of histological objects in a greyscale image exhibit intensity gradients. This means that when binary thresholding candidate regions, subsequent erosion and dilation-based resegmentations are more likely to track gray-scale intensity levels in the original image than blind set-theoretic segmentation operations on binary images. This observation highlights a limitation of the underlying method, namely, that once the segmentation mask of the target histological object is provisionally segmented, changes subsequently made and translated back to the images may not conform to all available information in the image. However, in some cases restricting oneself to information in an image may not, even in principle lead to a histological model. For example, uneven staining may fail to reveal the full extent of the cytoplasm in the sample; in which case a model-based resegmentation solution can be used to factor out those regions in an image that need to be treated differently than the rest.

It is also perhaps useful here, to give at least some indication of empirical methods and metrics envisaged for quantifying, validating and measuring our segmentation solutions. For example, given cell nuclei are easier to segment than their associated bodies of cytoplasm, these can be used as a gold-standard reference measuring how well the segmented cytoplasm conforms to the histological model. In this case a necessary condition is that segmented nuclei form part of some overlapping body of cytoplasm, so one possible measure is the number of region pairs that form a part-whole relation divided by all possible overlap cases. These simple examples show (i) that the underlying physical model should guide the abstraction and (ii) the danger of abstracting and working with generic cases too quickly, where empirical constraints restricting valid resegmentations may be missed.

6 Conclusions and Future Work

We have shown how DM provides the means to model cellular and tissue structure in digitised histological images. Segmentation and resegmentation satisfying a histological model can be achieved by a set of operations on regions that satisfy a set of constraints on pairs of regions. These constraints can be encoded as a set of graphs in which topological and set-theoretic operators lead from one vertex to another. The method is generic and can be applied to any domain where it is required to segment digitised images into regions satisfying specific sets of mereotopological relations.

Several directions for future work can be suggested. First, the set of operators and associated graphs can be extended to cover all the standard topological and set-theoretic operators, including the discrete exterior, boundary, the (absolute) complement and the symmetric difference. Second, various different metrics can be defined on the conceptual neighbourhoods and their graphs, allowing optimisation of segmentation models and prediction of the most likely path to take through the graphs from a given state to a segmentation goal. These could be based on what proportion of JEPD relations reached at each step can lead to a valid model, or probability measures determined from a statistical analysis of the data sets, taking into account *a priori* and empirically derived properties such as tissue type and morphological shape and size.

Acknowledgments. This work was supported by the EPSRC through funding under grant EP/M023869/1, "Novel context-based segmentation algorithms for intelligent microscopy".

References

1. Galton, A.: The mereotopology of discrete space. In: Freksa, C., Mark, D.M. (eds.) COSIT 1999. LNCS, vol. 1661, pp. 251–266. Springer, Heidelberg (1999). doi:10.1007/3-540-48384-5_17

2. Galton, A.: Discrete mereotopology. In: Calosi, C., Graziani, P. (eds.) Mereology and the Sciences: Parts and Wholes in the Contemporary Scientific Context, vol. 371, pp. 293–321. Springer, Cham (2014). doi:10.1007/978-3-319-05356-1_11
3. Landini, G., Randell, D.A., Galton, A.: Discrete mereotopology in histological imaging. In: Claridge, E., Palmer, A.D., Pitkeathly, W.T.E. (eds.) Proceedings of the 17th Conference on Medical Image Understanding and Analysis, pp. 101–106 (2013). ISBN 1-901725-48-0
4. Landini, G., Randell, D.: The complexity of cellular neighbourhoods. Fifth International Symposium on Fractals in Biology and Medicine, Locarno, Switzerland (2008). Riv. di Biol./Biol. Forum, **101**(1), 129–158 (2008)
5. Randell, D., Landini, G.: Discrete mereotopology in automated histological image analysis. In: Proceedings of the Second ImageJ User and Developer Conference, Luxembourg, pp. 151–156 (2008)
6. Randell, D., Landini, G., Galton, A.P.: Discrete mereotopology in automated histological image analysis. IEEE Trans. Pattern Anal. Mach. Intell. **35**(3), 568–581 (2013)
7. Bloch, I.: Spatial reasoning under imprecision using fuzzy set theory, formal logics and mathematical morphology. Int. J. Approx. Reason. **41**, 77–95 (2006)
8. Randell, D.A., Cui, Z., Cohn, A.G.: A spatial logic based on regions and connection. In: Proceedings of the Third International Conference on Knowledge Representation and Reasoning, pp. 165–176 (1992)
9. Li, S., Ying, M.: Generalised region connection calculus. Artif. Intell. **160**(1–2), 1–34 (2004)
10. Cui, Z., Cohn, A.G., Randell, D.A.: Qualitative simulation based on a logic of space and time. In: Proceedings of the AAAI 1992, pp. 679–684 (1992)
11. Sioutis, M., Condotta, J.-F., Salhi, Y., Mazure, B., Randell, D.A.: Ordering spatio-temporal sequences to meet transition constraints: complexity and framework. In: Chbeir, R., Manolopoulos, Y., Maglogiannis, I., Alhajj, R. (eds.) AIAI 2015. IAICT, vol. 458, pp. 130–150. Springer, Cham (2015). doi:10.1007/978-3-319-23868-5_10
12. Chen, J., Cohn, A.G., Liu, D., Wang, S., Ouyang, J., Yu, Q.: A survey of qualitative spatial representations. Knowl. Eng. Rev. **30**(1), 106–136 (2015)
13. Gurcan, M.N., Boucheron, L.E., Can, A., Madabhushi, A., Rajpoot, N.M., Yener, B.: Histopathological image analysis: a review. IEEE Rev. Biomed. Eng. **2**, 147–171 (2009)
14. Sirinukunwattana, K., Pluim, J.P.W., Chen, H., Qi, X., Heng, P.-A., Guo, Y.B., Wang, L.Y., Matuszewski, B.J., Bruni, E., Sanchez, U., Böhm, A., Ronneberger, O., Cheikh, B.B., Racoceanu, D., Kainz, P., Pfeiffer, M., Urschler, M., Snead, D.R.J., Rajpoot, N.M.: Gland segmentation in colon histology images: the glas challenge contest. Med. Image Anal. **35**, 489–502 (2016)
15. Li, S., Ying, M.: Region connection calculus: its models and composition table. Artif. Intell. **145**(1–2), 121–146 (2003)
16. Weidenbach, C., Dimova, D., Fietzke, A., Kumar, R., Suda, M., Wischnewski, P.: SPASS version 3.5. In: Proceedings of the 22nd International Conference on Automated Deduction, pp. 140–145 (2009)
17. Ruifrok, A.C., Johnston, D.A.: Quantification of histochemical staining by color deconvolution. Anal. Quant. Cytol. Histol. **23**(4), 291–299 (2001)

A 2D Morphable Model of Craniofacial Profile and Its Application to Craniosynostosis

Hang Dai[1(✉)], Nick Pears[1], and Christian Duncan[2]

[1] Department of Computer Science, University of York, York, UK
{hd816,nick.pears}@york.ac.uk
[2] Alder Hey Craniofacial Unit, Liverpool, UK
Christian.Duncan@alderhey.nhs.uk
https://www-users.cs.york.ac.uk/~nep/research/LYHM/

Abstract. We present a fully automatic image processing pipeline to build a 2D morphable model of craniofacial saggital profile from a set of 3D head surface images. Subjects in this dataset wear a close fitting latex cap to reveal the overall skull shape. Texture based 3D pose normalization and facial landmarking are applied to extract the sagittal profile from 3D raw scan. Fully automatic profile annotation, subdivision and registration methods are used to establish dense correspondence among sagittal profiles. The collection of sagittal profiles in dense correspondence are scaled and aligned using Generalised Procrustes Analysis (GPA), before applying Principal Component Analysis to generate a morphable model. Additionally, we propose a new alternative alignment called the Ellipse Centre Nasion (ECN) method. Our model is used in a case study of craniosynostosis intervention outcome evaluation and the evaluation reveals that the proposed model achieves state-of-the-art results. We make publicly available both the morphable model with matlab code and the profile dataset used to construct it.

Keywords: Morphable · Craniofacial · Fully automatic · Craniosynostosis

1 Introduction

In the analysis of head shape, the sagittal profile is often the most revealing and informative dimension to look for deviations from population norms and it is often useful, in terms of visual clarity and attention focus, for the clinician to examine shape from such a canonical viewpoint. Therefore, we have developed a novel image processing pipeline to generate a 2D morphable model of craniofacial saggital profile from a set of 3D head surface images. Other profiles are of course useful, as is a full 3D morphable model of the entire craniofacial region, but these are beyond the scope of this paper.

A morphable model is constructed by performing some form of dimensionality reduction, typically Principal Component Analysis (PCA), on a training dataset

M. Valdés Hernández and V. González-Castro (Eds.): MIUA 2017, CCIS 723, pp. 731–742, 2017.
DOI: 10.1007/978-3-319-60964-5_64

of shape examples. This is feasible only if each shape is first re-parametrised into a consistent form where the number of points and their anatomical meaning are made consistent. Shapes satisfying these properties are said to be in dense correspondence with one another. Once built, the morphable model provides two functions. Firstly, the it is powerful prior on 2D profile shapes that can be leveraged in fitting algorithms to reconstruct accurate and complete 2D representations of profiles. Secondly, the proposed model provides a mechanism to encode any 2D profile in a low dimensional feature space; a compact representation that makes tractable many 2D profile analysis problems in the medical domain.

Contributions: We propose a new pipeline to build a 2D morphable model of craniofacial sagittal profile. A new pose normalisation scheme is presented called *Ellipse Centre- Nasion* (ECN) normalisation. Extensive qualitative and quantitative evaluations reveal that the proposed normalisation achieves state-of-the-art results We use our morphable model to perform craniosynostosis intervention outcome evaluation on a set of 25 craniosynostosis patients. For the benefit of the research community, we will make publicly available the sagittal profile dataset, and our 2D morphable model with matlab code.

Paper Structure: In the following section, we discuss related literature. Section 3 discusses our new pipeline used to extract sagittal profiles and construct 2D morphable models. The next section evaluates several variants of the constructed models both qualitatively and quantitatively, while Sect. 5 illustrates the use of the morphable model in intervention outcome assessment for a population of 25 craniosynostosis patients. A final section concludes the work.

2 Related Work

In the late 1990s, Blanz and Vetter built a '3D morphable model' (3DMM) from 3D face scans [1] and employed it in a 2D face recognition application [2]. Two hundred scans were employed (young adults, 100 males and 100 females). Dense correspondences were computed using a gradient-based optic flow algorithm - both shape and colour-texture is used. The model is constructed by applying PCA to shape and colour-texture (separately).

There are very few publicly available morphable models of the human face and, to our knowledge, none that include the full cranium. The Basel Face Model (BFM) is the most well-known and widely used face model and was developed by Paysan et al. [3]. Again 200 scans were used, but the method of determining corresponding points was improved. Instead of optic flow, a set of hand-labelled feature points is marked on each of the 200 training scans. The corresponding points are known on a template mesh, which is then morphed onto the training scan using underconstrained per-vertex affine transformations, which are constrained by regularisation across neighbouring points [4].

Other deformable template methods could be used to build morphable models, such as the well-known method of Thin Plate Splines (TPS) [17] or the work of Li et al. (2008). Their global correspondence optimization method solves simultaneously for both the deformation parameters as well as the correspondence positions [5]. Myronenko and Song (2009) consider the alignment of two point sets as a probability density estimation [6] and they call the method Coherent Point Drift(CPD), and this is remains a highly competitive template morphing algorithm.

Template morphing methods need an automatic initialisation to bring them within the convergence basin of the global minimum of alignment and morphing. To this end, Active Appearance Models (AAMs) [7] and elastic graph matching [8] are the classic approaches of facial landmark and pose estimation. Many improvements over AAM have been proposed [9,10]. Recent work has focused on global spatial models built on top of local part detectors, sometimes known as Constrained Local Models (CLMs) [11,12]. Zhu and Ramanan [13] use a tree structured part model of the face, which both detects faces and locates facial landmarks. One of the major advantages of their approach is that it can handle extreme head pose and we exploit this directly in our model building pipeline.

Another relevant model-building technique is the minimum description length method (MDL) [18], which selects the set of parameterizations that build the 'best' model, where 'best' is defined as that which minimizes the description length of the training set.

3 Model Construction Pipeline

Our pipeline to build a 2D morphable model is illustrated in Fig. 1. It employs a range of techniques in both 3D surface image analysis and 2D image analysis and has three main stages: (i) Profile extraction: The raw 3D scan from the *Headspace dataset* undergoes pose normalization, preprocessing to remove redundant data, and profile detection to find the sagittal profile; (ii) Dense correspondence establishment: A collection of sagittal profiles are reparametrised into a form where each sagittal profile has the same number of points joined into a connectivity that is shared across all sagittal profiles. Furthermore, the semantic or anatomical meaning of each point is shared across the collection; and (iii) Similarity alignment and statistical modelling: The collection of sagittal profiles in dense correspondence are subjected to Generalised Procrustes Analysis (GPA) to remove similarity effects, leaving only shape information. The processed meshes are statistically analysed, typically with PCA, generating a 2D morphable model expressed using a linear basis of eigenshapes. This allows for the generation of novel shape instances.

Each profiles is represented by m 2D points (y_i, z_i) and is reshaped to a $2m$ row vector. Each of these vectors is then stacked in a $n \times 2m$ data matrix, and each column is made zero mean. Singular Value Decomposition (SVD) is applied from which eigenvectors are given directly and eigenvalues can be computed from singular values. This yields a linear model as:

$$\mathbf{x_i} = \bar{\mathbf{x}} + \mathbf{P}\mathbf{b_i} = \bar{\mathbf{x}} + \sum_{i=1}^{k} \mathbf{p^k} b_i^k \tag{1}$$

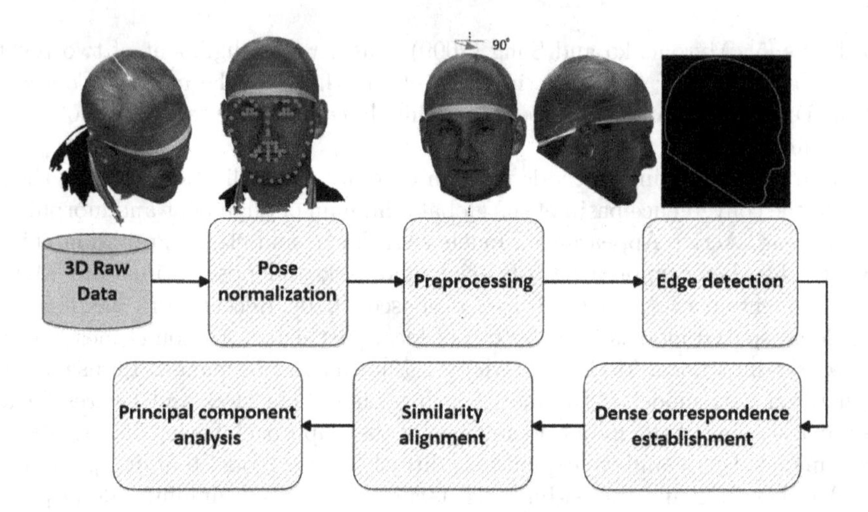

Fig. 1. The pipeline for 2D morphable model construction.

where $\bar{x}$ is the mean head profile shape vector and $\mathbf{P}$ is a matrix whose columns $\mathbf{p}^k$ are the eigenvectors of the covariance matrix (after pose alignment), describing orthogonal modes of head profile variation. The vector $\mathbf{b}$ holds the shape parameters $\{b^k\}$, that weight the shape variation modes which, when added to the mean shape, model a shape instance $\mathbf{x}_i$. The three main stages of the pipeline are described in the following subsections.

3.1 Profile Extraction

Profile extraction requires three stages, namely (i) pose normalisation, (ii) cropping and (iii) edge detection. Each of these stages is described in the following subsection.

Pose Normalisation. Using the colour-texture information associated with the 3D mesh, we can generate a realistic 2D synthetic image from any view angle. We rotate the scan over 360° in pitch and yaw (10 steps of each) to generate 100 images. Then the Viola-Jones face detection algorithm [14] is used to find the frontal face image among this image sequence. A score is computed that indicates how frontal the pose is. The 2D image with the highest score is chosen to undergo 2D facial landmarking. We employ the method of Constrained Local Models (CLMs) using robust Discriminative Response Map Fitting [15] to do the 2D facial image landmarking. The CLMs are trained using data from the Biwi Kinect Head Pose Database, which is equipped with ground truth rotation angles (pitch, yaw and roll). Then the trained system is used to estimate the three angles for the image with facial landmarks. Finally, 3D facial landmarks are captured by projecting the 2D facial landmarks to 3D scan. By estimating the rigid transformation matrix T from the landmarks of a 3D scan to that of

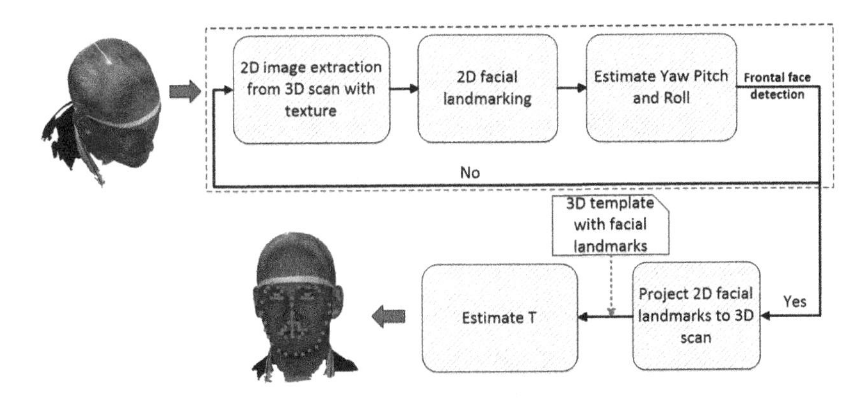

Fig. 2. 3D pose normalization using the texture information

a template, a small adjustment of pose normalization is implemented by transforming 3D scan using T^{-1}.

Cropping. 3D facial landmarks can be used to crop out redundant points, such as the shoulder area and long hair. The face landmarks delineate the face size and its lower bounds on the pose normalised scan, allowing any of several cropping heuristics to be used. We calculate the face size by computing the average distance from facial landmarks to their mean. Subsequently a plane for cropping the 3D scan is generated by moving the cropping plane downward an empirical percentage of the face size. We use a sloping cropping plane so that the chin area is included, but that still allows us to crop close to the base of the latex skull cap at the back of the neck to remove the (typically noisy) scan region, where the subject's hair emerges from under the cap (see Fig. 2).

Edge Detection. The scan is rotated 90° to reveal the head profile and we can generate a 2D image of this profile by orthogonal projection. Then the Canny edge detector [16] is employed to find the edges of this 2D image. The threshold is chosen to be the average of the set of pixels that excludes those that are white space (i.e. off the profile).

3.2 Dense Correspondence Establishment

To extract profile points using subdivision, we have an interpolation procedure that ensures that there is a fixed number of evenly-spaced points between any pair of facial profile landmarks. However, it is not possible to landmark the cranial region and extract profile model points in the same way. This area is smooth and approximately elliptical in structure and so we project vectors from the ellipse centre and intersect a set of fitted cubic spline curves, starting at the nasion, and incrementing the angle anticlockwise in small steps (we use one degree) over a fixed angle.

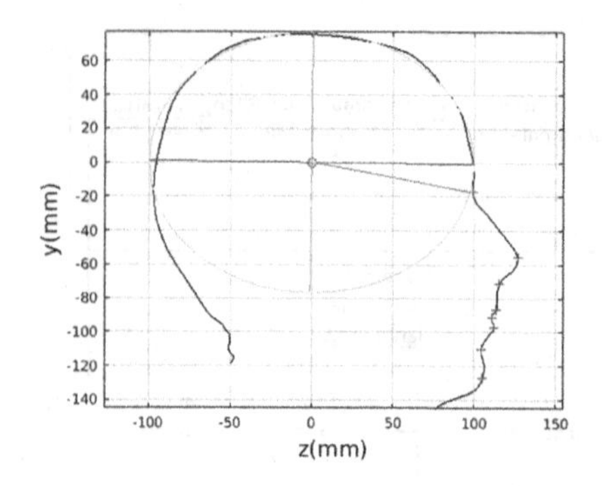

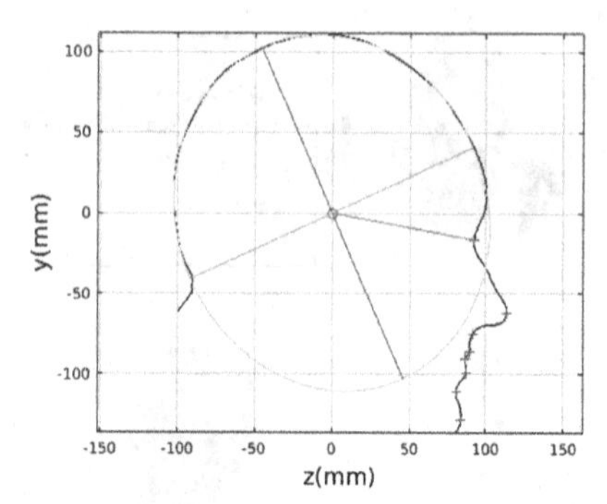

Fig. 3. Head tilt pose normalisation based on ellipse centre and nasion position. The extracted head profile is shown in blue, red crosses show facial landmarks and the ellipse fitted to the cranial profile is shown in cyan. Its major axis is red and its minor axis green. (Color figure online)

As well as using subdivision points directly in model construction, we form a model template as the mean of the population of subdivided and aligned profiles and we use template deformation on the dataset. The resulting deformed templates are re-parameterised versions of each subject that are in correspondence with one another. In this paper, we apply Subdivison, Thin Plate Splines (TPS) [17] nonrigid ICP (NICP) [4], Li's method [5], Coherent Point Drift (CPD) [6] and Minimum Description Length (MDL) [18] to the proposed pipeline for comparative performance evaluation.

3.3 Profile Alignment

A profile alignment method is needed before PCA can be applied to build the 2DMM. We use both the standard GPA approach and a new *Ellipse Centre - Nasion* (ECN) method. Ellipse fitting was motivated by the fact that large sections of the cranium appeared to be elliptical in form, thus suggesting a natural centre and frame origin with which to model cranial shape. One might ask, why not just use GPA over the whole head for alignment. One reason is because variable facial feature sizes (e.g., the nose's Pinocchio effect) induce displacements in the cranial alignment, which is a disadvantage if we are primarily interested in cranial rather than facial shape. We use the nasion's position to segment out the cranium region from the face and use a robust iterative ellipse fitting procedure that rejects outliers.

Figure 3 shows examples of the robust ellipse fit for two head profiles. The centre of the ellipse is used in a pose normalisation procedure where the ellipse centre is used as the origin of the profile and the angle from the ellipse centre to the nasion is fixed at $-10°$. We call this Ellipse Centre - Nasion (ECN) pose normalisation and later compare this to GPA. The major and minor axes of the extracted ellipses are plotted as red and green lines respectively in Fig. 3.

Figure 4 shows all the profiles overlaid with the same alignment scheme. We noted regularity in the orientation of the fitted ellipse as is indicated by the clustering of the major (red) and minor (green) axes in Fig. 4 and the histogram of ellipse orientations in Fig. 4. A minority of heads (9%) in the training sample have their major ellipse axes closer to the vertical (brachycephalic).

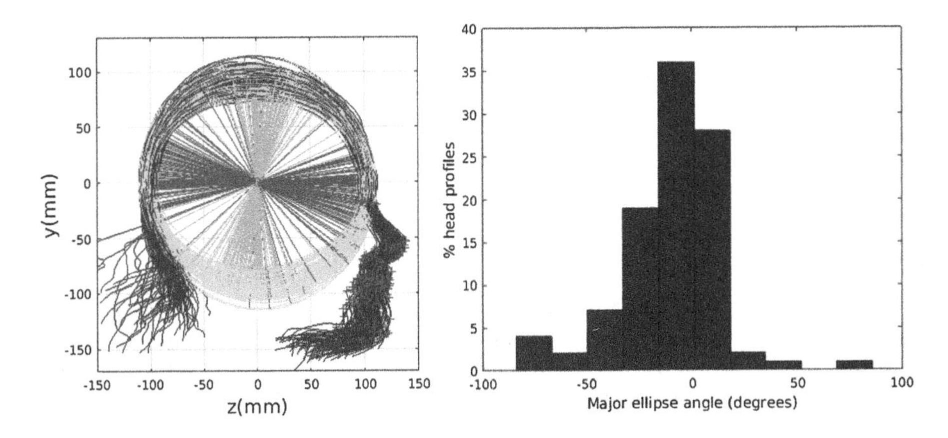

Fig. 4. (1)All training profiles after ECN normalisation; (2)Major axis ellipse angles with respect to an ECN baseline of $-10°$: median angle is $-6.4°$ (2sf). (Color figure online)

4 Morphable Model Evaluation

We built four main 2DMM variants using 100 adult males from the Headspace dataset and animated shape variation along the principal components. The four model variations correspond to full head, scale normalised and unscaled, and cranium only, scale normalised and unscaled.

As an example, when ECN is used (Fig. 5, 1st row), the following three dominant (unscaled) modes are observed: (i) *Cranial height* with *facial angle* are the main shape variations, with small cranial heights being correlated with a depression in the region of the coronal suture; (ii) The overall size of the head varies: surprisingly this appears to be almost uncorrelated with craniofacial profile shape. This was only found in the ECN method of pose normalisation; (iii) The length of the face varies - i.e. there is variation in the ratio of face and cranium size. The second row of Fig. 5, shows the model variation using GPA alignment for comparison.

For quantitative evaluation of morphable models, Styner et al. [19] gives detailed descriptions of three metrics: compactness, generalisation and specificity, now used on our *scale-normalised* models.

Compactness: This describes the number of parameters (fewer is better) required to express some fraction of the variance in the training set. As illustrated in Fig. 6, the compactness using ECN is superior to that of GPA, for

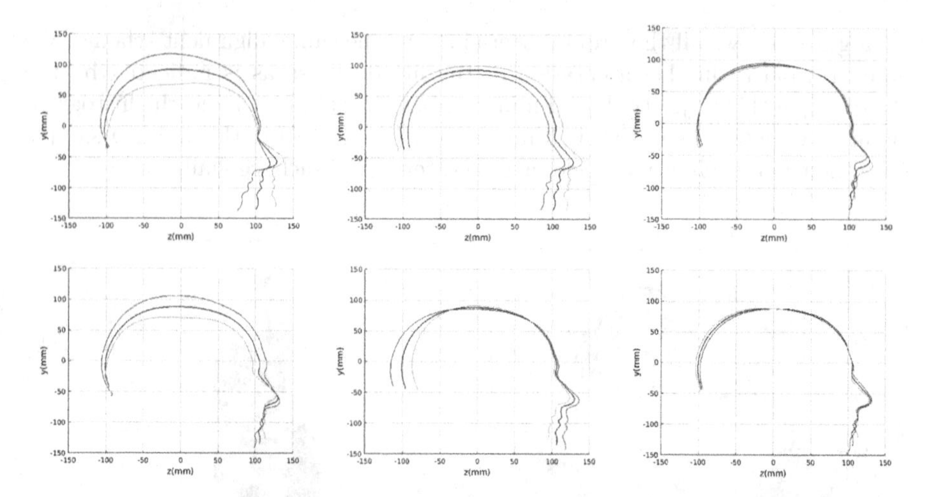

Fig. 5. 1st row: The dominant four modes (left: mode 1; right: mode (3) of head shape variation using automatic profile landmark refinement and ECN similarity alignment. Mean is blue, mean + 3SD is red and mean − 3SD is green. 2nd row: GPA similarity alignment. (Color figure online)

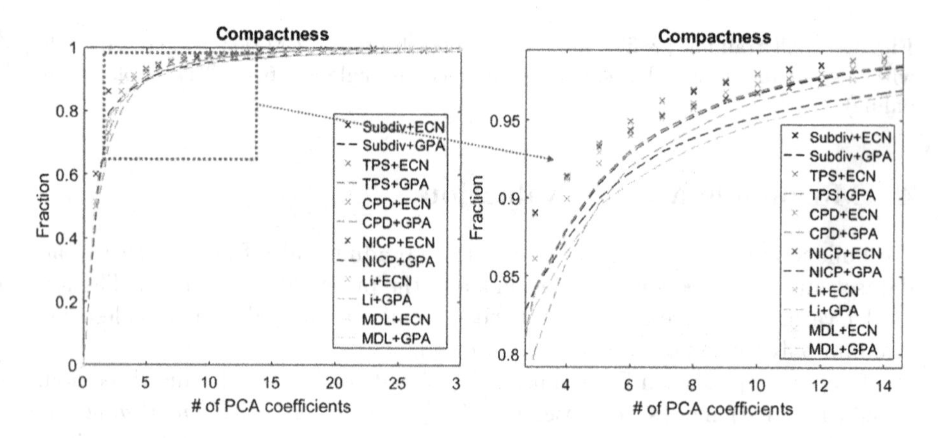

Fig. 6. Compactness

the same correspondence method. Among these methods, subdivision, TPS and MDL, all aligned with ECN are able to generate the most compact models.

Specificity: Specificity measures the model's ability to generate shape instances of the class that are similar to those in the training set. We generate 1000 random samples and take the average Euclidean distance error to the closest training shape for evaluation, lower is better. We show the specificity error as a function of the number of parameters in Fig. 7. Across all correspondence methods with GPA, it gives better specificity against all correspondence methods with ECN. This suggests that GPA helps improve the performance of modelling the underlying shape space. NICP with GPA captures the best specificity.

Generalisation: Generalisation measures the capability of the model to represent unseen examples of the class of objects. It can be measured using the *leave-one-out* strategy, where one example is omitted from the training set and used for reconstruction testing. The accuracy of describing the unseen example is calculated by the mean point-to-point Euclidean distance error, the lower the better. Generalization results are shown in Fig. 7 and for more parameters, the error decreases, as expected. NICP with GPA performs better in terms of Euclidean distance once less than 7 model dimensions are used. Between 7 and 20 model dimensions, TPS with ECN outperforms other methods. When more than 20 model dimensions are used, CPD with GPA has the best generalization ability. Overall, GPA is able to help more successfully model the underlying shape against ECN for the same correspondence method, thereby generating better reconstructions of unseen examples.

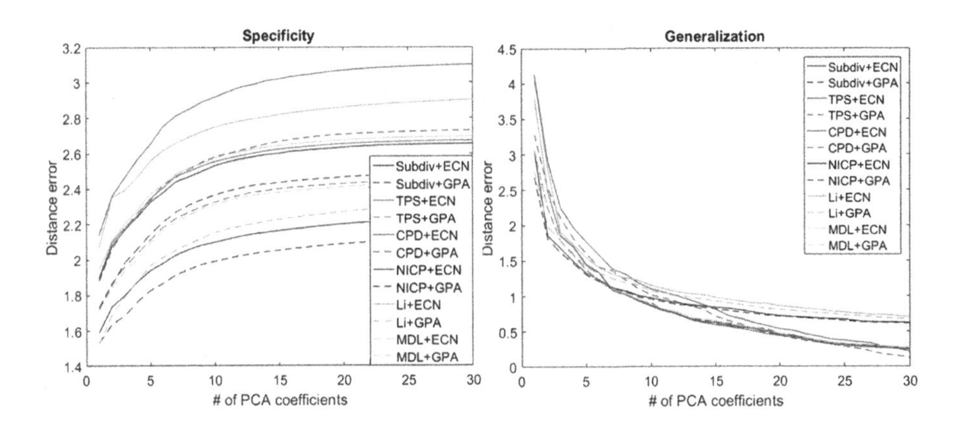

Fig. 7. Left: specificity; right: generalization.

5 Craniosynostosis Intervention Outcome Evaluation

Craniosynostosis is a skull condition whereby, during skull growth and development, the sutures prematurely fuse, leading to both an abnormally shaped head and increased intracranial pressure. We present a case study of 25 craniosynostosis patients (all boys), 14 of which have undergone one type of corrective procedure called *Barrel Staving* (BS) and the other 11, another corrective procedure called *Total Calvarial Remodelling* (TCR). The intervention aim is to remodel the patient's skull shape towards that of an adult and we can employ our model in assessing this.

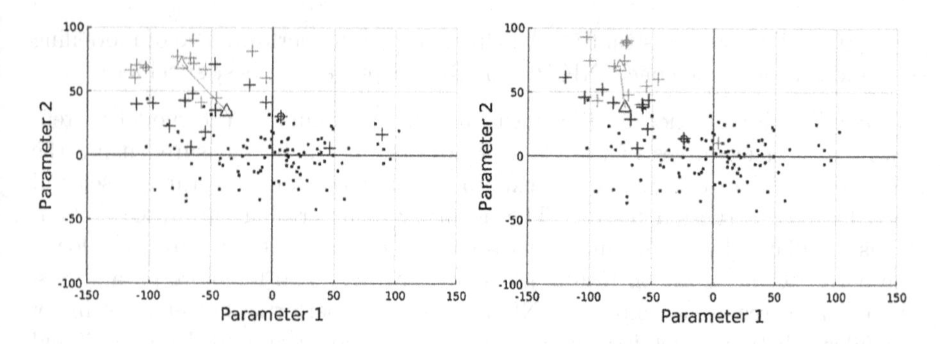

Fig. 8. Patient cranial profile parameterisations, BS (left) and TCR (right) intervention: pre-operative (red crosses) and post-operative (blue crosses) in comparison to the training set (black dots). The circled values represent an example patient (Color figure online)

We build a *scale normalised, cranium only* (to the nasion) 2D morphable model (2DMM) using 100 male subjects, without cranial conditions. Note that both facial structure and overall scale are now irrelevant and that major cranial shape changes are not thought to occur after 2 years old. The patients' scale normalised profiles are then parameterised using the model, indicating the distance from the mean cranial shape along in terms of the model's eigenstructure. The comparisons of pre-operative and post-operative parametrisations show the shapes moving nearer to the mean of the training examples, see Fig. 8.

For the BS patient set, the Mahalanobis distance of the mean pre-op parameters (red triangle in Fig. 8) is 4.670, and for the mean post-op parameters (blue triangle) is 2.302. For shape parameter 2 only (the dominant effect), these figures are 4.400 and 2.156. For the TCR patient set, the Mahalanobis distance of the mean pre-op parameters (red triangle in Fig. 8) is 4.647, and for the mean post-op parameters (blue triangle) is 2.439. For shape parameter 2 only these figures are 4.354 and 2.439. We note that most of this change occurs in parameter 2, which corresponds to moving height in the cranium from the frontal part of the profile to the rear. In these figures, we excluded one patient, who preoperatively already had a near-mean head shape (see red cross near to the origin in Fig. 8, so any operation is unlikely to improve on this (but intervention is required in order to relieve potentially damaging intracranial pressure).

It is not possible to make definitive statements relating to one method of intervention compared to another with these relatively small numbers of patients. However, the cranial profile model does show that both procedures on average, lead to a movement of head shape towards the mean of the training population. An example of analysis of intervention outcome for a BS patient and a TCR patient are given in Fig. 9. The particular example used is highlighted with circles on Fig. 8 to indicate pre-op and post-op parametrisations.

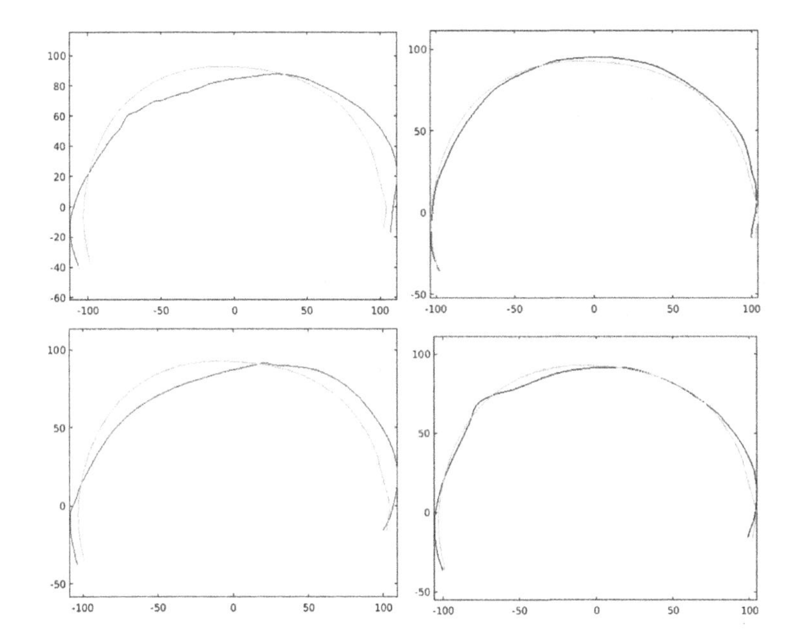

Fig. 9. 1st row: Pre-op and post-op profiles for a BS patient; 2nd row: Pre-op and post-op profiles for a TCR patient. The red and blue traces show the extracted sagittal profiles of the patient pre-operatively and post-operatively respectively, whilst the green shows the mean profile of the training set. (Color figure online)

6 Conclusions

We have presented a fully automatic general and powerful sagittal profile modelling pipeline. Alignment using the Ellipse-Centre Nasion method was introduced. The proposed model has been demonstrated to capture profile shape and assess the intervention outcomes. ECN builds more compact sagittal profile models when compared to GPA. Subdivision, TPS and MDL with ECN are recommended for a more compact sagittal profile model, while NICP with GPA is recommended to capture more specificity. NICP with GPA is able to generate better reconstructions of unseen profiles when fewer than 7 model dimensions are used. If using between 7 and 20 model dimensions, TPS with ECN is recommended for a better generalisation ability. When more than 20 model dimensions are used, CPD with GPA builds a model with better ability to reconstruct unseen examples.

Acknowledgments. The authors wish to thank the Royal Academy of Engineering and Leverhulme Trust for funding the University of York Senior Research Fellowship that primed this work. The Headspace project was funded through the Quality Improvement, Development and Innovation Scheme (QIDIS) from the National Commissioning Group (NCG) between 2011 and 2014. The authors also wish to thank Rachel Armstrong, the Headspace project coordinator.

References

1. Volker, B., Vetter, T.: A morphable model for the synthesis of 3D faces. In: Proceedings of the 26th Annual Conference on Computer Graphics and Interactive Techniques, pp. 187–194 (1999)
2. Volker, B., Vetter, T.: Face recognition based on fitting a 3D morphable model. IEEE Trans. Pattern Anal. Mach. Intell. **25**(9), 1063–1074 (2003)
3. Paysan, P., Knothe, R., Amberg, B., Romdhani, S., Vetter, T.: A 3D face model for pose and illumination invariant face recognition. In: Sixth IEEE International Conference on AVSS 2009, pp. 296–301 (2009)
4. Amberg, B., Romdhani, S., Vetter, T.: Optimal step Nonrigid ICP algorithms for surface registration. In: IEEE Conference on CVPR 2007. IEEE (2007)
5. Li, H., Sumner, R.W., Pauly, M.: Global correspondence optimization for nonrigid registration of depth scans. Comput. Graph Forum **27**(5), 1421–1430 (2008)
6. Myronenko, A., Song, X.: Point set registration: coherent point drift. IEEE Trans. Pattern Anal. Mach. Intell. **32**(12), 2262–2275 (2010)
7. Cootes, T.F., Edwards, G.J., Taylor, C.J.: Active appearance models. IEEE Trans. Pattern Anal. Mach. Intell. **23**(6), 681–685 (2001)
8. Wiskott, L., Krger, N., Kuiger, N., Von Der Malsburg, C.: Face recognition by elastic bunch graph matching. IEEE Trans. Pattern Anal. Mach. Intell. **19**(7), 775–779 (1997)
9. Sauer, P., Cootes, T.F., Taylor, C.J.: Accurate regression procedures for active appearance models. In: British Machine Vision Conference (2011)
10. Tresadern, P.A., Sauer, P., Cootes, T.F.: Additive update predictors in active appearance models. In: British Machine Vision Conference (2010)
11. Smith, B.M., Zhang, L.: Joint face alignment with nonparametric shape models. In: 12th European Conference on Computer Vision (2012)
12. Zhou, F., Brandt, J., Lin, Z.: Exemplar-based graph matching for robust facial landmark localization. In: 14th ICCV (2013)
13. Zhu, X., Ramanan, D.: Face detection, pose estimation, and landmark localization in the wild. In: Computer Vision and Pattern Recognition (2012)
14. Viola, P., Jones, M.J.: Robust real-time face detection. Int. J. Comput. Vis. **57**(2), 137–154 (2004)
15. Asthana, A., et al.: Robust discriminative response map fitting with constrained local models. In: Proceedings of the IEEE Conference on Computer Vision and Pattern Recognition (2013)
16. Canny, J.: A computational approach to edge detection. IEEE Trans. Pattern Anal. Mach. Intell. **6**, 679–698 (1986)
17. Bookstein, F.L.: Principal warps: thin-plate splines and the decomposition of deformations. IEEE Trans. Pattern Anal. Mach. Intell. **11**(6), 567–585 (1989)
18. Davies, R.H., Twining, C.J., Cootes, T.F., Waterton, J.C., Taylor, C.J.: A minimum description length approach to statistical shape modeling. IEEE Trans. Med. Imaging **21**(5), 525–537 (2002)
19. Styner, M.A., et al.: Evaluation of 3D correspondence methods for model building. In: Taylor, C., Noble, J.A. (eds.) IPMI 2003. LNCS, vol. 2732, pp. 63–75. Springer, Heidelberg (2003)

A Comparison of Texture Features Versus Deep Learning for Image Classification in Interstitial Lung Disease

Alison O'Neil(✉), Matthew Shepherd, Erin Beveridge, and Keith Goatman

Toshiba Medical Visualization Systems Ltd., Edinburgh, Scotland
alison.j.oneil@gmail.com

Abstract. Interstitial lung disease (ILD) is a multifactorial condition that is difficult to diagnose. High-resolution computed tomography (CT) is commonly the imaging modality of choice, as it enables detection and mapping of distinctive pathological patterns. The distribution of these patterns gives clues as to the correct histological diagnosis. This paper compares two approaches to detecting these complex patterns: "man-made" features, based on classical handcrafted texture descriptors, and "machine-made" features, built with deep learning convolutional neural networks (CNNs). The two paradigms are evaluated on scans from 132 subjects, derived from two public databases of high resolution ILD CT images and associated expert annotations. Five specific tissue patterns are included: healthy, emphysema, fibrosis, ground glass opacity, and micronodules. The subjects are divided into training, validation and test groups. On the validation data the best handcrafted solution achieves a class assignment accuracy of 76.0%, compared with the best deep learning accuracy of 79.0%. For the test group, which was not used during development and only tested once, the handcrafted method achieves 65.5%, compared with the CNN accuracy of 69.9%. The results indicate that deep learning CNNs can outperform traditional texture measures, even on a low-level texture classification task such as this.

1 Introduction

In this paper, man-made features are pitted against machine-made features. The target application is tissue classification in subjects with suspected interstitial lung disease (ILD). Distinctive ILD pathological patterns are observed on high-resolution CT scans; the goal is to identify the five pathology types illustrated in Figure 1. Their presence—and distribution within the lungs—gives clues as to the histological diagnosis [6, 20], of which there are many possibilities.

Texture recognition is one of the lowest levels of image abstraction. It is known that first-layer learned convolutional filters appear similar to commonly used analytic solutions, such as Gabor filters [21]. Hence, it may be expected that handcrafted solutions based on such analytic functions, and deep learning solutions would have comparable performance. Whilst a range of methods have been

M. Valdés Hernández and V. González-Castro (Eds.): MIUA 2017, CCIS 723, pp. 743–753, 2017.
DOI: 10.1007/978-3-319-60964-5_65

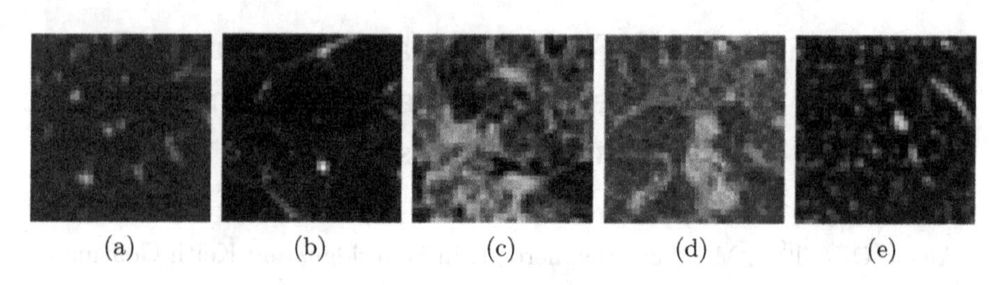

Fig. 1. Examples of the five different lung patterns. From left to right: (a) Healthy (b) Emphysema (c) Fibrosis (d) Ground glass (e) Micronodules

published addressing ILD classification, many on the publicly available MedGift dataset [3], direct comparison is difficult due to the variability in the choice of training/test/cross-validation data splits, and in the choice of evaluation metrics. In this paper, we perform a direct comparison of first-order intensity metrics, a selection of more complex texture measures, and two contrasting deep learning solutions.

An obvious first approach to classifying pathological patterns is to apply one or more of the many known image descriptors, and observe whether the pathologies may be separated on the basis of their respective descriptor responses. In the *Adaptive Multiple Feature Method*, Uppaluri *et al.* used classical texture features for the classification of six lung patterns: honeycombing, ground glass, bronchovascular, nodular, emphysema-like, and normal [17]. The features consisted of simple intensity-based metrics (e.g. mean, variance, skewness, kurtosis) as well as run-length features [4], co-occurrence matrix features [7], and a fractal feature (geometric fractal dimension [14]). The resulting classifier performed as well as a group of trained observers who had been given the primary diagnosis. The "classical" handcrafted approach in this paper comprises a similar set of texture features.

More recently, deep learning techniques have become the focus of research. For such an imaging task, convolutional neural networks are an obvious fit. Fairly simple feedforward (supervised) neural nets have been demonstrated for patchwise classification by Li *et al.* [13] and by Anthimopoulos *et al.* [2]. Both used very small nets, perhaps reflecting the nature of the texture application: Li *et al.* used a single convolutional layer with 7×7 filters, while Anthimopoulos used five convolutional layers consisting of very small filters (2×2). These two networks constitute the deep learning approaches in this paper.

Other deep learning approaches include transfer learning, which was explored by Shin *et al.* [15], who used the well-known AlexNet and GoogLeNet CNNs, pre-trained on non-medical images, before fine-tuning on the MedGift data. Gao *et al.* [5] also used a variant of AlexNet to perform *slicewise* detection. Finally, Restricted Boltzmann Machines (RBM) have been demonstrated by Li *et al.* [12] and van Tulder and de Bruijne [18,19]. Van Tulder and de Bruijne showed that *mixing* discriminative and generative learning gave better performance than either in isolation, although on balance the RBM appeared to have worse performance than the feedforward nets developed by other authors.

2 Methods

2.1 Datasets

Two publicly available ILD databases are employed:

MedGift ILD Database [3]**:** This consists of 93 ILD scans, which were collected at the University Hospitals of Geneva from patients undergoing high-resolution thorax CT. Patients had a histological ILD diagnosis, and radiographic images consistent with the diagnosis. In order to reduce the radiation dose, the scans were acquired with large slice spacing compared to the in-plane slice resolution, i.e. a slice spacing range of 10–15 mm compared with 0.4–1 mm within-slice resolution. The slices are partially annotated by a radiologist, with 2D regions of interest for sixteen pathological lung patterns, for which only four have significant numbers of examples (emphysema, fibrosis, ground glass opacity, and micronodules). Healthy examples are also included, however these were excluded from this study since they are annotated on patients with a history of ILD.

CT Emphysema Database [16]**:** This database is derived from high-resolution CT scans of a study group of 39 patients (9 never-smokers, 10 smokers, and 20 smokers with COPD). This database was used for healthy examples, and to supplement the "emphysema" class. The scans were acquired in an exploratory study carried out at Gentofte University Hospital in Denmark. The data has a within-slice resolution of 0.78 mm. A set of 61×61 pixel patches is provided, extracted at regions of normal tissue, centrilobular emphysema, and paraseptal emphysema. Normal tissue samples were marked on non-smoking subjects. Each patch is labelled with its leading diagnosis, based on the consensus of an experienced chest radiologist and a CT experienced pulmonologist. The patches for healthy tissue and centrilobular emphysema are used in this paper, since the paraseptal regions in this dataset were too small to extract patches. For the purpose of extracting smaller patches, sub-regions within the patches were manually outlined and labelled with the designated class, i.e. excluding non-parenchymal tissue, large blood vessels, and tissue which clearly belonged to a different class.

2.2 Patch Extraction

Prior to algorithm development, the data were randomly split into approximately 40-20-40 proportions for the training, validation and test sets, respectively, so as to approximately balance the number of patients in each class. Then, patches of 20×20 mm, with 80% overlap, were extracted at 1 mm/pixel resolution. Patches had to be fully contained within the labelled ground truth regions (see Fig. 2). Details of the resulting datasets are summarised in Table 1.

2.3 Texture Features: The Man-Made Feature Approach

Traditional approaches to classification problems such as this tend to require the selection of appropriate image-derived features, which are then passed to

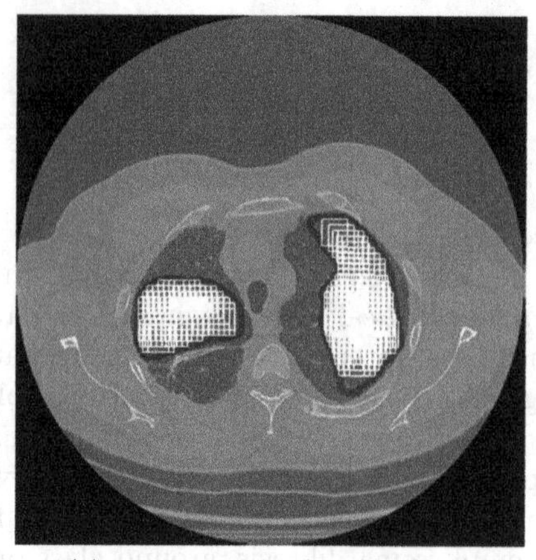

(a) MedGift slice (emphysema)

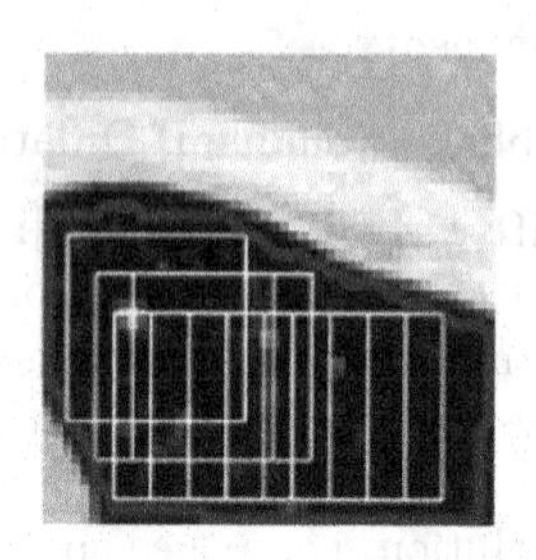

(b) Emphsyema patch (healthy)

Fig. 2. Images showing how patch extraction is performed in each of the two databases. The ground truth region is outlined in red, and the extracted patches are outlined in white. The patches are all fully contained within the ground truth region, and overlap by 80%. (Color figure online)

Table 1. Distribution of data in terms of patients and (patches).

Data subset	Training	Validation	Test
Healthy	3 (569)	1 (250)	4 (737)
Emphysema	4 (1880)	2 (448)	6 (836)
Fibrosis	12 (6212)	7 (2925)	10 (839)
Ground glass	9 (2510)	5 (587)	10 (1747)
Micronodules	7 (27107)	2 (665)	4 (1622)
Total	34 (38278)	15 (4875)	33 (5781)

some form of classifier (e.g. a support vector machine). Since it is difficult to know a priori which features are most useful—or indeed whether they are useful at all—such workflows also usually include a feature selection or dimensionality reduction step. The features chosen here are a combination of first-order intensity metrics, and well known image texture metrics:

1. First-order intensity metrics
   (a) Mean, standard deviation, skew, kurtosis, minimum, maximum, median, percentiles, interquartile range.
2. Texture metrics
   (a) Greylevel co-occurrence matrix (GLCM) [7]
   (b) Greylevel runlength matrix [4]
   (c) Laws' texture filters [11]
   (d) Neighbourhood grey tone difference matrix (NGTDM) [1]
   (e) Fractal dimension [14]

Feature selection was performed using a combination of the ReliefF algorithm [10] to detect useful features, and cross-correlation to exclude useful features that were very strongly correlated with a higher ranked feature. Three configurations were tested: (a) First-order features alone, (b) Texture features alone, and (c) both types of feature.

Two classifiers were compared: a Random Forest classifier and a Support Vector Machine with a quadratic kernel. Implementations from the MATLAB Statistics and Machine Learning Toolbox were used.

2.4 Deep Learning: The Machine-Made Feature Approach

Two contrasting convolutional neural network (CNN) architectures are presented in this paper, based on the original networks by Li *et al.* [13] and Anthimopoulos *et al.* [2]. The Li architecture is shallow: just a single convolutional layer with a large kernel size. The Anthimopoulos architecture achieves depth by using very small kernels of 2×2 (right at the lower size limit), and layering these together. Both networks have previously reported results on the MedGift data, however both used different divisions of training and test data, and different supplementary data.

Data Augmentation: The training data was augmented by rotating and flipping four ways each. In all, the volume of training data was increased sixteenfold. The data was then rebalanced to achieve equal numbers from each class, by randomly selecting patches equal to the number of patches in the smallest class.

CNN Architectures: In this paper, smaller patches are used than either of the networks were originally designed for (i.e. 20×20 pixels compared with 32×32 pixels). Also, a lower resolution is employed: 1 mm versus variable resolutions ranging between 0.4 mm and 1.0 mm in [13], and 0.4 mm in [2]. To address this, the relevant parameters of the architectures were re-tuned empirically using the validation data as follows. Firstly, the receptive field for each network was reduced to compensate the decrease in pixel resolution, by reducing the kernel size from 7×7 for the Li architecture to 4×4, and by reducing the number of layers in the Anthimopoulos architecture from 5 to 3; this gave both architectures an effective receptive field size of 4×4 pixels. To compensate for the smaller patches (20×20 versus 32×32), and to protect against over-fitting, the number of dense layer nodes in the Li architecture was reduced from 100/50 to 40/20, and the k factor in the Anthimopoulos architecture was reduced from 4 to 3. It was found to be further beneficial to adopt max pooling across the whole patch in the Li architecture, as used in the Anthimopoulos architecture. This may be expected, since assigning importance to the geographical location of features in a patch has little meaning in the context of texture. Max pooling did better than average pooling for both networks. The leaky ReLU activation units used in the Anthimopoulos architecture were found to give no advantage over standard ReLU units, and furthermore gave no advantage when inserted into the Li architecture, so all units were set to ReLU. Batch normalisation was trialled and found to be ineffective. The modified architectures are shown in Fig. 3.

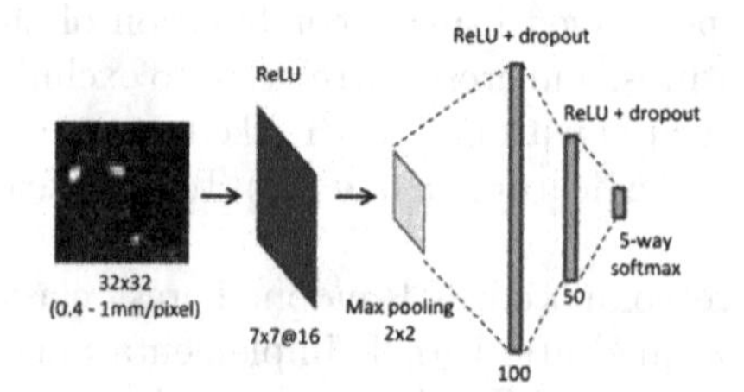

(a) Li *et al.* (276,605 parameters)

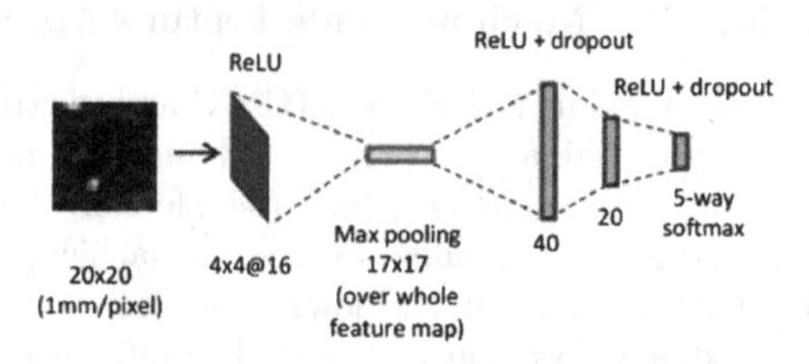

(b) Modified Li *et al.* (1877 parameters)

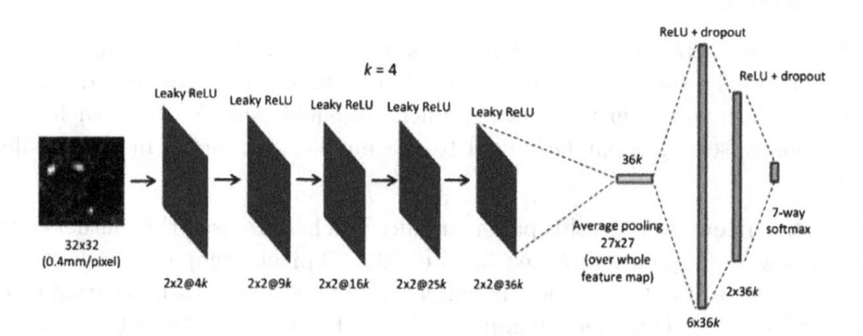

(c) Anthimopoulos *et al.* (470,989 parameters)

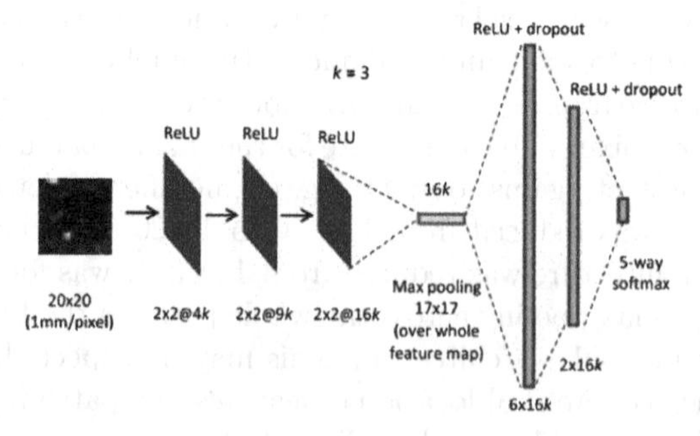

(d) Modified Anthimopoulos *et al.* (48,956 parameters)

Fig. 3. Images showing the two CNN architectures. The original architectures are shown alongside the modified networks which were used in this paper.

Training: Kernel and node weights were initialised using the method by He *et al.* [8]. The Adam adaptive gradient-descent algorithm [9] was used for learning, with its default settings of learning rate 0.001, and beta parameters 0.9 and 0.999, to optimise the categorical cross-entropy. Training was performed using mini-batches of 128 patches. The networks were trained for 10 and 20 epochs respectively for the Anthimpopoulos and Li networks; further epochs gave no further improvement on the validation set.

Ensembles: An ensemble of CNNs was trained using each architecture, by randomly shuffling the data to achieve different optimisations of the architecture parameters.

Implementation: The CNNs were implemented in Python using Keras, a neural networks library built on top of Theano. For fast throughput, parameter searches and final test runs were run on machines from the Amazon Web Services (AWS) cloud computing services. The AWS machines were p2.xlarge (1 NVIDIA K80 GPU, 4 vCPUs, 61 GiB RAM). The NVIDIA K80 is a dual GPU card, and for the experiments 10 of these machines were run in parallel, to give a total of 20 GPUs running the experiments. In total, the experiments ran for 60 h, which represented 600 h of processor time. Each experiment took approximately 15–20 min (including data loading, data pre-processing and validation/test accuracy computation).

3 Results

Results are reported in Table 2 for all three datasets: training, validation, and test. The training and validation metrics are computed after tuning the networks to the validation set, without sight of the test set. The test set metrics are computed using the same tuned hyperparameters, but with re-training of the network parameters using the combined training and validation data.

Weighted accuracy is chosen the evaluation metric, giving equal weight to the five classes. This is fairer, since the distribution of patients and classes is not representative of the normal population, nor necessarily of typical hospital referrals.

To measure the inherent variation in classifier accuracy, the training for each classifier was run ten times, and the mean and standard deviations reported. For the intensity and texture classifiers, the order of the *features* was randomly shuffled before each run. For the CNNs, the order of the *training data samples* was randomly shuffled before each run.

On the test set, the deep learning solutions outperform the intensity features and the traditional texture measures. This is slightly different to the validation set results, where the more complex Anthimopoulos *et al.* deep learning model remained the best, but the simpler Li *et al.* model did worse than the other solutions, including the intensity-based method.

It was a little surprising that the results on the test set were so much worse than those on the validation set. The confusion matrices (see Fig. 4) reveal that

Table 2. Weighted accuracy mean (%) and standard deviation (% in brackets) for all classifiers on all datasets. Figures are computed on the basis of 10 training runs for each classifier.

Classifier	Training	Validation	Test
Random forest (intensity)	96.6 (0.0)	72.4 (0.3)	61.5 (0.2)
Random forest (texture)	97.0 (0.1)	68.7 (0.5)	65.8 (0.3)
Random forest (all)	98.1 (0.0)	74.3 (0.2)	65.5 (0.2)
SVM (intensity)	95.1 (0.1)	74.7 (0.3)	64.1 (0.1)
SVM (texture)	97.9 (0.0)	73.2 (0.3)	64.2 (0.1)
SVM (all)	**98.2 (0.1)**	76.0 (0.2)	64.9 (0.2)
Modified li	85.0 (0.8)	68.5 (3.3)	67.1 (1.1)
Modified li (ensemble of 5)	88.2 (0.5)	72.1 (0.6)	68.7 (0.4)
Modified anthimopoulos	93.6 (0.7)	78.8 (0.9)	68.6 (1.2)
Modified anthimopoulos (ensemble of 5)	95.4 (0.3)	**79.0 (1.1)**	**69.9 (0.9)**

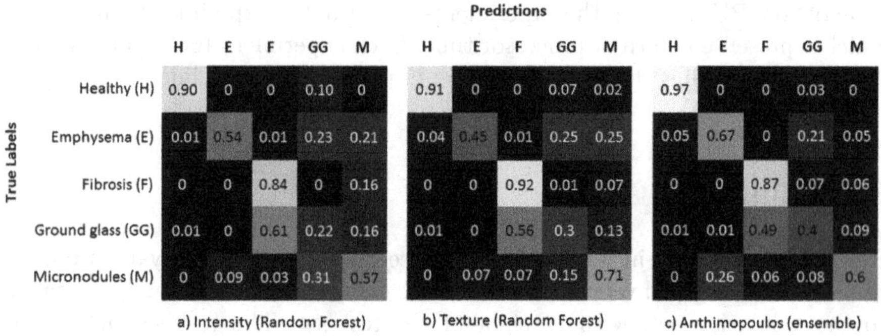

Fig. 4. Normalised confusion matrices for the best classifier in intensity, texture and deep learning respectively.

the ground glass examples are being confused by all classifiers with fibrosis. Closer inspection of some cases in the (previously unseen) test set reveal that this set contains some patients with unusually severe ground glass pathology compared to the other datasets, which might easily be confused with bright fibrosis regions to a classifier which relies heavily on intensity features to separate the two classes. Figure 5 shows a case of mild ground glass opacity from the validation set versus a case of severe ground glass opacity from the test set.

The emphysema class was often confused with ground glass, and also (at least in the intensity and texture classifiers) with the micronodules class—this trend was reversed in the CNN, which classified micronodule examples as emphysema. This is perhaps surprising, but may point to an inconsistency between the subject populations and ground truth labelling strategies in the two databases, since far more of the data from the emphysema database was allocated to the test set than to either the training or validation sets, and the ground truth for the emphysema database included much milder examples of emphysema.

In interstitial lung disease patterns, simple mean intensity is a powerful feature. The distributions of each data set are shown in the box plots in Fig. 6; these support the observations above that the test set had a significantly different intensity distribution to the training set for at least three of the classes. It appears that the classifiers which relied too heavily on intensity features performed worse on the test sets, where there were a wider range of pathology severities present, whereas the "texture features only" and the deep learning solutions were more robust to this.

In hindsight, perhaps the data sets should have been stratified in order to give each set a representative sample of patients showing the range of pattern

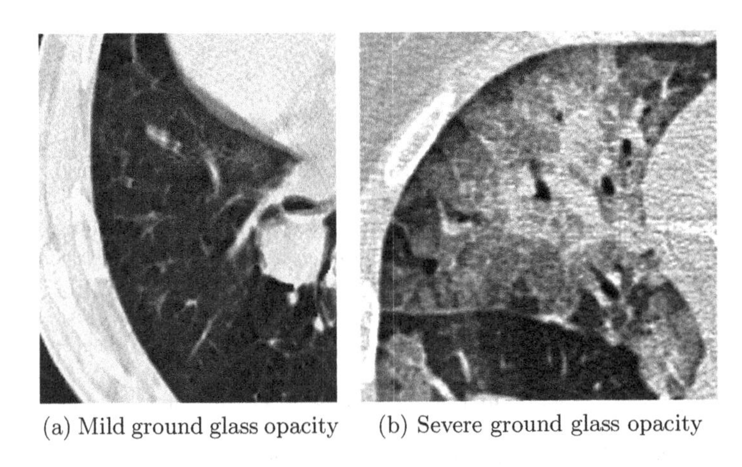

(a) Mild ground glass opacity (b) Severe ground glass opacity

Fig. 5. Images showing the range of attenuation of ground glass opacity.

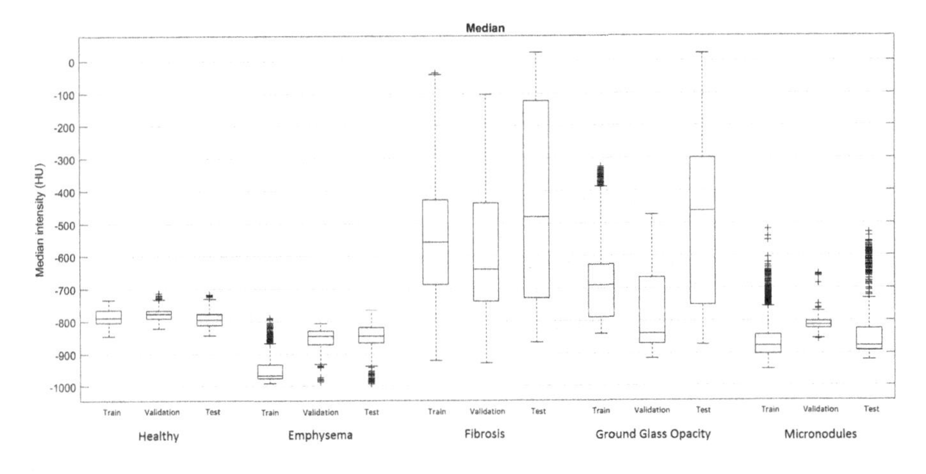

Fig. 6. Box plots showing the range in patch mean intensity (HU) for each lung pattern in each data set.

severities, rather than allocating patients randomly. Of course this is a symptom of insufficient training data, which does not cover the range of presentation variation seen in clinical practice.

4 Conclusion

Deep learning solutions have proven their worth for many problems, including in medical image analysis. In this paper a comparison was performed between man-made, traditional handcrafted features and machine-made features with deep learning to classify interstitial lung disease. The results demonstrate that even for relatively low-level problems, such as texture classification, a well-designed deep learning architecture with sufficient capacity can outperform traditional model-based approaches. The next step would be to reproduce this result on a larger dataset with more patients.

References

1. Amadasun, M., King, R.: Texural features corresponding to textural properties. IEEE Trans. Syst. Man Cybern. **19**(5), 1264–1274 (1989)
2. Anthimopoulos, M., Christodoulidis, S., Ebner, L., Christe, A., Mougiakakou, S.: Lung pattern classification for interstitial lung diseases using a deep convolutional neural network. IEEE Trans. Med. Imaging **35**(5), 1207–1216 (2016)
3. Depeursinge, A., Vargas, A., Platon, A., Geissbuhler, A., Poletti, P.A., Müller, H.: Building a reference multimedia database for interstitial lung diseases. Comput. Med. Imaging Graph. **36**(3), 227–238 (2012)
4. Galloway, M.M.: Texture analysis using gray level run lengths. Comput. Graph. Image Process. **4**(2), 172–179 (1975)
5. Gao, M., Xu, Z., Lu, L., Harrison, A.P., Summers, R.M., Mollura, D.J.: Multi-label deep regression and unordered pooling for holistic interstitial lung disease pattern detection. In: Wang, L., Adeli, E., Wang, Q., Shi, Y., Suk, H.I. (eds.) MLMI 2016. LNCS, vol. 10019, pp. 147–155. Springer, Cham (2016). doi:10.1007/978-3-319-47157-0_18
6. Grenier, P., Valeyre, D., Cluzel, P., Brauner, M.W., Lenoir, S., Chastang, C.: Chronic diffuse interstitial lung disease: diagnostic value of chest radiography and high-resolution CT. Radiology **179**(1), 123–132 (1991)
7. Haralick, R.M., Shanmugam, K., Dinstein, I.: Textural features for image classification. IEEE Trans. Syst. Man Cybern. **3**(6), 610–621 (1973)
8. He, K., Zhang, X., Ren, S., Sun, J.: Delving deep into rectifiers: surpassing human-level performance on ImageNet classification. CoRR, abs/1502.0 (2015)
9. Kingma, D.P., Ba, J.L.: Adam: a method for stochastic optimization. In: International Conference on Learning Representations, pp. 1–15 (2015)
10. Kononenko, I., Šimec, E., Robnik-Šikonja, M.: Overcoming the myopia of inductive learning algorithms with RELIEFF. Appl. Intell. **7**(1), 39–55 (1997)
11. Laws, K.I.: Textured image segmentation. Technical report, University of Southern California, January 1980
12. Li, Q., Cai, W., Feng, D.D.: Lung image patch classification with automatic feature learning. In: Conference Proceedings: Annual International Conference of the IEEE Engineering in Medicine and Biology Society, pp. 6079–6082 (2013)

13. Li, Q., Cai, W., Wang, X., Zhou, Y., Feng, D.D., Chen, M.: Medical image classification with convolutional neural network. In: 13th International Conference on Control, Automation, Robotics, and Vision, pp. 844–848, December 2014
14. Lopes, R., Betrouni, N.: Fractal and multifractal analysis: a review. Med. Image Anal. **13**(4), 634–649 (2009)
15. Shin, H.C., Roth, H.R., Gao, M., Lu, L., Xu, Z., Nogues, I., Yao, J., Mollura, D., Summers, R.M.: Deep convolutional neural networks for computer-aided detection: CNN architectures, dataset characteristics and transfer learning. IEEE Trans. Med. Imaging **35**(5), 1285–1298 (2016)
16. Sørensen, L., Shaker, S.B., De Bruijne, M.: Quantitative analysis of pulmonary emphysema using local binary patterns. IEEE Trans. Med. Imaging **29**(2), 559–569 (2010)
17. Uppaluri, R., Heitmman, E.A., Sonka, M., Hartley, P.G., Hunninghake, G.W., Mclennan, G.: Computer recognition of regional lung disease patterns. Am. J. Respir. Crit. Care Med. **160**(2), 648–654 (1999)
18. van Tulder, G., de Bruijne, M.: Learning features for tissue classification with the classification restricted Boltzmann machine. In: Menze, B., et al. (eds.) MCV 2014. LNCS, vol. 8848, pp. 47–58. Springer, Cham (2014). doi:10.1007/978-3-319-13972-2_5
19. Van Tulder, G., De Bruijne, M.: Combining generative and discriminative representation learning for lung CT analysis with convolutional restricted Boltzmann machines. IEEE Trans. Med. Imaging **35**(5), 1262–1272 (2016)
20. Wang, J.S., Cherng, J.M., Perng, D.S., Lee, H.S., Wang, S.: High resolution computed tomography in assessment of patients with emphysema. Respir. Care **58**(4), 614–622 (2013)
21. Yosinski, J., Clune, J., Bengio, Y., Lipson, H.: How transferable are features in deep neural networks? Adv. Neural Inf. Proc. Syst. 27 (Proc. NIPS) **27**, 1–9 (2014)

A Novel High-Throughput Multispectral Cell Segmentation Algorithm

Jenia Golbstein(✉), Yaniv Tocker, Revital Sharivkin, Gabi Tarcic, and Michael Vidne

NovellusDx, Jerusalem Bio-Park 5th floor, Hadassah Medical Center Campus, Jerusalem, Israel
{jenia,yaniv,revital,gabi,michael}@novellusdx.com
http://www.novellusdx.com/

Abstract. An increasingly common component of molecular diagnostics is the analysis of protein localization at the single-cell level using fluorescent microscopy. Manually extracting quantitative data from a large population of cells is unreasonably time-consuming and existing automatic systems are rather limited.

Here we present an integrated image analysis software system for high-throughput segmentation of cells and corresponding nuclei. The system is composed of robust image enhancement, followed by multispectral identification of putative cells using a statistical model (Gaussian Mixture Model) approach, followed by cross-spectral watershed that effectively segments clustered cells, and finally, a rule based refinement using statistical morphological attributes of the cells.

The robustness and accuracy of the system have been tested on artificial fluorescent beads, as well as on hand segmented and visually inspected images. Lastly, we compare our algorithm to state-of-the-art systems and show it does better on most performance parameters.

To date, the system has been used to accumulate data from over 300 million segmented cells each expressing specific set of genomic alterations.

Keywords: Cancer · Functional genomics · Live-cells · Gaussian mixture model · Watershed · Cross-spectral image segmentation

1 Introduction

1.1 Intracellular Protein Localization Is Central for Functional Genomics

Sequencing based methods to tailor the treatment of cancer to individual patients are at the cutting edge of Personalized Medicine and have come a long way in the past few years [11,15]. Nonetheless, the response rates to targeted therapies and the progression free survival (PFS) have remained low even when the treatment is genetically tailored [7]. Accurate diagnostic and predictive technologies can

M. Valdés Hernández and V. González-Castro (Eds.): MIUA 2017, CCIS 723, pp. 754–766, 2017.
DOI: 10.1007/978-3-319-60964-5_66

provide more information on how to correctly treat the patient from the available arsenal of targeted therapies [16], thereby improving these results [17].

While sequencing detects all mutations, it necessarily remains blind to their functional significance. It is able to identify a relatively small number of mutations that have been established as "driver" mutations - confers growth advantage to the cancer cell and is positively selected in the cancer's development. But, the function of the vast majority of mutations are not characterized. While the majority are neutral "passenger" mutations (which are not selected, do not confer clonal growth advantage, and therefore do not contribute to cancer development), there are many driver mutations that sequencing cannot recognize as significant in cancer growth [11].

NovellusDx developed a novel, live-cell based, platform to diagnose cancer and recommend an optimal, tailored treatment to the specific patient's cancer. The NovellusDx platform monitors the activity of the vast majority of the oncogenic signaling pathways (pathways through which 'information and instruction' are communicated to the cells) before and after administration of a panel of drugs and drug combinations in vitro. The system is based on quantifying the amount of reporter proteins in different compartments of the cells and therefore requires a very high-throughput and robust system to segment fluorescent microscopy images into constituent cells and sub-compartments.

1.2 Assay for the Functional Annotation of Genetic Mutations

The main principle behind this technology is the fact that the activity level of signaling pathways is constituted by translocation of key proteins from the cytoplasm of the cells into the nucleus of the cells, as seen in Fig. 1. Therefore, by tracking the ratio of these key proteins inside the nucleus versus the cytoplasm we are able to monitor the activity of major oncogenic signaling pathways. Comparing nucleus to cytoplasmic levels requires a system that is able to analyze thousands of cells and segment them to their cytoplasms and nuclei compartments in high efficiency and accuracy and under differing conditions.

2 Method

The system, as seen in Fig. 2, is composed of three main modules: preprocessing, nuclei segmentation, and cell segmentation. The results of the nuclei segmentation are used for the cross-spectral watershed algorithm in the cell segmentation. Below we expand on each of the main modules.

2.1 Preprocessing

The raw images were obtained using the following microscopy system: automated NIKON Ti-Eclipse microscope coupled with an Andor Zyla 4.2 PLUS sCMOS camera and a LED-based SOLA light source. The microscope uses a 10x CFI Plan-Apo objective to maintain high-throughput of the imaging step (numerical

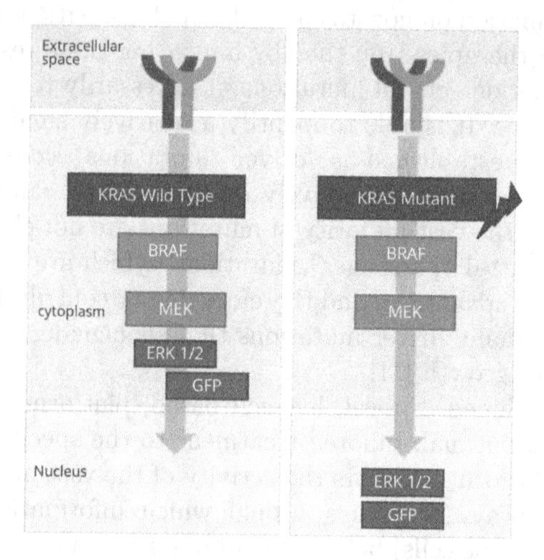

Fig. 1. Schematic overview of the biological principle of the NovellusDx assay. The system is designed to functionally annotate unknown mutations by tracking the translocation of downstream reporters from the cytoplasm to the nucleus. In the left panel, a simplified cell and signaling pathway. The KRAS gene that is being analyzed is in the wild-type form therefore it is not inducing a translocation of ERK into the nucleus and we expect to find the GFP spread throughout the cell evenly. This is in contrast to the right panel where the KRAS gene has a driver mutation which is activating the pathway and causing excessive translocation of ERK into the nucleus.

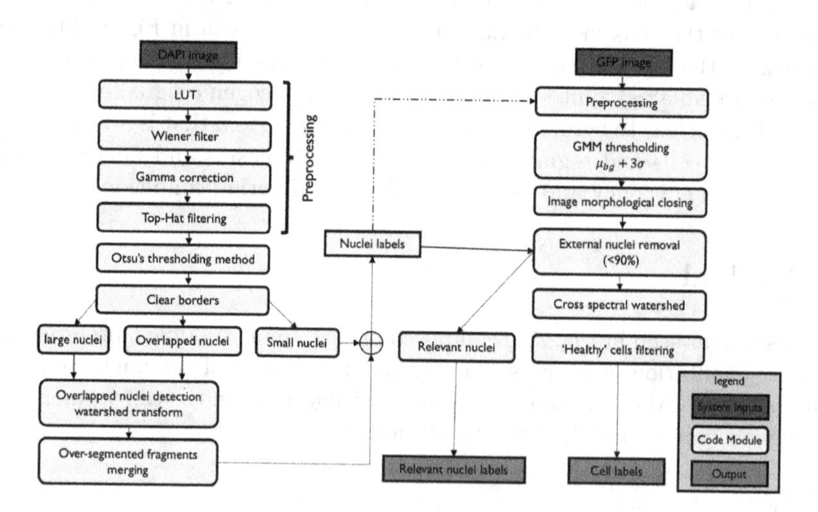

Fig. 2. Schematic overview of the code. The system receives two inputs: the DAPI images of the nuclei and the GFP image of the reporter protein. Due to the higher spatial segregation of the nuclei, we start by finding all nuclei, and use them to seed the watershed module of the reporter protein.

aperture 0.45 for high signal-to-noise ratio), and a laser-based auto-focus system (PFS3) for rapid, accurate, identification of the relevant focal plane. The camera has a 13.3×13.3 mm chip for maximal FOV (field of view) using this objective while maintaining relatively high image resolution of 4.2 Megapixel (2048×2048). It also has an 82% QE (quantum efficiency) for increased sensitivity. The images were acquired in two wavelengths: DAPI channel image with peak excitation and emission at 340–380 ms and 435–485 ms correspondingly; GFP channel image with peak excitation and emission at 512–552 ms and 565–615 ms correspondingly. Cell line used in the images - HeLa CRM CCL-2. After image acquisition, an automated system initializes the preprocessing step.

First, we use a Look-Up Table (LUT), subtracting the minimal value and dividing the difference between the maximal value and the minimal value. This is used to normalize the image values between 0 and 1 and allows utilization of the full dynamical range [3].

Subsequently, a 2D-Wiener filter [13] is used to enhance the images. The filter uses a pixel-wise adaptive Wiener method based on statistics estimated from a local neighborhood of each pixel. The Wiener filter minimizes the mean square error between the estimated random process and the desired process [2].

The third step of preprocessing is a Gamma correction [14] - raising the value of each pixel by a constant power, γ. Gamma correction is a common nonlinear operation used to decode luminance in imaging systems. It is used to optimize the usage of bits when encoding the image [3]. The optimal values of γ were obtained by testing the accuracy and sensitivity of the system to estimate the radius of artificial florescent beads, as seen in Fig. 3.

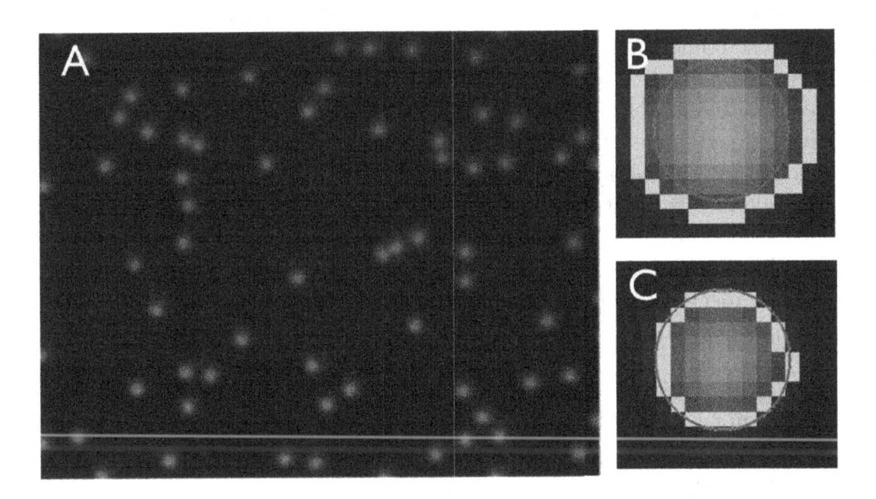

Fig. 3. A - a small portion of the artificial beads view of field showing about 50 florescent beads. B and C zoom in on one bead with the segmented border in green and the true radius of the bead in red. B - shows γ values that are too high and cause the radius estimation to overshoot while C shows the result of the optimal γ - 2.5 for nuclei segmentation and 1.2 for cytoplasm segmentation. (Color figure online)

A common noise in this type of images are large circular shapes of higher intensity due to dust particles diffracting the light, cell debris, coating of surface, or floating cells that were detached from the well bottom. Therefore, in the final step of the preprocessing in order to remove such noises, we used a Top Hat Filtering [4]. The top hat filter is based on neighborhood ranking, but it uses the ranked value from two different size regions. The brightest value in a circular interior region is compared to the brightest value in a surrounding annular region. If the brightness difference exceeds a threshold level, it is kept (otherwise it is removed).

2.2 Nuclei Segmentation

The image of the nuclei has higher SNR (compared to the image of the cells), and since most of the nuclei are more isolated from each other due to the physical extent of the surrounding cytoplasm therefore, we start the segmentation process with the nuclei. The nuclei will later on be used to seed the cell segmentation.

In fluorescence microscopy images, DAPI staining is often used to mark the cells nuclei. DAPI is a fluorescent stain that binds strongly to A-T rich regions in the DNA and is therefore bound to the nuclei which are rich in DNA [10].

In the first step we categorize the pixels to cells and background. We use the Otsu's algorithm [9] to perform image thresholding. Otsu's method is an iterative thresholding algorithm that calculates the optimum threshold separating the two classes so that their combined spread (intra-class-variance) is minimal. The algorithm correctly segments most of the isolated nuclei, but fails to resolve the ambiguity caused by nuclei that are overlapping or adjacent to each other.

To resolve the issue of adjacent nuclei, we first designed a filter to separate the putative'fused' nuclei from the well separated ones. The filter uses morphological features such as the area of the objects (relatively larger in fused nuclei), and their compactness (relatively uncompacted in fused nuclei). By separating the putative merged nuclei we avoid unnecessary processing of well segmented nuclei and reduce computational complexity.

After we find the putative merged nuclei objects, they are passed through a marker-based Watershed transform [1] to find the boundaries of an overlapping nuclei. With this method, we use bright objects to "mark" at least a partial group of connected pixels inside each nuclei cluster to be segmented. We also have to mark the background.

The Watershed algorithm is based on topology maps; whenever raindrops fall, they will be caught in basins, and the idea is to find the separatrix between the basins. This is a known method to separate two objects that were mistakenly classified as one. The Watershed line is found by "raising water" from each basin and the marked line is where the water meets.

The result of the Watershed transform are often over-segmented, due to initialization with too many basins. In order to undo this side effect, we used the suggested merging algorithm from [12] for over-segmented nuclei, i.e., a nucleus which has been segmented into more than one object due to watershed algorithm. Let N be the total number of segmented objects found by the watershed algorithm. Let A_{min} be the minimum size of a nucleus in the image. Two objects are considered merged if they belong to the same object before the application of the watershed algorithm. The merging process first finds the smallest of object within each iteration and then uses the following checking process to update the segmentation until no more adjacent objects can be merged. The checking process is implemented as follows:

1. If the size of an object is less than A_{min}, it is merged with its smallest adjacent neighbor. Isolated objects with sizes less than A_{min} are discarded.
2. If the size of an object is greater than A_{min} compactness values are calculated: for each fragment of the adjacent object, and for the adjacent object as a whole. If the merged object's compactness is the smallest, these fragments are merged.

2.3 Cell Segmentation

Cells are marked by GFP (Green fluorescent protein) gene which is fused to a reporter that reports the activity of a single signaling pathway. In light of this fact, segmenting cell bodies is a more challenging task than segmenting the nuclei. This is because of several factors:

1. When a signaling pathway is activated the location of the GFP-fusion reporter changes, most commonly by translocation from the plasma membrane or cytoplasm into the nucleus.
2. Cell bodies have more varied geometrical forms and can take many shapes.
3. Cell bodies are larger and tend to touch each other forming large clusters of cells.
4. Cell bodies have a large extent and a varied Z extent making the edges of the cells dimmer and harder to detect correctly.
5. The GFP amount expressed in the cells varies widely and we need to be able to segment even cells that are very weakly expressing GFP. This in turn causes the SNR to be very low compared to the nuclei case. This also brings about a problem of auto-fluorescence of cells that did not express GFP and should be excluded.
6. Dying cells are visible and usually seem brighter. We exclude them from segmentation.

These considerations mean that we need an extremely sensitive algorithm that uses side information to correct errors and remove false positives.

The main concept in this algorithm is to classify each pixel of the image, either to the background or to the foreground (to a cell). In order to do this, we first fit the histogram of pixel intensities with a Gaussian mixture model (GMM) [6]. We used a mixture of two Gaussians,

$$p(x) = \frac{a_1}{\sqrt{2\pi\sigma_1}} \exp\left(-(x-\mu_1)^2/(2\sigma_1)\right) + \frac{a_2}{\sqrt{2\pi\sigma_2}} \exp\left(-(x-\mu_2)^2/(2\sigma_2)\right), \quad (1)$$

under the assumption that the background has a Gaussian distribution that should be somewhere near zero, and that the pixels which belong to cells also come from a Gaussian distribution, but have higher mean intensity. Once the histogram fits the model, we can easily classify each pixel to one of the two sub-populations: background and cells. Subsequent pixel classification, we use morphological closing to close any "holes" in the cytoplasm of the cells where the pixels in a small neighborhood were mis-classified. Therefore, we get continuous objects after applying the image closing. The morphological close operation is a dilation followed by an erosion, using the same structuring element for both operations.

Since most of the cells in the field of view (FoV) are not expressing the investigated mutations, we find the *relevant nuclei* using a logical AND operation between nuclei labels and the cells black and white image. This way, we only need to consider the few hundreds of nuclei in areas where cells did express GFP rather than the thousands of cells in each well. The *relevant nuclei* we find will serve us as the seed to the cross-spectral watershed [8] algorithm for cell boundaries detection.

The cross-spectral watershed algorithm is similar to the simple watershed algorithm applied to separate nuclei. The only difference is in the initialization method of the algorithm. Instead of taking local maxima in the objects, we take the eroded nuclei as the basins for the watershed of the cells. The erosion is applied to ensure that each basin inside the corresponding cell (for example, that is not touching a border of a cell). The advantage of using the nuclei as the basins is that we start with the correct number of cells, and that we can ignore issues of intracellular luminance changes which might have caused several minima in each cell resulting in over-segmented cells.

Lastly, the cells and nuclei that were found go through a statistical rule based rejection of cells based on biological considerations. Specifically, these are based on different cell features, such as the ratio of nucleus area to total cell area, max/min size of cells, max/min size of nucleus, which we have accumulated.

3 Evaluation Results

In order to test the efficiency, accuracy and robustness of the proposed method, we tested the accuracy of system by: (1) having human experts (PhD in biology with 5+ years of fluorescent microscopy experience) evaluate the segmented images of an internal HeLa CRM CCL-2 dataset, (2) compare the results of the system on unseen benchmark datasets of U2OS and NIH3T3 cells from [5].

Our internal dataset is composed of images of transfected cells (according to the assay described in the introduction) of 10 96-well plates. These plates encompass a large variety of genetic mutations being transfected into the cells, as well as a large variety of external conditions such as light source fluctuations, user variability and more. All cells were treated and imaged at the NovellusDx laboratory. In Fig. 4 are examples of the segmentation results overlaid on the preprocessed images. We bring here results of outlier images (low-signal, high-signal, dense/sparse seeding, round/spread cells). The system shows remarkable accuracy as judged by the expert evaluation in a wide range of work parameters.

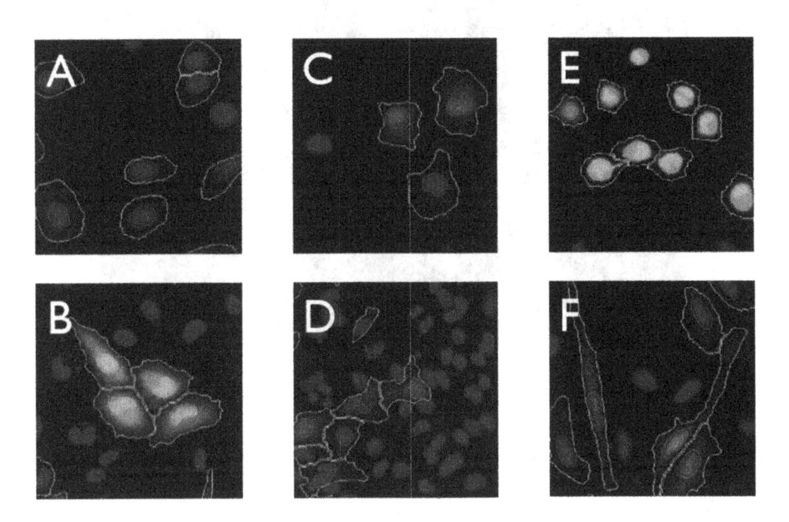

Fig. 4. Accuracy of the system in varied array of conditions. A & B low and high levels of signal. C & D sparse and high density of cell seeding. E & F round and spread cells due to different mutations introduced in the transfection

In Fig. 5 we show the results of the DAPI stained nuclei segmentation of the system on the benchmark dataset of [5]. In red, are the manually segmented borders of the nuclei and in green the automatically generated segmentation of the system. It is important to note that the segmentation required no changes at all even though the images were scanned in different microscopy systems and are of different cell-lines.

In [5] the authors provide a table summarizing the performance of 8 segmentation algorithms on the two datasets. In Table 1 we reproduce the main performance criteria of the table together with the performance of the system presented here on the same datasets. RI(Rand index) measures the fraction of the pairs where the two clusterings agree. The Rand index ranges from 0 to 1, with 1 corresponding to perfect agreement. Let S be a (binary) segmented image and R be a (binary) reference image. Let i and j range over all pairs of pixels where i not equal to j, then each pair falls into one of four categories: (a) $R_i = R_j$ and $S_i = S_j$, (b) $R_i \neq R_j$ and $S_i = S_j$, (c) $R_i = R_j$ and $Si \neq S_j$, (d) $R_i \neq R_j$ and $S_i \neq S_j$. If we let a, b, c, d refer to the number of pairs in its corresponding category, then the Rand index is defined as:

$$RI(R, S) = \frac{a + d}{a + b + c + d} \tag{2}$$

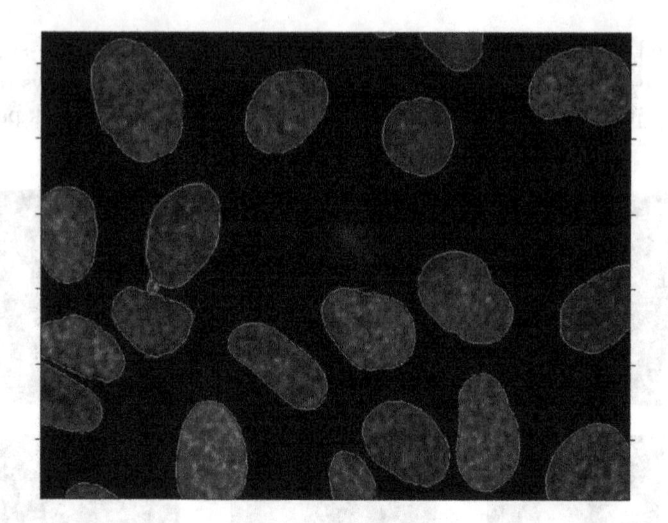

Fig. 5. Accuracy of the system compared to manually segmented nuclei. The red borders are the manual benchmark and the green borders are the automatically generated segmentations without any changes to the algorithm. (Color figure online)

Table 1. Comparison of Segmentation Algorithms. Result of various segmentation approaches are compared against the hand-segmented standard. Each entry contains two values corresponding to the statistic for two datasets used, U2OS and NIH3T3 cells, respectively. The table is reproduced from [5] and the last row was added to show the systems performance. Note, this dataset has only images of nuclei (no cell bodies) which is the more challenging task.

Algorithm	RI	JI	Split	Merged	Spurious	Missing
AS manual	95%/93%	2.4/3.4	1.6/1.0	1.0/1.2	0.8/0.0	2.2/3.2
RC threshold	92%/77%	2.2/2.1	1.1/1.0	2.4/2.4	0.3/1.9	5.5/22.1
Otsu threshold	92%/74%	2.2/2.1	1.1/0.8	2.4/2.1	0.3/1.7	5.6/26.6
Mean threshold	96%/82%	2.2/1.9	1.3/1.4	3.4/5.1	0.9/3.1	3.6/4.8
Watershed (direct)	91%/78%	1.9/1.6	13.8/2.9	1.2/2.4	2.0/11.6	3.0/5.5
Watershed (gradient)	90%/78%	1.8/1.6	7.7/2.6	2.0/3.0	2.0/11.4	2.9/5.4
Active masks	87%/72%	2.1/2.0	5.5/5.0	2.1/1.5	0.4/3.9	10.8/31.1
Merging algorithm	96%/83%	2.1/1.9	0.7/2.5	2.1/3.0	1.0/6.8	3.3/5.9
NovellusDx	**97%/92%**	**2.2/1.9**	**1.2/3.8**	**1.8/3.5**	**0.4/6.9**	**0.5/6.7**

Based on the same definitions for a, b, c, d, the *JI* (Jaccard index) is defined as:

$$JI(R,S) = \frac{a+d}{b+c+d} \tag{3}$$

The Jaccard index is not upper-bounded, but higher values correspond to better agreement.

split is a measure of over-segmentation (1 nucleus segmented as 2 or more); *merged* is a measure of erroneously clustering several nuclei as one nucleus; *spurious* is a measure for noise erroneously segmented as nucleus; *missing* is the measure of missing nuclei. More details about the performance matrices can be found in [5]. Even though the system presented here was not calibrated to the new datasets that were obtained from different cell-lines and different optical conditions, our system outperforms the reference algorithms in most criteria.

One of the most challenging tasks in the segmentation of florescent images is the overlap and aggregation of cells into clusters. Figure 6 shows such an example of a cluster of 9 adjacent cells and the resulting segmentation using the cross-spectral watershed algorithm. It is easy to see that the 9 transfected cells are well segmented.

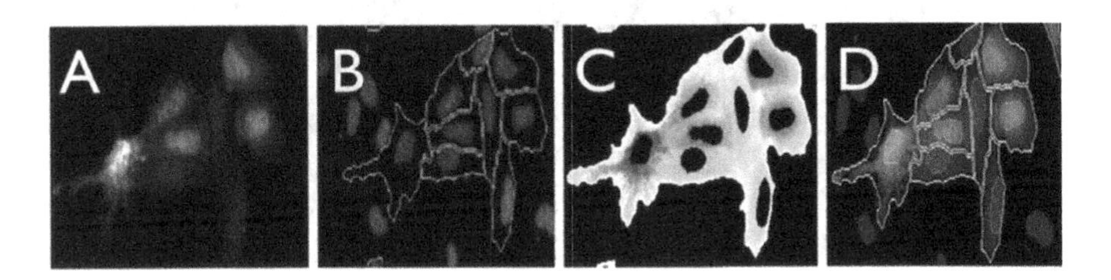

Fig. 6. A - An example of a cluster of cells, consisting a variety of cells. B - Corresponding DAPI image with the segmented cell boundaries. Note that there are more visible nuclei in the DAPI image than cells in A. C - The input of the cross spectral watershed algorithm: clustered cells (gray object) and their relevant nuclei (in black inside the cells). D - The final segmentation of the clustered cell as done by the watershed transform and the corresponding nuclei combined into a one RGB image where the R (red) initialized with nuclei perimeters, G (green) initialized with the preprocessed GFP image and B (blue) initialized with the preprocessed DAPI image. (Color figure online)

In order to test the robustness of the system to external variables, we scanned the same well in three different exposure times to simulate changing conditions of the light source and/or changes to the efficiency of the GFP transfection. We scanned the GFP channel in 600 [ms], 700 [ms], and 800 [ms]. The images were then analyzed in the system and the results of the three different segmentations are overlaid (red, yellow and white respectively) on the 800 [ms] preprocessed image in Fig. 7. One can see that there is complete agreement between the different conditions. When calculating the root mean square error (RMSE) between the borders of the segmented 600 [ms] image and the segmented 800 [ms] image we find it is smaller than 5% indicating the strong agreement in the location of the cell borders.

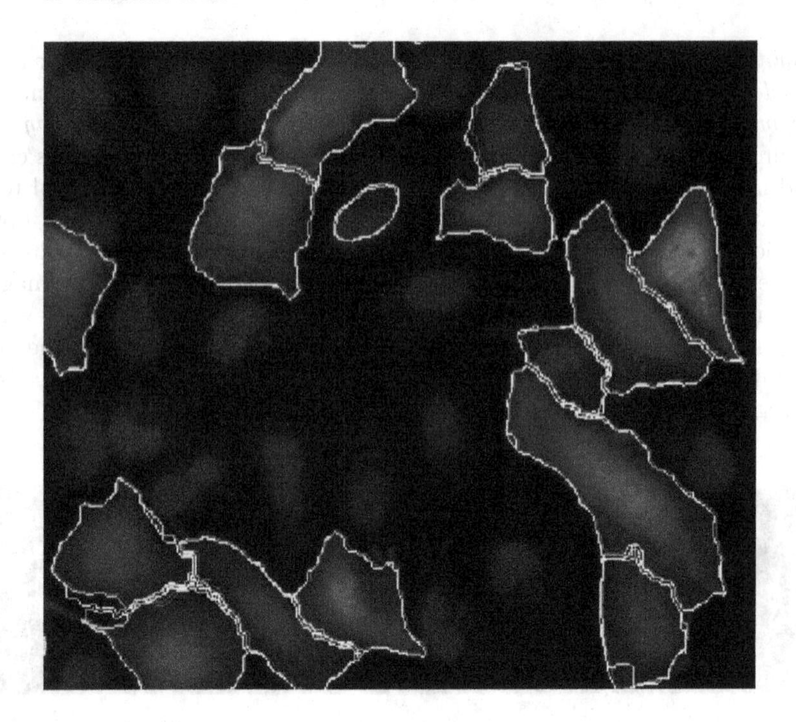

Fig. 7. Robustness to overall brightness. The sensitivity to overall brightness (due to expression level, light source variability, etc.) was tested by imaging the same cells in three different exposure times: 600, 700, 800 [ms]. The resulting segmented cells are represented with red, yellow and white respectively overlaid on the image showing the code is invariant to exposure length of these scales (RMSE of the area differences is less than 5%). (Color figure online)

4 Conclusion

We have shown that the cross-spectral watershed algorithm is effective in cell segmentation for a vast variety of cells. To date, the system was used to segment 2,677,378,332 nuclei and 498,291,752 transfected cells under 14,603 unique genetic conditions effecting the size, shape and other parameters of the cells. On average 300 cells can be segmented in 1 min on a single core running Ubuntu operating system and the MatLab code.

When compared to state-of-the-art systems, the system outperforms the benchmarks on most evaluation criteria

The system we introduced here is also the kernel of a live-cell segmentation algorithm. In it, the system introduced here segments the still frame images in a time-lapsed movie. The results of each still frame are the inputs to a tracking algorithm that determines the unique ID of each cell, it's movements, cell death and cells divisions.

We introduced here novel methods and improvements of the up-to-date algorithms for cells and nuclei segmentation. Specific emphasis was devoted to make this algorithm as accurate as possible and insensitive to environmental changes, all together in parallel with efficient MatLab code writing to be able to get results of over thousands of cells within a few minutes. The robustness of the system was examined with a different nuclei images data-set from [5] and has been proved to be effective there as well. The live-segmentation algorithm could be used as a tool that provides an opportunity to observe and measure cell-cycle progression of individual cells in a large population.

Acknowledgments. The authors would like to gratefully acknowledge the entire NovellusDx team which run the biological experiments and the deep discussions with regards to biological parameters to exclude cells. The authors would also like to thank Prof Zohar Yakhini for his careful review and insightful comments.

References

1. Bleau, A., Leon, J.L.: Watershed-based segmentation and region merging. Comput. Vis. Image Underst. **77**(3), 317–370 (2000)
2. Brown, R.G., Hwang, P.Y.: Introduction to Random Signals and Applied Kalman Filtering. Wiley, New York (1997)
3. Brown, C.M.: Fluorescence microscopy-avoiding the pitfalls. J. Cell Sci. **120**(10), 1703–1705 (2007)
4. Bright, D.S., Steel, E.B.: Two-dimensional top hat filter for extracting spots and spheres from digital images. J. Microsc. **146**(2), 191–200 (1987)
5. Coelho, L.P., Shariff, A., Murphy, R.F.: Nuclear segmentation in microscope cell images: a hand-segmented dataset and comparison of algorithms. In: Proceedings/IEEE International Symposium on Biomedical Imaging: from Nano to Macro IEEE International Symposium on Biomedical Imaging, vol. E6, pp. 518–521 (2009). doi:10.1109/ISBI.2009.5193098
6. Duda, R.O., Hart, P.E.: Pattern Classification and Scene Analysis. Wiley, Hoboken (1973)
7. Le Tourneau, C., et al.: Molecularly targeted therapy based on tumour molecular profiling versus conventional therapy for advanced cancer *SHIVA*: a multicentre, open-label, proof-of-concept, randomised, controlled phase 2 trial. Lancet Oncol. **16**(13), 1324–1334 (2015)
8. Lindblad, J., Wahlby, C., Bengtsson, E., Zaltman, A.: Image analysis for automatic segmentation of cytoplasm and classification of Rac1 activation. Cytometry **57A**, 22–33 (2004)
9. Otsu, N.: A threshold selection method from gray level histogram? IEEE Trans. Syst. Man Cybern. **SMC–8**, 62–66 (1978)
10. Tarnowski, B.I., Spinale, F.G., Nicholson, J.H.: DAPI as a useful stain for nuclear quantitation. Biotech. Histochem. **66**(6), 296–302 (1991)
11. Vogelstein, B., et al.: Cancer genome landscapes. Science **339**(6127), 1546–1558 (2013)
12. Chen, X., Zhou, X., Wong, S.T.C.: Automated segmentation, classification, and tracking of cancer cell nuclei in time-lapse microscopy. IEEE Trans. Biomed. Eng. **53**(4), 762–766 (2006)
13. Lim, J.S.: Two-Dimensional Signal and Image Processing. Prentice Hall, Englewood Cliffs (1990). Equations 9.26, 9.27, and 9.29

14. Poynton, C.: Digital Video and HDTV Algorithms and Interfaces. Morgan Kaufman Publishers, San Francisco (2003)
15. Turski, M.L., et al.: Genomically driven tumors and actionability across histologies: BRAF-mutant cancers as a paradigm. Mol. Cancer Ther. **15**, 533–547 (2016)
16. Friedman, A.A., Letai, A., Fisher, D.E., Flaherty, K.T.: Precision medicine for cancer with next-generation functional diagnostics. Nat. Rev. Cancer **15**, 747–756 (2015)
17. Chapman, P.B., et al.: Improved survival with vemurafenib in melanoma with BRAF-V600E mutation. N. Engl. J. Med. **364**, 2507–2516 (2011)

Unsupervised Superpixel-Based Segmentation of Histopathological Images with Consensus Clustering

Shereen Fouad[1(✉)], David Randell[1], Antony Galton[2], Hisham Mehanna[3], and Gabriel Landini[1]

[1] School of Dentistry, Institute of Clinical Sciences, University of Birmingham, Birmingham, UK
{s.a.fouad,d.a.randell,g.landini}@bham.ac.uk

[2] Department of Computer Science, University of Exeter, Exeter, UK
apgalton@ex.ac.uk

[3] Institute of Head and Neck Studies and Education, University of Birmingham, Birmingham, UK
h.mehanna@bham.ac.uk

Abstract. We present a framework for adapting consensus clustering methods with superpixels to segment oropharyngeal cancer images into tissue types (epithelium, stroma and background). The simple linear iterative clustering algorithm is initially used to split-up the image into binary superpixels which are then used as clustering elements. Colour features of the superpixels are extracted and fed into several base clustering approaches with various parameter initializations. Two consensus clustering formulations are then used, the Evidence Accumulation Clustering (EAC) and the voting-based function. They both combine the base clustering outcomes to obtain a single more robust consensus result. Unlike most unsupervised tissue image segmentation approaches that depend on individual clustering methods, the proposed approach allows for a robust detection of tissue compartments. For the voting-based consensus function, we introduce a technique based on image processing to generate a consistent labelling scheme among the base clustering outcomes. Experiments conducted on forty five hand-annotated images of oropharyngeal cancer tissue microarray cores show that the ensemble algorithm generates more accurate and stable results than individual clustering algorithms. The clustering performance of the voting-based consensus function using our re-labelling technique also outperforms the existing EAC.

Keywords: Superpixel segmentation · Consensus clustering · Histology · Histopathology · Image analysis

1 Introduction

The automatic segmentation of digitised histological images into regions representing different anatomical or diagnostic types is of fundamental importance

M. Valdés Hernández and V. González-Castro (Eds.): MIUA 2017, CCIS 723, pp. 767–779, 2017.
DOI: 10.1007/978-3-319-60964-5_67

for developing digital pathology diagnostic tools. Superpixel segmentation is an advanced method to group image pixels with similar colour properties into atomic regions to simplify the data in the pixel grid [1]. Recently, superpixel methods have been combined with pattern recognition techniques for image segmentation (e.g. [2]) where certain features (e.g. colour, morphology) are extracted from the superpixels and then fed to pattern recognition procedures that assign each superpixel to an expected histological class. Supervised methods are built from labelled training sets to predict the classes of novel unlabelled data and they require access to ground truth reference images for the training. In contrast, unsupervised approaches (clustering analysis) do not require pre-labelled training sets for their learning but instead rely on certain similarity measures to group data into separate homogeneous clusters. In histopathological imaging analysis, clustering is particularly useful as an exploratory tool as it can provide information about hidden anatomical or functional structures in images.

Clustering algorithms use different heuristics and can be sensitive to input parameters, i.e. repeatedly applying different clustering methods on the same dataset often yields different clustering results. Furthermore, a given clustering algorithm may give rise to different results for the same data when the initialisation parameters change. Consensus Clustering (CC) [3] methods have addressed this issue by combining solutions obtained from different clustering algorithms into a single consensus solution. In unsupervised learning, this enables more accurate and robust estimation of clusterings when compared to single clustering algorithms. CC is often performed in two main steps, (a) the cluster ensemble generation, and (b) the consensus function, which finds a consensual opinion of the ensemble. CC techniques have proved efficient in a variety of practical domains; their application to histological image segmentation is, however, relatively new.

In this work, we investigate the use of CC in the context of superpixel-based segmentation of haematoxylin and eosin (H&E) stained histopathological images. We suggest a multi-stage segmentation process. First, we use the recently proposed Simple Linear Iterative Clustering (SLIC) superpixel framework [1,4] to segment the image into compact regions. Colour features from each dye are extracted from the superpixels and used as input to multiple base clustering algorithms with various parameter initializations. The generated results (denoted here as partitions) pass through an ensemble selection scheme which generates a more effective ensemble based on partitions diversity. Two consensus functions are considered here, the Evidence Accumulation Clustering (EAC) [5] and the voting-based consensus function (e.g. [6,7]).

Unlike supervised methods, labels resulting from unsupervised techniques are symbolic (i.e. labels do not represent a meaningful class), and consequently an individual partition in the ensemble includes clusters that do not necessarily have labels that correspond to other clusters in different partitions of the ensemble. In the voting-based consensus function the label mismatch is defined as the problem of finding the optimal re-labelling of a given partition with respect to a reference partition. This problem is commonly formulated as a weighted bipartite matching formulation [6,7], and it is solved by inspecting whether data patterns in two partitions share labels more than with other clusters. In this paper, we present an alternative simple, yet robust, implementation for generating a consistent labelling scheme among the different partitions of the ensemble. Our approach considers the space occupied by each individual cluster in an image and exploits the fact that pairs of individual clusters from different partitions would match when their pixels largely overlap in a segmented image.

2 Related Work

SLIC [1] is an advanced superpixel method that generates compact, mostly uniform superpixels by agglomerating pixels based on colour similarity and proximity in the image plane. Achanta et al. [4] conducted an empirical comparison of SLIC with other state-of-the-art superpixel algorithms, which revealed the superiority of SLIC in terms of performance and speed. They also showed that SLIC is easy to use and implement, low in computational cost and requires fewer parameters than other algorithms. All these features are potentially useful for automatic segmentation of large, complex and variable histopathological images. SLIC superpixels have been used before to facilitate and improve unsupervised segmentation of histopathological images. SLIC was applied in [2] as a pre-processing step to decrease the complexity of large histopathological images. Colour descriptors of the generated regions were then used in an unsupervised learning formulation of the probabilistic models of expected classes using the Expectation-Maximisation (EM) [8].

Consensus Clustering (CC) methods have emerged for improving robustness, stability and accuracy of unsupervised learning solutions. Contributions in this field include the EAC [5] and voting-based algorithms. A comprehensive survey of existing clustering ensemble algorithms is presented in [3]. The voting-based literature utilizes different heuristics in attempting to solve the labelling correspondence problem. This problem is commonly formulated as a bipartite matching problem [6], where the optimal re-labelling is obtained by maximizing the agreement between the labelling of an ensemble partition with respect to a reference partition. The agreement is estimated by constructing a $K \times K$ contingency table between the two partitions, where K is the number of clusters in each partition[1]. Each entry of the contingency table holds the number of cluster label co-occurrences counted for the same set of objects in the two partitions.

There have been previous publications on CC in unsupervised histopathological segmentation, but to the best of our knowledge, its application to superpixel-based segmentation remains unexplored. Simsek et al. [9] defined a set of high-level texture descriptors of colonic tissues representing prior knowledge, and used those in a multilevel segmentation where they used a cluster ensemble to combine multiple partitioning results. Khan et al. [10] proposed ensemble clustering for pixel-level classification of tumour vs. non-tumour regions in breast

[1] The two partitions should contain the same number of clusters, K.

cancer, where random projections of low-dimensional representations of the features and a consensus function combined various partitions to generate a final result.

3 Unsupervised Superpixel-Based Segmentation with Consensus Clustering

3.1 Dataset and Preprocessing

Our data consisted of H&E stained tissue images (paraffin sections) of human oropharyngeal cancer processed into tissue micro arrays (TMAs), prepared at the Institute of Cancer and Genomic Sciences, University of Birmingham, UK. H&E is the commonest staining method used in routine diagnostic microscopy; haematoxylin primarily stains nucleic acids and nuclei in blue/violet while the eosin counter-stain primarily stains proteins in the intra- and extra-cellular compartments in pink. TMAs are usually used for the analysis of tumour markers of multiple cases (cores) in single batches where there is a need to identify various components in the samples. Samples were digitised using an Olympus BX50 microscope with a x20 magnification objective (N.A. 0.5, resolution 0.67 µm) and a QImaging Retiga 2000R camera and a tunable liquid crystal RGB filter (Surrey, BC, Canada).

Tissue core images were $\approx$3300 $\times$ 3300 pixels (inter-pixel distance of 0.367 µm). Fifty five images were used for the analysis (ten for training and forty five for testing), which provided the range of variations in tissue distributions typically found in this type of histological material (2.3 to 98.8% of epithelium tissue component and 25.5 to 83.2% of background out of the whole image).

As a preprocessing step, colour deconvolution [11] was applied to the H&E image I to separate the RGB information into haematoxylin-only and eosin-only images. With this procedure, up to three dyes (in our case, H&E) can be separated into 'stain' channels. This can be applied when the colours of the dyes on their own are known and combine as light-absorbing dyes. In the case of two-dye stains, a third component is a residual channel of the deconvolution process. The results of the colour deconvolution can be combined into a "stain" RGB image here denoted I^* to better represent the dye absorption of the different tissue types. In I^* the R, G and B channels now hold the light transmittance of the haematoxylin, eosin and residual images, instead of containing the RGB components. The feature extraction discussed in Sect. 3.3 is applied to this image I^*.

3.2 Superpixel-Based Segmentation

The SLIC segmentation spits-up the original image I into a set of superpixels held in a binary image S. The superpixels tend to be compact and relatively uniform. They are formed by pixel grouping based on colour similarity and spatial proximity. In detail, a k-means algorithm [12] is used to cluster a five-dimensional vector consisting of the 3 components of a pixel colour in CIELAB space and the pixel spatial coordinates. A special similarity measure is then exploited, replacing the standard Euclidean distance, which weighs the distance in the colour and spatial domain. This measure weighs the relative importance between color similarity and spatial proximity in the five-dimensional space. Furthermore, it allows the size and compactness of the resulting superpixels to be adjusted, providing some control over the number of superpixels generated.

In our experiments, we used the recently proposed jSLIC [13], a Java implementation of SLIC that is faster than the original (in [14]). Unlike the original, jSLIC avoids computing the same distances between data by exploiting precomputed look-up tables. Borovec et al. showed that the jSLIC is able to segment large images with intricate details into uniform parts, which is particularity useful for complexity-reduction problems (as is the case here). The authors also defined a function f that compromises between superpixel compactness and the alignment of object boundaries in the image. This is expressed as: $f = m \cdot z^2$, where m is the initial superpixel size and z is a regularisation parameter which affects the superpixel compactness. The value of z lies within the range [0,1], where 1 yields nearly square segments and 0 produces very 'elastic' superpixels. To ensure an effective segmentation, we performed a cross validation procedure for the configuration of these two parameters, as discussed in the Experiments and Evaluation section.

3.3 Feature Extraction

Colour features are known for their relevance in visual perception and are exploited here for the discrimination of superpixels into different histological regions. Our H&E images contain at least three types of regions that uptake dyes differently: (1) stratified squamous epithelial tissue (a 'solid' tissue with densely packed cells which appear more darkly stained than the rest), (2) connective stroma, which is less cellular and contains abundant extracellular matrix, blood vessels, inflammatory cells, and sometimes glandular tissue, and (3) background areas, often appearing white or neutral grey. First the colour descriptors for each superpixel in image S are computed but instead of referring to the original I, these are extracted from the data in image I^*, so they become 'stain features' that quantify the distribution of the stain uptake in the superpixels. We used eleven measures for each stain (mode, median, average, average deviation, standard deviation, minimum, maximum, variance, skew, kurtosis and entropy) for each of the three colour deconvolution components (haematoxylin, eosin and the residual channel), forming a vector of thirty-three colour descriptors per superpixel.

3.4 Consensus Clustering (CC) Frameworks

The CC framework exploited here involves three main steps (1) creation of an ensemble of multiple cluster solutions, (2) selection of an effective sub-set of cluster solutions based on their diversity measure, and (3) generation of a final partition via the so-called consensus function. A clustering algorithm takes the set $X = \{x_1, x_2, ..., x_n\}$ of n superpixels as an input, and groups it into K clusters (epithelium, stroma and background regions) forming a data partition P. Note that x_i is characterized here by the 33-dimensional colour features described in the previous section.

Ensemble Generation and Selection. First, a number of q clustering results are generated for the same X, forming the cluster ensemble E, where $E = \{P_1, P_2, ...P_q\}$. To this end, we used five different clustering algorithms and ran each of those multiple times while varying their parameters. There are two factors that influence the performance of this approach: one is the accuracy of the individual clusters (P_i) and the other is the diversity within the ensemble E. Accuracy is maintained by tuning a set of effective clustering methods to obtain the best set of results. Regarding the diversity of E, it was shown in [15] that a moderate level of dissimilarity among the ensemble members (E) improves the consensus results. For this, we studied the diversity within E, using the Rand Index (RI) similarity measure [16], and created a more effective sub-set of cluster solutions to represent the new ensemble, denoted here as E'. This new ensemble was obtained by pruning out significantly inconsistent partitions as well as identical or closely-similar partitions.

Given clustering solutions P_i in the original ensemble E, in order to decide whether P_i is included in E', we measure how well P_i agrees with each of the clustering solutions (P_j) contained in E, where $i = 1, \cdots q$, as follows:

$$\text{similarity}(P_i, E) = \frac{1}{q-1} \sum_{j=1}^{q} \text{RI}(P_i, P_j), \tag{1}$$

where ($P_i, P_j \in E$) and ($i \neq j$). The RI counts the pairs of points (in our case superpixel pairs) on which two clusterings agree or disagree and it is computed as:

$$\text{RI}(P_i, P_j) = \frac{TP + TN}{TP + FP + TN + FN}, \tag{2}$$

where TP and TN are the number of pairs correctly grouped in the same, and different clusters, respectively. FP is the number of dissimilar pairs assigned to the same cluster and FN is the number of similar pairs grouped in different clusters. The RI lies between 0 and 1, where 1 implies the two partitions agreeing perfectly and 0 that they completely disagree. We defined two thresholds T_1 and T_2 that correspond to the minimum and maximum accepted levels of diversity among the partitions. If P_i exhibits an acceptable level of diversity with respect to the rest of the population in E (i.e. similarity$(P_i, E) \geq T_1$ and similarity$(P_i, E) \leq T_2$) then it is considered as an eligible voter and is added to the new ensemble E'. If the opposite applies then the partition is excluded from the new ensemble. The total number of selected partitions in E' is denoted here as q', where $q' \leq q$. E' is formed as follows,

$$E' = \{P_i \,|\, \text{similarity}(P_i, E) \in [T_1, T_2]\}. \tag{3}$$

The next step consists of finding the consensual partition, denoted here as P^*, based on the information contained in E'. For this we use two consensus functions described below.

Evidence Accumulation Consensus (EAC) Function [5]. This method, denoted here as EAC-CC, considers the co-occurrences of pairs of patterns in the same cluster as votes for their association. In particular, the algorithm maps the q' partitions in E' into an $n \times n$ co-association matrix M. Each entry in M is defined as $M_{ij} = u_{ij}/q'$, where u_{ij} is the number of times the pattern pair (i, j) are grouped together in the same cluster among the q' partitions. The more frequent a pair of objects appear in the same clusters, the more similar they are. Note that M is needed here because of the label correspondence problem occurring among partitions of E'. M can now be viewed as a new similarity measure among the data patterns and it comprises real numbers ranging from 1 (perfect consensus among partitions) down to 0 (no association). The consensus cluster P^* is obtained by applying an appropriate similarity-based clustering algorithm on M (e.g. the hierarchical agglomerative clustering algorithm [17]). The final clustering output here (P^*) is represented in another image, namely S'. Although the interpretation of the results of the EAC are intuitive, it has a quadratic complexity in the number of patterns, $O(n^2)$.

Voting-Based Consensus Function. This method, denoted here as Vote-CC, utilizes a majority voting technique to find the P^* that optimally summarizes E', first, however, it solves the problem of labelling correspondence among different partitions in E'. Here we propose a simple re-labelling algorithm using imaging processing tools to match the symbolic cluster labels between the different partitions in E'. The method finds the optimal re-labelling of a given partition P with respect to a reference fixed partition P'. P' is selected from E' as the one with highest RI with respect to the ensemble (see Eq. 1).

As we are dealing with images, we first assign the labels resulting from the P' and P to the corresponding regions (or superpixels in this case) located in the binary segmented image S. The labelled regions are displayed in K unique colours in two images denoted here as IMG' and IMG for P' and P, respectively. For example, superpixels with cluster assignments of '1', '2' and '3' in P will be represented in IMG as blue, red and green, respectively. We assume that the number of clusters ranges from 1 to K and the partitions in E' group the data (superpixels) into three clusters (epithelium, stroma and background regions). However, due to the label mismatching problem a pair of correlated clusters from different partitions may be assigned different labels. Our target is therefore to permute the labels, so the cluster labels in P are in the most likely agreement with the labels in P'.

To this end, individual clusters displayed in images IMG' and IMG, denoted here as $k_{p'}$ and k_p, are visualized in two binary images $IMG'_{k_{p'}}$ and IMG_{k_p}, respectively. Note that $k_{p'} \in P'$ and $k_p \in P$. The algorithm then estimates the degree of overlapping between $IMG'_{k'_p}$ and IMG_{k_p} , to assess the similarity between the individual clusters ($k_{p'}$ and k_p). The similarity is obtained using the Jaccard Index (JI) [16], defined as the ratio between the pixel-counts of the intersection and union of $IMG'_{k_{p'}}$ and IMG_{k_p} as follows:

$$\mathrm{JI}_{(IMG'_{k_{p'}},IMG_{k_p})} = \frac{|IMG'_{k_{p'}} \cap IMG_{k_p}|}{|IMG'_{k_{p'}} \cup IMG_{k_p}|}, \tag{4}$$

JI values range from 0 (denoting no matching between $IMG'_{k'_p}$ and IMG_{k_p}, and hence between $k_{p'}$ and k_p) to 1 (denoting perfect matching). For every label $k_{p'} \in P'$ we compute $\mathrm{JI}_{(IMG'_{k_{p'}},IMG_{k_p})}$ obtained against all $k_p \in P$. Then, we find the maximum JI value which gives the most similar cluster in P to k'_p. If k'_p and its highest similar k_p have different labels then the match is achieved by swapping the labels in the original image IMG and therefore the labels in P. The procedure then stores the swapped labels as well as their corresponding JI in two variables. These are needed in order to track whether a label pair of $(k_p, k_{p'})$ has already been swapped in a previous iteration. If true, then swapping $k_{p'}$ and k_p is only performed if they have higher JI value than before (i.e. the swapped pair of $(k_p, k_{p'})$). The process is repeated until all labels in IMG have been inspected against the ones in IMG', and therefore clusters in P are matched with P'. Note that P' remains unchanged throughout the re-labelling process. The procedure is summarized in Algorithm 1 and it has a complexity of $O(K^3)$. The now aligned labels for all the partitions are combined into a final consensus partition P^* via a majority voting technique. In exceptional cases, where the number of votes are equal we select the vote of the partitions that produce the highest total similarity (RI) with respect to the ensemble E' (Eq. (1)). As before, P^* will be represented in image S'.

The idea of cluster re-labelling based on a similarity assessment has been proposed before in relation to voting-based consensus methods. However, those approaches are implemented based on inspection of the labels of data points (i.e. samples as abstract objects with no shape or size) while our re-labelling captures the similarity in a different way, based on the overlap of the superpixels, which in turn represent image regions with their own shapes and sizes.

4 Experiments and Evaluation

The effectiveness of the proposed methodology—CC applied to superpixel-based segmentation—was evaluated in the context of clustering accuracy obtained against five standard clustering approaches: (1) ***k*-means** [12], a centroid based algorithm, (2) **Unsupervised Learning Vector Quantization (LVQ)** [18], LVQ algorithm for unsupervised learning, (3) **EM** [8], a distribution based method (4) **Make Density Based (MDB)** [19], a density based algorithm, and (5) **Agglomerative Hierarchical Clustering (AH)** [17], a pairwise distance based approach. These algorithms were chosen to include a range of different clustering strategies to ensure diversity in the ensemble.

All imaging procedures and machine learning algorithms were implemented on the ImageJ platform [20] using the WEKA data mining JAVA libraries [21]

```
Algorithm 1: Label matching algorithm for the Vote-CC method

Input: P, P', S, n, K
Output: Labels matched for P with respect to P'

for (s=1 to n) do
  Assign label of P'_s to superpixel S_s and save in image IMG'
  Assign label of P_s to superpixel S_s and save in image IMG
end for
for (k'_p=1 to K) do
  for (k_p=1 to K) do
    Threshold k'_p in IMG', convert to mask and save in IMG'_{k'_p}
    Threshold k_p in IMG, convert to mask and save in IMG_{k_p}
    Compute JI_(IMG'_{k_p'}, IMG_{k_p}) using eq.(4)
  end for
  MaxJI = max{JI_(IMG'_{k_p'}, IMG_{k_p})}, where k_p = {1 ··· K}}
  if (k_p' ≠ k_p)
    SwappedLabels = (k_p', k_p)
    JISwappedLabels_(k_p', k_p) = MaxJI
    if (k_p, k_p') ∉ SwappedLabels
      Swap k_p' and k_p and save result in IMG
    else if (k_p, k_p') ∈ SwappedLabels and JISwappedLabels_(k_p', k_p) > JISwappedLabels_(k_p, k_p')
      Swap k_p' and k_p and save result in IMG
    end If
  end If
end for
Assign the new labels in IMG to partition P
```

running on an Intel R core(TM) i7-4790 CPU running at 3.60 GHZ, with 32 GB of RAM and 64-bit Linux operating system. All the algorithms were quantitatively evaluated by comparing their results with forty five gold-standard H&E stained images (denoted here as R) from oropharyngeal cancer TMAs. A set of R images were obtained by manually labelling them into epithelium, stroma and background areas by one of us (GL) with a background in Oral Pathology.

We used three well-known clustering measures [16] to evaluate the algorithm results: (1) **The Rand Index (RI)** was used to compare the final consensus clustering solution given in image S' with their corresponding reference partition given in the gold-standard image R and it is estimated as, $\mathrm{RI}(S', R)$ (see Eq. (2)), where TP, TN, FP, or FN were calculated by considering the overlapping superpixels of S' and R (as explained before). (2) **F1-score** that is defined as: $2 \cdot \frac{precision \cdot recall}{precision + recall}$, (3) **Jaccard Index (JI)** that is defined as: $\mathrm{JI} = \frac{|S' \cap R|}{|S' \cup R|}$,

In all experiments, (hyper)parameters of jSLIC and CC methods were tuned via a cross-validation procedure on a training set of ten additional images. For the superpixel segmentation, the regularisation parameter z and the initial superpixel size m were tuned over the values of (0.2, 0.3, 0.4) and (40, 50, 60), respectively. We found that the optimal values were at 0.3 and 60 for z and m, respectively. The number of clusters was fixed to three in all experiments, corresponding to three most distinct types of content: epithelium, stroma and background regions. The ensemble of cluster solutions was generated by running the five aforementioned clustering algorithms multiple times with various parameter settings. The number of seeds in k-means and EM algorithms were chosen randomly from the range [10, 300]. Learning rates in the LVQ algorithm were set at the values of 0.05, 0.07, 0.09, 0.1 and 0.3. The AH algorithm was used with Complete and Mean link types. The ensemble generation process yielded a total of thirty one clustering solutions, stored in the initial pool of cluster solutions, E. The diversity selection strategy was applied to form another better performing ensemble E'. For this, we assigned the values of 0.5 and 0.9 to the diversity acceptance thresholds T_1 and T_2, respectively.

Table 1 presents a quantitative comparison of the EAC-CC and Vote-CC methods with five individual clustering approaches (mentioned above). For each of the individual clustering algorithms, we selected the result of the best performing run (out of the multiple runs) then we evaluated its mean RI, F1-score, JI and the standard deviations across the forty five images. Figure 1 provides a visual comparison of our output against the clustering methods. For display purposes we randomly selected one clustering output (out of the multiple runs) to represent the performance of the individual clustering approaches.

The results show that EAC-CC and Vote-CC following jSLIC segmentation produce the most accurate results out of the individual clusterings tested (81% and 82%, respectively). The accuracy of the Vote-CC comes very close to the one in EAC-CC. However, Vote-CC significantly outperformed EAC-CC in execution time. This is due to EAC-CC having a large complexity of the order $O(n^2)$ (in our case, n reached

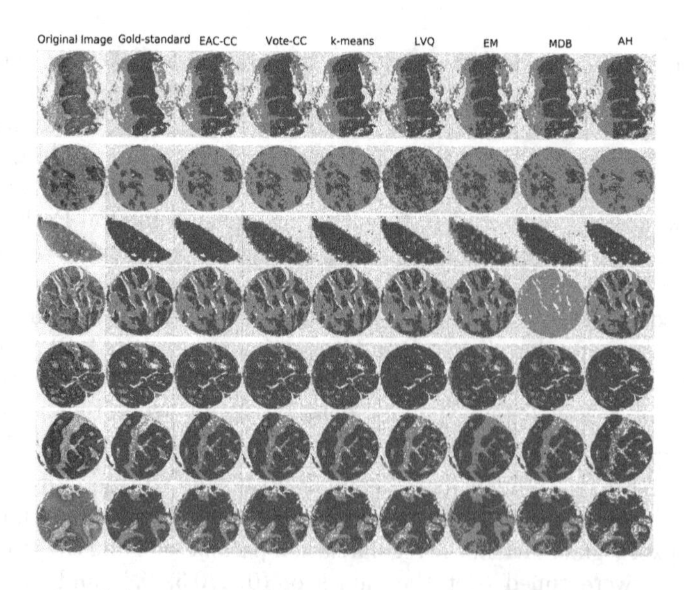

Fig. 1. Examples of tissue regions detection in seven H&E images. From left, the original image, gold-standard, EAC-CC, Vote-CC and the individual clustering methods after superpixel segmentation. Black, white, magenta and green colours correspond to the segmentation lines, background, epithelium and stroma regions, respectively. (Color figure online)

Table 1. Performance evaluation of the EAC-CC and Vote-CC frameworks compared against five individual clustering approaches in terms of mean RI, F1-score and JI along with standard deviations (±) across the forty five images. The best results (Vote-CC method) are marked in **bold font.**

Measure	EAC-CC	Vote-CC	k-means	LVQ	EM	MDB	AH
RI(±)	0.81	**0.82**	0.78	0.77	0.76	0.79	0.72
	±(0.05)	**±(0.05)**	±(0.15)	±(0.11)	±(0.16)	±(0.07)	±(0.12)
F1-score(±)	0.74	**0.75**	0.71	0.72	0.69	0.73	0.68
	±(0.08)	**±(0.09)**	±(0.15)	±(0.12)	±(0.14)	±(0.09)	±(0.15)
JI(±)	0.71	**0.72**	0.69	0.61	0.66	0.68	0.60
JI(±)	±(0.10)	**±(0.10)**	±(0.18)	±(0.12)	±(0.20)	±(0.13)	±(0.17)
Time (ms)	748.95	**31.02**					

up to 5000 in some images) while the complexity of Vote-CC is $O(K^3)$ (with $K = 3$). The results also reveal that CC methods result in greater consistency in performance over individual clustering methods as illustrated by lower standard deviations of the RI and F1-scores. This consistency can be seen visually by comparing the results in Fig. 1. In particular, despite the apparent satisfactory clustering results obtained by the single algorithms across most images, they all failed to perform well in some cases (e.g. notice the unstable performance of the LVQ, MDB and AH in the seven examples depicted in Fig. 1).

5 Conclusion

We presented a method of tissue segmentation of histopathological images using superpixels and Consensus Clustering (CC), a combination that, to our knowledge, has not been applied before in quantitative microscopy. Our approach decreases the spatial complexity of images while retaining important information about their contents which are essential for enabling automated pre-screening and guided searches on histopathological imagery. The proposed method performs an unsupervised detection of image regions that correspond to three classes of interest: epithelium, connective and background areas. A superpixel segmentation was initially performed that was followed by a CC technique which combines the 'opinions'of several clustering algorithms into a single, more accurate and robust result. Our work exploited two CC functions, the EAC and the voting-based. For the latter, we introduced a label matching technique which imposed consistency to the different base clustering outcomes. The method is easy to understand and implement and specially tailored for unsupervised imaging segmentations. Qualitative and quantitative results tested on a set of forty five hand-segmented H&E stained tissue images verified that the CC methods outperform the individual clustering approaches in terms of the accuracy of the results and consistency. Furthermore, the voting-base CC using our re-labelling technique outperforms the EAC in terms of execution time.

Acknowledgments. This work was supported by the EPSRC through funding under grant EP/M023869/1 “Novel context-based segmentation algorithms for intelligent microscopy”.

References

1. Achanta, R., Shaji, A., Smith, K., Lucchi, A., Fua, P., Susstrunk, S.: SLIC superpixels. Technical report, EPFL no. 149300 (2010)
2. Borovec, J.: Fully automatic segmentation of stained histological cuts. In: 17th International Student Conference on Electrical Engineering, Prague, pp. 1–7 (2013)
3. Vega-Pons, S., Ruiz-Shulcloper, J.: A survey of clustering ensemble algorithms. Int. J. Pattern Recogn. **25**(03), 337–372 (2011)
4. Achanta, R., Shaji, A., Smith, K., Lucchi, A., Fua, P., Susstrunk, S.: SLIC superpixels compared to state-of-the-art superpixel methods. IEEE Trans. Pattern Anal. Mach. Intell. **34**(11), 2274–2282 (2012)
5. Fred, A.L., Jain, A.K.: Combining multiple clusterings using evidence accumulation. IEEE Trans. Pattern Anal. Mach. Intell. **27**(6), 835–850 (2005)
6. Topchy, A.P., Law, M.H.C., Jain, A.K., Fred, A.L.: Analysis of consensus partition in cluster ensemble. In: 4th IEEE International Conference on Data Mining, pp. 225–232 (2004)
7. Dudoit, S., Fridlyand, J.: Bagging to improve the accuracy of a clustering procedure. Bioinformatics **19**(9), 1090–1099 (2003)
8. Dempster, A.P., Laird, N.M., Rubin, D.B.: Maximum likelihood from incomplete data via the EM algorithm. J. Roy. Stat. Soc. B **39**(1), 1–38 (1977)
9. Simsek, A.C., Tosun, A.B., Aykanat, C., Sokmensuer, C., Gunduz-Demir, C.: Multilevel segmentation of histopathological images using cooccurrence of tissue objects. IEEE Trans. Biomed. Eng. **59**(6), 1681–1690 (2012)
10. Khan, A.M., El-Daly, H., Rajpoot, N.: RanPEC: random projections with ensemble clustering for segmentation of tumor areas in breast histology images. In: Medical Image Understanding and Analysis (MIUA), Swansea, pp. 17–23 (2012)
11. Ruifrok, A.C., Johnston, D.A.: Quantification of histochemical staining by color deconvolution. Anal. Quant. Cytol. Hist. **23**(4), 291–299 (2001)
12. Hartigan, J.A., Wong, M.A.: Algorithm AS 136: a k-means clustering algorithm. J. R. Stat. Soc. Ser. C Appl. Stat. **28**(1), 100–108 (1979)
13. Borovec, J., Kybic, J.: jSLIC: superpixels in ImageJ. In: Computer Vision Winter Workshop, Praha (2014)
14. The jSLIC-superpixels plugin. http://imagej.net/CMP-BIA_tools
15. Hadjitodorov, S.T., Kuncheva, L.I., Todorova, L.P.: Moderate diversity for better cluster ensembles. Inf. Fusion. **7**(3), 264–275 (2006)
16. Hubert, L., Arabie, P.: Comparing partitions. J. Classif. **2**(1), 193–218 (1985)
17. Defays, D.: An efficient algorithm for a complete link method. Comput. J. **20**(4), 364–366 (1977)
18. Kohonen, T.: Learning vector quantization. In: The Handbook of Brain Theory and Neural Networks, 2nd edn, pp. 631–634. MIT Press (2003)
19. Ester, M., Kriegel, H.P., Sander, J., Xu, X.: A density-based algorithm for discovering clusters in large spatial databases with noise. In: 2nd International Conference on Knowledge Discovery and Data Mining, pp. 226–231, August 1996
20. Rasband, W.S.: ImageJ. US National Institutes of Health, Bethesda, Maryland, USA (1997). http://imagej.nih.gov/ij/
21. Frank, E., Hall, M., Witten, I.H.: The WEKA Workbench, 4th edn. Morgan Kaufmann, Burlington (2016)

A Non-integer Step Index PCNN Model and Its Applications

Zhen Yang, Yanan Guo, Xiaonan Gong, and Yide Ma(✉)

School of Information Science and Engineering,
Lanzhou University, Lanzhou 730000, Gansu, China
ydma01@126.com

Abstract. In this paper, based on the Simplified pulse coupled neural network (SPCNN) model, a non-integer step index PCNN model is proposed to solve "the mathematic coupled firing" phenomenon in classical PCNN. Method: A time parameter is introduced into SPCNN model, which make each iteration step value of SPCNN not an integer number any more, thus to emulate an analogue time system more closely. This model is used to accomplish two different tasks, detect micro-calcifications in mammograms and detect noise in natural image. The experimental results show that the model performers better in micro-calcifications detection. Furthermore, it is effectively to use this model in image noise reducing.

Keywords: PCNN · Non-integer step index model · Micro-calcifications detection · Mammogram · Image de-noising

1 Introduction

A bio-inspired neural network was developed by Eckhorn in light of synchronous dynamics of neuronal activity in cat visual cortex [1,2]. It was soon recognized as having significant applications in image processing, therefore, large number of modifications were introduced to Eckhorn's model to make it more suitable for image processing, and these models were collectively known as pulse coupled neural networks [3]. Since 1994, Johnson et al. modified Eckhorn's model to full mathematical description for image processing [3–5], there have been many modified PCNN models aim to achieve lower computational complexity while retaining the essential visual cortical property, the intersecting cortical model (ICM) [6–8], unit-linking PCNN model [9,10] and spiking cortical model (SCM) [11] are some of the representative models.

PCNN has developed rapidly in the image processing field during last decade, such as image segmentation [12,13], image fusion [14–16], pattern recognition [17], feature extraction [18,19] etc. Ma et al. [20] give more detailed applications of PCNN in their literatures. However, there is no standard rules for the setting of PCNNs parameters. On one hand, many researchers turned to the study of simplification of PCNN model, Kinser proposed the SPCNN [6] in 1996, it

M. Valdés Hernández and V. González-Castro (Eds.): MIUA 2017, CCIS 723, pp. 780–791, 2017.
DOI: 10.1007/978-3-319-60964-5_68

contains only one weight matrix and use a contrast threshold, which maintains the mechanisms of the full PCNN model while performs similar types of computations. Ekblad et al. [7] used the ICM in object detection. Zhan et al. [11] proposed a SCM model which simplified internal activity to a single equation while inherits both linking and feeding inputs from the full PCNN, therefore this model is less time consuming. On the other hand, some related research about automatically set values of the PCNN parameters were studied. Kuntimad and Ranganath [21] analyzed the determination of the minimum value for dynamic threshold which guarantees that each neuron pulses exactly only once during a pulsing cycle, and the authors also gave the range of linking coefficient for perfect segmentation. Karvonen [22] chose the linking coefficient based on the class distributions of the segments with different mean value obtained from training synthetic aperture radar images. Berg et al. [23] improved PCNN neurons for image segmentation through evolution algorithm. Ma and Qi [24] built an automated PCNN based on genetic algorithm. Chen et al. [25] proposed an automatic parameter setting method for the simplified PCNN.

However, up to now, most of the previous methods based on the external characteristic of PCNN while ignoring the shortage resulted from the discretization of the neuron firing time. In 2012, Deng and Ma [26] deduced the expressions of the firing time and analyzed the firing period of neurons, and revealed the "mathematics coupled firing" phenomenon of PCNN. Previous methods have not involved any analyzation of the time parameter of PCNN, they just emphasized on a discrete parameter n rather than a continuous parameter t, which may not be exact in some cases.

In this paper, a Non-integer step index PCNN model is proposed to emulate the analogue time system more closely. Based on this view, we attempt to build a direct relation between the firing characteristic and the time parameter, and use the improved model to detect micro-calcifications in mammograms, for further study, this model is also used to detect noise in natural image.

The rest parts of this paper are divided into six sections. Section 2 describes full PCNN briefly and mainly focuses on the derivation of the Non-integer step index model and the analyzation of its firing characteristic. Section 3 presents the proposed mammograms analysis approach in detail. Experimental results are discussed in Sect. 3, furthermore, some examples of image noise detection are given in this section. Conclusions are discussed in Sect. 4.

2 PCNN Models

2.1 Full PCNN Model

The discrete model of PCNN derived by Lindblad et al. [8] are described as follows:

$$F_{ij}[n] = e^{-\alpha_f} F_{ij}[n-1] + V_F \sum_{kl} M_{ij,kl} Y_{kl}[n-1] + S_{ij} \tag{1}$$

$$L_{ij}[n] = e^{-\alpha_l} L_{ij}[n-1] + V_L \sum_{kl} W_{ij,kl} Y_{kl}[n-1] \tag{2}$$

$$U_{ij}[n] = F_{ij}[n](1 + \beta L_{ij}[n]) \tag{3}$$

$$E_{ij}[n] = e^{-\alpha_e} E_{ij}[n-1] + V_E Y_{ij}[n] \tag{4}$$

$$Y_{ij}[n] = \begin{cases} 1, if U_{ij}[n] > E_{ij}[n-1] \\ 0, else \end{cases} \tag{5}$$

The (i, j) neuron N_{ij} embedded in a two dimensional array of neurons contains two main components: feeding input F_{ij} and linking input L_{ij}. Each neuron communicates with its neighboring neurons through the synaptic weights M and W respectively, and retains its previous state altered by decay factor $e^{-\alpha_f}$ and $e^{-\alpha_l}$. Only the Feeding input F_{ij} receives the input stimulus S_{ij}. The feeding input and the linking input are combined in a second order fashion modulated through linking strength β to create the internal activity U_{ij}. Then the internal activity is compared with a dynamic threshold E_{ij} to yield an output Y_{ij}, and then judge whether the neuron N_{ij} fires or not, i.e. Y_{ij} equals one or zero. The threshold is dynamic in that when the neuron N_{ij} fires the threshold then would increase by amplitude V_E immediately. Otherwise, the dynamic threshold would decay by factor $e^{-\alpha_e}$ until the neuron fires again. The parameters n denotes the discrete iteration time, V_F and V_L are normalizing constants denote the amplitudes of feeding input and linking input respectively.

2.2 Non-integer Step Index PCNN Model

The PCNN is a discrete time form of a biological system that operates in analogue time, however, the digital discretization processing of PCNN induces the neurons mathematic coupled firing. The direct effect is that the resolution of PCNN is decreased in image high components. In order to make a compromise between resolution and computational complexity, a non-integer step index PCNN model is proposed in which each neuron is scheduled to pulse along with quasi continuous time. This model can be described as Eq. 6 to Eq. 8:

$$U_{ij}[t + \Delta t] = e^{-\alpha_F \Delta t} U_{ij}[t] + S_{ij}(1 + \beta V_L \sum\nolimits_{kl} W_{ij,kl} Y_{kl}[t]) \tag{6}$$

$$E_{ij}[t + \Delta t] = e^{-\alpha_E \Delta t} E_{ij}[t] + V_E Y_{ij}[t] \tag{7}$$

$$Y_{ij}[t + \Delta t] = \begin{cases} 1, if U_{ij}[t + \Delta t] > E_{ij}[t] \\ 0, else \end{cases} \tag{8}$$

Compared with PCNN model, the non-integer step index PCNN model is actually an improvement over the digital system. All the parameters have the same meaning as mentioned above, Δt is a non-integer value which is used as the step index in each neurons iteration. In order to give an overview of the proposed model, analysis the of firing mechanism of this model is given as follows: For simplicity, Eq. 6 is reduced to Eq. 9:

$$U_{ij}[t+\Delta t]=e^{-\alpha_F \Delta t}U_{ij}[t]+S_{ij} \tag{9}$$

Assume the initial state of dynamic threshold $E_{ij}[0]$ is zero, and the initial state of $Y_{ij}[0]$ is zero too, then $Y_{ij}[1]=1$. Solving the equation of dynamic threshold shown in Eq. 7, which is discussed as follows:

If let $t=0$, according to Eqs. 7 and 8:

$$E_{ij}(\Delta t)=e^{-\Delta t\times\alpha_E}\times E_{ij}(0)+V_E\times Y_{ij}(0)=0 \tag{10}$$

$$Y_{ij}(\Delta t)=step\left(S_{ij}-E(\Delta t)\right)=1 \tag{11}$$

The neuron firing for the first time; As time increasing, when $t=\Delta t$, a new iteration is going on,

$$E_{ij}(2\Delta t)=e^{-\Delta t\times\alpha_E}\times E_{ij}(\Delta t)+V_E\times Y_{ij}(\Delta t)=V_E \tag{12}$$

$$Y_{ij}(2\Delta t)=step\left(S_{ij}-V_E\right) \tag{13}$$

When $t=2\Delta t$,

$$\begin{aligned}E_{ij}(3\Delta t)&=e^{-\Delta t\times\alpha_E}\times E_{ij}(2\Delta t)+V_E Y_{ij}(2\Delta t)\\&=e^{-\Delta t\times\alpha_E}V_E+V_E Y_{ij}(2\Delta t)\end{aligned} \tag{14}$$

If let $t=3\Delta t$,

$$\begin{aligned}E_{ij}(4\Delta t)&=e^{-\Delta t\times\alpha_E}\times E_{ij}(3\Delta t)+V_E Y_{ij}(3\Delta t)\\&=e^{-\Delta t\times\alpha_E}E_{ij}(3\Delta t)+V_E Y_{ij}(3\Delta t)\\&=e^{-2\Delta t\times\alpha_E}\times V_E+e^{-\Delta t\times\alpha_E}\times V_E\times Y_{ij}(2\Delta t)+V_E Y_{ij}(3\Delta t)\end{aligned} \tag{15}$$

Assume the neuron is fired for the second time at $t_2=n_2\Delta t$,

$$\begin{aligned}&E_{ij}(t_2)=e^{-\Delta t\times\alpha_E}E_{ij}\left((n_2-1)\Delta t\right)+V_E Y_{ij}\left((n_2-1)\Delta t\right)\\&=e^{-(n_2-2)\Delta t\times\alpha_E}V_E+e^{-(n_2-3)\Delta t\times\alpha_E}V_E Y_{ij}(2\Delta t)+\ldots+V_E Y_{ij}((n_2-1)\Delta t)\end{aligned} \tag{16}$$

According to the neurons firing condition, $E_{ij}(t_2)=S_{ij}$, so the equation can be simplified as $e^{-(n_2-2)\Delta t\times\alpha_E}V_E=S_{ij}$, thus $t_2=\Delta t-\frac{1}{\alpha_E}\ln\frac{S_{ij}}{V_E e^{\Delta t\times\alpha_E}}$ finally, the neuron firing period $t_m=n_m\Delta t$ is deduced as Eq. 17.

$$t_m=\Delta t-\frac{1}{\alpha_E}\ln\frac{S_{ij}}{V_E e^{\Delta t\times\alpha_E}+e^{n_2\Delta t\times\alpha_E}+\cdots+e^{n_{m-1}\Delta t\times\alpha_E}} \tag{17}$$

Equation 17 shows the firing time t_m is inversely proportional to the neurons stimulation S_{ij}. As shown in Fig. 1, when the value of Δt is set differently, the ladder shape curves of firing time of the improved model show different appearance, for each case, the ladder width is proportional to the pixel's gray value. Furthermore, from (a) to (b), the smaller value the index Δt is, the denser ladder the firing time has. A special case is that when $\Delta t=1$, the ladder graph is just as same as that of the discrete full PCNN model. In order to make

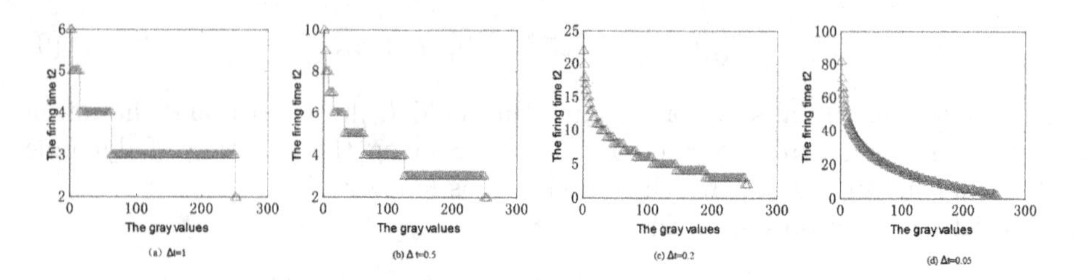

Fig. 1. The firing time t_2 of all gray values with the given Δt.

a comparison between the non-integer step index model and the discrete full PCNN model, the ladder shape curves of firing time of two models are plotted in Fig. 2, (a) denotes the curve in theory, (b) denotes the curve of discrete models, while (c) represent the curve of the improved model with $\Delta t = 0.02$. From these ladder shape curves, it is clear that the non-integer step index model is more closely to emulate an analogue time system, the resolution of which is higher, especially in the image high contrast regions, as indicated in the dashed boxes.

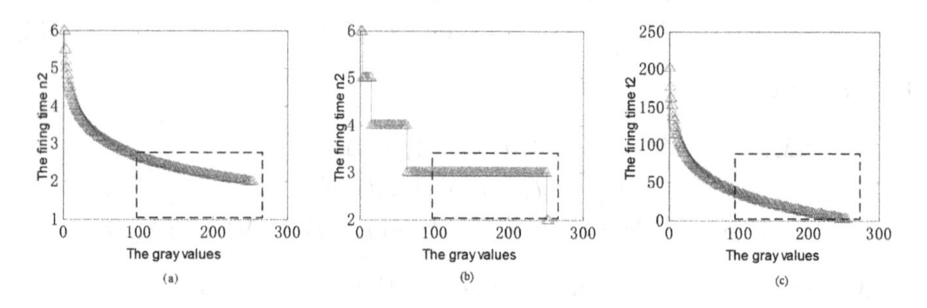

Fig. 2. The firing time t_2 of all gray values for different models.

3 Experimental Results and Discussion

3.1 Micro-Calcifications Detection

Breast cancer is a crucial public health problem in the world, as the causes of this disease is still unknown, it seems impossible for adopting effective prevention measures. Early detection is important for improving breast cancer prognosis and mainly focus on micro-calcifications detection. However, it's difficult to detect micro-calcifications automatically as they are very small in size, they don' have uniform geometric feature and their low contrast compare with the background. Computer aided diagnosis (CAD) systems will provide an opportunity for radiologists to improve their diagnosis. Conventional CAD systems mainly include four parts, i.e. image pre-processing (noise suppression, contrast enhancement between the region of interest (ROI) and background), ROIs selection, feature extraction and classification [27–31]. Using PCNN in mammograms

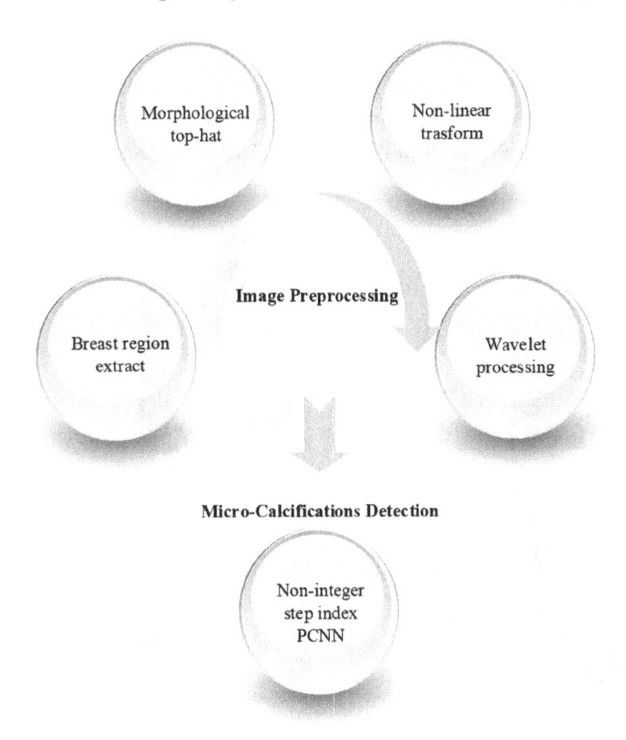

Fig. 3. General flow chart of the proposed method.

processing mainly focused on mass detection [32–35], while micro-calcifications have the similar characteristic with noise, it is not proper implement full PCNN directly in micro-calcifications detection as they don't have enough resolution in high frequency components [36]. Thus, a non-integer step index model is used in micro-calcifications detection. The proposed method does not involve any prior training and trials, which can be quite suitable for clinical diagnosing. To increase the efficiency of the classification and prediction process, it is necessary to enhance the quality of the input mammograms before PCNN process. The pre-processing includes breast region extraction, sub-image enhanced by a top-hat transform which based on a non-flat structuring element [37], further enhancement of the sub-image is implemented by using the exponential contrast enhancement function, subsequently, wavelet transform is used to suppress background and extremely high frequency components. Finally, non-integer step index PCNN is used to detect micro-calcifications. A general flow chart of the proposed method is shown in Fig. 3. In the following experiments, two databases acquired from patients with abnormal pathologies are used for performance evaluation. The first data set is mini Mammographic Image Analysis Society (MIAS), which contains 23 mammograms with micro-calcifications, and the second data set is from Japanese Society of Medical Imaging Technology contains 15 mammograms, 11 of which with micro-calcifications, while the remaining are normal, these images are digitalized with a resolution of 2510×2000, for each mammogram, only a hand draw illustration is provided by radiologists [38]. For each data set, a pre-processing is applied as stated early, the outputs of non-integer step index PCNN model for mini MIAS and JSMIT data set are shown in Figs. 4 and 5 respectively.

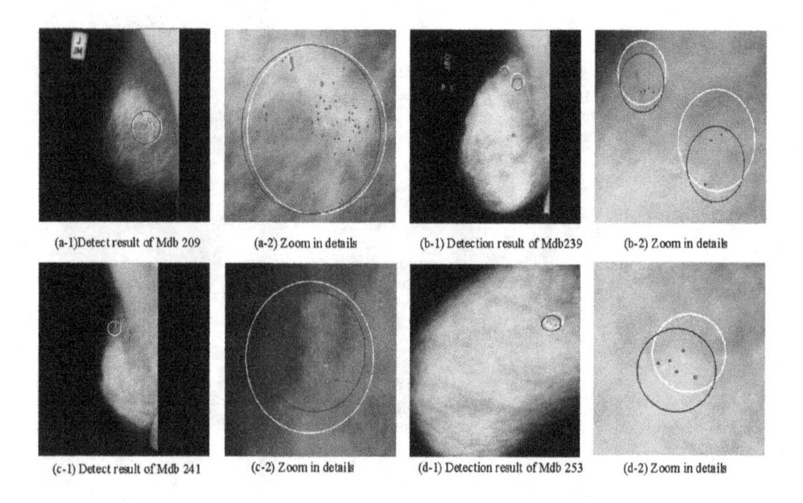

(a-1)Detect result of Mdb 209
(a-2) Zoom in details
(b-1) Detection result of Mdb239
(b-2) Zoom in details
(c-1) Detect result of Mdb 241
(c-2) Zoom in details
(d-1) Detection result of Mdb 253
(d-2) Zoom in details

Fig. 4. The test results of mini MIAS database, as show in (a-1), (b-1), (c-1), (d-1), while (a-2), (b-2), (c-2), (d-2) are enlarged to show details of the former respectively. The green circles represent the ground truth provided by radiologists, the blue ones represent the diseased area judged by our system. The white points enclosed by red circles are micro-calcifications detected by the non-integer index step PCNN model. (Color figure online)

To investigate the robustness of the proposed model, it is necessary to compare our methods with the latest algorithm. Accuracy, the receiver operating characteristic (ROC) curve, the area under the curve (AUC), the number of True Positives (TP) and False Positives (FP) are conventional evaluation indexes [39]. Table 1 denotes the discrimination power of different methods when use the same data mini-MIAS.

Table 1. Performance comparison between rivals and the proposed method.

Method	TP rate (%)	FP rate (%)	Accuracy (%)	AUC
Ref. [40]	90.9	47.5	82	0.7698
Ref. [41]	N/A	N/A	N/A	0.8294
The proposed	91.397	0.035	93.182	0.9739

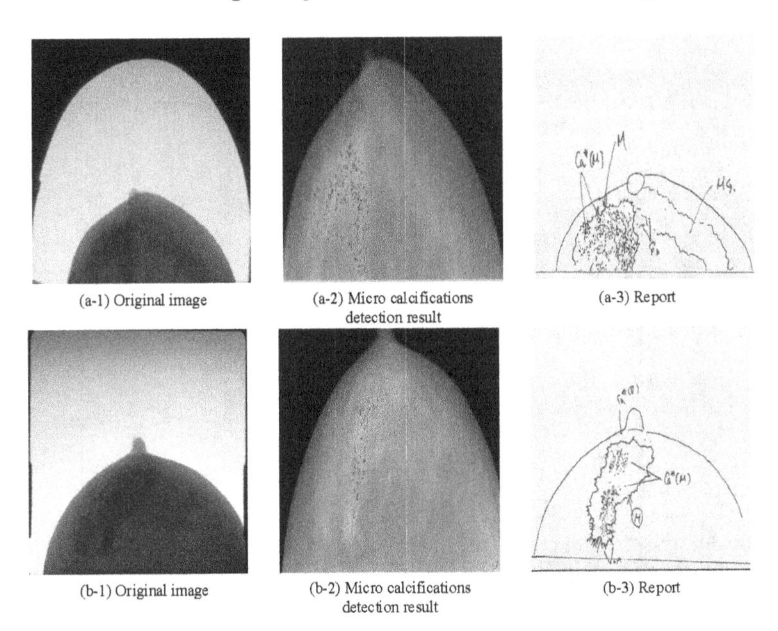

Fig. 5. The test result for database from Japanese Society of Medical Imaging Technology. (a-1) and (b-1) denote original abnormal case, (a-2) and (b-2) denote the output of the improved model, while the third column represent the hand drawn report provided by radiologists.

A comparative performance analysis based on ROC curve of various classifiers is shown in Fig. 6.

The experimental results demonstrate that the non-integer step index PCNN outperforms other systems. The proposed approach generates a sensitivity of 91.397% with a specificity of 99.965%.

3.2 Image De-noising

In this part, the non-integer step index PCNN is applied in noise detection, three standard images, Lena, aerial and dollar are used to test the proposed model, for each image, we add 10% pepper and salt noise to it. In order to make a comparison, median filtering, morphology filtering and the non-integer step index PCNN model are used to de-noising respectively. Figure 7 shows "Lena" processed by three different algorithms. In Table 2, the peak signal to noise ratio (PSNR) for each image is calculated before and after de-noising by three methods.

As shown in Fig. 7, it is clear that the proposed method has a better performance in image de-noising while maintain the image details more properly than comparing algorithms. Table 2 shows the PSNR for each image after de-noising, obviously, the proposed method is performing better.

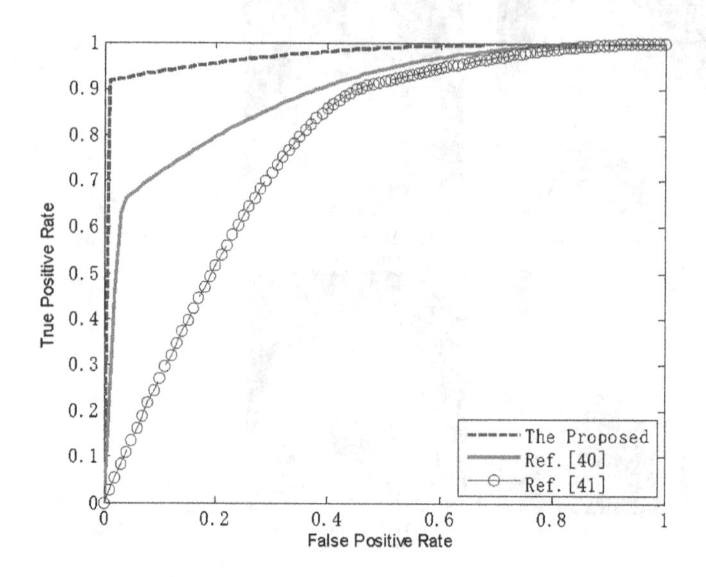

Fig. 6. ROC curve of different methods.

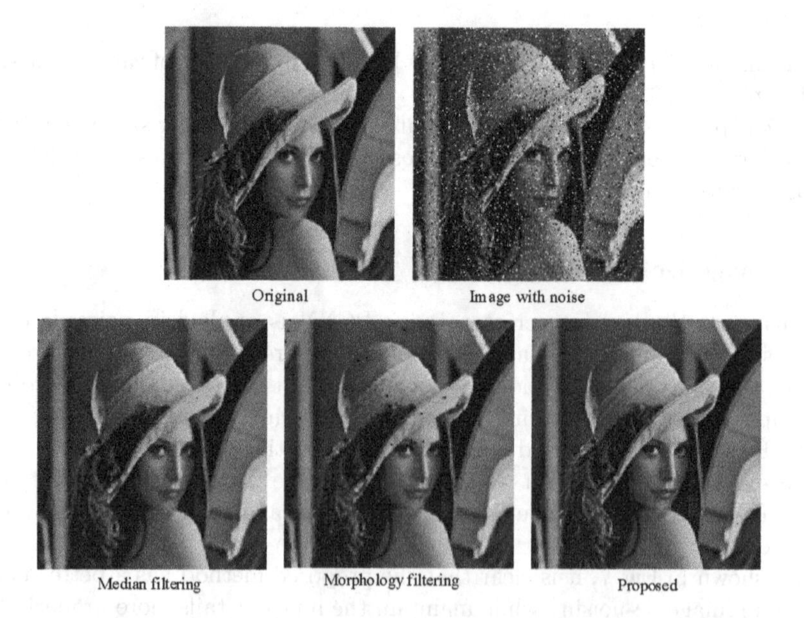

Fig. 7. Image de-noising method for Lena.

Table 2. PSNR for each image before and after de-noising by three methods.

Original	Noise	Median filtering	Morphology filtering	The proposed
Lena	15.2490	29.4658	27.7636	34.1173
Aerial	15.0324	26.2494	24.9549	27.7725
Dollar	14.8590	19.5410	19.3202	20.4268

4 Conclusions and Future Works

The non-integer step index PCNN produces an improvement in classification accuracy to the problem of micro-calcifications detection. Non-integer step index PCNN model is a quasi-continuous time system, which is more closely emulate an analogue time system, the resolution of the improved model is higher than discrete full PCNN. Thus, it is very suitable to solve the noise or noise like signal detection problems. The method gives a better classification accuracy than the traditional methods. The future work mainly focuses on testing much more data set from local hospital, meanwhile, other applications of our model should be considered.

References

1. Eckhorn, R., Reitboeck, H., Arndt, M., Dicke, P.: Feature linking via synchronization among distributed assemblies: simulations of results from cat visual cortex. Neural Comput. **2**, 293–307 (1990)
2. Reitboeck, H.J., Eckhorn, R., Arndt, M., Dicke, P.: A model for feature linking via correlated neural activity. In: Haken, H., Stadler, M. (eds.) Synergetics of Cognition, pp. 112–125. Springer, Heidelberg (1990)
3. Johnson, J.L., Ritter, D.: Observation of periodic waves in a pulse-coupled neural network. Opt. Lett. **18**, 1253–1255 (1993)
4. Johnson, J.L.: Pulse-coupled neural nets: translation, rotation, scale, distortion, and intensity signal invariance for images. Appl. Opt. **33**, 6239–6253 (1994)
5. Ranganath, H., Kuntimad, G., Johnson, J.: Pulse coupled neural networks for image processing. In: Proceedings of the IEEE Southeastcon 1995. Visualize the Future, pp. 37–43 (1995)
6. Kinser, J.M.: Simplified pulse-coupled neural network. In: Aerospace/Defense Sensing and Controls, pp. 563–567 (1996)
7. Ekblad, U., Kinser, J.M., Atmer, J., Zetterlund, N.: The intersecting cortical model in image processing. Nucl. Instrum. Methods Phys. Res. Sect. A: Accel. Spectrom. Detect. Assoc. Equip. **525**, 392–396 (2004)
8. Lindblad, T., Kinser, J.M., Lindblad, T., Kinser, J.: Image Processing using Pulse-Coupled Neural Networks. Springer, Heidelberg (1998)
9. Gu, X.: Feature extraction using unit-linking pulse coupled neural network and its applications. Neural Process. Lett. **27**, 25–41 (2008)
10. Gu, X., Zhang, L., Yu, D.: General design approach to unit-linking PCNN for image processing. In: Proceedings of the 2005 IEEE International Joint Conference on Neural Networks, IJCNN 2005, pp. 1836–1841 (2005)

11. Zhan, K., Zhang, H., Ma, Y.: New spiking cortical model for invariant texture retrieval and image processing. IEEE Trans. Neural Netw. **20**, 1980–1986 (2009)
12. Ranganath, H.S., Bhatnagar, A.: Image segmentation using two-layer pulse coupled neural network with inhibitory linking field. GSTF J. Comput. (JoC) **1**(2) (2014)
13. Dongguo, Z., Chao, G., Yongcai, G.: Simplified pulse coupled neural network with adaptive multilevel threshold for infrared human image segmentation. J. Comput.-Aided Des. Comput. Graph. **2**, 010 (2013)
14. Cheng, S., Qiguang, M., Pengfei, X.: A novel algorithm of remote sensing image fusion based on Shearlets and PCNN. Neurocomputing **117**, 47–53 (2013)
15. Wang, N., Ma, Y., Zhan, K., Yuan, M.: Multimodal medical image fusion framework based on simplified PCNN in nonsubsampled contourlet transform domain. J. Multimedia **8**, 270–276 (2013)
16. Das, S., Kundu, M.K.: NSCT-based multimodal medical image fusion using pulse-coupled neural network and modified spatial frequency. Med. Biol. Eng. Comput. **50**, 1105–1114 (2012)
17. Zhuang, H., Low, K.-S., Yau, W.-Y.: Multichannel pulse-coupled-neural-network-based color image segmentation for object detection. IEEE Trans. Ind. Electron. **59**, 3299–3308 (2012)
18. Nie, R., Zhou, D., He, M., Jin, X., Yu, J.: Facial feature extraction using frequency map series in PCNN. J. Sens. **501**, 204608 (2015)
19. Hoang, T.-T., Nguyen, N.-H., Nguyen, X.-T., Bui, T.-T.: A real-time image feature extraction using pulse-coupled neural network. Int. J. Emerg. Trends Technol. Comput. Sci. (IJETICS) **1**, 117–185 (2012)
20. Ma, Y., Zhan, K., Wang, Z.: Applications of Pulse-Coupled Neural Networks. Springer, Heidelberg (2010)
21. Kuntimad, G., Ranganath, H.S.: Perfect image segmentation using pulse coupled neural networks. IEEE Trans. Neural Netw. **10**, 591–598 (1999)
22. Karvonen, J.: Baltic sea ice SAR segmentation and classification using modified pulse-coupled neural networks. IEEE Trans. Geosci. Remote Sens. **42**, 1566–1574 (2004)
23. Berg, H., Olsson, R., Lindblad, T., Chilo, J.: Automatic design of pulse coupled neurons for image segmentation. Neurocomputing **71**, 1980–1993 (2008)
24. Ma, Y., Qi, C.: Study of automated PCNN system based on genetic algorithm. J. Syst. Simul. **18**, 722–725 (2006)
25. Chen, Y., Park, S.K., Ma, Y., Ala, R.: A new automatic parameter setting method of a simplified PCNN for image segmentation. IEEE Trans. Neural Netw. **22**, 880–892 (2011)
26. Deng, X., Ma, Y.: PCNN model automatic parameters determination and its modified model. Acta Electron. Sinica **40**, 955–964 (2012)
27. Yu, S., Guan, L.: A CAD system for the automatic detection of clustered microcalcifications in digitized mammogram films. IEEE Trans. Med. Imaging **19**, 115–126 (2000)
28. Pal, N.R., Bhowmick, B., Patel, S.K., Pal, S., Das, J.: A multi-stage neural network aided system for detection of microcalcifications in digitized mammograms. Neurocomputing **71**, 2625–2634 (2008)
29. Dheeba, J., Albert Singh, N., Tamil Selvi, S.: Computer-aided detection of breast cancer on mammograms: a swarm intelligence optimized wavelet neural network approach. J. Biomed. Inform. **49**, 45–52 (2014)
30. Tsai, N.C., Chen, H.W., Hsu, S.L.: Computer-aided diagnosis for early-stage breast cancer by using wavelet transform. Comput. Med. Imaging Graph. **35**, 1–8 (2011)

31. Yu, S.-N., Li, K.-Y., Huang, Y.-K.: Detection of microcalcifications in digital mammograms using wavelet filter and Markov random field model. Comput. Med. Imaging Graph. **30**, 163–173 (2006)
32. Hassanien, A.E., Kim, T.-H.: Breast cancer MRI diagnosis approach using support vector machine and pulse coupled neural networks. J. Appl. Logic **10**, 277–284 (2012)
33. Xie, W., Li, Y., Ma, Y.: Breast mass classification in digital mammography based on extreme learning machine. Neurocomputing **173**, 930–941 (2016)
34. Ali, J.M., Hassanien, A.E.: PCNN for detection of masses in digital mammogram. Neural Netw. World **16**, 129 (2006)
35. Hassanien, A.E., Ali, J.M.: Digital mammogram segmentation algorithm using pulse coupled neural networks. In: 2004 IEEE First Symposium on Multi-agent Security and Survivability, pp. 92–95 (2004)
36. Yang, Z., Dong, M., Guo, Y., Gao, X., Wang, K., Shi, B., et al.: A new method of micro-calcifications detection in digitized mammograms based on improved simplified PCNN. Neurocomputing **218**, 79–90 (2016)
37. Wirth, M., Fraschini, M., Lyon, J.: Contrast enhancement of microcalcifications in mammograms using morphological enhancement and non-flat structuring elements. In: Proceedings of the 17th IEEE Symposium on Computer-Based Medical Systems (CBMS 2004), p. 134 (2004)
38. Guo, Y.N., Dong, M., Yang, Z., Gao, X., Wang, K., Luo, C., et al.: A new method of detecting micro-calcification clusters in mammograms using contourlet transform and non-linking simplified PCNN. Comput. Methods Progr. Biomed. **130**, 31–45 (2016)
39. Wang Lei, M.Z., Yuan, X.: The research on computer aided detection of micro-calcifications in mammogram. Zhejiang University (2014)
40. Civcik, L., Yilmaz, B., ÖZBAY, Y., Emlik, G.D.: Detection of microcalcification in digitized mammograms with multistable cellular neural networks using a new image enhancement method: automated lesion intensity enhancer (ALIE). Turk. J. Electr. Eng. Comput. Sci. **23**, 853–872 (2015)
41. Shin, S., Lee, S., Yun, I.D.: Classification based micro-calcification detection using discriminative restricted Boltzmann machine in digitized mammograms. In: SPIE Medical Imaging, pp. 90351L–90351L-6 (2014)

Segmentation of Overlapping Macrophages Using *Anglegram* Analysis

José Alonso Solís-Lemus[1(✉)], Brian Stramer[2], Greg Slabaugh[1], and Constantino Carlos Reyes-Aldasoro[1]

[1] School of Mathematics, Computer Science and Engineering, City, University of London, London, UK
jose.solis-lemus@city.ac.uk

[2] Randall Division of Cell & Molecular Biophysics, King's College London, London, UK

Abstract. This paper describes the automatic segmentation of overlapping cells through different algorithms. As the first step, the algorithm detects junctions between the boundaries of overlapping objects based on the angles between points of the overlapping boundary. For this purpose, a novel 2D matrix with multiscale angle variation is introduced, i.e. *anglegram*. The anglegram is used to find junctions of overlapping cells. The algorithm to retrieve junctions from the boundary was tested and validated with synthetic data and fluorescently labelled macrophages observed on embryos of *Drosophila melanogaster*. Then, four different segmentation techniques were evaluated: (i) a Voronoi partition based on the nuclei positions, (ii) a slicing method, which joined the clumps together (junction *slicing*), (iii) a partition based on the following of the edges from the junctions (*edge following*), and (iv) a custom self-organising map to fit to the area of overlap between the cells. Only (ii)-(iv) were based on the junctions. The segmentation results were compared based on precision, recall and Jaccard similarity. The algorithm that reported the best segmentation was the junction slicing.

Keywords: Segmentation · Overlapping objects · Macrophages · Self-organising maps

1 Introduction

The migration of cells is of great importance in many biological processes, one of them is within the immune system. Macrophages are one of the cells of the immune system that settle in lymphoid tissues and the liver, which serve as filters for trapping microbes and foreign particles [1]. Cell migration is an essential biological process that ensures homeostasis in adults, where an unbalanced migratory response results in human disease [2]. The model organism *Drosophila melanogaster* can offer complementary insights into how macrophages integrate cues to migration [3]. It has been shown that interactions amongst the cells'

M. Valdés Hernández and V. González-Castro (Eds.): MIUA 2017, CCIS 723, pp. 792–803, 2017.
DOI: 10.1007/978-3-319-60964-5_69

structures appear to anticipate the direction of migration [4], thus, accurate cell segmentation could provide information for specific cells for biological studies.

Segmentation of cells in fluorescence microscopy is a widely studied area [5], with many approaches ranging from thresholding techniques [6], to active surfaces [7]. In recent years, techniques like adaptive active physical models [8] and multilevel sets [9] have been used to address the problem of cells that overlap in cervical cancer images. Other techniques like self-organising maps (SOM) have also been used for biomedical image segmentation [10].

Junctions are commonly acquired by looking for extrema in the curvature of the image gradient [11,12]. In this work, a novel approach to find junctions is proposed for boundaries of overlapping objects, whose intersections would correspond to the junctions acquired. The junctions detected would later be used as the basis for completing a segmentation of the overlapping cells. Four methods are presented, of which three use the information from the junctions detected.

2 Materials

2.1 Macrophages Embryos

Fluorescently labelled macrophages were observed in embryos of the model organism *Drosophila melanogaster*. The nuclei were labelled with GFP-Moesin, which appeared red, whilst the microtubules were labelled with a green microtubule probe (Clip-GFP) [4]. RGB images of dimensions $(n_h, n_w, n_d) = (512, 672, 3)$ and $t = 541$ time frames were acquired. The images have two layers of fluorescence. Figure 1 shows one representative time frame. The green channel illustrates overlap that makes an accurate segmentation of the cells complicated.

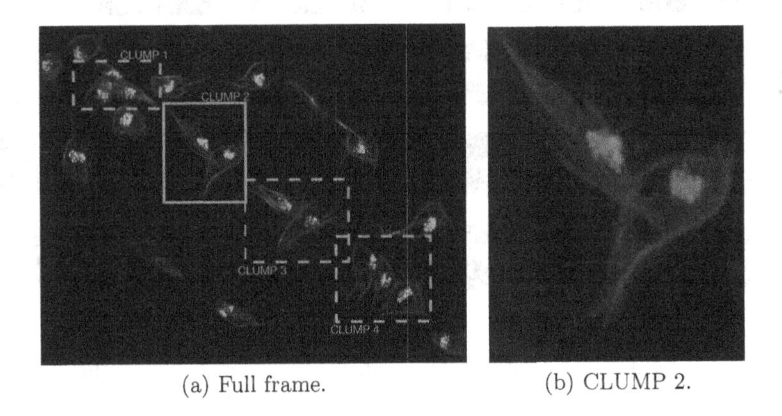

(a) Full frame. (b) CLUMP 2.

Fig. 1. Example of cell overlapping in a single frame. (a) Presents the full frame with (red) squares highlighting all regions where instances of overlapping cells (clumps) are shown and labelled for easy reference. (b) Detail of CLUMP 2, present in (a). (Color figure online)

2.2 Synthetic Data

In order to assess the limitations of the junction detection methodology, images of pairs of synthetic overlapping ellipses with varying angles and separation distances were generated ($n = 142$). Let $\mathcal{E}(\phi, \mathbf{x}_0) = \{\mathbf{x}_{\phi,\mathbf{x}_0}(t) : t \in [0, 2\pi]\}$, be the ellipse defined by the equation $\mathbf{x}_{\phi,\mathbf{x}_0}(t) = R(\phi)(a\cos(t), b\sin(t))^T + \mathbf{x}_0$, which is rotated with respect to the x-axis by ϕ degrees and whose centre is located on position $\mathbf{x}_0$. The pairs of ellipses constructed in this work differ both in angle and position with three conventions taken into consideration: (i) presetting the values of the axes (a, b); (ii) defining a *central* ellipse $\mathcal{E}_0 = \mathcal{E}(0, \mathbf{x}_0)$, common to all pairs; and (iii) the difference in position would only be made by moving the ellipses in the x-axis. Thus, the pairs of ellipses can be defined in terms of the differences to $\mathcal{E}_0$, namely the angle and distance from the centre (ϕ, Δ). A set of pairs of ellipses was generated in MATLAB® to test the method at different values of (ϕ, Δ) ranging ϕ from 0 to 90° and Δ from 0 to 160 pixels with increments of 10. Images of size $(n_h, n_w) = (256, 512)$ were generated with $\mathbf{x}_0 = (128, 128)^T$ and axes $(a, b) = (120, 53)$ that contained an overlapping of $\mathcal{E}_0$ and $\mathcal{E}_{\phi,\Delta}$. Disregarding the images where there was no overlap present in the generated ellipses, a total of 142 images was generated. Figure 2 contains a subset of the ellipses tested. Cases where there was no overlap were ignored from the analysis.

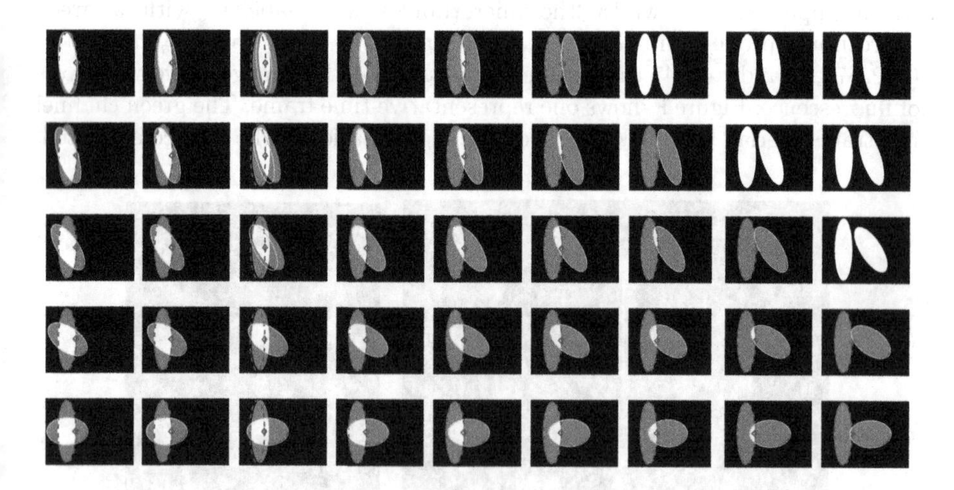

Fig. 2. Overview of the range of pairs of ellipses investigated. The pairs presented on this image represent a sample of the ellipses that were tested by the method presented. The boundary of the central ellipse $\mathcal{E}_0$ is highlighted in blue while the second ellipse's boundary is presented in red. (Color figure online)

3 Methods

In this work, a *clump* will be understood as a cluster of two or more overlapped objects. A clump was detected when two or more nuclei on the red channel were detected within a single region on the green channel. Segmentation of the green channel was performed by low-pass filtering with a 5×5 Gaussian filter, following with a hysteresis thresholding technique [6]. A morphological opening with a disk structural element ($r = 3$) was performed to smooth the edges and remove noise. Segmentation of the red channel followed the same methodology. Then, the number of nuclei per region was counted to determine the presence of *clumps*.

3.1 Junction Detection

Let $\mathcal{B}$ define the boundary of a clump, then for each of the ordered points $\mathbf{p}_i = \mathbf{x}_i \in \mathcal{B}$, the inner angle of the point is defined as follows,

Definition 1 (Inner angle of a point). *The inner angle of a point* $\mathbf{p}_i \in \mathcal{B}$ *in the boundary is the angle* $\theta_{i,j}$ *adjacent to the point, and measured from the jth previous point* $\mathbf{p}_{i-j}$ *to the following jth position* $\mathbf{p}_{i+j}$.

Figure 3 shows examples of the calculation of an *inner angle* for a given point in the boundary. By visual inspection, it can be noticed that the inner angle of a junction would be greater than 180° for a number of separations j. This number of separations will be referred to as the *depth* of the junction. Thus, the method consists of computing the inner angle $\theta_{i,j}$ at every point $\mathbf{p}_i \in \mathcal{B}$, and on every separation j. The **anglegram matrix** $\Theta = ((\theta_{i,j}))$ is defined as the values of the inner angles of each point i and per separation j, Fig. 3(c).

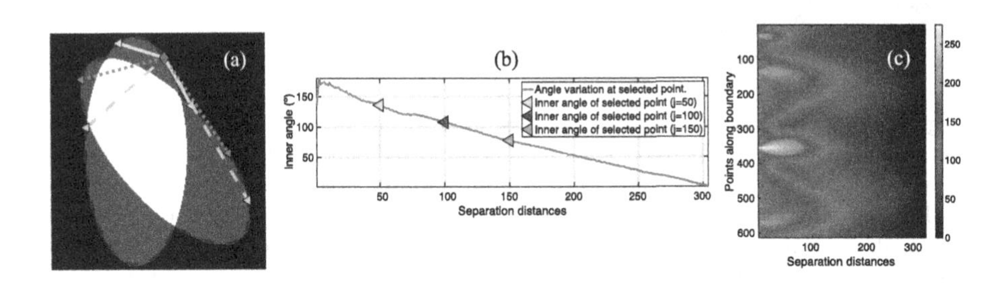

Fig. 3. Representation of *inner point angle* calculation and generation of *anglegram matrix*. (a) Represents a synthetic clump with its boundary outlined (blue, dotted), where a point (magenta ◇) in the boundary will have various inner point angles per separation j. All the inner point angles for the highlighted point are displayed in (b). (c) Shows the *anglegram matrix*, where each row represents the graph displayed in (b) for each boundary point. (Color figure online)

The local maxima on a projection over the horizontal dimension of the anglegram is related to the position of the junctions of the boundary and the depth of the junction. Each row, $\Theta(i,:)$, corresponds to the inner angles of point $\mathbf{p}_i$, therefore taking a summary of the rows would yield a measurement of the general inner angles of each point. For this work, the maximum intensity projection $\hat{\theta}_{\max}$, Fig. 4(b), was compared with mean, median and area under the curve, but maximum provided the best results (data not shown due to space limitations). To account for quantisation errors in the boundaries extracted from the clumps, an averaging filter of size 5×5 was applied to the anglegram matrix, Θ, before the calculation of $\hat{\theta}_{\max}$. The local maxima of the 1D projection were found by using the function `findpeaks` from MATLAB®, which identifies local maxima of the input vector by choosing points of which its two neighbours have a lower value. Due to quantisation noise in $\hat{\theta}_{\max}$, the parameters `MinPeakDistance` and `MinPeakHeight` were set to empirically consistent values. First, `MinPeakDistance`, which restricts the function to find local maxima with a minimum separation, was set to 25. Furthermore, the parameter `MinPeakHeight` was set to $\text{mean}(\hat{\theta}_{\max}) + 0.75 \times \text{std}(\hat{\theta}_{\max})$.

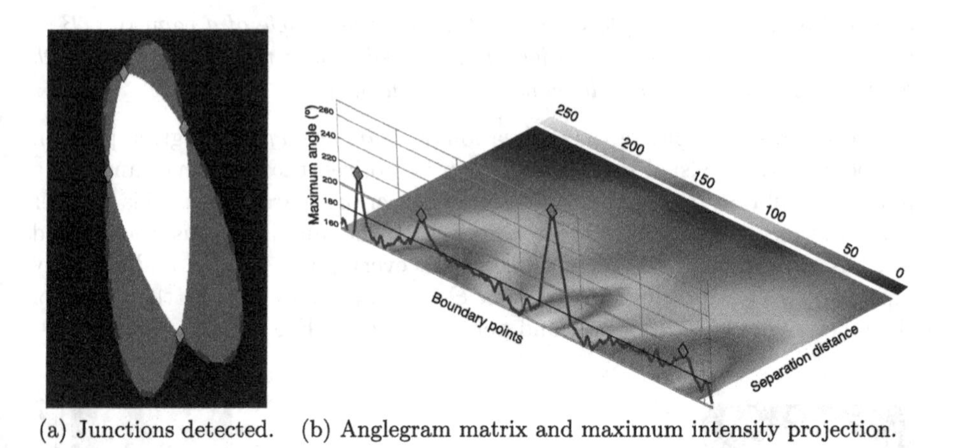

(a) Junctions detected. (b) Anglegram matrix and maximum intensity projection.

Fig. 4. Junction detection on overlapping objects through the maximum intensity projection of the anglegram matrix. The junctions detected on a synthetic pair or ellipses is shown in (a), where the boundary of the clump is represented as a dotted line (blue) as well as the junctions (magenta $\diamond$). The definition of $\hat{\theta}_{\max}$ is represented in (b), where the anglegram matrix Θ is displayed in a plane and $\hat{\theta}_{\max}$ is represented along the boundary points. Detection of junctions are shown with $\diamond$ markers (magenta). (Color figure online)

3.2 Segmentation of Overlapping Regions

This section describes the comparison of methodologies to segment clumps into overlapping cells. Initially, as a benchmark, a simple partition of the clump based on Voronoi partitioning [13] was developed. Then, the three methods, which incorporate the information from the junctions into a segmentation output were used. The methods differed in the way the junctions' information was incorporated into a complete segmentation. Junction Slicing (JS) and Edge Following (EF) involved the explicit use of the junctions' position, while the proposed self-organising map (SOM) fitting involved the information of junctions into creating a custom SOM that adapts to the overlapping section of the data. In this work, only the cases where two junctions were found were examined in detail. A diagram showing all methods presented and the data flow is presented in Fig. 5.

Voronoi Partition. This method was included as a lower-bound benchmark for comparisons against with the proposed methods. The results should be the worst as no information from the green channel is used. The image area was partitioned, using Voronoi tesselations [13]. The partition of the space was based on the centroids of the detected nuclei from the image's red channel. The clumps detected on the green channel were divided based on the Voronoi partition.

Junction Slicing (JS). This method partitioned the clump with the line that joined two junctions. For each junction detected, each of the two adjacent segments of the boundary of the clump would correspond to one of the different objects within the clump. Since the points in the boundary are ordered, starting at one point $\mathbf{p}_1$ and moving alongside $\mathcal{B}$ in a clockwise manner, then the segment that appeared before a detected junction would correspond to one cell, whereas the segment that appeared after the junction would correspond to the other cell. For cases where only two junctions were found, the problem of selecting which pair of junctions will be joined becomes trivial. However, considering a case like the one presented on Fig. 4(a), where four junctions would appear, different combinations of the boundary segments could yield different *candidates* of segments.

Edge Following (EF). In order to obtain the edge information, the Canny algorithm [14] was used on the green channel of the image. The algorithm consists of finding the local maxima of the image gradient. In this work, the parameter of the standard deviation was set to $\sigma = 1$. The trend of the two adjacent segments leading to the junction was defined by approximating the tangent line of the boundary at the junction point. The definition of the tangent line was taking an average slope of the secant lines leading up to the detected junction. The tangent line was extended, and a region of interest (ROI) was defined by a triangle where the approximated tangent line goes along the vertex and the adjacent angle corresponds to 20° to each side of the tangent line (Fig. 5). The ROI defined for each of the adjacent line segments was then intersected with the edge information of the image, resulting in a set of binary line segments, which were labelled. Labelling of the binary line segments allowed for individual analysis of each line. Each line detected was analysed in terms of its orientation and size, preserving the one that has the most similar orientation to the extended line segment. Binary line segments with a change in direction were split by removing the strongest corners, detected through the corner detection algorithm by Harris and Stephens [11]. The lines found by both ROIs on each junction were then used as new coordinates to add to the boundary of the corresponding cell.

Self-organising Maps (SOM) Fitting. This work proposes an alternative implementation of the self-organising maps [15] that adapts itself to the overlapped area. For this SOM, a custom network was defined, as well as the input data and additional rules to the definition of the step-size parameter, α. Let a Network $\mathcal{N} = (\mathcal{V}, \mathcal{L})$, where $\mathcal{V} = \{\mathbf{m}_i = (x_i, y_i) \in \mathbb{R}^2 : i = 1, \cdots, n_v\}$ are nodes assigned to positions in the plane and $\mathcal{L}$ are some edges linking the some of the nodes in $\mathcal{V}$. Each node $\mathbf{m}_i \in \mathcal{V}$ has an identifier, position, and a *speed* parameter, related to the movement of each node. The input data was determined by the positions and normalised intensity values of the image, i.e. $(\mathbf{x}_t, I(\mathbf{x}_t))$. Values in $I(\mathbf{x}_t)$ that were selected by an Otsu's threshold [16] and were located within a bounding box that contains the junctions. Given an input, the algorithm proposed by Kohonen [15] follows two basic steps: identifying the closest node in the network to the input, shown in Eq. (1), and update the positions of the nodes inside a neighbourhood, determined by a distance n_e to the winner node $\mathbf{m}_c$, (2),

$$\mathbf{m}_c(t) = \underset{c \in \{1, \cdots, n_v\} S}{\arg\min} \|\mathbf{x_t} - \mathbf{m}_i(t)\|_2^2 \tag{1}$$

$$\mathbf{m}_i(t+1) = \begin{cases} \mathbf{m}_i(t) + \alpha_t\left(\mathbf{x}_t - \mathbf{m}_c(t)\right), & (i, c) \in \mathcal{L} \text{ and } \operatorname{dist}(\mathbf{m}_i, \mathbf{m}_c) \leqslant n_e \\ \mathbf{m}_i(t) & \text{, otherwise} \end{cases}, \tag{2}$$

where $\operatorname{dist}(\mathbf{m}_i, \mathbf{m}_j)$ refers to the distance from node i to node j in the shortest path determined by the edges $\mathcal{L}$. In this work, the parameter α_t was determined the intensity level of the image, I, and the speed parameter of the node. The proposed formula for the parameter α_t is shown in Eq. (3),

$$\alpha_{t,i} = \alpha_0 \times (0.2 + I(\mathbf{x}_t))^4 \times \operatorname{speed}(\mathbf{m}_i), \tag{3}$$

where $\operatorname{speed}(\mathbf{m}_i)$ is 0.1, or 1, depending on where the node resides in the topology. The network was defined by taking a subset of the boundary points in $\mathcal{B}$ in a ring topology, and then adding two networks in a grid topology to each side of the line joining two junctions. The three networks are independent from each other. Thus, $\operatorname{speed}(\mathbf{m}_i) = 0.1$, if $\mathbf{m}_i$ was located in the boundary of the clump, and $\operatorname{speed}(\mathbf{m}_i) = 1$, if it was one of the grid networks. The assumption is that the network taken from $\mathcal{B}$ would be closer to the actual cell, and therefore it should not move abruptly, whereas the networks inside the clump will adjust and adapt to the shape of the overlapping area between the cells. In order to finalise the network final state into a segmentation, the external network was taken as a new clump and it was partitioned by the same line used in the junction slicing (JS) method. Finally, the area formed by the inner network that adapted to the overlapping section of the cell was dilated with a 5×5 square element and then attached to both partitions of the new clump. The right column of Fig. 5 displays the main steps of the SOM fitting method described.

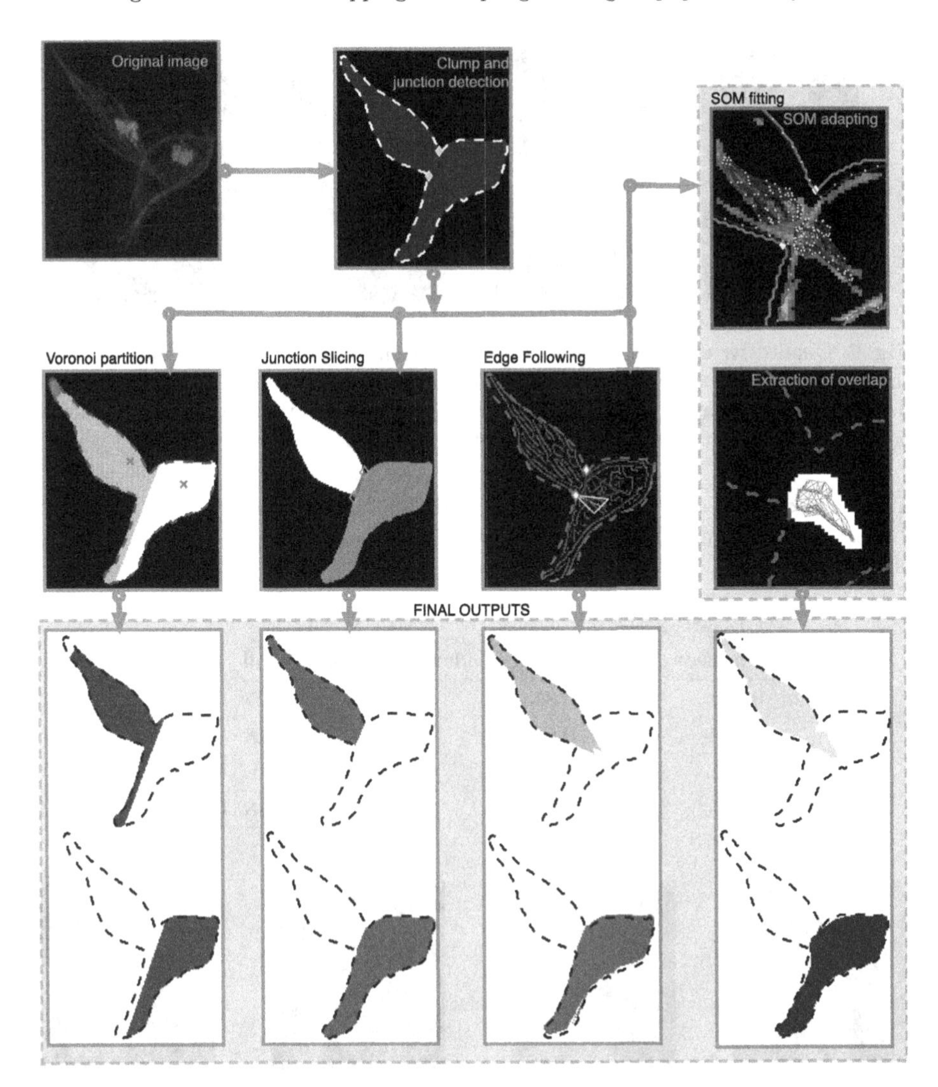

Fig. 5. Illustration of all the methods developed and the workflow to obtain results. Top left shows the detail of CLUMP 2 in the original frame. Clumps are detected and the boundary was extracted. With the boundary information, the *anglegram* was calculated and the junctions were detected (top, middle). On the second row, a diagram to the methods were presented. From left to right, the Voronoi partition, Junction Slicing (JS), Edge Following (EF) and SOM fitting. Bottom row shows the outputs from each method for both cells within the detected clump.

4 Results

The junctions that were correctly detected on the synthetic data had a range of angles (based on the corresponding value in vector $\hat{\theta}_{\max}$) [188.64–328.4] degrees;

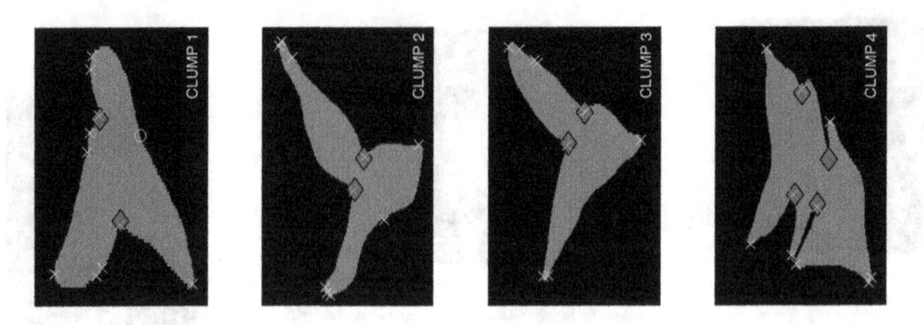

Fig. 6. Qualitative comparison of junction detection via anglegram (magenta ◇) versus the Harris corner detector (green ×). The strongest 10 corners from the Harris detector per clump are displayed. Only CLUMP 1 has a missing junction (cyan ○), it should be noticed how difficult detection of the junction would be. (Color figure online)

Table 1. Comparison of mean values of Precision, Recall and Jaccard Index for clumps 2 and 3 over 10 frames. This table summarises the results in Fig. 7. Highest results are highlighted.

	CLUMP 2			CLUMP 3		
	Precision	Recall	Jaccard Index	Precision	Recall	Jaccard Index
Voronoi	0.906	0.925	0.843	0.872	0.868	0.771
JS	**0.970**	0.953	0.926	**0.974**	**0.948**	**0.925**
EF	0.964	**0.983**	**0.948**	0.938	0.950	0.896
SOM	0.965	0.951	0.919	0.973	**0.948**	0.923

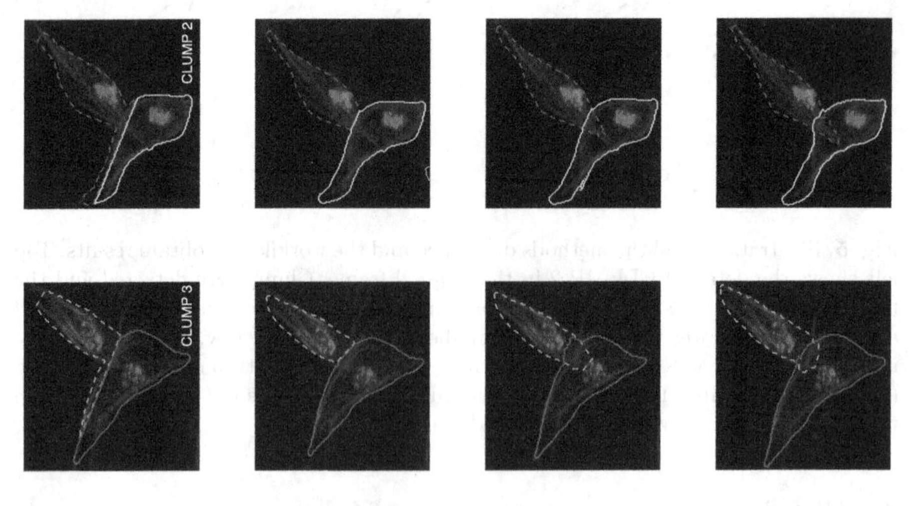

Fig. 7. Qualitative comparison of different segmentation methods in one of the frames. From left to right, the segmentation results for the Voronoi method, Junction Slicing (JS), Edge Following (EF) and SOM fitting are presented. Top and Bottom rows represent the results for CLUMP 2 and CLUMP 3 respectively.

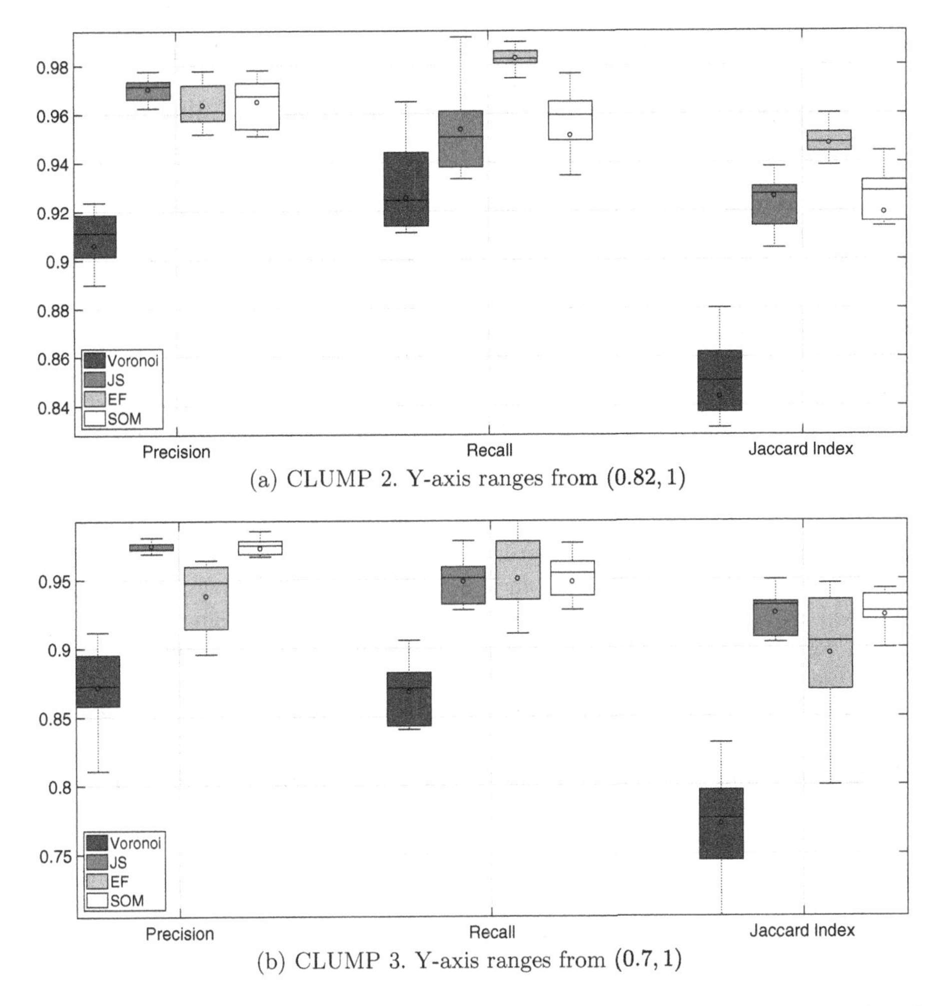

(a) CLUMP 2. Y-axis ranges from $(0.82, 1)$

(b) CLUMP 3. Y-axis ranges from $(0.7, 1)$

Fig. 8. Comparison of Precision, Recall and Jaccard Index for all methods of segmentation of overlapping in clumps 2 and 3. Horizontal axis correspond to the box plots from the different methods and their summarised performance in the metrics computed. Three groups corresponding to Precision, Recall and Jaccard Index contain four box plots; which, from left to right, correspond to Voronoi, JS, EF and SOM methods. Table 1 summarises the information on this image. (Color figure online)

whilst the missed junctions had a range of [162–191.96] degrees. This indicates that very wide angles, close to a straight line are easy to miss. The overlap region of [188–192] deserves a further investigation outside the scope of this paper. For overlapping clumps, junctions detected by the anglegram algorithm were compared qualitatively against the Harris corner detector [11], Fig. 6.

The results obtained for one frame of overlapping cells examined in detail are shown in Fig. 7, which are clumps that look similar throughout the images.

Finally, a comparison with manually segmented ground truth was performed. In order to have the best results shown for each of the methods, the input clumps used were taken from the ground truth images. The Jaccard Similarity Index [17], recall and precision [18] statistics were computed for both clumps on all the frames and all the methods described, box plots of the results are shown in Fig. 8 and summarised in Table 1. Qualitative comparisons are also provided for some examples of the ten images and some of the clumps not analysed in detail, as well as for the SOM outputs segmenting the overlapped area in the clumps.

5 Discussion

Preliminary work using thresholding techniques [6], active contours [19] or multilevel set methods [9] did not provide satisfactory segmentation of the overlapping cells, as these techniques could only detect clusters of overlapping objects without distinction between each of the individual objects (data not shown). In this paper, a method to segment overlapping cells through the analysis of the boundary of the clump was proposed. Its main advantage is to present a way to find relevant junctions from a boundary. Figure 6 shows the junctions detected by the anglegram would not require further processing to select the useful corners, unlike the Harris algorithm outputs. Consistent with synthetic tests, limitations were observed in CLUMP 1, where a junction was missed by both methods. This limitation depends on cell positions and is transferred to the underlying segmentation methods. Table 1 shows a better performance from all three junction-based methods compared to the Voronoi partition. Furthermore, the percentile box sizes in Fig. 8 show that the EF method (yellow) is less consistent than the SOM method (white).

Current experimental results demonstrate the promise of this method to produce correct segmentations of overlapping cells, Fig. 7. Further experimentation is ongoing applying these techniques to cases where there are more than two main junctions detected, such as the one presented in Fig. 4(b) and cases where more than two cells are present in the clump. Future work will consider the extension of the anglegram matrix as a prior to a probabilistic modelling of the position of the junctions.

References

1. Martinez, F., Sica, A., Mantovani, A., Locati, M.: Macrophage activation and polarization. Front. Biosci. **13**, 453–461 (2008)
2. Pocha, S.M., Montell, D.J.: Cellular and molecular mechanisms of single and collective cell migrations in drosophila: themes and variations. Annu. Rev. Genet. **48**, 295–318 (2014)
3. Wood, W., Martin, P.: Macrophage functions in tissue patterning and disease: new insights from the fly. Dev. Cell **40**(3), 221–233 (2017)
4. Stramer, B., Moreira, S., Millard, T., Evans, I., Huang, C.Y., Sabet, O., Milner, M., Dunn, G., Martin, P., Wood, W.: Clasp-mediated microtubule bundling regulates persistent motility and contact repulsion in Drosophila macrophages in vivo. J. Cell Biol. **189**(4), 681–689 (2010)

5. Maška, M., Ulman, V., Svoboda, D., Matula, P., Matula, P., Ederra, C., Urbiola, A., España, T., Venkatesan, S., Balak, D.M., et al.: A benchmark for comparison of cell tracking algorithms. Bioinformatics **30**(11), 1609–1617 (2014)
6. Henry, K., Pase, L., Ramos-Lopez, C.F., Lieschke, G.J., Renshaw, S.A., Reyes-Aldasoro, C.C.: PhagoSight: an open-source MATLAB® package for the analysis of fluorescent neutrophil and macrophage migration in a zebrafish model. PLoS ONE **8**(8), e72636 (2013)
7. Dufour, A., Shinin, V., Tajbakhsh, S., Guillén-Aghion, N., Olivo-Marin, J.C., Zimmer, C.: Segmenting and tracking fluorescent cells in dynamic 3-D microscopy with coupled active surfaces. IEEE Trans. Image Process. **14**(9), 1396–1410 (2005)
8. Plissiti, M.E., Nikou, C.: Overlapping cell nuclei segmentation using a spatially adaptive active physical model. IEEE Trans. Image Process. **21**(11), 4568–4580 (2012)
9. Lu, Z., Carneiro, G., Bradley, A.P.: An improved joint optimization of multiple level set functions for the segmentation of overlapping cervical cells. IEEE Trans. Image Process. **24**(4), 1261–1272 (2015)
10. Reyes-Aldasoro, C.C., Aldeco, A.L.: Image segmentation and compression using neural networks. In: Advances in Artificial Perception and Robotics, CIMAT, pp. 23–25 (2000)
11. Harris, C., Stephens, M.: A combined corner and edge detector. In: Proceedings 4th Alvey Vision Conference, pp. 147–151. Alvety Vision Club (1988)
12. Lindeberg, T.: Junction detection with automatic selection of detection scales and localization scales. In: ICIP, vol. 1, pp. 924–928 (1994)
13. Okabe, A., Boots, B., Sugihara, K., Chiu, S.N., Kendall, D.G.: Spatial Tessellations Concepts and Applications of Voronoi Diagrams. Wiley, Hoboken (2008)
14. Canny, J.: A computational approach to edge detection. IEEE TPAMI **8**(6), 679–698 (1986)
15. Kohonen, T.: The self-organizing map. Neurocomputing **21**(1), 1–6 (1998)
16. Hannah, I., Patel, D., Davies, R.: The use of variance and entropic thresholding methods for image segmentation. Pattern Recogn. **28**(8), 1135–1143 (1995)
17. Jaccard, P.: Étude comparative de la distribution florale dans une portion des Alpes et des Jura. Bull. Soc. Vaudoise Sci. Nat. **37**, 547–579 (1901)
18. Fawcett, T.: An introduction to roc analysis. Pattern Recogn. Lett. **27**(8), 861–874 (2006)
19. Caselles, V., Kimmel, R., Sapiro, G.: Geodesic active contours. Int. J. Comput. Vis. **22**(1), 61–79 (1997)

New Disagreement Metrics Incorporating Spatial Detail – Applications to Lung Imaging

Alberto M. Biancardi(✉) and Jim M. Wild

Polaris (IICD Department) and INSIGNEO, The University of Sheffield, Sheffield, UK
{a.biancardi+miua, j.m.wild+miua}@sheffield.ac.uk

Abstract. Evaluation of medical image segmentation is increasingly important. While set-based agreement metrics are widespread, they assess the absolute overlap, but fail to account for any spatial information related to the differences or to the shapes being analyzed. In this paper, we propose a family of new metrics that can be tailored to deal with a broad class of assessment needs.

Keywords: Disagreement · Evaluation metrics · Segmentation · Medical images · Lung MRI

1 Introduction

As computer-supported segmentation of medical images becomes increasingly commonplace, evaluating the outcomes plays an even more important role – for instance, for validation purposes (especially on large datasets) or for performance comparison. The typical goal of these evaluations is the assessment of the agreement, or disagreement, expressed as a measure of their spatial overlap. The most common approaches, across several applicative domains [1–4], are based on the quantification of the spatial agreement by means of set operations (Dice similarity, Jaccard index, etc.). However, these approaches assume that the assessed elements are independent among themselves (as entailed by the definition of these similarity assessments itself, being based on set operations), while, in the imaging domain, the segmentation-region elements (pixels, voxels, etc.) are characterized by their spatial location and this location introduces a correlation among the set elements. In this paper, a new family of metrics that quantify various aspects of the spatial differences between two regions is presented. We demonstrate the use of the proposed metrics amongst set-based techniques in the analysis of lung MRI.

2 Set-Based Measurements

Following [5, 6], let a scalar (medical) image be represented by a function defined on a regular grid $I : \mathcal{G} \rightarrow V$. Typically the elements of $\mathcal{G}$ are indexed by a subset of $\mathbb{Z}^n$, where n is the image dimensionality, and V is a subset of $\mathbb{Z}$ or $\mathbb{Q}$. Additionally we

M. Valdés Hernández and V. González-Castro (Eds.): MIUA 2017, CCIS 723, pp. 804–814, 2017.
DOI: 10.1007/978-3-319-60964-5_70

assume that consistent spatial locations are assigned to all of the elements of $\mathcal{G}$ and that, therefore, a metric is defined between grid element pairs.

We define a binary image as an image with two possible values:

$$b : \mathcal{G} \rightarrow [0, 1] \tag{1}$$

As we are going to analyze only single-label segmentations, a segmentation is represented by a binary image and the subset of $\mathcal{G}$ having a value of 1:

$$S = \{x : b(x) = 1, x \in \mathcal{G}\} \tag{2}$$

The function b, by definition, induces a partition of $\mathcal{G}$ by means of its two inverse images; therefore, the segmentation background, i.e. $b^{-1}(0)$, will be simply denoted by S', being the complement of S.

Given two segmentations T and R, typical set-based assessments of spatial overlap are defined by computing the cardinality of selected subsets of $\mathcal{G}$. If T is a test segmentation and R is a reference segmentation, for which its values have been considered to be in accordance with the expected outcome (or ground truth), then the customary confusion matrix can be expressed as:

	R	R'
T	$TP = T \cap R$	$FP = T \cap R'$
T'	$FN = T' \cap R$	$TN = T' \cap R'$

The performance parameters can then be expressed either in term of set operations, where there is no assumption of truth (see also Sect. 4), or based on the confusion matrix cardinalities:

Measure	Set-based	Truth-based
Dice	$\frac{2\lvert T \cap R \rvert}{\lvert T \rvert + \lvert R \rvert}$	$\frac{2TP}{2TP + FN + FP}$
Jaccard	$\frac{\lvert T \cap R \rvert}{\lvert T \cup R \rvert}$	$\frac{TP}{TP + FN + FP}$
Sensitivity or *recall*	$\frac{\lvert T \cap R \rvert}{\lvert R \rvert}$	$\frac{TP}{TP + FN}$
Specificity	$\frac{\lvert T' \cap R' \rvert}{\lvert R' \rvert}$	$\frac{TN}{FP + TN}$

3 Spatial Impact of the Image Domain

Even if the set-based measurements are sometimes referred to as assessing the spatial overlap, the extent to which the actual spatial characteristics of the two segmentations under evaluation are assessed is limited: only the exact overlap of the voxels is tested, while any level of proximity is lost. Additionally every element is given the same weight, regardless of possible constrains brought forth by the specific application where the evaluation takes place.

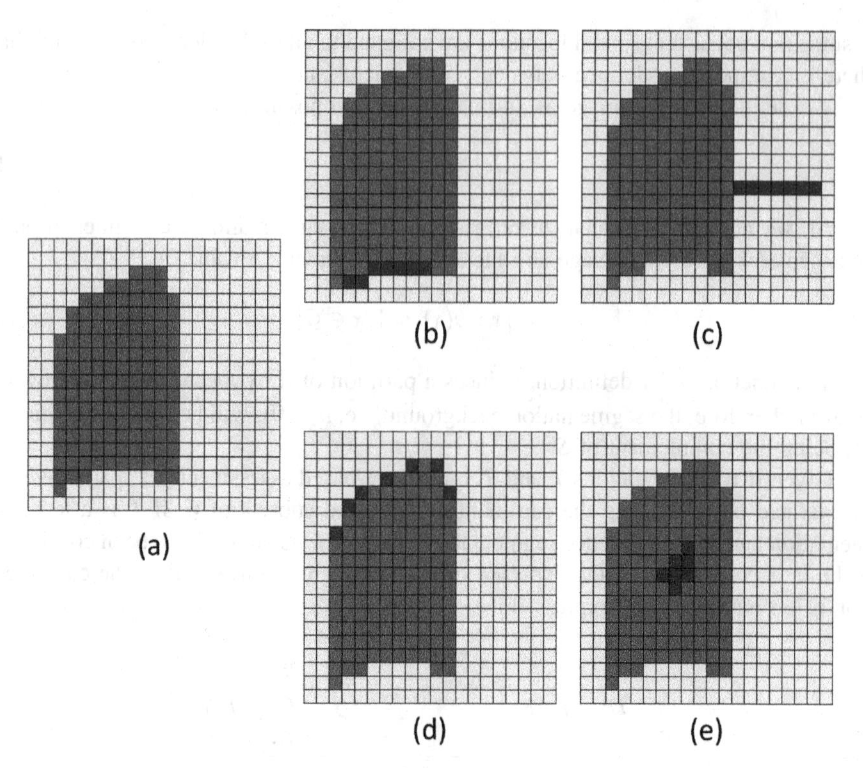

Fig. 1. Two groups of hypothetical segmentations having the same set-based agreement with the reference region (a). Regions (b) and (c) have additional pixels (i.e., false positives), marked in red/dark gray; regions (d) and (e) are missing pixels (i.e., false negatives), marked in orange/light gray (Color figure online)

Figure 1 shows the impact that the spatial location of the segmented voxel plays when assessing the dissimilarity between two segmentations, assuming that no a-priori knowledge is available as regards the region to be segmented. The reference region *R*, shown in (a), is assumed to be the ground truth while regions shown in (b) and (c) have 7 additional elements with respect to *R* (false positives highlighted in red). It is easy to see, that, no matter what cardinality-based measurement is chosen, the four regions will have the same outcome measure of agreement, regardless of the position of the red pixels:

	(b)	(c)
Dice	$\frac{2\lvert R\rvert}{\lvert R\rvert+7+\lvert R\rvert} = \frac{2\lvert R\rvert}{2\lvert R\rvert+7} = \frac{280}{287} = \mathbf{97.56\%}$	
Jaccard	$\frac{\lvert R\rvert}{\lvert R\rvert+7} = \frac{140}{147} = \mathbf{95.24\%}$	
Sensitivity	$\frac{\lvert R\rvert}{\lvert R\rvert} = \mathbf{100\%}$	
Specificity	$\frac{\lvert R'\rvert-7}{\lvert R'\rvert} = 97.30\%$	

Analogously, regions (d) and (e) have 7 pixels missing from the reference region (a), false negatives highlighted in orange. The lack of any spatial insight in the evaluation produces 4 values, for these image examples, that are always the same:

	(d)	(e)
Dice	$\frac{2(\vert R\vert-7)}{\vert R\vert-7+\vert R\vert} = \frac{2(\vert R\vert-7)}{2\vert R\vert-7} = \frac{266}{273} = \mathbf{97.44\%}$	
Jaccard	$\frac{\vert R\vert-7}{\vert R\vert} = \frac{133}{140} = \mathbf{95.00\%}$	
Sensitivity	$\frac{\vert R\vert-7}{\vert R\vert} = \frac{133}{140} = \mathbf{95.00\%}$	
Specificity	$\frac{\vert R'\vert}{\vert R'\vert} = 100\%$	

It is worth mentioning that, when dealing with medical images, the cardinalities of all the subsets that include segmentation complements (T', R') are somewhat arbitrary, being easily affected by crop operations that leave the segmented regions untouched.

4 Roles

Before proceeding, it is important to highlight a key difference between the two possible scenarios where an agreement assessment takes place: (i) comparison with respect to a reference region (typically called ground truth), and (ii) comparison between two regions with equal standing. Sub-figures 1*a* and *e* demonstrate this difference. If 1a is the reference region and we want to assess the agreement of *1e* with *1a*, then clearly *1e* is in error and, possibly, by an important one as the region has a hole in the middle of a supposedly filled region[1]. On the other hand, if *1a* and *1e* are equally reliable (we call it the *symmetric* case), then the hole in the middle might be part of the correct result, but we cannot infer it from the data that we have available. In the following sections, the discussion will assume the existence of a reference region; a preliminary analysis of the *symmetric* scenario is presented in Sect. 8.

5 Initial Considerations for New Metrics

Based on the previous considerations, the key aspect we would like to introduce in our metrics is the acknowledgment of and a grading of the different spatial positions where the disagreements occur. For instance, we would like to switch from a cardinality-based disagreement as in Eq. 3 (or, re-written to loop over all the image-domain elements, Eq. 4) to a disagreement metric where the disagreement is weighted by a spatially dependent function w as in Eq. 5 (and the normalization is scaled by function n)

[1] Notice that a specular reasoning still holds if the roles of *1a* and *1e* are swapped - with *1e* being the reference and *1a* being the assessed region.

$$dis(T,R) = \frac{|T \Delta R|}{|R|} \tag{3}$$

$$dis(T,R) = \frac{\sum |S_{ij} - R_{ij}|}{\sum R_{ij}} \tag{4}$$

$$new\_dis(T,R) = \frac{\sum w(i,j,R)|S_{ij} - R_{ij}|}{\sum n(i,j,R)} \tag{5}$$

As the new disagreement measures in (5) are still based on the cardinality of $\mathcal{G}$ subsets, they will clearly satisfy the conditions of a metric as long as the weighting functions are strictly positive.

6 A Family of Disagreement Metrics

A convenient family of metric-defining weighting functions can easily be built by using the signed Euclidian distance transform (SEDT) [6, 7] of the reference region, where the internal elements are given a positive value, while those outside the region are given a negative value, as shown in Fig. 2. It is worth highlighting how this approach, thanks to its use of the SEDT, provides several potential advantages (e.g., with respect to the perceptual-based approach of [8, 9]):

- it maps the n-dimensional image domain to a single dimension;
- it overcomes the need to account for the pixel/voxel size;
- it makes it possible to structure the weighting function according to problem-specific (anatomical) sizes, expressed in real word lengths.

Therefore, the weighting function can be written as $w(i,j,R) = a(d_R(i,j,R))$ where the metric-proxy $a : \mathbb{R} \rightarrow \mathbb{R}_0^+$ defines the amount of disagreement according to the signed distance from the reference region border(s) and provides ample freedom in expressing the wanted grading of the disagreement. The following equations exemplify this relationship between disagreement location and its measurement. The metric-proxy a_1 (Eq. 6) provides an example of a grading that is proportional to the distance from the region borders; the metric-proxy a_2 (Eq. 7) is designed to tolerate errors up to 10 mm from the border and then flag anything further up; the metric-proxy a_3 (Eq. 8) highlights errors near the border and discounts the other discrepancies:

$$a_1(x) = |x| \tag{6}$$

$$a_2(x) = (x/10)^4 \tag{7}$$

$$a_3(x) = e^{-(x/10)^4} \tag{8}$$

Figure 2 shows the signed distance transform computed on the sample reference region in Fig. 1a, and the resulting disagreement values for all the other regions in Table 1, according to different choices of function a (Eqs. 6, 7 and 8) and having the corresponding normalization function n defined so that a complete disagreement with the reference region is graded at 100%[2]. All the results are computed assuming a pixel size of 3 mm by 3 mm. The set-based disagreement is always 5% for all Sub-figs. 1b to e.

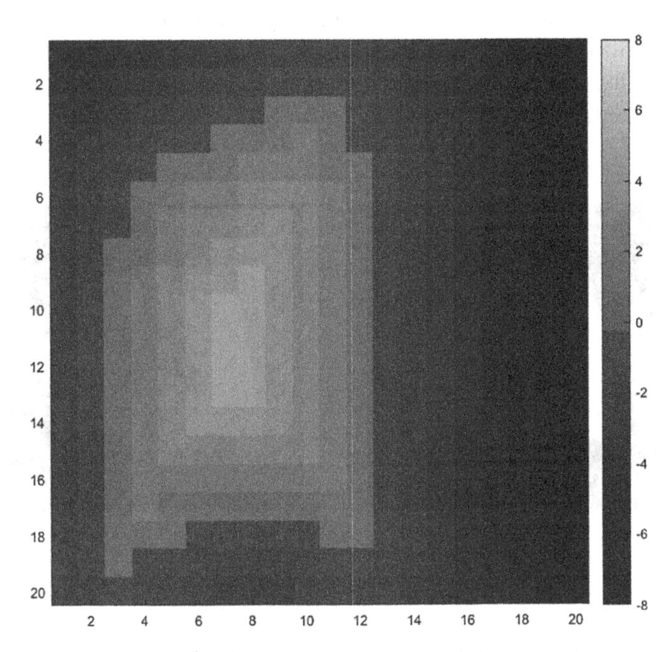

Fig. 2. The signed distance transform of the reference region R (shown in Fig. 1a).

Table 1. Disagreement performance of the 4 hypothetical segmentations of Fig. 1 with respect to the example metrics

Figure 1 region	Metric proxies		
	a_1	a_2	a_3
b	1%	0.01%	7%
c	7%	24%	3%
d	1%	0.01%	7%
e	7%	9%	3%

7 Application to Lung Imaging

In order to demonstrate the effectiveness of the new metrics in supplying meaningful summaries of disagreements' spatial distributions, they were applied in the evaluation of different thresholding levels of a chest anatomical scan, acquired on a GE HDx 1.5T

[2] The upper level of disagreement is arbitrary. The amount of 100% was chosen to be comparable with the set-based formulation.

MR scanner[3] (3D spoiled gradient echo sequence, $1.5625 \times 1.5625 \times 5\ \text{mm}^3$ voxel size). Three threshold levels (th_3, th_5, th_7) were computed as the lowest values of a multi-threshold Otsu algorithm [10, 11] with 3, 5, and 7 clusters, respectively. The image had previously been segmented manually to produce the reference segmentation. Representative coronal slices of the reference segmentation and the three thresholded regions are shown in Fig. 3. As the number of clusters increases, the threshold values decrease, causing the resulting regions to exclude areas with denser tissue such as vessels and airway walls.

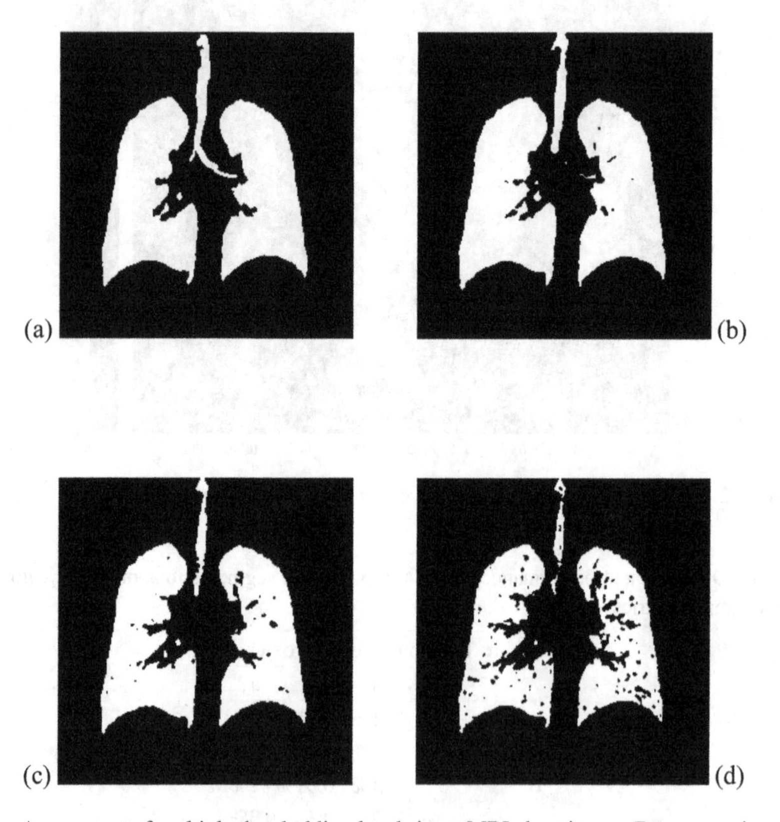

Fig. 3. Assessment of multiple thresholding levels in an MRI chest image. Representative slices of (a) the ground truth segmentation, (b, c, d) thresholded images at values th_3, th_5, th_7.

Table 2 reports the values for the set-based disagreement and for the metric-proxies a_2 and a_3. While the threshold th_7 is too low according to any metric, the set-based disagreement is unable to summarize the slight differences between th_3 and th_5. By considering all the values from a_2 and a_3, it is straightforward to acknowledge that, if one is limited to simple thresholding, a tradeoff must be chosen – as th_3 has a better

3 GE Healthcare, Milwaukee, IL, USA.

performance in the inner parts of the lungs, whereas th_5 is considerably better at capturing the lung borders.

Table 2. Disagreement values for the shapes in Fig. 2a and b.

Threshold	Set-based	Metric proxies	
		a_2	a_3
th_3	4.99%	0.05%	3.40%
th_5	5.11%	0.64%	3.29%
th_7	14.8%	2.49%	9.32%

8 Region Assessment Without Reference

Let us now consider two regions, *S* and *T*, neither of which is the reference region (ground-truth). In these cases, the set-based overlap metrics of regions *S* and *T* can be interpreted as measuring the size of the region where there is agreement, the intersection, against an estimation of the reference region size – the size average for Dice, the union for Jaccard. Similar normalization approaches can be used as estimates for the set-based disagreement metric (3):

$$dis_{Dice} = \frac{2|S\Delta T|}{|S| + |T|} \tag{9}$$

$$dis_{Jaccard} = \frac{|S\Delta T|}{|S \cup T|} \tag{10}$$

If we are to take into account the spatial component, then a different estimate is required and the natural solution for the ground truth estimation appears to be a shape that is the average of *S* and *T* [12]. In our example, a shape interpolation [13] was performed using the *itksnap* tool [14, 15]. With a proper estimate *R* of the reference region, a spatially aware disagreement can be expressed both as individual disagreements for each region (i.e., *S* with *R* and *T* with *R*) and as combined disagreement as the sum of the respective individual disagreements:

$$dis_S = new\_dis(S, R) \tag{11}$$

$$dis_T = new\_dis(T, R) \tag{12}$$

$$dis_{new} = dis_S + new_T \tag{13}$$

Clearly, there will be no disagreements deep inside there reference region because *R* is an average of the two regions. Nonetheless, the use of metric-proxies like a_3 can prove useful in summarizing the disagreement behavior around the region borders. This is well demonstrated by a simple example, where the disagreement between two shapes – a disk and a star, are investigated. Their set-based disagreements are 36.5% and 30.8%

for dis_{Dice} and $dis_{Jaccard}$. respectively. Figure 4 shows the two shapes, the interpolated shape and the SEDT of the interpolated shape. Table 3 reports the values of the metric proxies. In the case when the reference ground truth is missing, metric proxies that limit the effect of large positive values, such as a_4 in Eq. 14, can be used to assess the individual disagreements and, together with the other metrics, gain an additional understanding of the way spatial disagreement is distributed:

$$a_4(x) = 1 - \frac{1}{1 + e^{-x}} \tag{14}$$

Table 3. Disagreement values for the shapes in Figs. 2a and b.

Shape	Metric proxies		
	a_1	a_2	a_3
Disc	3.5%	9.4%	141.8%
Star	3.5%	9.4%	181.6%

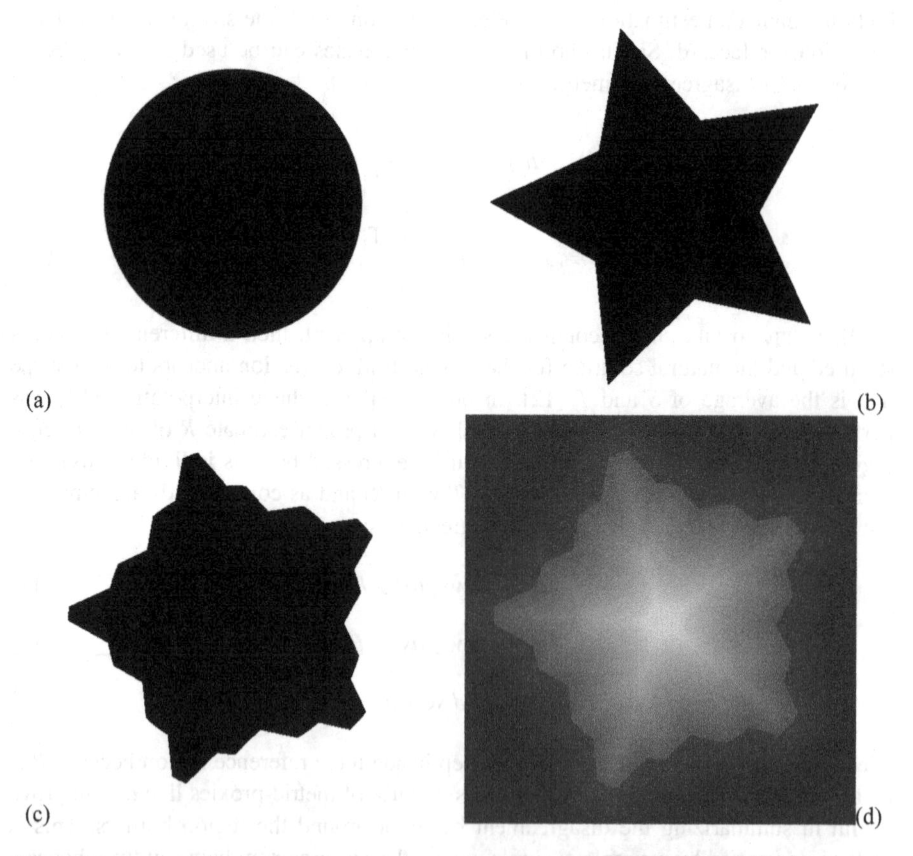

Fig. 4. Example of disagreement assessment without ground truth: (a) and (b) the shapes to be evaluated, (c) the interpolated shape as ground truth estimation, (d) the SEDT of c.

In our example, the a_4 disagreement for the star is 234.6% and for the disc is 490.8%. The higher number for the disk, in combination with a smaller a_3 value, tells us that the disk has much more disagreement on the outside of the reference region and that this disagreement is rarely far from the reference border.

9 Conclusions

Assessment of image regions plays an important role in computer-supported analysis because of the various outcomes relying on those regions for their computations. A new family of image disagreement metrics was introduced; these metrics can be easily adapted to the specific anatomical sizes under analysis and give a much richer summary of where the disagreement occurs when compared to set-based disagreement metrics. Preliminary applications show the potential usefulness of these additional spatial insights; however, further aspect can be investigated. Future work will study the relationship between these metrics and boundary/surface-based metrics such as the Hausdorff distance [16] and evaluate the extension to multi-object scenarios.

Acknowledgements. This work was supported by NIHR grant NIHR-RP-R3-12-027 and MRC grant MR/M008894/1.

The views expressed in this publication are those of the author(s) and not necessarily those of the NHS, the National Institute for Health Research or the Department of Health.

References

1. Cuingnet, R., Prevost, R., Lesage, D., Cohen, L.D., Mory, B., Ardon, R.: Automatic detection and segmentation of kidneys in 3D CT images using random forests. In: Ayache, N., Delingette, H., Golland, P., Mori, K. (eds.) MICCAI 2012. LNCS, vol. 7512, pp. 66–74. Springer, Heidelberg (2012). doi:10.1007/978-3-642-33454-2_9
2. Van Leemput, K., Bakkour, A., Benner, T., Wiggins, G., Wald, L.L., Augustinack, J., Dickerson, B.C., Golland, P., Fischl, B.: Automated segmentation of hippocampal subfields from ultra-high resolution in vivo MRI. Hippocampus **19**(6), 549–557 (2009)
3. Sun, S., Bauer, C., Beichel, R.: Automated 3-D segmentation of lungs with lung cancer in CT data using a novel robust active shape model approach. IEEE Trans. Med. Imaging **31** (2), 449–460 (2012)
4. Korez, R., Ibragimov, B., Likar, B., Pernuš, F., Vrtovec, T.: A framework for automated spine and vertebrae interpolation-based detection and model-based segmentation. IEEE Trans. Med. Imaging **34**(8), 1649–1662 (2015)
5. Johnson, H.J., McCormick, M.M., Ibáñez, L.: The Insight Software Consortium (2017). The ITK Software Guide. http://itk.org/ItkSoftwareGuide.pdf
6. Felzenszwalb, P.F., Huttenlocher, D.P.: Distance transforms of sampled functions. Theory Comput. **8**, 415–428 (2012)
7. Maurer, C.R., Qi, R., Raghavan, V.: A linear time algorithm for computing exact Euclidean distance transforms of binary images in arbitrary dimensions. IEEE Trans. Pattern Anal. Mach. Intell. **25**(2), 265–270 (2003)

8. Villegas, P., Marichal, X.: Perceptually-weighted evaluation criteria for segmentation masks in video sequences. IEEE Trans. Image Process. **13**(8), 1092–1103 (2004)
9. Gavet, Y., Fernandes, M., Debayle, J., Pinoli, J.C.: Dissimilarity criteria and their comparison for quantitative evaluation of image segmentation: application to human retina vessels. Mach. Vis. Appl. **25**(8), 1953–1966 (2014)
10. Otsu, N.: A threshold selection method from gray-level histograms. IEEE Trans. Syst. Man Cybernet. **9**(1), 62–66 (1979)
11. Liao, P.S., Chen, T.S., Chung, P.C.: A fast algorithm for multilevel thresholding. J. Inf. Sci. Eng. **17**(5), 713–727 (2001)
12. Raya, S.P., Udupa, J.K.: Shape-based interpolation of multidimensional objects. IEEE Trans. Med. Imaging **9**, 32–42 (1990)
13. Albu, A.B., Beugeling, T., Laurendeau, D.: A morphology-based approach for interslice interpolation of anatomical slices from volumetric images. IEEE Trans. Biomed. Eng. **55**(8), 2022–2038 (2008)
14. Zukic, D., Vicory, J., McCormick, M., Wisse, L.E., Gerig, G., Yushkevich, P., Aylward, S.: nD morphological contour interpolation. Insight J. (2016). http://hdl.handle.net/10380/3563
15. Yushkevich, P.A., Piven, J., Cody Hazlett, H., Gimpel Smith, R., Ho, S., Gee, J.C., Gerig, G.: User-guided 3D active contour segmentation of anatomical structures: significantly improved efficiency and reliability. Neuroimage **31**(3), 1116–1128 (2006)
16. Taha, A.A., Hanbury, A.: Metrics for evaluating 3D medical image segmentation: analysis, selection, and tool. BMC Med. Imaging **15**(1), 29 (2015)

Unsupervised Segmentation of Cervical Cell Nuclei via Adaptive Clustering

Srishti Gautam(✉), Krati Gupta, Arnav Bhavsar, and Anil K. Sao

Indian Institute of Technology, Mandi, India
{srishti_gautam,krati_gupta}@students.iitmandi.ac.in,
{arnav,anil}@iitmandi.ac.in

Abstract. Owing to the uncertainties of manual screening, automated diagnostic tools can aid in improving the reliability of cervical cancer diagnosis. Due to pre-cancerous changes, cell morphology is altered and hence it plays an important role in the screening process. Therefore, segmentation of the cellular parts is important for the classification of cells as normal/abnormal. This paper focuses on segmentation of nuclei in Pap smear images using contrast based adaptive versions of mean-shift and SLIC algorithms followed by an intensity weighted adaptive thresholding. The proposed method is evaluated on Herlev dataset. The performance of the proposed method is compared with state-of-the art clustering based method. The results show that the approach is effective in segmenting images having inconsistent contrast.

Keywords: Cervical cell segmentation · Pap smear · Cervical cancer screening · Nucleus segmentation · Mean shift · SLIC

1 Introduction

Cancer of the cervix is the most common cause of death among women in the developing countries [1]. For instance, among women in India with 24% of the total cancer-related deaths in 2015 [2]. The typical way of diagnosis involves the microscopic observation of Pap smear samples [3]. Due to the uncertainties in manual diagnosis, it is recommended to use automated screening systems [4]. The morphological characteristics of nucleus and cytoplasm yield information about the malignity of each cell [5]. Traditionally, such traits are employed in the classification. Hence, the first step in screening involves identifying region of interests (nucleus and cytoplasm) in Pap smear images by segmentation process. However, typically the segmentation of nucleus is relatively more reliable than that of cytoplasm. Further, it is noted that the nucleus-level analysis is valuable in cervical cancer screening [6]. Hence, instead of extracting features from relatively unreliable segmentation of cytoplasm, employing only the nucleus regions can be more effective in grading the abnormalities.

Considering the importance of the problem, some methods have been reported in this area. Phoulady et al. [6,7] used adaptive multilevel thresholding and ellipse fitting, but on a different problem of segmenting overlapping

M. Valdés Hernández and V. González-Castro (Eds.): MIUA 2017, CCIS 723, pp. 815–826, 2017.
DOI: 10.1007/978-3-319-60964-5_71

cells. Cheng and Hsu [8] used HSV color space and color clustering for the same. Genctav et al. [9] multi-scale hierarchical segmentation algorithm. Similar to our proposed method, Agarwal et al. [10] uses mean-shift segmentation but with constant bandwidths. Some supervised methods have also been used for this purpose, for example, multi-scale convolutional networks [11], deep CNN [12] and SVM [13]. [14] uses patch-based fuzzy C-means (FCM) clustering technique. On the over-segmented image obtained from FCM, a threshold is applied to classify the FCM cluster centers as nucleus and background. [15] uses superpixel-based Markov random field (MRF) framework Bora et al. [16] uses Wavelet and Haar transform along with MSER (maximally stable extremal region).

In this paper, we propose a new unsupervised approach for nucleus segmentation in Pap smear images based on an adaptation of popular approaches for superpixel segmentation viz. mean-shift (MS) [17] and the SLIC [18] methods. This approach can be divided as a three-step process. The images are first pre-processed to reduce noise and accentuate the differences between nucleus and background. We then apply superpixel estimation to group visually similar pixels into local regions. Due to a variable contrast between nucleus and background, we propose adaptive variants of well-known superpixel estimation methods such as mean-shift and SLIC, wherein the parameters in these methods important for the segmentation performance vary for each image depending on its contrast. This step yields a superpixel image with pixels grouping in small local regions. Finally, we propose a novel intensity weighted adaptive thresholding on the superpixels to single out the nuclei from cell and background. The thresholding approach considers the observation that nuclei are generally darker than rest of the cell contents. The algorithm is validated on Herlev dataset [19] with 917 images. The results obtained show an improvement over those in [14]. Further, it is noted that the accurate segmentation of abnormal cells is more important than normal ones because of no tolerance for false negatives in medical imaging. Hence, we also report separate segmentation performance for normal and abnormal cells (Table 1).

Table 1. Sample cervical cells with total number of cells in Herlev dataset.

Classes	**Abnormal**			**Normal**			
	Superficial Squamous (nsup)	Intermediate Squamous (nint)	Columnar (ncol)	Light dysplasia (ldys)	Moderate dysplasia (mdys)	Severe dysplasia (sdys)	Carcinoma in situ (cis)
Sample cells							
Total no. of cells	74	70	98	182	146	197	150

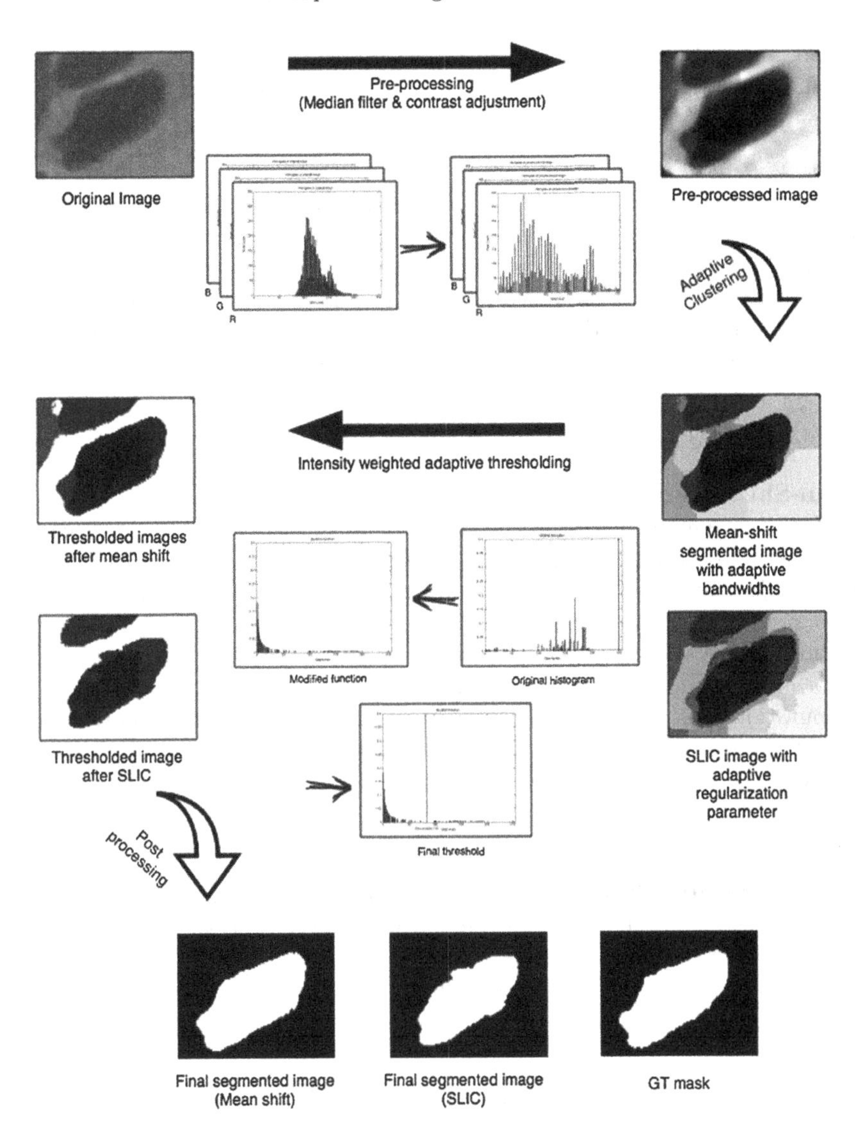

Fig. 1. Overview of the proposed method.

2 Proposed Approach

2.1 Pre-processing

The Pap smear images often have a poor contrast and hence a poor discrimination between the background, cytoplasm and nuclei due to inconsistent staining. Hence, a contrast enhancement operation is employed. Prior to this, we filter the image with a 7×7 median filter to uniformly smoothen it, so as to reduce any spurious noise that may affect the local contrast. The contrast of each channel is then adjusted by mapping the initial intensities to new values such that some percent of the data is saturated at low and high intensities of the input image.

$$I_{out} = l_{out} + (h_{out} - l_{out}) * ((I - l_{in})/(h_{in} - l_{in}))^{\gamma} \quad (1)$$

where, l_{in} and h_{in}, and l_{out} and h_{out} are the input and output limits, respectively for the image I and γ value shapes the mapping curve between the input and output elements. We have considered the value of γ as 1. We note that the choice of preprocessing approach is not very strict, and other well-known preprocessing methods could also be used, e.g. histogram equalization or CLAHE [20], depending on the desired contrast enhancement.

2.2 Adaptive Bandwidth Mean-Shift Segmentation

Mean-Shift Segmentation. The mean-shift algorithm [21,22] is an unsupervised, iterative mode detection algorithm in the density distribution space commonly used in computer vision problems such as filtering, tracking and segmentation. It is a non-parametric mode finding procedure. The mean-shift procedure takes a smoothing window around an observation. Then the mean-shift vector is computed by taking kernel-weighted average of the observations within the smoothing window and the window is shifted along the mean-shift vector. This computation is repeated until convergence is obtained at a local density mode.

In the mean-shift procedure-based image segmentation, each pixel is associated with a significant local mode of the joint density in color and spatial domain located in its neighborhood.

Adaptive Bandwidth Selection. For segmentation, two kernel-weighted windows are used in mean-shift algorithm. One is the spatial bandwidth h_s, which affects the smoothing and connectivity of segments and the other one is the range bandwidth h_r, which affects the number of segments. The kernel used can be expressed as,

$$K_{h_s,h_r}(x) = \frac{C}{h_s^p h_r^q} k\left(||\frac{x^s}{h_s}||^2\right) k\left(||\frac{x^r}{h_r}||^2\right) \quad (2)$$

where $p = 2$ for spatial domain with features on pixel locations and $q = 3$ for LAB color space features and x^s and x^r denote the variables corresponding to the spatial and range (color) domain respectively. Traditional mean-shift uses constant spatial and range bandwidths for all types of images.

The Pap smear images have considerable difference in terms of contrast as can be seen in Fig. 2. Hence range bandwidth must be decided adaptively based on the contrast of the image so that the final segmentation is independent of the contrast value. This ensures that in the images having low contrast, nucleus segment does not merge with the background segment (Fig. 2) and in the images having high contrast, nucleus and background segments do not over-segment and appear as one connected component which further helps in thresholding (Fig. 3).

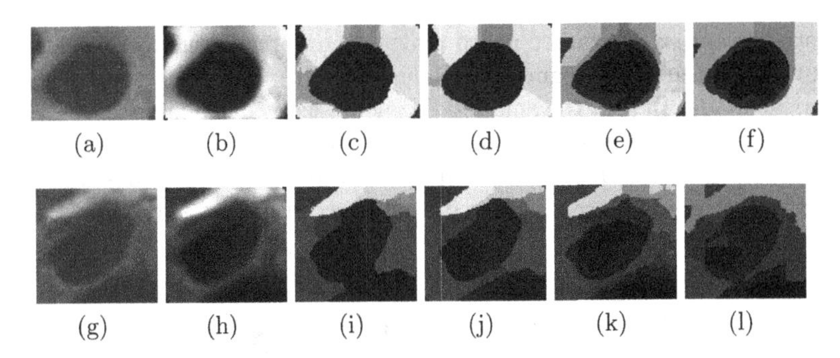

Fig. 2. Adaptive bandwidth. (a), (g) high and low contrast images respectively, (b), (h) pre-processed image ($c_i = 0.88, 0.62$ respectively), (c), (i) mean-shift segmentation with constant bandwidths ($h_r = 15, h_s = 10$), (d), (j) mean-shift segmentation with adaptive bandwidths ($h_r = 17.6, h_s = 10$; $h_r = 12.5, h_s = 10$ respectively), (e), (k) SLIC with constant regularization($m = 0.01$), (f), (l) SLIC with adaptive regularization($m = c_i * 10$).

The range bandwidth decides the number of segments. As h_r increases, only the features with high color contrast survives [17]. For the high contrast images, high range bandwidth works well because there is enough discrimination between the nuclei and background and the high range bandwidth will help in obtaining connected segments for both nuclei as well as the background. But for the low contrast images, we need low range bandwidth so that the nuclei and background segments do not merge. We have considered standard deviation of the images as a measure of contrast. The h_r for an image i is calculated as,

$$c_i = \frac{\sigma(I_i)}{\max\limits_{1 \le j \le n} \sigma(I_j)} \tag{3}$$

$$h_r = c_i * s \tag{4}$$

where c_i is the contrast value for the image i, $\sigma(I_i)$ is the standard deviation of image i, n is the total number of images in the dataset and s is the scaling constant whose value is varied in the range of 15–30 with a step size of 1. The range 15–30 is chosen to ensure meaningful segmentation by the mean-shift method. Such parameter values may vary with the dataset used, and can depend on typical size and intensity variation of the ROI (nucleus) in the dataset. We note that a single value, once chosen from the range, is used for all images in the dataset, and best results across all such values are reported.

Figure 2 shows the effect of the adaptive bandwidth mean-shift on a high and a low contrast image and also compares the results with using constant bandwidths. We have kept spatial bandwidth, h_s, fixed according to the image size [17]. This is due to the fact that the dataset consists of images with extremely varying sizes, smaller with around 80×80 pixels and larger with around 250×250 pixels. Based on the observed image size variation in the dataset, for the images greater than 150×150 pixels, h_s is chosen as 25, else a value of 10 is used. The smallest segment size parameter also plays an important role in the mean-shift segmentation because the segments smaller than this parameter are merged with the nearest mode. Hence, the smallest segment size is decided based on the size of the smallest feature in the image which in our case is the smallest nuclei and is fixed as 30. It is noted that the values of free parameters discussed above are selected empirically, but can be tuned easily according to the dataset.

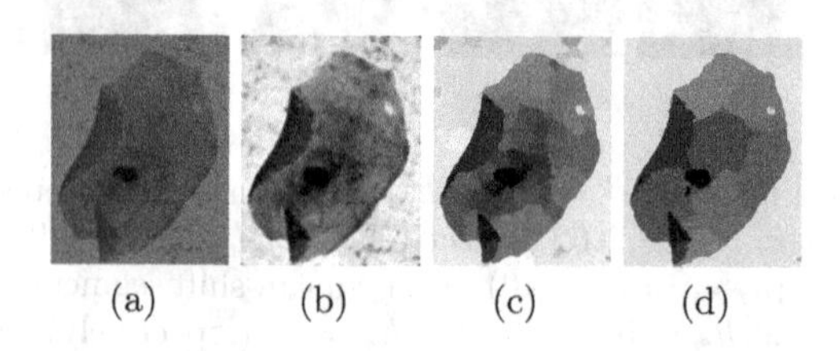

Fig. 3. (a) Original image, (b) pre-processed image, (c) mean-shift segmentation with constant bandwidths ($h_r = 15, h_s = 10$), (d) mean-shift segmentation with adaptive bandwidths ($h_r = 19.7, h_s = 25$).

2.3 Adaptive regularization SLIC

The simple linear iterative clustering [18] method can also be used for segmenting an image in visually consistent local regions. Typically, semantically meaningful objects (e.g. nucleus) are over-segmented into multiple similar sized regions. The approach starts by dividing the image domain into a regular grid $S = \sqrt{N/k}$ pixels apart, where N are the total no. of pixels in the image and k corresponds to the no. of approximately equally sized superpixels. Hence $S \times S$ is the initial superpixel size. The centers are initialized with each grid center. Each pixel is associated with the nearest cluster center with a search area of $2S \times 2S$. Cluster means are updated iteratively according to the distance measure D,

$$D = \sqrt{d_c^2 + \left(\frac{d_s}{S}\right)^2 m^2}, \tag{5}$$

$$d_c = \sqrt{(l_j - l_i)^2 + (a_j - a_i)^2 + (b_j - b_i)^2}, \tag{6}$$

$$d_s = \sqrt{(x_j - x_i)^2 + (y_j - y_i)^2} \tag{7}$$

where $[l, a, b]$ is the pixel's color in CIELAB color space, $[x, y]$ represents its position and m is the regularization parameter which allows to weigh the relative importance between color and spatial proximity.

Adaptive Regularization. Similar to the adaptive mean-shift, because of the extensive contrast variation, one parameter does not fit all of the images. For the images having low contrast, we suggest that the color similarity must weigh more than the spatial proximity to avoid the merging of similar colored nucleus and background into one segment. Similarly, for the images having high contrast, spatial proximity must weigh more than the color similarity so that the nucleus does not get extremely over-segmented. Hence, the regularization parameter, m, is selected adaptively on the basis of contrast as,

$$m = c_i * s \tag{8}$$

where c_i is inherited from Eq. 3 and s is the scaling constant whose value is varied in the range of 10 to 30 with a step size of 1, and best results are reported. When the contrast is high, m will be high, hence more weight will be given to the spatial proximity. Similarly, when the contrast is low, m will be low and more weight will be given to the color similarity. The comparisons of adaptive and non-adaptive SLIC are shown in Fig. 2.

2.4 Intensity Weighted Adaptive Thresholding

In this stage, the superpixels computed by the mean-shift and SLIC approaches are to be labeled either as nucleus or as background. In cervical cells, as the abnormality increases, the size of the nucleus increases [4]. Hence, our thresholding method must be size independent. Further, it has been observed that the darker pixels in the image correspond to the nuclei and as they start becoming lighter, they correspond to the background.

Keeping these observations in mind, we define a new function $P_r(r_q)$ in place of the original histogram of the image, where the input r_q represents the intensity values in the image and the output value represents its weight. $P_r(r_q)$ is defined as follows: For all the modes obtained from the mean-shift segmentation, the weight of the intensity values of the modes is set as 1 and all the other intensity values (not present in the segmented image) are set as 0. This ensures the size independence of the modes i.e. the smaller and bigger size of nuclei in normal and abnormal images, respectively, will not have any contribution in decision of the threshold because of the equalized weights. The function $P_r(r_q)$ is then multiplied by an exponential probability density function, as

$$Q_r(r_q) = P_r(r_q) * \left(\frac{1}{\mu} e^{\frac{-r_q}{\mu}}\right) \tag{9}$$

where μ is the mean of the distribution. We have taken μ as a vector consisting of values from 1 to 256. This ensures that the darker pixels are given more preference. $Q_r(r_q)$ is now normalized and passed on further for Otsu thresholding approach [23].

Because the contribution of lighter pixels is less due to the exponential weighing, the optimal threshold k^* shifts towards the darker pixels. This ensures that only the pixels which relatively more dark (typically the nuclei pixels), fall into one class. The results with this intensity weighted adaptive thresholding demonstrate striking difference in the final segmented mask obtained as compared to traditional Otsu's thresholding as shown in Fig. 4.

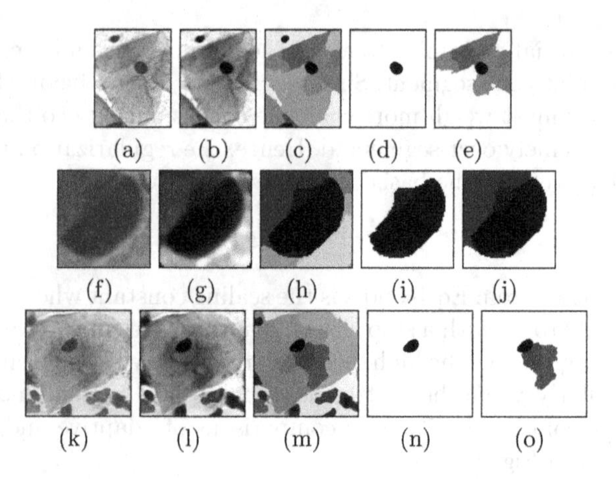

Fig. 4. Thresholding. (a), (f), (k) Original image, (b), (g), (l) pre-processed image, (c), (h), (m) mean-shift segmentation, (d), (i), (n) intensity weighted adaptive thresholding, (e), (j), (o) traditional Otsu's thresholding.

3 Experiments and Results

3.1 Data Description

We have evaluated our algorithm on Herlev Pap smear dataset [19]. It consists of 917 cell images whose description in the increasing order of abnormality is given in Table 1. Each image in Herlev dataset consists of a single nuclei.

3.2 Evaluation Measures

We have quantified the segmentation using pixel-based precision and recall.

$$\text{Precision} = \frac{A \cap B}{A} \quad \text{Recall} = \frac{A \cap B}{B} \tag{10}$$

where A denotes the pixels classified as nucleus in the segmented image and B denotes the ground truth nucleus pixels. Precision denotes the fraction of pixels classified correctly from all of the pixels that we classify as nucleus, and recall represents the fraction of ground truth nucleus pixels we are able to find.

We also use F-score which considers both precision and recall. It reaches its best value at 1 and worst at 0 and is defined as harmonic mean of precision and recall.

$$\text{F-score} = 2 * \frac{\text{Precision} * \text{Recall}}{\text{Precision} + \text{Recall}} \tag{11}$$

3.3 Results

The visual results of the algorithm on some examples are shown in Fig. 5, which indicate that the approach is able to localize the nucleus well.

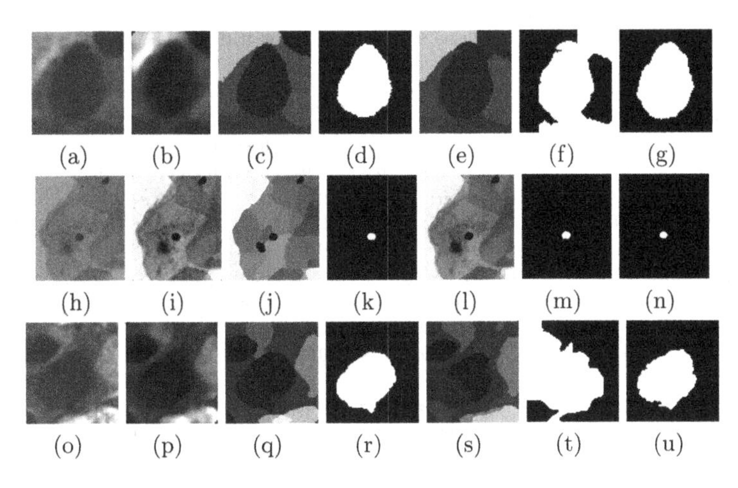

Fig. 5. (a), (h), (o) Original image, (b), (i), (p) pre-processed image (c_i = 0.63, 0.56, 0.55 respectively), (c), (j), (q) Mean-shift segmentation ($h_r = 12.6, 11.2, h_r = 11.14$, $h_s = 10, 25, 10$, respectively), (d), (k), (r) final segmentation by mean-shift, (e), (l), (s) SLIC superpixels, (f), (m), (t) final segmentation by SLIC, (g), (n), (u) ground truth.

The quantitative results of the proposed approaches on Herlev dataset are provided for each class in Table 2, along with the comparison with a contemporary method [14], which also uses a clustering based approach. Here, we just compare with [14] as it is the only known method providing metrics specified in Table 2 for different classes separately. It can be observed that all the variants of the proposed approach are outperforming the method of [14] for most of the classes. The results in Table 2 also demonstrate the positive effect of channel selection for the superficial and intermediate classes.

In Table 2, we also compare our approach with the non-adaptive counterparts (for various trials) followed by Otsu's thresholding. We observe that our approach greatly outperforms the latter. This highlights the importance of adaptiveness that we propose.

In Table 3, we provide the results after combining the cell classes into normal and abnormal categories, for different parameter settings. We can observe an improvement in F-score for both normal and abnormal classes over [14]. In addition, we note that accurate segmentation of abnormal cells is more important, if there is little tolerance for false negatives. Based on this observation, one can choose the parameters accordingly.

We also provide the overall Zijdenbos similarity index (ZSI) [15] comparisons with various contemporary methods such as FCM [14], basic thresholding, MRF and RGVF methods as reported in [15] in Table 4. We note that the MRF and RGVF approaches are more sophisticated methods involving constrained energy minimization, while the method in [14] and the thresholding method are relatively simpler clustering based approaches (similar to ours). Thus, among the simpler methods, the proposed variants show a clear improvement. Having said that, we acknowledge the better performance of the MRF and RGVF approaches.

Table 2. Class-wise nucleus segmentation results for Herlev dataset.

		nsup	nint	ncol	ldys	mdys	sdys	cis	Average
Our method (MS)	Precision	0.74	0.92	0.74	0.87	0.87	0.88	0.87	0.85
	Recall	0.93	0.96	0.88	0.84	0.88	0.89	0.84	0.88
	F-score	0.82	0.94	0.81	0.86	0.88	0.89	0.85	**0.86**
Our method (MS with best channel thresholding)	Precision	0.84	0.93	0.74	0.87	0.87	0.88	0.87	0.85
	Recall	0.92	0.98	0.88	0.84	0.88	0.89	0.84	0.89
	F-score	**0.87**	**0.95**	0.81	0.86	0.88	0.89	0.85	**0.87**
Our method (SLIC)	Precision	0.84	0.85	0.86	0.72	0.85	0.59	0.86	0.79
	Recall	0.89	0.87	0.90	0.91	0.98	0.89	0.84	0.91
	F-score	0.86	0.86	0.88	0.81	0.91	0.71	0.88	**0.85**
Our method (SLIC with best channel thresholding)	Precision	0.83	0.92	0.86	0.72	0.85	0.59	0.86	0.80
	Recall	0.92	0.98	0.90	0.91	0.98	0.89	0.84	0.92
	F-score	**0.87**	**0.95**	0.88	0.81	0.91	0.71	0.88	**0.86**
FCM [14]	Precision	0.95	0.98	0.88	0.80	0.81	0.79	0.70	0.85
	Recall	0.75	0.82	0.78	0.86	0.88	0.88	0.88	0.85
	F-score	0.84	0.89	0.83	0.83	0.84	0.83	0.78	**0.84**
Mean shift & Otsu (non-adaptive)	Precision	0.78	0.65	0.67	0.51	0.28	0.05	0.76	0.61
	Recall	0.97	0.97	0.97	0.99	0.99	0.98	0.96	0.97
	F-score	0.87	0.78	0.79	0.67	0.44	0.09	0.85	**0.64**

Table 3. Nucleus segmentation results comparison for abnormal and normal classes.

Classes	Our results (MS) F-score ($h_r = c_i * 20$)	Our results (MS) F-score ($h_r = c_i * 30$)	Our results (SLIC) F-score ($m = c_i * 10$)	Results of [14] F-score
Normal	0.86	0.88	0.90	0.85
Abnormal	0.87	0.84	0.83	0.82

Table 4. Pixel-based ZSI comparison

Segmentation method	Proposed (MS)	Proposed (SLIC)	FCM [14]	Threshold [15]	MRF [15]	RGVF [15]
ZSI of nuclei	0.86	0.86	0.80	0.78	0.93	0.93

3.4 Channel Selection for Thresholding Based on Prior Knowledge

It is evident from the literature that the normal intermediate and normal superficial cells are larger that the rest of the classes [19] and it is observed that the nuclei of these classes are more compact and circular. Hence, we suggest a further improvement for these two classes to achieve improved results. We first identify the superficial and the intermediate images based on their size. Our adaptive thresholding is then applied separately on the RGB channels of the superpixel images. The channel which yields the least anti-circular region after thresholding is selected as the final segmented output. The anti-circularity, a_c is defined as, $a_c = P^2/4\pi A$ where P is the perimeter and A is the area of the region. Note that this assumes that single cell images are available, which can be obtained via a coarse cell detection stage.

4 Conclusion

This paper reports on unsupervised segmentation algorithms for nuclei in Pap smear images. The proposed approaches exploit the flexibility of parameters of the mean-shift and SLIC approaches and adapt these based on the image contrast. An intensity weighted adaptive thresholding is used after the superpixel computation to get the final nuclei segments. The adaptive thresholding considers that nuclei are generally darker in color and vary in size with the variation in grades of abnormality. The algorithm is tested on Herlev dataset and is shown to outperform a contemporary clustering based approach. The results also demonstrate the importance of the adaptation proposed in the individual models.

References

1. Sreedevi, A., Javed, R., Dinesh, A.: Epidemiology of cervical cancer with special focus on india. Int. J. Womens Health **7**, 405–414 (2015)
2. Dey, S.: Cervical cancer back as top killer among women. TNN, Times of India, 4 May 2016, 03.46 AM IST
3. Chang, C.-W., Lin, M.-Y., Harn, H.-J., Harn, Y.-C., Chen, C.-H., Tsai, K.-H., Hwang, C.-H.: Automatic segmentation of abnormal cell nuclei from microscopic image analysis for cervical cancer screening. In: 2009 IEEE International Conference on Nano/Molecular Medicine and Engineering (NANOMED), pp. 77–80, October 2009
4. Bengtsson, E., Malm, P.: Screening for cervical cancer using automated analysis of pap-smears. Comput. Math. Methods Med. **2014** (2014)
5. Lassouaoui, N., Hamami, L., Nouali, N.: Morphological description of cervical cell images for the pathological recognition
6. Phoulady, H.A., Zhou, M., Goldgof, D.B., Hall, L.O., Mouton, P.R.: Automatic quantification and classification of cervical cancer via adaptive nucleus shape modeling. In: 2016 IEEE International Conference on Image Processing (ICIP), pp. 2658–2662, September 2016
7. Phoulady, H.A., Goldgof, D.B., Hall, L.O., Mouton, P.R.: A new approach to detect and segment overlapping cells in multi-layer cervical cell volume images. In: 2016 IEEE 13th International Symposium on Biomedical Imaging (ISBI), pp. 201–204, April 2016
8. Cheng, F.H., Hsu, N.R.: Automated cell nuclei segmentation from microscopic images of cervical smear. In: 2016 International Conference on Applied System Innovation (ICASI), pp. 1–4, May 2016
9. Gençtav, A., Aksoy, S., Önder, S.: Unsupervised segmentation and classification of cervical cell images. Pattern Recogn. **45**(12), 4151–4168 (2012)
10. Agarwal, P., Sao, A., Bhavsar, A.: Mean-shift based segmentation of cell nuclei in cervical pap-smear images. In: 2015 5th National Conference on Computer Vision, Pattern Recognition, Image Processing and Graphics (NCVPRIPG), pp. 1–4, December 2015
11. Song, Y., Zhang, L., Chen, S., Ni, D., Lei, B., Wang, T.: Accurate segmentation of cervical cytoplasm and nuclei based on multiscale convolutional network and graph partitioning. IEEE Trans. Biomed. Eng. **62**(10), 2421–2433 (2015)

12. Song, Y., Zhang, L., Chen, S., Ni, D., Li, B., Zhou, Y., Lei, B., Wang, T.: A deep learning based framework for accurate segmentation of cervical cytoplasm and nuclei. In: 2014 36th Annual International Conference of IEEE Engineering in Medicine and Biology Society, pp. 2903–2906, August 2014
13. Tareef, A., Song, Y., Cai, W., Feng, D.D., Chen, M.: Automated three-stage nucleus and cytoplasm segmentation of overlapping cells. In: 2014 13th International Conference on Control Automation Robotics Vision (ICARCV), pp. 865–870, December 2014
14. Chankong, T., Theera-Umpon, N., Auephanwiriyakul, S.: Automatic cervical cell segmentation and classification in pap smears. Comput. Methods Prog. Biomed. **113**(2), 539–556 (2014)
15. Zhao, L., Li, K., Wang, M., Yin, J., Zhu, E., Chengkun, W., Wang, S., Zhu, C.: Automatic cytoplasm and nuclei segmentation for color cervical smear image using an efficient gap-search MRF. Comput. Biol. Med. **71**, 46–56 (2016)
16. Bora, K., Chowdhury, M., Mahanta, L.B., Kundu, M.K., Das, A.K.: Automated classification of pap smear images to detect cervical dysplasia. Comput. Methods Prog. Biomed. **138**, 31–47 (2017)
17. Comaniciu, D., Meer, P.: Mean shift: a robust approach toward feature space analysis. IEEE Trans. Pattern Anal. Mach. Intell. **24**(5), 603–619 (2002)
18. Achanta, R., Shaji, A., Smith, K., Lucchi, A., Fua, P., Süsstrunk, S.: SLIC superpixels compared to state-of-the-art superpixel methods. IEEE Trans. Pattern Anal. Mach. Intell. **34**(11), 2274–2282 (2012)
19. Jantzen, J., Norup, J., Dounias, G., Bjerregaard, B.: Pap-smear benchmark data for pattern classification. In: NiSIS, pp. 1–9 (2005)
20. Vyavahare, A.J., Thool, R.C.: Segmentation using region growing algorithm based on clahe for medical images. In: 4th International Conference on Advances in Recent Technologies in Communication and Computing (ARTCom 2012), pp. 182–185, October 2012
21. Comaniciu, D., Meer, P.: Mean shift analysis and applications. In: Proceedings of 7th IEEE International Conference on Computer Vision, vol. 2, pp. 1197–1203 (1999)
22. Mehnert, A., Moshavegh, R., Sujathan, K., Malm, P., Bengtsson, E.: A structural texture approach for characterising malignancy associated changes in pap smears based on mean-shift and the watershed transform. In: 2014 22nd International Conference on Pattern Recognition (ICPR), pp. 1189–1193, August 2014
23. Otsu, N.: A threshold selection method from gray-level histograms. Automatica **11**(285–296), 23–27 (1975)

Feature Detection and Classification

Simultaneous Cell Detection and Classification with an Asymmetric Deep Autoencoder in Bone Marrow Histology Images

Tzu-Hsi Song[1(✉)], Victor Sanchez[1], Hesham ElDaly[2], and Nasir Rajpoot[1]

[1] Department of Computer Science, University of Warwick, Coventry, UK
T-H.Song@warwick.ac.uk
[2] Addenbrookes Hospital, Cambridge, UK

Abstract. Recently, deep learning approaches have been shown to be successful for analyzing histopathological images. In general, cell detection and classification are separated and assigned into two different networks, resulting in increased computational complexity for training the deep netwrok. Here we propose a novel deep autoencoder structure for classification with detection. This novel network uses one deep autoencoder to detect the positions of cells and classify types of cells simultaneously. In addition, the proposed network can efficiently detect the cells with irregular shape. The performance of the proposed method is shown to be similar to that of conventional deep learning approaches for detection and close to that of conventional deep learning approaches for classification.

Keywords: Cell detection · Cell classification · Bone marrow cancer · Digital pathology

1 Introduction

Bone marrow trephine biopsy is an important hematopathological specimen and is particularly useful in investigating specific bone marrow diseases, like myelofibrosis neoplasms (MPNs). It offers a complete assessment of bone marrow architecture and of the pattern of distribution of any abnormal infiltrate [1]. Moreover, bone marrow trephine biopsy also provides the information of cell morphology and cytology, which is potentially useful for precise diagnosis. In bone marrow diseases, the hematopoietic cells proliferate abnormally and quantitative measurement of cells is an essential diagnostic tool for pathological measurements that help distinguish different types of bone marrow diseases in routine clinical diagnosis. However, inter-observer variability among hematopathologists usually occur because quantitative estimation is largely subjective and manual quantification is tedious, time intensive and expensive [2,3,6,7].

In order to address these challenges, automated cell detection and classification are important steps in the quantification of cell level analysis but pose

M. Valdés Hernández and V. González-Castro (Eds.): MIUA 2017, CCIS 723, pp. 829–838, 2017.
DOI: 10.1007/978-3-319-60964-5_72

technical challenges, especially due to the complicated cellular characteristics of histopathological images. Most current cell detection methods are based on exploiting low-level hand-crafted features [7], such as color, edge, contextual information, and texture; they rely on the shape of nuclei and stability of features. However, in bone marrow trephine specimens, as can be seen in Fig. 1, there are a large number of different types of cells mixing, which have variable sizes, shapes and textures. Hence, we need to develop algorithms for automated detection of bone marrow cells that take into account this variability.

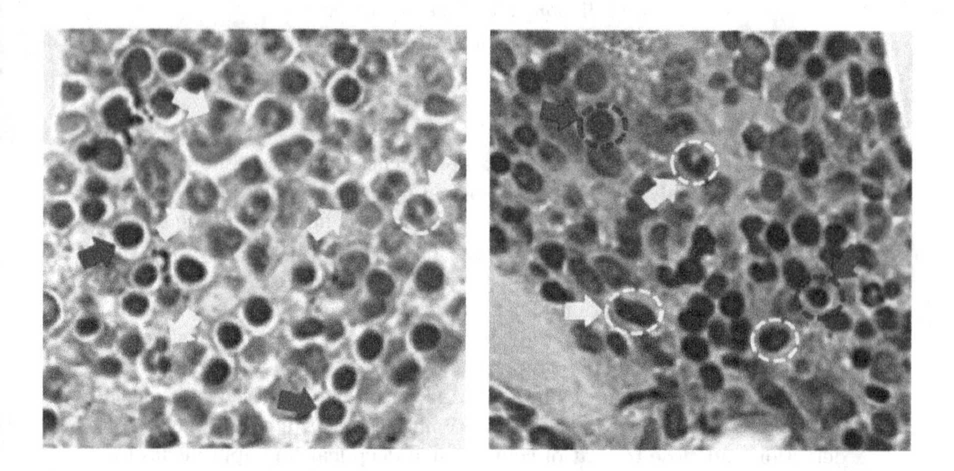

Fig. 1. Examples of bone marrow hematopoietic stem cells. We observe that there are different types of cells densely mixing and randomly distributed in bone marrow tissue. Green and yellow arrows point out erythroid cells and myeloid cells, respectively. Some specific shaped myeloid cells are shown in yellow circles. Otherwise, some myeloid cells have similar shape and intensity features as erythroid cells (green circles and yellow circles). (Color figure online)

In this paper, we present an autoencoder network for simultaneous detection and classification of erythroid and myeloid cells. We build on [8] by adding another deep autoencoder. This network can simultaneously process and generate the predicted results of detection and classification. The proposed network also takes less time on training the network than other conventional deep learning approaches while maintaining the accuracy of the detection and classification.

2 Related Works

Accurate detection and classification of different types and shapes of nuclei or cells are difficult tasks becasue the properties of histopathological images are complicated by many different factors, such as the morphological and textural variability of individual nuclei or cell. In recent years, deep learning approaches have been shown to be successful for the analysis of histopathological images.

Deep learning (DL) is a hierarchical learning approach for discovering hidden feature representations in natural images, and DL methods avoid the traditional design and computation of features and directly learn high-level features from training data; they are full feed-forward in terms of feature extraction [5,6,10]. DL methods have been shown to be more efficient and robust than the classical machine learning approaches. One of the major DL approaches is Autoencoder, which is a unsupervised learning network [4,5,10]. Autoencoder is an encoder-decoder architecture where the encoder network represents pixel intensities modeled via lower dimensional attributes, while the decoder network reconstructs the original pixel intensities using the low dimen-sional features. There are many different types of improved autoencoder frameworks, such as sparse autoencoder (SAE), denoising autoencoder (DAE), and convolutional autoencoder (CAE) [11], having their own regularization terms. These autoencoder-based networks can enhance the robustness of features and effectively avoid over-fitting in training the network. These unsupervised learning approaches are widely and successfully applied in histopathology image processing and analysis. For instance, Xu *et al.* [6] use the stacked sparse autoencoder with a softmax classifier to learn a high-level representation of nuclear and non-nuclear objects for detecting and classifying the nuclear region. Su *et al.* [12] use a denoising autoencoder to reconstruct the shape information for cell segmentation. In addition, our previous work [8] proposes a connect network to convert the input images to the probability response for nuclei detection.

3 The Proposed Method

The proposed synchronized deep learning method is driven from our previous work hybrid deep autoencoder (HDAE) [8] and designs a new connect layer structure to integrate HDAE network with another autoencoder network for simulatenously detecting and classifying dffierent types of nuclei or cells. In this section, we describe the construction of our proposed synchronized deep autoencoder structure.

3.1 Hybrid Deep Autoencoder (HDAE)

At first, we need to introduce our previous work, HDAE network, for nuclei detection [8]. The HDAE model extracts high-level feature representation from input images and probability maps and then connects the relative features. In other words, the HDAE model converts input images to probability responses to identify the positions of centroids of nuclei. For detecting irregular-shape nuclei, the HDAE utilizes curvature Gaussian model to construct irregular shape labels corresponding to the irregular-shape nuclei into training dataset. The HDAE provides the similar ability of nuclei detection and even better performance of detecting irregular-shape than other approaches [8]. Formally, given a set of samples $X = x_1, \ldots, x_n$ and a set of probability maps $Y = y_1, \ldots, y_n$, the formulation of sections in HDAE network are shown as:

$$\begin{cases} h_i^I = f(W_{l^I} x_i + b_{l^I}) \\ h_i^C = f(W_{l^C} h_i^I + b_{l^C}) \\ \hat{y}_i = f(W'_{l^P} h_i^C + b'_{l^P}), \end{cases} \tag{1}$$

where W and b represent the weight matrix and bias terms in the encoder, while $W^{'}$ and $b^{'}$ denote the same in the decoder. h^I and h^C represent the high-level hidden representations in the input and connect networks, respectively. l^I, l^C and l^P are the number of layers in the input, connect and probability networks. $f(\bullet)$ is the activation function, which is set as a sigmoid function. The loss function uses mean square error and is shown as:

$$Loss_{HDAE} = \sum_i \|y_i - \hat{y}_i\|^2, \tag{2}$$

3.2 Synchronized Asymmetric Deep Autoencoder (Syn-ADAE)

In general, we obtained the predicted detected center points of nuclei through HDAE and then use another deep learning method, such as CNN or SSAE network, to process these results for cell classification. However, it needs to train another deep learning approach and takes much time on it. Here we would like to simplify this conventional framework of detection and classification. We adopt a network architecture idea from the paper, which proposed a concatenated asymmetric deep learning structure for capturing the distorted feature and noise estimation [9]. According to the properties of autoencoder, input data are used to compare the reconstructed data to train the parameters of network via backpropagation, and this makes the network is symmetric. In [9], the authors build an asymmetric deep denoising network to estimate the noise and reconstruct clear features from input distortion features. It seems like one network has two different abilities. According to this idea, we attempt to construct a novel network that can process detection and classification at the same time.

Before developing an asymmetric network, we recall the architecture of networks for detection and classification. Our previous work focuses on nucleus detection and has described in Sect. 3.1. Moreover, for classification, a normal autoencoder network provides high-level features and then uses the softmax classifier to identify different classes of cells. When we train these two networks, the same input patches are applied to obtain the parameters of detection and classification separately. i.e. we can image that these two networks have the same part to capture high-level representations. In order to process detection and classification simultaneously, we would like to keep the same section and integrate the different parts of these two networks as a whole component or layer. In previous work [8], we build a connect network to connect input and probability network. Here we would like to redesign the connect network to achieve the abilities of detection and classification. First, we obtain the parameters of softmax classifier by training the input network of classification. Given a set of class maps $C = c_1, \ldots, c_n$, the formulation of input network with softmax classifier is shown as:

$$\begin{cases} h_i^I = f(W_{l^I} x_i + b_{l^I}) \\ \hat{c}_i = f(W_{SMC} h_i^I), \end{cases} \tag{3}$$

where W_{SMC} is the weight matrix of softmax classifier. We use the predicted class map $\hat{c}_i$ and the labelled class map c_i to calculate the loss function for obtaining the weight W_{SMC}. Second, in HDAE network, we use the probability maps as input patches to train the probability response network. Here we combine class maps with corresponding probability maps instead of original probability maps. It is because the input data have to include the information of detection and classification. Here we set $z_i = y_i \times c_i$ and the formulation of probability network is:

$$\begin{cases} h_i^{PC} = f(W_{l^{PC}} z_i + b_{l^{PC}}) \\ \hat{z}_i = f(W'_{l^{PC}} h_i^{PC} + b'_{l^{PC}}), \end{cases} \tag{4}$$

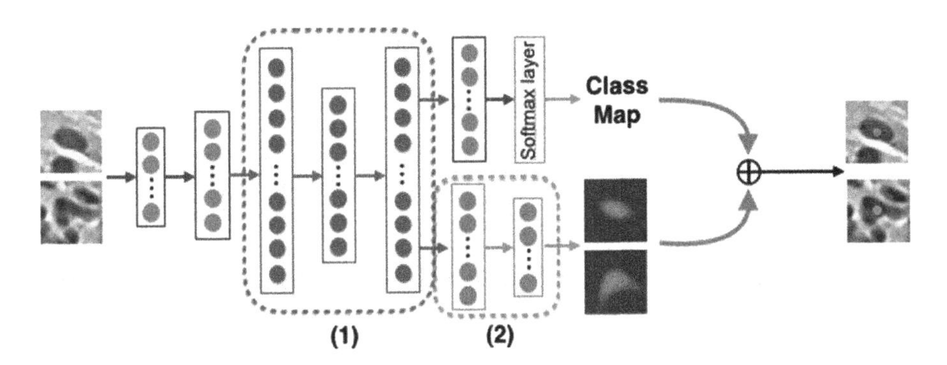

Fig. 2. The overview of synchronized asymmetric deep autoencoder (syn-ADAE): (1) asymmetric connect layer consists of two parts of high-level hidden representation (2) decoder layer uses the probability maps combining with corresponding class or label to obtain the parameters.

After we obtain above networks, we would like to rebuild the connect network for integrating these networks. According to (3) and (4), we obtain the high-level feature representation h^I from input images and h^{PC} from the class-probability maps. Here we need to create another three different networks to set the initial parameters of this connect network. We use the first autoencoder to obtain the weight matrix W_{AC_1} and W'_{AC_1} by using h^I being input patches for classification, while h^I and h^{PC} are used to train the second network to get the weight matrix $W_{AC_2\text{-}1}$ and $W_{AC_2\text{-}2}$ and bias term $b_{AC_2\text{-}1}$ and $b_{AC_2\text{-}2}$. The second network is a two-layer neural network and is derived from the same concept of the connect network in HDAE network. The last network is the proposed asymmetric autoencoder network. In [9], the asymmetric network concatenates two autoencoders. Here we adopt the paralleled structure to build the proposed asymmetric connect network. From the first and second networks, we have the pre-training parameters for detection and classification and then we use them to initialize the proposed connect network. It is to efficiently avoid the influence between detection and classification. Here we set the initial weight matrix of the asymmetric network as:

$$W_{AC_3\_1} = \begin{bmatrix} W_{AC_1} & 0 \\ 0 & W_{AC_2\_1} \end{bmatrix}, W_{AC_3\_2} = \begin{bmatrix} W'_{AC_1} & 0 \\ 0 & W_{AC_2\_2} \end{bmatrix}, \quad (5)$$

$$b_{AC_3} = \left[b_{AC_2\_1}\; b_{AC_2\_2}\right]^T, \quad (6)$$

and adjust them in fine-tuning by training the asymmetric connect network:

$$\begin{cases} h_i^{AC_3\_1} = f\left(W_{AC_3\_1} h_i^I + b_{AC_3}\right) \\ h_i^{AC_3\_2} = f\left(W_{AC_3\_2} h_i^{AC} + b_{AC}\right), \end{cases} \quad (7)$$

Then we calculate the loss function between $h_i^{AC_3\_2}$ and $\left[h_i^I\; h_i^{PC}\right]$ to optimize these parameters. Finally, we use the above parameters we obtained from input network, class-probability network, and asymmetric connect network to construct the whole asymmetric autoencoder network and then do the fine-tuning. Figure 2 shows the architecture of syn-ADAE network. The formulation of the loss function of syn-ADAE network is shown as:

$$Loss_{syn-ADAE} = \sum_i \|z_i - \hat{z}_i\|^2 + \sum_i H\left(c_i, \hat{c}_i\right), \quad (8)$$

where $H\left(c_i, \hat{c}_i\right)$ is the cross-entropy cost function to estimate the classification loss energy and defined by:

$$H\left(c_i, \hat{c}_i\right) = -\left[c_i log\left(\hat{c}_i\right) - \left(1 - c_i\right) log\left(1 - \hat{c}_i\right)\right], \quad (9)$$

For detecting cell centroids, we find the local maxima in the probability map. In order to avoid over-detection, a threshold is introduced, which is defined as a fraction of the maximum probability value found on the probability map. All local maxima whose probability values are less than the threshold are not considered in the detection. Moreover, we also consider the neighboring elements of class map around each predicted center point to identify which class this center point belongs to. It is because the classification section in the proposed syn-ADAE network could be affected by the detection section. For reducing the influence of detection section, we consider the neighboring region of class map around each detected point. However, we need to strictly define the size of considering range for accurately identify the class of center points. In all experiments, we set the radius = 12 pixels for detection and 5 pixels for classification.

4 Experimental Results

In our experiments, for training the autoencoder-based networks, we use training cell patches of size 29 × 29. We also consider the hematoxylin (H) stain channel to obtain training features for detecting the cells. Here we use color deconvolution [13] to generate the H-channel image patch. We initialize the input training cell patch to $29 \times 29 \times 4$ in RGB and H-channel and collect all training cell patches from 40× magnification bone marrow trephine biopsies. Our training set is comprised of 5,248 cell patches from 52 bone marrow images and we use two-fold cross validation to train and test our proposed network. We use an input network with two hidden layers and one softmax classifier, a probability label network with one hidden layer, and a connect network with two hidden layers. In input network, the input vector is represented as a column vector of pixel intensities with size $(29^2 \times 4) \times 1$. The first and second hidden layers of input network have 1600 and 400 hidden units. respectively. Moreover, the label network has 841 input uints and 400 hidden units. The number of neurons of each hidden layer is set to be less than that of the previous hidden layer for emphasizing high-level feature representations. For constructing the asymmetric connect network, we use the number of neurons is equal to the sum of the number of neurons in high-level representation of input network and class-probability network. Here the aysmmetric connect network has 400 input units from the output of input network, 800 hidden units, and 800 output units corresponding to the high-level feature representations of input and label networks. In order to optimize the parameters of whole syn-ADAE network, at first, we obtain the initial parameters by training two autoencoder networks that look for the relationship between similar feature representations of input and class-probability. We use these parameters to pre-train the asymmetric connect network and then fine-tune the whole proposed network with dropout. In the post-processing, local maxima and a threshold value are used to remove low probability responses and identify the centroid of all types of cells. Then we consider the neighboring predicted points surrounding the detected center points to identify final class.

Figure 3 shows the results of the proposed syn-ADAE network. Here we define that a predicted point is true positive when the detection and classification outcomes of this point are as same as ground truth. Moreover, It demonstrates that the syn-ADAE network can efficiently detect and classify different types of bone marrow hematopoietic stem cells. Moreover, we compare the quantitative performance of the syn-ADAE network with other deep learning approaches. At first, we need to confirm that the proposed syn-ADAE network has similar ability of nuclei detection to the other detection models, such as LIPSyM [14], SSAE [6] and SC-CNN [15], and our previous HDAE network because the proposed method is derived from the previous work and autoencoder structure. Table 1 shows comparative results in terms of precision, recall and accuracy between the proposed model and other detection approaches. It shows that our approach achieves similar performance to HDAE network and has better performance than other detection models in detection section. It also means the proposed network in detection section efficiently reduces the influence from the classification section in asymmetric structure and maintains the ability of detecting various nuclear shapes and precisely reflecting the locations of centroids of various cells. In classification section, for fair comparison, we use the same method to detect the center points and different types of autoencoder structures for classification. Here we

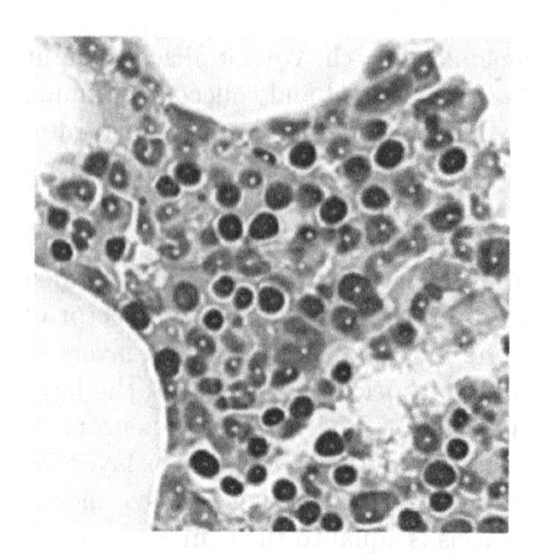
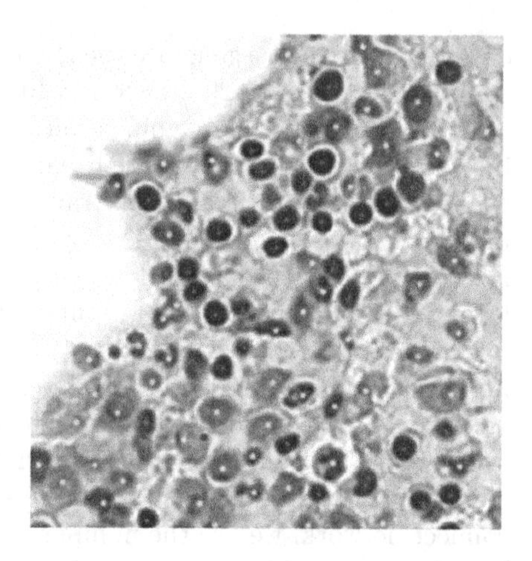

Fig. 3. Results of detection and classification via syn-ADAE network. Green and yellow dots represents the erythroid and myeloid cells, respectively. Red dots denote the false positive of detection and blue dots are false positive of classification. (Color figure online)

adopt sparse autoencoder (SAE), SSAE and convolution neural network (CNN) for comparing the classification performance with the proposed method. The number of hidden layer in SSAE network is set to 6, which is the same number of layer in the proposd method. CNN network includes 3 convolutional layers, 3 max-pooling layes, one full connected layer and one softmax layer. Here it notes that predicted points are used in quantitative measurement of classification as these points match with the ground truth of detection. Table 2 shows that our proposed network provides better performance than other frameworks. It means that the proposed network also provides similar ability of classification to other deep learning approaches but has less perfomance than CNN in precision and than SSAE in recall. This situation exists in both detection and classification sections becasue the detection and classification sections in proposed network are affected by each other, and the optimized parameters can not fit both two strategies but find a balance between them to maintain sufficient functionalities of dectection and classification as individual ones.

In addition, we compare the time of training among these frameworks of detection and classification. Table 3 shows our proposed network takes fewer time for training. It means the proposed network provides lower computational complexity and need to train less number of parameters then other conventional frameworks. However, in Fig. 3, there are a few incorrect detection or classification results. Here it is important to note that if the detected point is incorrect, it is unnecessary to predict the labelled class for this incorrect point. These incorrect outcomes in detection or classification occur because that (1) the cell has faded stain color or the size of cell is larger than the size of the training patch; (2) the thresholding value is affected by class labels and not efficient to remove

Table 1. Comparing the results of detection between the proposed model and other methods

Methods	Precision	Recall	F1-score
LIPSyM [14]	0.7267	0.6514	0.687
SSAE [6]	0.8733	0.719	0.7887
SC-CNN [15]	**0.9517**	0.9118	0.9313
HDAE [8]	0.9273	**0.9702**	**0.9483**
Proposed model	0.9129	0.9641	0.9378

the noise points in whole dataset; (3) some myeloid cells have similar shape and intensity features to erythroid cells, and this affects that the proposed network is uncertain of these two types of cells in classification.

Table 2. The comparison of classification results among proposed model and other approaches

Methods	Precision	Recall	F1-score
HDAE+SAE	0.8267	0.8108	0.8187
HDAE+SSAE	0.8461	**0.9037**	0.874
HDAE+CNN	**0.8605**	0.8922	0.8761
Proposed model	0.859	0.896	**0.877**

Table 3. Comparing the training time between conventional frameworks and the proposed methods

Methods	Training time
HDAE+SAE	5–6 h
HDAE+SSAE	7–8 h
HDAE+CNN	6–7 h
Proposed model	**4–5** h

5 Conclusions

We presented a novel synchronized deep autoencoder network to perform nuclei detection and classification simultaneously. The proposed network combines the hybrid autoencoder network [8] and a deep autoencoder network with an asymmetric connection network. The experimental results show that the proposed syn-ADAE method provides the ability of detection and classification as good as conventional deep learning frameworks and efficiently takes less training time than other approaches.

References

1. Bain, B.J., Clark, D.M., Wilkins, B.S.: Bone Marrow Pathology. Wiley, Hoboken (2011)
2. Wilkins, B.S., Erber, W.N., Bareford, D., et al.: Bone marrow pathology in essential thrombocythemia: interobserver reliability and utility for identifying disease subtypes. Blood **111**, 60–70 (2008)
3. Thiele, J., Imbert, M., Pierre, R., Vardiman, J.W., Brunning, R.D., Flandrin, G.: Chronic idiopathic myelofibrosis. who classification of tumours: tumours of haematopoietic and lymphoid tissues, pp. 35–38. IARC Press, Lyon (2001)
4. Zhang, G., Zhong, L., Huang, Y.H., Zhang, Y.: A histopathological image feature representation method based on deep learning. In: 2015 7th International Conference on IEEE Information Technology in Medicine and Education (ITME), pp. 13–17 (2015)
5. Baldi, P.: Autoencoders, unsupervised learning, and deep architectures. J. Mach. Learn. Res. **27**, 37–50 (2012)
6. Xu, J., Xiang, L., et al.: Stacked sparse autoencoder (SSAE) for nuclei detection on breast cancer histopathology images. IEEE Trans. Med. Imaging **35**(1), 119–130 (2016)
7. Irshad, H., Veillard, A., Roux, L., Racoceanu, D.: Methods for nuclei detection, segmentation, and classification in digital histopathology: a review-current status and future potential. IEEE Rev. Biomed. Eng. **7**, 97–114 (2014)
8. Song, T.H., Sanchez, V., EIDaly, H., Rajpoot, N.: Hybrid deep autoencoder with curvature Gaussian for detection of various types of cells in bone marrow trephine biopsy images. ISBI (2017, in Press)
9. Lee, K.H., Kang, S.J., Kang, W.H., Kim, N.S.: Two-stage noise aware training using asymmetric deep denoising autoencoder. In: 2016 IEEE International Conference on Acoustics, Speech and Signal Processing (ICASSP). IEEE (2016)
10. Schmidhuber, J.: Deep learning in neural networks: an overview. Neural Netw. **61**, 85–117 (2015)
11. Arpit, D., Zhou, Y., Ngo, H., Govindaraju, V.: Why regularized autoencoders learn sparse representation? Stat **1050**, 29 (2015)
12. Su, H., Xing, F., Kong, X., Xie, Y., Zhang, S., Yang, L.: Robust cell detection and segmentation in histopathological images using sparse reconstruction and stacked denoising autoen-coders. MICCAI **9351**, 383–390 (2015)
13. Khan, A.M., Rajpoot, N.M., Treanor, D., Magee, D.: A non-linear mapping approach to stain normalization in digital histopathology images using image-specific colour deconvolution. IEEE Trans. Biomed. Eng. **61**(6), 1729–1738 (2014)
14. Kuse, M., Khan, M., Rajpoot, N., Kalasannavar, V., Wang, Y.F.: Local isotropic phase symmetry measure for detection of beta cells and lymphocytes. J. Pathol. Inform. **2**(2), 2 (2011)
15. Sirinukunwattana, K., Raza, S., Tsang, Y.W., Snead, D., Cree, I., Rajpoot, N.: Locality sensitive deep learning for detection and classification of nuclei in routine colon cancer histology images. IEEE Trans. Med. Imaging **35**, 1196–1206 (2016)

Glomerulus Classification with Convolutional Neural Networks

Anibal Pedraza[1], Jaime Gallego[1], Samuel Lopez[1], Lucia Gonzalez[2], Arvydas Laurinavicius[3], and Gloria Bueno[1(✉)]

[1] University of Castilla La Mancha, Ciudad Real, Spain
{Anibal.Pedraza,Jaime.Gallego,Samuel.Lopez,Gloria.Bueno}@uclm.es
[2] Hospital General Universitario, Ciudad Real, Spain
lmgonzalez@sescam.jccm.es
[3] Vilnius University Hospital Santariskes Clinics and Vilnius University, Vilnius, Lithuania
Arvydas.Laurinavicius@vpc.lt

Abstract. Glomerulus classification in kidney tissue segments is a key process in nephropathology to obtain correct diseases diagnosis. In this paper, we deal with the challenge to automate the Glomerulus classification from digitized kidney slide segments using a deep learning framework. The proposed method applies Convolutional Neural Networks (CNNs) classification between two classes: Glomerulus and Non-Glomerulus, to detect the image segments belonging to Glomerulus regions. We configure the CNN with the public pre-trained AlexNet model, and adapt it to our system by learning from Glomerulus and Non-Glomerulus regions extracted from training slides. Once the model is trained, the labelling is performed applying the CNN classification to the image segments under analysis. The results obtained indicate that this technique is suitable for correct Glomerulus classification, showing robustness while reducing false positive and false negative detections.

Keywords: Glomerulus classification · Digital pathology · Nephropathology · Convolutional neural networks

1 Introduction

Whole-slide scanner allows the creation of high-resolution images from the tissue slides. This technology has opened the possibility to apply image processing and machine learning techniques to these Virtual Slides (VS) to automate and/or enhance traditional pathology processes in what is known as Digital Pathology field. This current trend is becoming a necessary tool for pathologists, and starts to assist them to perform a fast and robust diagnosis through interacting with high-resolution images or, in some cases, to completely automate quantitative evaluations. Bueno *et al.* [1] give a general survey of digital Pathology trends.

Deep Learning techniques based on Neural Networks, represent an important area inside of Digital Pathology field. These machine learning methods,

M. Valdés Hernández and V. González-Castro (Eds.): MIUA 2017, CCIS 723, pp. 839–849, 2017.
DOI: 10.1007/978-3-319-60964-5_73

such as Convolutional Neural Networks (CNN), are becoming central in VS analysis, since they give promising results in difficult tasks like regions detection and classification [2]. CNN allows a relatively simple configuration thanks to its unsupervised feature generation, the availability of greater datasets to train the models, and the possibility to import pre-trained models to partially set-up the neural networks. The main drawback of these techniques is the huge amount of computational power necessary to correctly train the networks in a reasonable time. Madabushi and Lee [3] show machine learning methods applied to different Digital Pathology challenges.

In this paper, we focus on Glomerulus classification based on deep learning framework, in the context of digital pathology applied to kidney tissue slides. Glomerulus classification is central in nephropathology field to get a correct medical diagnosis. These networks of capillaries serve as the first stage in the filtering process of the blood that retains higher molecular weight proteins in blood circulation in its formation of urine. Therefore, any pathological change associated to them inside a tissue section, like the number of Glomeruli or their sizes, is essential to detect diseases from the patients.

Glomeruli structures visualized in a VS present high variability in terms of size, shape and colour. The causes of this different appearance among samples can be various: the relative position of the Glomerulus inside the renal section, the heterogeneity in immunohistochemistry staining or the presence of internal biological processes.

- Size and shape: In a healthy kidney before sectioning, Glomeruli present a spherical shape with fixed size, but the aspect can be modified due to the presence of medical diseases. For instance: Glomeruli can present a swell aspect under hypertension [4] or diabetes [5] conditions. After sectioning, the presence of this pathologies affect the appearance inside the VS. Besides, the different Glomeruli sizes observed could vary depending on where the cross-section was taken with respect to each Glomerulus sphere.
- Colour: In our configuration, we use PAS (Periodic Acid Schiff) stain in tissue sections, which gives a purple-magenta colour to the slides. The amount of stain present in each slide will determine the colour intensity of the segments under analysis. Since this process is not perfect, each slide can present different intensities. Moreover, the presence of medical diseases can vary the amount of stain present in the Glomeruli under study.

All this variability related to the Glomerulus appearance, and in general to VS images, increases the difficulty to automate the classification using a digital pathology framework. Therefore, any proposal devoted to this task must be robust in front of these aspects.

1.1 Previous Work

In the last years, many authors have been working in digital pathology developing techniques to automate quantitative analysis tasks. Methods devoted to classify and detect Glomeruli in nephropathology field have become of main importance in diseases diagnosis and research. The classification methods of the literature can be divided into two groups according to the necessity or not to define features to characterize the structures: Handcrafted features approaches and unsupervised features approaches.

- Handcrafted features: These classification methods try to find specific measurable attributes in the image that can be utilized to correctly identify the structures. In [6], the authors propose to use a genetic algorithm for edge patching by using a canny edge detector to get discontinuous edges of glomerulus and its final detection and classification. The main problem of this technique is that borders detected by means of canny edge detector are not robust and is prone to errors under colour intensity variations. In [7,8], the authors use Rectangular-Histogram of Gradients (R-HOG [9]) as feature vector to classify Glomeruli. The main problem of these techniques resides in the fact that R-HOG has rigid block division that leads to keep considerable number of false positives. In order to improve the R-HOG framework, Kato *et al.* [10] suggest the Segmental HOG (S-HOG) as potential candidate descriptor for glomerulus objects detection, where the block division is not rigid, and uses nine discretized oriented gradients and Support Vector Machine (SVM [11]) as a supervised learning classifier. In [12], an analysis of renal microscopic images is applied by using a detection of borders improved with Convex Hull algorithm [13]. This method detects and classifies Glomeruli for the measurement of both, diameter and Bowman's Space thickness. Finally, as an example of Glomeruli classification in 3D dataset, Zhang *et al.* [14] propose a Glomerulus detection and classification using 3D Magnetic Resonance Imaging (MRI) by means of Hessian based Difference of Gaussians (HDoG) detector.
- Unsupervised features: These techniques avoid the necessity to define handcrafted features to perform the classification. Deep Learning methods such as CNNs belong to this group. They learn by themselves the best features that suit the problem, so that the scientists are only responsible of providing the better (and more populated) dataset as possible, as well as choosing the network architecture. Kothari *et al.* [15] perform a great review of how the application of this kind of techniques is the most suitable one nowadays than anytime, pointing out also the challenges and future opportunities that are involved.

Regarding the application of these techniques to specific problems, some of the most interesting works that have been published focus on different medical areas: Wang *et al.* [16] use a cascaded approach for mitosis detection in breast cancer that combines a CNN model and handcrafted features (morphology, colour, and texture features). In [17], the authors deal with mitotic figures classification combining manually designed nuclear features with learned features extracted by CNN. Xu *et al.* [18] explore the use of very deep CNN to automatically classify diabetic retinopathy by configuring a network with 18 layers. Finally, Havaei *et al.* [19] apply CNN architecture to automatic brain tumour segmentation in MRI images that exploits both, local features as well as more global contextual features simultaneously, in a CNN cascade configuration.

Table 1 shows a summary of the different experiments performed in the publications mentioned above, specifying the kind of problem, network architecture (and any particularity if necessary to be mentioned), the employed dataset and the obtained results. The revision of every publication stated before is out of scope of this article and therefore we focus on those works related to CNN.

Table 1. Previous work experiments and results

Year [Reference]	Application	Dataset	Features	Classification	Performance
2013 [17]	Mitotic figures	5 WSIs with 35 ROIs and 226 samples	Handcrafted (Colour, texture, shape) + CNN (LeNet based)	SVM	0.659 (F-score)
2014 [16]	Mitotic figures	35 HPFs	Handcrafted (Morphology, Intensity, Texture) + Custom CNN (2CV + FC)	Ensemble	0.735 (F-score)
2016 [2]	Nuclei segmentation	141 ROIs with 12,000 samples	AlexNet based CNN	Softmax	0.83 (F-score)
	Epithelium segmentation	42 ROIs with 1,735 samples	AlexNet based CNN	Softmax	0.83 (F-score)
	Tubule segmentation	85 ROIs with 795 samples	AlexNet based CNN	Softmax	0.84 (F-score)
	Lymphocyte detection	100 ROIs with 3,064 samples	AlexNet based CNN	Softmax	0.9 (F-score)
	Mitosis detection	311 ROIs with 550 samples	AlexNet based CNN	Softmax	0.53 (F-score)
	Invasive ductal carcinoma	162 WSIs	AlexNet based CNN	Softmax	0.765 (F-score)
	Lymphoma classification	374 samples	AlexNet based CNN	Softmax	97.0% (Acc)
2017 [19]	Brain tumour segmentation	BRATS2013	Custom CNN (4CV)	Softmax	0.84 (Dice)
Proposed	Glomeruli classification	10,600 ROIs from 40 WSIs	CNN - AlexNet (pre-trained)	Softmax	**0.999** (F-score)
Proposed	Glomeruli classification	10,600 ROIs from 40 WSIs	CNN - AlexNet (from-scratch)	Softmax	0.992 (F-score)
Proposed	Glomeruli classification	10,600 ROIs from 40 WSIs	CNN - GoogleNet (from-scratch)	Softmax	0.994 (F-score)

2 Materials and Method

This section explains how a glomerulus dataset have been built from Whole Slide Images (WSI) and the use of Convolutional Neural Networks to develop a model able to classify patches in two classes: glomerulus and non-glomerulus.

2.1 Dataset

The proposed goal for this problem is to be able to discern whether a patch from a WSI could be classified as a glomerulus structure or not. The available data was a collection of 40 WSIs in SVS format from collaborating institutions.

The samples used to obtain the collection were stained by PAS staining, which is mainly used for staining structures containing a high proportion of carbohydrate macromolecules (glycogen, glycoprotein, proteoglycans), typically found in e.g. connective tissues, mucus, the glycocalyx, and basal laminae. Many laboratories use a battery of stains for evaluating renal disease, e.g., H&E, PAS, trichrome, and silver stains. The silver stain accentuates collagenous structures, e.g., in the glomerulus, the mesangial matrix and the glomerular basement membrane. The PAS stain also accentuates matrix and basement membrane constituents, as does the blue or green component on the trichrome stain. In certain circumstances the trichrome stain demonstrates immune deposits as fuchsinophilic (red) structures. Regarding the utility, the PAS stain is used to identify pathological disorders in which weakness or improper functioning of basal membranes are present. Kidney glomerular diseases are an example of this kind of pathologies [20].

Tissue samples from biopsies were digitized with a scanner Aperio ScanScope XT at Hospital General Universitario de Ciudad Real at 20x and 40x with no compression, into format *SVS*[1].

The process of labelling was performed manually using the Aperio ImageScope tool, as shown in Fig. 1. As a result, the coordinates of the different glomerulus detected by the pathologists were available, so that the glomerulus patches could be cropped from the images.

Following this process, around 700 glomeruli and non-glomeruli structures were obtained. As shown in Fig. 2, these samples present high variability in terms of shape, size and texture.

Since this number of samples may seem barely enough to be fed into a Convolutional Neural Network, a data augmentation technique was decided to be applied. A combination of rotations in 0°, 90°, 180° and 270° and vertical flip were performed, so actually the amount of samples was increased in a factor of 8. The final number of samples per class (glomerulus and non-glomerulus) was established in 5300. Figure 2 shows some examples of patches from both classes.

[1] *SVS* is a semi-proprietary file format consisting on a single-file pyramidal tiled *TIFF*, which can be opened with Aperio ImageScope software, by Leica, and by ImageJ or Fiji via the Bio-Formats plugin, or the individual *TIFF* files can be extracted.

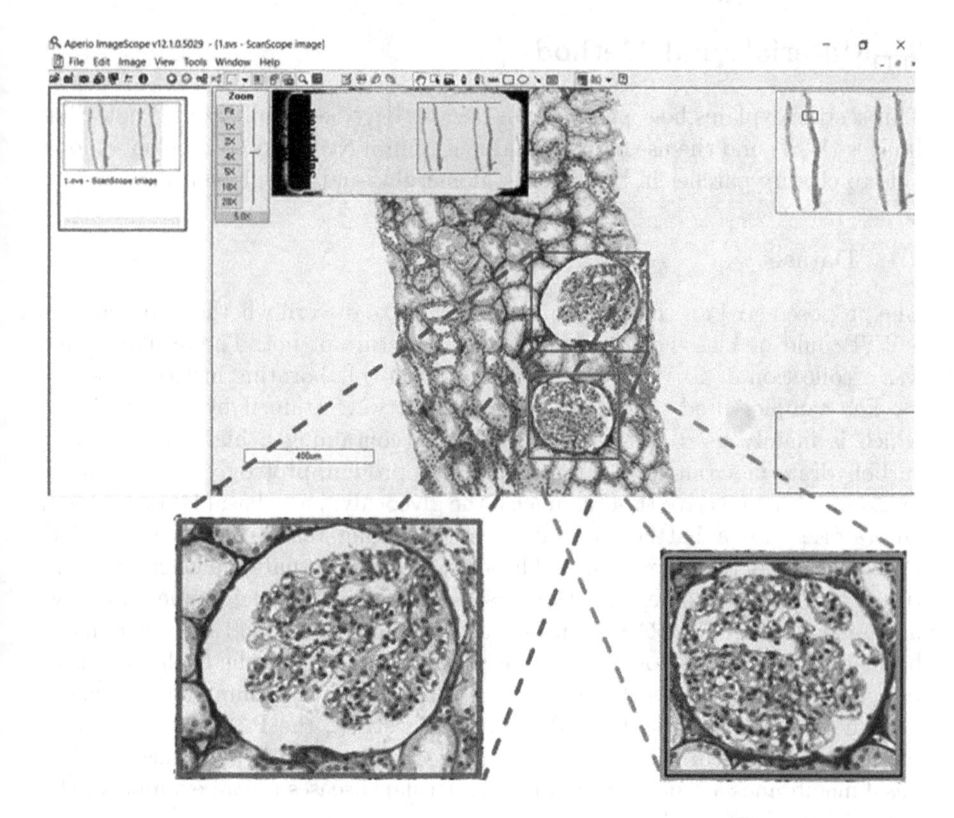

Fig. 1. Glomerulus examples labeled using Aperio ImageScope tool. Biomarker: Sirius Red (Color figure online)

2.2 CNN Training

The approach followed in the proposed method is based on Convolutional Neural Networks, focusing in specifically designed deep models to be used with images. Two different approaches have been covered. First, a technique called fine-tuning has been used. That is, taking a previously trained model that has learned good general features and apply that to our own dataset, modifying barely the weights to be adapted to this problem. The model used was a pre-trained AlexNet model from the ImageNet challenge [21]. The architecture of this network is shown in Fig. 3. Given that, it was possible to achieve good results with only a few epochs/iterations. After that, the same network was employed without any previous training, what is called to train a network "from-scratch". In order to increase the variability of architectures the GoogleNet [22] architecture was employed also with the later technique.

The training process was carried out using two different sets of parameters, depending whether the network was pre-trained or not. For the pre-trained network the learning rate is started out with an initial of 0.001, decreasing it with drop factor of 0.1 with a period of 8. For back propagation, the Stochastic

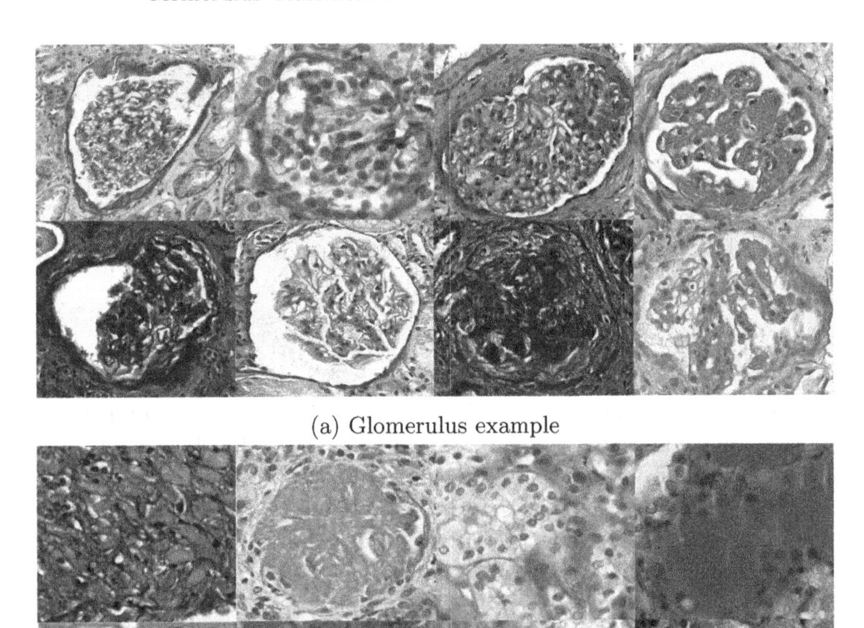

(a) Glomerulus example

(b) No glomerulus example

Fig. 2. Example of image patches from dataset

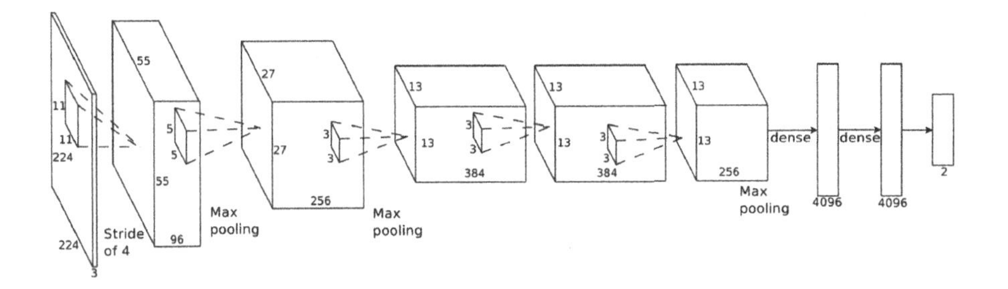

Fig. 3. AlexNet CNN architecture layers

Gradient Descent was used, establishing 0.004 for L2-Regularization. With the previous parameters, the training reaches the best results with just 10 epochs (after that number the loss value and accuracy do not improve). It takes around 17 min to perform the training process with this kind of network (i.e. time for training one-fold in cross-validation), using the GPU NVIDIA GTX 960 Ti with 6

GB of VRAM. On the other hand, training from-scratch models requires slightly different parameters to obtain the best convergence in terms of accuracy and loss values. For these experiments, and initial value of 0.01 was employed for the learning rate decreasing it with drop factor of 0.1 with a period of 16. For back propagation, the Stochastic Gradient Descent was used, establishing 0.004 for L2-Regularization. Due to a higher number of epochs, it takes close to 33 min to perform the training process with this technique and parameters, using the same hardware as described above.

2.3 Validation

To perform the experiments, 10-fold cross-validation has been applied, with 9 folds for training and the remaining one in each iteration for testing. Hence, joining the results from test folds, the whole dataset is validated.

In order to measure the performance, the F-score (also known as F-measure) has been chosen. This is one of the most commonly used metrics to obtain the performance of a binary classification problem, such as glomeruli, nuclei, or tumour detection, in which, for a given sample, is determined whether it belongs to the specified class or not. T. Fawcett performs a comprehensive review of this topic in [23]. Here, the most important concepts are summarized.

Equation 1 states the general expression that rules the F-score, in terms of True positive (TP), True negative (TN), *Type 1* error (False positive, FP) and *Type 2* error (False negative, FN), of a given confusion matrix:

$$F_\beta = \frac{(1+\beta^2)\cdot \mathrm{TP}}{(1+\beta^2)\cdot \mathrm{TP} + \beta^2 \cdot \mathrm{FN} + \mathrm{FP}} \tag{1}$$

This expression can be stated in terms of Precision (P, defined in Eq. 2) and Recall (R, defined in Eq. 3), as shown in Eq. 4

$$P = \frac{\mathrm{TP}}{\mathrm{TP} + \mathrm{FP}} \tag{2}$$

$$R = \frac{\mathrm{TP}}{\mathrm{TP} + \mathrm{FN}} \tag{3}$$

$$F_\beta = (1+\beta^2)\cdot \frac{P \cdot R}{(\beta^2 \cdot P) + R} \tag{4}$$

The F-score can be understood as a weighted measure of false positives and negatives, varying the balanced between them with the β value, so that it is given β times more importance to recall than precision. Some common values to β are 0.5, 1 and 2. For this problem the selected value is 1, so the statement employed to calculate the measure in our experiments is the one shown in Eq. 5.

$$F_1 = 2 \cdot \frac{P \cdot R}{P + R} \tag{5}$$

3 Results

Once the training process have been performed for the selected models and techniques, the test folds are used to build the confusion matrix. For example, Table 2 shows the results obtained in pre-trained AlexNet.

Table 2. Testing confusion matrix for pre-trained AlexNet

	Glomerulus	Non glomerulus
Glomerulus	5295	5
Non Glomerulus	0	5300

On average, the accuracy for this approach in the 10-fold cross-validation is 99.95%, with and standard deviation of 0.06.

The detailed results of every experiment that has been carried out are shown in Table 3.

Table 3. Experiment results

Architecture	Technique	F-score
AlexNet	pre-trained	0.999
AlexNet	from-scratch	0.992
GoogleNet	from-scratch	0.994

As the performance of the trained CNNs is nearly 100%, it becomes interesting to examine some patches that have been wrongly classified, in order to extract useful information about the features of those samples. The specific five glomerulus samples that have been detected as non-glomerulus in the experiment stated in Table 2 are shown in Fig. 4.

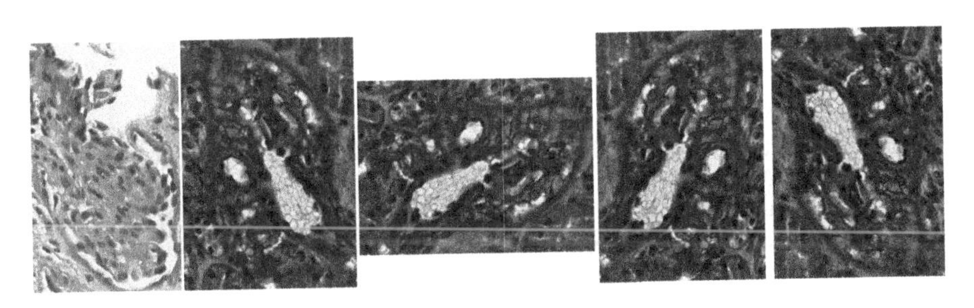

Fig. 4. Patches that have been classified as non-glomerulus

As it is observed, four out of five samples belong to the same image, since these examples are related with the data augmentation that have been performed to the dataset. This is interesting in terms of how the data augmentation can introduce orientation invariance to the model, since, as it is observed, whether a sample is learned as glomerulus (or non-glomerulus in this case), this behaviour remains the same for every variation of the same image.

Also, it is observed that the pre-trained approach slightly improves the performance of the model (and the computational time employed), but it is stated that from-scratch deep CNN models are also suitable to be used in this problem with a reduced dataset as the one that was employed.

4 Conclusions and Future Work

Rather than customized architectures with few layers, the work that has been presented has relied on deep models specifically designed to be used with images, applying different techniques to determine whether a pre-trained model can significantly improve the results. Therefore, models that have been designed for large-scale image classification problems can also be used with smaller datasets to specific tasks, such as glomeruli classification, in this case.

Regarding future work, the final aim is to perform detection and classification of candidate structures straightforward from full WSIs, so that the fully detection workflow is performed. To developed this technique, there would be two main approaches: the first one is to apply a sliding window throughout the image, applying the previously trained CNN model to every extracted patch; the other one is R-CNNs (Region-based Convolutional Neural Networks) one of the most promising Deep Learning techniques to be applied for this kind of problems. In the latter, a new training process is performed, where CNN architecture network variants are used to learn how to extract suitable random region proposals and applying them over the network to determine whether they belong to one of the classes that have been trained.

Acknowledgments. This project has received funding from the European Union's FP7 programme under grant agreement no: 612471. http://aidpath.eu/

References

1. Bueno, G., Fernandez-Carrobles, M.M., Deniz, O., Garcia-Rojo, M.: New trends of emerging technologies in digital pathology. Pathobiology **83**(2–3), 61–69 (2016)
2. Janowczyk, A., Madabhushi, A.: Deep learning for digital pathology image analysis: a comprehensive tutorial with selected use cases. J. Pathol. Inform. **7**, 29 (2016)
3. Madabhushi, A., Lee, G.: Image analysis and machine learning in digital pathology: challenges and opportunities. Med. Image Anal. **33**, 170–175 (2016)
4. Hughson, M.D., Puelles, V.G., Hoy, W.E., Douglas-Denton, R.N., Mott, S.A., Bertran, J.F.: Hypertension, glomerular hypertrophy and nephrosclerosis: the effect of race. Nephrol. Dial. Transplant. **29**(7), 1399–1409 (2014)
5. Rasch, R., Lauszus, F., Thomsen, J.S., Flyvbjerg, A.: Glomerular structural changes in pregnant, diabetic, and pregnant diabetic rats. Apmis **113**(78), 465–472 (2005)

6. Ma, J., Jun, Z., Jinglu, H.: Glomerulus extraction by using genetic algorithm for edge patching. In: IEEE Congress on Evolutionary Computation (2009)
7. Hirohashi, Y., Relator, R., Kakimoto, T., Saito, R., Horai, Y., Fukunari, A., Kato, T.: Automated quantitative image analysis of glomerular desmin immunostaining as a sensitive injury marker in spontaneously diabetic torii rats. J. Biomed. Image Process **1**(1), 208 (2014)
8. Kakimoto, T., Okada, K., Fujitaka, K., Nishio, M., Kato, T., Fukunari, A., Utsumi, H.: Quantitative analysis of markers of podocyte injury in the rat puromycin aminonucleoside nephropathy model. Exp. Toxicol. Pathol. **67**(2), 171–177 (2015)
9. Dalal, N., Triggs, B.: Histograms of oriented gradients for human detection In: International Confernce on Computer Vision and Pattern Recognition, vol. 1 (2005)
10. Kato, T., Relator, R., Ngouv, H., Hirohashi, Y., Takaki, O., Kakimoto, T., Okada, K.: Segmental HOG: new descriptor for glomerulus detection in kidney microscopy image. BMC Bioinform. **16**(1), 316 (2015)
11. Boser, B. E., Guyon, I.M., Vapnik, V.N.: A training algorithm for optimal margin classifiers. In: Proceedings of the Fifth Annual Workshop on Computational Learning Theory. ACM (1992)
12. Kotyk, T., Dey, N., Ashour, A.S., Balas-Timar, D., Chakraborty, S., Ashour, A.S., Tavares, J.M.R.: Measurement of glomerulus diameter and Bowman's space width of renal albino rats. Comput. Methods Programs Biomed. **126**, 143–153 (2016)
13. Graham, R.L.: An efficient algorithm for determining the convex hull of a finite planar set. Inf. Process. Lett. **1**(4), 132–133 (1972)
14. Zhang, M., Wu, T., Bennett, K.M.: A novel Hessian based algorithm for rat kidney glomerulus detection in 3D MRI. In: SPIE Medical Imaging. International Society for Optics and Photonics (2015)
15. Kothari, S., Phan, J.H., Stokes, T.H., Wang, M.D.: Pathology imaging informatics for quantitative analysis of whole-slide images. J. Am. Med. Inform. Assoc. **20**(6), 1099–1108 (2013)
16. Wang, H., Cruz-Roa, A., Basavanhally, A., Gilmore, H., Shih, N., Feldman, M., Madabhushi, A.: Mitosis detection in breast cancer pathology images by combining handcrafted and convolutional neural network features. J. Med. Imaging **1**(3), 034003 (2014)
17. Malon, C.D., Cosatto, E.: Classification of mitotic figures with convolutional neural networks and seeded blob features. J. Pathol. Inform. **4**(1), 9 (2013)
18. Xu, K., Zhu, L., Wang, R., Liu, C., Zhao, Y.: Automated detection of diabetic retinopathy using deep convolutional neural networks. Med. Phys. **43**(6), 3406 (2016)
19. Havaei, M., Davy, A., Warde-Farley, D., Biard, A., Courville, A., Bengio, Y., Larochelle, H.: Brain tumour segmentation with deep neural networks. Med. Image Anal. **35**, 18–31 (2017)
20. Agarwal, S.K., Sethi, S., Dinda, A.K.: Basics of kidney biopsy: a nephrologist's perspective. Indian J. Nephrol. **23**(4), 243 (2013)
21. Krizhevsky, A., Sutskever, I., Hinton, G.: Imagenet classification with deep convolutional neural networks. In: Advances in Neural Information Processing Systems, pp. 1097–1105 (2012)
22. Szegedy, C., Liu, W., Jia, Y., Sermanet, P., Reed, S., Anguelov, D., Rabinovich, A.: Going deeper with convolutions. In: Proceedings of the IEEE Conference on Computer Vision and Pattern Recognition, pp. 1–9 (2015)
23. Fawcett, T.: An introduction to ROC analysis. Pattern Recogn. Lett. **27**(8), 861–874 (2006)

Paediatric Frontal Chest Radiograph Screening with Fine-Tuned Convolutional Neural Networks

Jonathan Gerrand[1,2](✉), Quentin Williams[1], Dalton Lunga[3], Adam Pantanowitz[2], Shabir Madhi[4], and Nasreen Mahomed[4,5]

[1] Council for Scientific and Industrial Research, Pretoria, South Africa
JGerrand@csir.co.za

[2] Biomedical Engineering Research Group in the School of Electrical and Information Engineering, University of the Witwatersrand, Johannesburg, South Africa

[3] Oak Ridge National Laboratory, Oak Ridge, TN, USA

[4] Medical Research Council: Respiratory and Meningeal Pathogens Research Unit, University of the Witwatersrand, Johannesburg, South Africa

[5] Department of Diagnostic Radiology, University of the Witwatersrand, Johannesburg, South Africa

Abstract. Within developing countries, there is a realistic need for technologies that can assist medical practitioners in meeting the increasing demand for patient screening and monitoring. To this end, computer aided diagnosis (CAD) based approaches to chest radiograph screening can be utilised in areas where there is a high burden of diseases such as tuberculosis and pneumonia. In this work, we investigate the efficacy of a purely data-driven approach to chest radiograph classification through the use of fine-tuned convolutional neural networks (CNN). We use two popular CNN models that are pre-trained on a large natural image dataset and two distinct datasets containing paediatric and adult radiographs respectively. Evaluation is performed using a 5-fold cross-validation analysis at an image level. The promising results, with top AUC metrics of 0.87 and 0.84 for the respective datasets, along with several characteristics of our data-driven approach motivate for the use of fine-tuned CNN models within this application of CAD.

Keywords: Computer aided diagnosis · Convolutional neural network · Chest radiograph screening · Fine-tuning

1 Introduction

Forecasts have observed that by the year 2030, the demand for medical physicians in developing countries will greatly outweigh the supply, with twelve times the anticipated number of available doctors required to deal with patient volumes [1]. In areas for which the burden of infectious diseases such as tuberculosis and pneumonia is high, automated chest radiograph reading with CAD could reduce the volume of patients visiting specialists.

M. Valdés Hernández and V. González-Castro (Eds.): MIUA 2017, CCIS 723, pp. 850–861, 2017.
DOI: 10.1007/978-3-319-60964-5_74

Traditionally, the development of a CAD-based system involves several phases of data processing before an automated screening can occur, for example: lung field segmentation, lung field sub-division and feature extraction [2,3]. Each of these processing stages is highly specialised, requiring the use of manual 'hand-crafted' approaches which are specific to a given reading task such as the detection of lung nodules or pleural effusion [4].

With the limitations posed by these manual processing techniques in mind, recent advances within the training and use of deep convolutional neural networks (CNN), have given rise to marked improvements in the field of medical image processing. These advances have redefined the state-of-the-art across multiple imaging modalities as seen within the literature [5–7]. This success can be seen to stem from progress made within the computer vision community in which large-scale datasets such as ImageNet [8] enable the data-driven training of very large classification models [9]. As medical imaging datasets cannot currently achieve the same scale and representative quality of natural-image datasets commonly used in computer vision, techniques such as transfer learning [10] and fine-tuning [11] have found wide use within the medical imaging community.

Using CNN-based approaches, automatic chest screening can become a purely data-driven process. That is, end-to-end system learning can occur without the need for multiple processing phases and expert feature engineering. This is desirable as it relaxes the need for expert domain knowledge and reduces CAD system complexity. In addition, posturing the CNN learning as a weakly-supervised task (in which only image-level class labels are provided) can remove the need for manual region-of-interest (ROI) labelling to be performed. While the use of holistic image labelling has been shown to marginally reduce the performance of a CAD system as compared to a ROI-based approach [6], the latter technique is costly and human intensive [12]. Furthermore, work within the computer vision community has shown the ability of weakly-supervised CNN models to perform accurate object localisation in addition to classification [13]. These techniques have been extended to the medical imaging domain [14], holding the potential for pathology identification and further clinician interrogation upon classification.

Several authors have investigated the use of CNNs for automatic chest radiograph reading. Within the works of [15,16], various layer outputs of popular computer vision CNN models were fed into a support vector machine (SVM) in order to detect the presence of radiographic chest pathologies and pulmonary nodules respectively. In both these works, the use of CNN features alone was insufficient to advance the state-of-the-art and required fusion with hand-crafted features to produce favourable results. In comparison, Cicero *et al.* [17] report high classifier accuracy (AUC = 0.964) when using the GoogleNet CNN model [18] trained with 35038 posterior-anterior (PA) chest radiographs to classify between normal and pathology cases. Most radiographic datasets do not approach the scale seen within this study, however, with similar attempts to train a CNN model using smaller public datasets, as seen within the work of [19], producing inferior performance.

In this paper, we evaluate the efficacy of using fine-tuning to train CNNs on two distinct radiography datasets for pneumonia and opacity detection. While our approach is similar to that of [20], in which fine-tuned CNNs were used in the classification between frontal and lateral radiographs, we believe the use of these techniques for the detection of smaller, locally discriminative pathologies inherent within radiograph screening presents a complex and relevant task whose investigation is motivated.

Here our contribution is three-fold: (1) We perform chest radiograph screening in a purely data-driven manner, excluding the use of traditional radiographic CAD processing stages such as lung segmentation and feature extraction. (2) We achieve compelling AUC and F1 metrics using our fine-tuned CNN models performing image-level classification on a paediatric and adult radiographic dataset respectively. (3) We present saliency maps generated for the evaluation of classifier performance and use these to illustrate the merits of a data-driven approach to radiographic screening.

2 Background

2.1 Convolutional Neural Networks in Medical Imaging

CNNs differ from other neural networks (NN) in that their architectures contain multiple convolutional layers. In deep CNN models, these layers can learn representations exhibiting a hierarchical order of complexity, with lower layers approximating general colour blobs and Gabor filters, and higher ones representing specific object-like features [11]. It is important to note that these features are self-learnt from the data, and when combined with training techniques such as data augmentation, add a measure of robustness to the classification procedure [19].

Both task-specific and popular computer-vision CNN architectures have been proposed for use in CAD systems. For task-specific approaches, models typically possess smaller input dimensions, several fully-connected layers for classification, and are generally shallower than their computer-vision domain counterparts. While these adaptations tailor the models to perform better at patch-based classification [21], they are often limited in their scope of use to a specific modality or dataset. In contrast to this, deeper and more complex models for natural image classification have been shown to perform favourably across multiple medical imaging modalities [14]. The popular, open-source, status of these models also allows for greater research community support, with versions pre-trained on large-scale datasets being publicly available.

2.2 Base Model Training and Fine-Tuning

The standard process of training a CNN typically involves randomly initialising the weights and biases for each of its layers using a chosen initialisation strategy, for example [22], and then adjusting these weights over successive forward and backward propagation steps. This process, which is often referred to as training the base CNN model, is suitable in the presence of a large dataset and can produce state-of-the-art classification results [23]. However, in data constrained environments with small datasets, such as within the medical imaging community, standard training results in over-fitting due to the high capacity which deep CNNs possess [12]. As a means to overcome this low data constraint, the technique of fine-tuning has been investigated in the medical imaging domain.

In fine-tuning, a CNN model is initially trained on a large *source* dataset (D_s), after which a variable number of its classification layers are re-initialised and the entire network (with all its previously learnt weights) is updated on a smaller *target* dataset (D_t) [11]. Importantly, D_s and D_t are disjoint and often dissimilar in content. As such, fine-tuning is believed to act as a regulariser in which the feature descriptors learnt on D_s possess sufficient representational capacity to extract efficient discriminative features from a smaller set of images in D_t.

The efficacy of fine-tuning was investigated within the works of [12,14]. The results obtained within these studies are complementary and report an improvement in generalisation and overall accuracy when fine-tuning was performed. Furthermore, the work of [14] extended these results across multiple medical imaging modalities in which fine-tuning was consistently seen to improve performance. The same authors also introduced the notion of "deep" fine-tuning for medical contexts and demonstrated that updating all the weights across a fine-tuned network further improved classification performance.

3 Methodology

3.1 Model Selection and Training Procedure

Our data-driven approach for radiograph screening is evaluated by first training two medical tailored CNNs using Xavier [22] initialisation. The models are based on the *GoogleNet* [18] and *Network-in-Network* (NIN) [24] architectures depicted in Fig. 1 and shall hence be referred to as GoogleNet-R and NIN-R. We also perform deep fine-tuning using base models pre-trained from ImageNet [8] and call these GoogleNet-FT and NIN-FT respectively.

3.2 Datasets and Data Preparation

PCV-13 Paediatric Radiograph Dataset. The PCV-13 chest X-ray database (frontal and lateral radiographs) for children enrolled in a case control study evaluating the effectiveness of pneumococcal conjugate vaccine (PCV) against presumed bacterial pneumonia in hospitalised HIV-uninfected South African children (i.e.: sites were Soweto, Cape Town and KwaZulu-Natal) is used [25]. This dataset contains 3905 frontal and 2374 lateral chest radiographs (JPEG format) of children 1–59 months of age. To make this dataset suitable for study, complete anonymisation at a patient level was performed. Ethical

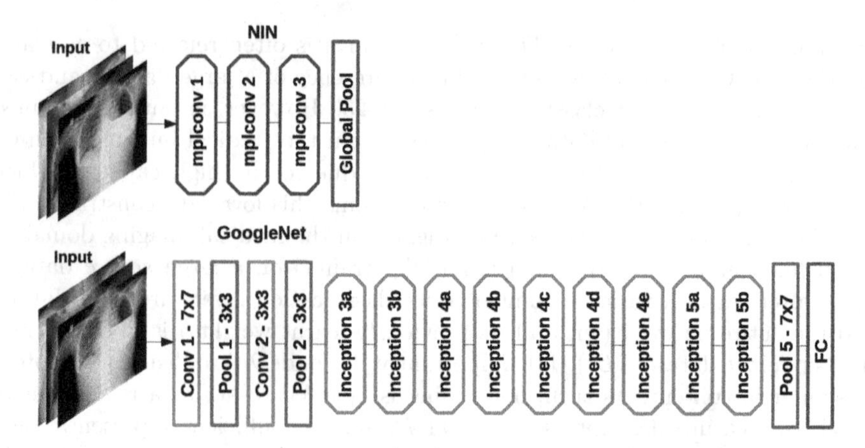

Fig. 1. Model architectures evaluated. Each *inception* and *mlpconv* module contains multiple sub-layers. As can be seen, the NIN architecture is shallower than that of GoogleNet

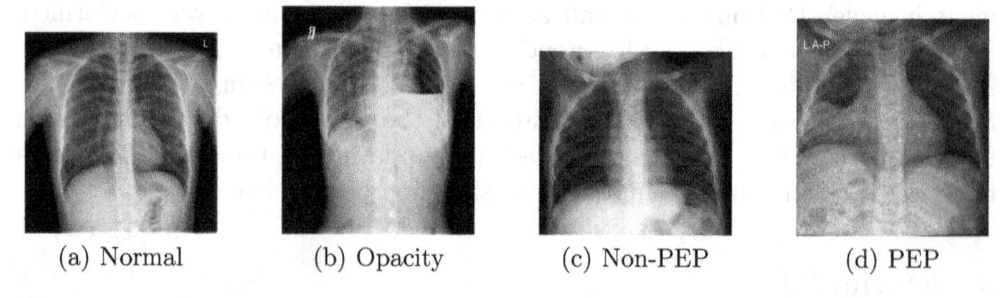

(a) Normal (b) Opacity (c) Non-PEP (d) PEP

Fig. 2. Example frontal chest radiographs and their respective class labels drawn from the Indiana OpenI (a), (b) and PCV-13 (c), (d) datasets.

clearance (certificate no. M1611133) from the medical human research ethics committee (MHREC) of the University of the Witwatersrand, Johannesburg, has also been granted.

A panel of two paediatricians trained in the WHO standardised chest X-ray interpretation methodology [26] was used to independently classify the frontal radiographs into three classes, with a third reader being used for discord readings to provide the reference standard. The final consensus human readings were: *Primary end-point pneumonia* (PEP) (n = 1394), *Other infiltrates* (OI) (n = 1751) and *No infiltration/consolidation/effusion* (Non-PEP) (n = 586). Using the WHO standardised classification [26], the remainder (n = 174) were determined to be uninterpretable. To pose our problem for radiographic pneumonia screening, we exclude the OI class from our study and perform binary classification on the PEP and No-PEP classes. An example of each of these classes can be seen in Fig. 2(d) and (c) respectively.

Table 1. Average number of datapoints per fold (5-fold cross-validation) after performing data augmentation on each training set

Dataset	Num. datapoints		
	Train	Validation	Test
PCV-13	11232	232	234
Indiana OpenI	16128	334	334

Indiana OpenI Radiograph Subset. Housed under the umbrella of the *OpenI* biological image and medical literature portal (https://openi.nlm.nih.gov/), the Indiana University chest X-ray collection comprises of 7470 fully anonymised PA and lateral chest radiographs (PNG format) obtained from the Indiana Network for Patient Care image archiving system. Associated with each radiograph is a set of Medical Subject Headings (MeSH) which provide a basis for pathology assignment to each image. Several MeSH terms can be present within one report, and thus a large proportion of the images can be classified under multiple pathologies. To disambiguate classification, we select two non-overlapping MeSH terms with the highest frequency of occurrence and make these our class labels. This gives rise to a subset of *normal* (n = 2696) and *opacity* (n = 840), with examples shown in Fig. 2(a) and (b). Further information surrounding the collection and anonymisation of the Indiana OpenI dataset can be found at [27].

Data Pre-processing. To prevent class imbalance, we sub-sample based on the lowest-volume class within each dataset. We then separate each dataset into 5 distinct folds at an image level (5-fold cross validation), with each fold containing a 60%:20%:20% split of training, validation and testing data respectively. As motivated within [19], several class-preserving data augmentation operations are then employed to artificially increase the amount of data within each fold's training set. These augmentations include slight rotation ($\pm 10°$), image cropping (90% corner-anchored crop) and horizontal reflection These augmentations increase the effective size of both datasets by a factor of 15. Finally, we use bilinear interpolation to rescale each image to a size of 224×224. The resulting dataset statistics are summarised in Table 1.

4 Results and Discussion

4.1 Experimental Setup and Training Parameters

During implementation, all experimentation is performed on a machine running Linux 14.04 with a single Nvidia 1080GTX (8 Gb) GPU card, 128 Gb ram and 28 CPU cores @2.4 Ghz. We use the Tensorflow [28] framework to train, evaluate and produce saliency maps for all our models.

In all training instances, we use the Adam optimiser [29] with a β_1 value of 0.9 and β_2 value of 0.999. The following additional model hyper-parameters are also employed; mini-batch size: 64 and weight decay: 0.0001. A dropout [30] coefficient of 0.4 is utilised for the GoogleNet models during training. Early stopping is used to limit model over-fitting and is determined from the F1-score of a model on a held-out validation set during training. Our initial learning rate selection is guided by the work of [19], with a grid search yielding a final learning rate of 0.015 across all training layers for base training. In fine-tuning we apply a learning rate of 0.015 for the re-initialised classification layer and reduce this by a factor of 10 for all other layers.

4.2 Evaluation Metrics

All models are trained using 5-fold cross-validation (CV) at an image level. During evaluation, batch (64 images) classification times for the GoogleNet and NIN models are determined as 263 ms and 131 ms respectively. The results obtained for chest radiograph classification on both the PCV-13 and Indiana OpenI datasets are reported in Table 2 where we provide both the average F1-score and area-under-curve (AUC) metrics obtained across all folds. We further produce receiver operating characteristic (ROC) curves for each dataset and display these in Fig. 3.

GoogleNet-R and NIN-R perform poorly across both datasets, with GoogleNet-R yielding the worst performance of all the models. This agrees with our expectations, as GoogleNet is an extremely deep model with a high capacity which can easily over-fit on the small number of data points within each dataset. In contrast to this performance, GoogleNet-TL is seen to produce superior results for both classification tasks (0.192 improvement for PCV-13 and 0.334 for Indiana OpenI). These increases are mirrored between the NIN-FT and NIN-R models and demonstrate the ability of fine-tuning to guide the data-driven training of CNN networks for chest radiograph classification. Furthermore, as illustrated within Fig. 4, fine-tuning of the models leads to far quicker convergence (and thus training times) compared to standard training.

When comparing the fine-tuning models on each respective dataset, GoogleNet-FT can be seen to produce the best F1-score for PCV-13 (0.789 vs. 0.775 for NIN-FT). However, for Indiana OpenI these results are not replicated

Table 2. Comparison of classification accuracies for both the PCV-13 and Indiana OpenI datasets. We use image-level 5-fold CV. Bold numbers indicate best performance.

Method	Dataset			
	PCV-13		Indiana OpenI	
	F1	AUC	F1	AUC
NIN-R	0.584	0.636	0.39	0.525
GoogleNet-R	0.597	0.674	0.09	0.507
NIN-FT	0.775	0.855	**0.768**	**0.842**
GoogleNet-FT	**0.789**	**0.873**	0.767	0.841

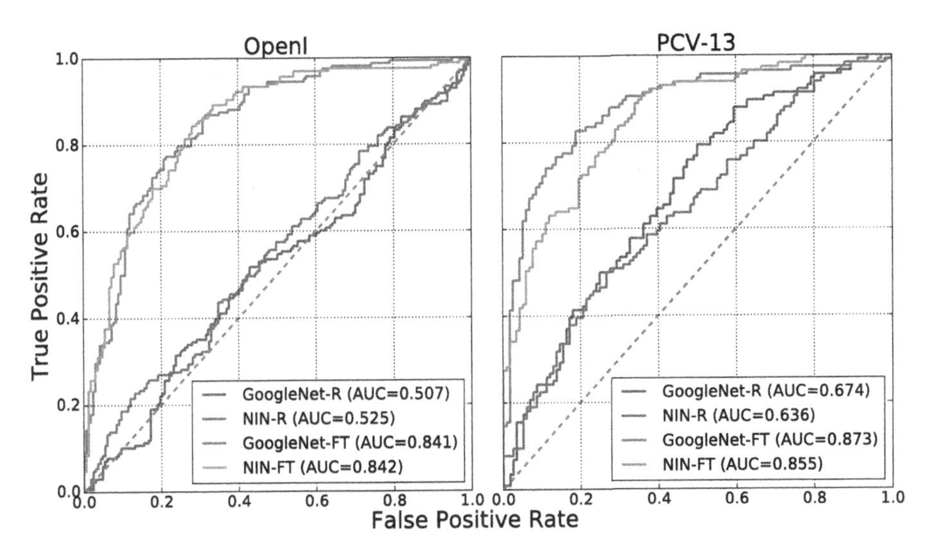

Fig. 3. ROC curves obtained during model evaluation. For both the OpenI and PCV-13 radiographic datasets, the fine-tuned (-FT) models significantly outperform their counterparts trained only on the target dataset (-R).

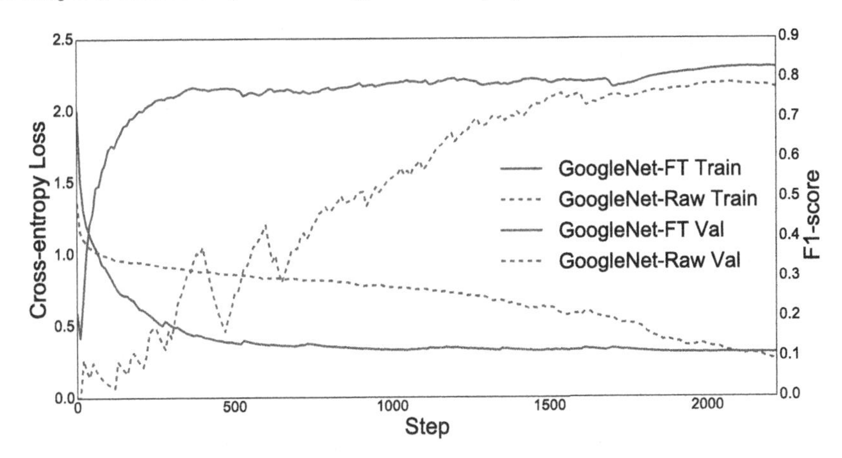

Fig. 4. Sample curves generated during training for the PCV-13 dataset. Here fine-tuning is seen to yield faster convergence as compared that of raw initialisation.

and the models are seen to perform nearly identically. Model capacity could be attributed to this once again; the OpenI dataset contains both frontal and lateral chest radiographs for each class label, increasing intra-class variability and effectively lowering the number of representative datapoints seen during training. This highlights the need for large, representative datasets for improved classifier performance.

When evaluated for their effectiveness in performing radiograph classification, both fine-tuned models show compelling performance. For the PCV-13

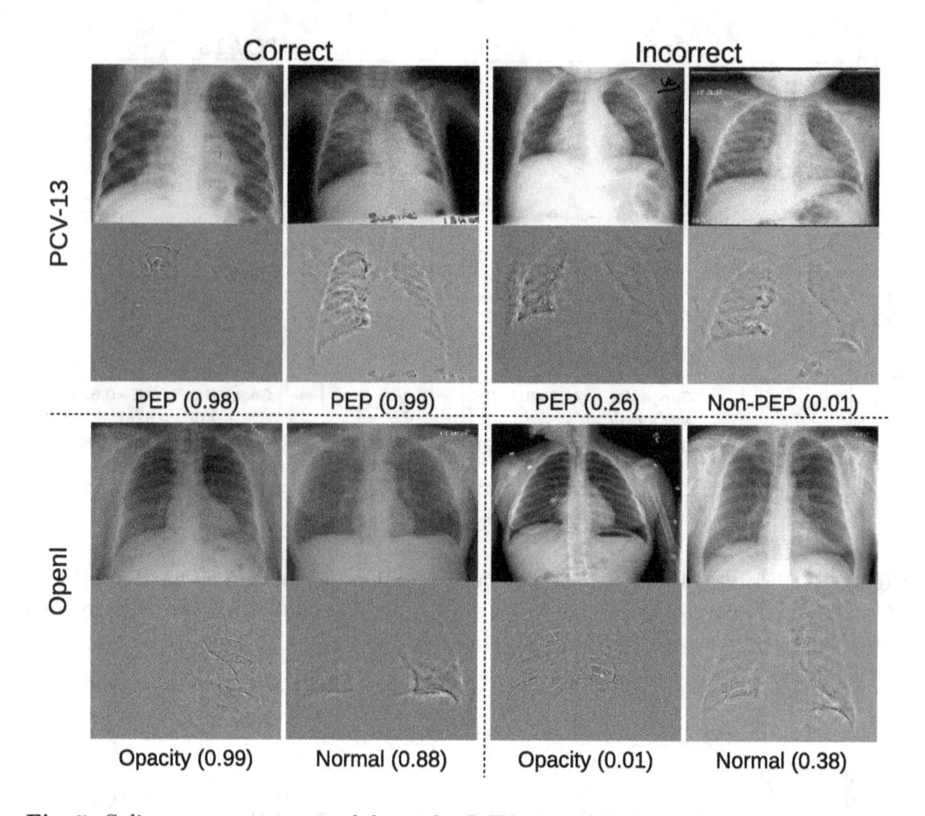

Fig. 5. Saliency maps generated from the PCV-13 and Indiana OpenI test sets using the GoogleNet-FT and NIN-FT models. Each map corresponds to the input radiograph placed above it, with darker regions in the map indicating discriminative areas of greater importance when performing a classification. The true class is provided below each image-map pair, with the classifier confidence for this class given in brackets.

dataset, the GoogleNet-FT model produces a sensitivity of 0.828 at a false-positive rate (FPR) of 0.19, while NIN-FT achieves a sensitivity of 0.826 at a FPR of 0.275 on the Indian OpenI dataset. The AUC scores obtained for each model are also comparable, highlighting the applicability of both architectures for use in frontal chest radiograph classification.

4.3 Saliency Maps

In order to further interrogate the performance of the trained models, we generate saliency maps from various dataset images using the *guided backpropagation* method outlined by [31]. These saliency maps visually represent the most discriminative parts of an input image for a given classification class, and can be thought of as visualising where the model places its attention when determining whether a pathology exists in the input.

We use the top performing GoogleNet-FT and NIN-FT models to generate the saliency maps presented in Fig. 5. Within these maps it can be seen that both models learn to focus on areas of the lung fields within each input image, which is an impressive feat seeing that no explicit labelling of these fields is ever provided. A review by a radiologist on sample saliency maps drawn from the PEP class of the PCV-13 dataset confirmed that certain areas of focus within these maps were discriminative for diagnosis, motivating the merits of our data-driven learning process in which multiple hand-crafted processing steps are not required.

4.4 Suitability for Chest Radiograph Screening

The results obtained during experimentation motivate the further study of fine-tuned CNNs for chest radiograph screening. The data-driven nature of this approach is advantageous as it reduces the number of processing steps needed for screening, relaxes labelling requirements, and robustly handles low dataset volumes. Furthermore, the rapid classification of radiographs is possible, facilitating screening applications in which real-time feedback between medical practitioner and patient can occur.

An implicit drawback in applying a data-driven approach for chest radiograph screening is the heavy reliance on larger data quantities to improve classifier performance. As such, while our future work seeks to explore training methods which will exploit current dataset volumes, there still exists a great need for publicly available, large radiographic datasets to facilitate research advances [19].

5 Conclusion

This work evaluates the efficacy of fine-tuning CNN models for use in chest radiograph screening. To this end, we apply a purely data-driven approach using two CNN models performing image-level classification on both paediatric and adult chest radiographic datasets. Using fine-tuning, we achieve appreciable AUC and F1 scores without the use of traditional processing stages such as lung segmentation and feature extraction. These results, along with saliency maps generated for model interpretation, motivate for the utilisation of fine-tuned CNN models in chest radiograph screening.

References

1. Schwab, K., Sala-i Martin, X., Brende, B., Blanke, J., Bilbao-Osorio, B., Browne, C., Corrigan, G., Crotti, R., Hanouz, M.D., Geiger, T., Gutknecht, T., Ko, C., Serin, C.: The Global Competitiveness Report. Economic World Forum (2014)
2. Campadelli, P., Casiraghi, E., Artioli, D.: A fully automated method for lung nodule detection from postero-anterior chest radiographs. IEEE Trans. Med. Imaging **25**(12), 1588–1603 (2006)

3. Carrillo-de Gea, J.M., Garca-Mateos, G.: Detection of normality/pathology on chest radiographs using LBP (2010)
4. Van Ginneken, B., Maduskar, P., Philipsen, R.H.M.M., Melendez, J., Scholten, E., Chanda, D., Ayles, H., Sánchez, C.I.: Automatic detection of pleural effusion in chest radiographs. Med. Image Anal. **28**, 22–32 (2016)
5. Roth, H., Lu, L., Liu, J., Yao, J., Seff, A.: Improving computer-aided detection using convolutional neural networks and random view aggregation, p. 17. arXiv preprint arXiv:1505.03046 (2015)
6. Gao, M., Bagci, U., Lu, L., Wu, A., Buty, M., Shin, H.C., Roth, H., Papadakis, G.Z., Depeursinge, A., Summers, R.M., Xu, Z., Mollura, D.J.: Holistic classification of CT attenuation patterns for interstitial lung diseases via deep convolutional neural networks. In: Workshop on Deep Learning in Medical Image Analysis (MICCAI). CIDI (2015)
7. Ciompi, F., de Hoop, B., van Riel, S.J., Chung, K., Scholten, E.T., Oudkerk, M., de Jong, P.A., Prokop, M., van Ginneken, B.: Automatic classification of pulmonary peri-fissural nodules in computed tomography using an ensemble of 2D views and a convolutional neural network out-of-the-box. Med. Image Anal. **26**(1), 195–202 (2015)
8. Deng, J., Dong, W., Socher, R., Li, L.J., Li, K., Fei-Fei, L.: ImageNet: a large-scale hierarchical image database. In: 2009 IEEE Conference on Computer Vision and Pattern Recognition, pp. 2–9 (2009)
9. He, K., Zhang, X., Ren, S., Sun, J.: Deep residual learning for image recognition. In: Proceedings of the IEEE Conference on Computer Vision and Pattern Recognition, pp. 770–778 (2016). Arxiv.Org
10. Sharif Razavian, A., Azizpour, H., Sullivan, J., Carlsson, S.: CNN features off-the-shelf: an astounding baseline for recognition. In: Proceedings of the IEEE Conference on Computer Vision and Pattern Recognition Workshops, pp. 806–813 (2014)
11. Yosinski, J., Clune, J., Bengio, Y., Lipson, H.: How transferable are features in deep neural networks? In: Proceedings of NIPS Advances in Neural Information Processing Systems 27, vol. 27, pp. 1–9 (2014)
12. Tajbakhsh, N., Shin, J.Y., Gurudu, S.R., Hurst, R.T., Kendall, C.B., Gotway, M.B., Liang, J., Member, S.: Convolutional neural networks for medical image analysis : fine tuning or full training? IEEE Trans. Med. Imaging **35**(5), 1299–1312 (2016). IEEE
13. Sermanet, P., Eigen, D., Zhang, X., Mathieu, M., Fergus, R., LeCun, Y.: Overfeat: integrated recognition, localization and detection using convolutional networks, pp. 1–16 (2013)
14. Shin, H.C., Roth, H.R., Gao, M., Lu, L., Xu, Z., Nogues, I., Yao, J., Mollura, D., Summers, R.M., Lu, M.G.L., Xu, Z., Nogues, I., Yao, J., Mollura, D., Summers, R.M.: Deep convolutional neural networks for computer-aided detection: CNN architectures, dataset characteristics and transfer learning. IEEE Trans. Med. Imaging **35**, 1285–1298 (2016). IEEE
15. Bar, Y., Diamant, I., Greenspan, H., Wolf, L.: Chest Pathology Detection Using Deep Learning with Non-Medical Training. In: IEEE International Symposium on Biomedical Imaging (ISBI), pp. 294–297 (2015)
16. Van Ginneken, B., Setio, A.A.A., Jacobs, C., Ciompi, F.: Off-the-shelf convolutional neural network features for pulmonary nodule detection in computed tomography scans. In: 2015 IEEE 12th International Symposium on Biomedical Imaging (ISBI), pp. 286–289 (2015)

17. Cicero, M., Bilbily, A., Colak, E., Dowdell, T., Gray, B., Perampaladas, K., Barfett, J.: Training and validating a deep convolutional neural network for computer-aided detection and classification of abnormalities on frontal chest radiographs. Inves. Radiol. **52**(5), 281–287 (2016)
18. Szegedy, C., Sermanet, P., Reed, S., Anguelov, D., Erhan, D., Vanhoucke, V., Rabinovich, A.: Going deeper with convolutions. In: 2015 IEEE Conference on Computer Vision and Pattern Recognition (CVPR), pp. 1–9 (2015)
19. Shin, H.c., Roberts, K., Lu, L., Demner-Fushman, D., Yao, J., Summers, R.M.: Learning to Read Chest X-Rays: Recurrent Neural Cascade Model for Automated Image Annotation (2016)
20. Rajkomar, A., Lingam, S., Taylor, A.G., Blum, M., Mongan, J.: High-throughput classification of radiographs using deep convolutional neural networks. J. Digit. Imaging **30**(1), 95–101 (2016)
21. Li, Q., Cai, W., Wang, X., Zhou, Y., Feng, D.D., Chen, M.: Medical image classification with convolutional neural network. In: 2014 13th International Conference on Control, Automation, Robotics & Vision Marina Bay Sands, Singapore, 10–12th December 2014 (ICARCV 2014) Th25.2. IEEE (2014)
22. Glorot, X., Bengio, Y.: Understanding the difficulty of training deep feedforward neural networks. In: Proceedings of the 13th International Conference on Artificial Intelligence and Statistics (AISTATS), vol. 9, pp. 249–256 (2010)
23. Krizhevsky, A., Sutskever, I., Hinton, G.E., (University of Toronto): ImageNet classification with deep convolutional neural networks (2012)
24. Lin, M., Chen, Q., Yan, S.: Network in network, p. 10. arXiv preprint https://arxiv.org/abs/1312.44001312.4400 (2013)
25. Madhi, S.A., Groome, M.J., Zar, H.J., Kapongo, C.N., Mulligan, C., Nzenze, S., Moore, D.P., Zell, E.R., Whitney, C.G., Verani, J.R.: Effectiveness of pneumococcal conjugate vaccine against presumed bacterial pneumonia hospitalisation in HIV-uninfected South African children: a case-control study. Thorax **70**(12), 1–7 (2015)
26. Cherian, T., Mulholland, E., Carlin, J., Ostensen, H.: Standardized interpretation of paediatric chest radiographs for the diagnosis of pneumonia in epidemiological studies variability in the interpretation of chest radiographs ! standardized method for identifying radiological pneumonia would facilitate read. Bull. World Health Organ. **83**(5), 353–359 (2004)
27. Demner-Fushman, D., Kohli, M.D., Rosenman, M.B., Shooshan, S.E., Rodriguez, L., Antani, S., Thoma, G.R., McDonald, C.J.: Preparing a collection of radiology examinations for distribution and retrieval. J. Am. Med. Inf. Assoc. **23**(2), 304–310 (2016)
28. Abadi, M., Agarwal, A., Barham, P., Brevdo, E., Chen, Z., Citro, C., Corrado, G., Davis, A., Dean, J., Devin, M., Ghemawat, S., Goodfellow, I., Harp, A., Irving, G., Isard, M., Jia, Y., Kaiser, L., Kudlur, M., Levenberg, J., Man, D., Monga, R., Moore, S., Murray, D., Shlens, J., Steiner, B., Sutskever, I., Tucker, P., Vanhoucke, V., Vasudevan, V., Vinyals, O., Warden, P., Wicke, M., Yu, Y., Zheng, X.: Tensorflow: large-scale machine learning on heterogeneous distributed systems. None **1**(212), 19 (2015). http://download.tensorflow.org/
29. Kingma, D., Ba, J.: Adam: a method for stochastic optimization, arXiv:1412.6980 [cs], pp. 1–15 (2014)
30. Srivastava, N., Hinton, G., Krizhevsky, A., Sutskever, I., Salakhutdinov, R.: Dropout: prevent NN from overfitting. J. Mach. Learn. Res. **15**, 1929–1958 (2014)
31. Springenberg, J.T., Dosovitskiy, A., Brox, T., Riedmiller, M.: Striving for simplicity: the all convolutional net, pp. 1–14 (2015)

Automatic Hotspots Detection for Intracellular Calcium Analysis in Fluorescence Microscopic Videos

David Traore[1(✉)], Katja Rietdorf[2], Nasser Al-Jawad[1], and Hisham Al-Assam[1]

[1] Applied Computing Department, The University of Buckingham, Buckingham, UK
1301722@buckingham.ac.uk

[2] School of Life, Health and Chemical Sciences, The Open University, Milton Keynes, UK

Abstract. In recent years, life-cell imaging techniques and their software applications have become powerful tools to investigate complex biological mechanisms such as calcium signalling. In this paper, we propose an automated framework to detect areas inside cells that show changes in their calcium concentration i.e. the regions of interests or hotspots, based on videos taken after loading living mouse cardiomyocytes with fluorescent calcium reporter dyes. The proposed system allows an objective and efficient analysis through the following four key stages: (1) Pre-processing to enhance video quality, (2) First level segmentation to detect candidate hotspots based on adaptive thresholding on the frame level, (3) Second-level segmentation to fuse and identify the best hotspots from the entire video by proposing the concept of calcium fluorescence hit-ratio, and (4) Extraction of the changes of calcium fluorescence over time per hotspot. From the extracted signals, different measurements are calculated such as maximum peak amplitude, area under the curve, peak frequency, and inter-spike interval of calcium changes. The system was tested using calcium imaging data collected from Heart muscle cells. The paper argues that the automated proposal offers biologists a tool to speed up the processing time and mitigate the consequences of inter-intra observer variability.

Keywords: Intracellular calcium signalling · Hotspots segmentation · Calcium change quantification · Cell parameters · Fluorescence microscopy

1 Introduction

Changes in the intracellular calcium concentration are a very critical and universal signalling mechanism used by cells [1, 2]. It is important for a myriad of processes. In the case of humans, life starts with an increase in the calcium concentration upon fertilisation of the egg [3]. Cell death [1, 4] is frequently caused by a prolonged increase in the intracellular calcium concentration. But between fertilization and death, changes in the calcium concentration regulate a plethora of processes in an organism, like memory formation, heartbeat, blood pressure regulation, bones and teeth development,

M. Valdés Hernández and V. González-Castro (Eds.): MIUA 2017, CCIS 723, pp. 862–873, 2017.
DOI: 10.1007/978-3-319-60964-5_75

blood clotting, hormone functions, cell division, muscle contractions, and antigen recognition in the immune system [1, 2]. Because a prolonged calcium increase is a death signal, it is very important to keep the intracellular calcium concentration tightly controlled in a narrow range, typically around 100 nM [1].

Further, a dysregulation of the normal calcium signalling processes may lead to several diseases like cardiac arrhythmias [5], neuronal disorders like Alzheimer's and Parkinson's diseases [5, 6], or immune dysfunctions [7].

Fluorescence imaging is a powerful and commonly used technique to study calcium signalling in living cells. Cells or tissues are loaded with a calcium sensitive indicator and changes in the intracellular calcium concentration can be recorded in real time by capturing a video of the fluorescence images on a microscope [8]. One downside of the technique is a labour-intensive and often subjective image analysis process. The proposal described in this paper aims to automatically detect and quantify intracellular calcium changes in these videos. In addition to automating the analysis by biologists, our data analysis framework also provides a fast and reliable tool for fluorescent video data analysis in cell phenotyping. The cells being monitored are assumed static. Issues addressed here are processing time, replicability of the data analysis, photo bleaching artefacts, and human-biased variations that are observed during the manual analysis of intracellular calcium imaging experiments.

All in all, this paper presents a fully automatic system to objectively detect active regions of interest(ROIs), which we refer to as hotspots, and to quantify their change in calcium concentration over the time period when they are imaged and saved in a series of fluorescent frames. The automated framework defines as hotspots any areas in cells whose the calcium concentration fluctuates and goes through the four key stages: (i) A pre-processing stage to enhance the contrast and increase the signal to noise ratio of the input video frames using median filtering and adaptive histogram equalization; (ii) A first level segmentation stage to detect candidate hotspots in cells based on changes in the fluorescence intensity using adaptive thresholding and morphological operations; (iii) A second level segmentation stage to fuse and identify the best hotspots from the entire videos by introducing the concept of hit-ratio of calcium fluorescence; (iv) An extraction and quantification stage of the calcium concentration over time for individual hotspots to calculate maximum peak amplitude, area under the curve, peak frequency, and inter-spike interval of calcium changes.

The rest of our paper is organized as follows: background and previous work are presented in Sect. 2. Section 3 gives an overview of the proposed system, focusing on key points. Section 4, presents and discusses the results of our approach. Finally, in Sect. 5 possible future work are highlighted followed by concluding remarks.

2 Background and Relevant ExistingWork

Fluorescent calcium indicators were first developed in 1982 by Roger Tsien and colleagues [9]. Since then, indicators with different characteristicshave been developed: colours, calcium binding affinities, or targeted to different intracellular organelles to name a few [8, 10]. Our framework was developed using calcium imaging data from a cardiomyocyte-like cell type, called pulmonary vein sleeve cells (PVCs), loaded with the calcium indicator Oregon Green BAPTA-AM [11]. Aberrant signalling processes in PVCs are important for the development of atrial fibrillation, the most common cardiac arrhythmia (heart rhythm disorder). In contrast to other cardiomyocytes, PVCs show a high level of localized spontaneous increases in their intracellular calcium concentration. The high level of this spontaneous activity in PVCs is one important factor in the development of atrial fibrillation. A better understanding of the processes underlying the calcium signalling in PVCs might provide new mechanisms to treat or prevent atrial fibrillation [11–14]. However, the spontaneous activity makes calcium imaging data from these cells very time-consuming to analyse, which prompted the development of the automated image analysis software presented here.

In a typical manual analysis of fluorescent video frames of a calcium imaging experiment, the readout is a file/spreadsheet giving the change in fluorescence for each hotspot over the length of the experiment. In order to compare the results from different experiments, these values are normalized [8]. First, the background fluorescence is subtracted from the initial readout. This is commonly done by setting one small background region in video frames through an interactive toolbox in Image J. Then, the background subtracted data need to be normalized to allow a comparison of the magnitude of the changes between different experiments. For that, the minimum fluorescence ($Fmin$) for each hotspot needs to be measured, and the fluorescence at any given time (F) will be normalized to ($Fmin$). The normalized ($F/Fmin$) is then used to quantitate parameters like the maximum amplitude of the change in fluorescence, the frequency of (calcium) increases or the area under the curve for the (calcium) transients, depending on which characteristic of the calcium changes is most informative for the experiment. This manual procedure raises concerns about the optimal estimation of the frame background value, because it does not use all of the cell-free area of a frame as a background, which should be the more accurate value. There are also concerns about the minimum fluorescence because in cells with spontaneous activity, it is very hard to estimate the minimum fluorescence by avoiding outliners giving too low an absolute minimum, or photo bleaching artefacts that may suggest fluorescence values lower than they should be. For all those reasons, an automated solution was needed to provide better estimations of background values and minimum fluorescence, and to shorten the analysis time.

Existing work on automatic cellular phenotyping in fluorescence microscopy is rather limited [15]. This is mainly due to the complexity of quantification and characterization of calcium signals through sequences of 2D grayscale images that tend to be of a low quality where cells boundaries are not clearly recognizable, and true signals are highly corrupted by noise. Recent publications investigated various image processing techniques to eliminate the noise effect in fluorescence microscopy images [16–18]. Some publications presented different techniques for fluorescence image segmentation [19, 20], while others conducted several cellular parameters analysis in fluorescence microscopy [21, 22].

One of the first automatic method to analyse calcium signalling through fluorescent images was described in 1999 to detect and measure calcium sparks inside skeletal muscle cells and cardiac myocytes through confocal line scan images based on double thresholding [22]. Later on, the limitations of the pioneering method to deal with highly noisy images led to the development of the two-phase greedy pursuit algorithm (TPGP) for the detection of calcium sparks through confocal single images based on wavelet transforms for noise removal and background subtraction [21]. Thus, various versions of 'home-built' analysis software of calcium signalling were available, but none come with a comprehensive manual, allowing them to be used by the wider calcium signalling community. Further, knowing that fluorescence microscopic images are generally affected by noise of type Gaussian, Poisson, or a mixture of both [15], state-of-art de-noising techniques were proposed based on two main principles: variance stabilization [16, 23] and patch-based techniques [18, 24]. In the former category, the general approach consisted of converting Poisson noise into Gaussian noise, while in patch-based techniques, researchers relied on the measurement of the similarity between small subdivisions of the images i.e. image patches to detect their random degradations prior to any segmentation procedure.

Alongside those techniques, interesting segmentation methods were also made available for vesicle segmentation in fluorescent stack images [19], tubule boundary segmentation in rat kidneys videos [20], and mid-point based nuclei segmentation in fluorescent images [25]. It is also the case for object-independent segmentation techniques such as Fuzzy set logics [26] and hierarchical merge tree [27], presented as alternative solutions to process fluorescence microscopy images. The Jayaprakas's ImageJ plugin [28], a less recent semi-automated method, allows both the manual setting of ROIs and the automatic measurement of their changes of fluorescence through an entire image stack. Data are output in graphics or exported into text or csv files for further analysis. But the plugin does not allow automated ROI segmentation.

To sum up, the literature in the area of intracellular calcium analysis based on fluorescence microscopic videos seems to be rich. However, this paper argues that the existing work on fully automated frameworks that take input videos and produce final measurements such as maximum peak amplitude, area under the curve, etc. of calcium changes over time, is very limited, and not available for use by the wider community.

3 The Proposed Solution for Calcium Signalling Analysis

As stated earlier, the proposed framework is based on four main tasks: pre-process individual video slides to increase their quality, identify regions with intracellular calcium changes (hotspots) by first-level segmentation of the input video frames, and detect calcium hotspots or best active ROIs by second-level segmentation. The final task of the proposed routine is the extraction and measurement of calcium signals in the detected hotspots. Figure 1 below highlights those stages.

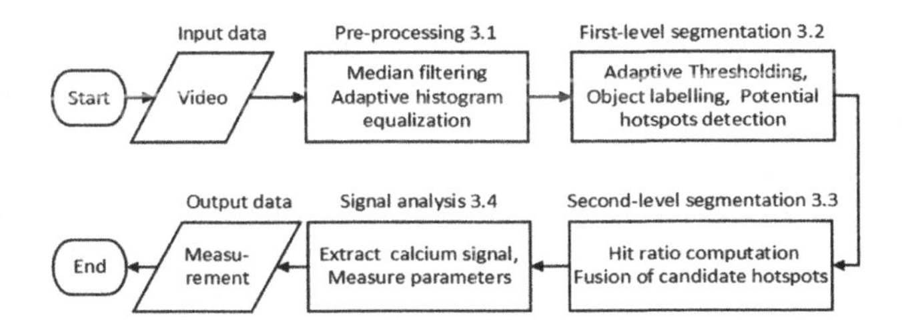

Fig. 1. Flowchart diagram of the proposed framework

3.1 Pre-processing by Median Filtering and Adaptive Histogram Equalization

Fluorescent video frames tend to be low-contrasted and corrupted by Gaussian noise. Therefore, we chose to perform median filtering over individual images to increase their signal to noise ratio. Then, we performed adaptive histogram equalization to enhance their contrast in such a way that changes of intracellular calcium concentration become different from the background pixels. Figures 2a, b, and c below demonstrate the effectiveness of the pre-processing in enhancing the quality of the fluorescent video frames.

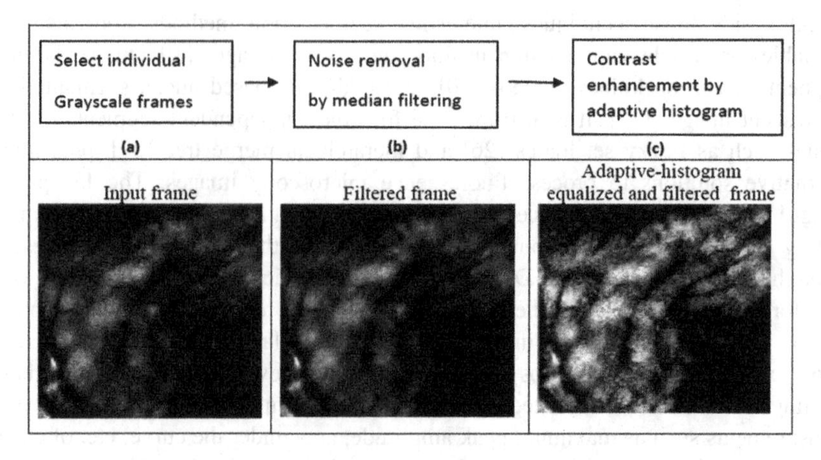

Fig. 2. Image quality increased by pre-processing

3.2 First Level Segmentation for Candidate Hotspots Detection

At this stage, we first aim to detect any changes in the intracellular calcium concentration expressed in the pre-processed video frames as transiently bright areas due to an increase in the fluorescence of the calcium indicator. By default any calcium change area is hotspot. Inside each frame, hotspots are detected based on a certain threshold pixel value T. To set the best T value, analysis of pixel intensity histograms of input frames suggested us that the fluorescence intensities of the hotspots correspond to the bright pixels that fall away from the mean of the frames. The ideal threshold T turned out to be equalling to $mean + 2standarddeviation\,(std)$ of the pixel intensities of individual frames. As a result, from a sequence of grayscale images input, the system generates a list of binary frames, where white pixels belong to objects changing their fluorescence, and dark pixels refer to background elements, that can be either inside or outside the cell. After that, morphological filters were applied on the binary output to transform the obtained connected components into closed-boundary objects in the following order: closing, holes filling, erosion, small objects removal, and image dilation with small structuring element of disk-square shapes. Figure 3 below demonstrates the thresholding approach adopted to detect intracellular calcium changes in individual fluorescent images.

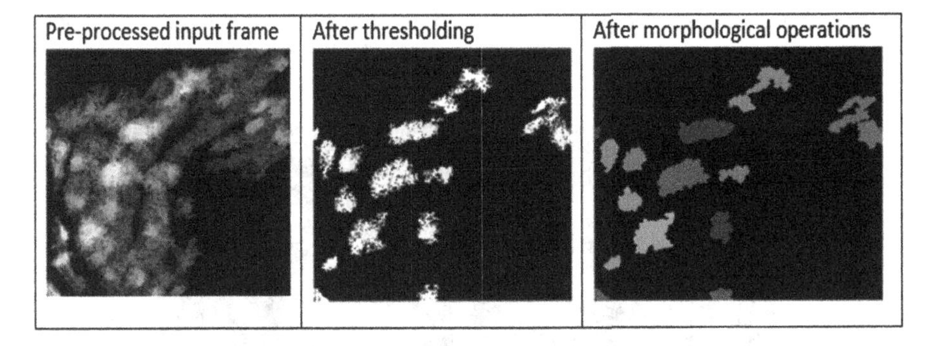

Fig. 3. Segmentation of calcium changes from fluorescent video frames

The first-level segmentation ends with assigning a set of geometrical properties such as centroid coordinates, area or object size, bonding box parameters, pixel element locations, and default hit-ratio value, etc. to each default hotspots representing intracellular calcium changes. From the average size of all detected objects, the system estimates a small significant interval where to look for potential active regions or hotspots. The best interval of size was found out to be between $\left[\frac{1}{4}, \frac{3}{4}\right]$ of the all objects average size, where to look for candidate calcium hotspots.

3.3 Second Level Segmentation for Frames Fusion Based on Hotspots' Hit-Ratio

The second level segmentation stage of our proposal for calcium signal analysis aims to identify the intracellular regions that show the most frequent calcium changes. They correspond to the candidate hotspots with highest hit-ratio parameters. The hit-ratio in this context represents the number of times a given cellular region appears or fluoresces in the entire video. For example, a hit-ratio of 50% means that a change in fluorescence appears in the same place in 50% of the frames. To calculate the hit-ratio of a candidate hotspot localised in a frame, the system counts the number of overlaps of this hotspot with others identified in the remaining frames. Overlapping of two objects is first assumed as an overlap of their bounding boxes (rectangle fitting), then confirmed by the comparison of their pixel elements location, elements whose union make the objects.

Once the hit-ratio computation is done, the system merges potential active regions into a single frame in which hotspots do not overlap. The fusion of candidate hotspots from individual frames is achieved based on two thresholds specified by the end user: the minimum hit-ratio, and the minimum distance between hotspots (we chose >15% and 10 pixels, respectively in this paper). At the end of the second level segmentation, the list of all hotspots is finalised and the system is ready to move on to the final stage of the framework which is to extract the measurements representing calcium changes over time from each hotspot. Figure 4 below presents a screenshot of 41 hotspots detected by the system through the fusion of 7352 default hotspots from 300 frames.

For visual purpose, hotspots are coloured according to their hit-ratio, the hottest ones corresponding to the most active regions, like highlighted in the figure.

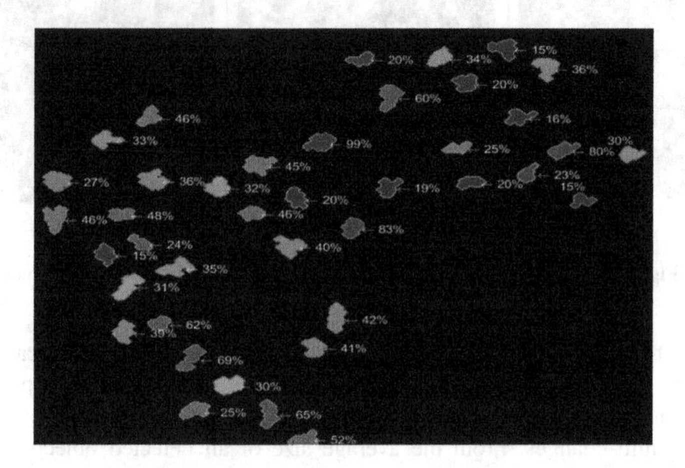

Fig. 4. Best active regions labelled with hit-ratio after video fusion

3.4 Extraction and Measurement of Calcium Signal Inside Hotspots

This final procedure stands for the retrieval of meaningful information from original input video frames. It is the most essential part of the automatic analysis of cellular calcium signalling as it provides biologists with traces illustrating the intracellular calcium changes and numbers to quantitate various parameters to perform statistical analysis for the understanding of the physiological implications of the experiments.

Our proposed software automatically detects active regions inside cells and sets hotspots by direct quantification of the fluorescence within a given area at a given time. These direct fluorescence readouts of a given hotspot correspond to the average intensity of the pixel points whose union makes the hotspot region in every original grayscale frame (mi). From the initial readout of the fluorescence inside hotspotsanother measurement called the corrected fluorescence (F) is computed.$F = mi - bg$, where bg is the background fluorescence in all areas of the frame that do not contain any cells such that $bg = image - imageCells$. The background value represents the average value of anything pixel that is neither part of a calcium-change event, nor a cell tissue component. The optimum parameter to represent an image containing all cells was found out to be $imageCells > mean - std/2$ of a frame's pixel intensities. In the same way, the system computes another calcium signal that is called normalized change of fluorescence ($F/Fmin$), where $Fmin$ corresponds to the minimum fluorescence value inside a hotspot after the background subtraction. $Fmin = imageCells + 1$. The ratio ($F/Fmin$) is the ultimate normalized version of the initial fluorescence readout of hotspots, as it allows effective comparison between different experiments. From the ratio ($F/Fmin$) signal, the system retrieves meaningful parameters about the intracellular calcium changes. For the experiments performed, these are the maximum amplitudes of peak in intracellular calcium changes of active regions, the area under the curve, the frequency of calcium transients, the inter-spike interval between two consecutive peaks of calcium concentration change, and the time of slope from peak to a baseline that is customizable, as illustrated in Fig. 5.

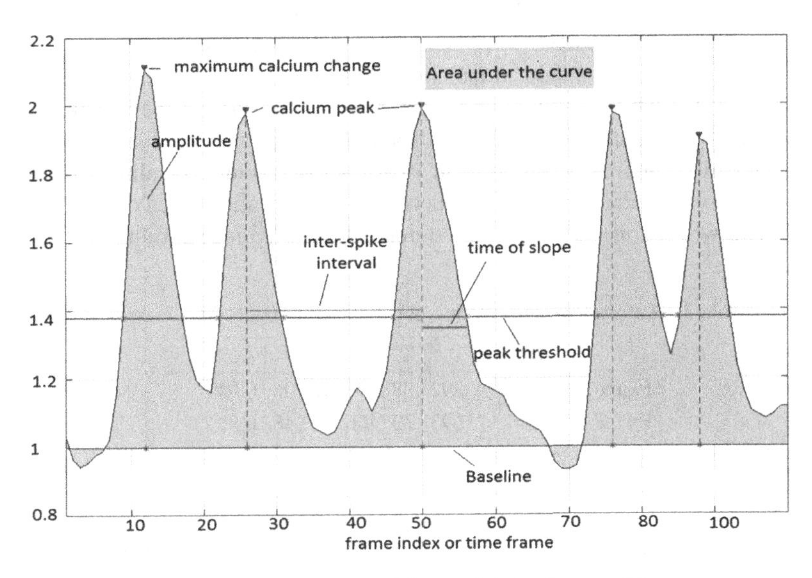

Fig. 5. Some parameters extracted from calcium change of fluorescence

4 Results and Discussions

Our proposal to automate the analysis of calcium imaging experiments based on fluorescence video imaging enables the automatic segmentation and measurement of intracellular calcium transients over time. Active hotspots can be filtered and monitored from portions or from entire input video files. The processes are fully automatic and rely on the video frames contents rather than users' biological knowledge or experience.

In details, for example, Table 1 below shows that the overall processing time of our proposal is proportional to the size of the input video file. Big files tend to take more time but compare to the manual procedure, the system processing time is impressive. Indeed, it takes approximately 1 day for biologists to complete the manual analysis of 1000 video frames, whereas our proposed pipeline does the job in less than 5 min.

Table 1. Results from a PC running Matlab 2014 on windows 7 OS with 4 GB of RAM

Video extension	Number of frames	Number of Ca^{2+} waves	Number of hotspots over potential ones	Processing time (in seconds)
.TIF	900	18,290	25/3,230	251
.MPG	1,194	17,195	20/2,355	303
.MOV	1,200	24,319	42/5,583	355
.MPG	1,499	19,158	15/4,381	390

Although, further testing of our proposed system for hotspots detection in fluorescence microscopy videos is still required, we proved that the automatic segmentation of intracellular calcium concentration in individual frames is about 80% accurate.

Here, the percentage of accuracy refers to the ratio between the sum of True Positive and True Negative (TP + TN) pixels over the sum of all Positives and Negatives pixels [29]. Indeed, when using hand-outlined images of intracellular calcium changes as gold standard to isolate calcium changes, like shown in Fig. 6, we were able to compare the automated segmented frames against the visual detection. Table 2 below illustrates the pixel classification we implemented initially over a random sample of 20 frames per fluorescent video files of calcium signalling experiments. Category Positive corresponds to the successful identification of calcium changes by our system, while True illustrates the reference from the ground truth.

Table 2. Automated segmentation performance through pixel classification

	TP	FP	TN	FN	Accuracy
Frame 1	7386	892	388186	25906	0.94
Frame 2	33423	3693	327406	57848	0.85
...	...	...	...	...	...
Frame 20	12649	1580	392998	15143	0.96

All in one, the comparison between biologist empirical methodology and our automatic approach establishes that our software can provide reliable results in shorter periods of time. The detection of intracellular calcium through individual frames is fast and accurate. The quantification of calcium concentration is easier and straightforward after ROIS detection.

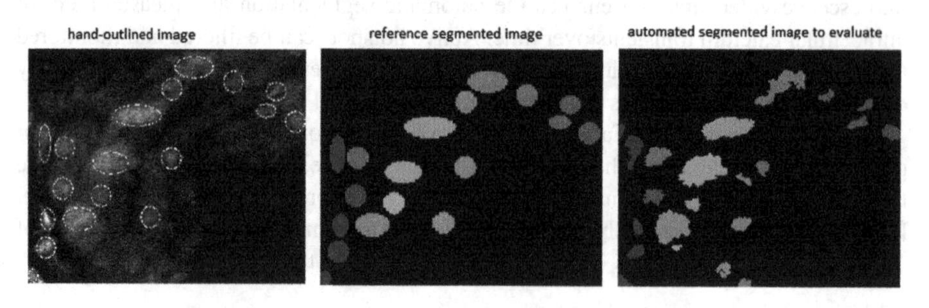

Fig. 6. Manual segmentation versus automated

5 Conclusion and Future Work

In this paper, we describe an automatic solution providing multi-level segmentation for the detection of changes in fluorescence in videos taken when using ion-sensitive fluorescent indicator dyes such as calcium indicators. Our proposal allows a fast and reliable detection of the hotspots to analyse and extract the changes of fluorescence over time. After the extraction of candidate hotspots from individual frames, we propose a new method based on the hit-ratio of fluorescence to identify the best active regions showing changes in the intracellular calcium concentration.

We also propose a normalized quantification of those changes over time for each hotspot. Further, the system enables the extraction of all major parameters researchers are typically interested in when analysing calcium imaging experiments, including some which are very time-consuming to perform manually. We have included the options of manual deleting hotspots which might not fit the criteria that researchers have identified. It is also possible to refine the hotspot detection by setting parameters such as minimum size and hit-ratio of hotspots as well as a minimum distance between them. The paper argues that obtaining the automatic measurements achieves a higher processing time compared with experts' manual ones. Higher accuracy of our proposed method still needs to be proven by further testing. However the automation of the entire procedure makes data analysis reproducible and easy comparable.

Despite those advantages, we acknowledge that our proposed system has got some limitations. Using videos of tissues, we currently cannot identity individual cells. Further, we currently cannot follow hotspots in case their position shift during experiment course, which might happen. It is the case for cardiomyocytes calcium signalling experiments where high contractions may lead cells to move slightly. In addition to those drawbacks, the system is currently designed to analyse data only from single-wavelength fluorescent indicators. Ideally the system should be extended for the use of multi-wave length fluorescent dyes such as the ratio metric indicators like fura-2. We also think the future system should also enable advanced comparison of cells' calcium activity by providing more useful parameters output. Finally, the handling of huge amount of data, and noisier raw images will have to be addressed.

References

1. Berridge, M.J., Lipp, P., Bootman, M.D.: The versatility and universality of calcium signalling. Nat. Rev. Mol. Cell Biol. **1**, 11–21 (2000)
2. Berridge, M.J., Bootman, M.D., Roderick, H.L.: Calcium signalling: dynamics, homeostasis, and remodelling. Nat. Rev. Mol. Cell Biol. **4**, 517–529 (2003)
3. Dumollard, R., Duchen, M., Sardet, C.: Calcium signals and mitochondria at fertilisation. Semin. Cell Dev. Biol. **17**(2), 314–323 (2006)
4. Zhivotovsky, B., Orrenius, S.: Calcium and cell death mechanisms: a perspective from the cell death community. Cell Calcium **50**(3), 211–221 (2011)
5. Berridge, M.J.: Calcium signalling remodelling and disease. Biochem. Soc. Trans. **40**, 297–309 (2012)
6. Cali, T., Ottolini, D., Brini, M.: Calcium signaling in Parkinson's disease. Cell Tissue Res. **357**, 439–454 (2014)
7. Feske, S.: Calcium signalling in lymphocyte activation and disease. Nat. Rev. Immunol. **7**, 690–702 (2007)
8. Bootman, M.D., Rietdorf, K., Collins, T., Walker, S. Sanderson, M.: Ca2+ sensitive fluorescent dyes and intracellular Ca2+ imaging. Cold Spring Harbor Protoc. 83–99 (2013)

9. Tsien, R., Pozzan, T., Rink, T.J.: Calcium homeostasis in intact lymphocytes: cytoplasmic free calcium monitored with a new intracellularly trapped fluorescent indicator. J. Cell Biol. **94**, 325–334 (1982)
10. Rodriguez, E.A., Campbell, R.E., Lin, J.Y., Lin, M.Z., Miyawaki, A., Palmer, A.E., Shu, X., Zhang, J.: The growing and glowing toolbox of fluorescent and photactive proteins. Trends Biochem. Sci. **42**(2), 111–129 (2017)
11. Rietdorf, K., Bootman, M.D., Sanderson, M.J.: Spontaneous, pro-arrhythmic calcium signals disrupt electrical pacing in mouse pulmonary vein sleeve cells. PLoS ONE **9**(2), 1–13 (2014)
12. Rietdorf, K., Masoud, S., McDonald, F., Sanderson, M.J., Bootman, M.D.: Pulmonary vein sleeve cell excitation-contraction-coupling becomes dysynchronized by spontaneous calcium transients. Biochem. Soc. Trans. **43**(3), 410–416 (2015)
13. Khaji, A., Kowey, P.R.: Update on atrial fibrillation. Trends Cardiovasc. Med. **27**(1), 14–25 (2017)
14. Heijman, J., Voigt, N., Nattel, S., Dobrev, D.: Cellular and molecular electrophysiology of atrial fibrillation initiation, maintenance, and progression. Circ. Res. **114**, 1483–1499 (2014)
15. Kervrann, C., Sorzano, C.Ó.S., Acton, S.T., Olivo-Marin, J.C., Unser, M.: A guided tour of selected image processing and analysis methods for fluorescence and electron microscopy. IEEE J Sel. Top. Sig. Process. **10**, 6–30 (2016)
16. Makitalo, M., Foi, A.: optimal inversion of the generalized anscombe transformation for Poisson-Gaussian noise. IEEE Trans. Image Process. **22**, 91–103 (2013)
17. Mustafa, N., Li, J.P., Khan, S.A., Giess, M.: Medical image de-noising schemes using wavelet threshold techniques with various noises. In: International Computer Conference on Wavelet Active Media Technology and Information Processing (ICCWAMTIP) (2015)
18. Tang, Y., Zhang, Y., Chen, Y., Gao, Y., Zhu, C.: Image denoising via expected patch log likelihood with Gaussian model identification. In: International Conference on Consumer Electronics (2016)
19. Thierry, P., Patrick, B., Jerome, B., Anatole, C., Sabine, B., Jean, S., Charles, K.: background fluorescence estimation and vesicle segmentation in live cell imaging with conditional random fields. IEEE Trans. Image Process. **24**, 667–680 (2015)
20. Gadgil, N.J., Salama, P., Dunn, K.W., Delp, E.J.: Jelly filling segmentation of fluorescence microscopy images containing incomplete labeling. In: International Symposium on Biomedical Imaging (ISBI), pp. 531–535 (2016)
21. Kan, C., Yip, K.P., Yang, H.: Two-phase greedy pursuit algorithm for automatic detection and characterization of transient calcium signaling. IEEE J. Biomed. Health Inf. **19**, 687–697 (2015)
22. Cheng, H., Song, L.S., Shirokova, N., González, A., Lakatta, E.G., Ríos, E., Stern, M.D.: Amplitude distribution of calcium sparks in confocal images: theory and studies with an automatic detection method. Biophys. J. **76**, 606–617 (1999)
23. Le Montagner, Y., Angelini, E.D., Olivo-Marin, J.C.: An unbiased risk estimator for image denoising in the presence of mixed Poisson-Gaussian noise. IEEE Trans. Image Process. **23**, 1255–1268 (2014)
24. Luo, E., Chan, S.H., Nguyen, T.Q.: Adaptive image denoising by mixture adaptation. IEEE Trans. Image Process. **25**, 4489–4503 (2016)
25. Gadgil, N.J., Salama, P., Dunn, K.W., Delp, E.J.: Nuclei segmentation of fluorescence microscopy images based on midpoint analysis and marked point process. In: Southwest Symposium on Image Analysis and Interpretation (SSIAI), pp. 37–40 (2016)
26. Li, B.N., Qin, J., Wang, R., Wang, M., Li, X.: Selective level set segmentation using fuzzy region competition. IEEE Access **4**, 4777–4788 (2016)
27. Ting, L., Mojtaba, S., Tolga, T.: Image segmentation using hierarchical merge tree. IEEE Trans. Image Process. **25**(10), 4596–4607 (2016)

28. Jayaprakash, B.: Time Series Analyzer (2007). https://imagej.nih.gov/ij/plugins/time-series.html. Accessed 15 June 2016
29. Gavet, Y., Fernandes, M., Debayle, J., et al.: Dissimilarity criteria and their comparison for quantitative evaluation of image segmentation: application to human retina vessels. Mach. Vis. Appl. **25**(8), 1953–1966 (2014)
30. Carraro, M., Bernardi, P.: Calcium and reactive oxygen species in regulation of themitochondrial permeability transition and of programmed cell deathin yeast. Cell calcium **60**, 102–107 (2016)
31. Huang, S.-Y., Chen, Y.-C., Yu-Hsun, K., Hsieh, M.-H., Chen, Y.-A., Chen, W.-P., Lin, Y.-K., Chen, S.-A., Chen, Y.-J.: Renal failure induces atrial arrhythmogenesis from discrepant electrophysiological remodeling and calcium regulation in pulmonary veins, sinoatrial node, and atria. Int. J. Cardiol. **202**, 846–857 (2016)
32. Treves, S., Jungbluth, H., Voermans, N., Muntoni, F. Zorzato, F.: Ca2+ handling abnormalities in early-onset muscle diseases: novel concepts and perspectives. In: Seminars in Cell & Developmental Biology (2016)
33. Conron, M., Young, C., Beynon, H.: Calcium metabolism in sarcoidosis and its clinical implications. Rheumatology **39**, 707–713 (2000)

Cervical Nuclei Classification: Feature Engineering Versus Deep Belief Network

Christoph Rasche(✉), Ciprian Ţigăneşteanu, Mihai Neghină, and Alina Sultana

Image Processing and Analysis Laboratory, University "Politehnica" of Bucharest, Bucharest, Romania
rasche15@gmail.com

Abstract. A database of 9405 cervical cells is introduced, which was collected from Pap-smear images: 1791 cells are pathologic cases (two types), the rest are healthy cases (three types). Their cell nuclei are classified using two methods: once with a traditional feature engineering approach using in particular iso-contours; and once with a Deep Belief Network made of Restricted Boltzmann Machines. The Deep Belief Network returns higher accuracy, but not in all classification tasks. The retrieval results show that nuclei information alone can be probably sufficient for a computer-assistive diagnosis of Pap-smear images.

Keywords: Cervical cancer · Pap-smear image · Feature extraction · Deep learning

1 Introduction

Cervical tissue is increasingly analyzed using liquid-based cytology (LBC), which offers higher smear quality and higher sensitivity than traditional staining methods such as Pap-smear cytology [11,19]. But the usage of the Pap-smear technique still persists and we therefore focus on an automatic analysis of such images. Our study is most comparable to Su et al.'s study on LBC images [17]. In that study, a set of ca. 20'000 cell patches was collected of which ca. 2'500 are affected. Those were then classified in two stages: in a first stage the cells were described with 41 features and classified into epithelial cells and other cells (or junk); in a second stage, 28 features were used to discriminate between normal and abnormal epithelial cells. In our study - using Pap-smear images -, we collect also ca. 20'000 cell patches, of which 1'791 cells are affected, and we also pursue a two-stage classification. But we use deep learning classification and compare it with feature engineering. Deep learning is convenient because it finds a good model completely automatically. Feature engineering has the theoretical advantage of greater 'structural specificity', but those specific features still need to be found. In this study we show that the use of iso-contours yields very good recognition rates.

Previous classification systems that were developed specifically for Pap-smear images, were trained with a relatively small set of 917 cells, and they pursued a

M. Valdés Hernández and V. González-Castro (Eds.): MIUA 2017, CCIS 723, pp. 874–885, 2017.
DOI: 10.1007/978-3-319-60964-5_76

feature extraction approach that is based on manual cell segmentations provided by a doctor [3,8,12]. Yusoff et al. use 508 cases [20]. Because finding a cell outline is relatively difficult in Pap-smear images, Plissiti and Nikou have focused only on nuclei [12], as they contain the majority of feature information to discriminate between benign and malign cells. Guo et al. also focus on nuclei information [4], but perform classification of the entire tissue and not of single nuclei.

But even finding nuclei in Pap-smear images is challenging. There are several attempts to use image segmentation techniques only [2,6,10,13,18], but they generally suffer from two downsides. One downside is that some of the techniques are extremely slow because they use propagation techniques; a popular example is the level-set method, which starts with an iso-contour and some other image aspects and then gradually evolves to an optimal status [1,7]. Another downside is that many have been developed only for a rather limited set of unaffected nuclei, which are generally small and dark. Affected nuclei however are large and bright, which makes detection rather difficult. In Pap-smear images those algorithms do not return good recall values. But this problem can be alleviated if one uses a classifier discriminating between nucleus and non-nucleus, in which case one can also provide structures to the classifier, that appear like a nucleus and then let the classifier confirm whether a nucleus is indeed present. We will show that in particular a Deep Neural Network is suitable at making this discrimination.

In total, three classifications tasks are pursued in this study:

1. Nucleus/Non-Nucleus Discrimination: a binary classification between nucleus and non-nucleus, as just mentioned. For that purpose ca. 10'000 image patches were collected that contained structures easily mistaken as nuclei.
2. Five-Type Classification: a classification into five different types of nuclei: we have collected three types of benign and two types of malign nuclei, as will be elaborated below.
3. Retrieval: a retrieval task of the individual cell types: that retrieval is supposed to return the most likely affected cases to the doctor (or cytologist), because it cannot be expected that the five-type prediction accuracy (of task no. 2) will be perfect.

In the following, the method section introduces the dataset, the Deep Belief Network and the feature description (Sect. 2). Then, the evaluation section reports the performances of the three types of classification tasks (Sect. 3). The discussion section argues that this 'direct' nuclei classification could be possibly sufficient for a computer-assistive diagnosis (Sect. 4).

2 Methods

2.1 The Cell Database

Our (digital) images of Pap-smear stains are obtained with a scanner by VENTANA iScan Coreo, at an optic zoom of 40. Cells from five subjects were collected and classified according to the Bethesda system (Table 1). Only one cytologist made the classification and the database would therefore profit from verification by other cytologists. Our collection contains two types of pathologic cells and three types of healthy cells. The two pathologic types are ASCUS and LSIL, which correspond to light and moderate progress in cancer development. The three healthy types are the superficial, intermediate and inflamated. The collected patches are of size 301×301, but because we focus on the nucleus only, the examples depicted in Fig. 1 show a zoom-in. A significant proportion of collected nuclei are not isolated but occur as part of overlapping cell agglomerates or other structures.

Table 1. Cell count per type.

Type	Grade	Count	# Subjects
ASCUS	Light	1154	5
LSIL	Moderate	637	5
Superficial	Normal	2693	4
Intermediar	Normal	3648	5
Inflamated	Normal	1273	5
Total Cells		9405	
Non-nuclei	Not appl	10743	5
Total Patches		20148	

In contrast to the Herlev database [8], our set of images does not contain pathologic cells that are in severe or even carcinogen stage. Those two cell types are however easier to discriminate from healthy cells than the cell types ASCUS and LSIL. Our database is therefore suitable for trying to detect the early stages of cancer development, which is the more pressing issue.

To carry out the first classification task - the nucleus/non-nucleus discrimination -, an additional 10743 patches are collected that do not contain a nucleus of interest in the patch center, see bottom row of Fig. 1 for examples.

2.2 Nucleus Description

To detect candidate nuclei we use our previously reported method, whose segmentation duration is shortest and whose detection recall is highest [14]. But as we use a nuclei/non-nuclei classifier, one could also use simpler methods that return only seed points, e.g. at H-minima [15] or at Maximally Stable Extremal Regions [9].

Previous methods of cervical cancer classification in Pap-smear images had gathered only 9 parameters about the nucleus [12]: area, brightness, shortest diameter, longest diameter, elongation, roundness, perimeter and the number of local maxima and minima. We also tried those 9 features, but did not nearly

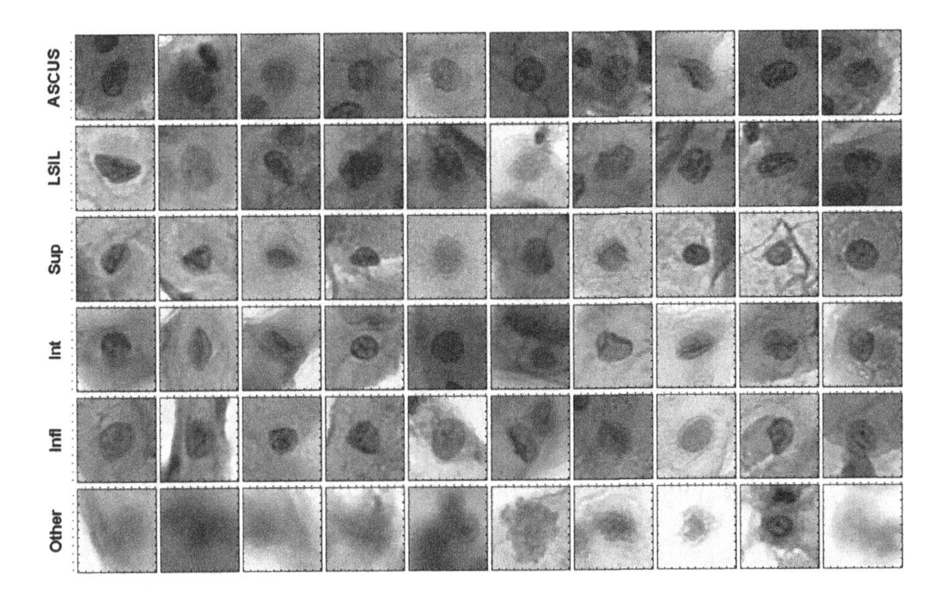

Fig. 1. Examples of collected nuclei. The first two rows are representatives of pathologic cases (ASCUS, LSIL); rows three to five are examples of healthy cells (superficial, intermediar, inflamated). The bottom row (other) shows structures similar to those cell nuclei or other nuclei types. The images show 101×101-pixel patches, but optimal classification with a Deep Belief Network can be achieved with patches of size 61×61 approximately.

achieve the classification accuracy as reported with our more elaborate feature description. Because cytologists observe carefully the 'granularity' of nuclei, we decided to focus on iso-contours for nuclei description (Fig. 2). It is pointed out, that this iso-contour description has nothing to do with active segmentation methods that merely start with an iso-contour: here iso-contours at different levels are taken without any further modification of them.

The first step toward that iso-contour description is to find the nucleus silhouette. For a patch containing a nucleus, the largest iso-contour of that patch is taken as the nucleus silhouette. In Fig. 2, the most outer iso-contour corresponds to that determined nucleus silhouette. As can be seen, this type of silhouette selection is not perfect: some silhouettes contain one or several protrusions. Three types of information are taken from such a localized nucleus: simple appearance, the structure of the silhouette and the structure of the inside.

Simple Appearance: Four attributes are determined: the nucleus area a, its mean intensity i_{mean}, the standard deviation of its pixel intensity values i_{std}, as well as the contrast, the range of intensity values i_{rng}.

Silhouette Structure: To describe the silhouette, the Fourier transform is applied to its radial signature. Studies on shape retrieval showed that this is the most

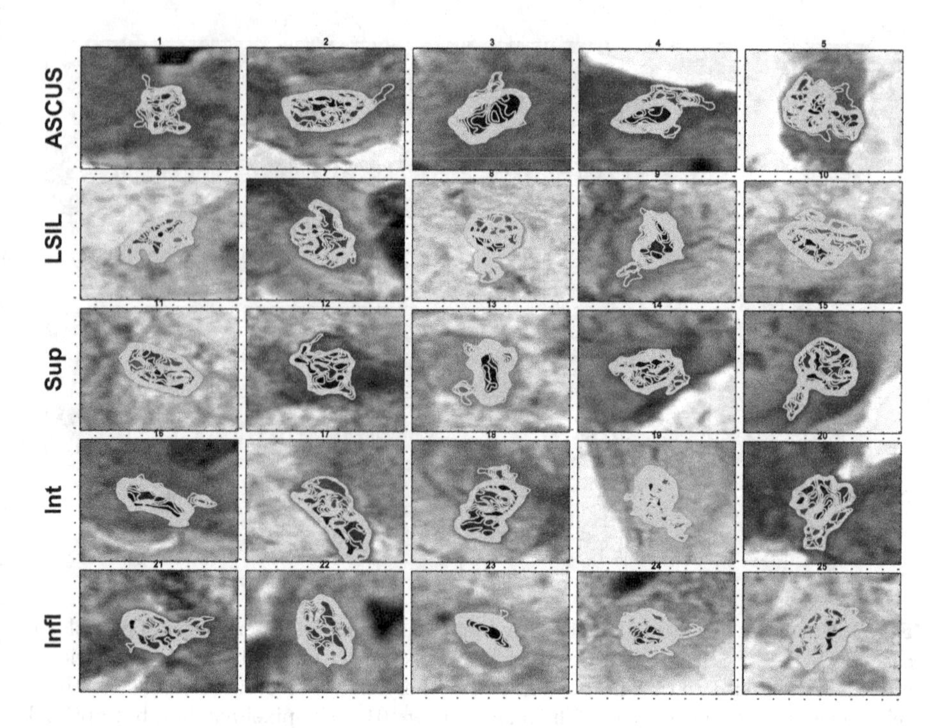

Fig. 2. Examples of iso-contours for nuclei. The silhouette iso-contour is selected by observing a number of geometric conditions. With the iso-contours inside that selected silhouette, a feature description is generated that analyses the sizes and relations between those iso-contours.

efficient shape description with regard to the spatial and temporal complexity [21]: shape descriptions that show a higher retrieval performance use excessive spatial and temporal complexity in comparison. More concretely, for a boundary $B(s)$ with arc length variable s, its radial signature $\rho(s)$ is determined, of which the first four (fast) Fourier descriptors are determined, f_1, f_2, f_3 and f_4. More Fourier descriptors did not improve performance significantly.

Inside Structure: The iso-contours inside the nucleus silhouette are determined, namely at a spacing value equal four, out of an intensity range of [0, 255]. That 'inside' count $\mathsf{n}_{\mathrm{iso}}$ can vary between a few - typical for the small healthy nuclei - to several tens - typical for the bloated nuclei, see Figs. 3 and 4; that count is therefore a parameter as well. For each inside iso-contour, its (average) radius R_i is calculated ($i = 1, .., \mathsf{n}_{\mathrm{iso}}$). To capture some structural aspects, two distributions are analyzed (right column in Figs. 3 and 4). One is the radius distribution, which is the sorted list of iso-contour radii R_i - in decreasing order - whereby the first value corresponds to the average radius of the nucleus silhouette, the last value corresponds to the smallest iso-contour. The other distribution is the luminance level of the iso-contours I_i, whereby the ordering is in accordance with the radius distribution. Those two distributions can be characteristic for

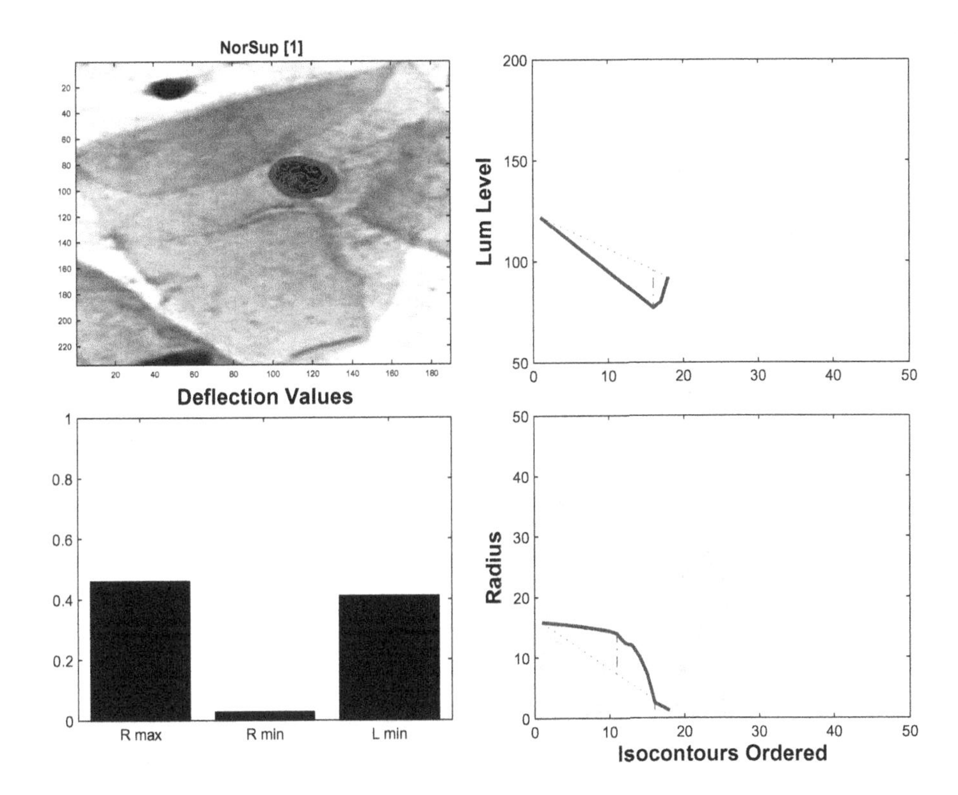

Fig. 3. Iso-contours for a normal nucleus (from the Herlev database). Upper right: Luminance level (intensity) against iso-contours ordered by $R(i)$. The dotted line oblique connects the first and last value linearly. The vertical dashed line indicates the linear decay; the negative deflection is marked by a dashed-dotted line. Lower right: Radii of iso-contours, sorted in decreasing order. Only the positive deflection is clearly visible. Lower left: deflection values δ_R^+, δ_R^- and δ_L^-.

different nuclei types (compare rights columns in Figs. 3 and 4). The goal is therefore to characterize those distributions with a few parameters. This is done with help of the linear decay between the first and last point of the distribution, for which we measure the amount of 'deflection' into the positive and negative range: if the decay occurs 'slower' than the linear decay, then it is a positive deflection; if the decay occurs 'faster' then it is a negative deflection. The amount of deflection corresponds to the maximal distance between the linear decay and the distribution. For the luminance distribution I_i only a negative deflection δ_L^- is possible, because the distribution is ordered according to the radius distribution R_i and the luminance decreases linearly due to sampling. For the radius distribution, both the positive and the negative deflection is determined and normalized by the range of radii, δ_R^+ and δ_R^-, respectively. In addition, we determine the minimum radius $r_{\min} = \min_i R(i)$ and the mean radius $r_{\text{mean}} = 1/\mathsf{n}_{\text{iso}} \sum_i R(i)$.

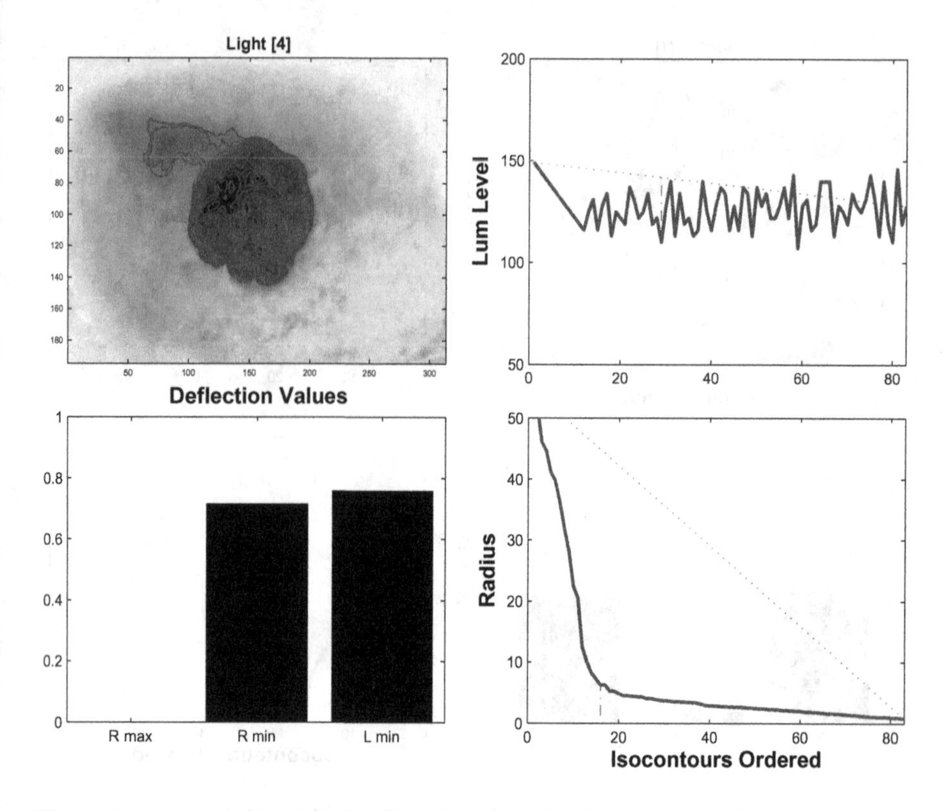

Fig. 4. Iso-contours of a lightly affected nucleus (Herlev database). The inside count n_{iso} for iso-contours is much larger.

In total there are 14 attributes and the following vector **n** is formed:

$$\mathbf{n} = [a, i_{\text{mean}}, i_{\text{std}}, i_{\text{rng}}, f_1, f_2, f_3, f_4, r_{\text{min}}, r_{\text{mean}}, \delta_R^+, \delta_R^-, \delta_L^-, \mathsf{n}_{\text{iso}}]. \quad (1)$$

2.3 Deep Belief Network (DBN)

We use a four-layer Deep Belief Network (DBN) as described in [5], written in Matlab: an input layer that takes the raw image - the pixel values; two hidden layers; and an output layer, the number of classes. This type of network has been used to classify a number databases and achieved competing prediction accuracies amongst Deep Network approaches, often with only two hidden layers (see [5] for details). The unit count of the input layer is now denoted as n_{inp} and will be $61 \times 61 = 3'721$ pixels for instance if it is a gray scale image. We made the general observation that the nuclei classification accuracy is maximal, if the first hidden layer has a unit count that is three to four times as large as the input layer n_{inp}; the second hidden layer has a unit count that is approximately one tenth of the second layer: $n_{inp} \times 4n_{inp} \times \frac{4}{10}n_{inp} \times n_{classes}$. Choosing more layers did not result in any performance gain. The batch size is 50; the learning rate is 0.2; the number of iterations 100 to 300; the other parameters are set as pointed out in [5]. Interestingly however, the method of drop out - as introduced in [16] for instance - did not improve but rather deteriorate performance and was therefore never used.

3 Evaluation

Figure 5 summarizes the classification performances for the three tasks, row-wise. In general we used five-fold cross-validation, but have not always used all five folds due to the slow learning rate for DBNs. For task no. 1, the binary classification, we report a ROC analysis. For task no. 2, the five-type classification, we report percentages and the confusion matrix. For task no. 3, we report retrieval measures. In more detail:

Task 1: Nucleus/Non-nucleus Discrimination. In this task one discriminates between the 9405 nucleus patches and the 10743 non-nucleus patches. Figure 5 upper left displays the ROC curve for one of the five folds as obtained with a DBN; the upper right displays the average area-under-curve (AUC) value, for gray-scale and color patches. Classification accuracies for iso-features are not shown because they are significantly lower. Those AUC values are obtained with a patch size of 61×61 pixels, for which the network's unit count is $3'721 \times 12k \times 2k \times 2$ for gray-scale patches, and $11'163 \times 30k \times 3k \times 2$ for color patches. For smaller and larger patch sizes, 51×51 pixels or 81×81 pixels, the AUC values were significantly lower (not shown).

The discrimination task was also tried with a sub-selection of the original patch, namely with its four principal cross-sections of an image: its two diagonals as well as its horizontal and its vertical cross-section that run through the center pixel of the image. For instance, for the original patch size (301×301), the input dimensionality would be 1204 for a gray-scale patch. In general, such sub-selections achieved the same prediction accuracy as for a DBN trained with a 61×61 pixel patch size, but only for cross-sections of larger patch sizes, namely for 201×201 pixel patches: in case of gray-scale patches a network unit count of $804 \times 3k \times 300 \times 2$ is used, in case of color the count is $2'712 \times 9k \times 900 \times 2$.

Task 2: Five-Type Classification. Now the five cell types are discriminated (ASCUS, LSIL, etc.). For the DBN, the architecture and its parameters remain the same as for task no. 1. To classify the 14-dimensional feature description vectors $\mathbf{n}$, a one-versus-all Support Vector Machine (SVM) is employed; the kernel function is a Gaussian function. Again, the DBN achieves higher prediction accuracy, but the feature classification also shows respectable prediction values (center left in Fig. 5). Color information again improves accuracy and does so more than in task no. 1. The confusion matrix for both classifiers looks very similar and only the one for the DBN is shown (Fig. 5, center right).

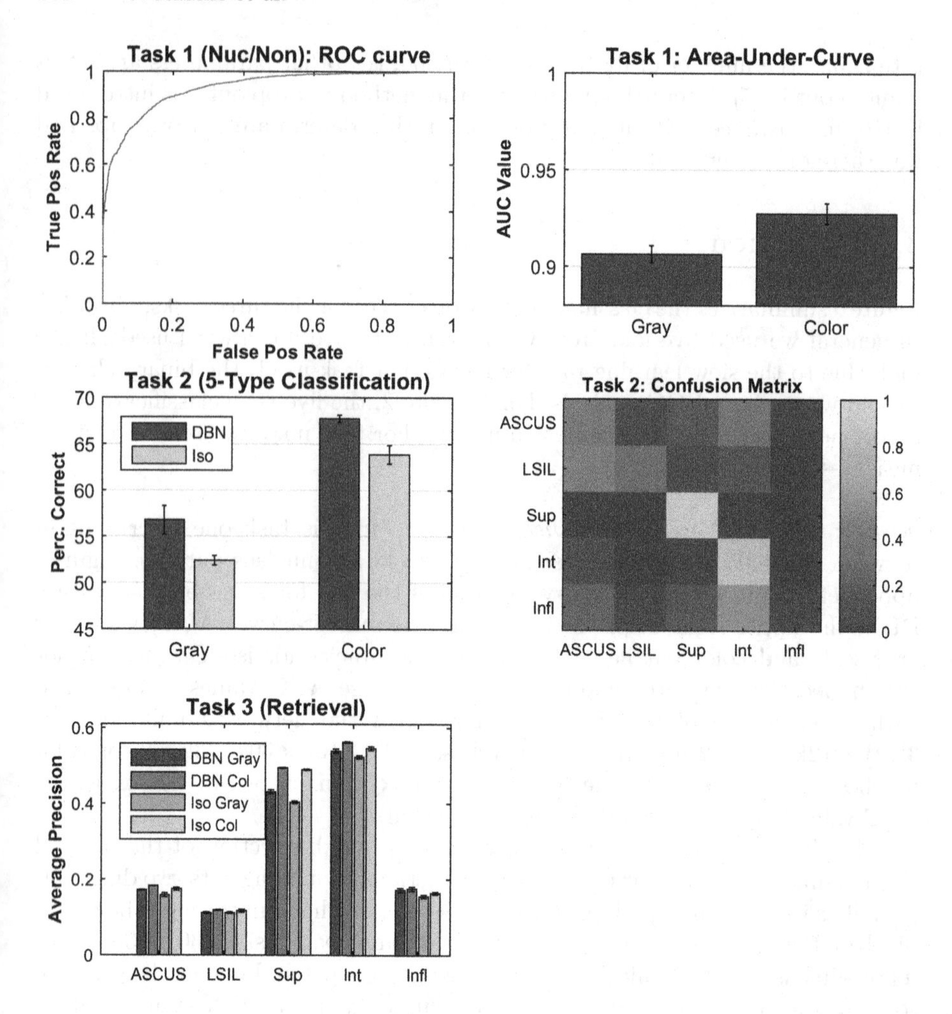

Fig. 5. Classification performances for the three tasks, based on five-fold cross-validation. Upper left: Task 1 (nucleus/non-nucleus discrimination): ROC curve for one fold for color patches (AUC value = 0.92). Upper right: Task 1: mean area-under-curve value, for both gray-scale and color patches; error-bars represent standard error of the five-fold cross-validation. Center left: Task 2 (Five-Type Classification): DBN: classification with the Deep Belief Network; Iso: feature description based on iso-contours, classified with a Support Vector Machine. Center right: Task 2: confusion matrix for the DBN for one fold. ASCUS: light cancer; LSIL: moderate cancer; see Table 1 for detailed labeling. Lower left: Task 3 (Retrieval): retrieval using the posterior values of the classifiers; average precision is the area under the precision-recall curve.

Task 3: Retrieval. A retrieval procedure is typically carried out with some sort of similarity measurement between pairs of items (i and j). If we intended to apply this principle to our case, this meant to either find a similarity measure for two pixel patches, or for its two vectors $\mathbf{n}_i$ and $\mathbf{n}_j$. As this is unlikely to produce best results, we directly exploit the power of classification algorithms: we use the exact same classifiers as above, but use the posterior values for retrieval, instead of choosing the class with maximal posterior value. More explicitly, instead of making a classification decision with the five posterior values per sample (p_m, $m = 1, .., \mathsf{n}_{\mathsf{classes}}$), the posterior values for *all* samples of *one* nucleus type are sorted in decreasing order (p_n, $n = 1, .., \mathsf{n}_{\mathsf{patches}}$); from that sorting one creates a precision-recall curve. As a performance measure, we take the commonly used area-under-curve (AUC) value, see lower left in Fig. 5.

The average precision value does however not tell us, if there are any hits amongst the first few retrievals. The retrieval of affected cases among the first few selected items is however important, because otherwise the doctor might as well analyze the entire image directly. We therefore verified for cancer types ASCUS and LSIL, that under the first 10 retrieval items at least one was present. This was always the case.

4 Discussion

The retrieval results with the first 10 items show that even at this stage of the development, a doctor can be aided in his analysis. A Pap-smear image can contain up to hundreds of thousands of cells - of which only few are affected; it can take a cytologist an hour to spot any of those affected cells. In that real-case scenario, our system's retrieval success might be ten times worse, meaning only one affected cell would be found under the first one hundred retrieved items: but even that would still help the cytologist to find malign cases much faster than by visual search of the entire image. This potential speed-up needs to be confirmed on novel cases and not on those that were used for training the system. In order to avoid the loss of nuclei during the nucleus/non-nucleus discrimination stage, one can bias the decision to show a higher recall at the cost of lower precision, a bias shift which should not deteriorate the retrieval performance substantially.

In this study, Deep Belief Networks have achieved a higher classification accuracy than the feature engineering approach. However for retrieval, the performance differences become almost marginal. In fact, for LSIL cases, the retrieval performance with iso-contours is the same as with a DBN. The feature engineering approach should therefore be continued and this study showed that the use of iso-contours is certainly promising, but whether such a description is also useful for liquid-base cytology needs to be tested. But one could also try other topological descriptions such as ridges and rivers, which have also been rarely used.

That the DBN is much better at the nucleus/non-nucleus task than the iso-contour description, can be explained by the circumstance that this is a discrimination between a target object and several distracter objects; in other words it is a discrimination between one class and multiple classes of very different structure, a task at which Deep Neural Networks have excelled by far the best to date. For the five-type classification however, there are only subtle structural differences between cell types and that is where feature engineering shows its greatest promise. In particular for the discrimination between LSIL and healthy types - for which there exist many confusions - one could build a more specific feature extraction process or even build a separate classifier.

For the nucleus/non-nucleus classification task (task no. 1), the optimal patch size for the DBN appeared to be just about the size of the nucleus, namely 61 × 61 pixels. For larger patch sizes the prediction accuracy decreased, which shows that the immediate surrounding cytoplasm probably does not carry information for a discrimination between cell types.

Acknowledgment. This work was fully supported by the Joint Applied Research Projects "Intelligent System for Automatic Assistance of Cervical Cancer Diagnosis", grant number: PN-II-PT-PCCA-2013-4-0202, funded by Executive Unit for Higher Education, Research, Development and Innovation Funding (UEFISCDI).

References

1. Cheng, J., Rajapakse, J.C.: Segmentation of clustered nuclei with shape markers and marking function. IEEE Trans. Biomed. Eng. **56**(3), 741–748 (2009)
2. Duanggate, C., Uyyanonvara, B., Koanantakul, T.: A review of image analysis and pattern classification techniques for automatic pap smear screening process. In: The 2008 International Conference on Embedded Systems and Intelligent Technology, 27–29 February 2008
3. Gençtav, A., Aksoy, S., Önder, S.: Unsupervised segmentation and classification of cervical cell images. Pattern Recogn. **45**(12), 4151–4168 (2012)
4. Guo, P., Banerjee, K., Stanley, R.J., Long, R., Antani, S., Thoma, G., Zuna, R., Frazier, S.R., Moss, R.H., Stoecker, W.V.: Nuclei-based features for uterine cervical cancer histology image analysis with fusion-based classification. IEEE J. Biomed. Health Inform. **20**(6), 1595–1607 (2016)
5. Hinton, G.E.: A practical guide to training restricted Boltzmann machines. In: Montavon, G., Orr, G.B., Müller, K.-R. (eds.) Neural Networks: Tricks of the Trade. LNCS, vol. 7700, 2nd edn, pp. 599–619. Springer, Heidelberg (2012). doi:10.1007/978-3-642-35289-8_32
6. Li, K., Lu, Z., Liu, W., Yin, J.: Cytoplasm and nucleus segmentation in cervical smear images using radiating GVF snake. Pattern Recogn. **45**(4), 1255–1264 (2012)
7. Lu, Z., Carneiro, G., Bradley, A.P.: Automated nucleus and cytoplasm segmentation of overlapping cervical cells. In: Mori, K., Sakuma, I., Sato, Y., Barillot, C., Navab, N. (eds.) MICCAI 2013. LNCS, vol. 8149, pp. 452–460. Springer, Heidelberg (2013). doi:10.1007/978-3-642-40811-3_57
8. Marinakis, Y., Dounias, G., Jantzen, J.: Pap smear diagnosis using a hybrid intelligent scheme focusing on genetic algorithm based feature selection and nearest neighbor classification. Comput. Biol. Med. **39**(1), 69–78 (2009)
9. Matas, J., Chum, O., Urban, M., Pajdla, T.: Robust wide-baseline stereo from maximally stable extremal regions. Image Vis. Comput. **22**(10), 761–767 (2004)
10. Moshavegh, R., Bejnordi, B.E., Mehnert, A., Sujathan, K., Malm, P., Bengtsson, E.: Automated segmentation of free-lying cell nuclei in pap smears for malignancy-associated change analysis. In: 2012 Annual International Conference of the IEEE Engineering in Medicine and Biology Society (EMBC), pp. 5372–5375. IEEE (2012)
11. Nance, K.V.: Evolution of pap testing at a community hospital-a ten year experience. Diagn. Cytopathol. **35**(3), 148–153 (2007)

12. Plissiti, M.E., Nikou, C.: Cervical cell classification based exclusively on nucleus features. In: Campilho, A., Kamel, M. (eds.) ICIAR 2012. LNCS, vol. 7325, pp. 483–490. Springer, Heidelberg (2012). doi:10.1007/978-3-642-31298-4_57
13. Plissiti, M.E., Nikou, C., Charchanti, A.: Combining shape, texture and intensity features for cell nuclei extraction in pap smear images. Pattern Recogn. Lett. **32**(6), 838–853 (2011)
14. Rasche, C., Oprisescu, S., Sultana, A., Radulescu, T.: Analysis of pap smear images with ISO- and edge-contours. In: IEEE 11th International Conference on Intelligent Computer Communication and Processing, Cluj-Napoca, RO, pp. 375–378 (2015)
15. Soille, P.: Morphological Image Analysis: Principles and Applications. Springer, Heidelberg (1999)
16. Srivastava, N., Hinton, G., Krizhevsky, A., Sutskever, I., Salakhutdinov, R.: Dropout: a simple way to prevent neural networks from overfitting. J. Mach. Learn. Res. **15**(1), 1929–1958 (2014)
17. Su, J., Xu, X., He, Y., Song, J.: Automatic detection of cervical cancer cells by a two-level cascade classification system. Anal. Cell. Pathol. **2016** (2016). Article no. 9535027
18. Sulaiman, S.N., Isa, M., Ashidi, N., Othman, N.H.: Semi-automated pseudo colour features extraction technique for cervical cancer's pap smear images. Int. J. Knowl.-Based Intell. Eng. Syst. **15**(3), 131–143 (2011)
19. Weintraub, J., Morabia, A.: Efficacy of a liquid-based thin layer method for cervical cancer screening in a population with a low incidence of cervical cancer. Diagn. Cytopathol. **22**(1), 52–59 (2000)
20. Yusoff, I.A., Isa, N.A.M., Othman, N.H., Sulaiman, S.N., Jusman, Y.: Performance of neural network architectures: cascaded MLP versus extreme learning machine on cervical cell image classification. In: 2010 10th International Conference on Information Sciences Signal Processing and their Applications (ISSPA), pp. 308–311. IEEE (2010)
21. Zhang, D., Lu, G.: Study and evaluation of different fourier methods for image retrieval. Image Vis. Comput. **23**, 33–49 (2005)

A New Method of Surgical Tracking System Based on Fiducial Marker

Shuaiyifan Ma and Zijian Zhao(✉)

School of Control Science and Engineering, Shandong University, Jinan, Shandong, China
austinma29@foxmail.com, zhaozijian@sdu.edu.com

Abstract. Now used commercial surgical tracking devicessuch as near-infrared optical tracking system is ofcomplex system, non-multi-target tracking and weak anti-interference ability. A low-cost and accurate surgical tracking system based on the artificial markers was proposed in our study for the real-time tracking of surgical tools. Bumblebee2 binocular vision camera system is used to capturing pictures for stereo vision process. We adopt ARTag fiducial marker system in our tracking system. A robust and fast ARTag detecting algorithm is proposed and this edge based approach is proved to be fast and robust with the conducted experiments. A relative algorithm for the registration of surgical tools with the test boarder is also proposed to realize and test the tracking system. Experiments are conducted to test the performance of the tracking system. We judge the tracking accuracy by the error of the distance between two near corner points, and the tracking stability by the location error of the same corner point with different tool postures. Results show that the mean error of the distance between two adjacent corners points is 0.40 mm, mean error with stable tip of different marker rotation is 0.12 mm, which is comparable to the common used optical tracking system.

Keywords: Surgery navigator · Binocular stereo vision · Spatial localization

1 Introduction

Surgical navigation system is the combination product of computer technology, stereotactic technique and image processing technology. It's now widely used in neurosurgery, orthopedic and ENT surgery applications [1–3]. Accurate and reliable tracking system is frequently used in surgical navigation system to perform robust and fast [3–9]. Navigation system based on optical is now widely researched. Reliable tracking system by detecting the targets assembled on the surgical tools [10, 11] has been proposed and researched for a long time. However, now used optical surgical navigation system usually adopts near-infrared tracking system [12, 13]. It is limited because the tracking devices are expensive and complex. For example, Chinese hospitals usually adopt the Stealth Station or Medtronic tracking system which is expensive and complex, blocking the development and employment of the surgical navigation system in China. Therefore, surgical navigation system need to more simple and low-cost to satisfy the situation of these developing countries. This development

M. Valdés Hernández and V. González-Castro (Eds.): MIUA 2017, CCIS 723, pp. 886–896, 2017.
DOI: 10.1007/978-3-319-60964-5_77

will help most of the hospitals in developing countries like China can purchase the new technology and take advantage of the system in surgeries. This is why we make efforts to find another easier solution of tracking system.

The Micron Tracker system [12] is a widely used simple surgical tracking system based on fiducial markers. As Fig. 1 shows, this system only consists of a binocular vision system and fiducial markers. By tracking the 3D position of two fiducial markers set on the surgical tools, the 3D position of the surgical tools can be calculated. The standard fiducial marker-object of Micron Tracker system consists of two black-white contrast patterns placed on a white background.

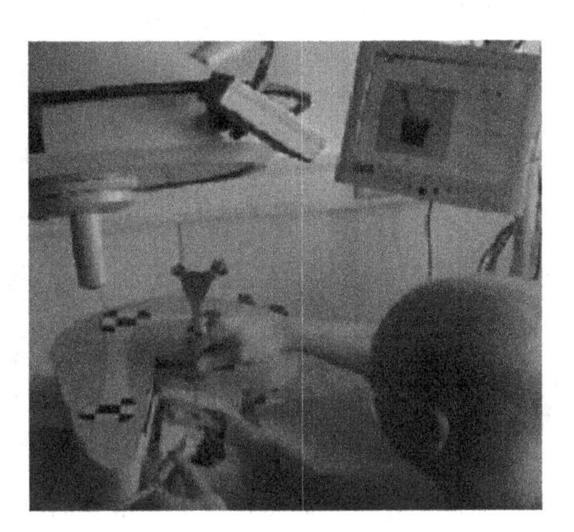

Fig. 1. Navigation system based of micron tracker

Though Micron Tracker system has a good performance on accuracy and speed, it has two drawbacks. The first is that it is sensitive to illumination, the second is that it failed to track two or more surgical tools because the fiducial marker. In this paper, we propose a fast tracking system method for surgical tools. Compared with the Micron System, we replace the micron marker system with ARTag marker system. The proposed method using ARTag marker system [14, 15] is designed to be robust against changing illumination conditions and occlusion of markers.

2 System

Three main components of the tracking system are: binocular camera system Bunblebee2 from Point Grey Research Inc.; computer and software; surgical tools with ARTag marker assembled on the surface. Binocular camera system acquires images, computer software extract the markers from the binocular images and get the marker IDs and 3D coordinate. Based on the process, the real-time tracking of the tools is achieved by transforming the image coordinate system to the world coordinate system. Detecting algorithm of ARTag markers is the crucial component of the whole tracking system. We proposed a detecting algorithm which is proved to be robust and fast with high accuracy.

2.1 Hardware

The tracking system based of ARTag marker system is shown as Fig. 2. The system consists of a Bumblebee2 binocular cam-eras with a resolution of 1024 × 768 and the frame per second value of the bunblebee2 is up to 20 fps. An ARTag marker assembled on the tool. A 1394 fire line is used to acquire image data from the Bumblebee2 and a computer driver is provided for secondary development.

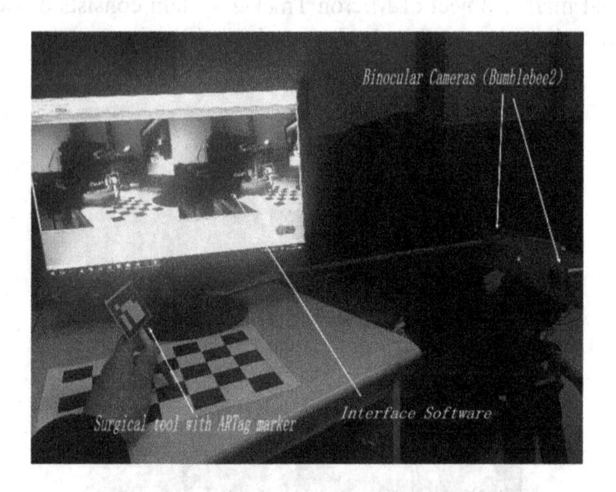

Fig. 2. Tracking system based on ARTag Marker System

As shown in Fig. 3. ARTag [16, 17] uses markers with a square border (black or white). But the interior region of an ARTag marker is filled with a 6 × 6 grid of black or white cells, representing 36 binaries '0' or '1' symbols. This 36-bit word is then processed in the digital domain. Furthermore, ARTag's id based markers do not require image matching with a library and therefore allow for a much faster identification.

Fig. 3. ARTag marker

The measurement range of the system and the accuracies of the measurements within this range must be determined when designing the optical system. And the geometric parameters like focal length of the two cameras, baseline length between two cameras, field angle and the image center are provided by the SDK of Bumblebee2. As well, Bumblebee2 also supplies rectified binocular images for more accurate 3D calculation.

2.2 Software

Figure 4 demonstrates the flow chart of the algorithm. The supplied SDK of Bumblebee2 enable us to acquire binocular images with preprocessing. We take advantage of the SDK to get processed images and then use our marker detecting algorithm proposed for the ARTag markers to find these ARTag markers in the left and right images. The next step is to recode the ARTag markers and match the marker pair and calculate the 3D coordinates of the corners of the markers. At last, get the 3D coordinate of tool tips.

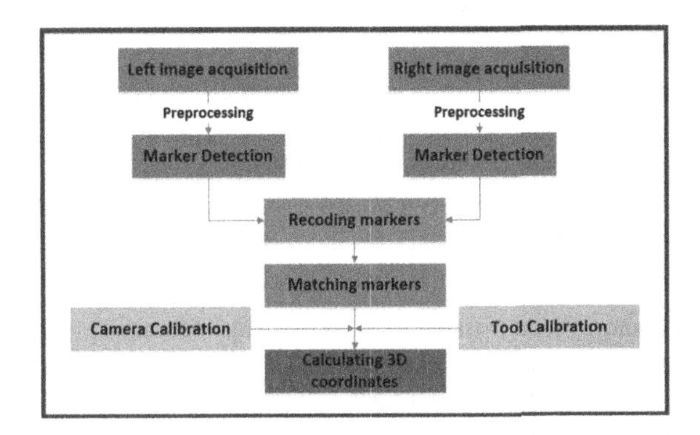

Fig. 4. System algorithm flow chart

Four main algorithms are integrated in the software of the system including algorithms for the acquirement of the left and right images, detecting the ARTag markers both in the left and right images, recoding the ARTag marker IDs, and 3D coordinate calculation. Before the real-time tracking of the surgical tools, these surgical tools need to be registered mainly by calculating the transformation between the ARTag marker and the tip of the tool as their relationship is stable.

During the tracking process, the image acquisition and preprocessing regarding the marker corner points on the surgical tools are performed by the left and right cameras, respectively. The system algorithm automatically searches all marker corner points, calculates their sub-pixel coordinates, decodes the marker ID, matches the markers from left and right images, performs the stereo matching of these points, and calculating the 3D coordinate of the tip of the surgical in the world coordinate system.

Marker Detecting

ARTag finds markers with an edge based approach. Edge pixels found by an edge detector. They are linked into segments, which in turn are grouped into quadrangles. The algorithm consists of three main steps. First, camera frames are divided by coarse grid and line segments are found by linking the detected edge pixels. And line segments are merged to obtain longer lines. Then, all detected lines are extended based on the gradient information, and get the lines of full length. At last, detected full lines are grouped into quadrangles as possible markers.

Line Detection
Camera frames are divided into small square regions by sampling grid. Local maxima of the intensity gradient are considered edge pixels, and the orientation of each edge pixel is calculated.

$$\theta = \tan^{-1}\left(\frac{\partial y}{\partial x}\right) \tag{1}$$

∂x, ∂y: x, y component of the gradient. Then RANSAC-grouper method is used to find straight line segments based on the edge pixels. Then two segments are merged if they meet two criteria as described below:

$$|\theta_a - \theta_b| \in \Delta_\theta, \quad \theta_{ab} \cong \theta_a \cong \theta_b, \quad L_{ab} \in \Delta_l \tag{2}$$

$\theta_a \theta_b$: Orientation of segment a and b, Δ_θ: orientation difference threshold, θ_{ab}, L_{ab} is the orientation and length of the connection line ab, Δ_l is the length threshold of connection line ab.

Line Extension
Usually there is a short piece missing at both ends at lines, thus we need to extend all lines to get the full length. Along the orientation of the line, we check the pixel near the end points. If it is compatible with the line's orientation, the point is added to the line. Then the next point is examined. At last, we get the extended lines and their end points. This step is conducted as Fig. 5 demonstrates.

the point of the merged line
the point passing the extension judgment
the point failing the extension judgment

Fig. 5. Extension judgment

Line Grouping
We obtain quadrangles by finding corner points by intersecting lines. Picking one line out of the set and find a corner by find intersecting line. After testing the whole lines, we get corner sequences and construct the possible markers.

Marker Decode
We need to get the marker ID by decoding the possible markers. There are three steps to finish this procedure. First, we should extract the possible markers to normal markers as it varies because of the warp perspective. As shown in Fig. 6.

The second is to get the binary image with Otsu operator. The last step is to decode the marker ID. As shown in Fig. 7, one marker possesses four possible direction pattern, thus we need to find the right possible pattern by calculating the Hamming distance. The one which possess the smallest value of Hamming distance is the right marker.

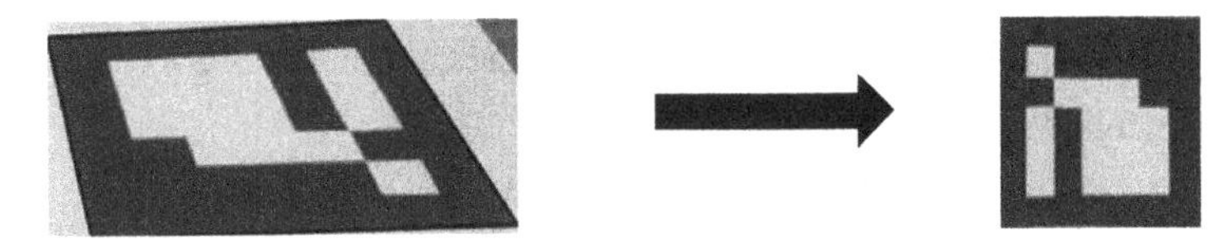

Fig. 6. Perspective transformation

Fig. 7. Four possible markers

Registration of Surgical Tools

Many methods can be used to acquire the position of thetip in the registration process of surgical tools [10–13]. We adopt a new method for the fast calculation of the coordinates of the tips in the registration process. In our registration, the location from the tip of the surgical tool to the world working coordinate system is fixed, and the surgical tool rotates around the tip. During the registration, the whole surgical tool including the ARTag marker and the tip must be in working area of the binocular camera system.

The coordinates of the tip Q should meet the following relation:

$$P_Q = R_Q \cdot P_Q + T_Q \tag{3}$$

Where R_Q and T_Q are the rotation matrix and translational matrix of the transformation from position U to position V in the surgical tool, respectively.

The coordinates of the tip Q can be obtained by transforming Eq. (3):

$$P_Q = (I - R_Q)^{-1} \cdot T_Q \tag{4}$$

Suppose that $R_Q = I - R_Q$, the angle of revolving to a specific coordinate axis may be zero in coordinate transformation. The coordinates of P_Q cannot be acquired by Eq. (4) because it makes R_Q in Eq. (4) non-full rank. To avoid this and improve the accuracy of the calculation, the surgical tool can be rotated to several positions. Suppose that the surgical tool rotates to N positions (N > 1), a respective calculation is made to acquire the rotation matrix R_{Qi} and the translational matrix T_{Qi} (i = 1, 2, 3… N−1) of two adjacent positions such that equation: $R_q^0 = \left[R_{q1}^{'T}, R_{q2}^{'T}, \ldots, R_{qn}^{'T}\right]^T$ and

$R_q^0 = \left[R_{q1}^{\prime T}, R_{q2}^{\prime T}, \ldots, R_{qn}^{\prime T}\right]^T$ can be obtained. Thus, the coordinates of P_Q can be obtained through the following equation:

$$P_Q = (R_q^{0^T} \cdot R_q^0)^{-1} \cdot R_q^{0^T} T_q^0 \tag{5}$$

Given that the positions of all light-emitting points of the surgical tool have a fixated rigidity relationship, the relative positions of the light-emitting points and the tip Q remain stable. Furthermore, the light-emitting points and the tip Q meet the same coordinate transformation. Therefore, the values of R_Q and T_Q can be obtained by the light-emitting points.

3 Experiments

In surgeries, navigation system help to supply the 3D coordinates of the surgical tools relative to the surgical area. Thus, the accuracy and real-time ability is the key parameters of the tracking system. Before the surgery, surgical tools need to be registered, to get the relative parameters for the real-time calculation of the tool tips in surgeries.

3.1 Accuracy and Reliability Regarding the Marker Corner Points

We judge the accuracy of the tracking system by calculating the corner points of the detected ARTag markers. One ARTag marker possess four corner points, the real distance between two near corner points is designed to be 40 mm, to reduce measurement error, the mean of 10 measurements with a Vernier caliper for each distance is taken as the final distance.

The calculated distances between the corners points of ARTag marker on the surgical tools can be taken as the standard for testing the accuracy of the tracking system designed in this paper, the distances between the corners points are calculated by acquiring the images of all surgical tools when they move. The calculated distances are then compared with the standard distances in Table 1 to determine the accuracy. As shown in Fig. 8, we take 30 sets of images with good light environment from different sides to be processed.

As shown in Table 1, the maximum mean square error for the distances between the corners points on the surgical tools obtained is 0.27 mm, whereas all other mean square errors are less than 0.27 mm, which are less than that of com-mon measurement equipment (0.35 mm) currently in use. All mean errors are less than 0.15 mm which meet the requirement of the system.

3.2 Precision Test of the Tracking System

We will test the tracking ability from two respects. One is the tracking stability, stationary tip with different rotation of the ARTag marker. The tool tip is on the same

Table 1. Distance between four corners points

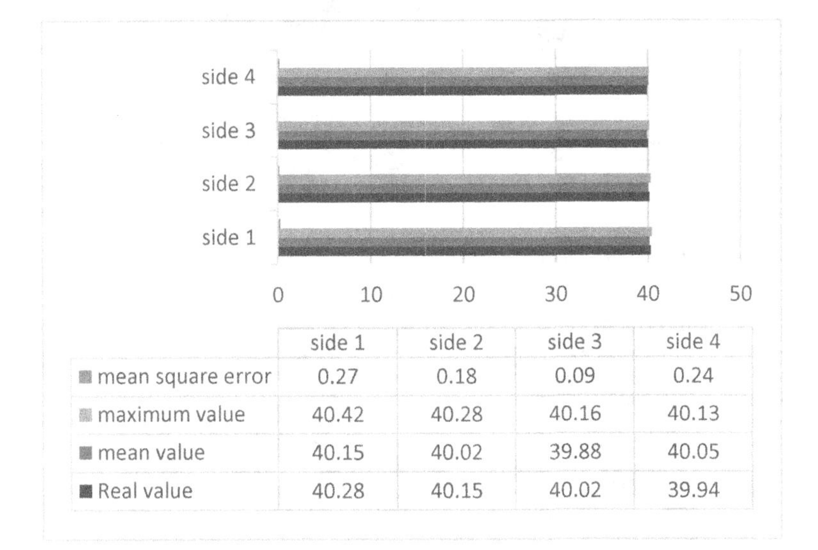

	side 1	side 2	side 3	side 4
mean square error	0.27	0.18	0.09	0.24
maximum value	40.42	40.28	40.16	40.13
mean value	40.15	40.02	39.88	40.05
Real value	40.28	40.15	40.02	39.94

Fig. 8. ARTag markers detecting

corner point and we calculate the location error while rotating the marker to different posture. Another is the tracking accuracy, tip on different corner points and we calculate the location error of the distance between the near two corner points.

Tracking Stability

Demonstrating as Fig. 9, we use the surgical tool stick on the same corner point and calculatingthe tip coordinate while rotating the tool with the marker different posture. A set of 20 rotation data is acquired. The location error data is demonstrated in Fig. 10.

Among the error data, the mean error is 0.1194 mm, maximum error is 0.3165 mm and mean square error is 0.1742 mm.

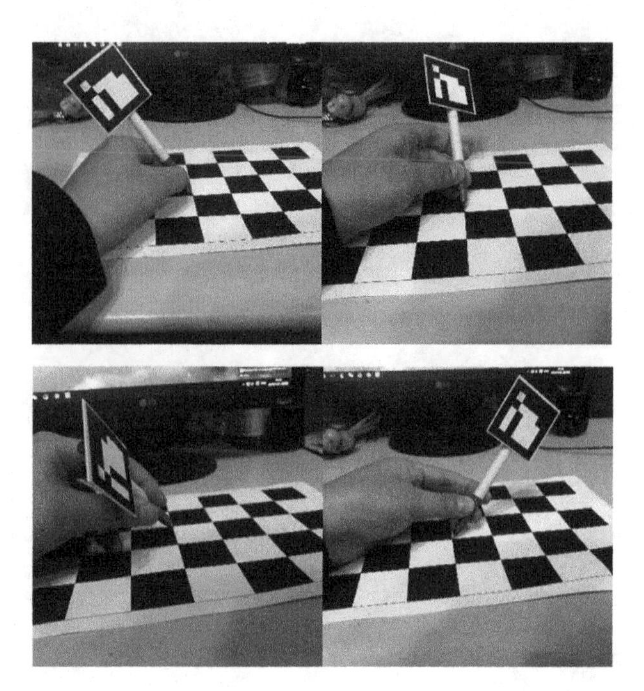

Fig. 9. Tracking the same tip with different rotation

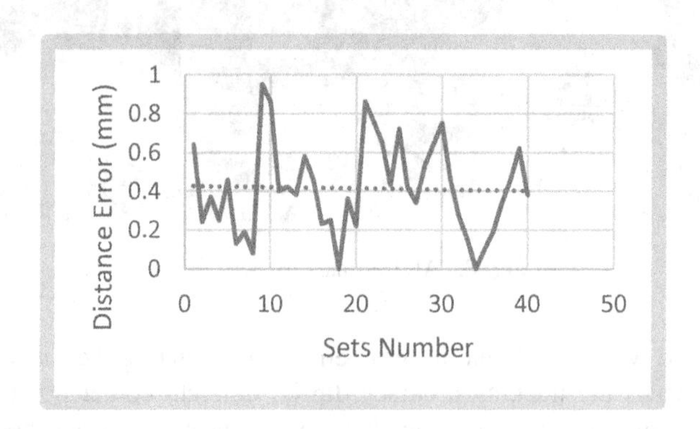

Fig. 10. Tracking stability

3.3 Tracking Accuracy

Figure 11 shows that these corner positions can be calculated by this tracking system. We choose the distance between the square grids on the board (40 mm) to get the value of the quantitative analysis of the tracking system.

We use the surgical tool stick the corner points from the lower left one to the top right one, 40 sets of distances between the corner points were calculated and compared to the real distance (40 mm). Then we obtained the whole error data as Fig. 11. The error data obtained is shown in Fig. 12.

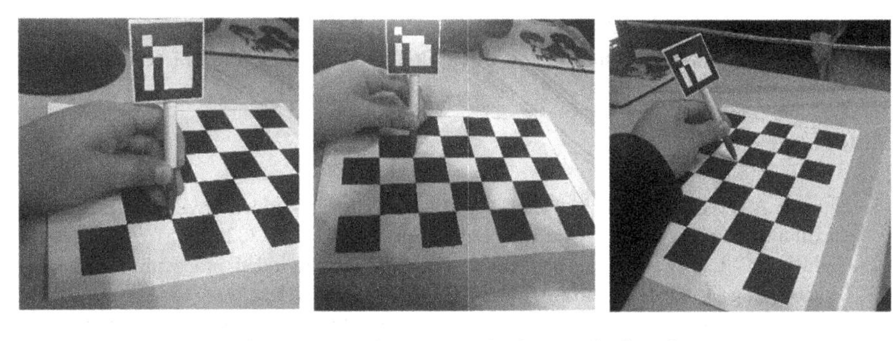

Fig. 11. Tracking and navigation on the board

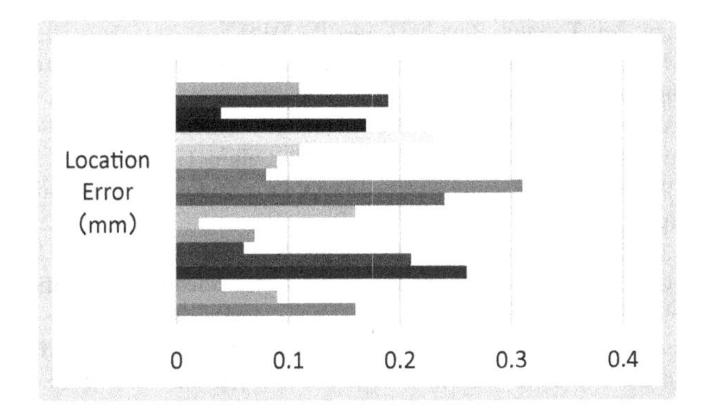

Fig. 12. Tracking accuracy

Among the distance error data, the mean error is 0.4006 mm, maximum error is 0.9500 mm and mean square error is 0.4458 mm.

The mean error of the tracking system is small and comparable to other tracking system, which shows the relatively high accuracy of tracking quality by our system built in this study.

4 Conclusion

The tracking system is the crucial component of CAS (Computer Assist Surgery). It provides the relative position and posture of the surgical tools to the surgery areas helping the surgeons to have a better surgery. Optical tracking system is not expensive and easy to conduct, what's more, it is also accurate and fast. A fast and accurate tracking system based on ARTag fiducial markers is proposed and proved to be practical for real applications in this paper. Also, with the marker features, we can track several tools though it may slow down the tracking real-time ability.

The test of our tracking system on the checkerboard acquire the tracking error which certificates that our tracking system is accurate to satisfy the needs of the navigation system in laboratory environment. However, more advance experiments are required to verify the accuracy of the tracking system in real surgical conditions.

References

1. Nijmeh, A.D., Goodger, N.M., Hawkes, D., et al.: Image-guided navigation in oral and maxillofacial surgery. Br. J. Oral Maxillofac. Surg. **43**(4), 294–302 (2005)
2. Galloway, R.L.: The process and development of image-guided procedures. Ann. Rev. Biomed. Eng. **3**, 83–108 (2001)
3. Williamson, T., Caversaccio, M., Weber, S., et al.: Image-guided microsurgery. In: Telehealth and Mobile Health, pp. 91–115. CRC Press (2015)
4. Rongqian, Y.A.N.G., Peifeng, G.U.A.N., Ken, C.A.I., et al.: A novel approach to synchronous image acquisition from near infrared camera in optical-surgery navigation system. J. Eng. Sci. Technol. Rev. **8**(3), 14–20 (2015)
5. Wenjuan, M.A.: Key Technologies of Infrared Surgical Navigation Instrument. Shanghai Jiao tong University, Shanghai (2010)
6. Peng, Z.H.A.O.: Study on the Method of Optimum Combination of Optics and Inertia in Surgical Navigation System. Tsinghua University, Beijing (2015)
7. Wittmann, W., Wenger, T., Zaminer, B., Lueth, T.C.: Automatic correction of registration errors in surgical navigation systems. IEEE Trans. Biomed. Eng. **58**(10), 2922–2930 (2011)
8. Simpson, A.L., Burgner, J., Glisson, C.L., Herrell, S.D., Ma, B., Pheiffer, T.S., Webster, R.J., Miga, M.: Comparison study of intraoperative surface acquisition methods for surgical navigation. IEEE Trans. Biomed. Eng. **60**(4), 1090–1099 (2013)
9. Liu, W., Ren, H., Zhang, W., et al.: Cognitive tracking of surgical instruments based on stereo vision and depth sensing. In: 2013 IEEE International Conference on Robotics and Biomimetic (ROBIO), pp. 316–321. IEEE (2013)
10. West, J.B., Maurer Jr., C.R.: Designing optically tracked instruments for image-guided surgery. IEEE Trans. Med. Imaging **23**(5), 533–545 (2004)
11. Brouwer, O.R., Buckle, T., Bunschoten, A., Kuil, J., Vahrmeijer, A.L., Wendler, T., Valdés-Olmos, R.A., van der Poel, H.G., van Leeuwen, F.W.: Image navigation as a means to expand the boundaries of fluorescence-guided surgery. Phys. Med. Biol. **57**(10), 3123 (2012)
12. Lenze, C., Chaudhry, N., Volling, P., et al.: Concept for a Navigated Micro Surgical Assistant System for Middle Ear Surgery. CURAC, München (2004)
13. Maier-Hein, L., Mountney, P., Bartoli, A., Elhawary, H., Elson, D., Groch, A., Kolb, A., Rodrigues, M., Sorger, J., Speidel, S.: Optical techniques for 3D surface reconstruction in computer-assisted laparoscopic surgery. Med. Image Anal. **17**(8), 974–996 (2013)
14. Wang, X., Kim, M.J., Love, P.E.D., et al.: Augmented reality in built environment: classification and implications for future re-search. Autom. Constr. **32**, 1 (2013)
15. Fiala, M.: Comparing ARTag and ARToolkit plus fiducial marker systems. In: IEEE International Workshop on Haptic Audio Visual Environments and their Applications, 6 p. IEEE (2005)
16. Khan, D., Ullah, S., Rabbi, I.: Factors affecting the design and tracking of ARToolKit markers. Comput. Stand. Interfaces **41**, 56–66 (2015)
17. Fiala, M.: ARTag, a fiducial marker system using digital techniques. In: IEEE Computer Society Conference on Computer Vision and Pattern Recognition, CVPR 2005, vol. 2, pp. 590–596. IEEE (2005)

Automated Detection of Barrett's Esophagus Using Endoscopic Images: A Survey

Noha Ghatwary[1,2(✉)], Amr Ahmed[1], and Xujiong Ye[1]

[1] University of Lincoln, Lincoln, UK
nghatwary@lincoln.ac.uk, noha.ghatwary@aast.edu,
{aahmed,xye}@lincoln.ac.uk

[2] Arab Academy for Science and Technology, Alexandria, Egypt

Abstract. Barrett's Esophagus (BE) is the predominant sign leading to esophageal adenocarcinoma (EAC) which is cancerous. Esophageal cancer has shown a very low survival rate in the last decade. Early detection of BE, and monitoring cells development is an effective way to control the progression of the cell transform into EAC in the lining of the esophagus. The examination of BE is done by using one the different endoscopy tools, the appearance of endoscopic imaging techniques created the opportunity for computer-aided detection (CAD) systems to develop more frequently. Such methods intend to help physicians by identifying abnormalities more accurately. The purpose of this survey is to discuss advances in the development of BE CAD systems as it has only started to grab attention recently. Starting with a brief introduction about endoscopy modalities used for esophageal examination. Then focusing on detection methods lately developed for BE detection. Remaining challenges are mentioned and some directions for future research are given.

Keywords: Barrett's Esophagus · Computer-Aided Detection (CAD) · Endoscopy · Esophageal cancer

1 Introduction

Esophageal cancer showed a significant increase in the past couple of years in the United states and Western European countries [29]. It is considered one of the main factor of death in industrial countries [5]. In 2012, Esophageal cancer has been counted as the sixth most common death from cancer [2] and recent studies ranked it as the top 18th cancer world-wide but with only a 5-year survival rate plan of 18.2% compared to breast cancer that has 89%, lung cancer has 55% stomach cancer 65%,... etc.

The main precancerous diseases leading to esophageal adenocarcinoma (EAC) is called Barrett's Esophagus (BE) [23]. BE is defined as the deformation that happens to normal cells in the lining of the lower part of the esophagus tube into metaplasia transformation with the presence of goblet cells [36]. The transformation of the cells in the esophagus can develop into something called

M. Valdés Hernández and V. González-Castro (Eds.): MIUA 2017, CCIS 723, pp. 897–908, 2017.
DOI: 10.1007/978-3-319-60964-5_78

dysplasia that can be divided into low grade dysplasia (LGD) and high grade dysplasia (HGD) (Also, known as early cancer stage). Once BE reaches its highest stage, and turns to be cancerous it is considered deadly with very low survival rate [7].

Endoscopic tools has been available for clinical examination for several decades. The improvement of the modalities facilitated the process of early detection was considered problematic [31]. BE examination can be done using different types of endoscopes as will be discussed in next section. BE can only be detected through endoscopy examination and the stage of cell deformation has to be confirmed by biopsy [11]. Regular follow-ups for infected patients using endoscopy is required to help decrease the chance of EAC by controlling the development of BE [20]. The physician needs to be an expert and well trained to the type of endoscopy used in the examination process to be able to detect the lesion as it usually overlooked because of its properties [27]. Moreover, it is considered challenging as the infection can be located in several separate cell regions within the esophagus [11].

One of the most popular tools that can assist and provide a second opinion to doctors is Computer Aided Detection (CAD) which provides auto-analysis of medical images to detect abnormalities [10]. It has been rapidly growing in the last several years as it proved its efficiency to accurately detect/diagnosis with less time for different types of diseases [8,32]. Therefore, early detection of precancerous esophageal abnormalities could make a valuable contribution in reducing death rate of EAC patients [3].

The benefit of CAD systems support became more popular now with the development of different endoscopy tools as it facilitated the option of studying closely the features of the targeted area. CAD systems for endoscopic videos have mainly draw attention in research fields related to detecting polyp, bleeding detection and cancer in colons and small intense [18]. The study of supporting BE analysis started to grab the attention recently as a result of the infection increase rate in the last couple of years. Early detection is important, for early intervention and higher recovery rate. However, the diagnosis are usually difficult due to the less apparent changes of the cells during the transformation stages and the random locations they appear at.

The main aim of this survey is to provide an overview of CAD systems focusing on recent methods reported in the literature to detect BE using different endoscopy modalities. To highlight the relevance of BE CAD systems we conducted an exhaustive search for publications dealing with technical papers published in reputable journal and conferences (Springer, Elsevier, SPIE, and IEEE) starting from 2009 till date focusing on the detection methods.

The rest of the paper is organized as follows. In Sect. 2, a review of the current available advanced imaging modalities used in BE evaluation. Followed by, Sect. 3 that reviews in details the key techniques available of CAD systems to analysis the captured images to detect BE. Problems of current CAD system and possible option to deal with them are discussed in Sect. 4. And finally, the paper is concluded in Sect. 5.

2 Endoscopy Modalities Used for Barrett's Detection

Endoscopy is a non-surgical process that examines the different cavities within a human body [12]. There exist several types of endoscopy procedures in the medical field based on examined area such as Colonoscopy, Thoracoscope, Neuroendoscopy, ...etc. For the examination of the Upper Gastrointestinal Tract (GI Tract) where the esophagus is located, the procedure is called Esophagoscope or Gastoscopy [18]. Several endoscopic technologies have been developed for a couple of decades that can examine different areas in GI.

Standard endoscopy white light endoscopy (WLE) (Fig. 1a) enables the image to be zoomed in up to 850,000 pixel [13], while the *High Definition-WLE* (HD-WLE) (Fig. 1b) and the *Magnification Endoscopy* magnified 115 times using optical magnifier. Although these three methods were an improvement for endoscopic technologies but they weren't effective alone where they sometimes require an advanced endoscopy tool to support the examination process [29]. Moreover, it cannot differentiate between the epithelium type, therefore, random biopsy sample takes place which might cause miss-leading diagnosis.

Chromoendoscopy (Fig. 1c) is another type of endoscope that injects a Methylene Blue (a type of dye) then selects the stained segment for the biopsy process [22]. The process seems simple but it requires an increased examination time. In addition to that, the pattern classification process showed instability and the experts required more training to be specialized in categorizing it. Moreover, the Methylene Blue has a risk in causing carcinogenesis and damage of DNA [22].

Wireless Capsule Endoscopy (WCE) is a non-invasive technology that exists in a pill shape which has the ability to approximately to capture an average of 50,000 images throughout an examination of 7–8 h where these images are sent and oriented remotely throughout the examination [26]. The problem with WCE is that its position cannot be controlled which might lead to uncertainty in its diagnosis. It is usually used to examine the colon and small intestine.

New technologies of endoscope were recently developed that were more effective towards detecting BE. *Narrow Band Imaging* (NBI) (Fig. 1d), has the ability to study the vascular pattern and mucosal by enhancing its surface resolution. NBI utilizes short wavelength of blue light that is supposed to absorbed by hemoglobin in the blood. Some studies showed NBI was superior than the WLE in detecting the HGD while other studies showed it disability in detecting neoplasia [29].

The *Optical Coherent tomography* (OCT) (Fig. 1e) uses light waves to capture the scattered coherent light for mucous. It has the ability to find the transformation of early dysplasia. Studies showed that OCT has an accuracy of 61% to 76% detection from endoscopists [24]. The results of OCT is recognized to be effective although it is not commonly used nowadays [28].

Confocal Laser Endomicrosciopy (CLE) (Fig. 1f) is regarded as one of the latest technologies used and it's an advanced version of the OCT [6]. It is a real-time endoscopic tool that allows both *imaging* and *patahlogoy diagnosis.* A blue laser light is used to concentrate on the mucosa while injecting a contrast agent called intravenous. The CLE captures 0.8 frames per second at a resolution of

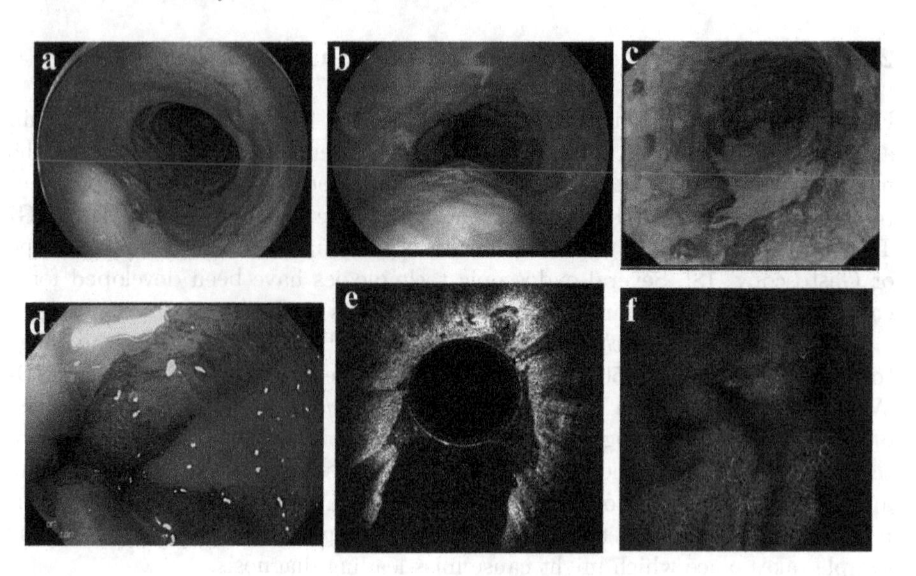

Fig. 1. Images of Barrett's Esophagus with different endoscopy tools; (a) WLE [37] (b) HD-WLE [34] (c) Chromoendoscopy [33] (d) NBI [15] (e) OCT [23] (f) CLE [11]

1024×1024 [4]. It was reported that an appropriate diagnosis of the histology grade using CLE might need less number of biopsies sample taken from the patient. It is also considered an important field that will grab attention for more research in the field of automatic classification [18].

Some of these advances as HD-WLE are intended to detect abnormal areas, while other modalities are more suitable for tissue characterization such as NBI and chromoendoscopy. Moreover, the CLE are used for histological confirmation [16].

3 Automated Detection of Barrett's Esophagus

CAD merges between the symptomatic images and the analysis of pattern recognition and image processing properties [9]. The benefit of utilizing CAD systems became more popular now with the development of different endoscopy tools as it facilitated the option to study closely the features of the targeted area. The study of CAD systems supporting BE analysis started to grab the attention recently as an outcome of the increased rate of infection in the last couple of years. Furthermore, the importance of early detection and the difficulty in diagnosis due to its similarity in the transformation stages and its appearance in random locations. The methods used for implementing an automatic detection system have a major effect on their performance.

In this section, we review the different techniques for BE detection based on the type of endoscopy used for examining the infected esophagus, the focus is on the key methods that have been reported recently in the literature.

3.1 WLE CAD Systems

The WLE is the standard type of endoscopy and used as the most regular method for endoscopy imaging [16]. An automatic detection of the region that is considered cancerous is presented by Yamaguchi *et al.* [37] that took the advantage of fractal dimension properties to apply the detection process. As a first step, the image is decomposed into components (Red, Green, Blue, Luminance). After that, regions from the image that clearly doesn't have cancer are clipped out to save more processing time aiming to reach a real-time model. Followed by re-sizing the images 1024×1024 to standardize the size for any image processed. Later, the image is decomposed into its DWT form and only the low-layer component is utilized dividing it into small non-overlapping blocks of size 128×128 to minimize the time of classification. More blocks are discarded that have a luminance value that is less than a total average value of luminance of all blocks. The remaining qualified blocks are exposed to DWT twice and divided into smaller sub-blocks. Each of these steps are applied to every component layer that was decomposed in the first step. The feature vector extracted for classification phase is finally calculated by multiplying the fractal dimension of each component layer by using box counting method extracted from for each sub-block. The block region is considered classified as cancer region if it has a very low fractal dimension value. The problem with this method that it is very time-consuming where a single image can undergo 3 min to reach a decision.

Ohura *et al.* [21] grabbed attention to the abnormal regions of early esophageal cancer after normalizing the input image to a given range, it is then converted from RGB to HSV color space. The Dyadic Wavelet Transform (DYWT) is then applied to the S and V layers and their low-level frequency components are fused together. Later on, contrast enhancement is applied to that fused image which is divided afterward into 16×16 non-overlapping blocks. The sum of the fractal dimension value is calculated as in their previously proposed work [37] to indicate either the block is normal or abnormal.

In all the WLE models/techniques discussed above, the evaluation was done through visual/qualitative approach only. The visual evaluation is done through visually comparing the detected region, by the proposed methods, with the annotation done by the experts (as a Ground-Truth). Hence, no quantitative evaluation results, through the common evaluation measures, were given.

3.2 HD-WLE

HD-WLE are a higher resolution endoscope that has been invented as a replacement to the standard endoscopy allowing a better visualization. Different clinical studies for BE detection using HD-WLE that showed to be efficient and allowed an improved procedure for targeted biopsy.

In order to implement a CAD system of early cancer detection in the esophagus, a study was represented by Setio *et al.* [27] that evaluated the texture feature that can be extracted from HD-WLE images for early esophagus cancer detection. The proposed study compared between the efficiency of Local Binary Pattern (LBP), Texture Spectrum, Histogram of Oriented Gradients (HOG), Dominant Neighbor Structure (DNS), Grey Level Co-occurrence Matrix (GLCM), Fourier feature and Gabor Features. After discarding the irrelevant texture tiles from the images as a pre-processing phase, the features were extracted. The results concluded that merging between the Gabor features and the Color features achieved 96.48% compared to the baseline of annotated accuracy and against the combination of other features. The method utilized the Principal Component Analysis (PCA) for reducing the features dimension and were classified using the SVM.

Based on this conclusion a CAD system has been proposed in Van der Sommen *et al.* [34,35]. The chosen features were used to detect and annotate infected lesion in the esophagus. The implemented method extracts the desired features and classifies them using SVM to allocate the region of interest. The dataset used consisted of 32 images from 7 different patients. Comparing the results with the specialist annotation the system was able to achieve 85.7% with a recall of 0.95 and precision 0.75. The model needed to increase its robustness and to have the ability to be real-time.

Later on, the previously stated method was used in [30] to automatically annotate the neoplastic lesions in Barrett's esophagus and compared to the annotations of 5 experts. The proposed method was tested on 100 images for 44 patients. The analysis of the proposed model was able to accomplish 0.86 sensitivity and 0.87 specificity. The results achieved were almost the same or less in comparison with the experts in annotation and detection as it is considered the ground truth. As a result, the study was considered a promising start where it was extended in [14] by changing the classifier from SVM to Random Forest to benefit from its properties. This change improved the results by 6% and 11% reaching a recall of 90% and precision of 75%.

3.3 NBI

One of the most-investigated endoscopy technique for enhancing imaging by improving the mucosal surface resolution is NBI [17]. Clinical studies proved that the NBI is more effective when used for detecting high grade dysplasia than for BE.

Boschetto *et al.* [5] suggested an automatic classification method that differentiates between normal and metaplasia (abnormal) regions. The method employed the super pixel method to cluster the image into regions by using Simpler Linear Interactive Clustering (SLIC) [1] method that mainly depends on K-mean clustering method to benefit from the simplicity and efficiency of computation time. Later on, eight different features are extracted from each super pixel. These features were divided as the mean of intensity values for the color channel of an image, then the mean intensity was extracted again but after applying three different filters which are Entropy filter, Range filter, and Top-hat filter. The last two values added to the feature vector were the contrast and homogeneity that are calculated from the GLCM texture method. Random forest classifier was used to classify between normal and metaplasia lesion from

116 NBI images of different patients. The method was able to achieve an overall accuracy of 83.9% accompanied with sensitivity of 72.2% and specificity of 87.3%. Since the method was mainly proposed as a proceeding step before classifying the type of metaplasia, therefore, the accuracy of the model needs to be higher than that and more accurate.

A study was proposed by Kage *et al.* [15] on NBI endoscopy images to prove the efficiency of utilizing automatic detection by classification systems with gastroscopy. The model extracted selected features from 326 Region-of-Interest (ROI) that were annotated by experts and classified as epithelium, cardia mucosa and Barrett's esophagus. The feature vector in the proposed work was composed of well known features (Co-occurrence matrices, Sum and Difference of histogram, Statistical geometrical, Gabor Filter) which lead to a high dimension vector that was reduced later by using the forward selection approach. The evaluation of the study has measured the performance of each selected feature separately and also by combining them altogether by using the Euclidian distance as similarity metric resulting in a range of accuracy between 85% and 92%. The best accuracy of classifying the BE individually was only 74%.

Based on a study proposed by Munzenmayer [19] of color texture analysis of medical images, Rajan *et al.* [25] applied several experiments using more the one classification method (SVM, KNN, Boosting) on images from various endoscopy modalities (WLE, NBI. Chromoendoscopy). The dataset here was classified into four categorize divided as Normal Squamous, Gastric Mucosa, Barrett's Esophagus (including intestinal metaplasia with goblet cells and Low grade dysplasia) and Hugh grade dysplasia. By down-sampling the endoscopy image and extracting features as in [19] the images were categorized as one of the four types. After applying the different classifiers, the accuracy for detecting BE had a varied range from 36.36% up to 89.17% according to the image and classifier used. The results of this method for each modalities on all the classifier is represented in details in Table 1.

4 Discussion

As mentioned in the previous section, there exist several methods aiming at assisting medical experts during the examination process. Although different approaches have been proposed in the literature, there exist some common drawbacks among these approaches. In this section, these issues and propose possible ways to cope with them will be discussed.

First of all, the limited number of dataset images is considered one of the main issues with most of the existing methods. The number of images varies significantly throughout the work as shown in Table 1. This is considered a problem especially with the experiments made on small sized dataset leading to doubtful results. Consequently, the method of evaluation is mostly done based on a Leave-one-out cross-validation (LOOCV). In order to achieve a more relevant evaluation, the dataset used for the experimental analysis should increase much more in size and the validation should be based on a separately divided data for

Table 1. An overview of the current CAD methods

Ref.	Endoscopy modalitay	No. of images and patients for evaluation	Classifier	Validation method	BE results
Van der Sommen *et al.* [30]	HD-WLE	100 images, 44 patients	SVM	LOOCV	Sensitivity 0.83 Sensitivity 0.83
				LOPOCV	Sensitivity 0.86 Specificity 0.87
Van der Sommen *et al.* [35]	HD-WLE	32 images, 7 patients	SVM	-	Accuracy 85.7% Recall 0.84 Precision 0.64
Janse *et al.* [14]	HD-WLE	100 images, 39 patients	Random forest	-	Recall 0.9 Precision 0.75
Boschetto *et al.* [5]	NBI	116 images	Ensemble	10 cross validation	Accuracy 87.3% Sensitivity 79.2% Sensitivity 87.3%
Kage *et al.* [15]	NBI	196 images, 30 patients	Euclidean distance	LOOCV	Accuracy 74%
Yamaguchi *et al.* [37]	WLE	10 images, 4 patients	-	-	-
Ohura *et al.* [21]	WLE	135 images, 23 patients	-	-	-
Rajan *et al.* [25]	WLE	125 images	SVM	10 cross validation	Accuracy 89.17%
			KNN		Accuracy 58.62%
			Boosting 1	10 cross validation	Accuracy 60%
			Boosting 2		Accuracy 65.83%
	NBI	122 images	SVM	10 cross validation	Accuracy 86.67%
			KNN		Accuracy 36.36%
			Boosting 1	10 cross validation	Accuracy 62.5%
			Boosting 2		Accuracy 51.67%
	Chromoendoscopy	150 images	SVM	10 cross validation	Accuracy 50.83%
			KNN		Accuracy 73%
			Boosting 1	10 cross validation	Accuracy 62.5%
			Boosting 2		Accuracy 67.5%

training and testing or a leave-one-patient-out (LOPOCV) which is considered much more convenient.

Secondly, the *time complexity* should be taken into consideration since most of the endoscopy is an on-spot examination method which requires a real-time processing of videos. All of the currently available methods apply their proposed technique on still images and consume a huge amount of time. For example

Yamaguchi *et al.* [37], proposed a method requires 3-minutes processing time for one image. However, an endoscopy tool generates N-frames per second, accordingly, the implemented CAD methods should take into consideration that the computational time must meet the video processing time to support the medical experts opinion through a real-time examination process.

Lastly is the *evaluation performance*, the *accuracy* that evaluates the outcome of the proposed CAD has to be compared to an available ground truth that has been given by more than one medical expert in the field. Moreover, the accuracy measured needs to be giving higher weight, and importance, to the BE class detection not as an overall system in-order to take into consideration the false positive or negatives produced by the system. Also, to assess the various BE CAD systems presented in literature against each other, the comparison needs to be made on the same dataset to have a fair evaluation of the results or the same endoscopy modality with almost same number of images.

5 Conclusion

CAD is a highly required technique for detection of early esophageal cancer. A notable amount of work has been done in this area over the last couple of years. Despite the fact that CAD acts as a second opinion, its performance needs to be improved to satisfy the needs of clinical purposes. Different methods have been developed to support an efficient CAD for detection of Barrets Esophagus to support decreasing the possibly of infected cells to turns into esophageal cancer.

The paper provided an overview of the recent advances of CAD systems focusing on BE/Early Esophageal cancer detection. Basic concepts related to main endoscopy modalities used for BE diagnosis by physicians has been described and the key techniques were reviewed and discussed. A comparison that highlighted the main classification methods, dataset and results of the review methods are summarized in Table 1.

Although significant development has been made over the couple of years, much work still needs to be done to develop more successful CAD systems. Further work on CAD BE systems is required to improve the efficiency and effectiveness, leading to an improved prognosis of early BE detection particularly accuracy, sensitivity, and specificity based on larger dataset size. The computation time also needs to be improved (aiming at real-time processing) to be compatible with used endoscopy frame processing time.

References

1. Achanta, R., Shaji, A., Smith, K., Lucchi, A., Fua, P., Süsstrunk, S.: Slic superpixels compared to state-of-the-art superpixel methods. IEEE Trans. Pattern Anal. Mach. Intell. **34**(11), 2274–2282 (2012)
2. Andrici, J., Eslick, G.D.: Epidemiology and risk factors for esophageal cancer. In: Saba, N.F., El-Rayes, B.F. (eds.) Esophageal Cancer, pp. 1–23. Springer, Cham (2015). doi:10.1007/978-3-319-20068-2_1

3. Arora, Z., Garber, A., Thota, P.N.: Risk factors for barrett's esophagus. J. Dig. Dis. **17**(4), 215–221 (2016)
4. Becker, V., Vieth, M., Bajbouj, M., Schmid, R., Meining, A.: Confocal laser scanning fluorescence microscopy for in vivo determination of microvessel density in barrett's esophagus. Endoscopy **40**(11), 888–891 (2008)
5. Boschetto, D., Gambaretto, G., Grisan, E.: Automatic classification of endoscopic images for premalignant conditions of the esophagus. In: SPIE Medical Imaging, International Society for Optics and Photonics, p. 978808 (2016)
6. Buchner, A.M., Wallace, M.B.: In-vivo microscopy in the diagnosis of intestinal neoplasia and inflammatory conditions. Histopathology **66**(1), 137–146 (2015)
7. Conteduca, V., Sansonno, D., Ingravallo, G., Marangi, S., Russi, S., Lauletta, G., Dammacco, F.: Barrett's esophagus and esophageal cancer: an overview. Int. J. Oncol. **41**(2), 414–424 (2012)
8. El-Dahshan, E.S.A., Mohsen, H.M., Revett, K., Salem, A.B.M.: Computer-aided diagnosis of human brain tumor through MRI: a survey and a new algorithm. Expert Systems Appl. **41**(11), 5526–5545 (2014)
9. Fu, Y., Zhang, W., Mandal, M., Meng, M.Q.H.: Computer-aided bleeding detection in WCE video. IEEE J. Biomed. Health Inform. **18**(2), 636–642 (2014)
10. Ghatwary, N., Ahmed, A., Jalab, H., et al.: Liver CT enhancement using fractional differentiation and integration. In: The 2016 International Conference of Signal and Image Engineering, vol. 1 (2016)
11. Ghatwary, N., Ahmed, A., Ye, X., Jalab, H.: Automatic grade classification of barretts esophagus through feature enhancement. In: SPIE Medical Imaging, International Society for Optics and Photonics, pp. 1013,433-1–1013,433–8 (2017)
12. Gotoda, T.: Endoscopic resection of early gastric cancer. Gastric Cancer **10**(1), 1–11 (2007)
13. Haringsma, J., Tytgat, G.N., Yano, H., Iishi, H., Tatsuta, M., Ogihara, T., Watanabe, H., Sato, N., Marcon, N., Wilson, B.C., et al.: Autofluorescence endoscopy: feasibility of detection of GI neoplasms unapparent to white light endoscopy with an evolving technology. Gastrointest. Endosc. **53**(6), 642–650 (2001)
14. Janse, M.H., van der Sommen, F., Zinger, S., Schoon, E.J., et al.: Early esophageal cancer detection using RF classifiers. In: SPIE Medical Imaging, International Society for Optics and Photonics, p. 97,851D (2016)
15. Kage, A., Raithel, M., Zopf, S., Wittenberg, T., Münzenmayer, C.: Narrow-band imaging for the computer assisted diagnosis in patients with barrett's esophagus. In: SPIE Medical Imaging, International Society for Optics and Photonics, p. 72,603S (2009)
16. Lee, M.H., Buterbaugh, K., Richards-Kortum, R., Anandasabapathy, S.: Advanced endoscopic imaging for barrett's esophagus: current options and future directions. Curr. Gastroenterol. Rep. **14**(3), 216–225 (2012)
17. Lee, M.M., Enns, R.: Narrow band imaging in gastroesophageal reflux disease and barrett's esophagus. Can. J. Gastroenterol. **23**(2), 84 (2009)
18. Liedlgruber, M., Uhl, A.: Computer-aided decision support systems for endoscopy in the gastrointestinal tract: a review. IEEE Rev. Biomed. Eng. **4**, 73–88 (2011)
19. Münzenmayer, C.: Color texture analysis in medical applications. Der Andere Verlag (2006)
20. Naini, B.V., Chak, A., Ali, M.A., Odze, R.D.: Barrett's oesophagus diagnostic criteria: endoscopy and histology. Best Practice Res. Clin. Gastroenterol. **29**(1), 77–96 (2015)

21. Ohura, R., Omura, H., Sakata, Y., Minamoto, T.: Computer-aided diagnosis method for detecting early esophageal cancer from endoscopic image by using dyadic wavelet transform and fractal dimension. In: Latifi, S. (ed.) Information Technology: New Generations. AISC, vol. 448, pp. 929–938. Springer, Cham (2016). doi:10.1007/978-3-319-32467-8_80
22. Olliver, J., Wild, C., Sahay, P., Dexter, S., Hardie, L.: Chromoendoscopy with methylene blue and associated DNA damage in barrett's oesophagus. Lancet **362**(9381), 373–374 (2003)
23. Qi, X., Sivak, M.V., Isenberg, G., Willis, J.E., Rollins, A.M.: Computer-aided diagnosis of dysplasia in barrett's esophagus using endoscopic optical coherence tomography. J. Biomed. Opt. **11**(4), 044,010 (2006)
24. Qi, X., Pan, Y., Sivak, M.V., Willis, J.E., Isenberg, G., Rollins, A.M.: Image analysis for classification of dysplasia in barrett's esophagus using endoscopic optical coherence tomography. Biomed. Opt. Express **1**(3), 825–847 (2010)
25. Rajan, P., Canto, M., Gorospe, E., Almario, A., Kage, A., Winter, C., Hager, G., Wittenberg, T., Münzenmayer, C.: Automated diagnosis of barrett's esophagus with endoscopic images. In: Dössel, O., Schlegel, W.C. (eds.) World Congress on Medical Physics and Biomedical Engineering. IFMBE, vol. 25/4, pp. 2189–2192. Springer, Heidelberg (2009). doi:10.1007/978-3-642-03882-2_581
26. Ramirez, F.C., Shaukat, M.S., Young, M.A., Johnson, D.A., Akins, R.: Feasibility and safety of string, wireless capsule endoscopy in the diagnosis of barrett's esophagus. Gastrointest. Endosc. **61**(6), 741–746 (2005)
27. Setio, A.A.A., van der Sommen, F., Zinger, S., Schoon, E.J., de With P.H.N.: Evaluation and comparison of textural feature representation for the detection of early stage cancer in endoscopy. In: Proceedings of the International Conference on Computer Vision Theory and Applications, (VISIGRAPP 2013), pp. 238–243 (2013). doi:10.5220/0004204502380243
28. Shahid, M.W., Wallace, M.B.: Endoscopic imaging for the detection of esophageal dysplasia and carcinoma. Gastrointest. Endosc. Clin. North Am. **20**(1), 11–24 (2010)
29. Singh, R., Yeap, S.P.: Endoscopic imaging in barrett's esophagus. Expert Rev. Gastroenterol. Hepatol. **9**(4), 475–485 (2015)
30. van der Sommen, F., Zinger, S., Curvers, W.L., Bisschops, R., Pech, O., Weusten, B.L., Bergman, J.J., Schoon, E.J., et al.: Computer-aided detection of early neoplastic lesions in barrett's esophagus. Endoscopy **48**, 617–624 (2016)
31. Streitz, J., Andrews, C., Ellis, F.: Endoscopic surveillance of barrett's esophagus. Does it help? J. Thoracic Cardiovasc. Surg. **105**(3), 383–387 (1993)
32. Tang, J., Rangayyan, R.M., Xu, J., El Naqa, I., Yang, Y.: Computer-aided detection and diagnosis of breast cancer with mammography: recent advances. IEEE Trans. Inf. Technol. Biomed. **13**(2), 236–251 (2009)
33. Trivedi, P., Braden, B.: Indications, stains and techniques in chromoendoscopy. QJM (2012). doi:10.1093/qjmed/hcs186
34. Van der Sommen, F., Zinger, S., Schoon, E.J., et al.: Computer-aided detection of early cancer in the esophagus using HD endoscopy images. In: SPIE Medical Imaging, International Society for Optics and Photonics, p. 86,700V (2013)
35. Van der Sommen, F., Zinger, S., Schoon, E.J., de With, P.H.N.: Supportive automatic annotation of early esophageal cancer using local gabor and color features. Neurocomputing **144**, 92–106 (2014)

36. Wang, K.K., Sampliner, R.E.: Updated guidelines 2008 for the diagnosis, surveillance and therapy of barrett's esophagus. Am. J. Gastroenterol. **103**(3), 788 (2008)
37. Yamaguchi, J., Yoneyama, A., Minamoto, T.: Automatic detection of early esophageal cancer from endoscope image using fractal dimension and discrete wavelet transform. In: 2015 12th International Conference on Information Technology-New Generations (ITNG), pp. 317–322. IEEE (2015)

Estimating Bacterial Load in FCFM Imaging

Sohan Seth[1(✉)], Ahsan R. Akram[2], Kevin Dhaliwal[2], and Christopher K.I. Williams[1]

[1] School of Informatics, University of Edinburgh, Edinburgh EH8 9AB, UK
sseth@inf.ed.ac.uk

[2] Queens Medical Research Institute, MRC Center for Inflammation Research, Pulmonary Molecular Imaging Group, University of Edinburgh, Edinburgh EH14 4TJ, UK

Abstract. We address the task of detecting bacteria and estimating bacterial load in the human distal lung with fibered confocal fluorescence microscopy (FCFM) and a targeted smartprobe. Bacteria appear as bright dots in the image when exposed to a smartprobe, but they are often difficult to detect due to the presence of background autofluorescence inherent to human lungs. In this study, we create a database of annotated image frames where a clinician has labelled bacteria, and use this database for supervised learning to build a suitable bacterial load estimation software.

1 Introduction

Fibered confocal fluorescence microscopy (FCFM) is a popular method for in vivo imaging of the distal lung, and has recently gained prominence in investigating the presence of bacteria using targeted *smartprobe* [1]. Smartprobes are specialized molecular agents introduced in the imaging area to make the bacteria fluoresce. FCFM imaging works by recording the number of emitted photons at each core of a optical fiber bundle. The photon counts are later translated into pixel intensities to achieve a 'smooth' image. Since the diameter of a bacterium is usually smaller than the width of the fibre core as well as the gap between two consecutive fiber cores, it appears as a high intensity dot in the image frame and, tends to 'blink' on and off in consecutive image frames due to movement of the apparatus.

Figure 1 shows examples of FCFM image frames without (top) and with (bottom) bacteria. In general, human lungs display a mesh-like structure due to autofluorescence of connective tissues (elastin) with or without the presence of smartprobe whereas bacteria appear as dots in the image frame when exposed to smartprobe (Fig. 1c). However, one can observe bacteria-like dots in the absence

S.S., A.A., K.D., and C.W. would like to thank Engineering and Physical Sciences Research Council (EPSRC, United Kingdom) Interdisciplinary Research Collaboration grant EP/K03197X/1 for funding this work. A.A. is supported by Cancer Research UK.

M. Valdés Hernández and V. González-Castro (Eds.): MIUA 2017, CCIS 723, pp. 909–921, 2017.
DOI: 10.1007/978-3-319-60964-5_79

of the smartprobe as well due to noise (Fig. 1a). Additionally, if bacteria are present, they are usually easy to detect when the elastin structure is not prominent (Fig. 1c-left), but it becomes more difficult to discriminate them from the background in the presence of elastin structure (Fig. 1c-right).

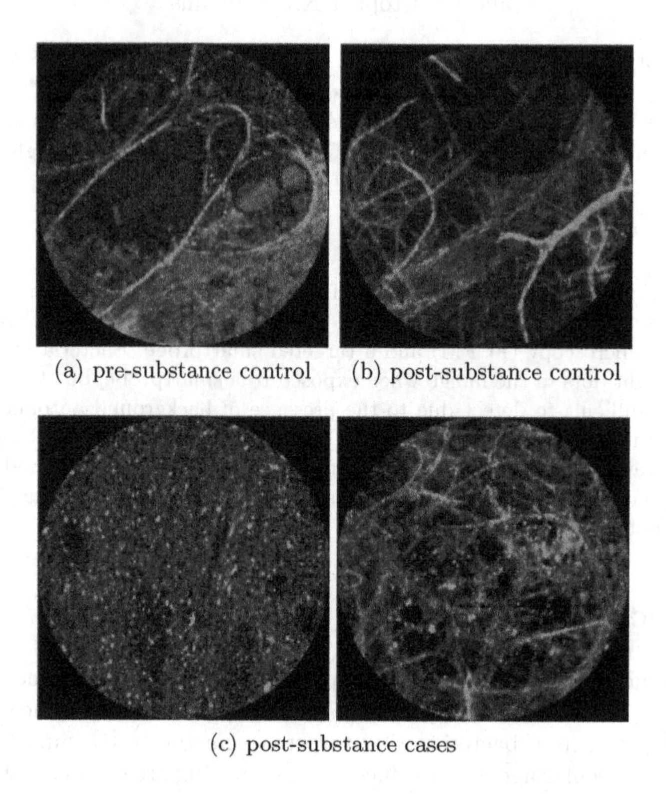

Fig. 1. FCFM image frames w/ or w/o smartprobe in control or case group.

We address the task of estimating bacterial load in a FCFM image frame. We consider a formal approach to the bacteria detection problem by explicitly annotating bacteria in image frames with the help of a clinician, and use this knowledge in a supervised learning set-up to learn a classifier that assigns a probability value to each pixel of the image of a bacterium being present. Estimating bacterial load can generally be framed as a *learning-to-count* [2] problem where we need to count the bacteria. However, although the learning-to-count framework usually bypasses the problem of detection before counting, we suggest detecting the object to allow the clinicians to see where the bacteria are appearing, ideally while performing the bronchoscopy. That said, our method bears resemblance with the established learning-to-count approaches with the major difference being that we learn a classifier to predict a probability value at each pixel whereas the other approaches learn a regressor to predict the 'count density' (intensity value) at each pixel (see Sect. 5).

Our final goal is to build a real time system to assist clinicians in estimating bacterial load while performing bronchoscopy. We suggest using a multi-resolution spatio-temporal template matching scheme using radial basis functions. Spatio-temporal analysis allows better capturing the 'blinking' effect, whereas multi-resolution analysis allows better discrimination between bacterial dots and elastin structure which may appear as series of dots. We use normalized intensity values around each pixel as features, which enables fast implementation of our method using 2D-convolution. We apply this method in estimating bacterial load in FCFM videos with and without bacteria (case and control), before and after applying the smartprobe, and show that we successfully infer low bacterial load in the control or the pre-substance videos and high bacterial load in the post-substance videos from the cases.

2 Dataset

2.1 Collection

In vivo imaging was performed in 6 patients (3 cases with gram-negative bacteria where a bacterial signal was detected and 3 cases with gram-positive bacteria (controls) where signal was not detected) where measurements were taken before and after administration of a gram-specific bacterial specific smartprobe (pre- and post-substance measurements respectively). Each measurement is a FCFM video (12 frames per second [4]) that were manually cleaned to ensure alveolar imaging by removing motion blur, air imaging, bronchi etc., and the remaining clean frames (~500 in each video) were used for this study.

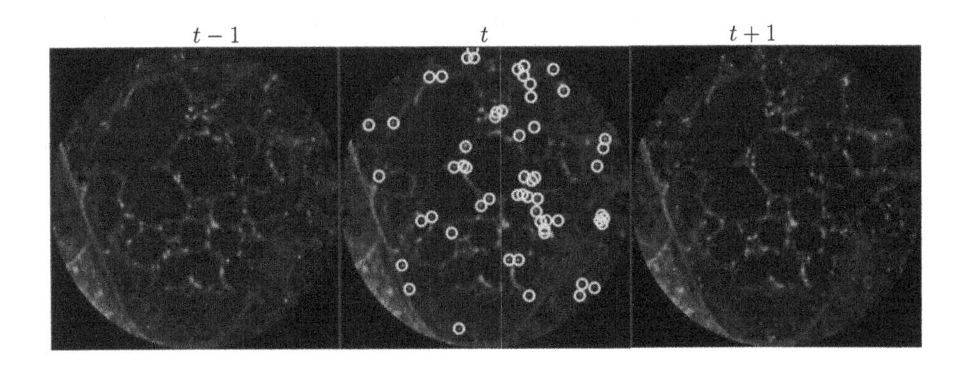

Fig. 2. FCFM image frame with bacteria annotated by a clinician in circles.

2.2 Annotation

The learning-to-count problem can be addressed in a variety of different annotation scenarios among which two widely used ones are the dot-annotation and the count-annotation [5]. While dot-annotation provides the location of where an object appear in the image, count-annotation only provides the number of objects in an image without explicitly revealing the locations. We use dot-annotations since they are more informative given the small size of the objects we are trying to count.

We chose 144 image frames from 12 videos such that 72 of them come from videos without bacteria (8 frames from 9 videos that are either in the control group or in the pre-substance group), and 72 of them come from videos with bacteria (24 frames from 3 videos that are in the post-substance case group). Along with the 144 image frames, the previous and next frames corresponding to those frames were extracted as well, and the clinician was allowed to toggle between the previous and next frame to annotate a bacterium in the current frame. Thus, a bacterium was identified in a spatio-temporal context. Figure 2 shows an example of annotated frame along with respective previous and next frames to demonstrate the blinking effect.

We observe that although we might encounter false positives, i.e., bacteria are annotated in either control group or pre-substance group, the clinician successfully annotates more bacteria in the frames where bacteria should exist, i.e., post-substance case group. For training purposes we only considered positive annotations from the post-substance case group.

3 Method

3.1 Preprocessing

To reduce the effect of noise and spurious intensity values we adjust the lowest and highest values of each image frame individually as follows: for each image frame, first, we set any intensity values below 1% quantile to 0, and next, we set any intensity values above 99% quantile to the respective 99% quantile values[1].

To allow supervised learning, we associate a feature vector $\mathbf{x}_p$ and label y_p to each pixel in the image and group the pixels in positive ($y_p = 1$) and negative samples ($y_p = 0$). We use the intensity values over a patch around a pixel as feature vector. However, we observe that image patches can vary significantly in contrast and therefore, we normalize them by the total intensity of the patch as follows: given an image patch $\{\tilde{x}^p_{ij}\}^w_{i,j=1}$ of size $w \times w$ around pixel p, we normalize the patch as $x^p_{ij} = \tilde{x}^p_{ij} / \sum_{ij}(\tilde{x}^p_{ij} + \epsilon)$ where $\epsilon = 1$ is added to suppress noisy image patch, i.e., $\tilde{x}^p_{ij} \approx 0$. Thus, our basic feature vector is $\mathbf{x}_p = (x^p_{11}, \ldots, x^p_{ij}, \ldots, x^p_{ww})$. $y_p = 1$ if p has been annotated by the clinician and 0 otherwise.

For positive labels ($y_p = 1$), we pool all image patches around the pixels annotated by the clinician. For negative labels ($y_p = 0$), we extract equispaced image patches over a grid (15×15 pixels apart). If any of the 'negative' image patches have a bacterium, they were assigned to the positive samples. Along with the original image patches, we performed data augmentation by rotating each image patch by 90°, 180° and 270° and adding them to the pool of samples. This

[1] This also increases the dynamic range, and thus helps the clinicians in annotating.

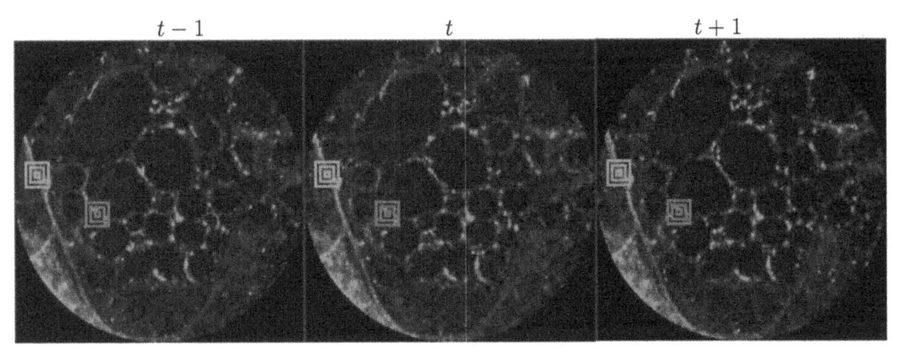

(a) Image patches around positive (red) and negative (green) annotations: the three boxes are of sizes 9×9, 27×27 and 45×45 pixels respectively

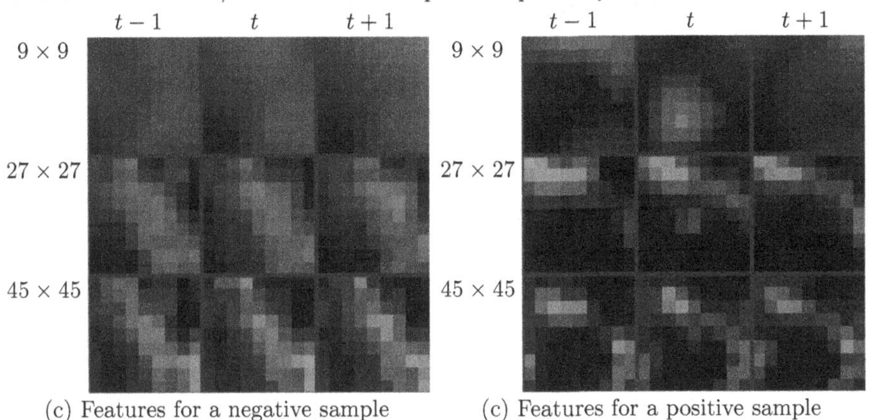

(c) Features for a negative sample (c) Features for a positive sample

Fig. 3. Illustration of spatio-temporal and multi-resolution feature extraction. (Color figure online)

results in about 18,000 positive samples and 540,000 negative samples. Notice that our classes are severely imbalanced: we use undersampling of the negative class to maintain a class balance while training a classifier.

For temporal analysis, we extract patches around the same pixel from the previous and the next image frame, normalize them individually and concatenate them to the feature vector from the current frame. For multi-resolution analysis, we extract larger image patches and 'downsample' them to the size of the smallest patch. This is done to allow equal importance over each resolution. These patches are normalized individually and concatenated to the feature vector. We 'downsample' the larger image patch, usually 3 or 5 times larger than the smallest image patch, by averaging over a 3×3 or 5×5 window around the pixel (in the patch) being downsampled. Figure 3 shows examples of positive and negative image patches. We also observe that (i) dot annotations can be noisy in the sense that the pixel with a bacterium might not be centered, and (ii) larger patch captures the context whereas smaller patch captures the object.

3.2 Supervised Learning

Our approach is to assign a probability at each pixel of a bacterium being present. Therefore, we essentially solve a classification problem from $\mathbf{x}_p$ to y_p. For simplicity and ease of implementation, we suggest *logistic regression* as a baseline method. However, we observe that it performs rather poorly. We then suggest *radial basis functions network* as an alternative, and show that it improves the performance significantly.

Logistic Regression: We solve a L_2 regularized logistic regression to learn a linear classifier, i.e.,

$$\min_{\mathbf{w},b} -\frac{1}{|\mathcal{P}|}\sum_{p\in\mathcal{P}} \left(y_p \log\sigma(\mathbf{w}^\top \mathbf{x}_p + b) + (1-y_p)\log(1-\sigma(\mathbf{w}^\top \mathbf{x}_p + b))\right) + \frac{\lambda}{2|\mathcal{P}|}||\mathbf{w}||^2$$

where σ is the sigmoid function and $\mathcal{P}$ is the set of all pixels in either positive or negative samples, and λ is the regularization parameter. We set $\lambda = 0.01$. Since we have class imbalance, we resample the negative samples to maintain class balance.

We test the performance of the baseline method with three different choices of features, (i) with 9×9 image patches extracted from the current frame, i.e., $\mathbf{x}_p \in \mathbb{R}^{81}$, (ii) with 9×9 image patches extracted from the current as well as previous and next frame, i.e., $\mathbf{x}_p \in \mathbb{R}^{81\times 3}$, and (iii) with 9×9 and 45×45 image patches extracted from the previous, current and future frames, i.e., $\mathbf{x}_p \in \mathbb{R}^{81\times 6}$. We expect (ii) to perform better than (i) since it captures the blinking effect, whereas (iii) to perform better than (ii) since it provides more contextual information around a bacterial dot.

Radial Basis Function Network: We learn a radial basis function network (RBF) [6] as a nonlinear classifier where the input vector $\mathbf{x}_p$ is first transformed through a set of nonlinear transformations, or radial basis functions, $\phi_i(\mathbf{x}_p)$ before being used as the input to the classifier, i.e., we solve the same problem as in logistic regression but replace the original feature vector $\mathbf{x}_p$ with the output of the radial basis functions $\{\phi(\mathbf{x}_p - \mathbf{c}_i)\}_{i=1}^{64}$ where $\mathbf{c}_i$ are centers of the radial basis functions. For ϕ we used Gaussian kernel with bandwidth set to median intersample distance. We chose the centers of the radial basis functions using k-means [4]. Since we have many fewer positive samples than negative, we chose 16 centers from the positive samples and 48 centers from the negative samples. Figure 4 shows examples of centers chosen from positive and negative samples. After choosing the centers, we perform logistic regression from 64 dimensional feature vector to the class label to learn the weight vector of the network. Notice that for selecting the centers we used the entire[2] negative set rather than undersampled set that we use for learning the weight vector.

We test the performance of the RBF network with two different choices of features, (i) with 9×9 image patches extracted from the current as well as

[2] After cross-validation split.

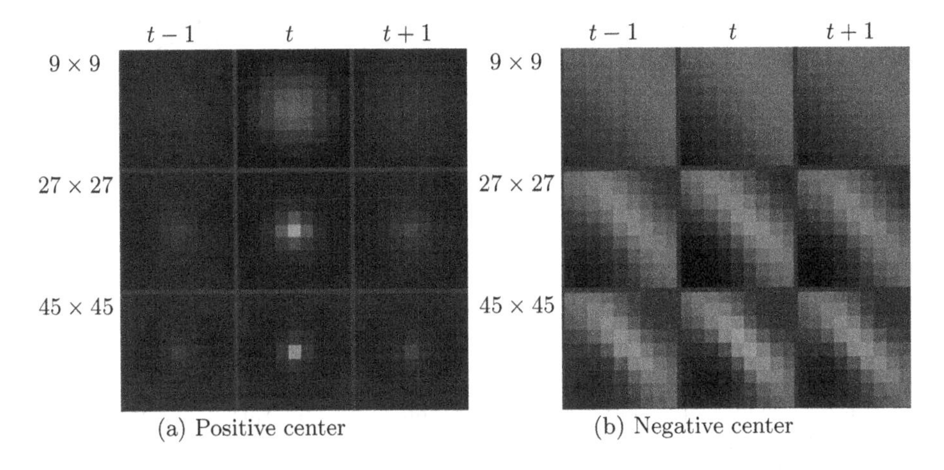

(a) Positive center (b) Negative center

Fig. 4. Examples of centers of RBF network.

previous and next frame, i.e., $\mathbf{x}_p \in \mathbb{R}^{81\times3}$, and (iii) with 9×9, 27×27 and 45×45 image patches extracted from the previous, current and future frames, i.e., $\mathbf{x}_p \in \mathbb{R}^{81\times9}$. We expect (ii) to perform better than (i) since it captures more contextual information around a bacterial dot. Additionally, we expect RBF network to perform better than linear logistic regression since it effectively uses 64 'templates' than one (weight vector in the linear classifier).

3.3 Postprocessing

The classifier returns a probability value at each pixel. These probability values are then thresholded, and pixels that exceed this threshold value are counted after non-maximum suppression [8] to estimate bacterial load.

3.4 Evaluation Method

Performance Metric: We compare different methods in terms of the precision-recall curve for varying thresholding of the probability map before non-maximum suppression. Given a contingency table of false positives, true positives, and false negatives, precision P and recall R are defined as follows.

$$P = \frac{\mathrm{TP}}{\mathrm{TP}+\mathrm{FP}}, \quad R = \frac{\mathrm{TP}}{\mathrm{TP}+\mathrm{FN}}.$$

Given the locations of pixels where a bacterium has been detected and the annotations by the clinician, we draw a disk of radius r around the annotations, and if a detection exists within the disk then it is declared to be a match as in [7].

- If a detection does not match any ground truth annotation, then it is declared to be a false positive.

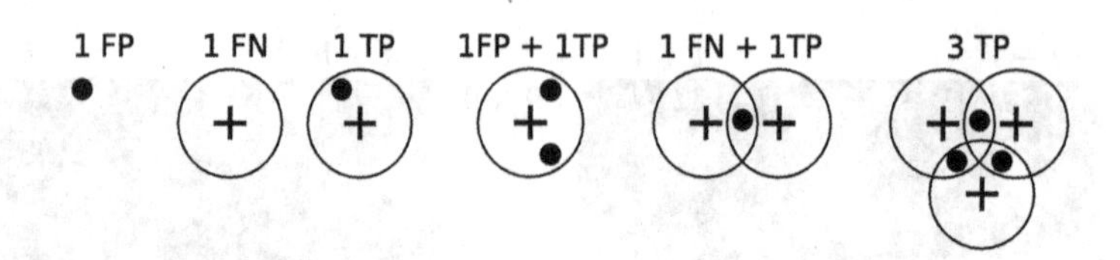

Fig. 5. Illustration of true positives, false negatives and false positives. + are ground truth annotations, • are detections and ∘ are disks of radius r around ground truth annotations.

- If a ground truth annotation does not match any detection then it is declared to be a false negative.
- If a detection exclusively matches a ground truth annotation and vice versa, then the detection is declared to be a true positive.
- If multiple detections exclusively match a ground truth annotation, then one of them is declared to be a true positive while the rest are declared to be false positives.
- If multiple ground truth annotations exclusively match a detection, then the detection is declared to be a true positive while the rest of the ground truth annotations are declared to be false negatives.
- If multiple detections and multiple ground truth match each other non-exclusively then their assignments are resolved greedily. Notice that the assignment might not be optimal.

Figure 5 illustrates these different situations. We set $r = 4$ pixels.

Precision Recall Curves: We utilize cross-validation to test the performance of the methods. To elaborate, we divide 144 image frames in 5 groups $\mathcal{C}_i, i = 1, 2, 3, 4, 5$. To learn the probability map for image frames in group $\mathcal{C}_i$ we train a classifier with positive and negative samples extracted from the remaining four groups $\mathcal{C}_j, j \neq i$. After repeating this process for each group, these probability maps are treated as the output of the classifiers, and the precision-recall curve is estimated by adding the false positive, false negative and true positive values over all image frames.

4 Result

4.1 Cross-Validation

Figure 6 shows the precision-recall curves from all learning methods and feature extraction strategies. We observe the following,

- RBF network performs better than linear logistic regression
- Using temporal information enhances performance. This can be seen from the performance of logistic regression with and without temporal information.
- Using multiple resolution enhances performance. This can be seen from the performance of RBF with and without multiple resolutions.

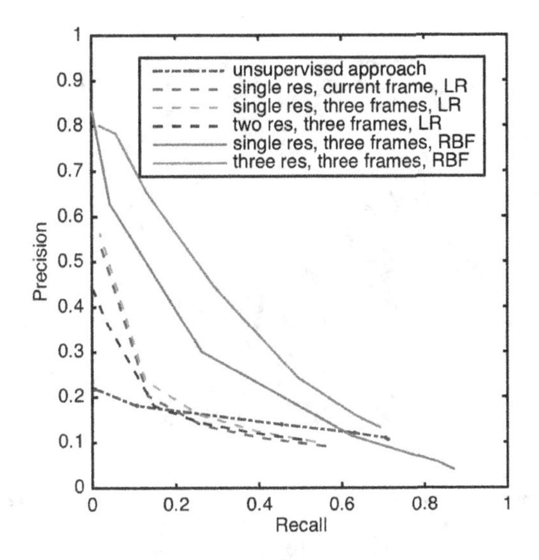

Fig. 6. Precision recall curves for different methods in detecting bacteria.

This follows the intuition behind the use of spatio-temporal multi-resolution analysis. Figure 7 compares annotations by the clinician and detections made by the learning algorithms on one of the validation image frame. We observe more true positives for RBF with multi-resolution spatio-temporal analysis.

4.2 Case vs. Control Study

We use the RBF network learned using spatio-temporal multi-resolution analysis from one of the cross-validation set to estimate bacterial load in each image frame of the entire dataset (12 videos). Figure 8 shows the estimated bacteria load. As expected, we observe that the estimated bacterial load shows significant change in the case group as opposed to control group. Notice that there might exist image frames in post-substance case videos which do not image a region with bacteria, thus resulting in low bacteria count.

5 Related Work

Estimating bacterial load has previously been addressed in an unsupervised learning set-up in our group where difference of Gaussians features were used to enhance the dots in the image. We assessed the precision-recall curve for the unsupervised approach in the context of the acquired ground truth annotations. This is presented in Fig. 6. We observed that the proposed supervised approach outperforms the unsupervised approach significantly. We also demonstrated the performance of the proposed method using fold change of bacterial load in control and case groups before and after the application of the smartprobe. In the

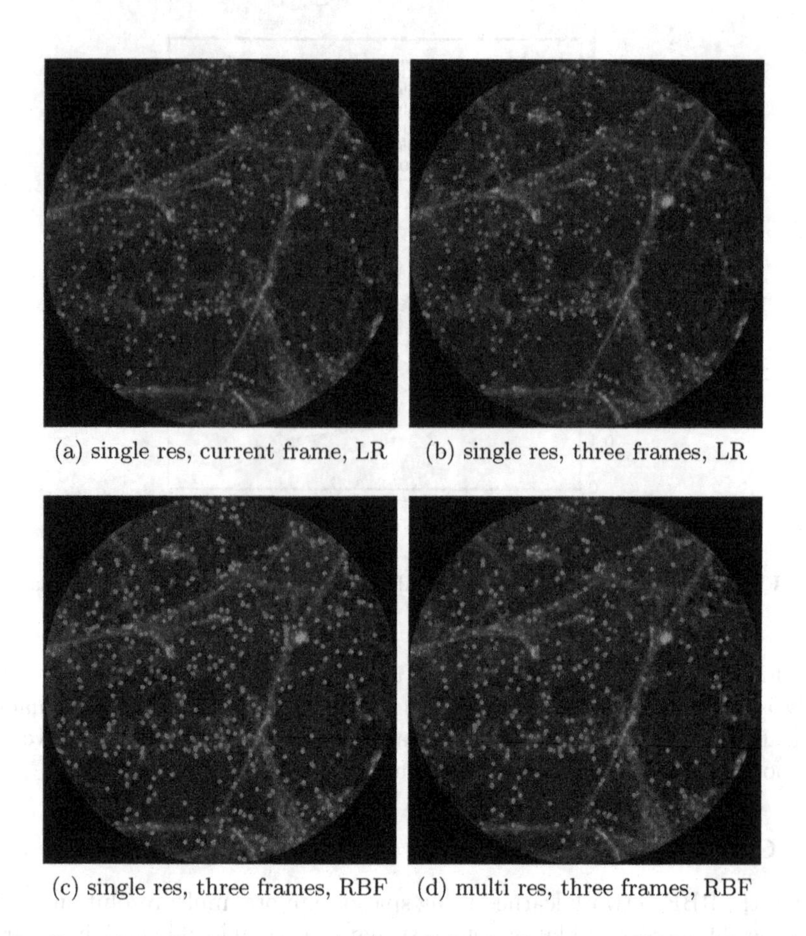

(a) single res, current frame, LR
(b) single res, three frames, LR
(c) single res, three frames, RBF
(d) multi res, three frames, RBF

Fig. 7. Ground truth annotations (red) and detected bacteria (blue) (Color figure online)

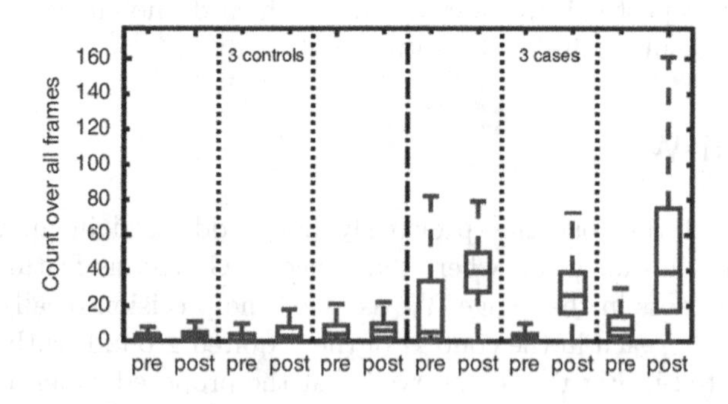

Fig. 8. Change of bacterial load in 3 controls and 3 cases, pre- and post-substance as assessed by RBF network with spatio-temporal and multi-resolution features.

control group, the estimated bacterial load did not change much when exposed to the smartprobe whereas in the case group the estimated bacterial load showed a significant change. Although the results showed the desired performance in case vs. control group in pre- vs. post-substance measurements, the method tended to overestimate bacterial load in an image frame.

Arteta *et al.* tackle the problem of object counting as density estimation, known as *density counting*, where integrating the resulting density gives an estimate of the object count [2]. The authors extract image patch based feature vector[3] $\mathbf{x}_p$ at each pixel p and construct a dictionary (512 elements) via k-means with l_2 distance. Each feature vector is then mapped to a binary vector $\mathbf{z}_p$ via one-hot encoding based on its smallest l_2 distance to the dictionary elements. The authors suggest learning a regression model from $\mathbf{z}_p$ to y_p. This, however, may lead to overfitting due to matching the density value at each pixel. Instead the authors suggest matching the densities such that they should match when integrated over an extended region which leads to a smoothed objective function, and is equivalent to a spatial Gaussian smoothing[4] applied to each feature vector and response vector before applying ridge regression. Essentially, the algorithm constructs a set of templates, and each template is assigned a probability value corresponding to the learned weight vector. Given a test image patch first the closest dictionary element is found, and the related probability value is assigned to the patch. Our approach is similar to [2] in the sense that we work with dot annotations. However, we learn a classifier instead of a regressor, and explicitly detect where each bacterium appears.

Following the work from [2], Arteta *et al.* address learning-to-count penguins in natural images [3]. The authors have access to around 500 thousands images, and each image has been dot-annotated by a maximum of 20 annotators. The authors use multi-task learning through convolutional neural network to, (1) separate foreground containing penguins from background, (2) estimate count density within foreground region, and (3) estimate variability in annotations in foreground region [3, Fig. 2]. We do not use a convolutional neural network due to lack of training images frames.

6 Discussion

We address the task of estimating bacterial load in FCFM images using targeted smartprobe. We create a database of annotated image frames where a clinician has dot-annotated bacteria, and use this database to train a radial basis function network for estimating bacterial load. We show that spatio-temporal features along with multi-resolution analysis can better predict the bacterial load since they capture the 'blinking' effect of the bacteria, and provide better contextual information about the bacterial dot. An attractive aspect of the suggested

[3] Contrast-normalized intensity values at each pixel in the patch after rotating the patch by the dominant gradient.

[4] The authors suggest using the width of the kernel to be greater than half of the typical object diameter.

method is that it can be implemented efficiently using convolution since it estimates the inner product between image patches to compute the outcome of the radial basis functions. We apply the suggested method in estimating bacterial load at each image frame of FCFM videos from control and case group. We observe significant fold change in the case videos before and after introducing smartprobe, which is not observed in the control group.

While annotating ground truth, it is highly likely that the annotator makes mistakes: (s)he can either falsely annotate a bacterium when it is noise, or simply misses annotating a bacteria due to their overwhelming numbers in each frame. These types of error are common in any annotation process, but it might have a more severe impact on learning since our objects are 'dots': while mis-annotation in other datasets such as penguin or crowd is mostly due to occlusion or objects being far away from the camera (for both cases the features of the object pixels are different from positively annotated objects), for bacteria datasets this is not true. Therefore, wrongly annotated bacterium can assign different labels to same feature vector.

We observe that even with more contextual information, elastin structure is often misinterpreted as bacterial dots. The proposed method can be extended further using *hard negative mining*, i.e., explicitly annotating dots which are misclassified as bacteria but actually part of the elastin structure. This definitely requires extra effort from the annotator, but should improve the performance of the classifier. A potential problem with this approach is that it needs to be revised when more diverse elastin structure becomes available, e.g., we do not have patients with granular structure (which arises from smoking). We would need case and control data for this situation in order to cover this scenario. We plan to explore these extensions and limitations as more clinical data becomes available, with the goal of building a robust clinical system.

References

1. Akram, A.R., Avlonitis, N., Lilienkampf, A., Perez-Lopez, A.M., McDonald, N., Chankeshwara, S.V., Scholefield, E., Haslett, C., Bradley, M., Dhaliwal, K.: A labelled-ubiquicidin antimicrobial peptide for immediate in situ optical detection of live bacteria in human alveolar lung tissue. Chem. Sci. **6**, 6971–6979 (2015)
2. Arteta, C., Lempitsky, V., Noble, J.A., Zisserman, A.: Interactive object counting. In: Fleet, D., Pajdla, T., Schiele, B., Tuytelaars, T. (eds.) ECCV 2014. LNCS, vol. 8691, pp. 504–518. Springer, Cham (2014). doi:10.1007/978-3-319-10578-9_33
3. Arteta, C., Lempitsky, V., Zisserman, A.: Counting in the wild. In: Leibe, B., Matas, J., Sebe, N., Welling, M. (eds.) ECCV 2016. LNCS, vol. 9911, pp. 483–498. Springer, Cham (2016). doi:10.1007/978-3-319-46478-7_30
4. Arthur, D., Vassilvitskii, S.: K-means++: the advantages of careful seeding. In: Proceedings of the Eighteenth Annual ACM-SIAM Symposium on Discrete Algorithms (2007)
5. Borstel, M., Kandemir, M., Schmidt, P., Rao, M.K., Rajamani, K., Hamprecht, F.A.: Gaussian process density counting from weak supervision. In: Leibe, B., Matas, J., Sebe, N., Welling, M. (eds.) ECCV 2016. LNCS, vol. 9905, pp. 365–380. Springer, Cham (2016). doi:10.1007/978-3-319-46448-0_22

6. Haykin, S.: Neural Networks: A Comprehensive Foundation, 2nd edn. Prentice Hall PTR, Upper Saddle River (1998)
7. Mandula, O., Šumanovac Šestak, I., Heintzmann, R., Williams, C.K.I.: Localisation microscopy with quantum dots using non-negative matrix factorisation. Opt. Express **22**(20), 24594–24605 (2014)
8. Neubeck, A., Gool, L.V.: Efficient non-maximum suppression. In: International Conference on Pattern Recognition, vol. 3 (2006)

Random Forest-Based Feature Importance for HEp-2 Cell Image Classification

Vibha Gupta(✉) and Arnav Bhavsar

School of Computing and Electrical Engineering (SCEE),
Indian Institute of Technology Mandi, Mandi, Himachal Pradesh 17005, India
vibha_gupta@students.iitmandi.ac.in, arnav@iitmandi.ac.in

Abstract. Indirect Immuno-Fluorescence (IIF) microscopy imaging of human epithelial (HEp-2) cells is a popular method for diagnosing autoimmune diseases. Considering large data volumes, computer-aided diagnosis (CAD) systems, based on image-based classification, can help in terms of time, effort, and reliability of diagnosis. Such approaches are based on extracting some representative features of the images. This work studies the selection of most distinctive features for HEp-2 cell images using the random forest framework. This framework provides a notion of variable importance for ranking features, which we use to select a good subset of features from a large set so that addition of new features to this subset does not increase classification accuracy. We perform various experiments to show the effectiveness of random forest in feature ranking as well as selection using three feature sets. We focus on using simple feature computation, and very less training data, and yet demonstrate high classification accuracy.

Keywords: Random forest · Class-specific features · HEp-2 cell image classification · Feature selection

1 Introduction

Antinuclear antibody (ANA) detection with an HEp-2 substrate, is used as a standard test to reveal the presence of auto-immune antibodies. If antibodies are present in the blood serum of the patient, their presence manifests in distinct nuclear staining patterns of fluorescence on the HEp-2 cells. Classifying these patterns using CAD systems has important clinical applications as manual evaluation demands long hours, fatigue, and is also highly subjective. Such CAD systems generally involve automated image analysis consisting of the following: (1) Feature extraction, (2) Classification. Various automated methods have been developed for HEp-2 cell image classification which typically compute large number of standard image-based features (e.g. morphology, texture, and other sophisticated features (Scale-invariant feature transform (SIFT), Stein's unbiased risk estimate (SURE), Local Binary Pattern (LBP), Histogram of oriented gradients (HOG) etc.)) and use some well-known classification methods (Support Vector machine (SVM), neural networks etc.).

M. Valdés Hernández and V. González-Castro (Eds.): MIUA 2017, CCIS 723, pp. 922–934, 2017.
DOI: 10.1007/978-3-319-60964-5_80

However, we believe that instead of using a large feature set, subsets of 'good' features from original set are sufficient to provide discriminative information regarding the data classes. Addition of any new feature in such a 'good' features set should not increase the classification accuracy further. Indeed, some directions for such a feature selection have been explored in the machine learning (ML) community [1]. These include: (1) Filtering methods that use statistical properties of features to assign a score to each feature. Some examples of scores are: information gain, chi-square test, fisher score, correlation coefficient, variance threshold etc. Theses methods are computationally efficient but do not consider the relationship between feature variables and response variable. They are potentially adhoc, and are usually applied before classification, (2) Wrapper methods explore whole feature space to find an optimal feature subset by iteratively selecting features based on the classifier performance. These methods are computationally expensive due to the need to train and cross-validate model for each feature subset combination, and require exhaustive search which grows exponentially with increase in original features. Some examples are: recursive feature elimination, sequential feature selection algorithms, genetic algorithms etc., (3) Embedded methods perform feature selection as part of the learning procedure, and are generally specific for given classifier framework. Such approaches are more elegant than filtering methods, and computationally less intensive and much less prone to over-fitting than wrapper methods. Some examples are: Random Forests for Feature Ranking, Recursive Feature Elimination (RFE) with SVM, Adaboost for Feature Selection etc. In addition, there are frameworks such as Principal Components Analysis (PCA), Linear Discriminant Analysis (LDA) which can be considered as performing feature selection (via dimensionality reduction). However, such methods project the features into a different subspace, while consider the selection process from the original feature space.

Based on the above discussion, we demonstrate the ability of Random Forest (RF) for selecting an important discriminative feature set and examine the contribution of each feature for the application of HEp-2 cell image classification. The choice of RF for this study is due to its various advantages over other classifiers such as, (1) From the feature selection perspective, RF inherently involves the notion of variable importance together with classification accuracy, which tells about how much the accuracy will degrade when a particular feature is removed from the feature set. By choosing the features which have higher ranks, an optimal subset can be found. (2) It has less parameters to tune (number and depth of trees), as compared to ensemble structures with popular classifiers (e.g. SVM, neural networks) where number of classifiers, choice of kernel and its parameters, cost function, number of hidden layers, number of nodes in the hidden layers etc. require tuning. In this work, we only tune the number of trees. (3) For a large dataset, it can be parallelized, and it can easily handle uneven data sets that have missing variables and mixed data (although this is not relevant in the current work). (4) It is known to perform well across various domains.

While the area of HEp-2 cell image classification has been somewhat explored, to the best of our knowledge, such a study on feature selection, yielding a good classification performance has not been reported. Random forest has only been utilized as classifier for this problem. Prasath at el. [2] utilized texture features such as rotational invariant co-occurrence (RIC) versions of the well-known local binary pattern (LBP), median binary pattern (MBP), joint adaptive median binary pattern (JAMBP), and motif cooccurrence matrix (MCM) along with other optimized features, and reported mean class accuracy using different classifiers such as the k-nearest neighbors (kNN), support vector machine (SVM), and random forest (RF). The highest accuracy is obtained with RIC-LBP combined with a RIC variant of MCM. In [3] authors utilized large pool of feature and used random forest as classifier. They have achieved a good accuracy, but the work uses the ICPR-2012 dataset, which is much smaller than the one which we use (ICPR-2014) for validation, and employs a leave-one-out cross validation, thus involving a large amount of training data. Li et al. [4] made use of the wavelet scattering network, which gives rotation-invariant wavelet coefficients as representations of cell images and achieved an overall accuracy of 90.59%. The work in [5] extracted various features (Shape and Size, Texture, Statistical, LBP, HOG, and Boundary) and used the random forest as a classifier, and achieved an accuracy of 92.85%. The feature selection ability of RF has been reported in some other domains such as gene selection [6], breast cancer diagnosis [7,8], and analysing radar data [9], which inspires us to explore it for the problem of HEp-2 cell image classification.

Unlike the existing works on HEp-2 cell classification, we also stress on the feature selection aspect of random forest. The selection of optimal feature subset can reduce the complexity of system and can also potentially increase the accuracy by removing features that have a negative effect. In addition to our study on feature selection, we focus on using simple feature definitions and very low training data (20%), and demonstrate state-of-the-art performance with the same. We believe that such a direction of feature selection is interesting, given that this is a emerging research area.

2 Methodology

In the following section we discuss about the random forest and its variable importance, the dataset which we have used and the different feature sets which we employ in this work. In this paper we use the terms ‘variable’ and ‘feature’ without distinction.

2.1 Random Forest

Random Forest [10] is a collection of decision trees where each tree is constructed using a different bootstrap sample from the original data. At each node of each tree, a random feature subset of fix size is chosen and the feature which yields maximum decrease in Gini index is chosen for split [10]. In the forest, trees are grown fully and left unpruned. About one-third of the samples, called out of bag samples (OOB), are the left out of the bootstrap sample and used as validation samples to estimate error rate. To predict the class of a new sample, votes received by each class are calculated and the class which has majority of the votes is assigned to the new sample. Figure 1 shows the general structure of random forest.

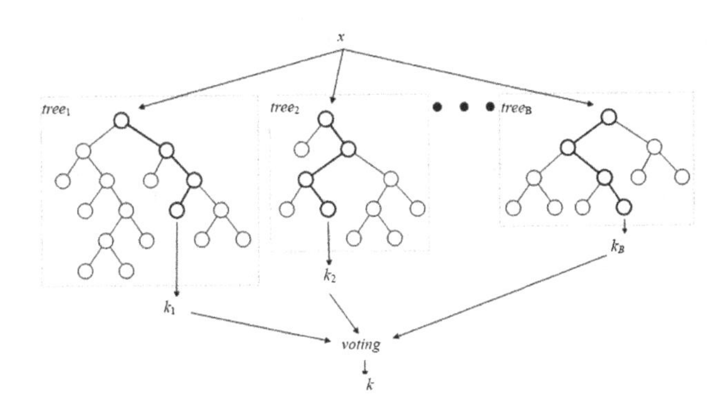

Fig. 1. Structure of random forest [8].

Random forest offers two different measures to gauge the importance of features, the *variable importance (VI)* and the *Gini importance (GI)* [10]. The average decrease in the model accuracy on the OOB samples when specific feature is randomly permuted gives the VI of that feature. The random permutation essentially indicates the replacement of that feature with a randomly modified version of it. VI shows how much the model fit decreases when we permute a variable. The greater the decrease the more significant is the variable. For variable X_j, the VI is calculated as follows: (1) Calculate model accuracy without permutation using OOB samples, (2) Randomly permute the variable X_j, (3) Calculate model accuracy with permuted variable together with the remaining non-permuted variable, (4) Variable importance is found out by taking difference of accuracies before and after permutation and is averaged over all trees.

$$VI(X_j) = \frac{\sum_{t=1}^{ntree} VI^t(X_j)}{ntree} \tag{1}$$

Gini importance (GI) indicates the overall explanatory power of the variables. The GI uses the drop in Gini index (impurity) as a measure of feature relevance. GI is a biased estimate [11] when features vary in their scale of measurement. Owing to this VI is more reliable for feature selection when subsampling without replacement is used instead of bootstrap sampling to construct the forest. Therefore, we consider the VI measure for variable ranking in this paper.

Finding all the important features which are highly related to response is useful for interpretation. Random forest provides the feature ranking based on their importance. We provide an examination how overall accuracy varies when the top-ranked features in some interval are chosen and how many features are actually required to achieve accuracy equal (or in some cases, even greater) to accuracy which had been obtained with all features. In result section, we will briefly discuss about the impact of top-ranked features on accuracy using three different feature sets.

2.2 Dataset Description and Feature Sets

Dataset: In this work, the publicly available dataset from the **ICPR** 2014 HEp-2 cell image classification contest is used which comprising more than 13,000 cell images [12]. The dataset consist of six classes termed as: Homogeneous (H), Speckled (S), Nucleolar (N), Centromere (C), Nuclear Membrane (NM), Golgi (G). In the dataset, each class consist of positive and intermediate images. The intermediate images are generally lower in contrast compared to the positive images. The dataset also includes mask images which specify the region of interest of each cell image. Figure 2 provides the details of examples for each class.

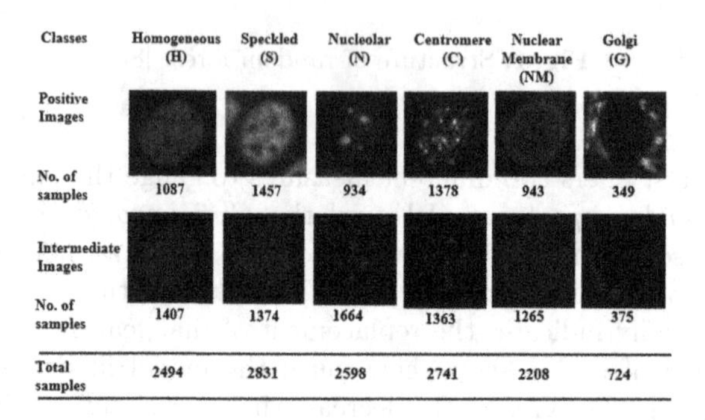

Fig. 2. Sample images (positive and intermediate) and number of examples of each class in dataset.

Feature Extraction: One of our intentions in this work is to demonstrate that even simplistic feature extraction (unlike the more sophisticated ones such as SIFT, HOG, SURF etc.), can also yield good classification performance. Our feature sets are constructed keeping this in mind.

1. Class-specific features: Motivated by experts knowledge [13] which characterizes each class by giving some unique traits, we define features based on such traits. As the location of these traits in each class may be different, features are extracted from specific regions of interest (ROI) for each particular class, computed using the mask images. Figure 3 shows the unique traits of each class. For example, in NM class useful information can be found in a ring which is centered on the boundary. Utilizing visually observed traits following features are extracted:

   1. Boundary area ratio: It is a area ratio in boundary ring mask.
   2. Inner area ratio: It is a area ratio in inner mask.
   3. Eroded area connected component: It gives the number of white pixels in inner mask.

   Due to space constrain, we do not describe all features elaborately here, and we refer the reader to our previous work [14]. Table 1 shows the feature list which are extracted for each class.

   These features, involve simple image processing operations such as scaler image enhancement, thresholding, connected components. And morphological operation. Our earlier work [14] explains the features in more detail which are extracted using the unique characteristics of each class. These feature are simple, efficient, and more interpretable. As listed in Table 1, the total feature definitions which are extracted from all classes is 18, and using various combination of threshold and enhancement parameters, a total 144 features are obtained.
2. Standard traditional features [3]: Features which include morphology (like Number of objects, Area, Area of the convex hull, Eccentricity, Euler number, Perimeter), texture (like Intensity, Standard deviation, Entropy, Range, GLCM) are extracted at 20 intensities equally spaced from its minimum to its maximum intensity. These extracted features are same as the features which were used in [3]. However, in [3], these features were applied on much smaller dataset.
3. Combination of standard and class-specific features: Five class-specific features (EACC, BAR, OAR, IAR, AOD) are added to standard features.

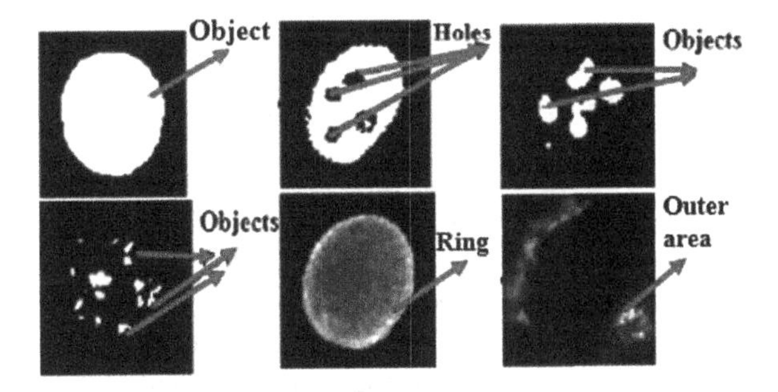

Fig. 3. Unique characteristics of each class.

2.3 Feature Selection Procedure

1. Training: Random forest (RF) internally divides the training data into 67%-33% ratio for each tree randomly, where 67% is used to train a model while 33% (out of bag) used for testing (cross-validation). The OOB error rate is

Table 1. Class-specific features for all classes.

Classes	Class-specific features
Homogeneous	Maximum object area (MOA), Area of connected component (ACC), Maximum perimeter (MP)
Speckled	Hole area (HA), Hole number (HN), Euler number (EN)
Nucleolar	Maximum object area (MOA), Average object area (AOA), Connected component (CC)
Centromere	Maximum object area (MOA), Average object area (AOA), Connected component (CC)
Nuclear membrane	Boundary area ratio (BAR), Inner area ratio (IAR), Eroded area connected component (EACC)
Golgi	Outer area ratio (OAR), Average object distance (AOD), Eroded area, connected component (EACC)

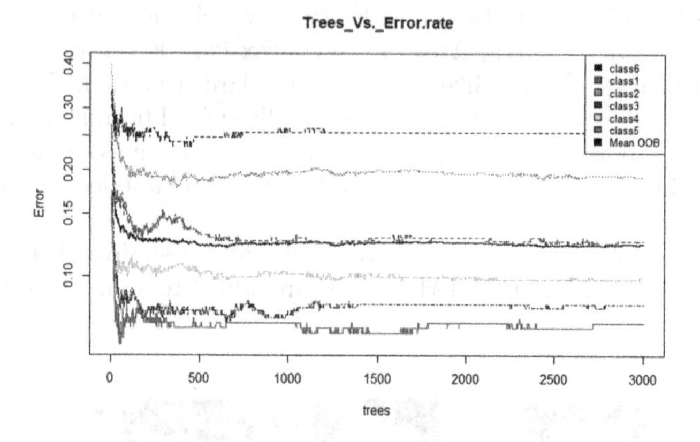

Fig. 4. Number of trees vs. error rate.

calculated using this 33% data only. To automatically decide the value of RF parameters such as number of trees and number of features used at each tree, the OOB error rate is generally considered [15]. In this work, we only decide the number of trees. For the number of features at each tree, we use a default value which is square root of the total number of features (based on the R package utilized for this study). Figure 4 illustrates the variation in OOB error rate with the variation in the number of trees. There are seven lines in graph where six correspond to each class and the black line corresponds to mean of OOB error rate of all class. The point where the OOB error reduces negligibly could be considered as a good point to fix the value of number of trees [15].

2. Feature selection: By using variable importance (VI) provided by RF (after cross-validation), a subset of good features is chosen to test the remaining 80% data. The procedure of selecting a good feature subset is given as follows:

(a) Top ranked features (in terms of VI) in sets of 10 (e.g. 10, 20, 30...) are used for testing.
(b) After some point, addition of more features do not increase the classification accuracy further.
(c) As there is no significant improvement after this point, the features upto this point is considered as good feature subset. This feature subset gives the accuracy which is equal to the accuracy obtained using all features.

3 Results and Discussion

In this section we discuss the experimental results obtained using the three different feature sets discussed above. An important point to note is that we use only 20% dataset from all the classes for training and remaining for the testing (80%). This is motivated by the fact that in realistic applications the amount of training data is low. We implemented the proposed approach using the R program language version R-3.1.2 with RF package. For more on usage and features of the R package please refer [16].

To determine the effectiveness of random forest for feature importance the experiment involves two steps: (1) Computing the variable importance (VI) of features, (2) Computing the performance of random forest using top ranked features. Figure 5 shows the ranking of feature given by VI in random forest for one of the feature set. After getting the ranking of features, we use top features in sets of 10 for computing the classification accuracy, and the same procedure is repeated for four random trials.

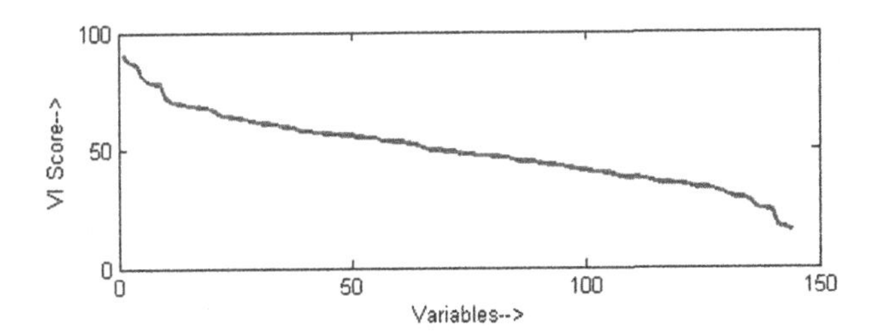

Fig. 5. Ranking of variables based on VI.

3.1 Evaluation Metrics

In all the experiments, the mean accuracy over all classes and false positive are calculated. The mean class accuracy (MCA) and false positive are defined as follows:

$$MCA = \sum_{i=1}^{N} \frac{CC(i)}{N_i} \tag{2}$$

where $CC(i)$, N_i are the classification accuracy and total samples of class i, and N is the total number of classes (here it is 6).

False Positive (FP): It is a proportion of negatives samples that are incorrectly classified as positive.

$$FP = 1/N \sum_{i=1}^{N} \frac{WC(i)}{FS_i} \tag{3}$$

where $WC(i)$ is the wrongly classification samples of class i and FS_i is the false samples for class i (for example, for class 1, the samples of other classes (2, 3, 4, 5 and 6) will be the false samples).

The overall accuracy (OA), which is the overall correct classification rate is also calculated for ease of comparison, is given as:

$$OA = 1/N \sum_{i=1}^{N} CC(i) \tag{4}$$

3.2 Performance Evaluation Using Three Feature Sets

Tables 2, 3 and 4 shows the accuracy details of three feature sets (Class-specific features, Standard traditional features, and combined features). To follow these tables, notions N_A, N_s and N_H are defined as:

N_A: All features.
N_s: Number of features that match the accuracy as with NA.
N_H: Number of features that yield the highest accuracy.

Each table provides the exact number for N_A, N_s, N_H and for their corresponding MAC, FP, OA for four random trails. From tables, it is clear that highest accuracy is slightly higher than accuracy obtained using all features. The reason behind is that importing features which is not important or correlated features negatively impact the accuracy. Importantly, about 50–60% features give the accuracy which equal to standard accuracy (using all features) in the first feature set. For the other feature sets, we note that (Tables 3 and 4) 15–20% features are sufficient to yield standard accuracy; less than 30% features are required to produce highest accuracy.

Table 5 specifies the best performing feature types in the experiment with the combined feature set. Note that the total of 977 features are essentially composed of parameter variations of 28 features types including the class-specific and traditional texture features. The best feature types among the traditional texture features are decided based on the absolute number of their parameter variations included in the top 230 selected features. For the class-specific feature types, as the original number of instances with parameter variations are quite low, the ranking is based on the relative fraction of their parameter variations included in the final selections. From the table, it is clear that the class-specific

Table 2. Accuracy details correspond to class-specific features.

	With all features			Features			Corresponding to highest accuracy		
	Overall accuracy	Mean accuracy over all classes	False positive (FP)	N_A	N_S	N_H	Overall accuracy	Mean accuracy over all classes	False positive (FP)
Trail 1	87.27	86.21	2.84	144	70	80	87.42	86.25	2.80
Trail 2	86.76	85.82	2.97	144	80	130	86.88	86.08	2.95
Trial 3	87.56	85.82	2.74	144	90	100	87.58	86.08	2.74
Trail 4	86.47	85.18	3.03	144	90	140	86.60	85.43	3.04
Average	87.01	85.75	2.89				87.12	85.97	2.88

Table 3. Accuracy details correspond to standard traditional features.

	With all features			Features			Corresponding to highest accuracy		
	Overall accuracy	Mean accuracy over all classes	False positive (FP)	N_A	N_S	N_H	Overall accuracy	Mean accuracy over all classes	False positive (FP)
Trail 1	90.76	88.86	2.02	932	180	300	90.98	89.24	1.99
Trail 2	91.02	88.94	2.0	932	120	260	91.39	89.55	1.93
Trial 3	91.42	88.69	1.87	932	90	280	92.03	89.58	1.75
Trail 4	90.42	88.19	2.09	932	110	150	90.89	88.89	1.99
Average	90.90	85.67	1.99				91.32	89.31	1.92

Table 4. Accuracy details correspond to combined feature set.

	With all features			Features			Corresponding to highest accuracy		
	Overall accuracy	Mean accuracy over all classes	False positive (FP)	N_A	N_S	N_H	Overall accuracy	Mean accuracy over all classes	False positive (FP)
Trail 1	92.49	91.84	1.70	977	90	230	92.93	92.58	1.61
Trail 2	92.83	92.25	1.65	977	130	200	93.04	92.73	1.61
Trial 3	93.01	91.92	1.56	977	90	190	93.55	92.87	1.46
Trail 4	92.57	91.82	1.67	977	70	120	93.14	92.69	1.57
Average	92.73	91.75	1.64				93.17	92.71	1.56

feature heavily contribute to the top ranks (in relative terms), while, a selection of texture feature types play an important role in absolute terms.

Table 6 details the accuracy obtained for positive and intermediate samples. It is clearly observed from the table that proposed framework produces comparable (in some cases good) performance even for individual experiments (positive and intermediate).

Table 5. Five best performing features.

Top 5 features	Absolute ranking		Relative ranking	
	Feature	Frequency	Feature	Frequency
1	Entropy	26	OAR	0.92
2	Range	24.5	AOD	0.92
3	Standard deviation	24.25	IAR	0.86
4	Eccentricity	15.5	BAR	0.86
5	Euler number	14	EACC	.78

Table 6. Accuracy details of positive and intermediate samples.

Trials	Positive samples			Intermediate samples		
	OA	MCA	FP	OA	MCA	FP
Trail 1	96.69	97.0	.79	92.67	90.53	1.63
Trail 2	96.24	95.74	.83	93.05	91.09	1.56
Trial 3	96.12	95.75	.88	92.43	90.74	1.69
Trail 4	96.79	96.92	.72	92.16	89.63	1.76
Average	96.46	96.35	.80	92.57	90.49	1.66

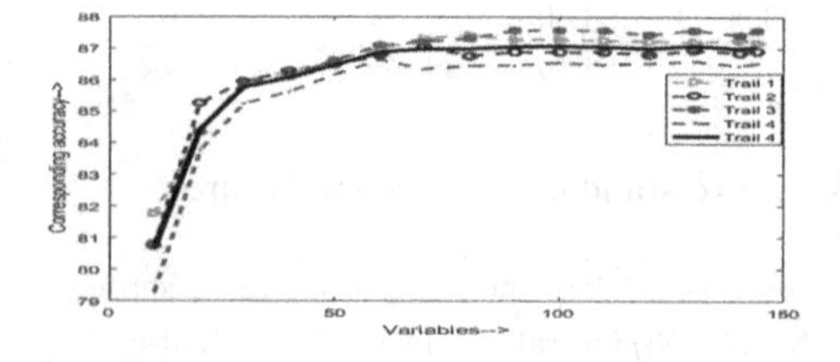

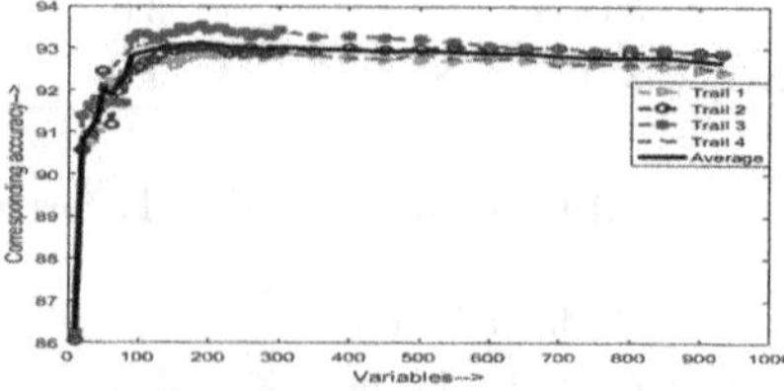

Fig. 6. Variation in accuracy using (a) class-specific feature set, (b) standard traditional feature set.

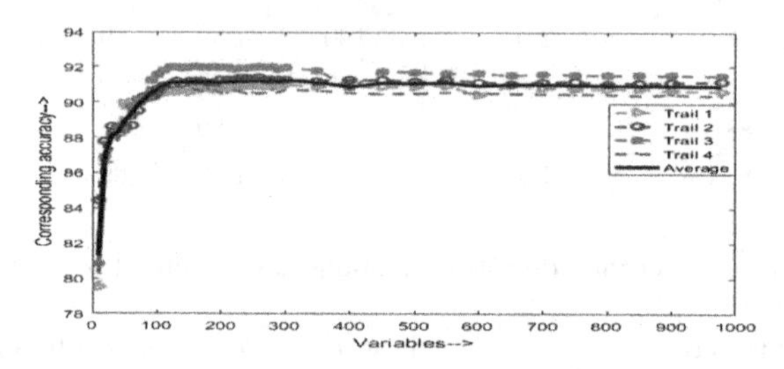

Fig. 7. Variation in accuracy using combined feature set.

Figures 6 and 7 show the variation in accuracy when top features are chosen (in sets of 10), each corresponds to one feature set. In each figure, the black line shows the average accuracy variation while others for different trails. It can also be noticed from Figs. 5, 6 and 7 that at some point accuracy is somewhat higher than the accuracy with all features.

3.3 Performance Comparison

Table 7 illustrates the performance comparison among various method, where in [2,4,5] utilized random forest for classification, [17,18] are recent methods using arguably, more sophisticated features, and the work of [19] uses deep learning.

For fair comparisons, the results below for the methods in [18,19], are those without data augmentation, as in our case (since the details for data augmentation are not clearly known). Indeed, with augmentation, these methods yield higher accuracy (mentioned in brackets in Table 7) than those obtained without

Table 7. Performance comparison.

Methods	Mean accuracy over all classes (%)	False positives (%)	Overall accuracy (%)
Prasath et al. [2]	94.29	NA	NA
Li et al. [4]	89.79	10.33	90.59
Agrawal et al. [5]	NA	NA	89.02
Ensafi et al. [17]	94.9	5.09	94.48
Manivannan et al. [18]	92.58 (95.21)	7.38 (4.8)	NA
Gao et al. [19]	88.58 (96.76)	NA	89.04 (97.24)
Proposed	92.71	1.56	93.17

argumentation. This experiment indicates that the proposed approach can yield good results in spite of relatively low training data.

In general, we observe that, considering the best results, the proposed approach outperforms most methods, especially in terms of false positive.

4 Conclusion

In this work, we explore random forest for feature selection for HEp-2 cell image classification. The notion of variable importance is used to select important features from a large set of simple features. Our experiments show that such a feature selection yields a significantly reduced feature subset, which can, in fact, result in accuracy higher than that using original large feature set.

References

1. Blum, A.L., Langley, P.: Selection of relevant features and examples in machine learning. Artif. Intell. **97**(1), 245–271 (1997)
2. Prasath, V.B.S., Kassim, Y.M., Oraibi, Z.A., Guiriec, J.-B., Hafiane, A., Seetharaman, G., Palaniappan, K.: HEp-2 cell classification and segmentation using motif texture patterns and spatial features with random forests. In: 23th International Conference on Pattern Recognition (ICPR). IEEE (2016)
3. Strandmark, P., Ulén, J., Kahl, F.: HEp-2 staining pattern classification. In: 2012 21st International Conference on Pattern Recognition (ICPR), pp. 33–36. IEEE (2012)
4. Li, B.H., Zhang, J., Zheng, W.-S.: HEp-2 cells staining patterns classification via wavelet scattering network and random forest. In: 2015 3rd IAPR Asian Conference on Pattern Recognition (ACPR), pp. 406–410. IEEE (2015)
5. Agrawal, P., Vatsa, M., Singh, R.: HEp-2 cell image classification: a comparative analysis. In: Wu, G., Zhang, D., Shen, D., Yan, P., Suzuki, K., Wang, F. (eds.) MLMI 2013. LNCS, vol. 8184, pp. 195–202. Springer, Cham (2013). doi:10.1007/978-3-319-02267-3_25

6. Moorthy, K., Mohamad, M.S.: Random forest for gene selection and microarray data classification. In: Lukose, D., Ahmad, A.R., Suliman, A. (eds.) KTW 2011. CCIS, vol. 295, pp. 174–183. Springer, Heidelberg (2012). doi:10.1007/978-3-642-32826-8_18
7. Paul, A., Dey, A., Mukherjee, D.P., Sivaswamy, J., Tourani, V.: Regenerative random forest with automatic feature selection to detect mitosis in histopathological breast cancer images. In: Navab, N., Hornegger, J., Wells, W.M., Frangi, A.F. (eds.) MICCAI 2015. LNCS, vol. 9350, pp. 94–102. Springer, Cham (2015). doi:10.1007/978-3-319-24571-3_12
8. Nguyen, C., Wang, Y., Nguyen, H.N.: Random forest classifier combined with feature selection for breast cancer diagnosis and prognostic (2013)
9. Hariharan, S., Tirodkar, S., De, S., Bhattacharya, A.: Variable importance and random forest classification using RADARSAT-2 PolSAR data. In: IGARSS, pp. 1210–1213 (2014)
10. Breiman, L.: Random forests. Mach. Learn. **45**(1), 5–32 (2001)
11. Strobl, C., Zeileis, A.: Danger: high power!-exploring the statistical properties of a test for random forest variable importance (2008)
12. Hobson, P., Percannella, G., Vento, M., Wiliem, A.: Competition on cells classification by fluorescent image analysis. In: Proceedings of 20th IEEE International Conference Image Processing (ICIP), pp. 2–9 (2013)
13. Foggia, P., Percannella, G., Soda, P., Vento, M.: Benchmarking HEp-2 cells classification methods. IEEE Trans. Med. Imaging **32**(10), 1878–1889 (2013)
14. Gupta, V., Gupta, K., Bhavsar, A., Sao, A.K.: Hierarchical classification of HEp-2 cell images using class-specific features. In: European Workshop on Visual Information Processing (EUVIP 2016). IEEE (2016)
15. Segal, M.R.: Machine learning benchmarks and random forest regression. Cent. Bioinform. Mol. Biostat. (2004)
16. Liaw, A., Wiener, M.: Classification and regression by randomforest. R News **2**(3), 18–22 (2002)
17. Ensafi, S., Lu, S., Kassim, A.A., Tan, C.L.: A bag of words based approach for classification of HEp-2 cell images. In: 2014 1st Workshop on Pattern Recognition Techniques for Indirect Immunofluorescence Images (I3A), pp. 29–32. IEEE (2014)
18. Manivannan, S., Li, W., Akbar, S., Wang, R., Zhang, J., McKenna, S.J.: An automated pattern recognition system for classifying indirect immunofluorescence images of HEp-2 cells and specimens. Pattern Recogn. **51**, 12–26 (2016)
19. Gao, Z., Wang, L., Zhou, L., Zhang, J.: HEp-2 cell image classification with deep convolutional neural networks (2015)

Automatic Quantification of Epidermis Curvature in H&E Stained Microscopic Skin Image of Mice

Saif Hussein[1,2(✉)], Sabah Jassim[1], and Hisham Al-Assam[1]

[1] Applied Computing Department, The University of Buckingham, Buckingham, UK
{Saif.hussein, sabah.jassim, Hisham.al-assam}@buckingham.ac.uk
[2] University of Baghdad, Baghdad, Iraq
saefrouef@yahoo.com

Abstract. Changes in the curvature of the epidermis layer is often associated with many skin disorders, such as ichthyoses and generic effects of ageing. Therefore, methods to quantify changes in the curvature are of a scientific and clinical interest. Manual methods to determine curvature are both laborious and intractable to large scale investigations. This paper proposes an automatic algorithm to quantify curvature of microscope images of H&E-stained murine skin. The algorithm can be divided into three key stages. First, skin layers segmentation based on colour deconvolution to separate the original image into three channels of different representations to facilitate segmenting the image into multiple layers, namely epidermis, dermis and subcutaneous layers. The algorithm then further segments the epidermis layer into cornified and basal sub-layers. Secondly, it quantifies the curvature of the epidermis layer by measuring the difference between the epidermis edge and a straight line (theoretical reference line) connecting the two far sides of the epidermis edge. Finally, the curvature measurements extracted from a large number of images of mutant mice are used to identify a list of genes responsible for changes in the epidermis curvature. A dataset of 5714 H&E microscopic images of mutant and wild type mice were used to evaluate the effectiveness of the algorithm.

Keywords: Skin layer segmentation · Epidermis layer quantification · Skin curvature quantification

1 Introduction

Although the Human Genome Project (HGP) determined the sequence of chemical base pairs, which make up human DNA and of identifying and mapping all of the genes of the human genome (approximately 20,500 genes), functionalities of the vast majority of genes are still under research or unknown [1]. The Mouse Genetics Project (MGP) is a large-scale mutant mouse production and phenotyping program aimed at identifying new model organisms of disease. The aim of the MGP is to produce over 20,000 mutant lines and the results are then translated into diagnostics, and treatments for diseases.

M. Valdés Hernández and V. González-Castro (Eds.): MIUA 2017, CCIS 723, pp. 935–945, 2017.
DOI: 10.1007/978-3-319-60964-5_81

Mammalian skin is a complex organ composed of a variety of cell and tissue types. It is the largest mammalian organ, and although apparently simple, it is a highly organized tissue comprised of the epidermis, underlying dermis containing connective tissue and a deeper subcutaneous adipose layer [2]. The skin can reveal evidence of inflammation, hyperplasia, connective tissue disorders and underlying metabolic changes resulting from local and systemic influences. As dermatology research requires a detailed understanding of skin structure and organization, which requires quantitative measurements, the automatic quantification of changes in skin structures has a variety of different applications for biological research. One requirement in this area is the accurate segmentation of the skin compartments followed by the assessment of various characteristics of those compartments. For example, changes in the curvature of the epidermis layer are often associated with many skin disorders, such as ichthyoses and generic effects of ageing [3]. Therefore, methods to quantify changes in the curvature are of a scientific and clinical interest. Arguably, this cannot be achieved without a reliable segmentation of different skin layers.

Traditional skin studies have utilized manual methods for the quantification of skin features, which is a challenging task due to the complexity of the analysis with big amount of datasets. However, there has been recently a move towards the automation of these techniques to improve accuracy and efficiency by reducing processing time and laboratory costs.

In previous works, one of the attempts to automate the analysis of skin layer in microscopic images was the introduction of the novel shapelet-based procedure for the epidermis boundaries identification and thickness measurement [4]. Later, a classification method was proposed to segment skin layers in images based only on their speckle information [5]. Another interesting work was a hybrid sequence segmentation and classification technique applied to split images into different channels by z-stack deconvolution, fitting model of skin layer and their classification into epidermis and dermis [6]. Recently high-definition optical coherence tomography was employed for grey imaging subsurface skin tissues to segment the epidermis layer with using weighted least square based edge-preserving smoothing method with weighted median filter followed by wavelet techniques an [7].

Although all those methods gave great results, they do not allow automatic quantification of epidermis curvature, which is a crucial feature for gene identification and skin disorders diagnostic. In addition to above there wasn't any of methods achieve sub-segmentation of the epidermis layer into cornified and basal layer.

In this paper, we propose an automated method to quantify and measure the curvature of the epidermis layer (cornified sub-layer). The ultimate aim is to use the measurements extracted from a large number of mutant mice to identify a list of genes responsible for significant alterations in the curvature of the epidermis.

The rest of the paper is organised as follows. Section 2 explains the proposed algorithm in details whereas Sect. 3 presents the method of identifying the interesting genes using reference range (RR). Section 4 presents the data set used to test the proposal algorithm as well as the result of interesting genes responsible for changes in the epidermis curvature. Finally the Sect. 5 summarises the whole paper by clarifying the aims of this work.

2 The Proposed Method

The proposed method consist of an algorithm that automatically computes the curvature along the epidermis layer and identify the genes associated with significant changes in the curvature. It is summarized with an illustration in Fig. 1, and works in 3 key stages.

1. Segments the skin image into its three layers, namely Epidermis, Dermis and Subcutaneous layers. And further segments the Epidermis layer into Cornified and Basal sub-layers [8].
2. Quantify the curvature of the epidermis layer by measuring the difference between the epidermis edge and a straight line (an artificial reference line) connecting the two far sides of the epidermis edge.
3. Using the curvature measurements extracted from a large number of skin images of mutant mice to identify a list of genes responsible for changes in the epidermis curvature.

The following subsections explain each of the key stages in details.

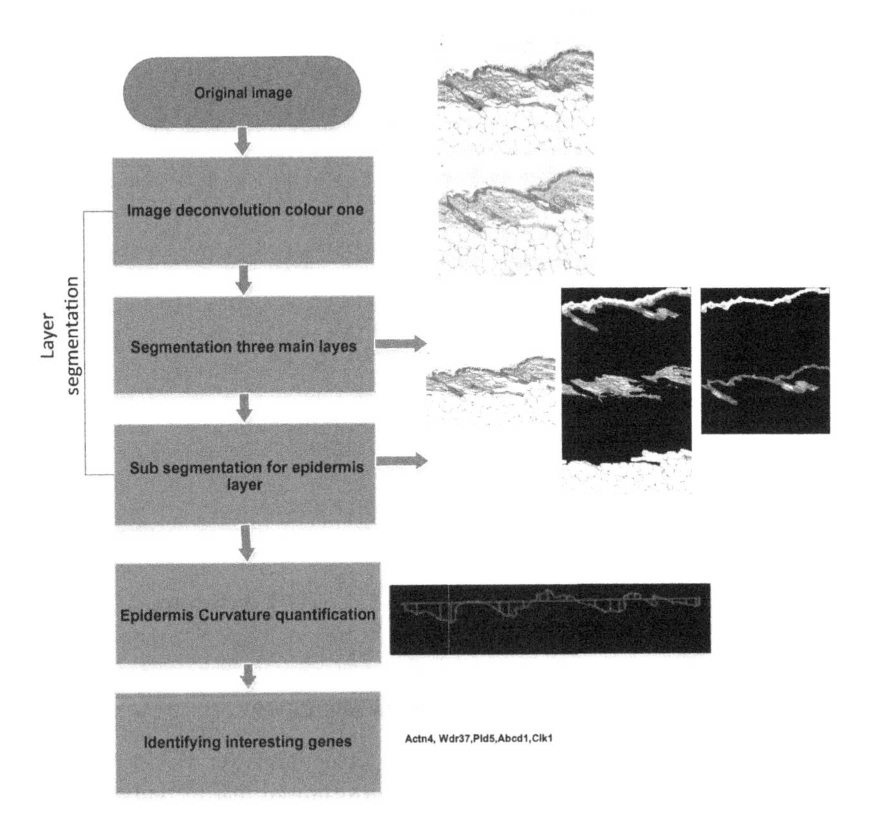

Fig. 1. The proposed algorithm to quantify the curvature of the Epidermis layer to identify the genes associated with significant changes in it

2.1 Layers Segmentation Algorithm

Segmentation is an important step in the algorithm because the quality of the results is impacted by the quality of the underlying segmentation. Our algorithm uses adaptive colour deconvolution techniques on the H&E stain images to separate different tissue structures as shown in Fig. 2 [8]. Then the algorithm uses a set of morphological operations with appropriate structural elements, such as image open, image close and image fill a gap. To further remove unwanted objects, logical operations, such as adding and multiplying images, have been applied to segment epidermis sub layers as illustrated in Fig. 3.

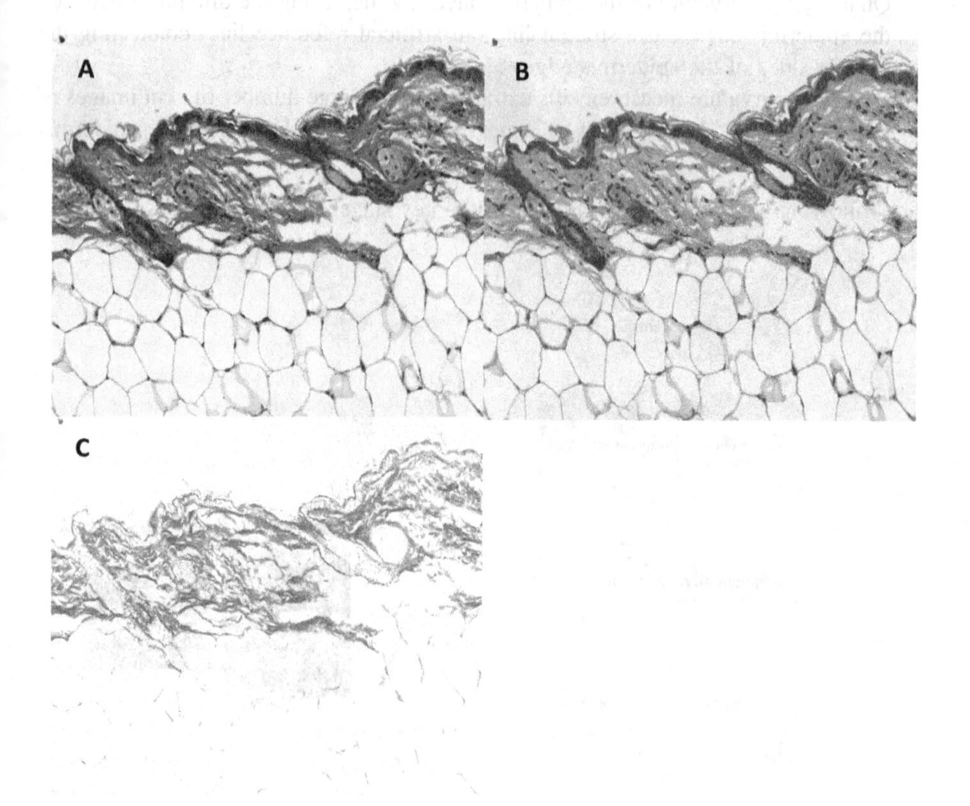

Fig. 2. Colour deconvolution (A) Original image. (B) Colour 2 deconvolution. (C) Colour 3 deconvolution. (Color figure online)

2.2 Quantifying the Curvature of the Epidermis Layer

Accurate quantification of the curvature of the epidermis layer would provide new insights into relevant skin disease. Several studies showed that the curvature of rete ridges can vary in ageing or obesity [9]. Other studies found several skin disorders are associated with changes in epidermal junction characterization, such as psoriasis [10]. The proposed method to quantify the curvature relies on measuring the difference

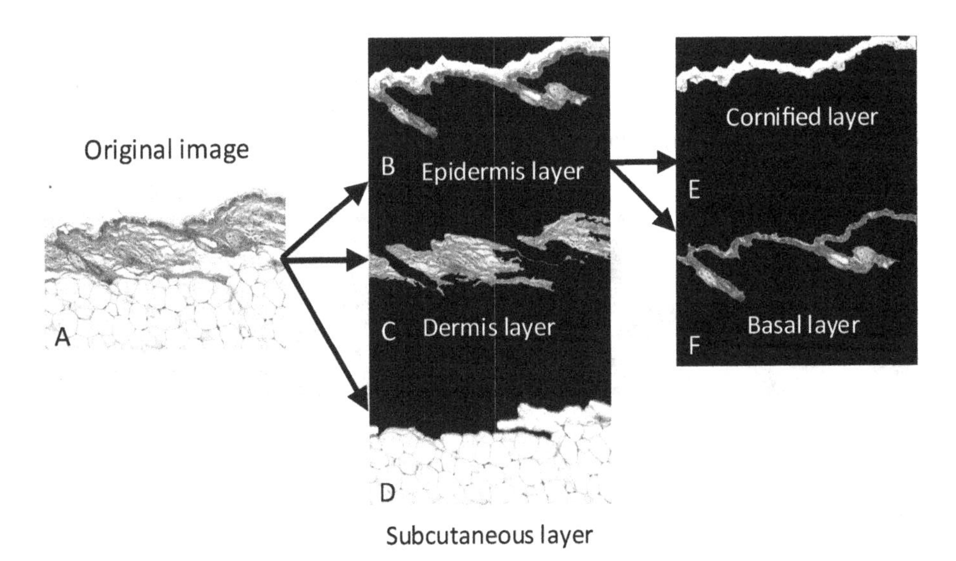

Fig. 3. Automatic method for layers and sub-layers segmentation (A) Original image. (B) Epidermis layer segmentation. (C) Dermis layer segmentation. (D) Sub-cutaneous layer segmentation. (E) Cornified sub layer segmentation. (F) Basal sub layer segmentation.

between the actual border of the epidermis and a straight reference line connecting the two far sides of the border. The curvature quantification stages can be summarized as follows.

- The border the cornified layers are first extracted from the binary mask of the layer obtained in the previous stage (Fig. 4B and C). This is simply done by enlarging the mask by one pixel from the top and the bottom followed by subtracting the original mask from it.
- Select the top border (Fig. 4D) and connect the far two sides using a theoretical reference line (the red line in Fig. 4E).
- Calculate the distances between the points on the actual top border with the reference straight line as explained below.
- The mean of all distances is used a measure to represent the curvature of the epidermis.

The distance from a line (the reference line) with equation

$$Ax + By + C = 0 \tag{1}$$

to the point (u, v) [11] is:

$$\text{Distance} = \frac{|Au + Bv + C|}{\sqrt{A^2 + B^2}} \tag{2}$$

The distance has been found between each points on the epidermis curve and the theoretical line (the red line on Fig. 4E), which was define by the Eq. (1). After that, small distances (less that a threshold) are ignored as they are more likely to represent a noise. The mean of distances is calculated for each image to represent the curvature measurement. It is worth mentioning that all images have more or the less the same orientations close to horizontal.

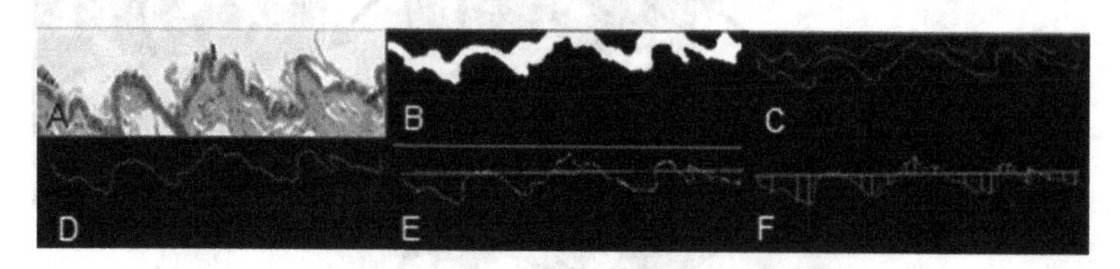

Fig. 4. Representative images for each step of the automated image analysis method to find the curvature in the cornified layer (NB: (A) Original image. (B) Cornified layer. (C) Border for the cornified layer. (D) Bottom curvature line. (E) Theoretical red line on bottom curvature line. (F) Distance between theoretical red line and the bottom curvature line.) (Color figure online)

3 Identifying the Genes Associated with Changes in the Curvature of the Epidermis

This paper follows a stringent protocol described by the MGP to identify interesting genotypes for further analysis [12], which can be summarised as follows. The reference range (RR) method is used to establish the distribution of the WT measurements for each genetic background. If 60% of the measurements obtained form images of a mutant mice falls outside the 95% confidence intervals (CI) of the RR range, the knocked-out gene for that mutant line is ladled as an interesting gene i.e. a gene that has led to significant changes to the epidermis curvature. The lower and upper 95% CI were computed by the following equations:

$$\mathit{Lower}\,95\%\,\mathit{CI} = -\mathrm{S} * 1.64 + \mathrm{M} \tag{4}$$

$$\mathit{Upper}\,95\%\,\mathit{CI} = \mathrm{S} * 1.96 + \mathrm{M} \tag{5}$$

Where the M is the mean of the WT measurements, and S is the standard deviation.

If the number of measurements from a mutant that are below the lower 95% CI was >60% or above the upper 95% CI was >60%, the gene of that particular mutant line is consider for future investigation by biologists.

4 Experiments

4.1 Dataset

The proposed method is tested on a dataset generated by the Wellcome Trust Sanger Institute (WTSI), which generates mouse genetic and phenotypic data, and distributes this data and resources to the scientific community. The program at WTSI to functionally annotate the mouse genome is illustrated. The primary phenotyping data generated by WTSI aims to discover genes involved in diseases. There are many other research projects that focus on the biological functions of genes in the mouse genetics area by WTSI, such as mouse behaviours, cancer and developmental genetics [13]. Data can be accessed via the mouse resources portal (http://www.sanger.ac.uk/mouseportal/). H&E stained skin from 16 week-old female mice. The testing data set has 5714 H&E stained image with 20× of magnification and dimensions 1444 × 908. The images contained 29 wild-type (WT) animals and 116-knockout animal selected randomly by WTSI. There were 2–3 slides available from each animal, and 6–10 images per slide (captured at the magnification and resolution above) were created.

4.2 Results

As it is not feasible to manually check the segmentation accuracy of 5714 images, we took random sample of 500 images. By close manual examination, we found that 18 images out of the 500 image were not segmented successfully as shown in Fig. 5 i.e. the segmentation algorithm achieved an accuracy about 96.4%.

As for the analysis results based on the RR method explained above, we identify 32 genotypes responsible for changes in the Epidermis curvature. Figure 8 shows 18 genotypes (e.g. Actn4/Actn4) in which the curvature measurements are bigger than the RR 95% CI upper threshold whereas Fig. 9 shows 14 genotypes (e.g. Wdr37) in which the curvature measurements are smaller than the RR 95% CI lower threshold.

Figure 10 shows an example of these interesting genotypes, particularly Actn4/Actn4 that increase the curvature in the epidermis layer and Wdr37 that is associated with a decrease in curvature. Figure 10 shows example of images, which have increased and decrease in the curvature of the epidermis layer of the H&E microscopic images in the mice skin.

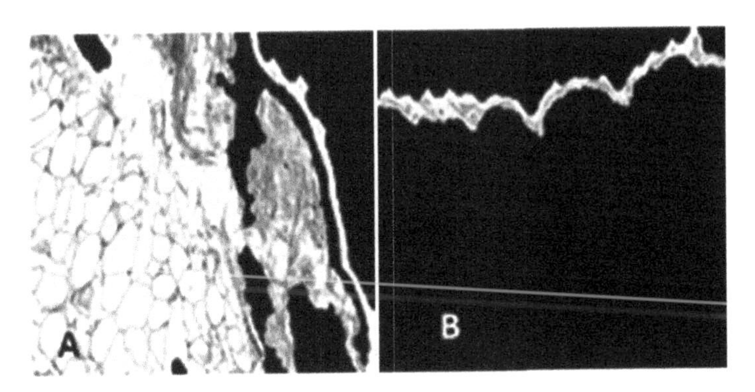

Fig. 5. Segmentation accuracy, (A) Bad segmentation. (B) Good segmentation

4.3 Curvature Accuracy Compared with Manual

Twenty five images was selected randomly from development data set (5714 images), then a manual technique was used to quantify the curvature in the epidermis layer to determine the truth distance values. Next the comparing operation has been done, which was clarified in Fig. 7, with the automatic result, which R^2 value, was 0.90, has used to measure the approach data to the fitted linear line ground truth value of the distance was found, by calculate the distance between the epidermis layer and the reference line, which was connected between a start point and end point of the epidermis layer as shown in Fig. 6, after that, the mean value was calculated for all distances values, for every single image, and then the comparison was applied between the mean value of the manual and the automatic method.

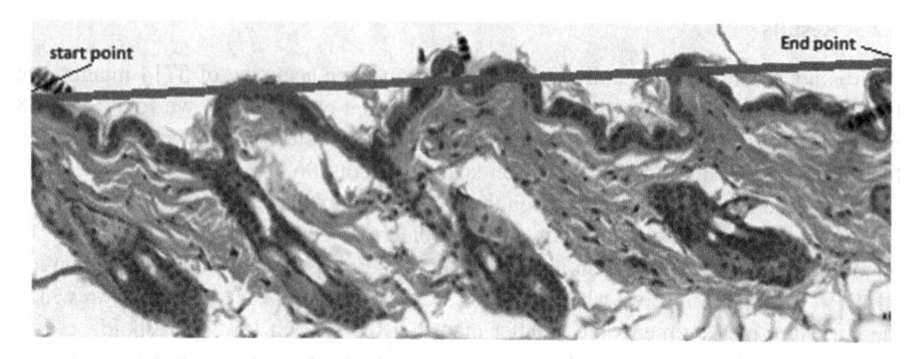

Fig. 6. Manual method for quantifying the curvature of epidermis layer

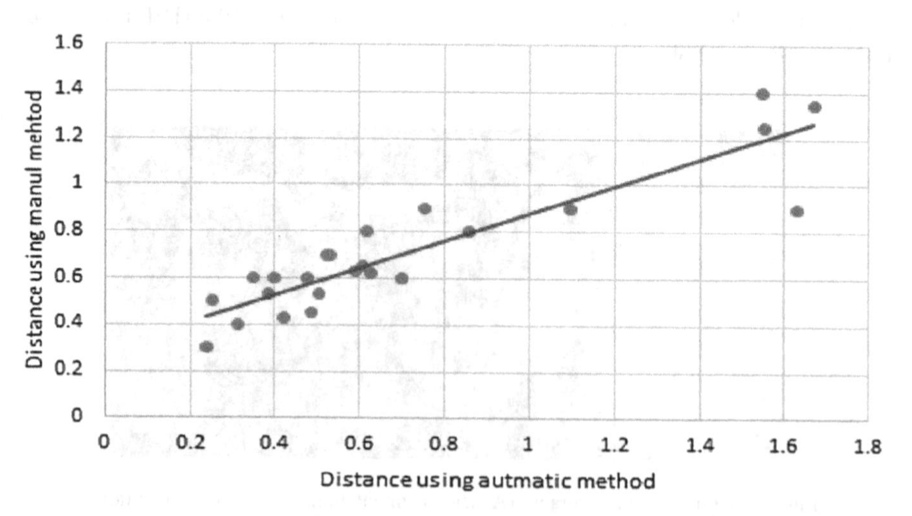

Fig. 7. Comparison between manual and automatic method

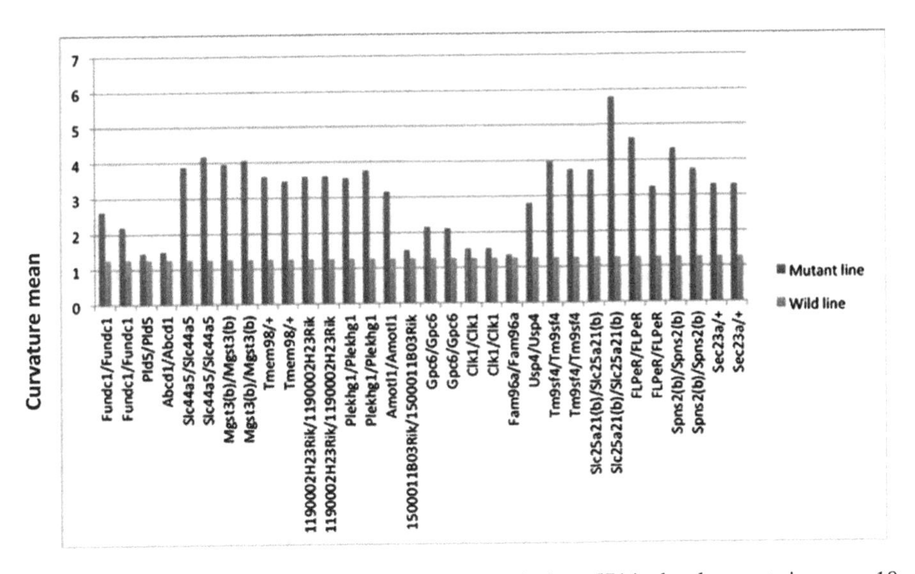

Fig. 8. Curvature in development data. From analysing 5714 development images, 18 genotypes showed curvature measurements bigger than RR 95% CI upper threshold.

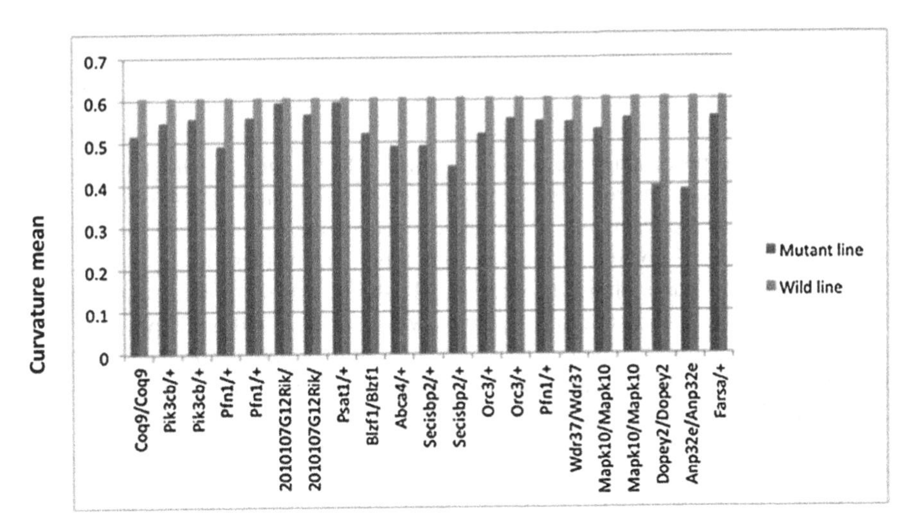

Fig. 9. Curvature in development data. From analysing 5714 development images, 14 genotypes showed curvature measurements smaller than RR 95% CI lower threshold.

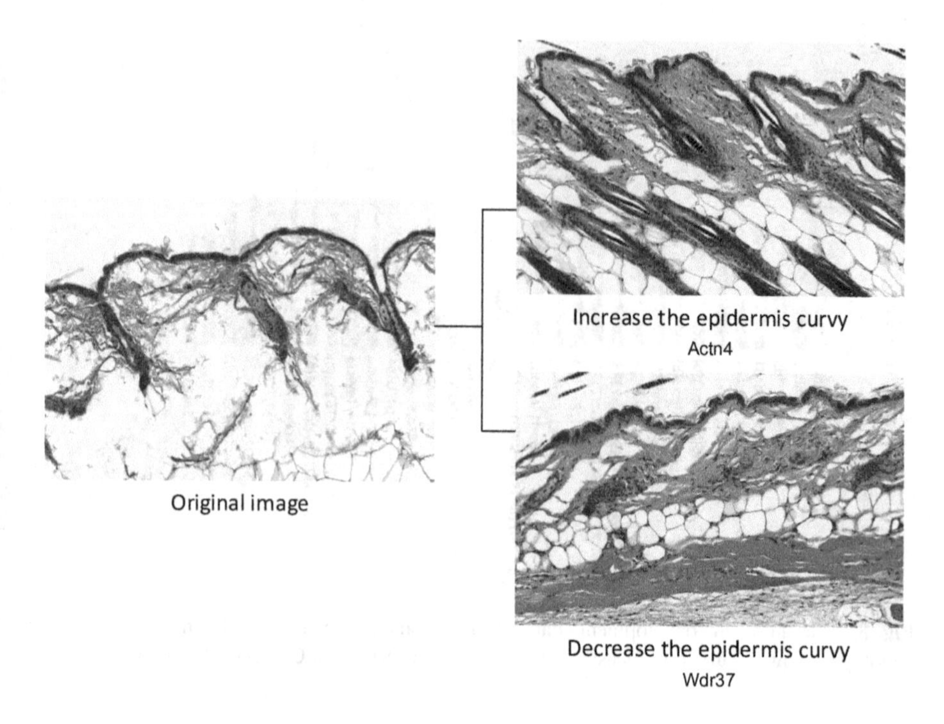

Fig. 10. Automatic estimation of interesting genes for curvature in epidermis layer.

5 Conclusion

Biological image processing and analysis provides techniques that help scientists to evaluate the effects of environmental exposure and physiological changes in a research context. These techniques may also be used to evaluate the effect of the treatments and drug efficiency in the context of the drug discovery. All of these analysis techniques could also be applicable clinically. In all applications, novel combinations of image processing/analysis techniques and pipelines will save time and are expected to produce more accurate results that will ultimately help or improve the speed and quality of dermatology and cosmetic treatments. This paper described a method for quantifying the curvature of the epidermis to help identifying genes responsible for changes in the curvature. Experiments on a large set of microscopic images of mutant and WT mice demonstrated the effectiveness of the proposal.

References

1. Collins, F.S., Jordan, E., Chakravarti, A., Gesteland, R., Walters, L.: New goals for the U.S. human genome project: 1998–2003. Science **282**, 682–689 (1998)
2. Zaidi, Z., Lanigan, S.W.: Skin: structure and function. In: Dermatology in Clinical Practice, Springer, London (2010)

3. Taylor, S.C.: Skin of color: biology, structure, function, and implications for dermatologic disease. J. Am. Acad. Dermatol. **46**, S41–S62 (2002)
4. Weissman, J., Hancewicz, T., Kaplan, P.: Optical coherence tomography of skin for measurement of epidermal thickness by shapelet-based image analysis. Opt. Express **12**(23), 5760–5769 (2004)
5. Ali, M., Hadj, B.: Segmentation of OCT skin images by classification of speckle statistical parameters. In: 2010 17th IEEE International Conference on Image Processing (ICIP) (2010)
6. Kurugol, S., Dy, J.G., Brooks, D.H., Rajadhyaksha, M.: Pilot study of semiautomated localization of the dermal/epidermal junction in reflectance confocal microscopy images of skin. J. Biomed. Opt. **16**(3), 3 (2011)
7. Li, A., Cheng, J., Yow, A.P., Wall, C., Wong, D.W.K., Tey, H.L., Liu, J.: Epidermal segmentation in high-definition optical coherence tomography. In: 2015 37th Annual International Conference of the IEEE Engineering in Medicine and Biology Society (EMBC) (2015)
8. Hussein, S., Selway, J., Jassim, S., Al-Assam, H.: Automatic layer segmentation of H&E microscopic images of mice skin. In: SPIE (2016)
9. Honda, K., Sakaguchi, T., Sakai, K., Schmedt, C., Ramirez, A., Jorcano, J.L., Tarakhovsky, A., Kamisoyama, H., Sakai, T.: Epidermal hyperplasia and papillomatosis in mice with a keratinocyte-restricted deletion of csk. Carcinogenesis **28**(10), 2074–2081 (2007)
10. Kurugol, S., Dy, J.G., Rajadhyaksha, M., Gossage, K.W., Weissmann, J., Brooks, D.H.: Semi-automated algorithm for localization of dermal/epidermal junction in reflectance confocal microscopy images of human skin. In: SPIE BiOS (2011)
11. Libby, J.: Teaching Practical Uses for Algebra, Geometry and Trigonometry. McFarland & Company, Jefferson (2017)
12. Bassett, J.H.D., Gogakos, A., White, J.K., Evans, H., Jacques, R.M.: Rapid-throughput skeletal phenotyping of 100 knockout mice identifies 9 new genes that determine bone strength. PLOS Genet. **8**, e1002858 (2012)
13. Liakath-Ali, K., Vancollie, V.E., Heath, E., Smedley, D.P., Estabel, J., Sunter, D., Di Tommaso, T., White, J.K., Ramirez-Solis, R., Smyth, I., et al.: Novel skin phenotypes revealed by a genome-wide mouse reverse genetic screen. Nature Commun. **5** (2014)

Author Index